The Encyclopedic Atlas of the Human Body

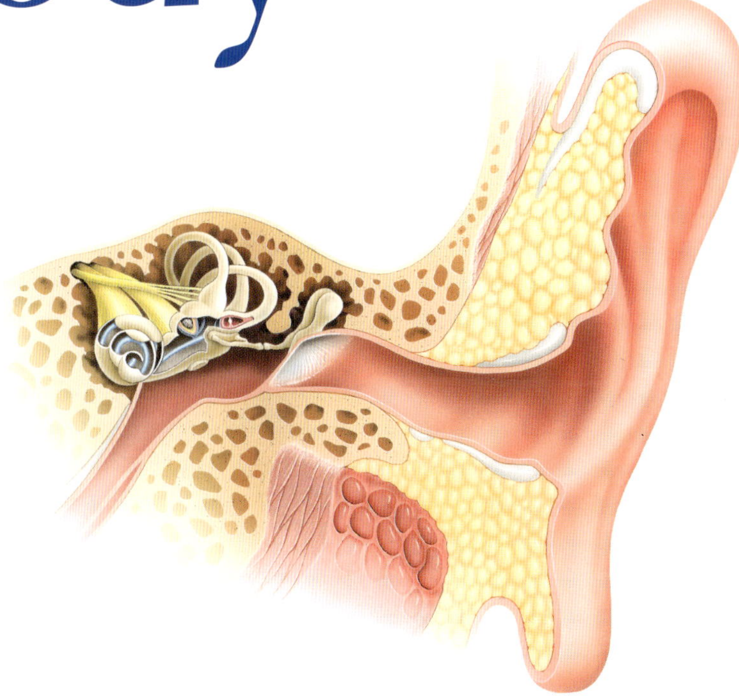

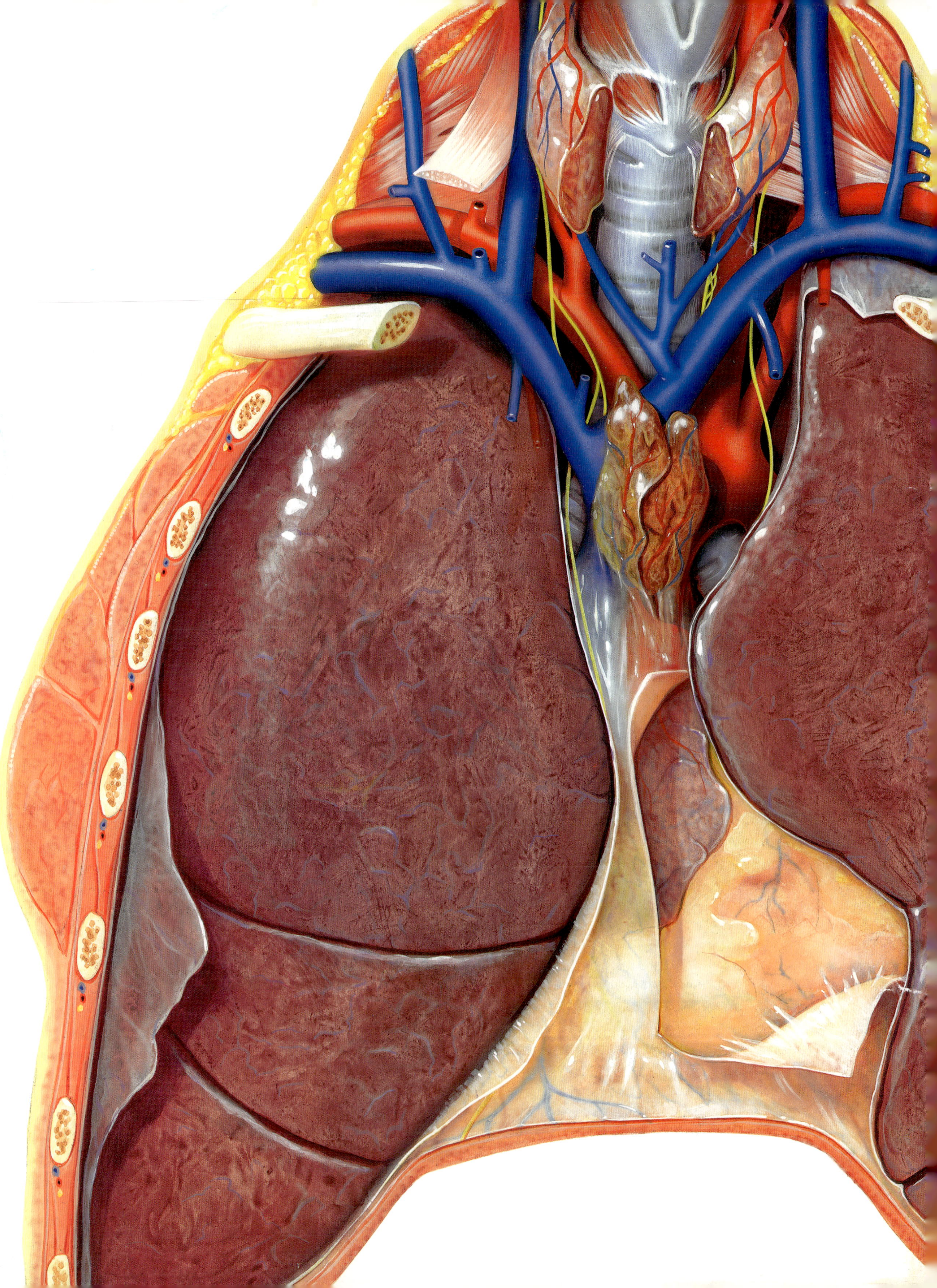

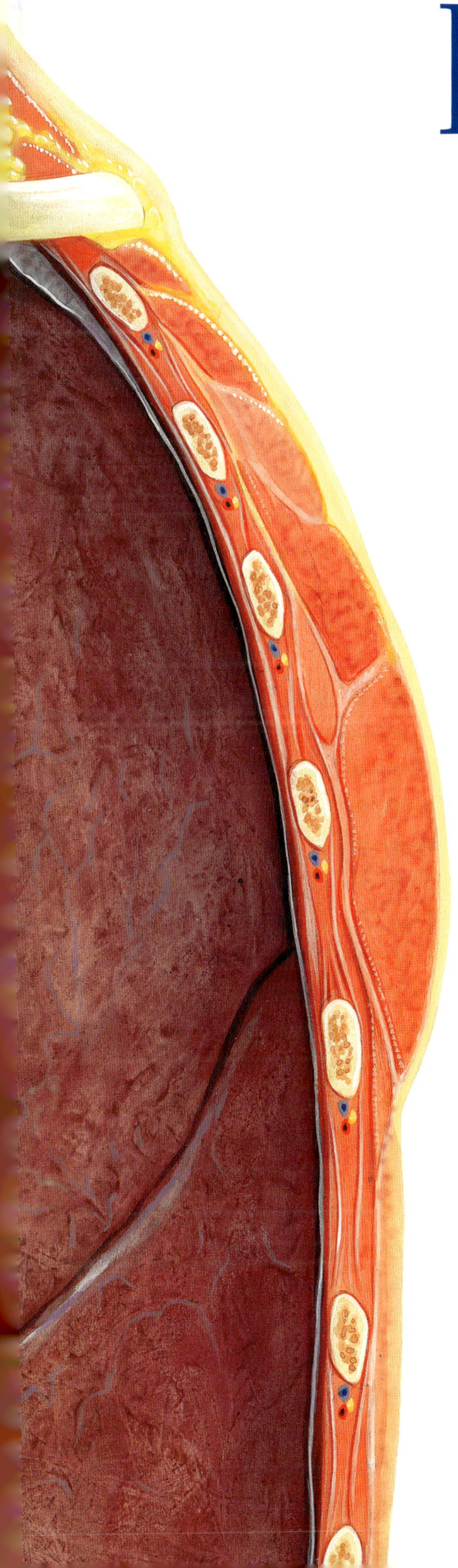

The Encyclopedic Atlas of the Human Body

A Visual Guide to the Human Body

RAINCOAST BOOKS
Vancouver

Publisher	Gordon Cheers
Associate publisher	Margaret Olds
Art director	Stan Lamond
Managing editor	Janet Parker
Chief illustration consultant	Dzung Vu MD MBBS DipAnat CertHEd
Illustration consultants	John Frith MBBS BSc(Med) DipEd MCH David Jackson MBBS BSc (Med)
Symptoms table text	Jenni Harman Melanie George MBBS DipPaed Annette Kifley MBBS Robyn McCooey BAppSci (Speech & Hearing) Sue Markham BAppSci (Phty)
Illustrators	David Carroll Peter Child Deborah Clarke Geoff Cook Marcus Cremonese Beth Croce Wendy de Paauw Levant Efe Hans De Haas Mike Golding Jeff Lang Alex Lavroff Ulrich Lehmann Ruth Lindsay Richard McKenna Annabel Milne Tony Pyrzakowski Oliver Rennert Caroline Rodrigues Otto Schmidinger Bob Seal Vicky Short Graeme Tavendale Jonathan Tidball Paul Tresnan Valentin Varetsa Glen Vause Spike Wademan Trevor Weekes Paul Williams David Wood
Labels	Thao Vu Jin-Oh Rhee
Cover design	Stan Lamond
Publishing assistant	Alan Edwards
International rights	Dee Rogers
Production	Bernard Roberts

NATIONAL LIBRARY OF CANADA CATALOGUING IN PUBLICATION DATA
The encyclopedic atlas of the human body.
Includes index.
ISBN 1-55192-747-0
1. Human anatomy—Atlases.
QM25.E53 2004 611'.0022'3 C2004-901961-9

Published in Canada by
Raincoast Books
9050 Shaughnessy Street
Vancouver, British Columbia
Canada V6P 6E5
www.raincoast.com

This publication and arrangement
© Global Book Publishing Pty Ltd 2004
Unit 1, 181 High Street, Willoughby,
NSW 2068, Australia
Phone: 612 9967 3100
Fax: 612 9967 5891
email: rightsmanager@globalpub.com.au

10 9 8 7 6 5 4 3 2 1

Printed in Hong Kong by Sing Cheong Printing Co. Ltd

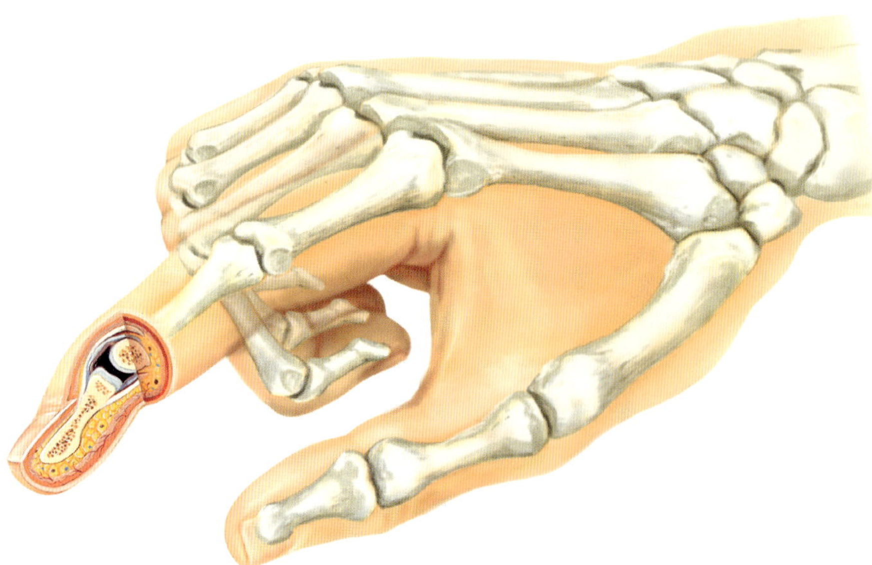

Consultants

CHIEF CONSULTANTS

Dr R. William Currie BSA, MSc, PhD is Professor of Anatomy and Neurobiology in the Faculty of Medicine at Dalhousie University, Halifax, Nova Scotia, Canada.

In his academic career, Dr Currie has taught all aspects of gross anatomy to medical, dental and health professional students. He is a pioneer and leader in his research on the protective role of heat shock proteins in the heart and the brain. He is a founding member of the editorial board of *Cell Stress and Chaperones.*

David Tracey BSc, PhD studied for his PhD in neuroscience at Stanford University and has worked as a neuroscientist in Paris, Munich, Melbourne and Canberra. In 1982 he joined the School of Anatomy at the University of New South Wales, where he teaches musculoskeletal anatomy. He became Professor of Anatomy there in 1995 and is Head of Department.

His research has included work on the anatomy and physiology of the spinal cord, and currently focuses on mechanisms of pain following nerve injury. He is a board member of the Federation of Australian Scientific and Technological Societies.

The Honorable Emeritus Professor Peter Baume AO, MD, BS (Syd), HonDLitt (USQ), FRACP, FRACGP, FAFPHM has been a professor since 1991. He was Head of the School of Community Medicine at the University of New South Wales until May 2000, and is currently with the Social Policy Research Centre within the University of New South Wales. He is Chancellor of the Australian National University.

He is a physician who holds a doctorate and several fellowships. He has been a consultant physician, a Senator for New South Wales (1974-91), Minister for Aboriginal Affairs, Minister for Health, Minister for Education and a Minister in Cabinet.

He is a Past President of the Public Health Association (New South Wales Branch), Chair of the Drug Offensive Council of New South Wales, a member of the Minister for Health Advisory Committee in New South Wales. He has published widely and was made an Officer in the Order of Australia in 1992. He is married with two children.

SENIOR CONSULTANT

John Frith MB, BS, BSc(Med), GradDipEd, MCH, RFD is a general practitioner and lecturer in general practice at the School of Community Medicine, University of New South Wales, Sydney. He graduated from the University of New South Wales in 1973, 1976 and 1994, and Sydney College of Advanced Education in 1988. His experience is in clinical and academic general practice and community health.

His professional memberships include Member of the Royal Australian College of General Practitioners and medical officer in the Royal Australian Naval Reserve. Other committee memberships include drug and alcohol education and prevention, health and safety in child care, and motor accidents compensation scheme.

He lectures in undergraduate general practice and medical ethics and law, and postgraduate primary health care, public health, and environmental health.

Dr Frith has contributed to publications and books on general practice and on public health.

CONSULTANTS

Kurt H. Albertine PhD is Professor of Pediatrics (Neonatology), as well as Adjunct Professor of Medicine (Pulmonary) and Neurobiology and Anatomy at the University of Utah School of Medicine in Salt Lake City, Utah, USA.

Dr Albertine received his bachelor's degree in biology from Lawrence University and a doctoral degree in human anatomy from the University of Chicago, Stritch School of Medicine. He then received postdoctoral training at the University of California, San Francisco, Cardiovascular Research Institute.

He has taught human gross anatomy for over twenty-five years, and is currently the course director for gross anatomy at the University of Utah.

One of his current scholarly projects is developing radiographic holograms for learning 3-dimensional human anatomy.

Laurence Garey MA, DPhil, BM, BCh, recently the Professor of Anatomy at the University of London, in the Division of Neuroscience of Imperial College School of Medicine at Charing Cross Hospital, London, is now the Professor of Anatomy in the Faculty of Medicine and Health Sciences at the United Arab Emirates University, Al Ain.

He qualified in medicine at Worcester College, Oxford, and St Thomas' Hospital, London, and obtained a doctorate in Oxford, based on research on the mammalian visual system. He worked in neuroanatomical research in Oxford, Berkeley, Lausanne and Singapore, before returning to London in 1990.

His research interest is on the structure, development and pathology of the human brain, especially the cerebral cortex. For a number of years he has been active in research on the pathophysiological basis of schizophrenia.

He has contributed to the *Oxford Companion to the Body*, and translated a number of biomedical science books from French, including *Neuronal Man* by Jean-Pierre Changeux (1985), *The Population Alternative* by Jacques Ruffié (1986), and *The Paradox of Sleep* by Michel Jouvet (1999). He has also translated (1994) from the German the famous *Localisation in the Cerebral Cortex* by Korbinian Brodmann, written in 1909.

Gareth Jones BSc (Hons), MB, BS (Lond), DSc (Univ West Aust), CBiol, FIBiol is the Professor and Head of the Department of Anatomy and Structural Biology at the University of Otago, Dunedin, New Zealand.

His main specialties are neurobiology—the organization and plasticity of synaptic connections in the brain; bioethics—issues related to the human body and human tissue; and anatomical education.

His recent book projects include *Medical Ethics*; *Speaking for the Dead*; *Universities as Critic and Conscience of Society*; and *Synapses in the Central Nervous System.*

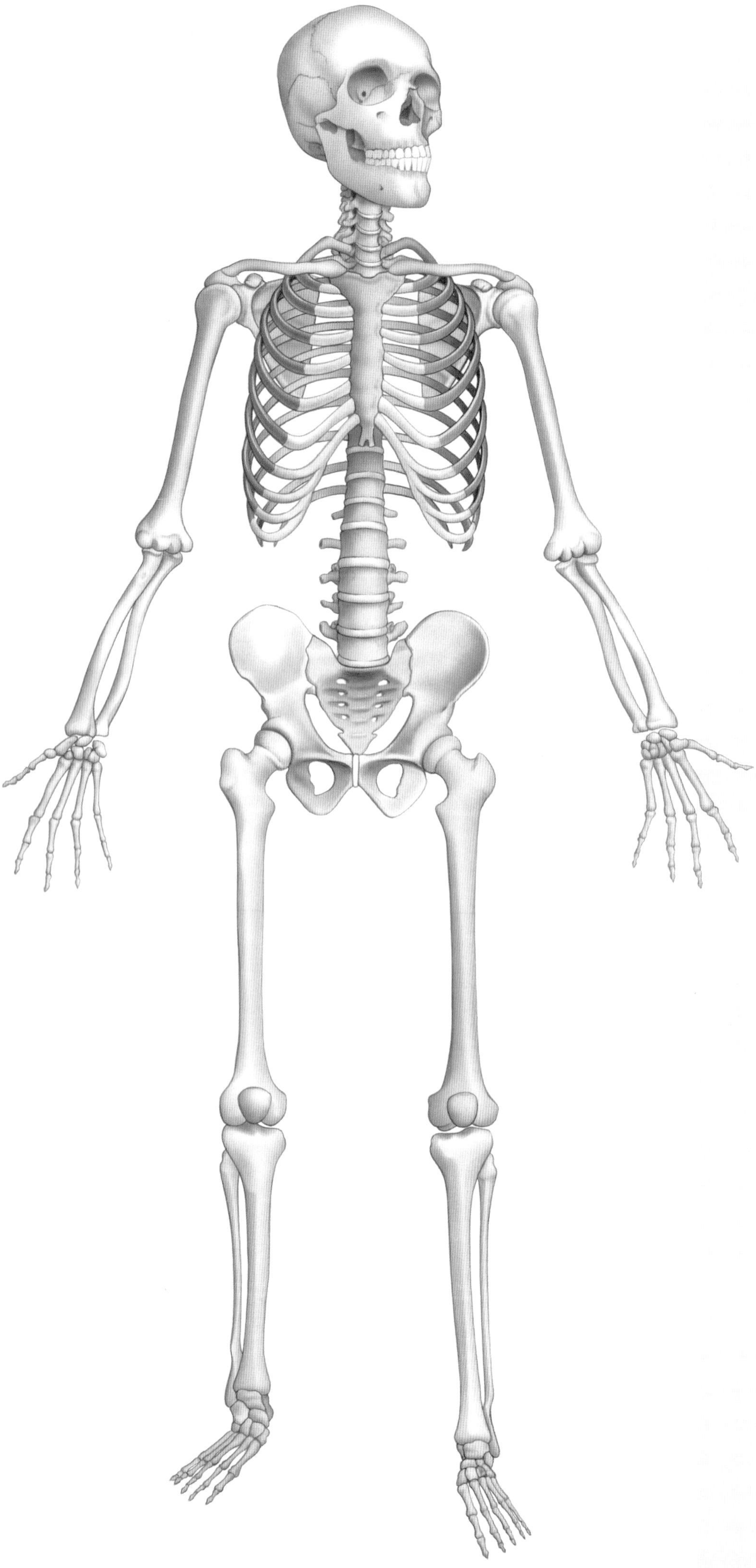

Contributors

Robin Arnold MSc is a lecturer in the Department of Anatomy at the University of Sydney. Her general professional interests include teaching gross anatomy, microscopy, medical history and dental histology. In the area of research her main interests lie in comparative mammalian female reproduction and atherosclerosis.

Ken Ashwell BMedSc, MB, BS, PhD graduated in medicine from the University of New South Wales in 1983. After a short time in clinical medicine he returned to research and teaching, undertaking a PhD studying processes in abnormal brain development and graduating from the University of Sydney in 1988. He has been teaching anatomy to medical, chiropractic and science students since 1984 and maintains an active involvement in research on brain development (both normal and abnormal) and brain evolution. He is the author of more than fifty scientific papers, two books and two book chapters and is currently Associate Professor in Anatomy.

Deborah Bryce BSc, MScQual, MChiro, GrCertHEd is a lecturer in the Department of Anatomy at the University of Sydney. Her special interests lie in teaching and learning in higher education and she has completed studies in this area at the University of NSW and University of Technology, Sydney. She has been central in the development of online teaching resources at the University of Sydney, including an online anatomy museum and anatomy glossary. Her particular area of teaching interest is musculoskeletal anatomy and the vertebral column.

Carol Fallows BA and **Martin Fallows** are joint founders of several consumer magazines in the areas of health and parenting. Both began their writing careers in the magazine industry and they share a passion for providing consumers with accurate up-to-date information. Carol is the author of several books on parenting including *The Australian Baby & Child Care Handbook* and *Having a Baby*. Martin is a freelance editor and publishing consultant.

John Gallo MB, BS (Hons), FRACP, FRCPA is a Senior Staff Hematologist at South Western Area Pathology Service and Liverpool Hospital, Sydney, Australia. He graduated in medicine from the University of Sydney in 1976 and, as part of his post-graduate training, was a Research Fellow at the University of Maryland Cancer Center in Baltimore, USA, in 1982–1983. His research interests were in the fields of chromosomal abnormalities in leukemia, and the cell cycle. After completing his specialist training he was in private clinical hematology practice in Sydney for ten years. His principal clinical interests are in Hodgkin's lymphoma and bone marrow disorders of the elderly.

Brian Gaynor MB, BS, FRACP, FRACGP, MRCGP (UK) DCH is a retired general pediatrician who is currently enjoying life as a part-time family doctor and part-time columnist on the internet. His experience in pediatrics has covered a range of aspects including hospital consultant, staff specialist in community pediatrics, superintendent of a large home for intellectually handicapped children, and he has also worked in remote areas, including Fiji, Iraq and Papua New Guinea. His writing career began as part of a midlife crisis when he started as a columnist for a Sydney daily newspaper, then progressed to being a feature writer for different newspapers, magazines and other publications.

Jenni Harman BVSc, BA (Hons I) is a freelance medical writer specializing in medical education. She trained and worked as a veterinary surgeon before completing an arts degree and working as a research assistant in a university education faculty. As a medical writer, she has specialized in continuing medical education for primary care physicians and specialists, as well as newspaper journalism. Her recent projects include the development of small-group discussion-based educational programs for doctors on asthma, cardiovascular disease and psychiatry, teaching resources for specialists, and fact sheets for patients.

Rakesh Kumar MB, BS, PhD graduated in medicine from the All-India Institute of Medical Sciences, New Delhi, and took up an appointment at the University of New South Wales, Sydney in 1977. He subsequently completed a PhD at the University of New South Wales and is now Associate Professor of Pathology. He is an enthusiastic teacher and is involved in courses for both medical and science students. He has a long-standing interest in chronic lung disease and his research focuses on mechanisms of inflammation of the airways in asthma.

Peter Lavelle MB, BS graduated in medicine from Sydney University in 1983 and practiced as a primary care physician for several years before becoming a full-time medical writer. He contributes regularly to the *Age* and the *Sydney Morning Herald* newspapers, the *Australian Doctor* and the *Medical Observer*.

Lesley Lopes BA Communications (Journalism) is a journalist and editor who began her career in publishing as a cadet newspaper reporter 15 years ago. She edited photographic and home entertainment magazines in Sydney for six years before working for a London publishing firm developing a range of children's and lifestyle publications. Lesley recently held the position of Features Editor of Australia's *Better Homes and Gardens* magazine.

Karen McGhee BSc is a Sydney-based freelance journalist. Her work has focused on the areas of science and the environment for more than a decade. She has written for newspapers (including the *Sydney Morning Herald*), magazines (ranging from *Time* to *Australian Geographic*), books and television documentaries for both Australian and overseas audiences.

Emeritus Professor Frederick Rost BSc (Med), MB, BS, PhD, DCP (London), DipRMS was born in London in 1934 and arrived in Australia in 1937. He was Professor of Anatomy at the University of New South Wales from 1974–95. Retired, Fred is now a freelance author, photographer and artist, and has honorary appointments at the University of New South Wales and Macquarie University. He is currently writing his fifth major book.

Elizabeth Tancred BSc, PhD is a Senior Lecturer in Anatomy at the University of New South Wales, with more than 20 years experience in teaching anatomy. Following a BSc with honours in anatomy, she completed a PhD in neuroanatomy in 1983. Her early research was in the area of visual neuroscience but she is now focused on the role of information technology in medical education. She is the author of several software packages for teaching neuroanatomy and cross-sectional anatomy, most notably, "BrainStorm: Interactive Neuroanatomy" which is used widely in universities throughout the world.

Dzung Vu MD, MB, BS, DipAnat., GradCertHEd is an orthopedic surgeon and clinical anatomist. He teaches clinical anatomy to medical students and candidates of specialist degrees, and is the author of many teaching videotapes and computer-assisted learning programs in anatomy. He was the recipient of the Vice Chancellor's Award for Excellence in Teaching at the University of New South Wales in 1992 and is an examiner of the Royal Australian College of Ophthalmologists. He is also a gifted medical illustrator and belongs to the Australian Institute of Medical and Biological Illustration, among many other national and international professional associations.

Phil Waite BSc (Hons), MBChB, CertHEd, PhD obtained a BSc in physiology followed by a PhD in sensory neurophysiology at University College, London. After emigrating to Australia, she held teaching and research positions at Monash University, Melbourne, and then at the University of Otago in New Zealand. She later graduated in medicine at Otago University and is now a Professor of Anatomy at the University of New South Wales, working on brain development and the effects of injury.

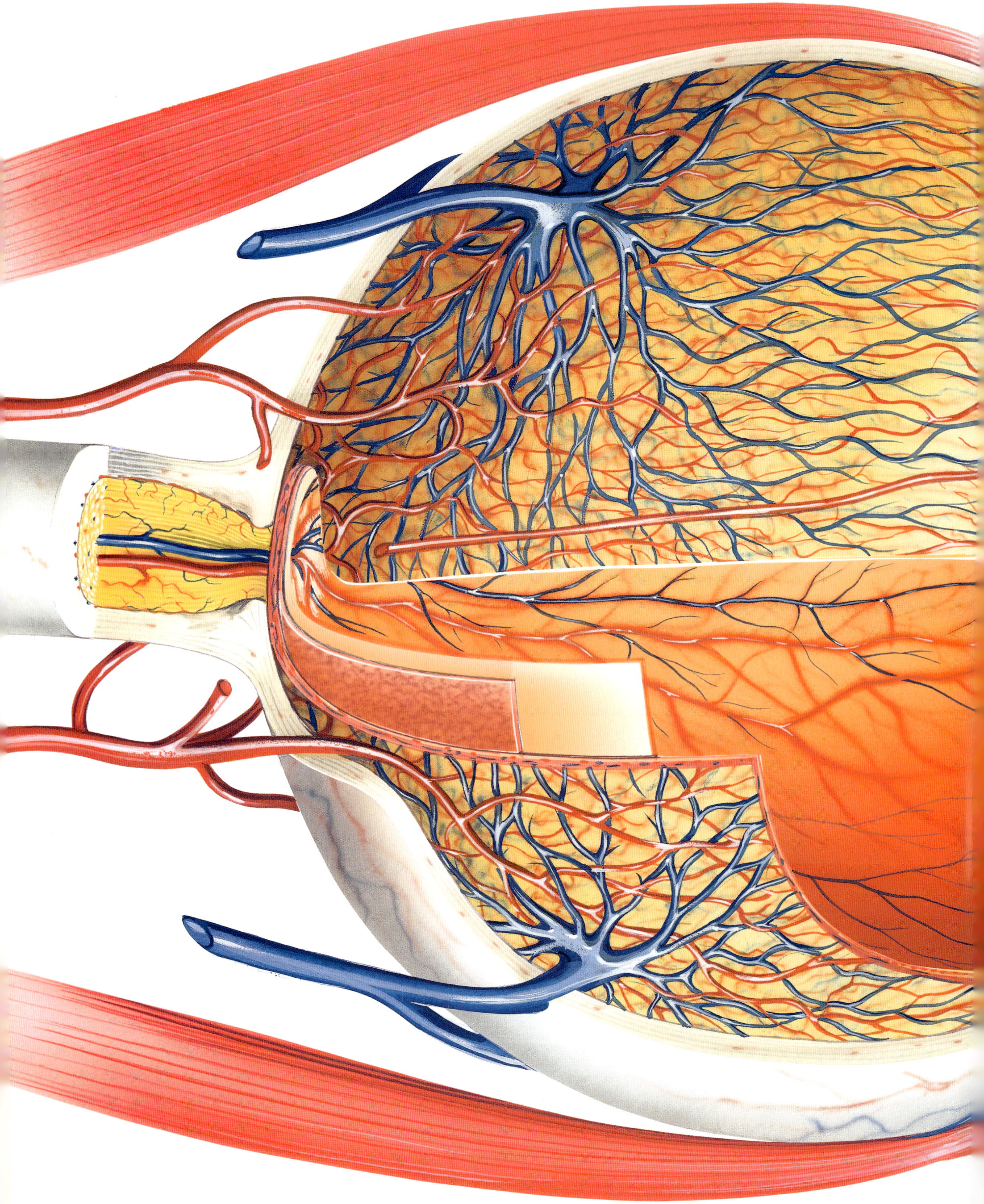

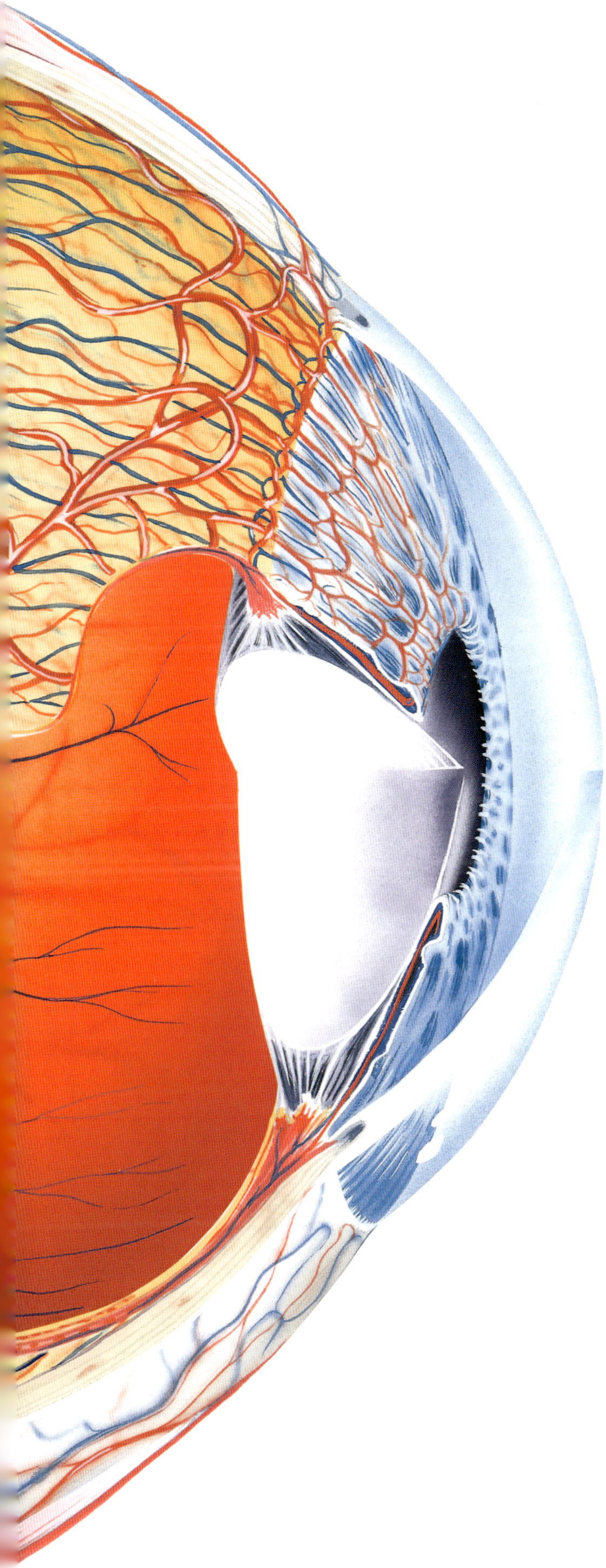

Contents

How This Book Works

The many facets of the human body are detailed in *The Encyclopedic Atlas of the Human Body*. This superb reference begins by discussing cells and tissues—the smallest yet most crucial elements in the human body. This is followed by sections on the major body systems (including the Skeletal, Muscular, Lymphatic, Nervous, Circulatory, Respiratory, Urinary, Digestive, Reproductive and Endocrine systems), giving an insightful overview of the body mechanisms and the interaction of the systems. Also included is a section on homeostasis and metabolism, which details many of the body's functions and capabilities.

Subsequent chapters deal with individual regions of the body, from the head and trunk through to the abdominal region and the hips, legs and feet. Each chapter discusses many of the illnesses, diseases, conditions and injuries pertaining to the particular region and its organs. There is a chapter dedicated to procedures, techniques and diagnostic tests, along with information on many of the medications commonly prescribed or available over-the-counter.

The Human Life Cycle is the focus of another chapter. All the stages and ages of life—pregnancy, infancy, childhood, adolescence, menopause and ageing—are outlined in detail, along with advice on the many treatments and procedures available for the illnesses, diseases or disorders that may be encountered along the way.

There is also an informative Symptoms Table—this guide lists many of the symptoms experienced in illness and disease, and advises when to seek medical advice. The final section, The Time of Your Life, discusses the developmental stages of each phase of life from birth to old age, and details the recommended health checks for each of these phases. An extensive index completes the volume.

Superb color illustrations provide a visual reference for many of the text entries, providing a snapshot of a region or organ. The illustrations are accompanied by captions and labels, which provide further information, and enhance understanding of the subject. In some cases, the illustration shows partial or complete removal of an organ in order to display another, and in such cases, this is noted with the caption. Helpful locator diagrams accompany many of the illustrations, to indicate where the organ lies in the body and its relationship with its surrounding organs and tissues.

Each entry has a heading, and where there is an illustration, it includes a caption and labels. Following discussion of each major body system, body region, organ or structure, the major illnesses, diseases and problems relevant to the subject are covered in detail. Many topics also feature SEE ALSOs, which provide a cross-reference to further information on the subject matter that can be found elsewhere in the book.

The Encyclopedic Atlas of the Human Body provides information on every aspect of the human body, explained in a clear concise style, and complemented by highly detailed illustrations. The aim of this volume is to provide accessible, useful and authoritative information—for everyone from school students to the medical profession—on the human body and how it works.

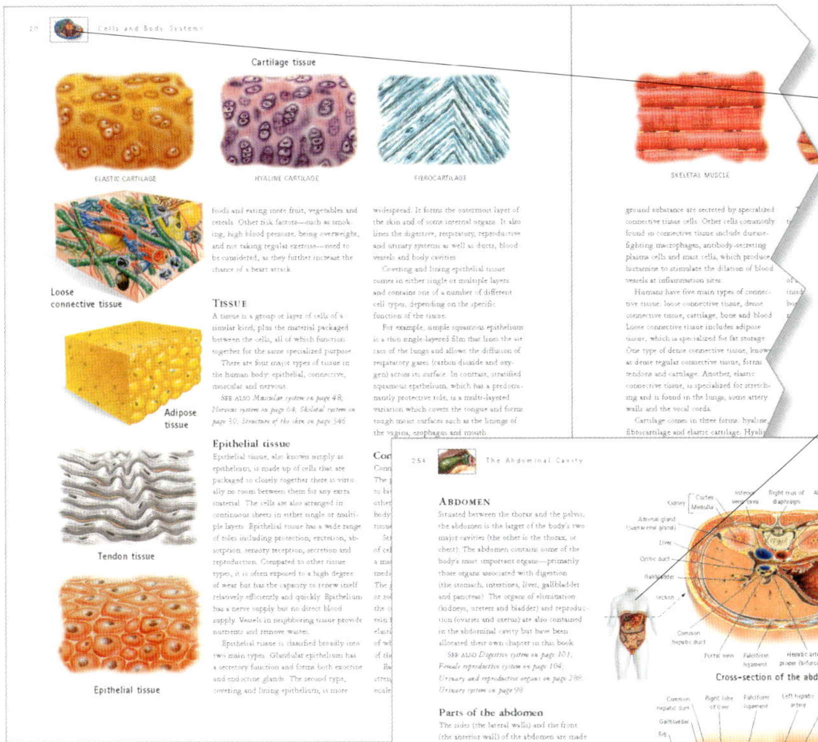

Running heads and chapter locators

A small icon of the main visual for each chapter appears next to the running heads to provide a handy reference to the chapters.

Locator diagrams

Locator diagrams are included to give an indication of where a body part is located in relation to the rest of the body.

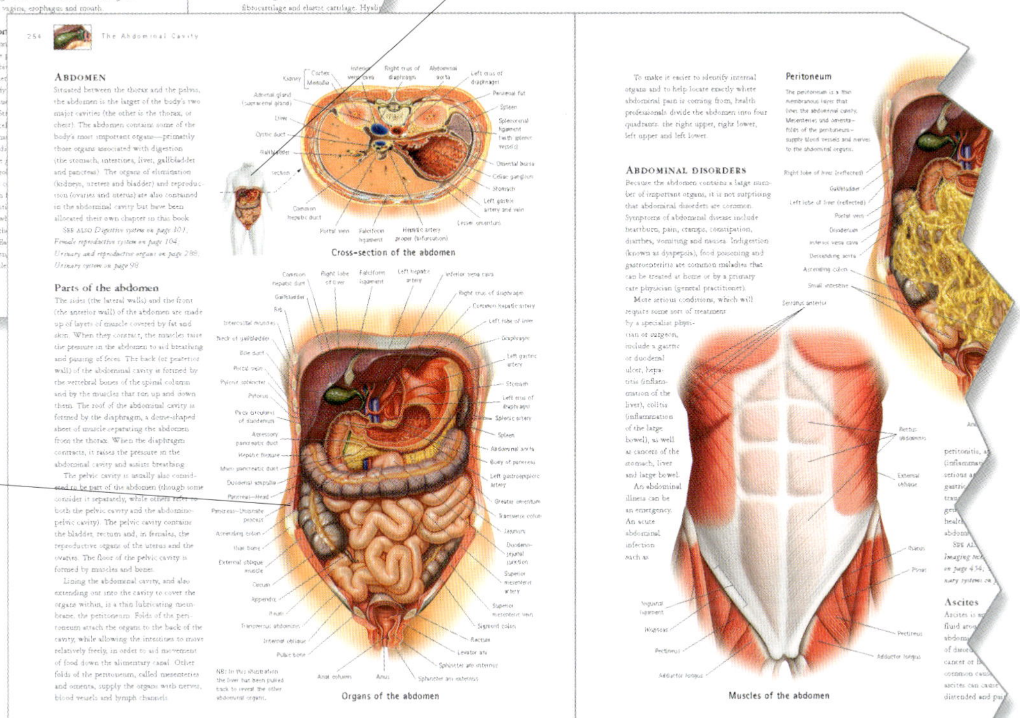

Illustrations

Illustrations show what is described in the text. They may show a whole body system, a single organ, microstructure, or the effects of damage or disease on a particular part of the body.

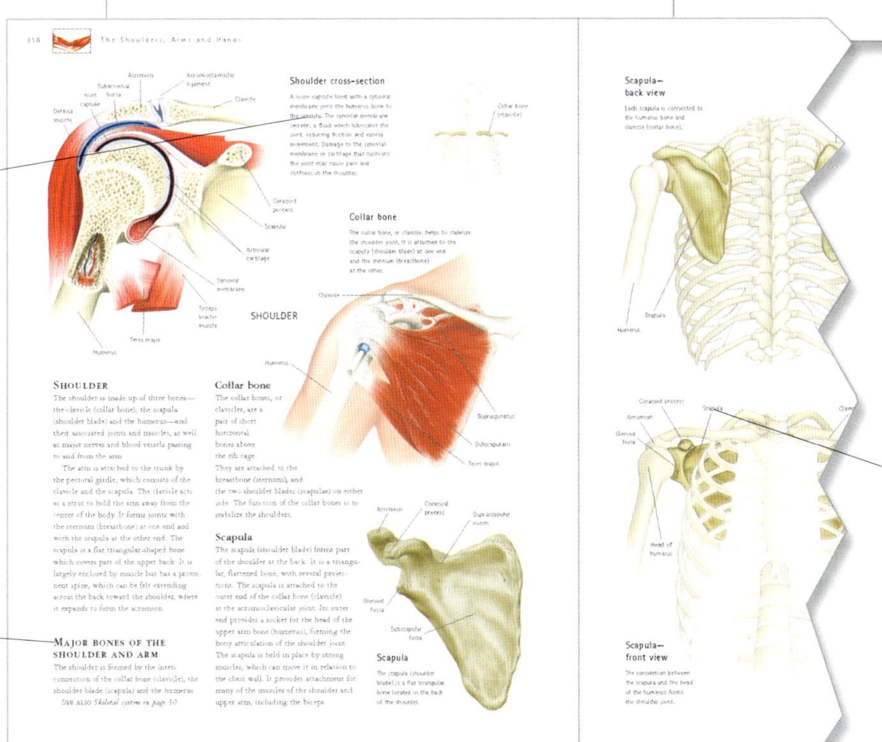

Captions

The captions to the illustrations make it easy to understand what the illustration is showing.

Labels

The labels on each illustration label particular elements that are relevant to the purpose of that illustration.

Headings

The headings mark the beginning of an entry.

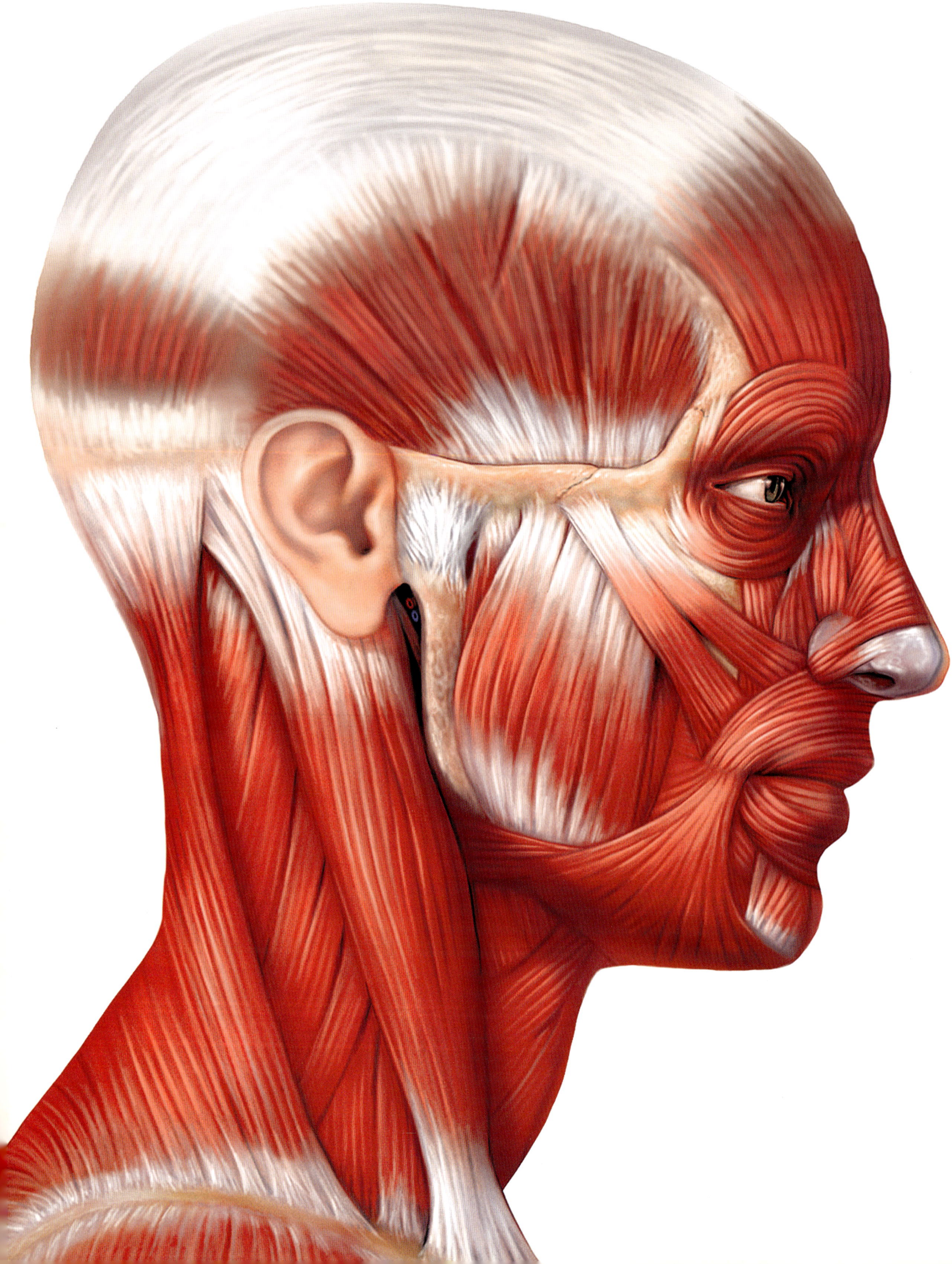

Introduction

Understanding the complexity of the human body has challenged humans since the beginning of time. With every discovery, yet another mystery is uncovered, and while we now have an incredible understanding of the human body, there is still much more to investigate and learn. Over the ages, as each milestone of discovery has led us to better understanding of the human body, so our desire to learn more has increased. Once the domain of academics and scientists, the human body is now a source of fascination to all, and this volume aims to provide easy-to-understand information to satisfy the thirst for knowledge. It will answer the questions many readers have relating to how the human body works and provide information on many of the illnesses encountered in current times, along with details on the latest diagnostic techniques and treatment options.

Discover the workings of the human body—how the body systems interact; where to find the major organs and structures; the function of many vital glands and organs; how each component contributes to the smooth operation of the body; and the diseases and disorders that can compromise the capabilities and function of each component. In this indispensable reference book, readers have this information at their fingertips.

Beginning with the foundation of every structure in the body, the cells and tissues, this book embarks on a journey of discovery—from the major body systems to the many regions of the body, then on to infectious diseases, the human life cycle and then diagnosis and treatment. Exquisitely detailed full color illustrations accompany many of the entries, providing a useful visual reference.

The human body is a fascinating subject, and as we all seek to be better informed about matters pertaining to our personal health, our concerns regarding disease, illness or complaints, or our natural curiosity regarding the various mechanisms of the body, *The Encyclopedic Atlas of the Human Body* will provide a wealth of information designed to answer these questions and many more.

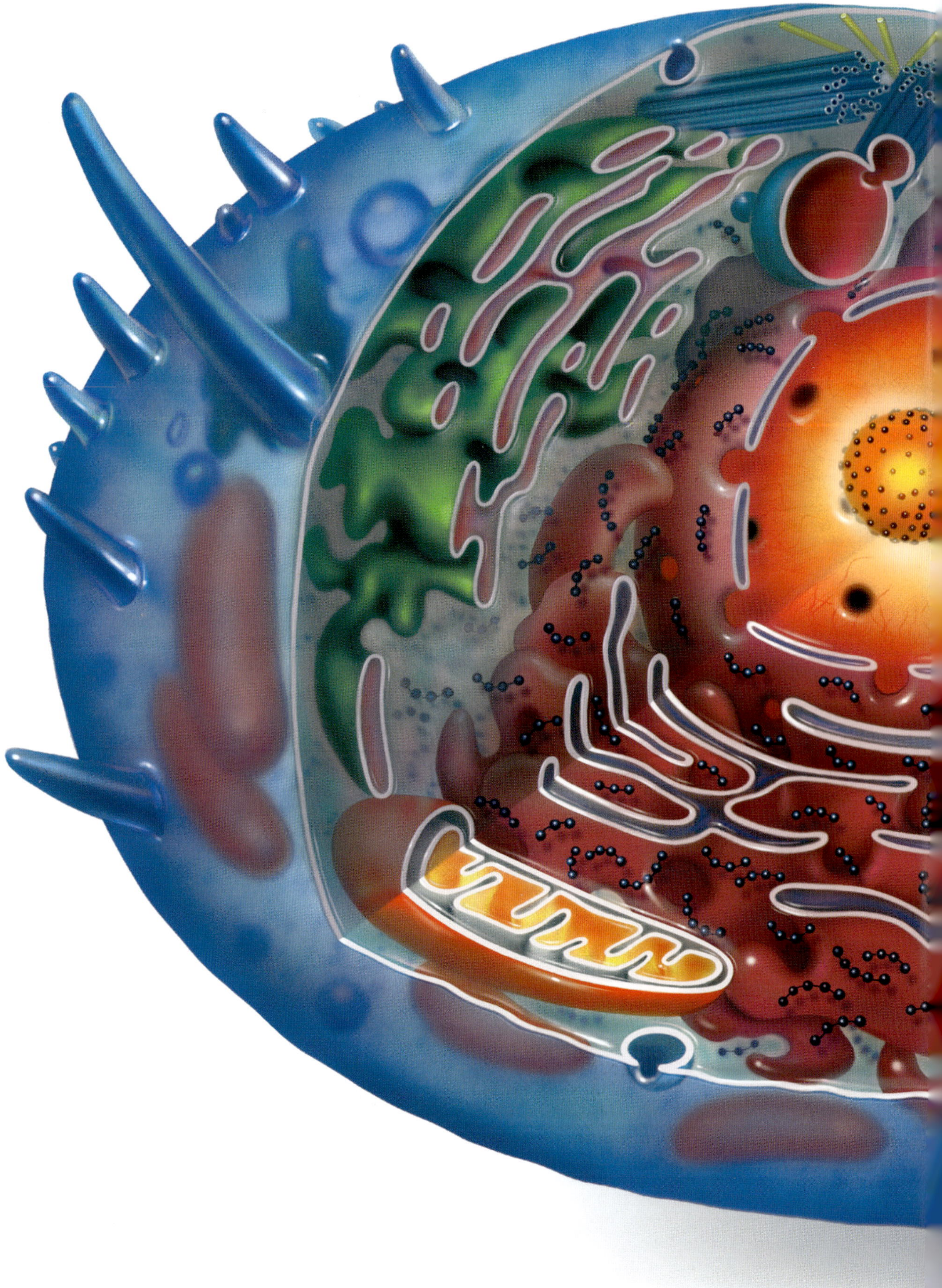

Cells and Body Systems

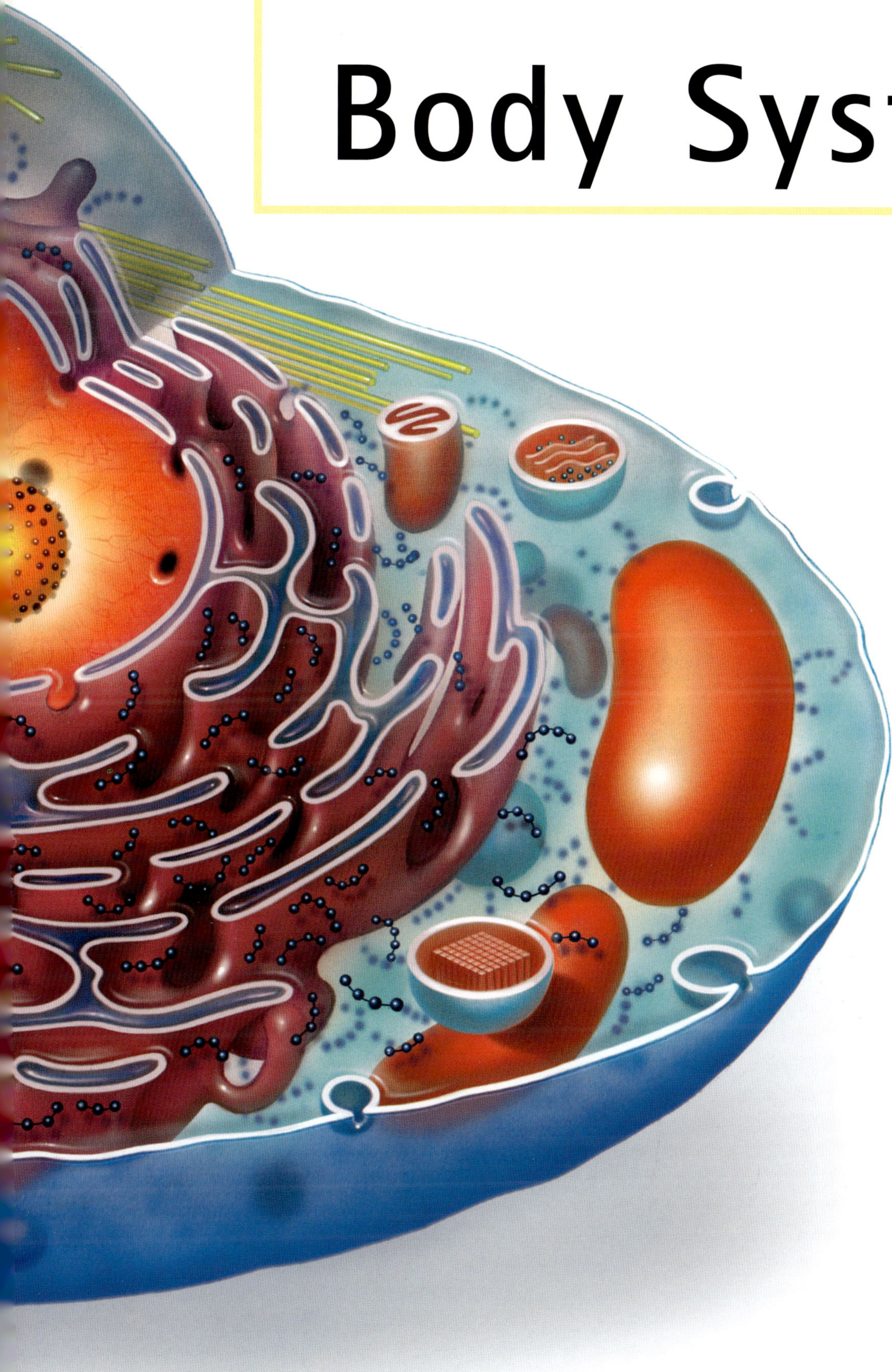

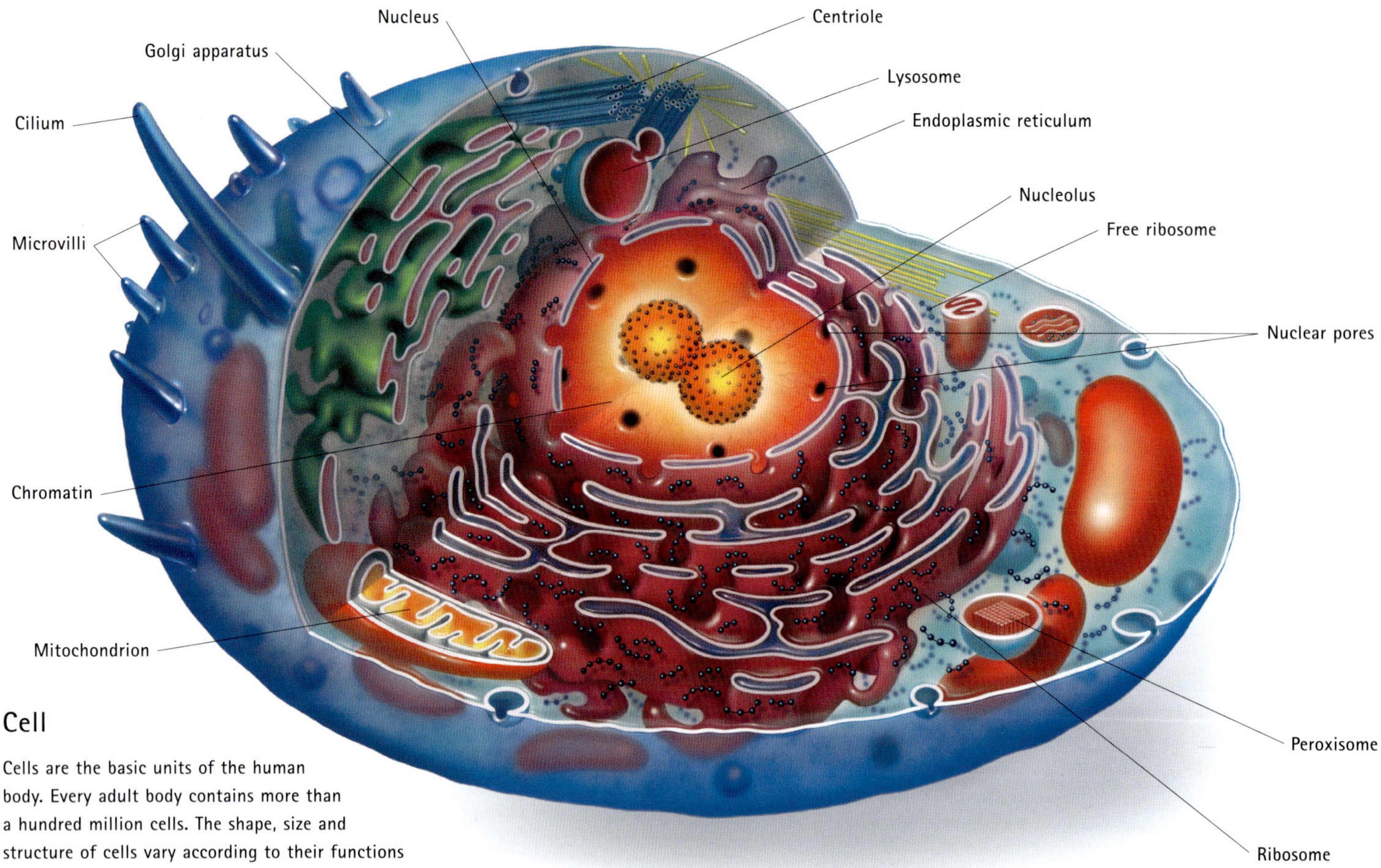

Cell

Cells are the basic units of the human body. Every adult body contains more than a hundred million cells. The shape, size and structure of cells vary according to their functions within the body.

CELLS AND TISSUES

The cell is the functional unit of all tissues and has the capacity to perform individually all the essential life functions. The various tissues of the body are composed of cells that specialize in a wide range of functions.

SEE ALSO *Ageing on page 394; Blood on page 82; Chromosomes on page 18; DNA on page 17; Fertilization on page 397; Genes on page 18; Lymphatic/immune system on page 55; Nervous system on page 64; Skeletal system on page 30*

CELLS

Cells are the basic structural units that make up all living organisms. Cells are organized into tissues, and tissues into organs. Typical cells are surrounded by a cell membrane. Within the cell membrane lies the cell's cytoplasm, which is a fluid containing many important structural units called organelles. Some organelles are concerned with the manufacture of protein (rough endoplasmic reticulum), the generation of energy (mitochondria), the packaging of manufactured products (Golgi apparatus) and the process of cell division (centrioles). The control of the cell's function is essentially directed by the nucleus, which contains genetic information in the form of DNA. In most of the larger and more complex cells, the nucleus is separate from the cytoplasm.

CELL DIVISION

Some types of cells undergo a process called mitosis, or cell division, whereby a mother cell divides into two daughter cells. This is particularly well seen in the cells lining the gut and the skin, where cells are continually being shed and must therefore be replaced. When cells are dividing continuously, they pass through a series of stages called the cell cycle. This cycle consists of four individual stages: a growth stage (G1), a synthetic stage (S), another growth stage (G2) and a cell division stage (M).

When cell division runs out of control and the daughter cells invade other tissues, a cancer results. Cancers usually arise in sites where cell division is already occurring rapidly, such as the skin, gut and airway lining. Most cells are formed through this process of cell division.

SPECIALIZED CELLS

Many cells are specialized to perform particular functions. For example, blood is made up of two broad types of cells (red and white blood cells) and specialized cell fragments called platelets.

Other specialized types of cells include nerve cells (neurons), which are electrically active and transmit information to other nerve cells across special junctions called synapses. Another type of electrically active cell is the muscle cell, which contracts in response to nervous or hormonal stimulus.

Skin cells are also highly specific. They are designed to be progressively shed, because of their continuous exposure to damaging effects of the external environment with its ultraviolet radiation, physical wear and tear, extremes of temperature and low humidity. To replace shed cells, the lower layer of the skin contains cells that divide rapidly.

Ovum

The ovum is the mature female germ cell. The ovaries of women of reproductive age contain about 800,000 immature ova (oocytes), each surrounded by a covering of specialized cells that secrete female hormones. Only about 400 of these will be ovulated. The rest degenerate sometime during the reproductive years or just after menopause.

Normal body cells are diploid, that is, they contain 23 pairs of chromosomes and divide by mitosis to produce daughter cells which also have 23 pairs of chromosomes. Germ cells (oocytes and sperm) are haploid, that is, they contain only one of each pair of chromosomes (23 single chromosomes). This is possible because oocytes are produced in the embryo by a special form of cell division called meiosis (reduction or division), which halves the number of chromosomes. Sperm are also produced by meiosis. When fertilization takes place and ovum and sperm join together, the chromosome number in the new individual is restored to the normal 23 pairs (46 single chromosomes).

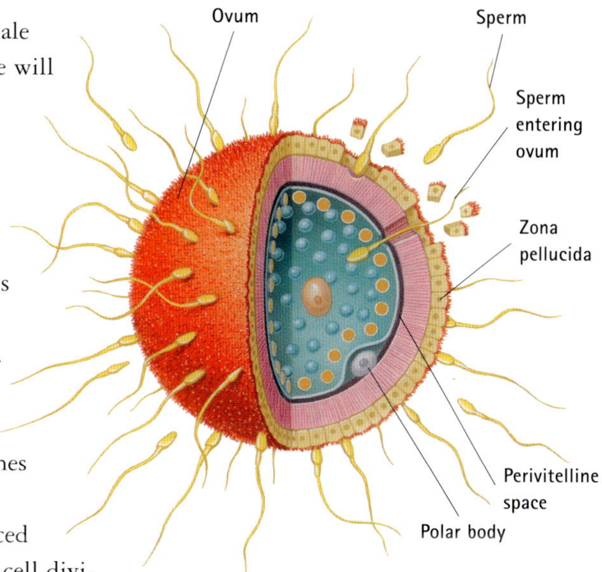

Zygote

A zygote is the cell created when a sperm enters and fertilizes an ovum, before it begins the process of division that will ultimately lead to the development of an embryo.

Zygote

In a normal pregnancy, the ovum (egg) will be fertilized in the fallopian tube within 48 hours of intercourse. This fertilized ovum, known as the zygote, begins to divide rapidly, first into two cells, then four, then eight, as it journeys down the fallopian tube, becoming a cluster of cells known as the blastocyst. About six days after fertilization the blastocyst implants itself into the uterine wall, where it will later develop into the embryo.

DNA

Deoxyribonucleic acid, commonly known as DNA, is a code for life found in almost every living organism. It is found in strands known as chromosomes in the cell nucleus. Each chromosome contains genes, that are blueprints of genetic information and are made up of segments of DNA which also contain the blueprints for making proteins and for replicating itself.

The structure of DNA was discovered in 1953 by Francis Crick and James Watson, who were awarded the Nobel prize for medicine in 1962 for their discovery. The DNA molecule is a double helix, which resembles a spiral ladder. The sides of the ladder consist of alternating units of phosphate and a sugar, deoxyribose.

Attached to the sugar units are the rungs of the ladder, which are made up of a combinations of bases. There are four bases: adenine, cytosine, guanine and thymine (A, C, G, and T). Each rung in the ladder consists of two bases. Because of chemical attractions, only a few combinations of bases are possible: A–T, T–A, C–G, or G–C. Lengthwise up and down the ladder, the bases form different patterns, for example ATCGAT.

Three of these bases join together to form a codon which encodes a single amino acid of a protein. The order of the bases in one strand (half) of the ladder determines the order of the bases in the other strand. For example, if the bases present in one strand are ATCGAT, the bases in the opposite strand would be TAGCTA.

Before a cell divides, the DNA duplicates. The ladder splits lengthwise, separating the bases of each strand. Then, with the help of special enzymes, the bases in each half ladder pick up their matching mates. The As attach to Ts, the Ts to As, the Gs to Cs, and the Cs to Gs. In this way, each new ladder becomes a duplicate of the original ladder. When the cell divides, the two new cells have identical DNA molecules.

DNA also determines the proteins a cell makes. It does this by encoding a messenger ribonucleic acid (mRNA) with information needed to make proteins in "cell factories"

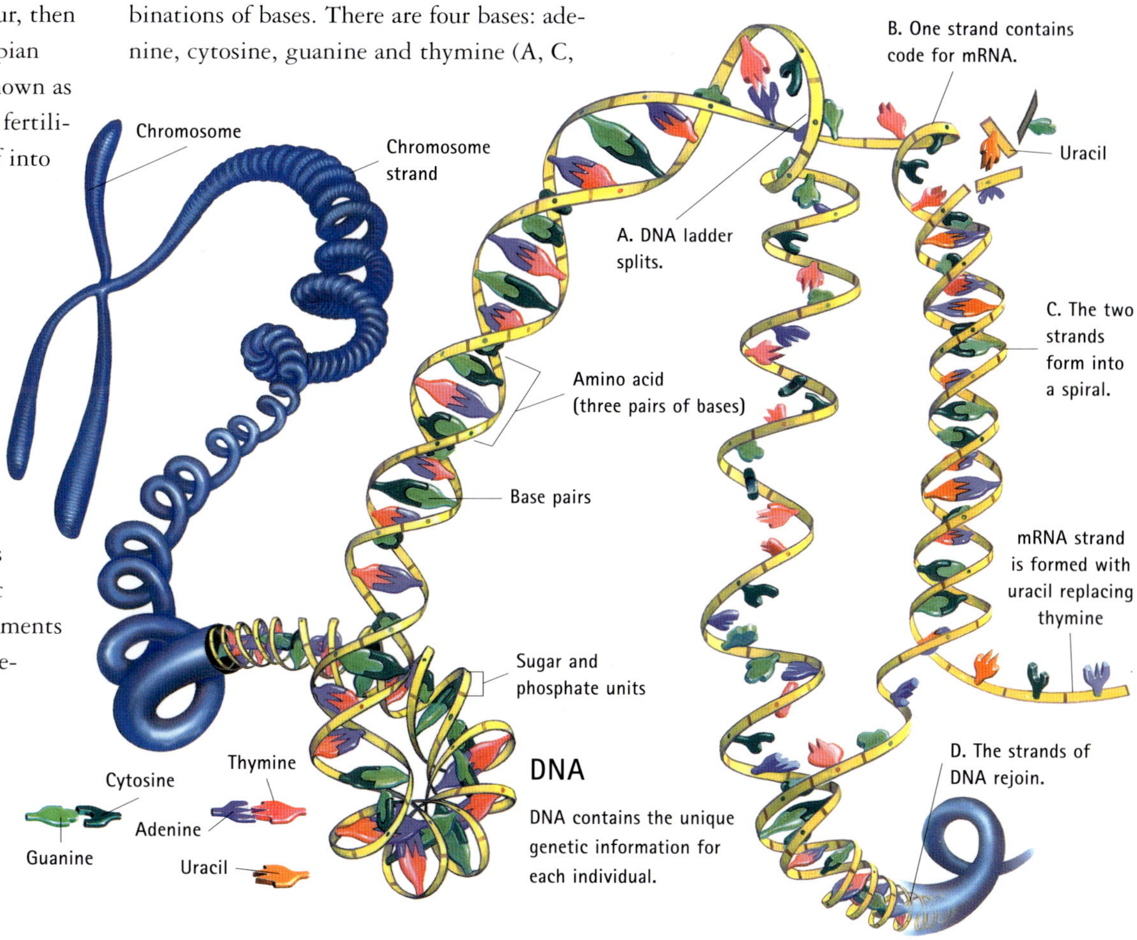

DNA

DNA contains the unique genetic information for each individual.

called ribosomes in the cytoplasm of the cell. The amino acid structure of each protein made by a ribosome corresponds to a particular sequence of bases in the DNA. If there is a mistake made during DNA replication and a sequence is altered (known as a mutation) the composition of a protein may also be changed. The result may be a genetic disorder. There are an estimated 4,200 diseases caused by genetic defects.

The Human Genome Project has mapped the entire genetic code of DNA. This amazing scientific achievement holds out the promise of an understanding of, and possibly a cure for, inherited disorders.

Chromosomes

The nucleus of each cell, except the sex cells, contains 46 chromosomes.

Genes

Genes are the units of genetic information, passed from parent to offspring and found on chromosomes in the nucleus of each cell. Humans have 23 pairs of chromosomes, and at conception each parent contributes one chromosome of each pair. These chromosomes are then copied into each cell in the body.

Chromosomes consist of deoxyribonucleic acid (DNA), with each gene being a section of DNA that instructs the cell how to make a particular protein. The instruction is contained in the sequence of nucleotide bases of the DNA, which codes the sequence of amino acids in the protein.

Humans have approximately 100,000 genes, with each of us having different combinations giving us our unique characteristics. Not all genes are active at any one time; gene expression can be inhibited or induced, depending on the function of the cell and the body's needs.

Because each person has a unique genetic makeup, DNA analysis ("DNA fingerprinting") can be used to for identifying individuals, as in forensic medicine. Since we inherit half our genetic makeup from each parent, the DNA of close family members contains more similarities than that of unrelated people.

Chromosomes

A chromosome is a thread-like structure in the nucleus of a cell. Every chromosome consists of a double strand of deoxyribonucleic acid (DNA), arranged in a helical shape. Each chromosome contains many hundreds of genes. The nucleus of every cell in the normal human body contains 46 chromosomes arranged as 23 pairs. The exceptions are the ova (eggs) and sperm cells, which have only 23 single chromosomes. At fertilization, the two sex cells fuse to form an embryo cell with 23 pairs of chromosomes. One pair of the 23 pairs of chromosomes are the sex chromosomes. In males, one of the two sex chromosomes is shorter and contains fewer genes than the other; this is the Y chromosome. The other longer sex chromosome is the X chromosome. Males have an X and a Y sex chromosome, while females have two X sex chromosomes. The Y chromosome is roughly a third the length of the X chromosome, and apart from its role in determining maleness, is otherwise genetically inactive. This confers a biological advantage on the female, because if a male inherits a recessive gene on the X chromosome, for example the gene causing hemophilia, there is no second X chromosome with a dominant gene to counteract it. By contrast, a female who inherits the recessive disease-causing gene on the X chromosome from one parent, will probably inherit the normal, dominant gene from the other parent, and so probably will not develop the disease.

The condition therefore is usually only seen in men (but is carried by women) and is termed sex-linked. Other examples of sex-linked conditions are red-green color blindness and night blindness.

Chromosomal abnormalities may arise through mutation of chromosomes or may be inherited. Some abnormalities are compatible with life, though usually the affected person has physical or metabolic abnormalities that may be severe. Trisomy 21, in which there are three number 21 chromosomes, not two, causes Down syndrome.

SEE ALSO *Fertilization on page 397*

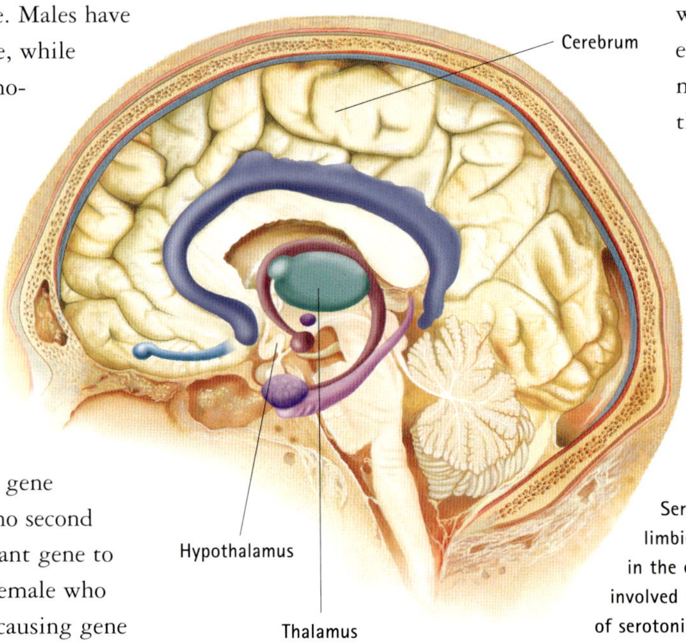

Serotonin

Serotonin is one of the key neurotransmitters in the limbic system, a collection of neural centers and tracts in the cerebrum, thalamus and hypothalamus that are involved in regulating mood and alertness levels. Low levels of serotonin have been linked to depression.

Amino acids

Amino acids are the building blocks of proteins. The body needs proteins in order to build cells and grow, and to maintain its metabolic functions.

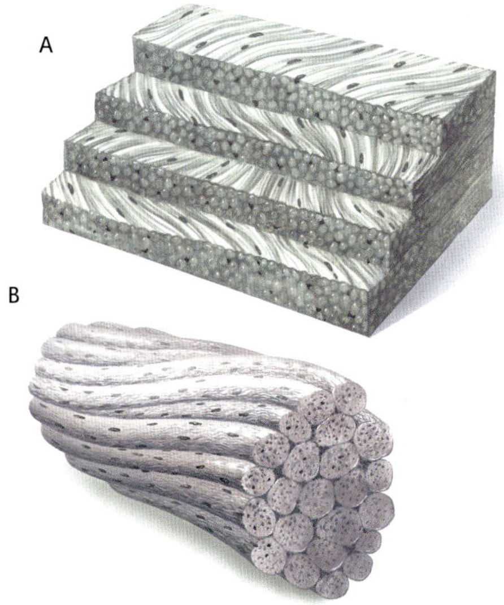

Collagen

Collagen is a tough, flexible protein, found in different arrangements, in structures such as ligaments (a), tendons (b), and the supporting capsules of internal organs (c).

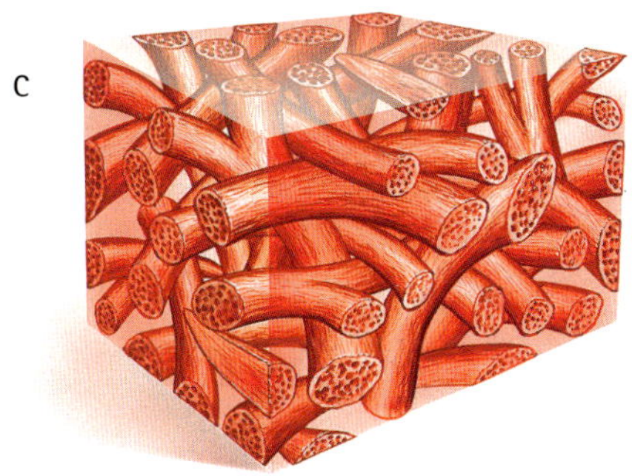

There are two groups of amino acids: non-essential (if we don't include them in our diet the body itself can make them) and essential (they must be included in the diet if we are to stay healthy).

Foods that provide rich sources of amino acids include meat, fish, poultry, egg white, milk, cheese, peas and beans.

Serotonin

Also called 5-hydroxytryptamine or 5-HT, serotonin is a chemical found throughout the plant and animal kingdoms. In the human body it is synthesized from the amino acid tryptophan (found in many foods) and has various roles. It stimulates muscle contractions in the intestine and blood clotting at the site of wounds. Blood vessels constrict when serotonin levels rise and dilate when levels fall.

Migraine pain is thought to be caused partly by blood vessels in the brain dilating due to low serotonin levels.

Serotonin also works in the brain as a neurotransmitter responsible for regulating moods. Low levels may trigger depression. The serotonin-specific reuptake inhibitor group of antidepressants help the brain optimize limited amounts of serotonin.

Collagen

Collagen is an important structural protein in the body. It is made up of chains of amino acids, with glycine, proline and hydroxy-proline being the most common. Collagen is often organized in long parallel bundles of fibers, forming connective tissue, which has a very high tensile strength (for example ligaments and tendons). Collagen may also be formed into sheets (such as mesenteries of the abdominal cavity).

Collagen diseases are a group of diseases (also known as connective tissue diseases), in which there is an attack by the body's immune system on the structural protein of the patient's body. One example is systemic lupus erythematosus.

Lipids

Lipids include fats and cholesterol. They are organic chemical substances of biological origin, and are usually esters (alcohol and acid compounds) of fatty acids. Lipids are characterized by not mixing with water. This gives them important physical properties, especially in the formation of cell membranes and the myelin sheaths of nerves. Lipids are widely distributed in the body, and form an important part of the diet.

CHOLESTEROL

Cholesterol is a fatty substance manufactured in the body and also absorbed from foods. It circulates in the blood in substances called lipoproteins (LDL). A high level of low-density lipoprotein is linked to atherosclerosis—a condition where cholesterol sticks to the inside walls of arteries and restricts blood flow. Arteries can become completely blocked, and where this happens in the arteries inside the heart the person will suffer myocardial infarction (heart attack). If arteries to the brain are blocked, a cerebral infarction (stroke) may result.

High-density lipoprotein (HDL) is believed to help remove cholesterol from the body by carrying it to the liver for processing and excretion.

International guidelines suggest that levels of LDL cholesterol in the blood should be below 200 milligrams per 100 milliliters (deciliter)—but levels differ with personal profile. In a small number of people, high cholesterol levels can be genetically inherited and the family's history plus a full medical examination will allow a doctor to consider all factors. Modifying the diet to exclude foods high in saturated fats can reduce levels of cholesterol in the blood. This means eating less meat, butter and other dairy

Cholesterol and lipoprotein

Cholesterol travels around the body in lipoprotein as free cholesterol and cholesteryl esters.

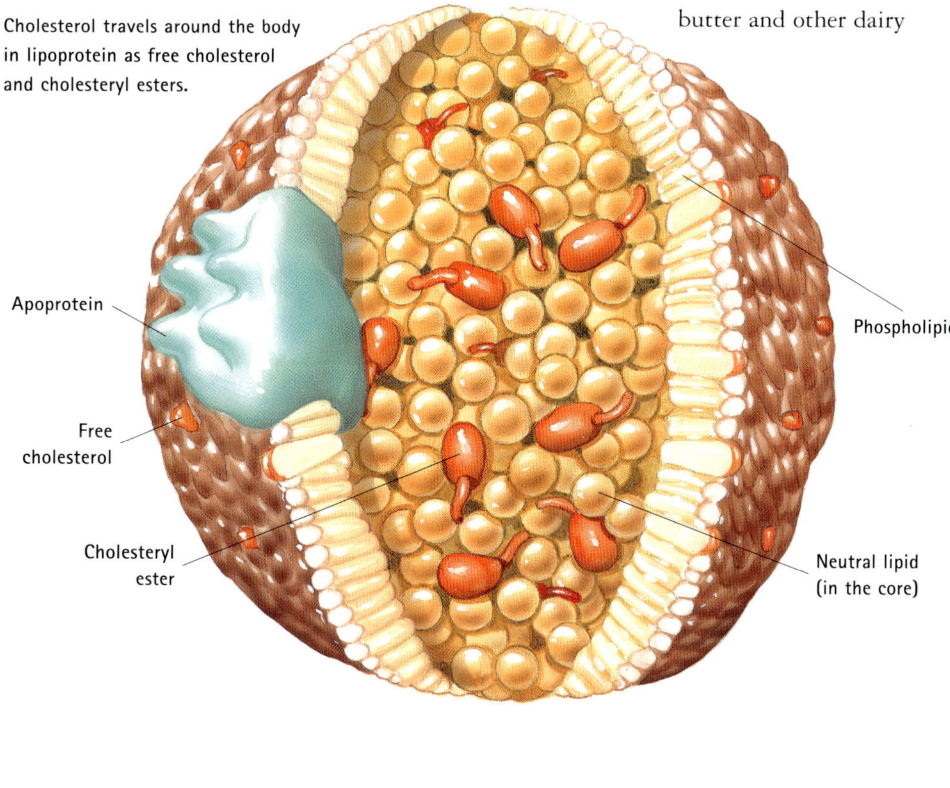

Cartilage tissue

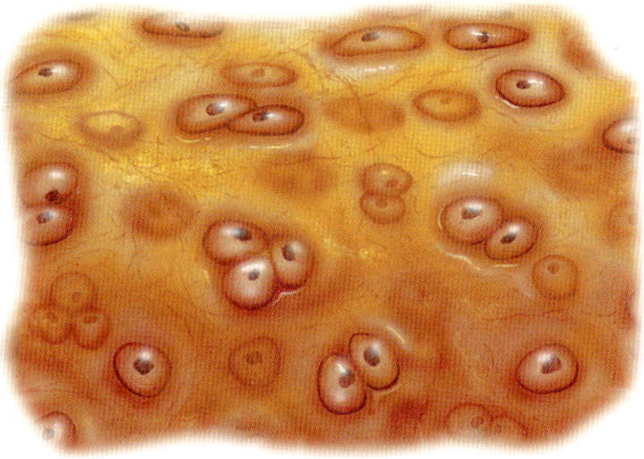

ELASTIC CARTILAGE

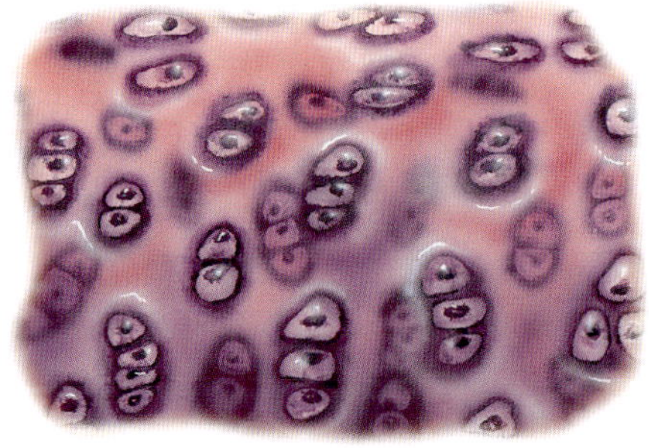

HYALINE CARTILAGE

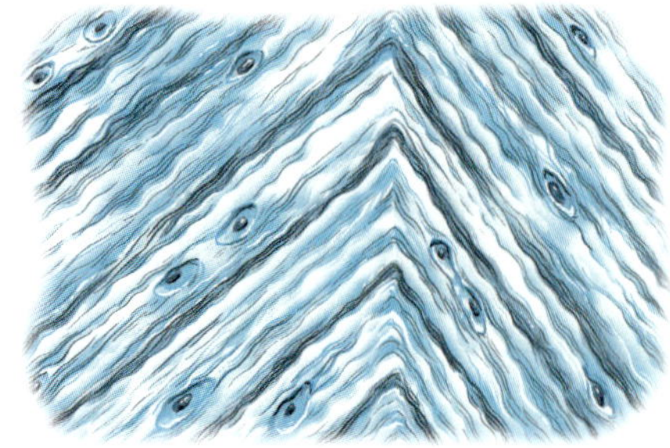

FIBROCARTILAGE

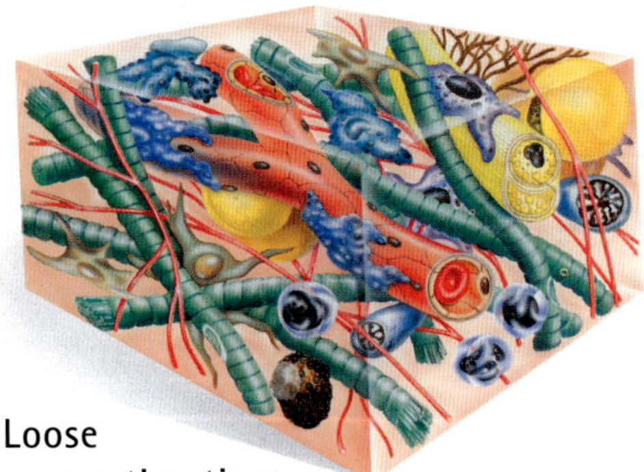

Loose connective tissue

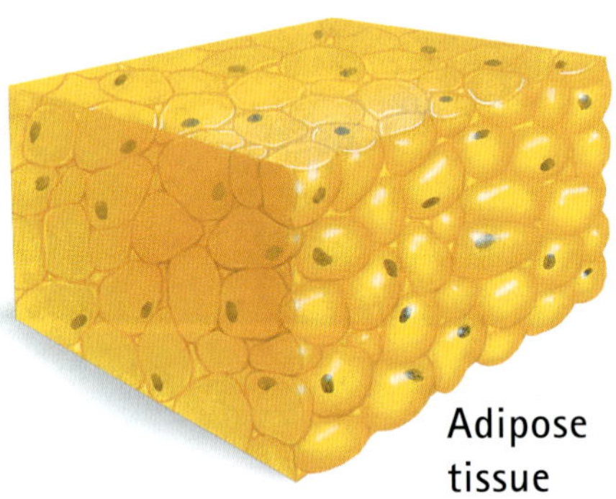

Adipose tissue

Tendon tissue

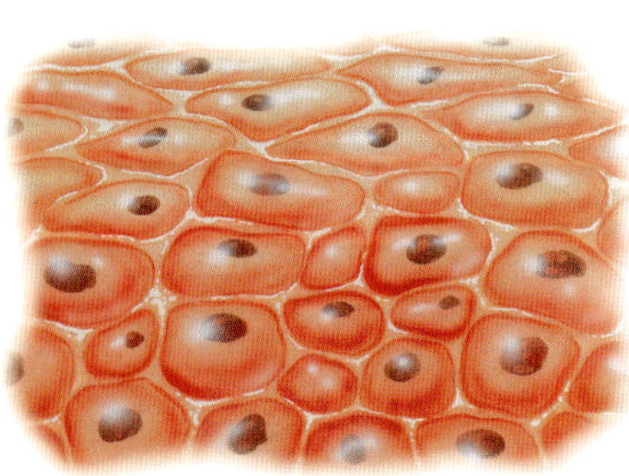

Epithelial tissue

foods and eating more fruit, vegetables and cereals. Other risk factors—such as smoking, high blood pressure, being overweight, and not taking regular exercise—need to be considered, as they further increase the chance of a heart attack.

Tissue

A tissue is a group or layer of cells of a similar kind, plus the material packaged between the cells, all of which function together for the same specialized purpose.

There are four major types of tissue in the human body: epithelial, connective, muscular and nervous.

SEE ALSO *Muscular system on page 48; Nervous system on page 64; Skeletal system on page 30; Structure of the skin on page 346*

Epithelial tissue

Epithelial tissue, also known simply as epithelium, is made up of cells that are packaged so closely together there is virtually no room between them for any extra material. The cells are also arranged in continuous sheets in either single or multiple layers. Epithelial tissue has a wide range of roles including protection, excretion, absorption, sensory reception, secretion and reproduction. Compared to other tissue types, it is often exposed to a high degree of wear but has the capacity to renew itself relatively efficiently and quickly. Epithelium has a nerve supply but no direct blood supply. Vessels in neighboring tissue provide nutrients and remove wastes.

Epithelial tissue is classified broadly into two main types. Glandular epithelium has a secretory function and forms both exocrine and endocrine glands. The second type, covering and lining epithelium, is more widespread. It forms the outermost layer of the skin and of some internal organs. It also lines the digestive, respiratory, reproductive and urinary systems as well as ducts, blood vessels and body cavities.

Covering and lining epithelial tissue comes in either single or multiple layers and contains one of a number of different cell types, depending on the specific function of the tissue.

For example, simple squamous epithelium is a thin single-layered film that lines the air sacs of the lungs and allows the diffusion of respiratory gases (carbon dioxide and oxygen) across its surface. In contrast, stratified squamous epithelium, which has a predominantly protective role, is a multi-layered variation which covers the tongue and forms tough moist surfaces such as the linings of the vagina, esophagus and mouth.

Connective tissue

Connective tissue is widespread in the body. The principal roles of connective tissue are to bind, support or strengthen organs or other tissues. It also functions inside the body to divide and compartmentalize other tissues and organ structures.

Structurally, connective tissue consists of cells linked together and supported by a matrix composed of protein fibers in a medium known as "ground substance." The ground substance can be in fluid, gel or solid form and is normally secreted by the cells of the connective tissue. The protein fibers come in three forms—collagen, elastin and reticular fibers—the proportions of which vary depending on the function of the tissue in which they are found.

Each, however, provides support and strength along with flexibility. The molecules that form the protein fibers and

Muscle tissue

SKELETAL MUSCLE

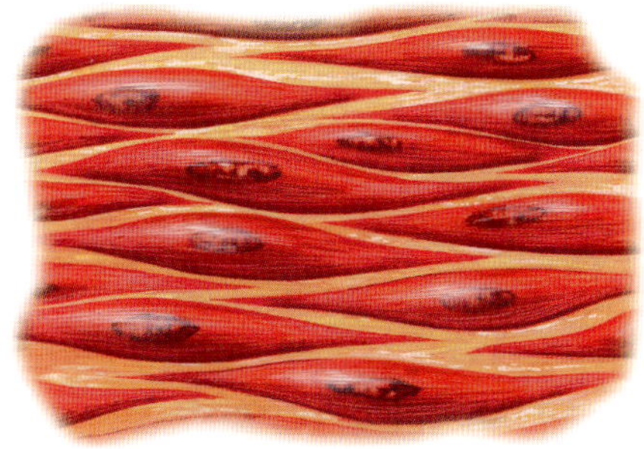
SMOOTH MUSCLE

CARDIAC MUSCLE

ground substance are secreted by specialized connective tissue cells. Other cells commonly found in connective tissue include disease-fighting macrophages, antibody-secreting plasma cells and mast cells, which produce histamine to stimulate the dilation of blood vessels at inflammation sites.

Humans have five main types of connective tissue: loose connective tissue, dense connective tissue, cartilage, bone and blood. Loose connective tissue includes adipose tissue, which is specialized for fat storage. One type of dense connective tissue, known as dense regular connective tissue, forms tendons and cartilage. Another, elastic connective tissue, is specialized for stretching and is found in the lungs, some artery walls and the vocal cords.

Cartilage comes in three forms: hyaline, fibrocartilage and elastic cartilage. Hyaline cartilage is the weakest and the most abundantly found cartilage in the entire body. It covers the bones where they form synovial joints. Fibrocartilage forms a component of some other joints, which usually have a rather limited range of movement. Elastic cartilage is the hard material under the skin, which can be felt in the external ear.

Muscle tissue

Muscle tissue accounts for appoximately 50 percent of total body weight in a healthy person. This tissue is made up of cells that are especially built for contraction. Muscle tissue also equips the body for movement, helps transport substances around the body and produces as much as 85 percent of the heat generated in the body.

Through sustained contractions, muscle tissue also helps stabilize the body's posture and regulates the volume of internal organs such as the bladder.

There are three main forms of muscle tissue: skeletal, cardiac and smooth. The muscle fibers, or myofibers, of skeletal muscle are long, cylindrical, arranged parallel to each other and have a striped appearance under the microscope. This is the sort of muscle tissue that produces movements inside the limbs. It is usually attached to bones and is termed voluntary because one normally has conscious control over it.

Cardiac muscle cells are similar in that they too have a striped appearance. They are, however, branched and operate outside of conscious control. For this reason, cardiac muscle is known as involuntary. This is the muscle found in the heart; it cannot regenerate after being destroyed.

Smooth muscle tissue is also involuntary but it has no striations. This is the type of muscle found in the walls of blood vessels, in airways and inside the eye. Compared to other muscle tissue types, it has quite good powers of regeneration.

Nervous tissue

Nervous tissue functions to sense changes both inside and outside the body, to analyze and interpret these sensory stimuli and to initiate a response. Nervous tissue forms the nervous system, which comprises the central nervous system (CNS) and the peripheral nervous system (PNS).

The CNS includes the brain and spinal cord and is responsible for sorting and responding to stimuli, generating thoughts and emotions and forming and storing memories. All other nervous tissue in the body is included in the PNS, which collects sensory stimuli and transports it into the CNS, and then transports messages controlling the body's response out to the muscles, glands and sense organs.

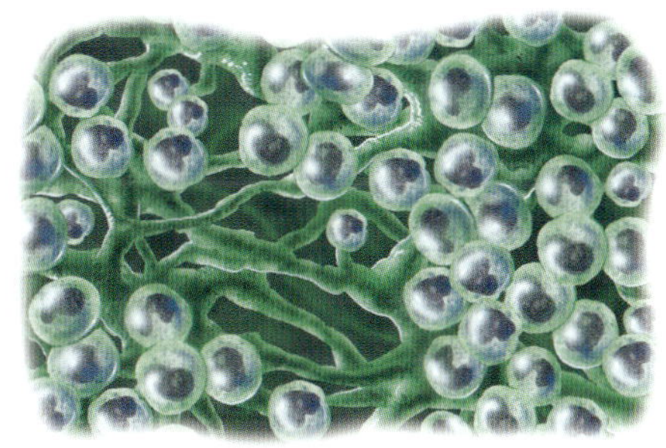
Lymphatic tissue

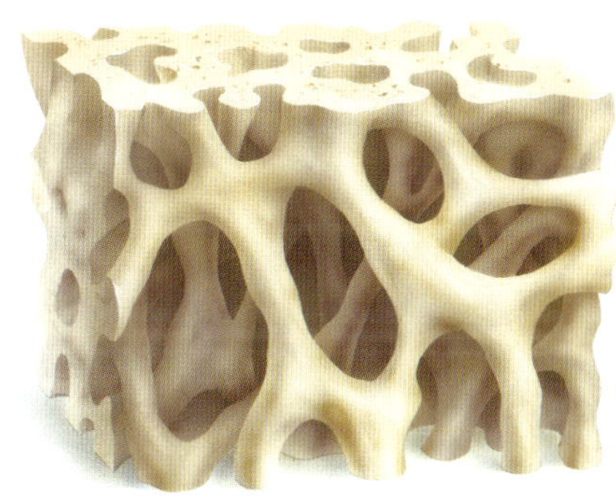
Bone

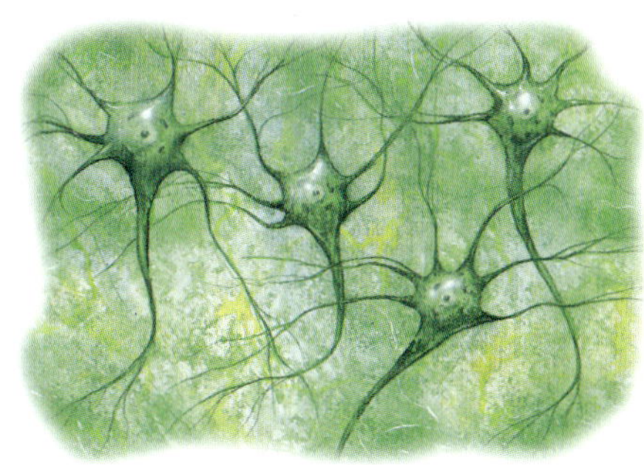
Neural tissue

Ligament tissue

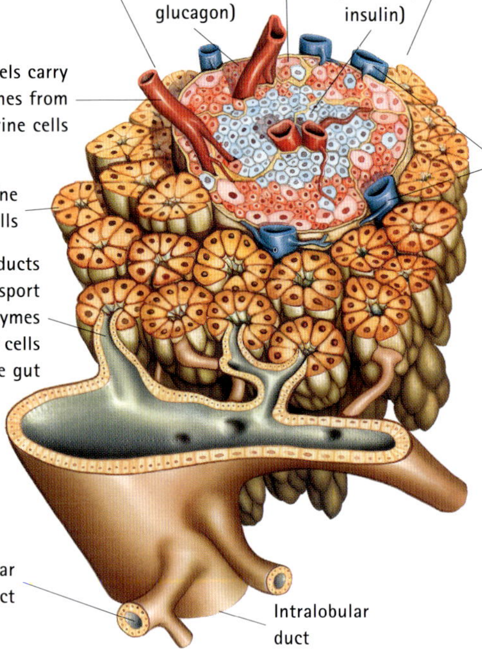

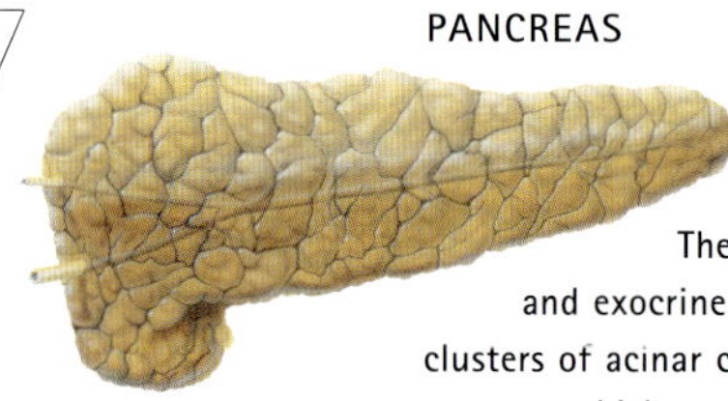

Exocrine and endocrine function—the pancreas

The pancreas is the only glandular organ that has endocrine and exocrine parts. Most of the organ is an exocrine gland comprising clusters of acinar cells with attached ducts. These cells produce digestive enzymes which are carried by the ducts to the gut. The endocrine cells are scattered throughout the pancreas. They are located in islets and secrete the hormones glucagon, insulin and somatostatin. As endocrine glands do not have ducts, a network of blood vessels transports the hormones out of the pancreas.

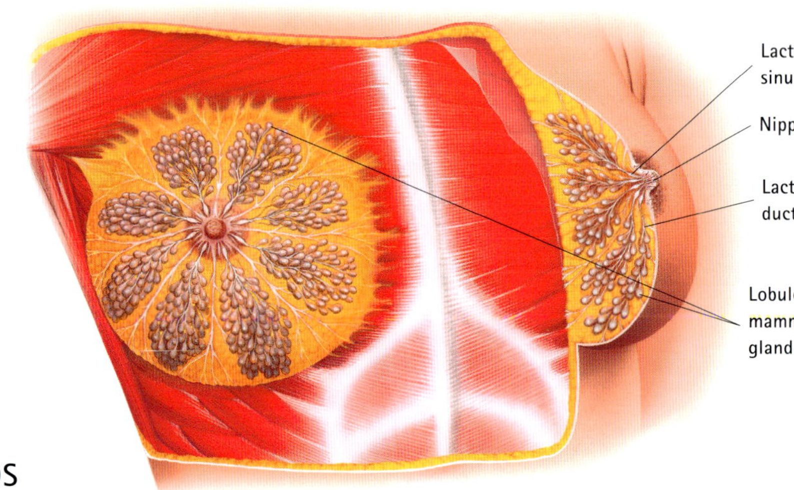

Exocrine glands

These glands drain their secretions to the surface of the body or into the gut, lungs or reproductive organs. They are comprised of secretory cells and ducts that carry the secretion to the body or gut surface. They include sweat and mammary glands and part of the pancreas.

GLANDS

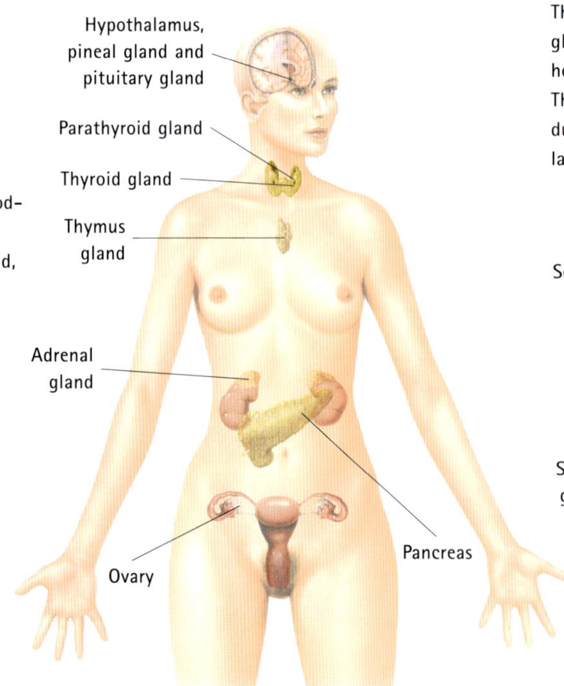

Mammary glands (exocrine)

The mammary glands (breasts) are modified sweat glands. They produce milk when stimulated by the hormone oxytocin, secreted by the pituitary gland. The mammary lobules secrete into the lactiferous ducts, which carry milk through a wider tube (the lactiferous sinus) to the nipple.

Endocrine glands

These glands secrete hormones directly into the bloodstream or into tissue spaces for circulation to other places in the body. They include the pituitary, thyroid, parathyroid, adrenal and pineal glands and part of the pancreas.

Pituitary gland (endocrine)

The pituitary is the main endocrine gland.

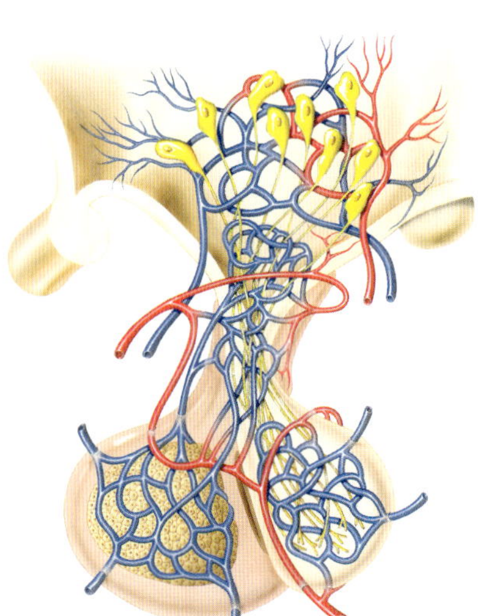

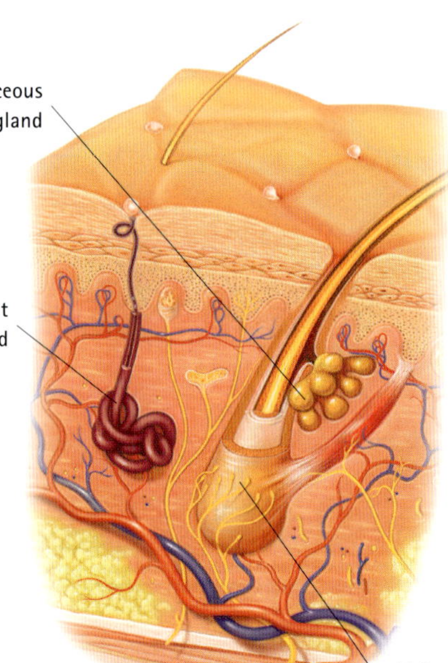

The skin (exocrine)

The skin contains two types of exocrine glands—the sweat and sebaceous glands—that secrete in different ways. Sweat glands release perspiration to the surface of the skin without losing any cellular material. The sebaceous glands which are mostly attached to hair follicles release an oily substance called sebum which moisturizes the skin and hair.

Sweat and sebaceous glands (exocrine)

The sweat and sebaceous glands of the skin are two examples of exocrine glands.

Nervous tissue is composed of two main types of cells: neuroglia and neurons. Impulses travel around the body via the neurons. Types of neuroglia include astrocytes, oligodendrocytes, Schwann cells, microglia, ependymal cells and satellite cells. The glial cells, which are smaller and more abundant than neurons, function to maintain the proper biochemical environment required by neurons, fight invading microbes and produce material that physically supports and protects the neurons. It is thought that, from about the age of 6 months, most human neurons lose the ability to reproduce themselves. As a result, when a neuron is destroyed completely it cannot be replaced.

Some types of damage can, however, be repaired. Under certain conditions, nerve fibers in the PNS can regenerate after injury—but not in the case of the brain or the spinal cord. Damage to nervous tissue in these locations usually has some sort of permanent impact.

Membrane

Membranes are thin and flexible sheets of tissue that line surfaces or divide spaces in the body. Three out of the four main types of membrane—mucous, serous and cutaneous—are composed of an epithelial cell layer overlaying a connective tissue layer. In synovial membranes the epithelial layer is modified as a mesothelial layer.

Mucous membranes form the lining of the hollow internal cavities of the body, notably the nose and mouth, the larger respiratory passages and the gut. A mucous membrane (mucosa) consists of several layers of cells, and is kept moist by secretions (mainly mucus) from glands found in or underneath the mucous membrane.

The cells on the surface are adapted to the functions of that particular organ. For example, the surface of the mucous membrane of the mouth has many layers of flattened cells which give some resistance to abrasion from food, whereas that of the respiratory passages carries hair-like cilia to sweep foreign particles away.

SEE ALSO *Muscular system on page 48; Nervous system on page 64; Respiratory system on page 95; Skeletal system on page 30; Structure of the skin on page 346*

Glands

Glands are a type of tissue that is made up of cells specialized for the production of a fluid secretion. These special secretions may have mineral salts, protein, fats or complexes of carbohydrates and proteins.

There are two types of glands: exocrine glands, which drain their secretions on to the surface of the body or into the interior of the gut, lungs or reproductive organs; and endocrine glands, which release their secretions directly into the blood stream or tissue spaces for circulation to some other places in the body.

Exocrine glands have a secretory part, which makes the secretory product, and a tubular duct, which carries the secretion to the body or the gut surface. It is the presence of the duct that distinguishes these glands from endocrine glands. In some types of exocrine glands, the cells release their secretion without any loss of cellular material, while in other types the entire cell becomes filled with the secretory product and is completely shed with the secretion.

Examples of exocrine glands include the sweat and sebaceous glands of the skin. Sweat glands are responsible for helping in the control of body temperature, while the fats of sebaceous gland secretions (sebum) prevent water loss from the skin and help to control the growth of microorganisms like bacteria and fungi. Mammary glands, found only in mammals, are a modified type of sweat gland. They produce milk, which is rich in many secretory products (protein, fats, carbohydrates, immune system proteins and vitamins).

Other types of exocrine glands include: the salivary glands of the oral cavity; the exocrine part of the pancreas, which produces digestive enzymes; the mucus-secreting cells of the lining of the gut and respiratory tract; enzyme secreting cells of the stomach and small intestine; and secretory cells lining the uterus. The pancreas is unique in that it contains both exocrine and endocrine glands.

Endocrine glands, or ductless glands, produce hormones which travel to all tissues of the body via the bloodstream. The main endocrine gland is the pituitary, located just below the brain and receiving commands from the hypothalamus in the brain by special nerve pathways and blood channels. The pituitary gland produces a number of important hormones with diverse functions. These include growth hormone (GH), which controls the growth of bones by affecting the cartilage at growth plates; thyroid stimulating hormone (TSH), which controls the production of thyroid hormones by the thyroid gland; follicle stimulating hormone (FSH); luteinizing hormone (LH); adrenocorticotropic hormone (ACTH); and prolactin.

Other endocrine glands are scattered throughout the body (such as the adrenal glands, thyroid gland, parathyroid glands, ovaries and testes). The adrenal gland is divided into two parts: an outer cortex and an inner medulla. The outer cortex produces corticosteroids, which control carbohydrate, lipid and protein metabolism, mineralocorticoids, which control the resorption of sodium in the kidney, and sex steroids, which play a minor role under normal conditions. The adrenal medulla produces epinephrine and norepinephrine (adrenaline and noradrenaline), released in times of acute physical or emotional stress.

The parathyroid gland produces a hormone that controls blood calcium. The ovaries and testes produce sex hormones (estrogen and testosterone), responsible for the bodily changes associated with sexual maturity (for example, growth of pubic hair, development of sexual organs and breasts). The placenta is also an endocrine gland, secreting chorionic gonadotropin.

The thymus is often called a gland, but is more accurately described as a lymphoid organ. During early childhood it produces a type of white blood cell known as the "T" lymphocyte. These cells are later distributed throughout the body, going to lymph glands (lymph nodes), the spleen and the gut wall.

The lymph glands are positioned along the lymphatic vessels—on both sides of the neck, in both armpits and on both sides of the groin, as well as the internal body cavities. They act as filters, inhibiting the spread of infection.

During a reaction to infection, the cells in a lymph gland multiply, and the gland becomes large and painful.

SEE ALSO *Adrenal glands on page 280; Breasts on page 222; Endocrine system on page*

110; Hypothalamus on page 136; Lymphatic/ immune system on page 55; Ovaries on page 308; Pancreas on page 277; Parathyroid glands on page 206; Pituitary on page 137; Sebaceous glands on page 348; Sweat glands on page 347; Testes on page 298; Thymus on page 251; Thyroid gland on page 205

HEALING

Healing is the restoration of structure and function of damaged tissues. Different processes are involved depending in the nature and extent of the injury, and the type of body tissue involved.

SEE ALSO *Diagnosis and treatment on page 430; Medication on page 438*

Inflammation

Inflammation is the body's natural response to tissue damage. It takes place after infections, burns, frostbite, or radiation exposure.

The process begins with dilation of blood vessels and increased blood flow to the area. The vessels become more permeable, allowing plasma to escape from the blood into the extracellular fluid. This produces swelling of the affected region. Leukocytes (white blood cells) also escape from blood vessels in the region, and release chemicals that may cause pain. These changes produce the classic signs of inflammation: heat, redness, swelling, and pain and tenderness.

As the inflammation progresses, leukocytes migrate in increased numbers from the bloodstream into the injured area. The inflammatory response proceeds until the dead tissue and invading organisms have been removed. The inflammation then resolves—the excess fluid is drained from the area by the lymphatics, the blood vessels constrict and become less permeable, and the swelling, heat and redness subside.

Scar

The growth of granulation tissue during wound healing may leave a mark or scar on the skin.

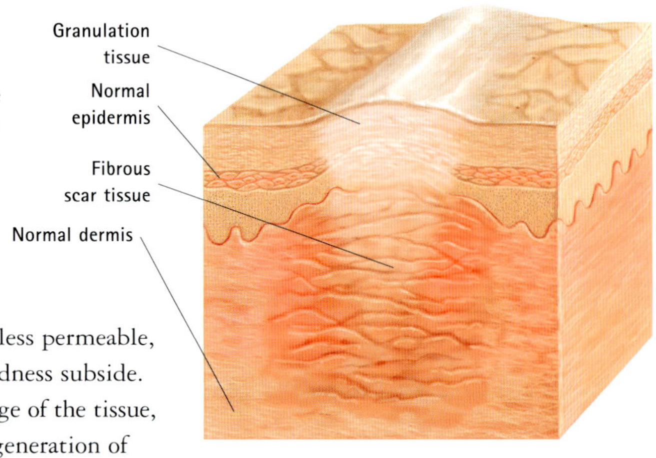

If there has also been damage of the tissue, then healing also involves regeneration of tissue—that is, the repeated division of surviving tissue cells to take the place of those that have been damaged. Not all tissue can regenerate (e.g. brain and nerve cells). However, if the damage is extensive, it may not be possible for regeneration to occur.

Surgical intervention sometimes may be necessary to aid the healing process. In some cases, dead bacteria, cells and tissue may collect to form pus, which may collect and form an abscess. This may need to be drained. If there are large areas of dead tissue, the dead tissue needs to be removed.

Wound healing

If an injury causes a wound such as a laceration, then healing takes place by a different mechanism. If the edges of the wound are close together, wound healing takes place by "primary intention." A blood clot forms in the wound which contracts, bringing the edges still closer together. From the edges, fibroblasts produce granulation tissue, which is gradually replaced with connective tissue. Meanwhile, epithelial tissues grow over the surface of the wound. If the edges of the wound are far apart, healing takes place by "secondary intention." Granulation tissue forms at the base and sides of the wound and "fills up" the wound until it reaches the level of the skin. The granulation tissue is gradually replaced with scar tissue.

Bone healing

Healing of bone is like that of other injured tissue, but uses special cells and materials. Immediately after a fracture, bone forming cells called osteoblasts begin to produce a tough binding material called callus, which knits the bones together. The bone is then remodeled as the callus is absorbed and replaced by true bone, which is gradually remodeled into its previous shape.

The process of bone healing is greatly helped if displacement of the ends of the fractured bone is minimal.

Several general factors can influence the healing process. The younger the person is, the faster healing takes place. Someone who is in good general health will heal faster than someone in poor general health or who is malnourished.

Sufferers from liver or kidney disease or diabetes do not heal as well as healthy individuals. Someone who is immunosuppressed or on cytotoxic, immunosuppressant or corticosteroid drugs may take longer to heal. Poor blood supply to damaged tissues also slows or prevents healing.

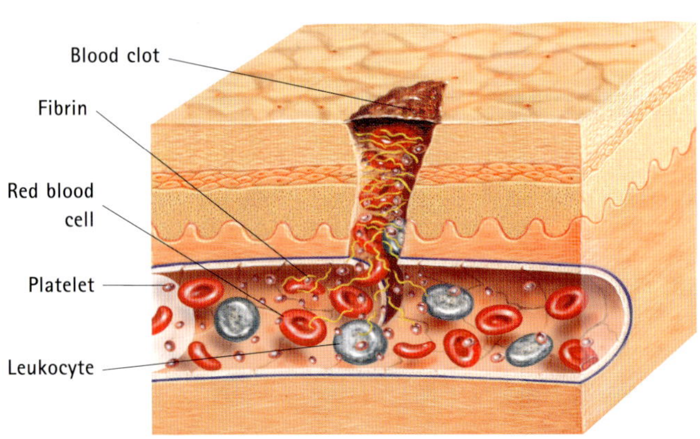

Healing

For a wound to heal, leukocytes (white blood cells) must first migrate from blood vessels to the injured tissues in order to remove dead cells and invading organisms. A blood clot comprising fibrin strands, platelets and red and white blood cells then forms to bring the edges of the wound together. Finally, fibroblasts produce new granulation tissue.

Fibrosis

Fibrosis is the formation of fibrous scar tissue. It normally follows after an infection, injury and inflammation. Too much scarring can cause a disorder, such as adhesions in the peritoneum following peritonitis, or keloid tissue, an overgrowth of scar tissue at the site of a skin injury. Scarring can also be a result of chronic inflammatory diseases such as interstitial lung disorders (a group of disorders characterized by scarring and thickening of the deep lung tissues) and hepatitis, which can cause cirrhosis, or scarring of the liver.

Scar

A scar, or cicatrix, is toughened fibrous tissue that develops while a burn, wound or surgical incision heals. A scar helps knit the wound together. Inside the body, scars rarely cause problems. People are more commonly concerned about scars on the skin. The extent to which an injury will leave a scar depends on where it occurs on the body, its size and depth. Scarring will also vary with a person's age, genetic predisposition to scarring, and skin characteristics.

There is a range of various treatments available to minimize scarring, for example dermabrasion, during which the surface of the skin is frozen with an aerosol spray and then abraded by mechanical means.

Diseases and Disorders of Cells and Tissues

Disorders of cells and tissues include cellular injury and death (due to hypoxia or ischemial infection) and abnormal growth (benign tumors and cancer). Generalized cellular damage may result from a number of disorders including radiation sickness and decompression sickness.

SEE ALSO *Genes on page 18*

Sclerosis

Sclerosis is the hardening or thickening of body tissue. Possible causes range from inflammation to the deposition of mineral salts to scarring. It is usually an abnormal and undesirable condition, often associated with disease. For example, the incurable and chronic illness known as progressive systemic sclerosis (also called scleroderma) is characterized by a thickening of the connective tissue causing debilitating changes to the skin, blood vessels and internal organs. In arteriosclerosis, the walls of the arteries thicken, calcify and lose their elasticity.

Abscess

An abscess is a collection of pus. The pus is made up of dead white blood cells, destroyed tissue and cells, and dead and live microorganisms (usually bacteria), which are all byproducts of inflammation and infection. It can form in an internal organ such as the large intestine, lung, liver or brain, when it is often the result of another debilitating disease, for example AIDS. Most abscesses are bacterial or fungal in origin, although they are sometimes caused by ameba (especially in the liver) or by the tuberculosis bacillus. These are carried to the internal organs through the bloodstream and once there can often cause inflammation and infection.

In healthy people abscesses more often occur in the soft tissues beneath the skin. The invading organism finds its way through the skin via an infected wound or bite, or the abscess may begin in a hair follicle, when it is known as a boil. Boils are most common in hairy sites such as the nostrils, the armpits, the back of the neck and between the legs and buttocks. An abscess below the skin is usually very painful, looks swollen and red, and feels hot to the touch. An abscess may also cause fever, sweating, tachycardia (rapid pulse rate) and malaise.

Antibiotics are often given, usually intravenously, although their usefulness is limited, as drugs do not readily penetrate past the abscess lining into the abscess.

The mainstay of treatment is surgical drainage. This involves making a cut into the abscess and providing a drainage route either through a drainage tube or by leaving the abscess open to the skin (although an abscess will often burst through the skin by itself). A dressing is then applied daily until the wound and infection have healed.

Hypoplasia

Incomplete or defective development of a body organ or tissue is known as hypoplasia. Potential causes are many and varied and may depend on the particular part of the body affected.

Cartilage-hair hypoplasia, in which the cartilage is affected leading to bone abnormalities, has a genetic cause. This condition is characterized by dwarfism. Enamel hypoplasia, which affects the teeth, can also be a genetic disorder. Pulmonary hypoplasia can occur when other organs have compressed the lungs during their development in the uterus. The cause of optic nerve hypoplasia remains unclear but it is sometimes blamed on substances taken during pregnancy. Sufferers of this disorder are missing 10–90 percent of the 1.2 million nerve fibers usually found in an optic nerve, impairing vision in the affected eye.

Hypoxia

Hypoxia is a shortage of oxygen in cells and tissues. There are many potential causes. It can result from a variety of disorders or substances, such as carbon dioxide or carbon monoxide, which reduce the blood's ability to transport and circulate oxygen throughout the body. It can be caused by environmental factors,

SKIN ABSCESS

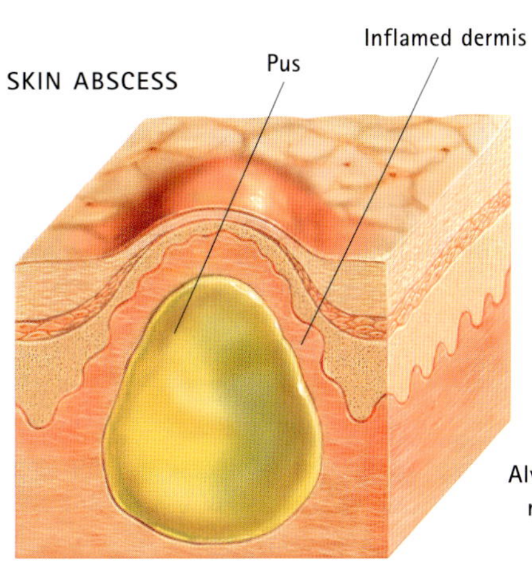

Abscess

An abscess is an infection that contains pus and causes inflammation of the tissues. Abscesses are common on the skin and teeth but also occur on internal organs such as the liver, lung or brain. In teeth, the pulp and nerves break down in the root canal. Treatment involves antibiotics and proceeding with root canal therapy.

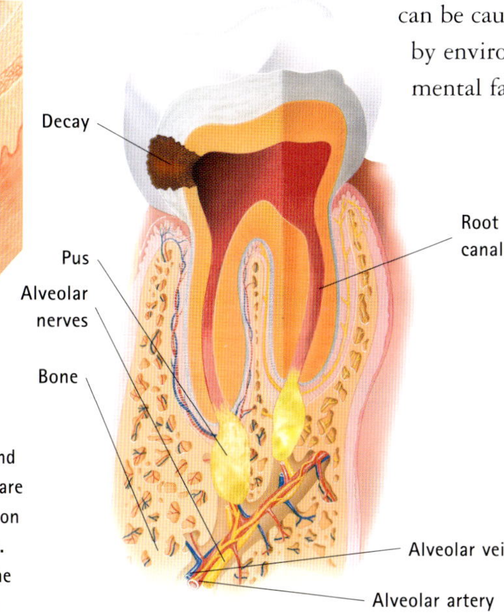

TOOTH ABSCESS (ADVANCED)

particularly high altitudes. Or it may be due to a disease or injury that affects a tissue's ability to use oxygen. Hypoxia can be isolated to a particular organ or it may be widespread in the body. The tissues that are most sensitive to reduced oxygen levels are the heart, brain, liver and blood vessels to the lungs. In severe or protracted episodes of hypoxia, permanent tissue damage due to cell death can occur.

The onset of the condition can often appear without warning. Early symptoms may include a rise in heart and respiration rates in response to falling oxygen levels in the blood. Another early symptom can be a sense of well-being—like an alcohol-induced high. Dizziness and mental confusion may follow. Treatment for hypoxia may require mechanical ventilation, drugs to stimulate respiration, or oxygen therapy.

Pilonidal sinus

Pilonidal sinus (or cyst) is a common disorder that occurs most frequently in hairy young males. The affected person has a minor congenital abnormality, a small skin sac (sinus) at the base of the spine in the cleft between the buttocks. It causes no problems, unless it is infected (pilonidal abscess), which usually causes pain and swelling and a discharge of pus.

The condition is treated with antibiotics, surgical drainage, and, when the infection has subsided, surgical removal of the sinus.

Tumors

A tumor (or neoplasm) is an abnormal growth or swelling of tissue. It may be cancerous (malignant) or noncancerous (benign). A malignant tumor is a neoplasm that grows and spreads throughout the body.

Malignant tumors can spread by extension into surrounding tissues, or beyond. They may spread to nearby lymph nodes via the lymphatic vessels, or to distant sites via blood vessels.

A benign tumor is a neoplasm that, if it grows, does so slowly, and does not spread or infiltrate other tissues of the body. The cells in benign tumors are similar under a microscope to the cells in the tissue they grow from. Cells in malignant tumors may look quite different. Not all cancers are tumors—cancers of the blood cells such as leukemia, for example, do not form growths.

Malignant tumors are more likely to be fatal (though in some circumstances benign tumors can cause death where their expansion destroys surrounding tissue, for example in the brain). But some malignant tumors, such as some skin cancers, are treatable, especially if diagnosed early. Others grow so slowly that the affected person may die of some other condition—this is true of many men with prostate cancer. Much depends on the site of the tumor—cancer of the colon, for example, may be slow growing but is often fatal because it only causes symptoms at a late stage and is often not detected until it has grown and spread. Some benign tumors may eventually turn malignant and need to be treated.

The aim of treatment of a tumor is to destroy as much of it as possible without destroying normal tissue. This can be achieved by surgery, chemotherapy, radiation therapy or combinations of all of these.

ADENOMA

An adenoma is a benign tumor of glandular tissue. Adenomas can occur in specialized glandular tissues such as the thyroid, pancreas or pituitary, or in organs and tissues that contain glandular tissue. They are common in the breast, ovaries and uterus, where glandular tissues are stimulated by hormones throughout a woman's reproductive life, and can also occur in the colon. As long as it remains benign, an adenoma does not pose any problems and needs no treatment. In some cases an adenoma can become cancerous. It is then known as an adenocarcinoma, and surgical removal is advised.

Adenomas of endocrine glands (those which produce hormones, like the pituitary gland, thyroid gland, adrenal glands and pancreas) can cause excessive hormone production, leading to disease. Pituitary adenomas, for example, can result in acromegaly or Cushing's syndrome. If two or more different endocrine glands are involved, the condition is called adenomatosis.

CYST

A cyst is an abnormal swelling that is sac-like in structure, with an outer wall of cells or fibrous tissue enclosing liquid or semi-liquid material. Cysts may contain a range of substances including blood, fat, pus or parasites. They are very common and can occur virtually anywhere in the body but are most often noticed in the skin, ovaries, breasts or kidneys.

Although usually benign, cysts can sometimes create complications. For example, symptoms caused by ovarian cysts (fluid-filled lumps on the ovaries) can include abdominal pain, abnormally heavy and irregular menstrual periods and increased growth of facial and body hair. If ovarian cysts rupture, they can cause severe pain, nausea, vomiting and shock. These cysts are usually benign in women during their reproductive years, but frequently are found to be malignant when they occur in young girls or post-menopausal women.

Hydatid cysts are formed by the larvae of the dog tapeworm. In humans they occur most often in the liver and tend to be more common where people have close associations with dogs and sheep. They can survive

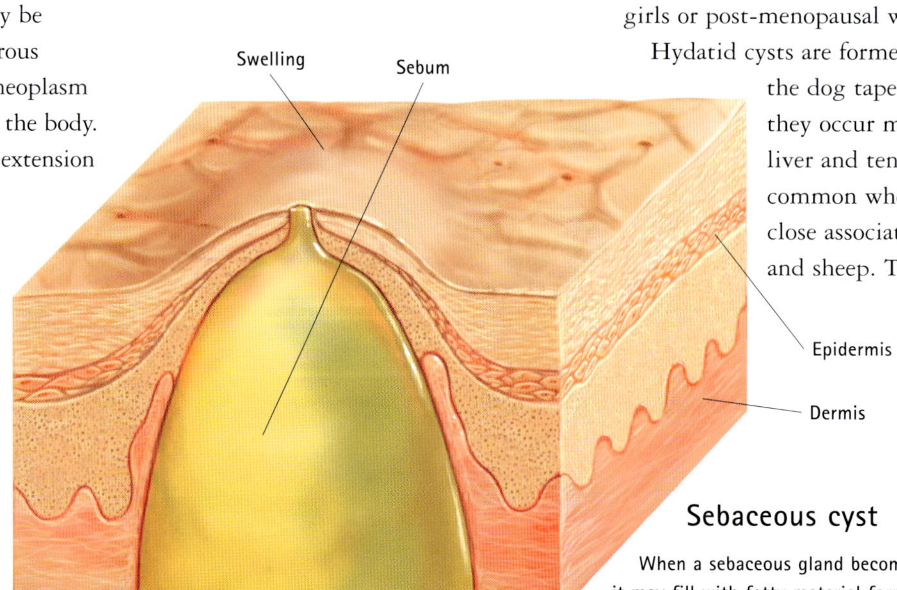

Sebaceous cyst

When a sebaceous gland becomes blocked it may fill with fatty material forming a cyst.

for many years growing very slowly without major consequences, but they have the potential to cause serious illness. They are usually treated with potent medications or removed surgically.

ANGIOMA

An angioma is an abnormal, though benign (non-cancerous), growth of small blood vessels. It may occur internally, or as a soft, reddish or purple mark on the skin. Cherry angioma, spider angioma and strawberry nevus are common examples. Treatment may not be necessary if the disfigurement can be concealed by cosmetics. If treatment is needed, a small angioma in the skin may be burned or cauterized, but larger ones usually need plastic surgery.

An angioma in the brain may bleed, causing subarachnoid hemorrhage or even a stroke. An angioma that bleeds in the digestive tract may cause anemia, black stools (melena), or vomiting of blood.

Ulcer

An ulcer is a region where the surface layer of the skin, or the lining of an internal organ such as the gut or airway, has been lost. There is often loss of underlying tissue, so that the region has a punched-out appearance, as if a cookie-cutter has been used.

Ulcers may occur in many areas of the body. On the skin, decubitus ulcers (bedsores) can occur following prolonged and unrelieved pressure on the skin, sometimes associated with poor circulation. Some skin cancers may develop ulcers in their central regions, as seen with rodent ulcers (basal cell carcinomas). Varicose ulcers develop around the ankles, due to problems with drainage of venous blood from the skin of the leg. The cornea of the eye may become ulcerated, with serious consequences for vision. In the mouth, aphthous ulcers may develop due to poor dentition, poor oral hygiene, excessive alcohol intake or cigarette smoking. Peptic ulcers develop in the upper gut, usually after the *Helicobacter pylori* bacterium has weakened and damaged the protective lining of the stomach or duodenum, allowing penetration by corrosive digestive juices. Multiple ulcers may occur in the colon and rectum in the condition known as ulcerative colitis.

Gangrene

This term refers to death of tissue, with secondary growth of bacteria that derive their nutrition by breaking down the dead tissue. In practice, gangrene is often used to refer to death of a large area of tissue (often as a result of loss of blood supply or in wounds infected by anaerobic bacteria), whether or not it has undergone significant bacterial decomposition.

Gangrene occurs most often in the extremities, for example in the foot or toe resulting from blockage of an artery, though it can also involve internal organs, such as with hernias. Fever, pain, darkening of the skin and unpleasant odor are common symptoms. Correcting the initial causes, medication and surgery may be required.

Gas gangrene

In this form of gangrene, dead tissue is invaded by bacteria called *Clostridium welchii*, which break it down and in the process release bubbles of gas into the tissue. Bacteria are frequently found in soil and include a variety of organisms that are able to secrete very powerful toxins. For example, the bacteria causing tetanus and the serious food poisoning known as botulism are members of this group. Despite its name, it is the toxins and the tissue-destroying enzymes released by the bacteria, not the gas bubbles, that make gas gangrene so dangerous.

Progressive breakdown of tissue, damage to blood cells and other effects of the toxins can eventually kill the patient.

Clostridia prefer a low-oxygen environment and therefore grow well in deep wounds with considerable tissue damage, especially if contaminated by soil and dirt. This is why gas gangrene is such a feared complication of wounds during war. It may also develop after injury or after surgery. Attention to wound care is essential to prevent the onset of gas gangrene.

Cancer

Cancer (also called malignancy) is a disease in which normal cells become abnormal and grow uncontrollably, often metastasizing (spreading from the site of origin to other sites). A leading cause of death in many countries, only diseases of the

ULCER

An ulcer forms when small parts of tissues or organs die, leaving painful, inflamed holes.

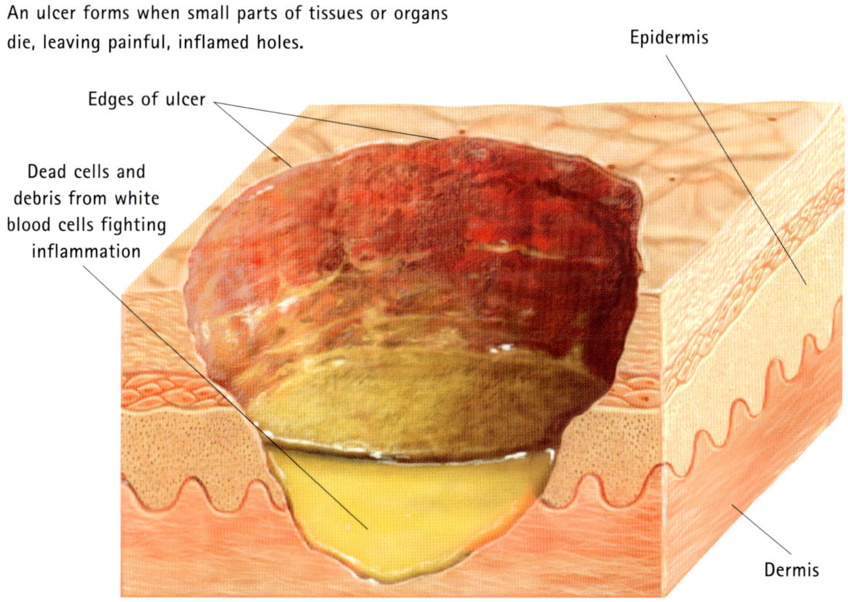

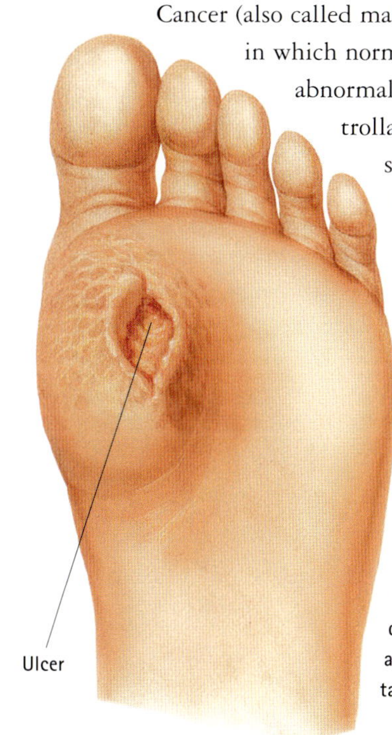

Foot ulcer

Diabetics often have poor circulation to the legs and feet and may develop ulcers in these areas. These ulcers often take a long time to heal.

heart and blood vessels kill more people than cancer in developed countries. It affects people of any age but is more common in middle and old age. There are over a hundred forms of cancer; most are named for the type of cell or the organ in which they arise. Almost any tissue in the body may become malignant, but the skin, the digestive organs, the lungs and female breasts are particularly prone to cancer.

There are three main classifications of cancer. Carcinoma is cancer of the epithelial tissue that forms the skin and the linings of the internal organs. Sarcoma is cancer of connective tissue, such as cartilage, muscle or bone. Cancers of the bloodstream (leukemia) and the lymph system (lymphoma) are a third category.

A cancer is, by definition malignant, that is, capable of spreading beyond its site of origin. (A benign tumor does not spread and is not malignant.)

Cancer spreads by infiltrating the tissue around it, or by distant spread to other parts of the body via blood or lymph vessels, or both. Frequently it spreads to the lymph nodes that drain the tissues in which the tumor has arisen; breast cancer, for example, tends to spread to the lymph nodes in the axilla (armpit). When a cancer spreads beyond the tissue of origin to a distant site and forms a tumor, the new tumor is called a metastasis, (also called a "secondary"), when there is more than one tumor, they are referred to as metastases (or "secondaries"). The cancer cells in a metastasis are the same as in the primary tumor.

A metastasis may also spread across the surface of a body cavity such as the peritoneum (the lining of the abdomen) or the pleura surrounding the lungs. Occasionally, metastases result from surgery and may be found in the scar of the wound through which a tumor has been removed.

Metastases are commonly found in the liver, lung, bone and brain. In such cases, they generally indicate a poorer expectancy or prognosis. Most cancers, if left alone, are fatal.

However, advances in diagnosing and treating cancer have improved greatly over the past few decades. About one-third of all persons who received treatment for cancer now recover fully or live much longer than they would have lived without undergoing treatment.

Skin cancer (including melanoma) is the most prevalent cancer for both men and women in developed countries. The next most common type among men is prostate cancer. Among women, breast cancer has the highest incidence after skin cancer. Lung cancer is the leading cause of death from cancer for both men and women. Brain cancer and leukemia are the two most common cancers affecting children and young adults.

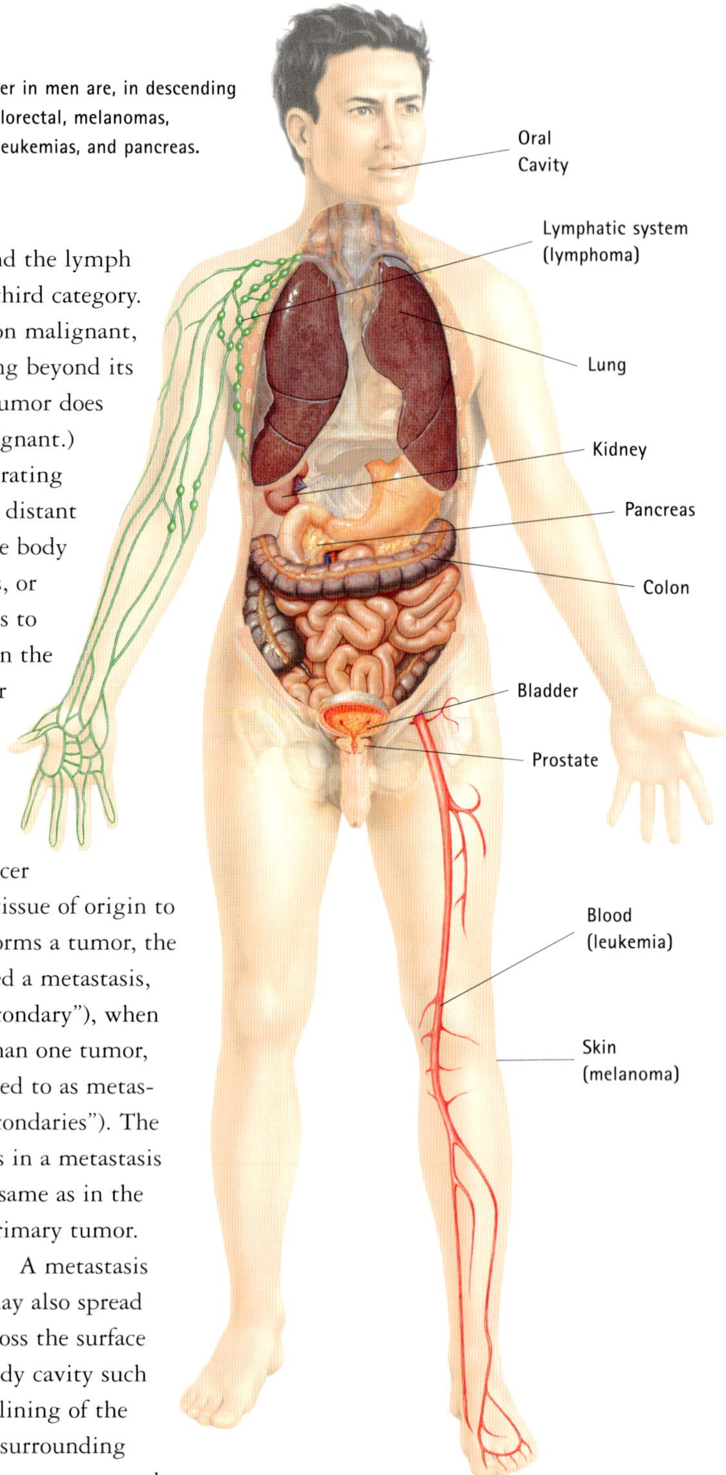

Male cancer sites

The most common types of cancer in men are, in descending order, prostate, lung, bladder, colorectal, melanomas, lymphomas, kidney, oral cavity, leukemias, and pancreas.

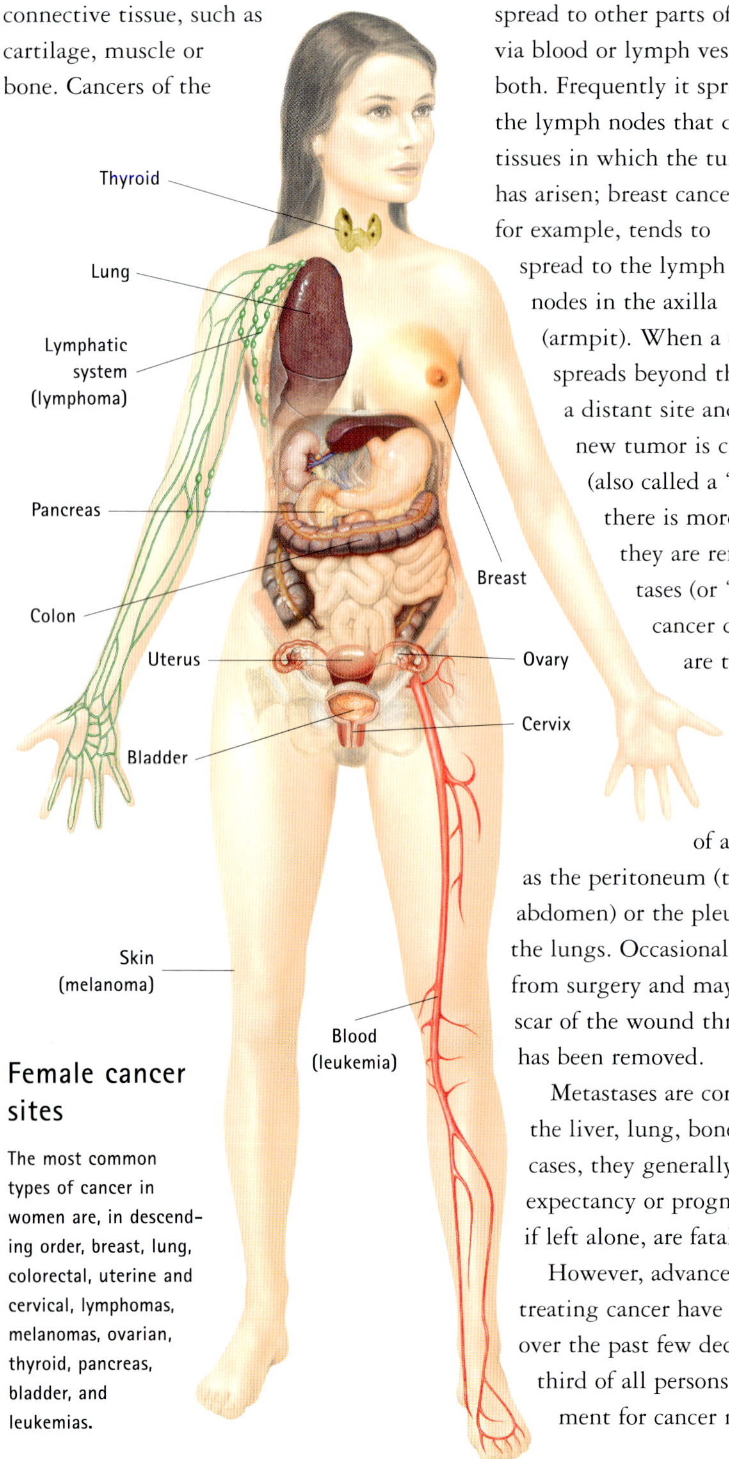

Female cancer sites

The most common types of cancer in women are, in descending order, breast, lung, colorectal, uterine and cervical, lymphomas, melanomas, ovarian, thyroid, pancreas, bladder, and leukemias.

The causes of cancer are not fully understood. Some cancers (including melanoma and cancers of the breast, ovary, and colon) tend to run in families. Others are known to be associated with certain risk factors; tobacco smoking is known to cause oral cancer and cancers of the larynx, lung, esophagus, pancreas, bladder, kidney, and cervix. Passive smoking, or exposure to smoke in the environment, increases the risk of lung cancer for non-smokers. A deficient diet may make cancer more likely. A high-fat diet may contribute to cancer of the breast, colon, uterus, and prostate. Obesity is associated with increased rates of cancer of the prostate, pancreas, uterus, colon and ovary, as well as increased rates of breast cancer in older women. Ultraviolet radiation from the sun, sunlamps and tanning booths may cause skin cancer, especially in those people with fair skin. Excess alcohol consumption may cause cancer of the mouth, throat, esophagus, larynx and liver. Chemicals and other substances in the workplace can cause cancer. Asbestos, nickel, cadmium, uranium, radon, vinyl chloride and benzene are examples of occupational carcinogens (cancer-causing substances). Radiation from medical x-rays can increase the risk of cancer.

In general, the sooner cancer is diagnosed and treated, the better the chance of survival. Regular medical check-ups and self-examination of organs such as skin, testicles (in men) and breasts (in women) will increase the chance of early detection. Depending on age and gender, screening tests may also be advised.

These include Pap tests (a smear taken of cells of the cervix), mammograms (x-rays of the breast) and colonoscopy (examination of the colon). If there are signs and symptoms that suggest the possibility of cancer, the physician may order tests such as x-rays, ultrasound, MRI or CAT scans, endoscopy, and blood tests. A biopsy may be taken for examination by a pathologist. The tests will allow a cancer to be "staged," that is, to be rated according to how far it has spread. Local lymph node involvement or distant metastases both indicate more advanced disease, with a correspondingly poorer outlook.

There are several approaches to treating cancer, which may be used alone or in combination with each other, depending on the type and location of the cancer, the stage of the disease, the person's age and general health. Surgery is frequently used to remove a primary tumor. The tissue around the tumor and nearby lymph nodes may also be removed during the operation.

Radiation therapy (radiotherapy) involves high-energy rays used to directly destroy or slow the growth of cancer cells. Chemotherapy is treatment with cytotoxic (cell-killing) drugs, introduced either directly into the tumor, or via the bloodstream. Hormone therapy and immunotherapy are used in certain cancers sensitive to these treatments. As cancer treatments affect normal cells as well as cancer cells, they can often cause unpleasant side effects such as nausea, skin rashes, loss of hair, and bone marrow suppression.

Prevention is important in managing cancer. Physicians recommend giving up smoking, avoiding smoke and other environmental carcinogens. A good diet will include foods that are low in fat, rich in vitamins A and C and high in fiber such as whole-grain cereals, fruits, and vegetables. Alcoholic beverages should be taken in moderation, and overexposure to the sun should be avoided, particularly by fair-skinned people.

SEE ALSO *Chemotherapy on page 446; Treating cancer on page 446; information on cancers of other parts of the body is included in the entry for the body part*

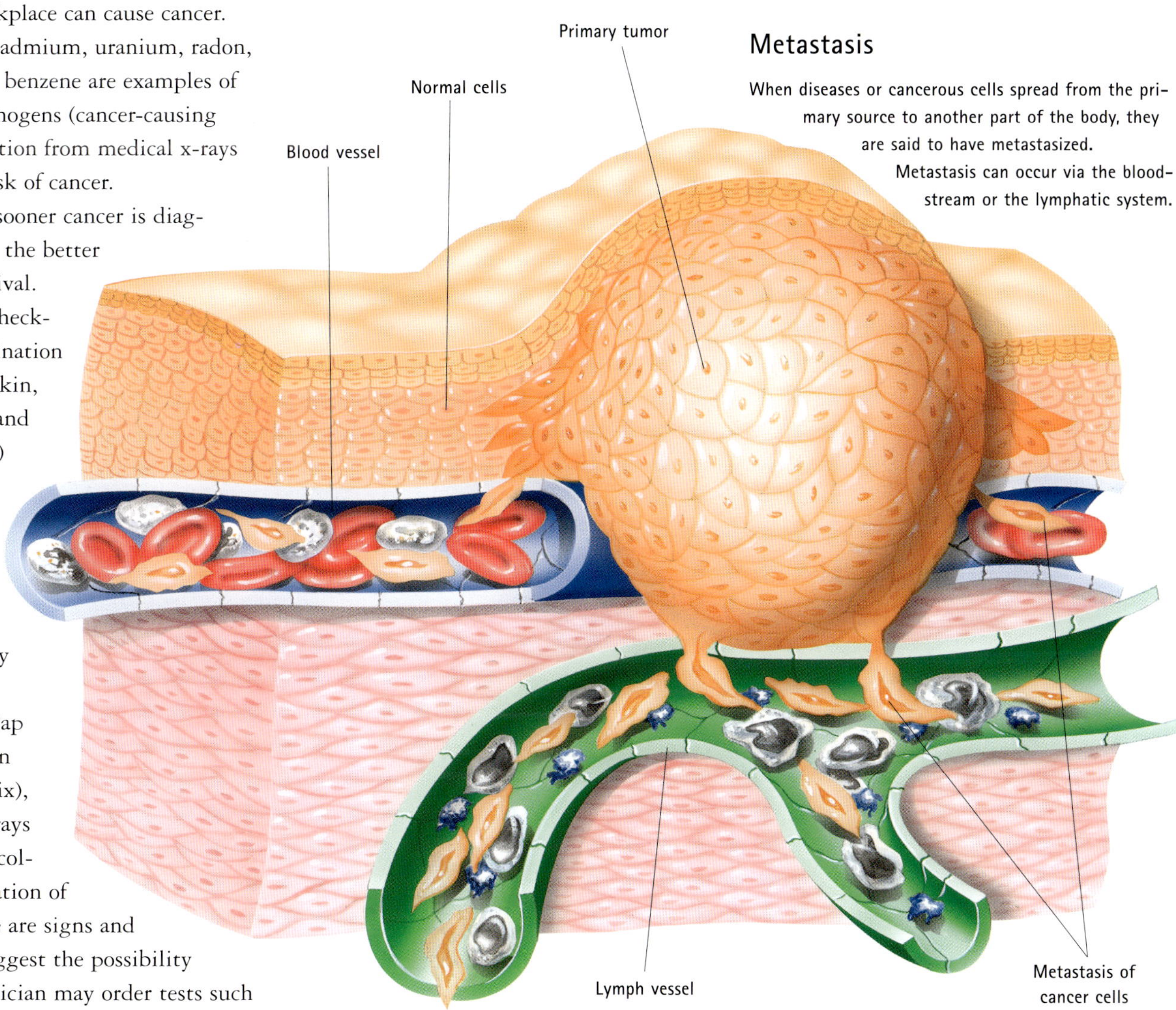

Metastasis

When diseases or cancerous cells spread from the primary source to another part of the body, they are said to have metastasized. Metastasis can occur via the bloodstream or the lymphatic system.

SKELETAL SYSTEM

The skeleton is the framework of the body and is usually described in two parts, the axial skeleton and the appendicular skeleton.

THE AXIAL SKELETON

The axis of the body is formed by the skull, the vertebral column (backbone) and the thoracic cage (chest).

SEE ALSO *Chest on page 220; Skull on page 124; Spine on page 211*

The skull

The skull forms the skeleton of the head. It consists of the cranium, the mandible (lower jawbone) and the hyoid bone at the base of the tongue. The top part of the skull (the cranial cavity) houses and protects the brain and part of the brain stem. The facial skeleton is the lower part of the skull that underlies the face. The upper jaw is fixed and formed by two bones called the maxillae.

Where the cranial cavity meets the facial skeleton, there are two orbits, or sockets, for the eyes. Underlying the nose is the nasal aperture of the skull that leads to the nasal cavity.

The base of the skull articulates with the first bone of the spine: the atlas vertebra. This joint allows the head to flex and extend (as in nodding) and to move sideways.

Many bones of the skull are hollow. The cavities inside them are called sinuses. They lessen the weight of the bone and give resonance to the voice.

The vertebral column

The vertebral column (backbone or spine) is a stack of small bones known as vertebrae. There are 7 vertebrae in the neck, 12 in the thorax (the chest region), and 5 in the lumbar region (the small of the back, behind the abdomen). The last two bones of the vertebral column, the sacrum and the coccyx, are formed by vertebrae which fuse after puberty (5 in the sacrum, 4 in the coccyx).

Vertebrae articulate with one another on intervertebral disks. These are flexible pads of cartilage that separate one vertebra from another. Each vertebra can only move a few degrees at its intervertebral disk, but the sum of all these individual movements gives great mobility to the vertebral column.

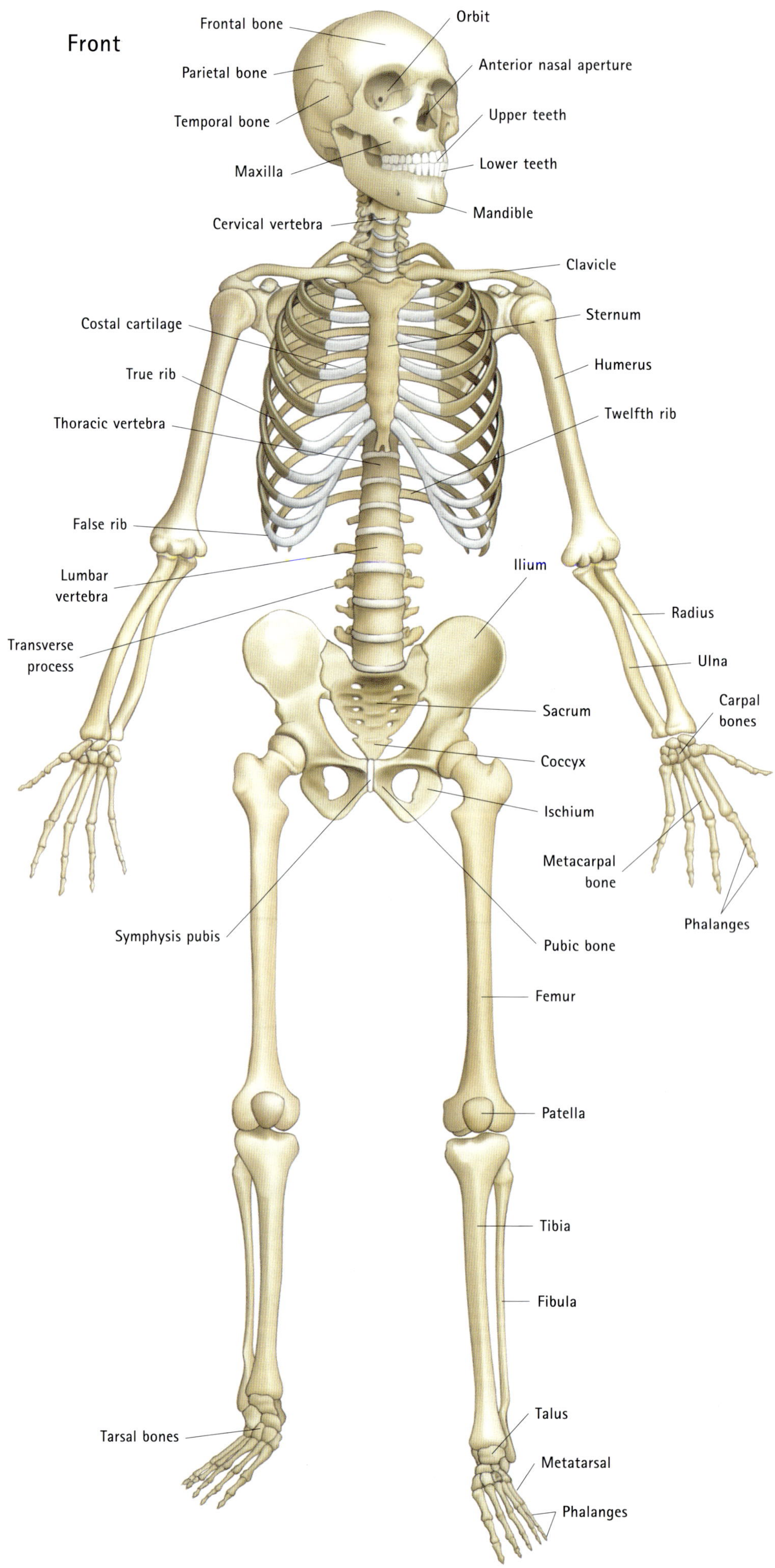

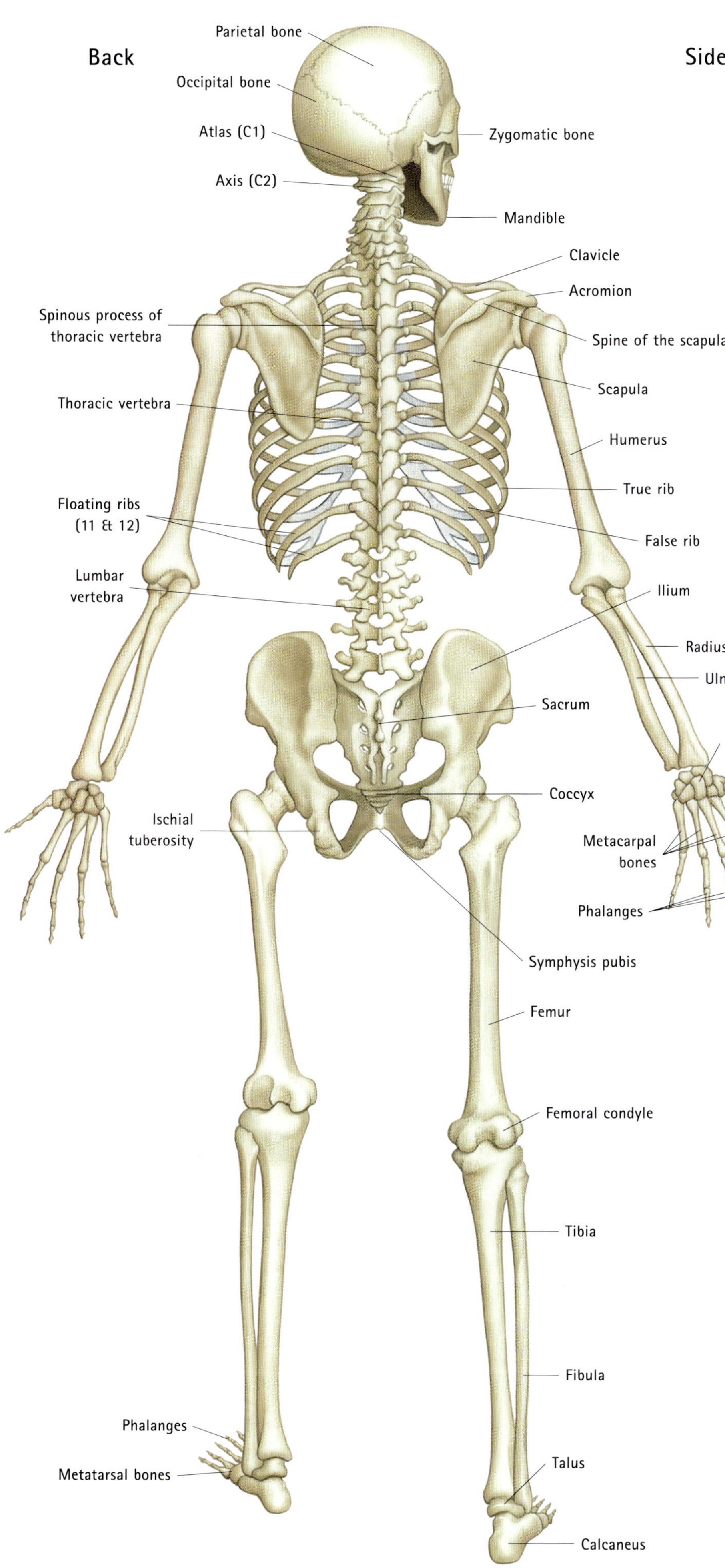
Back
Parietal bone
Occipital bone
Atlas (C1)
Axis (C2)
Zygomatic bone
Mandible
Clavicle
Acromion
Spinous process of thoracic vertebra
Spine of the scapula
Scapula
Thoracic vertebra
Humerus
True rib
Floating ribs (11 & 12)
False rib
Lumbar vertebra
Ilium
Radius
Ulna
Sacrum
Carpal bones
Coccyx
Ischial tuberosity
Metacarpal bones
Phalanges
Symphysis pubis
Femur
Femoral condyle
Tibia
Fibula
Phalanges
Metatarsal bones
Talus
Calcaneus

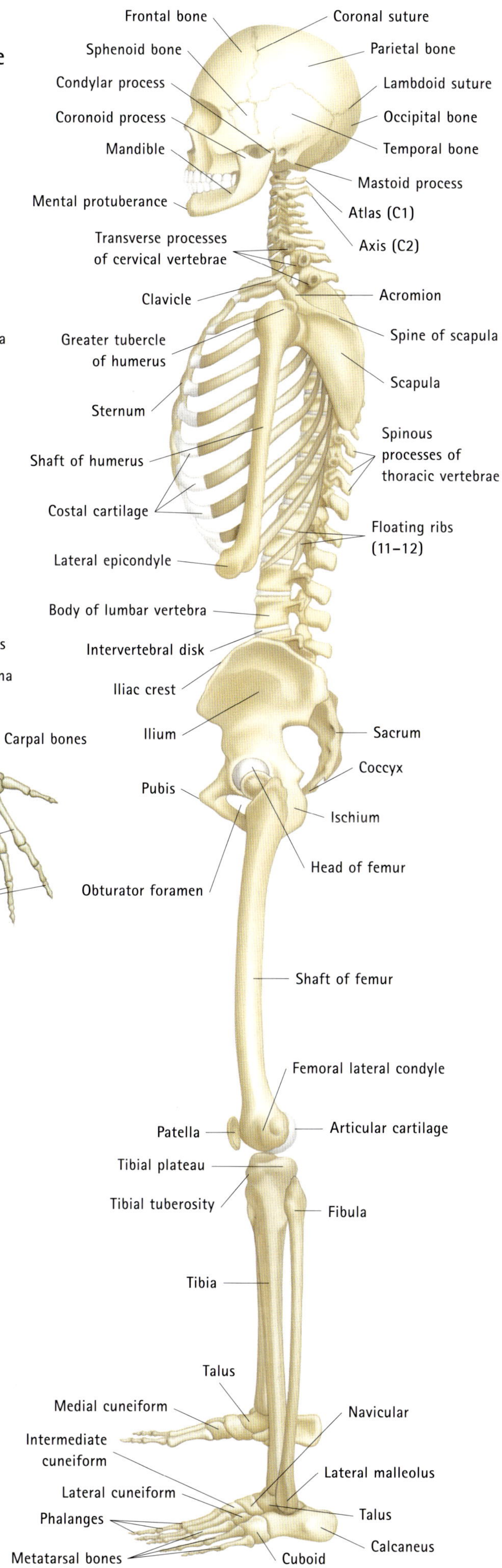
Side
Frontal bone
Coronal suture
Sphenoid bone
Parietal bone
Condylar process
Lambdoid suture
Coronoid process
Occipital bone
Mandible
Temporal bone
Mastoid process
Mental protuberance
Atlas (C1)
Transverse processes of cervical vertebrae
Axis (C2)
Clavicle
Acromion
Greater tubercle of humerus
Spine of scapula
Scapula
Sternum
Spinous processes of thoracic vertebrae
Shaft of humerus
Costal cartilage
Floating ribs (11–12)
Lateral epicondyle
Body of lumbar vertebra
Intervertebral disk
Iliac crest
Ilium
Sacrum
Coccyx
Pubis
Ischium
Head of femur
Obturator foramen
Shaft of femur
Femoral lateral condyle
Patella
Articular cartilage
Tibial plateau
Tibial tuberosity
Fibula
Tibia
Talus
Medial cuneiform
Navicular
Intermediate cuneiform
Lateral malleolus
Lateral cuneiform
Talus
Phalanges
Calcaneus
Metatarsal bones
Cuboid

BONE

Bone is made up of different layers, the outermost of which is the periosteum. Beneath this is the compact or cortical bone which surrounds the spongy or cancellous bone. At the core is the bone marrow.

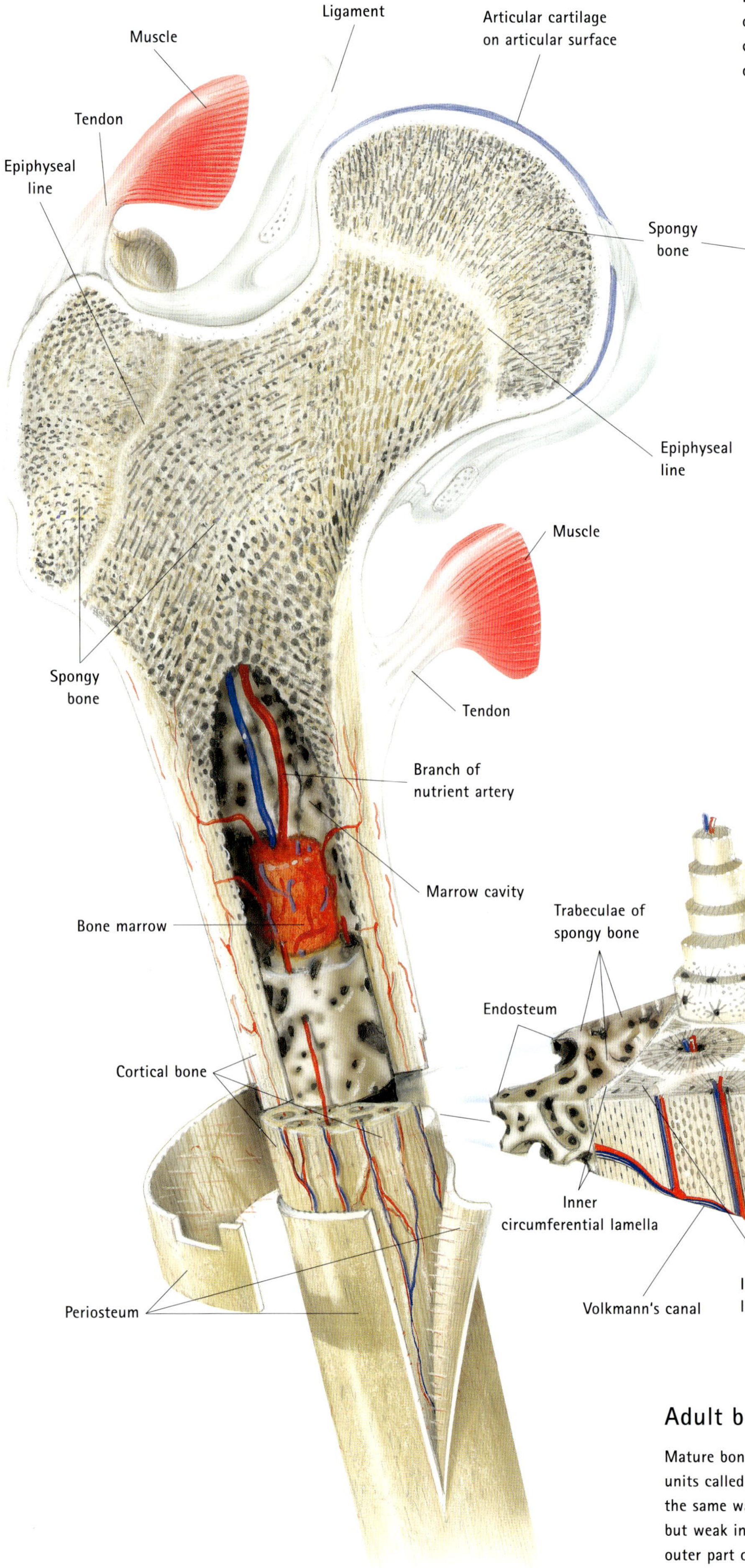

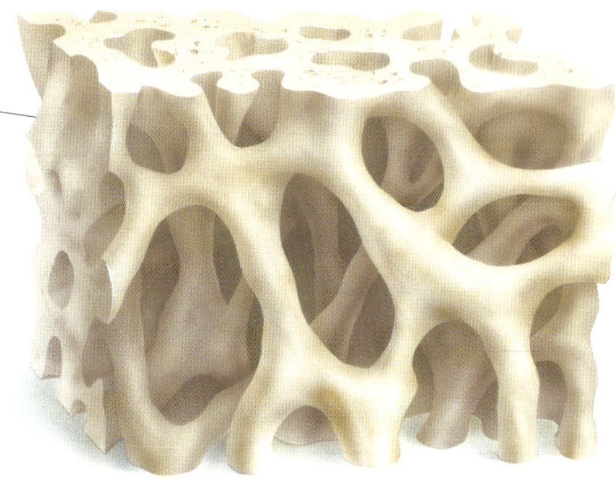

Spongy (cancellous) bone

Spongy or cancellous bone consists of a lattice work system of bony spikes called trabeculae, arranged in different directions. It is light and strong enough to withstand stress from many directions. An example is the head of the femur.

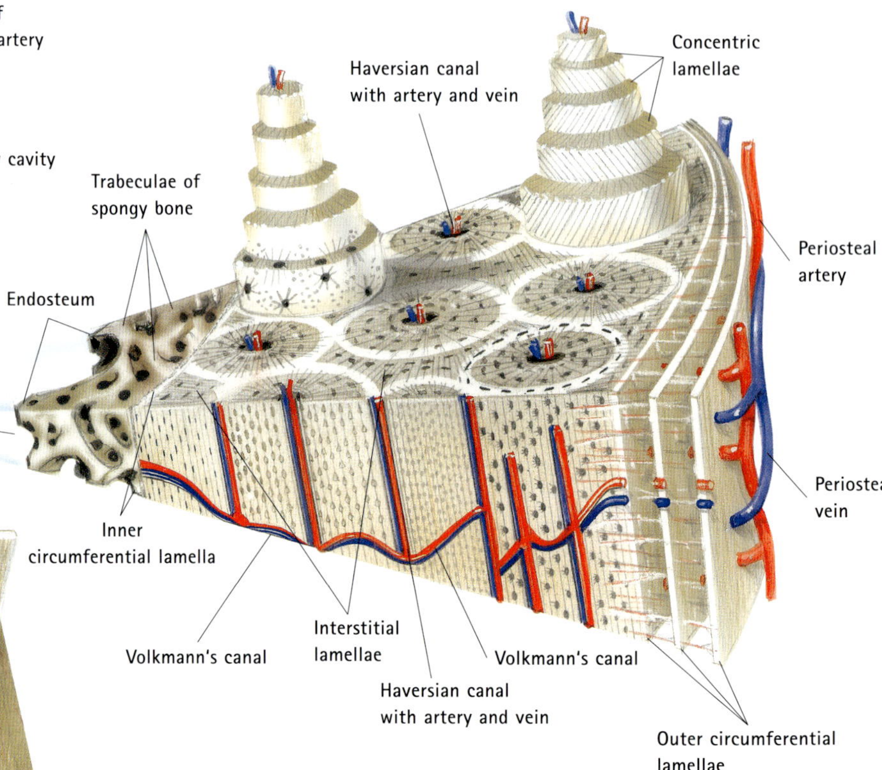

Adult bone

Mature bone consists of densely packed rod-shaped units called osteons (or Haversian systems) all aligned the same way. Cortical bone is strong along this axis but weak in any other direction. An example is the outer part of the shaft of the femur.

When you carry a heavy weight, the vertebral column is turned into a rigid pillar by the contraction of muscles at the back of the column. When you bend forward to lift a load, the force applied to the intervertebral disk can be large enough to damage the disk.

The thoracic cage

The thoracic cage, or chest, is made up of the 12 thoracic vertebrae, the ribs and the sternum. The top 7 true ribs extend from the vertebrae at the back and curve around the front to the sternum, connecting to the sternum by extensions of cartilage called the costal cartilages. The next 3 ribs, the false ribs, do not extend all the way around—their costal cartilage fuses onto the cartilage of the last true rib. The final 2 ribs, the floating ribs, do not reach the front. The rib cage protects the heart and lungs. The ribs, moved by the intercostal muscles, are involved in breathing.

Appendicular skeleton

The appendicular skeleton consists of the bones of the limbs and the shoulder and pelvic girdles, the bones that support and attach the limbs to the axial skeleton. The upper and lower limbs are similar in their composition. The shoulder girdle of the upper limb corresponds to the hip or pelvic girdle in the lower limb. The long bone of the upper arm is the humerus; in the lower limb the long bone is known as the femur. The forearm and the lower leg both have 2 long bones; the wrist has 8 bones, the ankle has 7. There are 5 bones in both the palm of the hand and the sole of the foot, with 14 bones making up the digits of both the hand and the foot.

The lower limb has to support the body weight and therefore is less flexible than the upper limb. The scapula (shoulder blade) can slide freely on the rib cage because it is attached to it by muscles. It is only stabilized at the front, by a strut known as the clavicle (collar bone). By contrast, the pelvic girdle is fixed to the axial skeleton where the 2 hip bones articulate with the sacrum at the base of the vertebral column. In front, the 2 hip bones join at the symphysis pubis. Each hip has an acetabulum, a deep socket that accommodates the head of the femur.

SEE ALSO *Pelvis on page 224; The hips, legs and feet on page 330; The shoulders, arms and hands on page 314*

Bones

Bone is the rigid, calcified tissue that makes up the skeleton. It supports the body and surrounds and protects its internal structures. It acts as a store of calcium, and houses the bone marrow in which blood cells are manufactured. Bones provide an attachment for muscles, which contract, allowing the body to move. There are 206 bones in the body in all.

SEE ALSO *Red blood cells on page 83; Tissues on page 20*

Bone structure

Being connective tissue, bone is composed of cells in a matrix. The major components of the matrix include mineral salts (mainly calcium phosphate), which provide hardness, and collagen fibers, which give strength.

Four types of cell are present: osteoprogenitor cells, osteoblasts, osteocytes and osteoclasts. Osteoprogenitor cells develop into osteoblasts, which form bone tissue. Osteoblasts mature into osteocytes, which maintain bone tissue. Osteoclasts, which occur on bone surfaces, are involved in the reabsorption of the matrix, required for bone development, growth and repair.

A typical mature long bone has a central shaft—the diaphysis—and ends known as epiphyses. Inside the diaphysis is the medulla, which contains yellow bone marrow, and is lined by the endosteum, a layer rich in osteoprogenitor cells and osteoclasts. The diaphysis meets the epiphysis at the metaphysis. This is the location, in an immature bone, of the cartilage layer from which lengthwise growth occurs. Most of the bone is protected and nourished by the periosteum.

Periosteum

Periosteum is a thin, fibrous membrane that covers all bone surfaces, except those that are involved in joints (which are covered by cartilage).

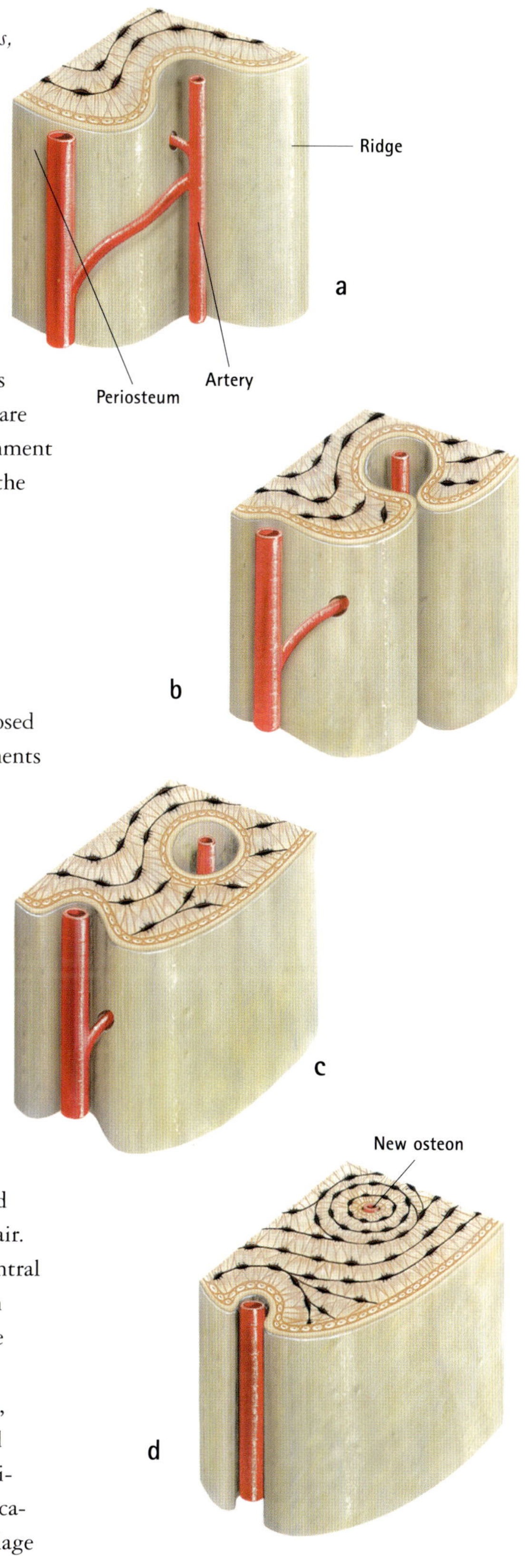

Bone formation

Bone grows in width as new bone is laid down in ridges (a) either side of a blood vessel. The ridges grow together and fuse, enclosing the vessel (b). More bone is laid down, diminishing the space around the vessel (c), and eventually an osteon is formed. The process continues, enclosing parallel blood vessels and causing the bone to become thicker (d).

Periosteum is made up of two layers. The inner layer comprises osteoblasts or bone-producing cells) which produce bone when a fetus or child is growing. Following a fracture, they produce new bone in order to mend the two ends of the broken bone together. Fibers from this inner layer also penetrate the underlying bone and help bind the periosteum to the bone. The dense outer layer contains nerves and blood vessels. Branches of these blood vessels penetrate the bone to supply the cells with nutrients.

Bone function

Although skeletons in museums give the impression that bones are brittle, living bones are much stronger than dry bones. Their structure is designed to best serve their mechanical functions and is continually remodeled throughout a lifetime.

The role of the bones in body movement is best seen in the limbs. The limb bones act as levers which are moved by the muscles attached to them in much the same way that the arms of a crane are activated by motors. The femur must be mechanically strong enough to bear the weight of a person's body but light enough to minimize the muscle force required to move that weight. A metal tube is much stronger against bending than a solid rod made out of the same amount of metal. Similarly, the shaft of a long bone, such as the femur, is made up of a cylinder of compact bone. In the center of the femur is spongy bone formed by bands of bone called trabeculae. The spaces between the trabeculae are filled up with bone marrow. This cylindrical arrangement is repeated within the compact bone, which is made up of tiny cylinders formed by concentric layers of bone surrounding blood vessels and nerves. The gaps between these microscopic cylinders are packed with thin plates of bone (laminae) running in all directions.

At the ends of the long bones, the trabeculae are aligned along the lines of stress applied to the bone. The body weight received by the head of each femur is transmitted down the neck of the bone to the shaft; the trabeculae there form arches along the lines of stress, like arches under a bridge.

The structure of a bone changes according to the stress applied to it. Exercise strengthens bones, while a prolonged state of inactivity weakens bone structure. When new bone is formed to bridge over a fracture, the trabeculae are rearranged along the line of stress. Weight bearing is therefore recommended during bone healing to promote the optimal development of new trabeculae.

Bone diseases and disorders

There are several tests for bone disease. Blood tests can show abnormally high or low levels of calcium, phosphorus and bone enzymes such as alkaline phosphatase. An x-ray can be used to detect fractures, tumors or degenerative conditions of the bone. A bone scan is a test that detects areas of increases or decreases in bone metabolism. This is done by determining how a radioactive isotope collects in the bone. It can be used to identify tumors or bone infection. Bone marrow biopsy uses a needle to extract bone marrow for examination. This test is used to diagnose leukemia, secondary bone tumors, anemia and infections.

SEE ALSO *Biopsy on page 435; Disorders of the bones and joints of the leg on page 331; Disorders of the foot on page 342; Disorders of the spine on page 215; Treating the musculoskeletal system on page 444; Traction on page 455*

Fractures

A break in a bone is called a fracture. Any bone in the body can be fractured, but some bones, because of their vulnerable positions (for example, the long bones of the arms and legs), tend to fracture more often than others. Most fractures occur as the result of injury or accident. Sometimes, a bone breaks following repeated minor strains. Some bones have a tendency to fracture easily because they are weak from a disease, osteoporosis for example.

There are several types of fractures. In a complete fracture, the two parts of the bone are completely separated. In an incomplete fracture, the two parts are partially separated. An incomplete fracture in a long bone in a child is often called a greenstick fracture. If there are more than two bone fragments at the fracture site, it is called a comminuted fracture. If the fractured bone has broken the skin and is exposed to the air, it is a compound or open fracture; if it hasn't broken the skin, it is a simple, or closed fracture. In an impacted, or compression fracture, the break occurs from extreme pressure on the bone. A stress fracture is a crack in a bone caused by repetitive and prolonged pressure on the bone, usually

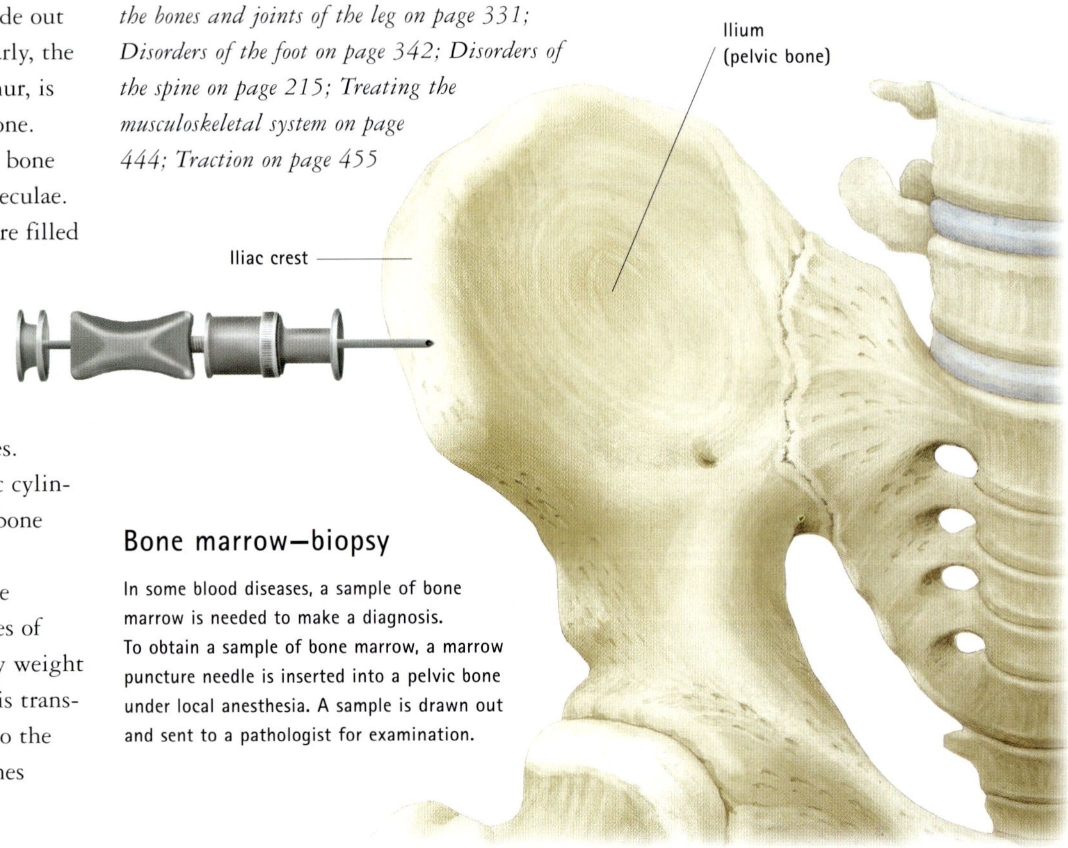

Bone marrow—biopsy

In some blood diseases, a sample of bone marrow is needed to make a diagnosis. To obtain a sample of bone marrow, a marrow puncture needle is inserted into a pelvic bone under local anesthesia. A sample is drawn out and sent to a pathologist for examination.

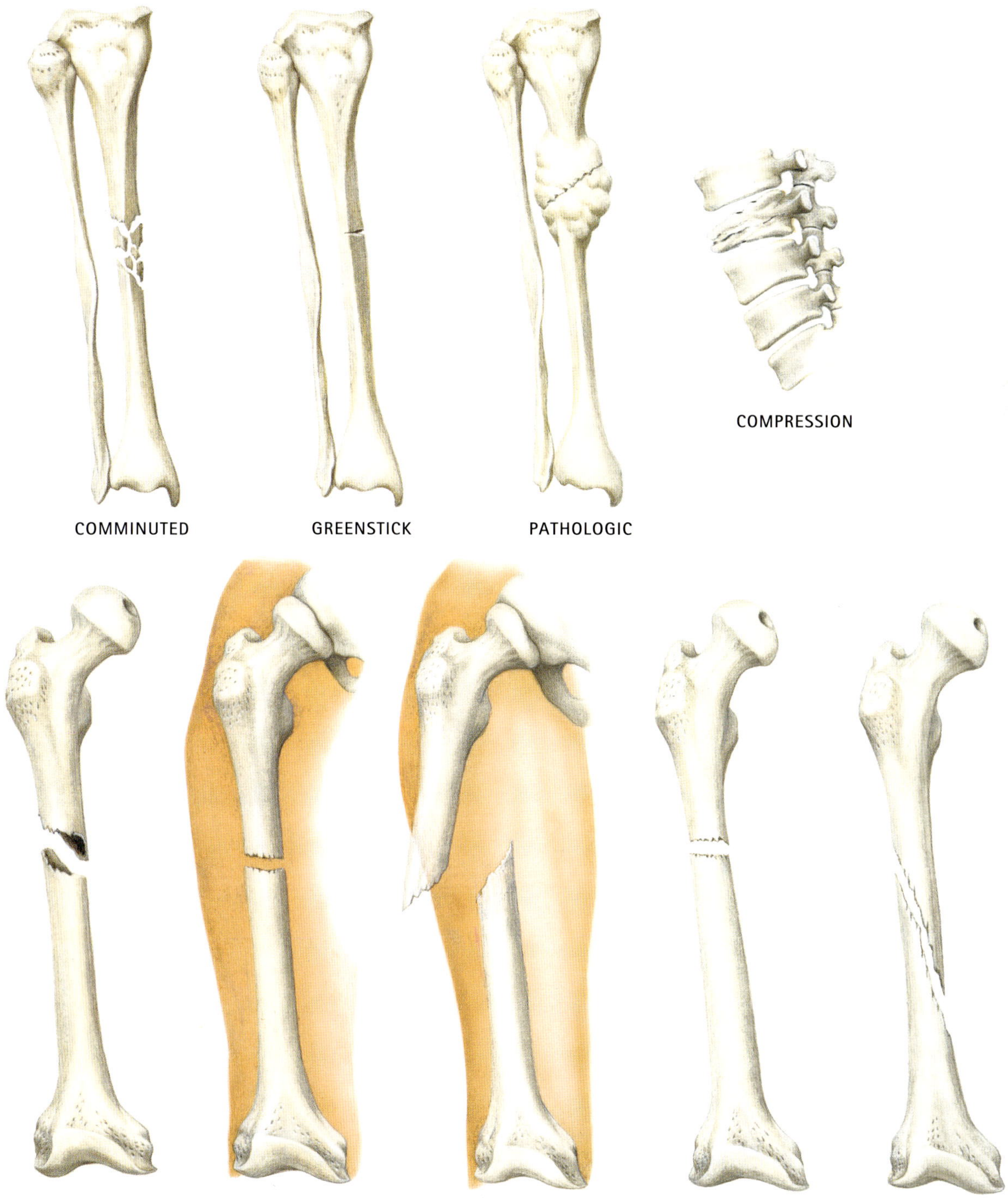

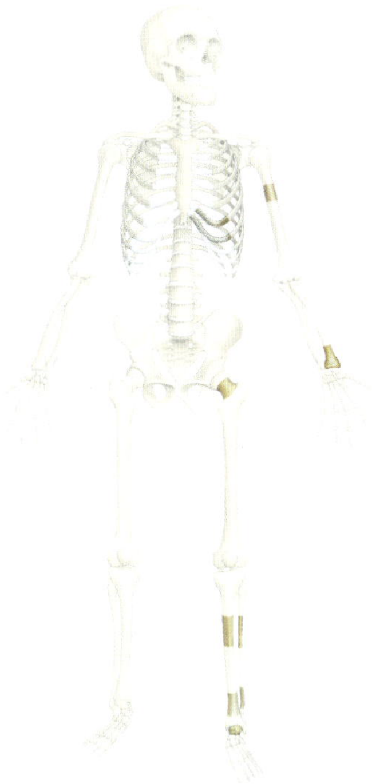

Common fracture sites

Common fracture sites include the arm, wrist, hip, leg and ankle.

Bone fractures

Fractures are named according to their appearance and characteristics.

from exercise. A pathologic fracture is one that occurs in bone that has been weakened or destroyed by disease such as osteoporosis or a tumor; the injury itself may be minor. A compression fracture is an example of this.

The symptoms and signs of a fracture are pain, swelling and deformity at the fracture site, weakness of movement, and inability to bear weight on the affected parts.

The arteries around the fracture may be damaged, causing bleeding or bruising at the site. In a limb especially, there may be loss of pulse, with cold white extremities. Nerves around the fracture sometimes become damaged; there may be numbness, tingling or paralysis in the extremity below the fracture.

A fracture can usually be seen clearly in an x-ray. The treatment of a fracture is to reduce (realign) the bone back into its normal position, then to immobilize the bone fragments to prevent any movement so they can join back together. Immobilization can be achieved by means of a plaster cast, traction (usually in hospital), or surgical insertion of rods, plates or screws. After the bones have healed (which may take from six weeks to six months), physical therapy and rehabilitation are required to ensure the restoration of complete mobility.

First aid is vital in the treatment of fracture. Follow the procedure for bleeding, and cover any open wounds. Do not move the person—especially if the injury is to the hip, pelvis, or upper leg—unless necessary. Do not try to straighten a bone or change its position, or give the person anything by mouth. Arrange emergency transport to the hospital immediately.

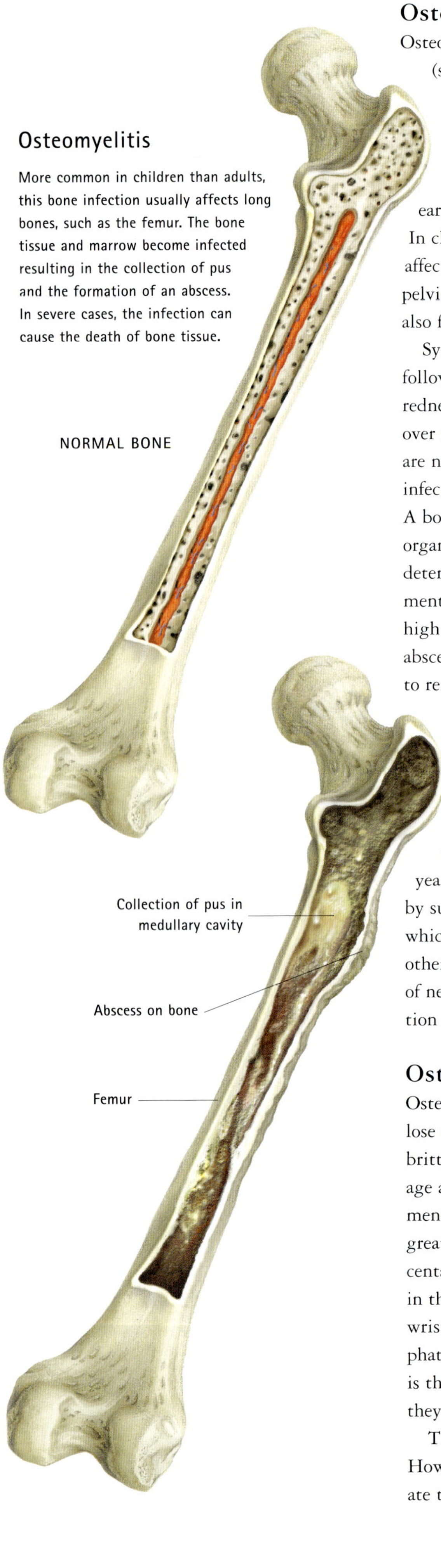

Osteomyelitis

More common in children than adults, this bone infection usually affects long bones, such as the femur. The bone tissue and marrow become infected resulting in the collection of pus and the formation of an abscess. In severe cases, the infection can cause the death of bone tissue.

Osteomyelitis

Osteomyelitis is an infection of bone. Acute (sudden onset) osteomyelitis, which occurs most often in children, is usually caused by a bacterium or fungus that has traveled from an infection elsewhere in the body—a boil, or an ear infection, for example—to the bone. In children, the long bones are usually affected. In adults, the vertebrae and the pelvis are most commonly affected. It can also follow injury, such as an open fracture.

Symptoms are fever and general illness followed a few days later by pain, swelling, redness, warmth and tenderness in the area over the infected bone. X-rays of the bone are normal in the early stages, but the infection will show up in a bone scan. A bone biopsy and culture will grow the organism causing the infection and thus determine what antibiotic to use. Treatment is with intravenous antibiotics in high doses for at least six weeks. If an abscess has formed, surgery will be needed to remove the abscess and drain any pus.

Slow onset (chronic) osteomyelitis results when bone tissue has died (become necrotic) following an acute infection. An opening to the skin (sinus) may form and drain pus, which may persist intermittently for years. Chronic osteomyelitis is treated by surgical removal of dead bone tissue, which is then replaced with bone graft or other material, which promotes the growth of new bone tissue. In some cases amputation is required.

Osteoporosis

Osteoporosis is a disorder in which bones lose their density and become weak and brittle. The condition gets worse with age and is seen most commonly in postmenopausal women. The bone loss is greatest in bones containing a large percentage of spongy (trabecular) bone, found in the vertebral column, the hips and the wrist. The bones lose calcium and phosphate, as well as the connective tissue that is the matrix of the bone. Consequently they become brittle and prone to fracture.

The cause of osteoporosis is unknown. However, some factors are known to accelerate the condition; they include a diet low in calcium, lack of exercise, cigarette smoking, excessive alcohol consumption and prolonged bed rest. In women, estrogen appears to protect against osteoporosis, but levels of estrogen fall after menopause and so from the age of 50 or so, osteoporosis is more common in women than men.

Some diseases can cause or accelerate osteoporosis. They include Cushing's syndrome (overactivity of the adrenal glands) and hyperparathyroidism (overactivity of the parathyroid glands). Liver, kidney or heart disorders can also accelerate osteoporosis. Other possible causes include eating disorders, such as anorexia, and some drugs, such as thyroid hormones and corticosteroid medication taken over a long period.

Typically the sufferer has no symptoms but gradually loses height over the years, finally becoming stooped with forward curvature of the thoracic spine (also known as dowager's hump, or kyphosis). Back pain can be severe and a back brace may be needed for support. Fractures, especially of the hip and wrist, are common in the elderly, and can happen without warning. Compression fractures of the vertebrae may compress the surrounding nerves and cause severe pain.

The diagnosis can be confirmed with x-rays and photodensitometry (a scanning technique that measures bone density). These tests show bones with diminished density, and may be used to predict future fractures. The main goal of management of osteoporosis is to prevent fractures. Various medications are available that will prevent further bone loss, improve bone density and decrease the risk of fractures. The biphosphonates, the most commonly used drugs to treat osteoporosis, act by inhibiting bone resorption by osteoclasts. Calcium and vitamin D supplements and adequate protein also assist in preventing further bone loss.

In postmenopausal women, hormone replacement therapy (HRT) is effective in preventing bone loss. However, recent studies have shown an increased risk of breast cancer in women taking long-term hormone replacement therapy and HRT is no longer recommended as long-term prevention for osteoporosis.

There are several drugs available, including the hormone calcitonin, which prevent loss of bone mineral and allow the bone to gain density. Calcitonin must be given daily by injection.

The best approach to the problem of osteoporosis is prevention. Adequate calcium in the diet (at least three to four glasses of milk per day or the equivalent), regular exercise, giving up tobacco and reducing alcohol consumption will prevent or slow the progression of the disease. Women should begin these measures in their teens.

Dowager's hump

Dowager's hump develops in older women with severe osteoporosis. The vertebrae in the spine may compress, so that the normal curve of the spine is exaggerated. This condition may progressively worsen.

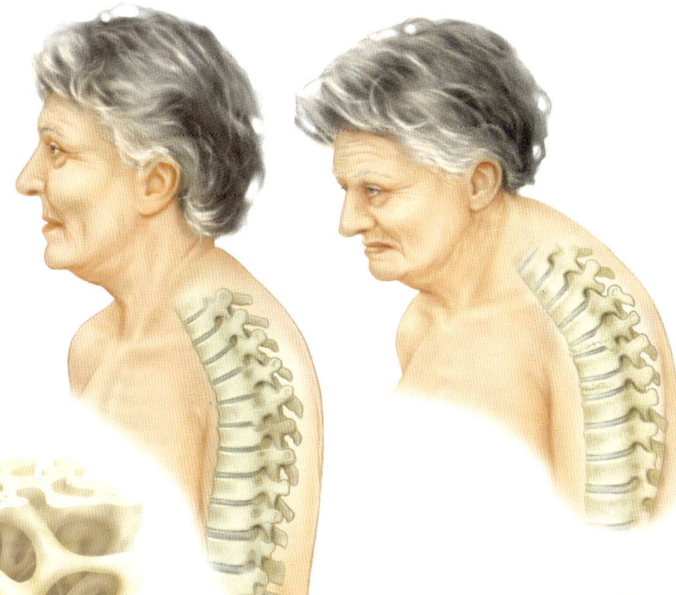

NORMAL BONE

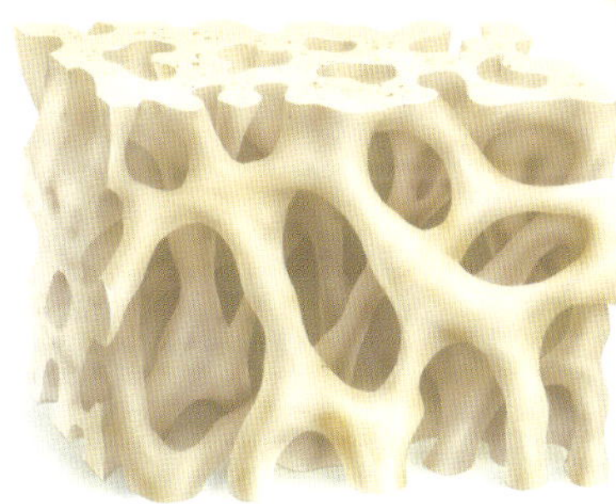

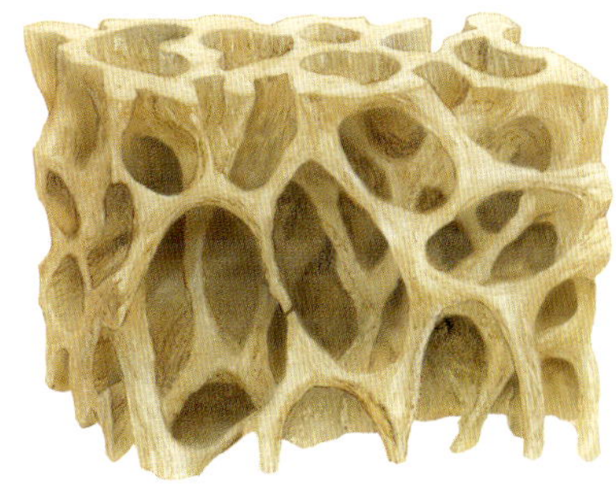

Osteoporosis

Osteoporosis occurs in both older men and women, however the incidence is higher for women. The condition occurs when bone mass is severely reduced, resulting in porous, weak and brittle bones which break easily.

Osteomalacia

Osteomalacia is softening of the bones caused by lack of vitamin D. It can be caused by poor dietary intake of vitamin D, poor absorption of vitamin D from the intestine, or too little exposure to sunlight, which is necessary for the formation of vitamin D in the body. Other causes include hereditary or acquired disorders of vitamin D metabolism, and kidney failure. In children, the condition is called rickets.

Common symptoms are aching bones, fractures and deformities, muscle weakness and spasms. The condition is diagnosed with x-rays of the bones, and blood tests for calcium, vitamin D and phosphorus. Treatment depends on the cause and may include vitamin D, calcium and phosphorus supplements.

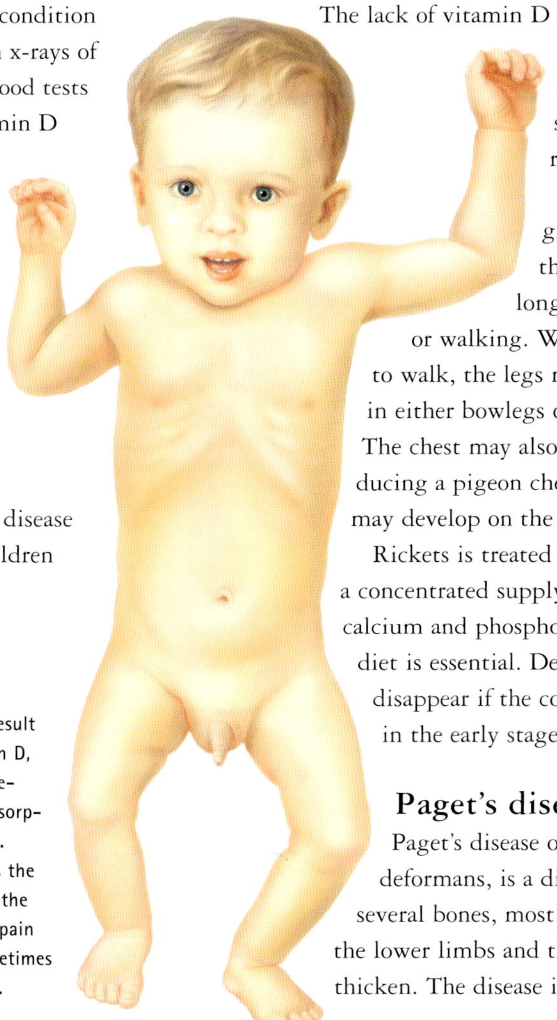

Osteomalacia

Osteomalacia is the result of insufficient vitamin D, either due to an inadequate diet or poor absorption by the individual. The condition inhibits the uptake of calcium by the body and this causes pain in the bones and sometimes also muscle weakness.

Rickets

Rickets is a bone disease of infants and children that is caused by a lack of vitamin D from either insufficient sunlight or inadequate diet. Occasionally, it may be caused by disorders of the kidney, liver or biliary system. The lack of vitamin D causes progressive softening and weakening of the bone structure that can result in deformity.

Infants with rickets grow more slowly than normal, and take longer to begin crawling or walking. When the infant starts to walk, the legs may bend, resulting in either bowlegs or knock-knees. The chest may also be deformed, producing a pigeon chest, and small knobs may develop on the ends of the ribs.

Rickets is treated by giving the child a concentrated supply of vitamin D, calcium and phosphorus; an adequate diet is essential. Deformities usually disappear if the condition is treated in the early stages.

Paget's disease of bone

Paget's disease of bone, or osteitis deformans, is a disorder in which several bones, most often the pelvis, the lower limbs and the skull, gradually thicken. The disease involves abnormally fast bone destruction and reformation where the new bone matter is structurally abnormal and fragile. This is a chronic, slowly progressive disease that mostly affects elderly people. It may be localized in one or two parts of the body or become widespread. The cause of the disorder is unknown, though recent research points to genetic causes or viral infection. It is known to be common in Europe, Australia and New Zealand.

The disorder (which has no connection with Paget's disease of the breast) often has no symptoms, and is generally discovered during examination or x-ray for some other complaint. In advanced cases, it can cause thickening of the skull, spinal curvature, barrel-shaped chest, bowing of the legs and leg pain. When the skull is affected the head may become enlarged, and hearing loss and blindness may occur if bone growth

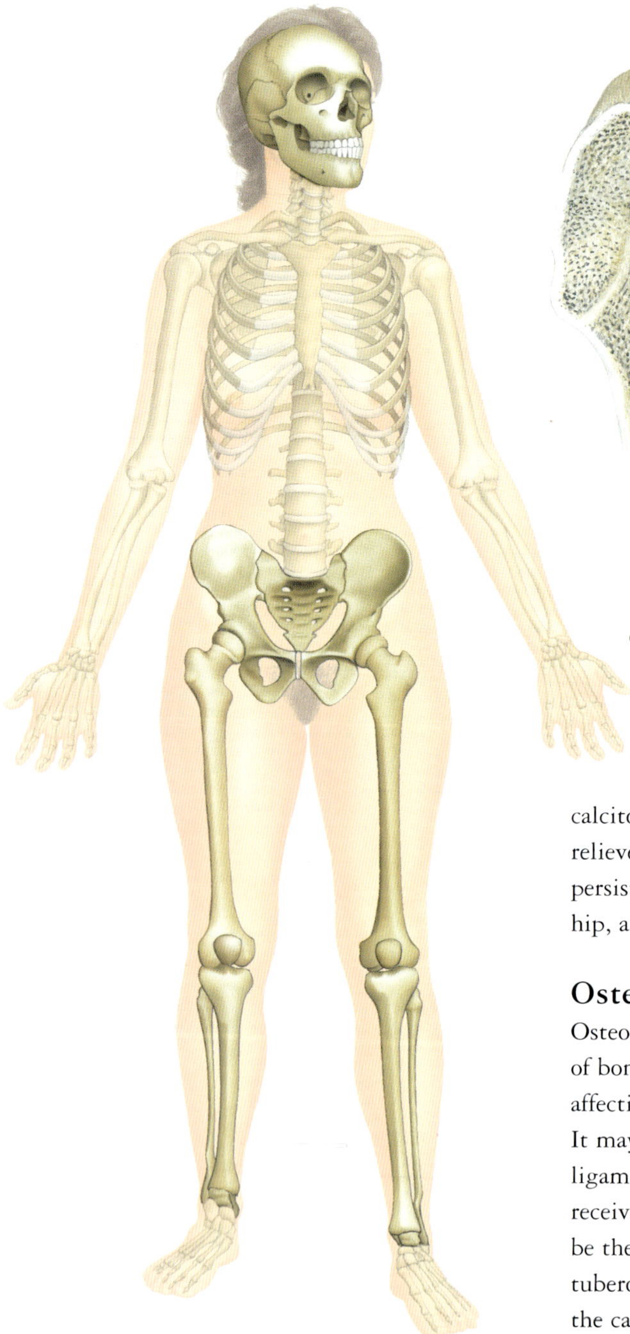

Bones affected by Paget's disease

In the advanced stages of this disease, the bone may be so weakened that even a light blow may cause a fracture. Paget's disease most often affects the pelvis, the lower limbs, and the skull.

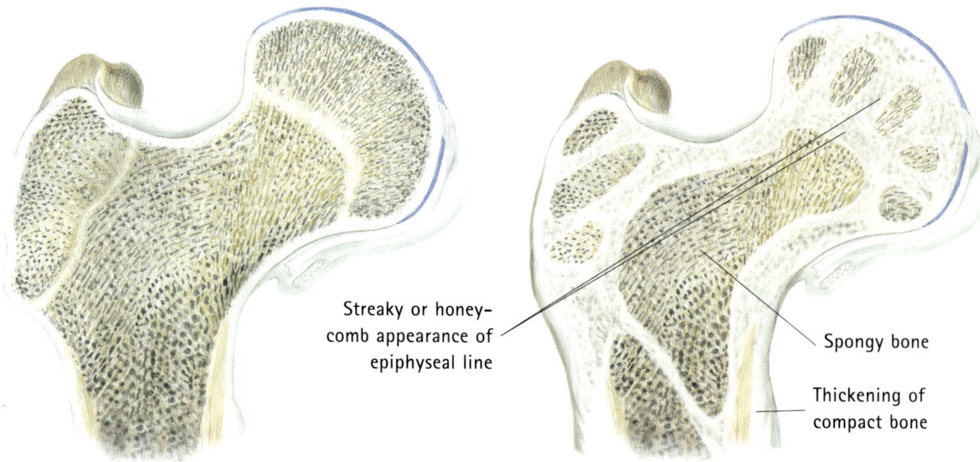

NORMAL BONE

Paget's disease of bone

In Paget's disease, the growth of new bone tissue is disrupted, resulting in bones becoming coarse and soft.

damages cranial nerves. Radiating sciatic pain in the lower extremities may be felt if the bones of the lumbar spine thicken. Bones affected by Paget's disease fracture more easily than normal bones. Osteosarcoma, a cancer of bone, may sometimes arise from the affected bones.

X-rays and bone scans help diagnose the condition, showing increased bone density and thickening. Blood tests show normal serum calcium with raised serum levels of the bone enzyme alkaline phosphatase. If treatment is required, anti-inflammatory drugs such as aspirin and ibuprofen may be used to relieve pain. Drugs such as calcitonin, which suppress bone loss and relieve pain, may be prescribed if symptoms persist. If there is extensive damage to the hip, a total hip replacement may be needed.

Osteochondrosis

Osteochondrosis is an abnormal condition of bone and cartilage formation in children, affecting growth plates at the ends of bones. It may occur at areas where tendons or ligaments attach to bones, or in areas that receive a lot of impact stress. This may be the hip (Perthes' disease), the tibial tuberosity (Osgood-Schlatter disease), or the calcaneus, the bone in the heel of the foot (Sever's disease).

In all cases, degeneration of bone and cartilage occurs; this is thought to be caused by an interruption to the blood supply of the bone, which is followed by spontaneous regeneration. The child feels pain and sometimes a lump at the site where the degeneration has occurred.

Treatment is aimed at protecting the bone and joint while spontaneous healing takes place, through bed rest and the use of appliances such as a brace, cast or splint. In most cases the bone will heal without any resulting deformity.

Periostitis

Periostitis is inflammation of the periosteum, the thin, fibrous membrane that covers bone surfaces. It may occur from overuse, especially along the medial side of the shin bone (a form of "shin splints"), in people who are physically active. Treatment involves rest and anti-inflammatory drugs. Periostitis may also occur in association with infection of bone (osteomyelitis) when it is caused by bacteria. In these cases, treatment is with intravenous antibiotics.

Bone tumors

Bone tumors (cancers) may be benign or malignant. Benign tumors include osteochondromas (the most common) and osteomas. They usually do not require treatment but can be removed for cosmetic reasons. Malignant bone tumors usually arise from primary cancers of the breast, lung, kidney, prostate or thyroid. Primary malignant bone tumors are rare and are more common in young men. These include osteosarcomas, Ewing's sarcoma, fibrosarcoma and chondrosarcoma. They require surgical removal followed by radiation therapy and chemotherapy. Often, an affected limb may need to be amputated.

EWING'S SARCOMA

Ewing's sarcoma is the second most common type of bone cancer. It occurs in children and adolescents (usually Caucasians) between 10 and 20 years of age. A person will experience pain and sometimes swelling at the tumor site, usually in the long bones of the arms and legs (especially the femur), or the pelvis

Osteochondrosis

Osteochondrosis occurs mainly in children, where there is degeneration of bone and cartilage. The bone may be moved out of shape causing a deformity. The hip is the most commonly affected.

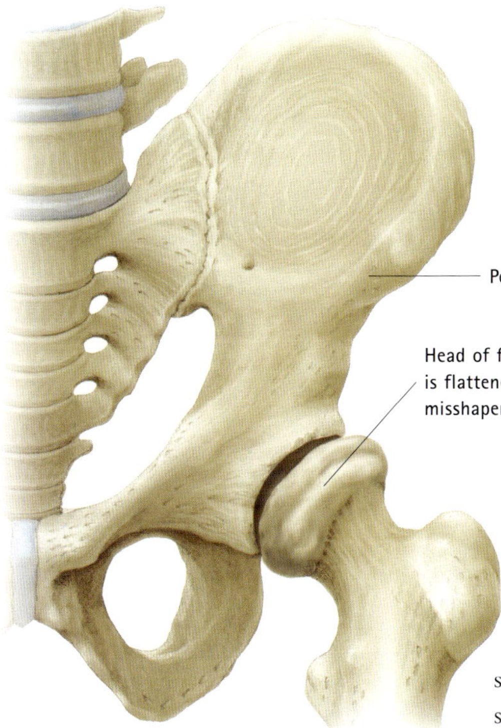

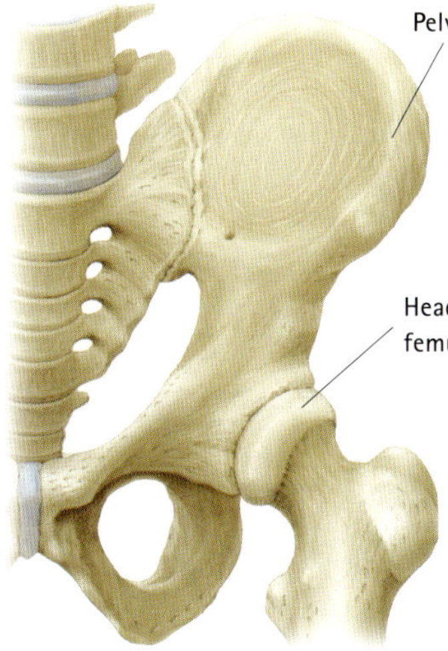

NORMAL PELVIS

Bone cancer

This bone tumor in the upper part of the shaft of the femur is a malignant osteosarcoma. It is treated with surgery, radiation therapy and chemotherapy. The leg may need to be amputated.

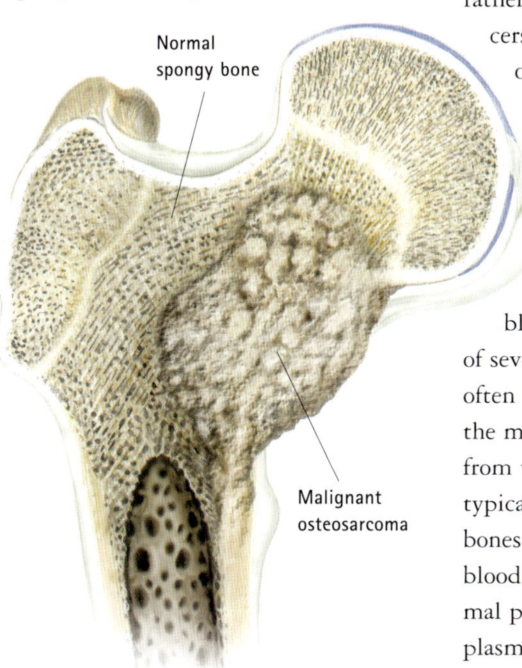

or ribs. The tumor grows quickly and spreads to other bones and the lungs.

Treatment involves a combination of intensive radiation therapy and chemotherapy, and (in some patients) surgical removal. If treated before it has spread, 60 percent of children with the tumor will survive.

MYELOMA

Myeloma is a tumor of plasma cells in the bone marrow, which results in pain and weakness of the bones, anemia and the production of an abnormal protein. It is the commonest tumor actually arising in bone, rather than spreading to bone as other cancers may do. It occurs most often in men over 65 years of age. There is an increased risk after exposure to high-dose ionizing radiation or pesticides. There have been recent reports of the possible role of a virus.

Myeloma may be detected by the finding of anemia and other blood abnormalities, or by the presence of severe, persistent bone pain. The pain is often in the spine or ribs and can be due to the myeloma itself or to fractures resulting from weakening of the bones. X-rays will typically show punched-out holes in the bones. If myeloma is suspected, testing of blood and urine is done to detect an abnormal protein produced by the malignant plasma cells. The diagnosis is confirmed by bone marrow biopsy.

The treatment of myeloma is initially pain control and treatment of complications, such as infection, high blood calcium or reduced kidney function. Then chemotherapy is given in the form of tablets or injections, usually as monthly cycles, for periods of up to 12 months. Myeloma is not curable but can be controlled for several years on average.

Recent advances in treatment include bone marrow transplantation and the use of the drug, thalidomide.

JOINTS

A joint is an area in the body at which two bones articulate. The bones may be fixed and immobile, such as the connections between the bones of the skull. These joints, (which are also called synarthroses) occur where two bones are fused or fixed together before or shortly after birth. Joints may be slightly mobile, such as the junction of the bones making up the front of the pelvis. These joints (also called symphyses) have a layer of cartilage between them and are held together by strong fibrous ligaments. Joints may also be freely mobile, such as the bones of the limbs. These joints (also called diarthroses) are held together and moved by muscles, ligaments, and tendons.

The bony surfaces of movable joints are covered with smooth cartilage. A thin fluid called synovial fluid, which is produced by the synovial membrane which lines the joint capsule, lubricates the joints.

There are several different kinds of mobile joints. In a ball-and-socket joint, such as the hip and the shoulder, free movement occurs in all directions. The elbow and the knee are essentially hinge joints, allowing movement mainly in one plane only. A saddle joint, such as the base joint of the thumb, allows sliding movement in two directions. A plane (or gliding) joint, such as those found between the carpal bones of the wrist allows only slight sliding movements. Pivot joints allow rotation about a single axis and are found between the first two vertebrae.

A mobile joint may have more than one type of joint movement; the elbow, for example, includes a hinge joint and a pivot joint.

SYNOVIAL JOINTS

Synovial joints are the most mobile joints in the body. The shape of articular cartilage surfaces in a synovial joint and the way they fit together determine the range and direction of the joint's movement.

Gliding joint

Aided by synovial fluid, the bones in the joint slide across each other in a limited movement.

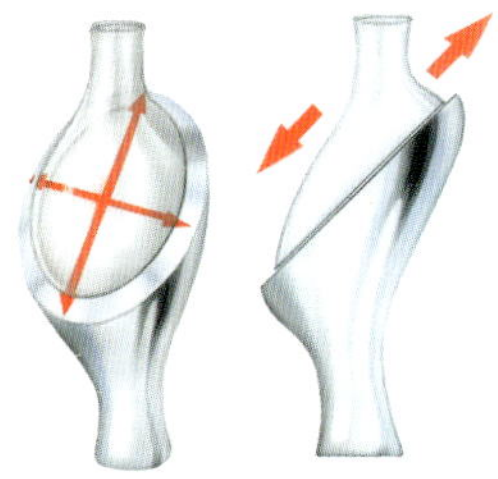

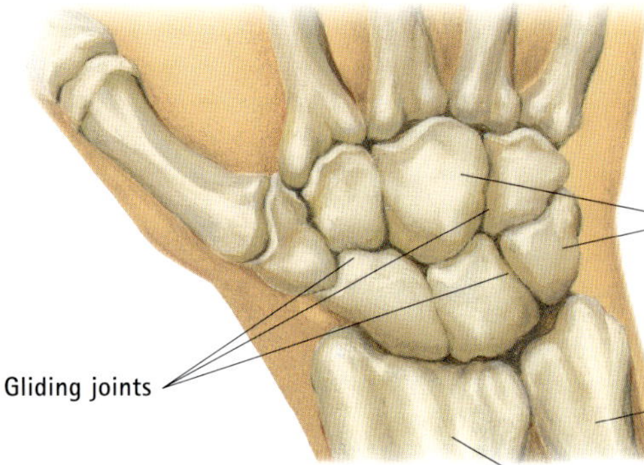

Ellipsoidal joint

This is a joint structure that allows movement in two directions, such as that which takes place at the wrist joint.

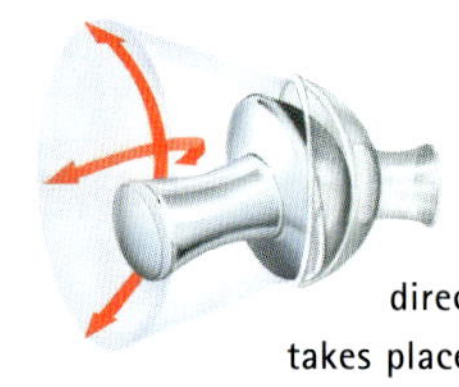

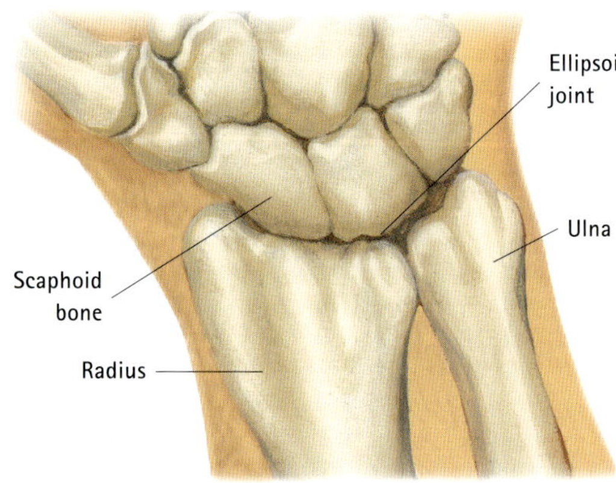

Saddle joint

A saddle joint allows almost as much movement as a ball-and-socket joint.

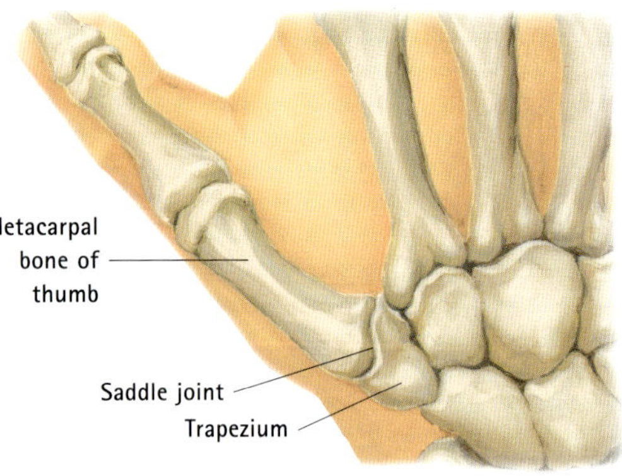

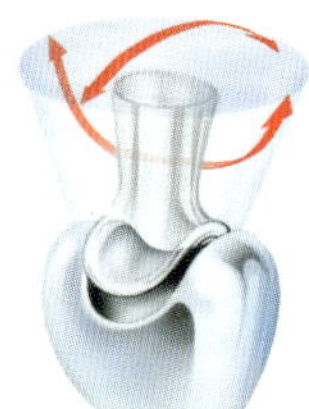

Hinge joint

As the name suggests, the joint can move in one plane only, such as the flexion (bending) and extension (straightening) of the elbow.

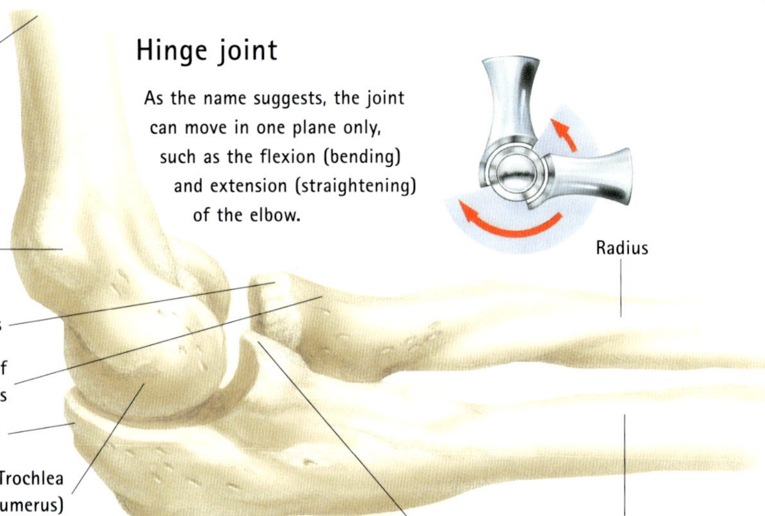

Ball-and-socket joint

A bone with a cup-like socket rests on the lubricated ball-like end of another bone. This type of joint gives the greatest freedom of movement.

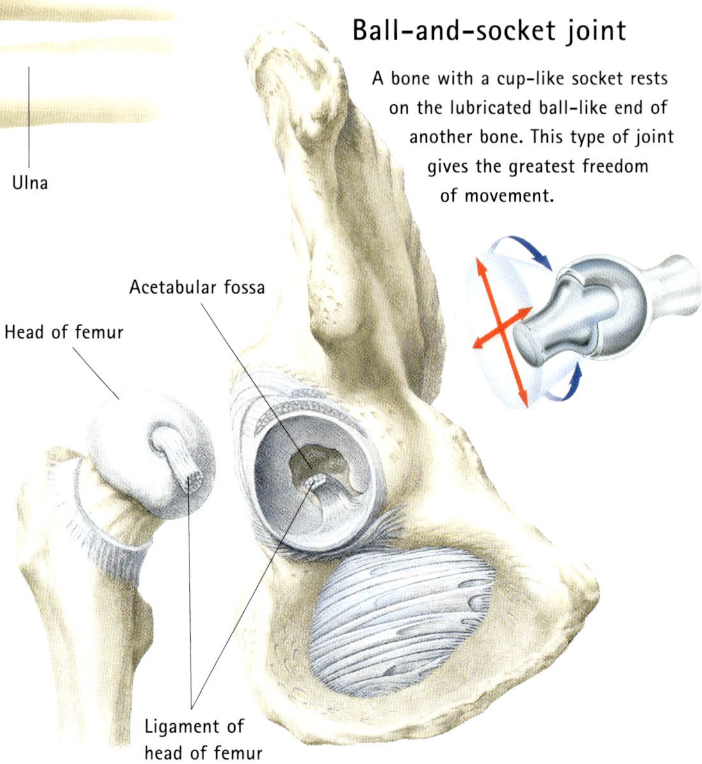

Pivot joint

The joint between the first and second cervical vertebrae rotates—this is known as a pivot joint.

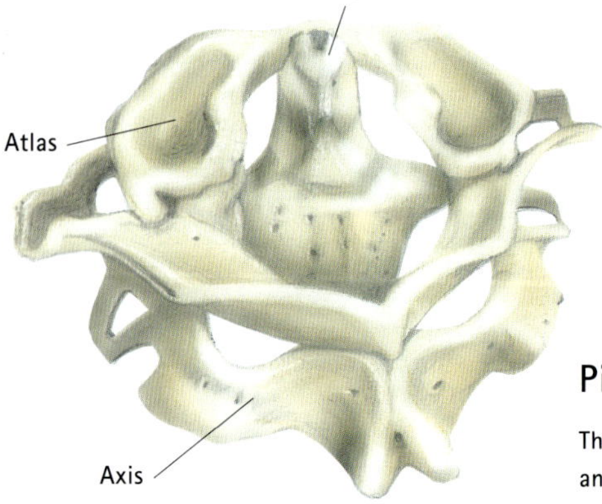

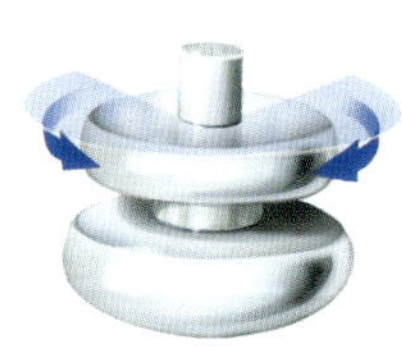

Where there is a series of joints, the total range of movement is the sum of the movements of the individual joints. The vertebral column is one such example of a chain of joints. The 8 carpal bones of the wrist allow a wide range of movement of the hand. The 7 tarsal bones of the ankle allow the foot to be tilted in many directions to negotiate uneven ground.

SEE ALSO *Ankle on page 340; Bones and joints of the leg on page 330; Elbow on page 321; Hand on page 324; Hip on page 336; Jaw on page 132; Knee on page 338; Spine on page 211; Tissues on page 20; Wrist on page 322*

SYNOVIAL JOINT

In a synovial joint, the ends of the bones are smooth, and are covered by articular cartilage that has an extremely low coefficient of friction.

The two bones are bound together by a capsule of fibrous tissue. The fibrous capsule is lined on the inside by a synovial membrane that secretes synovial fluid to lubricate the joint and nourish the cartilage.

The joint is reinforced by ligaments. Some ligaments are just thickened areas of the capsule itself, while others are attached to the bones. The cruciate ligaments of the knee joint, for example, are very strong fibrous cords that are separate from the capsule of the knee joint.

Capsules and ligaments are both made up of fibers of connective tissue. While the fibers in capsules are randomly arranged, those in ligaments are densely packed and run parallel in one direction. This arrangement gives ligaments a shiny appearance and great tensile strength in the direction of the fibers. In some injuries, bones break before ligaments rupture.

Joints can also be investigated by arthroscopy, a technique in which a small tube is inserted into the joint to transmit images of the interior to a screen; arthroscopy can be used to guide surgical procedures within the joint, such as taking a biopsy of synovial membrane.

When a joint is sprained, the ligaments may be stretched or ruptured. When a joint is immobilized for a long period of time, such as in a cast, the capsule and ligaments contract and become stiff, reducing the range of movement of the joint when the cast is removed. Physical therapy and stretching exercises often help the joint regain its mobility.

Cartilage

Cartilage is a tough, semi-transparent, elastic, flexible connective tissue consisting of cartilage cells (chondrocytes and chondroblasts) scattered through a glycoprotein material strengthened by collagen fibers. The exterior part of cartilage is covered by a dense fibrous membrane called the perichondrium. There are no nerves or blood vessels in cartilage, and when damaged it does not heal readily.

Cartilage has several functions. It covers the surfaces of joints, allowing bones to slide over one another, thus reducing friction and preventing damage; it also acts as a shock absorber. It forms part of the structure of the skeleton in the ribs, where it joins them to the breastbone (sternum). Cartilage is found in the tip of the nose, in the external ear, in the walls of the windpipe (trachea) and the voice box (larynx) where it provides support and maintains shape. In an embryo, the skeleton is formed of cartilage which is gradually replaced by bone as the embryo grows. Cartilage is known as elastic cartilage, fibrocartilage or hyaline cartilage, depending on its different physical properties.

A "torn cartilage" commonly refers to a disorder of the knee. A section of cartilage

TISSUE TYPES IN A SYNOVIAL JOINT

In a synovial joint, cartilage acts as a cushion between the bones. Ligaments reinforce the joint. Spongy bone tissue forms most of the bone with compact bone underlying the cartilage.

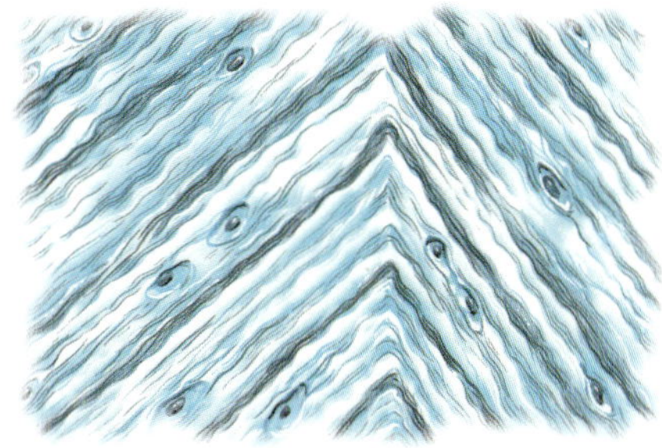

Fibrocartilage

Fibrocartilage contains large amounts of collagen, making it both resilient and able to withstand compression. It is found between the bones of the spinal column, hip and pelvis.

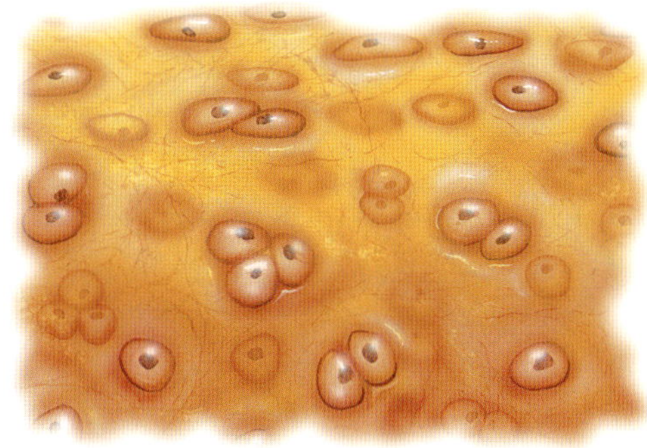

Elastic cartilage

Elastic cartilage is strong but supple cartilage containing proteins called elastin and collagen embedded in ground substance. Elastin gives it a distinctive yellow color. Elastic cartilage makes up the springy part of the outer ear, and also forms the epiglottis.

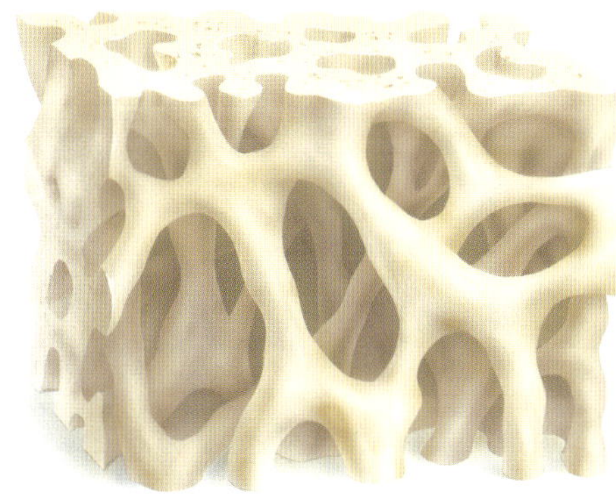

Spongy bone tissue

The air pockets and branching structure of spongy bone make it both light and strong.

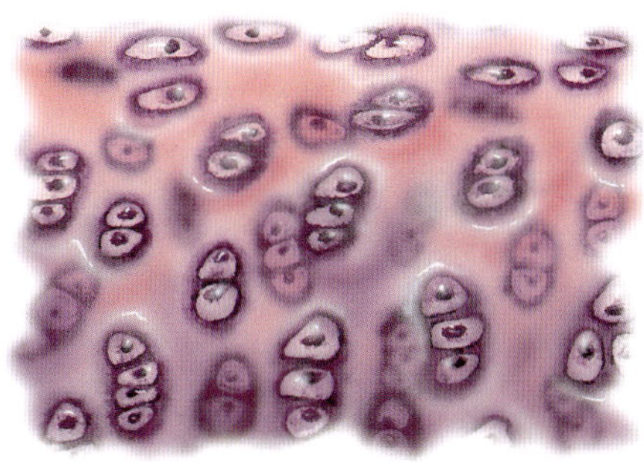

Hyaline cartilage

Hyaline cartilage contains collagen fibers. It forms the skeleton in the embryo and remains as a thin layer on the end of bones that form joints.

pad inside the knee joint known as the meniscus tears, and may move around in the knee joint, causing pain, swelling, and preventing weight being placed on that knee. A minor tear can be treated effectively with rest and a firm bandage around the knee. More serious tears require the surgical removal of part or all of the cartilage. This can often be done through an arthroscope—a tube that is passed into the knee joint, allowing visualization of the interior of the joint.

Costochondritis (Tietze's syndrome) is a painful inflammation of the cartilage of the ribs (commonly the third or fourth rib). It causes pain in the chest wall, which may be mistaken for cardiac pain. The cause is often unknown, though it may be the result of trauma (such as a blow to the chest), unusual physical activity or an upper respiratory infection. It usually clears up in a short time with rest and mild anti-inflammatory medications such as aspirin, acetaminophen (paracetamol) or ibuprofen.

Chondrosarcoma is an uncommon cancer that may arise from cartilage associated with bone or outside of bone.

The affected person notices a painful lump, usually in the long bones of the limb, pelvis or ribs. The diagnosis is confirmed with x-ray, CAT or MRI scans, biopsy and laboratory examination. The treatment is surgical removal of the tumor. If the tumor is slow growing and detected and treated early, the chances of survival are good.

Synovial membrane

Synovial membrane is found in synovial joints, bursae and tendon sheaths. It lines the non-cartilage areas of synovial joints, such as the joint capsule and exposed bony surfaces. The membrane is very vascular. It appears pink, smooth and shiny and may exhibit folds and fringes. Accumulations of fat are found in the membrane in some joints (for example, fat pads in the knee and elbow joints).

Specialized cells of the membrane (synoviocytes) both produce and reabsorb synovial fluid, a fluid that is critical to the lubrication of joint surfaces, and nourishment of the cartilage. These cells also remove debris from the joint cavity and may initiate an immune response to foreign material in the joint.

Synovial fluid

The fluid contained within synovial joints lubricates the joint surfaces, helps to reduce friction and provides nourishment for the joint cartilage. It is normally a clear, pale yellow, viscous fluid, and is only present in small amounts. Its composition is similar to blood plasma except that it contains hyaluronate and lubricin, which are essential for viscosity and lubrication. Monocytes, lymphocytes and macrophages are also found in low numbers in the synovial fluid. Synovial fluid may be aspirated from a joint and checked for the presence of red blood cells, inflammatory cells, infectious agents and crystals (for example, gout).

Ligaments

Ligaments (from the Latin ligamentum, meaning a band or tie) are tough, white, fibrous, slightly elastic tissues. They support many internal organs, including the uterus, the bladder, the liver and the diaphragm, and they also help in shaping and supporting the breasts. Their main function is to support and strengthen joints, preventing excessive movement that might cause dislocation and breakage of the bones in the joint.

A ligament may be damaged or torn if a joint, through injury or accident, is moved into a position it was not designed for. The tear may be a complete tear of all the strands of the ligament or a partial tear, where only some of the ligament strands are torn. A tear in a ligament is sometimes called a sprain.

Sprains most often occur in the ankle and knee joints, but may also occur in the fingers, wrist, shoulder and the spine.

Sprains should be treated by the application of cold compresses, immobilization of the joint, elevation of the joint to allow fluid around the joint to drain away, and the application of a bandage or splint. Nonsteroidal anti-inflammatory drugs (NSAIDs) may be given for pain relief and to aid healing.

Ligaments tend to heal very slowly because of their poor blood supply. Healing usually takes 7–10 days for mild sprains and 3–5 weeks for severe sprains. If a ligament is completely torn or severed from its point of attachment to the bone then surgery may be required.

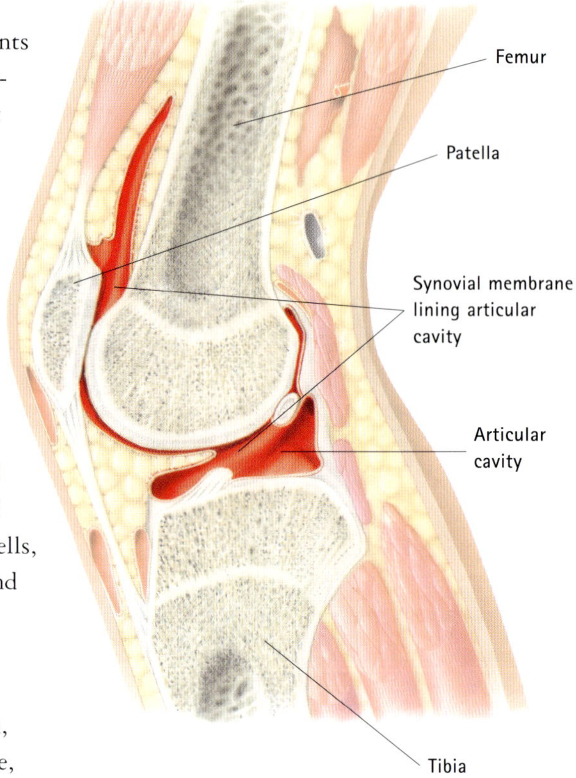

Synovial membrane

This type of membrane lines the non-cartilage areas of synovial joints such as the knee. Specialized cells in the membrane produce fluid that is critical for the lubrication of joint surfaces and the nourishment of cartilage.

Joint diseases and disorders

Joints are exposed to extreme stress and daily wear, making them prone to injury. Disorders affecting the bones, synovial membrane, or cartilage in a joint will also affect mobility and movement.

Displacement of the bones at a joint is called dislocation, and is more common in some joints, such as the shoulder, than in others which are more stable, such as the elbow. In some cases the dislocation is present from birth; this is called a congenital dislocation. This is common at the hip joint ("clicky hip") and, if not treated, will prevent the joint from developing normally.

Inflammation in the joints is called arthritis. There are two common types of arthritis, with different causes.

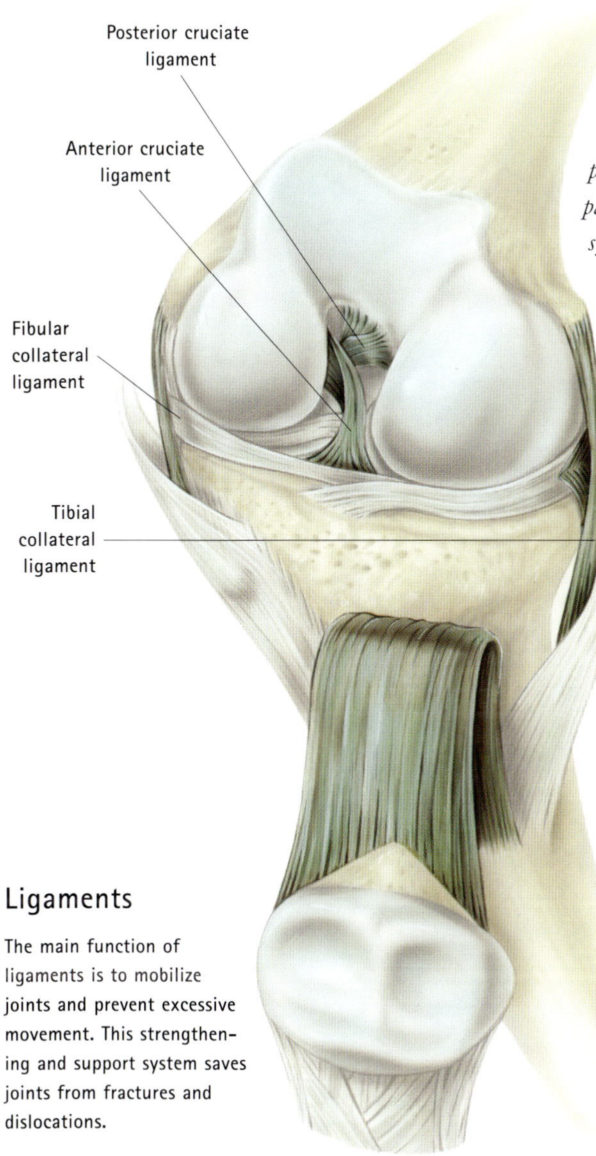

Ligaments

The main function of ligaments is to mobilize joints and prevent excessive movement. This strengthening and support system saves joints from fractures and dislocations.

Ligament microstructure

The direction of fibers within ligaments is related to the stress that is applied to them. The considerable interweaving increases structural stability and resilience.

SEE ALSO *Disorders of the ankle on page 341; Disorders of the hip on page 337; Disorders of the knee on page 339; Disorders of the shoulder on page 321; Disorders of the wrist on page 325; Treating the musculoskeletal system on page 444*

Arthritis

Arthritis is inflammation of one or more joints, causing redness, swelling, pain and sometimes loss of joint mobility. There are many different kinds of arthritis. Arthritis may result from wear and tear on the joints (osteoarthritis) or from active joint diseases such as gout or rheumatoid arthritis. It may also be a symptom of a generalized disease, such as connective tissue disease.

Symptoms of arthritis include joint pain (arthralgia), joint swelling, early morning stiffness and reduced joint movement. There may be warmth and redness of the skin around a joint, and more general symptoms such as unexplained weight loss and fever. Swelling is often due to a fluid collection called an effusion. The joint may be tender, and painful when moved. Treatment depends on the underlying cause. Rest and exercise, physical therapy, drug treatments and surgery all play a role in managing it.

OSTEOARTHRITIS

Osteoarthritis is the most common form of joint disease, where progressive deterioration of cartilage is accompanied by the formation of bony spurs and growth of dense bone at the margins of the joint. This condition does not necessarily involve the inflammation common in other forms of arthritis, many of which occur as a result of infection or accident trauma.

Estimates are that over 40 percent of the adult population of the USA and the UK have symptoms which show under x-ray. Joints are formed where two bones meet. Synovial joints have cartilage on the adjacent surfaces that cushions the adjoining surface of each bone, reducing friction on movement and protecting from shock. Synovial joints are lubricated by synovial fluid and enclosed within a fibrous capsule. Heavy use and the passage of time causes wear and tear on the cartilage coating bone ends, sometimes eroding it completely together with the underlying bone surface.

Symptoms of osteoarthritis have been found in skeletons of Neanderthal man and are consistently found in humans over the age of 70, but can occur much earlier. It is a condition common in all vertebrate animals, including birds and fish and all mammals except bats and sloths, which spend their lives hanging upside-down placing little weight on their joints.

Ageing of the cartilage normally begins in early adult life and most commonly affects joints of the hip, knee, spine and hands. This degenerative damage is irreversible. Gradually the cartilage becomes less well lubricated and less effective as a shock absorber, with increased friction and pain on movement. There may be increasing stiffness and discomfort, particularly on rising.

Osteoarthritis in older individuals most commonly affects the major weight-bearing joints, such as the hips and knees. The degree of pain and stiffness involved can vary widely from person to person.

Although osteoarthritis is a more common ailment in older age groups, factors other than age are the primary cause. Sites of sports injuries in youth, an injured knee for example, often become the first reported site of osteoarthritic effects such as limited mobility, pain and stiffness. Such damage can occur to weight-bearing joints of large-framed or overweight people early in life, yet go unnoticed until distinct symptoms are felt. Genetic factors may also be involved and twice as many women suffer as men.

Symptoms include swelling and pain at the joint in response to activity or a change in position, limited flexibility, and a condition known as Heberden's nodes, which are bony lumps, particularly noticeable at the joints nearest the ends of the fingers, and are thought to be genetic in origin. There may be tenderness at joints, and x-rays may show a change in shape and a reduction in the thickness of cartilage in the joint. Diagnosis may include blood tests.

Sufferers often complain of a deep ache in the center of the joint, worsened by use and relieved by rest. This pain may become constant and severe enough to interfere with sleep. Pain and discomfort may respond to analgesics and to corticosteroids if there is accompanying inflammation. Pain-relieving therapies include the application of heat by taking hot baths or applying heat pads, or the application of cold by the use of ice packs. Further damage to the joint may be preventable through taking exercise to strengthen supporting muscles and improve flexibility and range of movement. Excess weight should be reduced where this is a contributing factor. A common and increasingly successful solution is joint replacement, with artificial hip, knee and finger joints being recommended in some cases.

RHEUMATOID ARTHRITIS

Rheumatoid arthritis (RA) is a chronic and progressive condition which inflames connective tissue throughout the body. It most

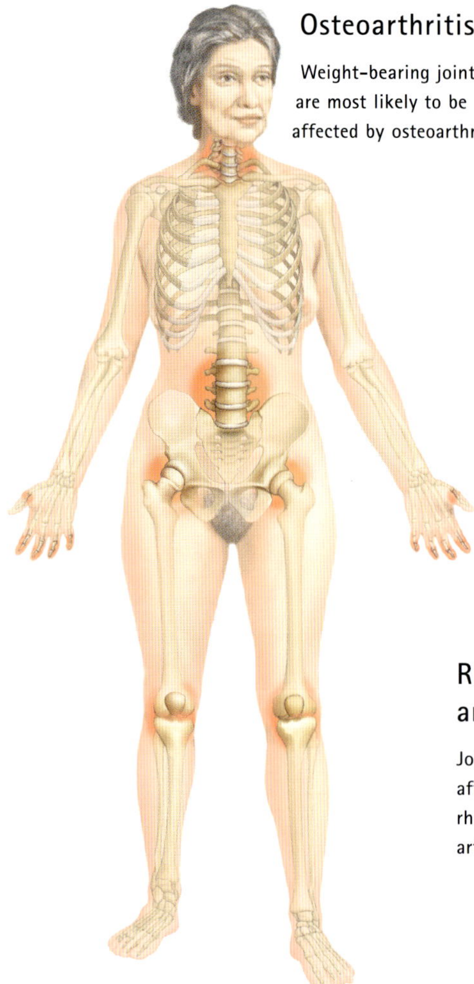

Osteoarthritis

Weight-bearing joints are most likely to be affected by osteoarthritis.

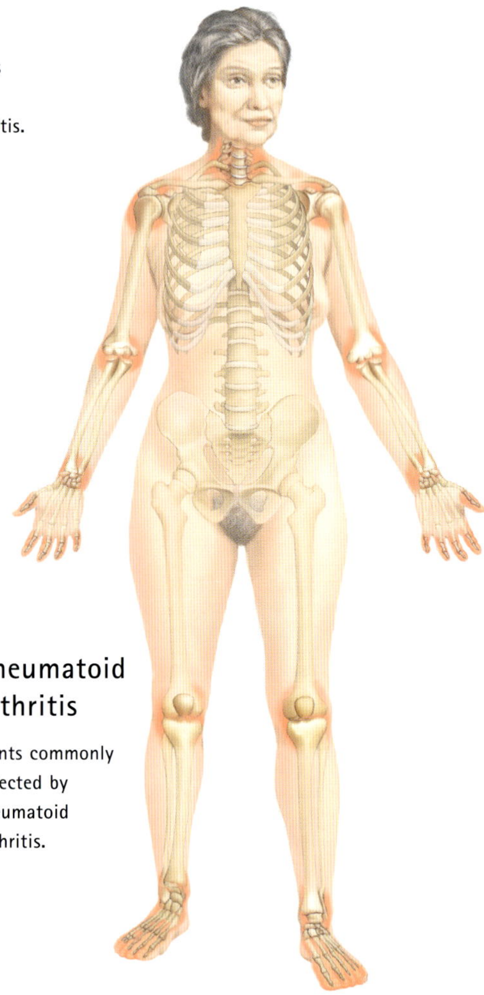

Rheumatoid arthritis

Joints commonly affected by rheumatoid arthritis.

Synovial deterioration

The cartilage that covers the ends of the bones in synovial joints, such as the knee, is lubricated by a fluid that is similar to blood plasma. It is produced by the synovial membrane and is essential to proper lubrication and function of the normal joint. There may be some deterioration of the membrane in osteoarthritis.

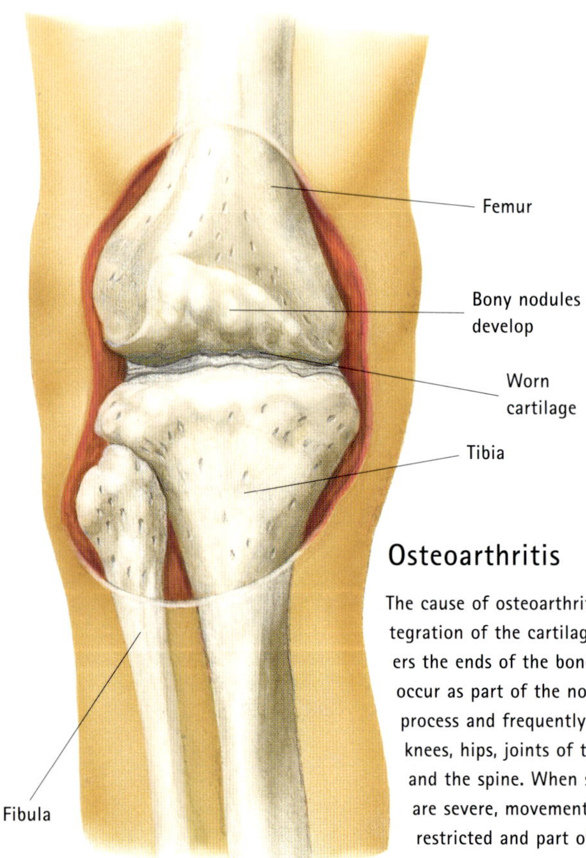

Osteoarthritis

The cause of osteoarthritis is disintegration of the cartilage that covers the ends of the bones. This will occur as part of the normal ageing process and frequently affects the knees, hips, joints of the big toes and the spine. When symptoms are severe, movement can be restricted and part of the affected bone may wear away.

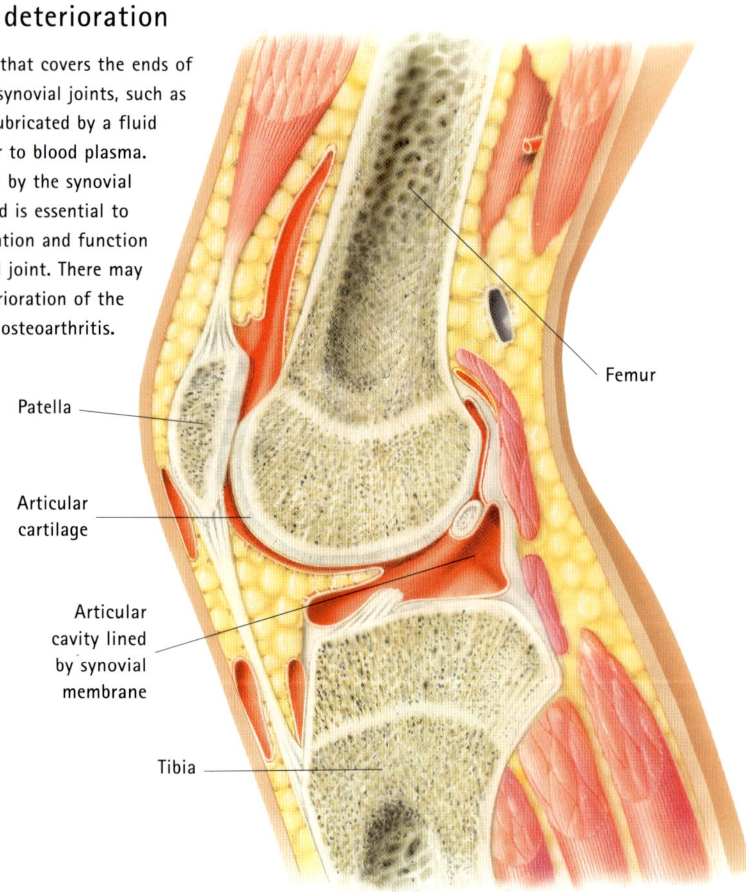

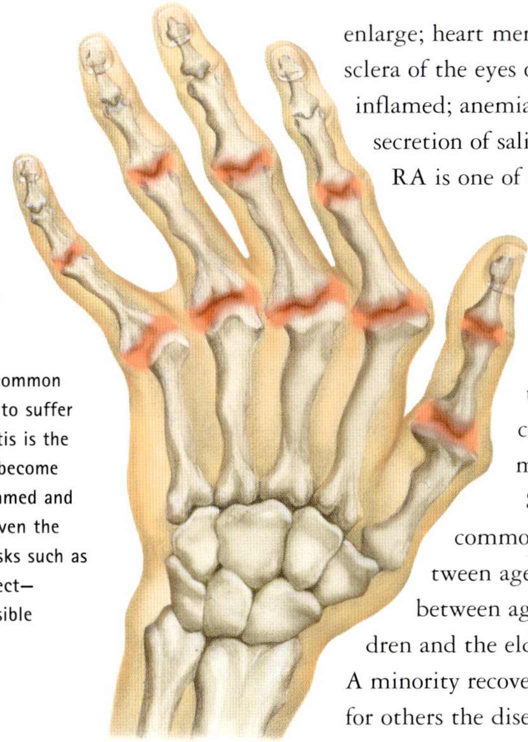

Rheumatoid arthritis

One of the most common areas of the body to suffer rheumatoid arthritis is the hands. The joints become stiff, painful, inflamed and swollen, making even the most simple of tasks such as picking up an object—difficult or impossible to do.

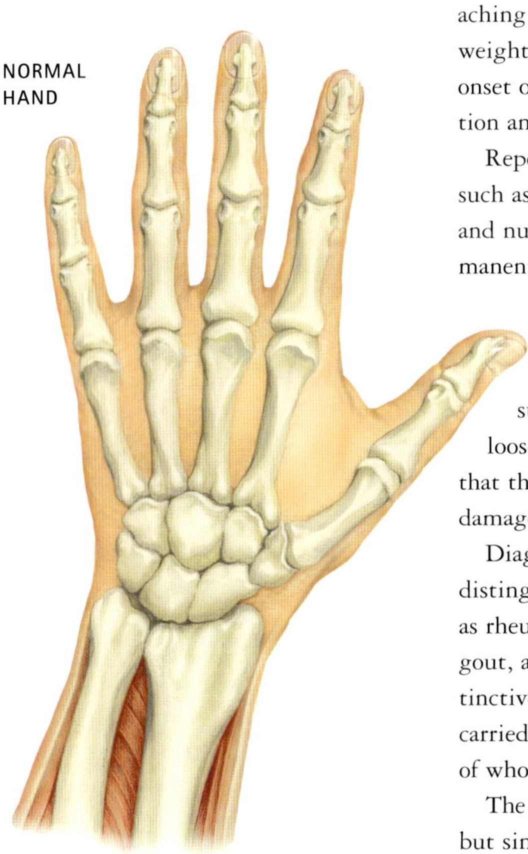

commonly affects both sides of the body simultaneously and involves the small joints of fingers, wrists, toes, ankles and elbows where the adjoining bone ends are enclosed in a membrane containing fluid for lubrication to ease movement.

The lungs and kidneys may also be affected by the disease; the spleen may enlarge; heart membranes, conjunctiva and sclera of the eyes or arteries may become inflamed; anemia, dry eyes and reduced secretion of saliva are also possible effects.

RA is one of many forms of arthritis, a group of around 100 disorders which may affect up to 20 percent of people in industrialized society, making it possibly the most common chronic cause of continuing or permanent disability.

Symptoms of RA most commonly arise in women between ages 35 and 55 and in men between ages 40 and 60, but children and the elderly can suffer attacks. A minority recover after only one attack, but for others the disease is progressive, needing continual treatment. The major effects are aching and stiff joints, fatigue and anemia, weight loss and wasting muscles, with the onset of the disease marked by inflammation and redness around the affected areas.

Repeated attacks may produce problems such as carpal tunnel syndrome, with pain and numbness in the hand and wrist; permanent swelling of finger joints, knuckles and wrist; inflamed tendons; tenosynovitis; and nodules under the skin of the arms. When attacks subside, joints may become excessively loose and mobile, and it is in this state that they are most susceptible to further damage through overuse.

Diagnosis involves a blood test used to distinguish RA from other conditions such as rheumatic fever, infectious arthritis and gout, and to detect the presence of a distinctive antibody called rheumatoid factor, carried by about 70 percent of people, some of whom will never suffer the disease.

The exact cause of the disease is unknown, but since the rheumatoid factor antibody is found in most sufferers, RA is thought to have an autoimmune mechanism, a reaction which causes the body's defense system to attack its own tissues. It is also thought that susceptibility to RA is a genetically inherited trait, with the disease being triggered by infection or possibly by environmental factors.

Treatment can involve rest, diet therapy, drugs and surgery, depending on individual symptoms. Rest may relieve pain and is recommended during attacks because movement aggravates the inflamed joints. Supports and splints can be used to immobilize joints for limited periods but regular use and movement of the joints is essential to prevent stiffness and preserve mobility. Certain foods can cause attacks in some people and paying careful attention to a balanced diet may help.

The drugs prescribed for symptomatic relief in rheumatoid arthritis are mainly non-steroidal anti-inflammatories (NSAIDs), including aspirin and ibuprofen. However, these drugs do not appear to alter the long-term course of the disease. They should not be taken by those with gastric ulcers as side effects may include digestive upsets as well as headaches, increased blood pressure and edema. There are also specific antirheumatoid drugs. These include penicillamine, gold preparations and methotrexate. These drugs all have potentially severe side effects, and close monitoring is required.

Lack of movement in joints and muscle weakness can create difficulties in walking and accomplishing everyday tasks. A range of aids is available, from orthopedic shoes to specially designed household appliances and hand tools. Fusion of small joints or replacement of hips or knees are options of last resort where other treatments have failed.

Still's disease, also known as juvenile rheumatoid arthritis (JRA), is a form of rheumatoid arthritis that affects children under 16 years of age.

Gout

Gout is an inflammation of the joints that is usually accompanied by the presence of excess uric acid (one of the body's waste products) in tissues in the body. If the uric acid level gets sufficiently high, needle-shaped crystals develop within a joint, leading to an inflammatory response in the joint. Gout most commonly occurs in the joint at the base of the big toe but it can affect other joints such as the hands, wrist, elbow and ankle. The attack usually starts suddenly, often at night, with the joint becoming red, swollen and very painful.

Gout is most common in middle-aged men, but can occur in women after menopause. The high levels of uric acid

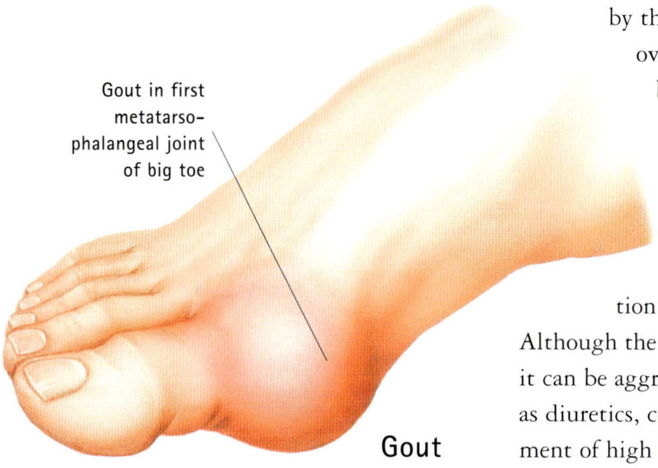

Gout

are basically due to either its overproduction or to not excreting enough, or a combination of both. Uric acid is formed from the breakdown of purines in the diet, which particularly come from offal, shellfish and some vegetables and fruits. Gout can also be aggravated by too much alcohol, which inhibits the excretion of uric acid by the kidneys. In rare cases, the overproduction of uric acid can be due to an inherited disorder of protein synthesis or to diseases that increase cell turnover.

The most common finding is reduced excretion of uric acid in the urine. Although the cause is often unclear, it can be aggravated by some drugs, such as diuretics, commonly used in the treatment of high blood pressure, and aspirin.

Besides being deposited in joints, crystals of urate may also be deposited in tissues around the joints and under the skin, for instance in the hands, elbows and around the ear. These deposits are called tophi; those under the skin can be felt as hard nodules. The presence of tophi in the joints, which can be seen on x-rays, can lead to arthritis and joint erosion. Crystals can also be deposited in the kidneys, where they are known as kidney stones. These may cause kidney damage, obstruction of urine flow, or painful renal colic as the stones are passed out of the body.

The diagnosis of gout can be made by measuring the uric acid levels in the blood, and by finding urate crystals within a joint. Gout is treated by anti-inflammatory drugs to reduce the pain and joint inflammation. Aspirin should not be used because it actually inhibits uric acid excretion. After the acute attack passes, long-term medication with drugs such as allopurinol can reduce the production of the uric acid. This can help to prevent future attacks, and if uric acid levels can be lowered sufficiently, may lead to resorption of some tophi and therefore prevent further joint destruction. High fluid intake is also helpful, especially for patients with kidney stones. If uric acid levels are untreated attacks usually become more common, and can lead to permanent joint damage and deformity.

Temporomandibular joint syndrome

Temporomandibular joint (TMJ) syndrome is a term for a range of problems in the joint and surrounding muscles of the jaw that may produce pain, discomfort, clicking noises, aching or tender muscles, locking or restricted movement of the joint, neck pain, apparent toothache and headaches.

The action of the temporomandibular joints is complicated, allowing movements in three directions—hinging to open and close, sliding backward and forward, and moving from side to side. Between the jawbone (mandible) and the temporal bone of the skull is a cartilaginous disk which

Temporomandibular joint syndrome

The temporomandibular joint has a complex action that allows movement up and down, side to side and backward and forward. Damage to cartilage in the joint or to the supporting muscles can result in temporomandibular joint syndrome. This may limit the action of the joint, cause clicking noises, or pain in the jaw muscles, neck or head.

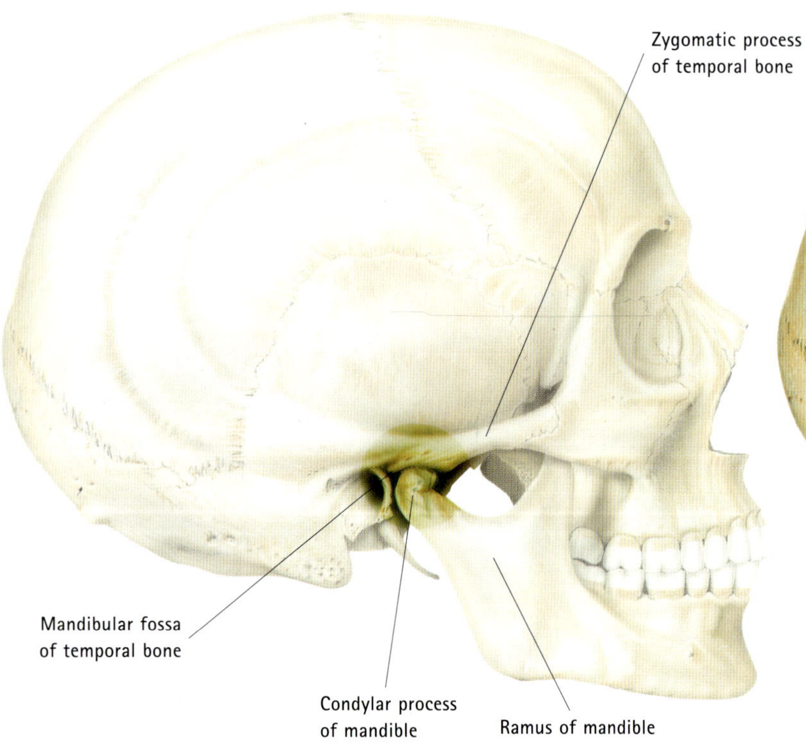

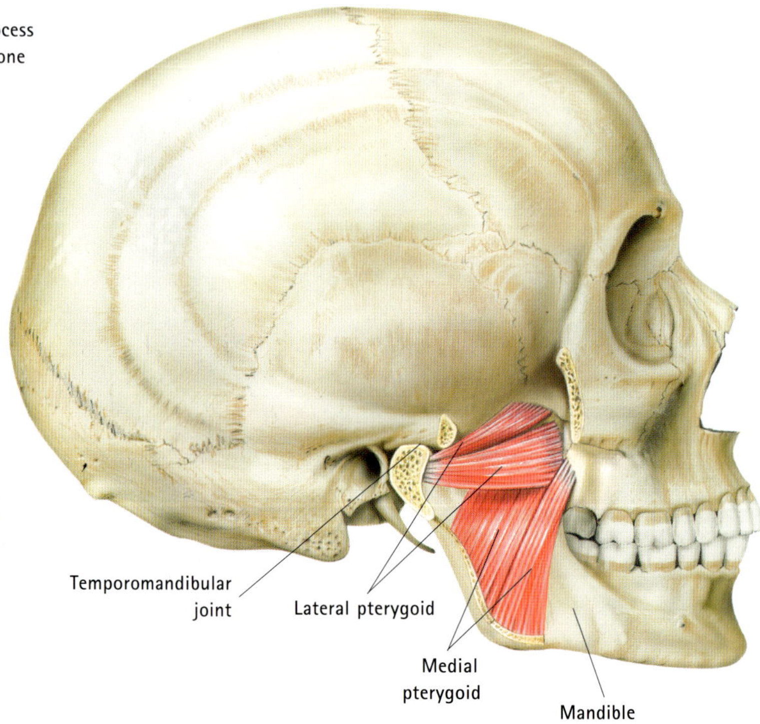

cushions the joint and can sometimes be displaced, as it is subject to great pressure during chewing.

Teeth-grinding (bruxism) during sleep can wear away biting surfaces, causing the teeth to meet unevenly ("bad bite"). This can cause jaw muscles (which must work together in a smooth and balanced way) to spasm, producing pain and headaches; the jaw may be displaced or lock in an abnormal position. Displaced disks rarely need surgery; teeth-grinding can be stopped by wearing a custom-made bite plate at night.

Osteoarthritis and rheumatoid arthritis can affect the joint and it is important to maintain mobility to avoid calcification of ligaments or fusion (ankylosis), which requires surgical correction.

In the absence of injury or other clear causes, muscular tension is often found to be the cause of TMJ syndrome. Rest and avoiding opening the jaw wide, even when yawning, may cure the problem, and reducing stress levels should also help.

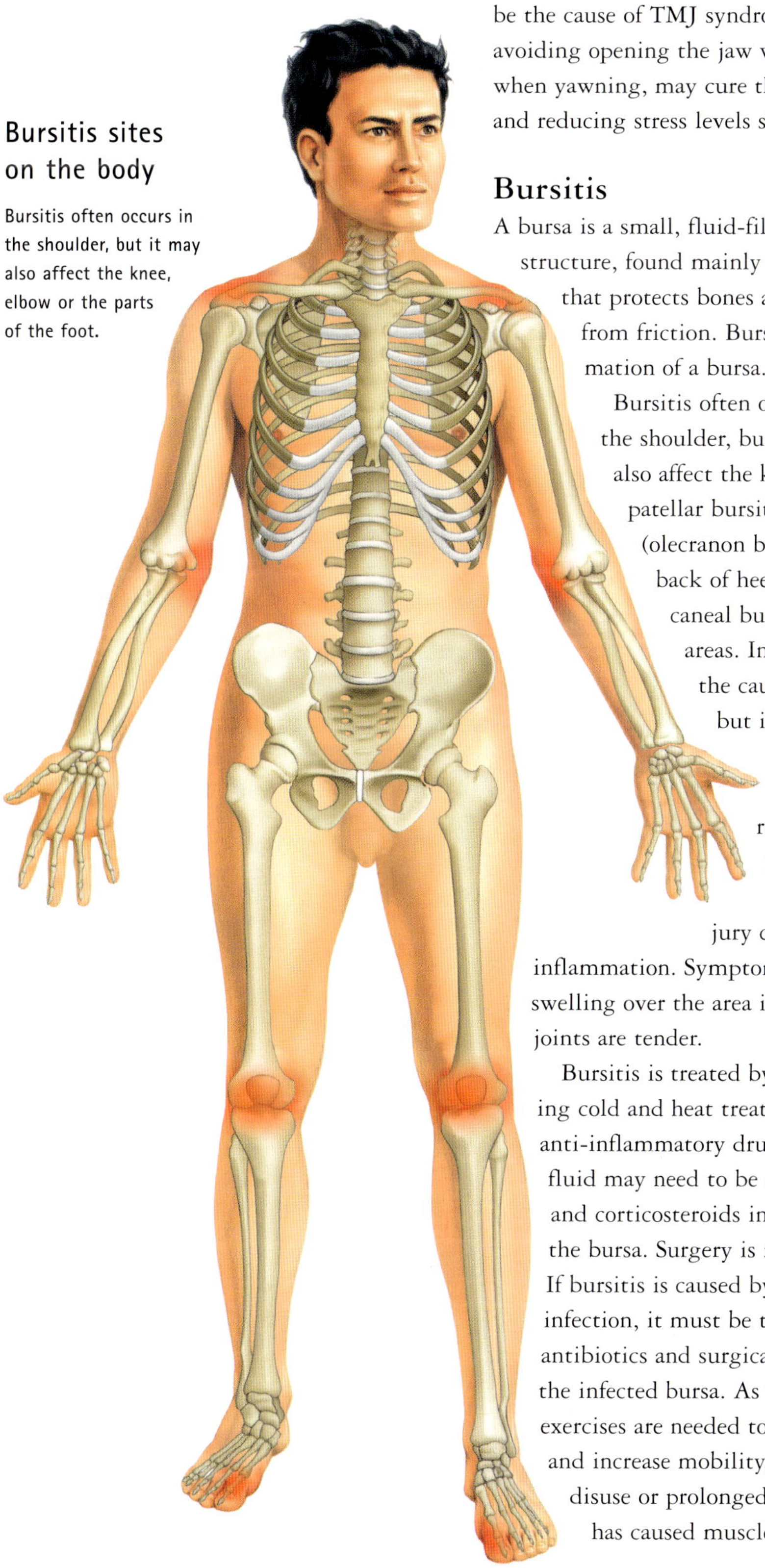

Bursitis sites on the body

Bursitis often occurs in the shoulder, but it may also affect the knee, elbow or the parts of the foot.

Bursitis

A bursa is a small, fluid-filled sac-like structure, found mainly around joints, that protects bones and tendons from friction. Bursitis is inflammation of a bursa.

Bursitis often occurs in the shoulder, but it may also affect the knee (infrapatellar bursitis), elbow (olecranon bursitis), the back of heel (retrocalcaneal bursitis) or other areas. In most cases the cause is unknown, but it can be caused by injury, infection and repeated friction. Repeated attacks of bursitis or injury can cause chronic inflammation. Symptoms are pain and swelling over the area involved. Nearby joints are tender.

Bursitis is treated by rest, alternating cold and heat treatments and oral anti-inflammatory drugs. Occasionally, fluid may need to be aspirated from, and corticosteroids injected into, the bursa. Surgery is rarely required. If bursitis is caused by bacterial infection, it must be treated with antibiotics and surgical drainage of the infected bursa. As the pain eases, exercises are needed to build strength and increase mobility, especially if disuse or prolonged immobility has caused muscle wasting.

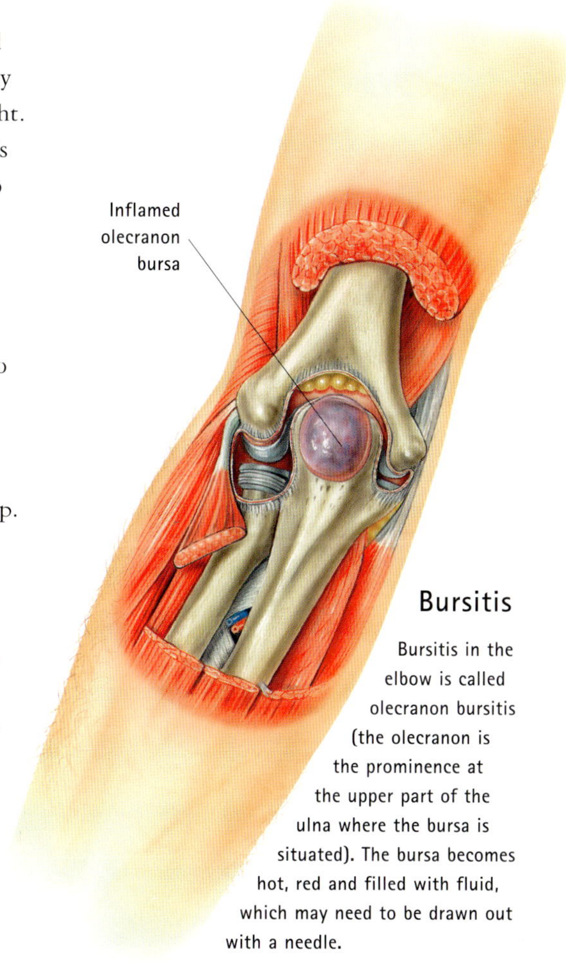

Bursitis

Bursitis in the elbow is called olecranon bursitis (the olecranon is the prominence at the upper part of the ulna where the bursa is situated). The bursa becomes hot, red and filled with fluid, which may need to be drawn out with a needle.

Frozen shoulder

The capsule of the shoulder joint can become inflamed and thickened, causing movements to gradually become more limited, a condition referred to as frozen shoulder. This requires pain relief; physical therapy can also be useful.

Synovitis

The cavities of freely movable joints are lined with synovial membranes, smooth, thin sheets of connective tissue that secrete a nourishing lubricant (synovial fluid) that helps bones move freely over other bones.

When a synovial membrane becomes inflamed, the condition is called synovitis. It can often cause an entire joint to become swollen and tender. Synovitis can occur with a bacterial infection, follow an irritation or trauma to the site such as a sprain or fracture, or be a complication of diseases such as gout and rheumatoid arthritis.

Treatment depends on the cause. In many cases time and rest are sufficient but severe or chronic synovitis may need treatment with analgesics or anti-inflammatory drugs.

MUSCULAR SYSTEM

The muscular system which brings about bodily movement includes the voluntary muscles of the body. These muscles range in size from the tiny muscles that wrinkle the forehead to the large muscles of the thigh. The voluntary muscle system does not include muscles like the cardiac muscle of the heart or the smooth muscles in the walls of internal organs such as the stomach, which are classed as involuntary muscles (not under conscious control).

There are about 700 muscles in the human body. Most of them have Latin names which may describe their location (brachialis, muscle of the arm), beginning and end (brachioradialis, running from the arm to the radius), shape (trapezius, shaped like a trapezium), location and shape (orbicularis oris, circular muscle around the mouth), organization (quadriceps, muscle with four heads) or function (dilator naris, dilator of the nostril).

When some muscles have the same name, qualifiers are added to distinguish between them. Of the two flexors of the thumb, the flexor pollicis longus is the long muscle running from the forearm to the thumb, while the flexor pollicis brevis begins in the wrist. The three muscles of the buttocks are named according to their size: gluteus maximus for the largest, gluteus medius and gluteus minimus for the medium and smallest.

Some muscles have fancy names; the buccinator (trumpeter) in the cheek is so-called because it blows air out of the mouth. Interestingly, the very small muscle that raises the upper lip and the nostril has one of the longest names: levator labii superioris alaeque nasi. Muscles vary greatly in size. The stapedius, which restricts movements of the eardrum, looks like a few millimeters of cotton thread. The gluteus maximus, on the other hand, forms the bulk of the buttock.

The organization of fibers also varies. In the common spindle-shaped muscles, all muscle fibers run from one tendon to another. In pennate (*penna*, meaning feather) muscles, fibers run obliquely down to the tendon, like a feather. Some muscles have several tendons at one end—for example, two in the biceps, three in the triceps and four in the quadriceps.

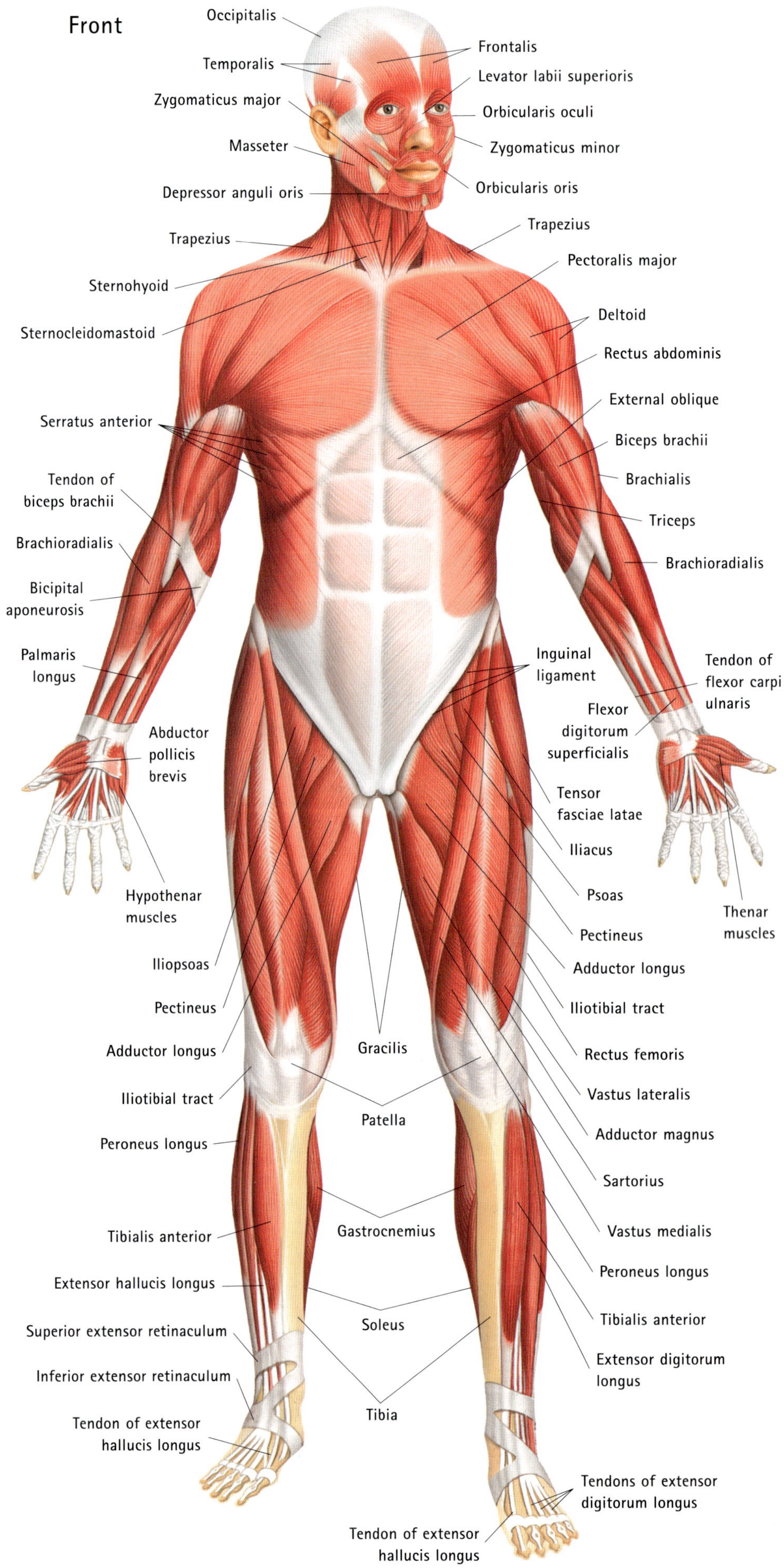

Back

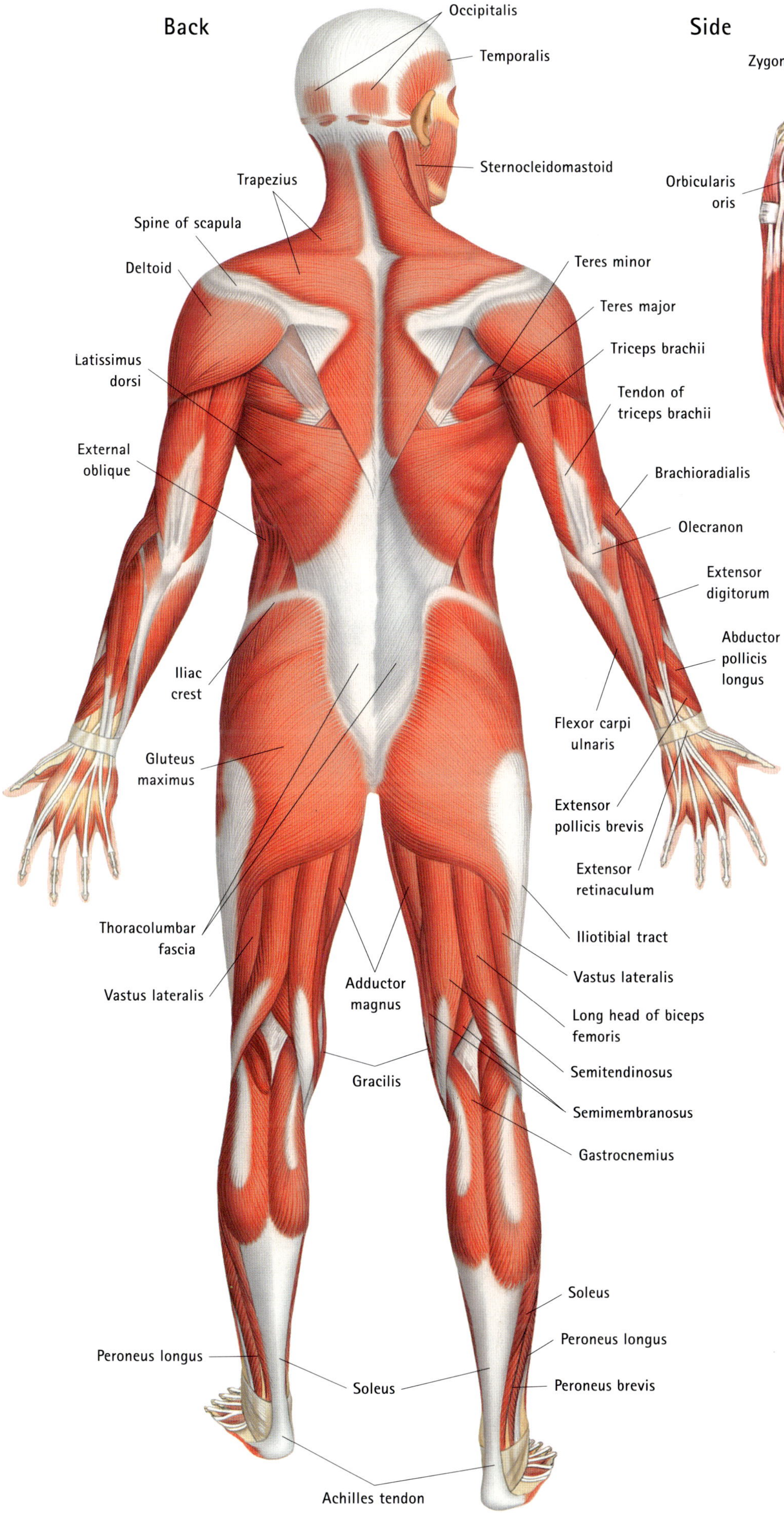
Occipitalis
Temporalis
Sternocleidomastoid
Trapezius
Spine of scapula
Deltoid
Teres minor
Teres major
Triceps brachii
Latissimus dorsi
Tendon of triceps brachii
External oblique
Brachioradialis
Olecranon
Extensor digitorum
Abductor pollicis longus
Iliac crest
Flexor carpi ulnaris
Gluteus maximus
Extensor pollicis brevis
Extensor retinaculum
Thoracolumbar fascia
Iliotibial tract
Vastus lateralis
Adductor magnus
Vastus lateralis
Long head of biceps femoris
Semitendinosus
Gracilis
Semimembranosus
Gastrocnemius
Soleus
Peroneus longus
Peroneus longus
Soleus
Peroneus brevis
Achilles tendon

Side

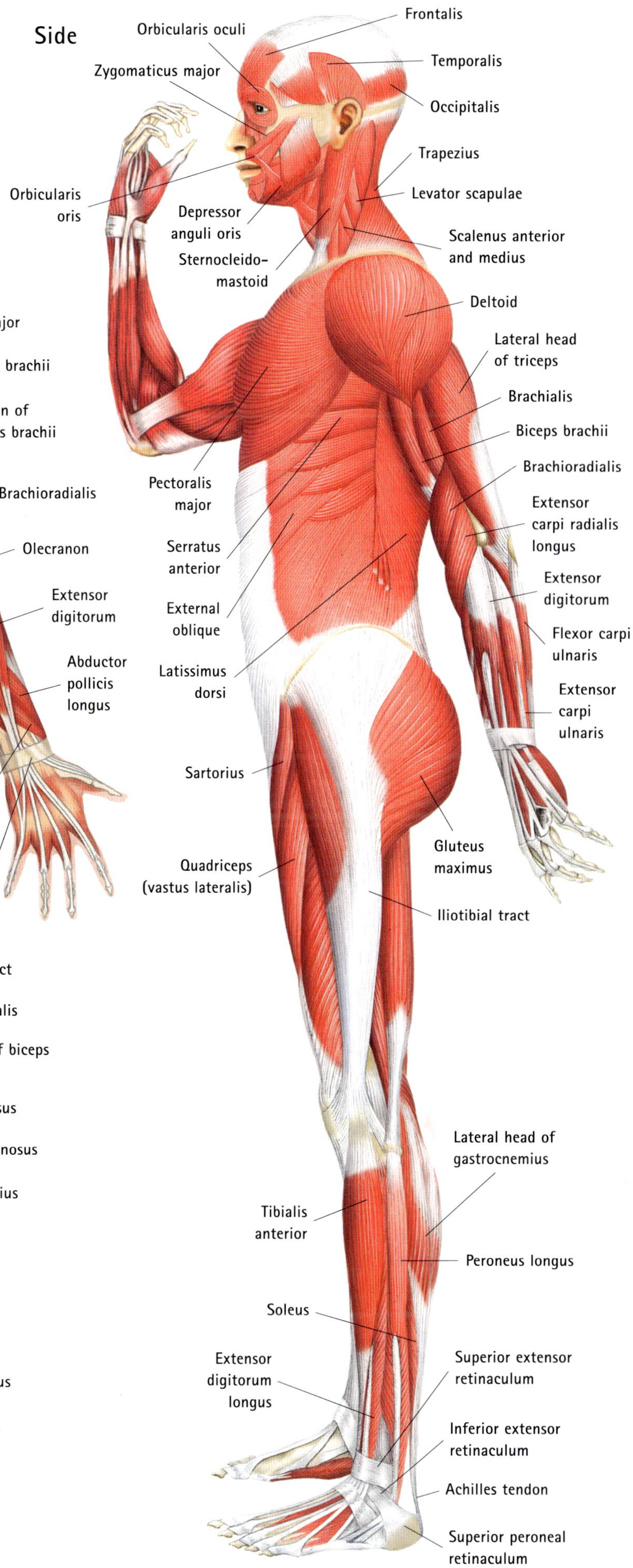
Frontalis
Orbicularis oculi
Temporalis
Zygomaticus major
Occipitalis
Trapezius
Orbicularis oris
Levator scapulae
Depressor anguli oris
Scalenus anterior and medius
Sternocleido-mastoid
Deltoid
Lateral head of triceps
Brachialis
Biceps brachii
Pectoralis major
Brachioradialis
Extensor carpi radialis longus
Serratus anterior
Extensor digitorum
External oblique
Flexor carpi ulnaris
Latissimus dorsi
Extensor carpi ulnaris
Sartorius
Gluteus maximus
Quadriceps (vastus lateralis)
Iliotibial tract
Lateral head of gastrocnemius
Tibialis anterior
Peroneus longus
Soleus
Extensor digitorum longus
Superior extensor retinaculum
Inferior extensor retinaculum
Achilles tendon
Superior peroneal retinaculum

MUSCLE TISSUE

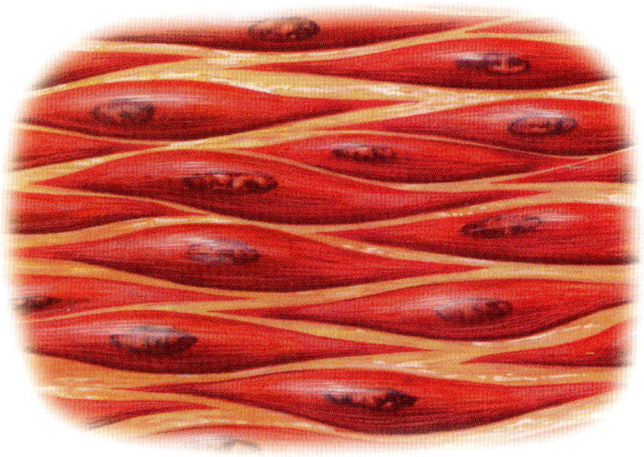

Smooth muscle tissue

Smooth muscle is controlled by the autonomic nervous system, and is found in the skin, the blood vessels, and the reproductive and digestive systems.

Skeletal muscle tissue

Skeletal muscles allow the body to move. They are voluntary muscles, controlled by the brain and spinal cord.

Cardiac muscle tissue

Cardiac muscle is the heart muscle, which contracts and relaxes rhythmically in an involuntary manner.

Muscle fibers attach either directly to a bone, or to a tendon which is fixed to a bone. The force produced by the contraction of the muscle fibers is transmitted to the bone by the tendon. Tendons are made up mainly of strong collagen fibers that run parallel in one direction and are tightly packed together to give maximum strength in the line of force of the muscle.

SEE ALSO *Heart on page 228; Muscles of the hand on page 324; Muscles of the leg on page 330; Muscles of the shoulder on page 316; Musculoskeletal column on page 198; Trapezius muscle on page 220*

Coordination of muscles

Each movement of a limb, however simple, is the result of a number of muscles working together. In kicking a football, the quadriceps is the prime mover, or agonist muscle. Its attachment on the femur is stabilized by muscles of the hip. The hamstring muscles, which have the opposite action to the quadriceps, begin to contract as the leg picks up momentum to control the force of the kick. The contraction of these muscles, which are called antagonist muscles, is maximal at the end of the kick to stop the movement.

The cooperation of the muscles in this example must be well orchestrated, each muscle having to contribute the right force at the right time. If the antagonist muscles contract too early, before the leg has gained enough momentum, the kick will be too weak. If they come in too late, the leg will extend too far and damage the knee joint.

The cerebellum of the brain is important in fine muscle (motor) control. Just watch a very young child attempting to throw a ball. The young cerebellum has not learned to program muscle actions. The arm flings out too far and the hand does not release the ball at the right time. Even movements that we take for granted, such as walking and running, are only possible after much training in the first years of life. More sophisticated muscle coordination, such as dancing, requires years of intensive training. When the cerebellum fails to work properly, either temporarily in a drunk person, or permanently in cerebellar diseases, even walking becomes difficult.

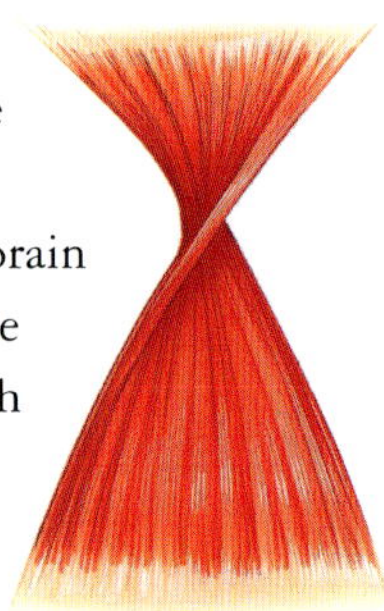

SPIRAL

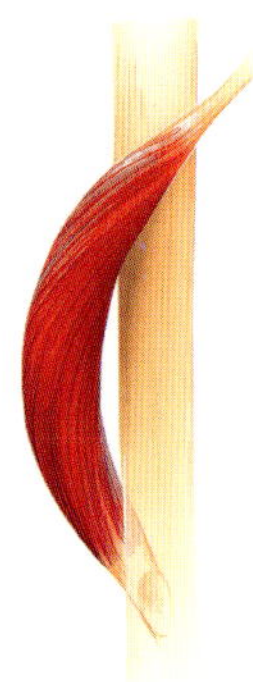

SPIRAL

RADIAL

QUADRILATERAL

STRAP

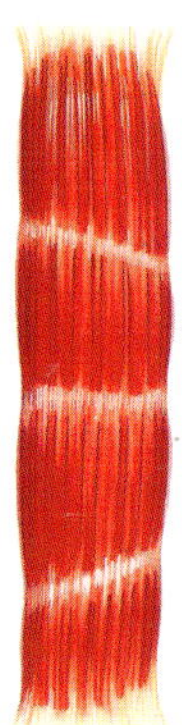

STRAP (With tendinous intersections)

CRUCIATE

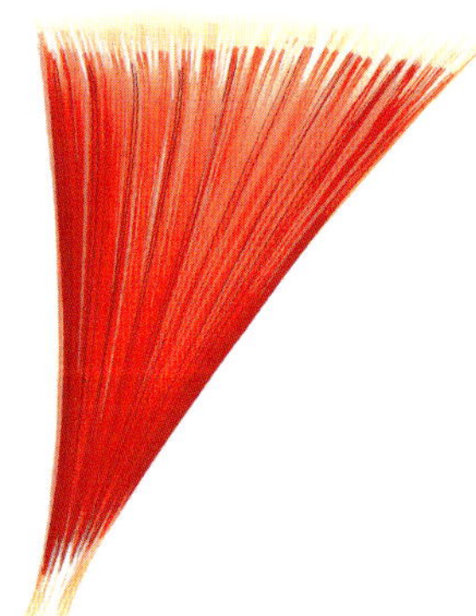

TRIANGULAR

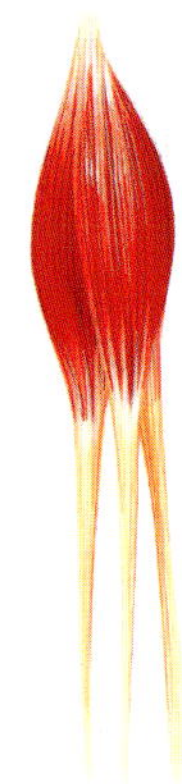

MULTICAUDAL

When a muscle is used repeatedly in lifting weight, it develops more strength and also enlarges because of the increase in diameter of individual muscle fibers. The best exercises to increase muscle bulk are those that make the muscles lengthen while they contract, such as slowing the fall of a weight.

SEE ALSO *Cerebellum on page 135*

Types of muscles

The human body has three types of muscle: cardiac, smooth and skeletal. Each has different characteristics.

Cardiac muscle is found only in the heart and consists of a network of branches of muscle fibers that do not contract voluntarily and are not under the control of the central nervous system. A heart muscle contraction originates with an electrical impulse in a natural pacemaker called the sinoatrial node, located within the heart itself. This node and the electrical conducting system that runs through the muscle control the heart rate. The heart is, however, supplied with nerves from the autonomic nervous system, but these nerves speed or slow the heart rate rather than originate contractions.

Smooth muscle is found in the digestive system, reproductive system, major blood vessels, skin and some internal organs. Smooth muscle contraction is involuntary and not under the control of the central nervous system (CNS), being influenced by the autonomic nervous system. Peristalsis—the regular and rhythmic contraction of smooth muscle in the gastrointestinal tract which propels food along the tract—is an example of the autonomous contraction of smooth muscle.

Skeletal muscle is the most prominent type of muscle and may account for up to 60 percent of the mass of the body. It is attached to the bones of the skeleton at both ends by tendons. It acts voluntarily: nerve endings that carry electrical impulses from the CNS control its movement. These nerve endings terminate at the cell membrane of the muscle fibers. The contraction of a muscle cell is activated by the release of calcium from inside the cell in response to electrical changes at the cell's surface. Skeletal muscle is usually found in bundles, forming characteristic shapes and sizes, depending on where it is located in the body and what bones and joints it moves. When skeletal muscle contracts, it usually thickens and shortens.

Most skeletal muscle is clearly visible below the surface of the skin and is responsible, together with the skeleton, for an individual's physique.

SEE ALSO *Autonomic nervous system on page 75; Digestive system on page 101; Heart on page 228; Nervous system on page 64; Reproductive system on page 104; Respiratory system on page 95*

Structure of muscles

Muscles have the property of being able to contract. They are composed of fibers, which are elongated cells containing tiny, threadlike structures made of complex proteins, myofibrils. Myofibrils consist of regularly arranged protein strands called myofilaments; the so-called "thick" myofilaments contain the protein myosin, while "thin" filaments contain the proteins actin, troponin and tropomysin. Thick and thin myofilaments lie side by side with their ends interlinked by chemical cross bridges. When stimulated, they slide along each other and the result is a muscle contraction.

UNIPENNATE

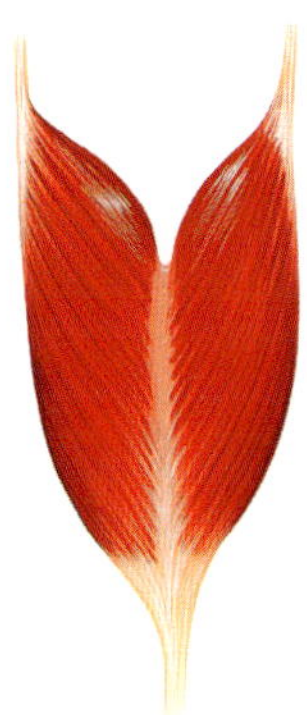

BIPENNATE

MULTIPENNATE

MUSCLES

Muscles can be classified based on their general shape—some muscles have mainly parallel fibers and others have oblique fibers. Muscle fiber arrangements may be very complicated. The shape and arrangement of the muscle fibers reflects the function of the muscle. For example, if its function is to move bone, its fibers will be aligned in the same direction. If its function is to support organs or soft tissues, fibers will be criss-crossed. If its function is to open and close an entrance, for example in the bowel or urinary tract, fibers will be arranged in a circular pattern.

FUSIFORM

BICIPITAL

TRICIPITAL

QUADRICIPITAL

DIGASTRIC

CIRCULAR

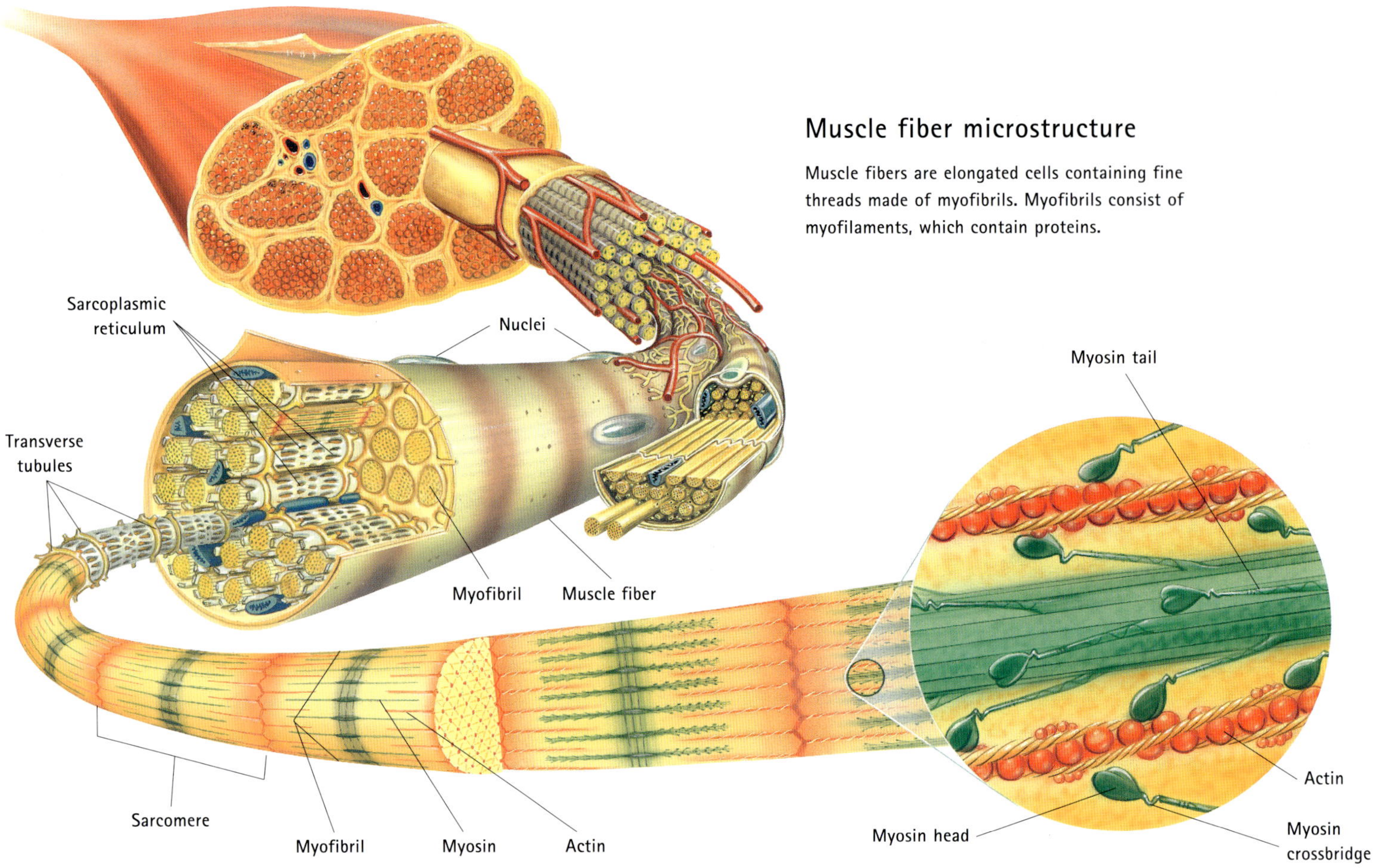

Muscle fiber microstructure

Muscle fibers are elongated cells containing fine threads made of myofibrils. Myofibrils consist of myofilaments, which contain proteins.

If the force of a muscle contraction is greater than the force that is resisting the contraction, the contraction is said to be isotonic. If, however, the resistance to contraction is equal to the force generated in muscle tissues, the muscle will not contract. This is called an isometric contraction.

There are two main classes of muscle fibers, called fast twitch (F) and slow twitch (S) fibers. F fibers generate more power and contract faster, but fatigue more quickly than S fibers. The relative proportion of the two types of fiber varies in different individuals and in different muscles. People tend to excel in sports that are suited to their predominant fiber type. For example, 95 percent of fibers in the gastrocnemius muscles in the legs of marathon champions are S fibers, compared to only 25 percent in champion sprinters.

Actions of muscles

Usually a muscle shortens when it is activated (or "contracts"), but not always. When holding a camera up in front of the eyes, for example, the muscles of the arm generate force against gravity to prevent the camera from falling, but do not change length. This kind of "contraction" is called isometric (meaning "same length") contraction. In the action of putting the camera down on a desk, the muscles that bend the elbow generate a force which is smaller than the pull of gravity on the camera. In this case the muscles of the arm are lengthened by the force of gravity and work to slow down the fall of the camera. Sometimes "contraction" is a poor term to describe muscle action.

The action of a muscle depends on its position in relation to the joint it works on. The biceps, crossing in front of the elbow, flexes the elbow; the triceps tendon crossing the back of the elbow straightens it.

The deltoid on the outside of the shoulder brings the arm out, away from the body (abducts the arm). The pectoralis major, running from the upper part of the arm toward the breastbone (sternum) pulls the arm in (abducts the arm).

However, most muscle actions are not as simple as these examples. The upper part of the pectoralis major, which runs from the clavicle to the arm, tends to raise the arm from a position of rest at the side of the body. Its lower part, which runs downward to the lower ribs, tends to pull the arm down when it is above the head. In muscles like this with different parts, the resulting movement depends on different levels of activity of different components.

SEE ALSO *Knee on page 338; Muscles of the leg on page 332; Muscles of the shoulder on page 321; Pectoral muscle on page 221; Trapezius muscle on page 220*

Muscle injuries and diseases

A pulled muscle, also known as a strained muscle, is a common term for a muscle that has been damaged by a sudden rupture of fibers within the muscle tissue. It is common in sporting and work-related injuries. The pulled muscle causes pain and stiffness that gradually improves over a number of days. It is treated with rest, ice packs and painkillers (analgesics).

Muscle cramps are painful, involuntary contractions of muscles experienced during exercise. They are caused by changes in the chemistry of muscle cells that occur during exercise brought on by lack of oxygen. They are treated with applications of ice or heat, gentle massage or physical therapy.

Muscles are subject to a variety of diseases. Muscles may become infected with bacteria such as *Staphylococcus*.

Disruption or damage to the blood supply may cause muscle tissue to die, a condition known as gangrene. Muscles may become paralyzed by injury or disease. Poliomyelitis and polyneuritis are viral diseases that result in paralysis and muscular wasting. Muscular dystrophies are hereditary diseases that are characterized by progressive muscular weakness and wasting.

Sometimes individual muscles or muscle groups may be affected by injury or disease; Bell's palsy is paralysis of all the muscles on one side of the face, caused by acute malfunction of, or damage to, the facial nerve that supplies them.

SEE ALSO *Diseases of the myocardium on page 235; Gangrene, Poliomyelitis and other individual disorders in Index; Shin splints on page 334; Sprain on page 341; Torticollis on page 203*

FIBROMYALGIA

Fibromyalgia is a common rheumatic condition consisting of painful muscles, body aches and pains, and sleep disorders. The cause is unknown; the muscles themselves are not weak, nor is there any sign of inflammation or disease in them. Fibromyalgia is usually mild and improves with treatment in the form of stretching exercises, application of heat and/or gentle massage, and taking of anti-inflammatory drugs, and, in some cases, antidepressant drugs.

DRUG TREATMENTS

Muscle relaxant drugs are commonly used to relieve painful muscle spasms that sometimes occur in stroke, in some muscle and rheumatic disorders, and in some skeletal muscle disorders.

TENDONS

A tendon is a glistening white cord of connective tissue that attaches muscle to bone. It is similar in structure to a ligament, which connects bone to bone. Tendons play a critical role in the movement of the human body by transmitting the force created by muscles to move bones. In this way, they allow muscles to control movement from a distance. The fingers, for example, are moved by tendons with force supplied by the forearm muscles.

Movement and support

The muscles of the shoulder joint, known as the rotator cuff muscles, provide important support for the joint, while allowing a wide range of movement.

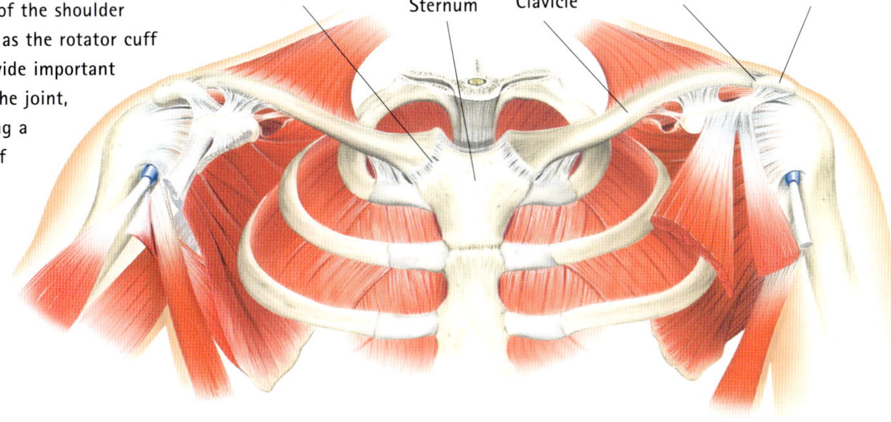

Like ropes, tendons are tough, fibrous and flexible. They are not, however, particularly elastic. If they were, much of the muscular force tendons are intended to carry would have dissipated before it had a chance to reach, let alone move, the bones.

Tendons are formed from the same components that make up other kinds of connective tissue, such as cartilage, ligament and bone. These components are collagen fibers, ground substance, and cells, which in the tendon are called fibrocytes. At the point where a tendon touches bone, the tendon fibers gradually pass into the substance of the bone and meld with it.

Some tendons run inside a fibrous sheath. Between the sheath and the tendon is a thin film of lubricant called synovial fluid. This arrangement helps tendons glide smoothly over surrounding parts.

SEE ALSO *Achilles tendon on page 341; Collagen on page 19; Tissues on page 20*

Diseases and disorders of the tendons

Inflammation of the tendon sheath is a painful condition known as tenosynovitis. Strain or trauma to a tendon sheath through repeated use, calcium deposits and high blood cholesterol levels are all potential causes. So too are diseases such as rheumatoid arthritis, gout or gonorrhea. Sometimes, during

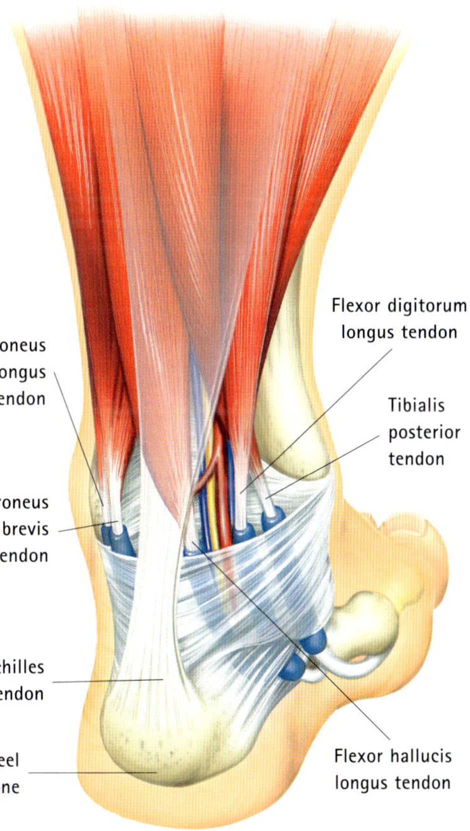

Tendons

Tendons are tough, fibrous tissues that join muscles to bones. In the hand, tendons link the fingers with the forearm muscles, allowing a full range of movement; while in the foot, the calf muscle is connected to the heel bone by the Achilles tendon.

Tendon microstructure

Tendons are constructed primarily of collagenic fibers arranged in a regular formation. This structure provides the strength needed to attach muscles to bones.

movement, a crackling noise occurs around the area of an inflamed tendon sheath.

Tenosynovitis is the underlying cause of two relatively common disorders responsible for pain in the hands and wrists—trigger finger and de Quervain's disease.

Inflammation of a tendon itself leads to a condition known as tendinitis. It is most commonly caused by overuse or a sudden overstretching of a tendon. Both situations can lead to small tears or ruptures in a tendon. As tendons undergo a gradual process of degeneration with age, they weaken as people get older and become more prone to tendinitis. This may be why tendinitis in the rotator cuff tendons, in particular, is more common with age. These tendons help create the flexibility and huge range of movement normally permitted in the shoulders. Catching a heavy object with the arm extended or carrying out repetitive overhead activities with the arms can lead to rotator cuff tendinitis.

Tendinitis is often an underlying cause of repetitive strain injury (RSI), a painful disorder involving the hands, wrists or arms, produced by excessive or repetitive motion, such as typing on a keyboard.

The elbow is commonly afflicted by two types of tendinitis—"golfer's elbow" or medial epicondylitis, and "tennis elbow" or lateral epicondylitis.

Swinging a golf club, chopping wood with an axe, pitching a baseball and any other activity that requires repetitive gripping, grasping and turning of the hand and bending of the wrist can cause golfer's elbow. This condition is characterized by pain on the inside of the elbow.

With tennis elbow, the pain occurs in the upper forearm on the outer side of the elbow. The pain is caused by repetitive grasping and twisting actions such as those involved in swinging a tennis racquet, painting a house or using certain tools repeatedly in the carpentry trade.

Over the last few years, tennis elbow has also started to appear in children who spend a lot of time playing hand-held computer games. Sufferers of tennis elbow not only experience pain, but they can more often then not, have difficulty actually straightening the forearm fully. Another common form of tendinitis, known as "jumper's knee," affects the patellar tendon of the knee. As its name suggests, it often affects people playing jumping sports such as basketball and netball and is caused by the repeated impact of the force on the knee tendon that occurs when the foot hits the ground after jumping.

There are other sites in the legs where tendinitis can develop and about which sportspeople should be particularly careful. Achilles tendinitis—an inflammation of the large Achilles tendon that stretches from the calf muscles to the back of the heel—is common in runners, particularly sprinters. Long-distance runners tend to be more inclined to develop a form of tendinitis called iliotibial band syndrome, which produces pain along the outside of the knee. It is also a common affliction of dancers, cyclists and football players.

SEE ALSO *Achilles tendinitis on page 342; de Quervain's disease on page 323; Rupture of the Achilles tendon on page 341; Trigger finger on page 325*

TREATMENT

Complete rupture of a tendon is rare and requires immediate medical attention and usually surgery. In the vast majority of cases, however, damage to tendons, or the sheaths surrounding them, responds well to ice treatments, elevation of the injury and rest, including immobilization of the affected area by strapping or even plaster, in severe cases. Non-steroidal anti-inflammatory drugs can help to reduce pain and swelling, and physical therapy will usually assist and accelerate the healing process. Steroid injections may sometimes be used in order to reduce severe pain and stiffness.

Occasionally, if tendinitis is failing to heal, surgery may be required. The extent of damage to soft tissues such as tendons can be assessed using the modern scanning technique known as magnetic resonance imaging (MRI).

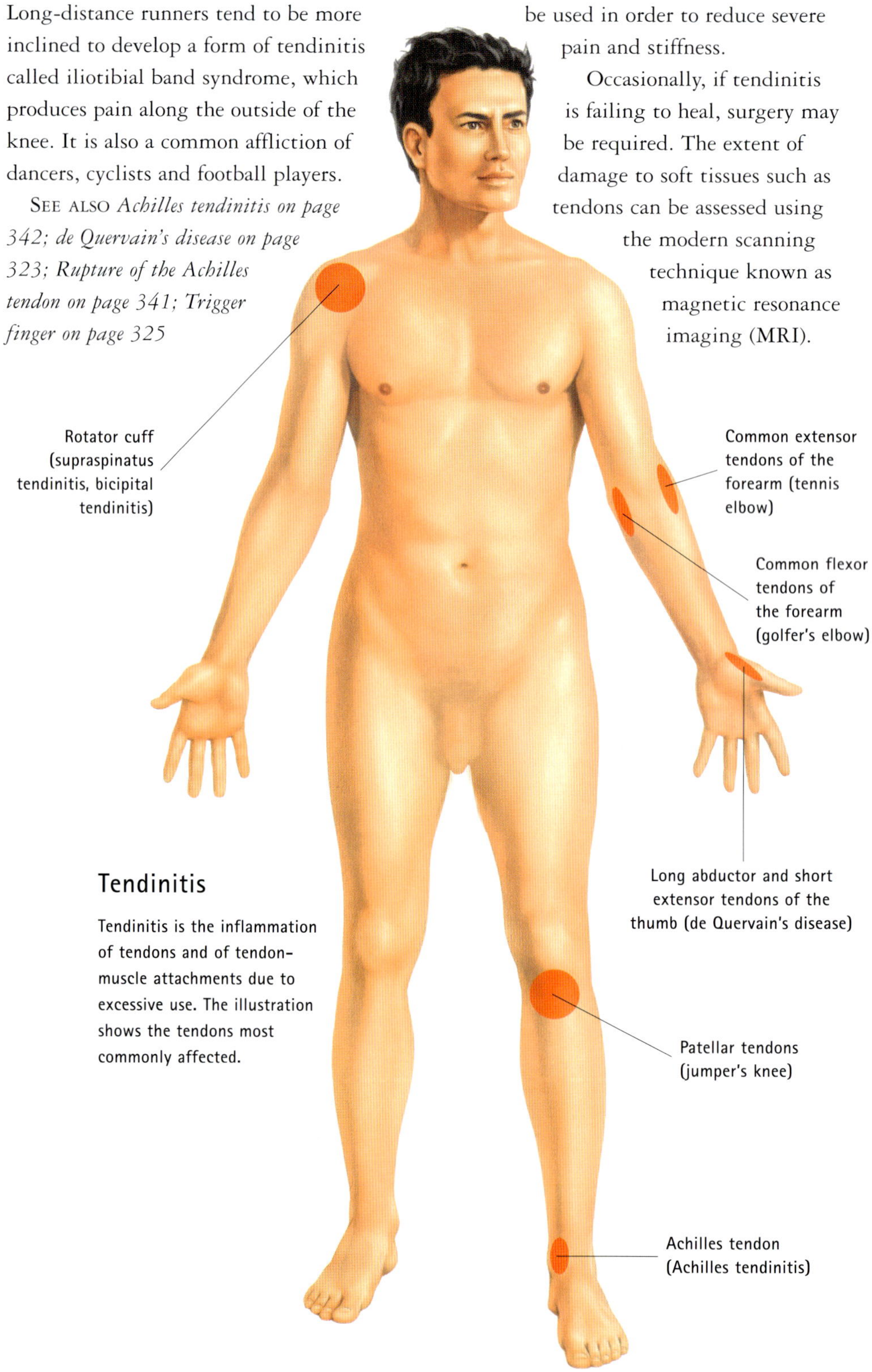

Tendinitis

Tendinitis is the inflammation of tendons and of tendon-muscle attachments due to excessive use. The illustration shows the tendons most commonly affected.

LYMPHATIC/IMMUNE SYSTEM

The lymphatic system is a complex network of vessels, aggregates of lymphoid tissue (lymph nodes) and lymphoid organs, and has two essential functions.

Apart from bringing back to the heart much of the interstitial fluid that bathes each and every cell of the body, it is also loaded with specialized white blood cells (also called lymphocytes) and macrophages. These are responsible for sweeping up foreign bodies or invaders, such as bacteria, viruses, and cancer cells.

SEE ALSO *Infectious diseases on page 364; White blood cells on page 59*

Lymph

The body's tissues are bathed in an almost-clear liquid called interstitial fluid, which filters out from blood vessels. It is collected (along with tissue and cellular wastes), drained away, cleaned in lymph nodes and recycled back into the blood by the vessels of the lymphatic system.

When the interstitial fluid is flowing through lymph vessels it is known as lymph. It contains mainly water, protein molecules, salts, glucose, urea and disease-fighting white blood cells. On its way to blood, lymph also carries an assortment of certain fats and fat-soluble vitamins collected from the digestive tract.

However, the lymphatic system does not have a pump. The movement of lymph through the entire body system is assisted by the actions of the skeletal muscle and breathing contractions and is slower than the movement of blood.

Lymph vessels

Unlike the complete circle of the circulatory system, the lymphatic system is a one-way system, which begins with a capillary network of blind-ended tubes. These capillaries absorb large molecules and particles (foreign bodies and nutritional elements), and converge to form gradually larger lymphatic vessels that carry the lymph toward the heart. Lymph vessels have valves to ensure one-way flow. Along the lymph vessels are found collections of lymphoid tissue or lymph nodes (frequently and colloquially called lymph glands).

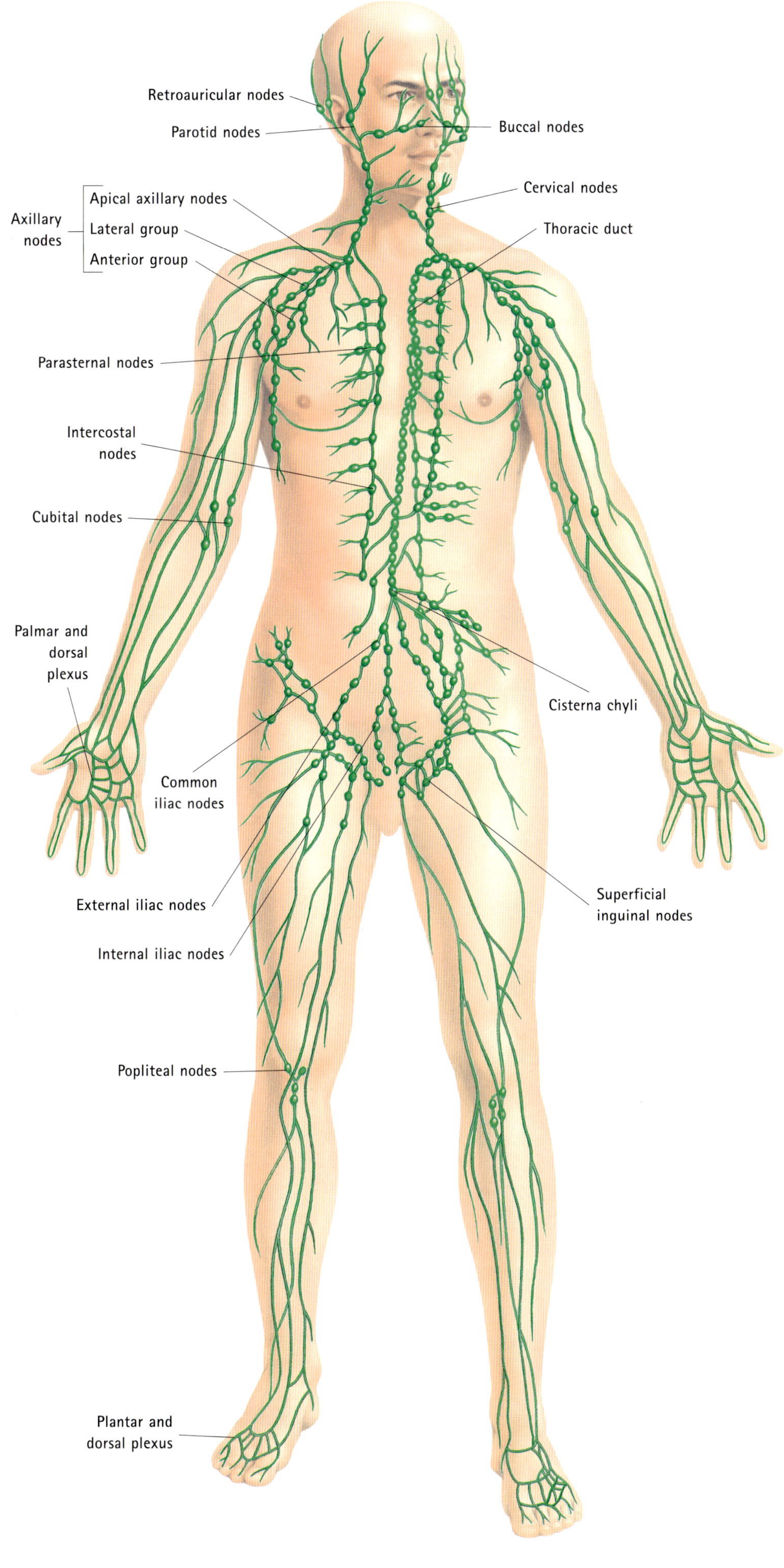

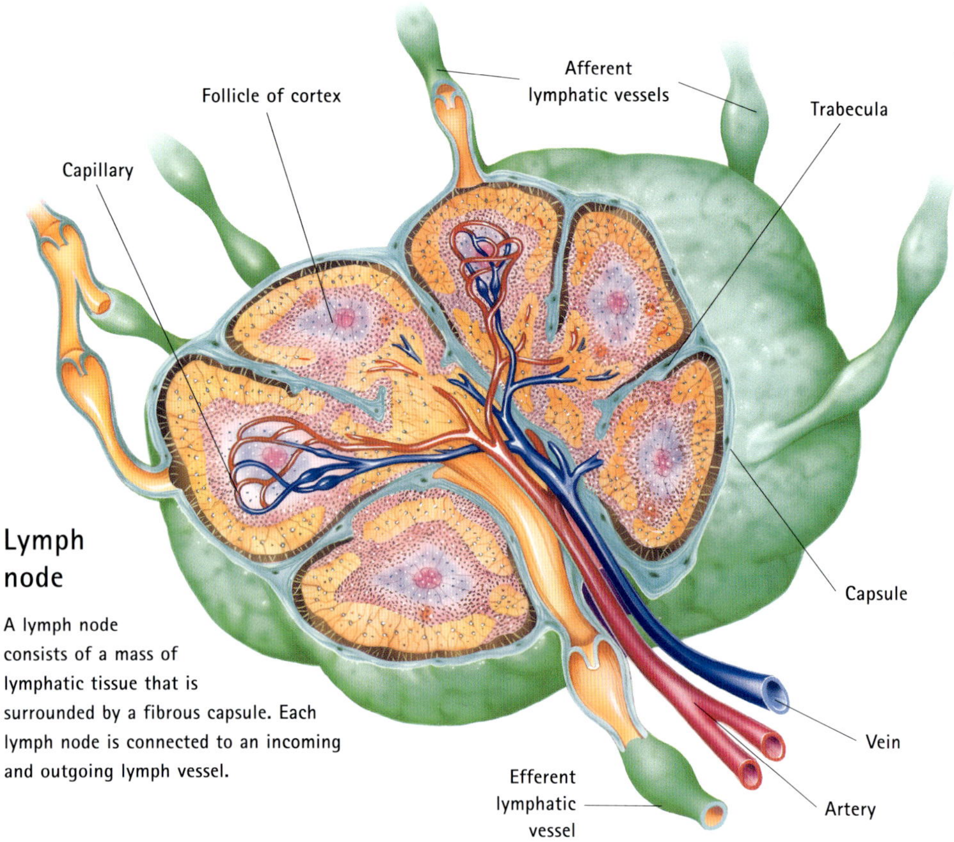

Lymph node

A lymph node consists of a mass of lymphatic tissue that is surrounded by a fibrous capsule. Each lymph node is connected to an incoming and outgoing lymph vessel.

Lymph nodes and lymphoid tissue

The lymph nodes are small pea-sized organs located in groups at the confluence of lymphatic vessels. Each node is connected to incoming and outgoing lymphatic vessels. Incoming lymph spreads throughout the node to pass through aggregations of lymphocytes and macrophages.

Lymph nodes not only filter out and destroy foreign bodies before they get into the circulation, they also activate the cloning of lymphocytes to produce specific antibodies, which is part of the immune response.

Lymph nodes can be compared to outlying fortresses along the main highway into a city. The battle against invaders begins in the outermost fortress. If the first fortress fails to contain the invasion, fighting will continue along the highway, fire and smoke progressing to fortresses closer in. In a similar fashion, inflammation traveling along major lymph vessels is sometimes visible as red streaks under the skin (lymphangitis).

SEE ALSO *White blood cells on page 59*

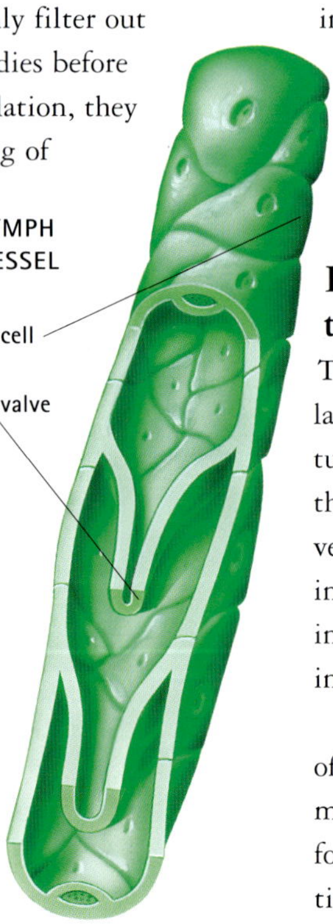

Lymph circulation

The lymph vessels contain numerous valves which prevent the backflow of lymph. The lymph is returned to the general circulation by two large lymphatic vessels which empty into the large veins at the base of the neck.

How lymph returns to the general circulation

The peripheral lymph vessels converge into large lymphatic trunks, which link up in turn to the right lymphatic duct and the thoracic duct, which empty into the large veins at the base of the neck. An overwhelming infection can spread along lymph vessels into the general circulation causing "blood infection" (septicemia), which can be fatal.

Lacteals are lymphatic vessels in the walls of the digestive system, which collect large molecules and lipids (chyle) extracted from food. They play a major role in the absorption of fats. They engorge after a meal and become visible as fine whitish streaks in the mesentery. The lacteals empty into the cisterna chyli, a sac below the diaphragm.

Lymphatic vessels from the head and neck pass through a collar of lymph nodes under the lower jaw, to end in chains of nodes along the internal jugular veins.

Lymphatic vessels of the upper limb ascend from the hand to the armpit (axilla). The nodes in the axilla, which can be felt along the upper arm and against the upper part of the rib cage, also receive lymph from the chest wall, back and the breast. Lymphatic vessels of the lower limb ascend from the foot to the inguinal lymph nodes in the groin. These nodes also drain the buttocks, the back and parts of the genitalia, and empty into the cisterna chyli.

Lymph from the internal organs of the thorax and abdomen drains into chains of lymph nodes along major arteries and the aorta. Lymph nodes draining the lungs are located around the bronchi and trachea. Lymph from the abdominal and pelvic viscera drains into lymph nodes along the iliac arteries and the aorta, and eventually into the cisterna chyli.

The cisterna chyli empties into the thoracic duct, which ascends through the thorax into the neck and empties into the junction of the left internal jugular and subclavian veins. The thoracic duct also collects the lymph from the thoracic organs, from the left upper limb and left half of the head and neck.

Lymph from the right half of the head, neck, and thorax, and the right upper limb, converges in a short right lymphatic duct which empties into the junction of the right internal jugular and subclavian veins.

LYMPHOID ORGANS

The lymphoid organs include the thymus, the spleen, and mucosa-associated lymphoid tissue.

SEE ALSO *Spleen on page 281; White blood cells on page 59*

THYMUS

The thymus lies in the upper part of the thorax, between the heart and the sternum.

Lymphocytes, which are manufactured in the lymph nodes and bone marrow, mature

Organs of the lymph system

The thymus gland is an important lymphatic organ in infancy, but gradually regresses after puberty. The spleen is the largest concentration of lymphatic tissue in the body; other concentrations of lymphatic tissue are found in the lymphatic nodules of the gut, and in the tonsils.

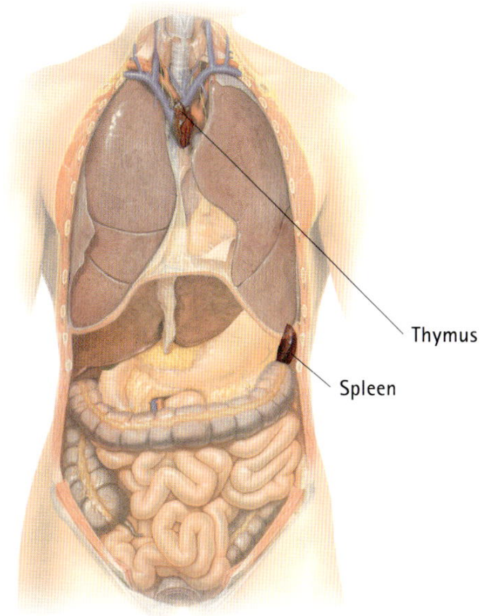

as they are pushed from the outer cortex of the thymus into the central part (medulla), and from there enter the circulation. Most of the lymphocytes in the thymus are T lymphocytes, which are able to recognize foreign-body antigens. B lymphocytes recognize only the body's own cells and antigens. The thymus also secretes hormones that regulate T cell production and function. When the thymus regresses after puberty, T cells continue to proliferate, thus maintaining an adequate number throughout life.

SPLEEN

The spleen lies under the left ninth, tenth and eleventh ribs, near the end of the pancreas. It has a rich network of blood capillaries and sinusoids, called the red pulp, and aggregates of lymphocytes around branching arteries, called the white pulp.

The spleen removes particles and aged red blood cells from the circulation.

The spleen also plays an important role in building up the immune response, functioning in a similar way to the lymph nodes.

MUCOSA-ASSOCIATED LYMPHOID TISSUE

Masses of lymphoid tissue are found in the linings (mucosa) of the respiratory system, urogenital tract and digestive tract. These mucosa-associated lymphoid tissues contain B and T lymphocytes and serve the same protective function as lymph nodes in the cavities of the body exposed to the external environment.

Tumors of the lymph nodes

Lymphomas (lymphatic tumors) often cause enlargement of the lymph nodes and are usually malignant. In late stage disease, cancer may spread from the nodes to other areas of the body. Treatment is usually with radiation therapy or chemotherapy.

SEE ALSO *Biopsy on page 435; Treating cancer on page 446; White blood cells on page 59*

HODGKIN'S DISEASE

Hodgkin's disease is a type of lymphoma (a cancer arising in lymph nodes), which was first described by Thomas Hodgkin in 1832. Lymphomas are a relatively common group of cancers (typically ranking fifth or sixth in frequency among both men and women) of which Hodgkin's disease makes up a variable proportion (approximately one-fifth in most economically developed nations). What makes Hodgkin's disease distinctive is that it usually develops in young to middle adult life, when most types of cancer are rare.

The exact cause of Hodgkin's disease development remains unknown, but there is a strong suspicion that certain viral infections may contribute to its emergence. Even the specific cell type that gives rise to Hodgkin's disease remains uncertain and it appears quite likely that more than one type of cell is involved.

Hodgkin's disease typically produces enlargement of the lymph nodes early in the disease. The patient may notice a swelling, for example in the neck, but the nodes are usually not painful. Quite often, patients with Hodgkin's disease develop symptoms such as fever, weight loss and night sweats as part of their illness, which are triggered by chemical signals released by the cancer cells and by the patient's response to the tumor. In addition, some patients develop complicating infections, because Hodgkin's disease is associated with suppression of the immune response, although the reason for this is not entirely clear.

The microscopic appearance of Hodgkin's disease is quite distinct from other types of lymphoma, which allows it to be diagnosed by examination of a sample of involved tissue. Unlike most other lymphomas, it usually does not spread far and wide at an early stage, but instead extends progressively from one group of nodes to the next, with later involvement of the spleen and other tissues. This relatively slow and orderly progression may be one reason why Hodgkin's disease is more responsive to treatment with anti-cancer drugs than many other varieties of lymphoma.

The extent of spread at the time of diagnosis is the most important determinant of the patient's likely response to chemotherapy. Also relevant is the specific variety of Hodgkin's disease that the individual has developed. With modern combination chemotherapy, the disease is controlled in the great majority of patients with early stage disease, and a substantial proportion can expect to be cured.

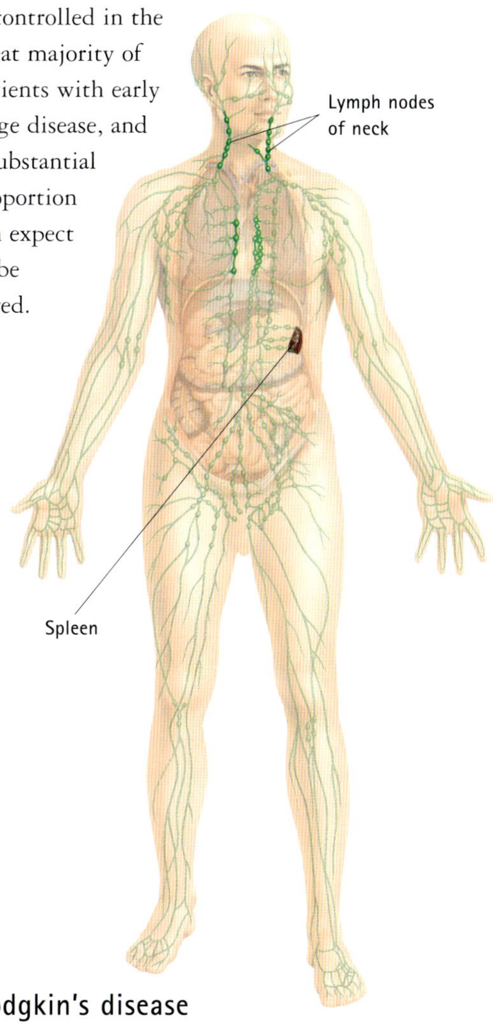

Hodgkin's disease

This type of lymphoma produces enlargement of the lymph nodes, often starting with the lymph nodes in the neck and spreading progressively to other nodes in the body, the spleen and other tissues. Symptoms include fever, night sweats and weight loss.

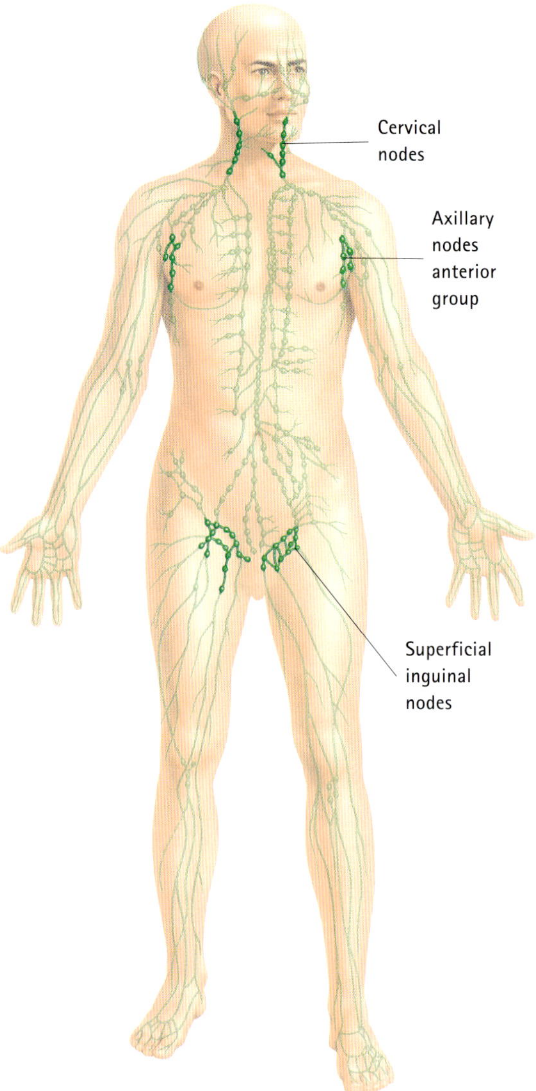

Non-Hodgkin's lymphoma

Non-Hodgkin's lymphoma is a group of malignant diseases that cause tumors of the lymph glands (nodes). The nodes of the neck, armpits, and groin area are most commonly affected. Enlargement of these nodes may be the first signs of disease.

NON-HODGKIN'S LYMPHOMA

Non-Hodgkin's lymphoma is a tumor of the lymph glands, which is distinct from Hodgkin's lymphoma (also called Hodgkin's disease). It is a group of malignant diseases arising in the lymph glands (also called lymph nodes) rather than spreading there, such as cancers do. Microscopic examination of a lymph gland by the pathologist accurately distinguishes between the two conditions.

Non-Hodgkin's lymphomas are more common than Hodgkin's disease, and the average age of patients is 50 years. The cause is generally unknown, although several possible causes have been identified. Exposure to high-dose ionizing radiation increases the risk, as do exposure to chemicals such as benzene and pesticides.

Chemotherapy drugs can lead to later development of lymphoma, which is also increased in patients with reduced immunity, either from medication or acquired immune deficiency syndrome (AIDS). Other viruses have also been associated with lymphoma, including Epstein-Barr virus, the cause of infectious mononucleosis (glandular fever).

The main symptoms of lymphoma are persistent fever, drenching night sweats and severe unexplained weight loss. Patients may notice enlarged lymph glands in the neck, under the armpits or in the groin area and there may be enlargement of the spleen. The diagnosis is made by biopsy of an enlarged lymph gland, using a fine needle or by a minor operation. Sometimes lymphoma causes a tumor that is not in the glands, such as lymphoma of the stomach or brain.

There is a variety of treatments for non-Hodgkin's lymphomas, depending on the particular type. Some lymphomas are very slow-growing and may not require treatment, particularly if the patient is elderly. If such a tumor causes symptoms, it can often be treated with mild chemotherapy tablets or radiation therapy to the affected area. Although such lymphomas are compatible with survival for years, they are not currently curable.

An important new development has been the production of antibodies directed against the cells causing lymphomas. These antibodies can kill the lymphoma cells without the side effects caused by most chemotherapy drugs. Other types of lymphoma are rapid-growing and require strong chemotherapy for treatment. The treatment consists of a combination of injections and tablets given on a cycle of three to four weeks for a total of six to nine treatments, generally given as an out-patient. Common side effects include nausea, hair loss and reduction of the normal white blood cells, leading to increased risk of infection. Injections are now available to stimulate the white cells to recover sooner, reducing the risk of infection. With such treatments, a proportion of aggressive lymphomas can be cured.

Currently, doctors are assessing higher doses of chemotherapy involving the use of the patient's own (autologous) marrow or stem cell transplants. In this procedure, normal cells of the patient are collected and stored in liquid nitrogen, so that they can be returned to "rescue" the patient following doses of chemotherapy so high that they usually destroy both lymphoma and healthy bone marrow cells.

BURKITT'S LYMPHOMA

Burkitt's lymphoma, also known as B cell lymphoma, is a tumor of the lymph glands. It arises from a type of white blood cell called a B lymphocyte, although it is a different type of lymphoma from a Hodgkin's lymphoma. It is often associated with the Epstein-Barr virus (EBV), which causes infectious mononucleosis (glandular fever).

Burkitt's lymphoma is usually first noticed as a painless but rapid swelling of the lymph nodes in the neck or below the jaw (though lymph nodes in other areas may be affected). The diagnosis can be confirmed by biopsy of the node. Treatment involves a combination of radiation therapy and chemotherapy. The disease is often curable if it is treated in the early stages.

IMMUNITY

Immunity is the body's ability to protect itself from disease. It is achieved via the immune system, a complex network of organs, cells and proteins that recognizes foreign substances (such as viruses, bacteria, fungi and other pathogens) in the body and destroys them. This process is called the immune response.

There are two broad parts to the immune system. One part is called the humoral immune system, so called because the immune response takes place in the body fluids (humors). When a foreign body, or antigen, is identified, proteins called antibodies are produced by B lymphocytes (white blood cells) in the blood and body fluids which then attack the antigens or render them more easily attacked by other white blood cells.

The second component is the cell-mediated immune system, involving the different types of T lymphocytes. Some ingest and destroy invading pathogens while other T lymphocytes destroy them directly. Antibodies are not involved.

The first time an individual is exposed to a pathogen, or antigen, there is a delay while the immune system responds and overcomes the pathogen. The next time the individual is exposed to that same pathogen, the response is much faster, and the individual may not actually develop the disease. This is due to "memory" B and T cells, which have been stored since the first encounter with that particular pathogen. The individual is then said to have immunity to that pathogen.

Immunity can be artificially induced by vaccination. A vaccine is a weakened or killed form of the pathogen. It causes the body to manufacture antibodies against the pathogen, so that in case is later exposed to the live form of the pathogen, the body can then launch a prompt and effective immune response.

Immunity is acquired through either active or passive means. Active immunity refers to those situations where the body itself has created the immunity, either as a result of past exposure to the disease, or because it has been vaccinated against it. In cases of passive immunity, the response has come from elsewhere, either from an injection of antibodies for example, or, in the case of the fetus, from the mother. Unlike active immunity, which has "memory" and can mount future responses to the same pathogen, passive immunity is usually temporary.

SEE ALSO *Immunization on page 447; Spleen on page 281; Thymus on page 251; White blood cells on page 59*

LYMPHOCYTES AND THE IMMUNE RESPONSE

Lymphocytes play a central role in the body's immune system. There are three major types of lymphocytes: natural killer cells (NK cells), B lymphocytes and T lymphocytes.

NK cells do not react to specific antigens like T lymphocytes, but kill a variety of target cells, including cancer cells.

T lymphocytes derived from the thymus can be either effector or regulator cells. Each effector T cell recognizes and is activated by one antigen. Activated T cells present antigen to antibody-producing cells to stimulate production of antibody to the particular antigen. Regulator T cells include T helper cells and T suppressor cells, which either facilitate or inhibit the immune response.

B lymphocytes, which come from bone marrow, produce specific antibodies. Each B cell is specialized in producing immunoglobulins or Ig (antibody) to a single antigen. There are five general classes of immunoglobulins: IgG, which is the most important serum globulin; IgM, which is the first antibody to appear in response to infection; IgA, which is found in secretions such as tears or saliva; IgE, which initiates allergic and hypersensitivity reaction; and IgD, which is found on the surface of B cells to help in binding antigens.

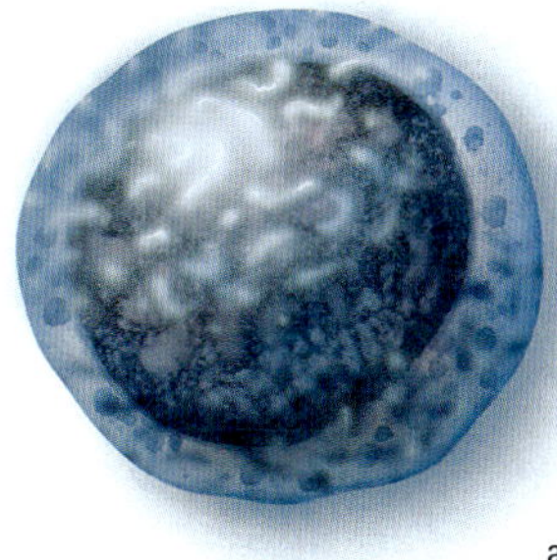

Lymphocytes

A lymphocyte is a type of white blood cell that plays an important role in the immune response. There are several types, including B cells, T cells and natural killer cells.

Humoral immune response

B lymphocytes (white blood cells) produce antibodies to help identify and eliminate invading antigens (carried by bacteria or viruses). They are helped in the body's defenses by circulating T lymphocytes and macrophages (scavenging white blood cells).

(a) Virus particles invade tissue through surface cells and multiply.

(b) Virus particles are consumed by macrophages.

(c) The macrophages break down the virus and present antigens to circulating T lymphocytes. These release proteins to recruit more T and B lymphocytes from nearby blood vessels and tissue to help defend the body.

(d) B lymphocytes divide into memory B cells (which remember the invading virus for future attacks) and plasma B cells which make antibodies specific to the invading virus.

(e) The circulating antibodies attach onto the virus particles.

(f) Macrophages primed to recognize the antibody consume the virus and break it down, saving the body from infection.

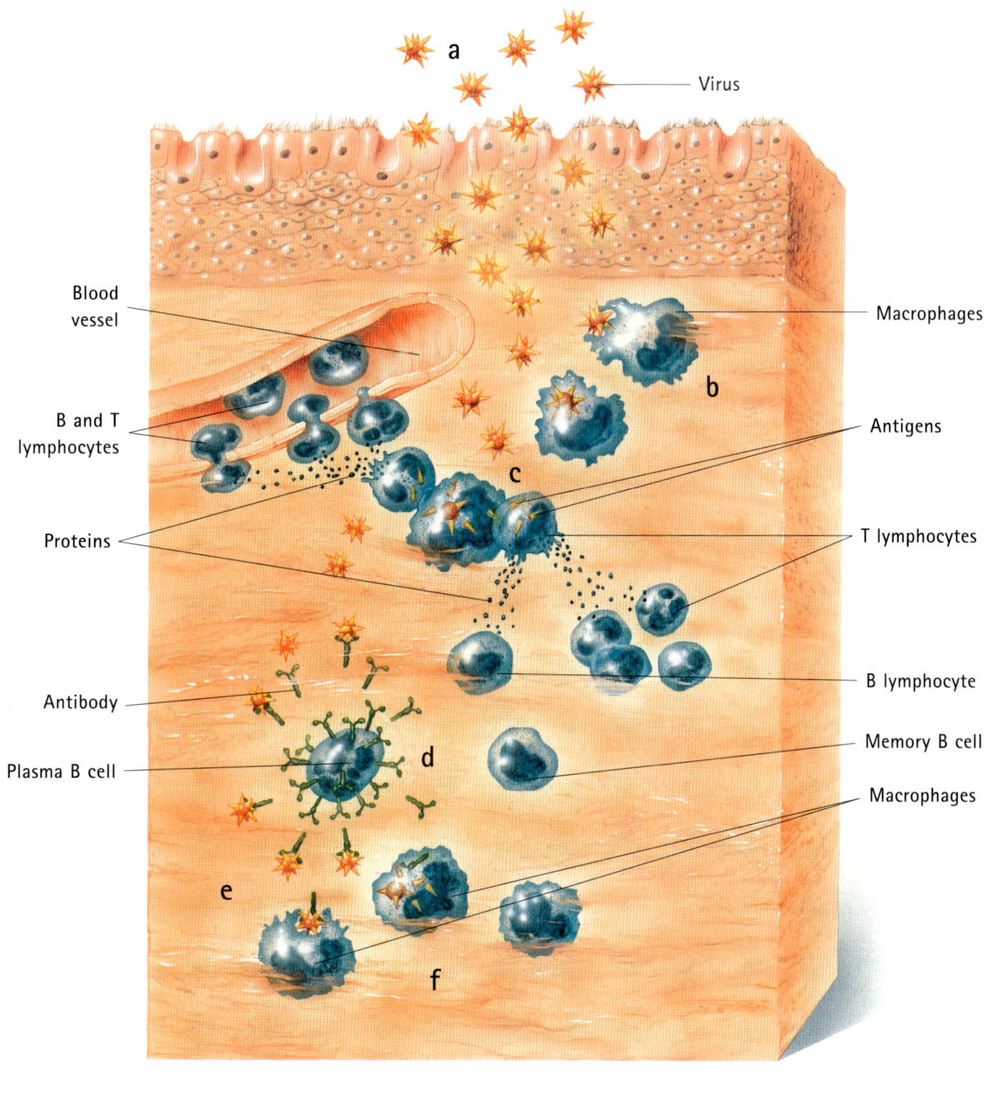

When a specific set of T and B cells are activated by an antigen, they multiply to increase production of the appropriate antibody as long as infection lasts. Some T and B cells, called memory cells, remain after the infection is over to ensure a quick response the next time the body is exposed to the same antigen.

Immune disorders

Autoimmune disease results when the immune system reacts to the normal cells and antigens of the body, producing autoantibodies that damage or destroy normal tissues. Common examples are psoriasis, rheumatoid arthritis, thyroiditis and systemic lupus erythematosus.

SEE ALSO *Antigen tests on page 438; Blood tests on page 436; Histamine on page 115; Immunization on page 447; Treating cancer on page 446; Treating infection on page 446*

Allergic disorders

Allergic disorders, which include the common conditions of hayfever and asthma, can be triggered by a number of factors.

Allergies

An allergy is a physical reaction to certain substances. In the allergic person, the immune system mistakenly identifies a substance as being harmful and mounts a defense against it. This unnecessary defense reaction is often excessively vigorous, and the antibodies manufactured to fight the substance have irritating or harmful effects, which constitute an allergic reaction.

Allergies tend to run in families but are also affected by environmental factors. They develop through exposure to substances, and a process called sensitization that can occur on first contact, or over a brief period or even through repeated exposure over several years. During this period, the immune system is activated to react against what is usually a relatively harmless substance.

Allergies can show up at any age but they often appear first in childhood, particularly contact allergies that are a reaction on first contact with the allergenic substance.

Asthma and hay fever, allergic rhinitis and sinusitis, cows' milk allergy and various other food allergies are well-known conditions. If you are a sufferer, contact with the offending substance will trigger a variety of unpleasant symptoms that could include skin rashes, itching or swellings, red and swollen eyes, runny nose, severe nasal inflammation, wheezing and shortness of breath. Sometimes a severe reaction can require immediate emergency treatment.

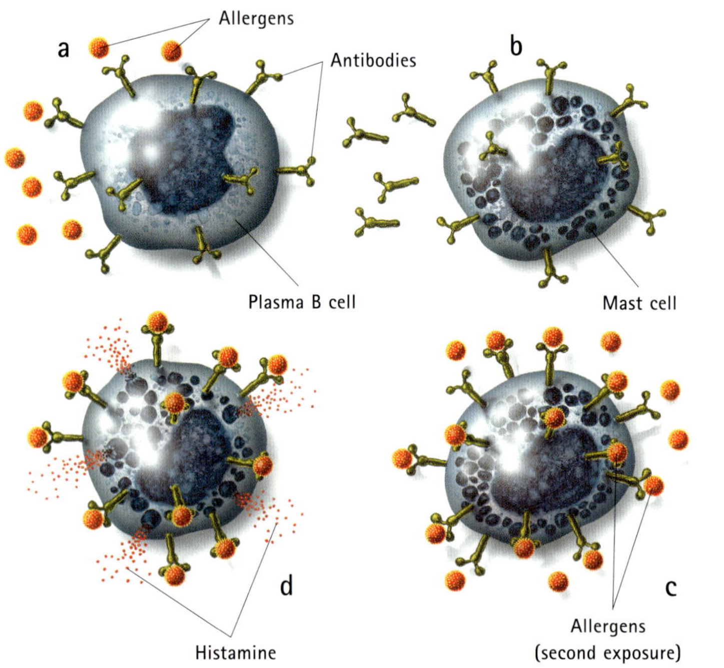

Allergic reaction

Exposure to invading allergens (that the body is sensitive to) leads to the release of histamine, which irritates tissues and causes symptoms such as sneezing and rash.

(a) On the body's first exposure to an allergen, plasma B cells produce antibodies.

(b) The antibodies attach to mast cells circulating in the body's tissues.

(c) The next time allergens enter the body they are captured by the antibodies on the mast cells.

(d) The mast cells respond by releasing histamine, a chemical that causes inflammation and the symptoms of allergy.

Angioedema

Angioedema is a severe allergic reaction, similar in many ways to urticaria. The chief difference is that urticaria affects the surface layers of the skin while angioedema affects the deeper layers. It may occur with or without urticaria. Angioedema is caused by an allergic trigger—this may be an insect bite or sting, food (shellfish, nuts, food additives or strawberries), exposure to animals or pollen, or a reaction to a drug such as penicillin.

In angioedema, the eyes, lips and skin around the eyes may swell markedly. If the swelling spreads to the throat, suffocation may occur. Mild cases of angioedema can be treated with antihistamine tablets. In more serious cases an intravenous injection of hydrocortisone or epinephrine (adrenaline) is given to reduce the swelling and remove the risk

Cell-mediated immune response

T lymphocytes (a type of white blood cell) are responsible for the delayed action of the cell-mediated response.

(a) Circulating mast cells ingest invading virus.

(b) Mast cells process the virus and present antigens to T cells.

(c) The T cells produce clones which each play a special role in the immune response: memory T cells remember the invading antigen for future attacks; helper T cells recruit B and T cells to the site of antigen attack; suppressor T cells inhibit the action of B and T cells; and killer T cells attach onto invading antigens and destroy them.

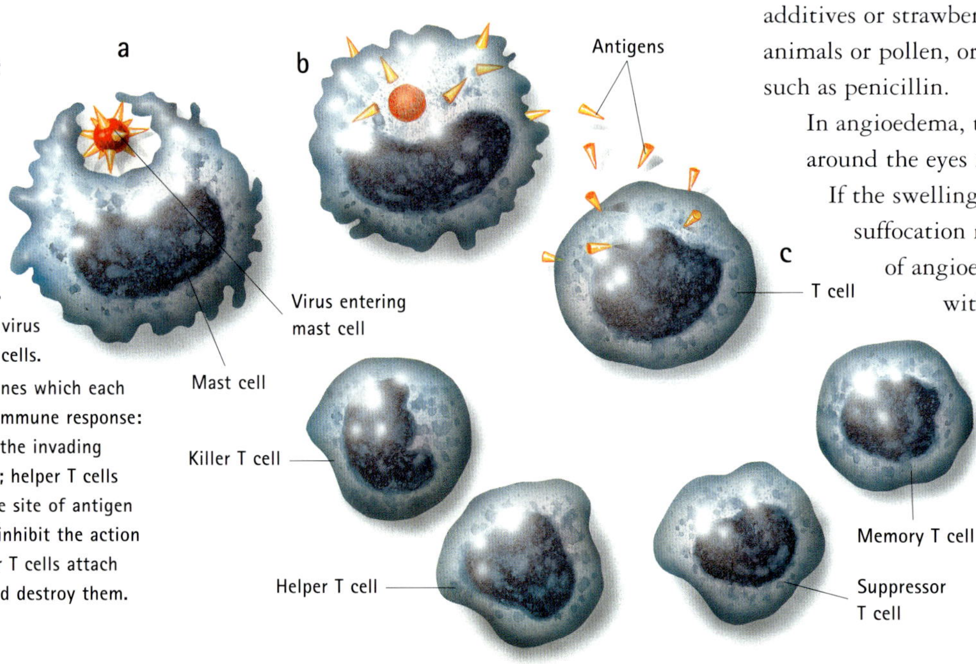

Immunodeficiency

The human immunodeficiency virus (HIV) compromises immunity by destroying T4 helper cells—a type of white blood cell. Here a T4 cell is under attack.

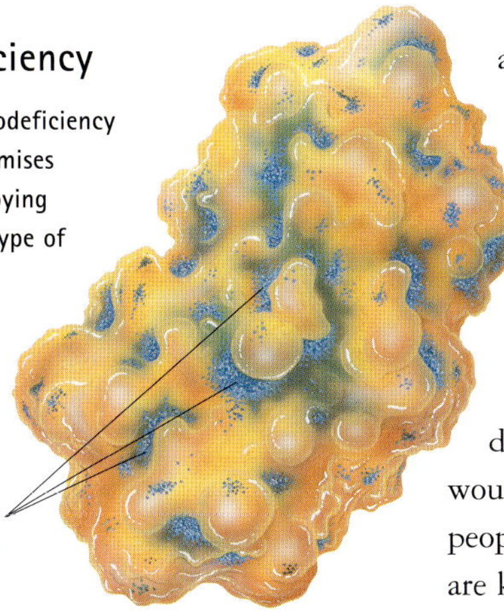

Human immunodeficiency virus particles

of suffocation. Any trigger factors known to bring on the condition should be avoided.

Anaphylaxis

Anaphylaxis is an immediate and violent reaction brought on by hypersensitivity to a particular substance. It is an extreme allergic response that occurs when a person comes into contact with a substance—an antigen—to which they are already allergic.

Antigens stimulate the body's immune system to fight them with specific antibodies and the release of histamine. The histamine within the body triggers the violent response known as anaphylactic shock. Capillaries enlarge and leak fluid into surrounding tissues; blood pressure collapses; the brain's oxygen supply is reduced; airways narrow and breathing is restricted; skin is pale and damp; there may be nausea and vomiting; there is a risk of heart failure and death. Anaphylaxis is often associated with insect bites or certain medications, such as penicillin.

Treatment often involves the injection of epinephrine (adrenaline) to restore blood pressure, and of other drugs such as steroids and antihistamines. People at risk often wear bracelets inscribed with their personal medical information and may carry their own adrenaline injection kits. First aid is the same as for a person who is in shock, and medical assistance is essential in all cases.

Immunodeficiency

When an individual's immune system has been damaged and is deficient, that person is said to have an immunodeficiency. Immunodeficiency can be congenital (the result of a genetic disease), or it can be acquired, as in the case of cancer, leukemia or infection with the human immunodeficiency virus (HIV). Congenital immunodeficiency is relatively rare and involves inborn defects in the production of B and T lymphocytes.

If the immune system is severely damaged, exposure to pathogens that would not normally cause disease in healthy people may cause serious infections; these are known as opportunistic infections. Examples include candidiasis (thrush); herpes simplex viruses, which can cause oral herpes (cold sores) or genital herpes; and *Pneumocystis carinii* pneumonia, which can cause a fatal pneumonia.

People with damaged immune systems are also prone to developing cancers such as Kaposi's sarcoma.

Immunodeficiency can be detected by measuring the levels of white blood cells in the blood. In HIV infection, for example, there is a fall in the number of white cells called T4 cells, which can be monitored. The T4 cell count gives an indication of how far the HIV infection is progressing.

Immunosuppression

Immunosuppression is inhibition of the immune system. It may occur as a side effect of drugs used to treat cancer, or in radiation therapy, when the bone marrow or other tissues of the immune system are damaged. It may also occur as a rare side effect of commonly used drugs.

Immunosuppression also refers to the use of certain drugs to prevent the body's immune system destroying its own tissues. This may be necessary, for example, after transplantation of organs or in autoimmune diseases. Corticosteroids and azathioprine are examples of immunosuppressant drugs.

Autoimmune disease

Autoimmune diseases develop when the body's immune system fails to recognize normal body tissues and attacks and destroys them as if they were foreign. The cause isn't fully understood, but in some cases is thought to be triggered by

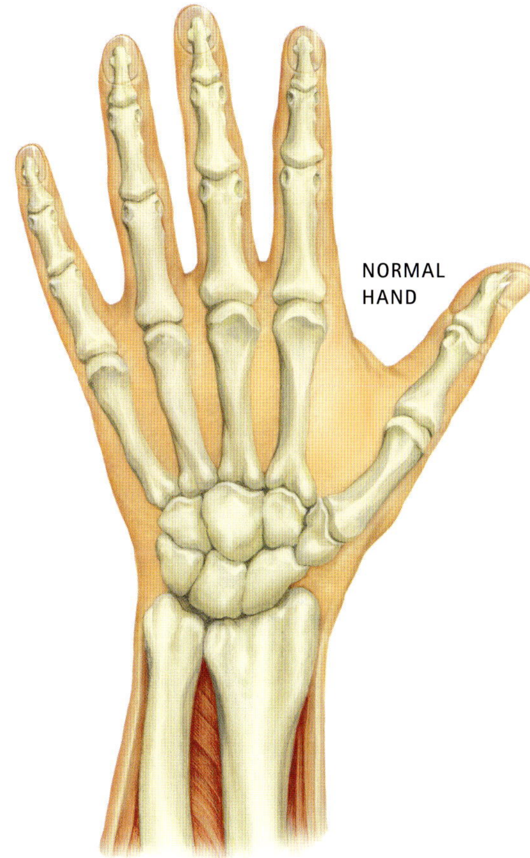

Autoimmune disease—rheumatoid arthritis

Rheumatoid arthritis is an autoimmune disease that inflames connective tissue throughout the body. It affects the small synovial joints of the fingers, wrists, toes, ankles and elbows, causing pain and swelling.

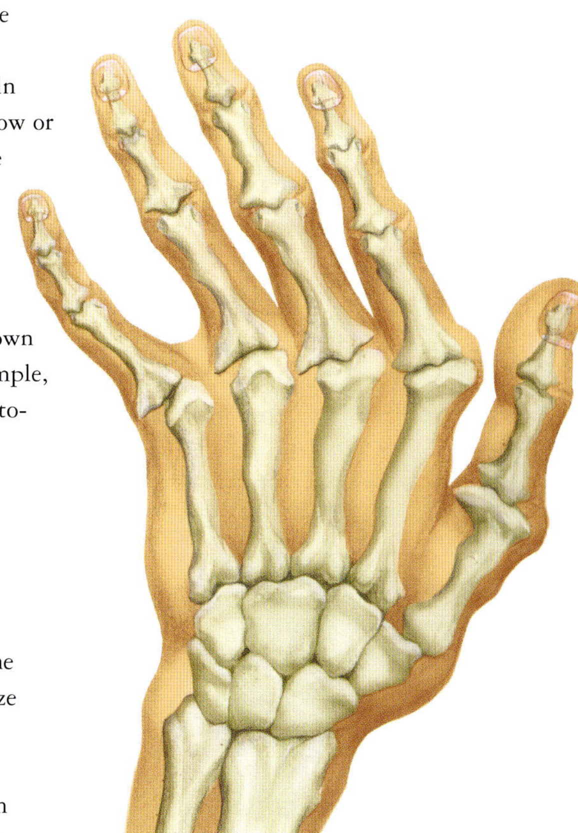

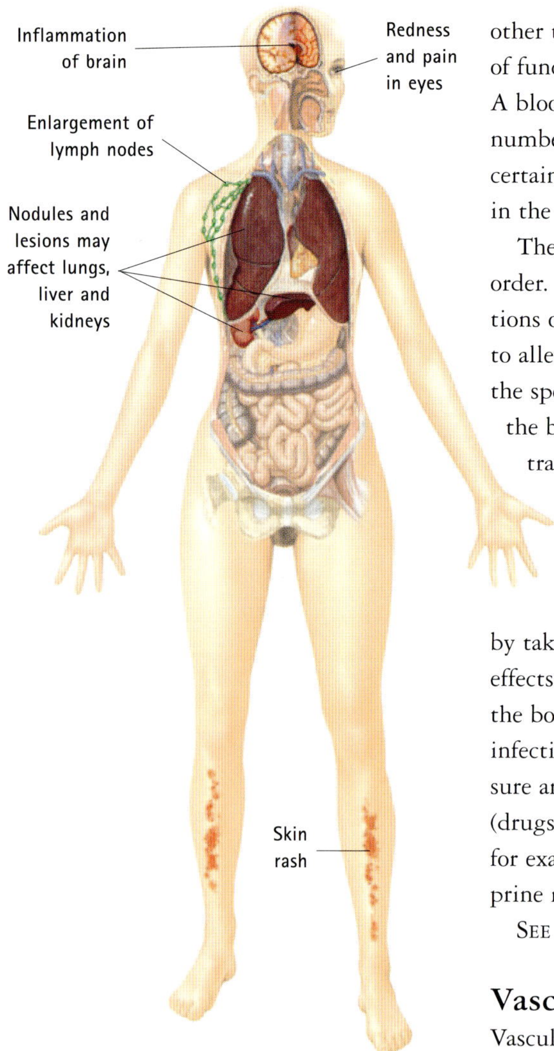

Sarcoidosis

This inflammatory disease can affect many parts of the body, but most frequently it is found in the lymph nodes, liver, spleen, lungs, skin and eyes.

exposure to microorganisms and drugs, especially in people with a genetic predisposition to the disorder. A single organ or multiple organs and tissues may be affected.

Examples of autoimmune diseases and the tissues they attack include: pernicious anemia (blood); Hashimoto's thyroiditis (thyroid); Addison's disease (adrenal cortex); diabetes mellitus (pancreas); rheumatoid arthritis (joints); systemic lupus erythematosus and dermatomyositis (connective tissues); and myasthenia gravis (muscles).

Symptoms of autoimmune disease are related to the lack of function of the organ or tissue involved. In addition, generalized symptoms common to all autoimmune disorders include tiredness and fatigue, dizziness, malaise and low-grade fever. The diagnosis is made from blood tests and other tests which indicate the degree of loss of function in the organ system involved. A blood cell count may show increased numbers of white blood cells. Levels of certain immunoglobulins and other proteins in the blood may be higher than normal.

There is no cure for an autoimmune disorder. Thyroid supplements, insulin injections or other supplements may be required to alleviate the symptoms, depending on the specific disease. Disorders that affect the blood components may require blood transfusions. Measures to assist mobility or other functions are sometimes needed for disorders that affect the bones, joints, or muscles.

Symptoms can often be controlled by taking corticosteroids. However, side effects such as osteoporosis (thinning of the bones), bruising, susceptibility to infections, diabetes and high blood pressure are common. Immunosuppressants (drugs that suppress the immune system), for example, cyclophosphamide or azathioprine may be used.

SEE ALSO *Diagnostic techniques on page 428*

Vasculitis

Vasculitis (angiitis) is a general term for a group of uncommon diseases causing inflammation of the blood vessels. Examples of vasculitis include Kawasaki disease, polyarteritis nodosa, Behçet's disease, Wegener's granulomatosis, Churg-Strauss syndrome, giant cell arteritis (temporal arteritis), Takayasu's arteritis, and Henoch Schönlein purpura. There is no known cure, but corticosteroids and immunosuppressants can help control the symptoms.

Lupus erythematosus

Lupus erythematosus is an autoimmune connective tissue disease in which the body's immune system attacks its own tissues. There are two forms: discoid lupus erythematosus (DLE), which affects only the skin, and systemic lupus erythematosus (SLE), which attacks joints and internal organs as well as the skin.

DLE is a chronic skin disorder that occurs most commonly in middle-aged women. It produces thickened, reddish patches on the face, cheeks and forehead. Sunlight makes the condition worse, so patients with DLE should wear hats and sunscreen to protect their skin. Corticosteroid skin creams are helpful. In severe cases, the antimalarial drug hydroxychloroquine may be beneficial.

In SLE, not only is the skin involved (in a similar way to DLE) but other organs are involved too. Generalized symptoms of fever, fatigue, weight loss and nausea are common. Arthritis usually develops in the fingers, hands, wrists and knees. A malar (butterfly-shaped) rash over the cheeks and bridge of the nose is a characteristic feature and may be made worse by sunlight. There may also be kidney, nervous system, blood, heart and lung damage.

No cure exists for SLE, but symptoms can be treated. Nonsteroidal anti-inflammatory medications (NSAIDs) are used in mild cases to treat arthritis. Corticosteroid creams are used to treat skin rashes. Antimalarial drugs (hydroxychloroquine) are sometimes prescribed for skin and arthritis symptoms. Severe or life-threatening cases may require the use of corticosteroid therapy, immunosuppressants (medications to suppress the immune system) or cytotoxic drugs, which block cell growth.

Scleroderma

Scleroderma is a disease characterized by increased deposition of fibrous tissue in the skin and other organs. This is associated with abnormalities of blood vessels, such as spasm precipitated by the cold, resulting in color change in the extremities (Raynaud's phenomenon). The fingers become swollen and small blood vessels may be prominent; gradually the skin over the fingers and face becomes thickened and tight. In severe cases the lungs, heart and kidneys can be affected.

Diagnosis is aided by the presence of autoantibodies in the blood. Treatment of scleroderma is difficult and the condition tends to progress gradually.

Sarcoidosis

Sarcoidosis is a systemic inflammatory disease of unknown origin, which may affect a variety of body organs. Most often it is detected as an incidental finding on a chest x-ray. It causes enlargement of lymph nodes in the chest and may cause shadows in the lungs. Later, symptoms such as persistent dry cough or shortness of breath may develop.

Other symptoms may include a painful red rash on the shins and redness and pain of the eyes. Less commonly, sarcoidosis may affect the kidneys, heart or brain. The inflammatory cells involved can produce an increased level of calcium in the blood and urine. Young women are most often affected.

Sarcoidosis is diagnosed by x-rays, blood tests and tissue biopsy. It needs to be distinguished from other causes of enlarged lymph glands, such as lymphoma and tuberculosis. The physician will often assess the function of various organs with further tests, such as an electrocardiograph or detailed lung function studies.

Often the condition is usually mild and may improve without any treatment, especially in children. In other cases treatment is required with anti-inflammatory medication. This can range from mild drugs, such as aspirin, to cortisone or powerful immunosuppressants in severe cases.

Chronic fatigue syndrome

Chronic fatigue syndrome (CFS), also known as myalgic encephalomyelitis (ME) or Tapanui flu (and also "yuppie flu"), is not fully understood. There is much debate as to whether this is a disease, but general agreement is that the syndrome does exist and the cause is unknown.

CFS may begin suddenly after an illness similar to flu or following surgery, accident, or bereavement. It affects the immune system and strikes more women than men. Said to exist where a person suffers chronic tiredness over at least six months and cannot gain relief through rest, the symptoms must be severe enough to prevent normal functioning at home and at work. Pain in muscles and abdomen, painful lymph nodes, and mild mental confusion are among the symptoms. Sufferers may develop sensitivity to certain foods, and to medications.

Diagnosis is made after excluding preexisting conditions such as psychiatric illness, depression, eating disorders, substance abuse and physical abnormalities.

Research has not found a cause, nor a cure, but it is commonly believed that stress seems to trigger the disease and worsen the symptoms. CFS can last for several years and some people never get better, although others recover in under a year. Treatment includes bed rest, avoiding stress, and prescription medications which help to alleviate anxiety.

Sjögren's syndrome

The second most common autoimmune rheumatic disorder after rheumatoid arthritis, Sjögren's syndrome is a syndrome of dryness of the eyes and mouth. It is caused by a failure of the lacrimal glands, which produce tears, and the parotid glands, which produce saliva. In Sjögren's syndrome (for reasons that are not yet understood) these glands are attacked by the body's own immune system, which produces antibodies against the gland tissues.

Sjögren's syndrome may occur on its own, or can even be associated with other connective tissue disorders such as rheumatoid arthritis, scleroderma, or systemic lupus erythematosus.

The symptoms of Sjögren's syndrome are dryness and grittiness of the eyes (sufferers complain of the sensation of "something in the eye," redness and burning of the eyes and sensitivity to light.

Dryness of the mouth causes difficulty in swallowing and talking, abnormal taste or smell, mouth ulcers and dental cavities. A physician will confirm the condition with a special test called Schirmer's test which measures the quantity of tears produced in 5 minutes. Salivary flow studies may also be performed.

The disorder cannot be cured, but the symptoms can be managed by using artificial tears for eye dryness, and methylcellulose or saline sprays for mouth dryness. In cases of severe underlying disease corticosteroid and immunosuppressive drugs may be prescribed.

Dermatomyositis

Dermatomyositis is a rare connective tissue disease that is characterized by inflammation and degeneration of the skin and the muscles throughout the body. It causes muscle weakness and a dusky red rash over the face, neck, shoulders, upper chest and back. Joint, heart and lung disease may occur. Treatment may include physical therapy and prednisone (a steroid hormone) or immunosuppressant drugs such as azathioprine and methotrexate.

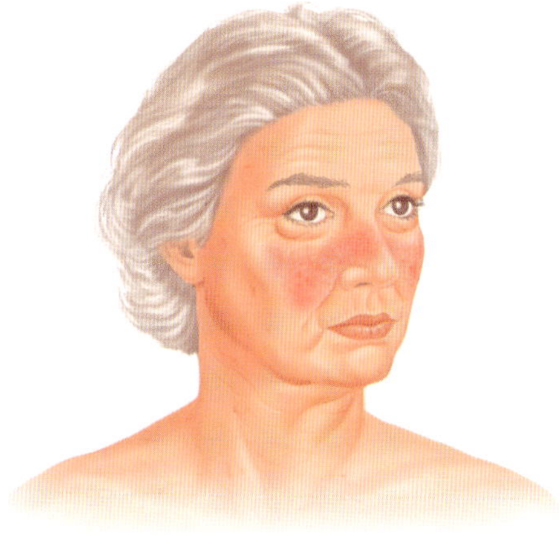

Lupus erythematosus

The body's immune system attacks its own tissues in this disease, producing a distinctive red rash on the face. Middle-aged women are most commonly affected.

Polymyositis

Polymyositis is an autoimmune inflammatory disease of muscle. It causes muscle weakness, especially around the shoulders and hips. The muscles ache and may be tender to the touch. Fatigue, weight loss and a low-grade fever are common. The cause is unknown. The condition is treated with physical therapy and corticosteroids or immunosuppressant drugs.

Myasthenia gravis

Most often an autoimmune disorder, myasthenia gravis may also be due to a tumor of the thymus gland (thymoma). Characterized by fluctuating but progressive weakness of the voluntary muscles, early symptoms can include double vision (diplopia) and lid-droop after prolonged reading, fading of the voice during a speech, difficulty swallowing the latter part of a meal and tiredness of the legs and arms. The weakness is reversed by resting the affected muscles, but the condition is slowly progressive.

Diagnosis is made when there is a rapid reversal of the weakness after the administration of certain drugs.

Treatment may be continued with drugs but the removal of the thymus gland will produce the greatest improvement. The occasional patient may require steroids or other immunosuppressive drugs. Myasthenia gravis may be associated with other autoimmune disorders. Small cell cancer of the lung and botulism can produce a somewhat similar myasthenia.

NERVOUS SYSTEM

Along with the endocrine and the immune systems, the nervous system is one of three systems concerned with coordinating the activities of the body. The nervous system receives information about the outside world and internal organs, determines the appropriate response to changes in both domains and responds rapidly. This information is processed and transferred by means of nerve cells firing electrical signals called action potentials, which can move along nerve fibers at speeds of up to 320 feet (100 meters) per second. The endocrine and immune systems respond more slowly because they depend on chemicals or cells released into the blood to communicate responses.

SEE ALSO *Autonomic nervous system on page 75; Brain on page 132; Memory on page 138; Nervous tissue on page 21; Senses on 140; Spinal cord on page 217; Spinal nerves on page 218; Touch on page 348*

DIVISIONS OF THE NERVOUS SYSTEM

The nervous system is divided into a central nervous system (CNS) and a peripheral nervous system. The CNS is made up of the brain and spinal cord, while the peripheral nervous system consists of all the nerves distributed throughout the rest of the body. The parts of the peripheral nervous system that control aspects of body function over which we have no voluntary control are called the autonomic nervous system. Although most of the constituent nerve cells of the autonomic nervous system are located peripherally, some are also located in the CNS. The autonomic nervous system is also under the control of parts of the brain such as the hypothalamus.

There are also many nerve cells located in the wall of the gastrointestinal tract (stomach and intestine). These cells coordinate and control the movement of the gut and the secretions of the gut glands, and also transport sensory information about conditions in the gut. Gut nerve cells may outnumber those in the spinal cord.

The brain and spinal cord consist of nerve cells and their processes, along with bundles of nerve fibers. Gray matter refers to the parts of the CNS where nerve cell bodies are concentrated; white matter refers to parts that have very few cells and many nerve fibers.

The CNS also contains many cells which are called glia. These fulfill diverse roles such as the formation of myelin, maintaining the correct concentrations of salts and chemicals in the spaces between the nerve cells, and providing surveillance against invading microorganisms.

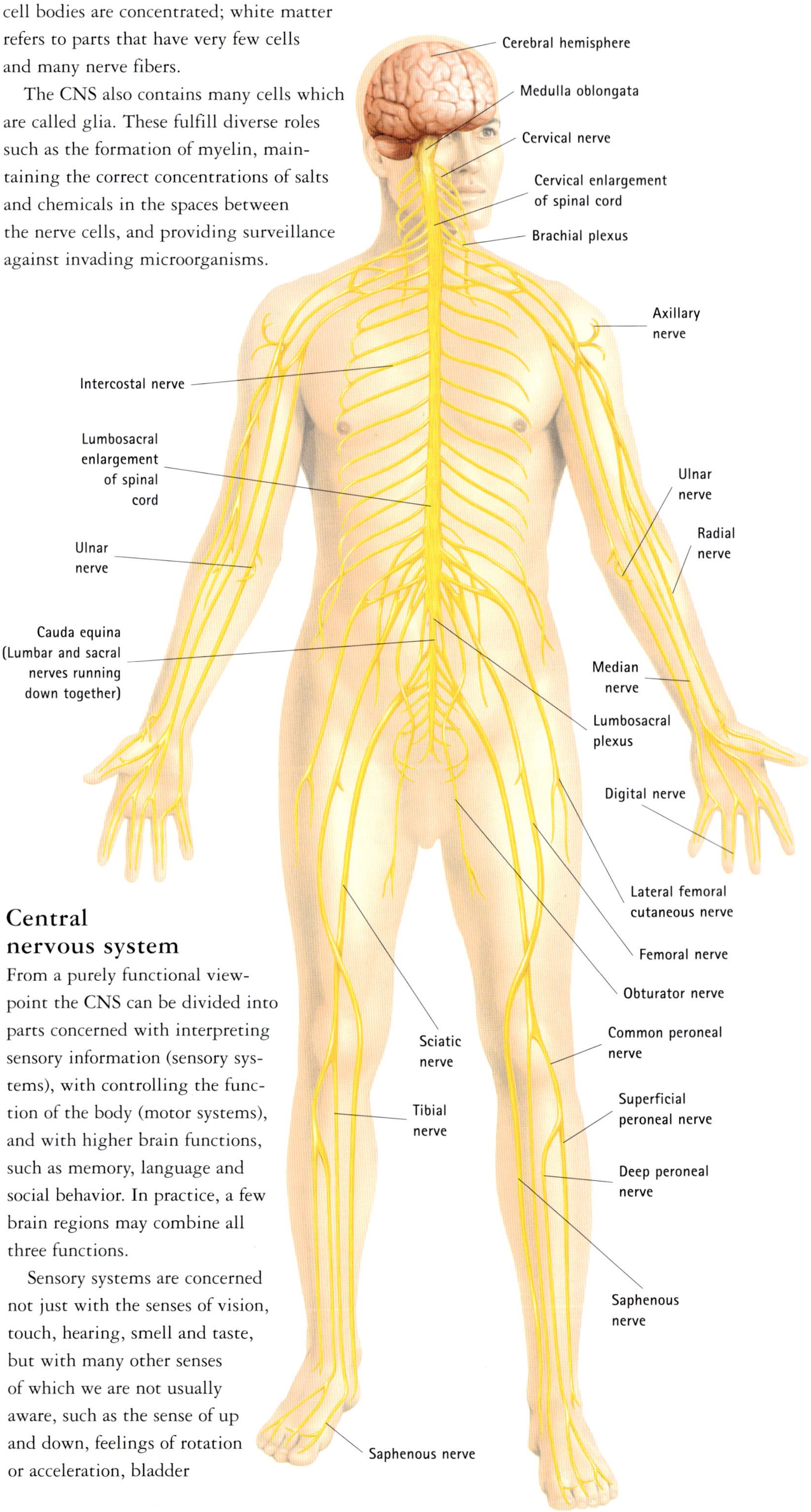

Central nervous system

From a purely functional viewpoint the CNS can be divided into parts concerned with interpreting sensory information (sensory systems), with controlling the function of the body (motor systems), and with higher brain functions, such as memory, language and social behavior. In practice, a few brain regions may combine all three functions.

Sensory systems are concerned not just with the senses of vision, touch, hearing, smell and taste, but with many other senses of which we are not usually aware, such as the sense of up and down, feelings of rotation or acceleration, bladder fullness, stomach distension, joint position and blood pressure. Motor (effector) systems are concerned with controlling our muscles and movement, and controlling many automatic functions, including sweating, blood pressure and gut movements.

Lobes of the brain

Fissures and sulci separate the lobes of the brain. The four lobes—frontal, parietal, temporal and occipital—are named after the skull bone that overlies each lobe.

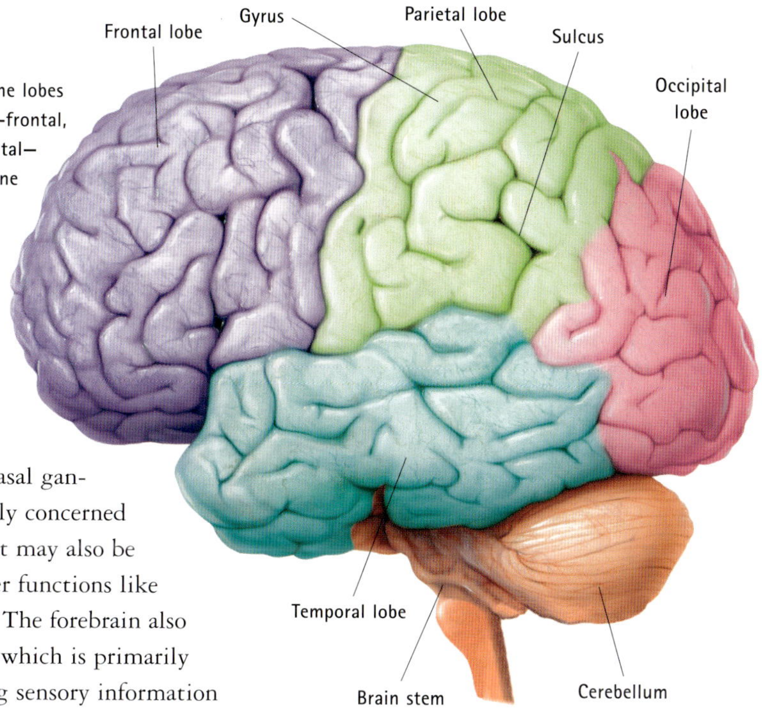

Anatomy of the brain

Anatomically, the brain is divided into three main regions—the forebrain, midbrain and hindbrain. The forebrain contains most of the brain substance and is capped by the highly folded cerebral cortex. The midbrain and hindbrain carry many fiber bundles that convey information to and from the upper parts of the brain. The hindbrain can be further divided into the pons, medulla and cerebellum. The midbrain, pons and medulla, collectively called the brain stem, contain nerve cell groups that control the muscles of the face and head and process information about touch on the face, taste and hearing.

Other important functions of the brain stem are the control of breathing, blood pressure and heart function.

The forebrain is a very complex and important region. At its surface, the cerebral cortex is broadly organized into the frontal lobe, the occipital lobe at the back, the parietal lobe in the middle and the temporal lobe below. Each lobe contains many different areas concerned with particular functions.

At the very front of the frontal lobe, the prefrontal area is concerned with the control of social behavior, motivation and planning. Behind the prefrontal area lie the premotor cortex, which plans motor actions, and the primary motor cortex, which sets those motor commands into action. The lower part of the left frontal lobe contains Broca's area, which is concerned with the expression of language. The occipital lobe contains a series of visual areas. The parietal lobe contains a primary somatosensory area, concerned with processing sensory information about touch, pain and joint position from the body and face. Further back in the parietal lobe lies a region that generates our sense of the spatial organization of the outside world. The temporal lobe contains areas concerned with hearing, smell and memory. At the junction of the temporal and parietal lobes lies Wernicke's area, concerned with understanding language. Deep inside the forebrain are groups of nerve cells called the basal ganglia, which are primarily concerned with motor control, but may also be involved in some higher functions like language and thought. The forebrain also contains the thalamus, which is primarily concerned with relaying sensory information to the cerebral cortex and controlling motor activity. Below the thalamus lies the hypothalamus, which serves as the interface between the brain and the autonomic nervous system and controls food and water intake, sexual function and body temperature, among other functions.

The spinal cord

The spinal cord acts as an intermediary between the peripheral nervous system and the brain. It extends from the base of the skull to a point about two-thirds of the way down the back, running through the vertebral canal. The spinal cord has many nerve fibers attached to it, arranged in sets and named according to their level on the cord. The cord itself has a central region of gray matter, which is divided into posterior (dorsal) and anterior (ventral) horns and an intermediate region. The gray matter is surrounded by white matter carrying ascending and descending fiber tracts.

At each level, the spinal cord receives sensory nerves, which convey information about touch, pain, temperature, muscle tension and joint position. This information may be used when it enters the spinal cord to control muscle tension or stimulate reflex responses, such as withdrawing a hand from a hot object, or it may be transmitted up through the white matter of the cord to the brain for conscious appreciation of the information. The spinal cord also gives rise to motor and autonomic nerve fibers,

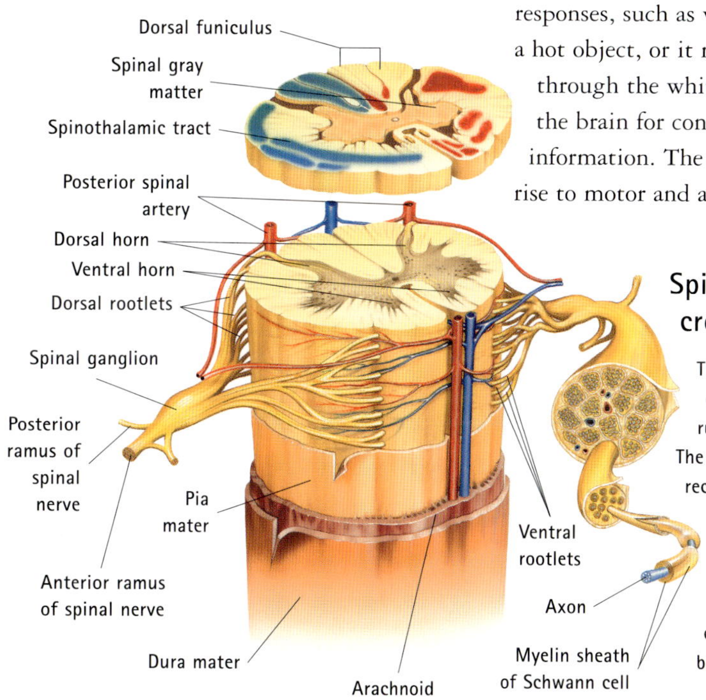

Spinal cord cross-section

The spinal cord is part of the central nervous system that runs down the vertebral canal. The central core of gray matter receives and processes sensory information, and sends signals to the muscles. The surrounding layer of white matter contains axons which communicate between the brain and spinal cord.

which control muscles and affect internal organs, respectively. The brain sends controlling signals through the white matter of the spinal cord to spinal motor neurons, so that we can consciously control our muscles.

DERMATOMES

Each of the 31 segments of the spinal cord gives rise to a pair of spinal nerves, which carry messages into and out of the central nervous system. These nerves branch into and service particular areas of the body. Ultimately, each ends up innervating a different region of the skin called a dermatome.

The location of dermatomes across the body forms a pattern that is significant when certain parts of the body require anesthesia. It indicates specific nerves that need to be blocked to cut sensations in a certain region. It is also relevant in spinal cord injuries—identifying dermatomes with abnormal or no sensations helps isolate the location of damage to spinal nerves or the spinal cord.

Peripheral nervous system

The peripheral nervous system consists of nerve fibers or axons that control muscle activity or carry sensory information to the spinal cord or brain stem. The motor axons come from the cell bodies located in the anterior horn of the spinal cord; the sensory axons arise from cell bodies in clumps or ganglia along the spinal cord. Many axons in the peripheral nervous system are coated with myelin (like axons in the central nervous system) produced by Schwann cells.

An important difference between the peripheral and central nervous systems is that peripheral nerves can repair and regenerate; central axons do not. This may be partly due to the behavior of Schwann cells after nerves have been injured.

Nerves

Nerves connect the brain and spinal cord (central nervous system, CNS) with peripheral regions such as the muscles, skin and viscera. Their role is to carry signals which provide us with sensations or control our muscles. Twelve pairs of nerves arise from the brain and are called cranial nerves; they include the trigeminal and facial nerves and the vagus nerve. Nerves arising from the spinal cord, between the vertebrae, are known as spinal nerves. There are 31 pairs of spinal nerves, supplying the trunk and limbs.

A special branch of the peripheral nervous system, the autonomic nervous system, is concerned with controlling muscles in the viscera, such as the gut, bladder, heart and blood vessels.

SEE ALSO *Autonomic nervous system on page 75; Brain on page 132; Cranial nerves on page 156; Nerves and blood vessels of the leg on page 334; Nerves and blood vessels of the shoulders and arm on page 322; Nervous tissue on page 21; Spinal cord on page 217; Spinal nerves on page 218*

Spinal cord

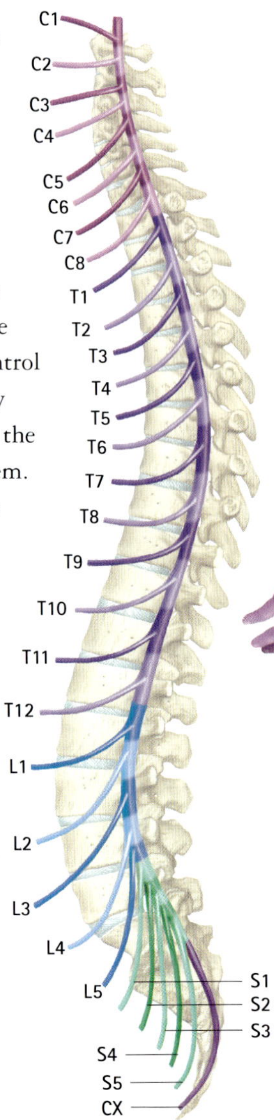

Nerve to skin link

The spinal nerves are numbered and correspond closely to the spinal vertebrae. Each pair of nerves supplies a specific dermatome (skin area) of the body. The face is supplied by branches of the trigeminal nerve (V1, V2 and V3).

DERMATOMES

The dermatomes are the areas of skin supplied by a single spinal nerve. However there is usually some overlap between adjacent dermatomes.

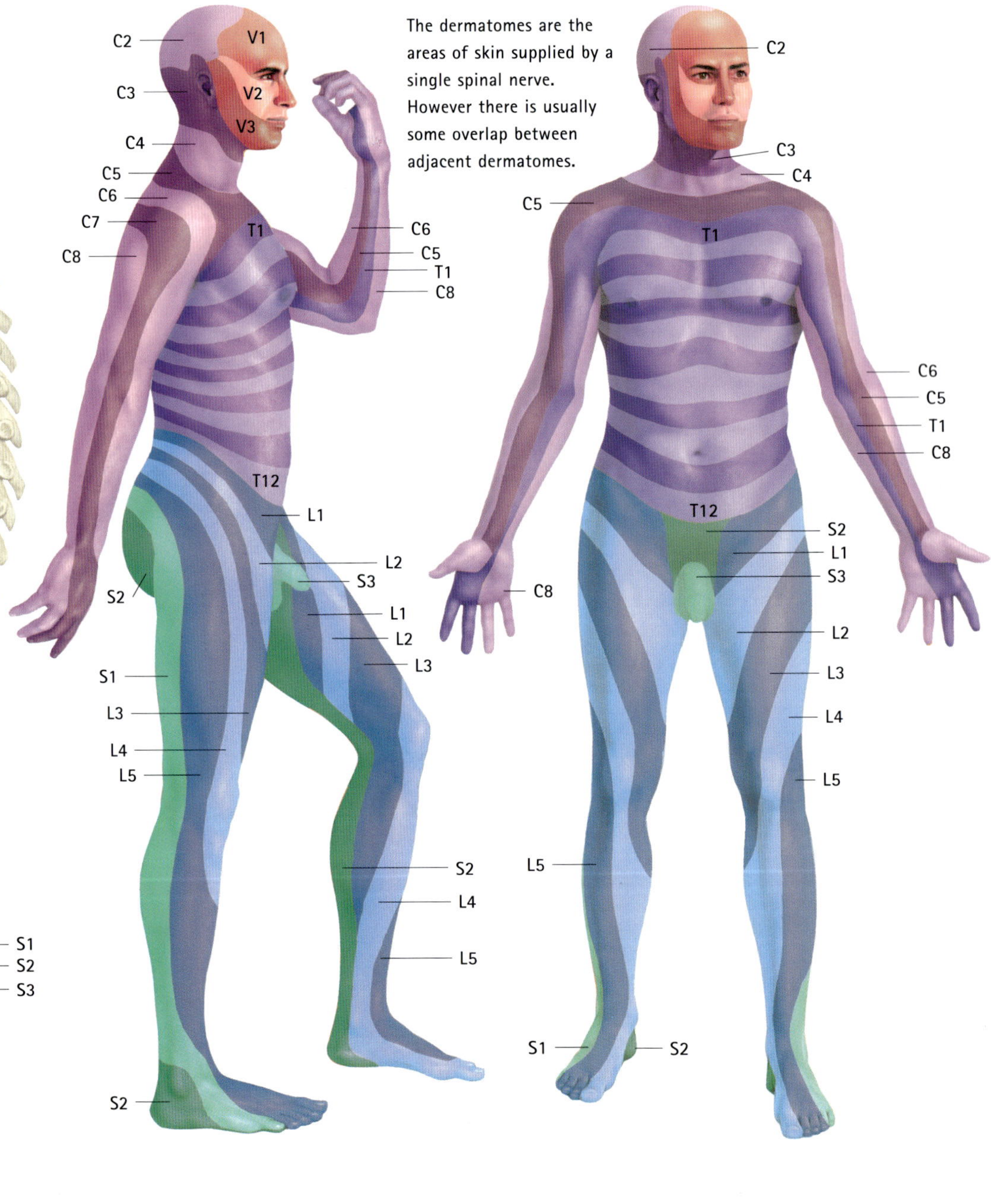

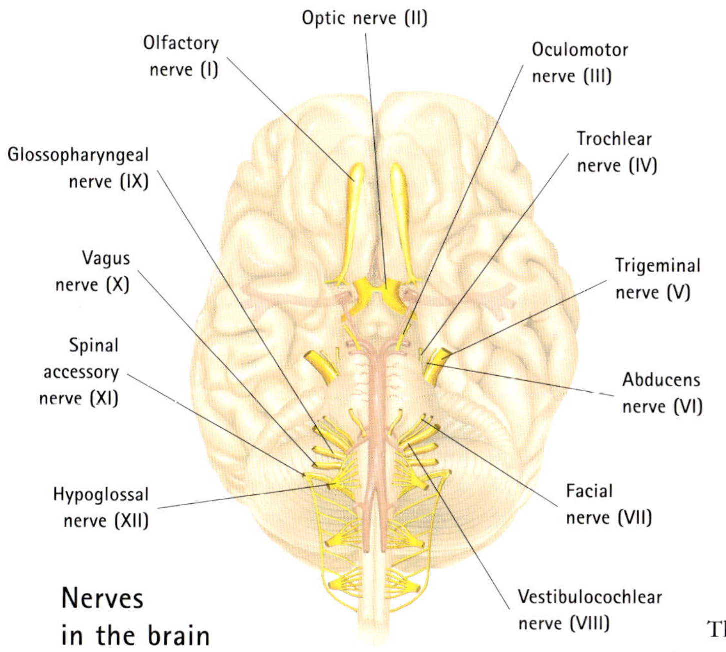

Nerves in the brain

There are 12 cranial nerves which are visible at the base of the brain. In addition to their motor and sensory functions, the cranial nerves serve the 5 systems for special senses.

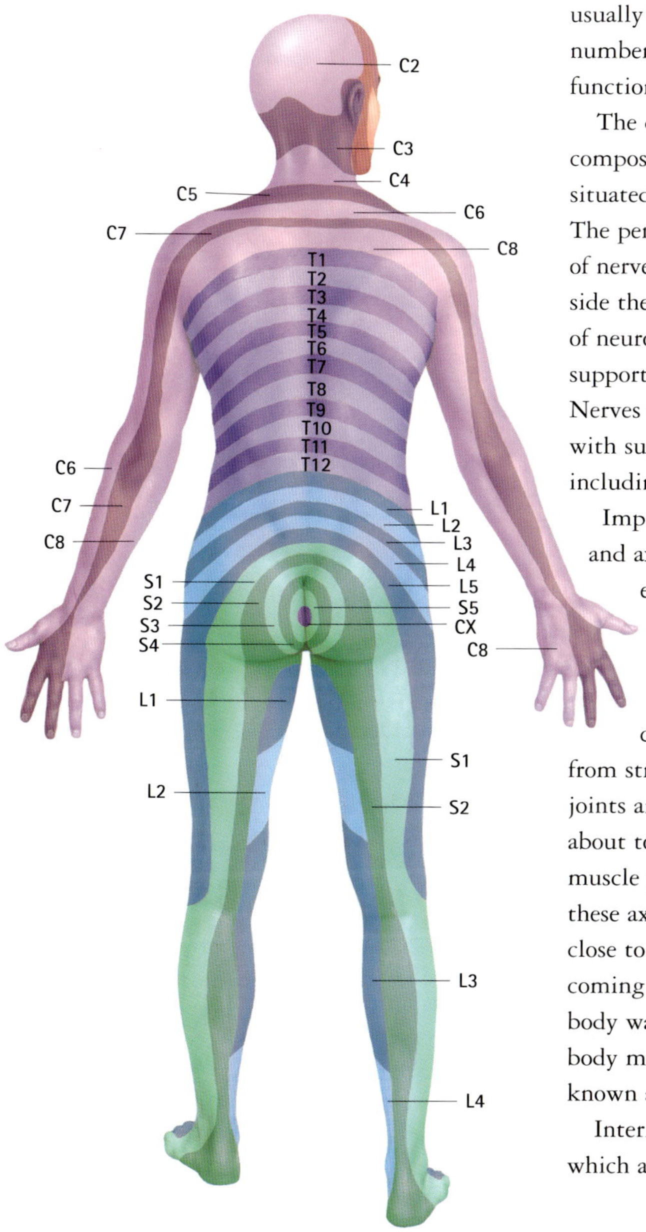

Neurons

Neurons, also known as nerve cells, are specialized cells involved in sensation, information processing and cognition, and control of muscle and gland activity. Neurons are the characteristic cells of the brain, spinal cord and nerves.

Every neuron consists of a cell body (perikaryon), a cell nucleus, contained within the cell body, and one or more narrow projections, known as processes. The processes, which radiate from the cell body, are of two types: dendrites, which convey impulses to the cell body; and axons, which normally convey impulses away from it. Neurons usually have a single axon, and a variable number of dendrites, depending on the function of the cell.

The central nervous system (CNS) is composed of nerve cells whose axons are situated in the brain, eye and spinal cord. The peripheral nervous system is composed of nerve cells whose axons are situated outside the CNS. The brain consists of a mass of neurons of various types, along with supporting cells (glia) and blood vessels. Nerves consist of bundles of axons, together with supporting cells (Schwann cells) including connective tissue.

Impulses are conveyed along dendrites and axons by a wave of chemical and electrical changes affecting the cell membrane of the dendrite or axon.

There are two types of axons, sensory and motor. Sensory axons carry signals coming into the CNS from structures such as the skin, muscles, joints and viscera. They carry information about touch, temperature, pain, joint and muscle position. The nerve cell bodies of these axons (sensory neurons) lie in ganglia close to the CNS. Motor axons carry signals coming from the CNS to muscles in the body wall, limbs and viscera, to control our body movements. Their nerve cell bodies, known as motor neurons, lie in the CNS.

Intermediate neurons (interneurons), which are the most numerous in the brain, connect sensory and motor neurons directly or by networks of cells. Cell networks are responsible for cognition and memory.

The signals carried by axons consist of short pulses of electrical activity, called action potentials, each about 0.1 volt and lasting about one-thousandth of a second. Action potentials in sensory axons start in the periphery (e.g. skin) with a stimulus (e.g. touch) which activates particular receptors (e.g. hair receptors). The action potentials then travel to the CNS where they activate other nerve cells, in sensory centers, to cause sensations. For motor outputs, action potentials start in motor cells in the CNS and travel out along the motor axons to muscles, causing the muscle to contract.

Most nerves contain sensory and motor axons, with signals going both to and from the CNS in the different axons. Because each axon is like a separate insulated wire it does not interfere with its neighbors. A few nerves are purely motor in function, such as the hypoglossal nerve controlling muscles in the tongue. Similarly, a few nerves are purely sensory, such as the sural nerve to the skin on the lateral side of the foot. Activity in a nerve can be blocked by a local anesthetic, which prevents action potentials from passing through the block. This can be useful for stopping pain signals, e.g. from the teeth during dental repairs.

Nerves also contain supporting cells called Schwann cells that envelop the axons. The combination of axon and Schwann cells is usually referred to as a nerve fiber.

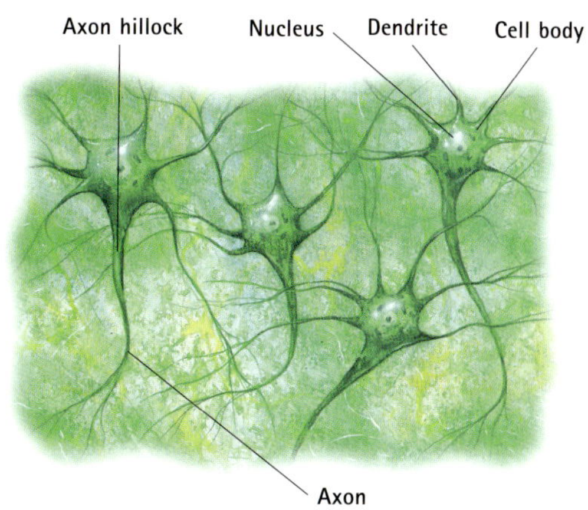

Neural tissue

Neural tissue processes neural data and conducts electrical impulses from one part of the body to another part. Ninety-eight percent of neural tissue is located in the brain and spinal cord.

A Schwann cell extends a short distance along the axon, with the following cell looking after the next section of axon, from one end to the other. Larger axons are surrounded by the Schwann cells, that wrap their cell membranes around the axon, making a special layer known as myelin. This acts as an electrical insulator, and is important for increasing the speed at which action potentials can travel (conduction velocity). These axons are called myelinated axons. Smaller axons do not have any myelin (unmyelinated axons). They are still encased by Schwann cells which helps to insulate them, but the cell does not wrap them in membrane, and their signals travel more slowly.

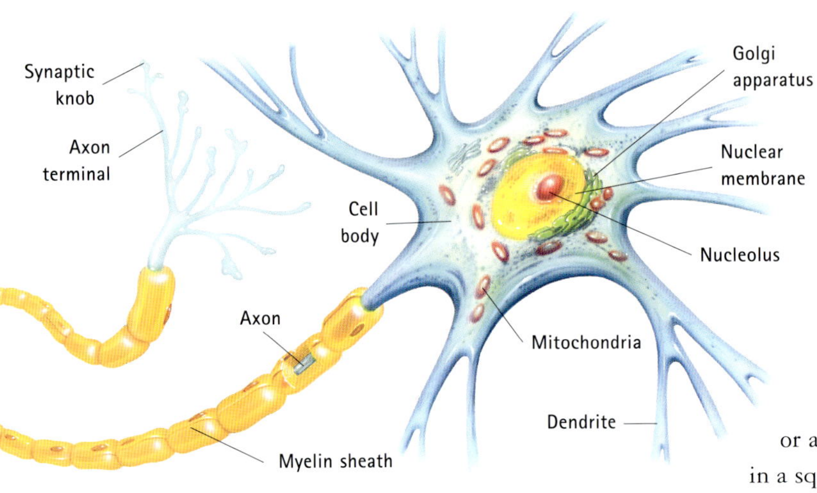

Neuron

Neurons are specialized cells found in the nervous system which conduct nerve impulses. Each neuron has three main parts: the cell body, the branching projections (dendrites) that carry impulses to the cell body, and one elongated projection (the axon) that conveys impulses away from the cell body.

In sensory axons there is a relationship between the size of an axon and the type of sensation it carries. Large myelinated axons carry information for fine tactile discrimination (e.g. texture of a surface), vibration and limb position. They can transmit signals at about 16–230 feet (5–70 meters) per second (it takes less than one-tenth of a second for action potentials to reach the spinal cord from the foot). Smaller unmyelinated axons carry signals about cold, warmth and pain, and a crude sense of touch. Being smaller and lacking a layer of myelin, their signals are carried more slowly, at about 3 feet (1 meter) per second. In these axons, action potentials from the foot, for example, take about 1 second to reach the spinal cord.

The activity traveling along a nerve can be recorded at certain sites by electrodes on the skin surface. Assessing the function of nerves in this way (nerve conduction testing) can be used to detect the slowing of conduction velocity which may occur when the nerve is injured or the myelin is damaged. This is similar to the evoked potential tests that can be used to detect slowing of conduction in central pathways, as with multiple sclerosis.

Reflexes, such as the knee jerk, involve a minimum of two neurons: a sensory neuron (which, in this case, detects stretching of the quadriceps muscle) and a motor neuron, which is stimulated by the sensory neuron and fires to initiate muscle contraction (in this case, contraction of the quadriceps muscle). Usually, there is at least one intermediate neuron in between the sensory neuron and the motor neuron.

The total number of neurons is greatest before birth, and decreases during life. Once lost, neurons do not regenerate, but cut axons in damaged nerves may regenerate.

Disorders affecting neurons may include injury, congenital defects, cancer, infections, vitamin deficiency and degeneration. Epilepsy is due to the inappropriate generation of impulses by neurons in the cerebral cortex.

Connective tissue

Nerves also contain several layers of connective tissue, which protect the nerve and carry blood vessels. Connective tissue surrounds individual axons and groups them into bundles (fascicles) as well as surrounding the whole nerve. Blood vessels to nerves have a special barrier, the blood-nerve barrier, similar to the blood-brain barrier. The blood-nerve barrier prevents certain chemicals in the blood from entering the nerve, thus providing further protection for the axons. When the blood supply to a nerve is restricted or there is pressure on a nerve, strange sensations, usually called paresthesias, can be felt in the area supplied by the nerve. These include tingling sensations, "pins and needles" and numbness. If such a sensation is because of tight clothing or a certain posture (e.g. sitting in a squatting position), it usually disappears when the pressure is relieved. Other paresthesias may need treatment, e.g. pressure to the median nerve at the wrist can give pins and needles in the thumb that may require surgery.

Synapses

Synapses are junctions between nerve cells (neurons), where impulses are passed from one cell to another, or from one cell to an

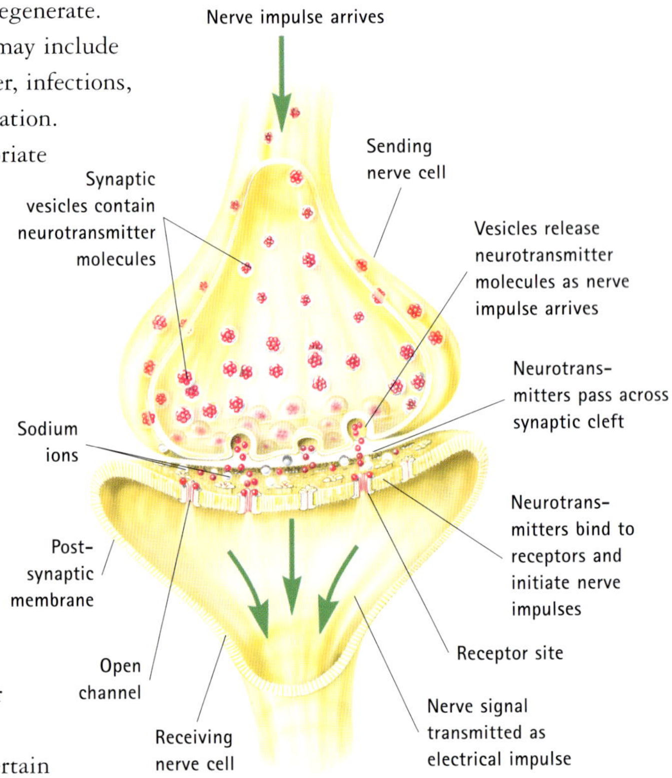

Synapses

A synapse is the junction between two nerve cells (neurons). Nerve signals are passed from cell to cell across the synaptic cleft by chemical molecules—neurotransmitters. The receiving cell has receptor sites specific to certain neurotransmitters. When the correct transmitters lock in place they open channels in the nerve membrane to let in sodium ions. This causes an electrical change which fires the receiving cell, passing on the nerve signal.

effector organ. Synapses usually occur between the axon terminal of one cell and a dendrite or cell body of another. The neural impulse is carried across the synapse, in one direction only, by chemical molecules known as a neurotransmitters. Common neurotransmitters include glutamate and serotonin. Low levels of neurotransmitters or disruption of their action can often cause neurological disorders.

REFLEXES

Reflexes are quick and automatic responses to stimuli, whereby a nerve impulse travels to a nerve center in the spinal cord. The nerve center then sends a message outward to a muscle or gland to effect a response without the person being consciously aware of it.

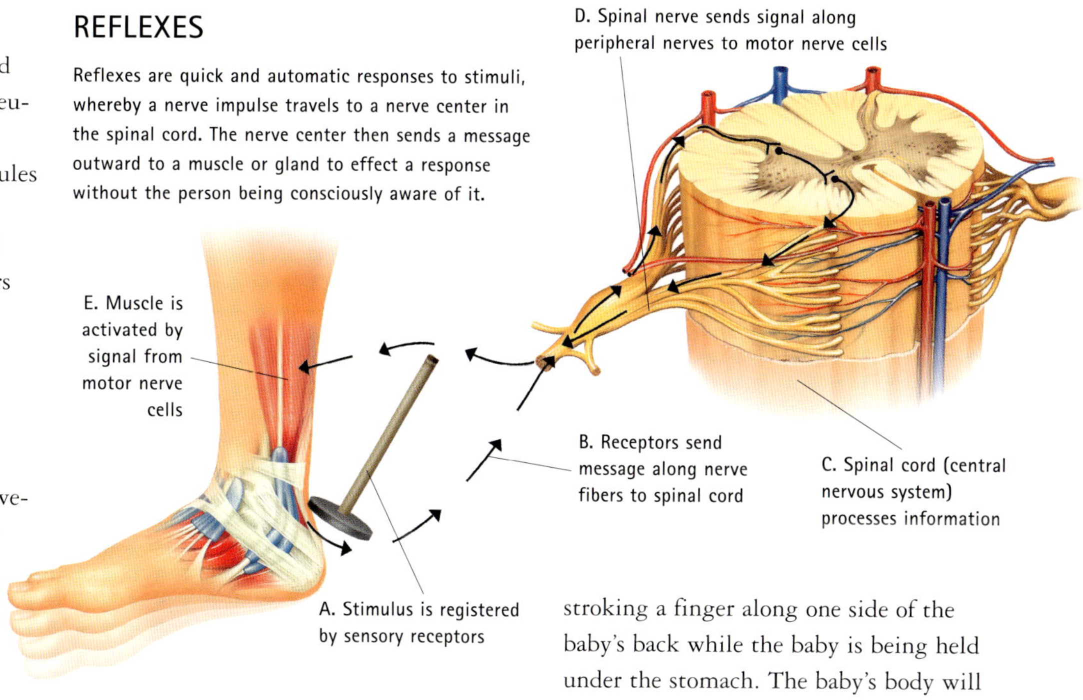

Reflexes

A reflex is an involuntary immediate movement or other response to an appropriate stimulus, which occurs unconsciously (without being willed). If you touch a sharp spike or a hot surface, you will automatically pull your hand away in a reflex action. There are many everyday reflexes that we take for granted.

When a bright light is shone into a person's eyes the pupils contract; a blink is a reflex action that protects the eye; a cough or a sneeze protects the lungs and the respiratory system, and there are many other reflexes. Four things happen when a reflex acts—reception, conduction, transmission and response. All of these reflexes involve the nervous system.

The well-known knee-jerk or patellar reflex is the sudden kicking out of the lower leg when the patellar tendon, just below the knee, is tapped sharply. This reflex is used to test for damage to the central nervous system and the peripheral nerves and also in recognizing thyroid disease. There are many other reflexes that can also be tested to evaluate the nervous system.

The concept of the conditioned reflex is attributed to the Russian physiologist Ivan Pavlov, who won the Nobel Prize in 1904 for his work. After noting that a hungry dog salivated at the sight of food, he rang a bell when the food appeared. After some time the dog would salivate at the sound of the bell when no food was present—thus the "conditioned" reflex.

There are three groups of reflexes found in a newborn. The first group, which are vital for survival, include the rooting reflex—when a baby's cheek is gently stroked it will turn its head and open its mouth on that side; the sucking reflex, which is triggered by pressure on the upper palate of the baby's mouth; the swallowing reflex; the gagging reflex, which is triggered if the baby swallows too much fluid; and the labyrinthine reflex, which makes the baby raise its head when lying on the stomach. Newborns also have protective reflexes; if a cloth is held over a baby's face the baby will brush the cloth away. Medication during labor, brain damage and prematurity can affect these survival reflexes and these will be tested if there is a doubt about the newborn's state of health.

The second group includes the stepping, crawling and Babinski reflexes. The stepping reflex, activated when a baby's foot touches the ground, is usually gone by the time a baby is two months old. The crawling reflex is displayed when a newborn is placed in a face-down position. The Babinski reflex occurs when the sole of the foot is stroked from heel to toe, resulting in the toes curling up and the foot turning in. It usually lasts until the baby is 2 years old. In older age groups, the Babinski reflex may be used as a diagnostic tool for disorders of the central nervous system.

The third group of reflexes are behavior patterns that may have stopped serving a function. These include the grasp reflex, which can also be found in the baby's foot; the Moro or startle reflex, which is used by doctors to test muscle tone; and the galant reflex, which is tested by gently stroking a finger along one side of the baby's back while the baby is being held under the stomach. The baby's body will bend like a bow, pulling the pelvis toward the side stroked. It indicates the development of the spinal nerves and will last until the baby is about 9 months old.

SEE ALSO *Newborn on page 412; Touch on page 348*

Proprioception

Proprioception is perception of one's own body, which is of vital for everyday activities. Activities such as standing, walking, and picking up objects require information on the position and movement of each part of the body and of muscle activity, which is provided by sensors in joints and muscles.

The sense of balance is provided by the semicircular canals and otolith organs of the inner ear. Visceral sensations include hunger, thirst, the feeling of food being swallowed, and the need to defecate or urinate. Other proprioceptive sensations include palpitations (awareness of the beating of the heart), and (on occasion) the need to breathe.

SEE ALSO *Balance on page 171; Senses on page 140*

Pain

Pain is an unpleasant experience associated with real or potential damage to the body. Pain usually serves as a signal that tissues are being damaged, and that the sufferer needs to move away from the painful stimulus as quickly as possible. It also urges rest and recovery from any damage that has been done, to allow healing to take place. There are two main types of pain. If pain does not outlast its cause, such as a burn or cut, then it is often referred to as acute pain. Such pain serves a warning function. For example, there are some rare people born without a sense of pain. While this might seem to be a blessing, such people die relatively young because they do not receive warning of potential damage, and may experience burns, fractures and joint damage.

Unfortunately, pain does not always serve a purpose. It may outlast the initial injury and the healing process. It may have a cause which cannot be removed, such as cancer or malfunctioning nerve cells. This is chronic pain and may involve hyperalgesia (increased sensitivity to painful stimuli), allodynia (pain caused by stimuli which would not normally be painful) and spontaneous pain, which has no obvious cause. Chronic pain is debilitating and difficult to treat.

SEE ALSO *Brain on page 132; Spinal cord on page 217; Spinal nerves on page 218; Touch on page 348; Treating the central nervous system on page 442*

PAIN RECEPTORS AND STIMULI

The sensory neurons that mediate pain sensation are known as nociceptors. They have endings or receptors in skin, muscles, joints and internal organs. These nerve endings are connected to the spinal cord or brain stem by nerve fibers which generally lack the myelin sheath which insulates most other sensory nerve fibers.

Nociceptors are activated by a range of potentially damaging (noxious) stimuli which may be mechanical (such as a pinch or cut), thermal (such as a burn) or chemical (such as exposure to acid). Once nociceptors are activated, signals are conducted along their nerve fibers to the spinal cord or brain stem, and from there to parts of the brain such as the thalamus and cerebral cortex.

Nociceptors have properties that depend on the tissue in which they are located. Nociceptors in the skin are readily activated by pinching or cutting, but internal organs, for example the liver or appendix, can be pinched with forceps without causing pain. This does not mean that the internal organs are entirely insensitive to pain. It is more likely that many of the nociceptors in the viscera are so-called "sleeping" nociceptors. These are only activated by mechanical stimuli when they have already been stimulated by chemicals produced when tissue is inflamed. Sensitization of nociceptors in this way contributes to the increased pain sensitivity felt following tissue damage.

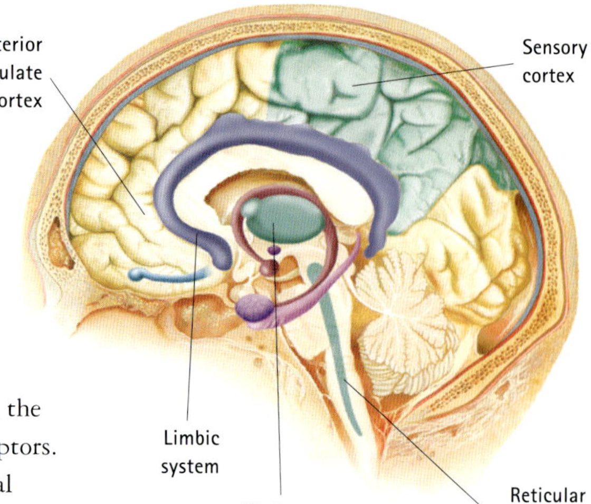

Processing pain

The brain controls our perception of pain in a number of different areas. Pain signals are relayed to the brain via the thalamus. The sensation and location of pain is registered in the sensory cortex and emotional responses are governed by the limbic system.

Most of us are familiar, for example, with increased sensitivity of the skin due to sunburn, or how ordinary movement becomes painful following a joint injury. Many analgesics, such as aspirin or acetaminophen (paracetamol), help relieve pain in damaged or inflamed tissues by preventing the production of chemicals which sensitize nociceptors.

TRANSMISSION AND RECOGNITION OF PAIN

Nociceptors form only part of a complex network of nerve cells that give rise to the sensation of pain. Nociceptors send signals to neurons in the spinal cord or brain stem. These neurons receive information not only from nociceptors, but also from mechanoreceptors. These sensory receptors transmit information about muscle length, skin pressure or joint angle—information which normally has nothing to do with pain. Signals from mechanoreceptors may not activate spinal neurons which deal with pain.

However, in chronic pain, the spinal neurons which deal with painful stimuli receive excessive input from nociceptors. This sensitizes the spinal neurons, which now overreact to inputs from nociceptors or even from mechanoreceptors. The body can thus become more sensitive to painful stimuli, and can often perceive harmless stimuli as painful as well. The changes in spinal neurons, which contribute to chronic pain, are often long-lasting and difficult to reverse.

Drugs that provide relief from acute pain include non-narcotic analgesics such as aspirin and acetaminophen (paracetamol), and for more severe pain, narcotic (opioid) drugs such as morphine, codeine and pethidine. There is a common misconception that using opioids for pain relief carries a risk of addiction, but this risk is negligible when they are used purely for the relief of pain in prescribed doses. Chronic pain is often more difficult to relieve than acute pain, but may be treated with antiepileptics and tricyclic antidepressants, non-steroidal anti-inflammatory drugs and opioids such as morphine and methadone.

Once pain signals have been processed in the spinal cord, they are transmitted to the thalamus and cerebral cortex. Relatively little is known about the role of the cortex in pain perception, but recent techniques of brain imaging have been very useful in showing that several areas of the cortex are involved, probably dealing with different aspects of pain. Objective aspects of a painful stimulus (intensity and location) usually appear to be represented in the somatosensory cortex, while emotional aspects (unpleasantness, associations with other events) are represented in other parts of the cortex, such as the anterior cingulate cortex. Decisions about what type of action to take in order to avoid the pain may involve the motor cortex.

People (including many doctors) tend to think that the pain they experience depends simply on the intensity of the painful stimulus. This is often not the case. While the nociceptors themselves are very simple, their connections within the spinal cord and brain are complex, and include circuitry which can suppress or even enhance pain sensation. Some people are very sensitive to painful stimuli, while others are quite insensitive. It is often thought that there are significant cultural differences in pain sensitivity, but such differences are probably about the expression of pain rather than the sensation itself.

Pain sensation can also depend on circumstance; someone injured in a football game or in the heat of battle may not notice even a severe injury; conversely pain may be exaggerated by fear or anticipation—an injection may be more painful if there is time to worry about it. Some of these differences in pain perception are due to the control that parts of the brain can exert on pain sensation.

TREATMENT OF PAIN

Drugs are perhaps the most common measure employed to relieve pain. These are not always effective, particularly for chronic pain, and people may turn to alternative therapies, such as chiropractic or acupuncture, for the relief of pain which conventional medical treatment has not been able to alleviate. People may be helped by these alternative treatments, although there is little firm clinical evidence to prove their efficacy.

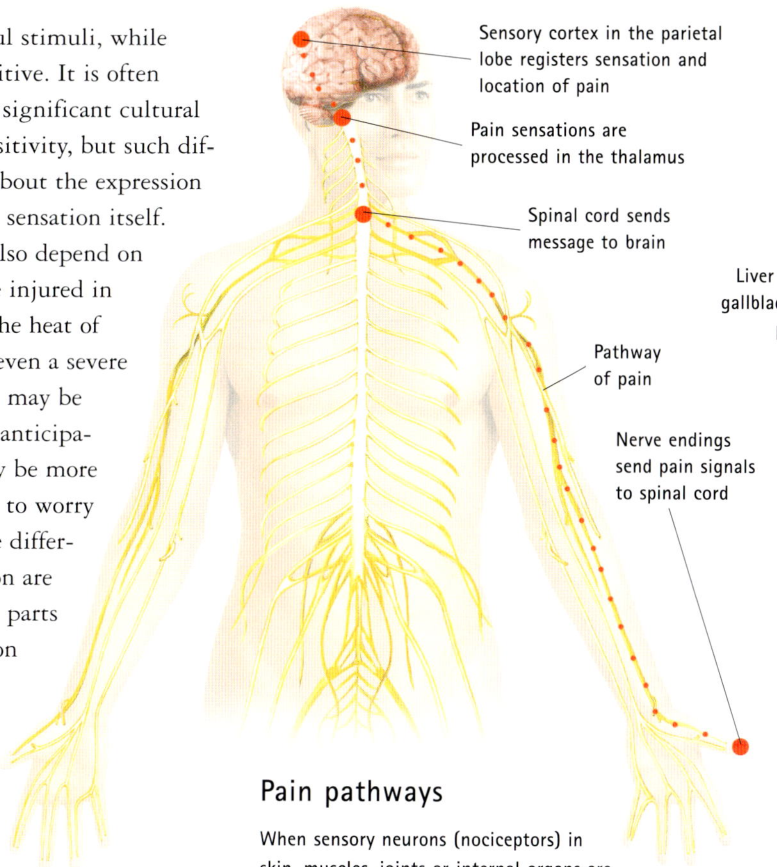

Pain pathways

When sensory neurons (nociceptors) in skin, muscles, joints or internal organs are activated by pain they send signals along nerve fibers to the spinal cord. From there the signals pass up the spinal cord to the brain stem and on to the thalamus and cerebral cortex where the sensation of pain is registered.

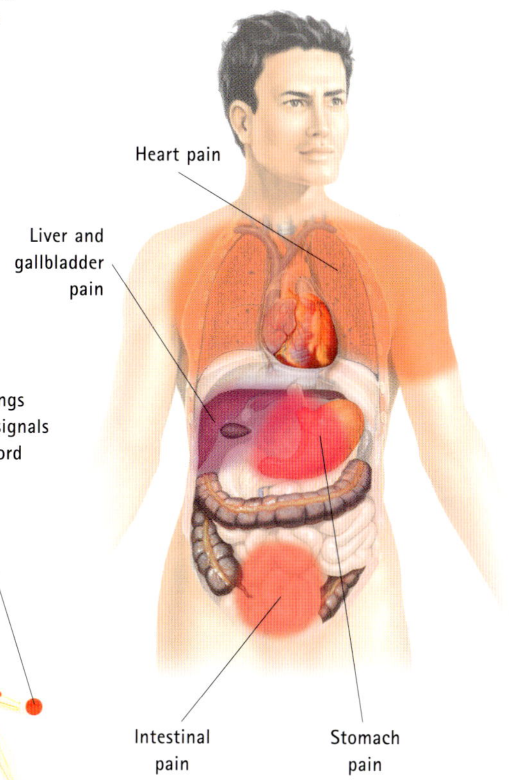

Referred pain

The pain from internal organs may be felt on the surface of the skin as well as internally. This is because the skin and internal organs may share the same pain pathways. When an organ sends pain signals to the spinal cord, the brain may perceive the signal as coming from an area of the skin.

DISORDERS OF THE NERVOUS SYSTEM

Diseases of the nervous system can be considered by category. Probably the two most important types of central nervous system disease in Western society are trauma and vascular disease. Injuries to the spinal cord may result in either complete or partial separation of the fiber tracts joining the brain to the lower spinal cord. Depending on the level of the injury, the patient may be paralyzed in both the lower limbs (paraplegia) or all four limbs (quadriplegia).

Vascular diseases of the brain can involve blood seeping into the brain from ruptured vessels or death of brain tissue due to the obstruction of brain arteries. Tumors may spread to the brain from tumors in other parts of the body or develop in the brain itself, usually from glial cells. Degenerative diseases of the brain include Alzheimer's disease and Huntington's disease.

Nerves can be damaged by mechanical injuries, such as fractures or stab wounds. Some nerves are particularly vulnerable. For example, the radial nerve can be injured in fractures of the arm, where the nerve passes close to the bone; the median nerve can be injured when the wrist is cut, for instance, in suicide attempts. Severing the nerve breaks the continuity of the axons and prevents them from transmitting signals to or from the periphery, thus loss of control of movement and loss of sensation occurs in the affected area. The part of the axon separated from the nerve cell body in the CNS dies, but the part connected to the cell body regenerates. As it grows at approximately $\frac{1}{25}$ inch (1 millimeter) per day, it may take many weeks to recover.

In order to grow back to the appropriate region, it is important for the separated ends of the nerve to be rejoined. Even very careful repair is not able to reconnect individual axons, and regeneration is often inaccurate, and recovery may be limited. Other nerves, such as the sural nerve in the leg, can be used for a nerve graft if the damage is extensive or part of the nerve is missing. Crush injuries to nerves usually recover better than injuries where the nerve is cut because the nerve is left in continuity after crushing, which can improve the extent and accuracy of the regeneration.

If nerve continuity is not re-established, the growing axons form a tangled ball of sprouts (neuroma) at the cut end of the nerve. This can occur after amputations and may be painful or contribute to phantom sensations that the limb is still present.

Diseases of peripheral nerves are usually called neuropathies, but if they involve inflammation the term neuritis may be used. Another term, neuralgia, refers to pain along a nerve. Nerve disorders may affect just one nerve, for instance in carpal tunnel syndrome and Bell's palsy. Neuropathies can involve many nerves, often symmetrically on the two sides of the body. Guillain-Barré syndrome is an example of an acute inflammatory neuropathy.

The optic nerve is unusual in that it is actually an extension of the brain, and therefore part of the CNS rather than a peripheral nerve. Inflammation of the part of the nerve close to where it leaves the eye is called optic neuritis.

A disease that affects only motor pathways, but involves both the CNS and peripheral nerves, is amyotrophic lateral sclerosis (motor neuron disease). The cause is unknown, but the disease leads to a gradual loss of motor neurons and their axons, with associated weakness and paralysis.

SEE ALSO *Bell's palsy on page 144; Carpal tunnel syndrome on page 323; Disorders of the spinal cord on page 219; Nerve conduction tests on page 430; Sciatica on page 336; Treating the central nervous system on page 442*

Neuritis

Neuritis is the inflammation of a nerve or group of nerves. When more than one nerve is subject to inflammation the disorder is sometimes referred to as polyneuritis. If the condition involves the root of a spinal nerve, it is termed radiculitis.

Symptoms of neuritis vary, depending on which nerves are affected and the severity of the condition. The impacts can range from strange but mild sensations, such as pins and needles, to paralysis. There may be pain (often in the form of a burning sensation), defective reflexes and either a loss of sensitivity or heightened sensitivity in the area supplied by the affected nerves.

The symptoms may not always point to nervous system involvement. In some cases, for example, muscles served by inflamed and dysfunctional nerves may weaken and degenerate over time. Similarly, problems in joints may be caused by neuritis. Low blood pressure is a possible outcome when neuritis affects the autonomic nervous system; in this case abnormal nerve signals lead to the failure of veins and leg muscles to return venous blood to the heart. This causes pooling of blood in the legs.

There are different potential causes for neuritis, including injury, nutritional deficiencies and disease. Toxins produced by the bacteria that causes diphtheria and leprosy can cause neuritis. Vitamin B deficiencies, associated with diseases such as beriberi and alcoholism, commonly cause neuritis. Other factors that can underpin the condition include diabetes mellitus, lung cancer, industrial poisoning and autoimmune diseases such as multiple sclerosis (MS).

Optic neuritis, which involves inflammation of the optic nerve serving the eye, can affect a person's vision. It occurs more commonly in women than men and normally afflicts one eye rather than both (although it often occurs in both eyes simultaneously in children). Optic neuritis is also known as retrobulbar neuritis. It can be caused by encephalitis or an infection that has spread from the sinuses or may arise after an injury to the eye area. In adults, the condition is often an early sign of MS. About 40 percent of people who develop optic neuritis later go on to develop MS.

The symptoms of optic neuritis usually appear over several days, often beginning with a blurring of vision in the affected eye, followed by a loss of color vision and the appearance of blind spots. Eye movements may be painful. Although effects on vision can be so extreme that a patient can barely detect light with the affected eye, optic neuritis usually gets better and full vision is restored. Treatment commonly involves anti-inflammatory medications but the condition often improves without intervention.

Vestibular neuritis is a condition involving inflammation of a vestibular nerve, which is integral to the sense of balance. The vestibular nerves convey messages about head movements from the inner ear to the brain. Neuritis in one of these nerves is usually caused by an infection due to a virus, possibly a member of the herpes family. The main symptom is usually dizziness. Hearing is not affected although a slight sensitivity to head movements can persist in some people for a few months, or even continue for years.

Neuroma

Neuromas are benign tumors formed from nerve cells, and are also called schwannomas. They may occur alone, or in multiples, along with pigmented patches on the skin (also known as neurofibromatosis or von Recklinghausen's disease).

Neuromas may occur between the bones in the ball of the foot (plantar or Morton's neuroma). This is due to compression of the nerves between the metatarsal bones where

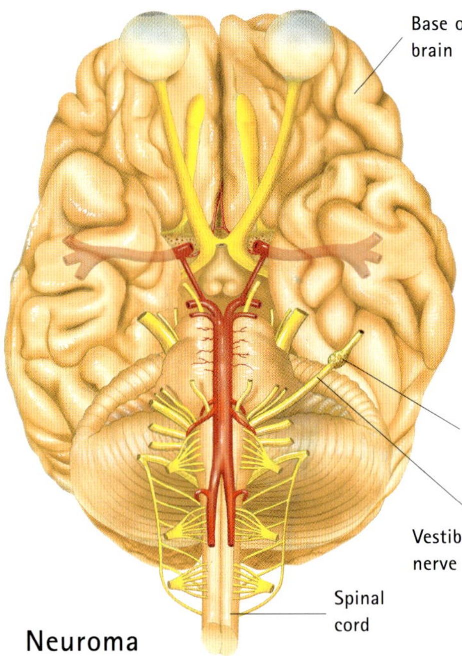

Neuroma

A neuroma is a benign tumor or new growth derived from nerve cells and fibers. An acoustic neuroma, growing on the vestibular cochlear nerve, can cause hearing loss and dizziness, and may damage other nerves close by.

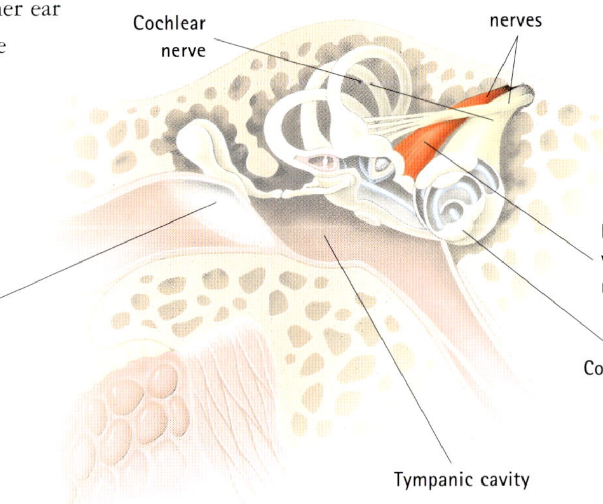

Neuritis

Neuritis is the inflammation of a nerve or a group of nerves. Vestibular neuritis affects the vestibular nerves in the ear and can cause dizziness.

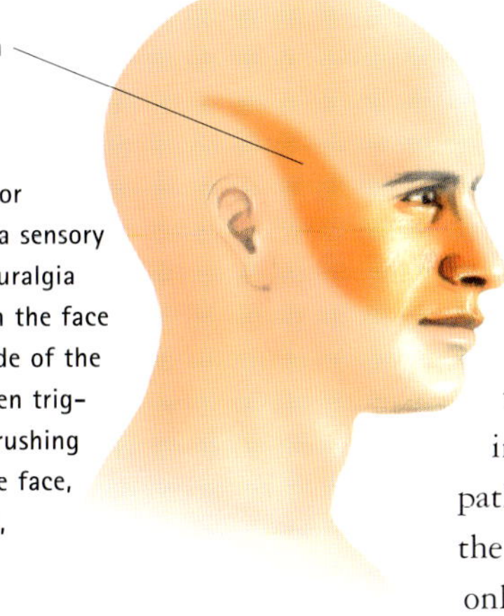

Neuralgia

Neuralgia is pain in or along the course of a sensory nerve. Trigeminal neuralgia causes severe pain in the face and cheek on one side of the face. The pain is often triggered by chewing, brushing the hair, washing the face, drinking cold liquids, and cold winds.

the toes join the foot, causing swelling, inflammation and a sharp burning sensation that may radiate to the toes. Surgery may be needed to remove severe tumors. Rest, ice packs and anti-inflammatory medications should offer pain relief.

Painful neuromas sometimes form on the stumps of amputated limbs where severed nerves grow back abnormally.

Acoustic neuromas are benign tumors of unknown cause that grow on the acoustic nerve in the ear canal and are sometimes associated with hereditary neurofibromatosis Type 2. They may cause hearing loss and ringing (tinnitus) in the affected ear, loss of balance, dizziness, pain and numbness. Large acoustic neuromas may cause complications by pressing on the skull. Surgical removal may result in total hearing loss in that ear and paralysis of facial muscles.

Neuralgia

Pain in or along the route of a sensory nerve is known as neuralgia. The pain is often sudden, severe and stabbing. Although any part of the body may be affected, the most common sites are the face, arms and chest. An infection such as tooth decay can trigger neuralgia. So too can pinching or pressure on a nerve, as occurs briefly with a knock on the "funny bone" at the elbow. Carpal tunnel syndrome, in which there is inflammation of soft tissues in the forearm, produces a more sustained form of neuralgia. A nervous system disorder can be responsible or associated with other conditions such as diabetes mellitus, some vitamin deficiencies and arthritis. In many cases, however, it is not always possible to identify an underlying cause for neuralgia.

One of the most common forms of neuralgia is trigeminal neuralgia (sometimes known as tic douloureux), involving the trigeminal nerve that carries sensory impulses to the brain from the jaws, nose, mouth and eyes. The sensations associated with this condition occur as very short but overwhelmingly painful repetitive stabs along the path of the trigeminal nerve on one side of the face. Pain may be short-lived, lasting only a few seconds, or can persist for minutes at a time. Trigeminal neuralgia also tends to occur in a series of episodes over several days, disappearing for a brief period and then returning.

It usually occurs in the middle-aged and elderly and is also common in people who have advanced multiple sclerosis.

Glossopharyngeal neuralgia is another form of neuralgia affecting the head region. In this case, however, the pain occurs deep inside the throat, starting near the tonsils and extending into an ear. Chewing and swallowing are frequent triggers. The cause is usually a blood vessel pressing on the nerves involved in swallowing.

A form of neuralgia known as postherpetic neuralgia sometimes occurs following episodes of shingles, a condition caused by the chickenpox virus. During an episode of shingles the virus multiplies and spreads along the nerves. Blisters and a rash form on the skin serviced by the affected nerve. In postherpetic neuralgia, pain along the nerve can persist for weeks or months after the shingles rash disappears.

Treatment depends on the cause, if one can be found. Sometimes the condition rights itself spontaneously. Analgesics such as ibuprofen and aspirin usually help to ease the pain. Pressure on affected nerves maybe relieved by using anti-inflammatory drugs to reduce tissue swelling. In severe and persistent cases of neuralgia, when drug treatments fail to bring relief, various surgical options are available.

Guillain-Barré syndrome

Guillain-Barré syndrome (also called acute inflammatory polyneuropathy) is an inflammatory disorder of the peripheral nerves. It starts suddenly (usually over days) with muscle weakness in the legs and arms, associated with numbness in the feet and hands. The weakness may spread to muscles in the face and trunk, and there may be difficulty swallowing or breathing. The syndrome usually follows an infectious illness and is thought to involve an autoimmune attack on the nerves. Damage to the nerves involves loss of the myelin sheath surrounding nerve fibers. This results in the loss of signals going along the nerves to control the muscles or provide sensation.

Guillain-Barré syndrome can be diagnosed by observing the rapid onset of weakness and the loss of reflexes and sensation on both sides of the body. Diagnosis is confirmed by lumbar puncture, from which cerebrospinal fluid is withdrawn for protein analysis;

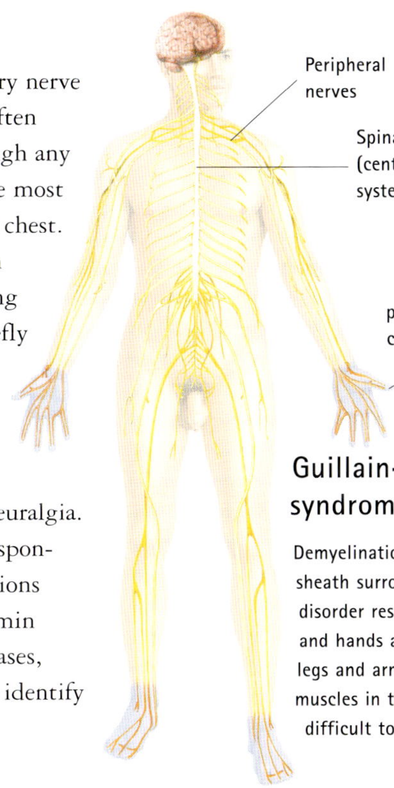

Guillain-Barré syndrome

Demyelination (loss of the myelin sheath surrounding nerve fibers) in this disorder results in numbness in the feet and hands and muscle weakness in the legs and arms. If weakness spreads to muscles in the face and trunk it may be difficult to breathe or eat.

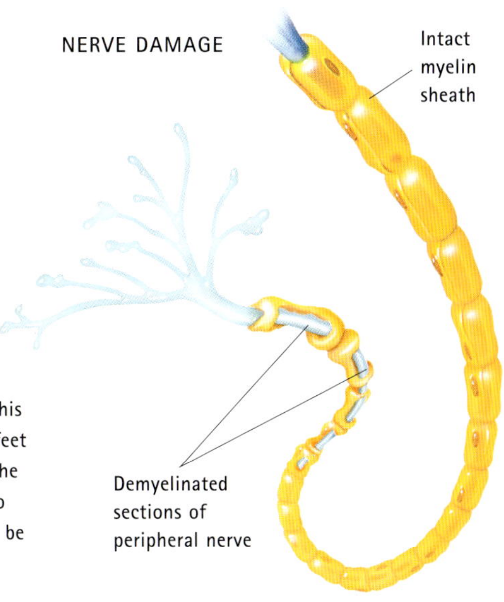

increased protein concentration indicates disease. If the disease progresses rapidly with serious complications, patients need to be hospitalized and monitored constantly. The airway must be kept clear and respiration supported artificially if necessary. Treatment may involve exchanging the patient's blood plasma for normal plasma, or replacing albumin and blood cells. Most patients recover but some may be left with muscle weakness and may require physical therapy and rehabilitation. If the symptoms are relieved within three weeks of their onset the outcome is usually good.

Tourette's syndrome

Tourette's syndrome involves the involuntary production of sudden sounds and movements, the results of multiple motor and vocal tics. The movements are purposeless and generally recur many times in a day but they subside during sleep. Sometimes there is meaningless repetition of swearwords as part of this syndrome. Symptoms may change frequently and differ in intensity, being worsened by stress and anxiety. In children the symptoms can even disappear just before a medical examination.

The condition generally starts in childhood or adolescence and affects three times as many males as females. It can sometimes be associated with learning difficulties or with hyperactivity. Children can sometimes suffer transient tics, which last for a relatively short period and disappear spontaneously; diagnosis of Tourette's syndrome relies partly on the presence of tics for more than a year.

Symptoms of Nervous System Disorders

Temporary or permanent disturbance of nerve pathways in any part of the body due to injury or disease can cause neurological symptoms such as loss of sensation, tingling, or involuntary movements.

Tremor

A tremor is an involuntary shaking of the body or part of the body, caused by a neurological disease that interferes with the nerve supply of certain muscles. Tremors are seen in a range of neurological diseases, such as Parkinson's disease or alcoholic brain damage, or as a side effect of drugs such as antipsychotics. If there is no discernible cause, it is known as essential tremor. A tremor may affect the whole body, or just areas such as the head, hands, arms or eyelids. It may occur sporadically, or at regular intervals. Often a tremor worsens with voluntary movement or emotional stress and disappears during sleep.

Treatment for tremors is directed at the underlying condition where possible. If the tremor prevents performing essential activities such as speaking or writing, drugs such as beta blockers, anticonvulsants or mild tranquilizers will help.

Paralysis

Paralysis is the loss of the ability to move a part of the body. It is caused by the inability to contract one or more muscles and usually results from injury to the brain, the spinal cord, a nerve or a muscle. It may be partial or full, and may be accompanied by the loss of sensation.

The kind and degree of paralysis differs according to whether the damage is to a peripheral nerve or to the central nervous system (brain and spinal cord). Damage to a peripheral nerve can cause loss of the ability to move a particular muscle or muscles, and a consequent wasting away of those muscles. Damage to the central nervous system, by comparison, produces weakness or loss of the use of a group of muscles, frequently affecting an entire limb. The muscles are stiff when moved (spasticity) but there is no wasting.

Paralysis of the cranial nerves in the head and neck results in paralysis of the facial, throat and eye muscles, usually causing

PARALYSIS

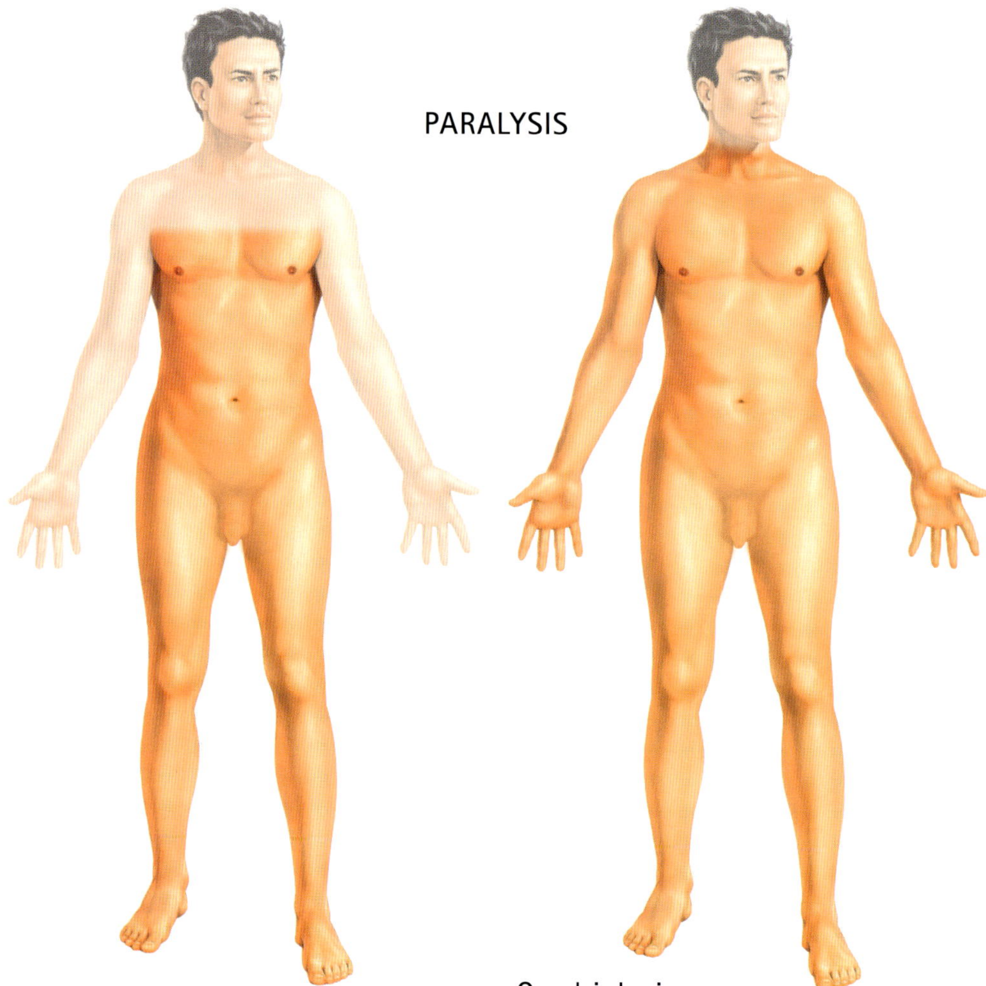

Paraplegia

Paraplegia is the result of injury or disease to the spinal cord between the T1 (thoracic) and L2 (lumbar) segments. It spares the arms but depending on the nerves damaged may involve the legs, pelvic organs and trunk.

Quadriplegia

Quadriplegia is the paralysis of all four limbs as a result of damage to the upper part of the spinal cord (cervical segments C1–C5). If any of the first three cervical segments (C1–C3) of the cord are damaged the injury is usually fatal, as the diaphragm, which helps us breathe, is paralyzed.

difficulty in speaking and swallowing, as well as blurred or double vision.

If there is partial or complete paralysis of both legs (and sometimes the trunk), the condition is known as paraplegia. If the arms are also affected, it is called quadriplegia (or tetraplegia). These conditions are caused by damage to the spinal cord, usually from trauma (most commonly road accidents and sporting injuries), but may also be caused by spinal cord tumors or birth defects. Neck injuries may result in quadriplegia, while an injury to the chest or lower back can result in paraplegia.

Hemiplegia is paralysis of the limbs on the same side of the body. It is usually the result of either a brain tumor or a stroke. The weakness or paralysis occurs on the side of the body opposite to the side of the brain which has been affected.

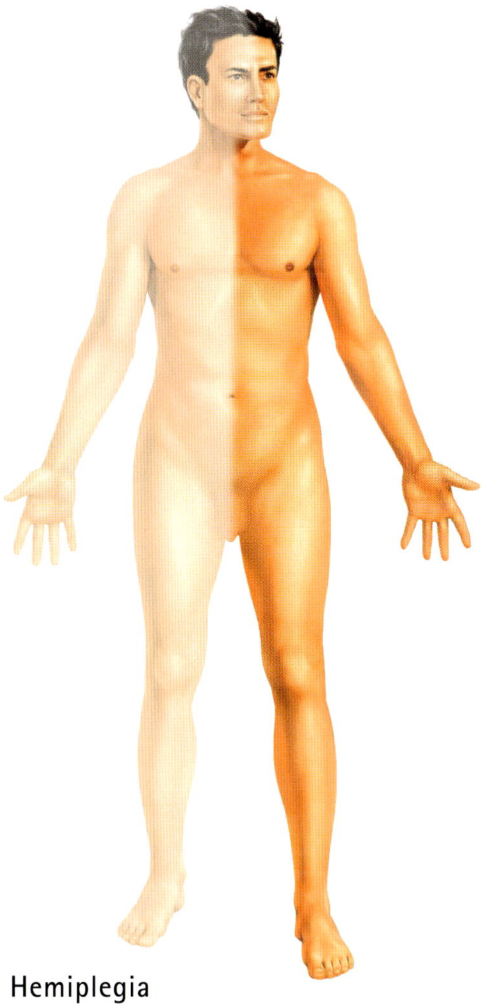

Hemiplegia

Hemiplegia is often the result of stroke (cerebral vascular accident) or a brain tumor. Injury to one side of the brain causes weakness or paralysis on the opposite side of the body, so when the right hemisphere is damaged, motor control on the left side of the body is affected.

Treatment of paralysis requires treating the underlying cause, where possible. Immediate treatment may mean hospitalization, including intensive care with artificial ventilation if breathing muscles are affected.

Surgery may be possible to limit further spinal cord damage or to remove bone fragments or a tumor if necessary. Peripheral nerve injuries can be helped by nerve transplants, orthopedic operations to immobilize a joint (arthrodesis), or the transplant of the tendon of a working muscle to aid paralyzed muscles. Physical therapy and rehabilitation play an important part in the treatment of paralysis. With rehabilitation, many lost functions can be compensated for or even restored. Passive exercises for paralyzed muscles can help to prevent contractures. Physical therapy will prevent joint stiffness.

Someone who has been in an accident and has neck pain or possible spinal cord injury should not be moved unless absolutely necessary. The injured person's neck should be immobilized with splints or pillows to prevent movement until an ambulance or other emergency service arrives.

Numbness

Numbness is the absence of sensation in a part of the body. It is usually due to damage or degeneration of a peripheral nerve, though it may also be caused by injury to the central nervous system (brain and spinal cord).

Total loss of sensation means all of the nerve is damaged. But if the nerve is partly injured, there may be partial numbness, accompanied by tingling (paresthesia).

Numbness may be temporary, as when the blood supply to a hand or foot is temporarily cut off because of the position a person is in, causing the hand or foot to fall asleep. When the blood supply is restored, normal sensation returns. Or the numbness may be permanent and irreversible. Conditions causing permanent nerve injury include diabetes, degenerative nerve diseases, local injury to the nerves under the skin, pressure on the nerves caused by a herniated disk, nerve damage from the effect of toxins (lead, alcohol or tobacco) or as the side effects of drugs. It is necessary to treat the underlying condition where possible. A numb hand or foot is prone to accidental injury; care must be taken to protect it from cuts, bumps or other injuries.

AUTONOMIC NERVOUS SYSTEM

This is the part of the nervous system concerned with controlling automatic bodily functions. The body's ability to maintain a constant internal environment, heart rate and blood pressure is called homeostasis. While many body systems contribute to homeostasis, the autonomic nervous system is the most important of them.

Most of the activities of the autonomic nervous system occur without our being aware of them, and so the system could also be called the involuntary nervous system. Nevertheless, some bodily functions can be influenced by conscious activity—the effect of relaxation therapy on blood pressure is an example of this.

While most of the activities of the autonomic nervous system are actions on body organs and tissues (such as motor functions) there are also many sensory nerves accompanying the autonomic motor nerves. These sensory nerves relay information about internal organs (such as the tension in the wall of a full stomach or blood pressure in parts of the cardiovascular system) back to the central nervous system (CNS).

This is essential for keeping the brain and spinal cord informed about changes in the body, and allows control procedures to be carried out to keep conditions in the body's interior in a relatively constant state.

SEE ALSO *Brain on page 132; Circulatory system on page 78; Homeostasis and metabolism on page 112; Nervous system on page 64; Reproductive systems on page 104; Spinal cord on page 217; Spinal nerves on page 218; Urinary system on page 98*

STRUCTURE

The autonomic nervous system consists of nerve cells (neurons) in the brain and spinal cord, their fibers which leave the central nervous system, nerve cells in the various body cavities, and also nerve fibers that are distributed in the internal organs. Collections of nerve cells in the body cavities are called ganglia. These ganglia are often embedded in networks of nerve fibers called plexuses, located near the heart and lungs, in front of the aorta in the abdomen, and in front of the sacral bone in the pelvis.

The autonomic nervous system differs from the somatic nervous system, which is concerned with voluntary control of the body's muscles. The autonomic nervous system has a series of two or more nerve cells between the CNS and the organ that is being controlled, while the somatic nervous system has only one nerve cell between the CNS and the muscle being controlled.

The somatic nervous system neurons have their cell bodies either in the brain or in the spinal cord and axons or nerve fibers that run directly to the muscle being controlled. Those autonomic nervous system neurons whose cell bodies lie inside the spinal cord or brain stem are called preganglionic nerve cells, while those neurons whose cell bodies lie in ganglia are called postganglionic nerve cells.

While both the somatic and autonomic nervous systems may use acetylcholine as a neurotransmitter, the sympathetic part of the autonomic nervous system also uses the chemical norepinephrine (noradrenaline), which is released from the postganglionic nerve cells onto smooth muscle and other target tissues.

Another difference between the somatic and autonomic nervous systems is that all the nerve fibers controlling voluntary muscles in the somatic nervous system have thick myelin sheaths, while usually only the preganglionic nerve cells of the autonomic nervous system have myelin sheaths. Since myelin sheaths around nerve fibers contribute to the rapid conduction of nerve impulses, it means that the somatic or voluntary nervous system acts on the muscles much quicker than the autonomic nervous system can influence the internal organs.

Divisions of the Autonomic Nervous System

Traditionally, he autonomic nervous system is divided into a sympathetic and a parasympathetic division. These are considered to be anatomically and functionally separate, but there are places in the body (such as the nerve supply of the pelvic organs) where the two may overlap.

Sympathetic and parasympathetic nervous systems have important complementary actions in the urinary and reproductive organs. The release of urine from the bladder is stimulated by the parasympathetic nervous system, which relaxes the sphincter muscle at the outlet from the urinary bladder and causes contraction of bladder wall muscle. Urination is inhibited by the sympathetic nervous system, which constricts the sphincter muscles around the outlet from the urinary bladder.

In the male reproductive organs, the parasympathetic nervous system causes erection of the penis by increasing the flow of blood into the cavernous spaces of the penis.

The system is also involved in ejaculation, the process of expulsion of semen from the penis. If a man experiences anxiety about his ability to perform sexually, then the sympathetic nervous system will be more active. This may cause difficulty in obtaining and maintaining an erection—a problem called impotence or erectile difficulty—and/or a too-rapid progression from arousal to ejaculation, known as premature ejaculation. This type of sexual dysfunction is often responsible for sexual problems in younger men, but in older men impotence is more likely to be due to vascular problems and diabetic damage to pelvic nerves.

Sympathetic division

The sympathetic division is often referred to as the "fight-or-flight" system. It comes into play during emergency situations when our bodies need extra energy to avoid or overcome danger. During emergencies we usually experience a pounding heartbeat, cold sweaty skin, rapid breathing and enlarged pupils—all produced by activation of the sympathetic nervous system. In addition, the sympathetic nervous system can cause changes such as increased blood pressure, a dry mouth, high blood sugar levels, and dilation of the small airways of the lung. The sympathetic nervous system increases the flow of blood to muscles and diverts blood away from the gastrointestinal tract.

The sympathetic nervous system also has an important effect on the control of body temperature. Some sympathetic nerve cells stimulate the activity of sweat glands in the skin to increase sweat production and lower body temperature by the evaporation of perspiration. Other sympathetic nerve cells control tiny smooth muscles in the skin, which when stimulated pull on the hairs of the skin causing them to stand up. This change is commonly called "goose bumps" and keeps a layer of relatively still air close to the skin to minimize heat loss. In the eye, the sympathetic nervous system causes enlargement of the pupil to increase the amount of light reaching the sensitive retina at the back of the eye.

Anatomically, the sympathetic nervous system consists of nerve cells in the thoracic and lumbar levels of the spinal cord as well as a long chain of nerve cells which lies alongside the backbone. This long chain—called the sympathetic trunk—gives off nerves to other plexuses and internal organs.

Some sympathetic nerve fibers go to the adrenal medulla, which is part of the endocrine system. These nerve fibers stimulate the adrenal medulla, which contains modified nerve cells, to release epinephrine (adrenaline) and norepinephrine (noradrenaline) into the bloodstream. These hormones are released in emergency situations where energy reserves need to be rapidly mobilized.

Parasympathetic division

The parasympathetic division is most active when the body is not under threat and when the person is largely at rest. Its main function is to conserve energy, restoring the internal body state to normal by promoting digestion and eliminating urine and feces from the body.

In the head, the parasympathetic nervous system increases the production of saliva by the salivary glands, while in the stomach and intestines it increases the activity of the smooth muscle and glands. The increased activity of the smooth muscle in the gut increases the movement called peristalsis, which helps to move food along the gut and break it up into a more readily digestible consistency.

The increased activity of glands in the gut increase the production of digestive enzymes to break down food, and the production of mucus and other fluid to enhance absorption and movement of food.

In the cardiovascular system, the parasympathetic nervous system lowers the heart rate and blood pressure and diverts blood from muscles to the gut to assist with digestion. In the lungs, it causes constriction of the small airways and increases the activity of the small glands in the airways, which will increase the amount of secretions.

In the eye, the parasympathetic nervous system closes down the pupil of the eye, decreasing the amount of light reaching the retina, and stimulates contraction of the ciliary muscle, which leads to bulging of the lens of the eye for close vision.

Diseases and Disorders of the Autonomic Nervous System

The autonomic nervous system can be affected in various diseases. In Raynaud's phenomenon, patients experience pallor and blue coloration of the ends of the fingers and toes. This may occur in young women with no apparent cause (Raynaud's disease), or it may accompany connective tissue disease or occupations involving the use of vibrating tools. Patients with this problem may develop ulcers on their fingertips and, in severe cases, develop skin infections. Treatment consists of keeping the hands and feet warm, but if this is not effective, surgically cutting off sympathetic nerves to the fingers may help.

Deliberate destruction of part of the sympathetic nervous system may also be helpful in some other conditions such as arterial obstruction due to atherosclerosis, excessive sweating, and frostbite. Patients with diabetes mellitus may have problems as a result of damage to the autonomic nervous system. Male patients may develop impotence, while both sexes may suffer from nocturnal diarrhea, problems maintaining proper blood pressure while standing and retention of urine in the bladder.

Sympathetic and parasympathetic divisions of the autonomic nervous system

The thoracic, abdominal and pelvic organs and tissues come under the dual influence of the sympathetic and parasympathetic divisions of the autonomic nervous system. A delicate balance between these two divisions helps maintain a fairly constant level of activity in these organs under normal circumstances. However, autonomic innervation extends beyond these organs to include the muscles of the eye, salivary glands, sweat glands, tiny muscles in the skin, and blood vessels.

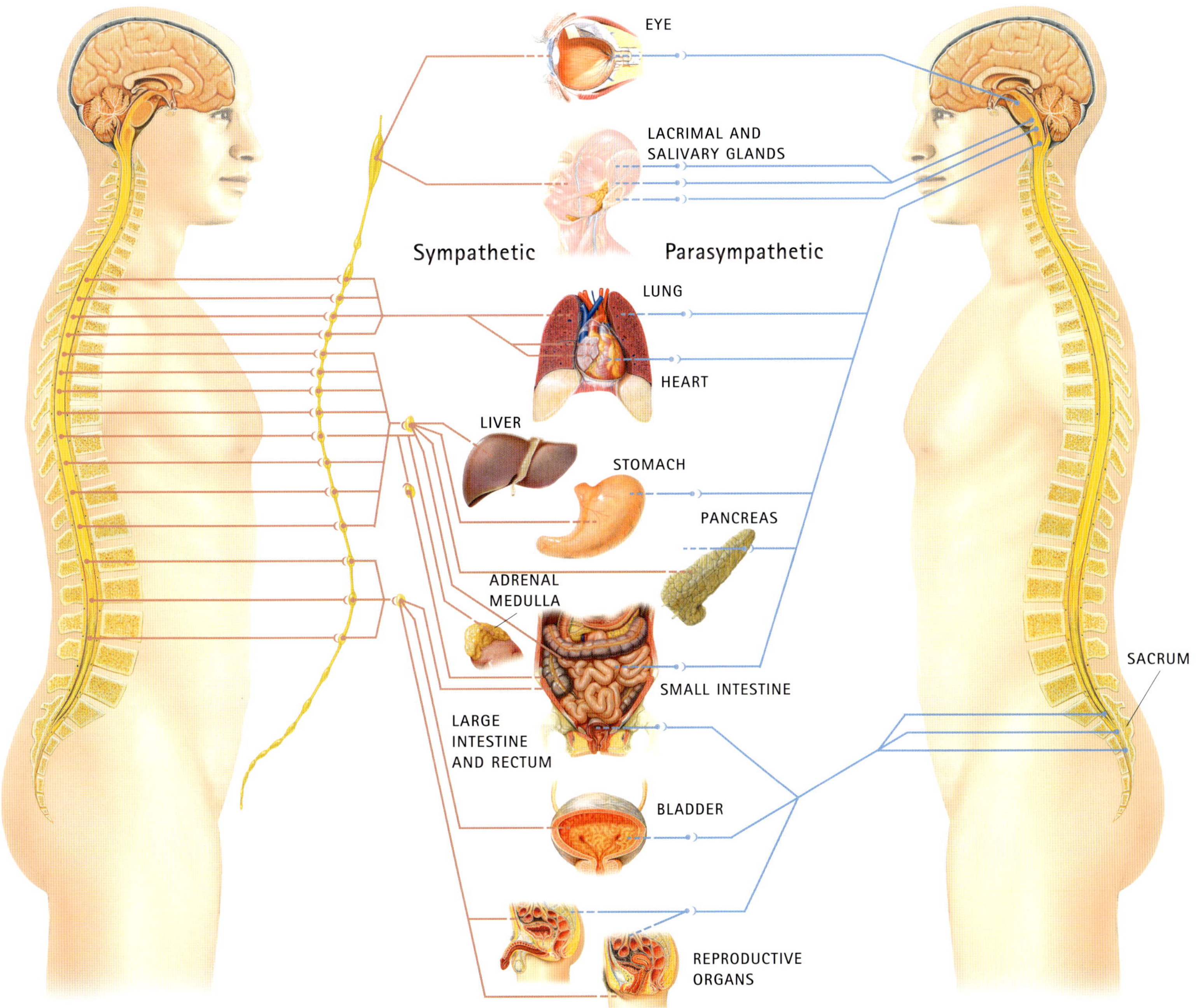

CIRCULATORY SYSTEM

The circulatory system includes the heart and blood vessels, which form a closed ring. The heart has four chambers. It pumps blood out from its two pumps—the left and right ventricles—and collects returned blood into its left and right atria. Blood is pumped out of the ventricles through arteries into a distribution network of tiny vessels, invisible to the naked eye, called the capillaries. After exchanging gases and nutrients with surrounding tissues, blood returns to the atria by the veins.

There are two separate circulations in the circulatory system, which are connected in series. Blood from the left ventricle is distributed to the capillaries throughout the body to deliver oxygen and nutrients to the entire body, and returns to the right atrium. This is the systemic circulation. Blood returning to the right atrium is depleted of oxygen and loaded with carbon dioxide. It then flows into the right ventricle where it is pumped into the capillary network in the lungs, from where, after exchanging carbon dioxide for new oxygen, blood returns to the left atrium. This is the pulmonary circulation. From the left atrium, blood flows to the left ventricle and the cycle continues.

The phase of the heart cycle when both ventricles contract is known as ventricular systole. The ventricles dilate during diastole to receive blood from the atria. In the last part of diastole, the atria contract to squeeze their contents into the ventricles.

SEE ALSO *Disorders of the heart on page 230; Heart on page 228; Respiratory system on page 95*

Anatomy of the heart

The heart lies in the midline of the thorax, between the lung, surrounded by a double-layered membrane called the pericardium. The heart looks like a pyramid lying on one of its sides, with the apex pointing forward and to the left side. The right and left atria are located at the back, and the right and left ventricles at the front.

A septum divides right and left atria, and right and left ventricles. Each atrium opens into a ventricle by an atrioventricular orifice, which is guarded by a valve to ensure that blood flows only in that direction. The valves consist of leaflets (cusps) located in

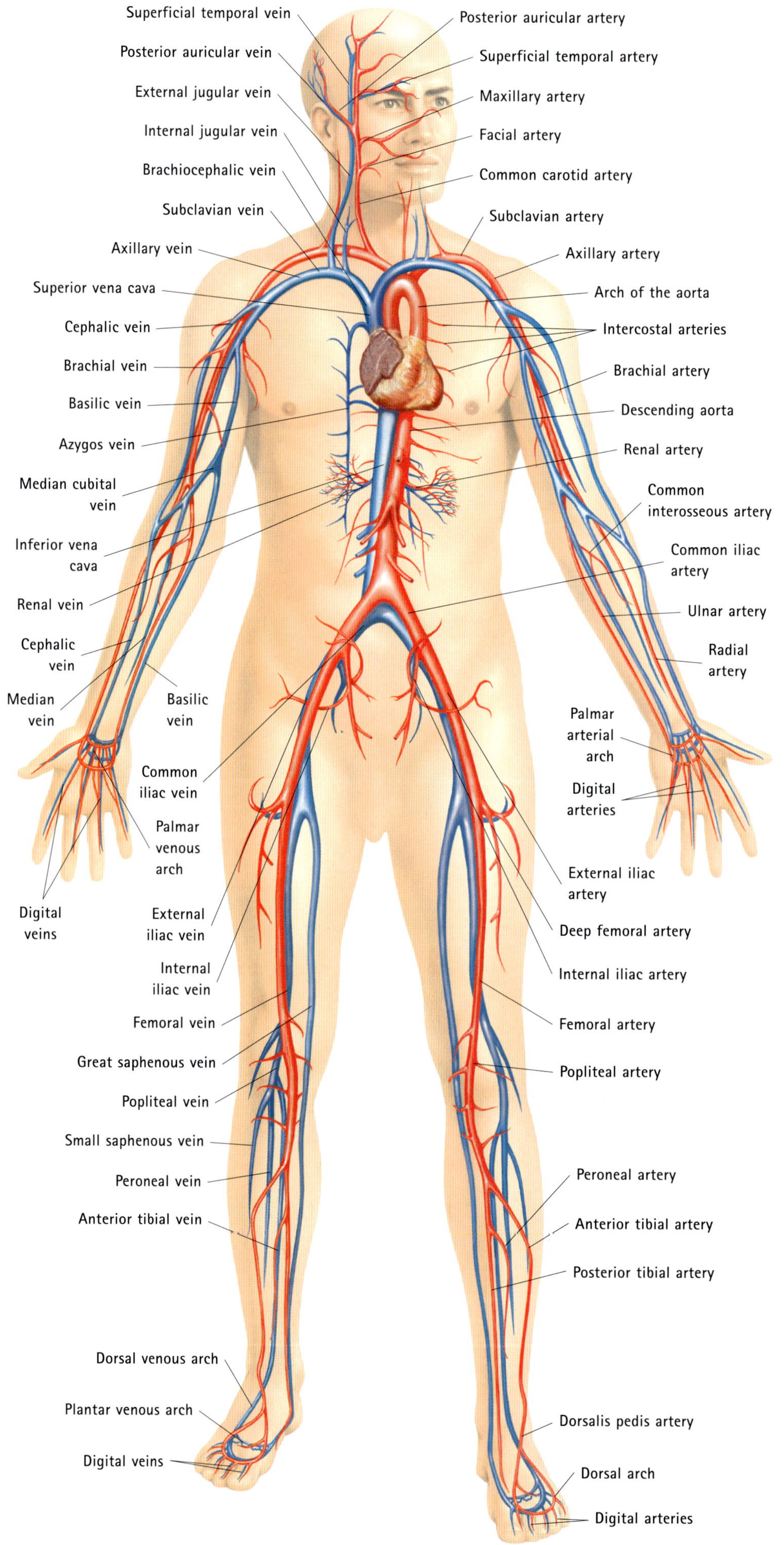

the ventricles and attached to the rim of the orifice. Blood can flow unimpeded into the ventricle, but when it tries to run backward into the atrium, it pushes the leaflets back toward the rim and closes down the orifice.

The right atrioventricular valve has three cusps and is called the tricuspid valve. The left atrioventricular valve is called the mitral valve because its two leaflets look like the split top of a bishop's mitre.

The function of the aortic and pulmonary valves is to prevent reflux of blood back into the ventricles. These valves are formed by three pockets attached to the inside wall of the aortic and pulmonary arteries. They are squashed against the arterial wall by outgoing blood and present no resistance to blood flow, but blood trying to return to the heart will fill up the pockets, causing the opening to close down.

Coronary arteries

The heart is only the size of the human fist, yet it is responsible for about 8 percent of the body's total oxygen consumption. Blood supply to the heart comes from the right and left coronary arteries, so-called because they wrap around the heart like a crown.

The right coronary artery runs in the groove between the right atrium and the right ventricle to the back of the heart, where it turns 90 degrees to run in the groove corresponding to the interventricular septum. The left coronary artery divides shortly after its beginning into the anterior interventricular artery, which runs in the groove of the interventricular septum at the front of the heart, and the circumflex artery, which runs in the groove between the left atrium and left ventricle.

Nerve supply

The heart can continue beating when taken out of the body because it has a built-in nerve supply. The sinoatrial node, located in the right atrium, is the pacemaker triggering the contraction of the right and left ventricles. The autonomic nervous system, by way of the cardiac plexus, modifies the intrinsic heart rhythm and adjusts heart rate and contraction power to the requirements of the body. In some cases, when the innervation of the heart malfunctions, an artificial pacemaker must be connected to drive the heart.

Pulse

As the heart beats, each beat causes a quantity of blood to be forced under pressure into the arterial system. These beats cause a pulse or shock wave that travels along the walls of the arteries. This pulse can be felt in several parts of the body by placing the finger over an artery. Arteries in which the pulse can normally be felt particularly easily are the carotid arteries (in the neck) and radial arteries (at the wrist).

Heart function

The ventricles of the heart pump blood around the body. Blood from the left ventricle delivers oxygen and nutrients to the whole body via the systemic circulation. Blood from the right ventricle is pumped into the lungs where it collects oxygen.

Heart front—cross-section

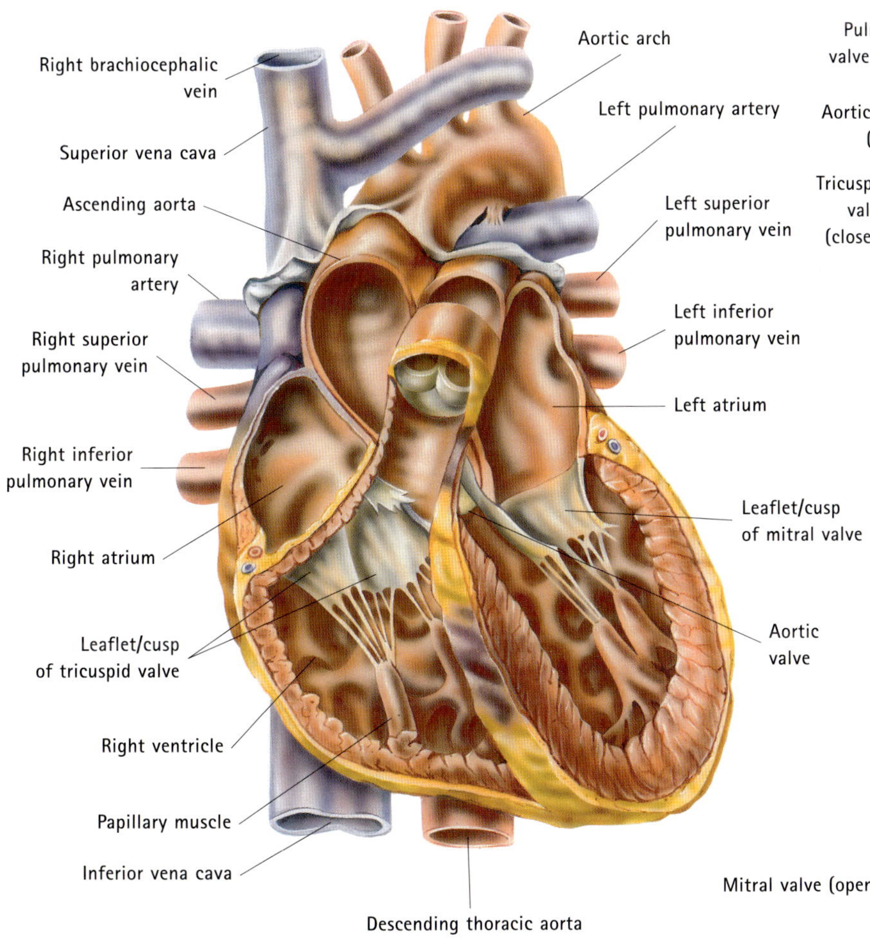

Heart valves

During ventricular systole (when the ventricles contract), the aortic and pulmonary valves open to allow blood to be pumped into the pulmonary and general circulatory system, while the mitral and tricuspid valves remain closed. During ventricular diastole (when the ventricles dilate), the aortic and pulmonary valves close while the tricuspid and mitral valves open to allow blood to pass from the atria into the ventricles.

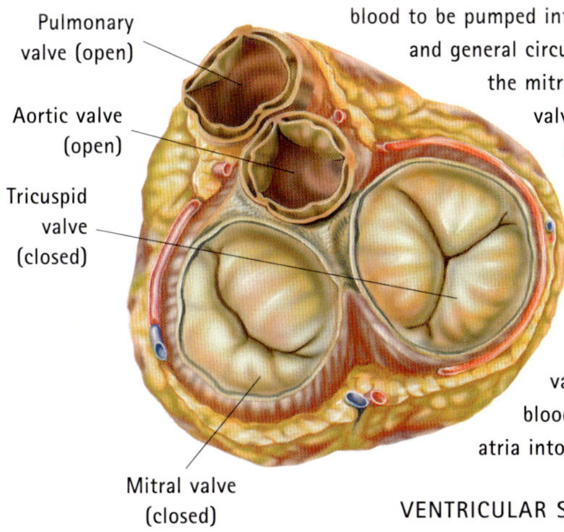

VENTRICULAR SYSTOLE

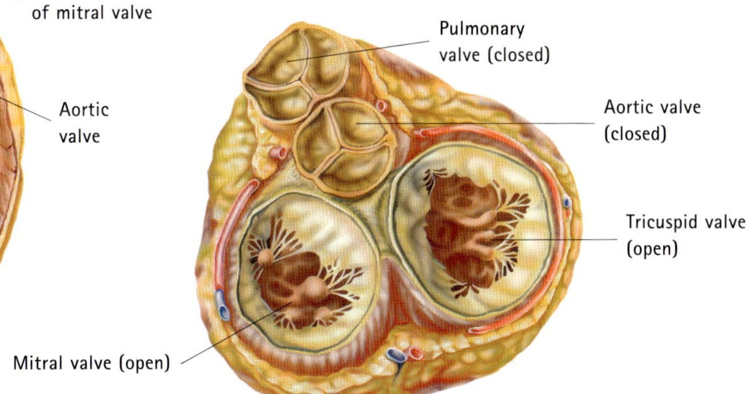

VENTRICULAR DIASTOLE

Pulse points

The pressure created by a beat of the heart can be felt quite easily where arteries are close to the skin.

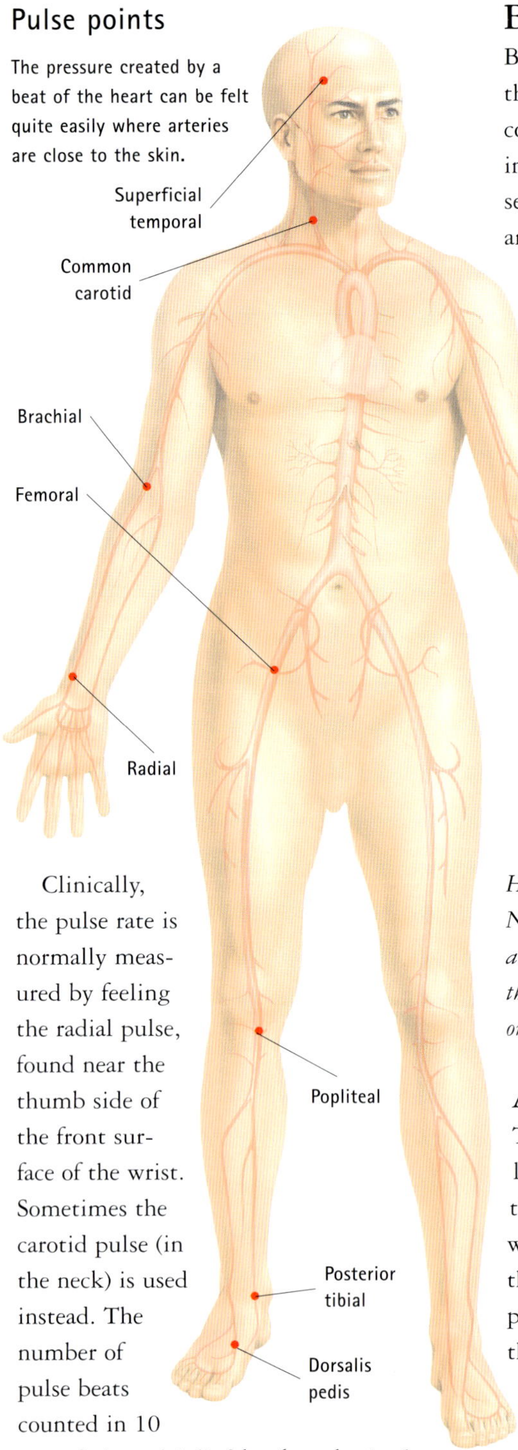

Clinically, the pulse rate is normally measured by feeling the radial pulse, found near the thumb side of the front surface of the wrist. Sometimes the carotid pulse (in the neck) is used instead. The number of pulse beats counted in 10 seconds is multiplied by 6 to obtain the rate in beats per minute. The pulse must, however, be felt for a longer period to check for irregular pulse rate. The strength or weakness of the pulse is also noted.

The pulse rate in healthy adults at rest is typically about 60–70 beats per minute. Athletes and those taking beta-blocker medication may have slower pulse rates, while children and babies have faster pulse rates. The pulse rate increases during exercise to increase the output of blood from the heart. The rate of the pulse is also increased by excitement.

Blood vessels

Blood vessels form an intricate system through which the blood circulates in a continuous cycle. The heart pumps blood into the aorta, a large elastic artery, which sends off branches to supply the head and arms, the internal organs and the lower limbs. Repeated branchings form thin-walled capillaries, across which oxygen and nutrients are transferred to all body cells and through which carbon dioxide is transferred away from the same cells. Specialized capillaries in the kidneys, known as glomeruli, allow waste to leave the body as urine.

After leaving the limbs and organs, blood is channeled into veins of increasing size, returning eventually to the heart to repeat the cycle.

SEE ALSO *Blood vessels of the neck on page 200; Cerebral arteries on page 132; Heart on page 228; Kidney on page 288; Nerves and blood vessels of the shoulder and arm on page 322; Nerve and blood supply of the leg on page 334; Pulmonary circulation on page 240*

Aorta

The aorta, a thick elastic tube, is the largest artery in the body. It arises from the left ventricle of the heart, arches upward, backward and to the left, then down the back of the thorax through the diaphragm and into the abdomen. From the thoracic aorta, arteries arise that supply the heart, head, neck and arms. From the abdominal aorta, arteries supply the abdominal organs, pelvis and legs.

Aortic aneurysm is the abnormal stretching of the walls of the aorta. Aortic stenosis is narrowing of the aortic valve, caused by valvular heart disease such as rheumatic fever.

Arteries

Arteries are flexible, thick-walled, tube-shaped blood vessels that carry blood away from the heart to the rest of the body.

The largest artery is the aorta, which channels blood from the heart to other arteries, and to the body's organs and other structures. Two small branches of the aorta, the coronary arteries, supply the blood to the heart muscle itself. The right and left carotid arteries carry blood to the two sides of the neck and head. Blood flows to the shoulders and arms through the right and left subclavian arteries. In the abdomen, the aorta divides into two large branches, the left and right iliac, supplying blood to the pelvic region. The iliac arteries then continue into the legs, where they are called the femoral arteries.

From the arteries, blood passes into very small blood vessels called capillaries, and from there to veins and then back to the heart. The heart pumps the blood through the pulmonary artery to the lungs, where it becomes oxygenated. The blood is then returned to the heart, where it is once again pumped out through the aorta.

Artery

The systemic circulation supplies the body with oxygenated blood. Arteries channel this blood from the heart to the rest of the body.

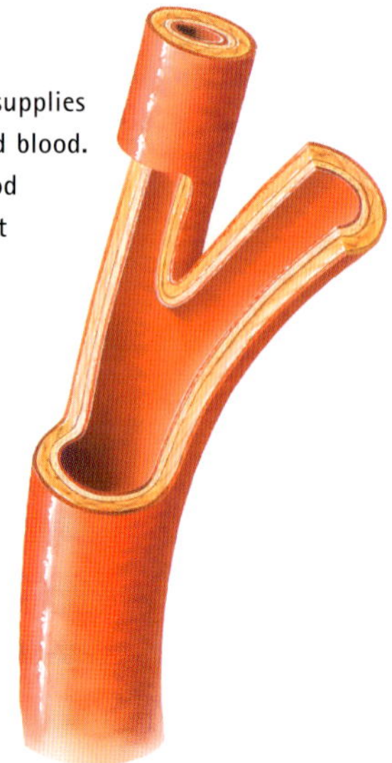

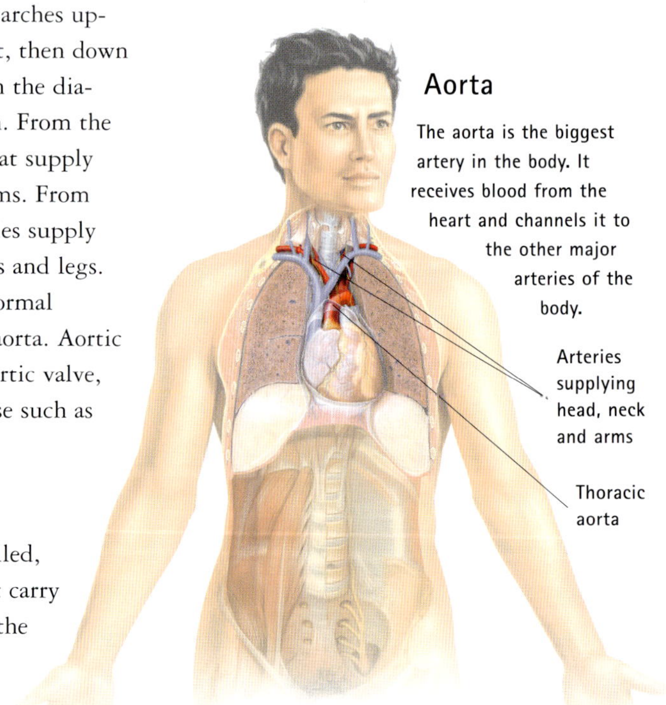

Aorta

The aorta is the biggest artery in the body. It receives blood from the heart and channels it to the other major arteries of the body.

Veins

Veins are thin-walled, low-pressure blood vessels that return blood to the heart. The smallest veins are the venules, which commence at the venous end of capillaries. Veins receive tributaries from other veins and progressively increase in size as they approach the heart. Many, but not all, veins contain one-way valves.

The two largest veins in the body are the superior and inferior vena cavae, which drain into the heart from above and below respectively. The brachial, basilic and cephalic veins drain the upper limbs. They drain into the axillary vein, which becomes the subclavian vein. The internal jugular vein, which drains the head and neck, joins the subclavian vein to form the brachiocephalic vein. The left and right brachiocephalic veins join to form the superior vena cava, which drains into the heart. The azygos vein, which drains the thoracic cavity, joins the superior vena cava just before the latter enters the heart.

The femoral vein drains the lower limb. It becomes the external iliac vein as it enters the trunk and is joined by the internal iliac vein from the pelvis to become the common iliac vein. The two common iliac veins join to form the inferior vena cava, which passes up the posterior abdominal wall. It is joined here by veins from the kidneys, gonads and back region, and just below the diaphragm by large veins from the liver. It passes through the diaphragm and almost immediately enters the heart.

Only a small number of veins in the trunk have valves; most of them are valveless. Blood in the trunk therefore flows according to pressure differences, making respiratory movements important in venous return. Inspiration creates a negative pressure within the thoracic cavity that not only draws air into the lungs but assists venous return to the heart.

Valves are common in veins within the limbs, where they assist the return of blood against the effect of gravity. They are particularly numerous in the lower limb. A large valve is also present in the lower end of the internal jugular vein, preventing the flow of blood back up toward the head and neck.

In the upper limb, most venous blood returns by the superficial veins which travel in the tissue just below the skin. In the lower limb, most venous blood returns by the deep veins which lie in compartments which contain muscles.

The contraction of nearby muscles compresses the veins and blood is forced by the valves in the direction of the heart. Blood then flows from superficial leg veins to the now empty deep veins, thus reducing pressure in the superficial veins.

Faulty valves in the veins communicating between the superficial and deep veins can lead to backflow of blood into the superficial veins when muscles contract. This results in the superficial veins becoming dilated and tortuous, otherwise known as varicose veins. Two of the superficial veins of the lower limb are named saphenous after the Greek word saphenes, meaning "obvious."

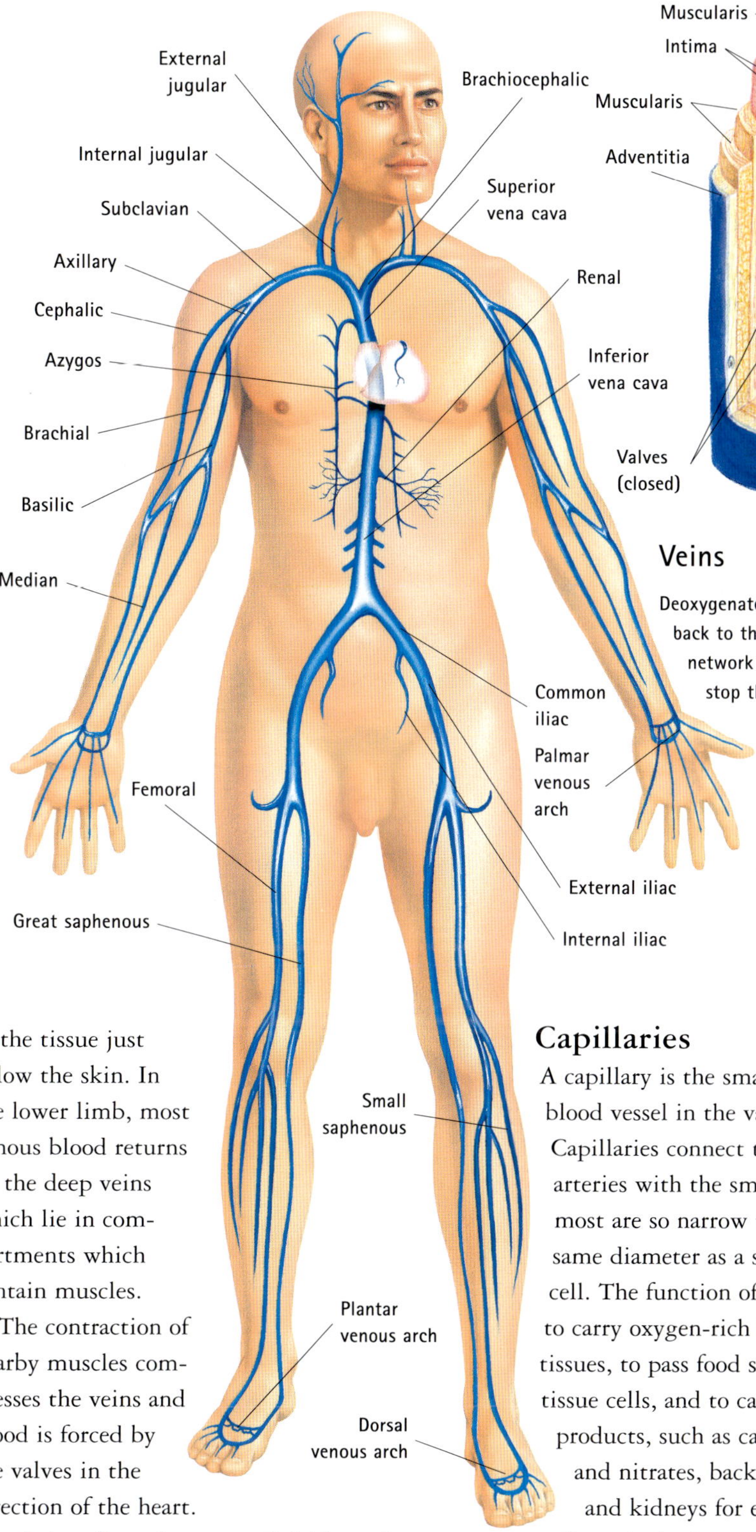

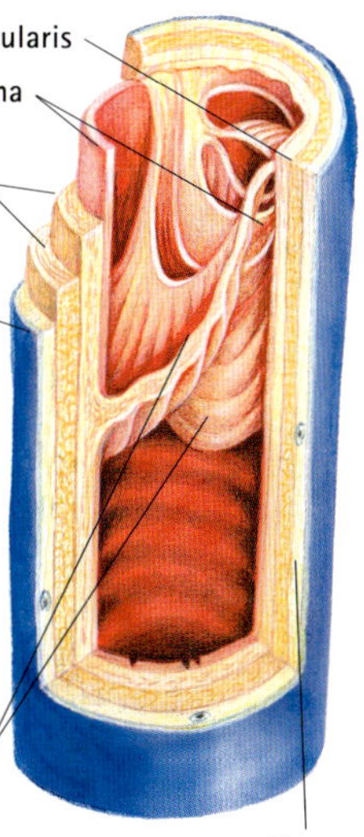

Veins

Deoxygenated blood travels back to the heart via a network of veins. Valves stop the blood from pooling in the veins, particularly where blood flow is fighting gravity (such as in the arms and legs).

Capillaries

A capillary is the smallest type of blood vessel in the vascular system. Capillaries connect the smallest arteries with the smallest veins; most are so narrow they have the same diameter as a single blood cell. The function of capillaries is to carry oxygen-rich blood to the tissues, to pass food substances to tissue cells, and to carry away waste products, such as carbon dioxide and nitrates, back to the lungs and kidneys for elimination from the body.

During inflammation of body tissues, the capillaries become more permeable. They allow white blood cells and proteins into the tissues to fight infections and stimulate inflammation. Capillaries can also become more permeable as a result of some allergic reactions. In serious reactions, such as anaphylaxis, the blood can lose volume and the person can go into shock.

Microangiopathy is a term used to describe a disorder of capillaries in which

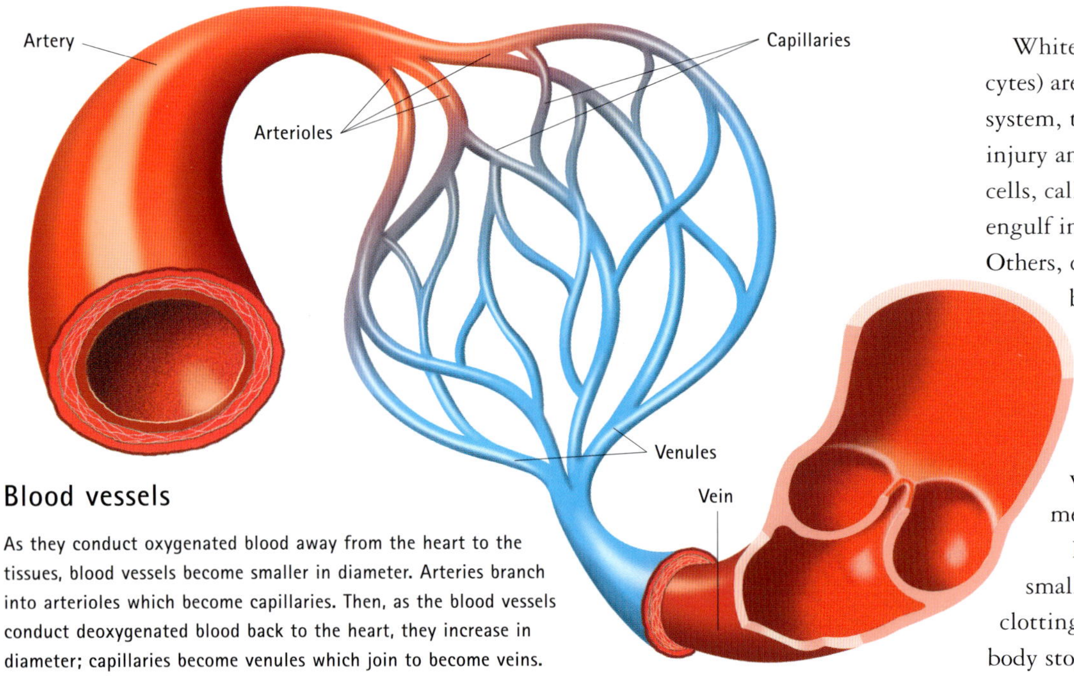

Blood vessels

As they conduct oxygenated blood away from the heart to the tissues, blood vessels become smaller in diameter. Arteries branch into arterioles which become capillaries. Then, as the blood vessels conduct deoxygenated blood back to the heart, they increase in diameter; capillaries become venules which join to become veins.

the capillary walls become so thick and weak that they bleed, leak protein, and slow the flow of blood. For example, diabetics may develop microangiopathy, which can cause thickening of the capillaries in the eye and the kidney.

Blood pressure

Blood pressure is the pressure blood exerts against the walls of the arteries. The amount of pressure depends upon the strength and the rate of the heart's contraction, the volume of blood in the circulatory system and the elasticity of the arteries.

Normal blood pressure varies from person to person, but usually normal blood pressure at rest is about 120/80 mm Hg. It is lower in children.

Systolic blood pressure, the top number, represents the maximum pressure in the arteries as the heart contracts and ejects blood into the circulation. Diastolic pressure, the bottom number, represents the minimum blood pressure as the heart relaxes following a contraction.

Blood pressure is measured using an instrument called a sphygmomanometer. A single blood pressure reading, unless it is very high or very low, should not be considered abnormal. Usually, several readings are taken on different days and the results are then compared.

SEE ALSO *Heart on page 228; Treating the cardiovascular system on page 439*

BLOOD

Blood is a suspension of red and white blood cells, platelets, proteins and chemicals in a straw-colored fluid called plasma. It is the means by which oxygen and essential nutrients are transported from the lungs and digestive tract to other parts of the body.

Blood transfers waste products to organs such as the kidneys and lungs, which eliminate them. It transports antibodies and white blood cells to the sites where they are needed to fight infection, and it transports heat from inner parts of the body to the skin to keep body temperature stable. A normal-sized adult has about 10 pints (nearly 5 liters) of blood. At rest, the heart pumps all of it around the body in about one minute.

The disk-shaped red blood cells are the body's means of transporting oxygen to the body's tissues. In a healthy person, each cubic millimeter (0.006 cubic inches) of blood contains between four and six million red blood cells. The oxygen is carried by an iron containing compound in the red blood cells called hemoglobin. When oxygenated, hemoglobin appears red—hence the red appearance of blood in the arteries. Once the red blood cells reach the tissues, the oxygen is exchanged for carbon dioxide, which the hemoglobin then carries back to the lungs, where it is exhaled.

White blood cells (also known as leukocytes) are the agents of the body's immune system, traveling via the blood to sites of injury and infection. Some types of white cells, called neutrophils and monocytes, engulf invading bacteria and small particles. Others, called lymphocytes, produce antibodies that destroy foreign organisms and are responsible for establishing ongoing immunity.

The average healthy person has between 4,000 and 10,000 white blood cells per cubic millimeter (0.006 cubic inches) of blood.

Platelets (or thrombocytes) are small fragments of cells responsible for clotting. Clotting is the normal way the body stops bleeding and begins the healing

Blood

Blood is composed of red blood cells, various types of white blood cells (leukocytes) and platelets in a solution of water, electrolytes, and proteins called plasma. About 40 percent (by volume) of blood is red blood cells. This illustration shows all the different types of blood cell—it does not accurately represent the proportions present in the blood.

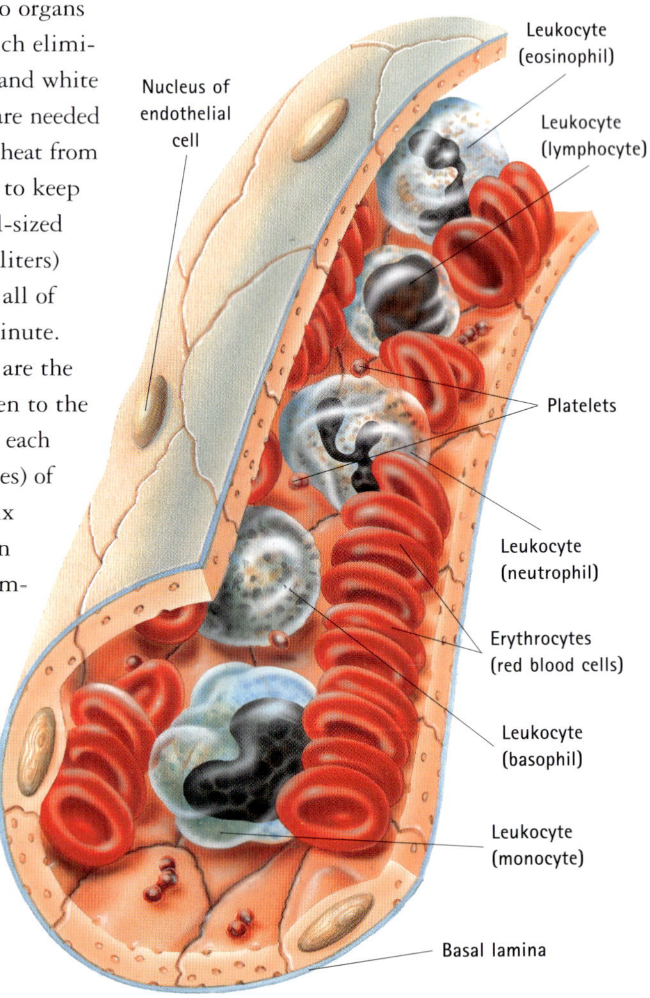

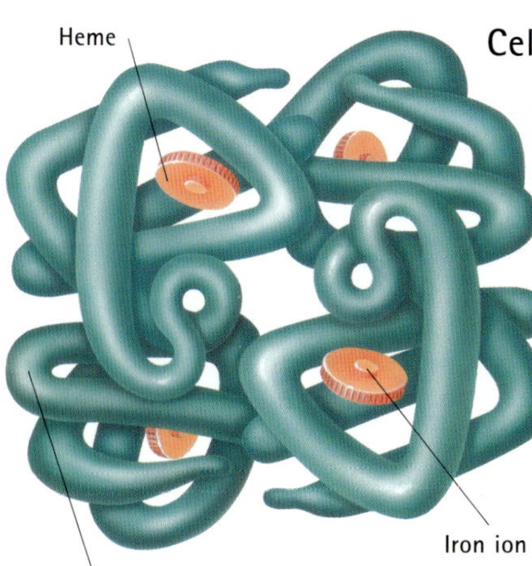

Cell content

Red blood cells contain hemoglobin, which carries oxygen to the tissues of the body. Hemoglobin contains the protein heme—a pigment that gives blood its color—and iron ions which bind with oxygen molecules.

RED BLOOD CELLS

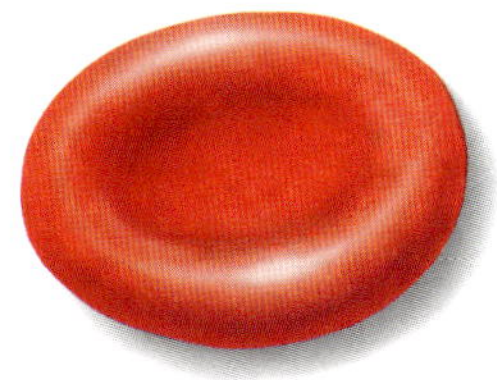

Red blood cells

The flat, plate-like shape of red blood cells enables them to bend and squeeze through small capillaries and also to stack together for ease of blood flow.

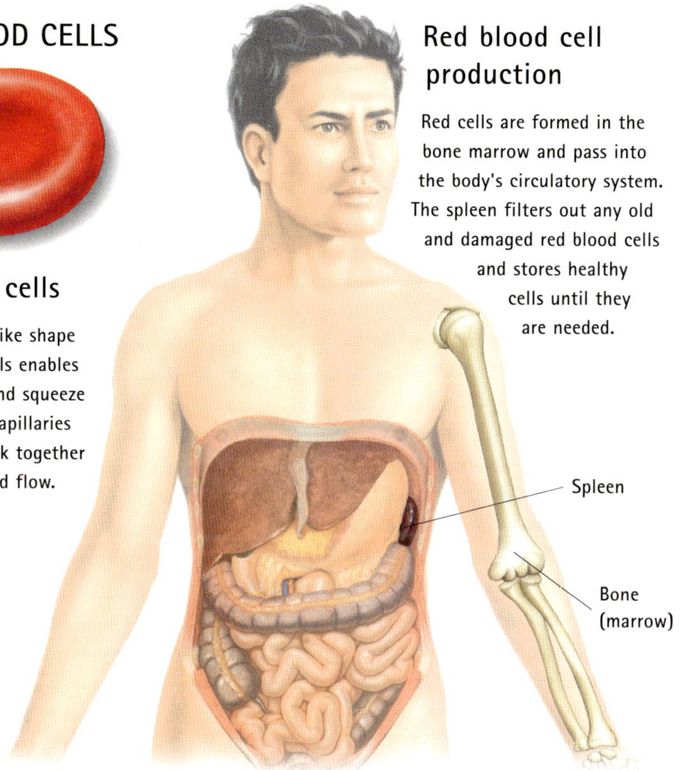

Red blood cell production

Red cells are formed in the bone marrow and pass into the body's circulatory system. The spleen filters out any old and damaged red blood cells and stores healthy cells until they are needed.

process. This process involves complex chemical reactions in the blood plasma.

In order to deliver oxygen and remove waste products, blood must reach the tissues. Two circulation systems maintain the body's blood supply. In the so-called systemic circulation, oxygenated blood is pumped from the heart through arteries to capillaries in tissues where the oxygen is removed, and then back to the heart via the veins. In the pulmonary circulation, deoxygenated blood flows from the heart to the lungs and then returns, oxygen-rich, to the heart.

SEE ALSO *Lymphatic/Immune system on page 55, Respiratory system on page 95; Spleen on page 284*

CLOTTING

Normal blood clotting, or coagulation, is a complex process involving as many as 20 different plasma proteins known as coagulation factors. These factors interact to form a reaction that leads to the production of fibrin, a protein that stops bleeding. In bleeding disorders, certain coagulation factors are deficient or missing. Such a disorder may be hereditary, such as hemophilia. Abnormal bleeding may also be caused by vitamin K deficiency, severe liver disease or prolonged treatment with anticoagulants.

A clot within a blood vessel is called a thrombus. A thrombus can block an artery and cause tissues downstream to die from lack of oxygen; it can also block a vein. The blood clot may dislodge and travel further upstream; it is then called an embolus.

Red blood cells

Red blood cells are the most numerous cells in the blood and their production rate by

Blood clot

This picture of a blood clot as seen by an electron microscope shows red blood cells trapped in a network of fibrin fibers.

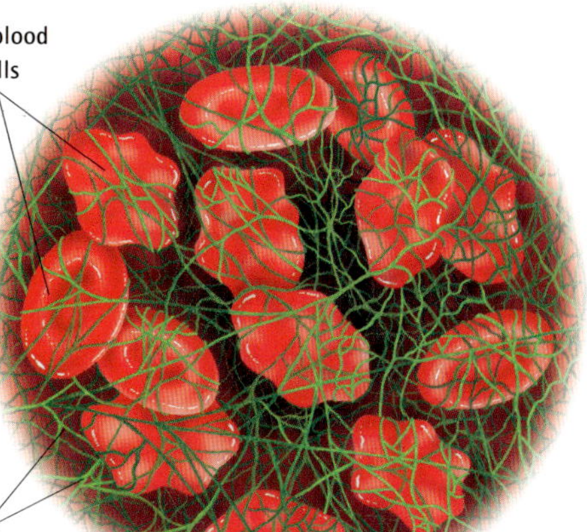

Blood clotting

Blood clots in three stages:

a. When a blood vessel is damaged, platelets and red blood cells spill into the damaged tissues.

b. Platelets increase in number and begin to attach to damaged surfaces. Strands of a protein called fibrin are formed.

c. Blood cells, platelets and strands of fibrin become enmeshed in a fibrous tangle called a clot.

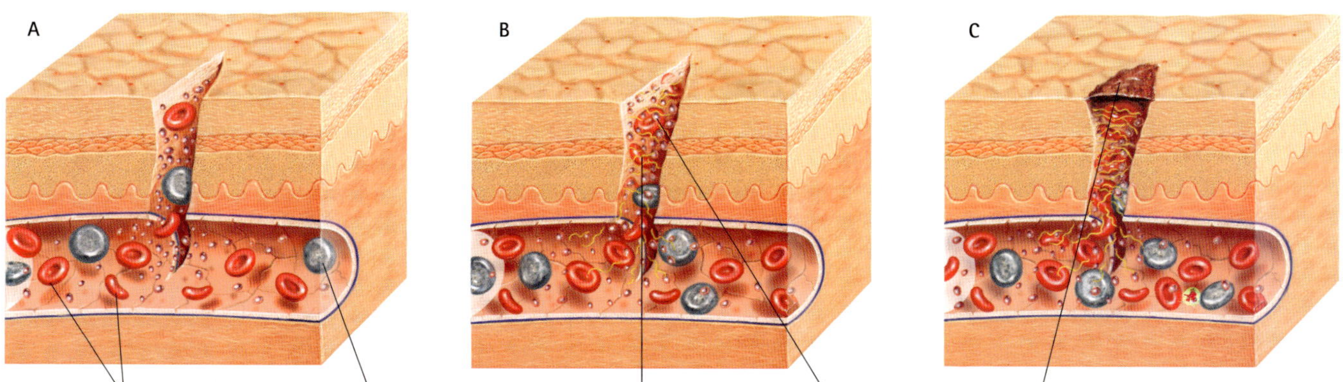

the bone marrow the highest in the body. They are highly specialized cells, devoted to the carriage of oxygen to the tissues and the removal of carbon dioxide. This allows the tissues to perform the metabolic processes required for the body to function.

Red cells are produced from early stem cells in the bone marrow, which multiply rapidly. The more mature red cells accumulate a red pigment, hemoglobin, which has special properties that ensure the efficient binding and release of oxygen. The fully mature red cell extrudes its nucleus and appears in the circulation full of hemoglobin. Its shape is that of a biconcave disk, which gives it the ability to deform and squeeze through narrow spaces, such as in the small capillaries. Red cells are stored in the spleen, which discharges them as needed. In emergencies, such as hemorrhage, some of the stored cells are released. After an average lifespan of 120 days the ageing red cell loses its ability to change shape and is destroyed in the spleen.

The number of red blood cells normally remains fairly stable, despite their constant production and destruction. The normal red cell count ranges between 4,000,000 and 6,000,000 per cubic millimeter, with the amount of hemoglobin in blood ranging between 14 and 18 grams per 100 milliliters in normal adults.

A deficiency of red cells results in anemia. There are many causes of anemia, which may result from diseases of the bone marrow, vitamin or mineral deficiencies affecting red cell production, diseases causing increased breakdown of red cells, or simple bleeding.

The most common cause of anemia is iron deficiency. Iron is a central part of the hemoglobin molecule, which is responsible for binding oxygen. It is obtained from the diet, red meat being especially rich in iron.

In industrialized countries, iron deficiency may be as a result of heavy periods in menstruating women, or from chronic gastrointestinal bleeding.

Sickle cell disease is an inherited abnormality of the globin gene. This disorder results in the production of abnormal sickle-shaped red cells. As well as causing anemia, sickle cell disease can result in the blockage of small blood vessels, which causes severe, painful secondary conditions.

WHITE BLOOD CELLS

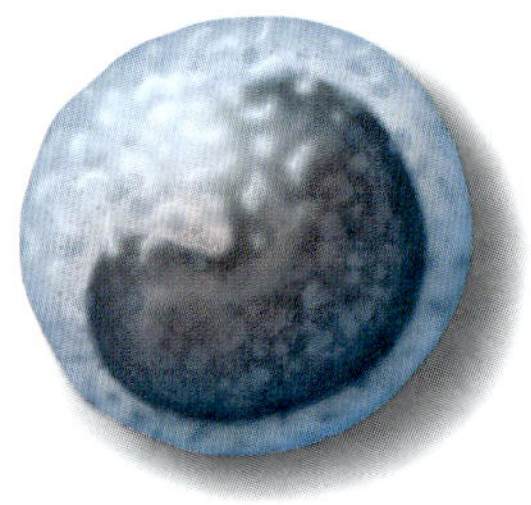

Monocyte

Monocytes circulate in the blood for 1–2 days before entering the body tissues to become macrophages.

Macrophage

Macrophages fight infection by engulfing foreign organisms and debris.

Neutrophil

The front-line defense against bacterial invasions, neutrophils engulf and destroy microorganisms.

Basophil

These cells release substances that increase the body's response to invading allergens.

Eosinophil

Eosinophils release enzymes which cause allergic reactions and kill some parasites.

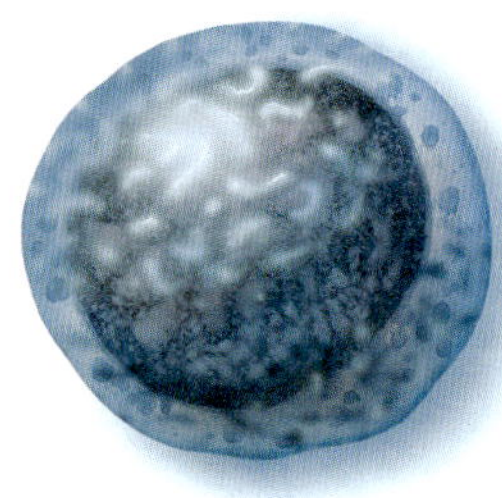

Lymphocyte

There are three types of lymphocyte. Natural killer cells and T cells attack foreign invaders directly; B cells make antibodies.

White blood cells

Throughout life the body faces ongoing assault from thousands of different viruses, bacteria and other microbes, as well as a range of invertebrate parasites. The first lines of defense—the skin and mucous membranes—keep most of these potential invaders out. The comparative few that make it through to deeper tissues to cause disease face a barrage of attacks from the white blood cells (WBCs). Known also as leukocytes, WBCs are produced in red bone marrow and lymphatic tissue. There are five major types: neutrophils, eosinophils, basophils, monocytes and lymphocytes.

Most WBCs can "squeeze" through very small spaces to sites of infection in the body, a process referred to as emigration. These spaces include those between the cells of capillary walls. Neutrophils are the first to arrive at the site of an invasion. There they release bacteria-killing enzymes and ingest microorganisms and foreign particles by a process called phagocytosis. The monocytes arrive a little later but in much larger numbers, becoming macrophages once they migrate from the blood into tissues. They kill offending organisms and clean up debris associated with infection. Eosinophils also consume and engulf foreign material; in addition they release enzymes associated with allergic reactions which kill certain parasitic worms and protozoa. Basophils are also involved in inflammation and allergic reactions. Once they enter infected tissue they become mast cells, which release substances such as histamine that intensify the body's inflammatory response to foreign matter, producing hypersensitivity reactions.

Lymphocytes are the white blood cells responsible for the body's ability to distinguish and react to foreign substances, such as bacteria, viruses and other microbes. They are part of the body's line of defense against invaders. Lymphocytes are small rounded cells that originate in the bone marrow from primordial cells (called stem cells). Once they have matured, they pass into the bloodstream where they travel to areas of lymphoid tissue, such as the spleen, lymph nodes, tonsils, and the lining of the intestines. There are three types of lympho-

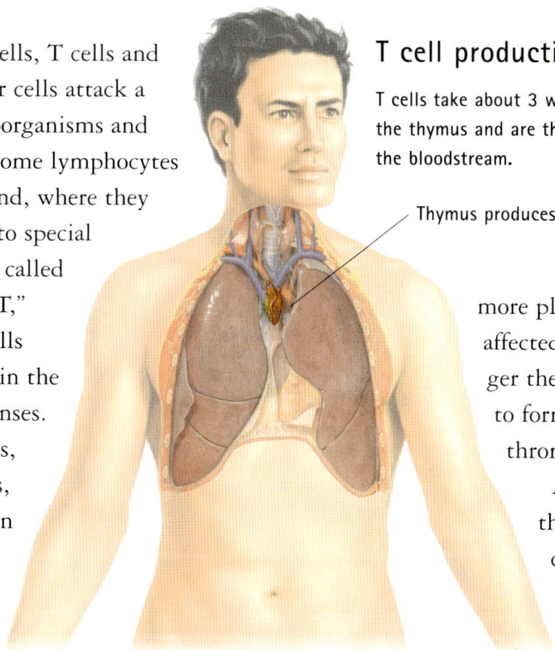

T cell production

T cells take about 3 weeks to develop in the thymus and are then released into the bloodstream.

cytes: natural killer cells, T cells and B cells. Natural killer cells attack a wide variety of microorganisms and certain tumor cells. Some lymphocytes enter the thymus gland, where they multiply and turn into special types of lymphocytes called thymus-derived, or "T," cells. The T and B cells are directly involved in the body's immune responses. T cells combat viruses, fungi and cancer cells, and also attack foreign tissue during organ transplants.

Before they can respond, T cells must become sensitized to a foreign invader. Different T cells learn to "recognize" different pathogens or harmful substances—there are literally millions of different T cells circulating in a healthy body.

The situation is similar with B cells, which deal particularly well with bacteria and are highly effective at deactivating their toxins. Unlike the T cells, B cells do not directly attack foreign material. Instead, they develop into plasma cells that secrete antibodies to do the work. T and B cells can circulate in the body for years. The other WBCs are, like suicidal soldiers, eventually killed by their combat efforts—particularly by the toxic material they engulf. Their life spans are usually only a few days but during times of infection they may survive for only a matter of hours.

An increase in the number of WBCs in the blood usually indicates inflammation or infection. Because each WBC type plays a different role, the percentage of each type present in the blood helps diagnose disease. Normally, WBCs number between 5,000 and 10,000 per cubic milliliter of blood; over 10,000 suggests infection. An elevated WBC count is called leukocytosis and an abnormally low count is called leukopenia.

Platelets

Platelets are small cell fragments in the blood that prevent bruising and bleeding. They adhere to small breaks in blood vessels and form a plug, which blocks the defect. Platelets release chemicals which attract more platelets to the affected area and trigger the clotting system to form a clot, or thrombus.

Aspirin blocks the release of chemicals by platelets and can cause easy bruising in some people. It is widely used to reduce the risk of thrombosis as it makes platelets less sticky or adherent.

Plasma

Plasma refers to the fluid component of the blood. It consists of water, salts and proteins and contains all the chemicals which circulate between body tissues, including glucose, hormones, enzymes and growth factors. Plasma generally contributes approximately half the blood volume, the rest being due to red and white blood cells.

The major plasma protein is albumin, which is synthesized by the liver and is important in maintaining the blood's osmotic pressure. The liver also produces some of the blood clotting proteins that are found in the plasma.

Another group of plasma proteins are the globulins, which include immunoglobulins produced by the immune system and are responsible for fighting infection.

Globulins can be divided into alpha, beta, and gamma subgroups. Alpha and beta globulins include fibrous and contractile proteins, transport proteins, and enzymes. Gamma globulins play a vital role in natural and acquired immunity to infection; they are manufactured by the immune system to help destroy or neutralize infection-causing bacteria. Gamma globulin (also known as antibodies) derived from blood of other humans can be used to induce temporary immunity to some diseases, for example hepatitis A.

Plasma is lost together with red and white cells during bleeding and may need to be replaced if loss is severe. Initial replacement can be with plasma substitutes, followed by the infusion of fresh frozen plasma. This is produced from donated blood, the plasma being separated by centrifugation and frozen until required.

In some diseases the patient's plasma may be exchanged with normal plasma or albumin solution as part of the treatment. The procedure is known as plasmapheresis and is performed using an apheresis machine.

Blood and Circulatory Disorders

The free circulation of blood round the body is vital to sustain all body systems. Any disruption to blood flow, pressure and content as a result of blood and circulatory disorders therefore has serious consequences.

SEE ALSO *Blood tests on page 436; Cerebral hemorrhage on page 144; Coronary artery disease on page 231; Disorders of the nerves and blood vessels of the leg on page 334; Portal hypertension on page 273; Stroke on page 142; Subarachnoid hemorrhage on page 144*

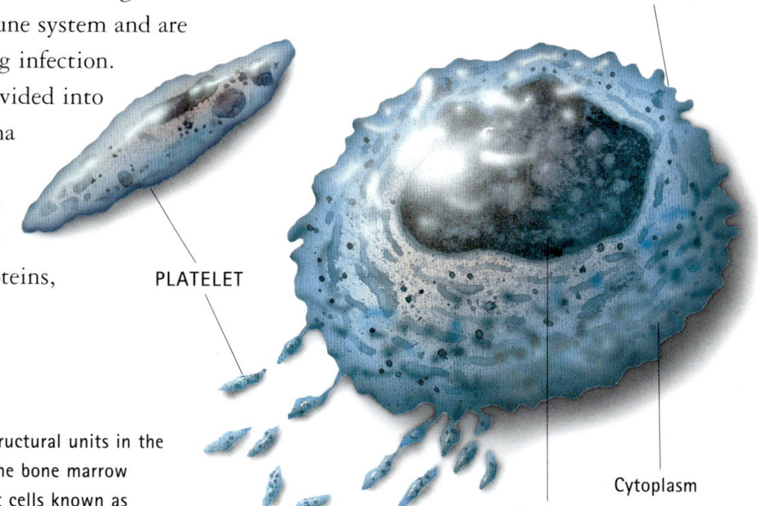

Platelets

Platelets are the smallest structural units in the blood. They are formed in the bone marrow from the cytoplasm of giant cells known as megakaryocytes.

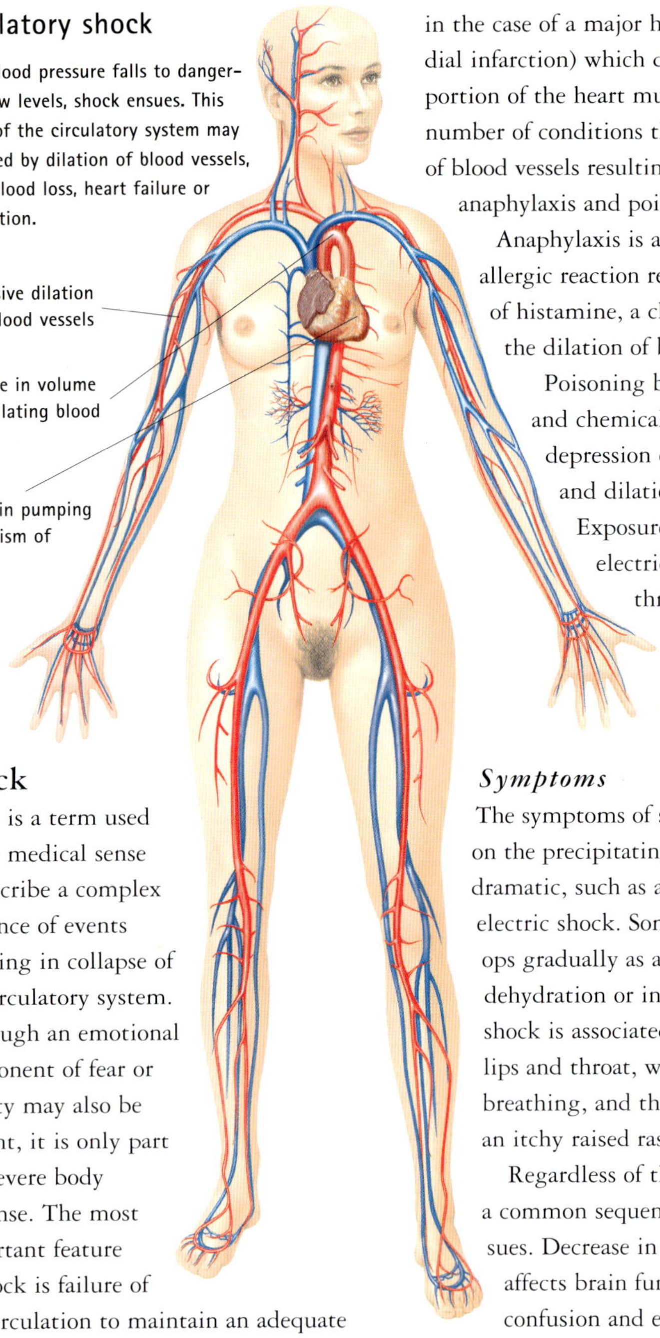

Circulatory shock

When blood pressure falls to dangerously low levels, shock ensues. This failure of the circulatory system may be caused by dilation of blood vessels, severe blood loss, heart failure or dehydration.

Shock

Shock is a term used in the medical sense to describe a complex sequence of events resulting in collapse of the circulatory system. Although an emotional component of fear or anxiety may also be present, it is only part of a severe body response. The most important feature of shock is failure of the circulation to maintain an adequate blood pressure to allow vital organs to continue to function.

Causes

Shock may result from a failure of the pumping ability of the heart, from a reduction in the circulating blood volume or from excessive dilatation of the blood vessels. The most common cause of shock is reduction in the blood volume, due to hemorrhage or severe dehydration. Another important cause of shock is septicemia, in which severe infection results in the release of chemicals called endotoxins, causing dilation of blood vessels. They may also impair the function of the heart. Heart diseases themselves can result in shock, particularly in the case of a major heart attack (myocardial infarction) which damages a large proportion of the heart muscle. There are a number of conditions that can cause dilation of blood vessels resulting in shock, including anaphylaxis and poisonings.

Anaphylaxis is a severe immediate allergic reaction resulting in the release of histamine, a chemical which causes the dilation of blood vessels.

Poisoning by a variety of drugs and chemicals may result in depression of cardiac function and dilation of blood vessels. Exposure to high-voltage electricity can cause shock through cardiac arrest and severe electrical burns, resulting in fluid loss.

Symptoms

The symptoms of shock depend initially on the precipitating factor. They may be dramatic, such as a heart attack or an electric shock. Sometimes shock develops gradually as a result of progressive dehydration or infection. Anaphylactic shock is associated with swelling of the lips and throat, wheezing and difficulty breathing, and the rapid appearance of an itchy raised rash.

Regardless of the initial symptoms, a common sequence of events then ensues. Decrease in the blood pressure affects brain function, resulting in confusion and even coma. The body mounts a protective defense of the circulation by releasing chemicals from the adrenal glands and activating the sympathetic nervous system. Chemicals such as epinephrine (adrenaline) increase the pumping action of the heart and constrict blood vessels. These chemicals are also responsible for symptoms of tremor and anxiety that accompany the shock syndrome.

The body's response may result in the person appearing pale, cold and clammy. While it is beneficial in maintaining the blood pressure, the compensatory response is only a temporary measure before organ function is compromised further. Constriction of blood vessels can reduce blood flow to vital organs, especially the kidneys. Also, the lack of blood supply to body tissues results in anaerobic metabolism with the accumulation of lactic acid. Ultimately shock can result in an irreversible situation, in which tissue damage has become so great that the condition becomes fatal.

Treatment

Treatment of shock relies on its early detection and institution of emergency measures, such as oxygen and epinephrine (adrenaline) in anaphylaxis. The maintenance of circulatory volume by intravenous fluids is essential and is combined with infusion of drugs to maintain blood pressure. Specific treatment, such as antibiotics for septicemia or antidotes for poisoning, is also given.

Raynaud's disease

In Raynaud's disease (also known as Raynaud's phenomenon), the smaller blood vessels go into spasm, especially those of the fingers and toes. The digits become progressively cold and white, then blue, and then turn red as they warm up again. Attacks may last from a few minutes to several hours and are usually set off by cold or emotional upset. In severe cases, ulcers and gangrene of the fingers or toes may develop. The cause is usually unknown, though it may occur as a complication of other conditions such as connective tissue disease, vascular disease, or trauma.

Treatment involves giving up smoking if the sufferer is a smoker, and avoiding the trigger factors of cold and emotional stress. The extremities, especially the hands and feet, should be kept warm and covered in winter. Any underlying cause should be treated. Vasodilators (arterial muscle relaxants) and sympathectomy (surgical excision of the sympathetic nerves supplying the arteries) may be tried in severe cases.

Cyanosis

Cyanosis is a bluish discoloration of the skin and mucous membranes (such as the lips). It is a sign that arterial blood is inadequately oxygenated. When it occurs in the whole of the body (central cyanosis) it is usually due to heart or lung disease.

Cyanosis often occurs in the terminal phases of a cardiac arrest, drug overdose, drowning or pneumonia, in fact whenever

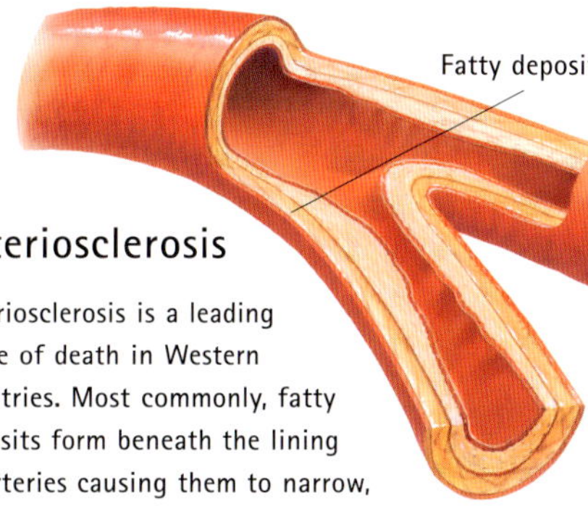

Arteriosclerosis

Arteriosclerosis is a leading cause of death in Western countries. Most commonly, fatty deposits form beneath the lining of arteries causing them to narrow, reducing the blood supply to the brain, heart, and other organs.

oxygenation of blood is not occurring. In rare cases it can be caused by abnormal hemoglobin (such as methemoglobinemia) and toxins such as cyanide.

Cyanosis of an area of the body, such as the feet or hands, can be caused by arterial disease. Arterial spasm in cold environments may cause cyanosis, which clears when the extremities are warmed. Arteriosclerosis may cause ischemia and cyanosis of extremities; angioplasty or amputation may be needed.

Hypertension

Hypertension is the term for high pressure (tension) in the arteries (i.e. high blood pressure). The average normal blood pressure at rest is about 120/80 mm Hg, and is usually between 100/60 and 140/90.

Although blood pressure varies from person to person and from time to time, someone is said to have hypertension when the blood pressure is 140/90 or above. High blood pressure is one of the causes of arteriosclerosis which in turn can cause heart attack, stroke, intermittent claudication and kidney failure. Hypertension can exist silently for decades; by the time it is diagnosed, the damage may have already been done. Hence, hypertension is sometimes called "the silent killer."

People with a family history of hypertension, who smoke, are overweight or obese, and consume a diet high in salt are at particular risk. Stress and excess alcohol consumption are also thought to be contributing factors. In 10 percent of cases there is a predisposing medical condition such as kidney disease or a tumor of the adrenal gland. The diagnosis of hypertension can be made by a primary care physician (general practitioner) using simple blood pressure measurements, repeated to confirm the diagnosis. More tests may be needed if the physician suspects an underlying disease is causing hypertension.

Because of its role in stroke, heart attack and kidney failure, it is important to bring hypertension under control. Preventive measures are the first line of treatment. Smoking and alcohol use should be stopped, or at least reduced, regular aerobic exercise and weight control introduced, and intake of salt and animal fats reduced. These measures can reduce blood pressure by about five points in 50 percent of sufferers. If this fails to control the hypertension, drug therapy is the next step. Drugs used for treating hypertension include beta-blockers, diuretics, alpha-adrenergic blockers, angiotensin-converting enzyme (ACE) inhibitors, and calcium channel blockers.

About 1 percent of people with hypertension have a severe form of the condition called accelerated or malignant hypertension. In these people, diastolic blood pressure exceeds 140 and is associated with headache, nausea and dizziness. This condition requires urgent hospital treatment to prevent stroke or brain hemorrhage.

Hypotension

Hypotension, (the clinical term for low blood pressure) is a condition in which the blood pressure is below the average measurement of 120/80 mm Hg.

Often a low blood pressure is of little significance, being a statistical variant from the average, and quite normal for a particular person. Low blood pressure may even indicate a prolonged life expectancy.

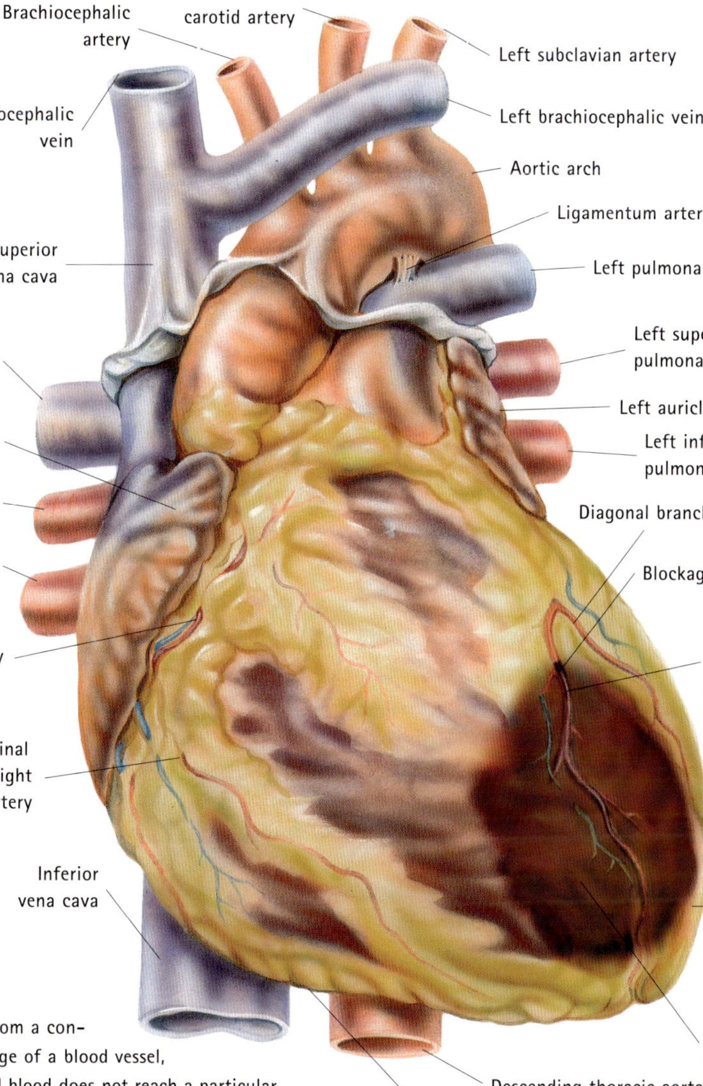

Ischemia

Ischemia results from a constriction or blockage of a blood vessel, so that oxygenated blood does not reach a particular part of the body. Obstruction of the arteries to the heart can lead to angina or a heart attack.

Anemia

In iron deficiency anemia the red blood cells (as shown here in a capillary) are smaller and paler than normal. In hereditary spherocytosis—an inherited form of hemolytic anemia—50 percent or more of red blood cells are replaced by abnormal, small round spherocytes.

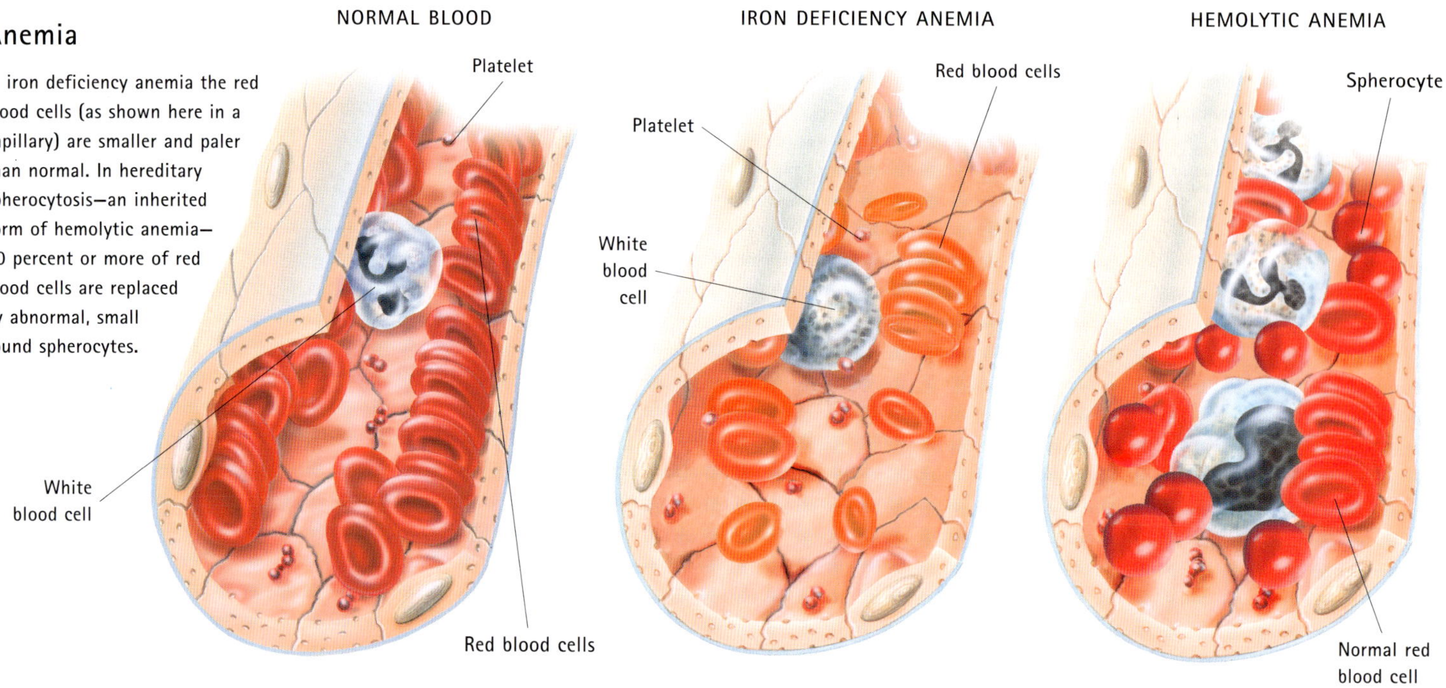

to gangrene. Prevention of these forms of thrombosis includes treating the risk factors for atherosclerosis, such as hypertension and high cholesterol; anti-platelet drugs such as aspirin are often prescribed.

Venous thrombosis occurs most commonly in the deep veins of the legs, resulting in pain and swelling due to blockages in the blood leaving the legs. More serious consequences can occur if the clot detaches from the vein wall and travels to the lungs (pulmonary embolism). Venous thrombosis can occasionally occur in the brain or the intestines, with serious effects. Thrombosis in the veins may be related to disease of the vein wall, such as varicose veins. However, there are generally other contributory factors, such as sluggish circulation (stasis) or an alteration in the composition of the blood, making it clot more readily. This is generally due to an increase in proteins of the coagulation system and/or a decrease in the protective antifibrinolytic system.

Deep vein thrombosis is most likely to occur during periods of immobility, such as after an operation. Apart from the resulting venous stasis, there are changes in the blood during surgery which make a person less likely to bleed and more likely to clot. Similar changes occur during pregnancy, also increasing the risk of venous thrombosis. It is increasingly recognized that some people inherit genes making them more susceptible to clotting. This tendency is known as thrombophilia and often explains why there is an increased incidence of clotting in some families, or why some pregnant women or some women taking the oral contraceptive develop thrombosis and others do not. The risk of venous thrombosis can be reduced by early mobilization after surgery and preventive treatment with anticoagulant drugs.

Hemorrhage

Hemorrhage is the technical term for bleeding. Bleeding can be a normal process, as with menstrual bleeding, but it is usually abnormal. The commonest cause of bleeding is trauma, such as from a cut. Bleeding is often a sign of damage to a blood vessel. This may be due to minor trauma, such as most nosebleeds; trauma can also cause subconjunctival hemorrhage under the white part of the eye.

Damaged blood vessels may be due to infection; cystitis, for example, can produce blood in the urine, and bronchitis can cause blood to appear in the sputum. Erosion of the vessel wall by acid can cause bleeding from peptic ulcers. Sometimes damage to the blood vessel is the result of a more serious disease, such as inflammation of the blood vessel (vasculitis) or a tumor invading it.

Bleeding can result from increased pressure in veins or arteries. An example of the former is bleeding from hemorrhoids and of the latter is subarachnoid hemorrhage, which is bleeding on the surface of the brain; this is often associated with high blood pressure.

Some types of bleeding can have serious or life-threatening effects, due either to the volume of blood lost or the site of the hemorrhage. Massive blood loss can result from major trauma, such as a car accident or, less often, from deficiencies of platelets or clotting factors. Bleeding that can be serious due to its location includes cerebral hemorrhage and bleeding before (antepartum), during (intrapartum) or just after (postpartum) childbirth.

Purpura

Purpura is a disease in which hemorrhages occur in the skin and mucous membranes. It manifests as small red spots on the skin. There are two main types: thrombocytopenic purpura, which involves a decrease in platelet count in the blood; and nonthrombocytopenic purpura, which does not involve any such decrease.

Causes of purpura include exposure to drugs or chemical agents; cancerous diseases such as leukemia; and infectious diseases such as rubella.

Anemia

In anemia, the red blood cells fail to provide sufficient oxygen to the tissues of the body. It can be caused by decreased amounts of hemoglobin in red blood cells, or decreased numbers of red cells in the blood. Signs of anemia include tiredness and weakness, pallor (especially in the hands and eyelids), fainting, breathlessness and rapid heartbeat. To diagnose anemia, a physician

RHESUS (RH) FACTOR

In pregnancy, problems may occur if the mother is Rh negative and the fetus is Rh positive. The red blood cells of the fetus may be destroyed by the mother's Rh antibodies, which can lead to hydrops fetalis (swelling in the fetus due to excessive fluid).

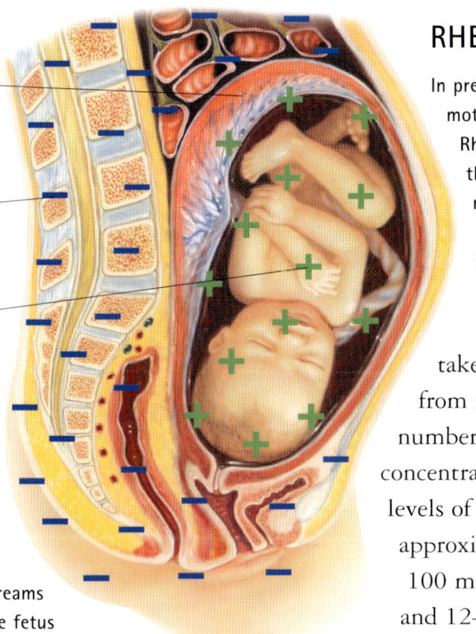

First pregnancy

Problems rarely occur during the course of the pregnancy, as the bloodstreams of the mother and the fetus do not mix.

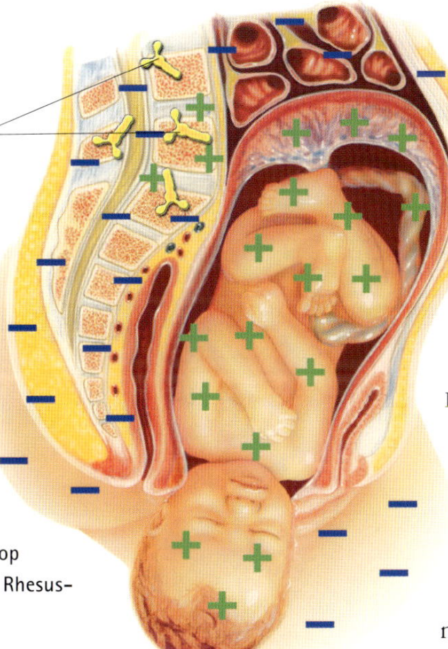

First pregnancy—delivery

During delivery, the baby's blood can leak into the maternal bloodstream; this causes the Rhesus-negative mother to develop antibodies which destroy Rhesus-positive blood cells.

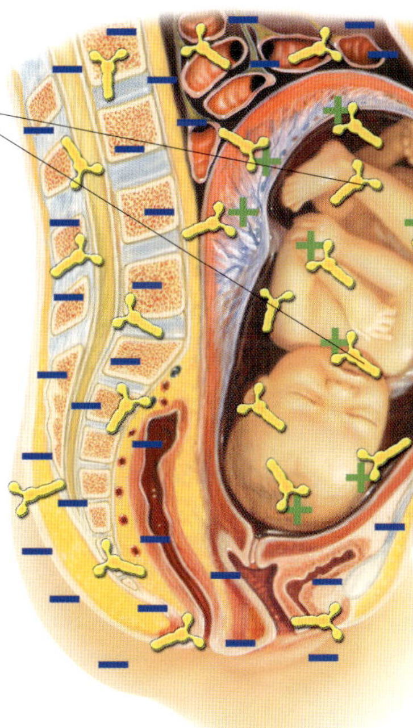

Second and subsequent pregnancies

If the mother does not receive an anti-D globulin injection following the first pregnancy (to halt the production of antibodies), her antibodies may attack the blood cells of subsequent Rhesus-positive babies.

takes a small sample of blood from the patient and counts the number of red blood cells and the concentration of hemoglobin. Normal levels of hemoglobin in the blood are approximately 14–17 grams per 100 milliliters (deciliter) for males and 12–15 grams per 100 milliliters for females.

The size and shape of red blood cells can give a clue as to the cause of anemia. In pernicious anemia, for example, the red blood cells are larger than normal, whereas in iron deficiency anemia they are smaller and paler than normal.

IRON DEFICIENCY ANEMIA

Iron deficiency anemia is caused by decreased absorption of iron, loss of iron from the body (usually from bleeding), or an increased need for iron. Malabsorption and/or poor nutrition may result in decreased absorption of iron and iron deficiency anemia. Premature babies often have low stores of iron at birth. Heavy menstrual bleeding or gastrointestinal disease with bleeding (such as bowel cancer) may deplete the body's iron stores. Pregnancy or rapid growth may place too much demand on the body's iron reserves. Treatment is to maintain an adequate iron intake through a well-balanced diet or supplements, and to correct the underlying cause.

PERNICIOUS ANEMIA

Pernicious anemia is the result of inadequate absorption of vitamin B_{12}, normally found in meat, fish and dairy products. It is caused by the absence of intrinsic factor, a chemical secreted by the stomach's lining which assists in the absorption of vitamin B_{12}. It may occur as an autoimmune disease or after stomach surgery. Pernicious anemia cannot be cured, but symptoms can be controlled with regular injections of vitamin B_{12}. Treatment needs to be continued for life.

APLASTIC ANEMIA

Aplastic anemia is caused by decreased bone marrow production of red blood cells. Platelet and white blood cell production is also diminished so that, as well as anemia symptoms, the patient suffers abnormal bleeding and reduced resistance to infections. The condition is most often caused by drugs, especially immunosuppressant drugs, anticancer drugs, chloramphenicol or chemicals such as benzene. It may also be caused by immunodeficiency or severe illness. The patient should be isolated in hospital to avoid infection, and may receive blood transfusions and bone marrow transplantation.

HEMOLYTIC ANEMIA

Hemolytic anemia is caused by hemolysis, the premature destruction of red blood cells. The cells may be destroyed because they are abnormal or because of antibodies that attack them; bone marrow cannot produce red blood cells fast enough to compensate. Along with other symptoms, there may be jaundice (yellow skin and eyes, dark urine) and an enlarged spleen (splenomegaly). The condition may be inherited (as in hereditary spherocytosis, sickle cell disease or thalassemia) or acquired, from blood transfusions or drugs. Acquired hemolytic anemia is treated by removing the cause. Immunosuppressant drugs may also be needed. Inherited forms of the condition are incurable, though symptoms can be controlled. Splenectomy (removal of the spleen) is sometimes needed.

Rhesus (Rh) factor

Rhesus (Rh) factor is named after an antibody produced in 1940 in the blood of a rabbit after immunizing it with red blood cells from a rhesus monkey. This antibody

LEUKEMIA

This malignant disease involves the rapid and uncontrolled proliferation of leukocytes (white blood cells) in the blood-forming organs. The white cell count is greatly elevated and large numbers of immature cells are found in the circulating blood.

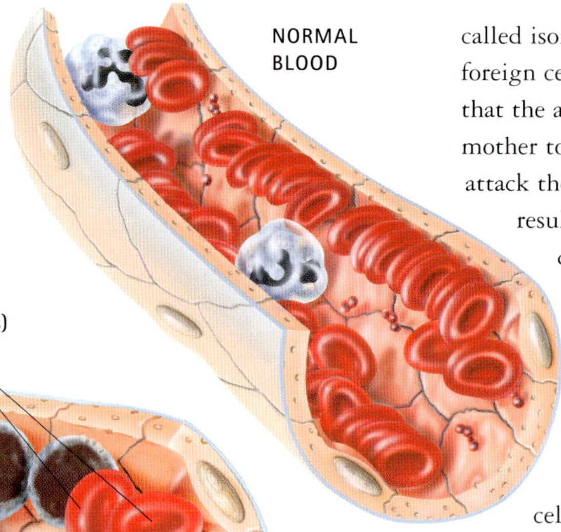

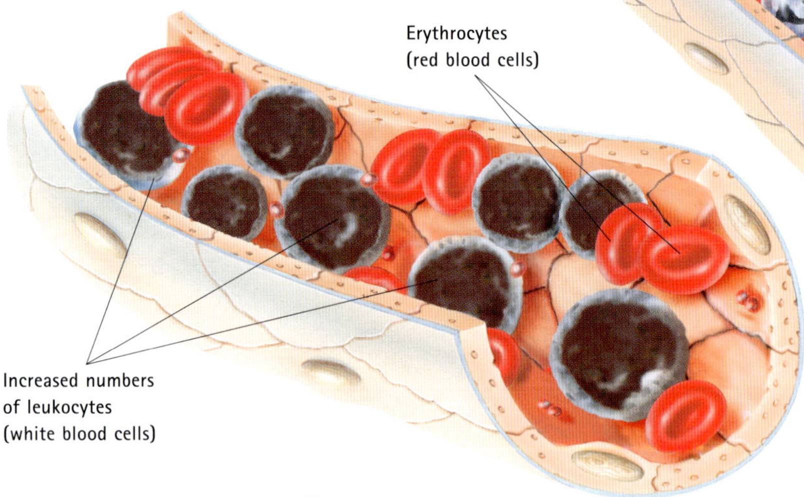

was shown to be the same as one found the previous year in the blood of a woman who had a stillborn child. The child was born with hydrops fetalis, or generalized swelling, due to severe anemia.

What had been discovered was a new blood group system, separate from the ABO system already known. Eighty-five percent of Caucasians are now known to have the Rh antigen on the cell membrane of their red blood cells and are Rh positive; the other 15 percent lack this antigen and are Rh negative. The presence or absence of an Rh factor is important when considering the compatibility of blood types, especially in pregnancy and blood transfusions.

When an Rh negative woman and Rh positive man have a baby, there is a possibility that the baby will be Rh positive and that its red cells will be destroyed by Rh antibodies. The Rh positive cells of the baby are introduced into an Rh negative woman via the placenta. This Rh incompatibility stimulates the woman's immune system to produce antibodies, a process called isoimmunization, which destroy the foreign cells or antigens. Of real concern is that the antibodies then travel from the mother to the baby, where they continue to attack the Rh positive red cells. This may result in increased physiological jaundice of the newborn, in neonatal anemia or, if severe, in hydrops fetalis. Immunization to Rh antigens can also result from the transfusion of Rh positive blood to an Rh negative recipient, leading to destruction of the transfused cells in a blood transfusion reaction.

The discovery of the Rh blood group system has resulted in safer blood transfusion, through testing for the Rh factor during crossmatching of blood. It has also led to measures to prevent Rh isoimmunization in pregnancy. Before the development of DNA sequence analysis, the Rh system contributed to the identification of blood or body fluids at the scene of crimes or accidents and was also used in testing for disputed parentage.

The Rh system consists of six major antigens, the most important being the D antigen, which results in the most severe isoimmunization. Initial exposure to the D antigen results in sensitization of the immune system; a second exposure produces a strong antibody response. It is now routine practice to test women during pregnancy

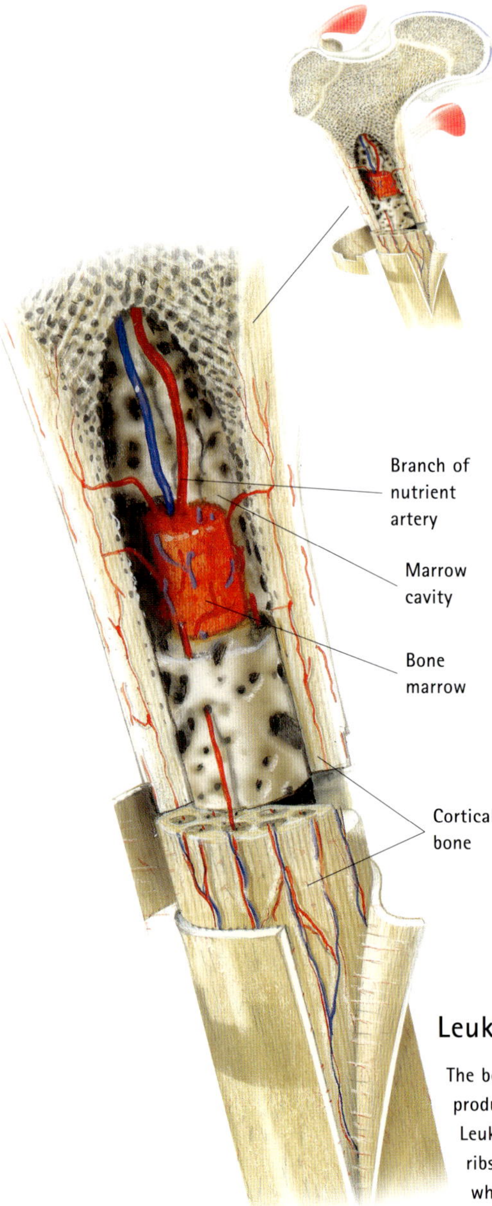

Leukemia—bone marrow

The bone marrow is the most important site for the storage and production of white blood cells (leukocytes) in the body. Leukocytes are produced mainly in the long bones, spine, skull, ribs, and sternum. In acute leukemia, the rapid expansion of white cells in the bone marrow can cause severe pain.

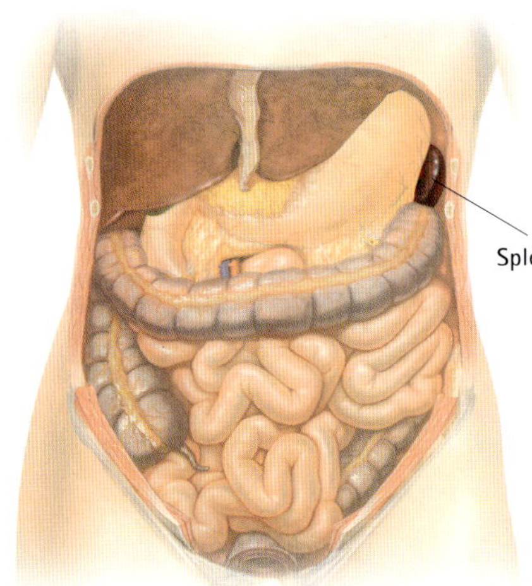

Leukemia—spleen

The spleen produces lymphocytes (one type of white blood cell) which are essential to the body's immune system. Overproduction of lymphocytes results in the most common form of leukemia—chronic lymphocytic leukemia. This can cause enlargement of the spleen.

for the Rh antigen. If a woman is Rh negative and her partner Rh positive, there is a possibility of isoimmunization. If this is the woman's first pregnancy, it is unlikely that she will become sensitized until delivery, when small numbers of fetal red cells enter the maternal circulation.

If she delivers an Rh positive baby, she will receive an injection of anti-D globulin to prevent her immune system reacting and forming Rh antibodies that could affect future pregnancies. Anti-D is also given to Rh negative women following an abortion, miscarriage or amniocentesis. The other means of preventing Rh isoimmunization is through the transfusion of Rh negative blood to Rh negative women of childbearing age. These precautions have resulted in a marked decrease in the incidence of Rh immunization and its effects on the newborn.

Hemolytic disease of the newborn (anemia caused by excessive destruction of red blood cells) does still occur, either due to episodes of sensitization not treated by anti-D globulin or due to other Rh or ABO antibodies. However, it is unusual now to see increasingly severe cases in successive pregnancies, as in the past, resulting in fetal death. Babies born with hemolytic anemia may need to be given blood transfusions after delivery.

Leukemia

Leukemia is a progressive, malignant disease arising from the white cells in the blood. It occurs in both adults and children, and is the most common type of malignant disease in children. The disease is characterized by an increase in white cells in the blood and in the blood-forming organs, the bone marrow and spleen. The leukemic white cells do not perform the normal function of fighting infection, but rather they interfere with the function of the normal white cells. They also interfere with the development of red blood cells and platelet cells in the bone marrow.

There are many types of leukemia, but they can be grouped according to the following criteria: duration (acute or chronic); the type of proliferating white cell; and the fluctuations in the number of abnormal cells in the blood.

Acute leukemia is a form of leukemia that develops rapidly, over weeks, and is quickly fatal if not treated. Patients develop infections, which are persistent and fail to respond to usual antibiotic treatment. The infections are generally due to bacteria and result in fever with or without local symptoms. Common sites of infection are the throat, gums, chest and skin. Acute leukemia also causes exceptional fatigue and shortness of breath, due to anemia from decreased production of red blood cells. Easy bruising and spontaneous bleeding, from the nose and gums or from cuts, results from interference with the production of platelet cells. The rapid expansion of white cells in the bone marrow can cause severe bone pain in the ribs and spine.

No single cause has been identified for acute leukemias and the cause of most cases is unknown. Exposure to high doses of ionizing radiation, such as in nuclear accidents like Chernobyl, resulted in increased cases of leukemia. However, it is not known whether low-level exposure to ionizing or other forms of radiation also increases the risk. The use of some toxic chemicals increases the occurrence of leukemia; benzene and its derivatives have the strongest association. Acute leukemia can also occur after the use of certain chemotherapy drugs recommended for other cancers, or after the use of certain drugs designed to suppress the rejection of kidney and other transplants.

There is a possibility that some drugs or chemicals affecting a woman during pregnancy can increase the risk of acute leukemia later developing in the child; these include alcohol, cannabis, benzene and pesticides. There is increasing interest in the possible role of viruses in causing leukemias in humans. Hereditary leukemia is rare but there is an increased incidence in conjunction with some genetic diseases, such as trisomy 21 (Down syndrome).

Acute lymphoblastic leukemia is the common form of childhood leukemia. It is derived from the type of white blood cells called lymphocytes. Blast cells are the immature cells from which the different types of blood cells normally develop. In acute lymphoblastic leukemia the blast cells that normally develop into lymphocytes do not do so and accumulate instead. Specific chromosomal or genetic abnormalities have now been identified in 60–75 percent of patients with acute lymphoblastic leukemia, giving hope for improvements in the understanding and treatment of this disease.

Acute myeloid leukemia is the common form of adult leukemia, whose incidence increases with age. It arises from an increase in myeloblasts in the bone marrow. At least six types of acute myeloid leukemia are recognized. Many are associated with specific chromosomal abnormalities and almost all patients have some chromosomal abnormality in their leukemic cells. In acute promyelocytic leukemia, this knowledge has led to a major breakthrough in treatment with the use of a form of vitamin A.

Treatment of acute leukemia involves combinations of chemotherapy drugs and supportive treatment to prevent infections and bleeding. The initial chemotherapy is administered by intravenous injection and in tablet form over several days. Often a plastic catheter is inserted into a large vein to make repeated injections more convenient. The purpose of the initial treatment is to achieve apparent eradication of the leukemic cells; this is known as a remission. Once remission has been achieved, it is followed by consolidation chemotherapy and, in the case of lymphoblastic leukemia, more prolonged maintenance chemotherapy.

The results of treatment have improved with the development of powerful antibiotics, platelet transfusions and the use of growth factors to stimulate recovery of the normal white cells. With current best methods of treatment, remission can be achieved in 70–80 percent of adults up to 60 years of age and 97–99 percent of children. Bone marrow transplantation in remission can achieve cure rates of 50–60 percent in adults and 75–80 percent in children.

Chronic, as differentiated from acute, leukemias are leukemias which run a more gradual and prolonged course. They are often detected by chance, such as through a routine blood test, in which the raised white cell count appears. These leukemias involve an increase in mature white blood cells.

The most common leukemia is chronic lymphocytic leukemia, in which there is an increase in the number of lymphocytes. This occurs mainly in men over 50 years of age. It can cause enlargement of the lymph glands and spleen. Often there is an increased susceptibility to bacterial infections, due to reduced production of antibodies by normal lymphocytes. However, some patients may be completely asymptomatic and may not require treatment for years.

Chronic myeloid leukemia has its peak incidence at 20–40 years and involves an increase in mature myeloid cells, or granulocytes, in the blood and bone marrow. Enlargement of the spleen is an important feature of this condition, as is the association with a chromosomal abnormality in the bone marrow cells, called the Philadelphia chromosome, after the city in which it was first described. This form of leukemia generally requires treatment and can be cured in young patients with bone marrow transplantation. In those not suitable for this procedure, the use of injections of interferons (agents that inhibit cellular growth) can decrease the number of Philadelphia-positive cells and prolong patient survival.

Thrombocytopenia

Thrombocytopenia is a reduction in the number of platelets circulating in the blood. Because platelets are essential for clotting, the main symptom of this disorder is abnormal bleeding, particularly into the skin where the blood forms bruises and small hemorrhages called petechiae, which appear as round purple-red spots. Mouth and nosebleeds are also common.

There is a range of possible causes of thrombocytopenia. One of the most frequent is an autoimmune disease called idiopathic thrombocytopenic purpura. This involves the production of antibodies against platelets by spleen and lymph tissue, resulting in the destruction of platelets in the spleen. As well as the characteristic bruising and skin hemorrhaging, this type of thrombocytopenia may also cause abnormal menstrual bleeding and sudden loss of blood in the intestinal tract. More children are affected than adults, mostly after viral infection, and often treatment is not necessary. In adults idiopathic thrombocytopenic purpura may develop into a chronic condition and does not usually follow viral infection.

Drug-induced non-immune thrombocytopenia is a reaction to certain drugs, some of which can damage bone marrow and slow the production of platelets. This is potentially fatal if it leads to bleeding in the brain or another vital organ. Other drugs may not affect the production of platelets but rather render them useless by ensuring that they cannot adhere to one another, a property that is required for blood clotting. Drug-induced immune thrombocytopenia is the development of antibodies to blood platelets, either as a result of the direct use of certain drugs or, in the case of an affected fetus, due to drugs taken by the mother during pregnancy. The condition may also occur with certain other diseases or infections, such as AIDS or leukemia or as a reaction to particular drugs. Treatment depends on the cause. Some forms of the illness will resolve themselves gradually over time. Others may require treatments such as a transfusion of platelets. Complications can include bloody stools and vomiting blood.

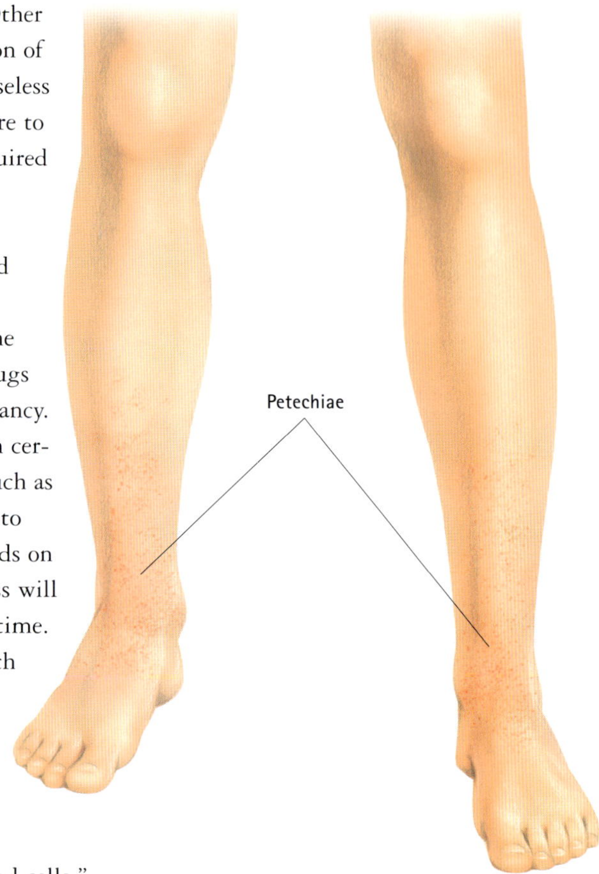

Thrombocytopenia

In this condition, reduced numbers of platelets in the blood inhibit clotting. This results in a distinctive rash of small red spots—usually on the legs—caused by abnormal bleeding into the skin. These spots are known as petechiae.

Polycythemia

Polycythemia, literally "many blood cells," refers to a group of blood conditions in which the numbers of cells in the blood increases. This may involve the red cells, white cells and/or the platelets. The most common and important form is an increase in the red cells, causing facial redness and increased viscosity of the blood. The affected person may have a sensation of fullness in the head and of itching, particularly after a hot bath or shower, which releases histamine from the white cells. The increased red cells and platelets result in a greater tendency to clotting in the veins and arteries, leading to deep venous thrombosis, heart attack or stroke.

Treatment is based on reducing blood viscosity through regular venesection, which is removal of blood from a vein—a needle is inserted into a large vein at the front of the elbow. Sometimes medication is required to control the increased cell production by the bone marrow.

Bacteremia and septicemia

Bacteremia and septicemia (also called blood poisoning) occur when bacterial infection has entered the bloodstream. Once infection has spread to the blood, it may be carried to other parts of the body. Symptoms include high fever, sweating, malaise and chills.

The difference between bacteremia and septicemia is one of degree; septicemia is more serious and usually indicates that the bacteria are multiplying in the blood and causing serious illness. In extreme cases of septicemia, abscesses may form in internal organs such as the liver or brain. Severe blood poisoning may be fatal. The condition requires urgent treatment with intravenous antibiotics in hospital.

Toxemia

Toxemia is a type of blood poisoning caused by toxins, or poisons. Many bacterial infections, such as bacterial dysentery, diphtheria, food poisoning and tetanus, cause toxemia. It is possible to immunize against some diseases that cause toxemia, such as tetanus or diphtheria.

Toxemia of pregnancy (also known as eclampsia or preeclampsia) is a serious condition that may occur in late pregnancy.

RESPIRATORY SYSTEM

In the process of metabolism, the human body consumes oxygen and produces carbon dioxide as a waste product. The respiratory system is designed to exchange the carbon dioxide accumulated in the blood for oxygen in the airways, which enters the lungs as air from the surrounding atmosphere. This air is breathed into the airways via the nose, pharynx, larynx, trachea and bronchi.

SEE ALSO *Circulatory system on page 78; Diaphragm on page 248; Epiglottis on page 202; Fetal development on page 398; Larynx on page 201; Lungs on page 237; Metabolism on page 116; Mouth on page 180; Pharynx on page 201; Ribs on page 192; Speech on page 187; Throat on page 200; Trachea on page 202; Treating the respiratory system on page 441*

Nose

The visible part of the nose is supported by cartilage and is only the front opening of the nasal cavity. The nasal septum divides the nasal cavity through the middle, and is formed by septal cartilage at the tip and by bone closer to the skull.

The inner surface of the nostrils are covered by skin with coarse hairs that trap dust from air that has been drawn in (inspired). The remainder of the nasal cavity is lined by mucous membrane with many blood vessels and mucus-secreting glands. Blood heat warms up inhaled air, while the moist sticky mucus traps more dust. Three curved structures known as conchae, jutting out from each side wall into the nasal cavity, increase the surface area for warming up inspired air.

A small area in the roof of the nasal cavity is supplied (innervated) by the olfactory nerve for smell.

Pharynx

The pharynx lies behind the nasal cavity, oral cavity and larynx and ends in the esophagus. Located in each side wall of the pharynx, just behind the nasal cavity, is the opening of the auditory tube, which connects the pharynx to the middle ear cavity. It equalizes pressure of the ear cavity with atmospheric pressure.

The larynx and speech

The larynx is a tube located in front of the pharynx and made up of a membrane reinforced by muscles. It begins with a cartilage—the epiglottis—which closes the entry into the larynx during swallowing to prevent food from passing into the airway. This tube has two segments separated from each other by a small gap. The top segment begins at the upper edge of the epiglottis and ends below it at the vestibular folds or "false vocal cords." The lower segment begins at the vocal cords and ends where it is attached to the cricoid cartilage. The cricoid cartilage is attached to the trachea below. The vocal cords are attached to the thyroid cartilage at the front and to the arytenoid cartilage at the back of the larynx.

We speak by blowing air from the lungs into the larynx, which vibrates the vocal cords. The arytenoids spin around a vertical axis, bringing the vocal cords close together or further apart. The thyroid cartilage tilts on the cricoid to control the tension of the vocal cords and the frequency of the vibration. The vibrations of the cords are converted into the different sounds of speech by the position of the tongue and lips.

The thyroid cartilage protects the front of the larynx. It has two plates (laminae) joined together at an angle, like the spine of an open book. The upper end of the angle is more prominent in males and known as the "Adam's apple." The thyroid cartilage is connected by membrane to the hyoid bone above and cricoid cartilage below. The larynx moves up and down when we swallow or when we want to sing with a vibrato effect.

Trachea, bronchi and lungs

The trachea is a stack of 15–20 C-shaped cartilages, connected to each other by fibrous tissue, forming a vertical gutter opening at the back. The trachealis muscle connects the ends of the Cs, closes up the gutter and turns it into a tube.

Food passing down through the esophagus bulges into the trachea through the trachealis. A large mass of food (bolus) stuck in the esophagus can block the air passage and choke the victim to death. If the larynx is obstructed by edema (swelling), as in smoke inhalation, tracheostomy may save life by creating a temporary opening in the trachea at the root of the neck.

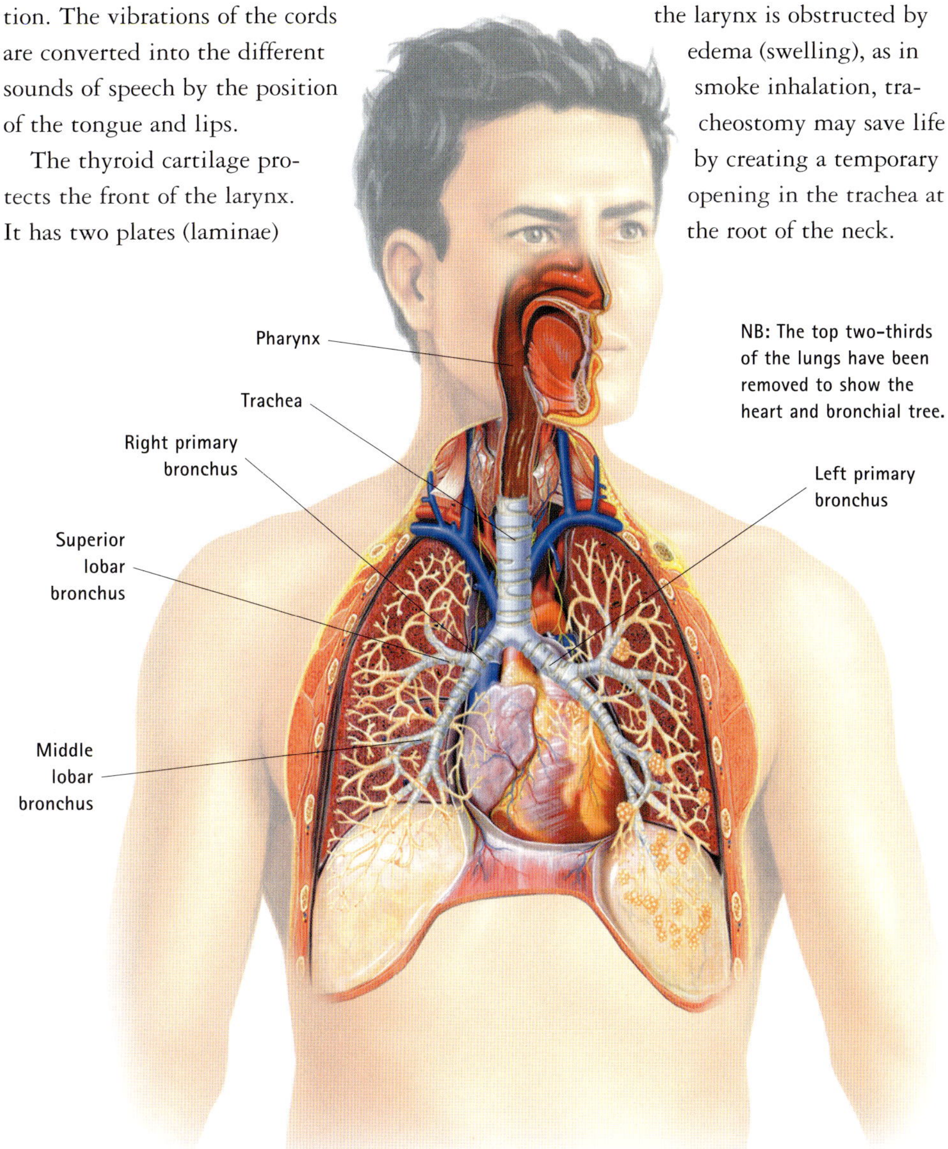

NB: The top two-thirds of the lungs have been removed to show the heart and bronchial tree.

Upper section of the respiratory system

The upper part of the respiratory system consists of the nose, the nasal cavity and the pharynx. The pharynx is shared by the respiratory and digestive systems.

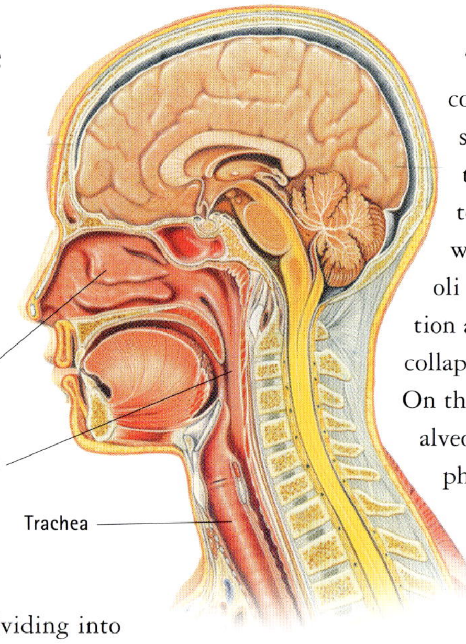

The trachea ends by dividing into the right and left main bronchi for the right and left lungs. The left lung is divided into upper and lower lobes by an oblique fissure. An additional horizontal fissure divides the right lung into three lobes, the upper, middle and lower lobes.

Inside each lung, the main bronchus divides first into lobar bronchi, then into smaller and smaller bronchi (bronchioles). The lining (epithelium) of the trachea and bronchi has hair-like structures known as cilia and contains many goblet cells and mucous glands which secrete mucus to trap dust. Dust trapped in mucus is moved upward by the cilia. The bronchial epithelium also contains lymphoid tissue that secretes antibody IgA. The bronchioles are entirely muscular and have no cartilage in their walls.

The bronchial tree ends in the air sacs, or alveoli, that branch out of the terminal bronchioles like bunches of grapes. Alveoli are separated from one another by inter-alveolar septa that greatly increase the surface area available for gas exchange.

The walls of the alveoli are thin and contain a network of capillaries. Blood coming from the pulmonary artery is depleted of oxygen and rich in carbon dioxide. On the other hand, inspired air is full of oxygen and low in carbon dioxide.

Gases move across the alveolar membrane: carbon dioxide passes into the alveolar air, and oxygen passes into the blood. The blood leaving the alveoli is again saturated with oxygen, and carbon dioxide in the alveoli is expelled into the atmosphere with expired air.

The alveolar walls also contain cells that secrete surfactant, a substance that reduces the surface tension of the alveolar wall, enabling the alveoli to expand in inspiration and preventing alveolar collapse during expiration. On the inner surface of the alveolar wall are macrophages which engulf inhaled bacteria, dust or carbon particles.

Pleura

Each lung is surrounded by a double-layered membrane called the pleura. Just imagine a slightly inflated balloon containing a teaspoonful of oil. Wrap the balloon around the lung, stretching it until it reaches the entrance of blood vessels and bronchus, and you have an image of the pleura. The half of the balloon in contact with the lung corresponds to the "visceral layer" of the pleura, the other half visible from outside corresponds to the "parietal layer" of the pleura. The parietal layer is attached to the inside of the rib cage. The small amount of fluid in the pleura allows frictionless movements of the lung against the rib cage.

Blood and nerve supply

Deoxygenated blood from the body is pumped by the right ventricle of the heart through the pulmonary arteries into the lungs. Branches of the pulmonary arteries follow the bronchial tree and end in the capillaries that surround the alveoli, where gas exchange occurs. Reoxygenated blood is collected by venules that converge into two pulmonary veins for each lung, and returns to the left atrium of the heart so that it can be pumped back around the body.

The airway and lungs receive oxygen and nutrients from the small bronchial arteries. Deoxygenated blood returns to the heart via the pulmonary veins.

Lymphatic vessels from the lungs eventually empty into the thoracic duct. The lymph nodes are located in the root of the lungs (where vessels and bronchi enter the lungs) and around the bronchi and trachea.

How do we breathe?

The rib cage is an airtight cylinder, with the diaphragm as its base, and the root of the neck as its top lid. The wall is made up of the ribs running from the thoracic vertebrae to the sternum, and connected together by three layers of intercostal muscles. When we breathe in, the ribs rotate around their

Muscles used for breathing

The diaphragm and intercostal muscles are involved in breathing. During inspiration these muscles contract to increase the front to back, side to side, and vertical size of the rib cage. Expiration is a passive process involving the recoil of the ribs and the return of the diaphragm to its normal position.

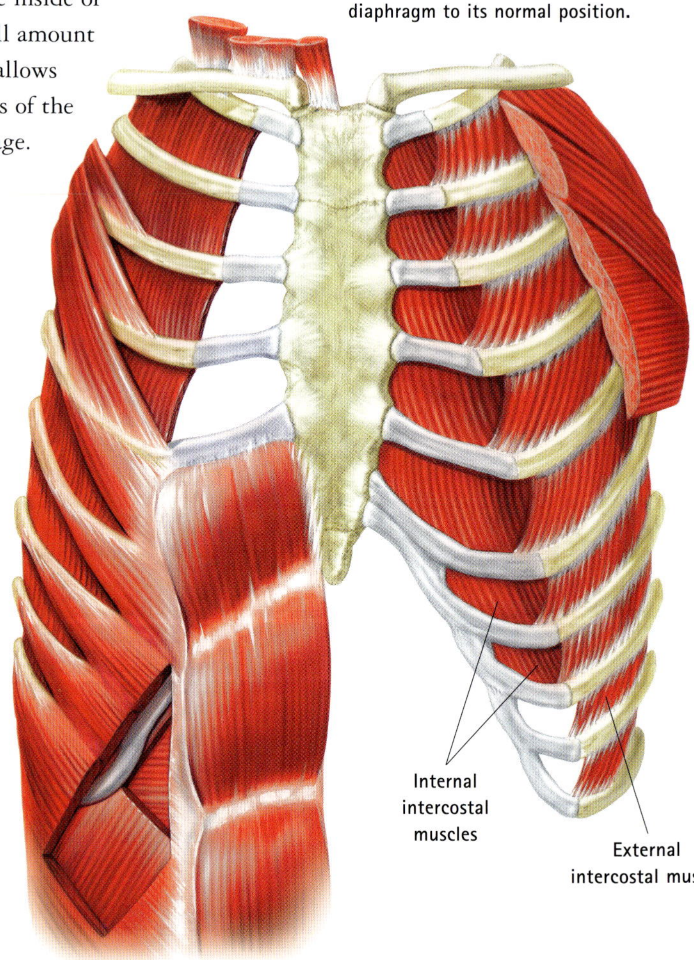

Bronchial tree

The bronchial tree provides a passageway for inspired air to pass into the alveoli where gas exchange occurs. Bronchi contain cilia (small hair-like structures) and mucus-secreting glands, which ensure the bronchial airways remain clear.

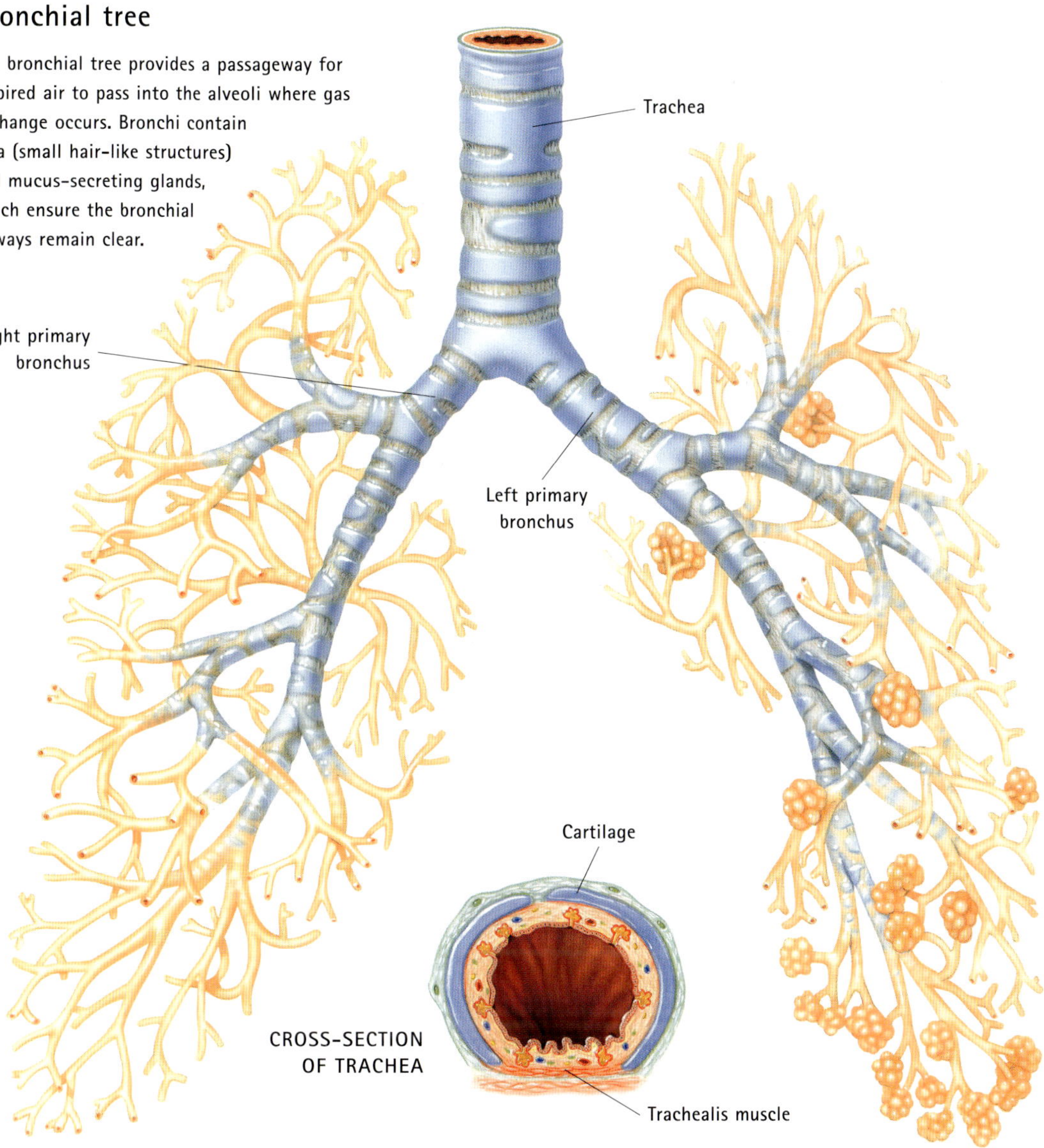

Gas exchange in the alveoli

The smallest bronchioles end in alveoli (air sacs). The sac walls are surrounded by small blood vessels which allow oxygen from the air to enter the bloodstream and carbon dioxide accumulated in the blood to pass into the alveoli to be breathed out.

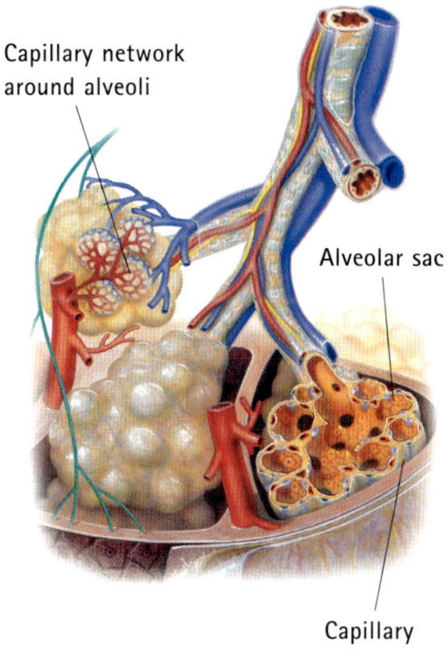

Gas exchange

The tiny alveoli in the lungs are the location for gas exchange. Oxygen from inspired air diffuses through the thin septa between each alveoli, to enter the bloodstream through the capillay network of the alveoli. At the same time, carbon dioxide from deoxygenated blood diffuses through the septa into the air sacs, ultimately to be expelled from the lungs during expiration.

vertebral articulations, and the sternum at their front ends is raised, increasing the front-to-back as well as the side-to-side diameters of the rib cage. The diaphragm moves downward, increasing the vertical size of the rib cage. The capacity of the thoracic cavity increases, resulting in a decrease in intrathoracic pressure to a value below atmospheric pressure. Air is sucked into the lungs through the airway.

Breathing out is a passive process—the recoil of the ribs as well as the lungs and the return of the diaphragm to a higher position squeeze the air out.

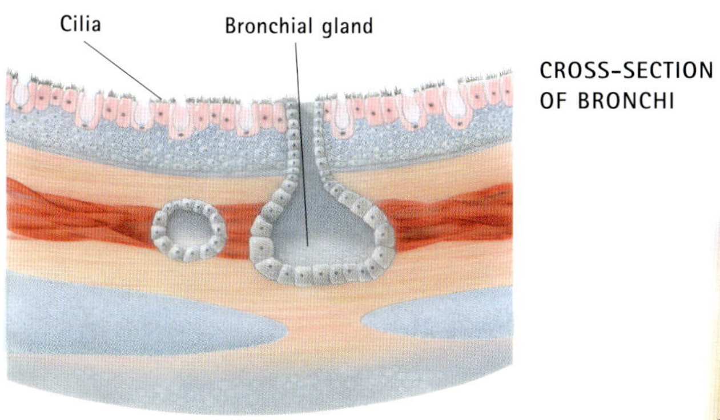

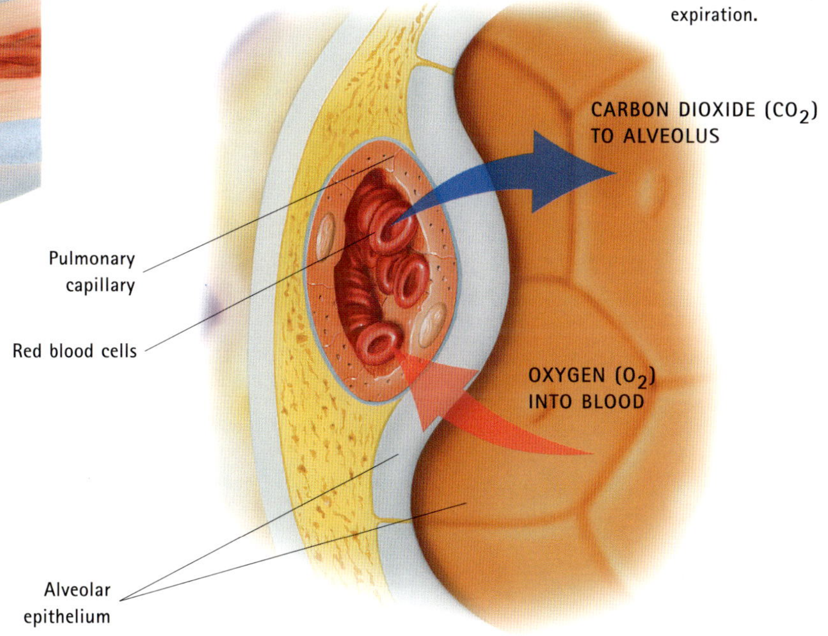

URINARY SYSTEM

The urinary system is essential to life. It excretes the waste products of metabolism and maintains the balance of water and electrolytes in the blood.

SEE ALSO *Adrenal glands on page 278; Dialysis on page 454; Electrolytes on page 114; Imaging techniques on page 431; Metabolic imbalances and disorders of homeostasis on page 117; Metabolism on page 115; Treating the digestive and urinary systems on page 444; Urinalysis on page 431; Urinary organs on page 286*

Blood and the internal environment

All cells and tissues of the body consume oxygen and nutrients and produce carbon dioxide and waste products. The respiratory system removes carbon dioxide from the body. Waste products are filtered out of the blood by the kidneys and excreted in the urine. An inorganic compound such as salt (sodium chloride) breaks down into sodium and chloride ions when dissolved in water. These ions allow the solution to conduct electricity and so are called electrolytes.

The cells of the body contain intracellular fluid and are surrounded by extracellular fluid that includes serum from the blood. All chemical reactions in the body and the functions of the nervous system depend on a constant concentration of electrolytes inside the cells and in the extracellular fluid. Sodium is the main electrolyte in extracellular fluid; potassium is the main electrolyte in intracellular fluid.

Changes in the proportions of ions in intracellular and extracellular fluid can have serious consequences. A high concentration of potassium in extracellular fluid can cause cardiac arrest; a low concentration may result in muscle paralysis. Hydrogen ions and bicarbonate ions determine the acidity of the blood, which influences the conduction of impulses in the nervous system and metabolic reactions of the body. The kidney is one of the main regulators of the balance of electrolytes and acids.

Kidneys

The kidneys are located on the back wall of the abdomen on either side of the vertebrae, enveloped by a renal capsule and enclosed in perirenal fat. Their position, below the last rib, explains why some kidney diseases cause pain at the angle between the lumbar vertebrae and the twelfth rib.

The right kidney lies behind the duodenum and is slightly lower than the left, which lies behind the pancreas and the stomach. The ends of the transverse colon, where it joins the ascending and descending colon, are located in front of the two kidneys. The adrenal glands cap the upper tip of each kidney.

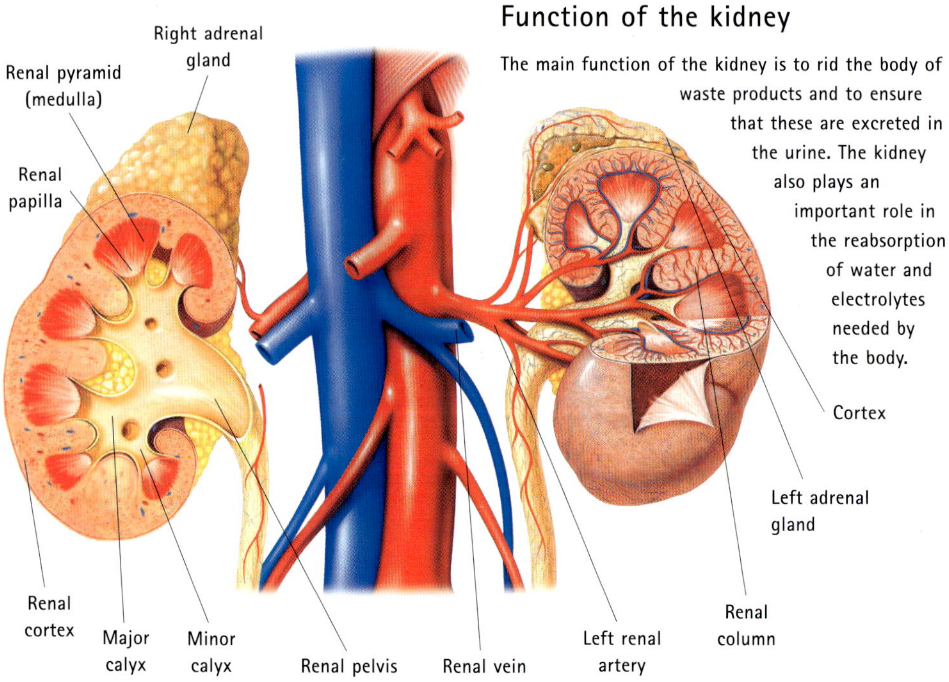

Function of the kidney

The main function of the kidney is to rid the body of waste products and to ensure that these are excreted in the urine. The kidney also plays an important role in the reabsorption of water and electrolytes needed by the body.

INTERNAL STRUCTURE OF THE KIDNEY

Each kidney has an outer part (cortex) and an inner part (medulla). The cortex contains filtration units (glomeruli and tubules). The medulla is made up of about a dozen renal pyramids, so-called because of their shape. The major component of each pyramid is a bundle of collecting tubules which collect the urine produced by the filtration units in the cortex.

The tips of the pyramids project into the sinus as small papillae, each of which is capped by a cup-like minor calyx. Minor calices join to form major calices which empty into the renal pelvis; this joins with the ureter.

Each renal artery branches off the aorta and divides into branches that enter the renal sinus to reach the medulla and cortex of the kidney. The veins accompany the arteries and end in the inferior vena cava.

THE FUNCTIONS OF THE KIDNEY

The kidney performs both excretory and endocrine functions.

Excretory function: the kidney filters the blood, removing metabolic waste products such as urea and any other unwanted substances. It maintains the balance of electrolytes in the blood and the water content of the body.

Each kidney has about a million microscopic functional units called renal tubules or nephrons. Each nephron begins with a glomerulus that is made up of a tuft of capillaries surrounded by the Bowman's capsule (a filtration membrane that is tightly wrapped around the capillaries of the glomerulus). Filtrate from the blood leaves the capsule and passes through the proximal and distal convoluted tubules of the nephron, where vitamins and various electrolytes are reabsorbed.

A part of the nephron (the loop of Henle) dips into the medulla of the kidney, and plays an important role in the reabsorption of water, sodium and potassium ions. Reabsorbed water and electrolytes enter the networks formed by the capillaries that leave the Bowman's capsule. Waste products and toxic substances are either not reabsorbed or are actively secreted into the urine.

The last part of the nephron, the collecting duct, is also the site of the reabsorption of water and electrolytes and plays an important role in adjusting the concentration of urine. The collecting ducts are bundled in the pyramids and open at the papillae.

The average urine output in the adult is approximately 1–1½ quarts (1–1.5 liters) a day. The amount and concentration of urine varies according to fluid intake and fluid loss through respiration, perspiration and fecal elimination. Only a small amount of very concentrated urine is produced when the body is dehydrated. People prone to urinary tract infections or kidney stones should maintain an adequate urine output to avoid the recurrence of infection or stone formation.

The filtration membrane in the renal glomerulus is a complex structure that can be damaged by the body's own immune reaction in glomerulonephritis. This disorder often develops after streptococcal infections or in some autoimmune diseases. In diabetes mellitus, thickening of the basement membrane of capillaries in the glomeruli disturbs the function of the glomeruli and causes renal failure.

The abuse of painkillers (analgesics) may cause severe damage to renal tubules. The tubules can also be blocked by hemoglobin released from muscles that have been destroyed by massive crush injury or excessive overheating in marathon runners. Damage to the renal tubules results in drastic changes in the composition of extracellular fluid. Renal failure may be fatal because concentration of some ions, such as potassium, can rise to lethal levels.

In late-stage kidney disease, life may be sustained by dialysis or artificial kidneys, both requiring close monitoring and constant intervention by a renal physician.

The function of the kidney extends far beyond simple filtration of the blood. Infection of the kidney is a common complication which used to result in death for bedridden patients. Even with recent advances in medicine, kidney failure is still an end stage of many general medical conditions.

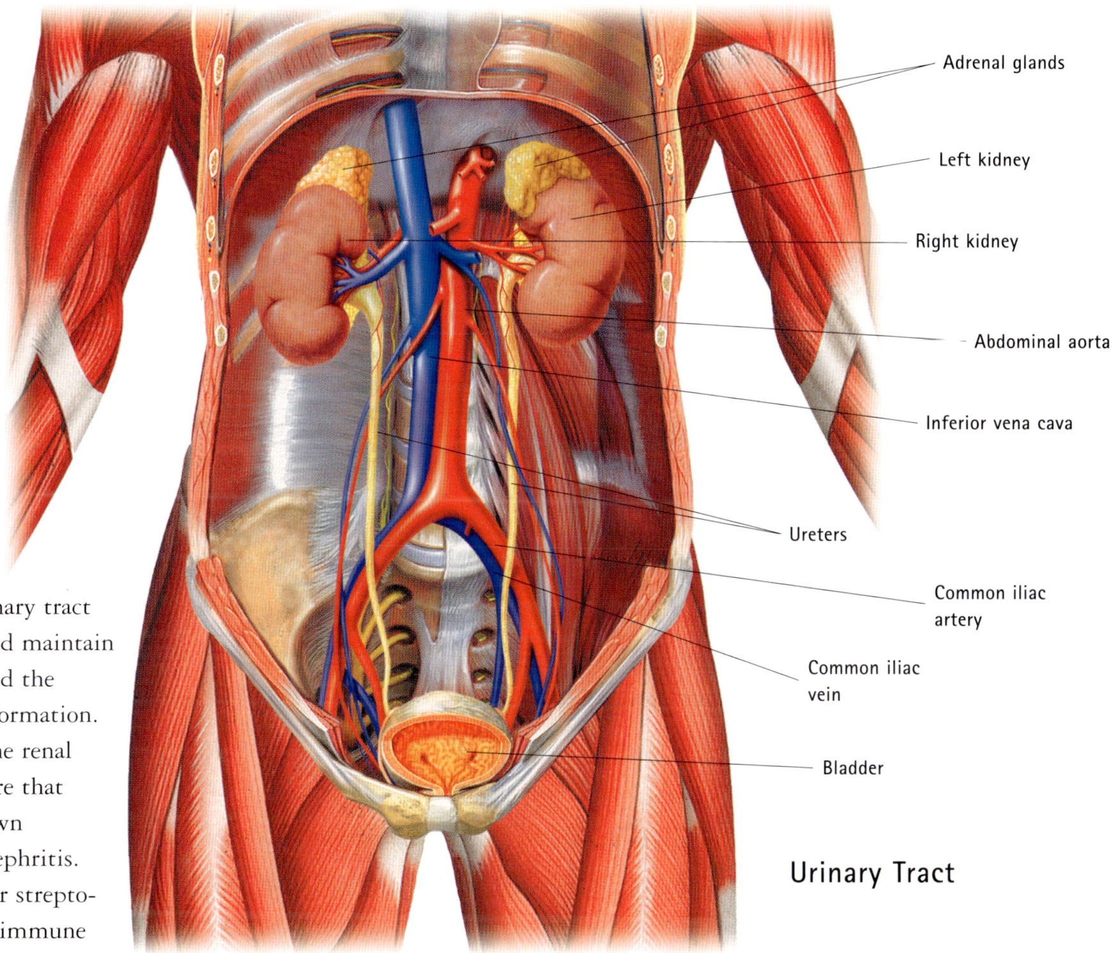

Urinary Tract

Endocrine function: the kidney's endocrine function involves the release of hormones into the blood: erythropoietin affects blood formation and hydroxycholecalciferol is involved in calcium metabolism.

Where the distal convoluted tubule joins the capillary in the glomerulus, specialized epithelium can sense blood concentration and endocrine secretion. This setup (the juxtaglomerular apparatus) also senses reductions in sodium concentration in the blood and responds by releasing renin, an enzyme that triggers a chain reaction to re-establish electrolyte balance.

Renin converts angiotensinogen in the blood into angiotensin I, which is then converted into angiotensin II by the lungs. Angiotensin II stimulates the adrenal cortex to secrete aldosterone, which acts on the nephron to increase the reabsorption of salt. This increases the output of diluted urine, restores the concentration of salt in body fluids, and raises blood pressure. Many antihypertensive drugs work by suppressing this renin-angiotensin-aldosterone system.

Ureter

The renal pelvis continues into the ureter, a muscular tube that carries urine to the bladder.

The ureter descends almost vertically along the line of the tips of the transverse processes of the lumbar vertebrae. At the hip bone it turns backward to enter the back of the bladder near the midline. The ureter is narrowest when it pierces the bladder wall. It runs obliquely for about ¾ inch (2 centimeters) in the bladder before opening to a slit-like aperture.

When the bladder is full, the increased pressure compresses the part of the ureter inside the bladder wall, preventing reflux of urine back to the renal pelvis. When there are bacteria in the bladder, reflux can spread infection to the kidney (pyelonephritis).

If the ureter runs perpendicularly through the bladder wall because of defective development, this "flap-valve" mechanism does not operate and kidney infection by reflux occurs frequently.

Male urinary system

The urinary system comprises the kidneys, the ureters, the bladder and the urethra. The male urethra is the passage for both sperm and urine.

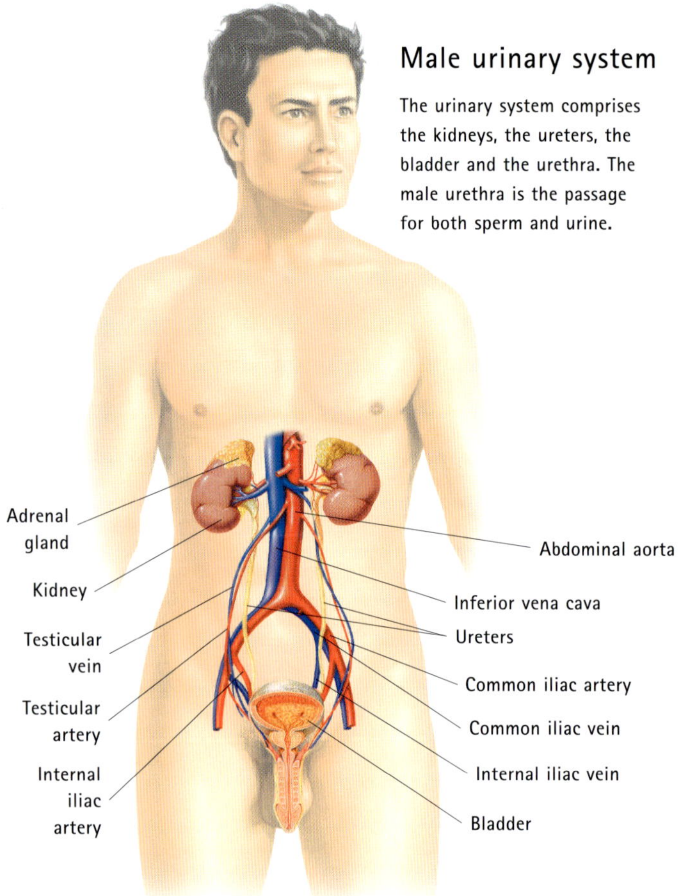

Female urinary system

The female urinary system is essentially the same as the male, except that the female urethra is much shorter. This provides bacteria with an easy route into the body, and is the reason why urinary tract infections are far more common in women than men.

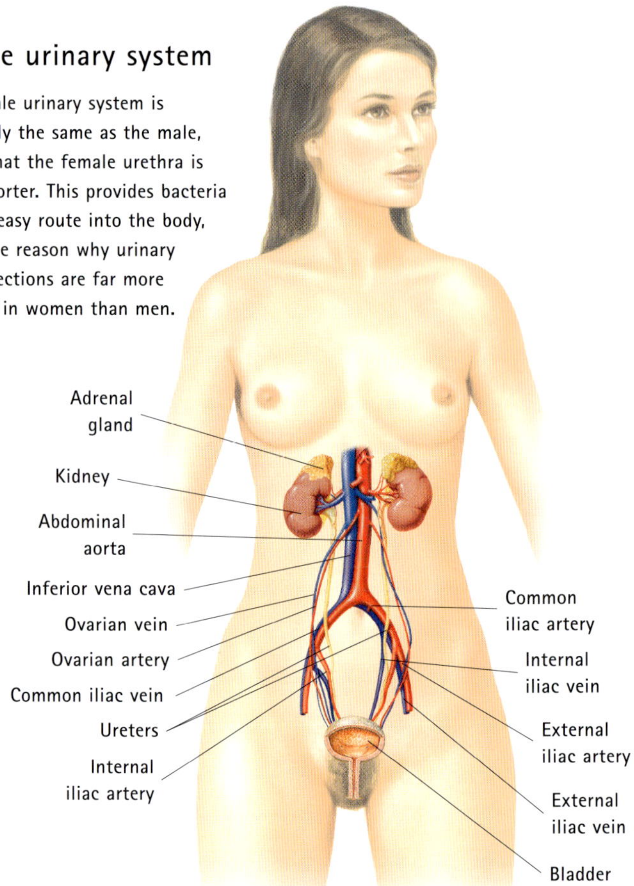

The urinary bladder

Urine is continuously produced by the kidney and stored in the bladder, a muscular sac that resembles an inverted pyramid. The urethra emerges from the neck (apex) of the bladder, which is located just behind the symphysis pubis. The rear surface of the bladder is in front of the rectum and receives the ureters. The openings of the two ureters and the urethra form the three corners of the triangular trigone.

The bladder is behind the pubis when empty; when full, it rises above the symphysis pubis and can be ruptured by a direct blow above the pubis. In males, the bladder rests on the prostate and is anchored to the pubis and to the side walls of the pelvis by ligaments attached to its neck.

The male urethra

The prostate, located below the apex of the bladder, rests on the pelvic floor, which is reinforced by a layer of muscles—the urogenital diaphragm. The urethra leaves the neck of the bladder and passes through the prostate and the membrane formed by the muscle layers before reaching the penis; it is usually divided into prostatic, membranous and penile parts.

The male urethra is the common passage for sperm as well as urine and is closely related to the reproductive system.

The urethra is a muscular tube reinforced by muscle fibers that extend like slings from the bladder around the upper part of the prostatic urethra. This sphincter is under autonomic control. Among the muscles of the urogenital diaphragm is another sphincter, which is under conscious control.

The full bladder sends a message to the reflex center in the sacral part of the spinal cord where it triggers a reflex contraction of the muscle of the bladder and causes the neck of the bladder to relax. This reflex is suppressed until there is an opportunity to relieve the bladder.

FEMALE URINARY SYSTEM

The structure and organization of the urinary system is similar in both sexes from the kidney down to the bladder. However, the position of the ureter within the pelvis is slightly different.

One example is the close proximity of the female ureter to the artery that supplies the uterus. During removal of the uterus (hysterectomy), surgeons have to be careful not to accidentally damage the ureter when they clamp and cut the artery.

Female bladder and urethra

The vagina lies between the rectum and the rear surface of the bladder, with the uterus lying on the top of the bladder. The urethra passes directly through the pelvic floor. As the female urethra is not involved in the reproductive system, it is very short and opens out in front of the entrance of the vagina.

Control of the closure of the urethra is less efficient in the female than in the male. Distortion of the muscle floor of the pelvis, or damage to the sphincters of the urethra by several pregnancies, often results in stress incontinence. Urine may leak with the slight raise in intra-abdominal pressure caused by a simple cough, for example.

The short urethra allows easy examination of the inside of the bladder using a cystoscope, but also provides an easy route of entry for bacteria. Thus infection of the lower portion of the urinary system (called urinary tract infection) is more common in females than in males, especially in young girls who have an even shorter and straighter urethra than adult women.

DIGESTIVE SYSTEM

The process of digestion breaks down foods into small, simple molecules for absorption and use as building blocks for human cells.

The design of the digestive system

The mechanical action of the digestive tract optimizes the chemical actions of enzymes. The entire tract is made up of smooth muscle fibers running in circular and longitudinal directions. In some places, the circular fibers are condensed into a thick ring (sphincter), which contracts to close down the cavity (lumen) within the tract. The circular fibers can contract sequentially along the tract (peristalsis) in order to knead the contents with digestive enzymes or to squeeze it along.

Digestive enzymes split the three main groups of food into their components: carbohydrates into glucose, proteins into amino acids, and lipids into glycerol and fatty acids. There are over two dozen amino acids, most of which are common to all animal species.

SEE ALSO *Amino acids on page 18; Lipids on page 19*

The digestive tract and glands

Food is bitten into mouthfuls by the incisor teeth, shredded by the canines, and ground by the premolars and molars. It is pushed around in the mouth and held between the molars by the tongue and the buccinator muscles in the cheeks. The top surface of the tongue is rough due to the presence of papillae which carry taste receptors.

Saliva in the mouth comes from three pairs of salivary glands. The largest, the parotid glands, are located just below the ears. The submandibular and sublingual glands are under the floor of the mouth. Saliva wets the food to facilitate chewing and swallowing. It also contains the enzyme amylase, which digests carbohydrates, and antibodies to help resist infection.

At the beginning of swallowing, food is rolled into a round mass (bolus) which is pushed to the back of the mouth by the tongue. Muscles of the pharynx then push the food bolus down the esophagus.

The esophagus runs vertically down behind the windpipe (trachea), passing through the diaphragm to end in the stomach. The opening of the stomach into the small intestine (the duodenum) is controlled by a thick bundle of circular muscle fibers, the pylorus.

Food is retained in the stomach where proteins are digested by hydrochloric acid and pepsin secreted by gastric glands. At intervals, the pylorus relaxes to empty the gastric content, now a half-digested mix known as chyme, into the duodenum.

The small intestine has three parts: the duodenum, jejunum and ileum. The lining of the duodenum and jejunum has many transverse folds called plicae circulares (circular folds). On each fold are numerous tiny projections called intestinal villi, giving the lining a velvety appearance. The plicae and villi greatly increase the surface area available for the absorption of nutrients. Glands at the root of the villi and in the submucosal layer secrete intestinal juice that protects the lining against acidity and digestive enzymes.

Most of the digestive process occurs in the duodenum due to the action of pancreatic enzymes. The pancreas is an elongated gland lying behind the stomach. Pancreatic secretions are collected by the main pancreatic duct which, together with the bile duct, enters the duodenum at a common opening, the duodenal ampulla. The pancreatic juice contains inactive pro-enzymes that are converted by the duodenal hormone (enterokinase) into active enzymes that digest carbohydrates, proteins, nucleic acids, and lipids.

The gallbladder is under the liver and connected to the bile duct. One of the functions of the liver is bile production. The gallbladder concentrates and stores this bile, which is discharged into the duodenum when chyme containing fat enters from the stomach. Bile emulsifies the fat to facilitate digestion by the pancreatic enzyme lipase. As the chyme reaches the jejunum, the products of digestion are ready for absorption. Small molecules enter the blood capillaries of the villi.

Larger molecules enter small lymphatic channels in the villi (lacteals), and flow along the mesenteric lymph vessels into the general circulation. Vitamin B_{12} is absorbed in the last portion of the ileum in the presence of intrinsic factor, a protein secreted by the stomach.

The ileum joins the large intestine, which consists of the cecum, colon, and rectum. The cecum—the blind-ended pouch of the colon—and the ileum meet at the ileocecal valve. The appendix is located near this valve. The contents become feces as they move along the parts of the large intestine: the ascending, transverse, descending, and sigmoid colon. The colon absorbs water and bile salts, and contains

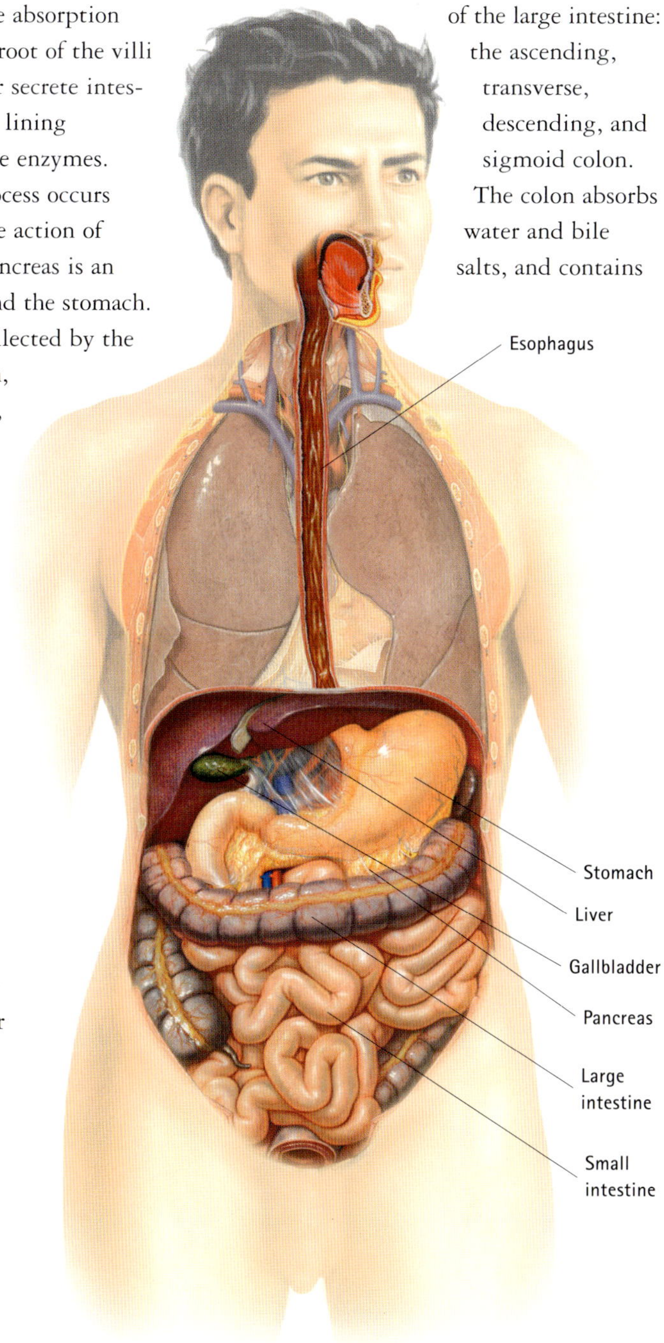

bacteria that synthesize some vitamins such as vitamin K and biotin. The epithelium (the lining) of the colon here contains numerous goblet cells and mucus-secreting glands to facilitate the movement of the feces.

The last parts of the large intestine are the rectum and anal canal, situated just in front of the sacrum. The anus is closed by an involuntary and a voluntary sphincter, which open up during defecation. When the rectum is distended by feces, the defecation reflex is initiated if the voluntary anal sphincter is intentionally relaxed. Veins in the lining of the rectum and anal canal facilitates the passage of feces.

SEE ALSO *Epiglottis on page 202; Esophagus on page 249; Mouth on page 178; Pharynx on page 201; The abdominal cavity on page 254*

Control of the activities of the digestive system

The mechanical and secretory activities of the digestive tract and its glands are controlled by neural and hormonal mechanisms. The autonomic nervous system sets the general level of activity of the digestive system; the parasympathetic system increases the contraction of the gastrointestinal wall and the secretion of the digestive glands. There are networks (plexuses) of autonomic nerves in the intestinal wall which coordinate muscle contraction, and the secretion of gastric acid may be initiated by the central nervous system via the vagus nerve.

A system of "local" hormones ensures that secretions begin and end at the right time. When the stomach is distended by food, its lower part secretes more gastrin, which increases the secretion of acid by gastric glands. The arrival of chyme in the duodenum initiates the release of hormones, which inhibit gastric secretion and stimulate the secretion of bile and pancreatic enzymes.

SEE ALSO *Autonomic nervous system on page 75; Nervous system on page 64; The abdominal cavity on page 254*

Peristalsis

Peristalsis consists of wave-like contractions in the muscular walls of the esophagus, stomach, intestines, ureters and fallopian tubes that propel the contents of the tube along. The walls of many tubular structures in the body are composed of smooth muscle whose contractions, like those of the heart muscle, are involuntary, that is, controlled by the autonomic nervous system. When food is swallowed, muscular contractions in the esophagus push the food downward into the stomach. There, peristalsis in the stomach mixes the chewed food with gastric juices and moves it through the pyloric sphincter to the small intestine, where peristalsis continues to move gut contents into the large intestine.

SEE ALSO *Autonomic nervous system on page 75; Esophagus on page 249; Muscular system on page 48; The abdominal cavity on page 254*

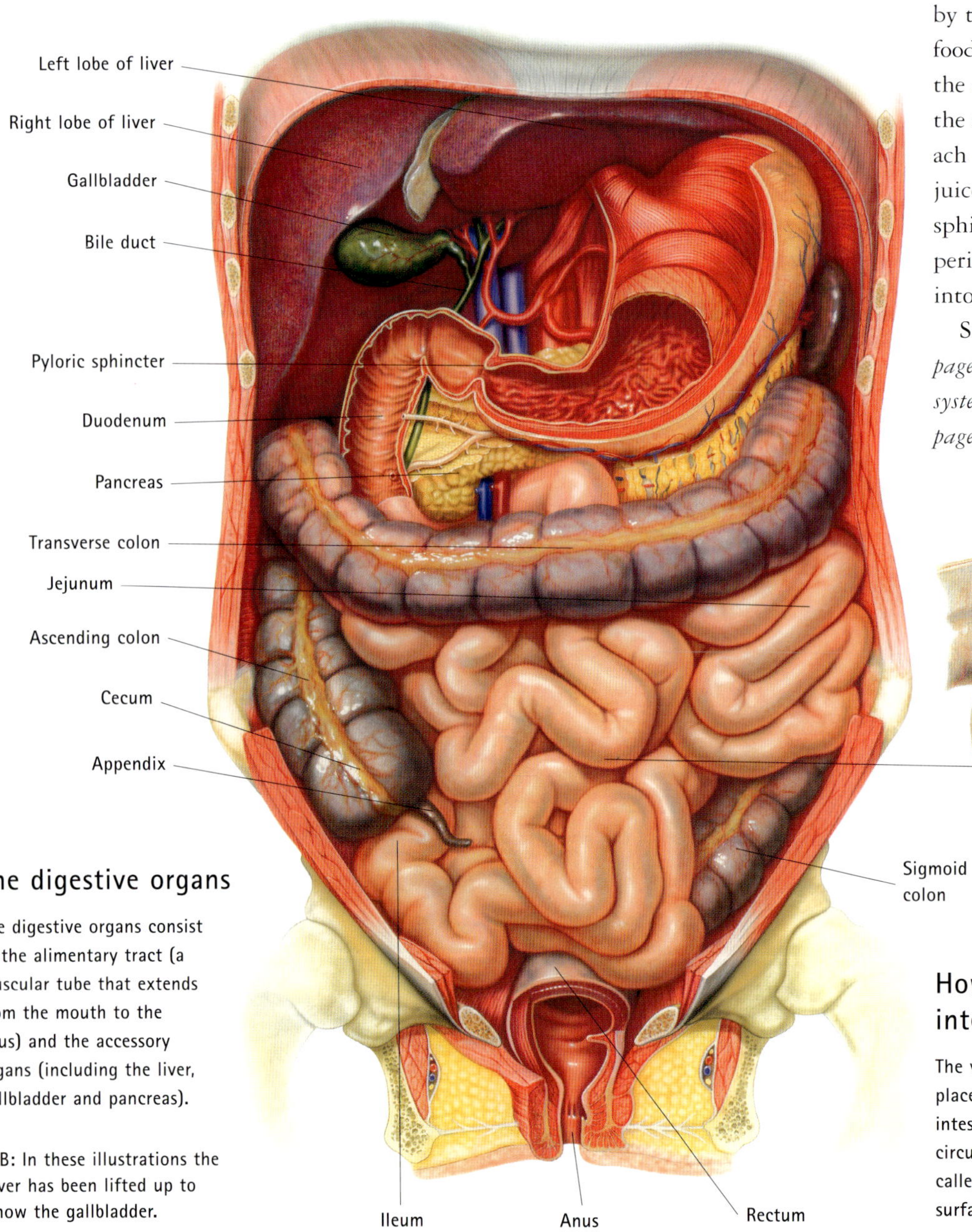

The digestive organs

The digestive organs consist of the alimentary tract (a muscular tube that extends from the mouth to the anus) and the accessory organs (including the liver, gallbladder and pancreas).

NB: In these illustrations the liver has been lifted up to show the gallbladder.

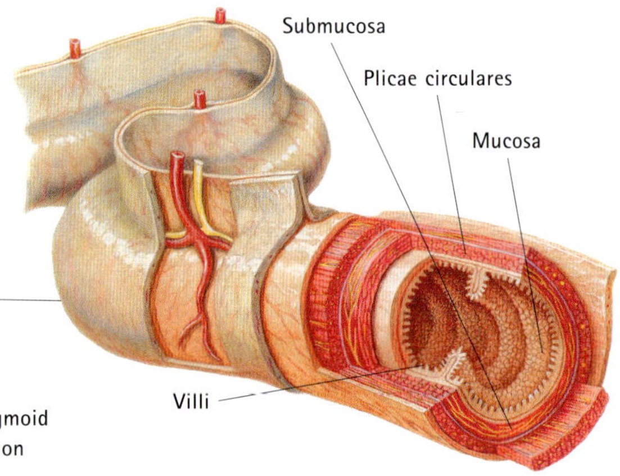

How the small intestines function

The vast majority of digestion and absorption takes place in the small intestine. The lining of the small intestine has many small transverse folds (plicae circulares) and numerous tiny finger-like projections called villi. The plicae and villi greatly increase the surface area available for the absorption of nutrients.

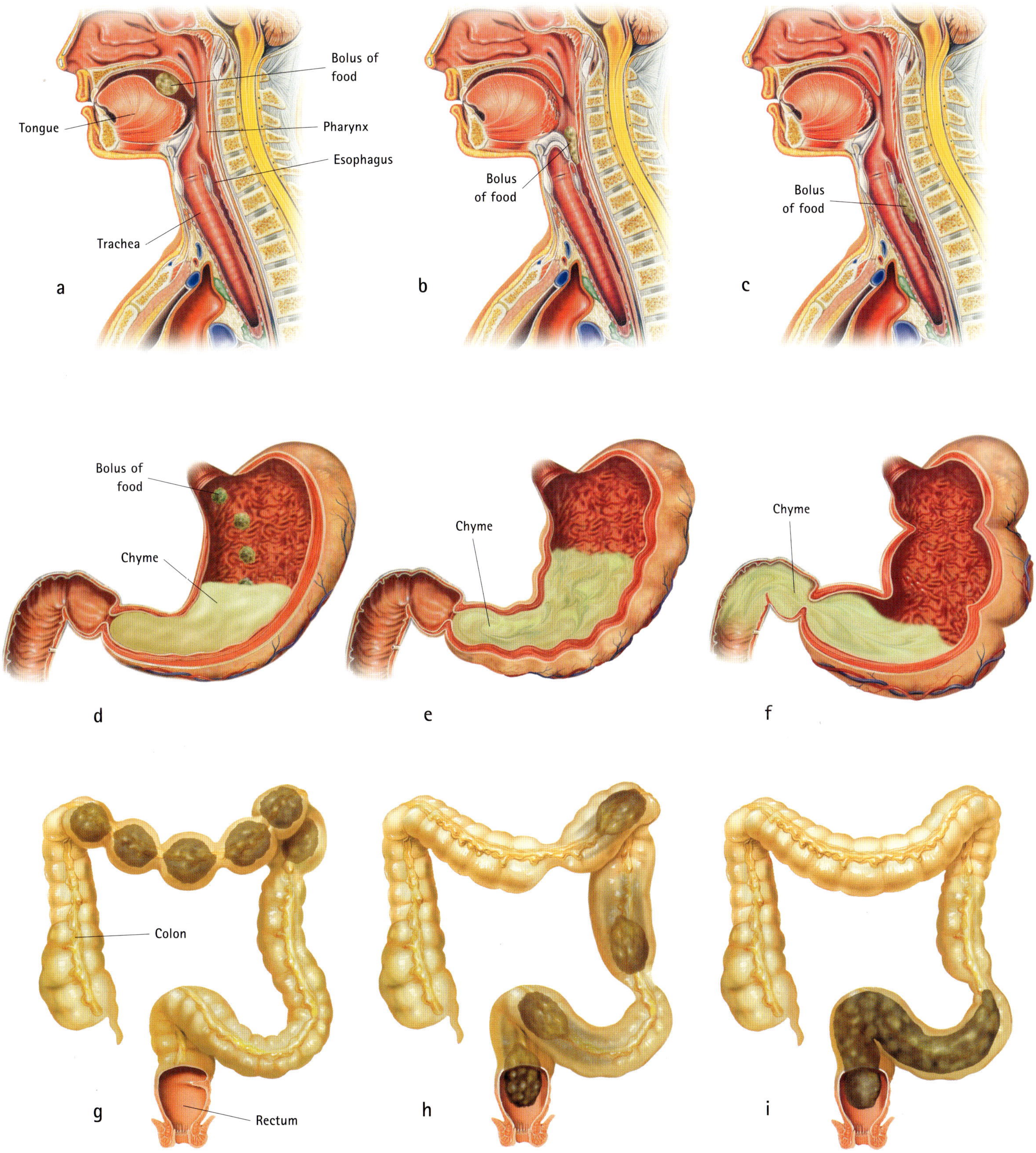

Peristalsis

The initial act of swallowing is under voluntary control. When swallowing begins, the tongue propels the bolus of food into the pharynx (a). Coordinated voluntary movement of the pharyngeal muscles permits entry of food into the upper esophagus (b). Once the bolus of food enters the esophagus, a peristaltic wave begins that travels toward the stomach, propelling the food before it (c). Food enters the stomach (d), triggering the release of gastric juices, which break down the contents (e), before gradually releasing the contents into the duodenum (f). In the colon, water and bile salts are absorbed before peristaltic contractions push the waste matter along to the rectum, where it is periodically expelled (g), (h) and (i).

REPRODUCTIVE SYSTEM

The male and female reproductive systems reach maturity during puberty.

Female Reproductive System

The reproductive system of the female includes two ovaries that produce ova (eggs) and female hormones, two fallopian (uterine) tubes that convey eggs to the uterus, the uterus itself, and the vagina, which is connected to the external genitalia.

SEE ALSO *Childbirth on page 408; Contraception on page 448; Disorders of pregnancy on page 407; Female procedures on page 462; Female reproductive organs on page 303; Fertility on page 396; Fetal development on page 398; Hormone replacement therapy on page 450; Hormones on page 112; Human sexuality on page 107; Infertility on page 397; Pregnancy on page 404; Puberty on page 421; Menopause on page 424; Menstruation on page 422; Sexual behavior on page 107; Sterilization on page 461*

Ovaries

The ovaries are almond-shaped organs in the side wall of the pelvis, situated just below the division of the common iliac artery. They are slightly flattened sideways, with a vertical axis measuring about 1½ inches (3.5 centimeters). All the ovary's blood vessels and lymphatics enter its upper part. Its lower part is bound to the corner of the uterus by a band of connective tissue, the ligament of the ovary.

The ovary is enveloped by peritoneum, the membrane that surrounds all organs in the abdomen. Inside its covering epithelium is a shell of cortex which encloses the medulla at its core. The bulk of the ovary is the supporting structure called the stroma. The cortex contains ova at different stages of development.

Unlike the testis, which produces sperm continuously, the ovary is endowed in embryonic life with a fixed number of ova. By birth, each ovary has about a million ova; by puberty, the number has been reduced to about a quarter of a million.

The ova begin as primordial oocytes, surrounded by a layer of flat cells called granulosa cells. With the onset of puberty, hormones from the pituitary gland stimulate the development of more mature primary and secondary follicles. The granulosa cells multiply and form the multi-layered theca interna that secretes estrogens; the surrounding stromal cells form the theca externa. A split appears in the theca interna and expands to form a fluid-filled cavity that pushes the oocyte to one side; the follicle is now a tertiary or Graafian follicle, visible as a small blister on the surface of the ovary.

Fallopian tubes

The fallopian tubes, or uterine tubes, are small tubes that begin as funnel-shaped passages with finger-like projections called fimbriae. The ovum released from a burst Graafian follicle is "captured" by one of the fimbriae and moved along to the uterine cavity by the action of cilia of the epithelium and the contraction of the tube.

Fertilization happens in the outer third of the fallopian tube if a sperm penetrates the ovum and its nuclear material fuses with that of the ovum. If the ovum is fertilized it will implant in the uterus and begin to develop into an embryo.

Uterus

The uterus looks like a slightly flattened, upside-down pear, with a slight constriction dividing it into two parts. The upper two-thirds is the body, the lower third is the cervix. The body rests on the bladder and has a very thick wall of smooth muscle. The uterus is flattened from the front to the back. The uterine tubes open into the two top angles and the angle below opens into the lumen of the cervix. The cervix is a cylindrical muscular tube that protrudes into the vagina. The lumen of the cervix is a spindle-shaped cervical canal.

The epithelium of the cervical canal secretes mucus that becomes thick, scanty and more acid after ovulation, a characteristic that can be used to determine the time of ovulation. The secretion is also thick during pregnancy, and forms the "mucus plug" which closes the cervical canal. The inner lining of the uterus, the endometrium, is highly vascular and contains numerous tubular glands. It is shed at the end of each menstrual cycle.

When the ovum is fertilized, the endometrium persists, and its glands secrete mucus rich in glycogen in preparation for implantation of the fertilized ovum. The part of the endometrium surrounding the developing embryo develops into the placenta.

During pregnancy, the uterus is distended to accommodate the developing fetus, extending beyond the umbilicus at full term. The muscular wall expels the fetus during labor and compresses the blood vessels to stop bleeding after expulsion of the placenta. The uterus is fixed in place by connective tissue from the pelvic wall together with vessels and nerves destined for the vagina and uterus.

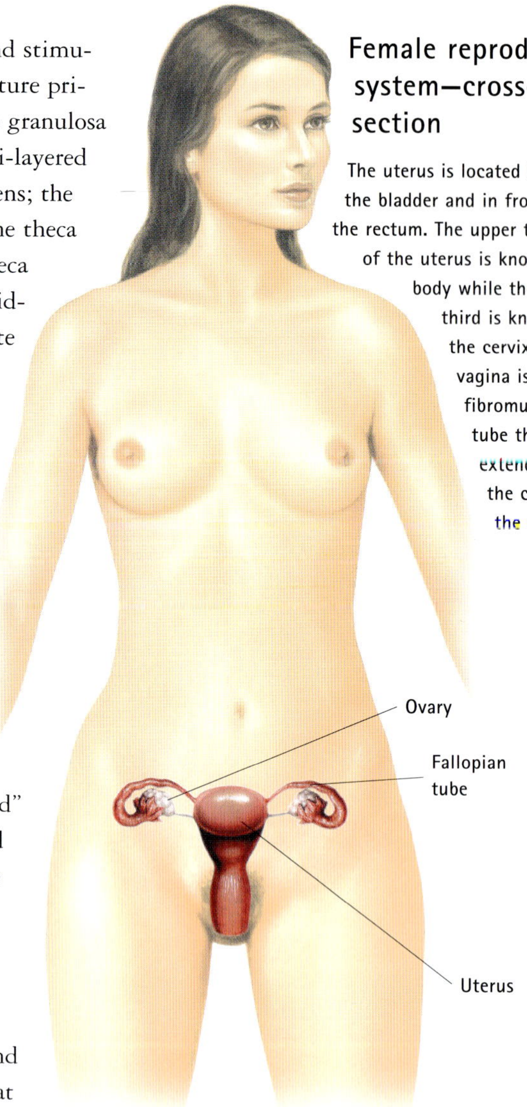

Female reproductive system—cross-section

The uterus is located behind the bladder and in front of the rectum. The upper two-thirds of the uterus is known as the body while the lower third is known as the cervix. The vagina is a long fibromuscular tube that extends from the cervix to the vulva.

Vagina

The vagina is a fibromuscular tube that runs from the cervix to the vestibule of the vulva. Normally, its front and back walls lie close together, but it is capable of much distension and elongation.

External genitalia

The external genitalia can also be called the pudendum or vulva. The lower end of the vagina continues into the labia minora. At their junction is a thin fibrous membrane called the hymen which, even when intact, allows the passage of menstrual blood. The labia minora may be pigmented and are devoid of hair. The space between them, the vestibule, receives the openings of the greater vestibular glands (Bartholin's glands).

The erectile bodies are comparable to those in the male genitalia. The equivalent of the corpus spongiosum is split into two masses flanking the vaginal opening. The corpora cavernosa occupy the same spot as in the male. Corresponding to the penis is the clitoris, which also has a prepuce, a small hood formed by the labia minora.

The skin outside the vaginal opening is thickly padded with subcutaneous fat and forms the labia majora. They join and continue over the symphysis as the mons pubis (mons veneris), a small mound raised by a thick underlying pad of fat. The mons pubis and labia majora are covered by hair, the distribution of which is controlled by female hormones.

Menstrual cycle

After puberty, the endocrine system has reached maturity, and the menstrual cycle becomes regular.

The menstrual cycle begins from the first day of bleeding. Under the influence of follicle stimulating hormone (FSH) from the pituitary gland, one or more follicles in the ovary mature, releasing estrogen into the blood, and stimulating the proliferation of the endometrium and its glands. At mid-cycle, about day 14, the surge of luteinizing hormone (LH) from the pituitary gland triggers ovulation and initiates the development of the corpus luteum, which secretes progesterone, maintaining the proliferation of the endometrium in anticipation of implantation of the fertilized ovum. Without fertilization, LH and FSH secretion ceases, the corpus luteum breaks down, and the levels of estrogen and progesterone drop. The endometrium then degenerates, dies, and sloughs off, causing menstrual bleeding which usually lasts about four days.

The pituitary gland, the master gland controlling the menstrual cycle, is under the influence of the hypothalamus and the limbic system, the part of the brain that controls emotions. Menstrual cycles can be disturbed in times of emotional distress.

Male Reproductive System

The reproductive system of the male includes two testes which produce spermatozoa (sperm) and male hormones, a system of ducts which convey sperm, glands which contribute secretions to semen, and the external genitalia, the scrotum and penis.

SEE ALSO *Circumcision on page 450; Contraception on page 448; Fetal development on page 398; Hormones on page 112; Human sexuality on page 107; Infertility on page 397; Male reproductive organs on page 297; Puberty on page 421; Sexual behavior on page 107; Sterilization on page 461; Urinary organs on page 286; Vasectomy on page 461*

Testes

The testes are two oval organs contained in the scrotum, the right one usually higher than the left by nearly ½ inch (about 1 centimeter). At the back, the testis is capped by the epididymis. The epididymis has a large head at the top, tapers toward its lower end, then makes a sharp turn back to become the ductus deferens (or vas deferens, hence the term vasectomy).

The testes are separated by a central partition. Surrounding each testis is a double-layered membrane, the tunica vaginalis. Each testis has a tough inelastic fibrous wall called the tunica albuginea which sends partitions inward to divide it into about 300 lobules. Each lobule contains coiled seminiferous tubules that produce sperm, and converge into a network which sends about 20 small ducts through the tunica albuginea into the epididymis. The ducts become larger and convoluted, forming the head of the epididymis, and gradually fuse into the ductus deferens, which runs back up toward the inguinal canal in the groin.

The tunica vaginalis is an extension of the peritoneal cavity in the embryo, and later separates. The germ cells in the seminiferous tubules—the spermatogonia—undergo division (meiosis) to become spermatids. Spermatids mature into sperm that have a head and a long tail. The head is capped by an acrosome which releases

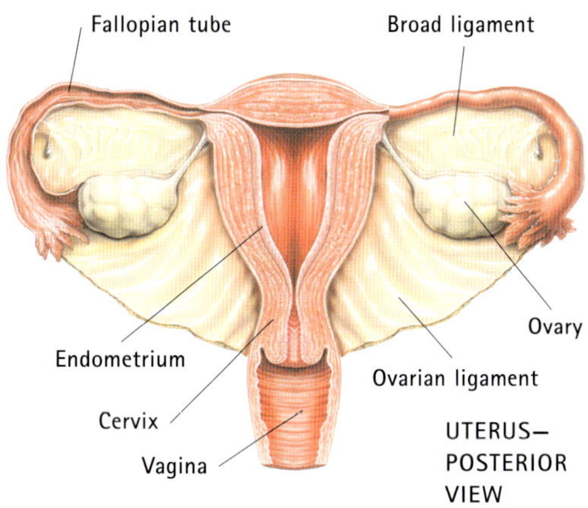

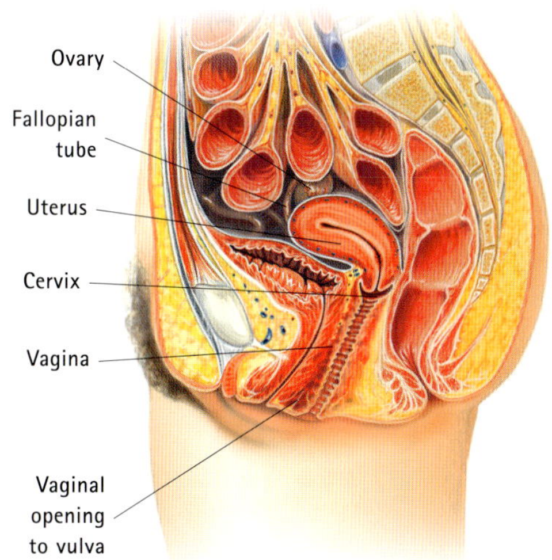

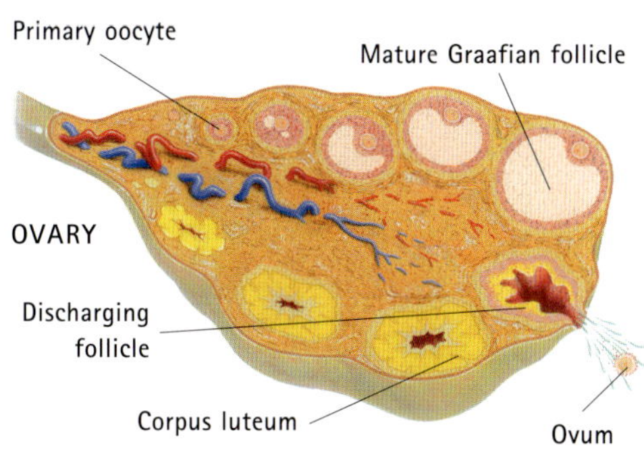

Female reproduction

During a menstrual cycle the ovary produces an oocyte (ovum) which enters the fallopian tube and travels toward the uterus. Fertilization (if it occurs) normally takes place in the outer third of the fallopian tube, and the fertilized ovum then implants into the inner lining of the uterus and develops into an embryo.

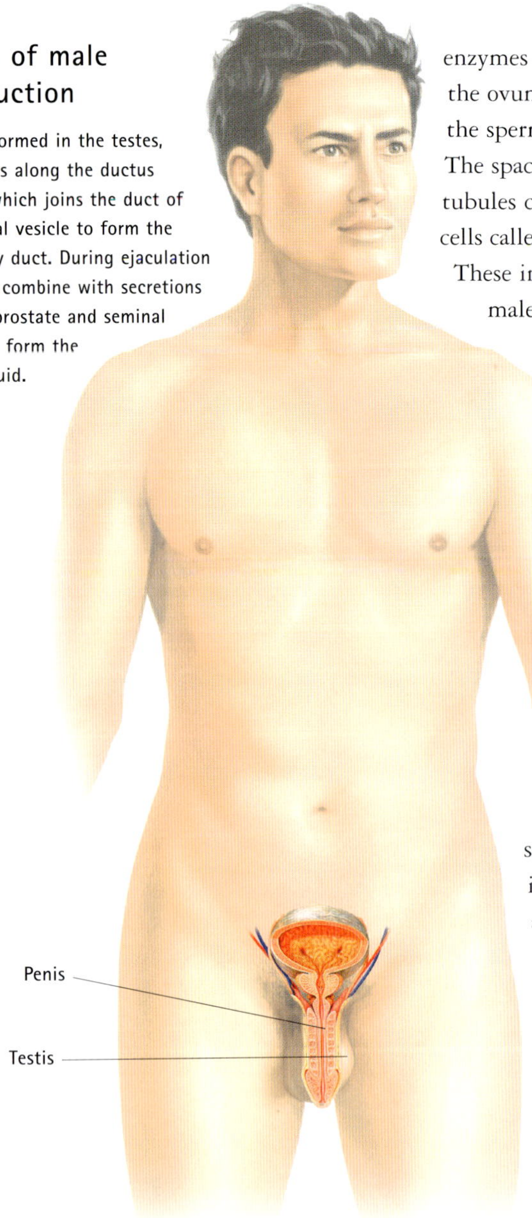

Organs of male reproduction

Sperm is formed in the testes, then passes along the ductus deferens which joins the duct of the seminal vesicle to form the ejaculatory duct. During ejaculation the sperm combine with secretions from the prostate and seminal vesicles to form the seminal fluid.

enzymes to help the sperm penetrate the ovum (egg). The tail propels the sperm like the tail of a tadpole. The spaces between the seminiferous tubules contain clumps of endocrine cells called interstitial cell or Leydig cells. These interstitial cells synthesize the male hormone testosterone, responsible for the development of sexual characteristics in the adolescent male and the functioning of the reproductive system.

Ductus deferens

The ductus deferens (or vas deferens), is a thick-walled muscular tube, which can be felt above the testis through the loose part of the scrotum. The ductus deferens is surrounded by the testicular artery, lymphatic vessels, autonomic nerve fibers and a plexus of veins called the pampiniform plexus. All these structures, enveloped by layers of connective tissue and muscle fibers, form the spermatic cord.

The spermatic cord runs upward to the level of the pubic tubercle on the pubic bone, passes through the inguinal canal, then turns sharply to enter the pelvic cavity. The ductus deferens then heads toward the back of the prostate where it expands into an ampulla, and joins the duct of the seminal vesicle to form the ejaculatory duct.

The muscle fibers in the spermatic cord (cremasteric muscle) pull the testes up in cold weather to maintain an optimal temperature for sperm formation. The pampiniform plexus fuses into a testicular vein that empties into the inferior vena cava on the right side, and the renal vein on the left side.

Seminal vesicle

The seminal vesicle is a single tube coiled upon itself into a pyramidal organ. It lies on the outer side of the ductus deferens. The fusion of its duct with the ductus deferens forms the ejaculatory duct, which penetrates the prostate gland to open into the prostatic urethra. Secretion from the seminal vesicle makes up about 60 percent of semen.

Prostate

The prostate is shaped like an inverted pyramid and lies under the bladder, with the apex pointing downward. The urethra emerging from the neck of the bladder runs vertically through the prostate, leaving it just in front of its apex.

The prostate contains two major groups of glands. The central zone surrounding the urethra contains periurethral glands and is the site of benign prostatic enlargement. The peripheral zone containing the main glands is usually the site where prostate cancer develops. All the glands open into the prostatiurethra and secrete the enzyme acid phosphatase, fibrinolysin, as well as some other proteins. Prostatic secretion makes up about 25 percent of semen.

After leaving the prostate, the urethra runs through the muscles of the urogenital diaphragm, and enters the penis. Here it receives the ducts of the paired bulbourethral glands. The secretion of these glands precedes the emission of semen and perhaps has a lubricating function.

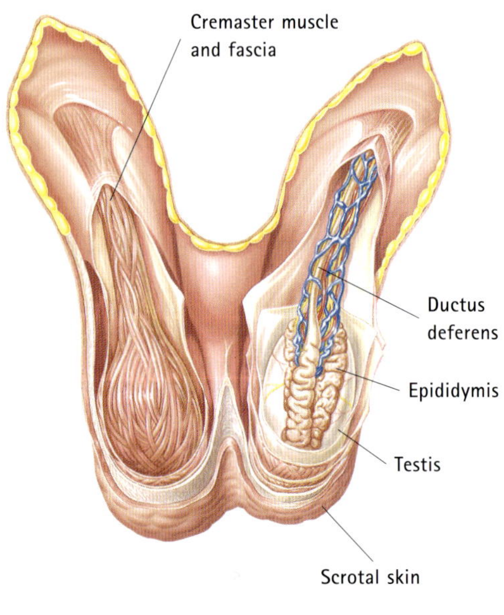

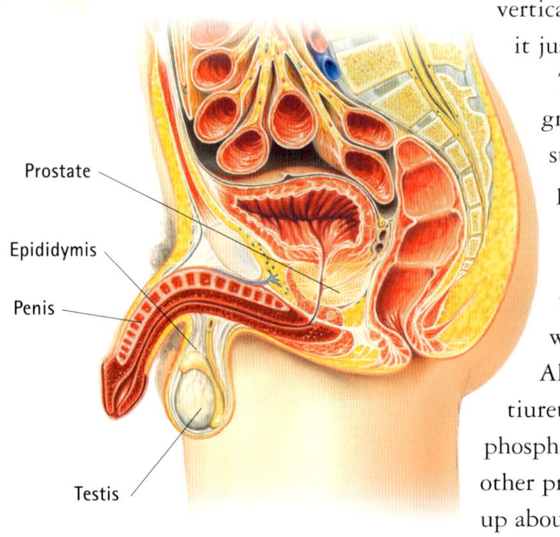

Male reproductive system—cross-section

The male reproductive system consists of the testes, the ductus deferens, the seminal vesicles, the prostate and the penis.

Penis

The penis is the organ of copulation and is made up of three cylinders of erectile tissue. The midline cylinder is known as the corpus spongiosum (the spongy body) which contains the urethra. Its front end flares out into the bulbous glans penis, which has in its center the opening of the urethra. On each side of the corpus spongiosum runs a corpus cavernosum (the cavernous body). All three cylinders are enveloped in a thick cylinder of tough connective tissue. The glans is covered by a hood of loose skin, which is called the prepuce, or foreskin. The foreskin can be retracted to the base of the glans.

The three cylinders are structured like a sponge, the interconnecting spaces containing blood. When the blood flows through them, they are flaccid. When the outflow of blood is prevented by closure of the veins, these cylinders become engorged with blood, thereby causing an erection. The connective tissue around the cylinders helps to build up pressure.

The mechanism of erection is controlled by the parasympathetic component of the autonomic nervous system and is the result of a very intricate interplay of many neurotransmitters.

Erection facilitates the entry of the penis into the vagina. When stimulation during intercourse reaches a threshold, the sympathetic system triggers a powerful emission of semen (this is known as ejaculation). The ejaculate, although it measures only about a teaspoon (5 milliliters), contains a few million sperm.

Fertility depends on the number of normal, healthy, motile sperm that is present in the ejaculate.

Human sexuality

A person's sexuality results from the combination of sexual attributes, behavior and tendencies. Sexual feelings are present from birth, and continue right through life. Sexual desire is influenced by many factors and its intensity may fluctuate greatly over a lifetime.

For women, sexual desire may be influenced by changes in hormone levels at different life stages (such as menopause), changes in roles and responsibilities, whether child-rearing or work outside the home, and economic and relationship factors. Men experience less fluctuation in hormone levels, but sexual desire and arousal is affected by many factors including economic problems, stress, work pressure and family situation.

Sexual feelings continue to be present as people age. Although sexual arousal may occur less often and men may find that erections are not as strong or as frequent, and women may suffer problems such as vaginal dryness, the need for, and the rewards from, an active sex life continues for many people to the end of their lives. Passing the menopause gives many women a greater sense of sexual freedom.

SEE ALSO *Hormones on page 112; Puberty on page 421; Menopause on page 424*

Sexual behavior

Sexual behavior is any activity that leads to sexual arousal; it may be solo, between two people or in a group. The sexual appetite begins in the brain in the hypothalamus which, together with the adrenal glands, is responsible for stimulating the gonads to produce estrogen and testosterone. These hormones combine with other factors to initiate sexual activity. In many societies, romantic love is associated with the sexual act.

Human sexual behavior can be divided into heterosexuality (a sexual relationship between a man and woman), homosexuality (a sexual relationship between people of the same sex) and bisexuality (sexual relationships with both sexes).

Social taboos have made the study of sexual behavior in industrialized countries difficult. Much of the recent information available comes from the Institute for Sex Research (also known as the Kinsey Institute) and from research in Sweden.

SEE ALSO *Contraception on page 448; Female reproductive organs on page 303; Hormone replacement therapy on page 450; Hormones on page 112; Hypothalamus on page 136; Pituitary on page 137; Puberty on page 421; Menopause on page 424; Sexually transmitted diseases on page 365*

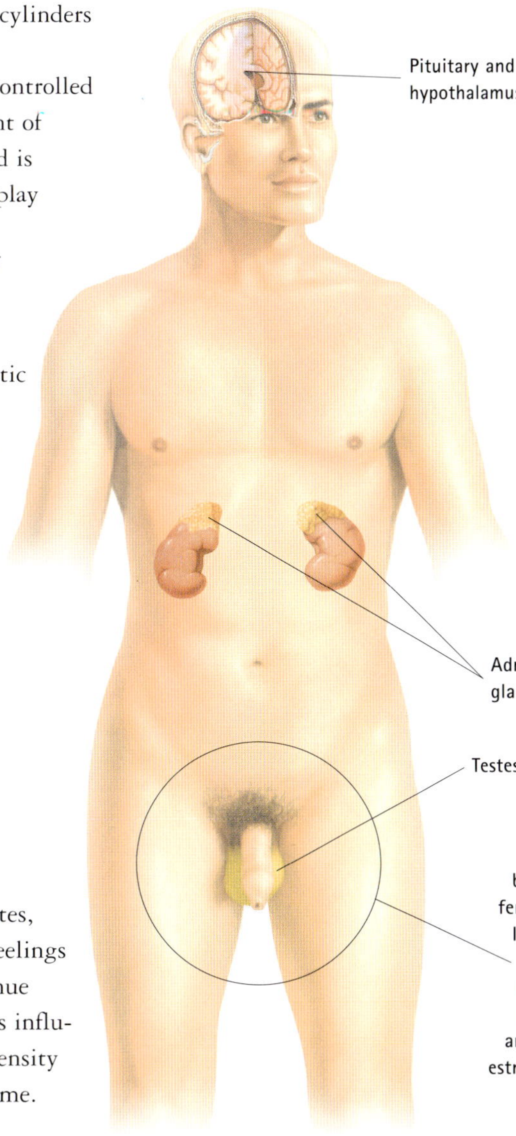

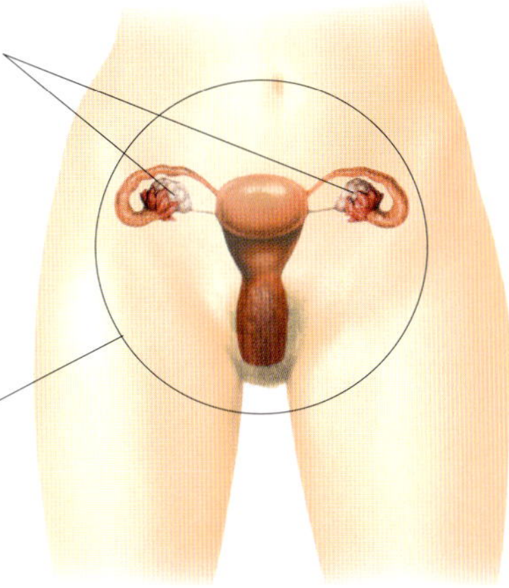

Sex hormones and behavior

The hypothalamus and pituitary gland in the brain, together with the adrenal glands, stimulate the gonads—testes in men and ovaries in women—to produce testosterone and estrogen. The release of these hormones modulates sexual appetite and activity.

Erection

Erection is a state in which the erectile tissue of an organ or body part has become distended and rigid by the accumulation of blood. Though the nipple and clitoris contain erectile tissue, the term usually refers to the distension of the penis.

The penis is comprised primarily of two cylinders of sponge-like vascular tissue. (A third cylinder contains the urethra, a tube that carries urine and the ejaculate.) After physical or psychological sexual stimulation, the cylinders become engorged with blood and the penis becomes erect and hard. This enables the male to insert the erect penis into the female's vagina during sexual intercourse. The blood is unable to drain out through the veins because they are temporarily closed by pressure from arterial blood in the spongy tissue of the penis.

Masturbation

Masturbation is the erotic stimulation of one's own, or a sexual partner's, genital organs for pleasure. It is a natural expression of human sexuality that begins in childhood and which can be an important part of self-discovery and satisfaction in a sexual relationship with a partner. Once the subject of taboos and superstitions, masturbation was said to cause numerous ill effects, ranging from hairy palms to insanity, none of which had any foundation in fact. Most people will masturbate at some time, some more often than others.

While children need to know the social etiquette of not masturbating in public, preventing them from masturbating can affect exploration of their own sexuality and lead to guilt and associated problems.

Sexual intercourse

Sexual intercourse, or coitus, is the union between a male and female whereby the woman is penetrated by the man's penis, usually resulting in ejaculation. It may also be defined more widely as any sexual contact between two individuals involving stimulation of the genitals of at least one of them. Using the first definition, sexual intercourse is an integral part of the reproductive process and basic to the existence of the human species (forgetting the development of technologies that allow reproduction without sexual intercourse).

Normal sexual intercourse begins with foreplay, when a couple arouse each other sexually by caressing, kissing and stimulation, preparing their bodies for sexual intercourse. Arousal increases blood supply into the penis, causing it to extend and become erect. It also increases blood flow to the woman's vulva and vagina, which becomes coated with a lubricating mucus; the clitoris becomes erect and the nipples more sensitive.

When the woman is ready for penetration the penis enters the vagina. The man, and often the woman, will usually experience orgasm during intercourse, not necessarily at the same time. The man usually experiences orgasm as he ejaculates. After ejaculation the penis becomes limp. After intercourse the bodies of both man and woman will return to normal, although it usually takes the woman's body longer.

Anal intercourse, in which the penis is inserted via the anus into the rectum of the other person, is occasionally used by heterosexual couples, but more frequently in homosexual relationships. As the anus is not designed to accommodate a thrusting penis, damage is more likely than in vaginal sex. There is also the possibility of infectious bacteria adhering to the penis after anal intercourse, thus hygiene is important. As autoimmune deficiency syndrome (AIDS) can be transmitted by unprotected anal intercourse (among other ways), a condom should always be used.

Oral sex is the term used to describe mouth and genital contact. It has been part of sexual activity for centuries, being depicted in many ancient Indian and carvings and also in the *Kama Sutra*, a famous Hindu love and sex manual from the second century BC.

Contraceptives can be used to prevent unwanted pregnancy. It is important to practice safe sex to protect both parties against sexually transmitted diseases.

Erogenous zones

An erogenous zone is any part of the body that, when stimulated, can produce sexual arousal or pleasure.

Erogenous zones in men are the external sex organs: the penis, particularly the glans, which is the sensitive head above the shaft; the skin of the scrotum and inner thighs; the buttocks and external skin around the anus; and in some men, the nipples. In women the clitoris, vulva, nipples, areolae, breasts and inner thighs are the main areas sexually responsive to stimulation.

Every part of the body can be sexually aware, and respond to stimulation, as our experience of sexual pleasure is largely mental. When injury or disablement affects the touch-receptive nerve endings in one

SEXUALITY

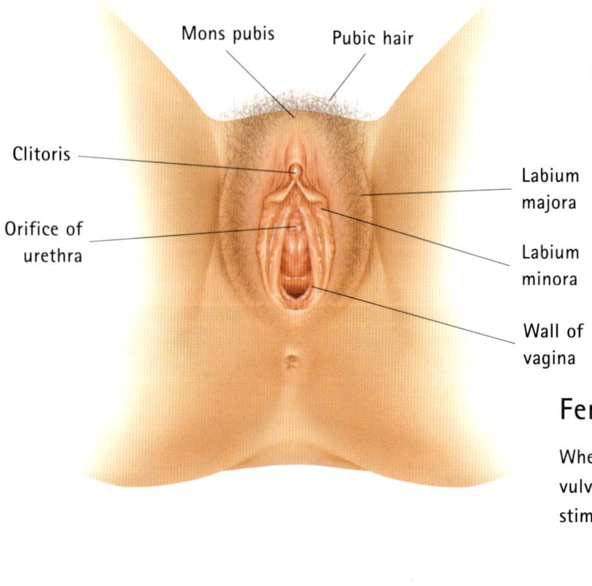

Male sexual arousal

Sexual stimulation causes the blood vessels of the penis to dilate. The increased blood supply makes the ordinarily flaccid penis swollen and rigid.

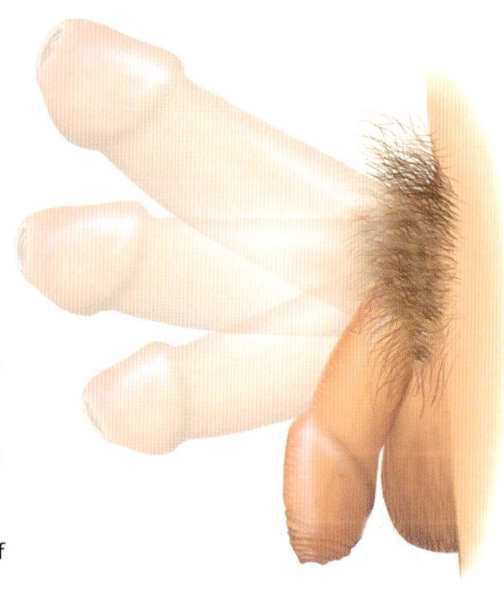

Female sexual arousal

When sexually aroused, blood flows to the woman's vulva and the clitoris becomes firm. Continued stimulation of the clitoris can lead to orgasm.

area, new areas can replace them, and relocate the traditional erogenous zones. The hypothalamus, which regulates the autonomic nervous system, receives nerve impulses generated by manual stimulation of the nipples and genitalia. It responds to sexual desire by releasing hormones that increase heart rate, blood circulation and respiration rates and dilate the pupils of the eyes, bringing about a state of sexual arousal or readiness for sexual intercourse.

Sexual arousal is a complex emotional and physical response to stimulation of the senses—sight, smell, hearing and touch. What a person finds stimulating will be a combination of individual preferences, personal history, memories and influences from culture and conditioning. A person's response at any one time can be influenced by their physical and emotional condition.

Under stress, such as financial pressures or relationship problems, sexual drive and response to normal stimulus may be very low. Tiredness, physical disorders and pain, mental attitudes and conditioning can also reduce or eliminate normal sexual arousal. The person may have no interest in sexual contact, although they may still need the physical and emotional reassurance given by the loving touch of a partner. Under these conditions, stimulation may not produce sexual arousal.

Sexual dysfunction

Not easy to define, a sexual dysfunction could be said to be any condition which results in dissatisfaction with performance, sensation or satisfaction during any part of a sexual interaction. Sexual dysfunctions include the following.

Anorgasmia is the inability to experience orgasm, either during sexual intercourse or sexual activity with a partner or during masturbation; this is mostly a female problem, although men may also suffer. It was originally termed frigidity. Sex therapists have discovered that the problem is mostly due to a lack of sexual knowledge or to religious or social prohibitions that have prevented the development of a sexual awareness through masturbation.

Dyspareunia or painful or difficult sexual intercourse in women is most commonly caused by insufficient arousal and lubrication of the vagina prior to penetration. Spending time in foreplay and ensuring that the woman is aroused will usually resolve the problem. Inflammation or irritation of the vulva can also cause dyspareunia and intercourse is best avoided until the condition is resolved. Pain during deep penile thrusting can be due to endometriosis; cervical infections or other infections surrounding the uterus, tubes or pelvic organs; or abdominal surgery. Vaginal dryness, often a problem for women after menopause, can usually be solved by hormone replacement therapy.

Impotence or problems with erection is one of the most common worries for men. Where impotence is due to lack of libido, the solution is often the same as for women who suffer anorgasmia. Damage to the blood vessels to the penis may cause impotence; diabetes, heart problems, smoking, trauma and radiation may also cause impotence, as may the drugs used to treat these conditions.

Libido problems include the absence, temporary loss or reduction of libido and can cause sexual difficulties. Natural differences in libido can create tensions in relationships, despite romantic love. Stress, fatigue and negative feelings about a partner can cause of temporary loss of libido which will return when, and if, the problems are solved. (As with anorgasmia, this was previously called frigidity in women.) Some medications can also be responsible for reduced libido.

Male "menopause" is considered by some medical authorities to be an emotionally triggered mid-life crisis, while others believe it is a physical condition resulting from changing hormone levels, in particular of testosterone.

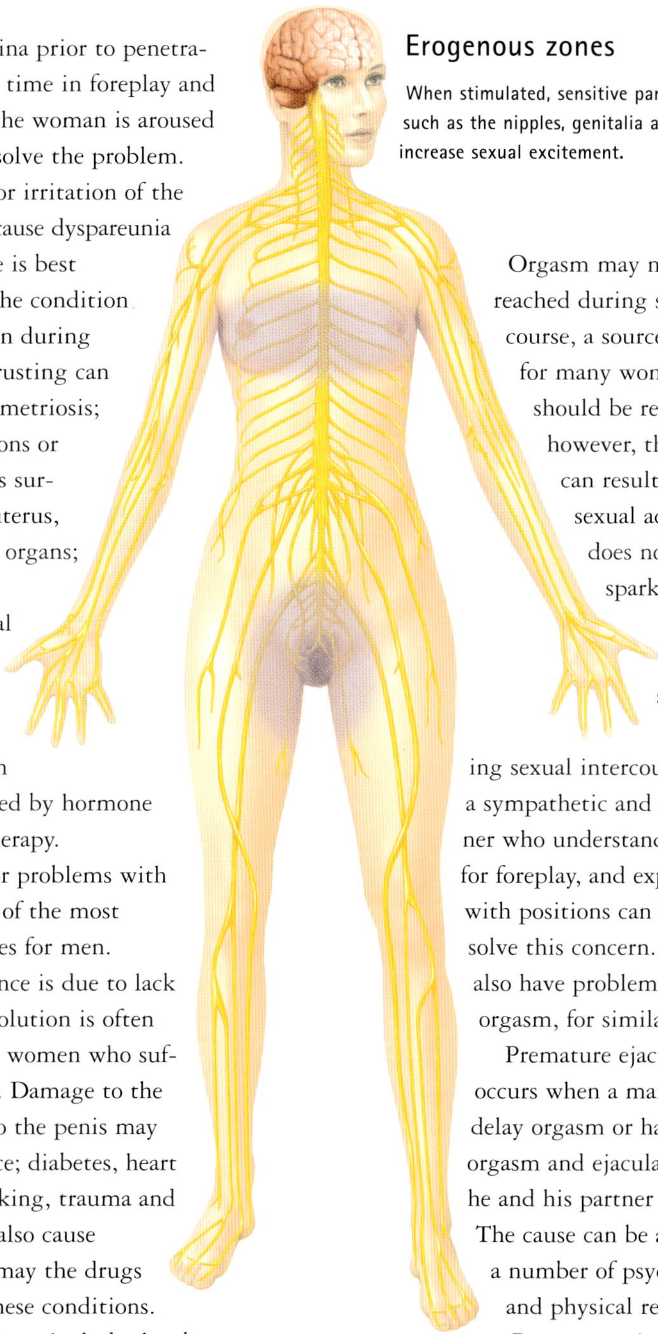

Erogenous zones

When stimulated, sensitive parts of the body such as the nipples, genitalia and inner thighs increase sexual excitement.

Orgasm may not always be reached during sexual intercourse, a source of concern for many women. It should be remembered, however, that orgasm can result from any sexual activity—it does not have to be sparked by sexual intercourse. Understanding and wanting sexual intercourse, a sympathetic and caring partner who understands the need for foreplay, and experimenting with positions can help to solve this concern. Men can also have problems with orgasm, for similar reasons.

Premature ejaculation occurs when a man cannot delay orgasm or has a partial orgasm and ejaculation before he and his partner are ready. The cause can be any one of a number of psychological and physical reasons. Premature ejaculation can often be solved with sexual therapy and a cooperative partner.

Retarded ejaculation occurs when a man cannot achieve orgasm. It is accompanied by emotional and physical frustration, discomfort and loss of interest in sexual intercourse. A sex therapist and an understanding partner can solve the problem.

Vaginismus is involuntary spasm of the muscles round the entrance of the vagina that causes pain during sex. A vaginal examination can also trigger this problem. An overwhelming fear of sex, often after a traumatic or violent sexual experience, is usually the cause. Sex therapy can often help, although treatment may have to continue for a prolonged period.

ENDOCRINE SYSTEM

The endocrine system, involved in coordinating the activities of tissues throughout the body, acts by means of organic chemicals called hormones.

Hormones, made of amino acids or steroids, are released from endocrine cells at specific times and in precise amounts to act on target organs often at some distance from their site of release. Hormones usually act by combining with special receptor sites on or inside target cells.

Endocrine organs (often called endocrine glands), secrete only into the bloodstream or body cavities, unlike exocrine glands which usually secrete their products onto the skin surface or the linings of the digestive and respiratory systems.

Endocrine glands

The endocrine glands secrete hormones that regulate growth, metabolism, sexual maturation and other important body functions.

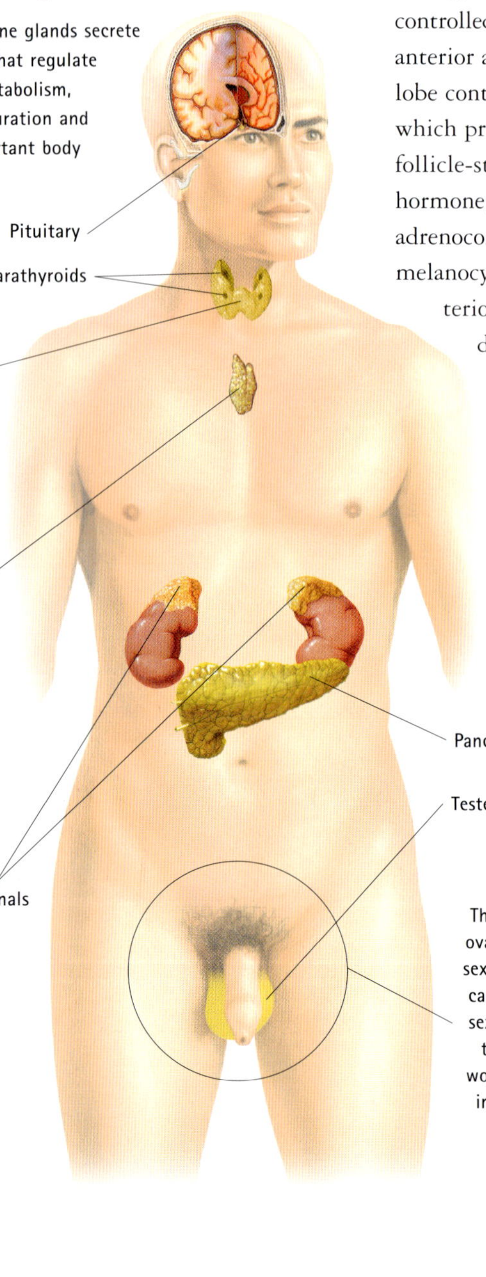

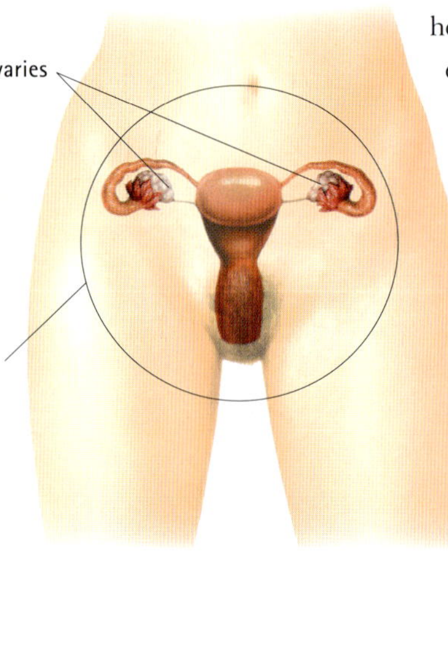

The testes and the ovaries produce the sex hormones which cause the differing sexual characteristics of men and women and play an important role in reproduction

Endocrine hormones affect the nervous system and many endocrine organs are stimulated or inhibited by nerve cells. The hypothalamus of the brain has an intimate connection with the chief organ of the endocrine system, the pituitary. This means that the endocrine and nervous systems share control of body functions: the nervous system usually controls activities occurring rapidly or in the short term, while the endocrine system controls slow or long-term changes.

The endocrine system (pineal, thymus, thyroid, parathyroids, adrenals, pancreatic islets, ovaries and testes) acts under the control of the pituitary gland. The ovaries and testes also function as endocrine glands, producing hormones which control sexual function and secondary sexual characteristics.

The pituitary, the central coordinator of the endocrine system, lies immediately below the hypothalamus of the brain and is closely controlled by it. The pituitary is divided into anterior and posterior lobes. The anterior lobe contains many different types of cells, which produce growth hormone, prolactin, follicle-stimulating hormone, luteinizing hormone, thyroid-stimulating hormone, adrenocorticotrophic hormone and melanocyte-stimulating hormone. The posterior lobe contains oxytocin and antidiuretic hormones, produced in the hypothalamus and transported to the pituitary within nerve fibers.

Growth hormone stimulates the growth of long bones, and is particularly important during childhood and early adolescence. Prolactin or lactogenic hormone acts on the mammary glands of the breast to stimulate and maintain the production of milk. Follicle-stimulating hormone (FSH) stimulates the production of eggs in women and sperm in men. Luteinizing hormone stimulates the release of eggs and the production of the hormone progesterone in women, and the secretion of testosterone in men.

Thyroid-stimulating hormone (TSH) stimulates the production of thyroid hormone and promotes its release into the bloodstream. Adrenocorticotrophic hormone stimulates the production of corticosteroid hormone by the adrenal gland. Oxytocin promotes contraction of the smooth muscle cells in the uterus and around the milk glands in the breasts. Antidiuretic hormone, or vasopressin, promotes the reabsorption of water from the urine in the kidney, thereby helping to control the concentration of salts in the blood.

The thyroid gland, under the influence of TSH from the pituitary, produces thyroxine and triiodothyronine, which act on cells throughout the body to increase energy production. Thyroid hormones also affect the developing brain; insufficient thyroid hormone during prenatal development can lead to mental deficiency (cretinism). Parafollicular cells or C cells, also found in the thyroid, produce calcitonin, responsible for reducing the concentration of calcium in the blood.

The parathyroid glands produce parathyroid hormone, which acts to raise the concentration of calcium in the blood and reduce the concentration of phosphate ions.

The adrenal glands lie on the upper end of each kidney. Each has an inner part (medulla) and an outer part (cortex). The adrenal medulla contains many modified nerve cells, which produce the hormones epinephrine and norepinephrine (adrenaline and noradrenaline). Epinephrine and norepinephrine are released in bursts during emergency situations or accompanying intense emotion (such as fright). These hormones act to increase the strength and rate of heart contraction, raise the blood sugar level and elevate blood pressure.

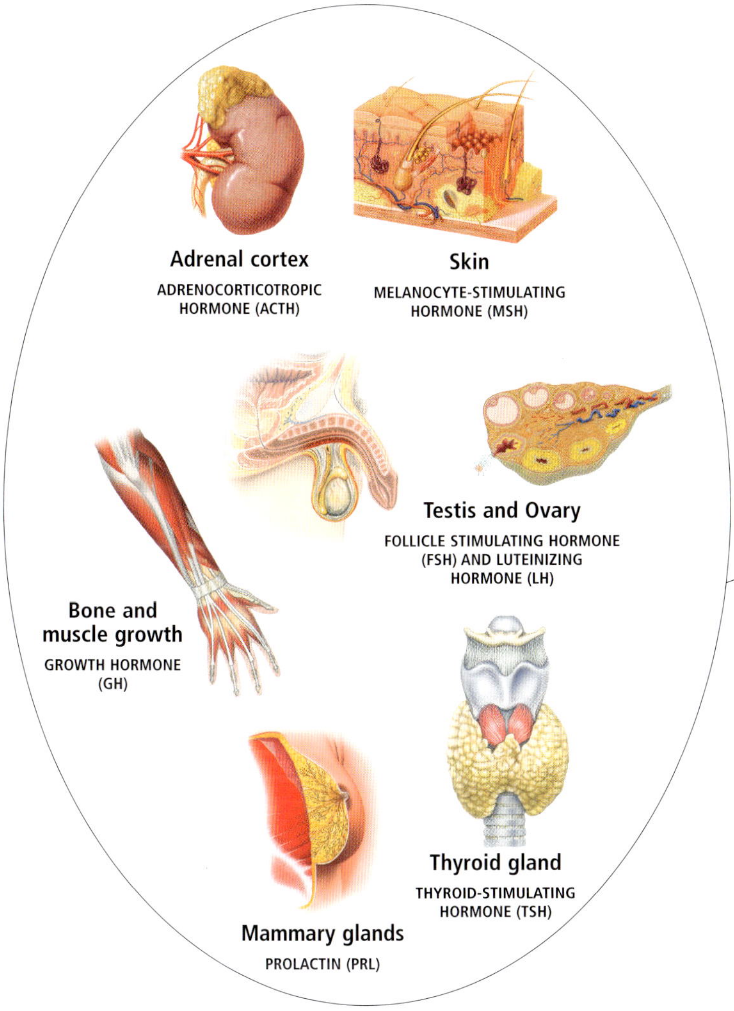

Pituitary gland

The pituitary gland exerts its effects on organs and tissues throughout the body by means of organic chemicals called hormones. The illustration shows the organs and tissues affected by the pituitary gland.

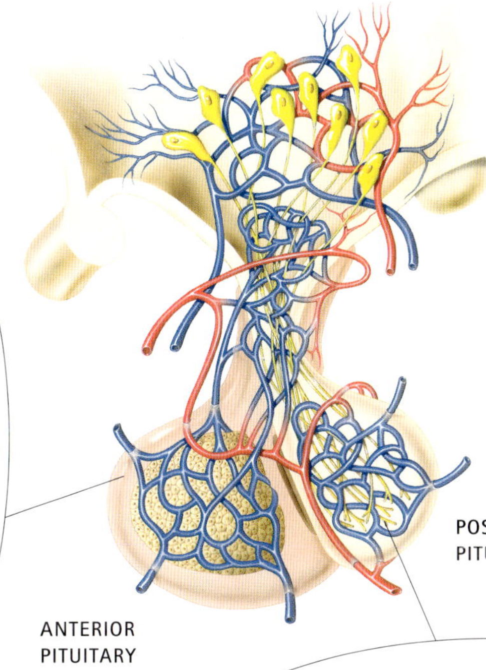

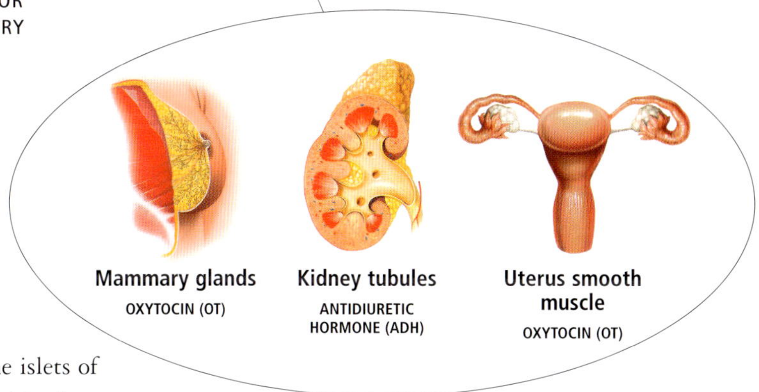

The adrenal cortex produces three main types of hormone: glucocorticoids, mineralocorticoids and sex steroids. Glucocorticoids, produced and released under the control of adrenocorticotrophic hormone (ACTH) from the pituitary, influence the metabolism of fat, protein and carbohydrates, promoting the breakdown of protein and the release of fat and sugars into the blood stream. Mineralocorticoids, such as aldosterone, stimulate the absorption of sodium in the kidney. The sex steroid produced by the adrenal cortex, dehydroepiandrosterone, has masculinizing effects if secreted in large amounts. Excessive amounts of glucocorticoids in the blood, usually due to medical treatment with high doses of the hormone, causes a condition known as Cushing's syndrome.

Insufficient production of corticosteroids produces Addison's disease. Excessive production of dihydroepiandrosterone in boys results in early puberty, while girls may develop an enlarged clitoris, and may be mistakenly raised as boys if clitoral enlargement occurs before birth.

The pancreatic islets (the islets of Langerhans) are located within the pancreas, an abdominal organ mainly concerned with producing digestive enzymes for the gastrointestinal tract. The islets produce hormones responsible for controlling blood sugar level. Insulin acts to lower blood glucose concentration, while glucagon acts to raise it.

The small pineal body is located inside the skull cavity, surrounded by the brain. It produces melatonin, whose concentration varies in tune with the body's circadian rhythm. The pineal gland probably has an effect on the ovaries and testes and may influence mood; its precise role is uncertain.

The ovaries produce estrogen and progesterone. Estrogen causes growth of the breasts and reproductive organs, among other functions; progesterone maintains the lining of the uterus in a state suitable to receive a fertilized ovum. Every 28 days estrogen and progesterone undergo cyclical changes in level, under the influence of FSH and luteinizing hormone from the pituitary.

During pregnancy, the placenta acts as an endocrine organ, producing a hormone to sustain the pregnancy (human chorionic gonadotrophin). The placenta also produces estrogen, progesterone and relaxin, along with human placental lactogen, which promotes milk production and fetal growth.

The production and release of hormones by glands such as the thyroid and adrenals is constantly regulated by a mechanism called feedback inhibition, in which high levels of the hormone tries to suppress its further production.

Other glands, such as the pancreatic islets and the C cells of the thyroid, detect changes in blood concentrations of sugars or calcium respectively, and modify their hormone production accordingly.

SEE ALSO *Adrenal glands on page 280; Glands on page 22; Homeostasis and metabolism on page 112; Ovaries on page 306; Pancreas on page 277; Parathyroid glands on page 206; Pituitary gland on page 137; Testes on page 296; Thymus gland on page 251; Thyroid gland on page 205*

HOMEOSTASIS AND METABOLISM

"Homeostasis" refers to the tendency to stability in the normal physiological state of the human body, while "metabolism" refers to all the physical and chemical processes that take place in the body.

SEE ALSO *Body systems in this chapter*

HORMONES

Hormones are chemical substances that are produced by endocrine organs, including the pituitary, thyroid, parathyroid, adrenal, pancreas, gonads (the testes in males and the ovaries in females) and the placenta (during pregnancy). They are released into the bloodstream and carried to other regions called target organs, where they alter the activity of target cells.

Hormones control a whole range of body functions including growth, metabolism and reproductive activity. They are divided into three main classes, according to their chemical structure: peptide and protein hormones, such as growth hormone and insulin: hormones made from amino acid tyrosine, such as epinephrine (adrenaline) and thyroxine; and steroid hormones, which include corticosteroids and the sex hormones.

The way hormones are produced depends on their type. Peptide and protein hormones are made from a mRNA (messenger RNA) sequence, like cellular proteins, and stored in the cell in membrane-bound vesicles until released. Thyroxine and epinephrine are made by specific chemical reactions within the thyroid and adrenal glands. Steroid hormones are made from cholesterol; they are not stored in vesicles but are found free in the cell cytoplasm.

Hormones work by combining with specific receptors in the target cells. These receptors are usually proteins and may be in the cell membrane (for peptide hormones) or in the cell cytoplasm or nucleus (for steroid hormones). Having bound to the receptor, the hormone may stimulate or inhibit specific metabolic pathways in the cell, for example, they may change the activity of an enzyme, or stimulate production of a new protein. Receptors for some hormones (for example, thyroxine) are present on many body cells and hence the effects of the hormone are widespread. For other hormones the receptors are only present on specific tissues and thus their effects are very localized.

While many hormones are carried in the blood, some cells are known to release secretions which act locally, generally referred to as paracrine secretions. Yet others, called autocrine secretions, can act on the same cell (that is, the secretory cell itself has receptors).

Hormones can have long-lasting effects, for instance growth hormone from the pituitary acts on a wide range of tissues to stimulate growth during childhood. Other hormones can cause very rapid changes, such as epinephrine (adrenaline). Produced in the adrenal medulla, it is important in the "fight or flight" response to stress and can cause an increase in heart rate, widening of the airways and release of glucose.

Besides regulating many body functions throughout our lives, some hormones are important at specific times, such as during pregnancy, childbirth and lactation. These include sex hormones like estrogen and progesterone, and hormones from the placenta (for example, chorionic gonadotropin) and pituitary (like prolactin).

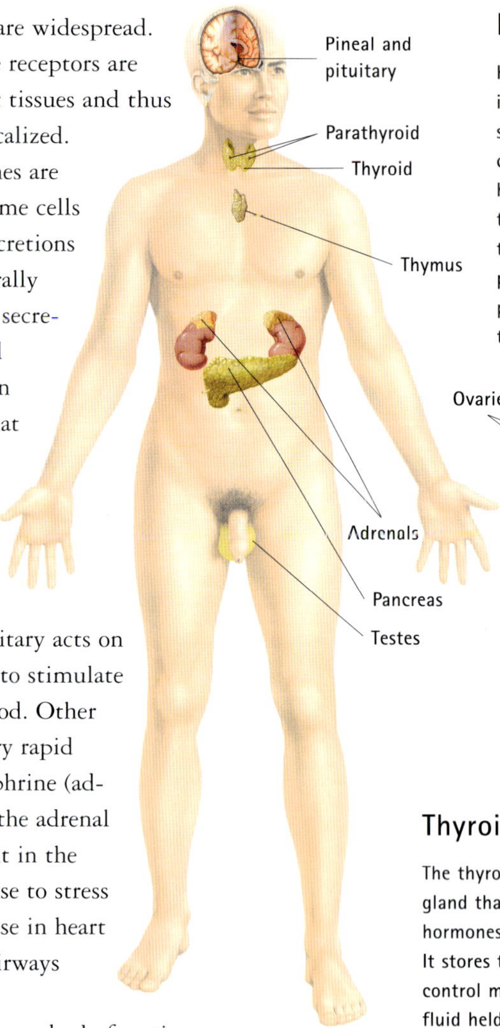

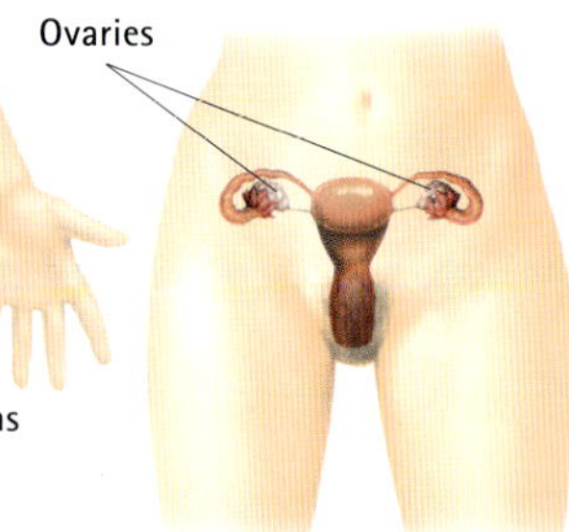

HORMONES

Hormones are produced in the endocrine glands, specialized organs whose cells produce and release hormones directly into the bloodstream. The thyroid gland, for example, is made up of cells producing and secreting thyroid hormones.

Thyroid microstructure

The thyroid is the only endocrine gland that does not release its hormones straight into the body. It stores thyroid hormones (which control metabolism) in colloid fluid held in the follicles of the gland. The hormones are then secreted gradually into the blood when needed.

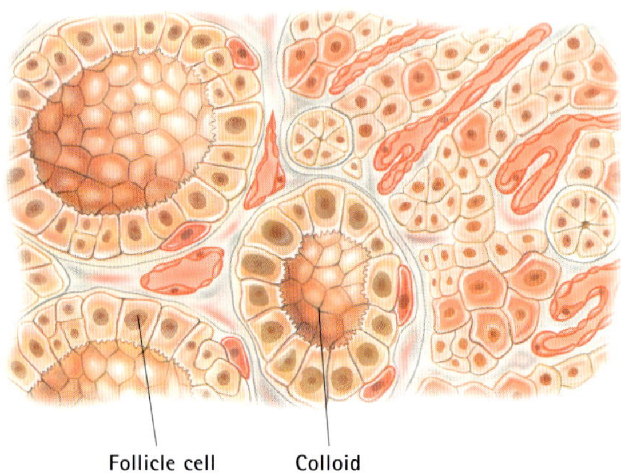

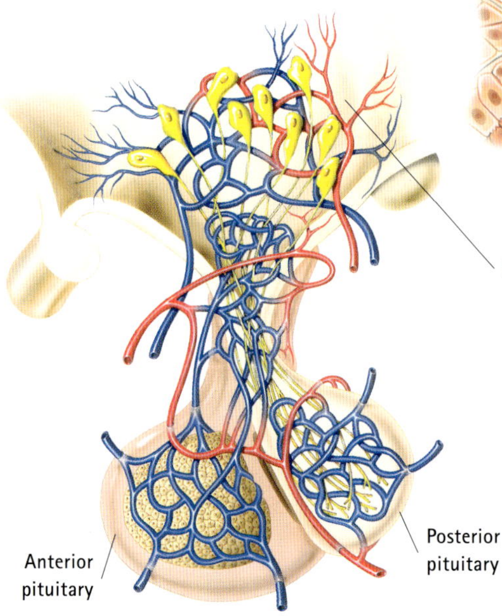

Pituitary gland

The pituitary gland is a small but very important gland in the base of the brain. It secretes hormones which control other endocrine glands elsewhere in the body. The pituitary gland itself is regulated by hormones secreted by the hypothalamus, which lies just above it.

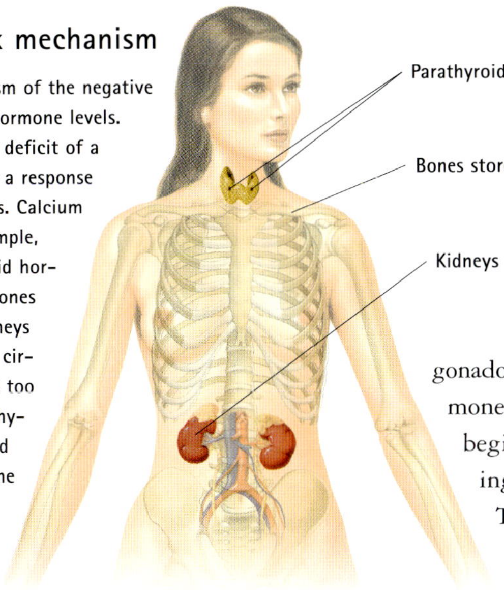

Negative feedback mechanism

The body uses the mechanism of the negative feedback loop to regulate hormone levels. In this system, an excess or deficit of a particular hormone triggers a response to normalize hormone levels. Calcium levels in the blood, for example, are controlled by parathyroid hormones which instruct the bones to store calcium or the kidneys to excrete it, depending on circulating levels. If levels fall too low, for example, the parathyroid registers the deficit and releases parathyroid hormone which tells the bones to release calcium, increasing blood calcium levels.

Hormone levels, and thus their impact on the body, are controlled in different ways, depending on the hormone, but a common mechanism is called negative feedback. This is a system whereby either an excess or deficit of a hormone prompts a response that results in that hormone's return to normal levels. Negative feedback loops operate for many hormones (such as the control of glucose by insulin), although several hormones often interact for the final outcome.

Hormone imbalance can occur if there is an excess or a deficiency of one hormone, which can often affect the function of others. Imbalances can also occur when there is sufficient hormone produced, but the target tissue is not able to respond properly. The inability to respond may occur because of changes in the receptors. The effects of hormone imbalance will differ depending on the particular hormone and its target area.

During menopause when menstruation has ceased, there are surges in pituitary hormones (FSH and LH) as a result of falling levels of ovarian hormones, because of the increasing unresponsiveness of the ovaries. These surges in pituitary hormones can result in hot flashes.

FEMALE AND MALE SEX HORMONES

The development of the gonad into either an ovary or testis occurs early in fetal life, and is dependent on the sex chromosomes (XX in females, XY in males).

During childhood, the ovaries and testes are relatively dormant. At puberty, the hypothalamus in the brain releases the gonadotropin-releasing hormone (GnRH), which begins the changes leading to sexual maturity. This stimulates the pituitary to release gonadotropins, luteinizing hormone (LH) and follicle-stimulating hormone (FSH) in both men and women.

These hormones stimulate the ovaries in females for follicle maturation and for ovulation, and the testes in males for sperm production and testosterone secretion. Both the ovary and testes secrete sex hormones; testosterone in males, and estrogen and progesterone in females. At puberty, sex hormones are involved in the development of the external genitalia as well as secondary sexual characteristics.

For the male, this includes the enlargement of the penis and testes, growth of body and facial hair, development of body musculature and growth of the larynx, causing deepening of the voice. For the female, the changes include development of the breasts, growth of body hair, and the beginning of menstruation (menarche).

After puberty has begun, the production of sex hormone is continuous in the male; in females, however, production occurs in cycles (menstrual cycles), which last until women reach menopause.

Insulin

Insulin is a hormone produced in the pancreas that affects the body's ability to use sugars. Produced in either abnormally high or abnormally low quantities, it can have a severe effect on the body's metabolism. Too much insulin leads to low blood sugar levels (hypoglycemia); too little insulin causes high blood sugar levels (hyperglycemia).

Normally, when food is digested, carbohydrates are broken down into sugars such as glucose and absorbed into the bloodstream. This triggers the secretion of insulin into the blood by clusters of cells in the pancreas known as the islets of Langerhans. The pancreas is part of the body's endocrine system, a network of glands that secrete various hormones into the blood to chemically regulate body functions. Insulin aids the absorption of glucose into body cells, for immediate use or storing.

Melatonin

Melatonin is a hormone secreted by the pineal gland, which is located deep inside the brain. Thought to be involved in the regulation of the body's sleep–wake "clock," melatonin production is highest during a person's normal sleeping hours and drops off as the body begins waking.

There is evidence that melatonin supplements may help some forms of insomnia

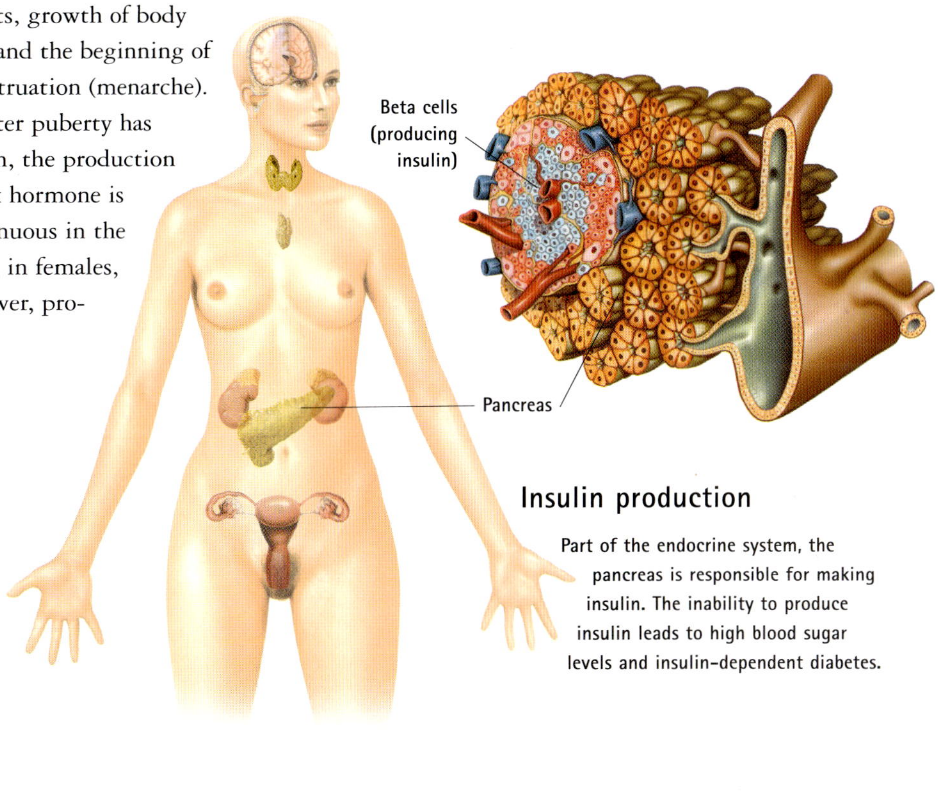

Insulin production

Part of the endocrine system, the pancreas is responsible for making insulin. The inability to produce insulin leads to high blood sugar levels and insulin-dependent diabetes.

and ease the symptoms of jet lag by speeding up the body's return to its regular usual sleep–wake cycle. Research investigating potential applications suggests melatonin supplements could be useful in birth control as well, boosting the immune system, treating cancer and depression and in the management of many other disorders.

Steroids

Steroids are a group of chemicals that comprise a large number of the body's hormones. They are synthesized from cholesterol mainly in the adrenal glands and gonads, and regulate a number of important body processes.

There are three main groups of steroids: glucocorticoids, mineralocorticoids and sex steroids (androgens and estrogens). The major glucocorticoid is cortisol, which is also known as the "stress hormone." It is released from the adrenal gland as a response to physiological stresses such as exercise, injury or infection, or psychological stresses, such as fear or depression.

Cortisol acts on cells in the body, and regulates their metabolism and activity. Cortisol prepares the body for action by mobilizing glucose stores and preventing excessive tissue reaction to injury; and this anti-inflammatory action comes into use when cortisol-type medication is prescribed for the treatment of autoimmune diseases. Mineralocorticoids help the body control salt and water balance. Androgens are produced by the adrenal glands; however, they require conversion in the testes to the main male hormone testosterone. The main female sex hormone (estradiol) is made from cholesterol by cells in the ovaries.

Corticosteroids

Corticosteroids, or corticoids, are steroid hormones produced in the outer layer (cortex) of the adrenal glands. The most important steroid hormones are aldosterone, which regulates the excretion of sodium and potassium salts through the kidney, and cortisol (hydrocortisone), which promotes the synthesis and storage of glucose and regulates fat distribution in the body. Production of the hormones is controlled by the pituitary gland via the hormone ACTH. Insufficient production of corticosteroid by the adrenal gland causes Addison's disease, excessive production causes Cushing's syndrome. The term corticosteroid is also used for some synthetic derivatives with similar properties to naturally occurring corticosteroids, used to treat allergies, rheumatic disorders and inflammation.

Androgens

Androgens are steroid hormones that produce male sex characteristics. The two main androgens are androsterone and testosterone. In men, they are secreted into the bloodstream by the testes (and to a lesser extent by the adrenal glands) under stimulation from the pituitary gland. In women, smaller amounts are secreted by the ovaries and adrenals. At the onset of puberty, production of testosterone in boys increases; facial and pubic hair develops, the larynx enlarges (so that the voice deepens), the penis and testes enlarge and there is an increase in body muscle strength.

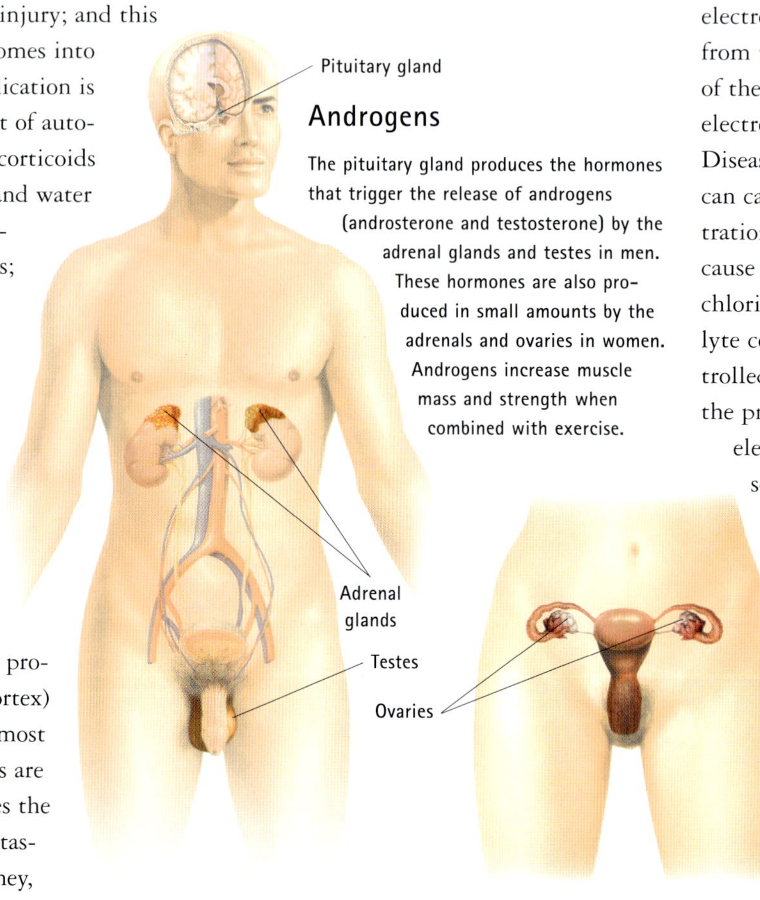

Androgens

The pituitary gland produces the hormones that trigger the release of androgens (androsterone and testosterone) by the adrenal glands and testes in men. These hormones are also produced in small amounts by the adrenals and ovaries in women. Androgens increase muscle mass and strength when combined with exercise.

Electrolytes

Electrolytes are ions in solution in the blood and other body fluids. They play an important role in all functions of the body. The major electrolytes are sodium, potassium, calcium, phosphate, chloride and bicarbonate. Derived from food, their concentration in the blood is regulated by the kidneys and lungs and can be measured by laboratory studies of serum (the clear liquid in plasma).

Different fluids in the body have different concentrations of electrolytes. For example, the concentration of sodium in serum is 142 millimoles/liter whereas the potassium concentration is only 4 millimoles/liter. Inside the cell, the concentration of sodium is 10 millimoles/liter, while the concentration of potassium is 160 millimoles/liter.

This difference in concentration is maintained by special ion pumps in cell membranes. Their function is to constantly move sodium ions out of the cell and potassium ions into the cell. The difference in concentration of electrolytes in different fluids is important for some metabolic processes, for example, the conduction of electrical impulses along nerve fibers.

The concentration of electrolytes in body fluids depends on an adequate intake of electrolytes in the diet, adequate absorption from the intestine, and proper functioning of the kidneys and lung, which also regulate electrolyte concentration in body fluids. Diseases that interfere with these processes can cause disturbances of electrolyte concentration. For example, kidney disease may cause the body to retain too much sodium, chloride, bicarbonate and calcium. Electrolyte concentrations in the body are also controlled by hormones, and diseases that affect the production of these hormones also cause electrolyte disturbances, which can severely affect the body's metabolism.

Different electrolytes play different roles in the body's metabolism. Sodium helps in the regulation of the body's water balance, and is also involved in maintaining normal heart rhythm, blood pressure and blood volume.

Too much sodium in the blood and body fluids (hypernatremia) or too little sodium (hyponatremia) can cause confusion, restlessness, anxiety, weakness, muscle cramps, edema and, in severe cases, stupor or coma.

Potassium assists in the regulation of the acid-base and water balance in the blood and the body tissues. It assists in protein synthesis from amino acids and in carbohydrate metabolism. It is necessary for the building of muscle and for normal body growth. Along with sodium and calcium, it maintains normal heart rhythm, regulates the body's water balance and is responsible for muscle contractions and nerve impulses.

Tissue cells usually have a high concentration of potassium, while blood usually has a low concentration. Too much or too less potassium in the blood can cause weakness and paralysis and can affect heartbeat.

Sodium and potassium levels in the blood and body fluids are regulated by aldosterone, a hormone secreted by the adrenal gland; it increases sodium resorption from the kidneys and promotes potassium loss by the kidneys. A tumor of the adrenal gland may cause excess secretion of aldosterone (hyperaldosteronism), causing hypernatremia, hypokalemia and fluid imbalance.

Calcium is a mineral component of blood which, along with phosphate, forms bone and teeth. Calcium also helps in the regulation of the heartbeat, transmission of nerve impulses, contraction of muscles and clotting of blood. Calcium and phosphate levels in the blood are regulated by parathyroid hormone, secreted by the parathyroid gland, which requires vitamin D to function. Calcium levels are also regulated by calcitonin, a hormone secreted by the thyroid gland.

Too much calcium often causes lethargy, delirium and seizures; too little calcium causes muscle spasms, twitching, cramps, numbness and tingling in extremities, seizures and irregular heartbeat.

Chloride is necessary in the maintenance of the body's acid-base and fluid balance. It is an essential component of the secretions of the stomach, aiding in digestion. Fluid loss due to excessive sweating, vomiting or diarrhea can cause a deficiency of chloride, resulting in excessive alkalinity of body fluids (alkalosis), low fluid volume (dehydration) and loss of potassium in the urine.

Bicarbonate acts as a buffer in blood and in body fluids, keeping the pH (acidity) to within a narrow range of 7.35–7.45. The concentration of bicarbonate is regulated by the kidney and lungs.

Histamine

Histamine is a chemical messenger found in all body tissues that reacts with histamine receptors on cell surfaces causing change in certain specific bodily functions. Histamine affects smooth muscles, such as those in the intestine, heart and lungs, causing them to contract; or can cause the smaller blood vessels to dilate resulting in a fall in blood pressure and release of lymphatic fluid into surrounding tissues, as when redness and swelling follows a sting or bite. Overproduction of histamine can occur when the body comes into contact with something to which it is allergic, for example. pollen in hay fever sufferers. Extreme cases can be life threatening.

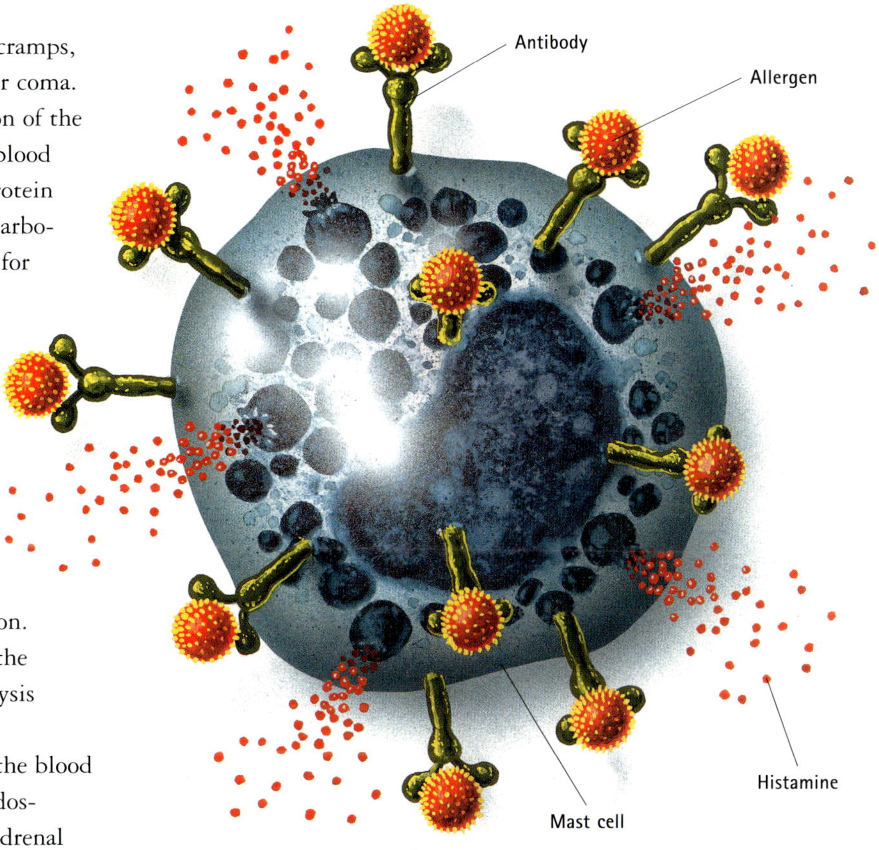

Histamine

When allergens (like pollen) enter the body, antibodies to the allergen trigger histamine release from circulating mast cells, causing symptoms such as runny nose and sneezing.

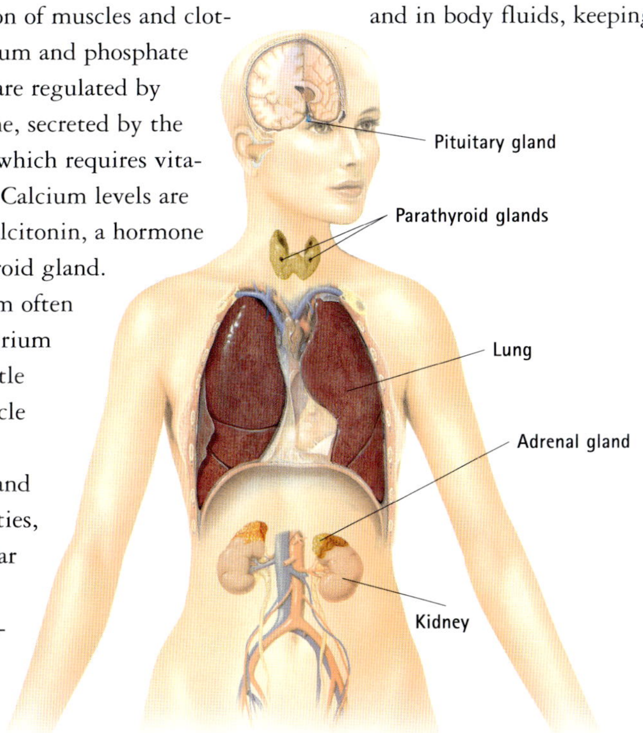

Electrolytes

The concentration of electrolytes in the blood is regulated by various organs. For example, the pituitary gland releases hormones which contribute to the regulation of the electrolyte balance; the adrenal glands regulate sodium and potassium; the parathyroid glands regulate calcium and phosphate; the kidneys and lungs regulate bicarbonate; and the kidney regulates chloride.

Metabolism

Everything that happens in the human body is dependent upon chemical reactions in the cells. These complex biochemical events or pathways that drive all of the body's systems are collectively referred to as metabolism. There are two fundamental phases or processes involved in metabolism—one is constructive and uses energy, the other breaks down compounds and creates energy.

During the building-up phase, called anabolism, simple molecules are used to create more complex molecules and substances. For example, amino acids are bonded together to build proteins and simple sugars are packaged up to produce polysaccharides. Anabolic reactions require energy and occur during the growth, repair and maintenance of body cells and systems. During the reverse process, catabolism, complex substances are broken down into simpler compounds. Catabolic reactions produce energy, which is stored in a substance known as ATP until required for use in anabolic reactions. Chemical digestion, during which food is broken down and energy released, is a catabolic process.

Substances that are crucial to the chemical reactions that help in running our bodies include enzymes (produced by the body itself) and nutrients, which are extracted from the foods we eat. Nutrients in food include carbohydrates, lipids, proteins, minerals and vitamins.

Basal metabolic rate is the lowest rate of energy use required by the body to sustain its essential functions, including respiration, blood circulation and temperature maintenance. Measured between 14–18 hours after the last meal, when the person is awake but at complete rest in a comfortable environment, the resulting measurement is an expression of the pace at which a particular person's body breaks down food. On an average, a healthy human uses approximately 1,500–2,000 kilocalories (6,000–8,000 kilojoules) per day in the maintenance of basic body functions.

A person's overall metabolism is regulated internally by hormones. Too much or too little of these can alter the metabolic rate and may cause serious health problems.

There are, however, many other factors that also affect metabolic rate. For example, our metabolism slows with age and increases with exercise. The metabolic rate of women tends to be lower than that of men, although it increases during pregnancy and lactation. Stress hormones raise the metabolic rate, as do other hormones such as testosterone. High temperatures, associated with illness, will also cause an increase in metabolic rate. The metabolic rate slows during sleep and when we are malnourished.

Temperature

The temperature of a normal healthy human body is assumed to be 98.6°F (37°C), but may range from 97.2–100°F (36.2–37.8°C); temperatures ranging from 95.9–101.2°F (35.5–38.4°C) have also been recorded in healthy individuals. Temperature can only be taken accurately by using a thermometer. Temperature can be taken in the mouth (orally) by placing a thermometer under the tongue; rectally, by inserting the thermometer into the rectum; or under the armpit (axillary). The rectal temperature is usually 1°F (0.5°C) higher than the oral temperature; the axillary temperature is usually 1°F (0.5°C) lower than the oral temperature. Temperature should always be taken by the same route for purposes of comparison in illness.

A child with a high fever, over 101°F (38.2°C), needs treatment to reduce their temperature in order to avoid the possibility of convulsions. This can be done by removing clothing, and bed clothes if necessary, and cooling the child with a tepid sponge. A tepid bath or a fan may help. Shivering is not the aim as this will raise the body's temperature. Acetaminophen (paracetamol) or ibuprofen may be given to children to help reduce fever, but never aspirin, as it may cause Reye's syndrome.

Temperature regulation

The human body's usual core temperature ranges from 97.2–100°F (36.2–37.8°C). The core body temperature varies according to the time of day and is also affected by the environment, physical activity, food intake and emotions. Typically, temperature is lowest early in the morning and highest in late afternoon, a normal part of the body's circadian rhythm.

Body temperature is the result of the balance between heat loss and heat production. Body heat is lost through radiation, which

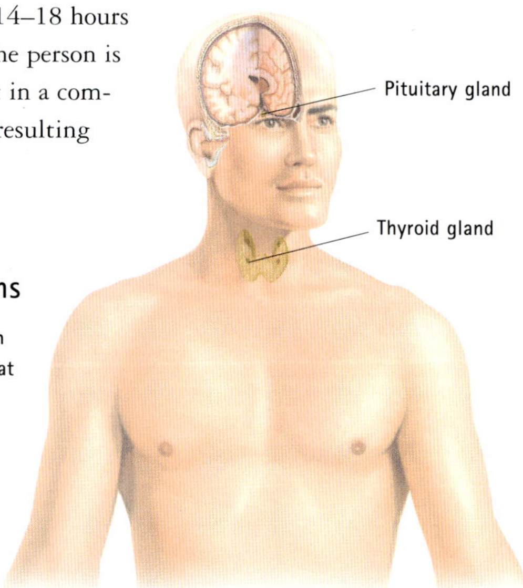

Metabolism—regulating mechanisms

Metabolism is the collective term for all the chemical processes that run our body. Thyroid hormones regulate the metabolic rate, so under- or over-activity of the thyroid can disrupt the body's metabolism. The pituitary gland controls the release of thyroid hormones.

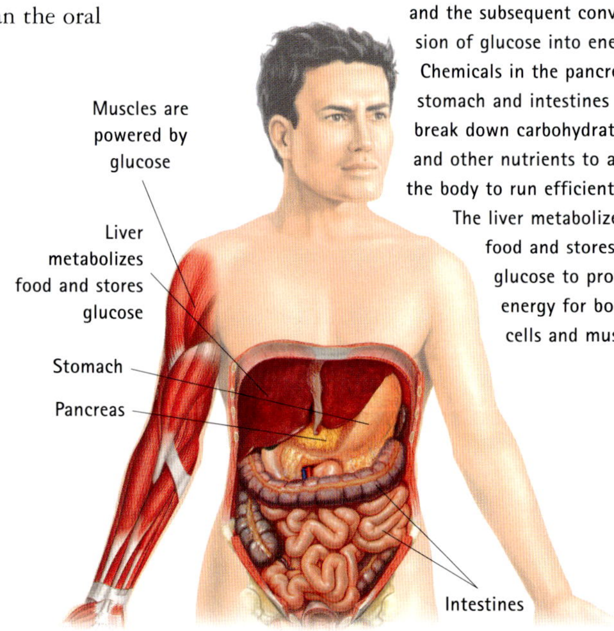

Chemical metabolism

One of the most important chemical processes in the body is the breakdown of carbohydrates into glucose, and the subsequent conversion of glucose into energy. Chemicals in the pancreas, stomach and intestines break down carbohydrates and other nutrients to allow the body to run efficiently. The liver metabolizes food and stores glucose to provide energy for body cells and muscles.

SKIN REACTION TO COLD

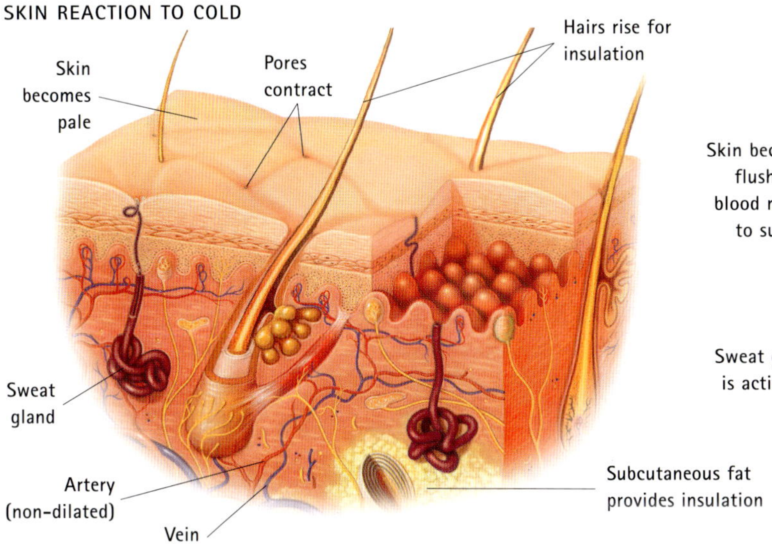

SKIN REACTION TO HEAT

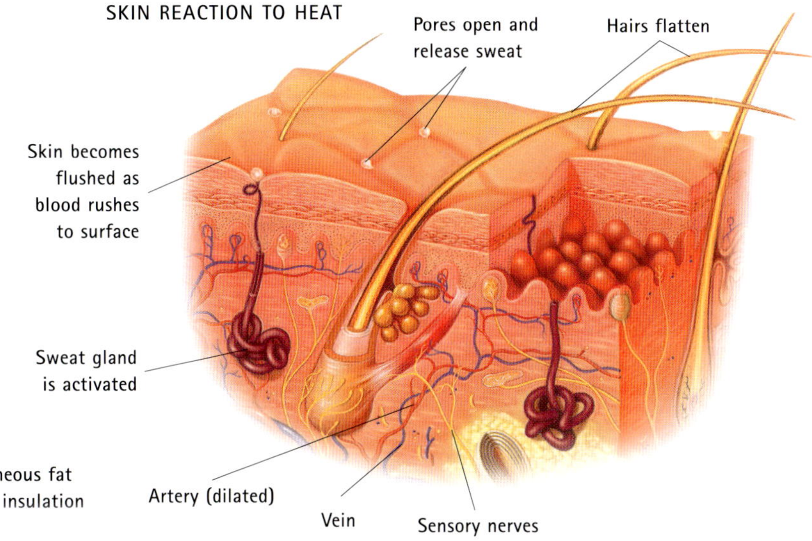

Skin and temperature

Both the dermal and epidermal layers of the skin are involved in regulating body temperature. When the body is hot, arteries dilate and blood flow to the skin increases, maximizing heat loss. The sweat glands are stimulated and release fluid, which evaporates and reduces the body's temperature. When the body is cold, the pores and arteries contract and the hairs of the skin stand up, providing an insulating layer that traps body heat close to the surface of the skin.

is affected by the rate of blood flow to the skin's surface; evaporation (the sweat glands can dissipate as many as 1,700 kilocalories (7,100 kilojoules) an hour; convection and conduction. The body's fat also provides the insulation which helps to maintain the body temperature when exposed to either hot or cold environments.

The part of the brain which is responsible for regulating body temperature is the hypothalamus. When the body is heated, heat loss is initiated by messages sent to the hypothalamus; the blood vessels dilate and the body sweats. If the body temperature is too high (hyperthermia) the skin may become dry, deep, fast breathing will follow, and the patient may experience headaches, nausea and unconsciousness. When the body is exposed to cold stress, the first symptom of hypothermia is pain. This is followed by numbness, mental confusion, lethargy and irregular heartbeat.

A sharp rise in body temperature (hyperthermia) can trigger heat-related illnesses such as heat exhaustion and heatstroke.

Hypothermia is a dangerous lowering of the body's temperature to below 95°F (35°C); there is a recorded case of survival after body temperature sunk to 57°F (13.9°C). Chilling is the first stage of cold injury; from this point body temperature can lower rapidly, leading to more serious cold injury. Frostnip is the next stage, when the skin blanches or loses its color; frostbite may follow and crystals of ice may form in the tissues.

Babies and children are vulnerable to extremes of temperature, particularly cold, because they have a greater body surface compared to their weight than adults. It is possible for babies to overheat when they are overdressed for the temperature. It is also possible for insufficient warm clothing to cause distress; this can happen where parents are overly concerned about sudden infant death syndrome and do not dress a baby warmly enough for sleep. The room temperature for a sleeping child should be kept around 68°F (20°C).

Metabolic imbalances and disorders of homeostasis

Disorders of homeostasis and metabolism encompass a large number of varied conditions, which include diabetes, tetany, heatstroke, and edema.

SEE ALSO *Body systems in this chapter; Diagnostic techniques on page 428; Medical therapies and medications on page 439*

Diabetes

Diabetes mellitus is a disorder caused by decreased production of insulin by the pancreas, or by a decreased ability of the body to use insulin. It is a serious, sometimes fatal, disorder and is a leading cause of death in Western societies. The cause of diabetes mellitus is not known, though it may run in families and is common in obese individuals.

Normally, food digested in the body releases glucose, a form of sugar, into the blood. This in turn causes beta cells in the pancreas to release insulin into the bloodstream. Insulin helps in transporting glucose from the blood into storage in liver and muscle cells; from here it can be later released into the blood for metabolism.

However, if the pancreas is producing insufficient amounts of insulin, or if there is a failure of the mechanism of transporting glucose into the cells (so-called "insulin resistance") then diabetes results.

TYPES OF DIABETES

There are two types of diabetes. Type I (insulin-dependent or juvenile-type diabetes) is caused by insufficient insulin being produced by the pancreas. It is usually found in persons under 25 years of age. Type II, sometimes called non insulin-dependent diabetes, maturity onset or adult-type diabetes, is more common. It is found in persons over age 40, and is usually caused by resistance to insulin.

In both types, there is excess sugar in the blood (hyperglycemia), which then needs to be removed by the kidneys. Symptoms of excessive thirst, frequent urination, and hunger develop. The metabolism of carbohydrates, fats, and proteins is altered. Fatty acids released from tissue throughout the body are converted by the liver into ketones,

Glaucoma

Glaucoma often occurs in diabetes. In glaucoma, pressure inside the eyeball rises, damaging the retina and causing loss of vision.

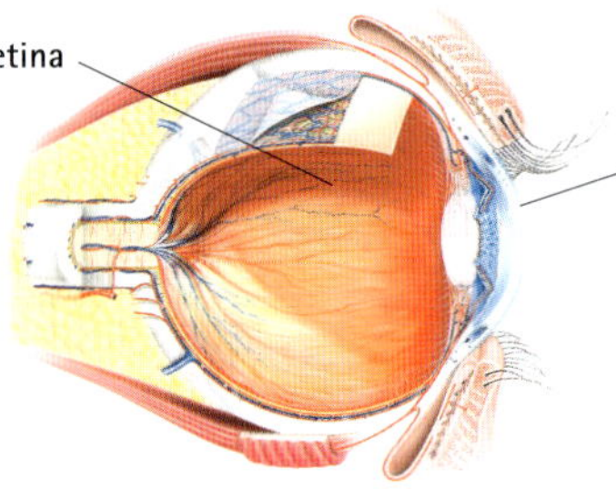

Diabetic retinopathy

Diabetes damages the small blood vessels in the retina of the eye. The damaged vessels can bleed, causing hemorrhages, or they can cause areas of the retina to die, resulting in loss of vision.

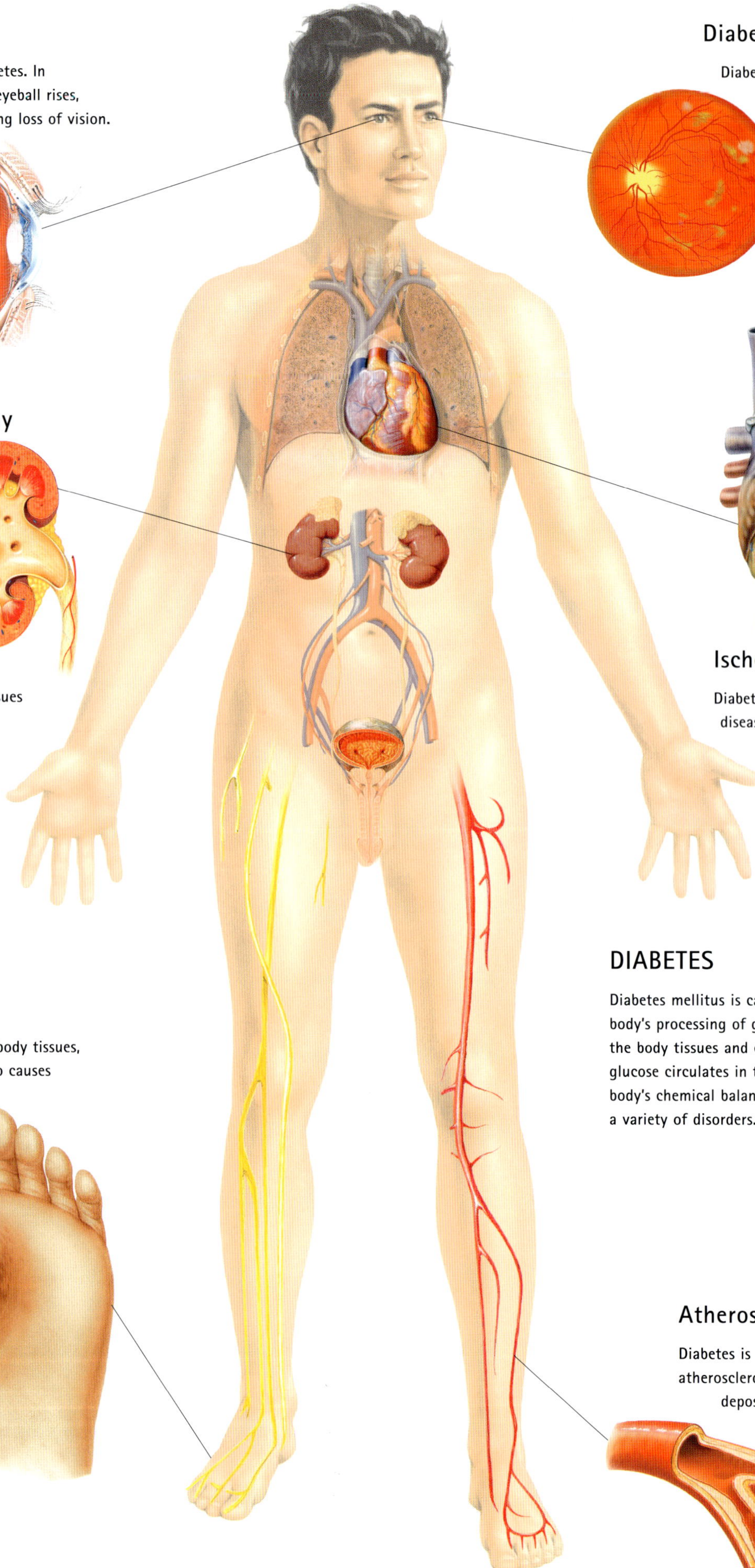

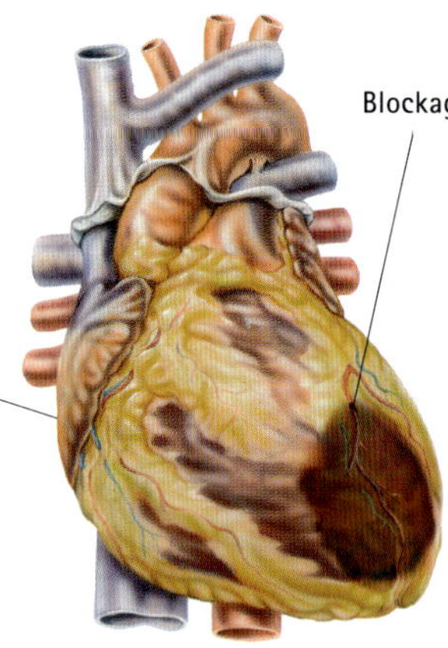

Diabetic nephropathy

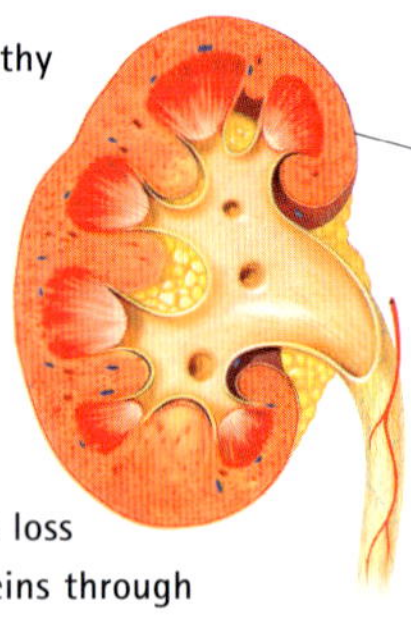

Diabetic nephropathy damages the glomeruli and small blood vessels of the kidneys, due to high levels of blood glucose. This results in the loss of necessary proteins through the urine, swelling of body tissues and eventually renal failure.

Ischemia

Diabetes can cause coronary artery disease, which in turn can cause death of cardiac tissue known as myocardial infarction.

DIABETES

Diabetes mellitus is caused by a disruption in the body's processing of glucose. Instead of reaching the body tissues and cells that need it, excess glucose circulates in the bloodstream upsetting the body's chemical balance. Over time, this can cause a variety of disorders.

Foot ulcer

Diabetes slows the healing of body tissues, damages blood vessels and also causes degeneration of the peripheral nerves (neuropathy). These two factors act together to cause foot ulcers in diabetics.

Atherosclerosis

Diabetes is one of the risk factors for atherosclerosis, a disease in which fatty deposits build up under the lining of the artery and block off its blood flow.

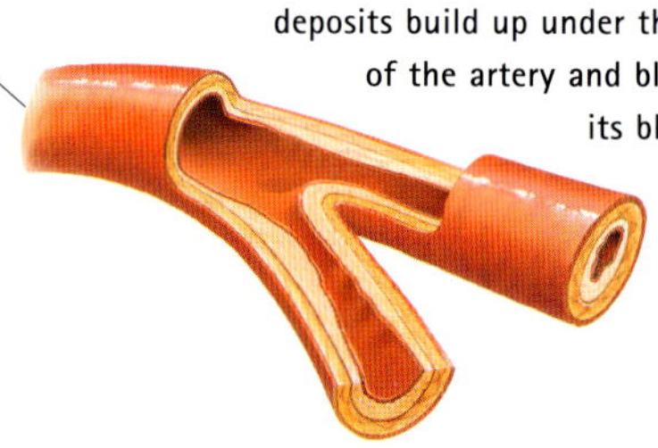

which enter the bloodstream causing the blood to become dangerously acidic (ketoacidosis).

A physician can diagnose diabetes by testing for sugar in urine and blood. A glucose tolerance test determines how well the body uses and stores sugar, by measuring blood glucose levels periodically for up to six hours after the patient swallows a glucose solution.

REGULATION OF DIABETES

Type I diabetics need daily injections of insulin (the hormone is not available in an oral form). Most diabetics self-administer their insulin by subcutaneous injection (below the skin) from one to four times per day, though some diabetics use a portable pump that delivers insulin directly to the body through an implanted cannula.

Careful regulation of activity and a strict diet are important to keep the levels of insulin and sugar in the blood within as normal a range as possible. Type II diabetes is usually treated with diet alone, or with diet plus oral antidiabetic drugs. In some cases, insulin treatment is needed. Weight control and regular exercise are also important. During periods of stress, such as surgery and infection, the insulin dosage may need to be temporarily increased.

IMBALANCE OF INSULIN AND GLUCOSE

Too much insulin or too little glucose in the diet causes hypoglycemia, or insulin shock. The signs of hypoglycemia are mild hunger, dizziness, sweating, and heart palpitations followed by mental confusion and coma. Diabetics can stop hypoglycemia by eating sugar, sweets or candy, or by injecting glucagon, a hormone that raises blood sugar. Diabetics should wear an identification card, tag or bracelet in case they need emergency care.

DIABETES AND PREGNANCY

Diabetes can be aggravated by pregnancy, so good management during pregnancy and labor is essential. Infants of women with poorly controlled diabetes are at risk for birth defects, but if the condition is well controlled, the risk is the same as for a nondiabetic mother. Gestational diabetes is diabetes that only appears during pregnancy. It usually becomes apparent during the weeks 24–28. In many cases, the blood glucose level returns to normal after delivery. However, there is a risk the mother may develop full-blown diabetes in the future.

LONG-TERM COMPLICATIONS

Long-term complications exist in diabetes mellitus as well. The longer the condition exists, the greater the complications. Atherosclerosis is the most serious, damaging small and large blood vessels especially in the retina of the eye (diabetic retinopathy) and in the kidney (diabetic nephropathy), causing blindness and kidney failure. Nerve degeneration resulting in the loss of sensation in peripheral nerves, and reduced resistance to infections are other complications. Foot problems caused by damage to blood vessels and nerves and lesser ability to fight infection can occur, so diabetics need to learn the techniques of foot care.

With good diabetic control, the onset of complications can be delayed. However, they cannot be prevented, especially in the case of Type I diabetes and long-standing Type II diabetes.

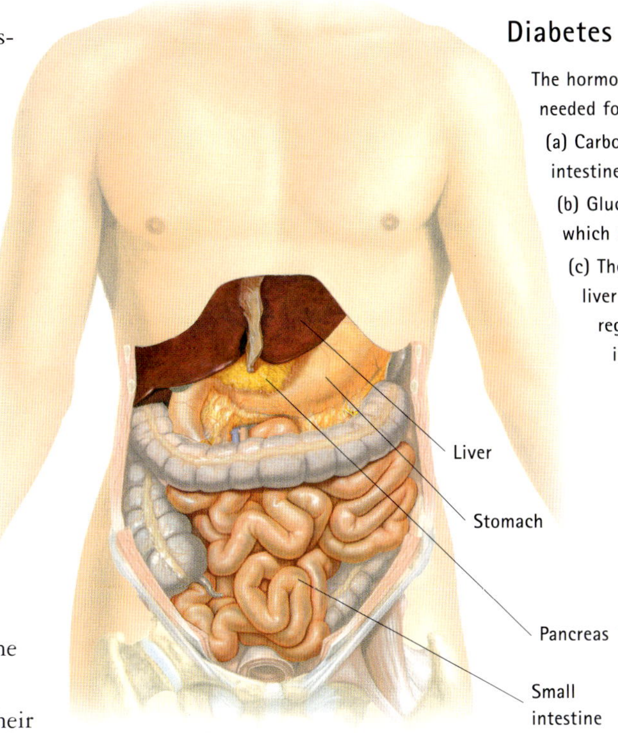

Diabetes and glucose metabolism

The hormone insulin helps the body process glucose—the sugar needed for energy by all body cells. In normal insulin production: (a) Carbohydrates are broken down in the stomach and small intestine and converted into glucose. (b) Glucose travels in the blood from the intestine to the liver, which releases it to body tissues for energy when needed. (c) The pancreas controls how much glucose is stored in the liver and how much is released into tissues. When the pancreas registers increased levels of glucose in the blood it secretes insulin. Insulin instructs the liver and other tissues to absorb more glucose, lowering blood sugar levels. In Type I diabetes, insufficient amounts of insulin are produced by the pancreas. In Type II diabetes, the body either produces too little insulin, or insulin receptors in liver cells and other tissues do not respond to the hormone. In both forms of diabetes too much glucose circulates in the blood, a condition known as hyperglycemia.

Ketosis

Ketosis is a condition in which there is a high concentration of a type of chemical called ketone bodies in the blood. Ketone bodies are produced during starvation, but ketosis is most often seen in the Western world in diabetics who have insufficient insulin to control their blood sugar concentration. This causes the blood to become dangerously acidic (ketoacidosis).

Common symptoms are thirst, frequent urination, nausea and weakness. The affected person is dehydrated, has rapid pulse and breathing, is confused or comatose, and has a fruity breath odor (indicating the presence of ketones). Ketoacidosis is a medical emergency requiring urgent treatment in hospital with intravenous insulin and fluid and electrolyte replacement. The condition has a 10 percent mortality rate.

Hyponatremia

An abnormally low concentration of sodium in the blood—known as hyponatremia—can cause some problems. Usually the condition occurs when too much water is consumed in relation to sodium, or the body fails to excrete excess water. Both situations can create an imbalance in the body.

People who are more likely to be susceptible to hyponatremia include the elderly, the very young and the mentally ill patients who may not be able to respond appropriately to their thirst, and people with certain kidney disorders. Symptoms of hyponatremia range from nausea, muscle cramps, weakness and fatigue to seizures and a loss of consciousness in severe deficiencies. In extreme cases the condition can be fatal.

Tetany

Tetany is a muscular spasm in the hands, feet and face, which is a symptom of a metabolic imbalance. This potentially life-threatening disease can be caused by abnormally low levels of calcium, potassium or magnesium in the blood or by an over-acid or over-alkaline condition of the body.

It is a painful condition in which the muscles of hands and feet cramp rhythmically and the larynx spasms, causing difficulty in breathing, nausea, vomiting and convulsions. There may also be sensory abnormalities such as an odd feeling in the lips, tongue, fingers and feet, general muscle aches and spasms of the facial muscles.

The condition may accompany poorly controlled hypoparathyroidism, hypophosphatemia, osteomalacia, renal disorders or malabsorption syndromes. If caused by abnormal calcium levels, tetany may be associated with a vitamin D deficiency. The treatment may restore metabolic balance.

Acidosis

Acidosis is an excess of acid in the body's fluids. Body metabolism works best in a narrow pH range—between 7.35 and 7.45 (pH is a measure of acidity). The lungs, kidneys and chemical buffers in the blood (such as bicarbonate) keep the pH at this level. But in illness, the pH can fall, causing acidosis. If uncorrected, it can lead to death.

In mild cases of acidosis the usual symptoms are agitation, headache and tachycardia (abnormally fast heartbeat). In more serious cases it causes lethargy, confusion, seizures, stupor and coma. The treatment of acidosis is to correct the underlying cause. If the acidosis is severe, the patient is given sodium bicarbonate intravenously.

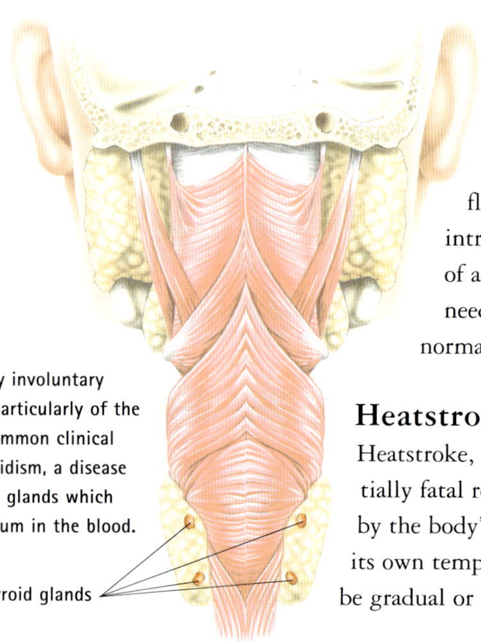

Tetany

Tetany is characterized by involuntary contraction of muscles, particularly of the hands and feet. It is a common clinical feature of hypoparathyroidism, a disease affecting the parathyroid glands which causes low levels of calcium in the blood.

Alkalosis

Alkalosis is an excess of alkali in the blood and other body fluids. Normally, the body's regulatory mechanisms keep the pH of the blood slightly alkaline, within the range of 7.35 and 7.45 (pH is a measure of acidity). But certain conditions cause it to rise above this level, causing alkalosis.

There are two types of alkalosis. Metabolic alkalosis occurs when there is a loss of acid from the body fluids. This generally happens during kidney failure, vomiting and intestinal obstruction, or due to the effect of some drugs. Metabolic alkalosis can also occur if there is excess absorption of alkali, as occurs sometimes after a blood transfusion or excess ingestion of bicarbonate in drug preparations.

The other type is respiratory alkalosis. This is caused by hyperventilation (rapid breathing) that results in excessive loss of carbon dioxide. Respiratory alkalosis can also be caused by head injury, lung disease, liver failure and salicylate (aspirin) poisoning.

Symptoms of alkalosis include tingling skin, muscle weakness and muscle cramps. The condition is confirmed by a blood test, which shows changes in blood pH, bicarbonate and oxygen and carbon dioxide levels.

Alkalosis is a serious condition. Hospitalization is necessary to correct the underlying cause and to facilitate replacement of lost body fluids. In some cases, intravenous administration of acidic compounds will be needed to restore the pH to normal levels.

Heatstroke

Heatstroke, or sunstroke, is a potentially fatal reaction to heat caused by the body's inability to regulate its own temperature. Its onset may be gradual or sudden with headache, weakness and nausea followed by mental confusion and high body temperature around 104–115°F (40–46°C). Shock, convulsions, brain damage, coma and death may follow.

Treatment is to reduce the temperature below 102°F (39°C) by moving the person to a cooler place, removing clothing where possible, and cooling with fans and ice packs to the head, neck, armpits and groin.

Dehydration

Dehydration—the depletion of the body's water content—is a life-threatening condition. Babies and children, who need more fluid relative to their body weight than adults, can become dehydrated—and be rehydrated—more easily.

The most common causes of dehydration are not drinking enough liquid (the deprivation of water is even more serious than the deprivation of food), vomiting, diarrhea (commonly associated with gastroenteritis), the use of diuretics (causing the kidneys to excrete excess water and salt), overheating and fever. Dehydration due to diarrhea is a major cause of death in children of developing countries. Dehydration may also be due to diabetes mellitus and Addison's disease.

Overheating or heat stress can even cause death in the elderly, babies and children who are not protected from high temperatures. Babies and children die every year when they are left in cars in summer.

On average the human body loses around 2.5 percent of its total body water each day, which is equal to 2½ pints (1.2 liters) in urine, expired air, perspiration and from the gastrointestinal tract. Drinking plenty of water before, during and after exercise helps prevent dehydration.

Symptoms of dehydration include thirst, sunken eyes (in a baby the fontanelle will also look sunken), dry mouth, infrequent dark urination (fewer wet diapers), lethargy and irritability. Skin tone will deteriorate so much so that when the skin is pinched it will not spring back as normal skin does. Blood in the stools, a high fever, extreme weakness or collapse require immediate medical treatment.

Since dehydration is almost invariably associated with some loss of salt (sodium chloride), rehydration also requires the

Controlling and maintaining fluid balance

The pituitary gland works with the kidneys to control water balance. When too much water is lost and dehydration occurs, the posterior part of the pituitary secretes larger than normal amounts of antidiuretic hormone (ADH) which make the kidneys retain fluid. The hypothalamus monitors water levels in the blood. When levels drop too low the hypothalamus instructs the pituitary to release antidiuretic hormone (ADH). This hormone makes the kidneys reabsorb water into the blood and decreases urine production. When the body is rehydrated, the hypothalamus tells the pituitary to slow ADH secretion, the kidneys keep less water in the blood and the bladder releases urine again.

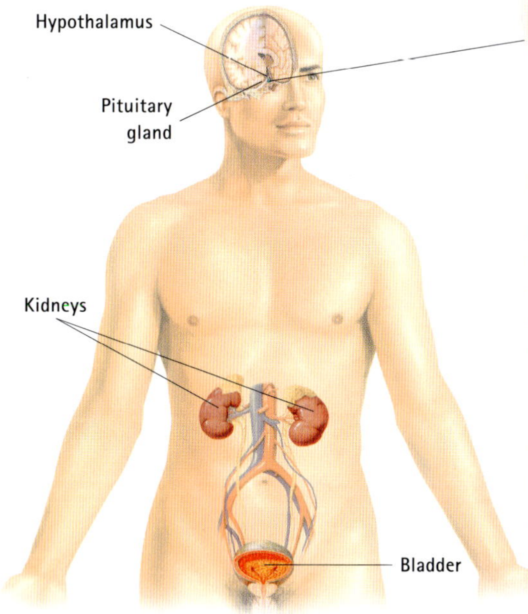

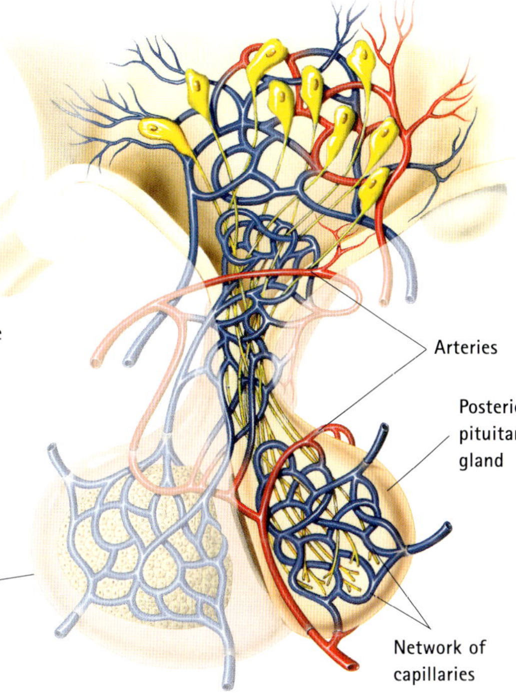

DEHYDRATION

Dehydration occurs when fluid levels in the body are depleted. Under normal conditions water accounts for around 70 percent of lean body mass. The symptoms of dehydration include thirst, sunken eyes, dry mouth and infrequent urination. The skin may also lose elasticity.

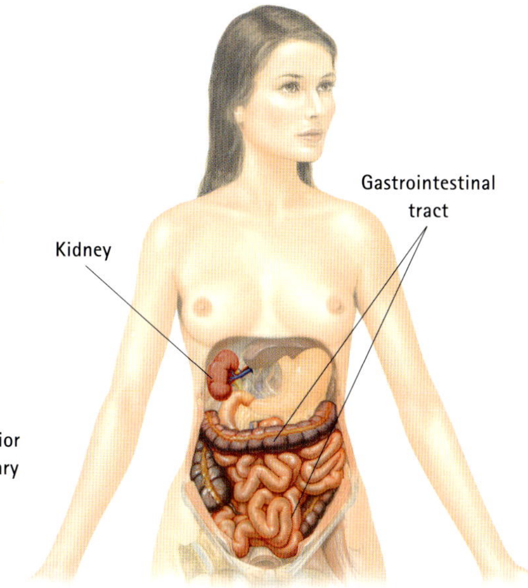

Causes of dehydration

Disorders of the gastrointestinal tract (diarrhea and vomiting), the kidney and the skin (such as sunburn) can all cause fluid loss and dehydration. Children are especially vulnerable to dehydration because water constitutes a higher percentage of their body weight (about 80 percent) than adults.

restoration of the normal concentration of salt in the body fluid. Mild cases can be treated with an oral rehydration fluid. But severe cases need medical aid so that the correct amounts of salts and water are given to restore the normal osmotic relationships between cells, allowing the kidneys to once again begin to work properly.

Fluid retention

Fluid retention is the abnormal accumulation of the fluid that surrounds cells and is a symptom of an underlying disorder of kidney function and fluid balance.

The balance of fluid in the body is regulated by the hypothalamus, which triggers the release of hormones to control the functions of the kidneys and bladder.

Normal fluid balance can be upset by excessive blood loss, and by various disorders. When sodium levels rise the blood volume also rises and the excess fluid results in edema and swelling in the feet and legs. Dietary salts can aggravate fluid retention, and heart patients particularly are advised to follow low-sodium diets. Hormonal changes during the menstrual cycle may cause retention of sodium and fluid and produce a bloated feeling. Heart, liver or kidney failure may also cause fluid retention.

In pregnancy, swelling caused by fluid retention is common and can usually be relieved by resting with the legs up; if it is severe, a doctor should be consulted.

Other causes include certain drugs used in the treatment of heart disease (notably calcium channel blockers), head injuries, stroke, and any surgery or disorder that interferes with normal drainage of lymphatic fluid and its return to the blood.

Diuretic drugs may be prescribed to increase urine output and decrease fluid volume throughout the body. It is advisable to consult a physician who will prescribe them if required, since they cause dilation of blood vessels and loss of potassium.

Edema

Edema is swelling caused by the build-up of fluid in the tissues. It can be localized (limited to a part of the body), or generalized (occurring throughout the body).

Localized edema may result from injury or infection, sunburn or varicose veins. Slight edema of the legs commonly occurs in warm summer months and during pregnancy.

More generalized edema can be caused by heart failure or by a lack of protein in the blood from cirrhosis of the liver, chronic nephritis or malnutrition. It can also occur as a result of toxemia of pregnancy (preeclampsia).

Edema is treated by correcting the underlying cause. Diuretic drugs, which make the kidneys eliminate excess salt and water, are often used to relieve the symptoms. Elastic stockings will help prevent edema caused by pregnancy, failure of lymphatic drainage, or varicose veins.

Pulmonary edema is a complication of heart disease in which the failing heart allows fluid to accumulate in the lungs, which eventually seeps into the air spaces (alveoli). This interferes with the exchange of oxygen and carbon dioxide, causing severe breathlessness. Pulmonary edema is a medical emergency, requiring immediate hospitalization and treatment with oxygen, diuretics, morphine and drugs for the underlying heart failure.

The Head

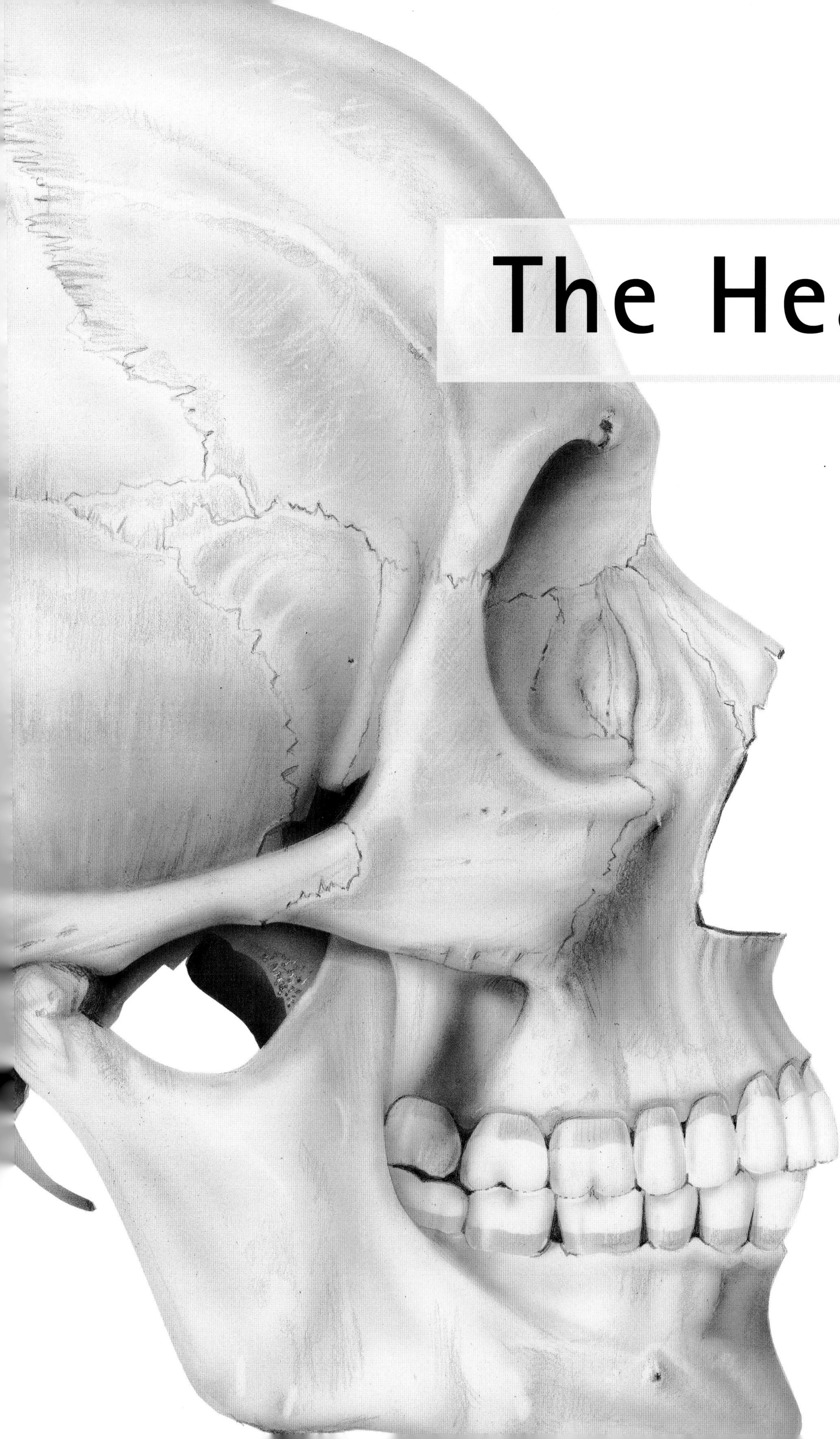

HEAD

The head contains the brain, encased in the cranium within the skull. There are five special senses in the head associated with cranial nerves: smell, vision, taste, hearing and balance (semicircular canals, inner ear).

The tongue is contained in the oral cavity (mouth). At the back of the oral cavity is the pharynx, which allows food to pass to the esophagus and air to pass into the trachea. Muscles in the neck allow the head to be flexed, extended and partially rotated.

SKULL

The skull forms the skeleton of the head, and is a part of the axial skeleton. It serves to protect the brain, eyes and inner ears; forms the upper and lower jaws; and provides attachment for muscles of the face, eyes, tongue, pharynx and neck.

The skull is the most complex bony structure in the body. With the exception of the lower jaw (mandible), the numerous bones of the skull are joined to each other by fibrous joints called sutures. Many of the skull bones develop separately from a membrane which covers the embryonic head.

At birth a significant amount of this membrane is still present between the partially formed bones, allowing them to slide over each other as the head passes through the birth canal. This means the bones are not crushed during the birth process. For descriptive purposes the skull is divided into two parts, the cranium and the face.

The skull provides a strong protective case for the brain and the sensory organs for sight, hearing, balance and smell. This protective value is based on several factors, including the convex shape of the part enclosing the brain, the strength of the interlocking arches of the face, the hardness of the outer bones, and the presence of spongy bone within to absorb impact.

The skull consists of a number of small, irregular-shaped bones. There are seven unpaired, midline bones: the ethmoid, frontal, hyoid, occipital and sphenoid bones and the mandible and vomer. There are 10 pairs of paired bones, one on each side: the lacrimal, maxillary, nasal, palatine, parietal and zygomatic bones, and the temporal bone and middle ear bones (the malleus, incus and stapes). There is also a variable number of sutural bones.

The forehead is formed by the frontal bone. Below that are the two nasal bones, with an opening below each. On either side are cavities for the eyes, called the orbits, so named because the eyes rotate in them. An arch of bone, the zygomatic arch, forms the skeleton of the cheek. The roof or vault of the cranium is formed by the frontal, parietal and occipital bones. The rear view of the skull is dominated by the occipital bone in the midline below, with the parietal bones above on each side. At the sides of the skull, the temporal bones contain the middle and inner ears, with the special sense organs for hearing and balance. The middle ear is unique in containing a chain of three tiny bones (ossicles) which relay sound vibrations from the tympanic membrane.

Strictly, the skull consists of the cranium and the mandible (lower jawbone) plus the hyoid bone at the base of the tongue, but the term "skull" is usually applied loosely to the cranium only.

SEE ALSO *Fetal development on page 398; Skeletal system on page 30*

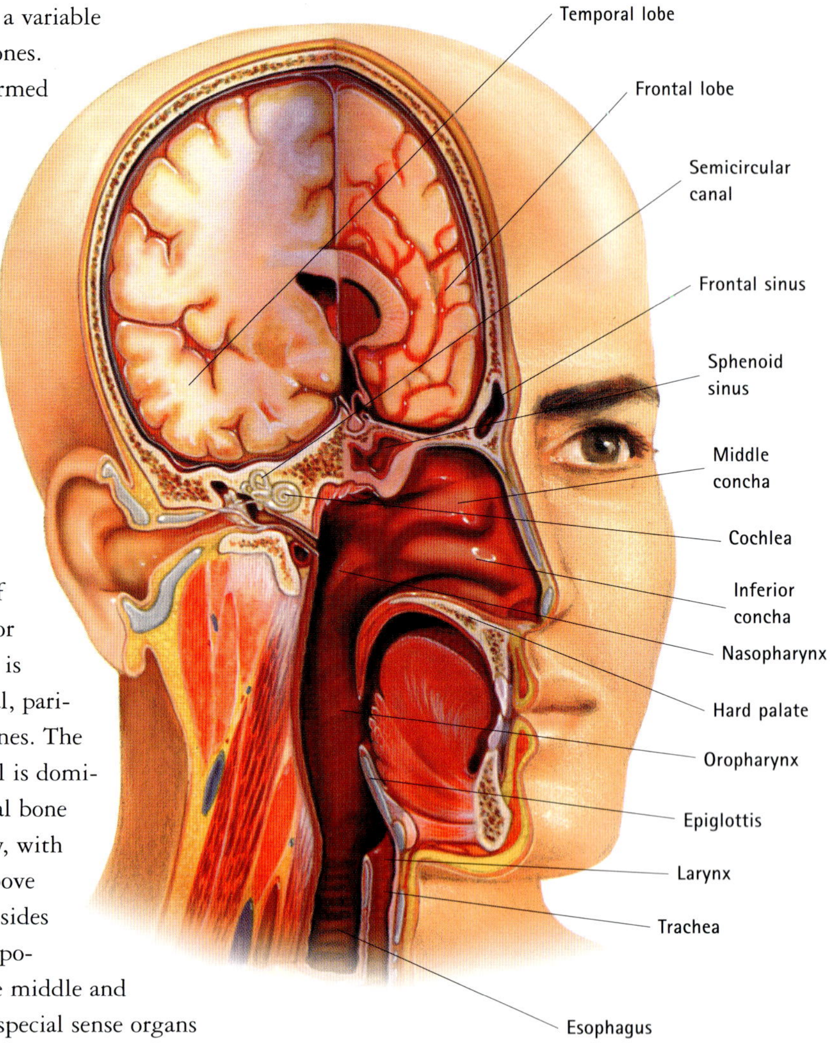

Head

The head contains the brain and special sense organs. Muscles in the neck enable the head to flex, extend and partially rotate.

Cranium

The cranium consists of two regions, one accommodating the brain and the other forming the skeleton of the face. The first part consists of a strong box, approximately ovoid in shape, supported by the spine. To this box is attached a series of arches, which form the skeleton of the face. From the side, the division of the skull into the larger, ovoid brain case and the smaller, approximately triangular skeleton of the face, is clear.

The bones of the cranium are linked together by "joints" known as sutures, found only in the skull. No active movement occurs. The adjacent bones have irregular, interlocking edges (rather like a jigsaw puzzle) bound together by fibrous connective tissue. Sometimes there are small sutural bones interpolated in sutures between the bigger bones.

The cranium is formed, from front to back, by the frontal bone, the paired parietal and temporal bones and the occipital bone with the sphenoid bone also forming part of the cranial joint. The lobes of the cerebral hemispheres of the brain are named according to the bone which overlies them: the frontal, parietal, temporal and occipital lobes. The temporal bone contains a system of spaces within it, which form the middle and inner parts of the ear. Each bone (except the parietal) also contributes to the floor of

Skull—side view

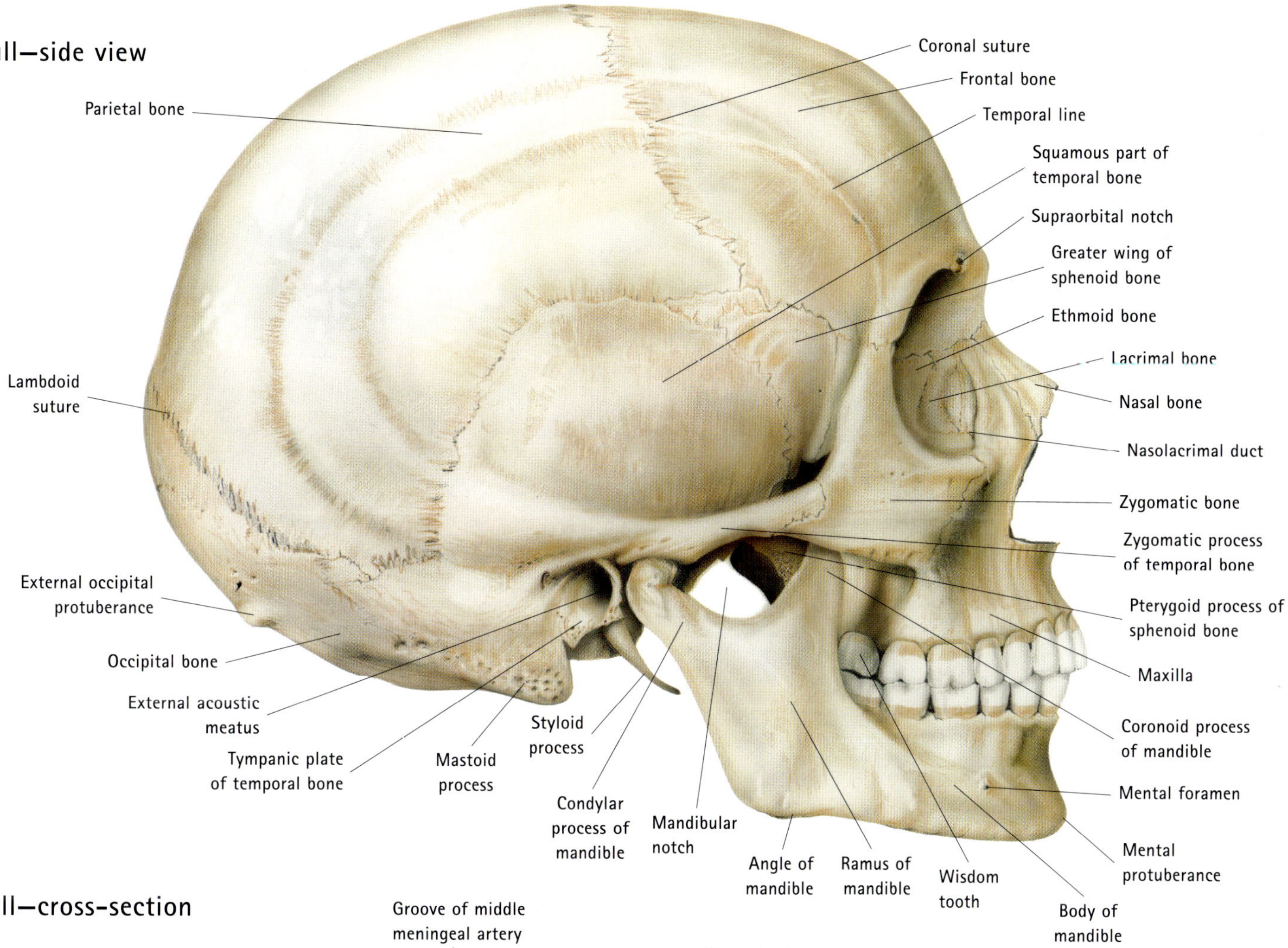

Skull—cross-section

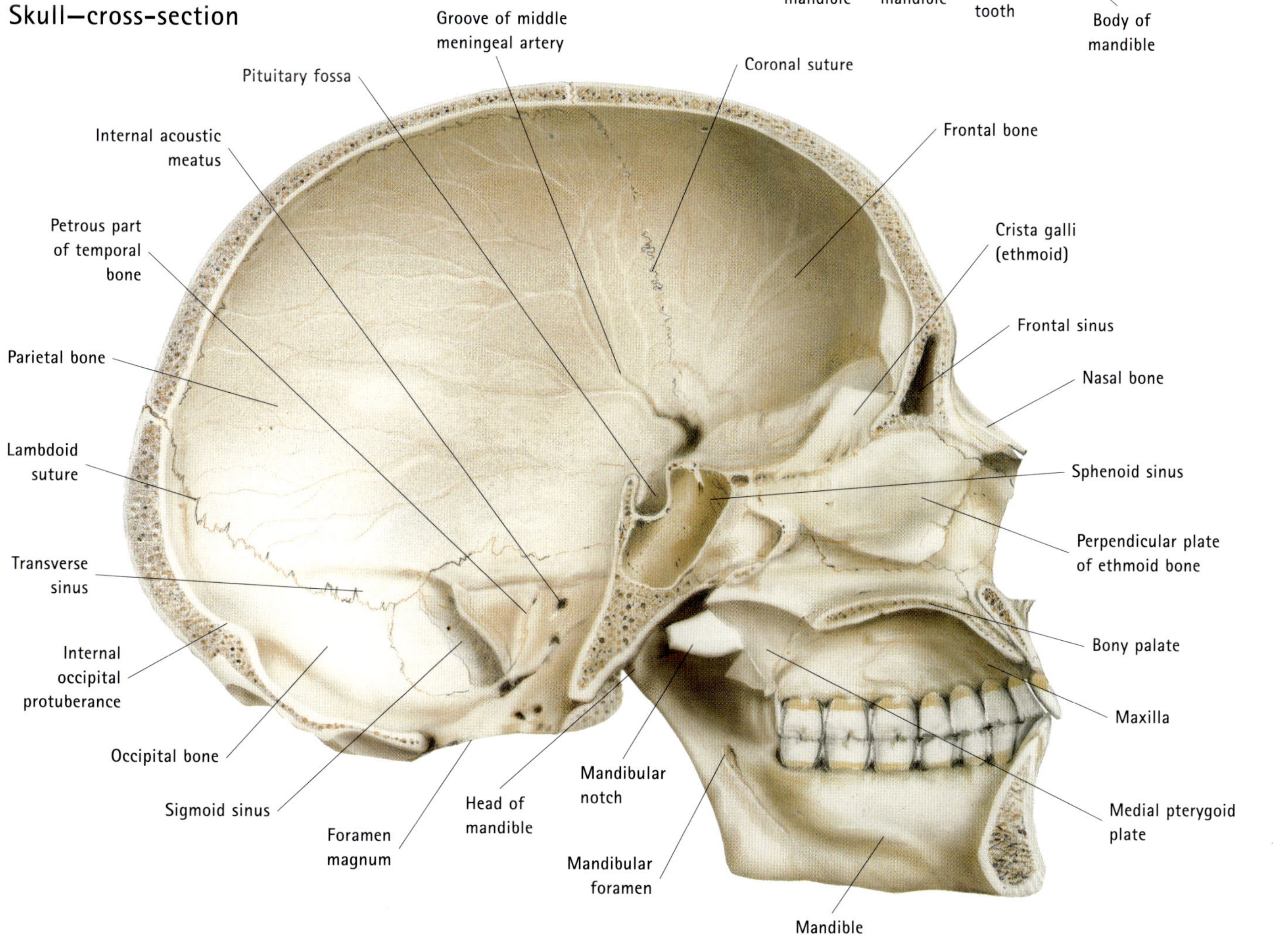

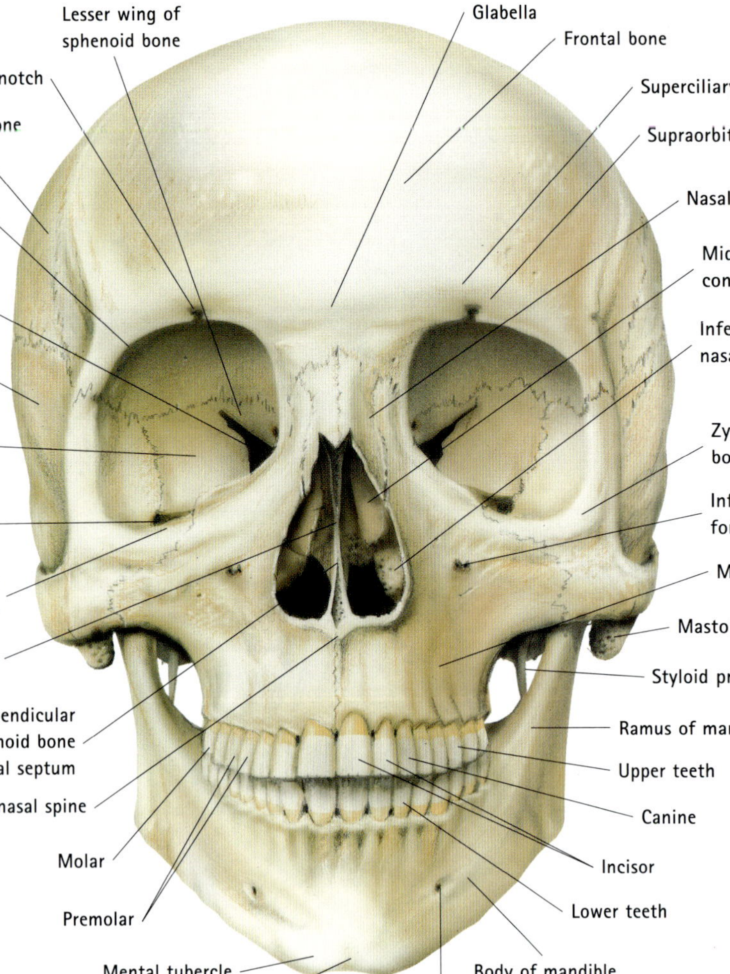

Skull—front view

Skull—back view

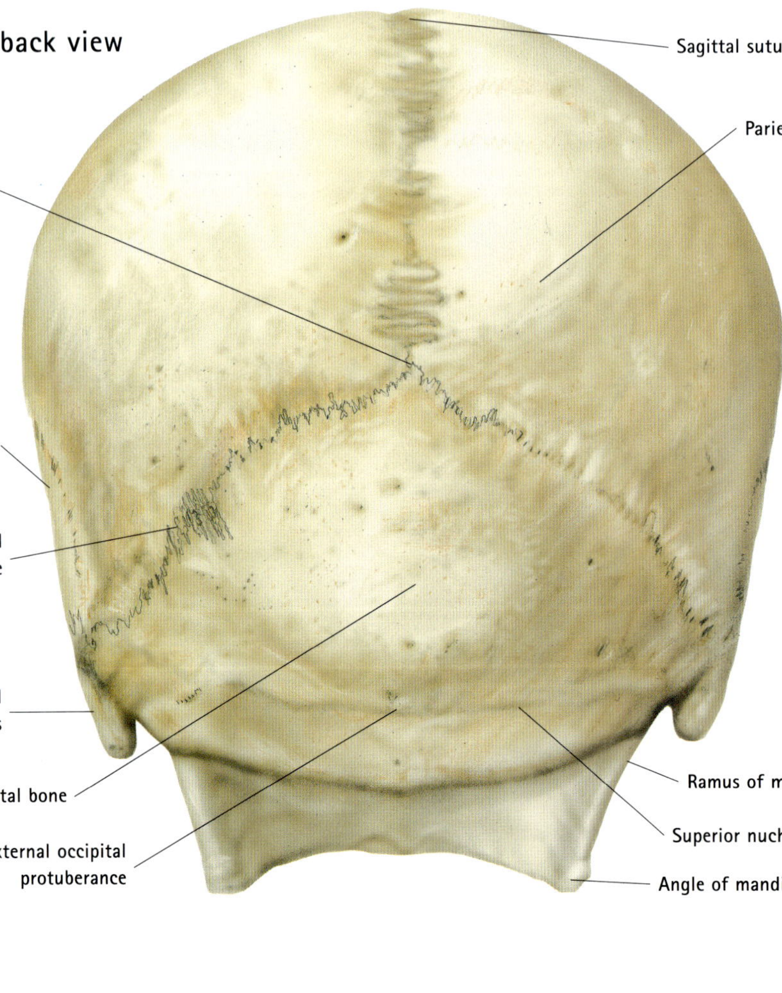

the cranium, which is divided into three terraces known as the anterior, middle and posterior cranial fossae.

The anterior fossa is occupied by the frontal lobes of the brain, the middle one by the temporal lobes and the posterior one by the cerebellum and brain stem. The floor of the cranium contains holes (foramina) through which the spinal cord, cranial nerves and blood vessels enter or leave.

The brain is separated from the skull by three layers of membrane, which are known as the meninges. The outermost layer, which is called the dura mater, is a tough, fibrous membrane which adheres to the inner surface of the skull. The middle layer, known as the arachnoid mater, lines the inner surface of the dura, to which it is loosely attached. The innermost layer, also called the pia mater, adheres to the brain and follows its contours. Between the arachnoid and pial layers is the fluid-filled subarachnoid space, which forms a cushion around the brain, providing it with buoyancy and protection.

At birth, the cranium is relatively large and the face is small. The teeth are not fully formed and the paranasal sinuses are rudimentary. There is no mastoid process. Between some bones of the vault, ossification is incomplete, leaving gaps containing fibrous connective tissue, particularly at the angles of the parietal bones. These gaps are called fontanelles. The anterior fontanelle, easily felt in the newborn child, lies between the separate halves of the frontal bone and the two parietal bones; about 1½ inches (3 centimeters) long and 1 inch (2 centimeters) wide, it closes during the second year. The posterior fontanelle at the apex of the occipital bone, between the two parietal bones, is closed at 2 months. There are several smaller fontanelles. Growth of the skull is rapid for the first 7 years, then slows until puberty, when there is another period of rapid growth.

Arteries which supply the skull and meninges, and veins draining blood from the brain, lie between the dura and the skull. An extradural hemorrhage occurs when these vessels are torn (usually in association with a skull fracture) and blood rapidly accumulates between the dura and the skull, putting pressure on the underlying brain. This is a potentially fatal condition which must be treated quickly, usually by making a hole in the skull to release the pressure. Subdural hemorrhage occurs when veins are ruptured while passing through the dura mater, causing blood to accumulate between the dural and arachnoid layers.

A subdural hemorrhage can occur from a bump on the head but is not as serious as an extradural hemorrhage because the blood accumulates at a much slower rate. The major arteries supplying the brain run through the subarachnoid space before entering the brain. Spontaneous subarachnoid hemorrhage may occur due to a rupture of congenital aneurysms in one of these arteries.

Facial bones

The bones of the front of the skull constitute the face. The frontal bone forms the forehead; the zygoma forms the cheek bone; the maxilla forms the upper jaw, palate and outer walls of the nasal cavity; and the mandible forms the lower jaw.

Jaw bones

The upper jaw and the bony roof of the mouth are formed by the maxillary and palatine bones. The mandible, the skeleton of the lower jaw, articulates with the temporal bones on each side at the temporomandibular joints.

SEE ALSO *Jaw on page 132*

Orbit

Each eye and its associated structures (muscles, nerves and blood vessels) occupies a cavity on the front of the skull know as the orbit. The walls of the orbit are relatively thin and fragile and can be easily fractured if a small object is inadvertently poked into the orbit, or by a blow to the front of the eye. The outer margin of the orbit, however, is thick and strong, and protects the eye from damage caused by larger objects. The eyelids protect and lubricate the eye. The lacrimal gland, which produces tears, is located just behind the upper eyelid.

Nose (nasal cavity)

The walls of the nasal cavity are formed by bone at the back and by the nasal cartilages at the front; the cartilages surround the nostrils. The cavity is divided into two halves by the nasal septum and its floor is formed

Skull—base

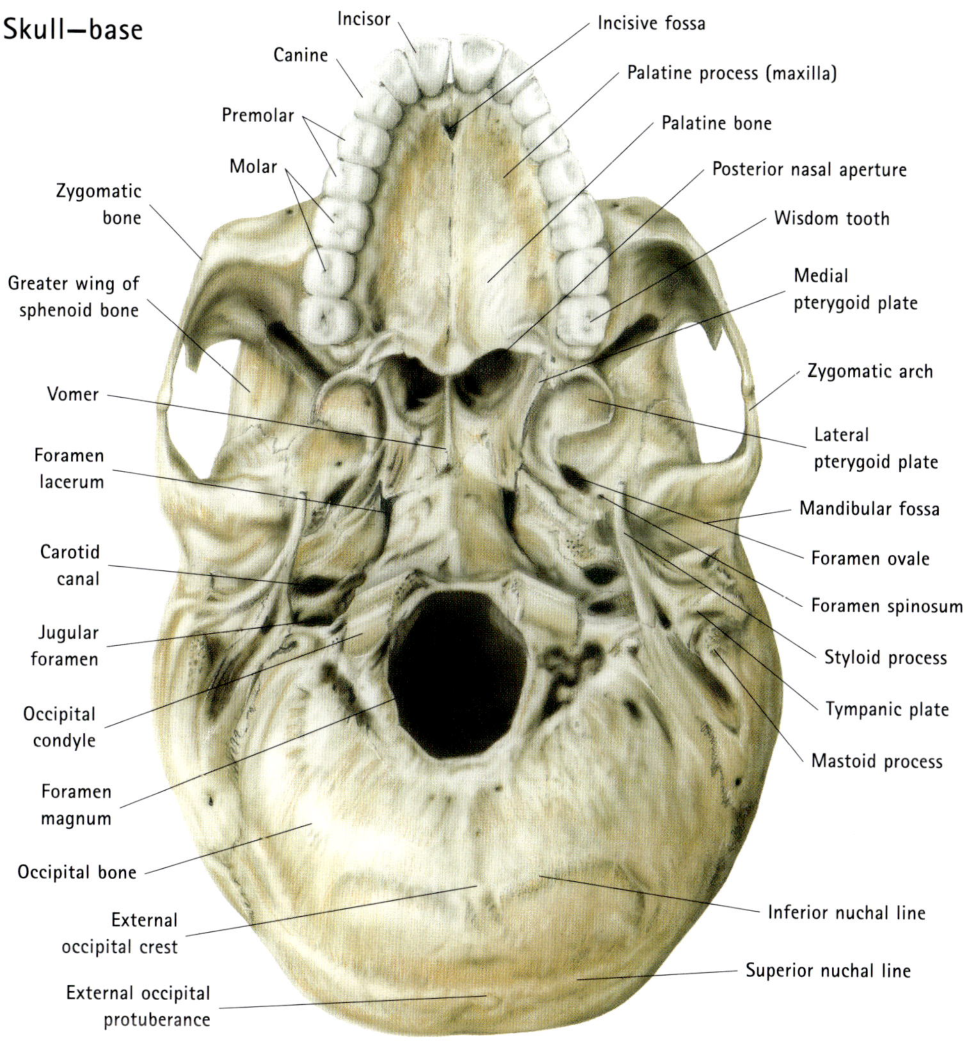

Skull—top

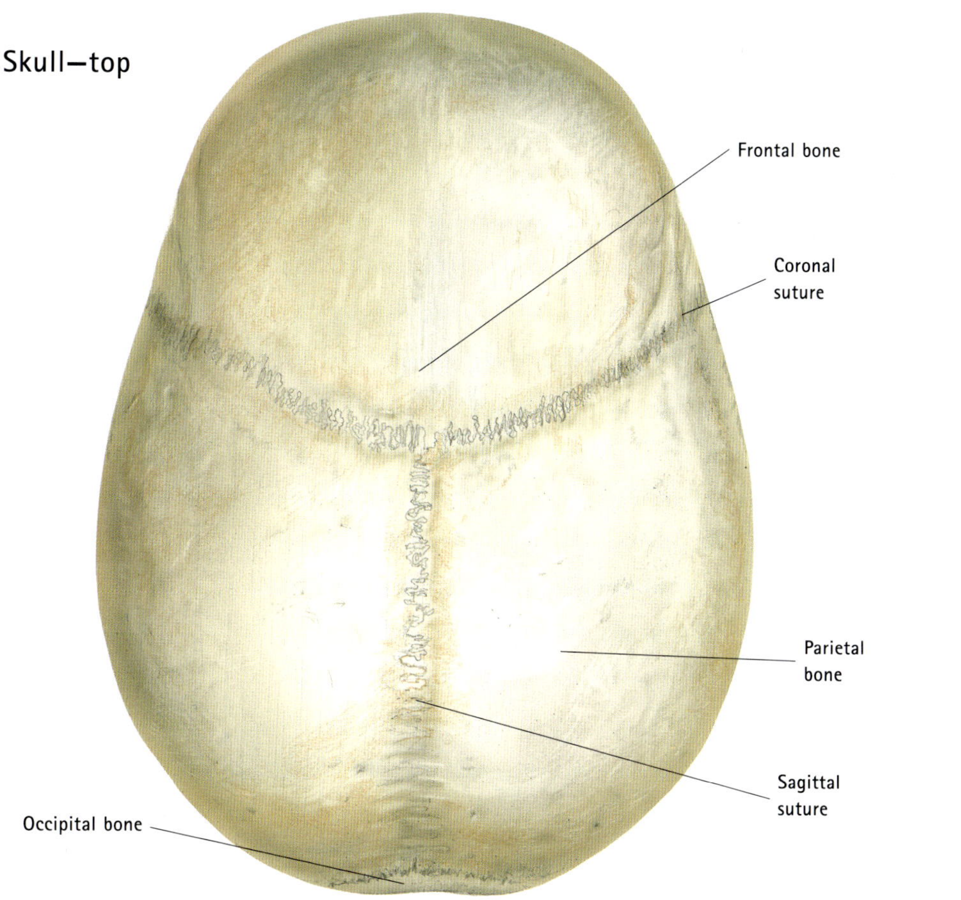

by the palate, the roof of the mouth. The inner lining of the nostrils is formed by hairy skin; the membrane at the roof of the nasal cavity contains specialized olfactory cells, which are responsible for detecting smells. The remainder of the nasal cavity has a specialized lining called respiratory epithelium, which has a very rich blood supply and numerous mucous glands to warm and moisten the air as we breathe.

In order to maximize the available surface area for this epithelium, the walls of the nasal cavity contain coiled inward projecting bones, known as conchae or turbinate bones. The nasal cavities open behind into a space known as the pharynx, through which air passes on its way to the lungs.

Paranasal sinuses

The nasal cavity also communicates with spaces known as the paranasal sinuses, which occupy the major bones near the nasal cavity (the maxillary, frontal, ethmoid and sphenoid sinuses, named after the bones in which they are located). These sinuses are also lined by respiratory epithelium and infection can spread to them from the nose, causing congestion and pain (sinusitis).

The maxillary sinuses, which lie to the sides of the nose, are the largest and most susceptible to infection, from either the nose or the upper teeth.

Oral cavity (mouth)

The oral cavity or mouth extends from the lips at the front to the throat (pharynx) at the back, and is bounded at the sides by the cheeks. Its roof is formed by the hard (bony) palate and behind by the soft (muscular) palate. Its floor is formed by muscle. Its major contents are the teeth and the tongue.

There are a total of 20 baby (deciduous) teeth, which usually appear between the ages of 6 and 24 months. From approximately 6 years of age onward, the deciduous teeth are gradually replaced by permanent or adult teeth. The mature adult has a total of 32 teeth. Each jaw contains 4 incisors (for cutting), 2 canines (for tearing), 4 premolars and 6 molars (for grinding).

The third molars, or wisdom teeth, are the last to appear and cause problems in those people whose jaws are not long enough to fit them comfortably.

Palatine bone

The palatine bone is an irregularly shaped bone at the back of the nasal cavity that forms part of the hard palate. It consists of a horizontal plate in the bony palate and a vertical plate that has three projections, or processes, which help form the floor of the eye socket, the outer wall of the nasal cavity and other adjoining parts of the skull.

Base of the skull

The base of the skull is supported by joints between the occipital bone and the uppermost bone of the spine (the atlas or first cervical vertebra). The joint with the atlas permits the head to make a nodding motion. There is a large hole in the occipital bone, the foramen magnum, above the spinal canal of the atlas vertebra, which allows for the passage of the medulla oblongata and meninges.

Skull injuries

Fractures of the bones of the skull can be associated with serious injury to the brain, meninges and the sensory organs for sight, hearing, balance and smell. A particular risk with skull fractures is intracranial hemorrhage, which may cause pressure damage to the brain. The most frequently fractured bones are the lower jaw and the nasal bones. The skulls of children are more elastic than those of the adult; blows to the head may result in serious injury to the brain and meninges without fracturing bones.

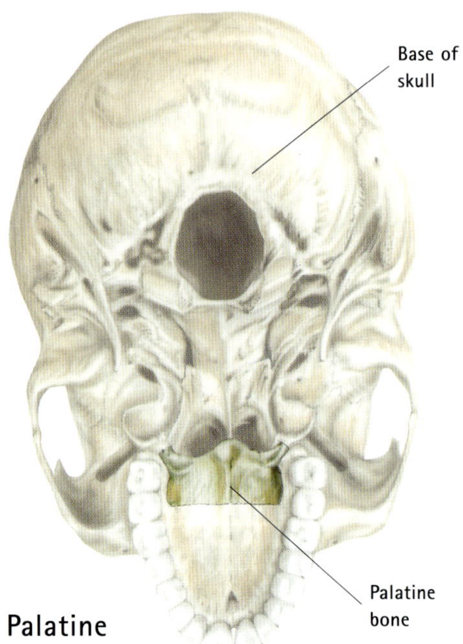

Palatine bone

Situated at the back of the nasal cavity, the palatine bone forms part of the hard palate.

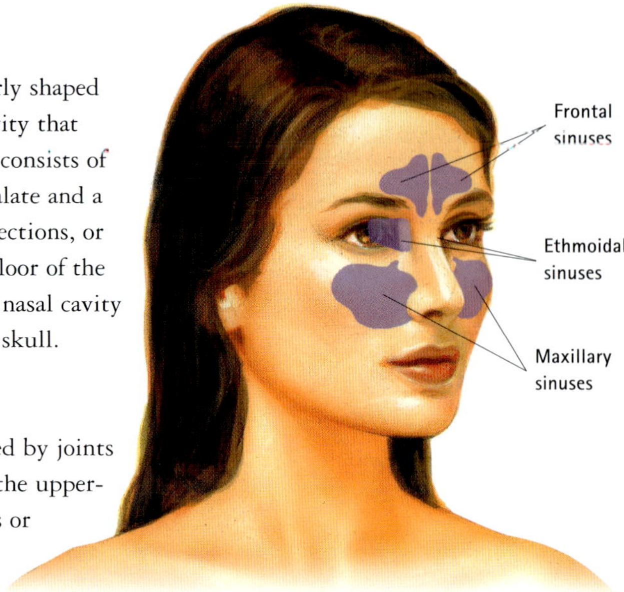

Paranasal sinuses

The paranasal sinuses consist of the frontal, ethmoidal and maxillary sinuses. Infection of the paranasal sinuses (sinusitis) may cause localized facial pain.

FACE

The face is made up of facial bones, muscles, skin, eyes, nose, jaws, cheeks and chin, as well as the nerves and blood vessels supplying these structures.

The facial bones are covered with muscles and skin. Variations in these features account for our individual appearance; it is thanks to these variations that we can recognize each other and tell one another apart. Much of what we think and feel is, more often than not, expressed in the face.

SEE ALSO *Muscular system on page 48; Skeletal system on page 30*

EVOLUTION AND DEVELOPMENT

During evolution from the prehuman *Australopithecus* to modern human *(Homo sapiens)*, the face became smaller compared to the overall size of the head. The brain and the cranium (braincase) tripled in volume, but the jaws became shorter and the teeth smaller. In consequence, the face receded beneath the forehead.

As a result, the modern human face exhibits an essentially vertical profile, in marked contrast to the protruding facial muzzle of the gorilla and the chimpanzee. As the jaws receded, they left the distinctive modern human features of a prominent nose and a sharply defined chin.

By the age of six years, the brain and the cranium have reached 90 percent of their adult size. But the face grows more slowly; at birth it is less than one-fifth the size of the braincase; by adulthood it has increased to nearly half. Facial dimensions increase most in depth, next in height (length), and least in width. Facial musculature

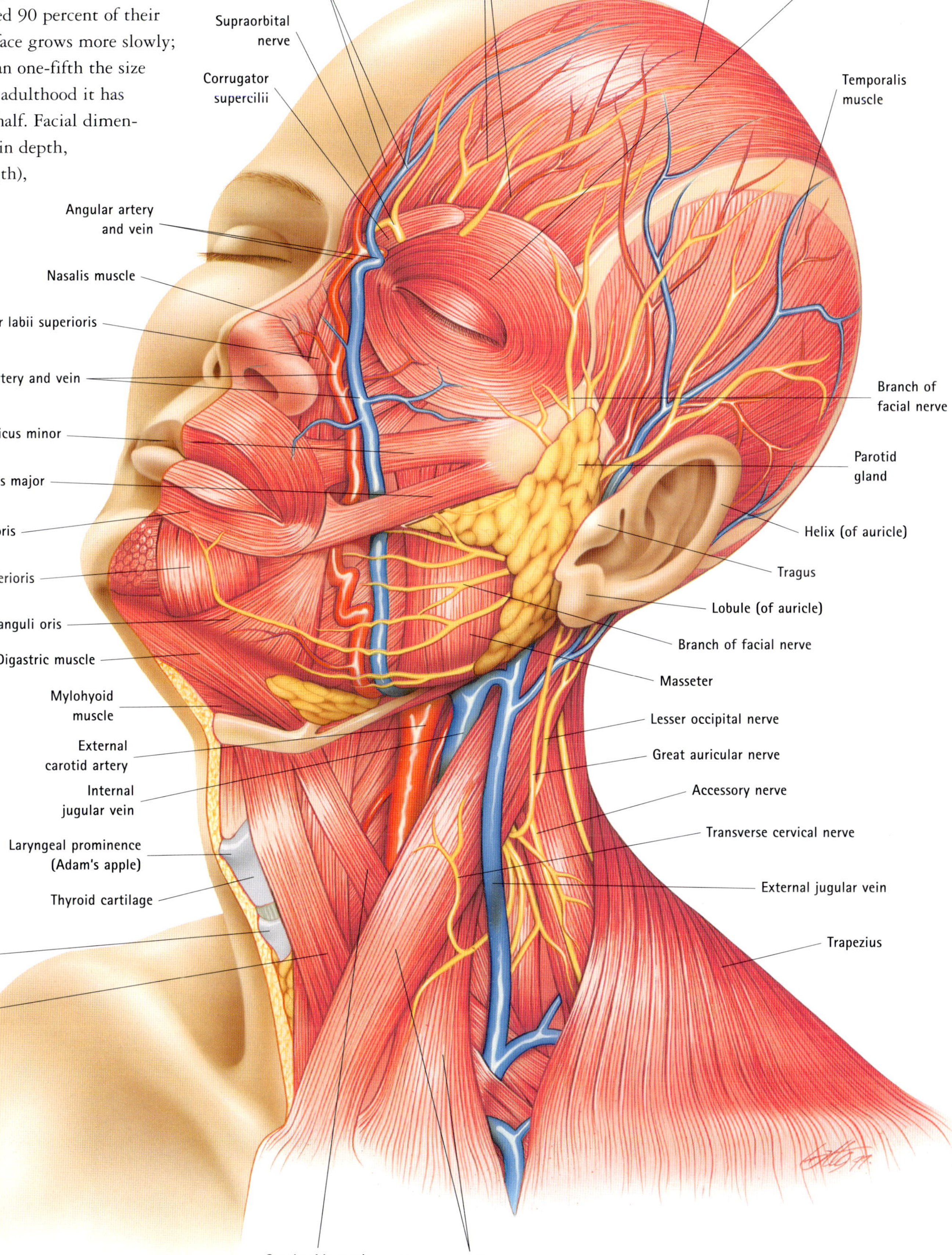

Face

A complex range of thoughts and emotions is expressed in the face—this requires an intricate system of muscles and nerves. The special sense organs (eyes, nose, ears and tongue) are all part of the facial structure.

increases and the nasal sinuses enlarge during adolescence, particularly in males.

PARTS OF THE FACE

The facial skeleton is made up of 14 bones. The frontal bone forms part of the forehead. The facial bones include the two nasal bones, forming the upper portion of the bridge of the nose; two lacrimal bones, which are located in each eye socket (orbit) next to the nose and close to the tear ducts; two maxillary bones (upper jaw); the mandible (lower jaw); the two palatine bones of the hard palate; the vomer, which, with a part of the ethmoid bone, makes up the nasal septum; and the two inferior turbinates of the nose.

The eyes are the organs of sight. Shaped like a ball, with a slight bulge at the front, each eye lies within a bony socket of the skull, protected from glare and dust by the eyelids, lashes and eyebrows. When the eyelids are closed as in sleep, the surface of the eye is covered; when awake, the eyelids blink roughly once every six seconds, washing the eye with salty secretions from the lacrimal (tear) glands, which are situated at the outer corner of the eye behind the upper eyelid. These secretions drain through the tear duct at the inner corner of the eye and into the nose; in certain emotional states, secretions from the lacrimal glands overwhelm the ducts and tears spill out over the eyelids. If an object suddenly moves too close to the eye, the eyelids automatically close.

The nose is a protuberance consisting of two cavities around a wall of cartilage called the nasal septum. The skeleton of the nose is cartilage at the tip, but bony closer to the skull. The nose functions as part of the breathing apparatus—filtering, warming, and moistening incoming air on its way to the lungs. The nose also contains olfactory nerve endings that detect smells. Just inside the nostrils grow short, coarse hairs which filter dust particles from the incoming air.

There are also a number of muscles in the face. There is a circular muscle around the mouth and one around each eye. Other muscles spread out over the face from the edge of the circular muscles.

The mouth is the opening between the maxillae and the lower jaw. It is used for ingesting food, breathing air and for making sounds, especially speech. Lips, which form the mouth's muscular opening, contribute to formation of words during speech and also help hold food in the mouth. They also help form facial expressions, such as smiling and frowning.

The sides of the mouth are formed by the cheeks. These are composed of muscle tissue covered on the outside by skin and on the inside of the mouth by mucous membrane. The cheeks play an important role in speech andhold food as it is chewed and swallowed.

The jaws are three bones making up the bony framework of the mouth. The two upper jaw bones (maxillae) are fixed, while the lower jaw (mandible) is moveable. By moving in opposition to each other, jaws can bite and chew food while eating to prepare it for swallowing.

Fixed to the bottom of the maxillae and the top of the mandible are the teeth, which are used for biting into and chewing food.

Facial nerves

The facial nerves, one on either side of the face, are the nerves that control the muscles of the face. They also supply the salivary glands below the mouth and the lacrimal glands in the eye, and carry taste sensations from the front two-thirds of the tongue. They are the seventh pair of the twelve pairs of cranial nerves. The facial nerves emerge from the brain stem, pass through the temporal bone in the skull (which houses the middle and inner ears), exit the skull through a small hole called the stylomastoid foramen, and then fan out over each side of the face anterior to (forward of) the ear.

Disorders of the facial nerves may be caused by fractures of the base of the skull, injuries to the face or middle ear, and birth or surgical trauma. The result is paralysis, weakness, or twitching of the face on the affected side; also, facial features lose their symmetrical arrangement, the mouth

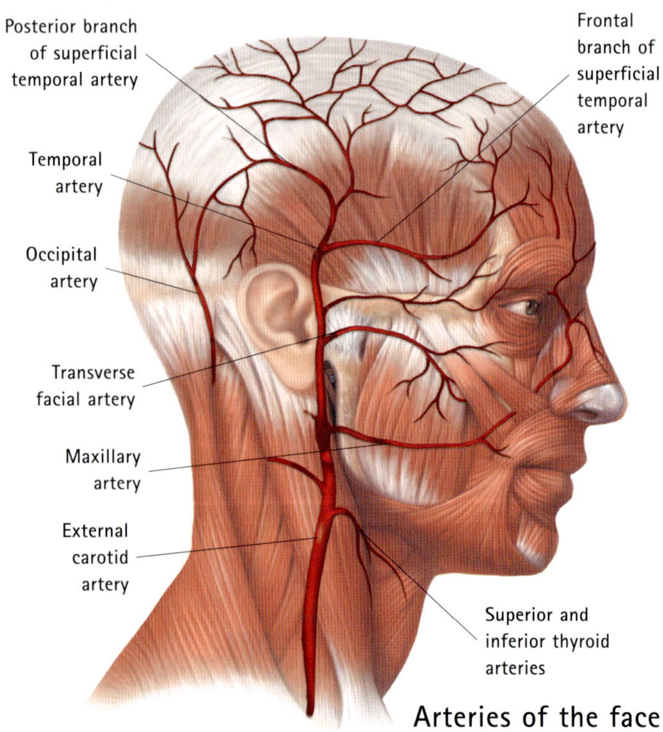

Arteries of the face

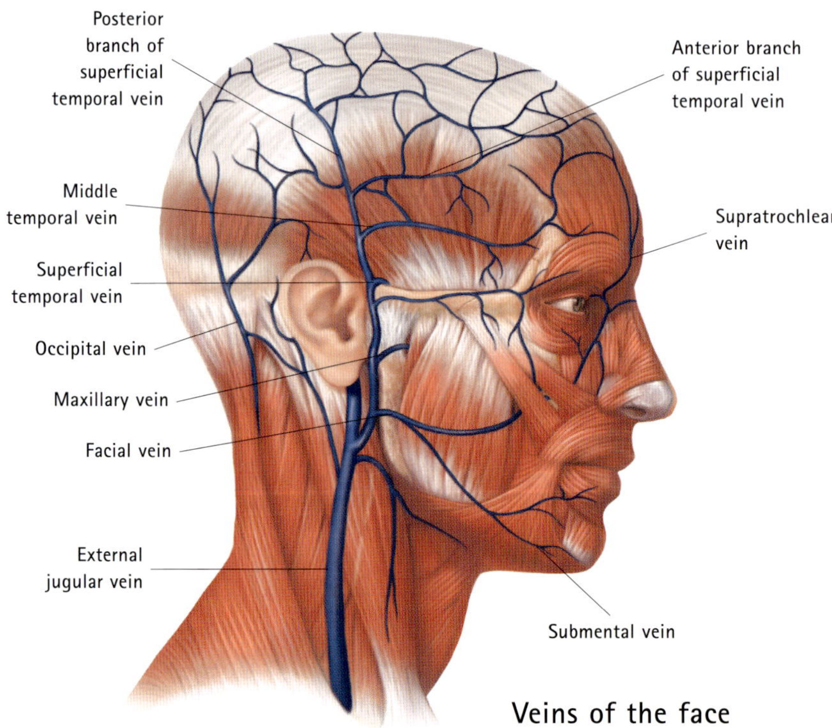

Veins of the face

droops at one corner, the eyelid may not close properly, and there may be dryness of the eye or mouth and loss of taste. When the condition occurs with no known cause it is called Bell's palsy. This condition may come on suddenly or develop over several days; often there is a preceding condition such as stress, fatigue or the common cold. In most cases it improves by itself over a few months. Prompt treatment with corticosteroid drugs may help recovery.

Nerves of the face

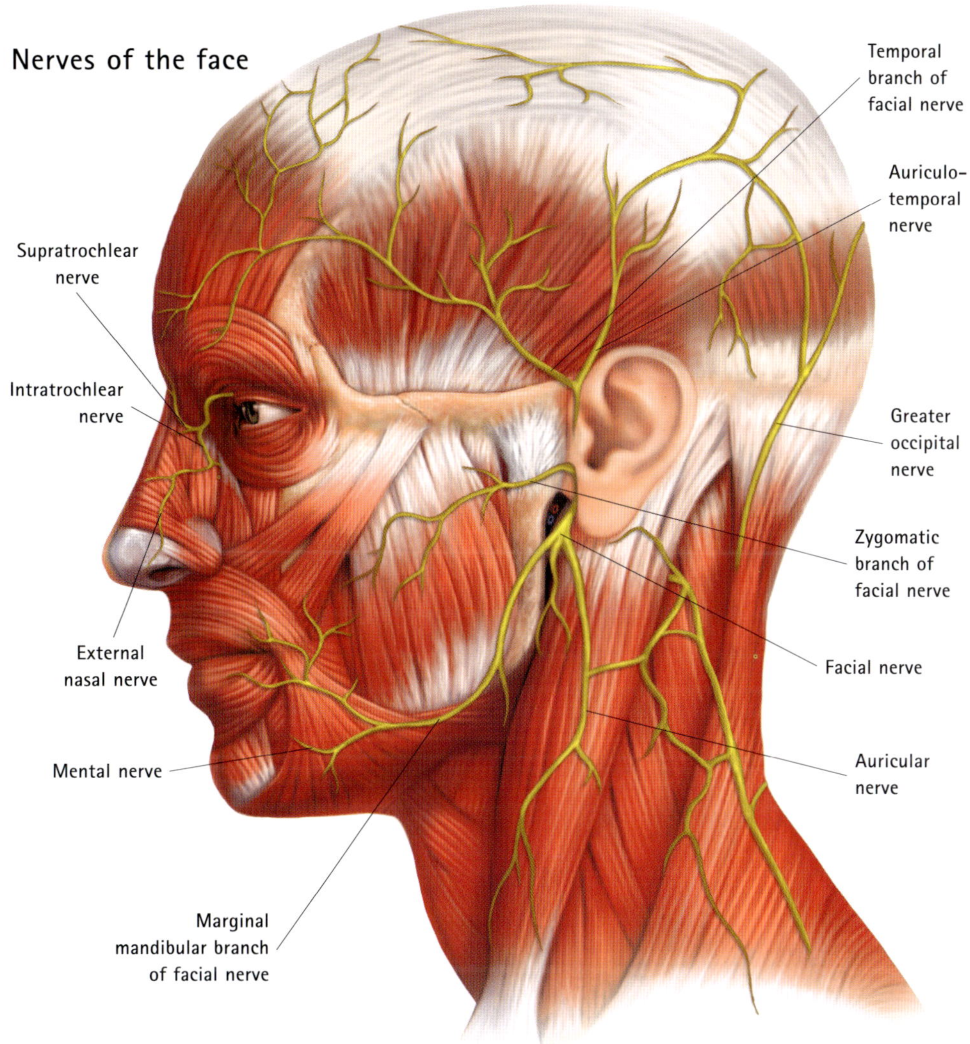

Disorders of the face

The face can be affected by a variety of problems. Diseases including infections of the skin such as herpes zoster (shingles), herpes simplex, tinea and impetigo can affect the face. Conditions of inflammation of the skin (dermatitis), such as eczema and psoriasis, can also occur. Acne may occur, especially during adolescence, and can cause unsightly scarring. Cancers of the skin such as squamous cell carcinoma, basal cell carcinoma and melanoma commonly occur on the face. Many systemic (generalized) diseases affect the face. Endocrine diseases such as Graves' disease, hypothyroidism, Cushing's disease or acromegaly may affect the features by altering the soft tissues or bones of the face. Muscle movements and facial expressions may be affected in neurological diseases such as myasthenia gravis, Parkinson's disease and Bell's palsy. Infections such as tuberculosis, leprosy and cellulitis can damage the tissues of the face and affect its features pemanantly.

A number of conditions can cause facial pain. Inflammation of the tempormandibular joint can cause aching pain over or around the jaw. Infection around a tooth can cause a throbbing pain on one side of face that worsens at night, when eating, or when touching a particular tooth. Sinus infection can cause pain or tenderness around the eyes and cheekbones which worsens when bending the head forward. It follows a recent cold or nasal allergy. Headache may be caused by migraine, tension or stress, meningitis, or high blood pressure (hypertension). Herpes zoster may also cause severe facial pain.

Fractures of the facial bones are common, and are usually due to a sporting injury or a blow to the face. There is severe pain at the injury, swelling and bruising of soft tissue around the fracture (including black eyes), and deformity if the fracture is complete and bone fragments separate enough. Often no treatment is needed, as the fracture heals spontaneously within six weeks.

If there has been displacement of bones, surgery may be needed to realign fractured bones and reconstruct normal facial contours. Fractures to facial bone can often be avoided by wearing protective face masks and headgear when playing contact sports.

Hairs are numerous in the skin of the face, especially in males. A condition called folliculitis, or infection of the hair follicles with staphylococcus bacteria (also known as beard rash), is common on the face. Characteristic yellow-white pustules surrounded by reddish rings form on areas of beard.

The condition can be treated by avoiding hot and moist conditions, using an antibacterial cream on affected areas. If the infection is severe, oral antibiotics will also help in treating the condition.

SEE ALSO *Plastic surgery on page 459; The skin on page 346; Treating infections on page 445; Acne, Bell's Palsy, Herpes zoster, Shingles, Squamous cell carcinoma, Treating cancer, and other individual disorders in Index*

Jaw

The jaw consists of two parts: a moveable lower jaw, formed by the mandible, and a fixed upper jaw, formed by the maxillae. The mandible is made up of a thickened body (which forms the lower border of the face); an alveolar part (which contains sockets for the lower teeth); and a ramus (which projects upward from each end of the body). The top of each ramus has a small rounded head, which fits into a socket on the base of the skull to form the temporomandibular joint. This joint, which is reinforced by a capsule and strong ligaments, allows gliding (backward, forward and sideways) movements, as well as a hinge movement, which occurs during the opening and closing of the jaw.

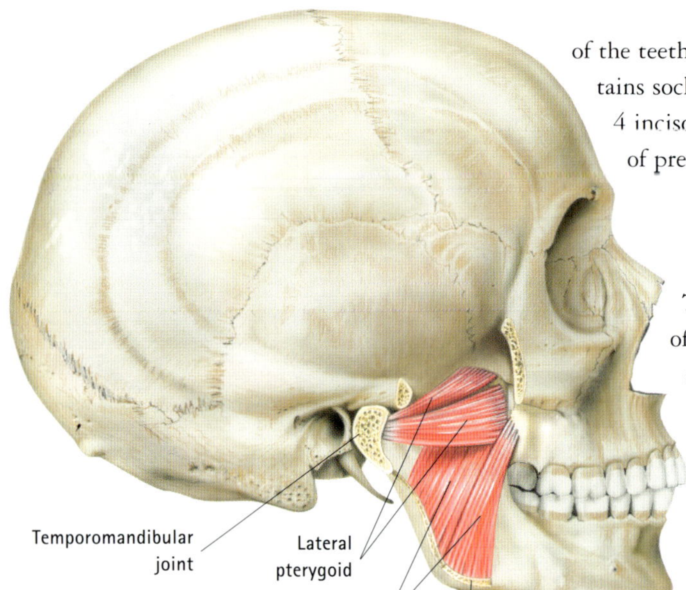

Deep muscles of the jaw

This important group of muscles around the lower jaw moves the mandible up, down, sideways, forward and backward, and is used in biting and chewing. Here the mandible has been cut away to show the pterygoid muscles of the jaw.

The movements of the temporomandibular joint are brought about by a group of muscles known as the muscles of mastication. These include the medial and lateral pterygoid, masseter and temporalis muscles. The medial pterygoid and masseter muscles cover the medial (inner) and lateral (outer) surfaces respectively of the ramus of the mandible and, with the temporalis muscle, act to elevate (close) the mandible. They can all be palpated when the teeth are clenched. The lateral pterygoid muscle is the major depressor (opener) of the mandible.

The upper jaw is formed by the maxillae. It has an alveolar part (which contains sockets for the upper teeth); a palatine part (which forms the hard palate in the roof of the mouth); and a hollow body (which forms part of the cheek). The cavity in the body is known as the maxillary sinus, and is a frequent site of minor infections (sinusitis) which can spread to it from the nose, or in some cases from the upper teeth.

The alveolar parts of the maxilla and mandible contain sockets for the teeth (alveolar sockets). The base of each socket contains a hole through which branches of the superior and inferior alveolar nerves and blood vessels reach the inside (pulp cavity) of the teeth. In an adult, each jaw contains sockets for 16 teeth, including 4 incisors, 1 pair of canines, 2 pairs of premolars and 3 pairs of molars.

Brain

The brain is the headquarters of the nervous system. As well as providing overall control of vital body functions, it enables us to perceive and respond to incoming sensory information, to think, speak and make decisions, and to carry out an entire range of purposeful and highly coordinated movements.

Its average weight is around 3 pounds (1.4 kilograms) and it lies mostly within the cranial cavity of the skull. It is made up of billions of nerve cells (neurons) and supporting cells (glia). Messages are transmitted from one part of the neuron to another electrically, and from one neuron to another by the release of chemicals.

The complexity of the connections between different neurons in the brain is almost incomprehensible, with some neurons commonly making 10,000 or more connections with other neurons. In the nervous system, neuronal cell bodies group together as gray matter and their processes group together as white matter.

The human brain can be divided into four main parts: the cerebrum, diecephalon, brain stem and cerebellum.

SEE ALSO *Autonomic nervous system on page 75; Endocrine system on page 110; Fetal development on page 398; Nervous system on page 64; Spinal cord on page 217*

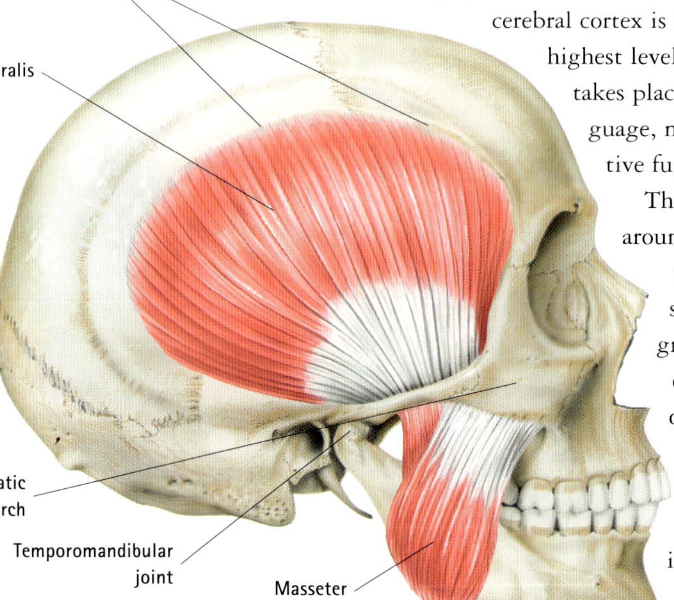

Surface muscles of the jaw—lateral skull

The masseter muscle is used in biting and chewing (mastication), and it is believed to be the strongest muscle in the human body based on force per unit of mass.

Cerebral arteries

The cerebral arteries are blood vessels in the head that transport blood (containing oxygen and other nutrients) to the cells of the hemispheres of the brain. Three cerebral arteries on each side arise near the base of the brain and give branches to deep structures before supplying the cerebral cortex.

The cerebrum

The largest part of the brain is the cerebrum, which is formed by the cerebral hemispheres. The hemispheres are joined together by a massive bundle of white matter called the corpus callosum and covered by the cer-ebral cortex. This is a sheet of gray matter (1.5–4 millimeters thick). The cerebral cortex is the site where the highest level of neural processing takes place, including language, memory and cognitive function.

The cortex makes up around 40 percent of the total brain mass, its surface area being so great in humans (just over 1 square yard, or approximately 1 square meter) that it is thrown into numerous folds in order to fit inside the cranial cavity.

While basic pattern formed by these folds is quite similar in all humans, the size and shape of some folds varies between individuals. The cortex covers the frontal, parietal, temporal and occipital lobes.

We know from studies on patients who have sustained damage to the cortex that the results of damage depend on which part of the cortex is affected. For example, the occipital lobe is involved in the perception of vision, the temporal lobe in memory, and the parietal lobe in the perception of touch and the comprehension of speech. The frontal lobe is not only important in movement but also has a large area devoted to thinking, behavior and personality.

Beneath the gray matter of the cerebral cortex is a thick mass of white matter, formed by fibers, which transmits information between different parts of the cortex or between the cortex and other parts of the brain. Embedded within the white matter of each hemisphere are some islands of gray matter known as the basal ganglia, which play a role in the control of movement and are affected in disorders such as Parkinson's disease or cerebral palsy.

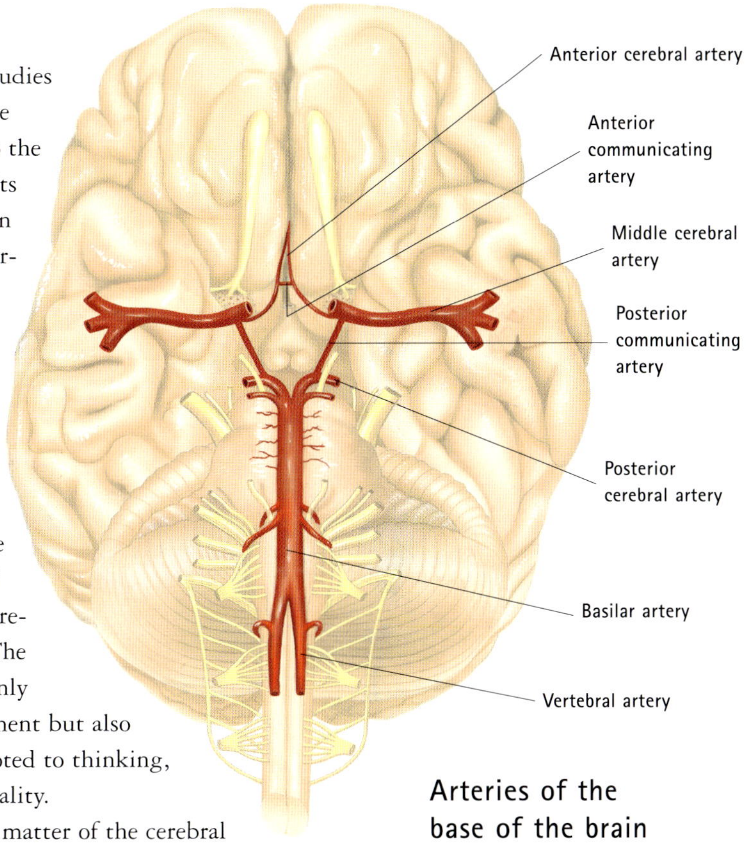

Arteries of the base of the brain

The brain stem

The brain stem, which is continuous with the spinal cord below it, consists of the midbrain, pons and medulla. Ascending pathways pass through the brain stem carrying sensory information from the spinal cord to the brain, and descending pathways carry motor commands down to the spinal cord.

The brain stem contains many important reflex centers which control vital functions such as heartbeat and respiration. It is also important in the regulation of consciousness levels—injury to the brain stem can result in prolonged loss of consciousness or death.

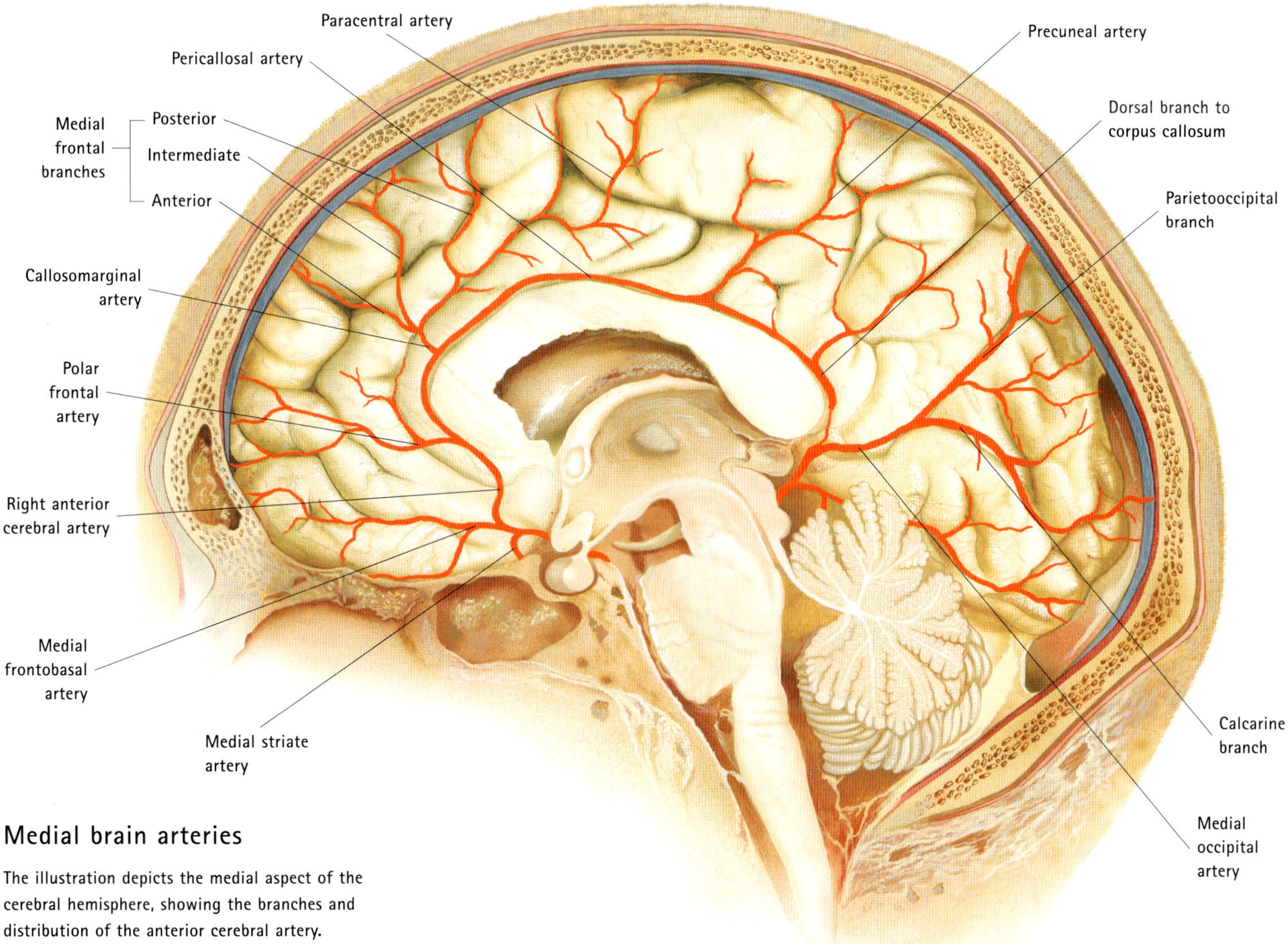

Medial brain arteries

The illustration depicts the medial aspect of the cerebral hemisphere, showing the branches and distribution of the anterior cerebral artery.

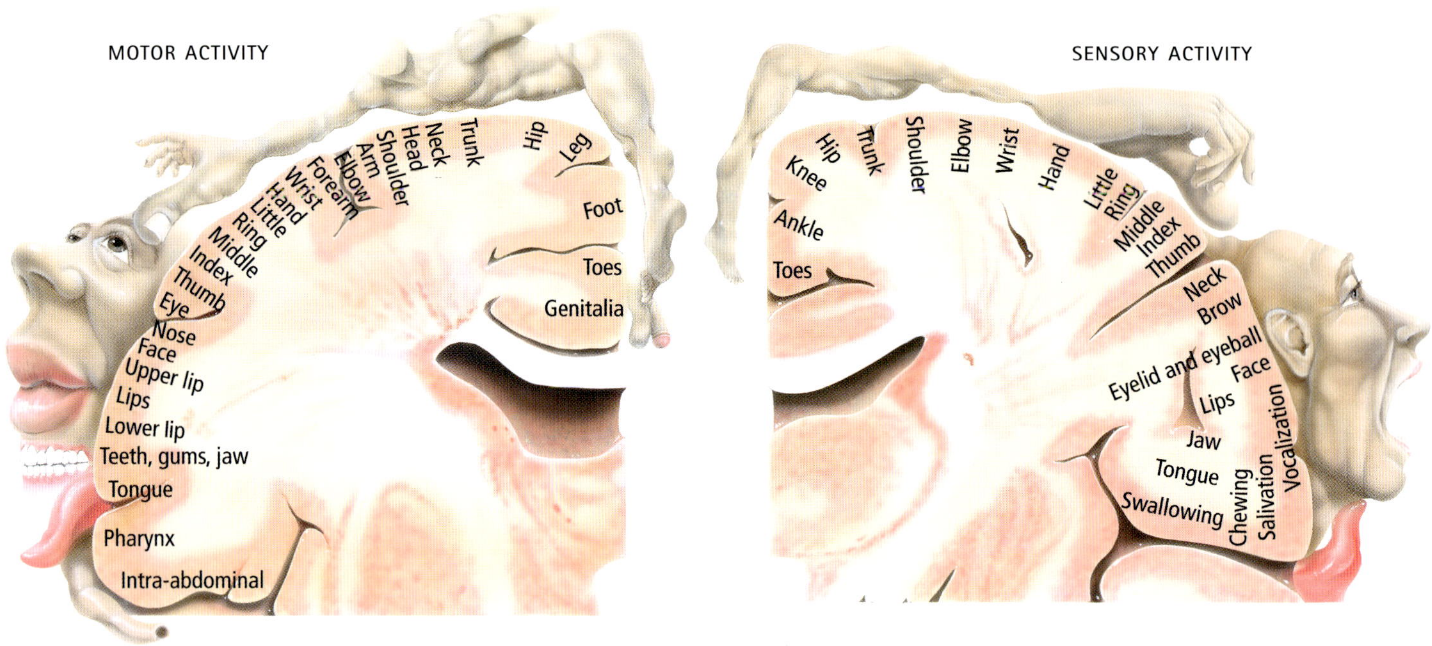

Organization of the motor and sensory areas of the cerebral cortex

The varying sizes of the body parts (top left) show the proportions of the precentral gyrus involved in motor activity in particular parts of the body, and (top right) the proportions of the postcentral gyrus involved in sensory activity in particular parts of the body.

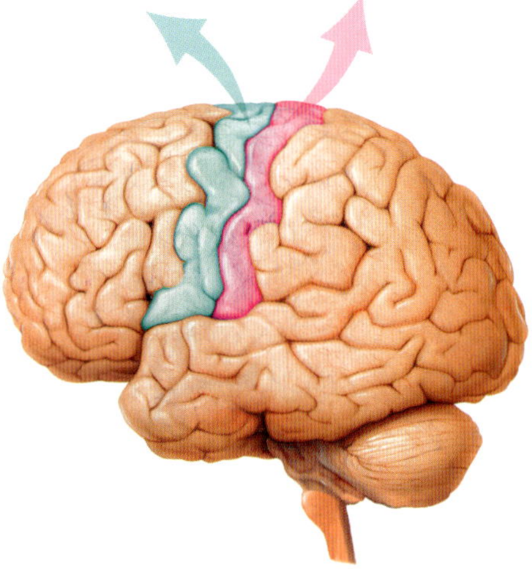

Functional areas

Particular regions of the cerebral cortex are associated with certain functions. For example, the precentral gyrus (motor cortex) is associated with the voluntary control of skeletal muscles, while the postcentral gyrus (sensory cortex) is associated with sensations from skin, muscles and joints.

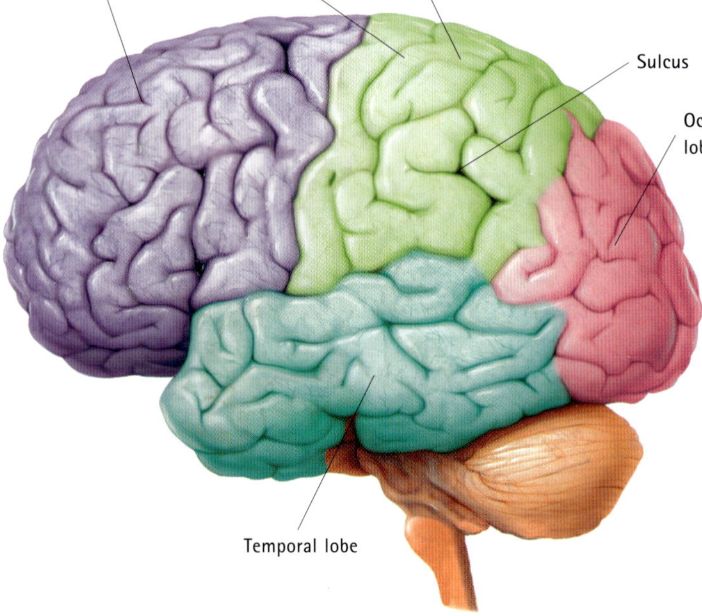

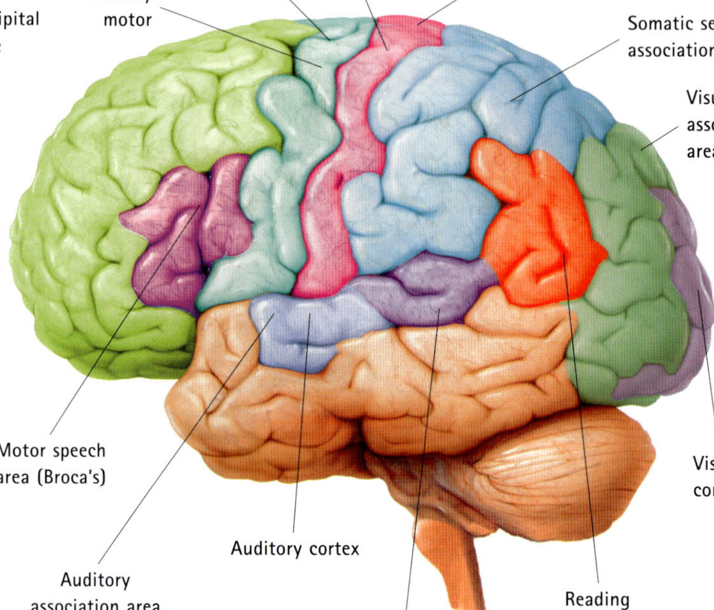

Lobes of the brain

The surface of the cerebral cortex (also called the cerebrum) is heavily folded. A ridge on the surface of the cortex is called a gyrus. A groove is called a sulcus if shallow, or a fissure if deep. Fissures and sulci divide the cortex into separate functional areas called lobes.

MIDBRAIN

The midbrain is deep inside the brain, below the cerebrum. It sits directly above the pons and, together with this structure and the medulla oblongata, forms the brain stem, the part of the brain attached to the spinal cord. The brain stem is the most primitive part of the brain and is involved in many body functions. The midbrain relays motor signals from the cerebral cortex to the pons, and sensory transmissions in the other direction, from the spinal cord to the thalamus. Cranial nerves III and IV, which service the eye muscles, start in the midbrain, making it important in eyelid, eyeball, lens and pupil movements. The midbrain is also called mesencephalon.

MEDULLA OBLONGATA

The medulla oblongata is the upward continuation of the spinal cord which forms the lower part of the brain stem. It contains pathways taking information between the brain and spinal cord and gives rise to the hypoglossal, accessory, glossopharyngeal and vagus nerves. It contains a central core of gray matter called the reticular formation, which is involved in regulating sleep and arousal, and in pain perception. The reticular formation also includes vital centers that regulate breathing and heart activity.

The cerebellum

The cerebellum (Latin for "little brain") is attached to the brain stem and resembles a cauliflower. Like the cerebrum, its surface is formed by a highly folded cortex. The cerebellum is important in the control of movement, particularly in the coordination of voluntary muscle activity and in the maintenance of balance and equilibrium. It is particularly sensitive to excess alcohol, and the effects of severe drunkenness (poor balance and coordination) to some degree mimic cerebellar disease.

The diencephalon

The diencephalon lies beneath the cerebral hemispheres and has two main structures, the thalamus and the hypothalamus.

THALAMUS

The thalamus is an ovoid structure composed of a group of nerve cells deep within the brain. There are two thalami, lying on either side of the third ventricle, a fluid-filled space in the midline of the brain. A large bundle of nerve fibers, known as the internal capsule, lies to the side of each thalamus.

There are several parts of the thalamus, which serve as sensory relay centers between so-called "lower" parts of the central nervous system and the surface of the brain, the cerebral cortex. For example, sensory information from the retina in the eye is

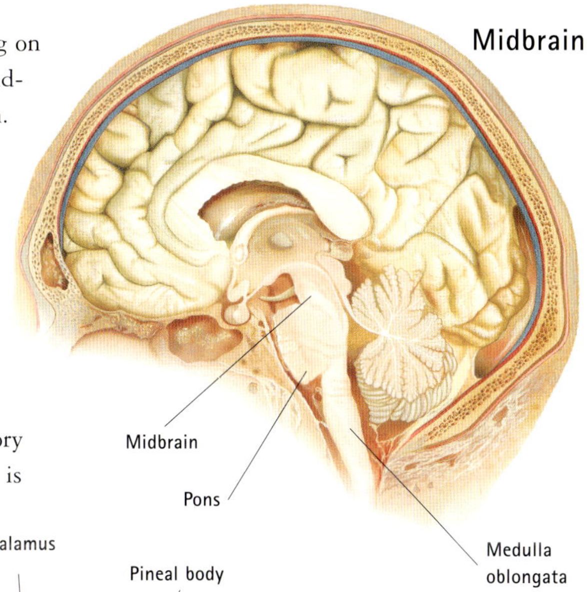

Midbrain

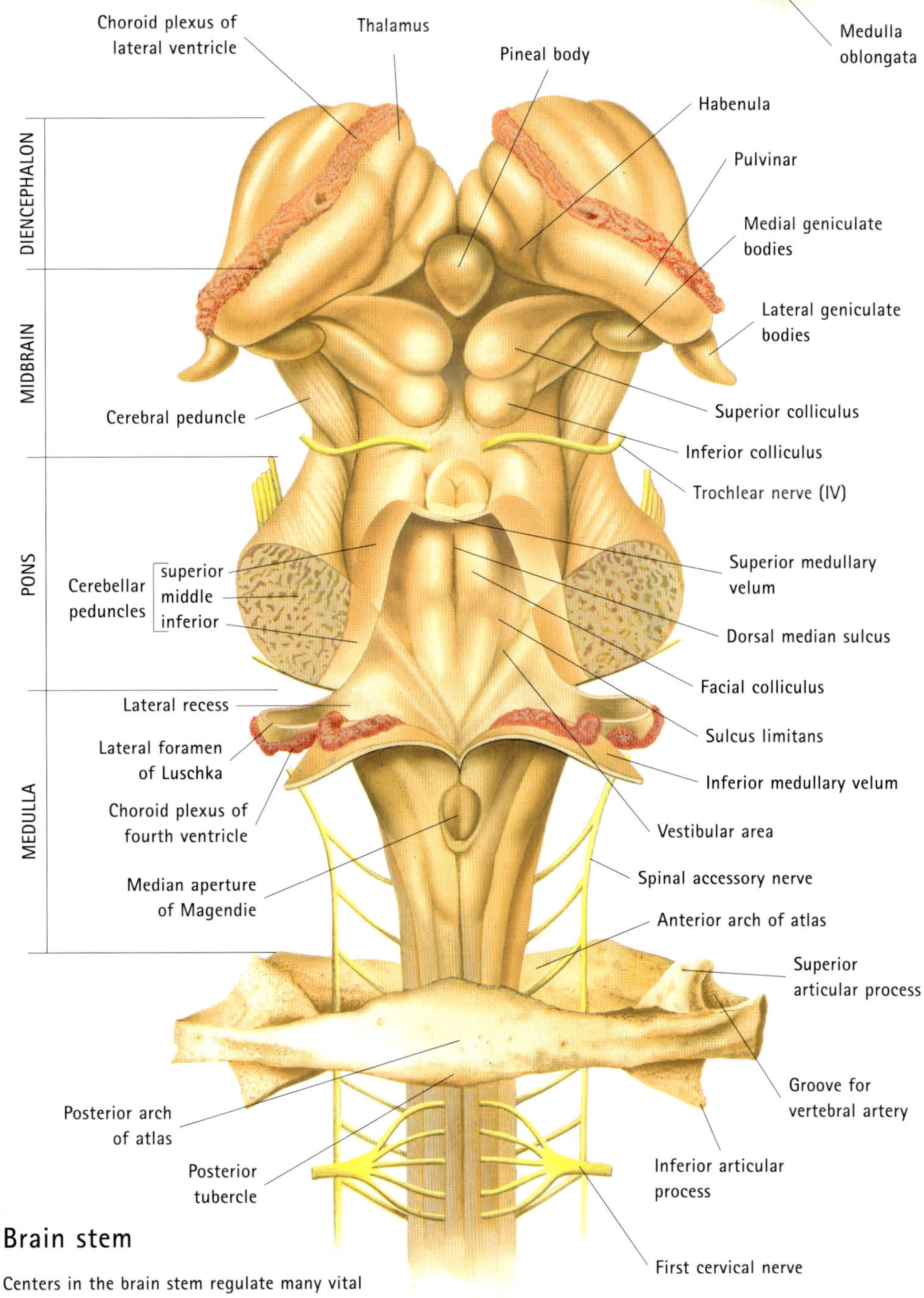

Brain stem

Centers in the brain stem regulate many vital functions including breathing, heart rate and blood pressure. Damage to it from stroke or other injury may result in death.

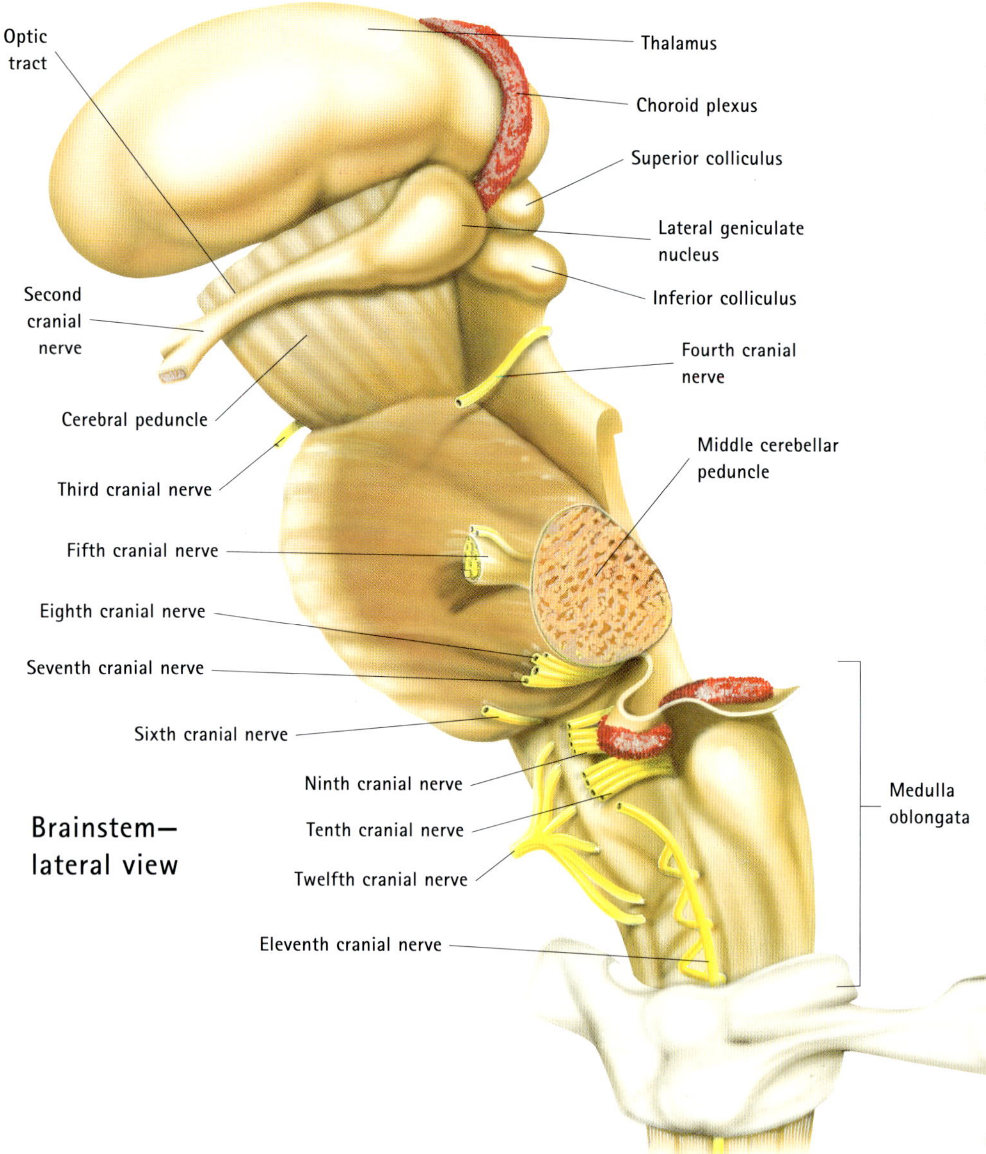

Brainstem—lateral view

relayed through the thalamus to the visual cortex at the back of the brain. However, the thalamus is not a passive relay station—some of the information processing necessary for sensory perception must occur within the thalamus.

Other parts of the thalamus are motor, or muscle control, relay centers. The thalamus is involved in two "motor loops," circuits by which nerve impulses are transmitted from the motor parts of the cortex to nerve cells in the lower parts of the brain, back up to the thalamus and thence returned to the cortex. One of these loops involves the brain stem and cerebellum; the other involves the basal ganglia, which are large groups of nerve cells lying to the side and in front of each thalamus.

There are also parts of the thalamus that are said to be "non-specific" and have connections with the cortex, for which no clear functional significance is available at present.

HYPOTHALAMUS

The hypothalamus is a small but vital region located at the base of the brain, which is essential for the maintenance of life. It contains specialized receptor cells that can detect changes in the properties of circulating blood (for example, temperature, hormone levels, osmotic pressure). The pituitary gland, or hypophysis, is attached to its exposed surface.

By regulating hormone production in the pituitary gland and through neural connections with other parts of the brain and spinal cord, the hypothalamus provides overall control of the autonomic nervous system, which coordinates activity in the body's internal organs. It contains centers which regulate the heart and blood pressure, body temperature, water balance, food intake, growth and sexual reproduction. It is also important in the expression of emotions such as fear, anger and pleasure.

THALAMUS

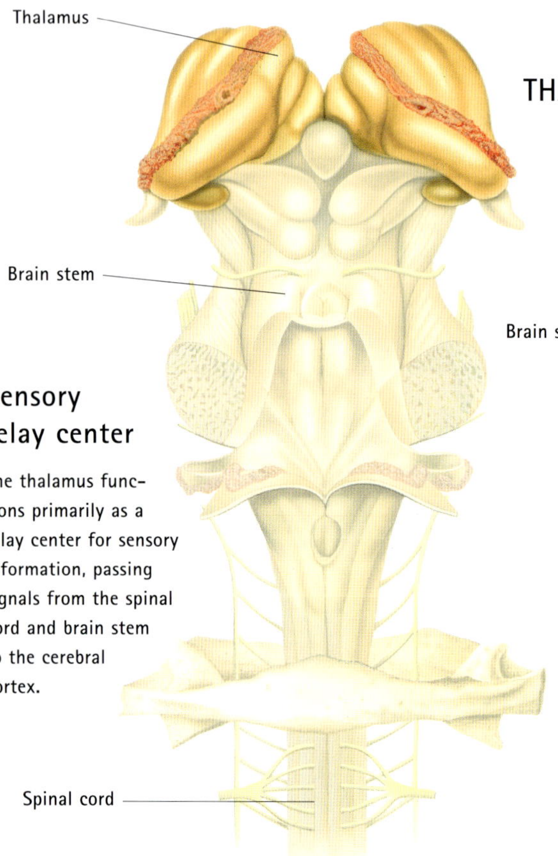

Sensory relay center

The thalamus functions primarily as a relay center for sensory information, passing signals from the spinal cord and brain stem to the cerebral cortex.

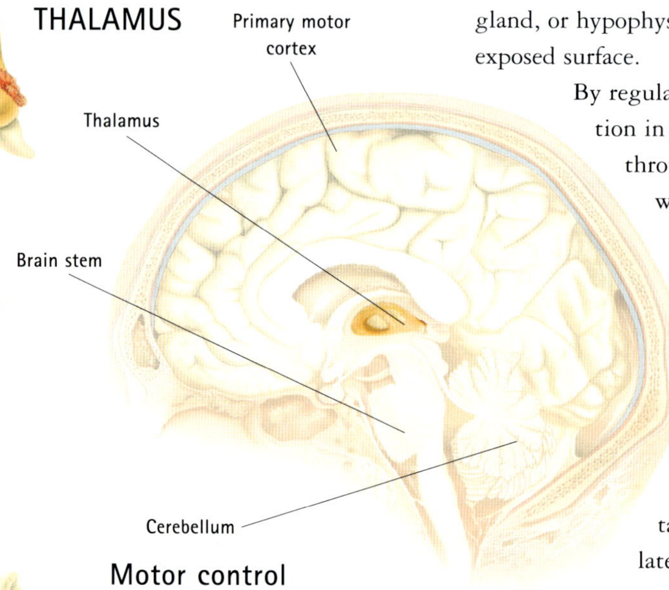

Motor control

The thalamus plays a role in motor control of the body. Some parts act as a motor relay center, sending information to the cerebellum and the motor cortex in the brain.

Pituitary gland

The pituitary gland (hypophysis) is a small organ lying immediately below the hypothalamus of the brain. It weighs only about one-sixtieth of an ounce (0.5 gram), but plays a very important role in the control of endocrine gland function throughout the body. Endocrine glands are those glands that secrete special chemicals called hormones into the bloodstream or body cavities.

The pituitary gland is divided into two basic parts, each of which has a different origin during embryonic life. The part toward the back of the gland is derived from the embryonic brain and is called the neurohypophysis (posterior pituitary). It is connected to the brain by a stalk called the infundibulum. The other part of the gland toward the front is derived from a pouch in the roof of the developing mouth (Rathke's pouch) and is called the adenohypophysis (anterior pituitary).

NEUROHYPOPHYSIS

The neurohypophysis is responsible for the release of two hormones: vasopressin (antidiuretic hormone—ADH) and oxytocin. Both hormones are polypeptides (chains of amino acids) that are made in the hypothalamus. Vasopressin increases water reabsorption from the urine as that fluid is being formed by the kidneys, thus making the urine more concentrated. The ultimate effect of this is to dilute the blood. Not surprisingly, vasopressin release is regulated in response to blood concentration, with special parts of the hypothalamus detecting changes in that concentration. Oxytocin causes contraction of smooth muscle cells in the uterus during childbirth and in the breasts during milk release (the milk ejection reflex).

ADENOHYPOPHYSIS

The adenohypophysis is also under the control of the hypothalamus, but is regulated by the release of hormones which flow in a special blood vessel system from the hypothalamus to the pituitary. The adenohypophysis releases many important hormones, with functions implied by their names. Growth hormone (GH) is important in the control of cartilage growth in long bones. Thyroid stimulating hormone (TSH) stimulates the production of thyroid hormone. Prolactin (lactogenic hormone) triggers the secretion of milk by the breasts (lactation). Adrenocorticotrophic hormone (ACTH or

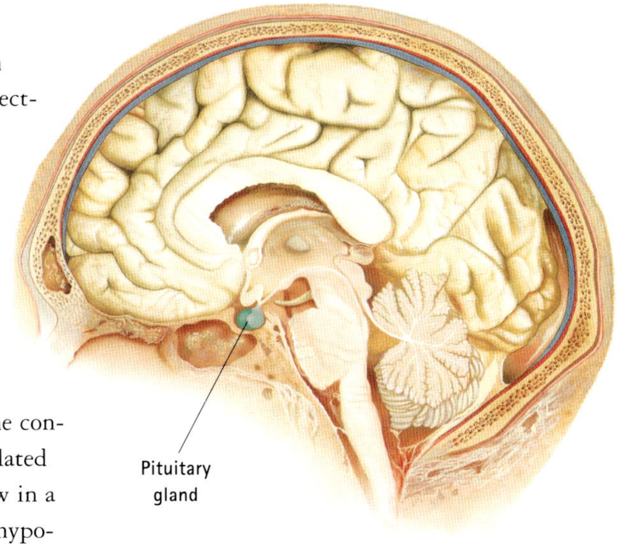

Pituitary gland

The pituitary gland is an endocrine gland located in a recess of the sphenoid bone at the base of the brain.

PITUITARY FUNCTION

The pituitary gland controls all the other endocrine glands in the body. It has two parts: the neurohypophysis (posterior pituitary), which secretes two hormones (vasopressin and oxytocin), and the adenohypophysis (anterior pituitary), which secretes hormones that control the thyroid and adrenal glands, and the follicles and corpus luteum in the ovaries.

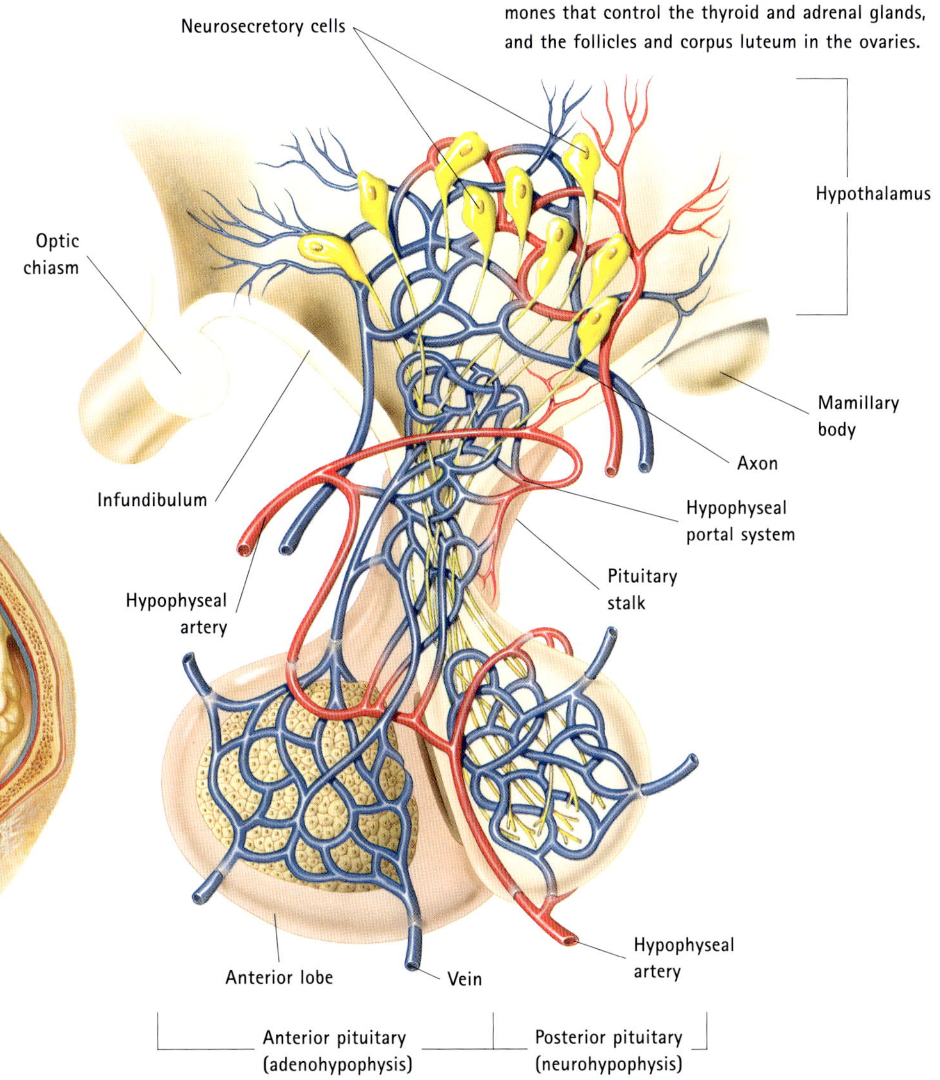

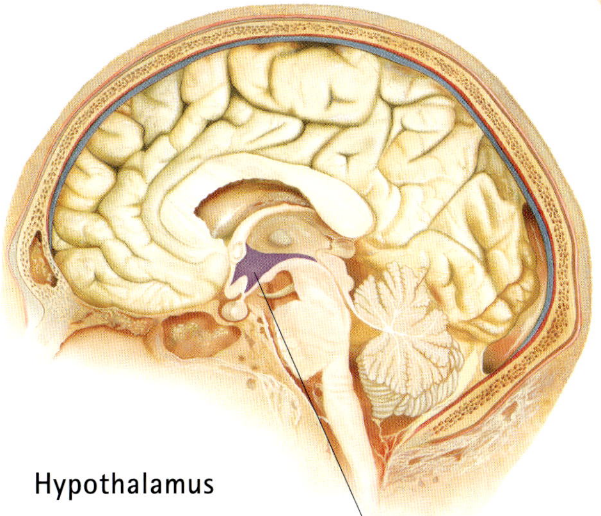

Hypothalamus

Although only a small part of the brain, the hypothalamus exerts an influence on a wide range of body functions.

corticotropin) stimulates the production of hormones (corticosteroids and sex hormones) from the outer part of the adrenal glands (adrenal cortex).

Follicle stimulating hormone (FSH) stimulates the development of egg follicles in a woman's ovaries and sperm cells in a man's testes. In the ovary, Luteinizing hormone (LH) stimulates the rupture of egg follicles, and the formation of a corpus luteum to produce progesterone during the latter half of a woman's menstrual cycle. The proper functioning of these hormones is essential to the correct growth, maturation and reproduction of an individual.

Limbic system

The limbic system is a collective term for a group of interconnected brain structures that are involved in behaviors associated with survival, including the expression of emotion, feeding, drinking, defense and reproduction, as well as the formation of memory. The term "limbic system" comes from the fact that the earliest parts of this system to be identified were observed to form a ring or "limbus" around the central structures of the brain (cingulate, parahippocampal and hippocampal gyri), but later definitions include a number of other structures. The key components of this system are the hippocampus, amygdala, septal area and hypothalamus.

The hippocampus is located deep within the temporal lobe and is continuous with the cortex on the inner part of the lower surface of that lobe. It is connected to other parts of the cerebral cortex, thalamus and hypothalamus and is essential in the formation of new memories.

The amygdala is located in the temporal lobe, just in front of the hippocampus. It is strongly linked to the olfactory (smell) system, hippocampus, cerebral cortex and hypothalamus, and is an important center for the expression of emotions.

The septal area, a small region of the inner surface of the brain, beneath the front of the corpus callosum, is linked to the hippocampus, amygdala and hypothalamus. It is thought to be a pleasure or reward center and is a focus of ongoing research investigating addictive behavior. The hypothalamus is a tiny but vital region that regulates the activity of the body's organs (viscera) through connections with other parts of the brain and through regulating the production of hormones. It is interconnected with all parts of the limbic system and is responsible for bringing about visceral changes associated with emotions, such as the increase in blood pressure, heart and breathing rate which occurs when scared or anxious, or blushing when embarrassed.

The behavioral and cognitive changes associated with emotional expression are brought about mainly through projections from the various limbic structures (hypothalamus, amygdala, hippocampus) to the cortex of the frontal and temporal lobes.

Ventricle

A ventricle is the term used for a small cavity or chamber either in the brain or in the heart. In the brain, there are four ventricles. These ventricals connect with each other, with the central canal of the spinal cord, and the subarachnoid space surrounding the brain. Specialized capillaries called choroid plexuses in the ventricles of the brain produce cerebrospinal fluid.

This clear liquid fills the ventricles and the other cavities with which they are connected to create a protective cushion for the central nervous system.

Memory

Memory is a cognitive process which allows humans to retain and retrieve information about previously experienced events, impressions, sensations and ideas. Humans acquire knowledge and store it as memory. The ability to learn or to reason is largely dependent on the ability to remember. For example, the ability to perform a simple task is based on remembering earlier experiences. In solving a problem or even simply recognizing that a problem exists, one is depending on memory.

Although not exactly understood, neuroscientists describe the storing of memory in three ways: sensory memory, which lasts from milliseconds to seconds; short-term memory, which lasts seconds to minutes; and long-term memory, which lasts days and even years.

Different storage mechanisms are said to exist for short-term and long-term memory. In short-term memory, a limited amount of information (from five to ten separate items) can be held for a few seconds, after which they must be transferred to long-term memory or they will be lost. We use short-term memory to remember a phone number after looking it in up in a directory, but only for as long as it takes to dial the number. The capacity of long-term memory

Limbic system

This "system" is a collective term referring to a group of elements in the brain. These structures are involved in behaviors associated with survival, such as feeding, defense and reproduction, and also govern emotional states and memory storage.

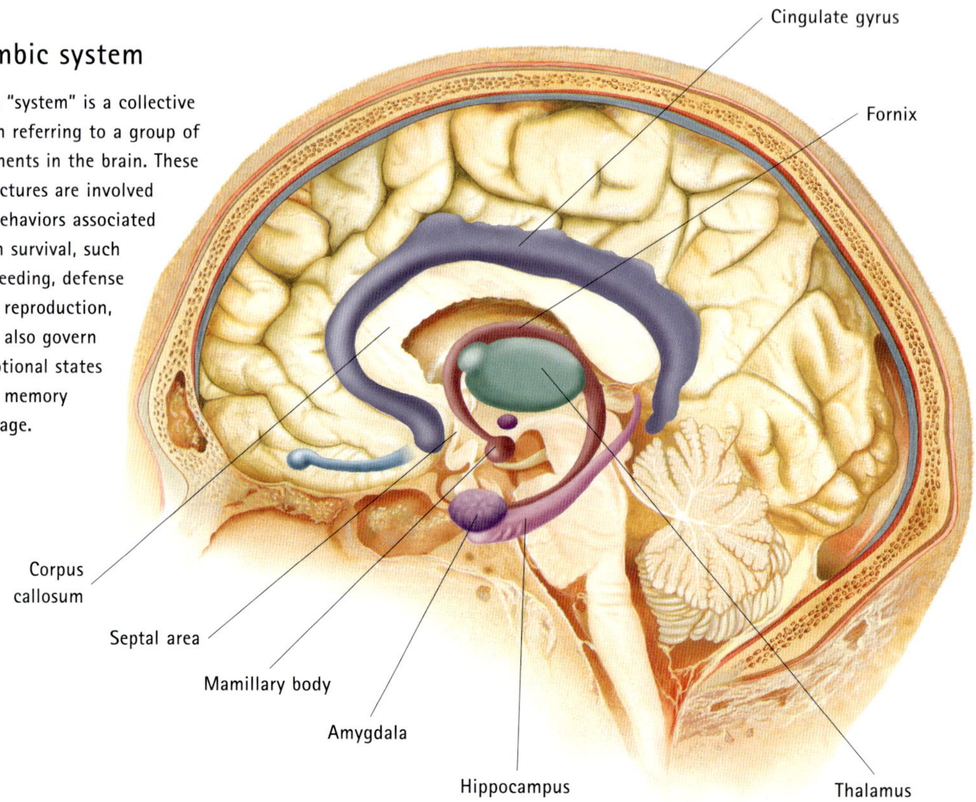

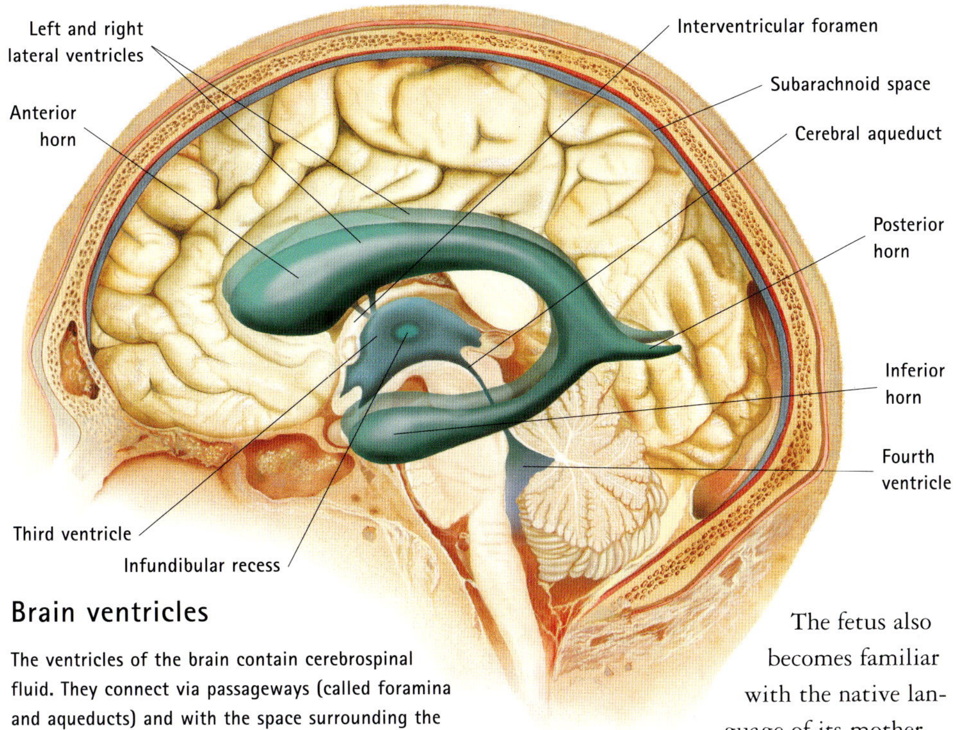

Brain ventricles

The ventricles of the brain contain cerebrospinal fluid. They connect via passageways (called foramina and aqueducts) and with the space surrounding the outside of the brain (the subarachnoid space). The cerebrospinal fluid acts as a shock absorber, cushioning the brain from mechanical forces.

is quite large—certain items, especially important events, can be stored for life.

Psychologists also divide memory into four different categories: recollection, recall, recognition and relearning. Recollection is the mental reconstruction of previous events from reminders that "jog" the memory. Recall is the unprompted active remembering of events or incidents from the past. Recognition is the identifying of a stimulus as being familiar from the past. Relearning is the ability to commit to memory with relative ease something that has been forgotten previously, the learning process being easier than the first time.

Memory is thought to be stored over wide areas of the brain rather than in any single location. However, the limbic system, notably the hippocampus on the medial side of the temporal lobe and some parts of the thalamus, are thought to be particularly important in the laying down of memories and in their recall when required. Injury to these areas of the brain results in amnesia, a disorder of memory.

Learning, and thus remembering, begins in the womb. Ultrasound observations of twins have shown the development of gestures and habits as early as 20 weeks gestation, which continue into early childhood. The fetus also becomes familiar with the native language of its mother. Taste is also learned *in utero*, as the baby becomes familiar with its mother's diet by inhaling and swallowing the amniotic fluid it lives in. Emotion too is learned; studies have found that babies whose mothers were depressed during pregnancy also exhibited symptoms of depression at birth.

PROBLEMS

Forgetting is normal, and apart from a rare few people, everyone forgets. Over time something which is not practised will be forgotten. The ability to forget, however, is also important to the process of learning. One is continually adjusting the ability to learn with the ability to forget in order to adapt to new learning experiences and learn new skills. Those who are unable to forget have been found to be extremely confused.

Certain situations, conditions or illnesses can affect memory. Depression, if it includes agitation and psychomotor retardation, makes it difficult for people to remember as well as they would normally, and such depression can lead to pseudodementia. Anxiety can reduce the ability to concentrate, for instance in an examination situation, and can affect memory. With age, the ability to learn new skills and to recall information from memory can diminish. This may be partly because the memory store is greater in the elderly than in younger persons, thus making recall more complicated. This has been described as age-associated memory impairment (AAMI). Boredom, tiredness, hearing and sight impairment, alcohol, drugs and pain can reduce the ability to remember and thus learn.

Infections, particularly of the brain (such as meningitis), can cause memory problems, as can an underactive thyroid gland, which will slow down the body's processes. Severe heart or lung disease will affect memory by reducing the supply of oxygen to the brain, and untreated, or poorly treated, diabetes with high or low levels of sugar in the blood can affect the brain's workings.

Dementia is the most serious cause of memory problems is dementia. Rarely a problem for people under 65 years of age (when it is usually associated with conditions such as Creutzfeld-Jakob disease), the risk of dementia increases with age. One in five people over the age of 80 will suffer from dementia, the most common cause being Alzheimer's disease.

SEE ALSO *Fetal development on page 398, Learning disorders on page 420; Alzheimer's disease, Anxiety, Creutzfeld-Jakob disease, Meningitis and other individual disorders in Index*

BRAIN WAVE ACTIVITY

Because nervous impulses are transmitted electrically, it is possible to record the activity of cells in the brain by placing electrodes on the scalp. The resulting graph, the electroencephalogram (EEG), shows wavelike patterns of activity, known as brain waves, which vary with different states of consciousness. Alpha waves indicate a relaxed awake state, beta waves are typical when we are mentally alert, theta waves are common in children but not adults, and delta waves appear during sleep. Brain wave activity becomes intensified during epileptic seizures. Prolonged absence of brain wave activity indicates brain death.

SEE ALSO *Electroencephalogram on page 430*

SENSES

The senses are the faculties that enable individuals to perceive changes in their external and internal environments. These changes are detected by sense organs—specialized

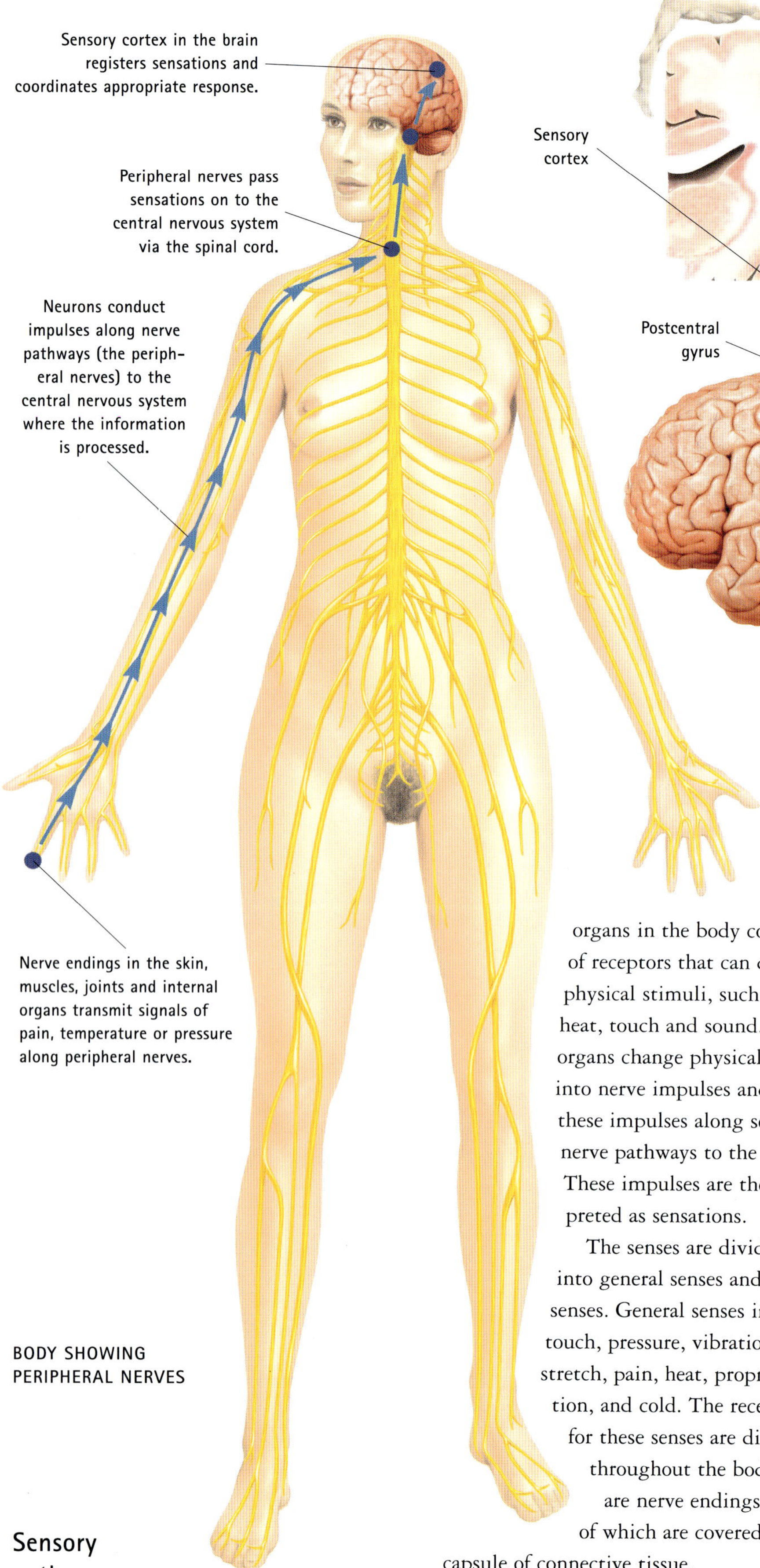

BODY SHOWING PERIPHERAL NERVES

Sensory pathways

Sensory receptors provide information on conditions inside and outside of the body. When stimulated, they pass information along the peripheral nerves to the central nervous system, where the signals are recognized as a sensation.

organs in the body consisting of receptors that can detect physical stimuli, such as light, heat, touch and sound. The organs change physical stimuli into nerve impulses and send these impulses along sensory nerve pathways to the brain. These impulses are then interpreted as sensations.

The senses are divided into general senses and special senses. General senses include touch, pressure, vibration, stretch, pain, heat, proprioception, and cold. The receptors for these senses are distributed throughout the body. They are nerve endings, most of which are covered by a capsule of connective tissue.

The special senses include olfaction (smell), vision, taste, hearing and equilibrium (balance). The receptors for these senses are found in very localized areas of the body and are more complex.

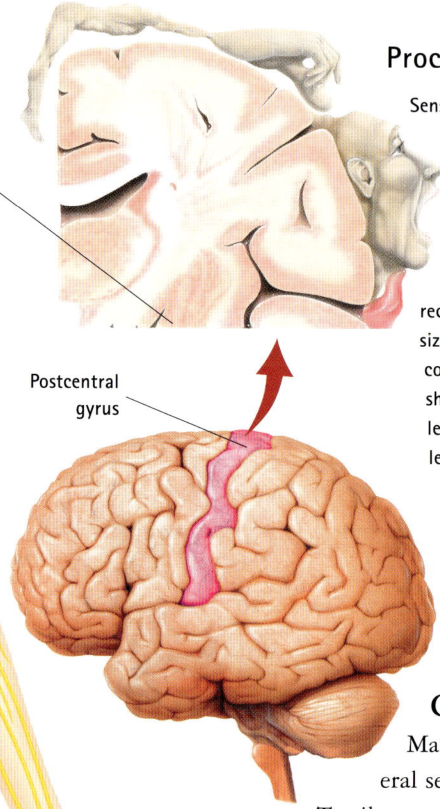

Processing centers

Sensory information from all over the body is processed in the somatic sensory cortex. Some parts of the body have a higher density of sensory receptors than others and send more information to this part of the brain. The number of nerve receptors in an organ (as opposed to its size) determines its share of the sensory cortex. This illustration, for example, shows that the lips take up an equivalent proportion of the cortex to the legs because of their greater sensitivity.

SEE ALSO *Nervous system on page 64; Spinal cord on page 217; Touch on page 348*

General senses

Many of the receptors of the general senses are found in the skin. Tactile receptors are mechanoreceptors, which can detect touch, pressure, vibration, tickle and itch. There are at least six different types of tactile receptors found in the skin, including Ruffini endings, Meissner's corpuscles, Krause's end bulbs, Pacinian corpuscles, and Merkel's disks.

Tactile receptors are particularly numerous in certain areas such as lips, fingertips, palms, toes, nipples, the glans of the penis and the clitoris. When a stimulus, such as a pinprick or vibration, is applied to one of these receptors, it generates a nerve impulse, which then travels along a sensory (or afferent) nerve to the spinal cord. The stimulus then travels up the spinal cord along special pathways through the thalamus to the cerebral cortex, where it reaches consciousness as a vibration, pinprick or other sensation according to the nature of the original stimulus and the brain's general level of consciousness.

Other types of receptors in skin include thermoreceptors and nociceptors. Other general sensory receptors are found deep in the tissues of muscles, tendons and joints. Proprioceptors give rise to sensations of weight, position of the body, movements of body parts such as limbs, and the position of joints. Proprioceptors include muscle spindles, Golgi tendon organs and joint receptors, which the body with the sensory information needed to coordinate muscle movements and maintain body position. Most of the information from proprioceptors does not reach consciousness.

Thermoreceptors (receptors that detect changes in temperature) are widely distributed in the body. Separate thermoreceptors detect heat and cold and can detect wide variations in temperature, ranging from freezing cold to burning hot. These receptors are especially numerous in the skin around the lips, mouth and anus.

Nociceptors (pain receptors) are also found in most tissues of the body. They may detect somatic pain (from the skin, muscles, tendons or joints) or visceral pain (from the internal organs of the body). These receptors respond to chemicals released by damaged cells, by high temperatures, by stretched muscle fibers and other stimuli.

Nerves carrying the impulses from visceral nociceptors may enter the spinal cord at the same place as nerves from a different and separate area of skin on the surface of the body. Pain is often felt in the skin area rather than in the affected organ; this is known as referred pain. Some nociceptors, known as "silent" nociceptors, do not respond to intense stimuli unless they are sensitized by chemicals released by cells in inflamed tissue.

(balance). The special sense organs are all found in the head and the impulses from these organs travel to the brain through the cranial nerves.

If a person is deprived of one or more special senses, other senses will often become sharper. For example, if people lose their sight, their sense of hearing may become more acute, or they may develop a heightened sense of touch.

Special senses

The special senses are smell, taste, hearing, sight, and equilibrium (balance). All the special sense organs are found in the head and nerve impulses from these organs travel to the brain via the cranial nerves.

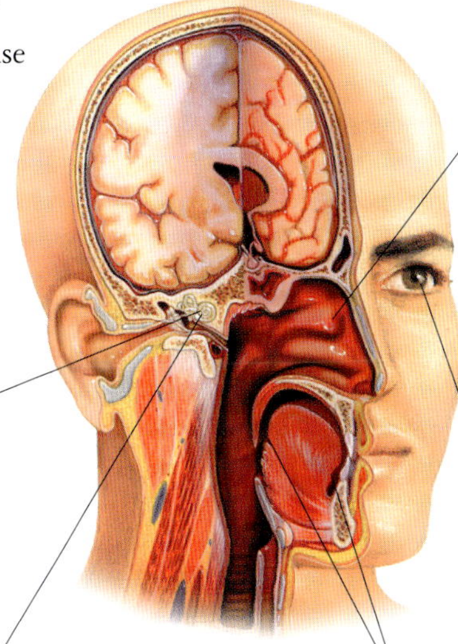

Special senses

The special senses are smell, vision, taste, hearing and equilibrium

SENSES

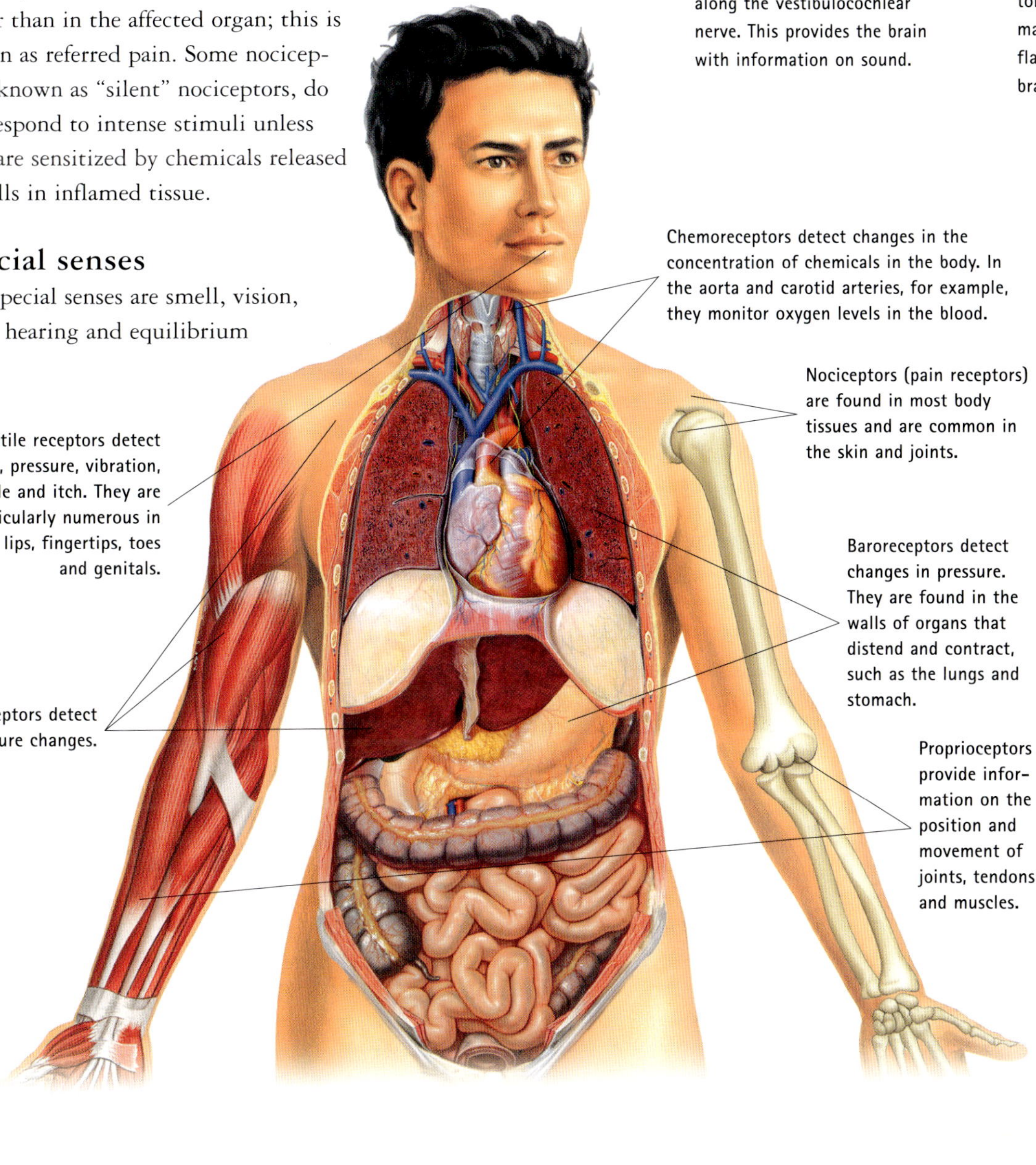

General senses

The general senses include temperature, pain, pressure, proprioception and vibration. The receptors for these sensations are distributed throughout the body. Some receptors provide information about the environment outside the body, some detect sensations in internal organs and tissues, others monitor the position and movement of joints, muscles and tendons.

NB: In this illustration the lungs and pleura have been cut to reveal the heart.

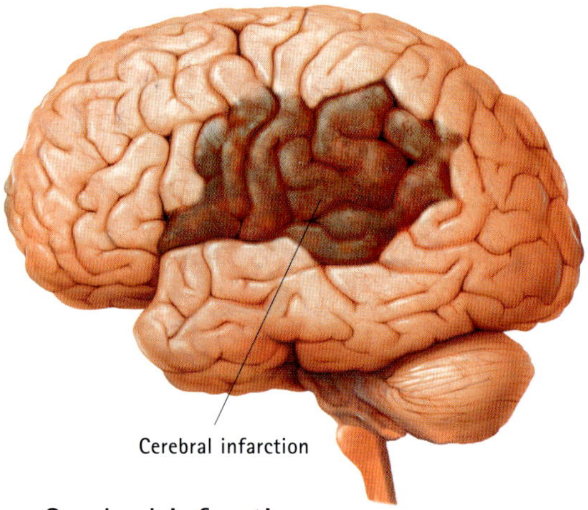

Cerebral infarction

Cerebral infarction is the death of brain tissue. This occurs when the brain is deprived of sufficient oxygen to keep the tissue alive and functioning normally. All the cerebral tissue may be affected if there is insufficient oxygen in the blood, for example following cardiac arrest. When localized, as in this illustration, infarction is usually due to the disease of a cerebral artery, which has blocked off the blood supply to that area.

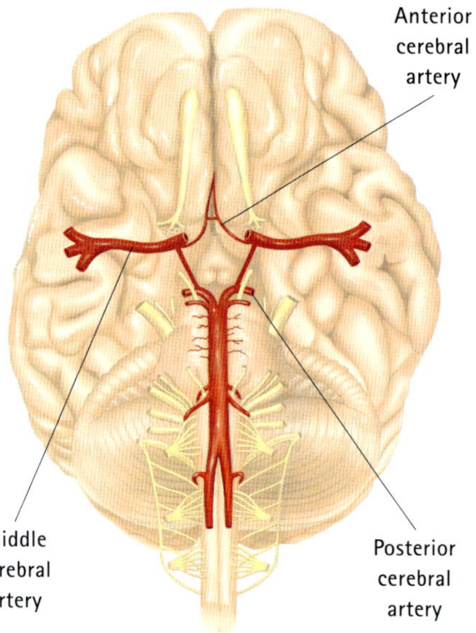

Cerebral thrombosis

Blockage of the cerebral arteries by a thrombus (blood clot) is a common cause of stroke. The clot may form in the cerebral arteries due to high blood pressure or arterial disease, or may form elsewhere and travel to arteries in the brain before blocking the blood supply.

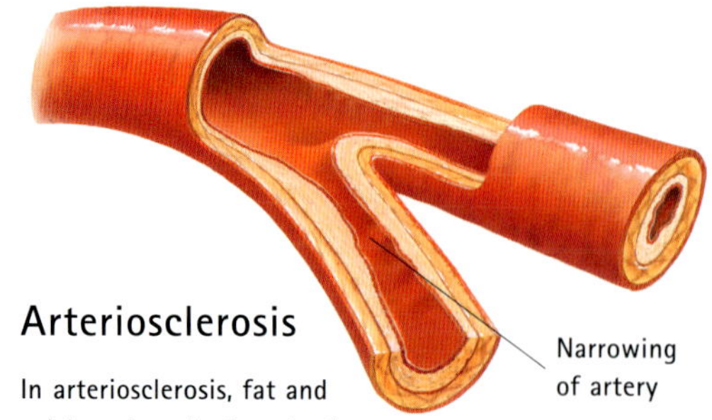

Arteriosclerosis

In arteriosclerosis, fat and calcium deposits form in the wall of an artery, gradually narrowing it until it blocks off the blood supply to an organ or tissues. This major cause of stroke is more common in people who smoke, have a high fat content in their diet, are overweight, who have a family history of stroke or heart disease, or have high blood pressure.

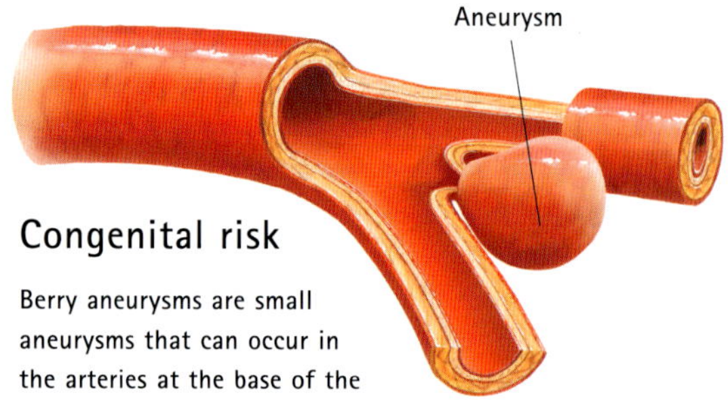

Congenital risk

Berry aneurysms are small aneurysms that can occur in the arteries at the base of the brain. These areas of weakness in the artery wall may burst, causing cerebral hemorrhage and stroke. A tendency to develop berry aneurysms may be inherited and people who have them may suffer a stroke at a relatively early age.

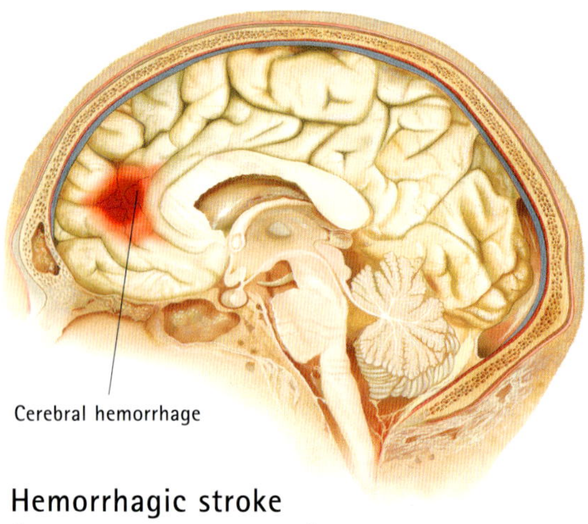

Hemorrhagic stroke (cerebral hemorrhage)

Sometimes blood may escape from an artery into the brain tissue (cerebral hemorrhage). The pressure of this bleeding destroys surrounding brain tissue, causing a stroke. Bleeding usually occurs at a section of artery that is damaged by arteriosclerosis. The diseased artery may weaken and balloon out, forming an aneurysm, which may burst. The risk of an aneurysm bursting is higher in people who suffer from hypertension (high blood pressure).

STROKE

Stroke is the term for loss of brain function arising from the death of brain tissue usually caused by arterial disease. It is one of the leading causes of death and illness in industrialized countries and particularly affects the elderly. Stroke usually cannot be cured, but preventive measures can be taken and, in some people, rehabilitation after a stroke may restore much of their former function.

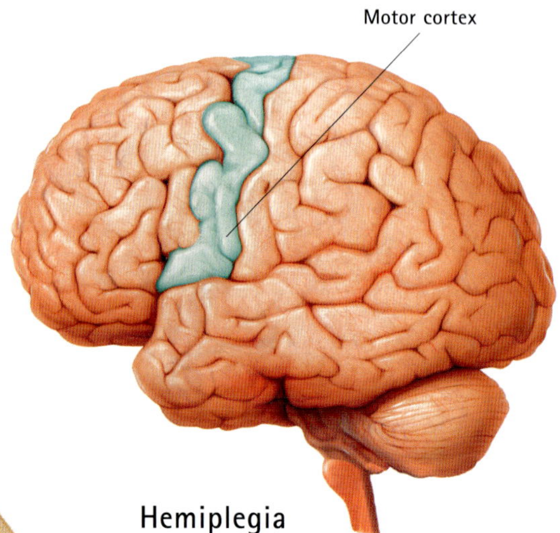

Hemiplegia

When a stroke damages the motor part of the cerebral cortex, the result is often hemiplegia, or paralysis of half of the body. Because of the way the nerve pathways are arranged in the brain and spinal cord, the part of the body affected is on the opposite side to the damaged part of the brain (i.e. when the left side of the brain is affected, the right side of the body is paralyzed).

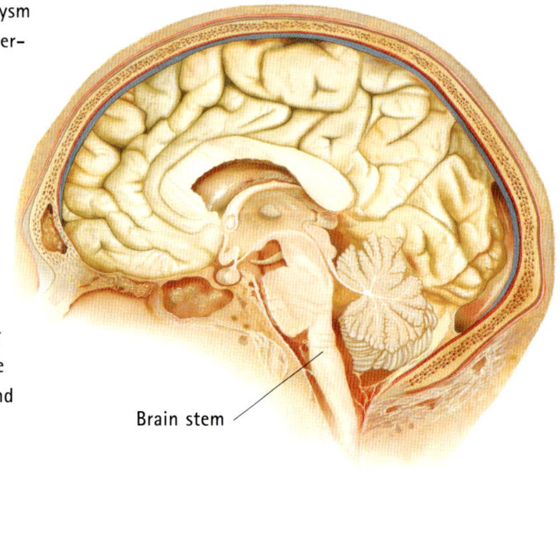

Brain stem

A stroke in the brain stem may result in coma or death, because the vital centers that regulate the basic functions of the body, such as breathing and blood pressure, are located here.

Disorders of the brain

Disorders of the brain range from mild intellectual disabilities to life-threatening conditions; the sudden rupture of blood vessels in the brain can quickly lead to death.

See also *Circulatory system on page 78; Electroencephalogram on page 430; Fetal development on page 398; Imaging techniques on page 431; Lumbar puncture on page 436; Treating the central nervous system on page 442*

Cerebrovascular accidents

Cerebrovascular accidents (also known as strokes) are caused by damage to the arteries which supply oxygenated blood to the brain and are characterized by a sudden loss of neurological function.

There are two main types of cerebrovascular accident. The first, more common type of cerebrovascular disorder involves occlusion or blockage of an artery, resulting in a lack of oxygen (ischemia) and consequent death (infarction) of the brain tissue supplied by that artery. The second type involves a sudden rupture (hemorrhage) of one of the brain's arteries, causing large quantities of blood to accumulate in the brain tissue (cerebral hemorrhage) or in the space surrounding the brain (subarachnoid hemorrhage). This leads to a sudden increase in pressure within the skull, causing severe headache, decreased consciousness and vomiting.

Stroke

Stroke is loss of brain function as a result of cerebral infarction (lack of oxygen and death of tissue in some part of the brain) as a consequence of interruption to the blood supply to the brain. It is a common condition, particularly in the elderly, affecting about one in 500 people. Stroke is the third largest cause of death in industrialized countries after heart disease and cancer.

Causes

There are several causes of cerebral brain infarction. Most commonly, a blood clot forms in one of the carotid arteries and obstructs blood flow to the brain. The clot forms in a section of the artery damaged by atherosclerosis (hardening of the arteries).

A blood clot may also form in the carotids, then dislodge and travel via the bloodstream to the brain, where it causes the infarction. Similar blockages may occur in any of the branches of the carotid or vertebral arteries, which supply blood to areas of the brain. Less commonly, stroke is caused by bleeding (hemorrhaging) from a diseased artery. This may occur in an artery affected by atherosclerosis or one with a congenital berry aneurysm, either of which may rupture. Bleeding into surrounding brain tissue damages the tissue and causes stroke.

Symptoms

The symptoms of stroke vary according to the part of the brain affected. Common symptoms are loss of movement (paralysis) of a body area, weakness, decreased sensation, numbness, loss of coordination, vision problems and difficulty in speaking.

Symptoms may occur suddenly, or develop gradually, fluctuating in severity. With hemorrhage there may be a sudden severe headache and discomfort with bright light.

A precursor to a stroke is often a transient ischemic attack (TIA), a "mini stroke" in which the deprivation of blood is not sufficient to cause permanent damage. The symptoms are the same as those of a stroke, but are temporary, and the affected person makes a full recovery, usually within 24 hours. The episode serves as a warning of the possibility of a complete stroke in the future.

Treatment

Sufferers of stroke are usually treated in hospital, and may require intensive care. The sooner the patient is taken to hospital the better the chances of effective treatment. A CAT (computed axial tomography) scan or MRI (magnetic resonance imaging) of the head may be used to rule out bleeding (hemorrhage) or other lesions and to define the location and extent of the stroke.

If there is bleeding or a blood clot, surgical removal of blood or blood clots from the brain cavity, or repair work at the source of the bleeding, may be possible. Recombinent tissue plasminogen activator (TPA) given within three hours of a stroke may dissolve the clot and improve neurologic outcome. Otherwise, treatment is centered around rehabilitation. Programs of speech therapy, occupational therapy, physical therapy and other measures are designed to recover as much function as possible.

Part of treatment involves minimizing the risk of future strokes. This includes reducing risk factors for atherosclerosis, such as smoking, high blood pressure, high levels of lipids in the blood and diabetes.

About a quarter of stroke sufferers recover most or all impaired functions, a quarter die of the stroke or its complications, and half experience long-term disabilities.

Transient ischemic attack

Transient ischemic attacks (commonly known as TIAs) are short episodes that result from the temporary obstruction of one of the small blood vessels carrying oxygen and other nutrients to the brain. The blockage is usually caused by a floating body (embolus) that gets lodged in one of these arteries, but frees itself after seconds or minutes. While the artery is blocked the brain tissue supplied by branches arising beyond the blockage is affected but, because it is only deprived of oxygen for a short time, function returns to normal once the floating body is freed and the blood supply to the tissue is restored.

The effects of a TIA can last from a few seconds or up to 10 minutes, but may last up to 24 hours. The effects depend on which artery has been blocked, and may include temporary disturbances of vision or speech, dizziness, and numbness and/or weakness of one or more of the limbs or the face.

People who are suffering from heart disease, hardening of the arteries (atherosclerosis) or high blood pressure (hypertension) are the most at risk of having a TIA, which often precedes or accompanies the development of a stroke. Treatment usually involves the use of blood-thinning (anticoagulant) medications.

Cerebral hemorrhage

A cerebral hemorrhage is a form of stroke that occurs when a blood vessel in the brain suddenly ruptures, releasing blood into the brain. Within a few minutes the patient will experience decreased consciousness, headache and vomiting. Other symptoms depend on which part of the brain is affected.

Subarachnoid hemorrhage

Subarachnoid hemorrhage is bleeding into the subarachnoid space over the surface of the brain. This occurs mostly from an aneurysm of an intracranial artery, which weakens its wall. The bleeding results in sudden severe headache with vomiting and temporary unconsciousness. It is a surgical emergency which can have serious consequences such as coma or death.

Encephalitis

Encephalitis is inflammation of the brain, causing swelling of brain tissue (cerebral edema), bleeding within the brain (intracerebral hemorrhage) and, sometimes, brain damage. Encephalitis is usually caused by one of a number of viruses. In some types of encephalitis, the virus may be transmitted to humans by mosquitoes or ticks, especially in rural areas. It can also follow other viral disorders such as measles, mumps, chickenpox, rubella, infectious mononucleosis (glandular fever) and coxsackievirus illnesses. Symptoms range from a mild illness with fever and tiredness to headache, stiff neck, vomiting and, in more serious cases, seizures, paralysis and drowsiness progressing to coma.

Laboratory analysis of blood and cerebrospinal fluid (taken via lumbar puncture) will confirm the presence of the virus. Electroencephalography and cranial magnetic resonance imaging (MRI) of the head may be needed to determine the actual extent of the infection.

Mild viral encephalitis is common and often requires no treatment other than bed rest and painkillers (analgesics). Severe cases are rare and usually require hospitalization and treatment with antiviral drugs such as acyclovir or amantadine, corticosteroids and drugs to control seizures, if needed.

Most cases recover fully within two to three weeks. A small percentage of cases suffer permanent brain damage—usually infants or the elderly.

Subacute sclerosing panencephalitis

A rare progressive disease, subacute sclerosing panencephalitis develops months or years after measles infection in childhood. It is thought to result from persistent infection of the central nervous system by altered forms of the measles virus. There is damage

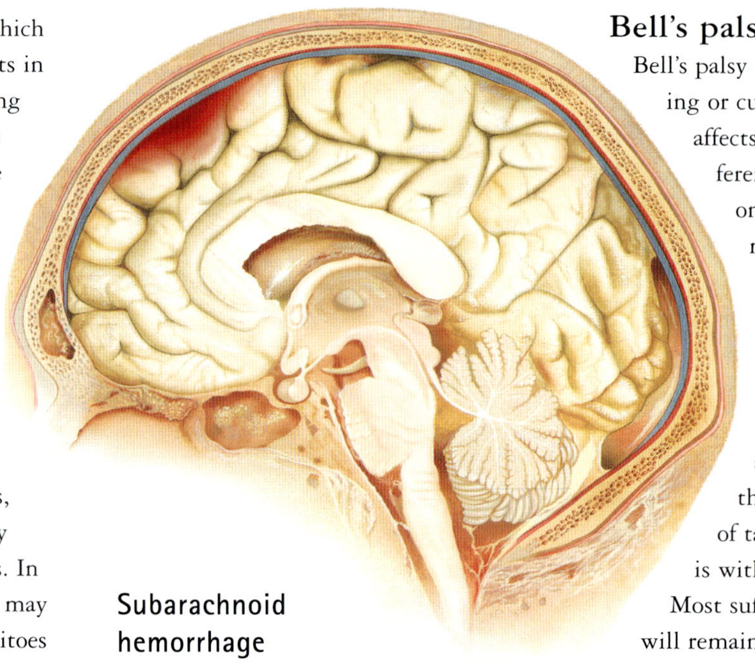

Subarachnoid hemorrhage

An aneurysm or the dilation of an intercranial artery may rupture and cause bleeding (hemorrhage) into the subarachnoid space.

to nerve cells and to the protective covering (the myelin sheaths) of nerves, and inflammation in the brain.

The sufferer develops seizures, spasticity and impaired mental function. Treatment can include physical therapy, occupational therapy, and drugs such as anticonvulsants and muscle relaxants.

Sydenham's chorea

Chorea is a neurological disorder characterized by involuntary, purposeless, spasmodic movements of the body. The most common types of this disorder are Sydenham's (or rheumatic) chorea, once known as Saint Vitus' dance, and Huntington's disease.

Sydenham's chorea is generally regarded as an inflammatory complication of Group A B-hemolytic streptococcal infection, which causes rheumatic fever. Sydenham's chorea occurs in 50 percent of children aged 5–15 who have had rheumatic fever and is more common in girls.

Triggered by emotional upset and episodes of crying, the spasms can range from mild to completely incapacitating. Attacks can last for several weeks and recurrence is usually recurring and frequent.

Bed rest is the best treatment option; and some patients can benefit from sedation and tranquilizers.

Bell's palsy

Bell's palsy is a paralysis caused by swelling or cutting of the facial nerve; it affects one side of the face. The sufferer is not able to close the eye on that side, or to contract the muscles controlling the forehead, mouth or cheek. The mouth droops and the face is distorted. The unaffected side retains normal function. The disorder may appear after a short period of pain and there may be loss of the sense of taste on the tongue. Treatment is with steroids and antiviral drugs. Most sufferers will recover, but a few will remain permanently impaired.

Intellectual disability

Formerly known as mental retardation, intellectual disability affects around 2 percent of the world's population. It can be defined as limited intelligence or cognitive potential. Those who have this condition have a reduced capacity to learn, to solve problems and possibly to perform other functions, depending on the degree of disability.

Intellectual disability is classified according to severity, and is usually measured by intelligence quotients (IQ) or scores.

Generally the average intelligence quotient is 100. The upper range of intellectual disability, known as mildly disabled, falls in an IQ range around the low 50s to high 60s. These people comprise the majority of intellectually disabled persons and they are able to learn basic academic skills with some difficulty and to be employed, usually in unskilled or semi-skilled jobs; they are also able to function independently. Those who are classified from the high 30s to mid 50s are described as moderately disabled. They are able to care for themselves, to live in a sheltered situation and to live in a home with supervision. The severely disabled fall into the low 20s to high 30s, and have slow motor development and limited communication skills. They may also have physical disabilities but will be able to care for their basic needs and contribute to their own care in work and living situations.

The last and smallest group are the profoundly disabled, with IQs below the low

20s. These people need full-time care because they have poor motor development and communication skills and are unable to care for themselves or to work except in highly structured activities.

Caused by events before, during or after birth, intellectual disability can be found in those with genetic disorders, such as Down syndrome; infectious disease, e.g. meningitis; metabolic disorders; fetal alcohol syndrome; physical malformations; poisoning, e.g. lead poisoning; trauma or injuries to the head; and malnutrition.

DIAGNOSIS

Children with more severe disabilities will usually be diagnosed before they reach their first birthday, often as a result of parental concerns. Mildly intellectually disabled children are sometimes not diagnosed until they are attending preschool or school, when it is the abstract areas of reasoning, problem solving and use of language which may indicate a problem.

A comprehensive evaluation is very important. Tests which are coordinated by a pediatrician or a child psychiatrist, in areas such as neurology, psychology, psychiatry, special education, hearing, speech, vision and physical therapy, may be necessary to determine the extent of intellectual disability.

TREATMENT

Intellectual disability cannot be cured, but early diagnosis may lead to early treatment and help the family to establish appropriate expectations for their child and to handle the stresses which this level of disability can bring to both the child and the family. Children diagnosed as intellectually disabled will benefit from an education that is tailored to their needs and from continued monitoring and evaluation. In many countries it was believed that mental health problems could be solved by deinstitutionalizing and that the ordinary health care system could give the same or better care than that provided by institutions. However, recent research has found that governments need to address the problem of the stress a disability causes to a family and the decrease in the quality of life for both the family and the disabled person that this stress can bring.

It has been found that the majority of mental health problems are not solved by deinstitutionalization and this is not a substitute for professional assistance.

Bell's palsy

One side of the face is paralyzed in Bell's palsy. The condition is due to injury or inflammation of the facial nerve. This nerve also controls the muscle in the eardrum that dampens loud noises; people with Bell's palsy may be abnormally sensitive to loud sounds.

Cerebral palsy

Cerebral palsy is a general term used to describe a group of disorders in which there is faulty development or damage to motor areas in the brain, which impair the brain's ability to adequately control movement and posture. The damage may result from disease, faulty growth, or injury, which may occur before, during, or shortly after birth. Rubella (German measles) in pregnancy, premature birth and brain damage due to a difficult delivery are common causes. Quite often, the cause may never be found; and the condition is never inherited.

The symptoms differ from one person to the next. Someone with cerebral palsy may have difficulty with fine motor tasks, have trouble maintaining balance and walking, or be affected by involuntary movements such as uncontrollable writhing motion of the hands, or drooling. In some persons, there is also intellectual disability, learning difficulties, slow growth, seizures, and hearing and vision problems. Usually, the condition is not apparent until the child is between one and two years old.

There is no cure for cerebral palsy. Treatment is aimed at helping affected people make best use of their abilities and may include physical therapy, speech therapy, and psychological support. Drugs to help control seizures and muscle spasms, special braces to compensate for muscle imbalance, and other mechanical aids may be required. The earlier treatment begins, the better a child will do. Many people with cerebral palsy can enjoy near-normal lives if their problems are properly managed.

Dysphasia

Dysphasia is an impairment of speech due to damage to the brain (usually the temporal or frontal lobes). It differs from dysarthria in that the speech impediment in dysphasia is due to damage to the parts of the brain that recognize and formulate speech, whereas dysarthria is an inability to articulate speech.

Dysphasia may be caused by stroke, transient ischemic attack (TIA), head trauma or Alzheimer's disease. The impairment may vary according to which part of the brain is affected. Damage to Wernicke's area in the temporal lobe (the interpretative center) results in trouble understanding speech and the written word. Damage to Broca's area in the frontal lobe (the motor speech

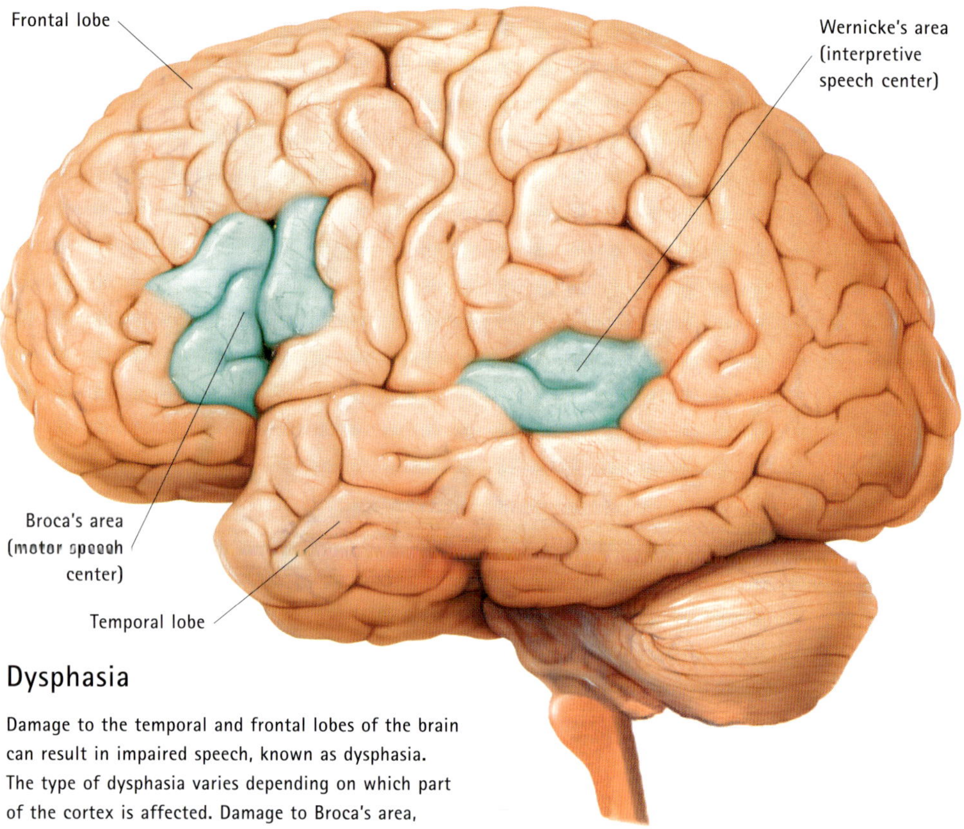

Dysphasia

Damage to the temporal and frontal lobes of the brain can result in impaired speech, known as dysphasia. The type of dysphasia varies depending on which part of the cortex is affected. Damage to Broca's area, for example, causes problems in the formation of words as this area controls motor function. Damage to Wernicke's area affects speech comprehension.

area) results in difficulty in mentally constructing written and spoken language. In total dysphasia, both comprehension and language formation are impaired. Speech therapy may improve the symptoms.

Aphasia

Aphasia is a disorder of the brain that results in the loss of speech or ability to understand language, including the ability to read and write. It is usually caused by brain disease, stroke or injury affecting the speech areas of the cerebral cortex. An aphasic seizure is a brain disturbance causing temporary speech loss.

Concussion

Concussion is a sudden alteration in levels of brain function following a blow to the head, often resulting in unconsciousness. It may be caused by a fall in which the head strikes against an object, or by a moving object striking the head. Concussion frequently occurs in contact sports, and also in auto, motorcycle or bike racing. Often, the injured person may not be aware of the problem; it may be teammates or observers who notice the confusion and disorientation. The injured person must be made to abandon the sport or activity, especially if there has been loss of consciousness.

Following the injury, the sufferer may have temporary retrograde amnesia; that is, for some time there will be no memory of events preceding the injury. There may be headache, difficulty in concentrating and focusing, nausea, vomiting, and depression.

Usually concussion is temporary and causes no permanent brain damage. However, if the concussion is severe, there may be prolonged unconsciousness and persistent confusion. The level of consciousness is the single most important indicator of the severity of a brain injury; the more severe the concussion, the longer the period of unconsciousness. In more serious cases, there may be convulsions, vomiting, a weakness of the muscles, and permanent brain damage, depending on the extent of the injury. Loss of memory of events following the concussion (anterograde memory loss) also signifies a serious concussion.

Concussed persons should seek medical treatment. A physician or neurologist will order x-rays of the head and neck to rule out the possibility of a skull fracture, and a CAT scan of the head if internal bleeding is suspected. If the injured person recovers and there are no signs of complications, rest at home and analegsics may be sufficient treatment. However, serious after-effects may be delayed and can appear 48 to 72 hours after injury, so a responsible person must watch the patient for serious symptoms.

The first 24 hours are the most critical. Danger signs include repetitive vomiting, unequal pupils, confused mental state or varying levels of consciousness, seizures, or the inability to wake up (coma). If these signs are present, urgent medical advice should be sought.

If there are no further signs or symptoms, the patient can rest in bed for a few days, after which time, normal activity may be resumed. However, sporting and athletic activities should be avoided for three months. A second or subsequent concussion is particularly dangerous, especially if it occurs before the symptoms of the earlier concussion have cleared. Even though the second injury may be milder than the first, together they have a compounding effect that may cause acute brain swelling leading to a rapid death.

To prevent concussion, protective head gear such as helmets should be worn when engaging in contact sports or any other activity that may result in head injury.

Subdural hematoma

Subdural hematoma is a blood clot on the surface of the brain, beneath its covering layer, the dura. It results from head injury, either from a direct blow or from sudden acceleration, as in whiplash injury. It may develop immediately (acute) or gradually (chronic), after injury.

The main sign of acute subdural hematoma is rapidly decreasing consciousness following a severe head injury, together with enlargement of the pupil on the same side as the injury. This is a surgical emergency requiring immediate drainage of the hematoma. If drowsiness is more gradual, a CAT scan of the brain is usually performed to confirm diagnosis.

Chronic subdural hematoma is more likely to occur in the elderly or alcoholic person following minor head trauma. The condition may cause rapidly developing dementia associated with headache. Surgery is not always required for the chronic forms.

Delirium

Delirium is a mental disorder marked by confusion and disorientation. A mental abnormality with a sudden onset, sometimes confused with dementia, delirium is not a disease and is potentially reversible.

The common symptom is confusion. People who are delirious may suffer hallucinations, be drowsy, be gripped by imaginary fears or impelled by sudden visions of catastrophe. They may be unable to recognize their surroundings, their speech may ramble and their memory of recent events may be poor or completely missing. They will have trouble concentrating and be unable to take in new information.

Delirium can result from a number of conditions or disorders, such as injury, intoxication, drug overdose, abrupt cessation of drug taking, dehydration and disease. It is quite common to find the person has deteriorated physically prior to the onset of delirium.

The most common causes are alcohol, illegal drugs and poisons, high fever, stroke, brain injuries and disorders, or severe physical injury. The treatment will be dictated by diagnosis of the exact cause. If delirious people seem likely to injure themselves through physical agitation, they may have movement restricted with padded restraints.

Dementia

Dementia is a condition in which there is a long-term loss of intellectual function, with little or no disturbance of consciousness or perception. It involves deterioration of memory and the ability to reason and changes in personality. It usually results from the degeneration of cells in the cerebral cortex of the brain.

Dementia generally occurs in elderly people and is becoming more common as the life-expectancy of the population increases. In the USA, about 1 in 15 people over the age of 65, and 1 in 3 people over the age of 85, suffer from dementia.

It is, however, not an inevitable consequence of ageing, as was previously thought. (The majority of people will not develop dementia in their old age.) It is a condition which develops in association with some diseases that happen to be more common in old age. By far the most common cause of dementia is Alzheimer's disease, which accounts for up to 70 percent of dementia cases and currently affects about 2 million Americans. The onset and progression of this form of dementia (also known as senile dementia) is gradual, and usually extends over a period of 5–10 years.

The affected person becomes increasingly forgetful, disoriented and confused. Intellectual functions, such as reading, writing and decision-making, gradually diminish. As the disease progresses, the person may also undergo a personality change, becoming aggressive, paranoid or depressed.

Alzheimer's disease is distinguished from other forms of dementia by characteristic changes which occur in the cerebral cortex, particularly the presence of dying cells, abnormal protein deposits (plaques), knotted fibers (known as tangles), and degenerate neurons.

The cause of Alzheimer's disease is not well understood but it is thought to be the result of a combination of genetic and as yet unidentified environmental factors. Some medications that enhance cholinergic transmission may improve memory in the early stages of Alzheimer's disease. However, they do not modify the steady worsening of the underlying pathology.

The second major type of dementia, multi-infarct or arteriosclerotic dementia, is caused by a number of episodes in which tiny blood vessels supplying the frontal and temporal lobes of the cerebral cortex get blocked due to cardiovascular disease, resulting in death of nerve cells (neurons)in these areas because of the lack of oxygen.

These episodes (mini-strokes) may remain undetected at the time but they have a

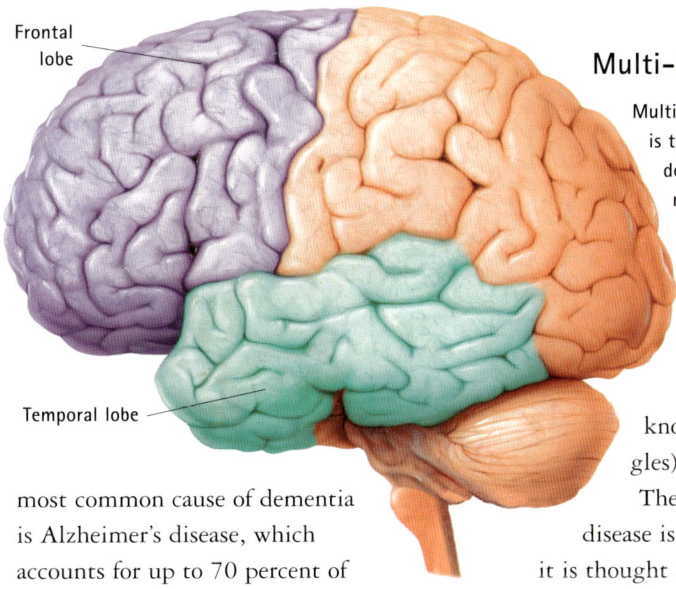

Multi-infarct dementia

Multi-infarct (arteriosclerotic) dementia is the second most common type of dementia. It is caused by repeated mini-strokes that block the blood supply to parts of the brain, eventually destroying brain tissue. The frontal and temporal lobes are most commonly affected.

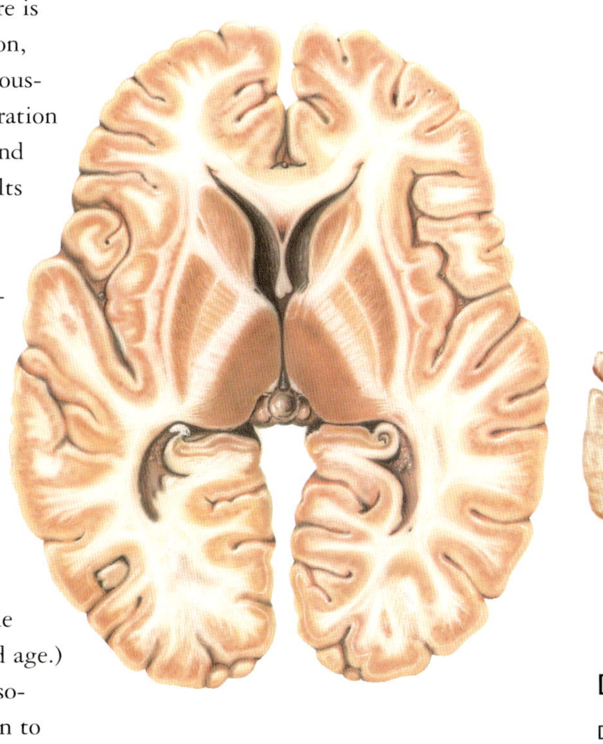

NORMAL BRAIN

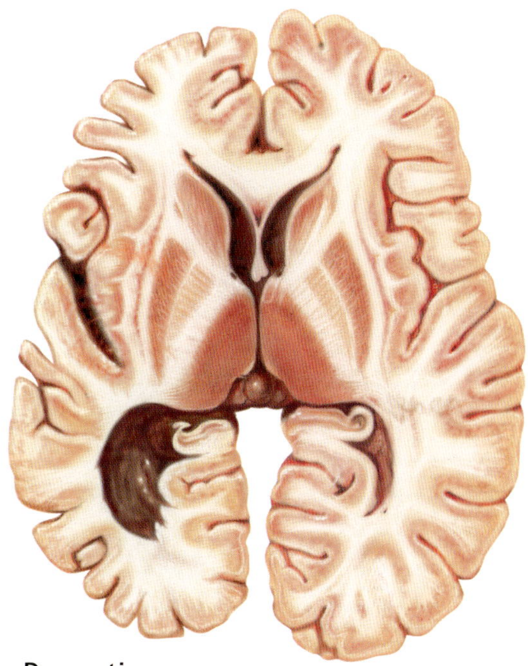

Dementia

Degenerative changes take place in the brain tissue in some types of dementia.

cumulative effect on the cortex and make up for 10–15 percent of dementia cases. The symptoms may sometimes be difficult to distinguish from those that occur in Alzheimer's disease, but they often tend to appear in a more step-by-step fashion (as the mini-strokes occur).

Other causes of dementia are numerous and include chronic alcoholism, degenerative diseases such as Huntington's disease and AIDS, nutritional and metabolic disorders, infections and tumors. Some of these disorders are treatable and recovery from the dementia can be expected but in most cases dementia is a chronic ongoing condition.

When a person suffers a temporary loss of intellectual function, such as when they have a very high fever or are withdrawing from alcoholism, they are said to be suffering from delirium, not dementia.

Alzheimer's disease

Alzheimer's disease is a common brain disease which results in dementia (confusion and loss of intellectual function). It usually occurs in elderly people but may occasionally appear during middle age.

This is the most common (but not the only) cause of senile dementia. After a person reaches the age of 65, the risk of Alzheimer's disease increases twofold with each five years of age. As the average life expectancy of the world population continues to rise, the number of those affected will continue to increase.

The symptoms of Alzheimer's disease appear very gradually, followed by a progressive deterioration over a period of five to ten years. Its major feature is increasing forgetfulness—initially forgetfulness of the names of objects, people and places as well as day-to-day events. The person begins to miss appointments and lose possessions. One of the most striking features in the early stages of the disease is that, although sufferers can't remember recent events, they may often be able to describe in great detail incidents or people from early in their life.

Unfortunately, the ability to recall even old memories fades as the disease progresses. Affected people gradually become more disorientated and confused, frequently getting lost or forgetting how to do simple tasks such as getting dressed or setting the table. Their intellectual capability diminishes—reading, writing, mathematical calculations and decision-making are all affected.

Eventually they undergo personality changes, neglecting personal hygiene and exhibiting uncharacteristic and sometimes bizarre patterns of behavior, such as paranoia or sexual indiscretions. They gradually become immobile, bedridden and very susceptible to infections.

The development of Alzheimer's disease is not considered a normal consequence of ageing—it is a real disease in which there are characteristic changes in both the structure and chemistry of the brain.

Alzheimer's sufferers lose up to 20 percent of their normal brain volume, with the shrinkage (cell death) occurring mainly in parts of the temporal, frontal and parietal lobes of the brain. These are the areas which are responsible for creating and storing memory and for intellectual processing. Parts of the brain dealing with motor and sensory functions (vision, touch and hearing) seem relatively unaffected. Examining affected tissue under a microscope shows three main characteristics: large spaces formed by dying cells; abnormal protein deposits, called senile plaques; and twisted, knotted bundles of fibers, known as tangles.

What causes these changes in the brain is yet to be understood. Less than 5 percent of cases have a proven genetic basis and these are almost invariably cases which have an early age of onset. For the vast majority it appears likely that the disease results from long-term exposure of genetically susceptible individuals to a combination of (as yet unidentified) environmental factors.

At present there is no known cure for Alzheimer's disease. Although several drugs have been shown to slow down the intellectual decline they cannot stop the progression of the disease.

Amnesia

Amnesia is the partial or complete loss of memory. It is often caused by brain damage due to trauma or disease, though it can sometimes be caused by psychological trauma. Memory loss is usually temporary and selective, being confined to one part of the affected person's experience, such as memory of recent events.

Causes of amnesia include Alzheimer's disease, head trauma or injury, seizures, general anesthetics, alcoholism, stroke or transient ischemic attack (TIA), drugs such as barbiturates or benzodiazepines, electroconvulsive therapy (especially if prolonged) as well as brain surgery.

There are a number of types of amnesia. Anterograde amnesia involves the loss of

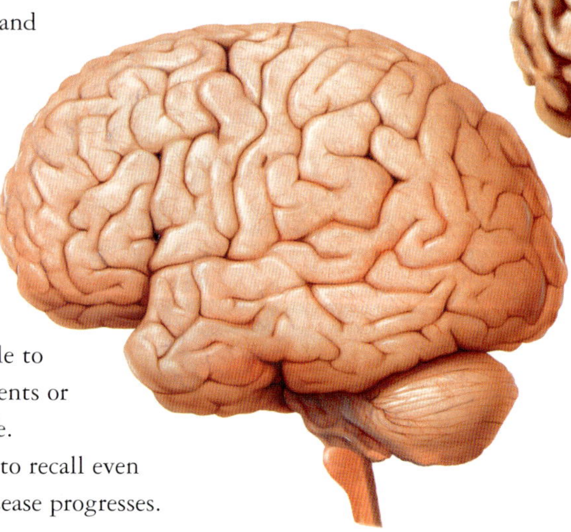

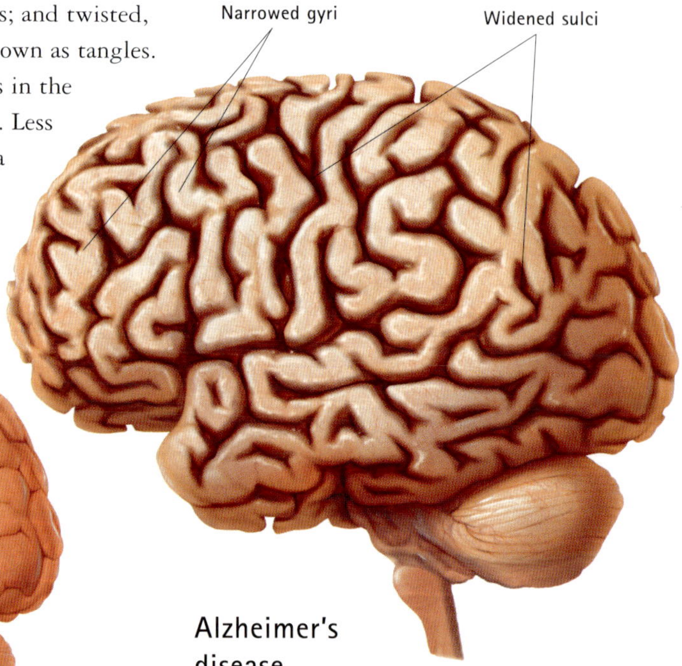

Alzheimer's disease

In Alzheimer's disease, the brain has shrunk slightly and the gaps between the folds of the cerebral cortex have widened.

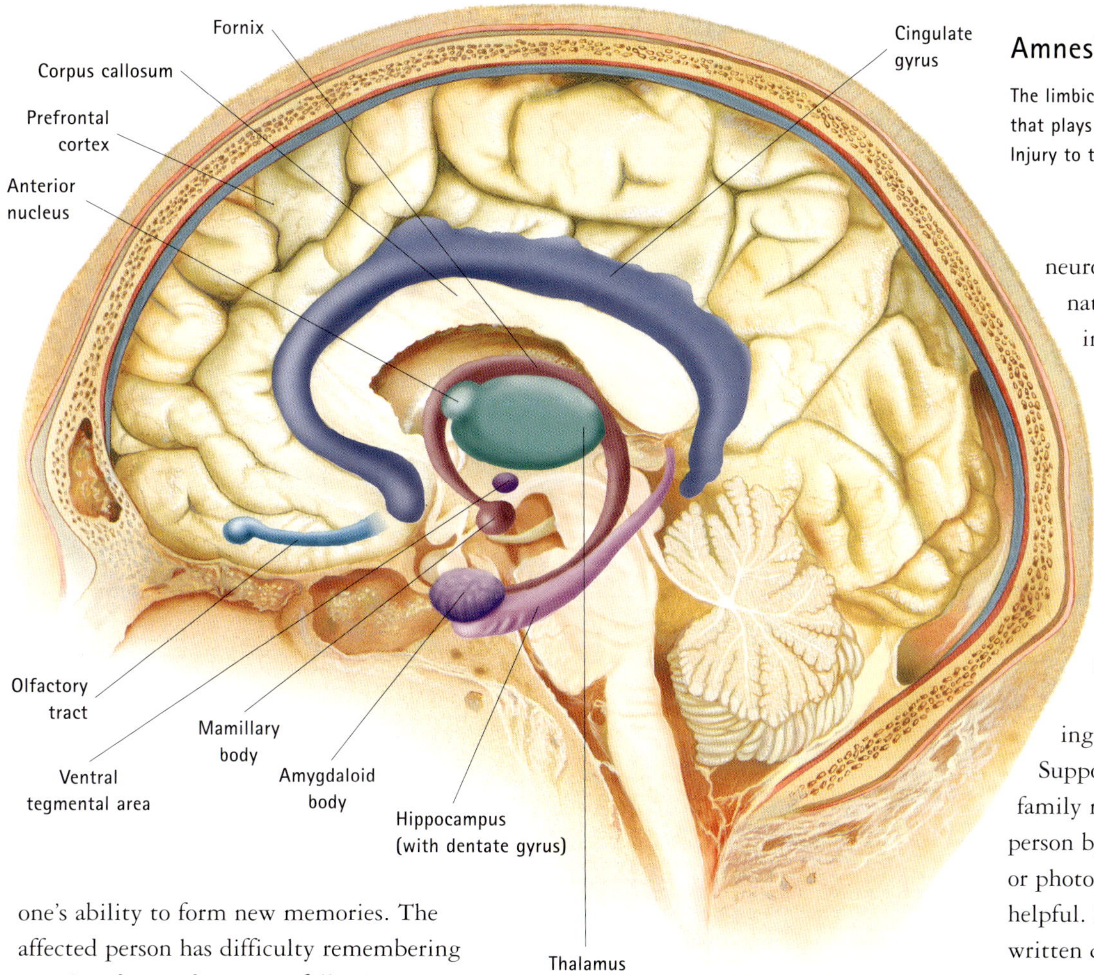

Amnesia

The limbic system, shown here, is part of the brain that plays an important part in processing memories. Injury to the limbic system can cause amnesia.

one's ability to form new memories. The affected person has difficulty remembering ongoing day-to-day events following an injury to the head, although they remember events prior to this. It may also affect alcoholics and usually leads to dementia.

In retrograde amnesia, the affected person has difficulty recalling events prior to head injury. Part or all of the memory loss may return in the days, weeks or months that follow. In a condition called transient global amnesia (TGA), the affected person, who is usually elderly, suddenly forgets how they came to be where they are. Their name and the names of family and friends can be recalled, but not the events leading to the attack. The condition is thought to be caused by unusual electrical activity in the temporal lobe of the brain. Recovery usually takes place in four to six hours. The condition does not cause permanent damage and requires no medical treatment.

In Wernicke-Korsakoff syndrome, memory loss is caused by a thiamine deficiency due to alcohol abuse. The affected person will have normal short-term memory, but will have difficulty acquiring new information and in remembering events that happened before the illness. It is a progressive disorder, and is usually accompanied by neurological problems such as uncoordinated movements and the loss of feeling in fingers and toes.

Hysterical amnesia (also known as fugue amnesia) is often temporary and is triggered off by a traumatic event that the mind of the affected person cannot cope with. Usually, the memory returns after a few days, though the memory of the traumatic event may remain incomplete.

Amnesia can be treated by reversing the underlying cause if possible. Support of the family is important. The family may need to orientate the affected person by providing familiar music, objects or photos, and relearning programs are also helpful. Medication schedules should be written down so they are not forgotten and lost. Nursing home and other extended care facilities may be needed if safety, nutrition or other basic needs are at risk.

Creutzfeldt-Jakob disease

Creutzfeldt-Jakob disease is a rare, degenerative, invariably fatal brain disorder that causes movement abnormalities and a rapid decrease of mental function. The disorder first appears about age 60 and progresses rapidly to loss of brain function similar to that of Alzheimer's disease. There may be muscle tremors, rigid posture and changes in coordination. It may occur spontaneously with no known cause.

In a small number of cases the disease is hereditary. In rare instances, it is acquired through exposure of brain or nervous system tissue during

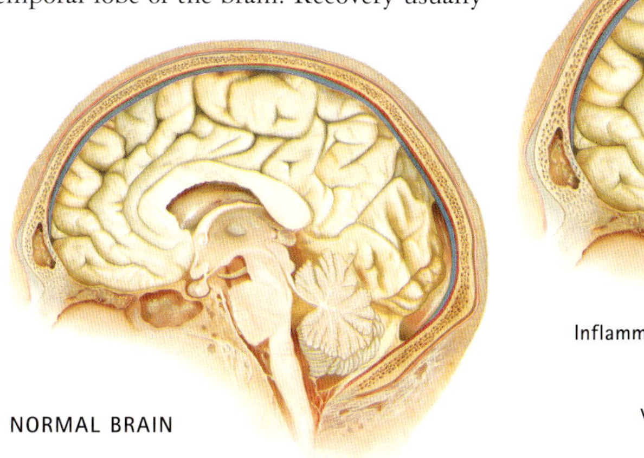

NORMAL BRAIN

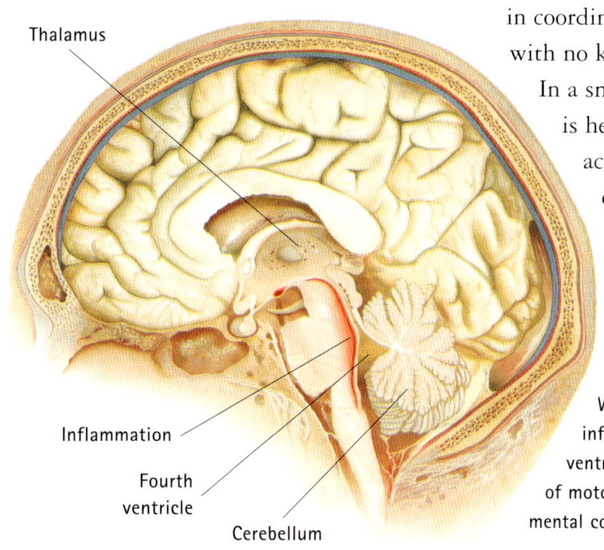

Wernicke-Korsakoff syndrome

Caused by a deficiency of thiamine, Wernicke-Korsakoff syndrome involves inflammation of the cerebellum, fourth ventricle and thalamus. This results in loss of motor skills, failing memory and general mental confusion.

medical procedures; it is thought to be transmitted via a viral-like protein called a prion. Adolescents who have received growth hormone derived from cadavers have contracted the disease; the use of synthetically manufactured growth hormone has meant contagion is no longer a problem. The disorder is fatal, usually resulting in death within a year.

Headache

Headache is a very common problem which can seriously interfere with a person's normal activities. It is not usually the result of another, more serious, underlying disease but severe or persistent headache, especially in children, requires medical investigation to exclude the presence of any such disorder.

By far the most common headaches are those referred to as tension headaches. Most persons experience occasional headaches of this type, often related to emotional stress or fatigue, but not necessarily triggered by tension. Tension headaches can also be triggered by caffeine, alcohol, certain foods, stress, fatigue, and skipping meals. The precise mechanics behind the development of these common headaches remain unclear. There can be associated tightening and tenderness of the muscles of the scalp, neck or jaw. Pain is usually moderate but may be severe with a sensation of the head being "gripped in a vise."

Episodic tension headaches usually respond to over-the-counter analgesics such as acetaminophen (paracetamol) or aspirin. Chronic tension headache can be associated with depression, anxiety, insomnia or other medical problems. Chronic or severe tension headaches should be investigated by a primary care physician, as they can be due to another cause, and there are other treatments that are available.

Vascular headaches make up a second large and important group, in which there is an associated dilation of the blood vessels supplying the head. The specific mechanisms involved are again poorly understood.

Migraine and its many variants are included in this category. Most migraine sufferers develop pain that usually affects one half of the head and lasts for several hours, often with throbbing and sometimes accompanied by nausea and vomiting. A number of precipitating factors for migraine headaches are recognized and several types of migraine can be distinguished. So-called hormonal headaches, associated with menstruation, pregnancy, oral contraceptive therapy and menopause, are forms of migraine headache.

Another important, although uncommon, type of vascular headache is known as cluster headache. They are of relatively short duration, but very severe attacks occur in clusters over periods of several weeks.

Headache is sometimes a manifestation of an underlying disease, although such headaches occur much less commonly than either tension or vascular headaches. Among the specific disease processes that may initially appear as headache are: inflammation of the blood vessels supplying the head and neck (known as cranial or temporal arteritis), which is of particular concern in older people; pain originating from inflamed nasal sinuses, inflamed teeth, osteoarthritis of the vertebrae in the neck, or injured or inflamed nerves; inflammation of the meninges covering the brain; very high blood pressure; and pain caused by masses within the cranium, including brain tumors.

Excessive self-medication for headaches can cause further headache, sometimes referred to as rebound headache. This is especially true of overuse of analgesics such as aspirin and acetaminophen (paracetamol).

The varieties of headaches that develop in children are similar to those in adults; both tension headaches and migraines commonly occur. Children's headaches can frequently be the first sign of underlying diseases (such as inflammation of eyes, ears, nose, head trauma, meningitis, and brain tumors), so severe or recurrent headaches should always be investigated by the family physician.

Epilepsy

Epilepsy is the disturbance in the normal electrical functions of the brain, causing seizures that may range from brief attacks of unusual behavior, a change in consciousness, erratic movements, or major seizures involving loss of consciousness.

The condition affects between 0.5–1 percent of people (mostly children) in Western countries, and usually begins between two and fourteen years of age.

HEADACHE

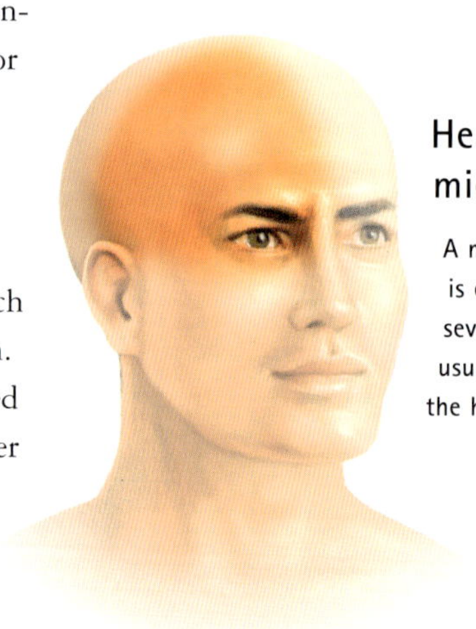

Headache—migraine

A migraine headache is characterized by severe, throbbing pain, usually on one side of the head.

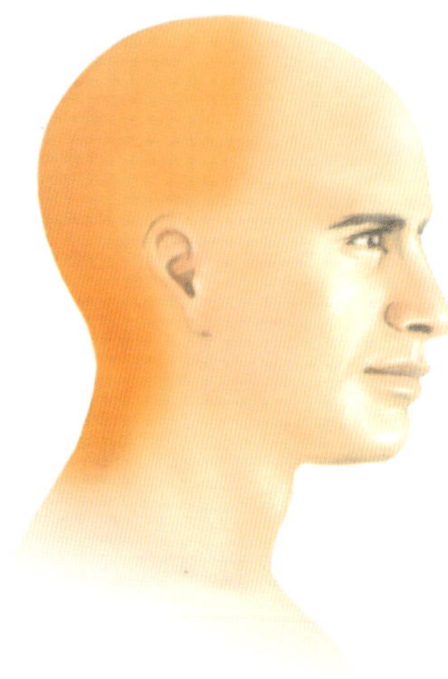

Headache—occipital neuralgia

Occipital neuralgia is a painful, stabbing sensation in the back part of the head (occiput).

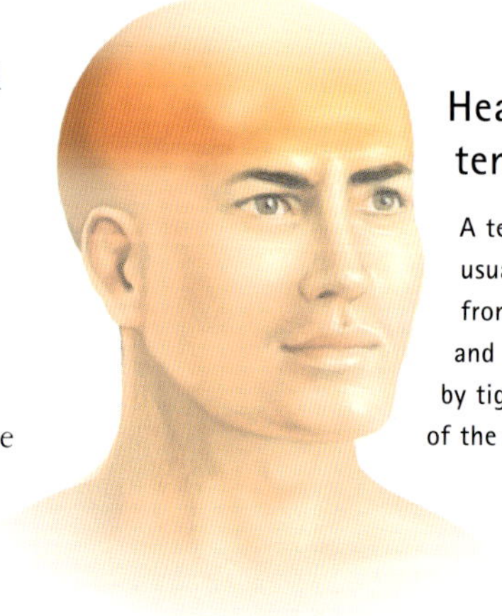

Headache—tension

A tension headache usually occurs in the front part of the head and is often accompanied by tightness in the muscles of the scalp, neck and jaw.

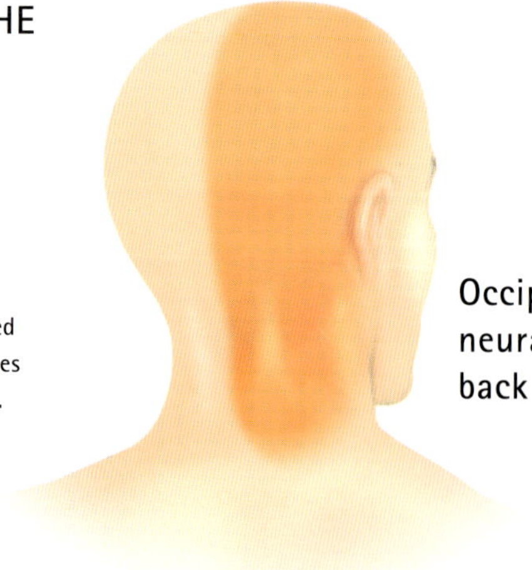

Occipital neuralgia—back view

The cause is usually unknown, though there might be a family history of seizure disorders. In about 25 percent of cases, there is an organic brain disorder such as a head injury, brain tumor, cerebral palsy, meningitis or encephalitis.

Seizures may occur in a generalized form (affecting all or most of the brain) or in a partial form (affecting only a portion of the brain). Generalized seizures cause loss of consciousness and include grand mal or tonic-clonic seizures (major seizures) and petit mal (minor) seizures (which occur mostly in children). Partial seizures that are termed focal result from a disturbance in the cortex; there may be abnormal movements or sensations, but the sufferer usually remains conscious. Complex partial seizures most commonly result from a disturbance in the temporal lobe.

Generalized tonic-clonic seizures begin with a sudden loss of consciousness. The person falls and the muscles become rigid (the tonic phase). As the abdominal muscles contract, forcing air from the lungs through the larynx, the person may give a shrill scream. Respiration ceases briefly, and the skin may turn blue. The lower limbs become extended and the upper limbs flexed. After about a minute, the clonic phase follows, consisting of jerking contractual movements of muscles in all four limbs. Breathing starts again, but is heavy and irregular, with frothing of saliva. Incontinence (loss of bowel and bladder control) is common and the sufferer may bite the tongue. Recovery occurs after 3–5 minutes. A period of confusion follows, when the person feels sleepy and may have a headache. Afterward, there is no recollection of the seizure or the confusion after it.

A petit mal seizure is much shorter than a grand mal seizure. It generally lasts less than 15 seconds and is characterized by loss of consciousness, but there are no involuntary movements and the person does not fall over. After the seizure, the person is alert and can resume their previous activity.

Other symptoms may also accompany the seizures; including headache, changes in mood or energy level, dizziness, fainting, confusion and memory loss. An aura (sensation such as a peculiar smell, vision or other sensation) may occur just before a generalized seizure. In some people, seizures may be triggered by hormone changes such as pregnancy or menstruation, illness, or by sensory stimuli such as lights, sounds and touch.

An electroencephalogram (EEG), which measures electrical activity in the brain, will confirm the diagnosis; it shows abnormal patterns of electrical activity and may show where the seizure is emanating from. A CAT or MRI scan of the brain may help to rule out an organic cause. Unless there is a reversible cause, epilepsy is considered to be a chronic, incurable condition. Nevertheless, anticonvulsant drugs can prevent most seizures and allow a near-normal life.

Grand mal seizures can be treated with phenytoin, carbamazepine, phenobarbital, valproic acid, or primidone. Petit mal seizures usually respond best to valproic acid, ethosuximide or clonazepam. Focal seizures or partial complex seizures are treated with phenytoin or carbamazepine.

If the sufferer has been free of seizures for a period (usually some years), the medications may be withdrawn gradually; many sufferers (especially children) will then stay free of seizures without medication.

Any circumstance that has triggered a seizure should be avoided and sufferers should wear a bracelet or pendant that identifies the condition. If someone is having a seizure, clear the area of objects that might get in the way. Do not try to restrain the person. Turn the person onto the side if vomiting occurs. When the seizure is over, keep the person on their side.

Status epilepticus is a serious, potentially life-threatening condition, usually defined as recurrent major convulsions that last for more than twenty minutes. Permanent brain damage or death can result if the seizure is not treated effectively; the longer the seizure lasts, the greater the danger. Treatment with intravenous anticonvulsants should be given as soon as possible.

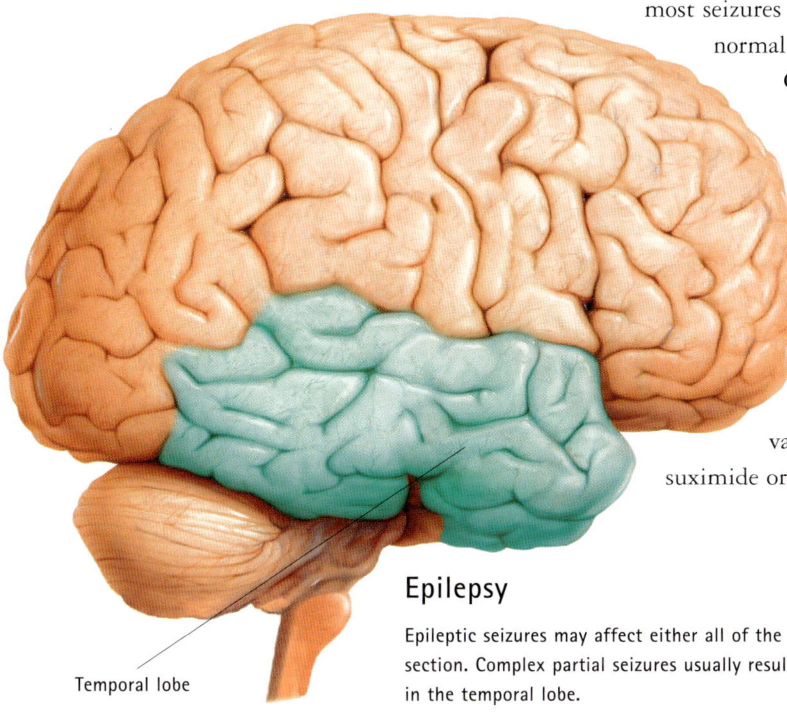

Epilepsy

Epileptic seizures may affect either all of the brain, or only a small section. Complex partial seizures usually result from a disturbance in the temporal lobe.

Convulsions

Convulsions or fits, epileptic attacks and seizures, are characterized by abnormal, often violent and uncontrolled spasmodic contractions and relaxations of the voluntary muscles. The eyes may roll, the teeth clench and the sufferer will twitch and shake. Convulsions can be a symptom of a disease and vary in severity, in some cases being accompanied by a loss of consciousness. Around 1 in 100 people will have a convulsion of some kind during their lifetime and about half of these will occur during childhood, most commonly between the ages of 3 months and 5 years. Seizures, particularly the first one, can be particularly worrying to parents of young children; however, they are rarely harmful.

Simple febrile convulsions (convulsions caused by a high fever) usually last less than a minute and are not repeated, though the child does have a slightly increased chance over children who never have a seizure of having subsequent attacks not associated with fever. Complicated febrile convulsions will last longer than 15 minutes and recur 2 or more times in 24 hours. Children who suffer from these have a greater risk of subsequent attacks and around 25 percent of cases will have a family history of seizures.

When people have convulsions, the most important thing is to ensure they do not injure themselves. They should be placed on a flat surface with the head to one side. It is not appropriate to put anything in the sufferer's mouth: it is more likely to cause harm. It is also not appropriate to attempt expired air resuscitation if children hold their breath in the early stages because this can cause damage as well. When the convulsion has passed, medical help should be sought immediately.

Reducing and controlling fever with tepid sponge baths and acetaminophen (paracetamol) or ibuprofen—not aspirin—may prevent a convulsion. A child who suffers convulsions may be prescribed anticonvulsant medicine; however, such medication can affect a child's ability to learn so it is prescribed with care. Most children outgrow febrile convulsions.

There are a number of other types of convulsion. They can occur because of scarred brain tissue following a head injury, or can be triggered by flashing lights. They can also be partial, only affecting one part of the body, or general, involving the entire body. Petit mal convulsions are not violent and show as a sudden cessation of activities. Epilepsy is diagnosed when an electroencephalogram (EEG) shows characteristic abnormalities. Prevention of these types of convulsion depends on the cause, and requires specialist advice.

Parkinson's disease

First described in 1817 by the British physician James Parkinson, Parkinson's disease is a disease of the central nervous system, characterized by gradual, progressive muscle rigidity, tremors and clumsiness.

The affected person suffers muscle stiffness and physical slowness, has a mask-like expression and an awkward or shuffling walk with a stooped posture, and talks in a slow, monotonous voice. Walking, talking, or performing other simple tasks becomes progressively more difficult. The symptoms are worsened by fatigue or stress. People with Parkinson's disease may become severely depressed. In later stages, mental deterioration and dementia may occur.

Parkinson's disease affects approximately two out of 1,000 people, and most often develops after age 50; it is one of the most common neurologic disorders of the elderly. The exact cause is unknown, but common to all is a progressive deterioration of the nerve cells in the part of the brain that controls muscle movement. As a result, there is a deficiency of a neurotransmitter (a chemical that relays messages across the nerve pathways) called dopamine that is normally found in this area of the brain. Without dopamine, the nerve cells cannot properly transmit messages. This results in abnormal firing of neurons and consequently, abnormal muscle movement. The disorder may affect one or both sides of the body, with varying degrees of loss of function.

Parkinsonism refers to those cases where Parkinson's disease occurs as a result of another disorder. It can also be caused by medications (especially the phenothiazine tranquilizers), and brain disorders such as post-influenza encephalitis and even by some slow virus infections.

Parkinson's disease is not curable. However, symptoms can be relieved or controlled by medications, which work by increasing levels of dopamine in the brain. They include levodopa, which is converted by the body to dopamine, and carbidopa, which reduces the side effects of levodopa.

These drugs can decrease tremors and reduce muscle rigidity, but they often have significant side effects. These may be controlled with antihistamines, antidepressants, bromocriptine, monoamine oxidase inhibitors, and other drugs.

Most people respond to medications, but to a variable degree. The therapeutic effect of the medications tends to wear off after a few years. Surgery to destroy parts of the nerve pathways in the brain responsible for tremors may reduce symptoms in some people. A new technique involves surgical grafting of dopamine-secreting neurons into the brains of Parkinson's sufferers. The long-term success of this technique has still not been established.

Regular rest and avoiding stress will help symptoms such as tremor. Physical therapy, speech therapy, social work, occupational therapy and other counseling services help the affected person to function normally.

Multiple sclerosis

Multiple sclerosis (MS) is a progressive and frequently debilitating disease of the central nervous system (CNS). It involves the ongoing destruction of the protective myelin sheaths around neurons in the brain and spinal cord. This interferes with the transmission of impulses by neurons, and disrupts signals sent throughout the CNS.

The cause of MS is unknown, although evidence suggests the trigger may be viral. Whatever the trigger, it is thought to stimulate an autoimmune response in which T cells mistakenly identify the myelin of the CNS as foreign. The T cells, which normally fight disease, then set up an immune response that leads to myelin destruction.

Just over 1 million people worldwide suffer from MS. Statistics reveal it to be far

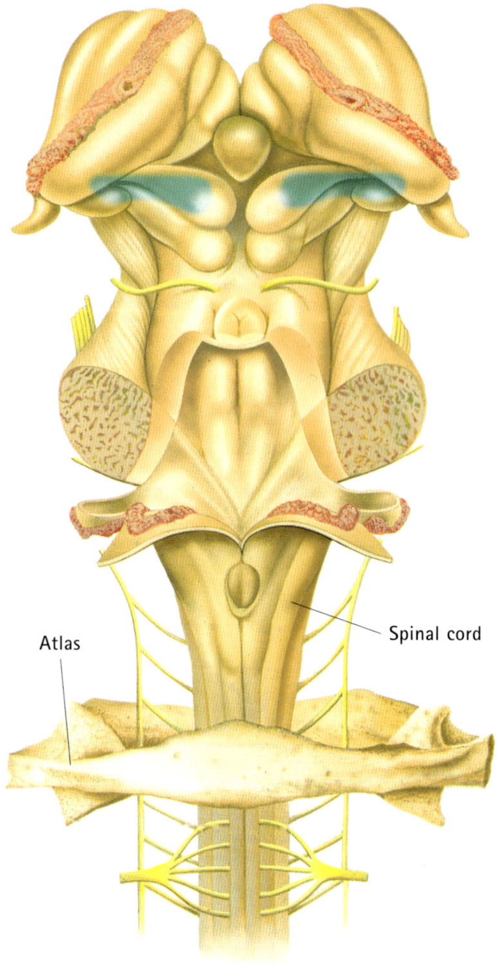

Parkinson's disease

Nerve cells in the substantia nigra (part of the brain stem) control muscle tone and movement. The progressive deterioration of these cells is the most common cause of Parkinson's disease, which causes muscle rigidity and tremors.

more prevalent in women than men, most common in people of northern European descent (particularly those with Scottish ancestry), extremely rare in people of Asian or African descent, and far less common in the tropics than in temperate areas.

Apart from ethnicity and ancestry issues, there are other indications that genetics may predispose a person to developing MS. Children of a parent suffering from MS are 30 to 50 times more likely to develop the disease than the general population.

MS is not fatal, although the life span of sufferers is usually reduced by about six years due to complications of the disease, with lung and kidney infections being the principal life-threatening risks. MS can also have a debilitating impact on a person's quality of life. Because the symptoms and progression of the disease vary widely between sufferers, however, it is hard to predict how an individual will be affected.

Between 20 percent and 35 percent have such a mild form of the disease that they hardly suffer. Up to 12 percent of patients fall into the category at the other end of the spectrum and suffer from a serious and particularly aggressive form of the disease which may involve paralysis and dementia.

The first signs vary and do not always immediately suggest a serious disorder. They usually appear some time between the ages of 15 and 40, but the disease most often strikes for the first time when the sufferer is in their twenties or thirties. Blurred or double vision is frequently one of the earliest indicators. Pain and involuntary movements in the eyes are also common in people who later develop MS. Extreme tiredness, clumsiness and tingling sensations are other early symptoms.

The disease takes years to progress, with sufferers experiencing periods of remission when symptoms apparently disappear, followed by relapses. Advanced-stage MS symptoms can include spasticity, poor coordination and vertigo, loss of bladder control, constipation, uncomfortable but short-lived sensations in the extremities, sexual problems including impotence in men, tremors in arms or legs and mental dysfunction that can range from simple memory lapses to impaired problem-solving abilities.

There is no cure for MS and symptoms are often difficult to treat, although there are therapies and medications available that can help make life more tolerable, including drugs which prevent muscle stiffness. Various drugs are available that help reduce the frequency of relapses in MS and help delay the eventual disability.

Amyotrophic lateral sclerosis

Amyotrophic lateral sclerosis (ALS) is a progressive, fatal disorder resulting in the loss of the use and control of muscles. Also called Lou Gehrig's disease (after the American baseball hero who suffered from it), ALS is a member of the class of disorders known as motor neuron diseases.

The condition is caused by gradual degeneration of nerve cells in the brain and spinal cord that control voluntary movement. As neurons shrink and disappear, the muscles under their control weaken and waste away. Symptoms include poor coordination, paralysis of hands and arms, twitching and cramping of muscles, and difficulty in speaking, swallowing and breathing. Symptoms usually do not develop until well into adulthood, often not until after 50 years of age. Mental faculties remain unaffected. The condition is confirmed by electromyography (EMG), which shows the nerve and muscle degeneration.

There is no cure for ALS, although one drug, riluzole, has been shown to prolong the survival of people suffering from the condition. Physical therapy, rehabilitation and use of braces or a wheelchair will assist the affected person to stay mobile as long as possible. ALS is usually fatal within five years after symptoms appear. The condition may run in families; if so, genetic counseling is advisable.

Hydrocephalus

Hydrocephalus ("water on the brain") is a condition in which there is an excess of fluid within or surrounding the brain. This fluid, known as cerebrospinal fluid or CSF, is produced at a constant rate within the ventricles (cavities) of the brain. It circulates from the ventricles into the space surrounding the brain (subarachnoid space) and from there it drains into the venous system. The total volume of CSF is replaced about three times per day so, if CSF circulation is blocked (e.g. by a tumor) or its drainage is defective, the CSF accumulates and exerts pressure on the brain. Occasionally, in some infants, the openings between the ventricles and the subarachnoid space fail to develop, and the resulting hydrocephalus causes the head to enlarge, because the skull bones have not yet fused. In adults, where the skull has fused and cannot expand any further, nearby structures can become compressed. CSF pressure can then

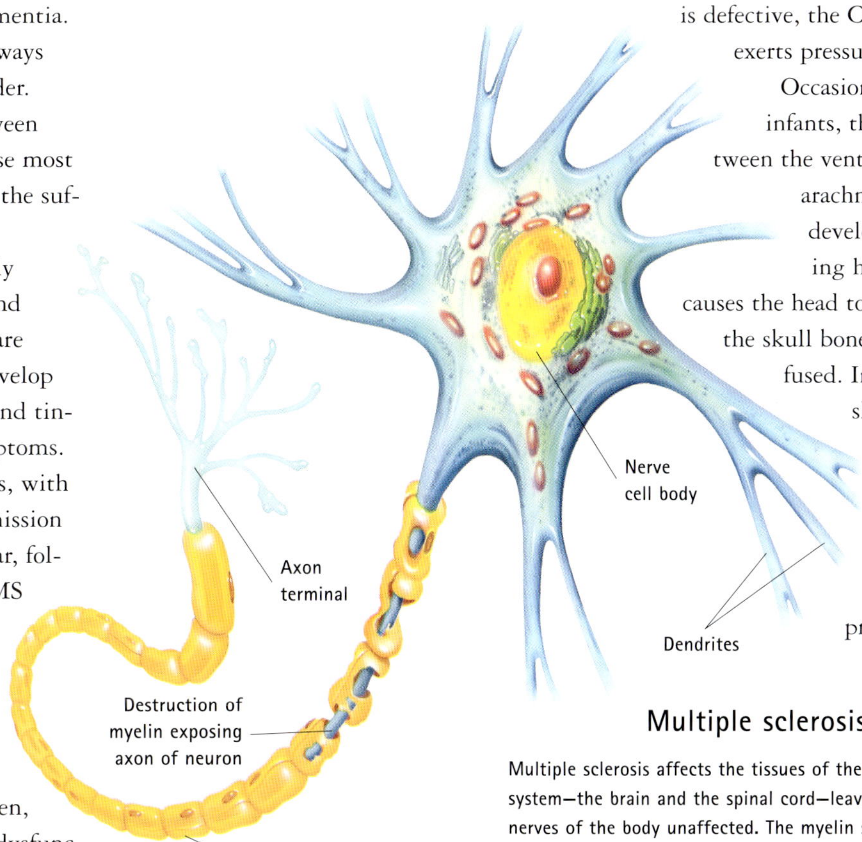

Multiple sclerosis

Multiple sclerosis affects the tissues of the central nervous system—the brain and the spinal cord—leaving the peripheral nerves of the body unaffected. The myelin sheath around neurons in the central nervous system is progressively destroyed. This leads to symptoms such as blurred vision, fatigue, poor coordination and tingling sensations.

build up inside the skull, which can affect consciousness and result in headache and vomiting.

Hydrocephalus is usually treated by placing a tube (shunt) into the ventricles, which enables the excess CSF to drain into the internal jugular vein in the neck.

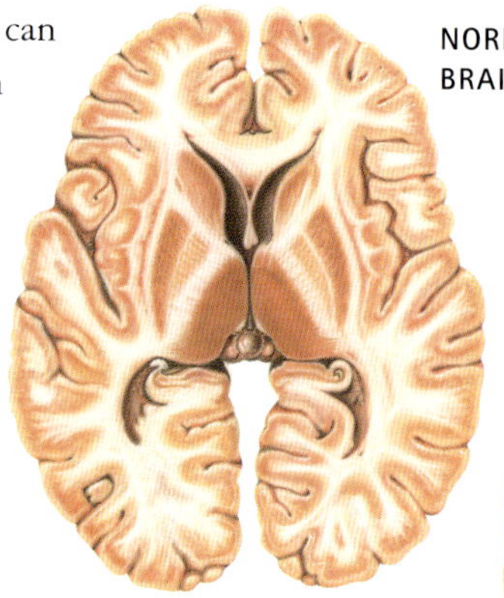

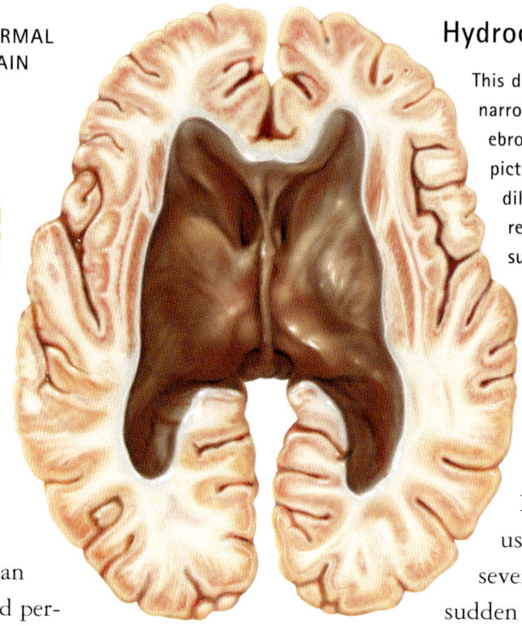

Hydrocephalus

This disorder is caused by blockage or narrowing of the pathways for cerebrospinal fluid (CSF). The brain pictured here shows the massive dilation of the lateral ventricles, resulting in compression of the surrounding brain tissue.

Diabetes insipidus

Diabetes insipidus is a rare condition causing pronounced thirst and the passage of large quantities of dilute urine. It is unrelated to diabetes mellitus. The most common type is so-called "central" diabetes insipidus, which is caused by a lack of antidiuretic hormone (ADH or vasopressin). This hormone is produced in the hypothalamus of the brain and controls the way the kidneys filter blood to make urine. Lack of this hormone (or failure of the kidneys to respond to the hormone), allows excess fluid to pass through the kidneys.

A person with diabetes insipidus must drink large quantities of water to compensate for the fluid loss in order to avoid dehydration. The lack of ADH secretion can be caused by damage to the hypothalamus as a result of surgery, infection, tumor or head injury. There is no cure for the condition; however, synthetic ADH injections will correct the hormone deficiency. ADH is also available as a nasal spray.

Diabetes insipidus can also be caused by a rare hereditary defect in the tubules of the kidney. The condition is called nephrogenic diabetes insipidus and produces symptoms similar to those of central diabetes insipidus. It may be treated with diuretic drugs.

Acromegaly

Affecting only about 1 in every 20,000 people, acromegaly is a rare hormonal disease that causes overgrowth of the body's bones, muscles and other tissues. It is caused by overproduction of an essential hormone called growth hormone (GH) by the pituitary gland.

In young children this can cause abnormal growth of long bones in the limbs. The condition is then called gigantism. If it happens in older children or in adulthood, the disorder is called acromegaly.

Acromegaly can cause the affected person's hands and feet to grow. The facial features coarsen; the jaw line, nose and forehead grow; the tongue grows larger and teeth get more widely spaced as the jaw grows larger. The voice gets deeper because of swelling of the larynx.

The cause of the overproduction of GH is usually a small benign tumor in the pituitary called a pituitary adenoma. As it grows, the tumor may also press on surrounding structures, especially the optic nerve nearby, so that vision may be affected. Changes to the menstrual cycle and abnormal production of breast milk are also common. Growth hormone can also have other unwanted metabolic effects, including diabetes, high blood pressure, gallstones and kidney stones.

Treatment of acromegaly is the surgical removal of all or part of the pituitary adenoma. Drugs such as bromocriptine and octreotide, which suppress growth hormone production by the pituitary, are often used as well.

Fainting

Fainting (syncope) is usually triggered by a severe and deep-seated pain, sudden shock or grief. It can usually be after a period of acute anxiety. Fainting may also be caused by prolonged standing, particularly in a crowd and especially if the person has not eaten before going out. Early pregnancy may be a contributing factor. Sudden blood loss can cause a fall in blood pressure followed by a fainting episode, as can the sight of severe trauma in others.

The experience of fainting can include a sudden drop in blood pressure, slowing of the heart, and pooling of blood in the extremities due to reflex dilation of their blood vessels. The person usually feels anxious, becomes clammy, and then falls unconscious to the ground. If the person is surrounded by a crowd of people and is not able to fall over, the brain becomes starved

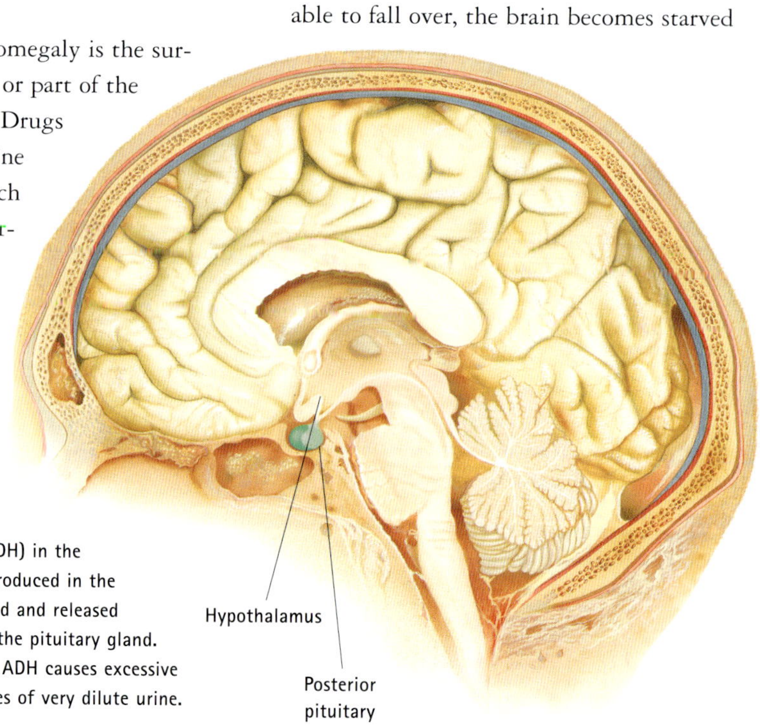

Diabetes insipidus

Diabetes insipidus is a disorder resulting from insufficient levels of antidiuretic hormone (ADH) in the body. ADH is normally produced in the hypothalamus then stored and released by the posterior part of the pituitary gland. Decreased production of ADH causes excessive thirst and large quantities of very dilute urine.

of sufficient blood supply and a seizure might occur. Once the head is at or below the level of the heart, consciousness rapidly returns. Lying the patient flat and elevating the legs hastens the process. Traditionally a fainting person would be given spirits of ammonia to inhale. However, by the time anyone can find a bottle of ammonia these days, the patient will have already returned to consciousness (also, ammonia can be very dangerous if inhaled).

Coma

Coma is a deep, often prolonged, state of unconsciousness. It may be caused by a disease (such as diabetes), liver or kidney failure, head injury, stroke, reaction to drugs or alcohol, or an epileptic seizure. It differs from sleep in that the subject cannot be roused by external stimulation.

An unconscious person should be laid on their side in the so-called recovery position. Emergency services should be called at once, as the comatose person has to be hospitalized and treated for reversing whatever is causing the coma. Meanwhile, airways must be kept open, and artificial respiration and cardiopulmonary resuscitation started if pulse and breathing are absent.

Meninges

Meninges are three thin, protective, continuous membranes surrounding the brain and spinal cord. The outermost membrane is a tough, fibrous layer called the dura mater. The middle membrane, the arachnoid, is a fragile network of collagen and elastin fibers with a cobweb appearance. The innermost membrane, the pia mater, is a layer of collagen and elastic fibers containing many blood vessels. Cerebro-spinal fluid fills the space between the arachnoid and pia mater.

SEE ALSO *Spinal cord on page 217*

Diseases of the meninges

The meninges can suffer from a number of problems. Tumors called meningiomas originate in the meninges, and though usually benign, these tumors may cause problems by placing pressure on the brain. Surgical removal is often successful. Meningitis is infection of the meninges.

Meningitis

Meningitis is an infection in the meninges, the membranes which cover the brain and spinal cord. It can be caused by fungi, protozoa, a virus or, in its most serious and potentially fatal forms, by a number of different types of bacteria.

Meninges in the brain

There are three layers of meninges: the fibrous outside layer (dura mater), the middle layer of collagen and elastin (arachnoid), and the inner layer, which contains many blood vessels (pia mater).

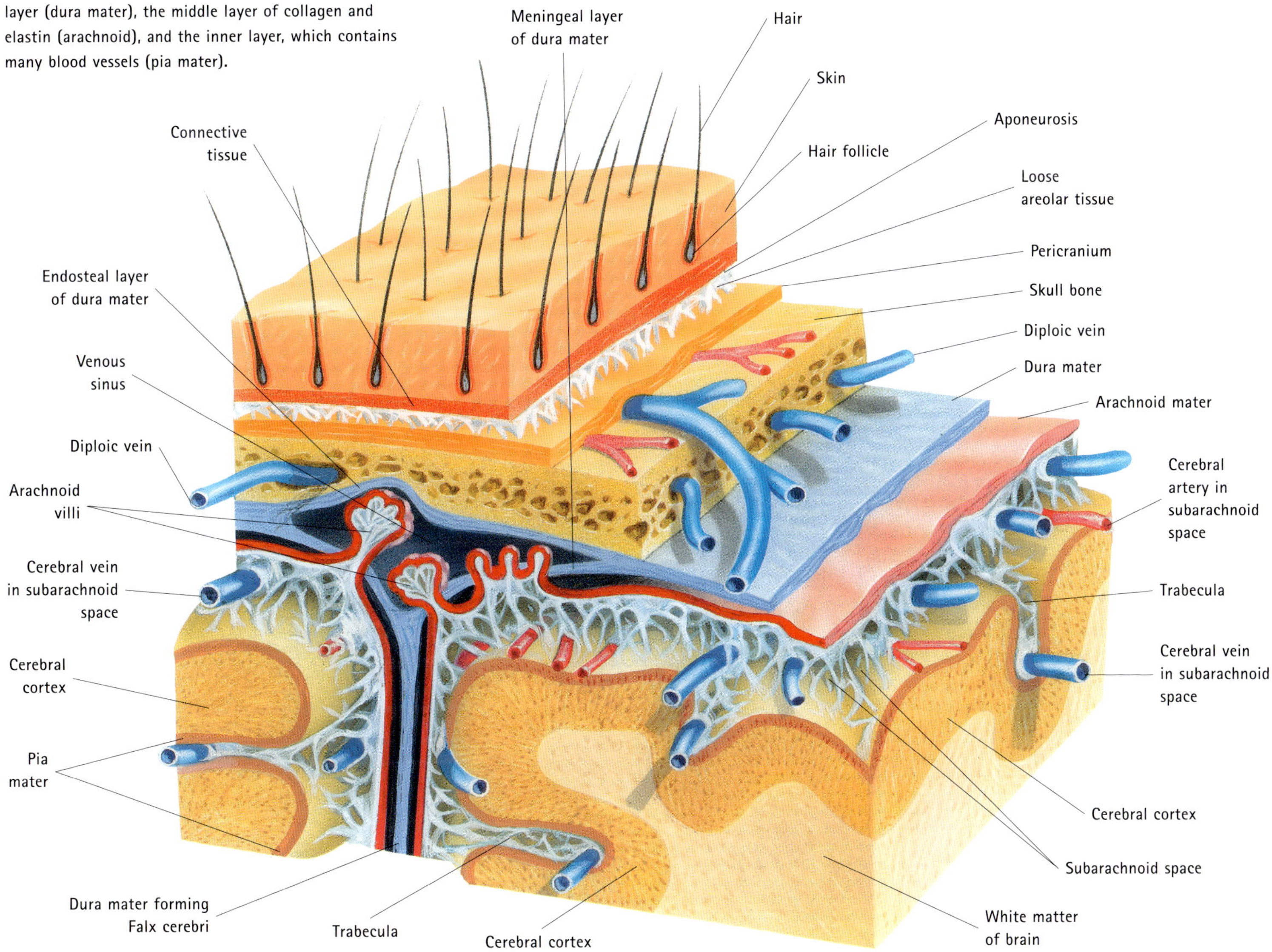

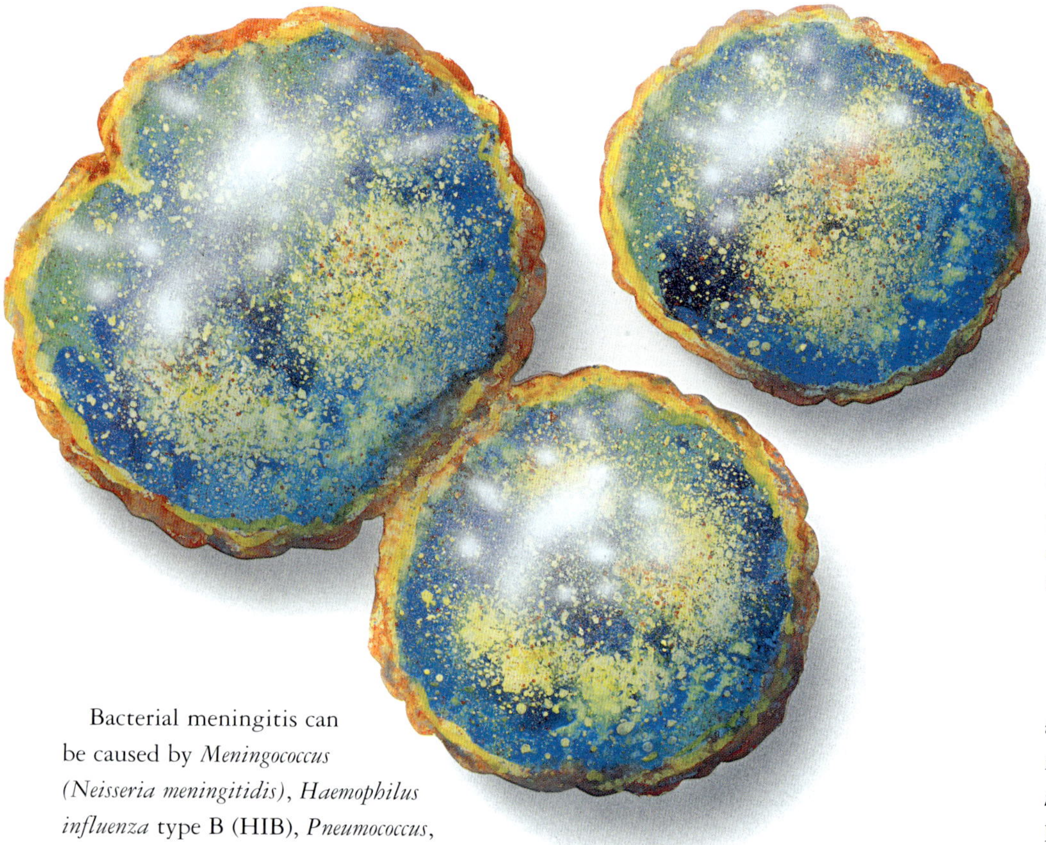

Meningitis bacteria

Meningitis can be caused by a number of different agents, including fungi, viruses or bacteria. The condition causes inflammation of the tissues (meninges) which encase the brain and spinal cord.

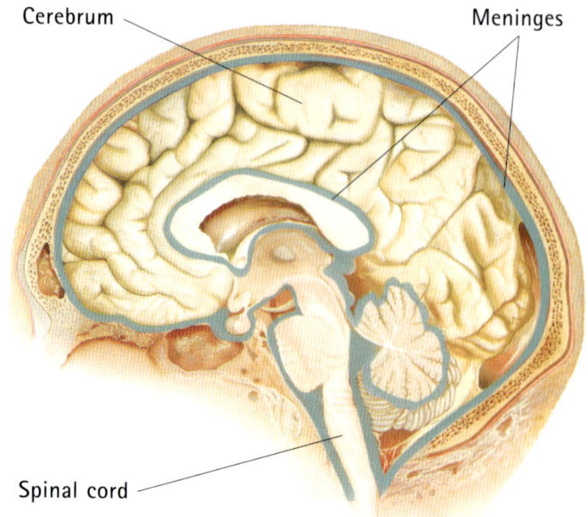

Meninges

The meninges (highlighted here in pale blue) are a set of three continuous membranes that cover and protect the brain and spinal cord.

Bacterial meningitis can be caused by *Meningococcus (Neisseria meningitidis)*, *Haemophilus influenza* type B (HIB), *Pneumococcus*, *Streptococcus* or *Staphylococcus*.

Meningococcal meningitis is found in all countries and primarily affects adolescents and children under age 10, as does the type of meningitis produced by *Haemophilus influenzae*. In adults, the most common cause is *Streptococcus pneumoniae*. It is an infection which can be only a very mild disturbance and therefore difficult to diagnose, or can make the sufferer extremely ill.

Bacterial infections of the middle ear or of another region of the body can be carried to the meninges via the blood. The bacteria then multiply quickly, causing the first symptoms to appear very rapidly.

The first symptom is usually vomiting, followed by a severe headache due to inflammation of the meninges and increased pressure of the cerebrospinal fluid. The neck may be very stiff, even arched and drawn backward in young children. Fluid may accumulate in the brain, causing coma and death unless relieved.

Other symptoms include moderate to high fever, headache, vomiting, collapse, convulsions, lethargy, inability to tolerate bright light, bulging fontanelle in children under two, and a purple rash all over the body. (This kind of purple rash is associated with Meningococcus infection, which is a very virulent form of meningitis.) In very young children, the fever may be the only sign until the child is suddenly critically ill.

The various forms of bacterial meningitis are spread through the secretions of the nose and throat, by coughing or kissing, but not by less intimate contact or in the air. Prolonged contact, such as between people sharing the same room or house, or children at the same daycare centers or school classroom, can cause infection and these people would be considered at risk in an outbreak.

A diagnosis is usually made by taking a sample of fluid from the spine and testing it for bacteria and abnormal chemical components. This is done in order to differentiate between meningitis and encephalitis, and also to establish which type of an organism is responsible for the infection. Early diagnosis of meningitis is essential in preventing death from the more serious bacterial forms of the disease.

Safe and effective vaccination is available against HIB, some strains of meningococcal meningitis (also known as *Neisseria meningitidis*) and forms of *Streptococcus pneumoniae*. HIB vaccines should be given routinely to infants with three doses before age 6 months, and a fourth between 12 and 18 months. Available vaccines against pneumococcal meningitis were, until recently, not effective in children under the age of 2 years, the major group at risk. However, a vaccine developed and approved for use in the USA offers hope for a dramatic improvement in protection against brain damage, hearing loss and death rates—currently 10 percent of those infected—for babies and infants up to the age of 5 years.

Adults may be vaccinated during outbreaks of the disease and should consider vaccination prior to travel in infected areas.

Cranial nerves

The cranial nerves provide innervation (distribution of nerves) to the muscles and sensory structures of the head and neck (including skin, membranes, eyes and ears). They also distribute nerves to the organs of the chest (trachea, bronchi, lungs and heart) and the upper part of the gastrointestinal tract. The twelve pairs of cranial nerves arise mainly from the brain stem.

SEE ALSO *Autonomic nervous system on page 75; Nervous system on page 64; Optic nerve on page 161*

CRANIAL NERVES

There are 12 cranial nerves that lead directly from the brain to various parts of the head. They control movements of the face, tongue, eyes and throat, and receive sensory input from the organs of hearing, sight, smell and taste.

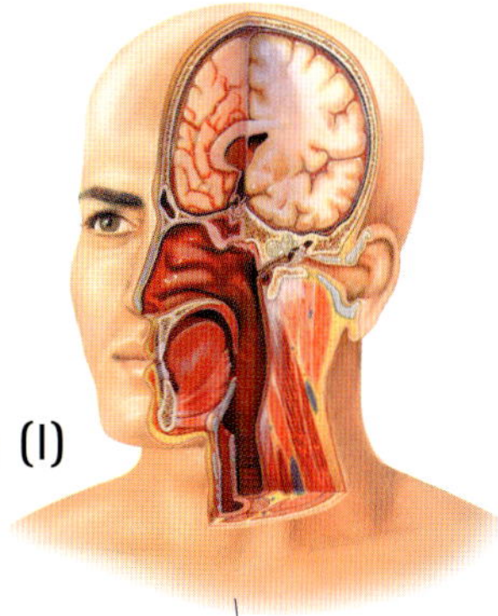

Olfactory nerve (I)

The first cranial nerve is concerned with the sense of smell. Nerve fibers starting in the mucous membranes of the nose carry messages to the cerebrum.

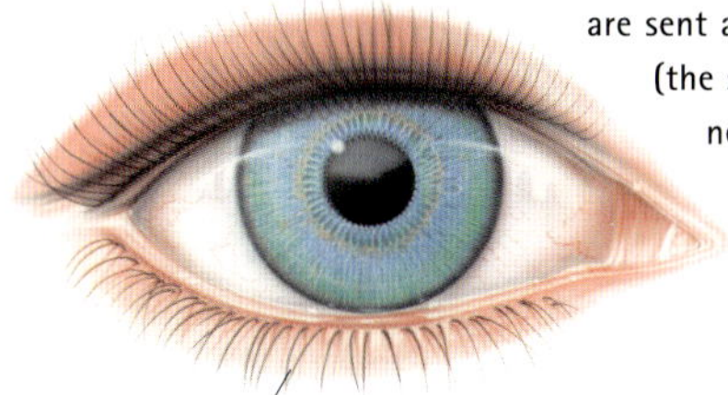

Optic nerve (II)

Visual impulses from the retina are sent along the optic nerve (the second cranial nerve) to the brain.

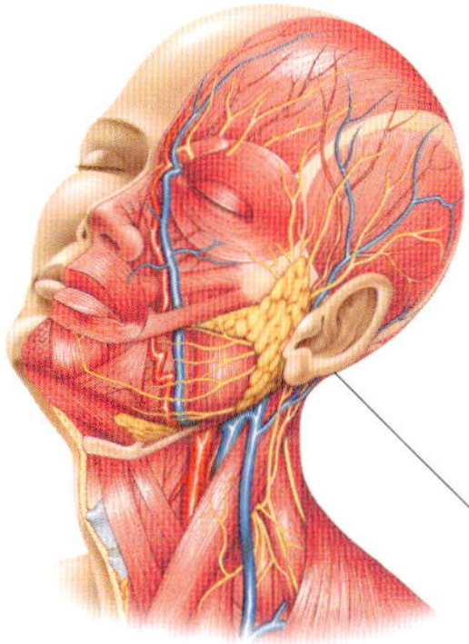

Trigeminal nerve (V)

The trigeminal nerve (fifth cranial nerve) has three sections: the ophthalmic, maxillary and mandibular divisions. They supply sensory fibers to areas such as the forehead, skin of the cheek and the muscles used for chewing.

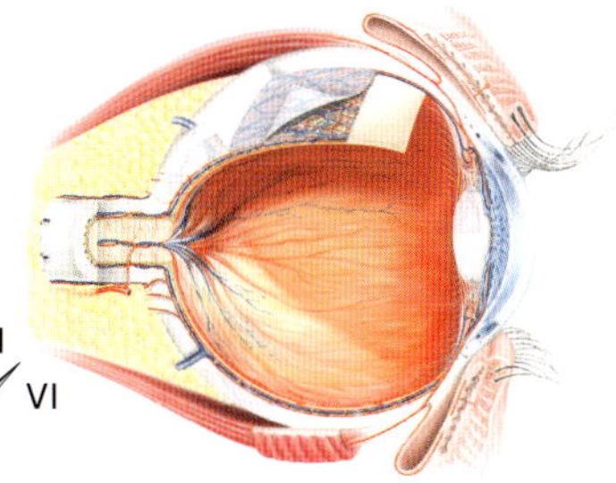

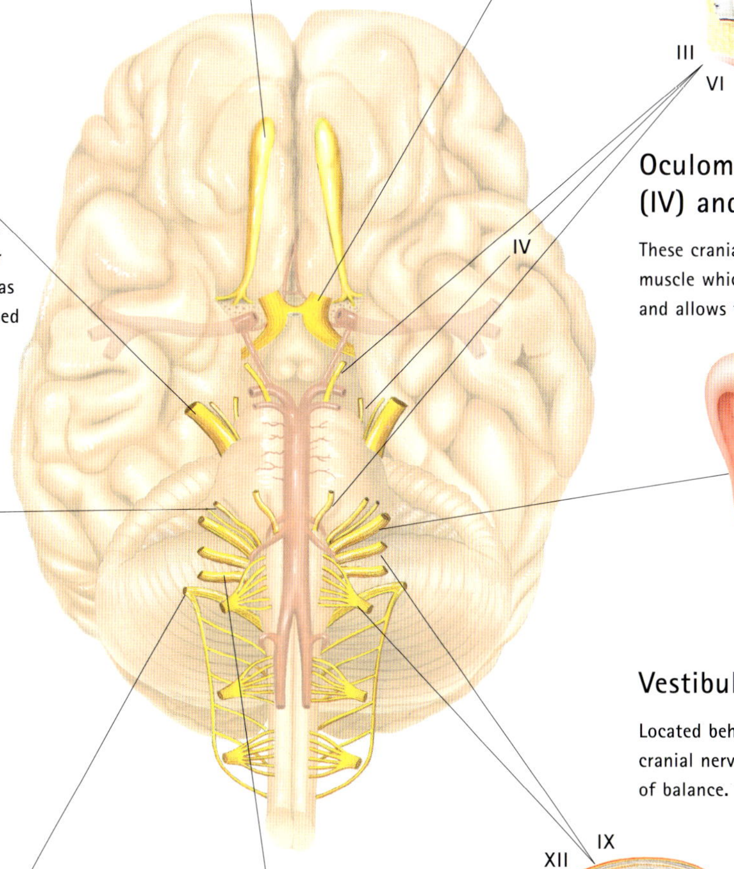

Oculomotor (III), trochlear (IV) and abducens (VI) nerves

These cranial nerves control movement of the muscle which moves the eyeball and eyelids, and allows focusing.

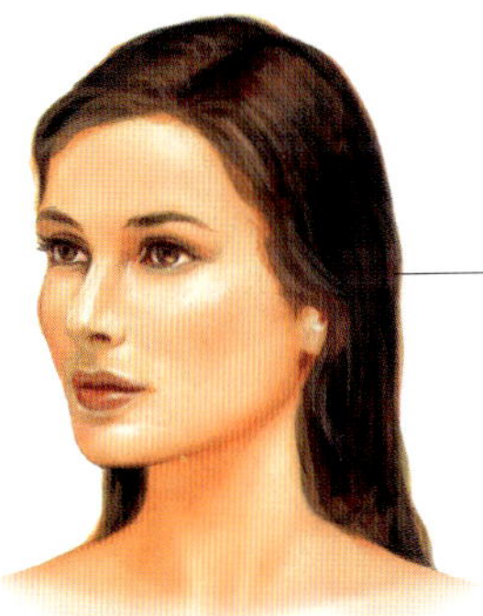

Facial nerve (VII)

The facial nerve is the seventh cranial nerve. It provides the motor fibers for facial expression. It is also responsible for the sensation of taste in the front part of the tongue.

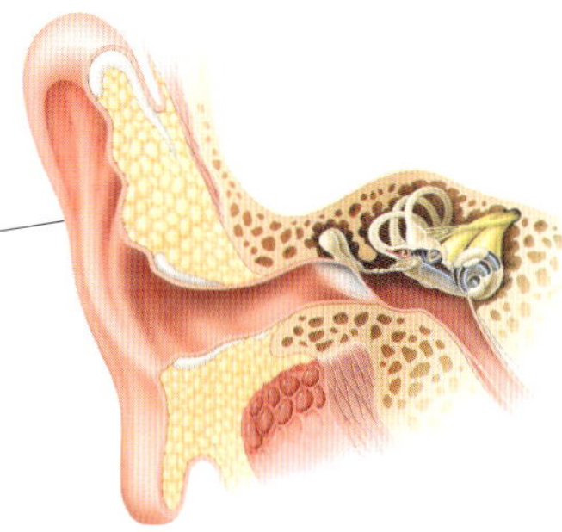

Vestibulocochlear nerve (VIII)

Located behind the facial nerve, the eighth cranial nerve carries impulses for the sense of balance.

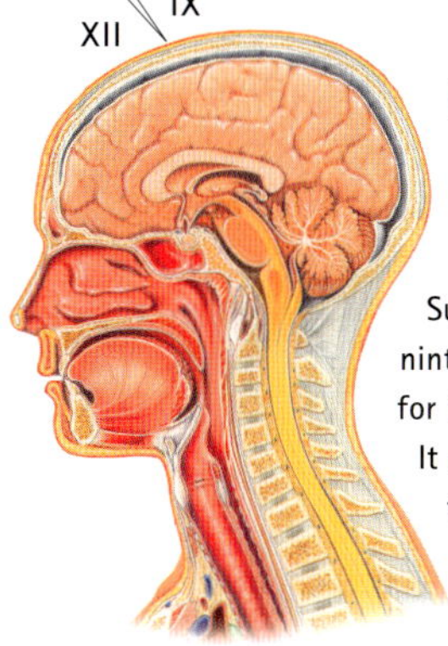

Glossopharyngeal (IX) and hypoglossal (XII) nerves

Supplying the carotid sinus, the ninth cranial nerve is responsible for the reflex control of the heart. It also supplies the back part of the tongue and the soft palate. The twelfth controls movement of the tongue.

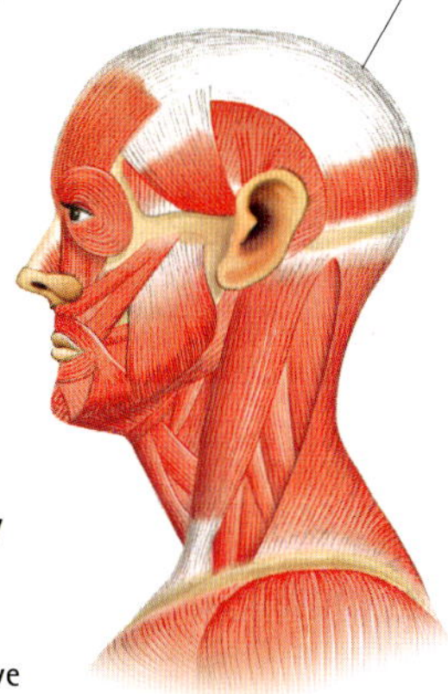

Spinal accessory nerve (XI)

The eleventh cranial nerve is primarily responsible for movement of the muscles of the upper shoulders, head, neck, and larynx and pharynx.

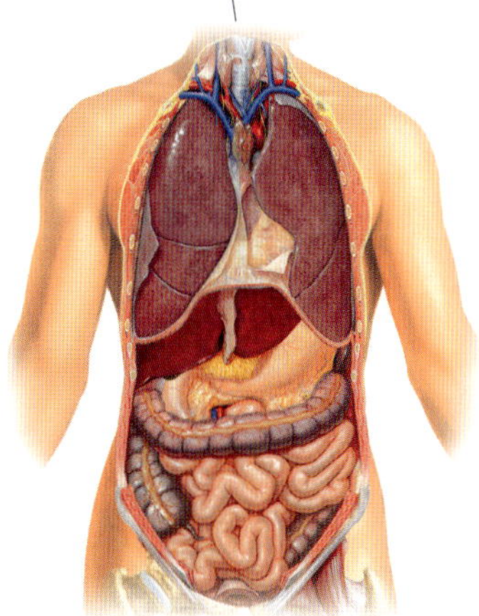

Vagus nerve (X)

The tenth cranial nerve is involved with functions such as coughing, sneezing, swallowing, speaking, secretions from the glands of the stomach, as well as the sensation of hunger.

EYE

The eye is the organ of sight: a complex, versatile and delicate structure. Every time we look at a scene, an image is formed on the retina of the eye and sent to the brain for analysis. The eyeball can be compared to a camera, which has three main parts: the camera body, the lens and the film.

The eyeball is made up of three layers. The outer layer consists of the sclera and cornea, the middle layer is the uvea and lens, and the inner layer is the retina.

SEE ALSO *Autonomic nervous system on page 75; Eye disorders on page 164; Fetal development on page 398*

Sclera and cornea

The eyeball is a sphere formed by a white layer called the sclera, the so-called white of the eye. It is composed of dense, tough fibrous tissue. Besides containing and protecting the optical parts of the eye, the sclera provides attachment for the muscles which move the eye. The front part of this sphere is cut off and replaced by a section of a smaller sphere. This section is more curved and formed by a transparent layer called the cornea.

The sclera gives the eyeball its shape and maintains a constant distance between the cornea and the retina at the back of the eye. It serves the same function as the camera body. When the eye is focused on infinity, the image should fall on the back of the eye. If the sphere of the sclera is slightly too large, images of close objects will fall exactly on the back of the eye but distant objects will not, and so the eye can only clearly see close objects, a condition known as near-sightedness or myopia.

The opposite condition is far-sightedness or presbyopia, occurring when the scleral sphere is too small.

The sclera has a hole near the posterior pole of the eyeball through which the optic nerve, which connects the retina to the visual area of the brain, passes.

Lying at the front of the eye, the cornea is the transparent part of the outer layer of the eyeball. The cornea is transparent because its fibers and cells are organized in a very orderly fashion. Opacity of the cornea results when this arrangement is disrupted by injuries and scarring.

Just like the camera lens, the cornea must have a perfect and adequate curvature, because it is the front element of the optic system of the eye. The cornea bends the light and works together with the lens to form an image on the retina. The surface of the cornea may be ulcerated through injury by foreign bodies or by infection with bacteria *(Streptococcus pneumoniae, Pseudomonas aeruginosa)* or viruses (*Herpes simplex*). Corneal ulcers will impair sight and may require corneal transplantation.

Examination of the inner parts of the eye is made through the cornea with the aid of an instrument known as an ophthalmoscope. Vision is very poor in the condition called keratoconus (the cornea, *kerato-*, is shaped like a cone, *conus*). In this condition, the image is distorted because the cornea does not have the profile of a sphere but is more pointed at the center.

In astigmatism, the image on the retina becomes distorted because the cornea is not exactly spherical.

The cornea is very sensitive to pain and temperature. We all experience that pain when a tiny speck of dust gets in our eyes.

The junction between the sclera and the cornea is called the limbus (meaning the rim) and contains Schlemm's canal, a circular channel which drains the fluid from the front part of the eye.

Uvea and lens

The uvea is the middle layer of the eye and has three parts: the choroid, ciliary body and

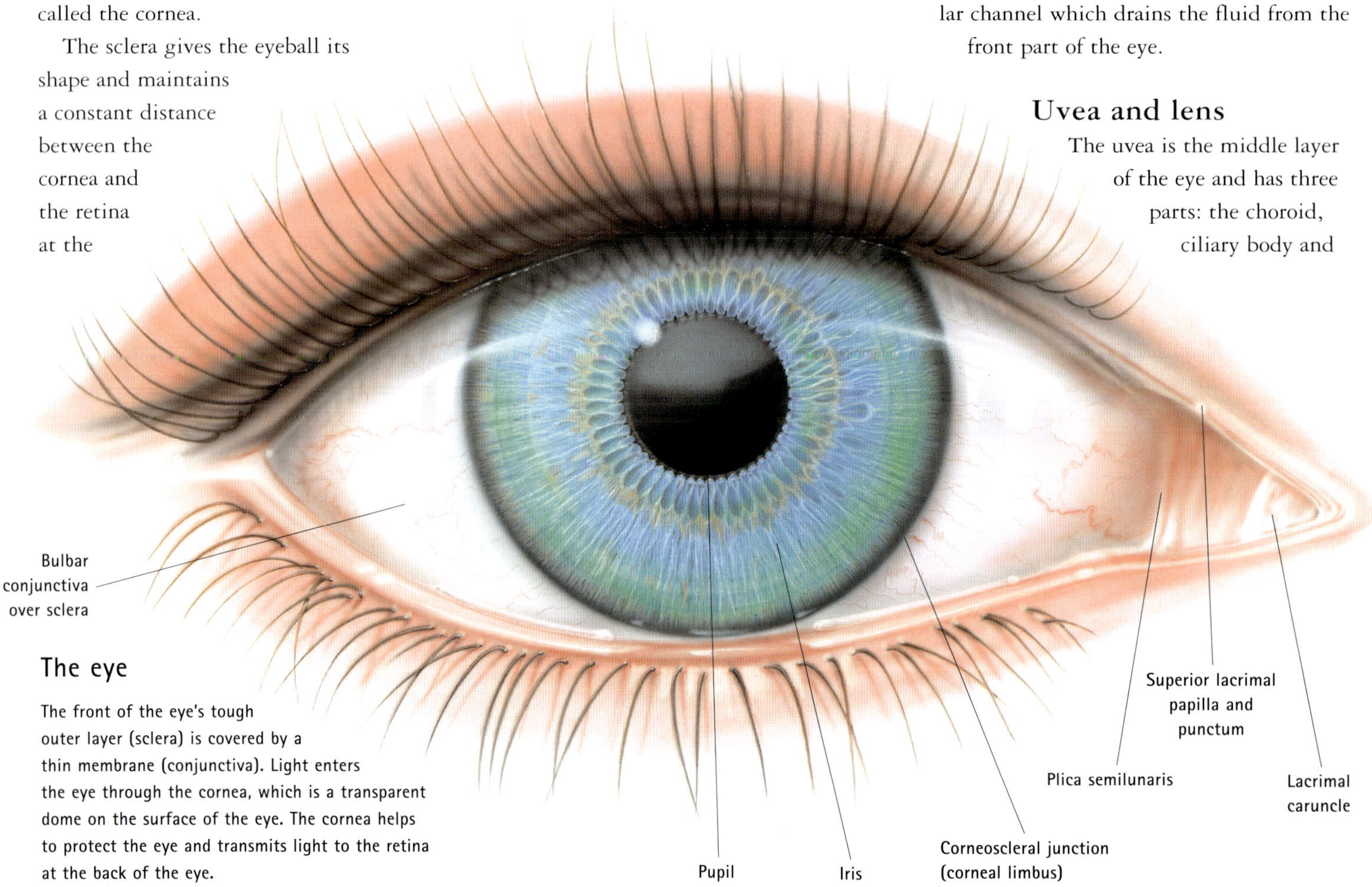

The eye

The front of the eye's tough outer layer (sclera) is covered by a thin membrane (conjunctiva). Light enters the eye through the cornea, which is a transparent dome on the surface of the eye. The cornea helps to protect the eye and transmits light to the retina at the back of the eye.

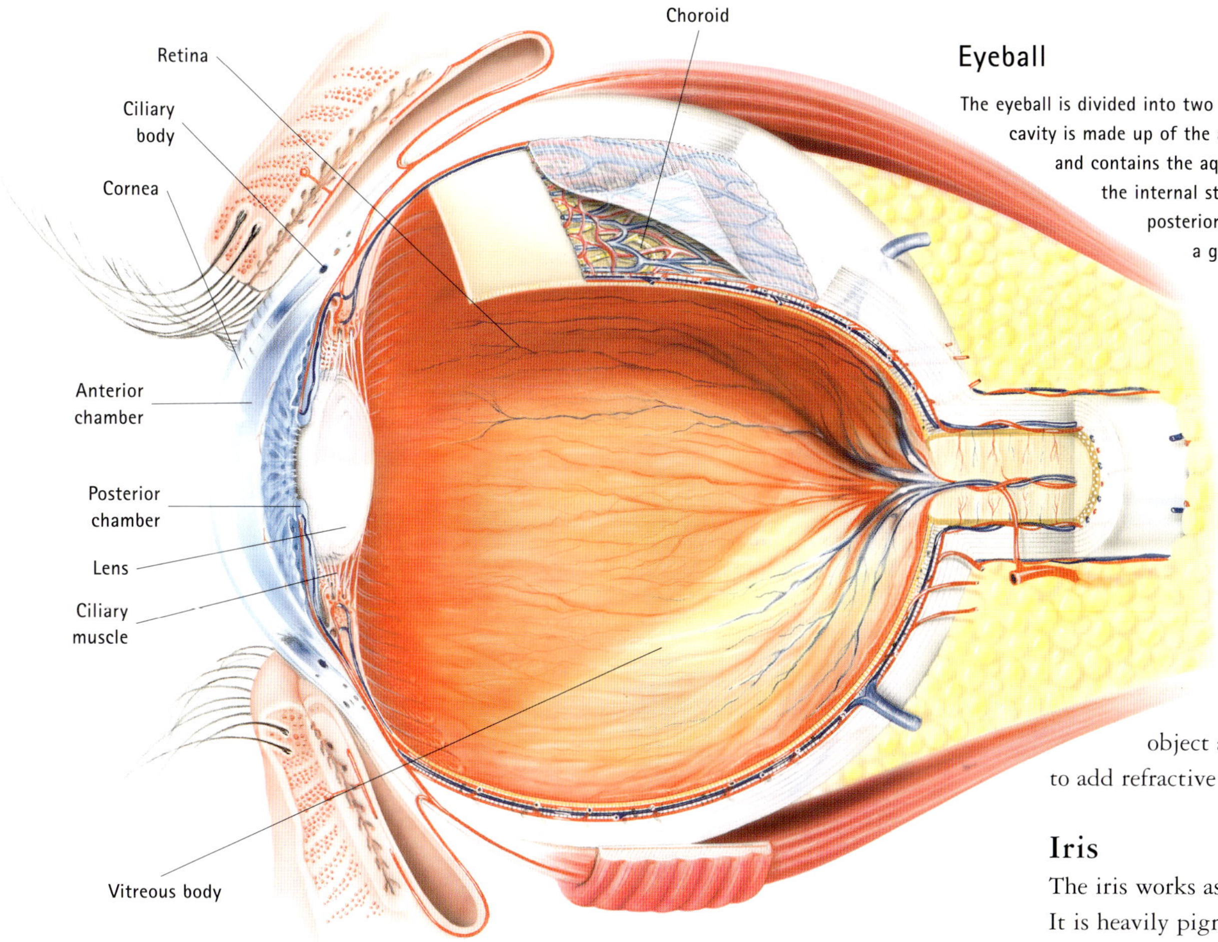

Eyeball

The eyeball is divided into two fluid-filled cavities. The anterior cavity is made up of the anterior and posterior chambers and contains the aqueous humor which nourishes the internal structures of the eyeball. The posterior cavity contains the vitreous body, a gel-like material that helps the eyeball maintain its shape.

iris. The choroid is the back part of the uvea. It contains several blood vessels and gives passage to nerves going to the cornea, ciliary body and iris.

The choroid continues forward as the ciliary body. If you were to cut the eye in two and look at the inside of the front half, you would see the ciliary body as a dark ring that gradually becomes thicker at the front. When the lens is removed, one can see that the ciliary body joins the iris, a flat ring that stretches down in front of the lens. The hole in the center of the iris is the pupil.

The ciliary body contains ciliary muscles. Tiny fibers that look like nylon strings run from the ciliary body to a ring near the equator of the lens. They form a suspension mechanism for the lens called the zonule.

The lens is a clear structure that looks and works as a magnifying glass. It does not bend the light as much as the cornea (i.e. its refractive power is not as great), but it can change its curvature. When the lens is in place in a normal eye that is looking into the distance, it is stretched and flattened by the pull of the zonule.

An experiment with a magnifying glass will help us understand how the eye focuses. Point the magnifying glass at a tree far away outside a window and collect the inverted image on a piece of white cardboard. Move the cardboard back and forth until the tree looks sharp. The image of the window frame (which is nearer to us) is blurred. To get a sharp image of the window, we must either move the lens further from the cardboard or add another lens in front of it to increase its refractive power.

The first conclusion is that the image of a close object is further from the lens than that of a distant object. That is why someone whose eyeball is slightly too large will not be able to see distant objects—their images fall in front of the retina. Corrective glasses must be concave to effectively reduce the refractive power of the lens of the eye to bring the image further back onto the retina.

The second conclusion is that when we want to get a sharp image of a nearer object without moving the lens, we have to increase the thickness of the lens. This is achieved in the eyeball by the action of the ciliary muscles. The ciliary muscles contract, causing a movement of the ciliary body, which slackens the tension on the zonule. The lens shrinks back by its elasticity, becomes thicker (more rounded) and the image of the object is pulled back to the retina. As we age, the lens loses its elasticity and fails to achieve the thickness required. We can no longer focus on a near object and need reading glasses to add refractive power to our lenses.

Iris

The iris works as the camera's diaphragm. It is heavily pigmented to block light and has circular muscle fibers around the pupil which contract in bright light to constrict the pupil. The iris also has contractile cells radiating out from the pupil. In darkness, these cells contract and the pupil is dilated.

The adjustment of pupil size and actions of the ciliary muscles is controlled by the autonomic nervous system, the part of our nervous system that operates without conscious effort. If this system is disrupted, the pupils will not constrict when a light is shone into the eyes.

The color of the eyes is determined by pigments in the iris; eye color is inherited. An albino person has no pigmentation in the iris so the eyes look pink due to the small blood vessels in the retina.

Pupil

The pupil of the eye is the central hole in the iris through which light enters into the eye. Tiny muscles in the iris control the size of the pupil.

In daylight, the pupil is typically about 1⁄10 inch (3 millimeters) in diameter. In dark conditions, it opens to about 1⁄3 inch (7 millimeters) in diameter to allow more light to enter the eye. Adaptation to dark also involves adaptation in the retina.

Arteries of the eye

The central artery of the retina enters the eye through the center of the optic nerve. It fans out into four main branches, each accompanied by veins. These arteries spread out to form a capillary network within the eye. The arteries can be adversely affected by conditions such as high blood pressure or diabetes.

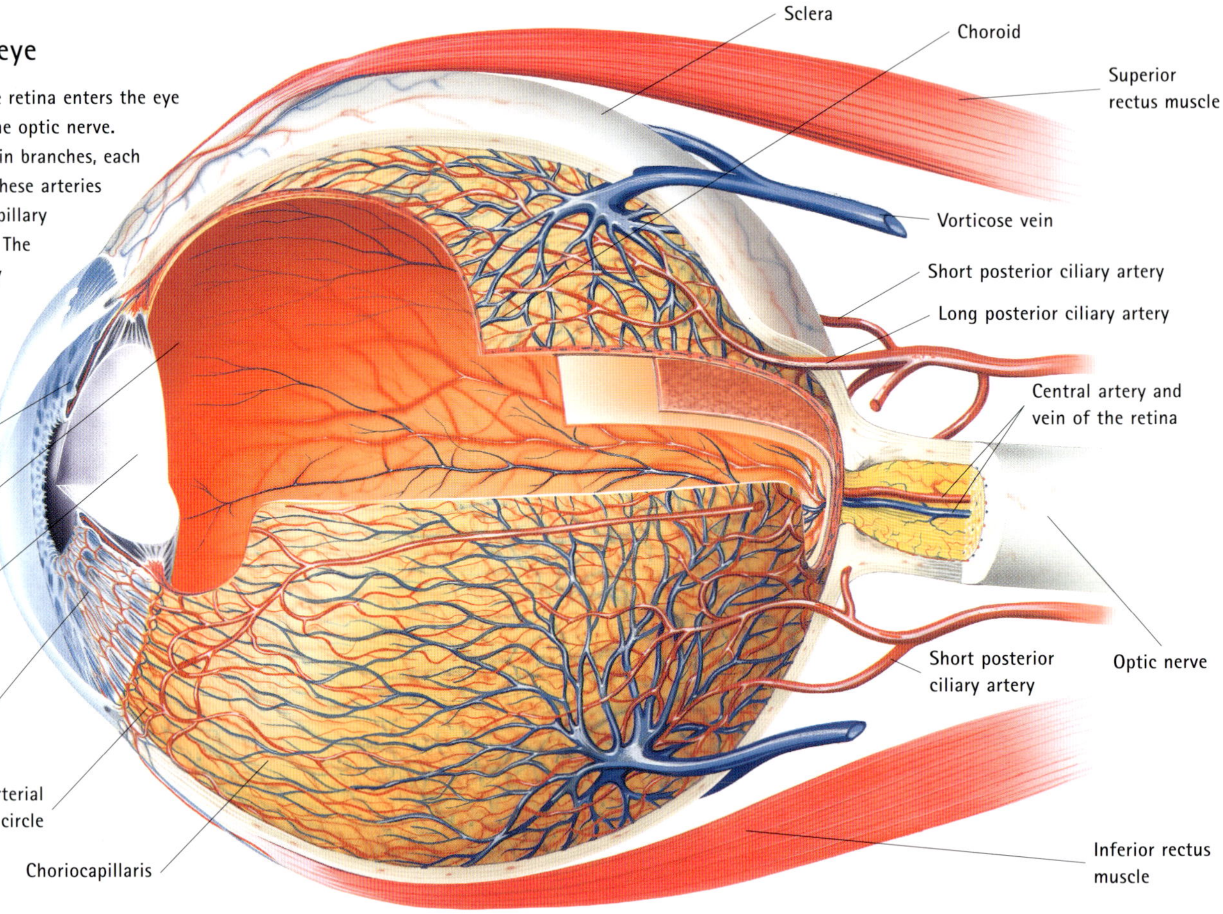

Retina

The retina is a light-sensitive layer at the back of the eye; it has photoreceptors that send visual information to the brain via the optic nerve. Light reaches the retina after passing through the cornea into the pupil, then through the optic lens and vitreous humor. The retina contains light-sensitive cells called rods and cones, which specialize in perceiving light intensity and color vision.

The optical axis of the eye is a straight line between the center of the lens and the center of its image on the retina. The point where all the nerve fibers from the retina converge into the optic nerve is known as the optic disk.

PHOTORECEPTORS

Light is converted into neural impulses by specialized cells called photoreceptors. These are classified into two types based on their shape: rods and cones. Cone receptors are not as sensitive to light as the rods but they can detect colors. Signals from individual cones are not mixed together much because only a few cones are connected together before the signal is sent to the brain. They are therefore responsible for detailed vision.

According to laws of optics, the image made by a lens is sharpest around the optical axis of the lens. The area of the retina around the optical axis is designed for high resolution perception with a high concentration of cones. At the central point, each cone is connected by one optic nerve fiber to one point on the cortex of the brain. This point-to-point projection to the brain ensures the highest resolution possible. While reading this page, you really only see clearly the words in the center of your visual field. You cannot recognize the words on the rest of the page.

Rod photoreceptors are more sensitive to light, but a large number of them are connected together before the sum of signals is sent to the brain. Thus they are good for sensing brightness and movement, not for color or detailed vision. The rods are almost nonexistent around the optical center of the retina, but they are the only type of photoreceptor in the periphery of the retina. That is why we can only see movements of a friend from the "corner of the eye," not the facial expression.

There is no light detection at the optic disk where there are no photoreceptors. Light falling on the optic disk will not be perceived by the eye, it is in the "blind spot."

When the photoreceptors are exposed to light, they lose an amount of a light-sensitive pigment called rhodopsin which has to be replaced by synthesis in the cell. After staring at a bright light, the eye is blinded for a short while because the photoreceptors are depleted of their rhodopsin.

CENTRAL ARTERY OF THE RETINA

The central artery of the retina enters the eyeball by running in the center of the optic nerve. As it emerges from the optic disk it fans out into four main branches accompanied by their veins. These branches run on the inside of the retina and spread out into a capillary network.

Observation of these blood vessels through an ophthalmoscope is not only important for the diagnosis of disorders of the eye but also tells the doctor much about the general condition of arteries in diseases such as high blood pressure or diabetes.

Vitreous body

The cavity behind the lens and its zonule is filled up by a viscoelastic gel which is called the "vitreous body" because it is as clear as glass. In old age, tiny particles may form in the vitreous body and these are visible as floating spots in the visual field because they cast shadows on the retina.

Anterior and posterior chambers

The space in front of the lens and zonule is divided by the iris into the anterior and posterior chambers. The ciliary body secretes into the posterior chamber a fluid called aqueous humor which passes through the pupil into the anterior chamber.

The lens and cornea do not receive any blood vessels and rely entirely on aqueous humor for nutrition. Aqueous humor is drained into Schlemm's canal in the anterior chamber, and from there into the veins.

The pressure in the eye, the intraocular pressure, depends on the balance between the rate of production and drainage of aqueous humor. This pressure is often measured when you have your eyes examined by your doctor or optometrist.

Optic nerve

The optic nerve connects the eye to the brain and is the second of the 12 cranial nerves. It contains the long processes (axons) of ganglion cells in the retina, the light-sensitive layer at the back of the eye. The retina and optic nerve grow out from the brain and are actually part of the central nervous system. The axons of the ganglion cells converge at the optic disk (blind spot), where the optic nerve leaves the eye. The nerve continues to the optic chiasm, where it joins the nerve from the other eye. There are about a million axons in each optic nerve, as well as blood vessels supplying the retina. Action potentials in the nerve are transmitted to the brain to provide visual sensations. Damage to the optic nerve can cause partial or complete blindness. For example, the optic nerve can get inflamed at the optic disk.

Eyeball

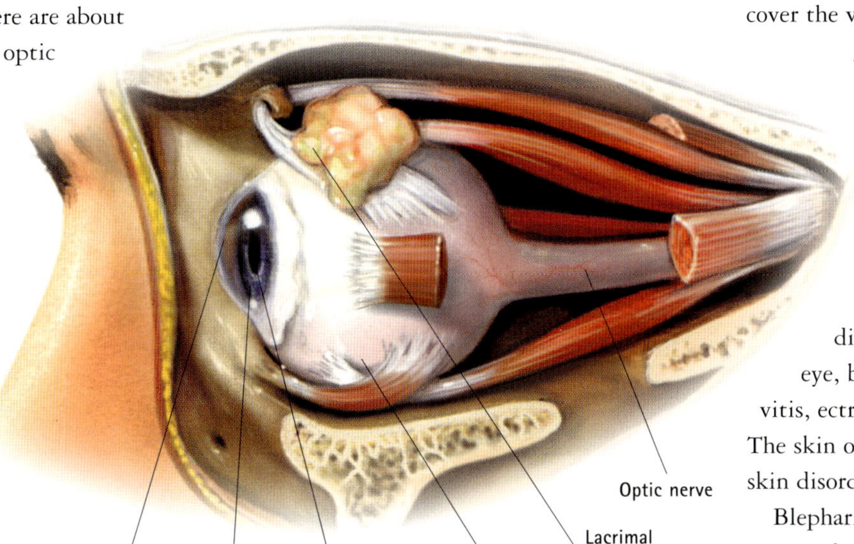

Movements of the eyeball

We move our eyes to look around us or to follow a moving object. Movement must be precise because the eye turns only a tiny angle to fixate on a person walking across the visual field a hundred yards (or meters) away. Moreover, movements of both eyes have to be coordinated to maintain stereoscopic vision.

The eyeball in its resting position is suspended by six muscles that run from the bones of the orbit (the cavity of the skull which accommodates the eyeball) to the eyeball. To simplify their actions, we can think of them as working in pairs to move the eyeball. For example, when the right eye looks to the right, the lateral rectus on the right side of that eye contracts and the medial rectus on the left side of the same eye relaxes to allow for the movement. Every movement of the eyeball involves actions of all six muscle. The neural mechanism of control is thus extremely complex.

Eye movement and innervation

The movement of the eyeball is controlled by six muscles that run from the bones of the orbit to the eyeball. These allow the eye to look up, down, left and right, providing a wide field of vision.

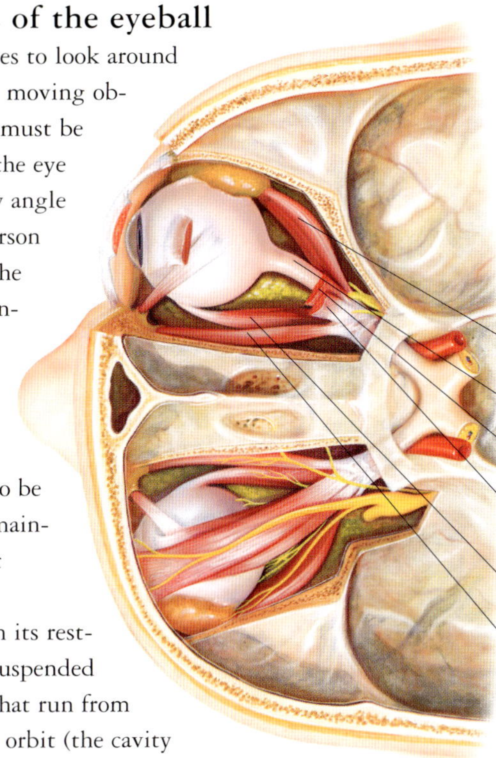

Eyelashes

The eyelashes are hairs that extend out from the eyelids. They have a protective as well as a cosmetic function. Trichiasis is a condition in which the eyelashes grow inward and rub against the cornea of the eye, causing irritation, watering of the eyes, and a feeling of a foreign body in the eye. It is usually caused by entropion, in which the eyelids curl inward. The condition is treated by removing the inturned eyelash or by surgical correction.

Eyelids

Eyelids are folds of skin that protect the front of each eyeball. When the eye is open, the upper eyelids retract above the eyeball; when closed, both upper and lower eyelids cover the visible area of the eye. The margins of the eyelids contain meibomian glands, which secrete oils that lubricate the eyelids. The conjunctiva, a mucous membrane covering the surface of the eyeball, also extends to cover the undersurface of the eyelids. Eyelid disorders include black (bruised) eye, blepharitis, chalazion, conjunctivitis, ectropion, entropion, ptosis and stye. The skin of the eyelids is also affected by skin disorders such as eczema.

Blepharitis is a disorder caused by an increased oil secretion from, or infection

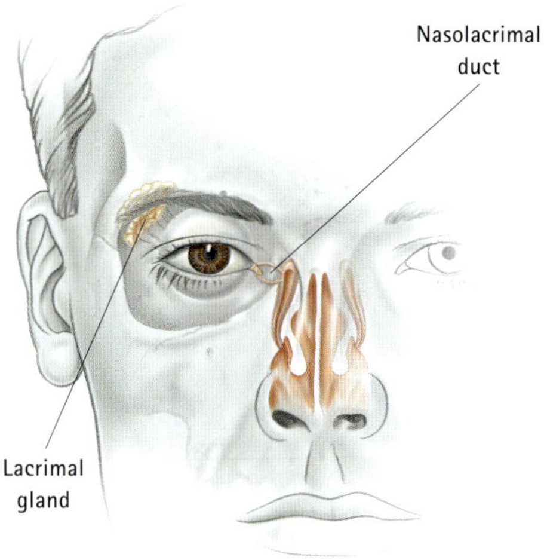

Lacrimal secretions

The lacrimal glands excrete a complex fluid (tears) to the eye surface. The fluid moves across the eye surface, lubricating the eyeball and keeping it moist.

of, the meibomian glands. The eyelids become red and inflamed; the treatment is to cleanse the eyelids regularly with a warm solution of salt water.

Ectropion of an eyelid is a condition where the eyelid (usually the lower eyelid) turns outward instead of remaining close to the eyeball. Common among the elderly, it is usually caused by the degeneration of the muscles of the eyelid.

Entropion of the eyelids is the opposite of ectropion: the eyelid curls inward toward the eye and may rub against the cornea, causing pain and redness of the eye. Both these conditions can be corrected with the help of minor surgery.

Lacrimal apparatus

Tears keep the eye moist as well as providing lubrication and protection against infection. They are secreted by the lacrimal gland which is located at the upper outer corner of the orbit.

Tears form a film over the eye and flow down across the cornea toward the inner corner of the eye to be collected by two tiny canals which open near the inner end of each eyelid. Tears go from these canals into the lacrimal sac and flow down the nasolacrimal duct, a small tube which opens into the nose. This is why we can taste bitter eye drops—they get into the nose and drip onto the back of the tongue. Movement of tears is facilitated by blinking. Tears overflow when we cry, when we have hay fever or when smoke gets into our eyes, because their production exceeds the draining mechanism.

Tears are essential in maintaining the integrity of the eye. The cornea may be ulcerated if it is dry. When their production is deficient, in conditions such as Sjögren's syndrome, patients have to instil artificial tears every few hours.

Sight

Sight is a process in which light received by the eye triggers nerve impulses in the brain to enable the perception of the shape, size, color, movement and position of objects. The ability to see clearly defined images with the correct color and intensity depends on the way in which rays of light pass through the eye and the resulting chain of physical, chemical and electrical reactions.

The amount of light entering the eye is controlled by the iris, the colored part of the eye which lies behind the transparent curved cornea at the front of the eye. The cornea focuses light rays through the pupil to the lens, another transparent structure responsible for the fine focusing of light. As the light rays travel through the eyeball they are refracted, or bent. The degree of refraction depends on the shape of the cornea and lens; the shape of the lens can be altered by surrounding muscles to allow the eye to focus on objects both near and far.

The light rays then pass through a jelly (vitreous humor) in the center of the eye to the retina, a highly sensitive layer of cells at the back of the eye. It is here that nerve impulses are generated for transmission to the brain via the optic nerve.

BINOCULAR VISION

Normal human vision is binocular, that is, the images from each eye are fused into one focused image. Each eye sees the same scene or object in a slightly different way because each sees it from a slightly different angle. Binocular vision is the ability to maintain focus on a scene or object with both eyes, blending the two pictures to create one image that is seen with depth. A person without perfect binocular vision may have difficulty visually estimating distance due to problems with the perception of depth.

REFRACTIVE ERRORS

For an object to be seen clearly, light rays must focus at a precise point on the retina. When this does not happen, due to the shape of the eyeball, it leads to vision known as refractive errors. These include shortsightedness (nearsightedness or myopia), longsightedness (farsightedness or hyperopia), presbyopia and astigmatism. Rather than being caused by disease or trauma, these focusing errors occur due to the eye's natural physical characteristics.

In longsightedness the eye has difficulty focusing on nearby objects but no problems

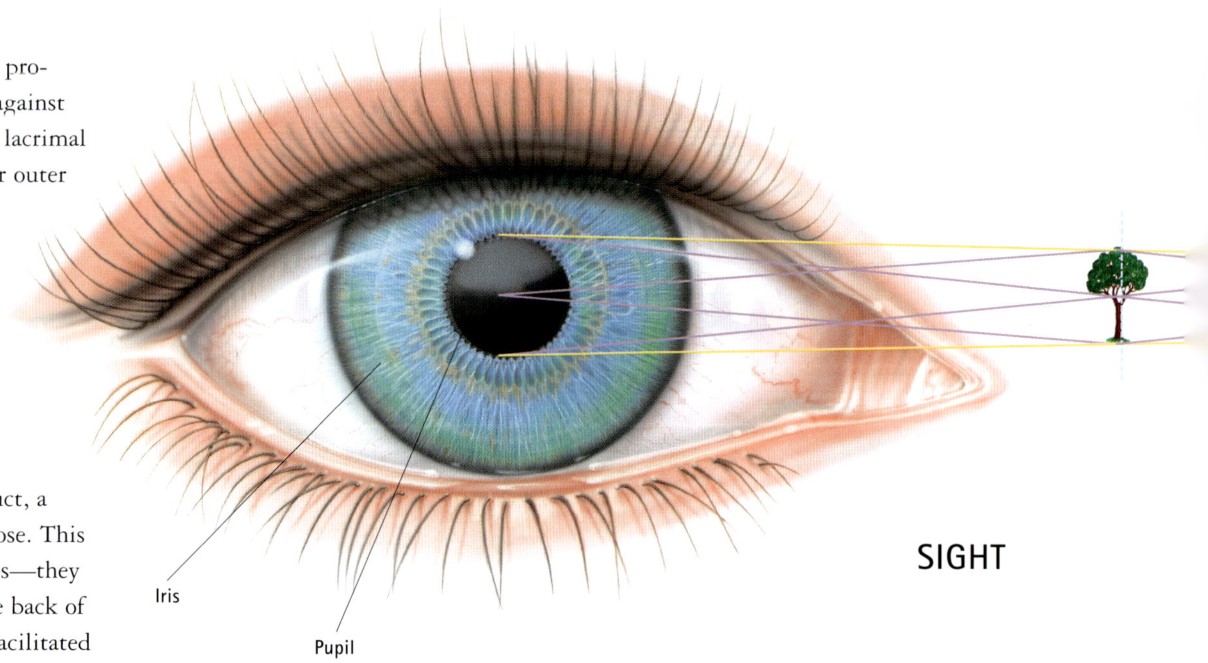

SIGHT

with objects far away. Light rays from close objects focus on a point beyond the retina instead of directly on it, because the eyeball is not deep enough, or because the lens is weak. The blurred vision due to longsightedness may lead to eyestrain and headaches.

Longsightedness caused by a decrease in the elasticity of the lens due to the normal ageing process is known as presbyopia. This makes it difficult to adjust focus for viewing close objects. Everybody suffers from presbyopia to some degree, mostly after the age of 45. Sometimes known as "old man's eyes," this condition cannot be prevented.

In shortsightedness the eye has difficulty focusing on objects far away but has no problem with those close up. This is because the eyeball is too deep or the lens too curved, causing light rays to bend too much so that visual images are focused in front of the retina instead of on it. This focusing error, like longsightedness, may be hereditary or may develop later in life as the lens becomes less elastic. An irregularly curved cornea or lens results in a condition known as astigmatism, where light also focuses in front of or behind the retina instead of on it. This can be hereditary or due to injury or disease and causes blurred or distorted vision.

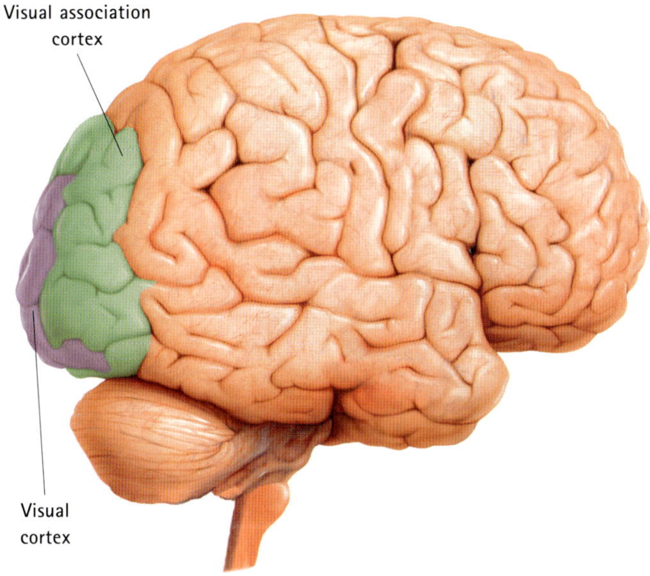

Visual cortex and visual association cortex

The visual cortex interprets and makes sense of the nerve impulses sent from the eyes via the optic nerves and thalamus. The visual association cortex processes more complex features of the visual stimulus, such as color and movement.

Visual pathways

The left and right eyes have slightly different, but overlapping fields of vision. The discrepancy between the images in the binocular field allows us to judge how far away an object is and its 3-D structure. Images are inverted, transposed and converted into nerve impulses. The impulses pass down nerve fibers to the optic nerves and through the optic chiasm to the lateral geniculate nuclei of the thalamus. These nuclei carry out some processing of the visual information and then send it to the visual cortex in the occipital lobes of the brain. Images are combined and interpreted in the visual cortex.

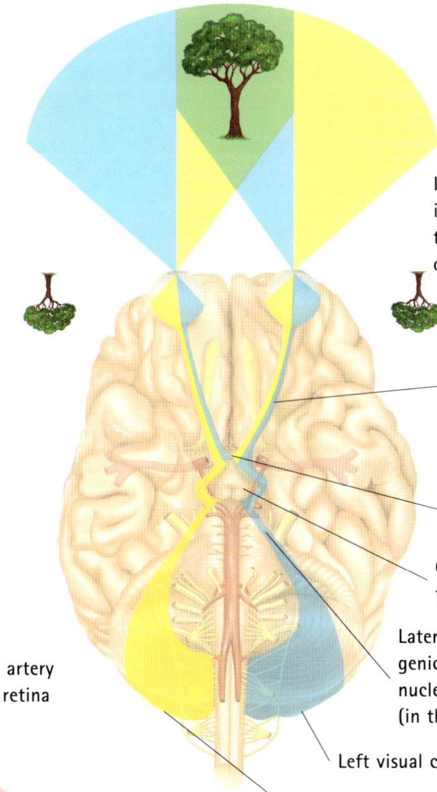

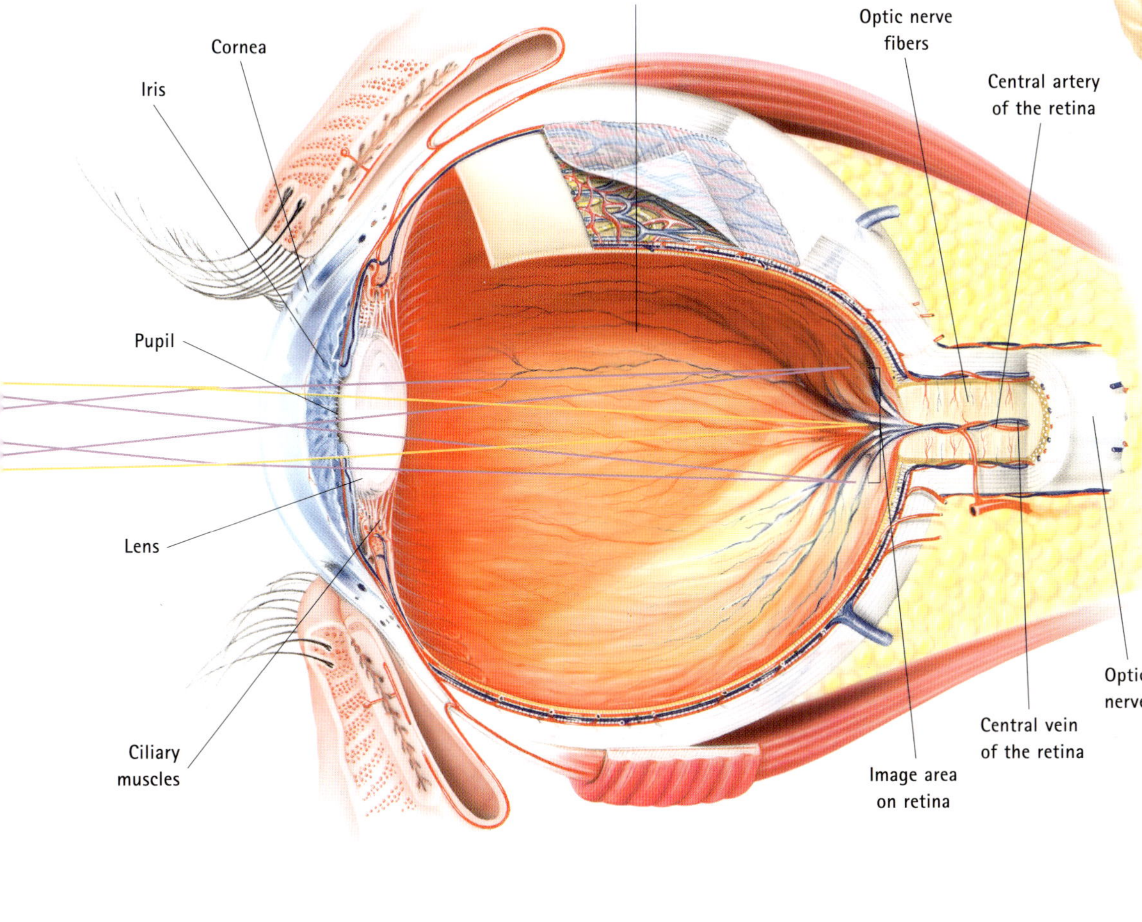

Field of vision

The structure of the eye is specially designed to bend and concentrate light rays to form a tiny image of a seen object on the back of the eye. This is then transported to the brain in the form of nerve impulses. Light rays entering the eye strike the cornea, which bends (refracts) the rays bringing them closer together. The rays then pass through the lens which focuses the rays on the back of the retina. The retina consists of a layer of light-sensitive cells—rods and cones. When stimulated by light, the rods and cones send electrical signals along the cells of the optic nerve.

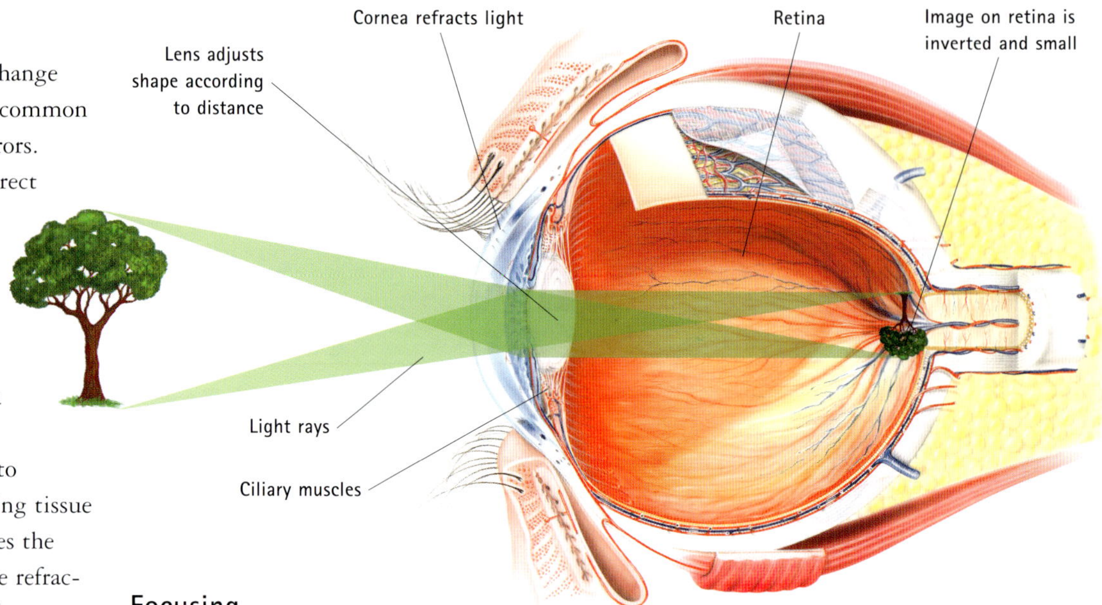

Focusing

The cornea refracts light rays as they enter the eye. The rays are brought together for focusing and cross over, creating an upside down image on the back of the retina. The shape of the lens is changed by the ciliary muscles according to whether an object is near-by or far away. This alters the angle of incoming light rays and focuses them on the retina.

TREATMENT OPTIONS

Glasses and contact lenses, which change the refractive power of the eye, are common forms of treatment for refractive errors.

Laser surgery may be used to correct or improve nearsightedness, long-sightedness and astigmatism. The most common form of laser therapy for treatment of refractive errors is known as laser insitu keratomileusis (LASIK), in which a thin flap of the cornea is surgically peeled back and a laser beam used to remove some of the cornea by heating tissue cells to bursting point. This changes the shape of the cornea according to the refractive error being treated, allowing the correct movement of light through the eye for focusing on the retina.

There is a chance, however, that best corrected vision—with the use of glasses or contact lenses—may be worse after laser surgery than it was before. Laser surgery may also be used to rid the eye of scar tissue and excess blood vessels caused by certain diseases and injuries.

Vision which is below the normal range and cannot be corrected by the use of glasses, contact lenses, surgery or other medical treatment is known as partial sight or low vision. It is often caused by damage to the macula, as well as by diseases and disorders such as diabetic retinopathy, glaucoma, cataracts and retinitis pigmentosa. Many people with partial sight may appear to be blind but may still be able to retain some usable vision.

Disorders of the Eye

The eye is a delicate organ, so the ability to see can be easily affected by foreign bodies in the eye, inflammation of or injury to eye tissues, and congenital defects.

SEE ALSO *Fetal development on page 398; Ophthalmoscopy on page 430, Tonometry on page 430*

Scleritis

Scleritis is inflammation of the sclera, the dense white fibrous covering of the eye beneath the transparent conjunctiva.

A mild nodular scleritis sometimes occurs in the superficial layers of the sclera; thought to be an allergy, it generally responds to corticosteroid eye drops.

A more severe and often painful form of scleritis is often associated with autoimmune diseases such as Crohn's disease, rheumatoid arthritis or other connective-tissue disorders. However, in some cases, no cause is found.

The symptoms encountered are severe eye pain, blurred vision, and sensitivity to light. Purple-red, inflamed areas may appear in areas of the white of the eye.

Scleritis may affect one or both eyes. The condition usually responds to treatment with corticosteroid eye drops or oral corticosteroids, but may recur.

If the underlying disease can be treated, the condition will improve. In some cases scleritis is chronic and progressive. Any vision loss that occurs during scleritis is usually permanent.

Episcleritis

Episcleritis is inflammation of the episclera, a thin layer of loose connective material that covers the sclera (the white of the eye). The cause of episcleritis is unknown, but it is often associated with diseases such as herpes zoster, rheumatoid arthritis, and Sjögren's syndrome. The affected eye becomes sensitive to light, and is painful and bloodshot. Episcleritis clears up in a few weeks without treatment, sometimes sooner with corticosteroid eyedrops.

Keratitis

Inflammation of the corneal surface is known as keratitis. This may be caused by chemicals (acids or alkalis) splashed into the eye; ultraviolet radiation (arc welding, sun exposure or snow blindness); bacterial, viral or fungal infections (especially herpes simplex virus type 1); or sensitivity to cosmetics, air pollution, or allergens such as pollen. Symptoms vary, but pain and an inability to tolerate light are usual.

Treatment will depend on the original cause. A temporary eye patch is often needed. Antibiotic or antiviral eye drops and ointments may be indicated for infection. Nonprescription eye drops containing topical corticosteroids should not be used as they may worsen the condition or perforate the eyeball. With early treatment, most types of keratitis can be cured. Severe cases may cause corneal scarring, and corneal replacement may be necessary.

Iritis

Iritis is inflammation of the iris, the colored ring in front of the lens and behind the cornea. Iritis causes pain, redness, photophobia, blurred vision, and a small, irregular iris. A foggy cornea may develop. It is related to various illnesses, including ankylosing spondylitis and collagen disease.

Treatment is with eye drops that dilate the pupil, such as atropine sulfate and homatropine, and corticosteroid drops.

Cataracts

Cataracts are cloudy spots which develop in the lens of the eye, sometimes due to injury but more often to age. As the cloudiness increases, vision becomes hazy and a halo may be seen around bright objects. As the cataract grows, vision worsens to the point where it is severely restricted and surgery is needed. Cataracts are very common—about 20 percent of people over 60 will be subject to them—but they are not painful.

Surgery involves the removal of the cloudy lens and implantation of a plastic or silicone lens. Certain types of cataract may respond to medication that dilates the pupil.

Astigmatism

An astigmatism is a lack of symmetry in the curvature of the cornea of the eye (the cornea is the transparent wall in front of the pupil and iris). It can also be a lack of symmetry in the crystalline lens. The result is that the patient sees an image which is blurred or smeared in one direction, either vertically, horizontally or obliquely.

The cause is not known, though some types of astigmatism may run in families. Astigmatism is rarely serious, and corrective lenses can be used if necessary. Surgery and laser treatment are used in some cases.

Myopia

Myopia (nearsightedness or shortsightedness) is a condition in which close objects are clearly visible while distant objects are blurred. It is caused by an abnormality of the eye in which the image is focused in front of the retina rather than directly on it. This may happen because the eyeball is too long, or because the lens is focusing the image too strongly. It usually develops in children and may run in families. Myopia is treated by wearing eyeglasses or contact lenses. Radial keratotomy, a surgery on the cornea can improve or correct the condition.

Retinoblastoma

Retinoblastoma is a potentially fatal malignant tumor on the retina, usually found in children under the age of five years. It may strike one or both eyes, giving the pupil a whitish glow and leading to impaired vision, pain and inflammation. Removal of the eye may be necessary in severe cases, though the disease may recur in the other eye. Other treatment options available include laser surgery, and a combination of radiation therapy and chemotherapy.

Retinopathy

Retinopathy is a non-inflammatory disease of the retina which manifests in a number of ways. Retinopathy of prematurity affects premature infants, causing extreme vision impairment or blindness due to rapid tissue production and retinal detachment. The exact cause is unknown and symptoms include white spots in the pupil, cross eyes, and cataracts. The condition is treated by surgical reattachment of the retina, or by reducing unwanted tissue via laser therapy or cryotherapy (a treatment involving the application of extreme cold).

Diabetic retinopathy is damage to the blood vessels nourishing the retina, causing cloudy vision and possibly blindness. The duration of diabetes is the major risk factor for diabetic retinopathy. Visual symptoms generally do not occur in the early stages of the disease. In more advanced disease, visual symptoms may include the sudden onset of blurred vision, loss of vision in either eye, decreased color perception and flashing

DISORDERS OF THE EYE

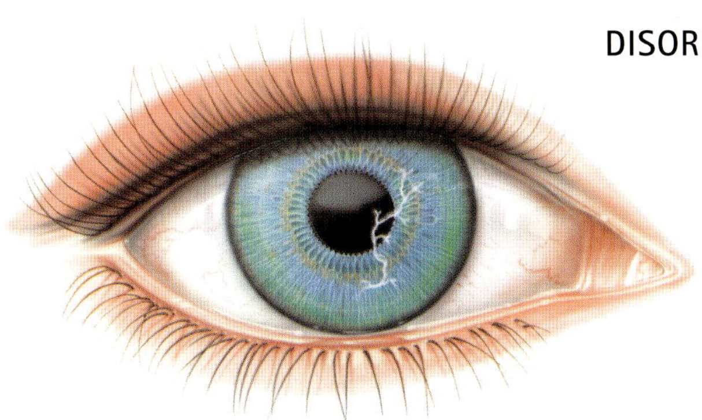

Dendritic keratitis

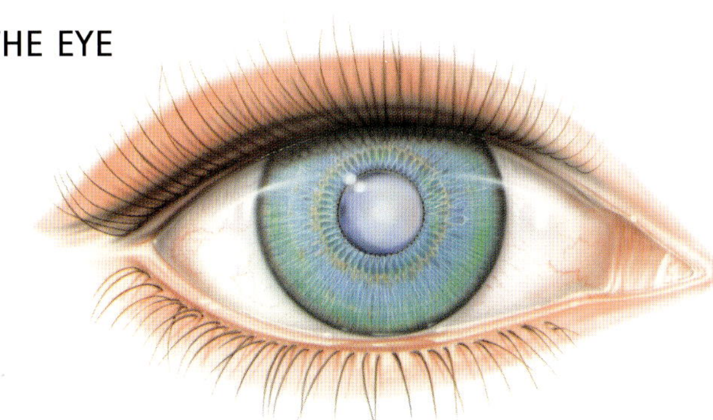

Cataract

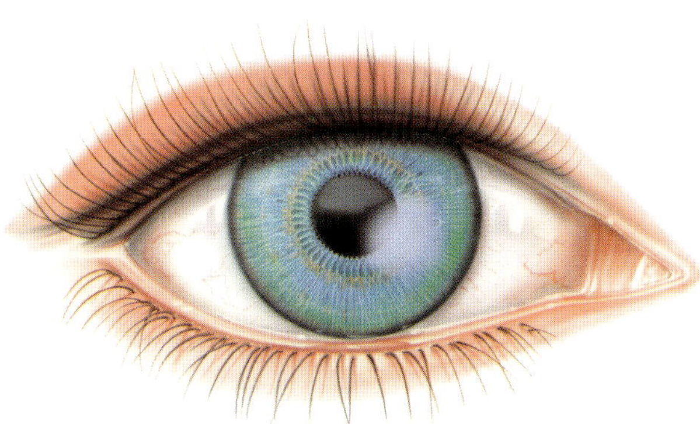

Diskiform keratitis

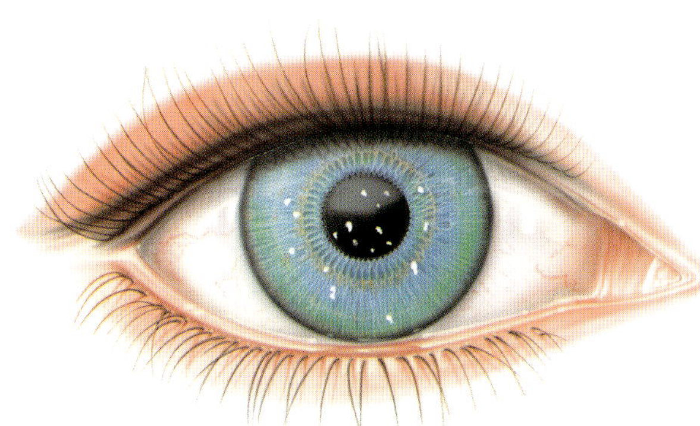

Punctate keratitis

lights or black spots. Laser, surgery may seal or destroy abnormal or bleeding blood vessels, though this may not improve vision.

Hypertensive retinopathy is caused by high blood pressure and may result in hemorrhaging, lesions, permanently impaired vision or blindness. Blood pressure must be controlled to arrest progression of the disease.

Macular degeneration

Macular degeneration is caused by impaired blood supply to the macula, which causes gradual vision loss. The macula is the area on the retina that provides fine visual acuity, used in driving, reading, watching television or activities that require focusing on very small objects. The disorder results in the loss of central vision only; peripheral visual fields are always maintained. It usually develops over a long period, often going unnoticed in its early stages. The cause is often not known.

Macular degeneration is a leading cause of blindness in Western countries and is common in the elderly. There is no treatment; though laser surgery in some cases can slow the progression of the disease.

Detached retina

Retinal detachment occurs when the retina of the eye, which contains light-sensitive cells, gets separated from the choroid, or middle layer of the eyeball. Separation may be partial or complete. The detachment usually occurs without any obvious cause, but some cases may be due to trauma, such as a blow to the head. The patient usually complains of progressively blurred vision.

Many of the nerve cells in the separated retina will die if they remain detached from the choroid, so it is important to reattach the retina as soon as possible. Patients must rest in a position such that gravity will encourage reattachment. Surgical treatments include draining fluid below the retina, fusing the retina to the choroid with lasers, electrical diathermy or very cold probes.

Retinitis pigmentosa

This rare degenerative disease of the retina involves the progressive breakdown of pigment, which stops the eye from responding to light and color in the usual way. Night blindness is often the first indication of the disease, followed by the loss of peripheral vision, a reduction in color sensitivity, deterioration of daytime vision and narrowing of the field of vision from the edges inward, or tunnel vision. In some instances the disease may result in blindness.

Retinitis pigmentosa is thought to be hereditary and there is currently no definitive treatment or cure.

Blindness

Blindness can be a total or partial inability to see. Total blindness means the inability to distinguish darkness from light. Partial blindness means having some sight in one or both eyes. A person may have partial sight can be declared blind—that is, blind according to a definition in law. This may be important in establishing a right to benefits, or training or employment allowances.

Blindness may be congenital (present at birth) or may be acquired through illness or injury. Industrial accidents are common causes of eye injury. Infections after injury can also cause permanent sight loss. A violent blow or accident can cause the retina to be detached from the inner layer of the eye called the choroid, with loss of sight. In snow blindness, vision is lost due to corneal damage caused by the sun's bright ultraviolet rays reflecting on the snow. Color blindness is the inability to see, or at least differentiate between, specific colors. Complete or partial blindness may also be due to injury to the neural pathways from the retina to the visual cortex.

Night blindness

Night blindness is a condition that can be caused by the disease retinitis pigmentosa. It can occur in people who have normal daytime vision but cannot see properly in reduced light. Some of the light-sensitive cells (rods) in the retina degenerate, and vision in low light gradually diminishes. The field of vision narrows, resulting in tunnel vision in the later stages; deterioration progresses to total blindness.

Night blindness is an inherited condition and there is no treatment to slow or reverse the damage to the retina. Ophthalmic examination may show early signs.

Color blindness

Color blindness is an inability to distinguish between certain colors. A total inability to see color—achromatic vision, where everything appears as black, white or shades of gray—is rare in humans, although 10 percent of males have impaired color vision.

Macular degeneration

This condition involves the breakdown of the cells in the eye that allow the perception of detail. Those with macular degeneration can compensate for this partial loss of vision by using a magnifying glass.

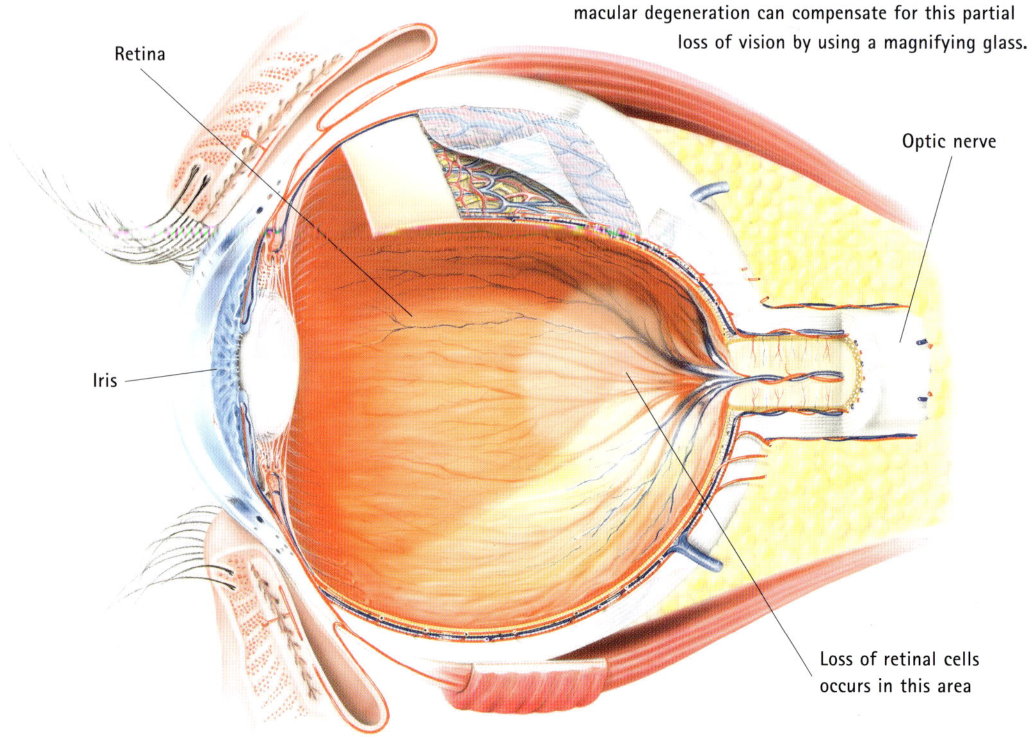

The ability to distinguish colors is the job of cone cells in the retina of the eye. There are three types of cone cells, each absorbing different wavelengths.

When one type is missing it may cause red blindness, with difficulty telling red from green, or blue blindness where it is difficult to distinguish between blue and yellow, or there may be various grades of difficulty, depending on the severity of the impairment.

Color blindness is primarily an inherited condition that tends to run in families, caused by a gene linked to the Y (male) chromosome. The gene can be carried by both males and females, but is more likely to be inherited by male children. Another factor is age—the lens darkens with age and colors become more difficult to distinguish. Some medications and various eye diseases can also affect color vision.

Children having learning difficulties at school, and anyone with a family tendency, should be tested for color blindness. People are often ignorant of their impairment until tested. There are no cures for color blindness and sufferers will be unable to perform certain jobs where normal color vision is vital.

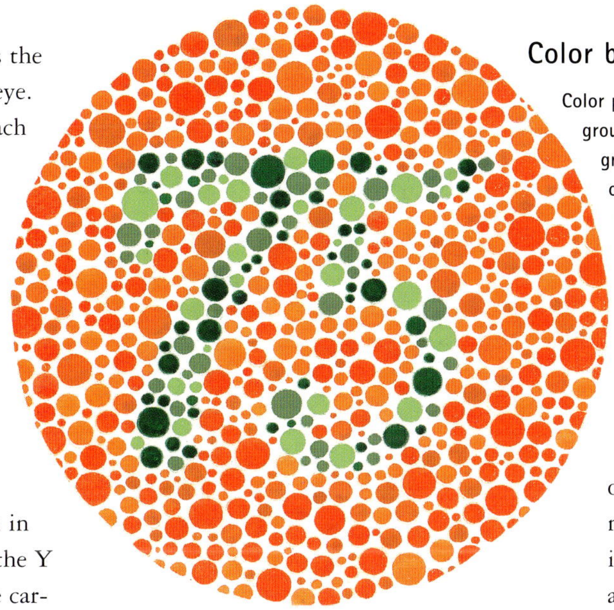

Color blindness

Color perception is the result of the stimulation of three groups of cone cells in the retina which react to red, green and blue. Color blindness occurs when one group of cones is missing or not functioning. Inability to distinguish between red and green (one of the most common forms of color blindness) occurs when red cones are missing. Tests can reveal difficulties in distinguishing between colors—anyone with red-green color blindness will not be able to see the number in this image.

Glaucoma

This is a group of eye diseases that can cause damage to the optic nerve. Glaucoma is typically, although not invariably, associated with increased pressure within the eyeball. If untreated, it causes progressive loss of vision and may ultimately lead to blindness. Glaucoma ranks with diabetic eye disease, macular degeneration and cataracts as a major cause of blindness today.

The most common form of glaucoma is open-angle glaucoma. In this form of the disease, the angle between the iris and the cornea, through which the nutrient fluid (the aqueous humor) drains, remains open. Despite the open angle, drainage of the fluid (into an area known as the canal of Schlemm) is inadequate, for reasons unknown. The resulting tendency toward accumulation of fluid in the eye causes pressure (known as the intra-ocular pressure) to rise above the normal range of 12–21 millimeters of mercury. However, the pressure may be normal in some forms of glaucoma.

In its early stages, open-angle glaucoma produces few symptoms and may not be recognized or treated. Over time the increased pressure causes death of the photoreceptor cells within the retina and their nerve fibers. This change first affects the peripheral (side) vision so that the patient progressively develops what is termed tunnel vision, but eventually sight may be completely lost. The exact mechanism by which elevation of intraocular pressure damages the retinal nerve cells is not yet well understood.

In closed-angle glaucoma, a blockage between the iris and cornea prevents the drainage of aqueous humor. Intraocular pressure increases rapidly and sharply and may result in severe pain. Other less common forms of glaucoma, such as congenital glaucoma and glaucoma secondary to injuries or other eye diseases, are also associated with noticeable symptoms. Although glaucoma is not yet curable, it can be controlled by a variety of measures, including drug treatment, and laser surgery or conventional surgery aimed at improving drainage.

For control of the disease to be effective, it is essential that the diagnosis is made as early as possible. This can be achieved by regular eye examination, especially for individuals at risk, such as persons over 60 and those with a family history of the disease.

Optic neuritis

Optic neuritis is inflammation of the optic nerve, causing sudden partial blindness in the affected eye. It may occur as the result of a viral infection, an autoimmune process or in multiple sclerosis.

The symptoms are a sudden loss of vision in one eye and pain on movement of the affected eye. Vision often returns to normal in a few weeks. If intravenous steroids are given, visual acuity may return earlier than

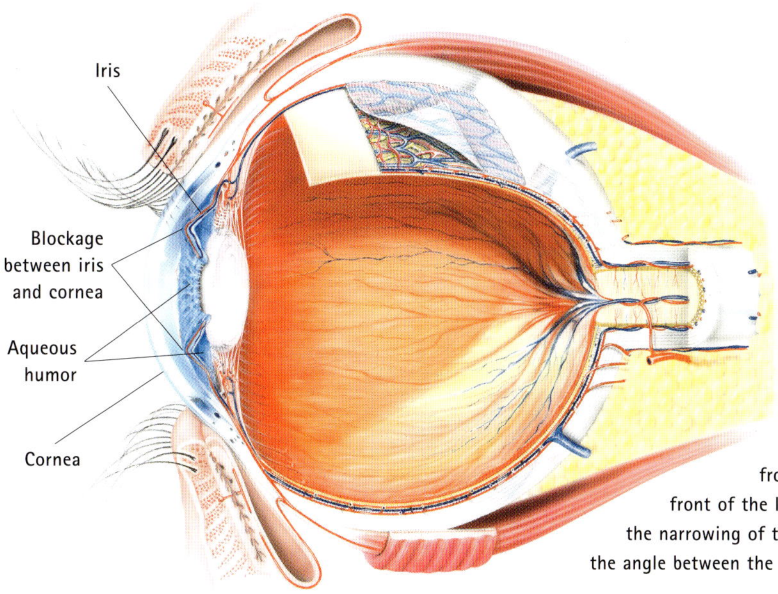

Glaucoma

The buildup of pressure in the eye may result in the loss of side vision, blurred or fogged vision and the appearance of colored rings or halos around bright objects. In closed angle glaucoma, the drainage of aqueous humor from the chamber in front of the lens is disrupted by the narrowing of the exit channel at the angle between the iris and cornea.

this. However, if the condition is due to herpes zoster or systemic lupus erythematosus (an autoimmune disease), then complete recovery is less likely.

Optic atrophy

Optic atrophy is wasting or degeneration of the optic nerve, the nerve that transmits the impulses for the sense of sight. It can occur in one or both eyes, and is the end result of some type of injury or damage to the optic nerve, such as severe head injury or anoxia (lack of oxygen). It may also occur as a hereditary disorder appearing in childhood. Optic atrophy usually results in blindness and there is no cure.

Papilledema

A serious eye disorder, papilledema involves inflammation and swelling of the optic disk, at the point where the optic nerve joins the eye. It is caused by increased pressure inside the brain, often due to a tumor, infections such as meningitis, or cerebral hemorrhage. Symptoms include blurred or double vision, headaches and nausea. Partial blindness can develop quickly. Papilledema requires immediate medical attention.

Floaters

Also known as muscae volitantes or "spots before the eyes," floaters appear as small shadows that seem to float over the fields of vision. Most common among the aged, they are caused by anything that breaks into the jellylike portion (vitreous humor) of the eyeball. Such breaks may be due to a few blood cells floating away from a broken capillary at the back of the eye, degenerative deposits, escaped fluid from blood vessels (exudates), and contraction of the vitreous gel and its detachment from the retina. These agents cause shadows on the retina, characteristically perceived by the patient as tiny shadows, usually trailing just behind the center of vision. They are most often noted while reading across a page. A sudden shower of floaters may indicate detachment of the retina, which may result in flashing lights and disturbed vision.

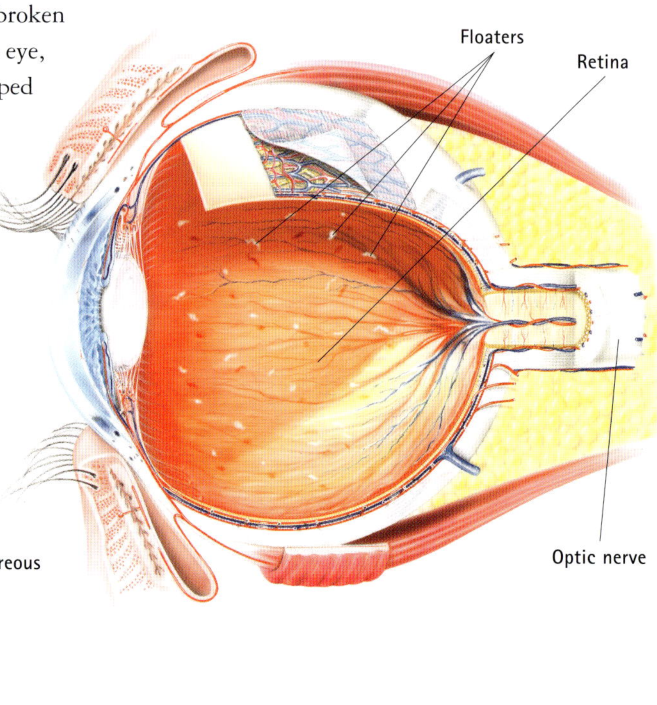

Floaters

"Spots before the eyes" are often caused by tiny organic particles moving about in the vitreous humor of the eye.

The majority of floaters are not serious and generally occur when the patient is tired. A major problem, however, can be the cause and early referral to an eye specialist is required to ensure that this is not the case.

Stye

A stye is the inflammation of one or more sebaceous glands on or under the eyelid due to bacterial infection.

Styes occur near the roots of eyelashes and look like pimples or boils. These red, swollen lumps fill with pus over 1–2 days, causing pain, watering eyes and blurred vision. Its best to let them on their own.

In the meantime, applying a warm, wet cloth for 10 minutes a few times each day may offer some relief. Antibiotics are used to treat persistent styes. Styes can be prevented by washing the hands before touching the eye area, particularly after treating acne and skin infections.

Chalazion

A chalazion (also known as a meibomian or tarsal cyst) is a small cyst that forms when a meibomian gland (a gland which secretes oils that lubricate the eyelids) becomes blocked with oil secretion. In most cases,

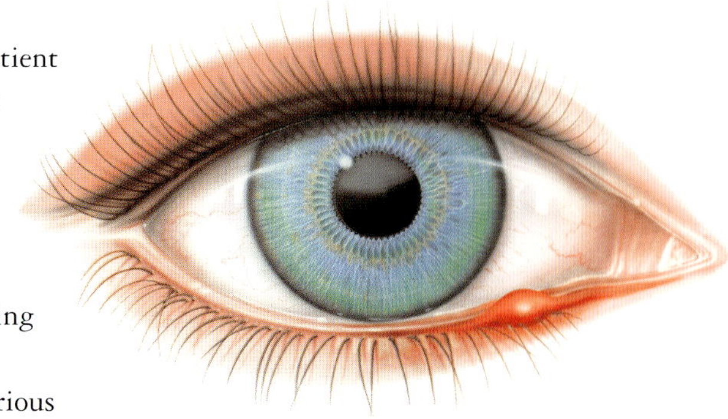

Stye

Bacterial infection of one or more of the sebaceous glands on or under the eyelid can cause inflammation, causing a stye.

the chalazion will disappear after a few months. It is harmless, but can be surgically removed if it becomes painful or unsightly.

Conjunctivitis

The conjunctiva is the membrane that lines the inner part of the eyelids and covers the whites of the eyes. Conjunctivitis, also called "pinkeye," is inflammation of the conjunctiva. The eye looks red and is painful and itchy with a watery discharge. The condition is common in childhood. Acute conjunctivitis may be caused by bacterial or viral infection, allergy or irritation. In bacterial conjunctivitis, the discharge from the eye is a yellow or greenish color. It accumulates during sleep, so the person wakes with the eyelids "stuck together." A warm washcloth applied to the eyes will remove the discharge. Antibiotic eye drops prescribed by a physician will cure the condition.

Viral pinkeye is usually associated with a more watery discharge and other viral "cold-like" symptoms. There is no treatment for viral conjunctivitis, though decongestant drops can relieve the symptoms.

Allergic conjunctivitis is frequently seasonal, occurring in the spring and summer, and the sufferer has typical allergy symptoms such as sneezing and runny nose. The eye is intensely itchy and the conjunctiva is swollen. Decongestant, antihistamine or corticosteroid eye drops will bring relief.

Conjunctivitis may be caused by irritation by dust, cosmetics, or smoke. Prompt, thorough washing of the eyes with large amounts of water will relieve the symptoms. If conjunctivitis is caused by a bacteria or virus, do not to rub the eye because the infection may be transmitted to the other eye. Anyone with conjunctivitis should wash their hands often and use their own towel so as not to transmit the disease.

Conjunctivitis may also be caused by the eye disorder trachoma or by other rare conditions such as rheumatic diseases and some inflammatory bowel diseases.

Double vision

Double vision, known clinically as diplopia, occurs when the movements of the two eyes are not properly coordinated.

The eyes are moved by a number of "extraocular" muscles in the bony cavity known as the orbit. Normally when we view an object, both eyes are directed toward the object. If the muscles moving one eye are weakened or paralyzed the patient will not be able to turn that eye in the correct direction. Consequently, the good eye looks at the object, the bad eye looks elsewhere, causing two quite different images to reach the brain which results in double vision.

Nystagmus

Nystagmus is involuntary circular movement of the eyes. It may occur normally, for example, when a person is looking out the window of a moving vehicle: the slow phase of the nystagmus occurs as the eyes drift while maintaining the gaze at the object outside, the quick phase occurs when the eyes dart back to their original position.

Nystagmus may be horizontal (side to side movements), vertical (in an up and down direction) or rotary (in a circular pattern). It may occur without any underlying disorder and usually does not affect vision.

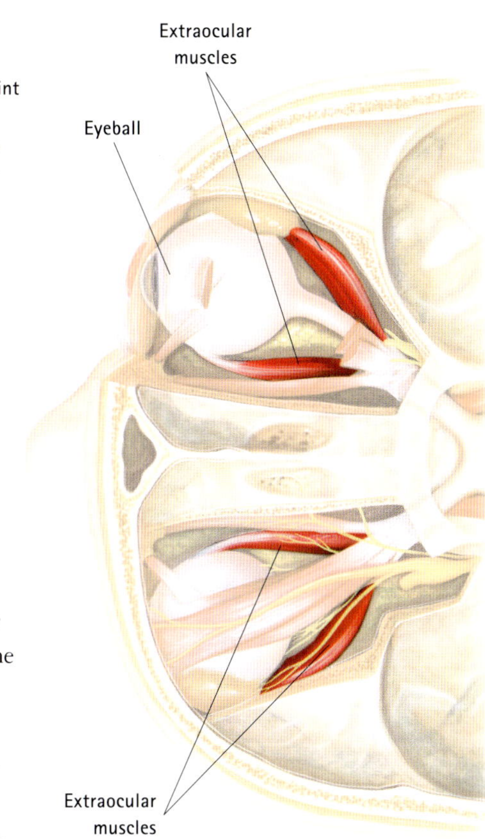

Squint

Three pairs of muscles control eye movement. A squint (misalignment of the eyes) occurs when any of the pairs does not function correctly. With a squint, one or both of the eyes may look inward (cross-eyed) or outward (wall-eyed).

Nystagmus may also be a symptom of disease; it can be the result of injury to the cerebellum in the brain, injury to the labyrinth of the inner ear, hereditary diseases or ingestion of toxins (poisons). The underlying condition will need to be identified and treated.

Horner's syndrome

Horner's syndrome is the result of damage to the sympathetic nerves which supply the face. It is characterized by a small pupil, drooping eyelid and loss of sweating on the same side of the face. It may be an indicator of more serious disease. It is most commonly caused by Pancoast's syndrome, when cancer in the apex of the lung invades the sympathetic nerves.

Squint

A squint (also known as strabismus) is a condition in which a person looks with partially closed or crossed eyes,primarily due to misalignment of the eyes. There are six muscles attached to each eye that work in pairs to allow movement; if muscle coordination is affected the eyes have trouble working in unison and can point in different directions.

One eye may turn inward (cross-eye) or outward (wall-eye) while the other looks straight ahead. This sends two different pictures to the brain, which may ignore one image or allow two images to be seen, resulting in double vision. When one eye is favored over the other, vision in the nondominant eye can be lost as a consequence and the sense of depth perception distorted.

A squint may be a congenital condition, or the result of a neurological disorder, cranial nerve and eye diseases, eye trauma or cerebral palsy. Farsightedness can be a contributing factor. The condition can usually not be prevented, but exercises to strengthen muscles around the eye often improve symptoms. In severe cases, the muscles of one or both eyes may be repositioned and shortened by surgery.

This may not totally cure the condition, but the squint will be significantly reduced. Squinting generally appears at a young age and early treatment is important to avoid permanent loss of vision in one eye.

Dacryoadenitis

Dacryoadenitis is inflammation of the lacrimal gland, which is situated above the eye. It is usually caused by bacteria from other infection such as conjunctivitis or upper respiratory tract infection, resulting in a blockage of the lacrimal duct,

Allergic conjunctivitis

In allergic conjunctivitis, the eye becomes intensely itchy and the conjunctiva are swollen. This condition is typically seen in conjuction with other allergy symptoms such as sneezing and runny nose.

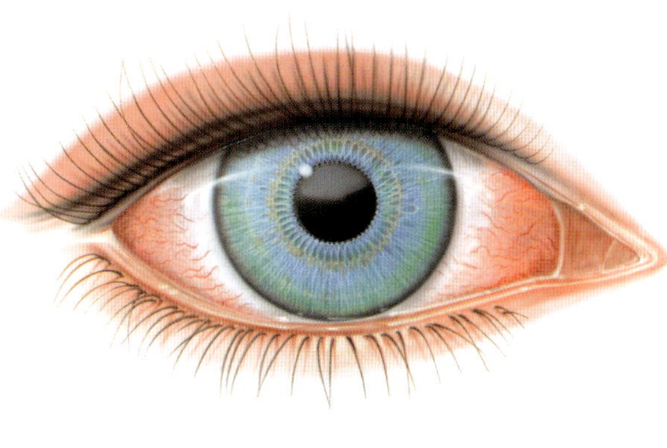

Bacterial conjunctivitis

Bacterial conjunctivitis causes a yellow or green discharge from the eye that builds up when the eyes are closed. This often causes the eyelids to stick together upon waking.

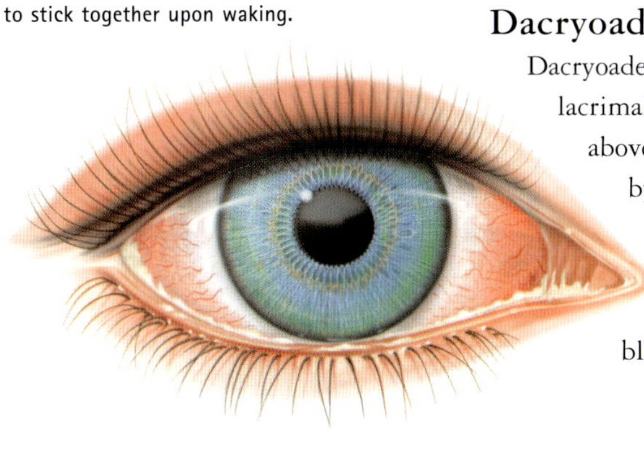

the duct that drains tears from the eye into the nasal cavity. It often occurs along with dacryocystitis. The eye is red, and the lacrimal gland is swollen and tender. The condition is treated with oral antibiotics.

Dacryocystitis

Dacryocystitis is inflammation of the tear (lacrimal) sac, caused by a blockage of the lacrimal duct, which normally drains tears from the eye into the nasal cavity.

It is most common in children, and may follow sinus or nasal infection, nasal polyps, eye injury or infection, especially conjunctivitis. It may also occur as an inherited condition, usually appearing in infants at 3–12 weeks of age.

The symptoms of dacryocystitis are pain, swelling and tenderness in the corner of the eye, with a discharge of pus and tears. Treatment is required to prevent infection spreading to the cornea, or permanent scarring of the tear duct (dacryostenosis). Irrigation of the infected ducts, massage of the tear duct and application of antibiotic ointments will usually cure the condition.

Severe cases may require surgery to dilate and probe the tear duct canal to clear the infection and prevent recurrences. Complete obstruction may require a surgical opening from the eye into the nasal passage.

Ear

The ear is the organ of hearing and balance. It receives sound waves traveling through the air and changes them first into mechanical vibrations and then into electrical nerve impulses that are sent to the brain and interpreted as sounds. The ear also senses the body's position relative to gravity, sending information to the brain that allows the body to maintain postural equilibrium.

The ear is positioned in a hollow space in the temporal bone of the skull. It is comprised of three separate sections: the outer ear; the middle ear; and the inner ear.

See also *Fetal development on page 398*

Outer ear

The purpose of the outer ear is to collect sound waves and guide them to the tympanic membrane (eardrum). The outer ear has three parts. The auricle (called pinna)

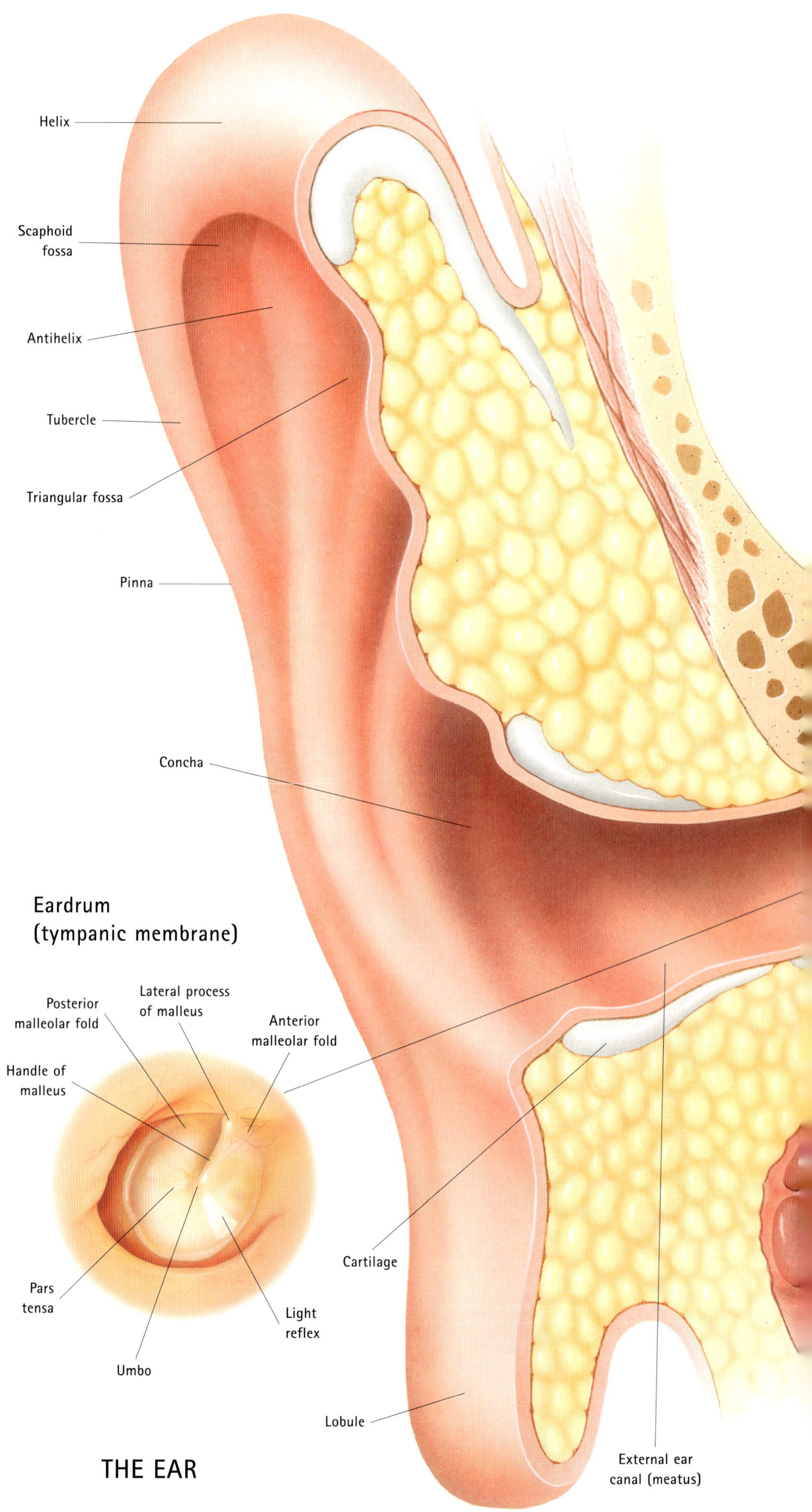

THE EAR

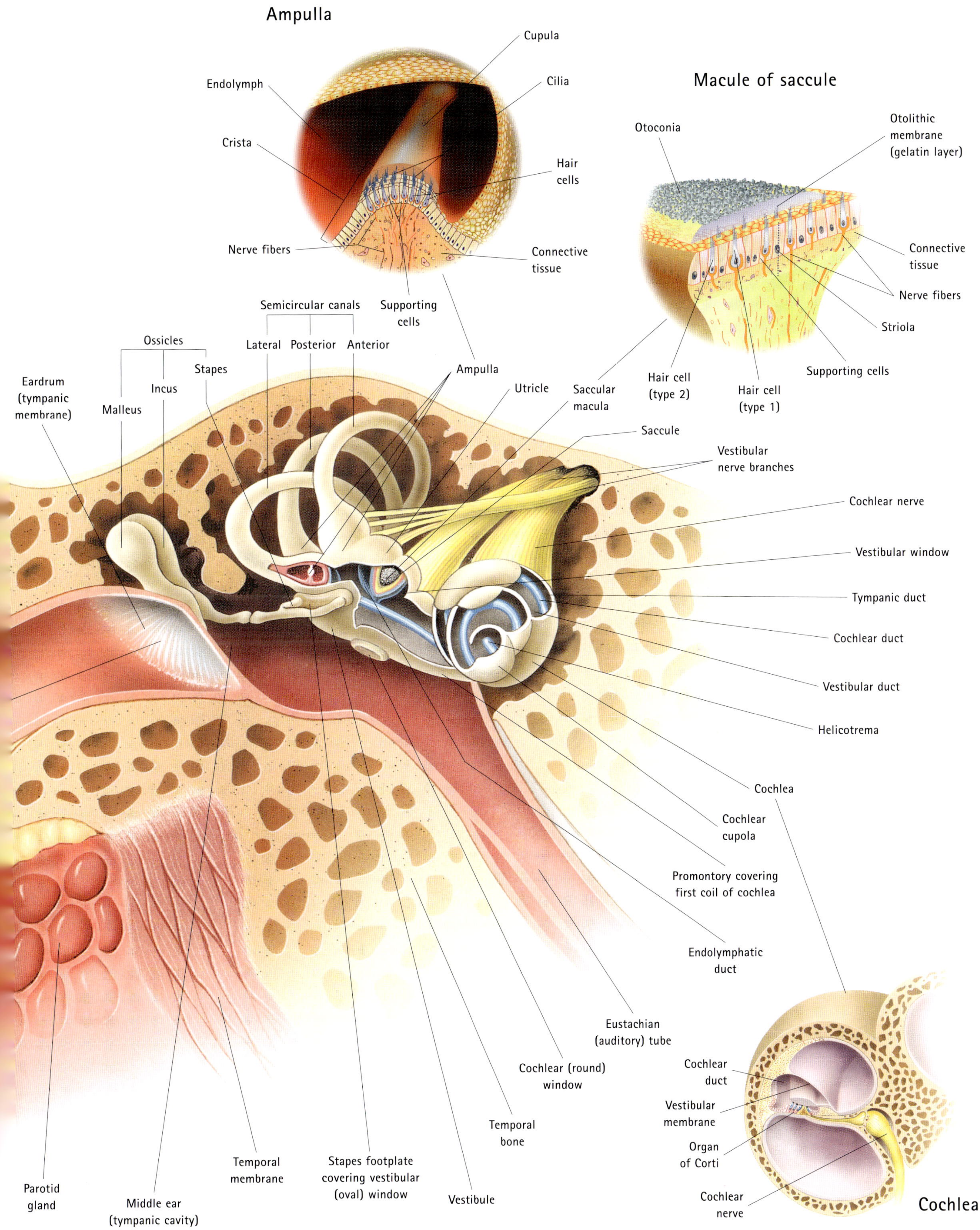
Ampulla
Cupula
Endolymph
Cilia
Crista
Hair cells
Nerve fibers
Connective tissue
Supporting cells
Macule of saccule
Otoconia
Otolithic membrane (gelatin layer)
Connective tissue
Nerve fibers
Striola
Supporting cells
Hair cell (type 2)
Hair cell (type 1)
Semicircular canals
Lateral
Posterior
Anterior
Ossicles
Malleus
Incus
Stapes
Eardrum (tympanic membrane)
Ampulla
Utricle
Saccular macula
Saccule
Vestibular nerve branches
Cochlear nerve
Vestibular window
Tympanic duct
Cochlear duct
Vestibular duct
Helicotrema
Cochlea
Cochlear cupola
Promontory covering first coil of cochlea
Endolymphatic duct
Eustachian (auditory) tube
Cochlear (round) window
Temporal bone
Vestibule
Stapes footplate covering vestibular (oval) window
Temporal membrane
Middle ear (tympanic cavity)
Parotid gland
Cochlear duct
Vestibular membrane
Organ of Corti
Cochlear nerve
Cochlea

funnels sound waves into the external acoustic meatus (or ear canal), a narrow canal that leads to the eardrum.

Eardrum

The eardrum (tympanic membrane) is a thin, semitransparent membrane, approximately ⅓ inch (9 millimeters) across that separates the external ear from the middle ear. Wax (called cerumen) secreted by glands lining the auditory canal, protects the eardrum from damage by dust and dirt.

Middle ear

The middle ear is an irregular-shaped, air-filled space, about ¾ inch (19 millimeters) high and ⅕ inch (5 millimeters) wide. It is spanned by three tiny bones: the malleus (hammer); the incus (anvil); and the stapes (stirrup), collectively known as the ossicles.

When sound waves strike the outer surface of the eardrum, they cause the tympanic membrane to vibrate. These vibrations are then mechanically transmitted through the middle ear by the ossicles (the malleus, incus and stapes), which relay them to a membrane that covers the oval window (the opening into the inner ear).

Connecting the middle ear to the throat is a short narrow passage called the eustachian tube, which helps to ensure equal air pressure on both sides of the eardrum.

Inner ear

The function of the inner ear is to convert the mechanical vibrations received from the ossicles of the middle ear into nerve impulses.

The inner ear is also the organ of equilibrium, or balance. It contains tiny organs that sense the body's relationship to gravity. A person knows which way is up because these organs send information to the brain about the body's position. Comprised of the utricle, the saccule and the three semicircular canals, they are collectively known as the vestibular organs.

COCHLEA

The inner ear contains a structure called the cochlea, a small, spiral-shaped structure containing fluid and special hairs that serve as sound sensors. The vibrations of the membrane covering the oval window cause waves to form in the cochlear fluid. These vibrations are picked up by the organ of Corti, the hearing organ housed in the cochlea. The organ of Corti contains tiny hair-like nerve endings anchored in a membrane that extends the length of the cochlea.

These nerve endings are highly specialized endings of the eighth cranial nerve, also called the vestibulocochlear nerve. When vibrations in the fluid reach them, these nerve endings fire off electrical impulses, which are sent along the nerve to the brain to be interpreted as sounds.

UTRICLE AND SACCULE

Both the utricle and the saccule are hollow sacs filled with a gelatinous fluid called endolymph. Fixed into each of their inner surfaces are tiny, hair-like structures, with their free ends projecting into the hollow space. Crystals of calcium carbonate, known as otoliths, lie over the hair cells; when the head is tilted, the otoliths change position. In shifting, they pass over the hair cells, which in turn generate impulses which are sent to the brain. The brain then triggers the body's reflex mechanisms to correct the position of the body.

SEMICIRCULAR CANALS

Head movements are detected deep inside the inner ear (labyrinth) by the semicircular ducts—three liquid-filled looped tubes in the bony semicircular canals arranged at right angles to each other. The end of each has a bulge (ampulla), containing sensory hairs which bend as the liquid in the canals moves. Nerve cells at the base of the hairs send signals to the brain about the head's movement. Each canal registers a slightly different movement: one detects the head nodding, another senses side-to-side shaking and the third perceives tilting.

Auditory centers

The vestibulocochlear nerve carries information on sound from the ear into the brain stem. From here, the information passes through the midbrain and on to the auditory cortex in the temporal lobe of the brain. This is where we recognize and interpret sounds. High frequency sounds activate one part of the cortex, low frequency sounds activate another part.

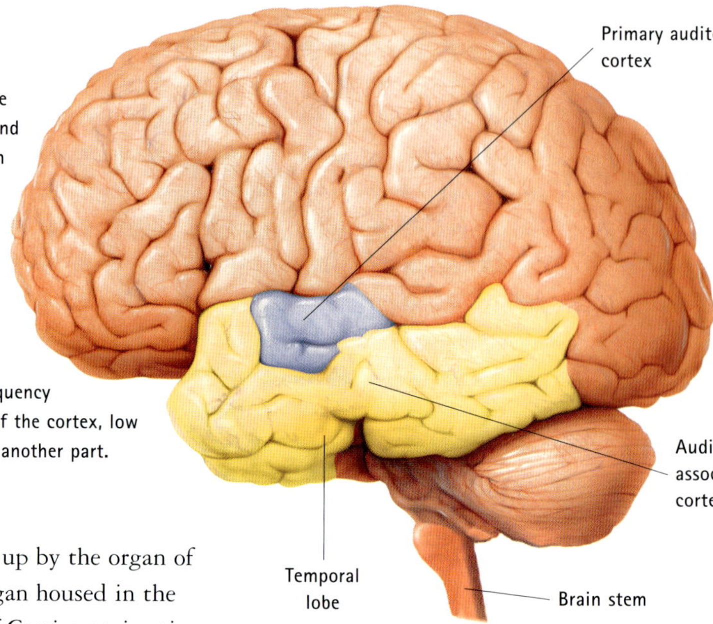

How we hear (opposite page)

(a) Sound waves enter the ear canal and hit the eardrum. (b) The eardrum vibrates and passes vibrations to the ossicles (three tiny bones in the middle ear). (c) The ossicles intensify the pressure of the sound waves and transmit vibrations to the oval window (a membrane that covers the entrance to the cochlea). (d) The vibrations pass into the cochlear spiral where fluid displaces tiny hair-like receptor cells in the organ of Corti. (e) These cells send nerve impulses along the cochlear nerve to the brain stem, midbrain, then the hearing center in the temporal lobe of the brain where sounds are interpreted.

Hearing

Hearing is the ability to perceive sound vibrations, which are transmitted in waves through gas, water or solid matter. Sounds can vary in frequency (be pitched high or low) and in intensity. Frequency is measured in hertz (cycles per second); sounds are audible from a low frequency of about 50 hertz to a high frequency of about 16,000 hertz. Intensity is measured in decibels, ranging from inaudible at zero to the devastatingly loud (above 140 decibels). Sound is conducted to the brain through the delicate and complicated mechanisms of the ear. Binaural hearing (i.e. with both ears) gives information about direction and distance of the source of sound.

The ear contains the sensory organ that perceives the sound through a process called

Auditory ossicles

The ossicles in the inner ear (the malleus, incus and stapes) are three tiny bones that act as an amplifier by magnifying the movement of the eardrum. This allows us to hear even very faint sounds. The malleus is attached to the eardrum; when the eardrum vibrates the malleus vibrates the incus, which in turn vibrates the stapes. The base of the stapes covers the oval window, a membrane that passes vibrations into the spiral of the cochlea.

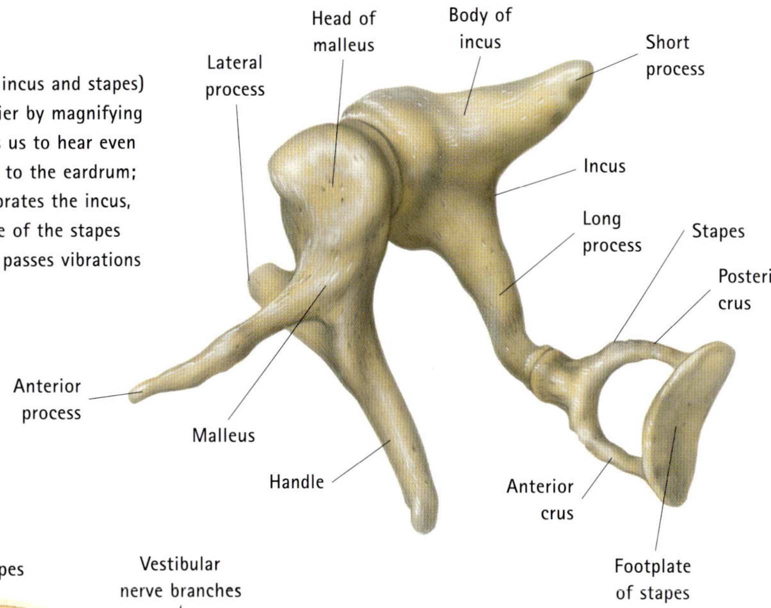

HEARING

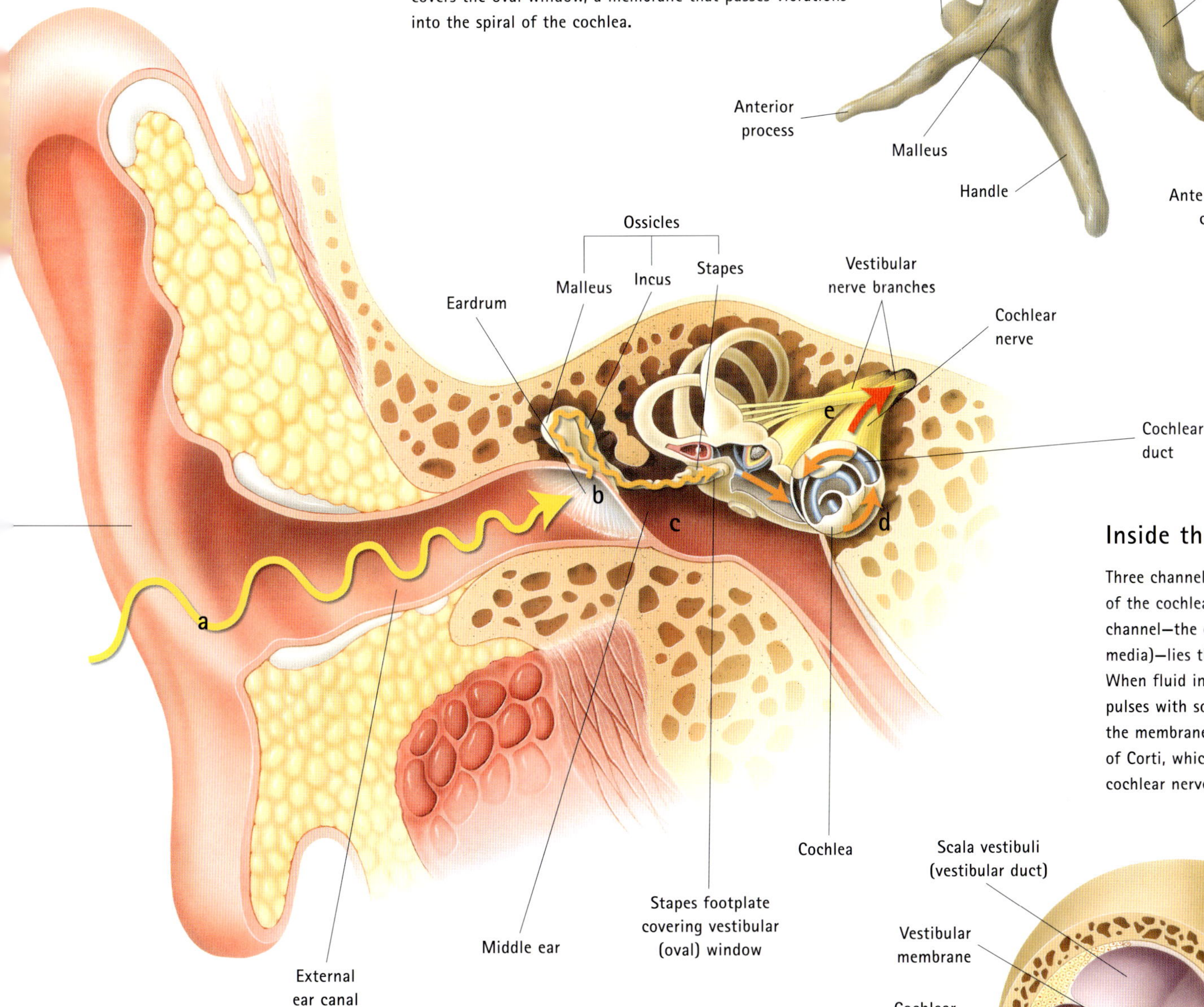

Inside the cochlea

Three channels run through the spiral of the cochlea. Inside the central channel—the cochlear duct (scala media)—lies the organ of Corti. When fluid in the central channel pulses with sound waves, it moves the membrane and hairs of the organ of Corti, which send signals to the cochlear nerve and on to the brain.

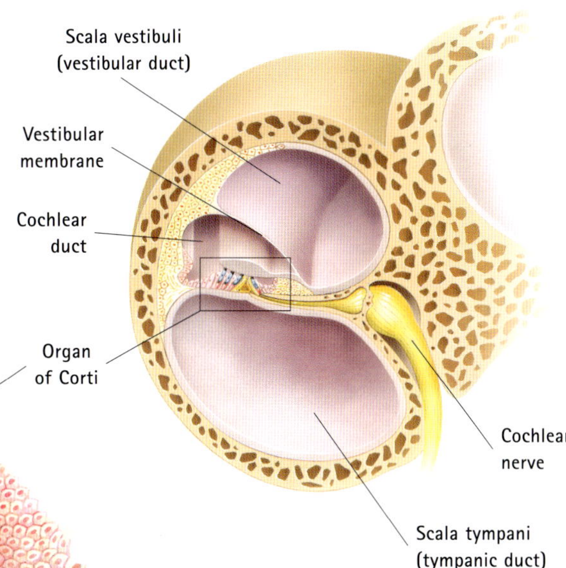

Organ of Corti

The organ of Corti consists of rows of cells with fine hair-like projections under a layer of membrane. Movement of the hairs provides the brain with information on the volume or intensity of sound. Soft sounds stimulate only a few cells; loud sounds stimulate cells in many of the rows. The cells stimulate the vestibulocochlear nerve carrying information to the brain.

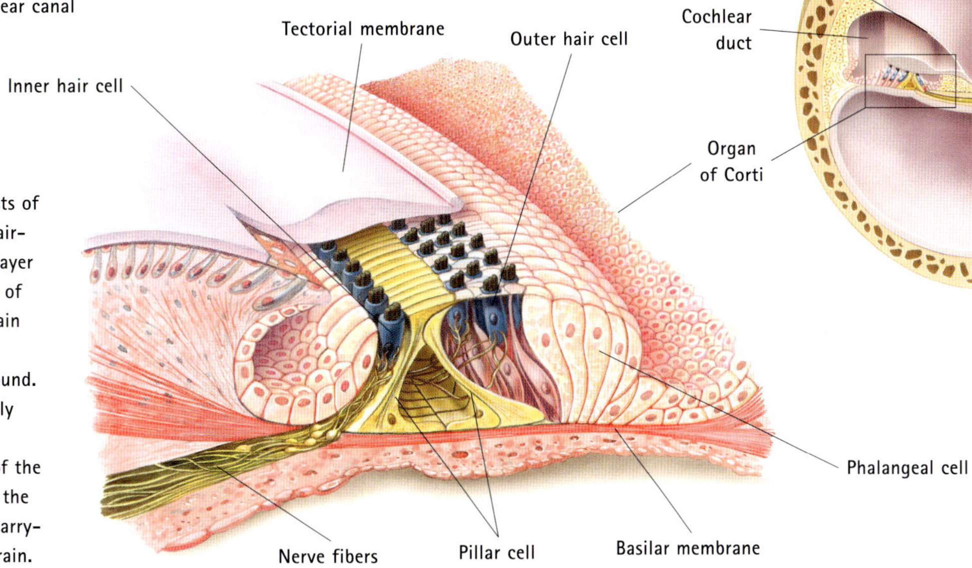

audition, and converts the mechanical energy of the sound wave into nerve impulses which are transmitted to the brain and interpreted. Hearing is present before birth, the inner ear developing in the fetus at around 9 weeks.

Airborne sounds or sound waves are captured by the auricle (outer ear), which directs them through the auditory canal to the eardrum (in the middle ear). From here waves are relayed by vibration of the ossicles to the organ of Corti, which is in the cochlea (the inner ear). There they disturb the cochlear fluid, exciting thousands of tiny hair cells which transduce mechanical energy to electrical impulses that are relayed via the cochlear nerve to the brain. Different hair cells relay high- and low-pitched sounds, those in the deepest part of the cochlea receiving the lowest pitches.

Conductive hearing is the reception of sound through the bones of the skull and is the main transmission route for one's own voice. Sound waves go direct to the cochlea, bypassing the middle ear. The loudness of sound is measured in decibels, and the lowest sound of a certain pitch that can be heard is expressed as the hearing threshold for that pitch. Sounds below 10 decibels (dB), the volume of a whisper, are difficult to hear; those above 140 dB will cause pain. Constant or regular exposure to sounds above 70 dB in volume can accelerate hearing loss; those above 100 dB may inflict permanent damage on a single exposure.

Hearing sensitivity has some protection from the acoustic reflex which, in response to loud noise, stiffens the chain of bones—the ossicles—in the middle ear to diminish the strength of vibrations transmitted to the inner ear. This protection, however, is not sufficient to prevent damage caused by repeated exposure, which permanently destroys hair cells in the cochlea.

Balance

Changes in body position are detected by different sensory receptors in the semicircular canals and otolith organs of the inner ear, the eyes, and the sensors in the joints and muscles that send messages to the brain. These organs, together with the nerves and muscles that control motor coordination and movement, maintain the body's balance.

Any disruption to these organs and pathways can lead to disturbances of balance. Inner ear infection (labyrinthitis) can affect the semicircular canals, causing disturbance of balance. Vertigo, in which balance is so severely affected that the room seems to be spinning around, may be a symptom of Menière's disease or some other ear disorder.

Lack of an adequate blood supply, and thus oxygen to the brain, may disrupt balance. Anemia, heart disease and circulatory disorders may cause dizziness, faintness and loss of balance.

Motion sickness is discomfort caused by repeated movements in cars, boats, airplanes and amusement rides such as carousels and roller coasters. The symptoms experienced include nausea and vomiting. The condition is more common in children; motion sickness usually disappears with age as the semicircular canals in the ear become less sensitive to movement.

BALANCE MECHANISM IN THE EAR

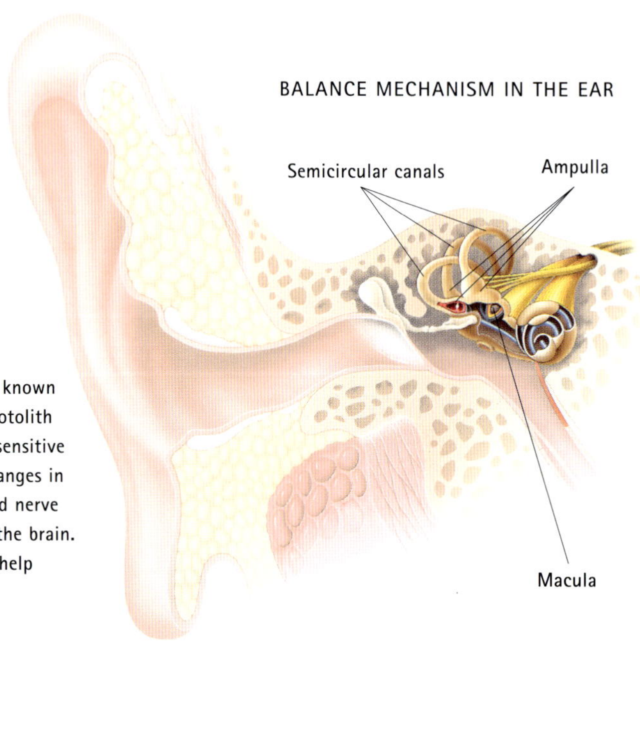

Balance

Specialized organs in the inner ear known as the semicircular canals and the otolith organs contain tiny hairs that are sensitive to the body's position in space. Changes in position excite the hairs which send nerve signals via the vestibular nerve to the brain. The brain uses this information to help balance the body.

Disorders of the ear

The three parts of the ear (the outer, middle and inner ear) are affected by disease or injury in different ways. Disorders of the outer ear are mainly related to disorders of the skin, glands and hair follicles in the outer ear canal. Infection and inflammation can affect the middle ear and (rarely) the inner ear.

Some people may experience some degree of hardened wax build-up in the ear. Wax in the outer part of the ear can be removed by wiping it away with a clean, damp washcloth. Never try to reach the wax in the ear canal with a cotton swab or a finger, as you may push the wax deeper into the ear canal. Instead, the wax must be removed by a physician who will syringe it out with warm water. The wax may be softened first either with a warm solution of olive oil or bicarbonate of soda, or with an over-the-counter solution for softening earwax.

After swimming or washing the hair, water may be trapped in the ear canal if there is wax in it, causing temporary deafness. If this happens, tip your head to one side and gently pull the external ear forward. The water will then flow out of the ear.

Children commonly push small objects into their ear. Parents should never attempt to remove anything lodged in the ear, but instead should seek professional assistance. Similarly, if an insect lodges in the ear, then medical assistance should be sought.

A perforated or "burst" eardrum is a painful injury which can lead to partial hearing loss and discharge of fluid or blood from the ear. It is caused by a sudden inward pressure on the eardrum such as that from an explosion, a foreign object being pushed into the ear, a slap or diving too deep when scuba diving. If left alone, the eardrum can often repair itself; surgery may be required if it does not heal. Those with a perforated eardrum can expect to have their hearing restored in one to two weeks, but in some cases healing may take up to two months.

Many conditions of the outer ear are treated with ear drops. Warm the drops by placing the container in a bowl of warm water (though antibiotic drops should not be warmed). Lie the person on their side with the affected ear uppermost. Pull back the earlobe to create as large an opening into the canal as possible. Rest the end of

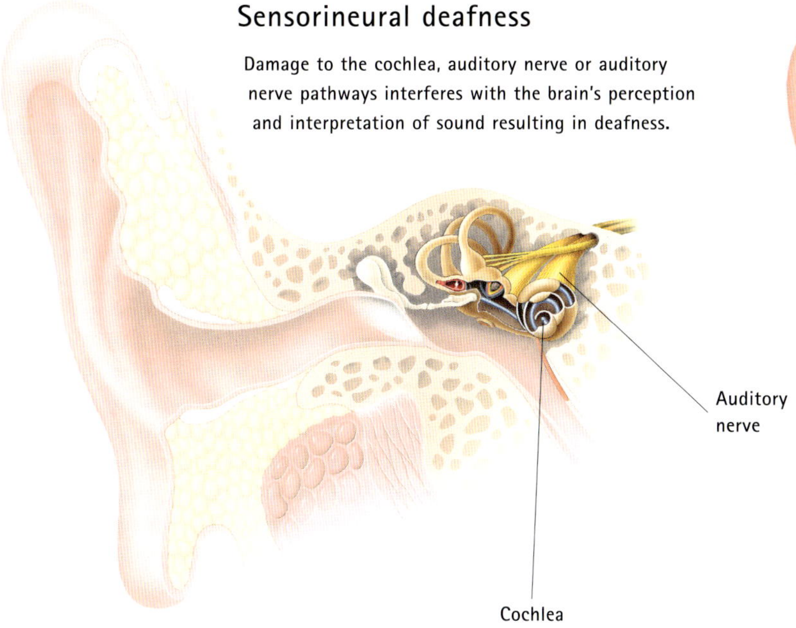

Sensorineural deafness

Damage to the cochlea, auditory nerve or auditory nerve pathways interferes with the brain's perception and interpretation of sound resulting in deafness.

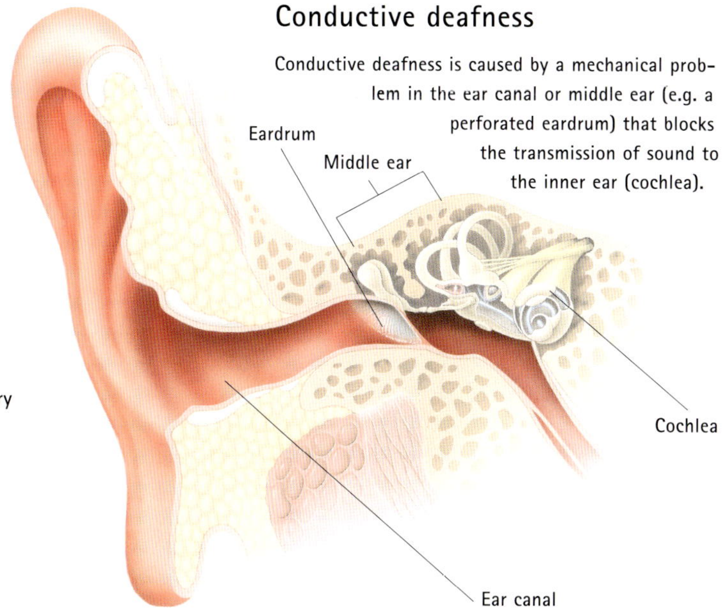

Conductive deafness

Conductive deafness is caused by a mechanical problem in the ear canal or middle ear (e.g. a perforated eardrum) that blocks the transmission of sound to the inner ear (cochlea).

the dropper over the ear opening and allow the drops to trickle gently into the ear. Place a small plug of cotton in the outer ear to prevent the drops from leaking out. The person should continue to lie in this position for about five minutes.

Unless the physician advises otherwise, do not attempt to administer ear drops if the eardrum is perforated or if a child has plastic tubes in the ears.

SEE ALSO *Audiometry on page 430*

Deafness

Deafness can be defined as an inability to hear. It can be partial, in which a person has an inability to hear in one ear, or inability to hear sounds at a certain frequency or below a certain volume (intensity). Or it can be profound, or a total inability to hear.

Unlike partially deaf people, people with a profound deafness have a hearing loss so severe that a hearing aid is useless. About 10 percent of people with deafness are profoundly deaf. There are two main types of hearing loss: sensorineural hearing loss and conductive hearing loss.

Sensorineural hearing loss is caused by damage to the cochlea or auditory nerves. This type of deafness occurs, for example, when tiny hair cells of the cochlea in the inner ear are damaged, with the result that sound reaching the cochlea is not adequately processed, and faulty nerve signals are sent to the brain. A hearing aid is often of no use to a person with a sensorineural loss.

Conductive hearing loss occurs when sound is not properly transmitted to the cochlea (the small, spiral-shaped structure in the inner ear) because of a blockage, disease or disorder of the middle ear. A person with a conductive hearing loss can generally benefit from a hearing aid.

Some experts recognize two other categories of deafness apart from conductive hearing loss. These two categories are mixed hearing loss and central hearing loss. Mixed hearing loss is generally caused by illness or injury in both the outer or middle ear and the inner ear. Central hearing loss can be caused by damage or injury to the neural pathways or parts of the brain that are concerned with hearing.

FACTORS AFFECTING HEARING LOSS

Babies will respond to sound soon after birth and their speech usually begins to develop at two to three months, words following from eight to twelve months. If these developments do not occur, a hearing problem may be the reason.

Hearing problems are particularly serious in childhood as the ability to hear affects the development of speech. Hearing enables the reception of sounds and recognition of language, and allows self-monitoring when those sounds are imitated, which is an essential part of the learning process.

Birth defects causing loss of hearing can be inherited, or can be the result of injury or disease. Lack of oxygen or Rh disease

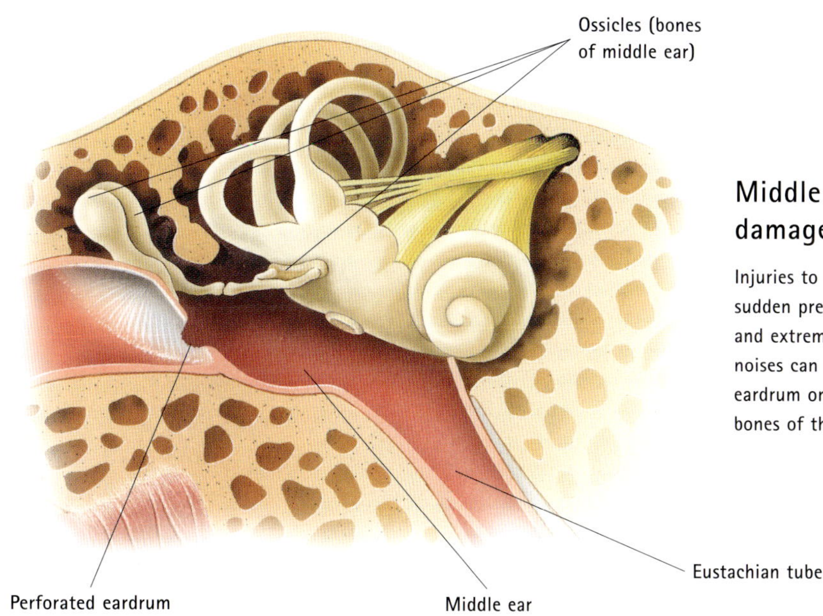

Middle ear damage

Injuries to the head, sudden pressure changes and extremely loud noises can rupture the eardrum or damage the bones of the middle ear.

(Rh incompatibility) can damage the auditory nerve and inner ear. The likelihood of a child suffering birth defects that affect hearing is greater where other family members were born with hearing defects or where there is exposure before birth to infection by rubella, cytomegalovirus, herpes, toxoplasmosis, or syphilis.

Down syndrome and other genetic problems can result in impaired hearing, as can exposure of the mother to certain antibiotics, quinine or radiation during pregnancy. Risk factors at birth are greater for babies with a low birth weight; those who need to spend more than ten days on a ventilator; those with an Apgar score below three; and those with high levels of bilirubin, indicating possible liver dysfunction.

Childhood illnesses including measles, mumps, meningitis and ear infections—all of which can cause scarring that restricts the movement of the ossicles—can permanently impair hearing. Head injury, exposure to very loud noises and side-effects of certain antibiotics are also causes of hearing loss.

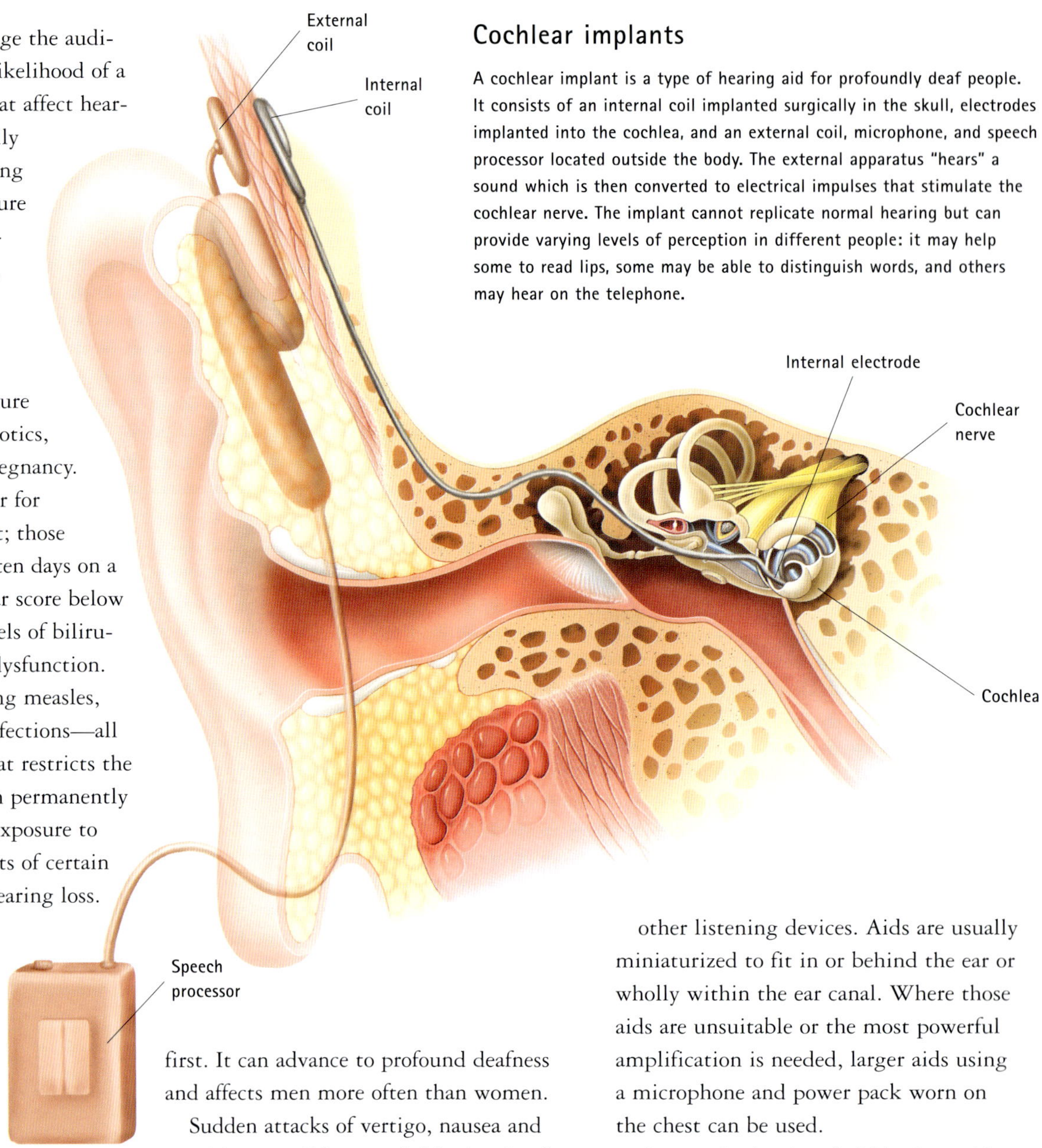

Cochlear implants

A cochlear implant is a type of hearing aid for profoundly deaf people. It consists of an internal coil implanted surgically in the skull, electrodes implanted into the cochlea, and an external coil, microphone, and speech processor located outside the body. The external apparatus "hears" a sound which is then converted to electrical impulses that stimulate the cochlear nerve. The implant cannot replicate normal hearing but can provide varying levels of perception in different people: it may help some to read lips, some may be able to distinguish words, and others may hear on the telephone.

Normal hearing can be affected temporarily in children by a blockage in the ear canal, which can be due to excessive ear wax, or a foreign body such as a bead or piece of food. Ear infections can infect or block the eustachian tube very easily in children as the tube is narrow. Hearing can also be lost temporarily after being exposed to loud noise.

Injuries to the head, blows, sudden pressure changes (for example, when divers rise too quickly to the surface) and extremely loud noises (such as explosions) can rupture the eardrum and damage the bones of the middle ear or the delicate inner ear. Repeated exposure to loud noise can also produce gradual hearing loss. Noise intensity greater than 85 decibels can destroy hair cells in the inner ear. Damage is permanent and may be accompanied by tinnitus, a high-pitched noise heard in one or both ears.

Age-related hearing loss (presbycusis) can begin in early adulthood and usually diminishes ability to hear high frequency sounds first. It can advance to profound deafness and affects men more often than women.

Sudden attacks of vertigo, nausea and vomiting, possibly preceded by hearing loss and tinnitus may indicate Menière's disease, which is a chronic disease of the inner ear. There may be pressure in one or both ears, and hearing ability gradually diminishes. The cause is still unknown, but vertigo can usually be treated temporarily with drugs, or more permanently, by surgery to the nerves of the inner ear.

HEARING AIDS

A hearing aid is an electronic device that improves hearing. The device consists of a microphone, amplifier, receiver and power source (usually batteries). It is adjustable for pitch (frequency response) and volume (saturation response) by the prescribing audiologist. Some aids have volume controls which the user can adjust, and switches for when using the aid with a telephone and other listening devices. Aids are usually miniaturized to fit in or behind the ear or wholly within the ear canal. Where those aids are unsuitable or the most powerful amplification is needed, larger aids using a microphone and power pack worn on the chest can be used.

A recently developed aid is the cochlear implant, which is surgically implanted in profoundly deaf people. The implant converts sounds to electrical impulses that are sent to electrodes implanted in the cochlea and thence to the brain.

A bone-conduction hearing aid is placed on a headband, which holds it firmly in contact with the head for the greatest possible sound conduction through the skull bone.

TREATMENT

Profoundly deaf people must communicate by means other than spoken language. There are a number of ways in which this can be accomplished, including sign language, lip reading (known also as speech reading), finger spelling, cued speech, manually coded language as well as writing, typing, gesture and mime. Each deaf person has a preferred method of communication that may change from situation to situation; for example they may prefer lip reading with some persons but not with others.

Deaf people face unique problems. Unemployment or underemployment is higher than in the general population and they may be discriminated against in the workplace. However, deaf people are employed in almost every type of job, and increasingly employers are hiring deaf people and adapting the workplace to accommodate them, for example, by providing sign language interpreters.

Deafness does not affect a person's intellect or learning ability. However, deafness may make the learning of language more difficult. Consequently, a hearing-impaired child's progress at school may be slower than that of a child who can hear. However, the hearing-impaired child who is taught lip reading and sign language at an early age is more likely to do well.

Dizziness

Dizziness is a feeling of light-headedness, unsteadiness or falling, accompanied by weakness and swaying, or a sensation of whirling rotation and general loss of balance.

Low blood pressure (hypotension) can cause dizziness as the brain may not receive enough blood and not enough oxygen. Orthostatic or postural hypotension occurs when a person stands up quickly and causes a temporary fall in blood pressure, enough to cause dizziness or sometimes fainting. Sitting, or lying down will make it easier for blood to reach the brain, increasing the flow and relieving the dizziness.

Infections, damage and tumors of the inner ear, brain disorders, head injuries, brain tumors, and medications used to relieve high blood pressure are all possible causes of dizziness.

Vertigo

Vertigo is a hallucination of movement, usually a feeling that the person or their surroundings is revolving. It may produce feelings of dizziness and confusion; intense vertigo may cause nausea and vomiting.

Vertigo may be caused by disorders of the inner ear, such as a blockage of the fluid in the semicircular canals, which regulate balance and detect movements of the head, or by disorders of the central nervous system. Vertigo is dangerous if experienced in high places and is an occupational hazard for divers and pilots, who are frequently in situations where there are no visual reference points. They may suffer spatial disorientation, becoming unable to properly judge their direction or speed of movement.

Labyrinthitis

Labyrinthitis is inflammation of the labyrinth—the fluid-filled semicircular canals in the inner ear that are responsible for monitoring the body's position and movement, and helping the brain to maintain equilibrium and balance. The condition is usually caused by a virus, but may also be caused by bacteria spreading from a middle ear infection (otitis media), meningitis or following an ear operation.

Symptoms of labyrinthitis include vertigo; dizziness, especially with head movement; loss of balance; nausea and vomiting; and sometimes temporary deafness and tinnitus (ringing in the ear).

Viral labyrinthitis is treated with anti-nausea (or anti-motion sickness) drugs and bed rest; recovery may take several weeks.

Bacterial labyrinthitis is treated with antibiotics. Surgical drainage of the ear may be necessary if the condition is associated with serious otitis media (commonly known as "glue ear").

Menière's disease

Menière's disease is a disorder of unknown cause characterized by periodic attacks of vertigo and hearing loss. The first symptom can be a ringing or hissing sound in the ears (tinnitus). This is followed by debilitating vertigo, nausea and vomiting that may last up to 24 hours. There may be a feeling of pressure or fullness in the ears. There is progressive hearing loss over time. The vertigo can be treated temporarily with drugs or more permanently by surgically cutting the nerves of the semicircular canals, the organs of balance within the inner ear.

Tinnitus

Tinnitus is the sensation of hearing a sound in one or both ears when there is no external noise from the environment. The sound may be a buzzing or ringing, a high pitched hiss or whine, or more complex sounds, and may be continuous or intermittent.

Tinnitus is a symptom of a number of possible conditions. It is generally associated with ear damage from exposure to loud noise, or with age-related hearing loss, and can also follow a middle ear infection. A rushing noise may indicate vascular problems. Certain drugs, after long periods of use, can affect the inner ear.

Tinnitus can also be caused by tumors on, or injury to, the vestibulocochlear nerve; problems with the temporomandibular joint; otosclerosis, a stiffening of the bones

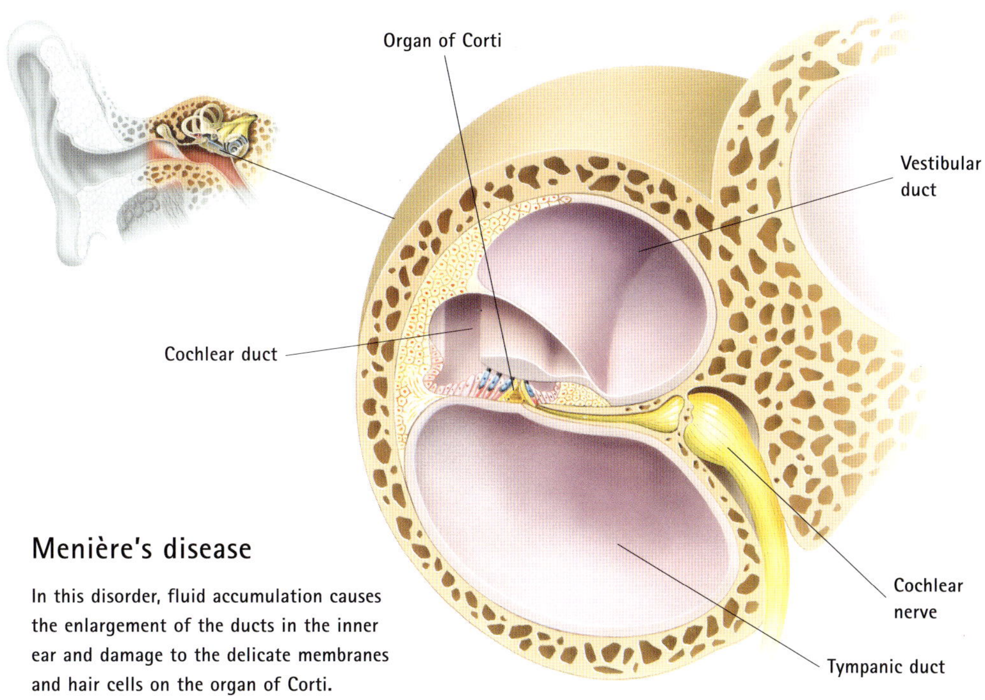

Menière's disease

In this disorder, fluid accumulation causes the enlargement of the ducts in the inner ear and damage to the delicate membranes and hair cells on the organ of Corti.

(ossicles) of the middle ear; head and neck injury; and Menière's disease. Sometimes tinnitus may arise with no identifiable cause. It may be so severe as to interfere with concentration or sleep.

There is no cure, but to prevent the condition worsening, loud noise and drug use should be avoided. The noise can sometimes be masked by "filling silence" with more pleasant sounds such as a radio or softly ticking clock; users of hearing aids often report that this makes the tinnitus less intrusive. Some people find that the symptoms are worse under anxiety or stress.

Mastoiditis

Inflammation of the air-filled spaces encased in the mastoid part of the temporal bone behind the ear can follow inadequately treated inflammation of the middle ear (otitis media). The acute phase can often be reversed with antibiotics. However, should the infection become chronic, the patient will complain of severe pain behind the ear with associated fever, local redness and swelling. Appropriate antibiotics chosen on the basis of a culture from the inflamed middle ear may control the infection, but more often an operation to drain the area is necessary. Diagnosis is made both clinically and by confirmatory x-rays or CAT scans.

Otitis media

Otitis media is inflammation of the middle ear. It is most commonly caused by the spread of bacteria from the throat into the middle ear via the eustachian tube. Usually there is an associated infection of the throat, such as the common cold or tonsillitis.

The sufferer, usually a child, develops fever, deafness in the affected ear, and severe earache, due to an accumulation of pus in the middle ear which can build up until it ruptures the eardrum, releasing the pus and relieving the earache. Diarrhea, abdominal pain, and vomiting accompany the symptoms—in infants, they may be more obvious than the earache.

Treatment is with painkillers and antibiotics. If the eardrum has ruptured or has been surgically opened and is discharging pus, the ear should be kept clean and dry until the eardrum has healed. Sometimes, pus under pressure may need to be released surgically via an incision in the eardrum (myringotomy). In some children, otitis media becomes recurrent and may become chronic (glue ear), when the insertion of a drainage tube becomes necessary. If untreated, chronic otitis media can cause deafness.

Otosclerosis

Otosclerosis is a disorder of the middle ear that leads to progressive deafness. It is caused by the gradual build-up of abnormal spongy bone tissue around one of the small bones (the stapes) in the middle ear. The

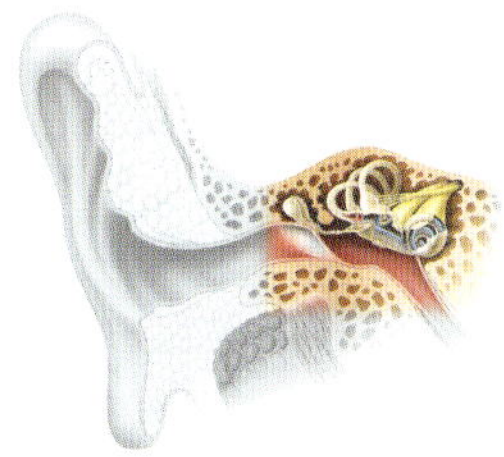

Otitis media

Otitis media is inflammation of the middle ear, which occurs frequently in infants and young children. Fluid accumulates in the ear behind the tympanic membrane. There is hearing loss, which ceases after treatment.

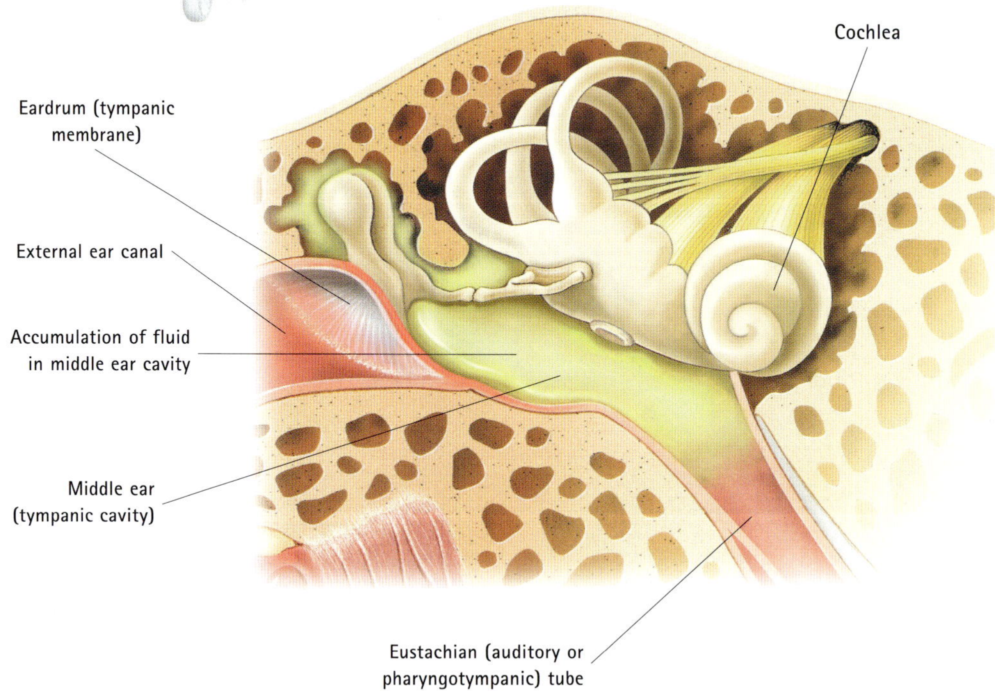

Glue ear

When children have glue ear (persistent otitis media), a grommet is inserted to allow drainage.

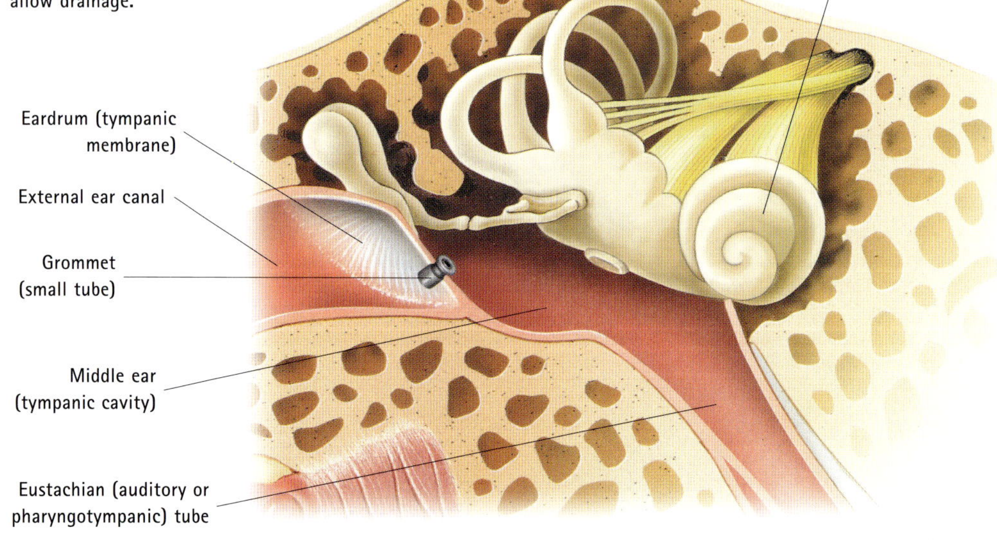

abnormal bone prevents the stapes from vibrating, thereby preventing the transmission of sound vibrations from the eardrum to the inner ear. The result is progressive deafness in the affected ear and, sometimes, ringing in the ear (tinnitus). Eventually both ears become affected. Otosclerosis is most commonly seen in women between the ages of 15–30 and may be triggered by pregnancy. Otosclerosis may also be heredity.

The condition is treated by a surgery called stapedectomy, in which the diseased stapes bone is replaced with a prosthesis, which can restore hearing. A hearing aid may be used as an alternative to surgery.

Otitis externa

Otitis externa is inflammation of the outer ear, i.e., the skin of the ear canal extending from the eardrum to the outside of the ear.

Symptoms of otitis externa include itching, ear pain that worsens when the earlobe is pulled, a slight discharge and (sometimes) deafness. The cause is a bacterial or fungal infection resulting from a number of causes, including swimming in dirty, polluted water (the condition is also called swimmer's ear), scratching the ear, using ear plugs for prolonged periods, or excessive sweating. It may also occur in people with eczema or in people with diabetes mellitus.

Otitis externa can usually be treated with ear drops that contain antibiotics to fight infection and cortisone drugs to control inflammation. Oral antibiotics may also be required in the case of severe infection. Any dead skin, pus or wax should be removed only by a physician.

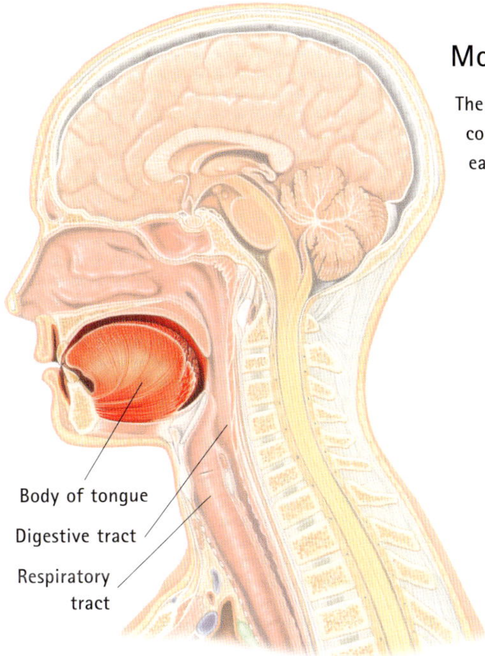

Mouth—entrance to body

The mouth is the entrance to the digestive tract and also connects to the respiratory tract. It plays a vital role in eating and speaking, and can also contribute to breathing.

Mouth

The mouth is the first part of the digestive tract. It consists of an outer vestibule, which lies between the teeth and the cheeks or lips, and an inner true oral cavity within the arches formed by the teeth.

The true oral cavity has a roof formed by the hard palate in the front and the soft palate at the back, which separate the mouth from the nasal cavity. The hard palate is bony; the soft palate is formed of muscle covered by mucous membrane. The prominent droplet-shaped fleshy structure which hangs from the rear edge of the soft palate is called the uvula. The floor of the oral cavity is made up of the tongue and the tissue between the tongue and the teeth. At the back, the oral cavity leads into the oropharynx, which is part of the throat.

The external opening and the lips are encircled by a muscle called the orbicularis oris, which allows the lips to be pursed for

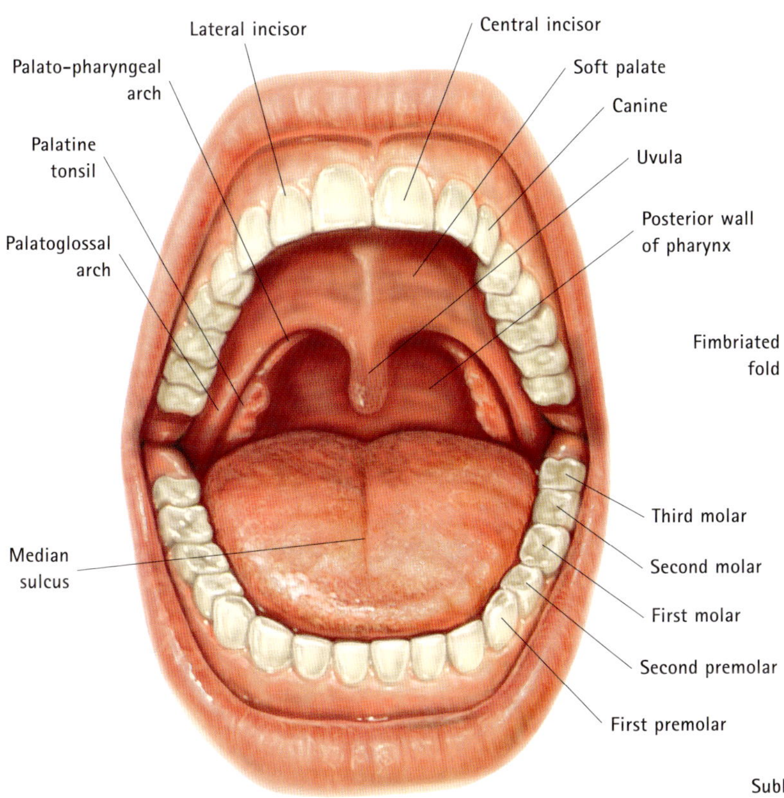

Structure of the mouth (with tongue down)

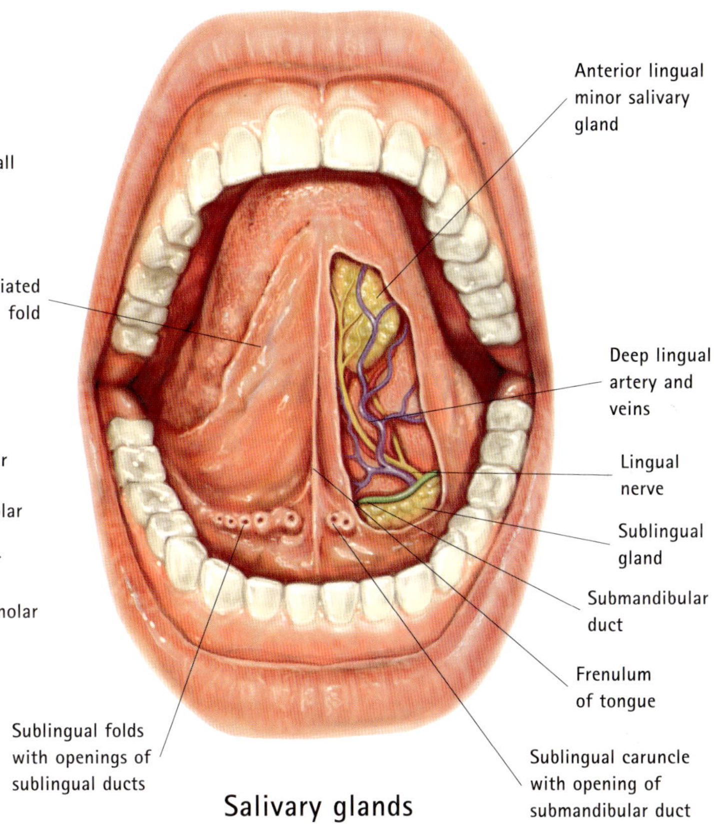

Salivary glands (with tongue up)

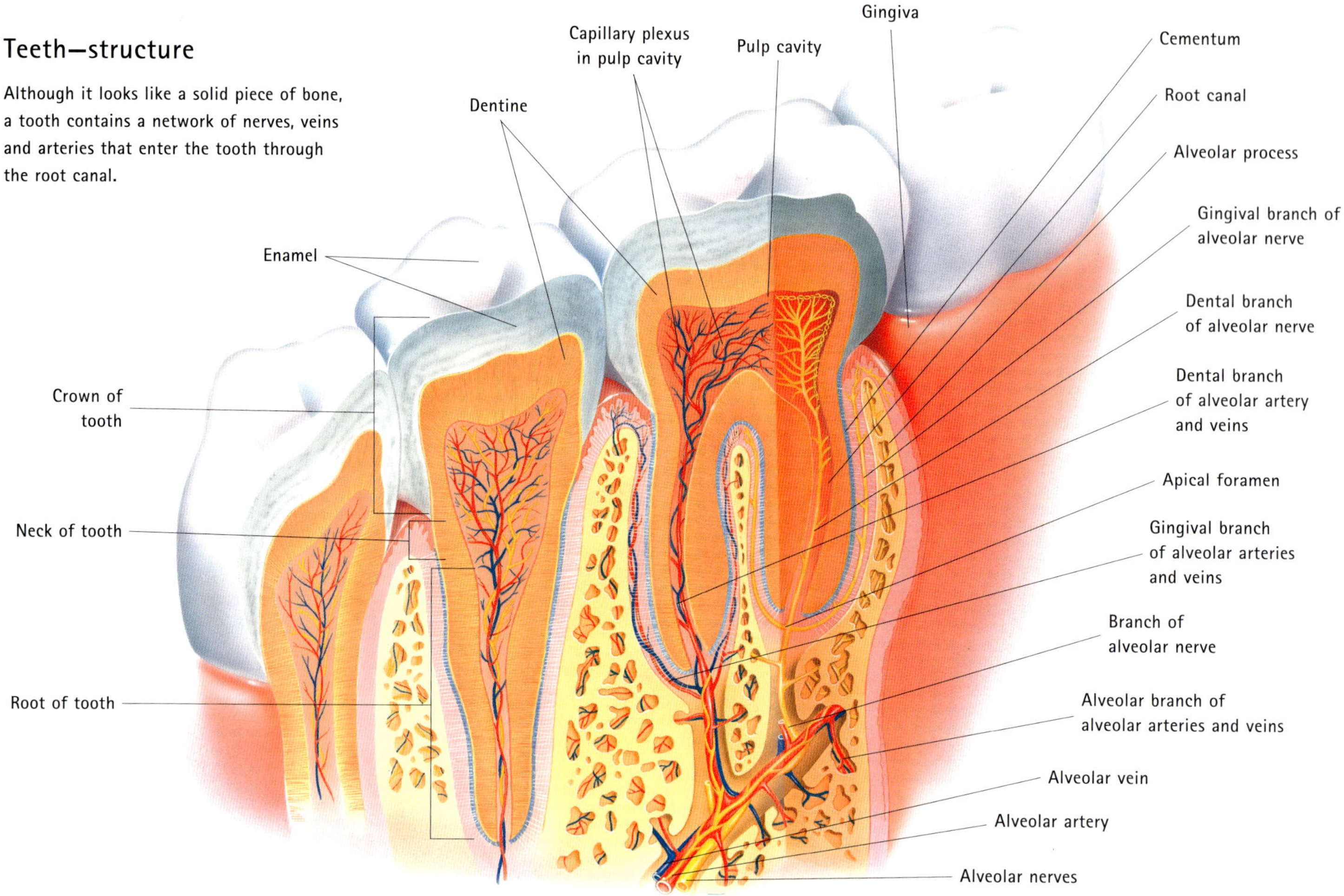

Teeth—structure

Although it looks like a solid piece of bone, a tooth contains a network of nerves, veins and arteries that enter the tooth through the root canal.

whistling and sucking on straws. Each cheek is formed by another facial muscle, the buccinator, whose fibers run forward into the orbicularis oris.

SEE ALSO *Autonomic nervous system on page 75; Digestive system on page 101; Teething on page 416; Visceral column on page 199*

Teeth

Teeth are calcified bone-like structures in both jaws whose role is to chew food, aid with speech and influence the shape of the face. The crown of the tooth, that part above the gum line, is covered with enamel, which is the hardest substance in the body. Under the enamel is dentine, slightly softer, which makes up the main part of the tooth. The dentine below the gum line is covered with cementum, a hard bony substance covering the roots of the teeth. Dentine is

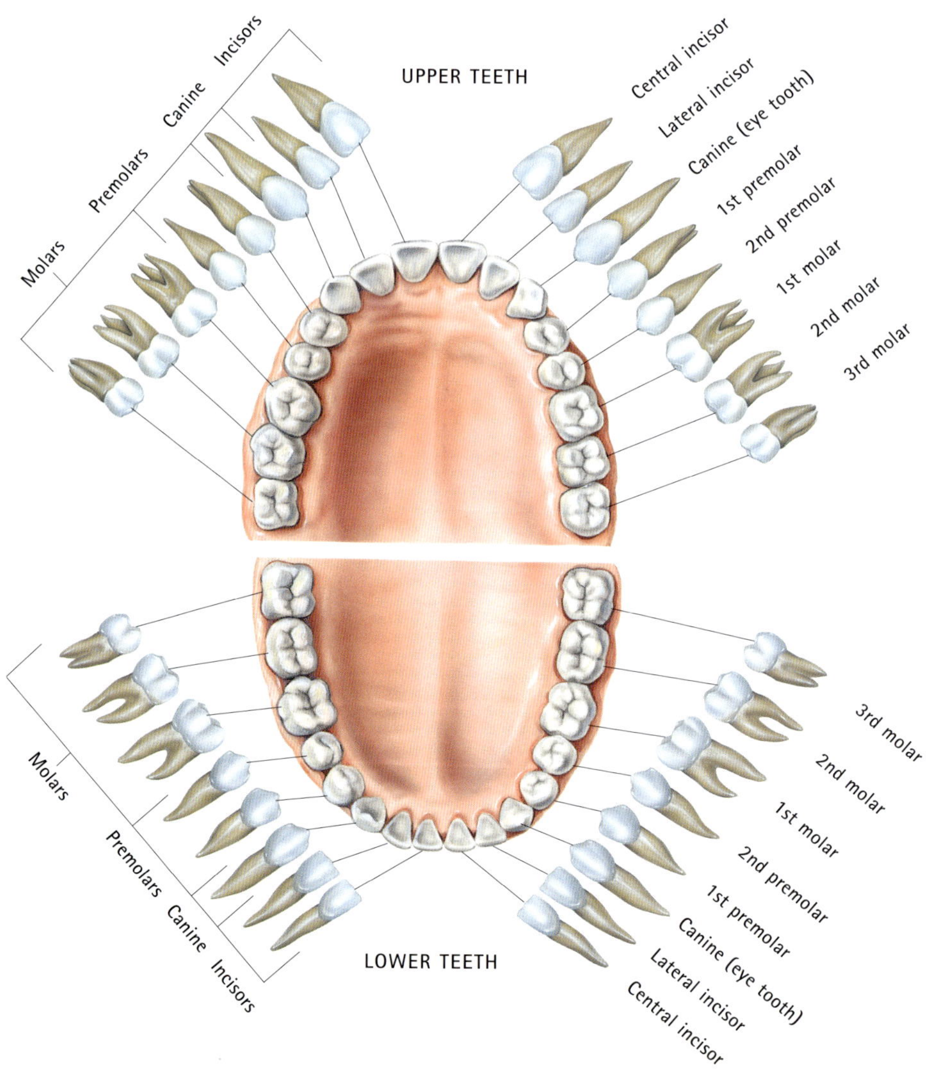

Teeth

Adults have 16 teeth in each jaw. The enamel that covers the teeth is the hardest substance in the human body.

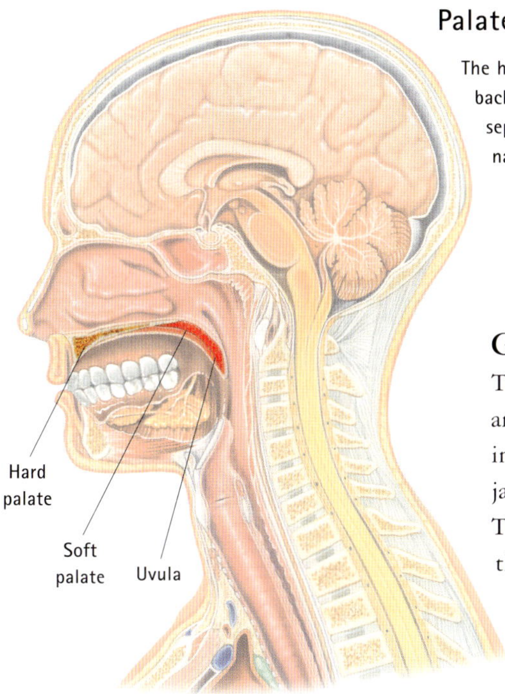

Palate

The hard palate extends back from the top teeth, separating the oral and nasal cavities.

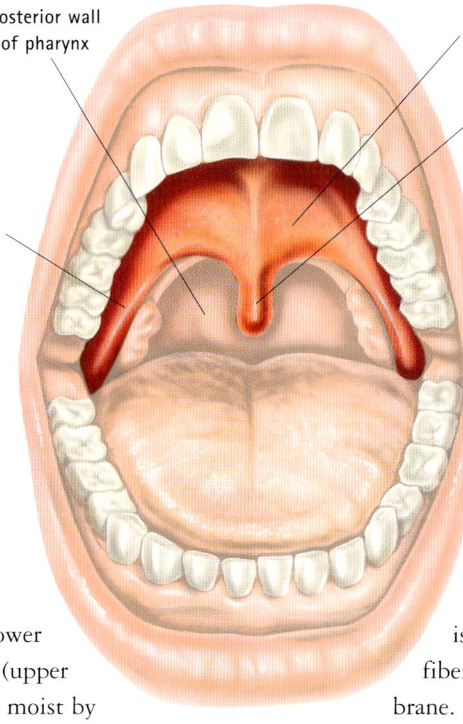

Palate—front view

The soft palate is mainly composed of muscle fibers and mucous membrane. The uvula is its most prominent feature.

a sensitive tissue, with millions of tubules running into the central pulp or nerve, which runs from the tip of the root into the center of the tooth. The cementum is surrounded by the periodontal ligament, which contains the fibers that anchor the tooth in the bone of the gum.

Humans develop two sets of teeth in a lifetime. The first set, 20 in number, are known as the deciduous, primary or baby teeth. The secondary or permanent set, containing 32 teeth, begins to replace the first set around the age of 7 years; there are 16 teeth in each jaw.

There are different types of teeth. The incisors, with sharp edges for biting, are at the front of the mouth; next to them, the canines have sharp points to tear food; at the back of the mouth the molars, together with the premolars which only appear in the second set of teeth, are used to grind food while eating.

A baby is born with the teeth already developing in the jaws. Occasionally a baby is born with some teeth already apparent. In most babies the teeth begin to appear between the ages of 5 months and 1 year; the average age for the appearance of the first four front teeth is 7 months. It takes nearly 20 years for the complete set of permanent teeth to be established, the final set of molars (often called "wisdom teeth") erupting usually in late adolescence.

Gums

The gums (gingivae) are the soft tissue covering the upper and lower jaws, inside the mouth. They extend from inside the lips, around and between the teeth, to the floor of the mouth (lower jaw) and the palate (upper jaw). The gums are kept moist by saliva and receive sensory nerves, similar to the skin. The gums are attached around the neck of the teeth, where food particles can become lodged and cause gum inflammation (gingivitis). Swelling, ulcers or discoloration of the gums may indicate more widespread disease and should be promptly investigated. A severe lack of vitamin C (scurvy) also leads to swollen and bleeding gums and loose teeth.

Palate

The roof of the mouth, separating the oral and nasal cavities, is called the palate. It comprises two sections—one hard, the other soft. Both are covered by mucous membrane containing numerous lubricating glands that keep the mouth and throat moist.

Much of the hard palate, extending from directly behind the top teeth, is formed by parts of the upper jaw bones, or maxillae. These normally fuse at the midline during fetal development. The posterior section of the hard palate is formed by the two L-shaped palatine bones of the skull. Ridges on the hard palate help with maneuvering food in the mouth during chewing and swallowing. When the mouth is closed, the tongue rests on the hard palate.

The soft palate is the fleshy structure that extends from the edge of the hard palate at the back of the mouth. Its own edge is like an incomplete curtain suspended between the back of the mouth and the beginning of the throat (pharynx). The soft palate is composed of muscle fibers and mucous membrane. A small cone-shaped projection, the uvula, hangs from it. Both the soft palate and uvula move up during swallowing or sucking to prevent food from entering the nasal cavity.

Tongue

The tongue is a muscular and sensory organ that is attached to the floor of the mouth. It has a dorsum or upper surface, a base attached to the floor of the mouth, a soft lower surface and a tip.

STRUCTURE AND FUNCTION

The tongue plays an important role in tasting, chewing and swallowing food, and in speech. Taste is sensed through the many taste receptors (taste buds) that are located on the dorsum of the tongue. The tongue's role during chewing is to move food around the mouth, pushing partially chewed food between the back teeth (molars). In swallowing, the tongue helps form a ball or bolus of food, which is gripped between the back of the tongue and the soft palate and squeezed back into the oropharynx.

During speech, the tongue makes contact with other structures in the mouth to help form consonants. Some consonants are formed by the tongue meeting the teeth (dental consonants like D and T), the soft palate (glottal consonants like G and K) or the hard palate (palatal consonants like N).

The dorsum of the tongue is studded with small projections called papillae, which are of three types. Filiform papillae are tiny cone-shaped elevations, which lack taste buds. Their role is to grip food and they are particularly well-developed in animals such as cats, which groom their fur with the tongue. Fungiform papillae are mushroom-shaped projections. They are less numerous than the filiform papillae, often contain taste buds and have a red appearance.

About two-thirds of the way back from the tongue's tip lies a V-shaped group of 7–12 vallate (also called circumvallate) papillae. These papillae have a central elevation, which is surrounded by a deep groove, much like a moat surrounding a castle. Taste buds located in the walls of the moat are continuously bathed in fluid, which clears food from the taste buds so that new taste stimuli can be tested. Other taste buds are located in nearby regions such as the palate, epiglottis and pharynx.

The pharyngeal part of the tongue behind the vallate papillae contains a lymphoid organ called the lingual tonsil. The lingual tonsil is involved in defending the body from microorganisms entering by the mouth.

The bulk of the tongue is made up of muscle, which can be divided into two main groups. The first group is called the intrinsic tongue muscles. They lie within the tongue itself and are responsible for changing its shape. The fibers of the intrinsic muscles are arranged in three directions: vertical, longitudinal and horizontal.

When vertical fibers contract they make the tongue thinner. When longitudinal fibers, which run the length of the tongue, contract, they make it shorter.

Horizontal fibers make the tongue narrower when they contract. Some people have a genetically determined ability to curl the tongue either upward or downward by controlling different groups of horizontal fibers separately.

Extrinsic tongue muscles, attached to the jaw, skull, palate and hyoid bones, are responsible for changing the position of the tongue. The hyoid bone is a small bone in the neck immediately below the jaw, which protects the airway from being crushed.

The extrinsic muscles can move the tongue forward (genioglossus muscle), backward (styloglossus muscle), upward (palatoglossus muscle) and downward (hypoglossus muscle). Most of the tongue muscles are supplied by the hypoglossal nerve, which is one of the cranial nerves relaying messages to the brain.

The underside of the tongue is soft and is kept moist by secretions from salivary glands. Beneath the tongue lie the openings of the ducts from the sublingual and submandibular salivary glands. The sublingual glands create a ridge on the floor of the mouth on each side of the tongue's base.

The midline ridge on the lower surface of the tongue is called the frenulum. On each side of this ridge lie paired deep veins of the tongue, which are visible through the thin surface layer.

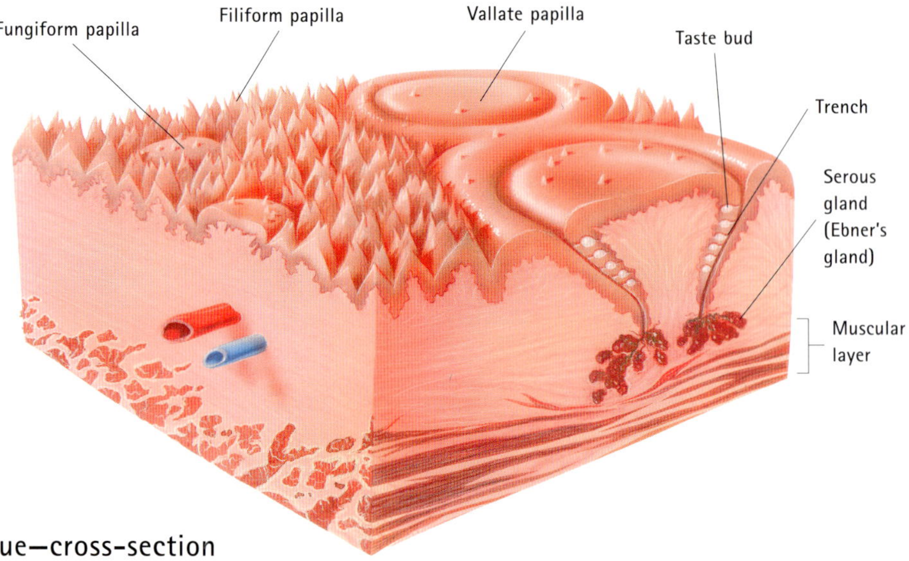

Tongue—cross-section

The tongue is a muscular organ which is used in chewing, swallowing and speech. Taste buds are located in the papillae, which are projections on the upper surface of the tongue.

TONGUE

The bulk of the tongue is made up of muscles. The intrinsic muscles lie within the tongue and are responsible for changing its shape. The extrinsic muscles are attached to the jaw, skull and palate and change the position of the tongue.

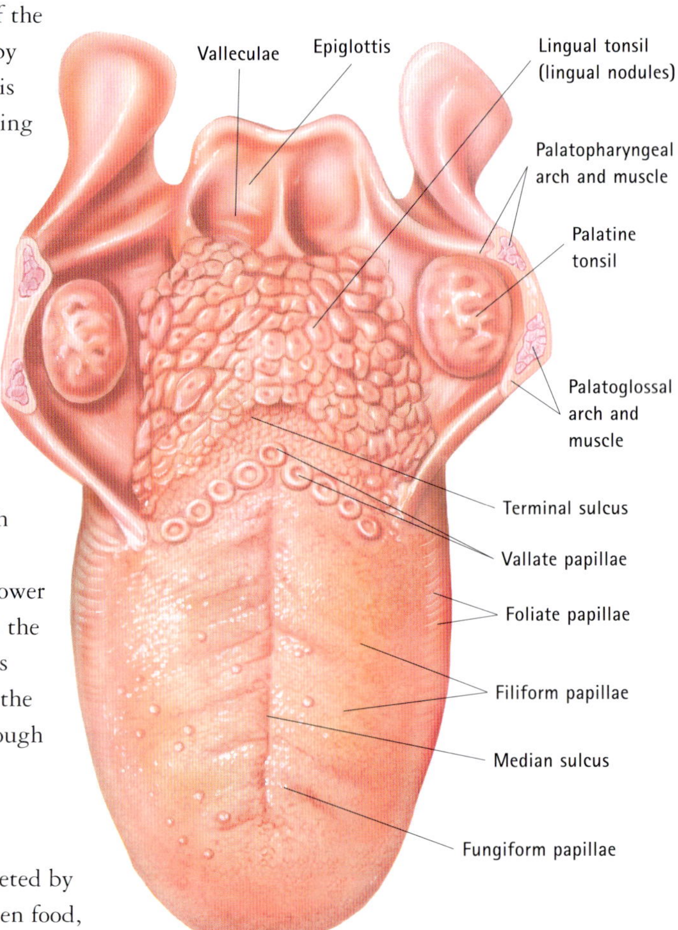

Saliva

Saliva is an alkaline fluid secreted by salivary glands that helps soften food, moistens the mouth and aids in digestion. Saliva consists of mucus, water, mineral salts, proteins and amylase. The mucus helps in swallowing the food, the water dissolves some of the components of food and helps in tasting it, and the amylase begins the digestion of carbohydrates.

Saliva also helps in keeping the mouth moist and clean. It also plays an important function in helping the body to retain control of its water balance. One of its other functions–the role of removing food debris—helps in reducing tooth decay (caries).

Salivary glands

The salivary glands are located around the beginning of the digestive tract. They produce saliva, a fluid that moistens food and enables food to be bound together into a mass called a bolus, thereby making chewing and swallowing easier. The moisture of saliva allows chemicals in the food to be dissolved and delivered to the taste buds for tasting. Saliva is also rich in digestive enzymes—chemicals that initiate the breakdown of food into simpler substances. The main enzyme in saliva is known as amylase, which begins the breakdown of starches into their constituent chemicals. This digestive process will continue until the food bolus reaches the stomach and the acid of that region reduces the activity of the amylase. Human saliva also contains substances such as lysozyme, and antibodies that control bacteria, thereby protecting the body against invasion by potentially disease-causing bacteria.

The salivary glands are divided into two groups. The major salivary glands are large structures that are easily seen with the naked eye. They consist of three pairs of glands: the parotid, submandibular and sublingual glands. The minor salivary glands are microscopic and are scattered around the mouth, palate and throat.

Human saliva is secretions of the submandibular gland, with contributions from the parotid, sublingual and minor salivary glands in decreasing order of importance.

The parotid gland is located in front of the ear. It gives rise to a duct that runs forward to open into the mouth opposite the second molar of the upper teeth on each side. A flap of mucosa is present at the point where the duct opens into the mouth, and may be felt with the tip of the tongue. The submandibular gland is located below the jaw on each side, about 1 inch (2–3 centimeters) in front of the angle of the jaw. It gives rise to a duct which runs forward a short distance and opens into the floor of the mouth under the tongue. The sublingual glands are small, and lie within ridges on the floor of the mouth beneath the tongue. They open by many small ducts into the floor of the mouth.

Salivary glands are mainly under the control of the nervous system, although some hormones may affect their function.

SALIVARY GLANDS

The salivary glands secrete saliva into the mouth. This fluid is needed to moisten food to ease swallowing and begins food breakdown in the preliminary stage of digestion. Infection of the salivary glands can cause swelling and pain.

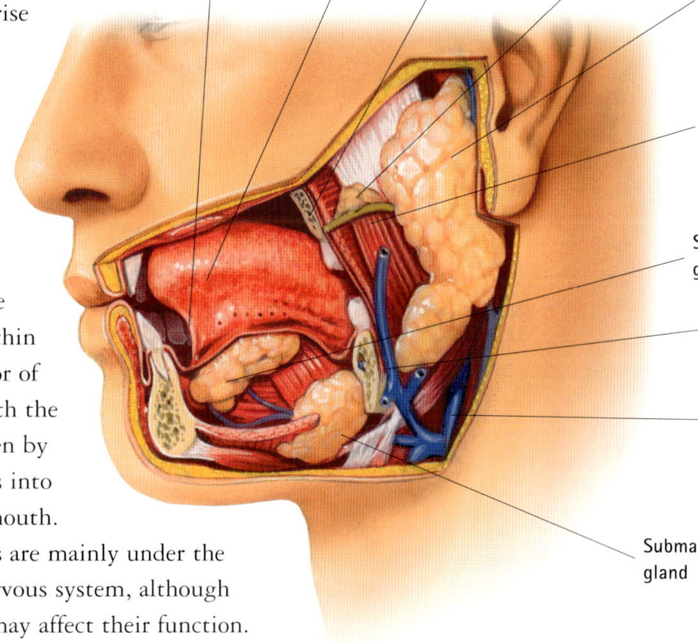

Salivary glands microstructure

There are three distinctive pairs of salivary glands—the parotid, sublingual and submandibular glands. Each pair has a unique cellular organization and produces saliva with slightly different properties.

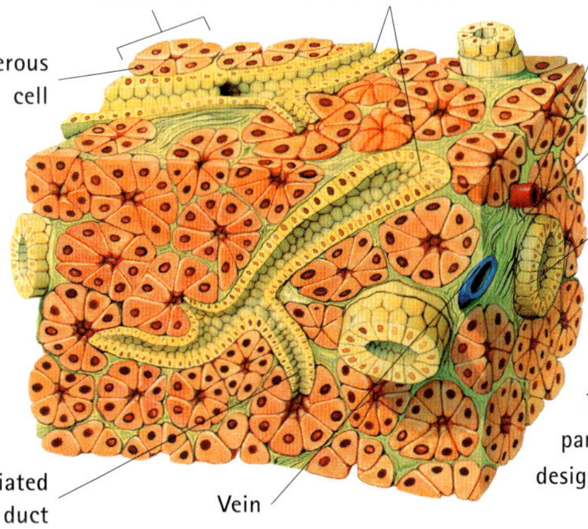

PAROTID MICROSTRUCTURE

The thin saliva produced by the parotids contains enzymes specially designed to break down starch.

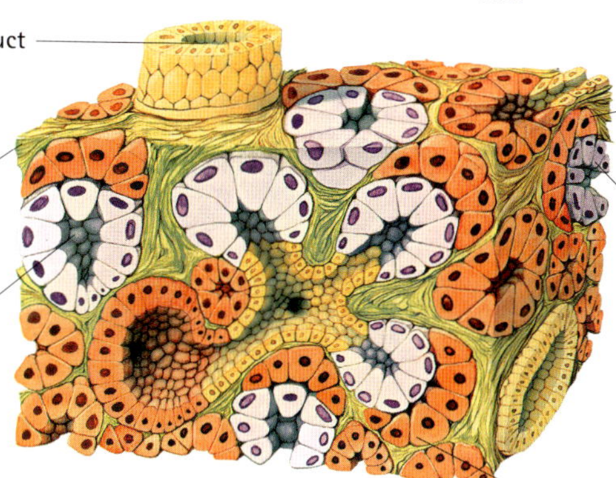

SUBMANDIBULAR MICROSTRUCTURE

The submandibular gland is comprised of a mixture of enzyme-producing serous cells and mucus-producing cells. Its saliva is predominantly water.

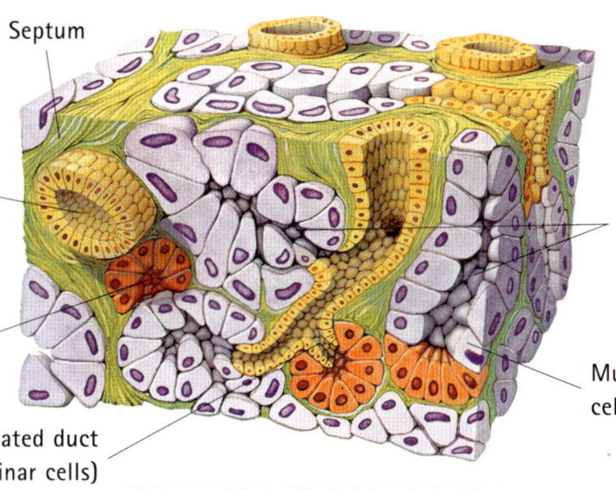

SUBLINGUAL MICROSTRUCTURE

This gland produces thicker, watery mucus, particularly in response to milk or cream, which helps to lubricate the mouth.

The two parts of the autonomic nervous system, the parasympathetic and sympathetic divisions, both contribute nerves to the salivary glands. The parasympathetic nerves are probably most important, because they provide the stimulus to release copious amounts of saliva that is rich in digestive enzyme. This occurs in response to the smell and sight of food, as well as to the presence of food in the mouth.

Sympathetic stimulation of salivary glands tends to produce a dry mouth (such as during states of anxiety or fear) with very little enzyme content in the saliva.

The salivary glands may be involved in disease. Infection with the mumps virus produces a characteristic enlargement of the parotid gland, which causes painful swelling in the face and cheek. Tumors may also appear in the salivary glands. Most tumors arise in the parotid gland, and about 80 percent of these are benign (non-invasive and less dangerous). About half of the tumors arising from the submandibular gland are benign, while the others are malignant (invasive and potentially fatal). Cancers of the salivary gland may be treated by surgical removal or radiation therapy.

Tonsils

The tonsils are lymphoid organs which lie under the surface lining of the mouth and throat. There are three sets of tonsils, named according to their position. The lingual tonsil lies on the back third of the tongue; the palatine tonsils lie on either side of the back of the tongue, between pillars of tissue which join the soft palate to the tongue; the pharyngeal or nasopharyngeal tonsils (adenoids) lie in the space behind the nose.

The tonsils are arranged around the entrance to the respiratory and digestive tracts to protect the body from bacteria and viruses which may enter from the mouth and nose. Tonsils produce lymphocytes, which cross into the mouth and throat.

Tonsillitis is inflammation of the tonsils, usually due to bacterial infection of the tonsillar tissue. If the infection spreads from the palatine tonsil to the space around the soft palate, a pus-filled peritonsillar abscess will form. This disease is known as quinsy and requires high-dose intravenous antibiotics and surgical drainage.

Adenoids

The adenoids are two glandular swellings at the back of the throat, above the tonsils, usually present in children before adolescence. Composed of lymphatic tissue, the adenoids are thought to assist the body in fighting throat infections. The adenoids are one of the first lines of defense against microscopic invaders entering the body via the nose and mouth.

Normally, adenoids grow slowly in size from the age of three until the age of five, when they shrink again, disappearing around puberty. But in some children who suffer repeated throat infections, they keep growing, becoming swollen and painful. Tonsils often also become enlarged. Eventually the adenoids may block the space between the nasal passages and the throat. Inflammation of the mucous membrane of the nose (rhinitis) and of the air sinuses behind the nose (sinusitis) may follow.

The adenoids can also block the opening of the narrow eustachian tube that connects the middle ear to the throat. When this happens, bacteria grow inside the middle ear and infection (otitis media) can develop. Repeated infections in the ear can turn into a chronic condition known as "glue ear" and can cause deafness.

The diagnosis is made by a pediatrician or primary care physician, who will examine the back of the child's throat, using a mirror with a light attached. Hearing should be tested and an x-ray of the sinuses may be required. Usually the adenoids themselves don't need to be treated, as they decrease in size around puberty. Any infections that arise must be treated with antibiotics.

If infections are very frequent or persist in spite of antibiotic treatment, then an operation to remove the adenoids may be necessary. This procedure is called an adenoidectomy and is performed under general anesthetic. Removal of the tonsils (tonsillectomy) is often performed at the same time. The child is in hospital for approximately three days and usually recovers rapidly.

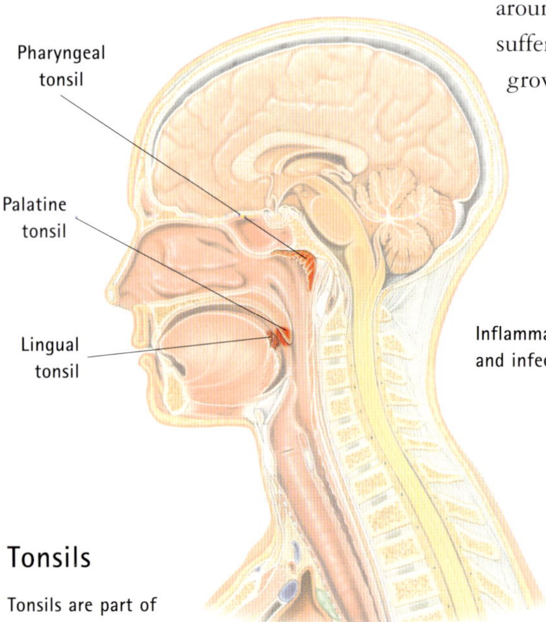

Tonsils

Tonsils are part of the lymphatic system and filter the circulating lymph of bacteria that may enter the body through the nose and the mouth.

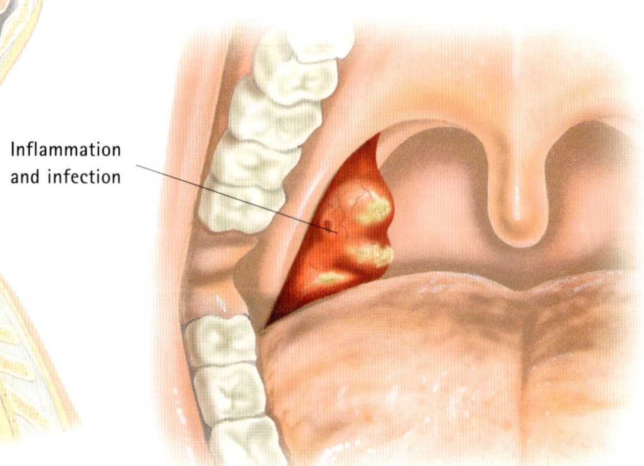

Taste

Taste, also known as gustation, is one of the five special senses. The organs of taste are the taste buds, specialized receptors made up of small clusters of cells called papillae, located on the surface and on the sides of the tongue, the roof of the mouth, and the entrance to the pharynx. At the top of each taste bud there is an opening called the taste pore. For food to be tasted it must be dissolved in a watery solution like saliva so that it can activate the receptors. Each person has 2,000–5,000 individual taste buds (women have more than men). Taste buds can

Nerve stimulation

Three cranial nerves that lead directly to the brain from different parts of the head and neck are involved in our sense of taste

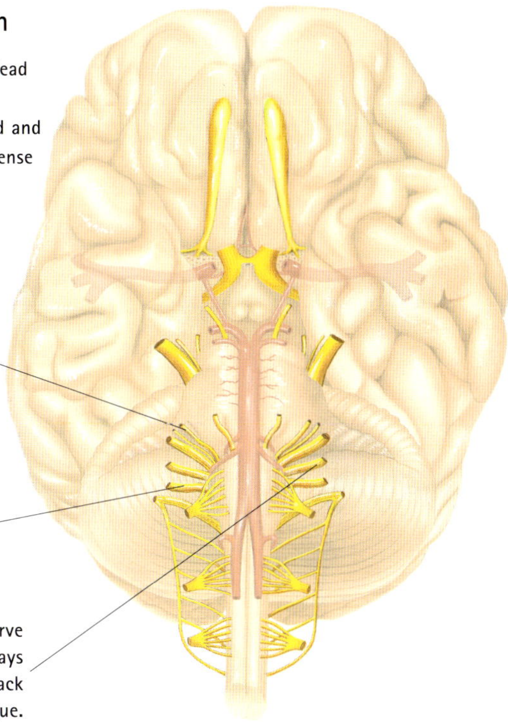

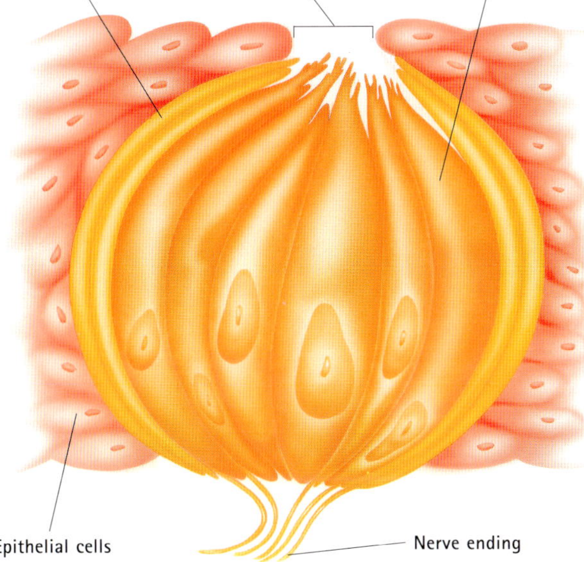

Taste buds

Taste buds are packed together in groups at various places on the tongue. These bundles of cells are sensitive to sweet, salty, bitter and sour flavors. Substances must be dissolved in a watery solution like saliva in order to activate the taste buds.

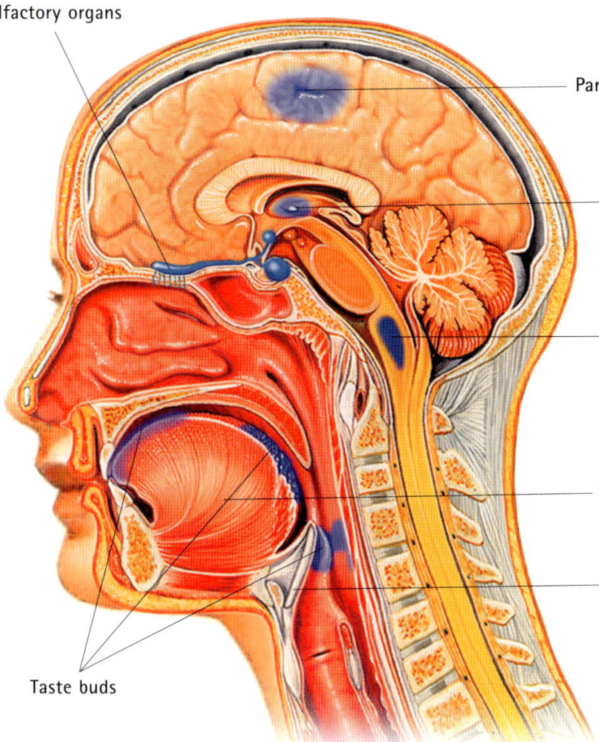

Taste pathways

Taste buds at the front and back of the tongue and in the throat send nerve impulses via the cranial nerves to the medulla in the brain stem. From here, the information passes to the thalamus and on to taste-receiving areas in the parietal lobe of the cerebral cortex where the taste is identified. The olfactory organs provide additional information vital for interpreting and appreciating different tastes.

TASTE

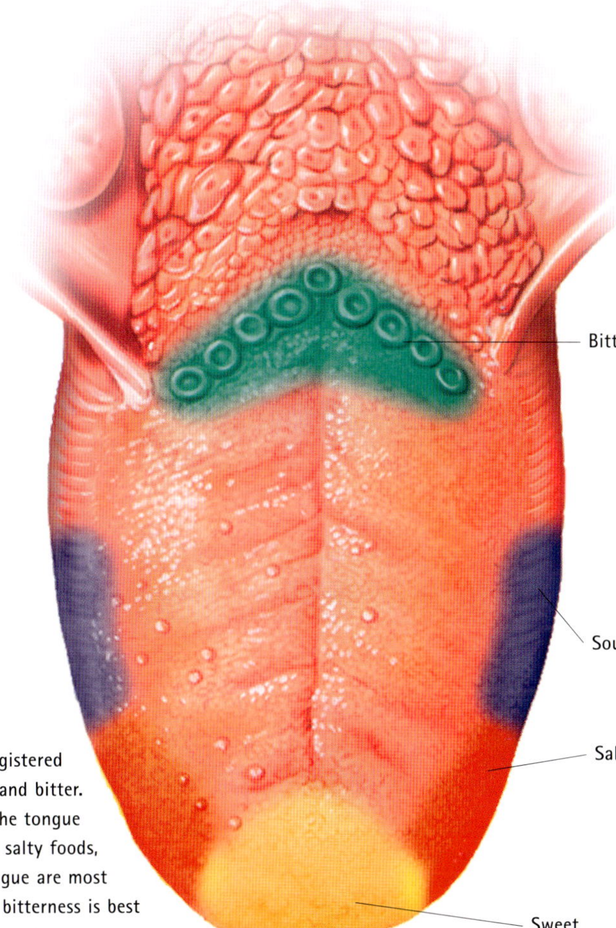

Tongue zones

There are four distinct tastes registered by taste buds: sweet, salt, sour and bitter. The taste buds at the front of the tongue are most sensitive to sweet and salty foods, the ones at the sides of the tongue are most sensitive to any sour taste, and bitterness is best registered at the back of the tongue.

discern among the four taste sensations: sweetness, sourness, saltiness and bitterness.

Historically, it was believed that each bud could only taste one type of taste—for example, those at the tip of the tongue were thought to detect sweetness, whereas those on the sides of the tongue would detect only saltiness and sourness.

Researchers now believe, however, that all taste buds are capable of detecting multiple combinations of the four basic taste types. Nevertheless, different areas of the tongue are not equally sensitive to all four tastes. The back of the tongue is more sensitive to bitter, the sides of the tongue to sour, the tip to sweet, and the tip and the sides to salt. Taste can be tested by applying a salty or sweet solution to the front two-thirds of the tongue.

The exact mechanism whereby a taste bud detects a particular taste is poorly understood. However, it is thought that when a food particle is dissolved in saliva and comes into contact with the taste bud, it causes a chemical reaction within the bud. The taste bud then sends a nerve impulse via a nerve fiber attached to the base of the bud, which joins nerve impulses from other taste buds and then travels via the cranial nerves to the brain.

Three different cranial nerves relay taste: the facial nerve relays taste impulses from the front two-thirds of the tongue, the glossopharyngeal nerve supplies the back of the tongue and the vagus nerve supplies taste buds in the throat. These impulses are then relayed through nerve pathways in the brain stem and the thalamus to a taste-receiving area in the anterior cerebral cortex where they are experienced as a particular taste.

The taste fibers follow a complicated route. The taste fibers of the facial nerve are distributed to taste buds in the front two-thirds of the tongue, predominantly along the lateral borders. The taste fibers travel with the lingual nerve but cross to the facial nerve in the chorda tympani, so that taste is impaired if the facial nerve is damaged above this junction with the chorda tympani (as is often the case in Bell's palsy). These fibers enter the brain stem in the sensory root of the facial nerve (nervus intermedius). The fibers are joined by taste fibers from the glossopharyngeal and vagus nerves and terminate in the nucleus of the tractus solitarius. The part of this nucleus that receives taste fibers is often referred to as the gustatory nucleus. Fibers ascend from here to reach the thalamus. Fibers then ascend from the thalamus to the cortical area for taste, which is located at the lower end of the sensory cortex in the parietal lobe. The taste area is adjacent to the sensory area of the cortex for the tongue and pharynx.

The taste fibers of the glossopharyngeal nerve are distributed to taste buds in the back of the tongue and the pharynx, while the taste fibers of the vagus nerve only supply the epiglottis. The vagal taste fibers are relatively unimportant because few continue into adult life. The vagal and glossopharyngeal fibers terminate in the gustatory nucleus. The fibers then ascend to the cortex from the gustatory nucleus in the same way as described above for the facial nerve.

Our state of consciousness, our cultural conditioning, our past experiences of taste and, in particular, the sense of smell, are all important in how we finally perceive a particular taste. About 80 percent of what we experience as taste is actually due to smell.

Taste abnormalities can be caused by conditions that affect the tongue and throat, the nasal passages, or the nerve pathways and brain. Conditions that can attenuate the sense of taste include the common cold, nasal infections, influenza, viral pharyngitis, mouth dryness, ageing (taste buds tend to diminish in number with age) and heavy smoking (which tends to dry the mouth).

SEE ALSO *Nervous system on page 64; Smell on page 192*

SPEECH

Speech and language depend on the function of many different organs and structures within the body. These can be divided into: nervous system elements, for planning language production and control of muscles; parts of the respiratory system and larynx, for the production of the raw sound (phonation); and the mouth area, for the modification of this raw sound to produce the vowels and consonants of speech (articulation).

Within the brain, usually in the left hemisphere, there are two specialized language regions known as Broca's area and Wernicke's area. Wernicke's area is concerned with the comprehension of language, while Broca's area is involved in the expressive aspects of language. These areas act through the cerebral cortex to control neuron activity in the brain stem and also to produce phonation and articulation. The cerebellum at the back of the brain also plays an important role in articulation.

Raw sound is produced by expelling air from the lungs through the larynx. Within the larynx are the two vocal cords, which can be separated during intake of breath and brought close together during speech. When the vocal cords are close together, and air is forced between them, they begin to vibrate, in the same way that two leaves held close together will vibrate when air is forced between them. This raw sound can be altered by increasing or decreasing the tension or length of the vocal cords.

Vowel sounds (such as "ah" or "ooh") are produced by modifying the shape of the expelled air above the vibrating vocal cords. This is done by moving the soft palate, tongue and lips in a coordinated fashion.

Most consonants are produced by temporarily stopping the airflow. Consonants are often named according to the part of the airway involved. Labial consonants (B, P) are produced by bringing the lips (labia) together; dental consonants (D, T) involve the tongue touching the teeth; nasal consonants (M, N) involve passing air through the nose; and glottal consonants (Q, G, K) involve the temporary closing of the back of the tongue (the epiglottis) against the soft palate.

Speech disorders

Speech disorders have been a concern since the beginning of recorded history. It is estimated that about 10 million Americans have a speech disorder of some kind. The major types include voice disorders (dysphonias), which can be caused by paralysis of the larynx, injury or a disease of the endocrine glands; and speech disorders, the most common being those that disrupt a child's ability to learn a language.

These disorders include unintelligible or absence of speech, lisping, and may be caused by genetic factors or by damage before, during or shortly after birth. Poor language

Vocal cords

The vocal cords, or vocal folds, are two folds of mucous membrane that vibrate to make sound when we speak. One end of each vocal cord attaches to cartilage at the front of the larynx. The other ends of the cords are attached to cartilages that can move freely, allowing the cords to vibrate and make sound. The folds are also very flexible—relaxing them will make low-pitched sounds and making them taut will produce high-pitched sounds.

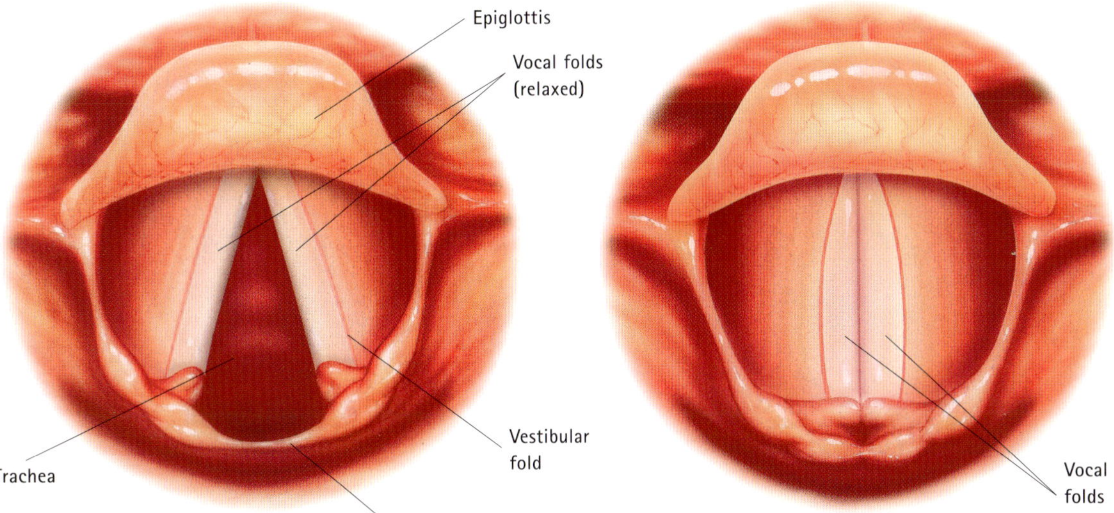

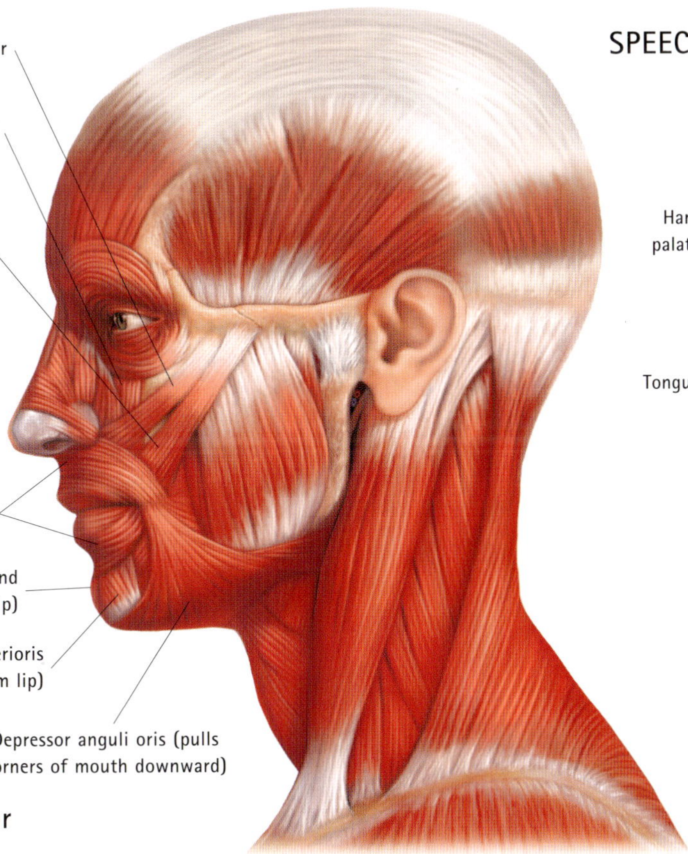

Lip movement for speech

The movement of the lips modifies the sounds that come from the larynx and vocal cords into speech. The lips need to join to make the sound "m," for example, need to touch the teeth (with the bottom lip) to make "f," and round to make "o." The accurate movements that speech requires are made possible by a complex arrangement of muscles around the mouth and cheek areas.

SPEECH

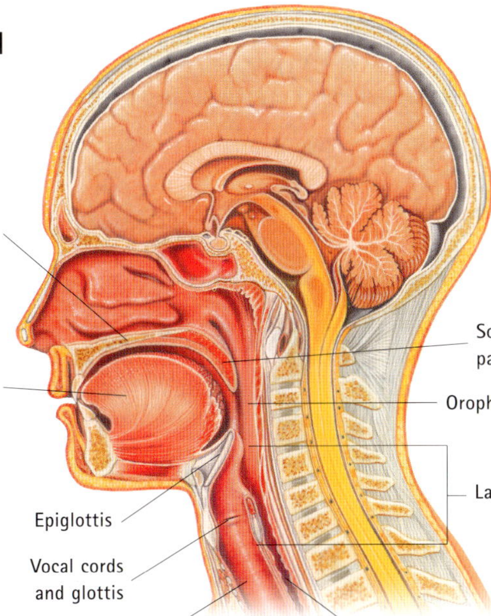

Producing sound

Producing sound, and turning it into speech, is a three-step process. First, the lungs expel air. Second, the vocal cords open and close to alter the air flow and cause vibrations, making a sound. Third, this sound is modified by muscles in the mouth and tongue—involving movement of the soft palate, tongue and lips.

Speaking and understanding speech

Two areas in the brain coordinate speech and our understanding of speech. Broca's area is involved in the expressive aspects of language. It gives instructions to the breathing muscles, the muscles of the larynx, pharynx, tongue and lips to regulate the airflow and vocalization needed for speech. Wernicke's area is involved in understanding and interpreting speech.

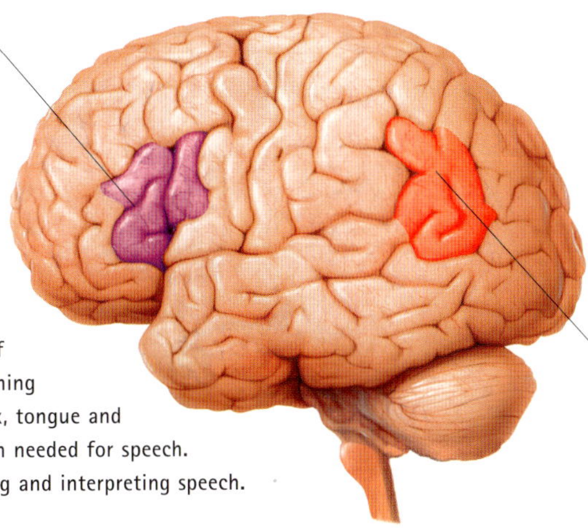

skills within the family, parental neglect or prolonged illness can also play a part.

Difficulties in articulation include cluttering (tachyphemia), lisping and stuttering or stammering (dysphemia). Physical problems can also affect speech. Injury sustained to the part of the brain related to language results in dysphasia; it may be the result of stroke or head injury.

Aphasia (speechlessness) occurs when the left side of the brain is damaged. It involves the loss of memory for the meaning of language and how it is produced. Sufferers may know what they want to say but be unable to say or write it. Together with treatment, the sufferer must also re-educate the parts of the brain that are still functioning normally.

Lisping includes a range of speech impediments, the most common being the pronunciation of the sibilant sounds of "s" or "z" as "th". This happens when the tip of the tongue is too far forward in the mouth, instead of against the hard palate. Lisping may be treated with speech therapy.

Stuttering or stammering (dysphemia) is the most common and obvious type of disturbed speech. Every child will stutter at some time, as will many adults. The condition is characterized by hesitant or jerky speech and an inability to pronounce or join syllables. While experts are uncertain of the causes, it is found more in males than females and may be caused by a combination of genetic and emotional problems. Treatment may include medication, aimed at blocking excess dopamine activity in the brain, or behavioral therapy focusing on changing speech mechanisms.

Tongue-tie (ankyloglossia) is easily corrected with surgery, but major defects of the tongue reduce the ability to articulate. Sufferers can be taught to speak despite these defects. Hypernasal speech, which results in a person "talking through the nose," can be caused by paralysis, congenital malformation, injury or palate defects. Treatment needs a thorough understanding of the causes. Symptomatic speech disorders, caused by lesions in the nervous system, are known as dysarthria. When speech development is limited by some mental disorders, it is known as dyslogia.

Hearing loss in early years results in distorted speech (audiogenic dyslalia); speech problems caused by defects to lips, teeth or mouth are known as dysglossia. Cerebral palsy, chorea, Parkinson's disease and other nervous disorders also affect speech.

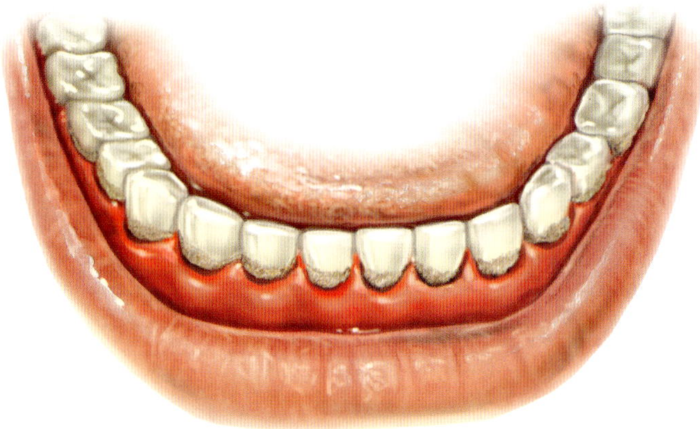

Gingivitis

The accumulation of food particles in the crevices between teeth and gums can cause gingivitis (inflammation of the gums). Symptoms include gum bleeding and swelling.

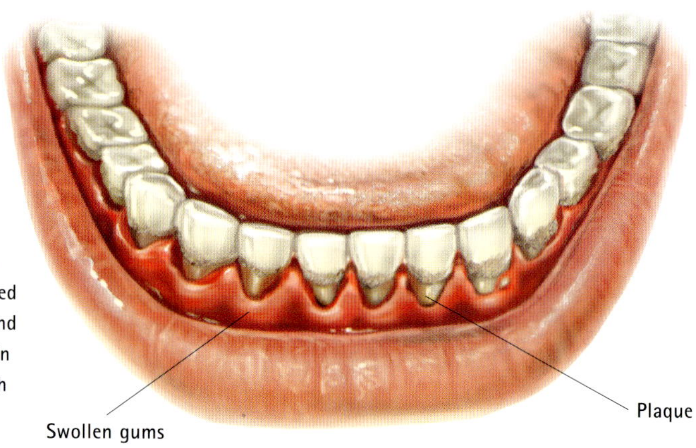

Periodontitis

In this condition, a painful inflammation of the tissues supporting the teeth is caused by bacteria, food particles and calcium deposits collecting in the spaces between the teeth and gum.

Disorders of the mouth

The oral cavity may be involved in a variety of infections due to viruses like herpes simplex and Coxsackie virus type A, or yeast. Infection with yeast can result in oral thrush or candidiasis, particularly in those people whose immune systems are functioning poorly due to chemotherapy for tumors or leukemia. Infection of the tooth pulp with pus-forming bacteria can eventually result in abscesses of the jaw.

Mouth ulcers may result from herpes simplex or from diseases such as aplastic anemia, in which the bone marrow stops producing red and white blood cells. Mouth ulcers may accompany leukemia, erythema multiforme and Stevens-Johnson syndrome.

The color and texture of the tongue is used by naturopaths to assess ill health. Red in the center of the tongue may indicate a stomach-related problem.

SEE ALSO *Dental procedures and therapies on page 457; Erythema multiforme, Herpes simplex Leukemia and other individual disorders in Index*

Gingivitis

Gingivitis is an inflammation of the gums. It commonly occurs when small particles of food get trapped between the tooth and gum, causing the build-up of bacteria in these areas. The gums become swollen and red, and bleed easily.

Chronic gingivitis is the result of poor oral and dental hygiene, with build-up of debris and plaque (tartar) around the teeth. It may also occur with poorly fitting dentures, or with tooth decay or abscesses.

If left untreated, gingivitis can lead to periodontitis, where the infection spreads to deeper tissues such as the tooth socket and bone. This causes bone loss, and leads to enlarging of the tooth socket, so that teeth become loose and may eventually fall out. Regular visits to the dentist and improved tooth brushing, with flossing between the teeth, are recom-mended in this condition.

Acute ulcerative necrotizing gingivitis is a severe inflammation caused by a gum infection. The infected areas bleed heavily and are very painful and ulcerated, and the breath smells foul. Again the inflammation can spread to deeper tissues and eventually inflammation will lead to the destruction of periodontal tissues and tooth loss. An urgent dental referral is recommended as soon as the inflammation is detected.

Periodontitis

Periodontitis is inflammation of the gums that leads to infection of the ligaments and bone supporting the teeth. It is usually caused by a build-up of plaque on the teeth due to poor dental hygiene. If untreated, the teeth become loose and fall out. Warning signs include swelling and a red-purple coloring of the gums, blood on the toothbrush, tenderness and bad breath. Bone structure can be damaged before the condition is realized as there is usually little or no pain. Treatment involves thorough cleaning by a dentist and perhaps surgical trimming of scar tissue and reshaping of bone. In severe cases, some teeth may need to be removed to stop the spread of infection.

Teeth problems

The most common problems experienced with teeth include abscesses and bad breath.

Abscesses occur when the pulp of a tooth becomes infected. The infection can spread into the tissues near the root tip, and a pocket of pus (an abscess) can form in the gum. Treatment begins with antibiotics.

Poor oral hygiene, gum disease, a dry mouth (xerostomia), tobacco products, various foods and medical problems are the main causes of bad breath. Once the cause has been determined, improving oral hygiene is the best method of keeping bad breath at bay.

Decay (caries) can be stopped or reversed in the early stages by appropriate use of fluoride and proper cleaning. More advanced decay can be repaired by removing damaged tissue and filling the tooth with amalgam or composite resin, which is tooth-colored. Other materials that can be used are glass ionomer cement, porcelain and gold.

There are many causes of unattractive blemishes on the teeth. Smoking, inherited conditions, childhood illnesses, inappropriate antibiotic treatment, injury and fluorosis, are some of the most common.

Wisdom teeth are the last of the back molars to erupt and play a valuable role if they are healthy and properly positioned. If the jaw is not large enough for proper eruption (partial eruption), the wisdom teeth may damage adjacent teeth, or a cyst may form, destroying surrounding structures, and they should be surgically removed.

Tartar

Saliva, scraps of food and other material such as calcium carbonate form tartar, a hard, yellow deposit on the teeth.

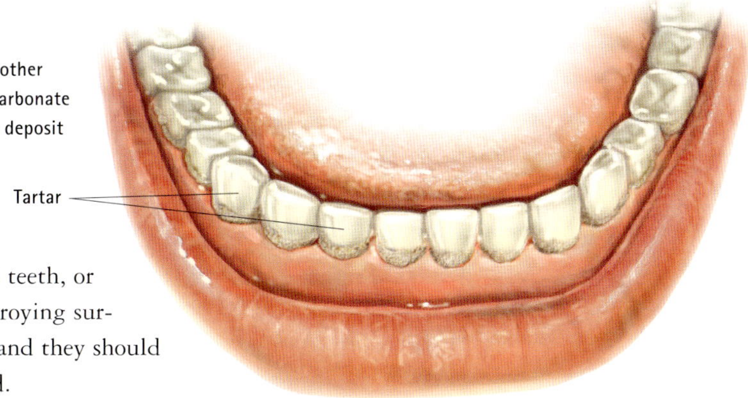

Tartar (dental calculus or dental plaque) is a hard yellowish film composed of calcium and food particles which is deposited on the teeth by the saliva. Found mostly at the line where the gum and tooth meet, and behind the lower front teeth and sides of the back teeth, it is not in itself responsible for tooth decay; rather it is the bacteria which lodge behind it that cause both decay and gum disease. Thorough brushing, preferably with an electric toothbrush, is essential to prevent build-up of tartar.

Teeth-grinding (bruxism) is the act of grinding or clenching the teeth, usually during sleep. It does not affect sleep—though it can damage teeth and cause face pain or headache. It seems to run in families and can occur at all ages. The most important treatment is to protect teeth and for this a mouth guard, as prescribed by a dentist, may be necessary.

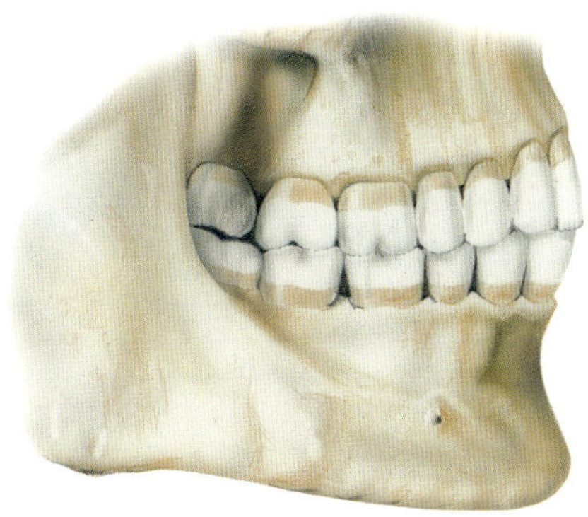

Normal bite

When teeth are correctly positioned, the middle lower teeth should abut against the back of the middle upper teeth.

Malocclusion (overbite)

When the maxilla (upper jaw) protrudes over the mandible (lower jaw), or in some cases when the mandible is underdeveloped, an overbite malocclusion occurs.

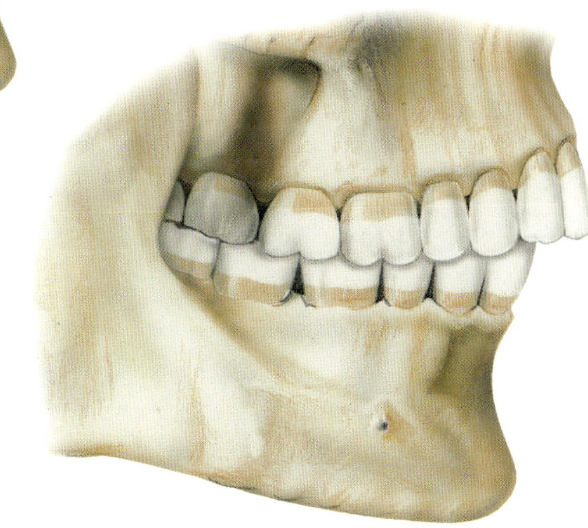

Malocclusion

Ideally, with the mouth closed, the middle lower teeth should abut against the back of the middle upper teeth. Any deviation of the teeth forward, backward or sideways will produce malocclusion. Although usually a problem with growing children—particularly if the teeth are too large for a small

Malocclusion (underbite)

When the mandible (lower jaw) juts out further than it should, the teeth become misaligned and malocclusion occurs.

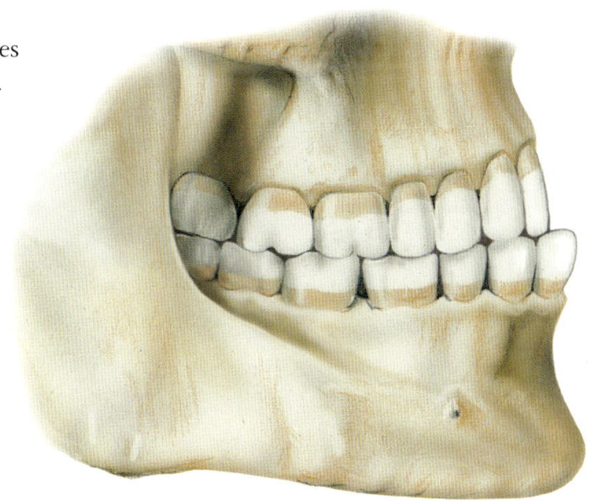

mouth—adults who have teeth removed from one side of the mouth may find that the remaining teeth drift sideways toward the gap and result in malocclusion. Other causes include unbalanced contractions of the muscles responsible for chewing, and prolonged thumb sucking. Tooth decay, inability to chew properly and tension headaches can all result.

Treatment is provided by orthodontists who, by bracing or banding the teeth and applying gentle but prolonged pressure, can eventually straighten the teeth. In older children and adults, wisdom teeth must occasionally be removed to allow adequate room for the remaining teeth.

Cleft palate and cleft lip

Cleft palate is a congenital abnormality in which there is an abnormal opening in the roof of the mouth. As the fetus develops, separate tissues from either side of the mouth fuse together to form the palate, upper lip and upper jaw. If the tissues do not fuse, an abnormal gap connecting the nasal passages with the mouth is the result. The gap is called a cleft palate; it can be closed by surgery, but a series of operations may usually be required.

Often, a cleft palate is accompanied by a similar gap in the upper lip—this is known as a cleft lip, or harelip. Cleft lip may occur on its own or with cleft palate. Normally, both sides of the upper lip should fuse in the first 35 days in the uterus. If it fails to fuse, a cleft lip is the result. The cleft may vary in size from a notch to a fissure that extends across the whole lip. Usually, it extends from the mouth up into the nostril. It can be on one side only (unilateral) or on both sides (bilateral).

The treatment for cleft lip is surgery, usually when the infant is about 10 weeks old or weighs 10 pounds (4.5 kilograms).

Both cleft palate and cleft lip affect females more often than males and may run in families. A cleft palate may occur with heart defects, and an abnormal face, and also learning problems in the congenital disorder called Shprintzen syndrome or the velocardio-facial syndrome).

Glossitis

Glossitis is inflammation of the tongue. Acute (sudden) glossitis often occurs in children; symptoms of acute glossitis include a painful, bright red, swollen tongue, which is sometimes ulcerated. There may be difficulty in swallowing, and the child may complain of an unpleasant taste in the mouth. Other mouth disorders such as gingivitis (inflammation of the gums) and stomatitis (inflammation of the mouth) also may be present. Treatment for acute glossitis is with antiseptic mouthwashes and an anesthetic solution to reduce pain.

Chronic (long term) glossitis is associated in an adult with chronic ill health, anemia, poor nutrition, vitamin deficiencies, tooth infections, smoking, alcohol consumption, and occasionally as a side effect of antibiotic drugs. Chronic glossitis can be treated by correcting the underlying cause.

Leukoplakia

Leukoplakia refers to a whitish-looking patch on the mucous membrane of the cheeks, gums or tongue that cannot be removed by scraping. It is most often the result of thickening of the surface layer of cells (the epithelium), and is a response to injury or chronic irritation. This may be a result of friction caused by rough edges on teeth, fillings, and ill-fitting dentures and crowns. It occurs most often in smokers (especially pipe smokers) and in users of chewing tobacco, but other irritants can also trigger leukoplakia.

A white or gray colored lesion usually develops slowly over a period of weeks or months. This area may be sensitive to touch, heat and spicy foods. In a few patients, the change is either pre-cancerous or represents an early cancer of the mouth, both of which can only be excluded by microscopic examination of a biopsy sample.

Other forms of this condition include hairy leukoplakia, which involves fuzzy patches on the tongue and occasionally other parts of the mouth. It is a symptom of AIDS. A rare form of leukoplakia of unknown cause has also been reported on the external genital area (vulva) of women.

Halitosis

Halitosis is an unpleasant, disagreeable or offensive breath odor. Poor oral hygiene is the leading cause of halitosis, though it may also be due to eating smelly foods such as onion or garlic, or to smoking. It may be a symptom of an underlying illness, such as alcoholism, throat infection, sinusitis, lung infection, gum disease, tooth abscess, or a foreign body in the nose (usually in children). Occasionally it may be due to more serious disease such as acute renal failure, chronic renal failure or bowel obstruction.

The condition is treated by improving oral hygiene (brushing teeth more frequently, rinsing with a mouth rinse, and visiting a dentist regularly) and by treating any underlying disorder.

Oral cancer

Oral cancer encompasses cancer of the lips, tongue, floor of the mouth, inside of the cheek, gums and palate. Oral cancers occur most commonly in the elderly, and more in men than in women, in people whose dental and oral hygiene is poor and who smoke cigarettes, cigars or pipes, or who are heavy users of alcohol. The cause is thought to be the irritant effect of these toxins.

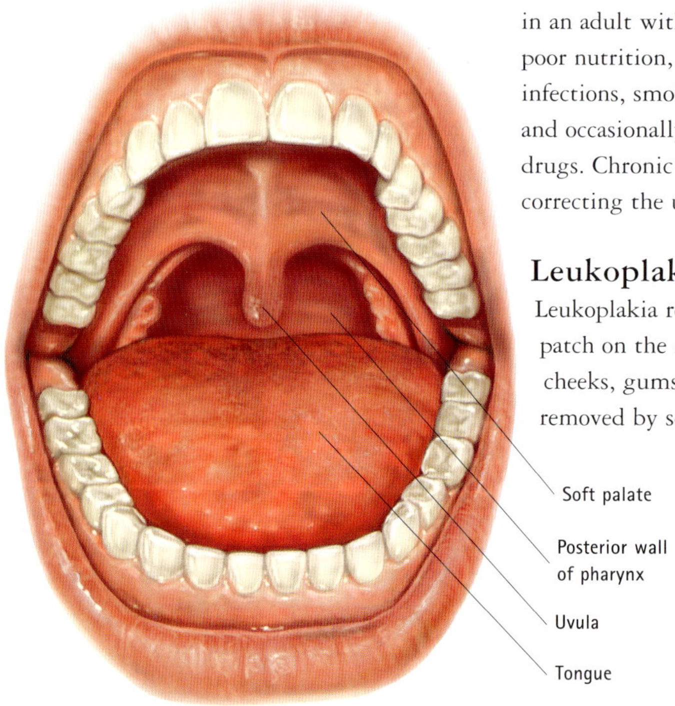

Glossitis

In glossitis the tongue becomes bright red and swollen.

Oral cancers are malignant squamous cell carcinomas. They begin as a small, painless lump or ulcer on the tongue, lips or elsewhere in the mouth. They grow and spread rapidly, ulcerating and bleeding, and as they become larger they cause difficulties in talking, chewing and swallowing.

The diagnosis is conformed by a biopsy of the cancer. Treatment is surgical removal, which may involve radical head and neck surgery if the cancer has spread to the lymph nodes in the neck. Surgery can sometimes be combined with radiation therapy. Early detection gives the best chances of a cure, but unfortunately most cancers are advanced by the time of diagnosis. About 50 percent of people with oral cancer will survive longer than 5 years.

Quinsy

Quinsy, also known as peritonsillar abscess, is a relatively uncommon infection of the tissue surrounding the tonsils that results in an abscess or collection of pus. Often caused by spreading infection from tonsillitis, its symptoms include fever, severe pain when swallowing, and difficulty in opening the mouth. The condition may be treated with antibiotics if caught in the early stages.

More severe conditions may require draining of the pus via a surgical incision. The infection can recur and if left untreated may spread to the mouth, neck, chest and lungs, causing life-threatening tissue swelling that can block airways. Quinsy most commonly affects older children and young adults.

Lip

Part of the face, the lips are the muscular borders of the mouth. They are covered in front with skin, and at the back by a mucous membrane that forms part of the lining of the oral cavity. The free edges of the lips are covered by a type of very thin skin called the vermilion border which, due to its dense nerve supply and thinness, is extremely sensitive to touch. The vermilion border itself is unpigmented, but blood from the blood vessels beneath shows through it, giving the lips their characteristic color.

The muscles of the lips are divided into two groups. The fibers of one group run concentrically around the mouth like purse strings and act to close it. The other muscles have fibers that run radially and are used to open the mouth. The lips serve to make a watertight and airtight closure of the mouth for drinking, swallowing and chewing. The muscular movements of the lips form an important part of the expressiveness of the face and, in infants, are essential for suckling. The lips and teeth form the embouchure (mouthpiece). The delicate surface of the border is especially liable to injury. Lips may also require sunscreen and moisturizers to protect against painful sunburn and chapping.

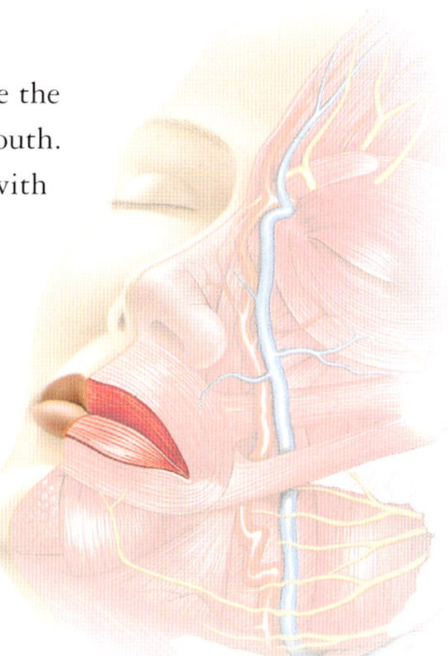

Lip

The lips play an essential role in speech by modulating sounds into recognizable words. They are an important part of the expressiveness of the face and contain many nerve endings, making them highly sensitive.

Nose

The nose is a part of the respiratory system. Its bony structure forms part of the skull.

The bones of the external nose consist, on each side, of a nasal bone and the maxilla. The framework of the nostrils is made of cartilage, while the nasal septum, which separates the nostrils in the midline, is part bone and part cartilage. Inside, the nose contains cavities that form part of the respiratory tract.

The nose serves to warm and humidify inhaled air, and to filter out particles of dust. Air is breathed into and out of the body through the nose, via the nostrils. The nostrils are guarded by hairs (known as vibrissae) whose function is to prevent entry by insects and larger particles of dust.

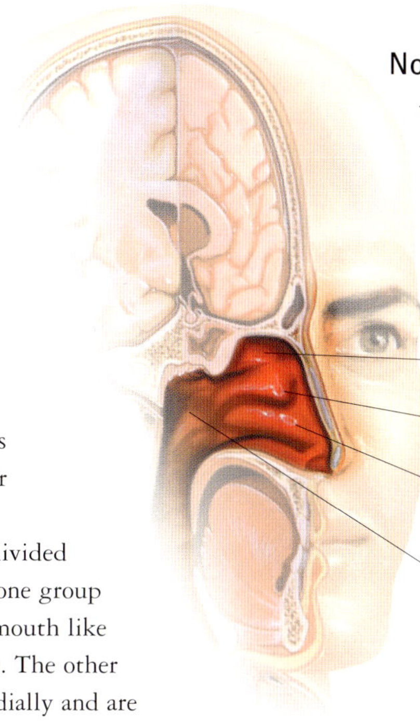

Nose

The main passageway for air entering the body is the nose. Air enters the nostrils and passes through the nasal cavities into the nasopharynx, trachea and down into the lungs.

The nostrils lead to the nasal cavities, one on each side, which in turn lead to the pharynx and thence to the voice box (larynx) and windpipe (trachea). The nasal cavities are also connected to the paranasal sinuses and receive drainage of tear fluid through the nasolacrimal ducts; the eustachian tubes connect the ears to the pharynx.

In each nasal cavity, three curved plates, called the conchae (turbinates), project from the side wall. These increase the surface area of the cavity, exposing inspired air to a greater amount of warm, moist surface due to a rich supply of blood vessels.

The bones surrounding the nasal cavities are the vomer and portions of the frontal, ethmoid, maxillary and sphenoid bones. The floor of the nasal cavities forms the roof of the palate. The two nasal cavities are separated by a partition called the nasal septum, which is commonly deviated to one side, thereby enlarging one nasal cavity at the expense of the other.

Most of each nasal cavity is lined by a specialized type of mucous membrane called respiratory mucosa. This is characterized by the presence of cilia, minute hairs that guard the nasal cavity, wafting foreign particles on a sheet of mucus toward the nasopharynx. In the upper region of each nasal cavity, the mucous membrane changes to olfactory mucosa, containing specialized nerve cells for the reception of smell. Glands of the mucous membrane produce a watery secretion that both protects the walls of the nasal cavities and is evaporated in order to humidify the inspired air.

Problems affecting the nose include the common cold (a viral infection predominantly affecting the nasal mucosa), hay fever (an allergic reaction to pollen particles in inhaled air) and bleeding. Besides injury, in most people nose bleeding may be associated with high blood pressure.

Smell

The sense of smell is one of the major senses. It responds to the chemical nature of airborne substances breathed into the nose, and also to odors from food and drink that reach the nose from the mouth and pharynx. It is extraordinarily sensitive to some volatile substances, such as methyl mercaptan, which is added to natural gas to give it an odor. Odors are sensed by special nerve cells in the lining of the nose. Sensations of smell are conveyed by the olfactory nerves to a part of the brain, lying above the nose, referred to as the rhinencephalon or limbic lobe. The senses of smell and taste act in unison to assess food.

On the roof of the nasal cavities, the mucous membrane contains specialized nerve cells for the reception of odors. This region of the mucous membrane is known as the olfactory epithelium or olfactory organ, and has a total area of about 1 square inch (5 square centimeters). The surface of the

Olfactory path

Odor molecules enter the nostrils and pass into the nasal cavity where they dissolve in olfactory mucosa in the nasal lining. They stimulate olfactory receptors which send signals to the olfactory bulb. From here, nerve signals travel along the olfactory tract to reach the olfactory cortex, the limbic system and the hypothalamus in the brain. This is where smells are identified and the body's response coordinated. Smell has the most direct pathway to the brain of all the senses.

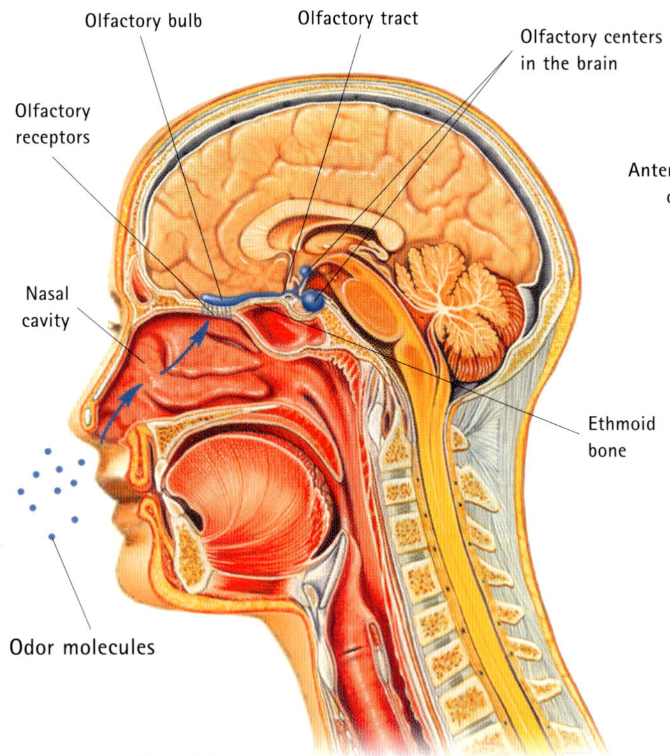

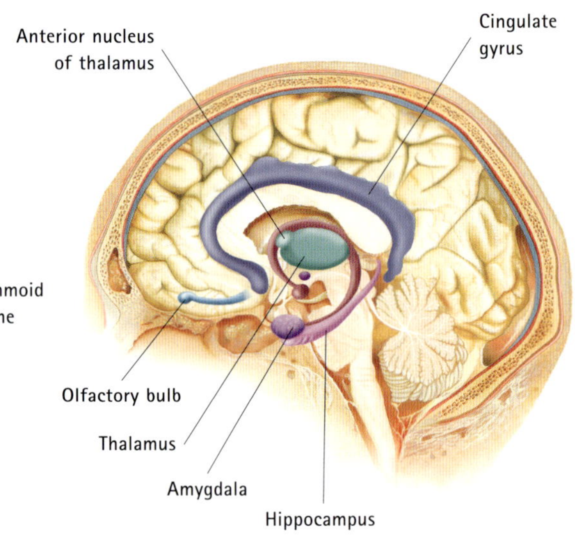

Smell and the limbic system

The olfactory bulbs are directly connected to the hippocampus and the amygdala in the limbic system, which is important for memory and emotion. This is why smells are evocative of past places and feelings. Some smells stimulate the limbic system to activate the hypothalamus and pituitary gland, which triggers the release of hormones associated with appetite and emotional responses including pleasure, fear, and sexual drive.

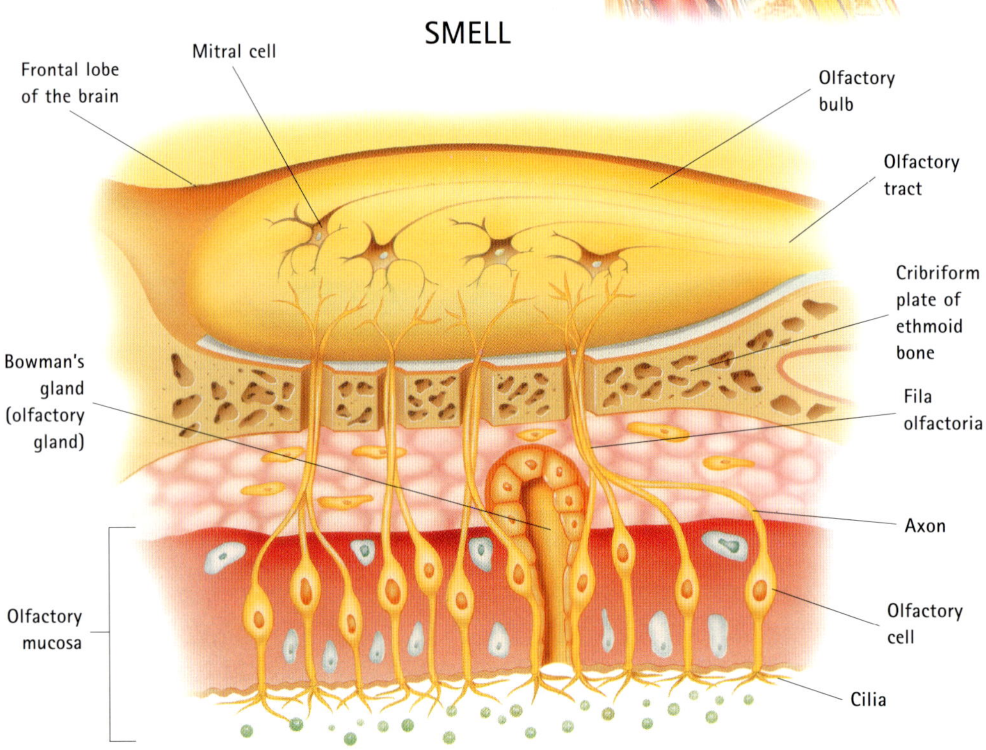

Olfactory apparatus

The olfactory receptors are located in the roof of the nasal cavity. A small area of mucous membrane (the olfactory epithelium) contains millions of nerve cells bearing cilia. When odor molecules are dissolved on the moist surface of the epithelium, they stimulate the nerve cells. Nerve impulses are transmitted along the nerve fibers through holes in the cribriform plate to the olfactory bulb under the frontal lobe of the brain.

epithelium is kept moist, to dissolve odors from passing air. The olfactory receptor cells, of which there are about 100 million, are modified neurons. A dendrite extends from each cell body toward the surface of the epithelium, where it terminates as a swelling termed an olfactory vesicle. This vesicle is specialized for the reception of smell.

The sensation is carried by an unmyelinated axon (nerve fiber) from each receptor cell into the connective tissue beneath the epithelium, where it joins with others to form bundles of olfactory nerve fibers, called the olfactory nerves. These are about 20 in number and pass through holes (foramina) in the cribriform plate of the ethmoid bone, to terminate in the brain at the olfactory bulbs.

The olfactory bulbs are ovoid structures forming forward extensions of the olfactory area of the brain. In the olfactory bulbs, axons from olfactory sensory cells converge and end (synapse) on dendrites of cells called mitral cells because of their shape (conical). Each mitral cell receives about 1,000 axons. These synapses between olfactory sensory cells and mitral cells are grouped to form conspicuous clusters known as glomeruli.

The axons of the mitral cells run back (posteriorly) to terminate in the cerebral cortex and adjacent parts of the forebrain, providing a pathway for conscious perception of smell. The pathway from the olfactory receptors to the cerebral cortex therefore has only one synapse (in the olfactory bulb), a more direct connection than that for any other type of sensation. Other cells (interneurons) in the olfactory bulb link one glomerulus with another, and also link the two olfactory bulbs across the midline. Axons from mitral cells and other cells of the olfactory bulb travel through the olfactory tract and terminate, either directly or through relay neurons, in two areas of the brain called the medial olfactory area and the lateral olfactory area.

Both the medial and lateral olfactory areas have neural connections to the hypothalamus, hippocampus and brain stem nuclei. These latter areas control automatic responses to smells, particularly feeding activities (such as salivation) and also emotional responses (such as pleasure, fear, and sexual drives). Loss of the sense of smell (anosmia) is most commonly due to blockage of the nose. Hallucinations of smell may be due to physical or psychological causes, including tumors of the temporal lobe.

SEE ALSO *Autonomic nervous system on page 75; Limbic system on page 138, Taste on page 184*

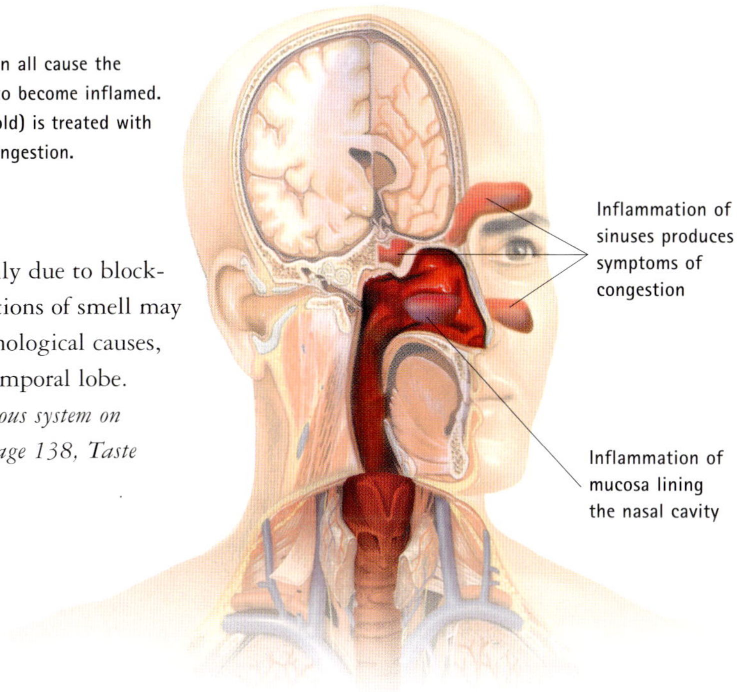

Rhinitis

Viruses, bacteria and allergens can all cause the mucous membranes of the nose to become inflamed. Acute rhinitis (or the common cold) is treated with decongestants to relieve nasal congestion.

DISORDERS OF THE NOSE AND SINUSES

The delicate mucous membranes of the nose are highly susceptible to irritation and inflammation, which cause excessive mucus production, Damage to the network of blood vessels within the nose often leads to nosebleed.

SEE ALSO *Allergies on page 60; Antihistamines on page 439*

Catarrh

Catarrh is a term generally used for an inflamed mucous membrane, which usually occurs either in the nose or in the throat, causing a discharge of mucus. It occurs as a symptom of various upper respiratory tract infections such as hay fever, rhinitis, laryngitis, or the common cold.

Rhinitis

Rhinitis is the inflammation of the mucous membranes lining the nose and is usually accompanied by an excessive production of mucus, which may be watery or thick. There may be difficulty breathing through airways which are restricted by swelling and mucus.

Infection with the common cold or other viruses, bacterial infections, irritation by smoke or airborne pollutants, and allergic reactions can all produce similar symptoms.

Symptoms can be treated with nasal sprays or oral drugs, which constrict the blood vessels in the swollen membranes and return them to near normal size, freeing up the airways. These drugs should be used for a few days only as their prolonged use may produce symptoms of congestion. If the mucous membrane is infected, nasal mucus

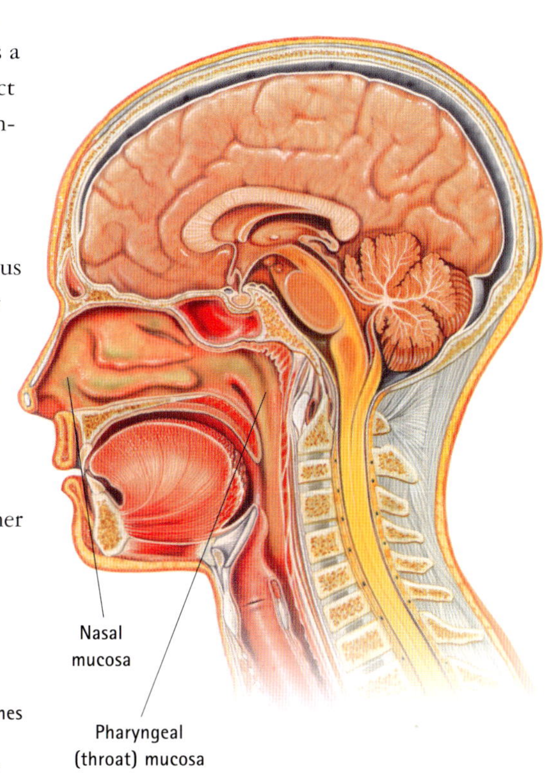

Catarrh

Catarrh is an inflammation of the mucous membranes of the nose or throat, leading to an increased discharge of mucus (runny nose, mucus in the throat).

may be yellow and pus-like and the nose will bleed easily. Antibiotics may be prescribed, depending on the cause.

Allergic rhinitis results from inhaling a substance to which the individual has been sensitized, whether or not there has been a previous allergic reaction. There is increased production of mucus, and there may also be redness and irritation of the eyes. Severe reactions can bring feelings of tightness in the chest and difficulty breathing. These changes are caused by the body's production of histamine, a substance used to help fight invading organisms.

Medication with antihistamines may counter the immediate symptoms; for serious reactions, corticosteroids may be needed.

Sinusitis

The paranasal sinuses are cavities in bones around the nose and eyes. These cavities are lined with mucous membrane. Under normal conditions mucus moves steadily from the sinuses into the nose. In sinusitis, the sinuses become inflamed and fill with mucus or pus, causing headaches, facial tenderness and minor breathing difficulties.

There are two types of sinusitis disorders, which affect 35 million people in the USA alone. Acute sinusitis is caused by a disorder that causes swelling of the membranes of the nose, such as a viral respiratory infection or allergic rhinitis. The swelling prevents fluid from draining out of the sinus normally, and infection with viruses, bacteria or fungi then follows. Swimming or immersion of the head in water may allow water and bacteria to enter the sinus, causing irritation and infection. Less commonly, dental infections such as a tooth abscess may infect the sinus.

One of the symptoms of sinusitis is headache, the location of which depends on the sinus(es) involved. There may also be pain in the front of the head or around the eyes, the forehead or cheeks, or in the roof of the mouth or teeth. The pain results from the accumulation of undrained fluid, which causes pressure within the sinus. There is often a thick yellow or yellow green nasal discharge and there may be fever and chills.

If the condition is due to bacterial infection, sinusitis is treated with antibiotics. Oral or nasal decongestants may help the sinuses to drain, although the use of nasal drops or sprays should be temporary, since long-term use can cause damage to the nasal lining.

If sinusitis is persistent or recurrent the condition is known as chronic sinusitis. It is less common than acute sinusitis and may be caused by a deviated nasal septum or other obstruction of the nose. Chronic sinusitis can also be treated with antibiotics (if infection) and steroid or cromoglycate nasal sprays (if allergic).

Hay fever

Hay fever or seasonal allergic rhinitis (also sometimes called pollinosis) is caused by extreme sensitivity to airborne pollen, and is common through spring and summer. Hay fever may affect up to 10 percent of the population and tends to run in families, as do many allergies. It can develop at any age. Pollen is a fine powdery yellow dust, the

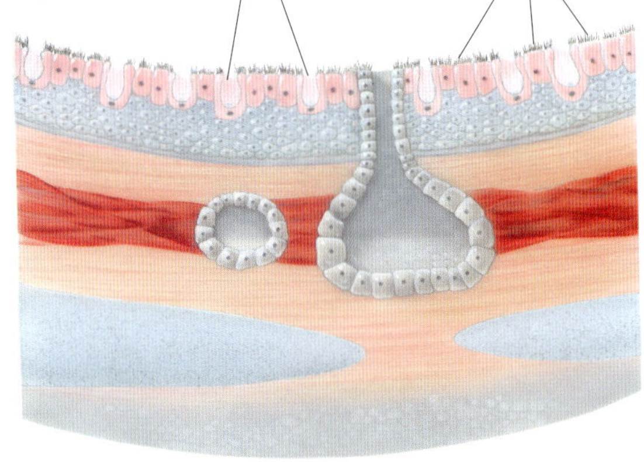

Mucous membrane

The lining of the nasal passages, trachea and lungs is covered in a sensitive mucosal lining that intercepts airborne particles. When pollen is inhaled and trapped in this lining it can trigger an allergic reaction—causing inflammation and mucus production.

HAY FEVER

Sensitivity to pollen in the air, particularly during spring, triggers an allergic reaction in those who suffer from hay fever. This often results in blocked sinuses, a runny nose, red eyes and sneezing.

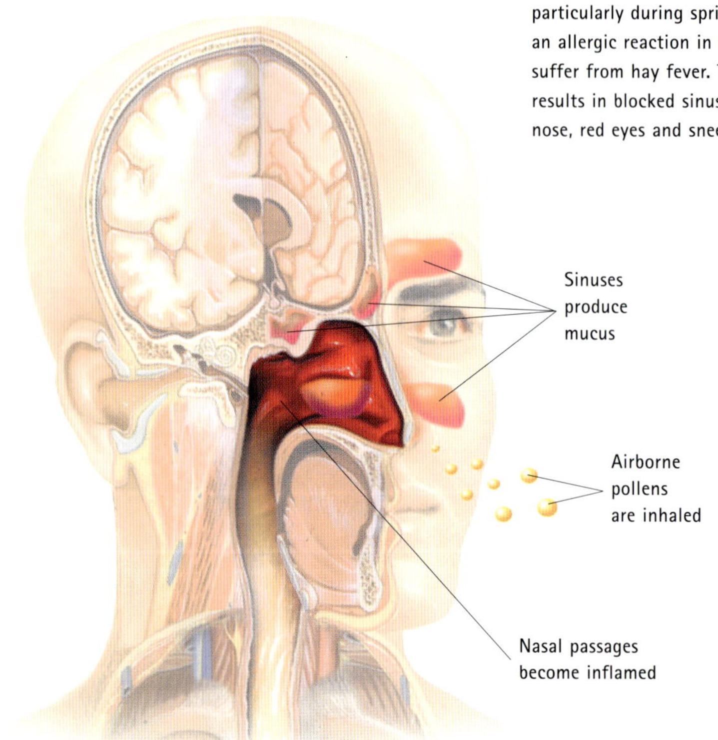

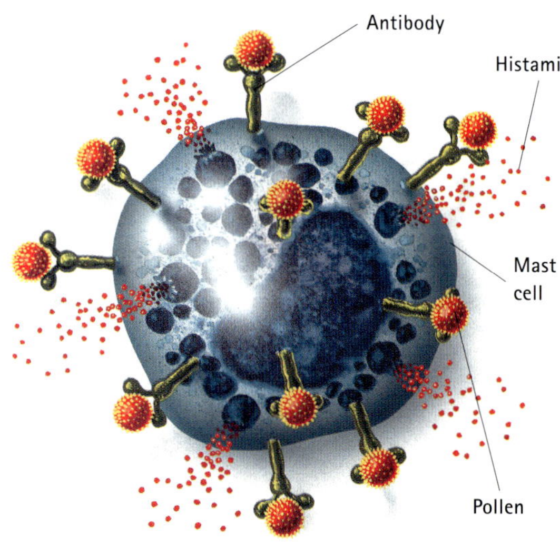

Allergic response

Each time pollens (allergens) enter the body of a hay fever sufferer, they are captured by antibodies attached to mast cells. These cells release histamine—an inflammatory substance which produces the symptoms of hay fever.

fertilizing agent of plants. Airborne pollen from grasses, weeds and trees is the major cause of hay fever. A mixture of pollen with pollution from vehicle exhaust gases can produce irritation similar to hay fever.

There are many airborne substances that can trigger hay fever. Besides pollen, molds and fungal spores can cause or worsen seasonal symptoms, and most are light enough to be carried hundreds of miles if the wind is right. This is what makes airborne allergens almost impossible to avoid. If you suffer from hay fever it helps to know which substances you are most sensitive to, and this can be established by observation and through specialized tests.

It is almost impossible to eliminate contact with airborne allergens in your local area, but avoiding heavy concentrations of whatever you are most sensitive to will certainly limit the amount of the allergens you are exposed to, which can decrease the severity of your allergic reaction.

When pollen or dust is inhaled, it is trapped in the lining of the trachea (windpipe), triggering a reaction in a sensitized or allergic person. The release of histamines then causes typical signs of hay fever: a runny nose, red and swollen mucous membranes in the nose and eyes, and sneezing.

Symptoms present at times when pollens are not may be caused by sensitivity to household dust mites, or animal hairs. This condition is called perennial rhinitis. Antihistamines (commonly used to control hay fever) block the action of histamine, which is responsible for the runny nose and other symptoms. The side effects of antihistamines include drowsiness, nausea and dryness of the mouth. Some of the newer antihistamines do not cause drowsiness; these can be used under medical supervision.

If the condition is severe, surgery may be required to control it. This may involve unblocking the sinus, or repairing a deviated septum or nasal obstruction.

Nasal polyps

Nasal polyps are caused by chronic infection or allergy in the nose (allergic rhinitis). They cause a nasal discharge and chronic stuffiness. Nasal polyps are easily removed with minor surgery under local anesthesia. They may also be treated with oral steroids.

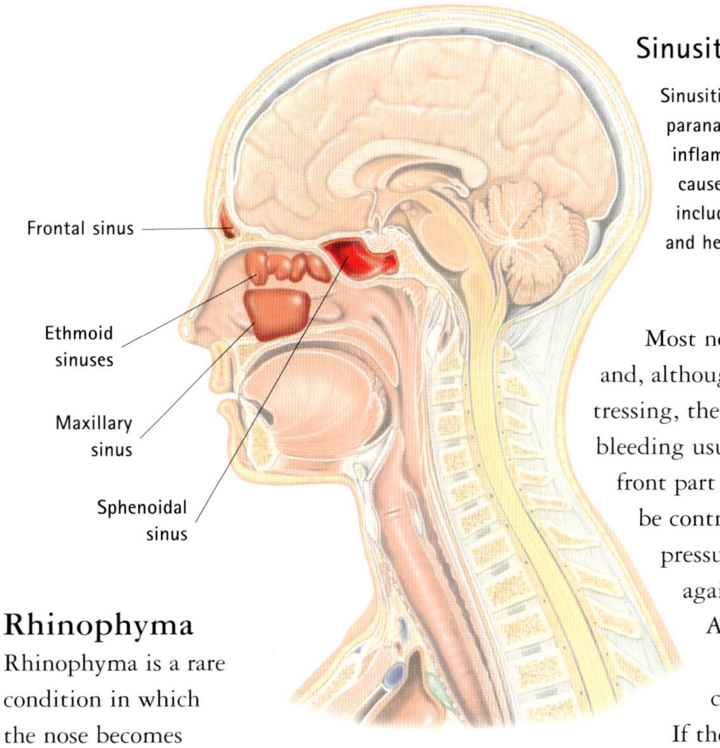

Sinusitis

Sinusitis occurs when the paranasal sinuses become inflamed. The mucus build-up causes a number of complaints including breathing difficulty and headaches.

Rhinophyma

Rhinophyma is a rare condition in which the nose becomes enlarged and red, with thickened skin and the appearance of veins near the surface. The nose takes on a bulb-like shape and oil-producing glands become enlarged. Once thought to be the result of excessive alcohol consumption, rhinophyma strikes just as many non-drinkers. It usually affects men and is associated with the skin disease acne rosacea. While antibiotics may be successful in reducing symptoms in the early stages, surgical reshaping of the nose is the only other treatment.

Nosebleed

A nosebleed (known as epistaxis) is common, and is due to rupture of the delicate blood vessels just beneath the mucous membrane covering the nasal septum. In this position the blood vessels are susceptible to damage from fingers or trauma.

Nosebleeds are common in childhood and are often related to upper respiratory infection, which results in sneezing or vigorous noseblowing. In children, the possibility of a foreign object in the nose needs to be considered.

Nosebleeds can also be caused by serious medical conditions. They can result from clotting disorders, similar to hemophilia, or a low number of platelet cells. Blood-thinning drugs, nasal polyps, tumors, or, occasionally, abnormally fragile blood vessels are other causes. Some believe that high blood pressure can cause nosebleeds.

Most nosebleeds are minor and, although alarming and distressing, they are harmless. The bleeding usually comes from the front part of the nose and can be controlled by firm finger pressure on the nostril against the nasal septum. An icepack over the bridge of the nose can be helpful also.

If the bleeding is severe or is not being controlled satisfactorily with these simple measures, medical attention must be sought. Such bleeding is treated with nasal packing or special balloon catheters, and the cause is then investigated.

Sneezing

Sneezing (sometimes also called a sternutation) is a sudden, noisy and spasmic exhalation of air through the nose and mouth. A sneeze is an involuntary reflex action triggered by stimulation of nerves in the mucous membranes of the nose or by overstimulation of the optic nerve by very bright light. Sneezing often accompanies respiratory infections such as the common cold where nasal membranes become swollen and inflamed. Inhaling tiny foreign particles, such as specks of ground pepper or talcum powder, can also lead to the sort of nasal irritation that elicits sneezing.

Allergens, such as mold or pollen, are among the most common causes of persistent sneezing. For people suffering from protracted allergy-related sneezing, there are several over-the-counter medications that may offer relief, such as antihistamines, decongestants and nasal sprays.

Interrupting or suppressing a sneeze is dangerous and can result in damage to abdominal muscles or to the middle ear as air is forced up the eustachian tube. Very occassionally, the pressure that results from stifling a sneeze can rupture an eardrum.

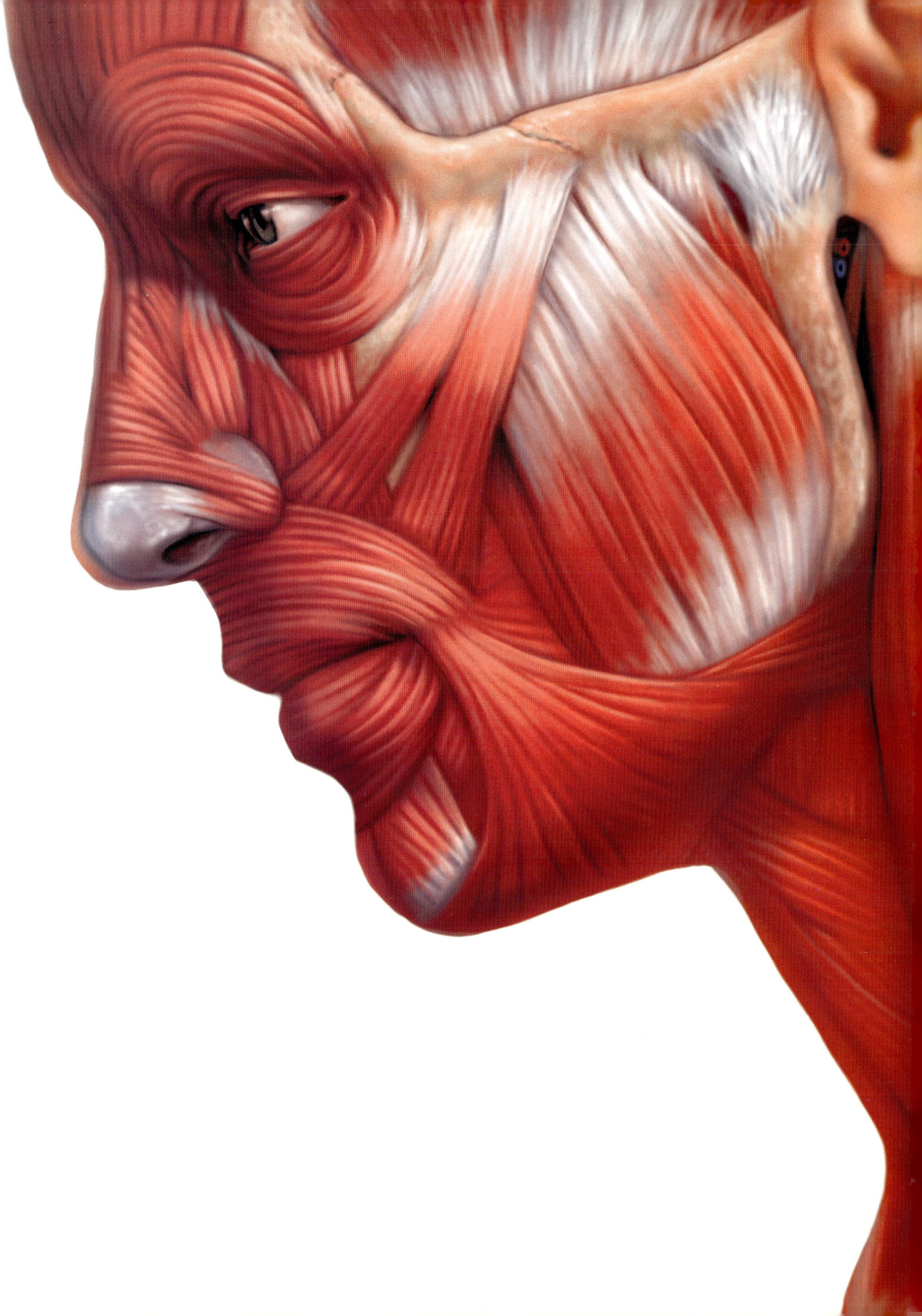

The Neck

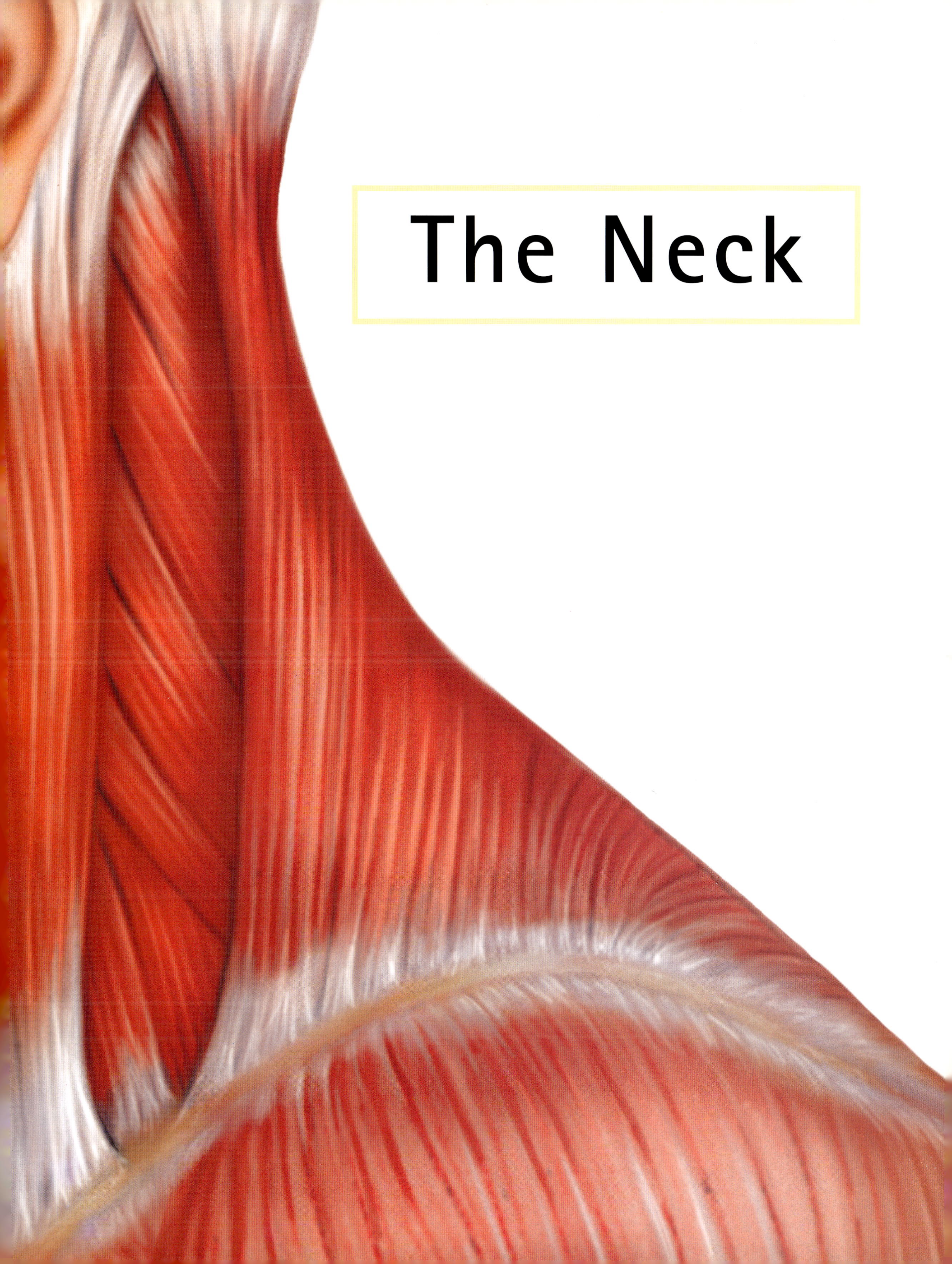

Musculoskeletal column

The neck supports and provides mobility for the head and contains a lage number of important structures in a relatively confined space: the spinal cord, protected by the vertebrae; major blood vessels to the brain and face; and passageways for food and air. Important nerves course through the neck and some arise from the neck region.

The neck can be divided into two major columns. At the back, the nuchal region comprises the vertebrae of the neck (cervical vertebrae) and their supporting musculature. In the front is a "visceral" column containing the larynx and trachea and behind them the pharynx and esophagus.

Surrounding the muscles and visceral structures in the neck are a series of connective tissue sheaths that can affect the spread of infection.

Compared to the thoracic and lumbar vertebrae found lower in the spine, the vertebrae in the neck are small as they carry very little weight, but the opening in the vertebrae for the spinal cord is relatively large. In all, there are seven cervical vertebrae, the upper two of which are specialized.

The first vertebra, the atlas, is a bony ring, and forms two joints (atlanto-occipital joints) with the base of the skull, allowing a nodding action.

The next vertebra, the axis, has an upward projecting bony element that forms a joint in the midline with the atlas. This, together with a joint on either side, allows a pivoting movement to occur, as in shaking the head when saying no. About 45° of rotation in the neck occurs at these joints alone (the atlantoaxial joint).

The remaining vertebrae have a typical vertebral pattern with a body in front, a bony arch behind and spines projecting backward (spinous processes) and to the sides (transverse processes).

The intervertebral disks that separate and cushion neighboring vertebrae from each other are relatively thick in the cervical region of the spine, permitting great freedom of movement.

SEE ALSO *Muscular system on page 48; Skeletal system on page 30; Spinal cord on page 217; Spine on page 211*

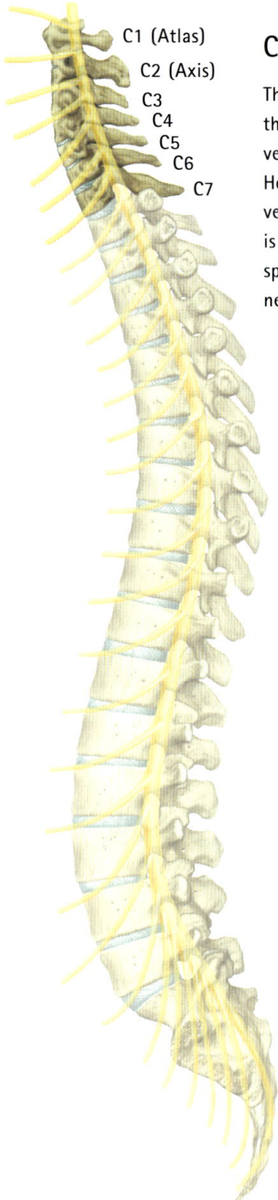

Cervical vertebrae

The 7 cervical vertebrae are smaller than the thoracic and lumbar vertebrae as they carry less weight. However, the opening in the vertebrae (the vertebral foramen) is comparatively large because the spinal cord is relatively thick at the neck and also in the lumbar regions.

Neck column

The front of the neck houses several important structures in a relatively small area. These include the larynx (voice box), the upper part of the trachea (windpipe), the thyroid gland, and the pharynx (throat) which leads to the esophagus. Running beside these are large blood vessels, one of which is the carotid artery.

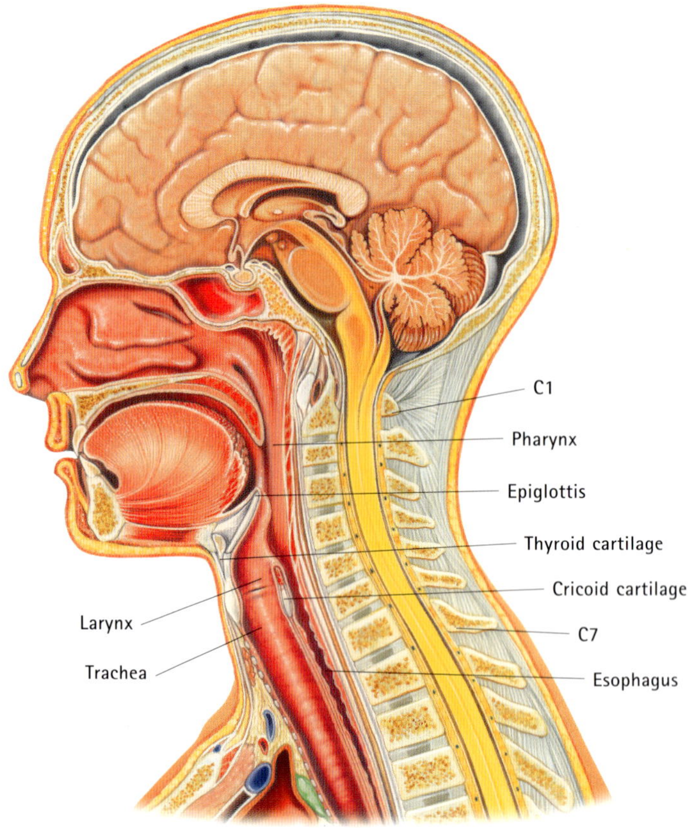

THE NECK

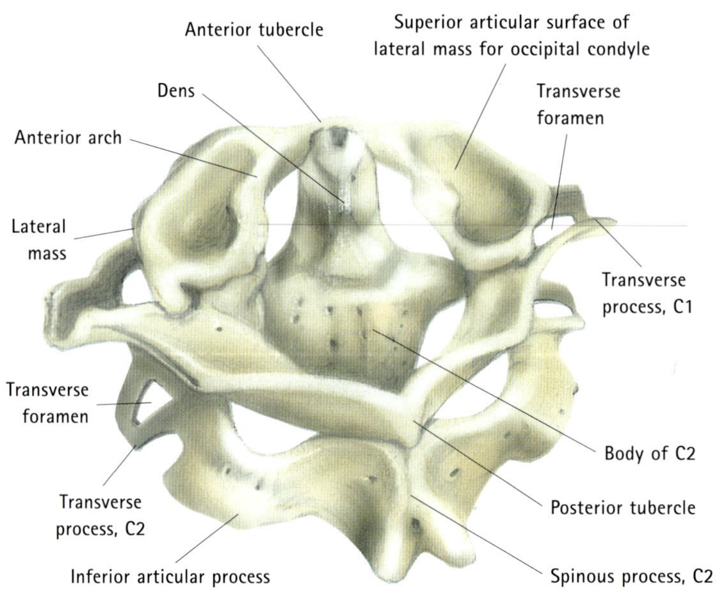

Atlas and axis

There are 7 vertebrae in the neck—the first 2 are more specialized than the others. C1, the atlas, supports the head and allows it to have a nodding movement. C2, the axis, articulates with the atlas and gives the head its ability to rotate.

BLOOD VESSELS

A feature of cervical vertebrae is that an artery (vertebral artery) passes through openings in the bony transverse processes. It ascends on the left and right, protected by the vertebrae, and supplies blood to the lower parts of the brain (cerebellum, brain stem and lower posterior part of the cerebral hemispheres).

MUSCLES

Muscles attach to the front, back, and sides of the vertebrae, producing forward, backward, and sideways movements. Those with an oblique orientation also produce rotation (turning). The largest musculature lies to the back. Some of these muscles are exclusively related to moving the head and neck (for example, splenius capitis and cervicis, semispinalis capitis and cervicis), while other muscles are related to moving the shoulder (for example, trapezius, levator scapulae) or raising the upper two ribs (the scalene muscles).

NERVES

The spinal cord in the neck region gives origin to nerves (cervical nerves C1–C8) that pass to the skin and muscles of the neck (C1–C4) and upper limb (C5–C8). They pass out through openings between the vertebrae (intervertebral foramen) and can be compressed here by bony outgrowths of the vertebrae (osteophytes), giving rise to symptoms such as tingling, numbness and weakness in the arm and hand.

The phrenic nerve is formed in the neck. It supplies the diaphragm, the muscle that contracts on inspiration. Damage to one or both phrenic nerves will usually affect breathing. Fracture of the cervical vertebrae may affect the spinal cord and/or cervical nerves. If damage occurs at spinal cord level C4 or higher it can be fatal, because the diaphragm is paralyzed.

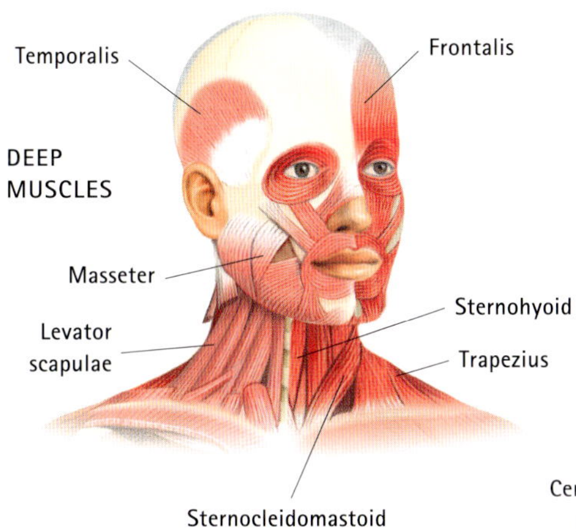

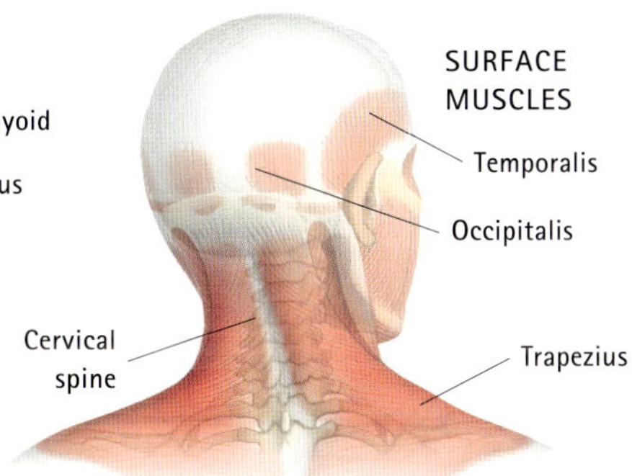

Musculoskeletal column

The various muscles of the neck have fibers arranged in a number of different directions, giving the head and upper spine great freedom of movement.

Visceral column

At the angle in the upper part of the front of the neck is a small, U-shaped bone called the hyoid bone and, suspended from this, the voice box (larynx). The larynx is part of the upper respiratory tract, and a cartilaginous framework serves to keep the airway open. The epiglottis, a piece of flexible cartilage at the inlet to the larynx, helps to prevent food and fluids from entering the airway when swallowing. The vocal cords, or vocal folds, are located within the larynx and control the size of the aperture between them (the rima glottidis). The folds lie close together and vibrate during speech, move wide apart in deep breathing, and may stop the flow of air completely (for example, when holding the breath).

The trachea commences at the lower end of the larynx and descends into the thoracic cavity to connect with the lungs. It can be

Neck cross-section

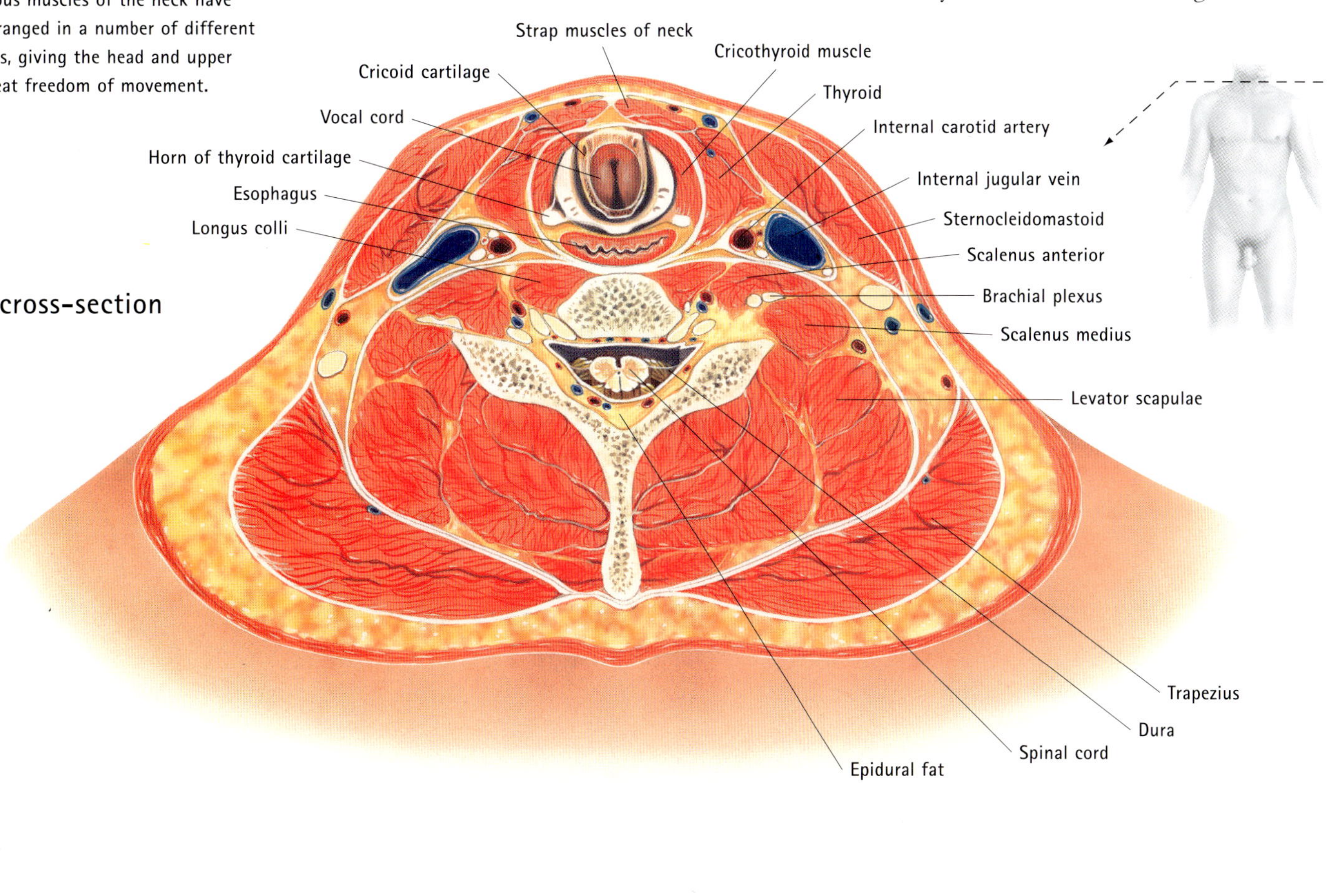

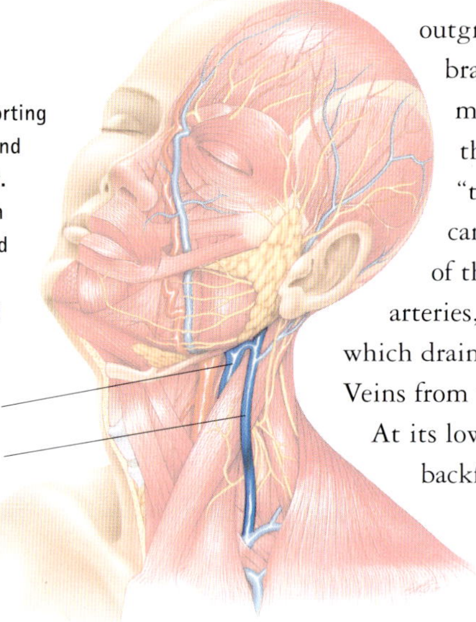

Jugular vein

The jugular veins are responsible for transporting blood from the head and neck back to the heart. Severing a jugular vein will lead to rapid blood loss and death if the wound is not attended to immediately.

felt, and moved from side to side, in the lower part of the neck. Wrapping around the upper trachea and extending onto the sides of the larynx and trachea is the thyroid gland. It is bound to the larynx and trachea by connective tissue and so moves with them on swallowing, a feature that allows an enlargement of the thyroid gland to be distinguished from other swellings in the neck (such as an enlarged lymph node, which does not move with swallowing).

The pharynx lies behind the larynx. It is connected below to the esophagus, which lies directly behind the trachea, and empties into it when food is swallowed. The pharynx and esophagus have muscular walls that contract in a milking action to move food toward the stomach.

SEE ALSO *Digestive system on page 101; Endocrine system on page 110; Respiratory system on page 95; Speech on page 187*

BLOOD VESSELS

On either side of the major structures of the visceral column—the pharynx, larynx, trachea and esophagus—are large blood vessels bound together by a connective tissue sheath (carotid sheath). Within the sheath, in the lower part of the neck, is the common carotid artery, which divides into internal and external carotid arteries at about the level of the laryngeal prominence. The internal carotid artery supplies the brain and the external carotid artery supplies the face and neck.

The pulse of the common carotid artery may be felt at the side of the larynx, and can be compressed against a bony outgrowth of the sixth cervical vertebrae (the carotid tubercle), stemming its flow. Carotid comes from the Greek word *karoo*, meaning "to put to sleep." Within the carotid sheath, on the outer aspect of the internal and common carotid arteries, is the internal jugular vein, which drains the blood from the brain. Veins from the face and neck drain into it. At its lower end is a valve that prevents backflow of blood toward the brain.

MUSCLES

In front of the visceral tubes and thyroid gland are thin strap-like muscles (infrahyoid muscles) that move the larynx and hyoid bones. Attached to the skin of the neck is a wide thin muscle, the platysma muscle, that is involved in facial expression, and tenses in anger. A large muscle that extends from the skull to the clavicle and sternum is the sternocleidomastoid muscle. It contracts when one lifts the head, and can be subject to spasm. The spinal nerves which supply the arm (C5 to T1) come together in the brachial plexus and separate into the nerves of the arm. They then pass obliquely down and outward, just below the skin, in the angle between this muscle and the clavicle. The outer margin of the neck is defined by the trapezius muscle, and the degree of development (bulk) of this muscle, can affect the shape and appearance of the neck.

NERVES

Behind and between the blood vessels and within the sheath is the vagus nerve, an important nerve supplying the larynx, trachea, pharynx, esophagus, heart, lungs and much of the gastrointestinal tract. Branches of the vagus (external laryngeal nerve and recurrent laryngeal nerve) pass to muscles controlling the vocal cords and may be damaged in thyroid surgery.

Throat

The throat is the front portion of the neck. Within it are the fauces, the opening that leads from the back of the mouth into the pharynx, and the pharynx itself, the cavity that connects the mouth, nose and larynx, which is situated behind the arch at the back of the mouth.

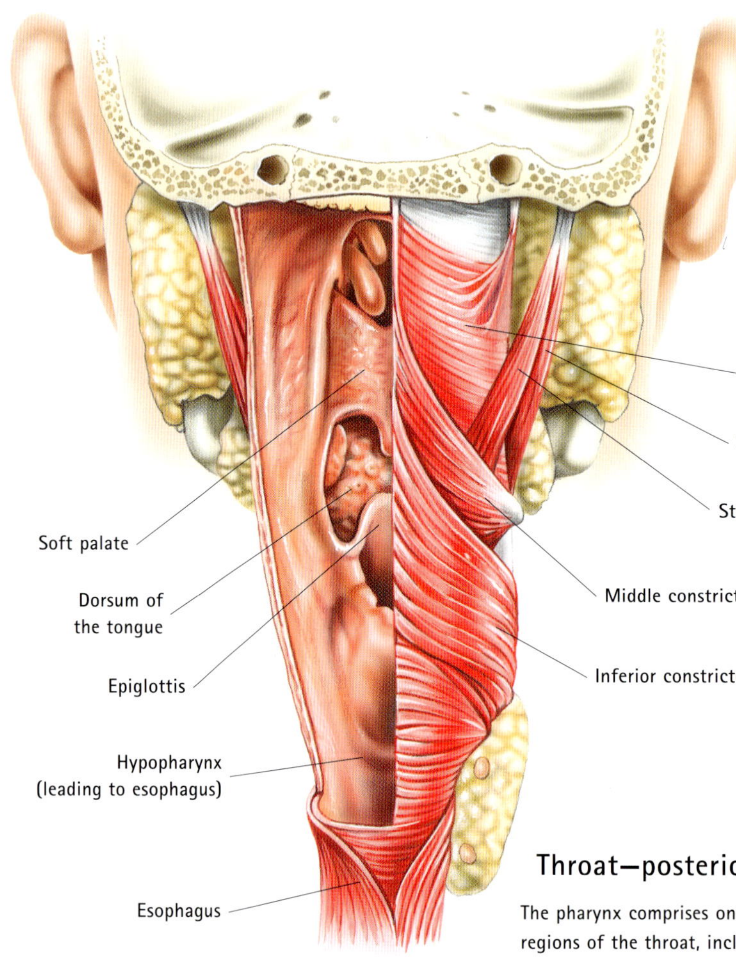

Throat—posterior view

The pharynx comprises one of the most important regions of the throat, including structures used in the processes of breathing, speaking and swallowing.

Pharynx

The pharynx is the common passageway for air, fluids and food entering the body. It comprises the nasopharynx (behind the nose), the oropharynx (behind the mouth) and the laryngopharynx (the voice box). The muscles that surround the pharynx are used in speech and swallowing.

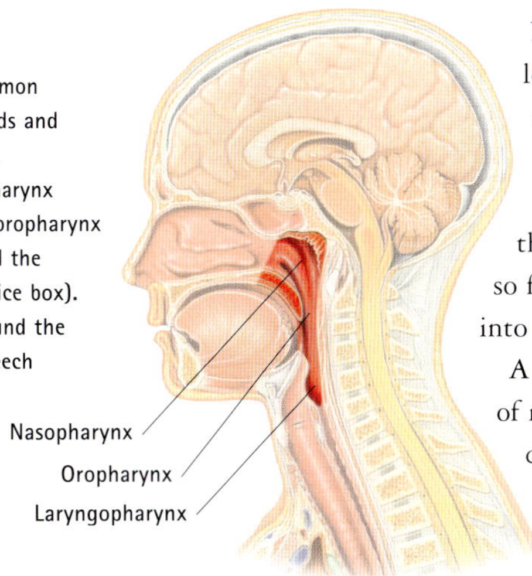

Pharynx

The pharynx is a vertically elongated tube that lies behind the nose, mouth and voice box (larynx). The pharynx has openings to all three of these regions and is a common passage for air, water and food. The pharynx is divided into three parts. From the top down, these are the nasopharynx, the oropharynx and the laryngopharynx.

The nasal part of the pharynx, or nasopharynx, is the space just above the soft palate that joins with the back of the nose. It contains the adenoids and the openings of the eustachian tubes on each side (which lead to the middle ear). The oropharynx lies at the back of the mouth and contains the tonsils (one on each side) and the back of the tongue. The laryngeal pharynx connects the back of the throat to the voice box, or larynx, and the esophagus (gullet). At the top of the laryngeal pharynx is the epiglottis, a flap of tissue that lies just behind the base of the tongue. From the epiglottis the laryngeal pharynx leads downward to the esophagus. A separate passageway leads to the larynx.

During the action of swallowing, muscles in the walls of the throat lift the pharynx, pushing food down to the esophagus, and closing the epiglottis over the trachea so food and liquids do not pass into the trachea.

A sore throat is a symptom of many disorders, including colds, diphtheria, influenza, laryngitis, pharyngitis, tonsillitis, measles, and infectious mononucleosis (glandular fever). Sore throats may be caused by bacteria such as *Streptococcus*, but most are caused by viruses; therefore, treating all sore throats with antibiotics (which do not cure viral infections) is inappropriate.

Larynx

The larynx (also called the voice box) is the part of the throat that leads from the pharynx to the trachea (or windpipe) and lungs. The larynx serves two main functions: to protect the airway to the lungs from inhalation of food and water; and to produce a source of air vibration for the voice. The larynx is composed of nine cartilages which provide strength for the airway and attachments for the various muscles, ligaments and membranes of the larynx.

The uppermost cartilage is the epiglottis, which lies immediately behind and below the tongue. The largest cartilage of the larynx can be felt at the front of the throat and is called the thyroid cartilage, because the thyroid gland lies in front of its lower part. This cartilage consists of two slightly curved plates, which meet in the midline at a prominent ridge. Above the ridge lies a notch, called the thyroid notch.

In males, the thyroid cartilage grows rapidly after puberty, resulting in an increased prominence of the ridge and notch of the thyroid cartilage (the Adam's apple). This enlargement elongates the vocal ligament,

Larynx—posterior view

The larynx is a triangular box composed of nine cartilages that are joined by ligaments and controlled by skeletal muscles. It serves as a passageway for air between the pharynx and the trachea, and provides a framework for the vocal folds (vocal cords). Muscles in the larynx close the air passage while food is pushed into the esophagus.

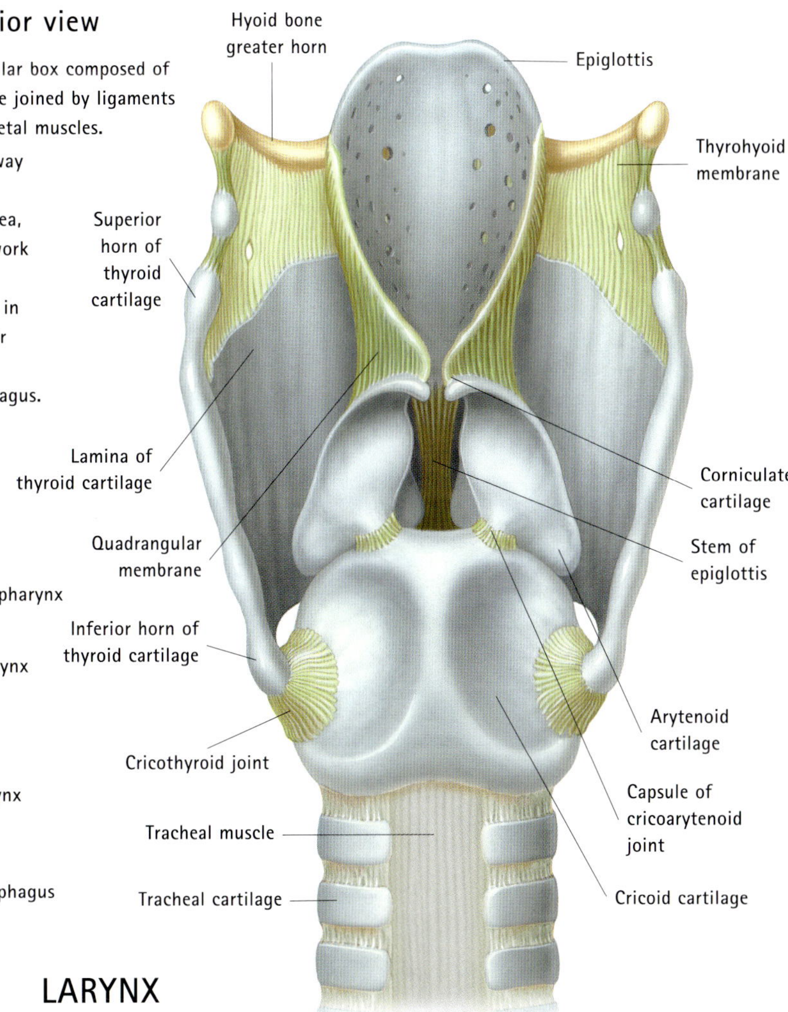

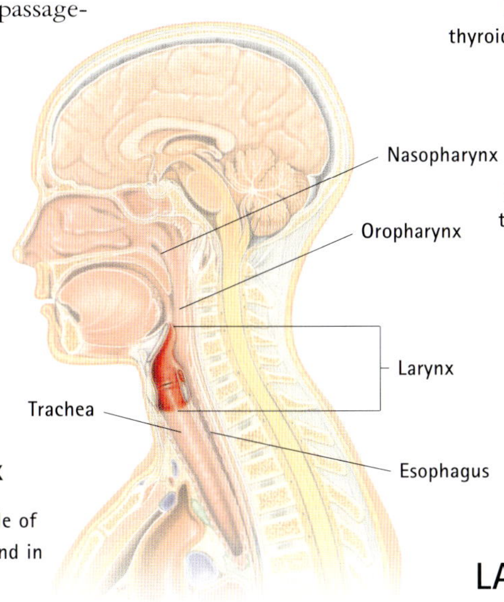

Position of the larynx

This passageway lies in the middle of the neck below the hyoid bone and in front of the laryngopharynx.

LARYNX

FUNCTIONS OF THE LARYNX

Swallowing

During swallowing the epiglottis folds down over the glottis to prevent food and drink passing into the airway.

Speaking

Exhaled air flowing through the larynx vibrates the vocal folds (vocal cords), producing sound. The tension and length of the cords determines the pitch of the sound.

Breathing

In breathing, the vocal folds are moved apart by laryngeal muscles.

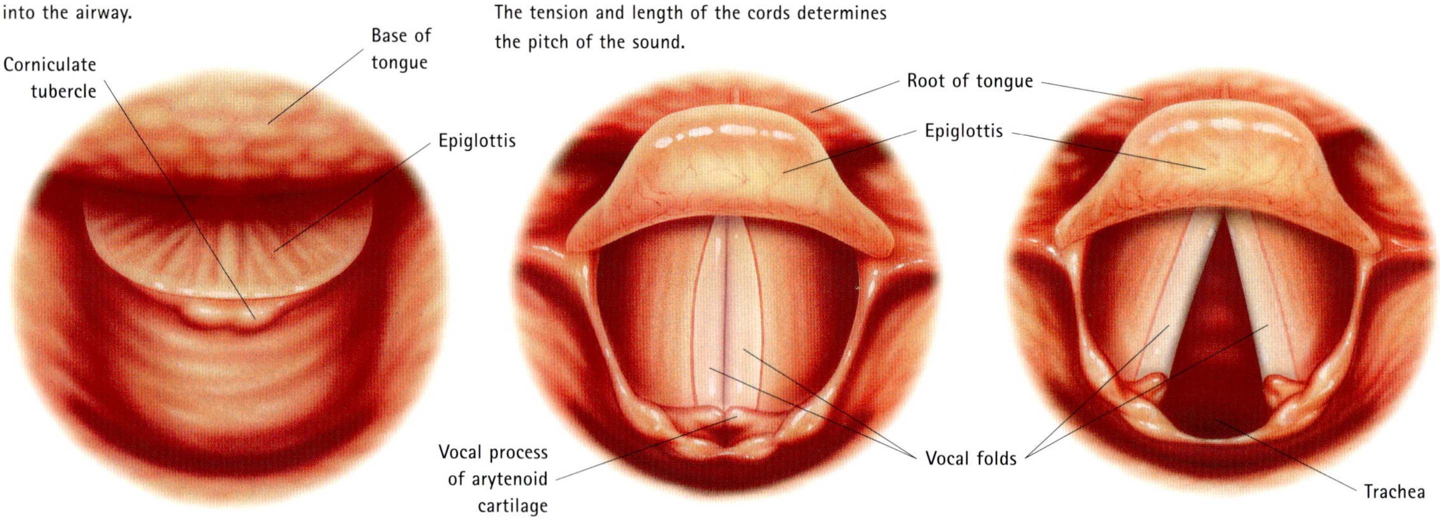

which is attached to the back of the thyroid cartilage and vibrates to produce the sound of the voice. A longer vocal ligament vibrates with a lower frequency (pitch), thus making men's voices lower in pitch than those of women and children.

Other cartilages of the larynx include the cricoid, which encircles the airway; the arytenoids, which are paired cartilages; and the tiny corniculate and cuneiform cartilages, which strengthen the folds of membrane around the laryngeal entrance.

Voice production depends on several key elements. The first is that the column of air above the larynx must be set vibrating. The larynx contains the vocal cords, which are the source of the sounds we produce. The vocal cords consist of a tent-shaped fold that extends upward on each side from the cricoid cartilage to attach to the thyroid and arytenoid cartilages. The paired upper free edges of this membrane (vocal ligaments) are covered by mucous membranes to form paired vocal folds. When these folds are brought together across the larynx, they can be set vibrating by forcing exhaled air between them, thus producing sound waves.

In speech, sound is modified by the complex action of muscles in the throat, larynx and mouth. The length and tension of the vocal ligaments can be adjusted to produce sounds of different pitch. Volume is determined by the force used when breathing air through the vocal cords.

Epiglottis

The epiglottis is a leaf-shaped flap of tissue in the throat that lies just behind the base of the tongue and over the opening of the voice box (larynx) and windpipe (trachea). It closes off the trachea when swallowing, preventing food and liquids from accidentally passing into the trachea, directing them instead into the esophagus.

Trachea

The windpipe (trachea), which measures approximately 3½–5 inches (9–12 centimeters) long and ⅝ inch (1.5 centimeters) wide, is a tube for the passage of air. It begins at the lower end of the voice box (larynx), and passes into the thoracic cavity where it terminates by dividing into the left and right main bronchi.

It is a fibro-elastic and muscular structure, reinforced by U-shaped cartilages. The cartilages prevent collapse of the airway, and the elastic fibers allow it to stretch and recoil with movements of the larynx (used in swallowing and speech) and diaphragm (used in breathing). The back of the trachea is flat. Here, the ends of the cartilage are bridged by transversely oriented muscle (trachealis muscle), whose contractions reduce the diameter of the airway. The esophagus lies against this surface and expands into the gap in the cartilage when food is swallowed.

The mucous membrane lining the trachea traps dust particles. Hair-like projections on the membrane (cilia), move the dust-laden mucus toward the throat where it may be swallowed or spat out. This clearing activity of the cilia is inhibited by smoking.

Since the larynx and trachea are the only air passage to the lungs, obstruction of the larynx (due to an allergic reaction, for example) may require surgical opening of the trachea (tracheostomy).

The cavity, or lumen, of the trachea is very small in infants and inflammation of the trachea (tracheitis) can cause severe breathing difficulties, and can result in a serious medical emergency.

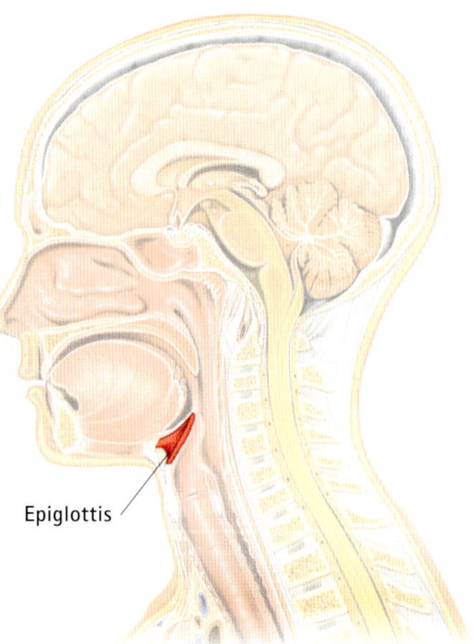

Epiglottis

The epiglottis is a lid made from cartilage that covers the entrance to the larynx and trachea.

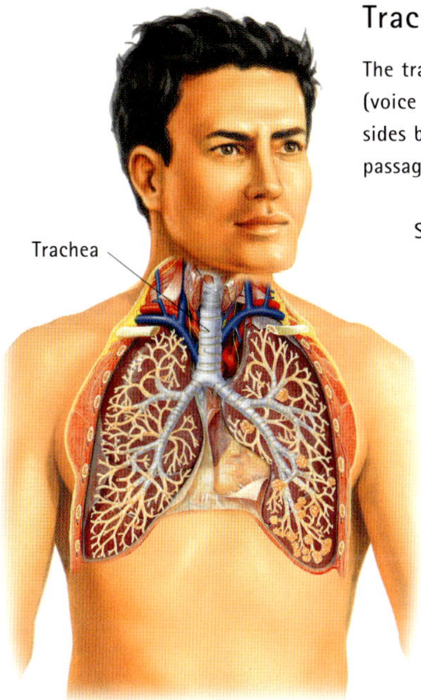

Trachea

The trachea (windpipe) is the air passage that connects the larynx (voice box) and the two bronchi. It is reinforced at the front and sides by a series of C-shaped rings of cartilage, which keep the passageway open. Gaps between the rings are occupied by muscle.

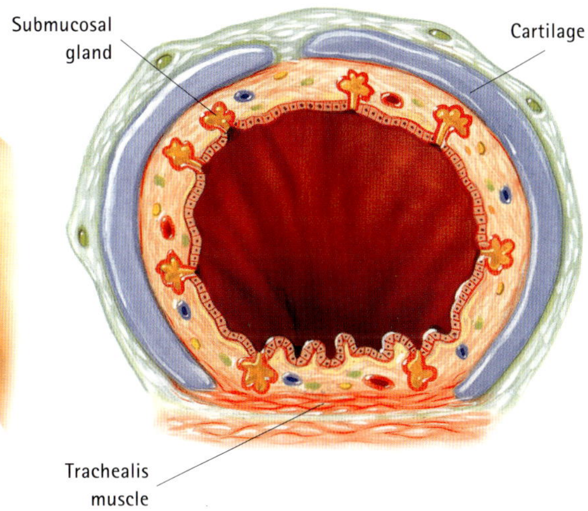

Disorders of the Neck

Disorders of the neck most commonly involve the cervical spine and supporting ligaments and musculature.

SEE ALSO *Laryngectomy on page 462; Tracheostomy on page 462; Treating the musculoskeletal system on page 444*

Whiplash

The neck is easily traumatized in whiplash injuries. Damage to ligaments, muscles and joints caused by the rapid movement of the head and forceful flexion of the neck may produce pain and instability.

Following severe trauma, the neck needs to be immobilized and assessed for spinal cord damage that can result in quadriplegia or may be fatal if damage occurs to the highest part of the spinal cord.

Treatment is usually protective support for the neck and back and sometimes the use of weights attached to the head or legs to stretch the injured muscles and relieve pressure on the nerves.

Branchial cyst

A branchial cyst is an abnormal cyst just in front of one of the muscles (the sternocleidomastoid) in one side of the neck. It is caused by defective embryonic development of the second branchial cleft in the neck.

Though present at birth, it is often not recognized until adolescence, when it tends to enlarge and become more noticeable. It may be left alone if it causes no symptoms, but if it makes an opening in the skin it may become infected or cause mucus to drain, so should be removed surgically.

Torticollis

Torticollis (or spasmodic torticollis) is a condition in which involuntary muscle spasms lead to prolonged contraction of the neck muscles. This causes the head to turn to one side, lean toward one shoulder or forward or backward, or to shake. Neck pain and headache may also occur. Symptoms increase gradually and usually plateau in 2–5 years. They may also spontaneously disappear in this time.

Torticollis may be inherited or may result from neck trauma or nervous system damage. Babies may suffer from torticollis at birth due to the positioning of the head and neck in the uterus. Treatment involves stretching exercises, massage, traction or surgery, depending on the cause.

Spondylosis

Spondylosis is osteoarthritis of the spine, due to natural ageing. It often occurs in the neck (cervical) region of the spine and is caused by the degeneration of the intervertebral disks. It can be aggravated by excessive activity or repeated injury. Abnormal bone growths (spurs) may occur, leading to compression of nerve roots and spinal cord. This can cause severe neck pain, radiating to the arms and shoulders, and loss of movement, function and sensation in parts of the body below the compression. There may be difficulty moving the head, buzzing in the ears, muscle weakness and loss of balance.

Treatment, if any, may include various exercises to strengthen the neck and maintain good head movement, plus anti-inflammatory medication, and hospitalization and traction for severe cases.

Whiplash

People involved in car accidents often suffer whiplash, as the sudden impact makes the head snap forward and then backward, damaging the muscles and tissues of the neck and overstretching the cervical vertebrae.

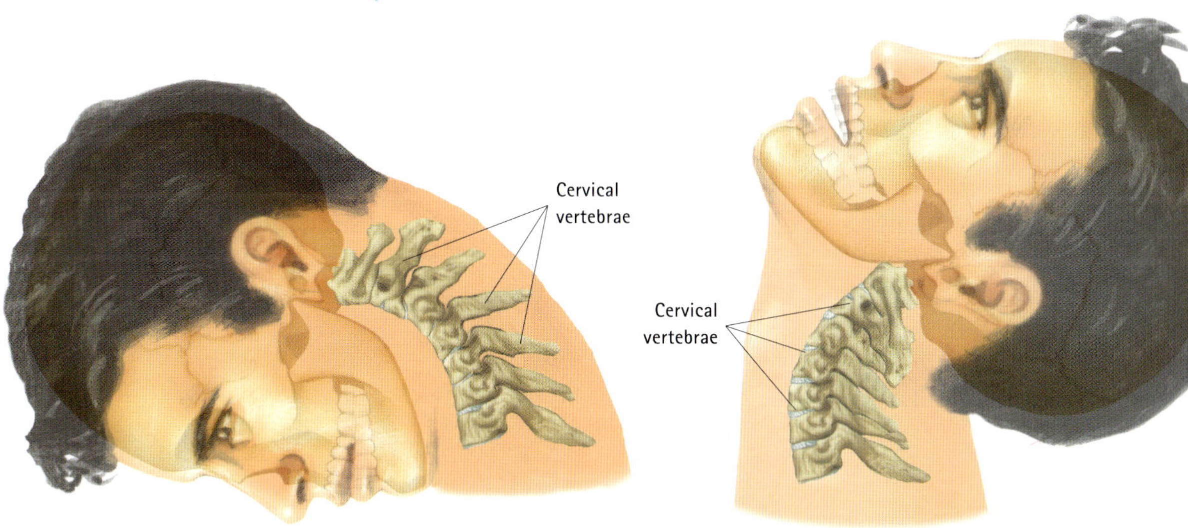

Cervical rib syndrome

Some people are born with a small extra rib known as a cervical rib, which is an appendage to the seventh cervical vertebra in the neck. It generally causes no problems, but in some cases it can compress the nerves in the neck, causing pain, numbness and tingling in the neck, shoulders, arms and hands and weakness in the arms and fingers. If these symptoms occur, the rib should be surgically removed. The condition is sometimes called thoracic outlet obstruction syndrome. A similar disorder may be caused by an injury to the neck (often while someone is unconscious or asleep). That disorder can be treated with physical therapy.

Cancer of the pharynx or larynx

Cancer may arise in the pharynx or larynx and, like mouth and tongue cancers, and is usually associated with heavy smoking and drinking. These cancers need to be treated early and aggressively, with surgery and irradiation, for the patient to have a good chance of survival. Spread of the cancer to nearby lymph nodes is often present at diagnosis.

Pharyngitis

Pharyngitis is an inflammation of the throat (pharynx). The symptoms of pharyngitis include a sore throat and discomfort or pain on swallowing. Acute pharyngitis may be caused by the viruses which cause such conditions as the common cold, laryngitis, infectious mononucleosis (glandular fever), tonsillitis and sinusitis. It may also be caused by the streptococcal bacterium, when the infection is known as strep throat. Other bacterial causes of acute pharyngitis include gonorrhea and mycoplasma.

When a physician examines the affected throat, it is seen to be red and swollen. The lymph nodes in the neck may be enlarged and tender and there may be a fever. If the cause is bacterial, antibiotics will cure the condition (antibiotics are ineffective in viral pharyngitis). Painkillers and decongestants will help the symptoms, which clear up in a week or so.

Chronic pharyngitis may be caused by smoking cigarettes, by drinking too much alcohol, or by postnasal drip resulting from chronic nasal or sinus inflammation. Treatment involves improving oral hygiene, giving up smoking and alcohol, and using antiseptic gargles.

Laryngitis

Laryngitis is an acute illness characterized by hoarseness due to an inflammation of the inner lining of the voice box (larynx). It normally lasts up to a week. During the acute phase talking should be limited. If a high temperature persists, a doctor should be consulted. Small children may develop croup (noisy difficult breathing) from swelling of the vocal cords and windpipe.

The most common form of laryngitis is an infectious condition caused by a virus. It may be associated with a bacterial infection or illnesses such as the common cold, influenza, bronchitis, pneumonia and upper respiratory infection. Allergies, trauma, laryngeal polyps and malignant tumors are among other causes. Fever and upper respiratory infection may accompany the characteristic hoarseness or loss of voice.

Diphtheria-type laryngitis is rare nowadays but if undiagnosed, it can be fatal.

Tubercular laryngitis is very painful, as is epiglottitis or bacterial inflammation of the epiglottis and larynx. The latter condition is a medical emergency; typically the patient is unable to swallow due to pain, and drooling and a high fever are other symptoms.

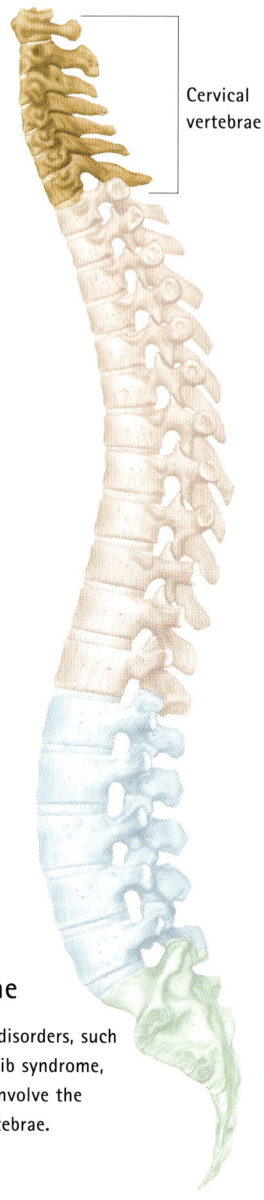

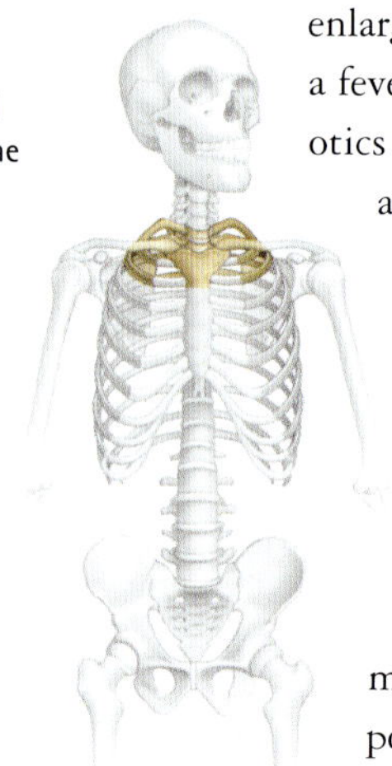

Cervical vertebrae

Many neck disorders, such as cervical rib syndrome, commonly involve the cervical vertebrae.

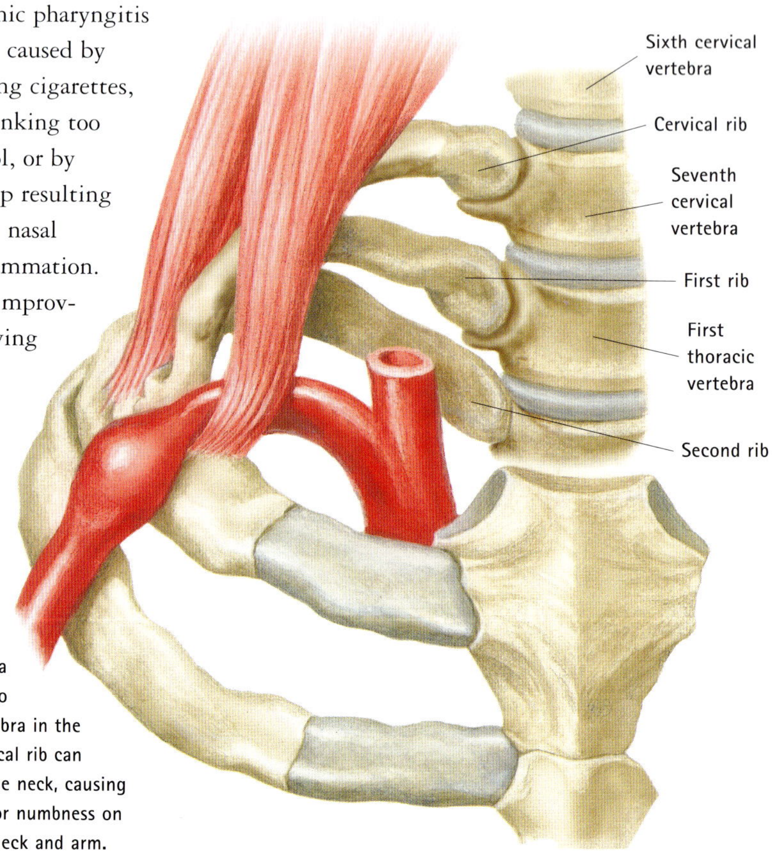

Cervical rib

Occasionally, a small extra rib is formed, attaching to the seventh cervical vertebra in the neck. This so-called cervical rib can compress the nerves in the neck, causing tingling, pain, weakness or numbness on the affected side of the neck and arm.

Chronic laryngitis can be caused by overuse of the voice, or frequent inhalation of chemical fumes and other irritants. For all types of laryngitis, the initial treatment is rest. The use of a humidifier may offer some relief from the discomfort felt in the throat. Associated upper respiratory infection may be alleviated with the help of an analgesic or decongestant. If the condition does not improve, medical advice should be sought.

Thyroid gland

The thyroid gland is one of the endocrine glands, which secrete hormones directly into the bloodstream or body cavities. The thyroid gland consists of two lobes joined together in the midline by a narrow bridge or isthmus. It is located in the neck immediately below and in front of the voice box (larynx). The thyroid gland is very well supplied with blood by a series of arteries and lies in close proximity to several important nerves which supply the larynx. These nerves must be carefully identified and protected during surgery on the thyroid.

The thyroid gland is made up of many follicles, which are spherical or polygonal structures consisting of cells arranged around a cavity filled with a gelatinous substance called colloid. The follicles of the thyroid gland make thyroid hormone and secrete it into the bloodstream.

Thyroid hormone is composed of two different substances: thyroxine (also called T4, or tetraiodothyronine) and triiodothyronine (T3). Most thyroid hormone is made up of thyroxine. An essential component of both substances is iodine, found in the diet.

Thyroid hormone has several functions, the main one being to determine the metabolic rate of the body tissues, that is, how fast the tissues of the body will use up oxygen and produce waste materials. An excess of thyroid hormone will speed up metabolism and a deficiency will slow it down. Thyroid hormone is also necessary for the normal growth and development of children—too little thyroid hormone will produce short stature and mental retardation.

The production of thyroid hormone is under the control of thyroid stimulating hormone (TSH), which is released from the pituitary gland. The presence of thyroid hormone in the bloodstream inhibits the production of TSH in a feedback loop control system.

Between the thyroid follicles are parafollicular cells (C cells), which are responsible for the production and secretion of another hormone, called calcitonin, which acts to reduce the concentration of calcium in the blood.

SEE ALSO *Endocrine system on page 110; Hormones on page 112; Metabolism on page 115*

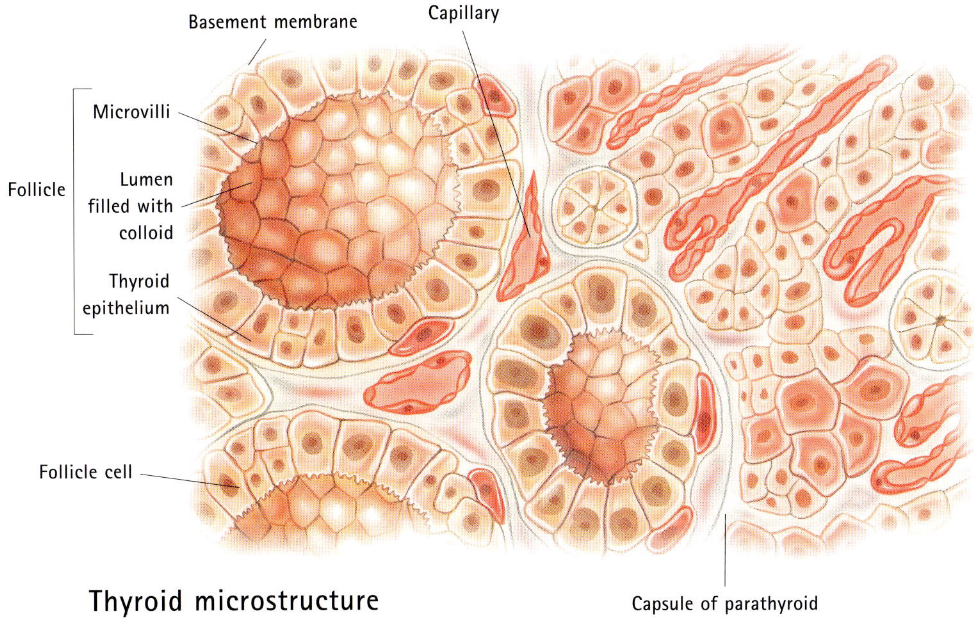

Thyroid microstructure

The thyroid gland comprises many follicles which make thyroid hormone and secrete it into the bloodstream. Each follicle consists of thyroid epithelial cells arranged around a cavity (or lumen) filled with colloid, a gelatinous substance. Thyroid hormones are stored within the colloid.

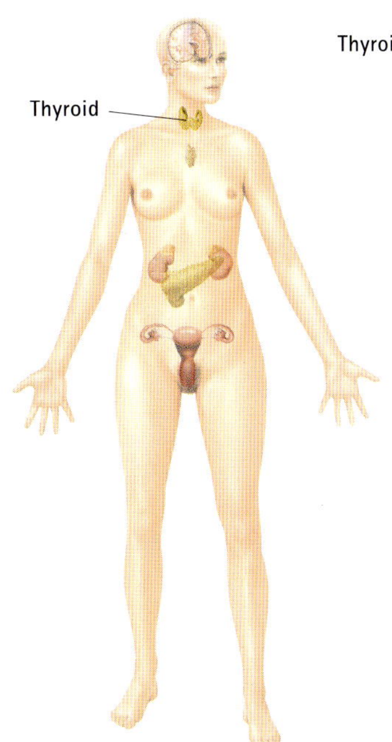

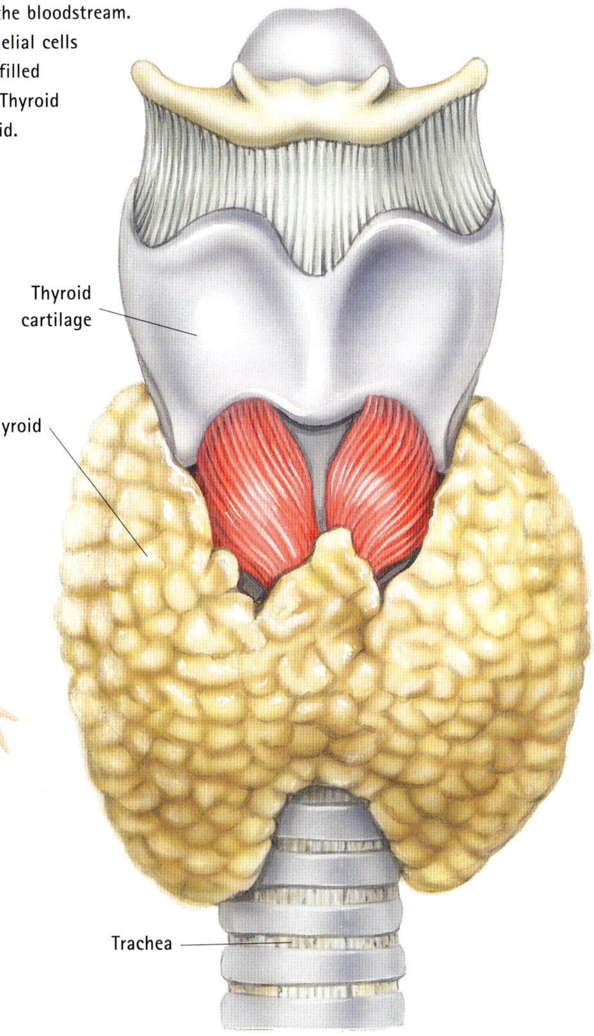

Thyroid

The largest of the endocrine glands, the thyroid is situated at the front of the trachea in the neck. The two lobes of the gland are joined by a narrow bridge (or isthmus).

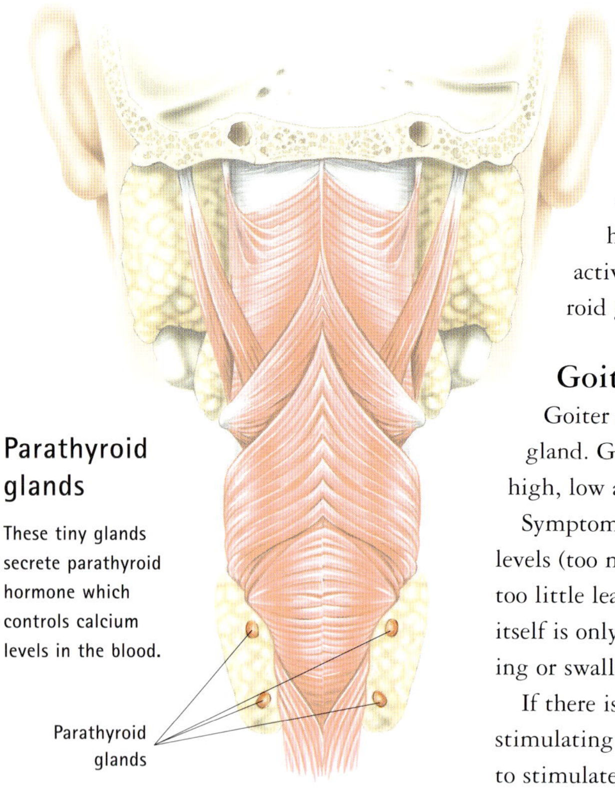

Parathyroid glands

These tiny glands secrete parathyroid hormone which controls calcium levels in the blood.

Parathyroid glands

The parathyroid glands are four (or occasionally three) small endocrine glands, which lie just behind the thyroid gland. The glands are only about the size of peas. Sometimes these glands are embedded within the thyroid gland itself; they may occasionally be found in the chest.

Each parathyroid gland comprises a fibrous tissue capsule and two types of cell called chief and oxyphil cells. Chief cells produce parathyroid hormone, which is involved in the control of calcium and phosphate concentrations in the blood.

A reduction in the levels of calcium in the blood stimulates the parathyroid gland to release parathyroid hormone. This hormone in turn stimulates the release of calcium from the bones by increasing the activity of cells called osteoclasts, which break down the mineral part of bone.

SEE ALSO *Endocrine system on page 110; Hormones on page 112*

Disorders of the thyroid and parathyroid glands

Tests of the thyroid gland and its function include blood tests to determine the concentrations of thyroid hormone and thyroid stimulating hormone in the blood, and isotope scans to determine the presence of "hot" or "cold" spots in the thyroid gland.

Tests of the parathyroid gland include a blood test to assay the level of parathyroid hormone. Both under- and overactivity of the thyroid and parathyroid glands can result in illness.

Goiter

Goiter is an enlargement of the thyroid gland. Goiters can be associated with high, low and normal thyroxine levels.

Symptoms of goiter depend on thyroxine levels (too much leading to thyrotoxicosis; too little leading to myxedema); the goiter itself is only a problem if it affects breathing or swallowing.

If there is insufficient thyroxine, thyroid stimulating hormone (TSH) will be released to stimulate the thyroid to produce more, and this can lead to enlargement of the gland (goiter). A common cause of goiter is a lack of iodine in the diet. With insufficient iodine, and low thyroxine levels, TSH production will increase to stimulate the thyroid to make more thyroxine. This type of goiter can be treated by addition of iodine to the diet.

Goiter can rarely be caused by excess stimulation of the thyroid gland from other causes, such as abnormally high levels of TSH occurring with pituitary tumors.

Hyperthyroidism

Hyperthyroidism (or thyrotoxicosis) occurs when overactive thyroid tissue secretes too much thyroid hormone into the bloodstream. The thyroid tissue may be overactive as a result of abnormal stimulation, as in Graves' disease; there may be an isolated overactive thyroid nodule or an entire overactive gland.

Symptoms of thyrotoxicosis include palpitations, nervousness, increased appetite with weight loss, poor tolerance of hot weather, increased sweating, increased frequency of bowel motions, menstrual problems, infertility and muscular weakness. An enlarged thyroid gland, increased heart rate, fine hair and warm moist skin may also occur. If blood levels of thyroid hormone reach dangerous heights during thyrotoxicosis, the patient may develop a condition known as thyroid storm. In this situation the patient will experience accentuated thyrotoxic symptoms, heart failure, grossly elevated fever and delirium.

Blood tests usually reveal elevated levels of thyroid hormone and the absence of TSH. If left untreated, hyperthyroidism can cause worsening angina, heart attacks, arrhythmias and heart failure, progressive weight loss and death.

Treatment is with antithyroid drugs to interfere with thyroid hormone production, radioactive iodine to destroy hyperactive thyroid tissue or surgery to remove excessively active thyroid tissue. The choice of treatment depends on the age and state of health of the patient and the size of the goiter.

Graves' disease

Named after the Irish physician Robert James Graves, Graves' disease is the most common (though not the only) cause of hyperthyroidism. Also known as toxic diffuse goiter, it is an autoimmune disease, in which the body's own immune system attacks and inflames the thyroid gland. Antibodies to their own thyroid gland can be detected in the blood of people who have Graves' disease.

In this condition, the thyroid gland swells in size (develops into a goiter) and secretes excessive amounts of thyroid

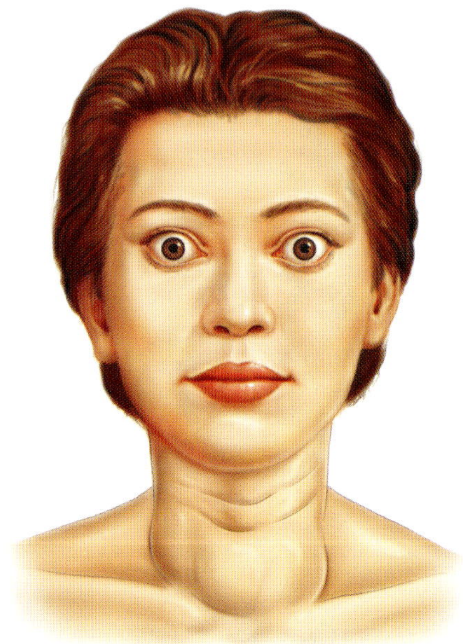

Graves' disease

Exophthalmos (abnormally protruding eyes, as if staring) is a common feature of this disease.

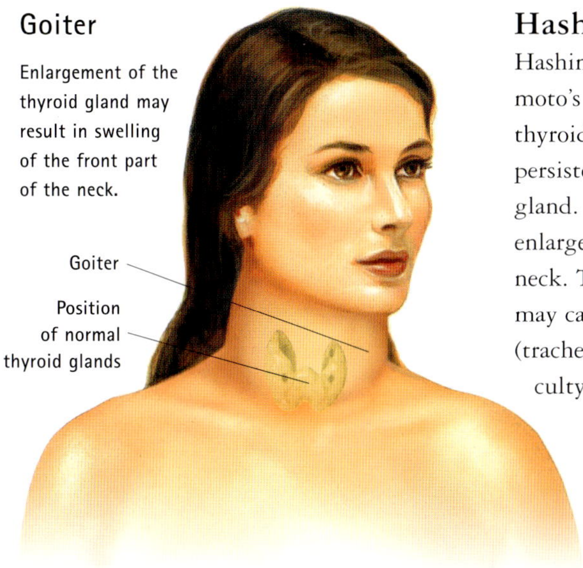

Goiter

Enlargement of the thyroid gland may result in swelling of the front part of the neck.

hormone into the bloodstream, causing thyrotoxicosis. The affected person experiences rapid heartbeat, tremor, increased sweating, weight loss (despite increased appetite), and weakness and fatigue.

A condition called exophthalmos often develops, in which the eyeballs protrude and the eyelids retract. This is caused by edema (fluid accumulation) in the tissues surrounding the eyeball in its socket. This eye protrusion may respond to treatment of the excessive thyroid activity, but may cause loss of vision if not treated promptly.

Graves' disease is more common in women than in men. It is usually inherited and tends to run in families. The condition is diagnosed by blood tests which show excess thyroid hormone in the blood, and a radioactive thyroid scan.

There are several treatment options, including treatment with drugs such as propylthiouracil, the surgical removal of part of the thyroid gland, or oral administration of radioactive iodine.

Hypothyroidism

In hypothyroidism (also called myxedema) there is a reduced level of thyroid hormone in the blood. In this condition the patient will experience a puffy thickening of the skin below the eyes, and in the skin of the lips, fingers and legs. The patient will also complain of lethargy, the slowing of thought processes, weight gain and the loss of hair. Hypothyroidism in adults is most commonly seen in Hashimoto's disease and in endemic goiter.

Hashimoto's disease

Hashimoto's disease (also known as Hashimoto's thyroiditis or chronic lymphocytic thyroiditis) is a slowly developing and persistent inflammation of the thyroid gland. Patients experience a painless enlargement of the thyroid gland in the neck. The enlargement of the thyroid gland may cause compression of the windpipe (trachea) and esophagus, resulting in difficulty in breathing and swallowing.

Similar to Graves' disease, it is an auto-immune disorder, affecting women more often than men, and tends to run in families, but unlike Graves' disease, the inflammation may result in the undersecretion of thyroid hormone into the bloodstream, or hypothyroidism.

Hashimoto's disease is usually slow to develop, and is often associated with other autoimmune endocrine disorders such as diabetes mellitus or Addison's disease. High levels of thyroid autoantibodies are almost always present.

Treatment of Hashimoto's disease usually requires lifelong replacement therapy with thyroid hormone to decrease goiter size and treat the hypothyroidism. Treatment may involve surgery to remove excess thyroid tissue that is compressing nearby structures.

Cretinism

In areas of the world where dietary iodine is inadequate, the fetus and neonate may not be able to produce sufficient thyroid hormone for normal brain maturation. In this condition, known as cretinism, the child will be intellectually disabled and have stunted growth. The face will usually be broad, emergence of the teeth will be delayed, and the tongue and mouth will tend to be large. Parts of the world where this may occur are usually mountainous. Fortunately, supplementation of the diet with small amounts of iodine can completely prevent the problem.

Cretinism may also arise in children with congenital absence of the thyroid gland or in those who have a genetic defect in the enzymes that make thyroid hormone. In these children the condition may be remedied by thyroid hormone supplementation. Early diagnosis and treatment are essential.

Cancer of the thyroid gland

Cancers may arise from the thyroid gland, and are more common if there has been irradiation of the head and neck earlier in life. Thyroid cancers are of several different types. Those found in young patients (papillary adenocarcinoma) are usually slow-growing and spread outside the gland relatively late in the course of the disease. There are good survival rates for these tumors (over 80 percent after 10 years).

At the other extreme are the so-called undifferentiated thyroid cancers, which usually appear later in life and invade surrounding tissue in an aggressive manner. Survival rates with this kind of cancer are quite low (10–15 percent after 10 years). Patients with thyroid cancers complain of a painless lump in the neck, which gradually increases in size. They may also experience difficulty in swallowing and hoarseness of the voice, particularly if the nerves to the larynx have been damaged.

Treatment is by surgery to remove the tumor and any involved lymph nodes, accompanied by radioactive iodine therapy. Aggressive tumors may need to be treated with chemotherapeutic agents. Tumors from the kidney, breast and lung sometimes spread to the thyroid gland, but they usually produce multiple lumps rather than one.

Hypoparathyroidism

Hypoparathyroidism is a rare disorder in which production of parathyroid hormone—which helps regulate blood calcium levels—is either reduced or nonexistent due to dysfunctional or absent parathyroid glands. It may be a congenital condition or, less commonly, acquired later in life (usually due to surgical damage to or removal of the parathyroids).

Hypoparathyroidism leads to very low blood calcium levels (hypocalcemia) which can trigger a nerve disorder called tetany, characterized by painful muscle spasms and twitches. People born with hypoparathyroidism may suffer from dry skin, hair loss and a susceptibility to yeast (*Candida*) infections. Without treatment, hypoparathyroidism in children can lead to impaired physical and mental development. It can be treated with calcium and vitamin D supplements, which need to be taken for life.

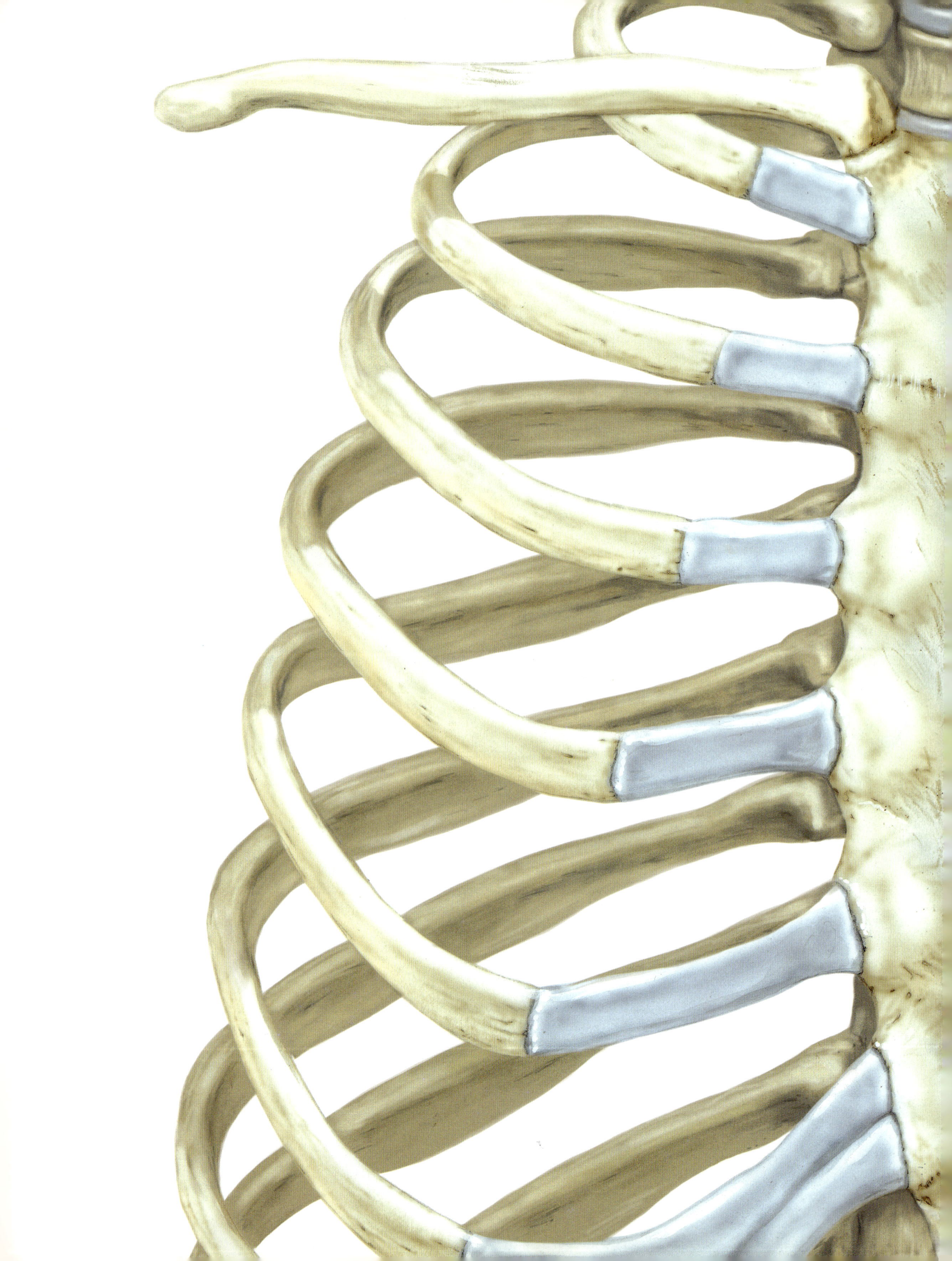

The Trunk

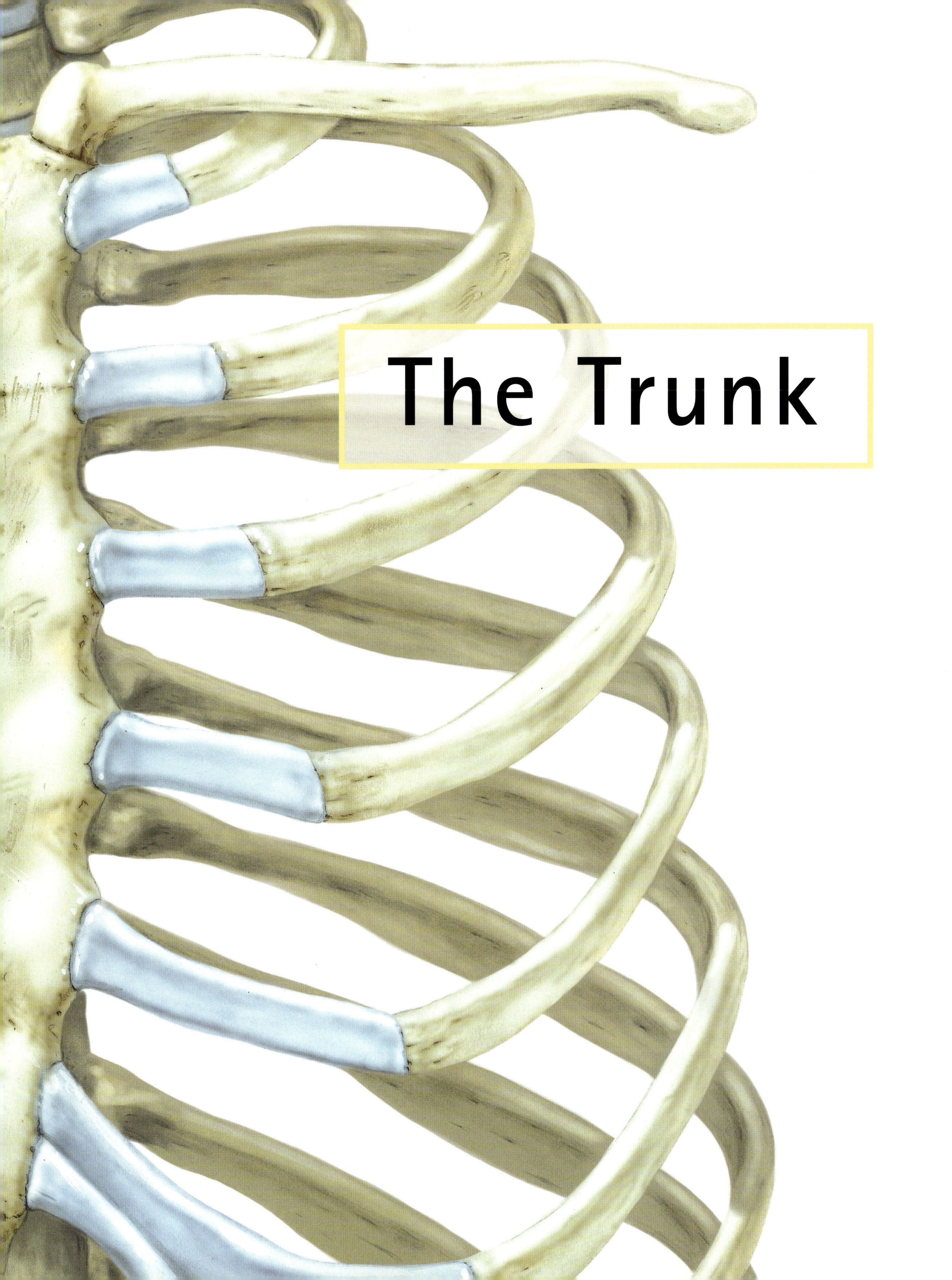

Back

The back is the part of the human body from below the neck to the lower end of the spine, just above the buttocks. The bones of the spine are known as the vertebrae and are joined to each other by disks (intervertebral disks) and joints (facet joints) to form the vertebral column; the spinal cord runs through a canal formed by the vertebrae. The spinal cord is composed of nerve tissue, and nerves branch off through spaces between vertebrae. The vertebrae are separated from each other and cushioned by the intervertebral disks, which are flexible. The disks bear weight and cushion the surfaces of the vertebrae as the spine moves. The vertebral column protects the spinal cord and spinal nerves, supports the weight of the body and head, and anchors the rib cage. It plays a major role in movement and posture, and provides an attachment point for muscles of the trunk, arms and legs.

SEE ALSO *Muscular system on page 48; Skeletal system on page 30; Spinal cord on page 217; Spinal nerves on page 218; Spine on page 211*

Causes of back pain

Low back pain is a common complaint in many societies because of the sedentary nature of daily work. Symptoms can be pain or numbness in the lower back and surrounding areas, commonly called lumbago. When pain extends down the buttocks and into the upper legs, it is associated with pressure on spinal nerves which form the sciatic nerve and is known as sciatica. Factors that place people at risk of back strain injury and low back pain include poor physical condition, loading unprepared muscles, poor posture, and lifting heavy weights.

Care of the back

Back care starts with good posture, whether sitting, standing or working. Good muscle condition plays a very large part, as poor posture often results in muscle weakness.

SEE ALSO *Disorders of the spinal cord on page 219; Disorders of the spine on page 215*

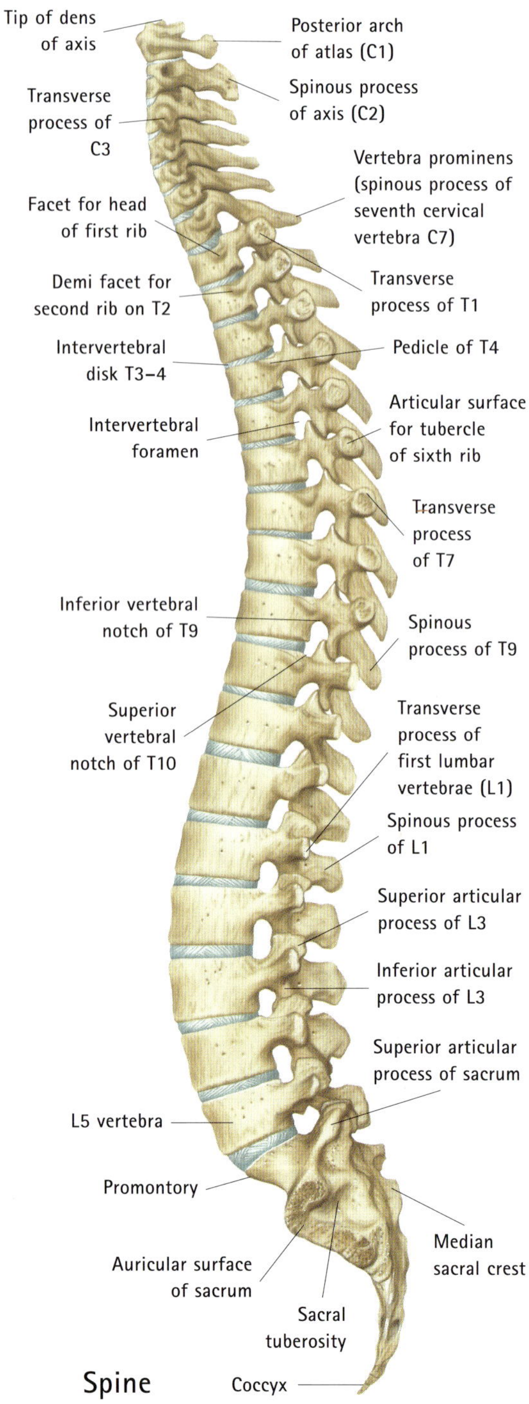

Spine

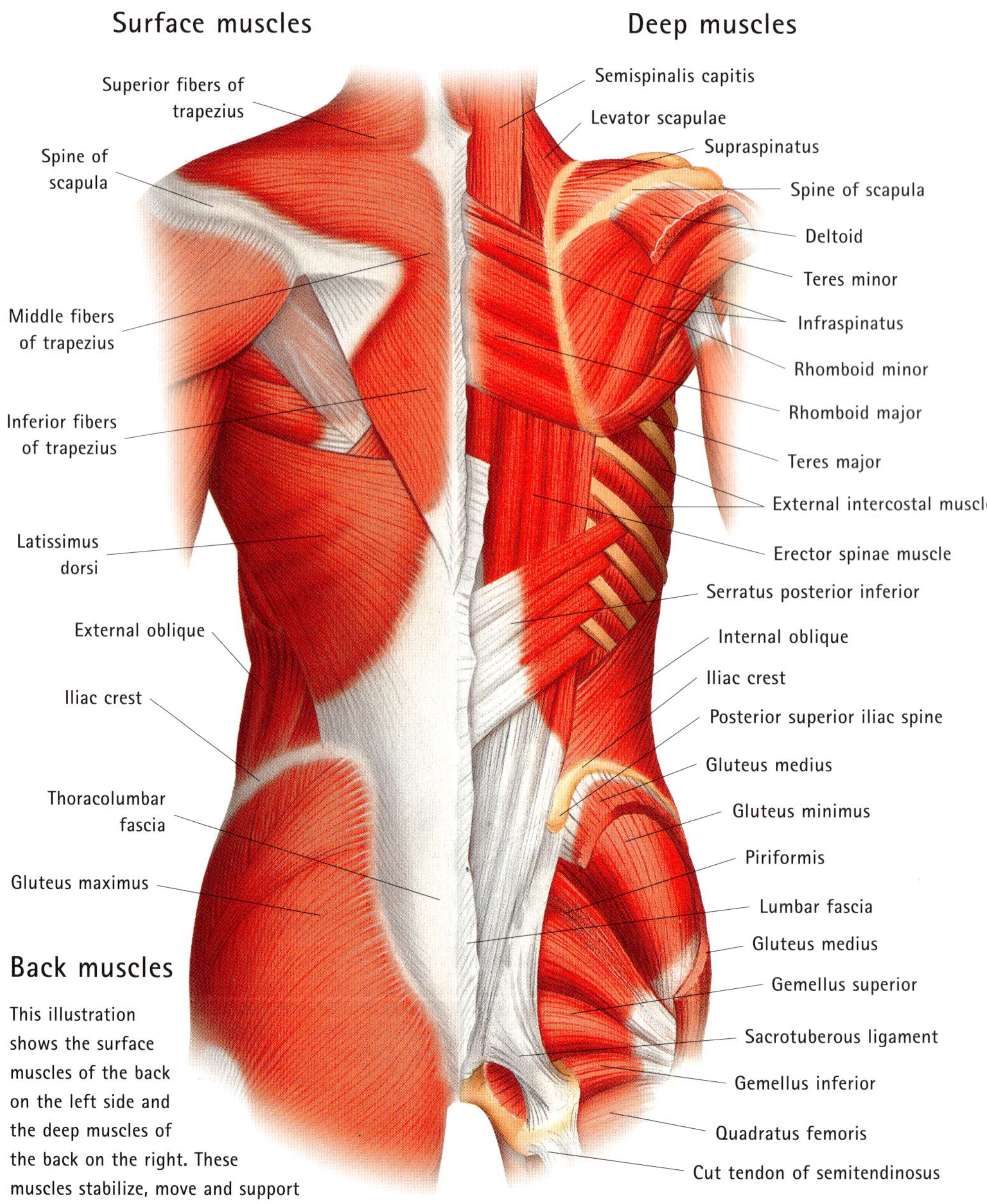

Back muscles

This illustration shows the surface muscles of the back on the left side and the deep muscles of the back on the right. These muscles stabilize, move and support the spinal column (the vertebrae).

SPINE

The spine, or vertebral column, extends down the midline of the back, from the base of the skull to the pelvis, and forms the central axis of the skeleton. It also functions to protect the spinal cord, a nervous structure located in a hollow canal (the vertebral or spinal canal) which runs down the center of the vertebral column. The spine must be firm enough to support the body weight but it also requires flexibility to allow bending of the trunk. These requirements are satisfied by its curved, segmented structure which is made up of 26 bones (vertebrae), which are separated from each other by pads of cartilage (intervertebral disks). These disks make up 25 percent of the length of the vertebral column in a young adult.

The spine is divided into five regions—cervical, thoracic, lumbar, sacral and coccygeal. The cervical region is formed by seven vertebrae, numbered 1–7 from the top down (C1–C7). The thoracic region is formed by twelve vertebrae (T1–T12), all of which have ribs attaching to their sides. The lumbar (lower back) region has five vertebrae (L1–L5). The sacrum, a single curved bone in the adult, actually develops as five separate vertebrae (S1–S5), which fuse to each other during early development to form a single bone. Similarly, the four vertebrae of the tail region fuse during development to form the coccyx, a rudimentary bone attached to the lower end of the sacrum.

Viewed from side on, the vertical column is not straight, but curved into an S-shape. At birth the vertebral column is bent forward into a C-shape, this forward curvature remaining in the adult in the thoracic and sacral regions (primary curvatures). In childhood two reverse curves appear, in the cervical and lumbar regions, to better balance the weight of the head and body. The cervical curvature appears when the child begins to lift the head and the lumbar curvature appears as the child learns to sit, stand and walk. Because the cervical and lumbar curvatures appear after birth, they are said to be secondary curvatures.

SEE ALSO *Disorders of the spine on page 215; Skeletal system on page 30*

Back pain

Lifting, bending or rotating can cause back injuries, especially if you are carrying a heavy load at the same time. Injuries are most common in the sites indicated.

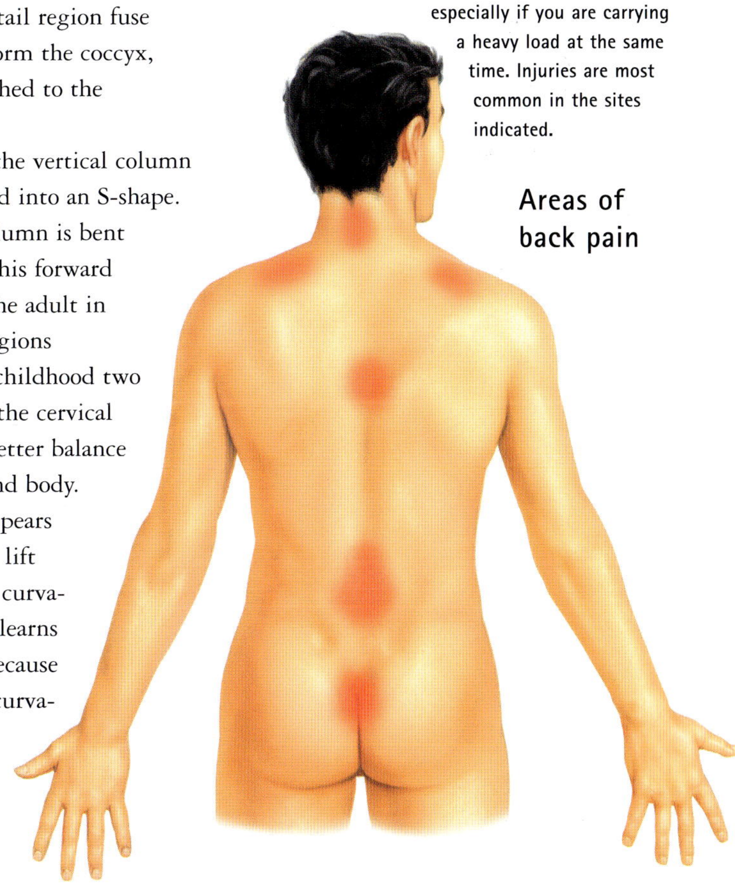

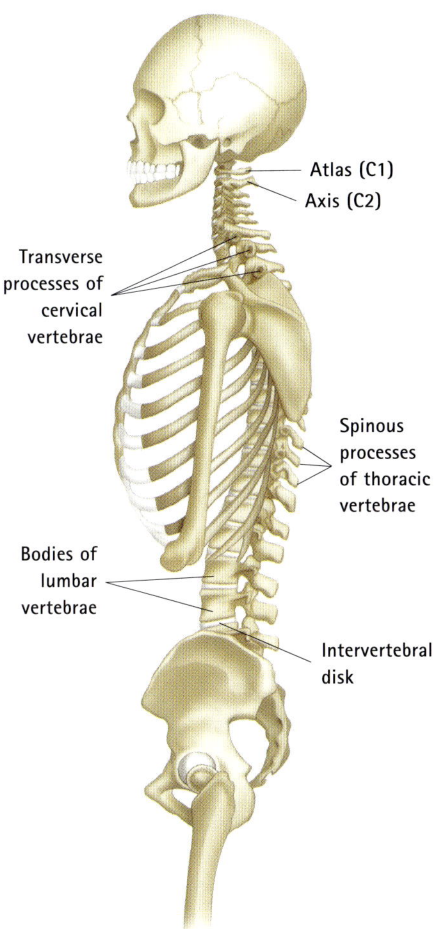

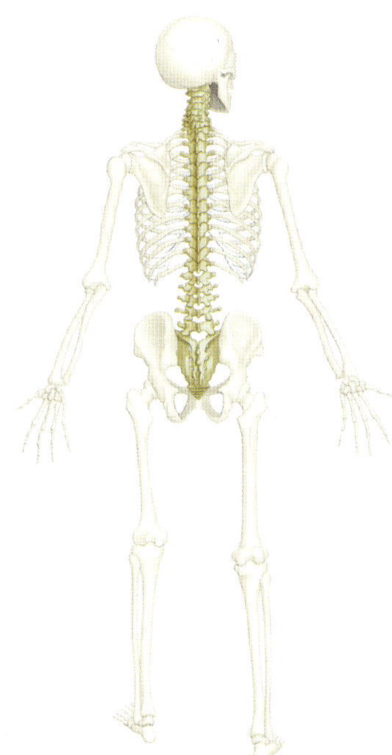

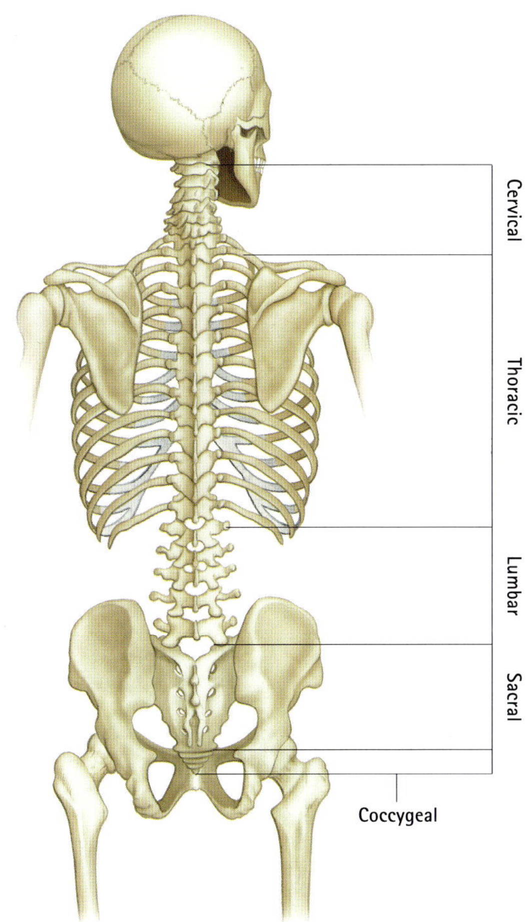

Spine

The spine forms the central axis of the skeleton and supports the weight of the head, neck and trunk. Its segmented structure provides flexibility, allowing the trunk to bend and twist.

Vertebrae

The vertebrae are the bony building blocks of the vertebral column (or spine). The individual vertebrae, separated by intervertebral disks, give the vertebral column both strength and flexibility. The vertebral column is comprised of seven cervical, twelve thoracic, five lumbar, five fused sacral and three to five fused coccygeal vertebrae.

All vertebrae have a similar structural pattern, although the size and shape of the individual features varies in different regions. Each vertebra consists of the following.

- A body at the front, which gets progressively larger from the top to the bottom of the column, because it is the weight-bearing part of the vertebra.
- An arch of bone, known as the vertebral arch, which attaches to the back of the body and surrounds a hole in the center called the vertebral foramen. The arch is usually divided on each side into a pedicle attaching to the body and a lamina at the back.
- A spinous process (spine), which extends backward from the arch. The spinous processes can be easily felt extending down the midline of the back. That of the C7 vertebra (the "vertebra prominens") is particularly prominent at the base of the neck and can be used as a landmark for counting other vertebrae.
- A pair of transverse processes, which extend outward from each side of the arch. The spinous and transverse processes function as levers during movements of the vertebral column.
- Two pairs of articular processes, which extend from the upper and lower surfaces of each side of the arch. These processes form joints with those of the vertebrae both above and below.

The atlas (C1) is atypical in that it does not have a body or spinal process. The upper and lower surfaces of the pedicles of each vertebral arch are notched, so that when two vertebrae sit together, the adjacent notches form an incomplete ring, which is known as an intervertebral foramen. The spinal nerves enter and exit through these holes from each side of the vertebral canal. Vertebrae may fracture as a result of trauma. Fracture or displacement may affect the spinal cord and associated nerves, and produce temporary or permanent loss of sensation and movement.

Movements of the vertebral column

The vertebral column is a flexible structure that acts as a single unit, in which large movements result from the sum of the many small movements occurring at the joints between the vertebrae. Movements of the vertebral column are flexion (to bend forward), lateral flexion (to bend sideways), extension (to bend backward), rotation (around its own axis) and circumduction (a combination of all these movements). The range of movement varies in different regions—the cervical and lumbar regions are the most mobile, with less movement possible in the thoracic region because of the presence of the ribs.

Joints

Each vertebra forms three separate joints with the vertebra above or below—a pair of facet joints and a single anterior intervertebral joint. A facet joint is formed on each side between the articular processes of the two vertebrae. The facet joints are synovial joints so they have a fluid-filled cavity between the bones, which are held together by a joint capsule. The adjacent cartilage-covered surfaces are able to glide on each other during movements of the vertebral column. The direction of orientation of the joint surfaces determines the type of movements that are permitted in different regions. For example, pure rotation of the vertebral column can only occur in the thoracic region. The anterior intervertebral joints, between the bodies of the vertebrae, are designed for strength and involve a pad of strong fibrous cartilage called the intervertebral disk.

Intervertebral disk

The intervertebral disks are flexible cartilaginous structures, which lie between adjacent vertebrae and make up approximately 25 percent of the length of the vertebral column in a young person. Intervertebral disks form joints between the bodies of the vertebrae, which serve to unite adjacent vertebrae and to permit movement between them. They also act as shock absorbers when force is transmitted along the vertebral column during standing and movement. The disks may be relatively thin and flat in shape (as in the thoracic region) or wedge-shaped and thicker, as in the lumbar and cervical regions.

The intervertebral disks are reinforced in front and behind by the anterior and posterior longitudinal ligaments. These are bands of ligamentous fibers, which attach to the bodies of the vertebrae and extend along the length of the vertebral column. The anterior longitudinal ligament prevents excessive extension, and is commonly injured in automobile accidents, when a car is struck from behind and the head is suddenly and forcibly thrown backward. This type of injury is known as whiplash.

Each intervertebral disk consists of two parts; a central region known as the nucleus pulposus, and a surrounding region called the annulus fibrosus. As its name suggests, the nucleus has a pulpy or gelatinous texture, and is easily able to change shape when pressure is placed upon it. The annulus is tough and fibrous, and is firmly attached to the vertebrae above and below. It consists of concentric rings of strong fibers, which pass in an oblique direction from one vertebra to the next. The fibers in adjacent rings are oriented at right angles to each other, allowing some movement to occur between the bones but also providing a strong bond between

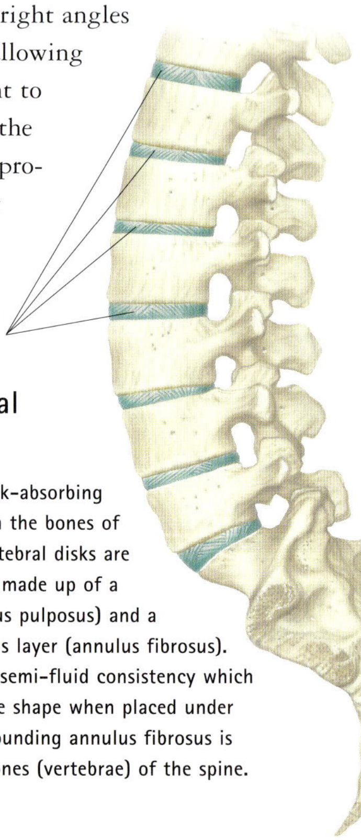

Intervertebral disks

These are the shock-absorbing structures between the bones of the spine. Intervertebral disks are flexible structures made up of a soft center (nucleus pulposus) and a surrounding fibrous layer (annulus fibrosus). The nucleus has a semi-fluid consistency which allows it to change shape when placed under pressure. The surrounding annulus fibrosus is attached to the bones (vertebrae) of the spine.

VERTEBRAE

The vertebrae of the spine form a strong yet flexible housing for the spinal cord.

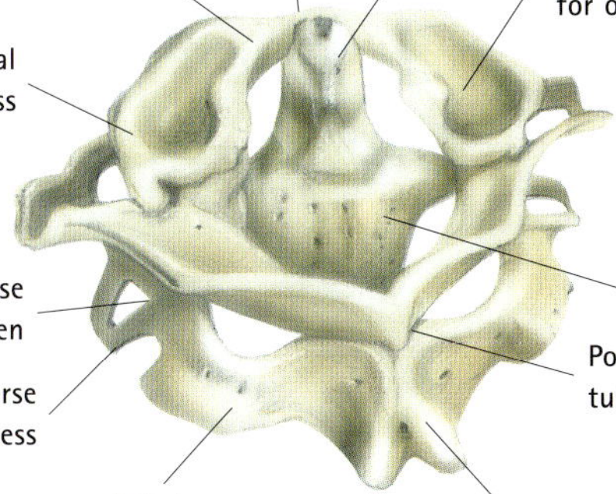

Axis and atlas

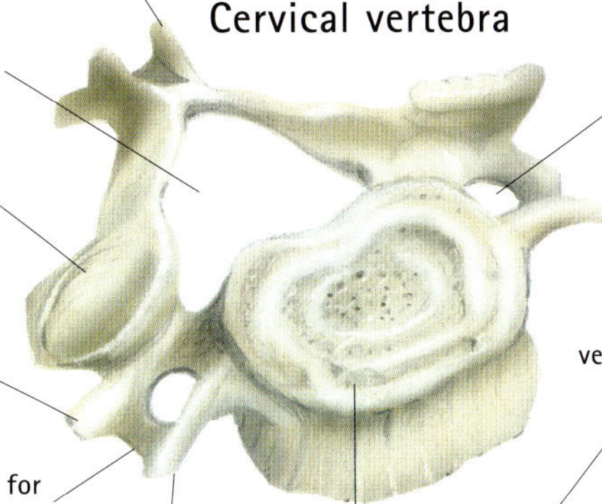

Cervical vertebra

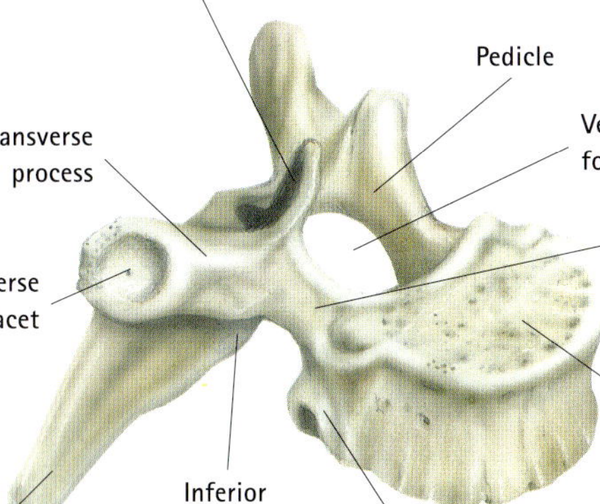

Thoracic vertebra

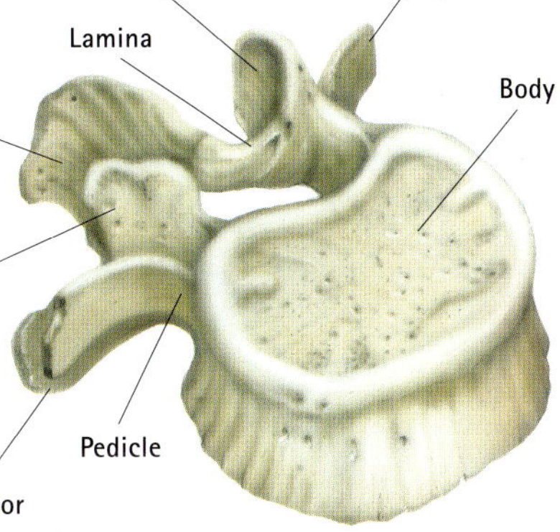

Lumbar vertebra

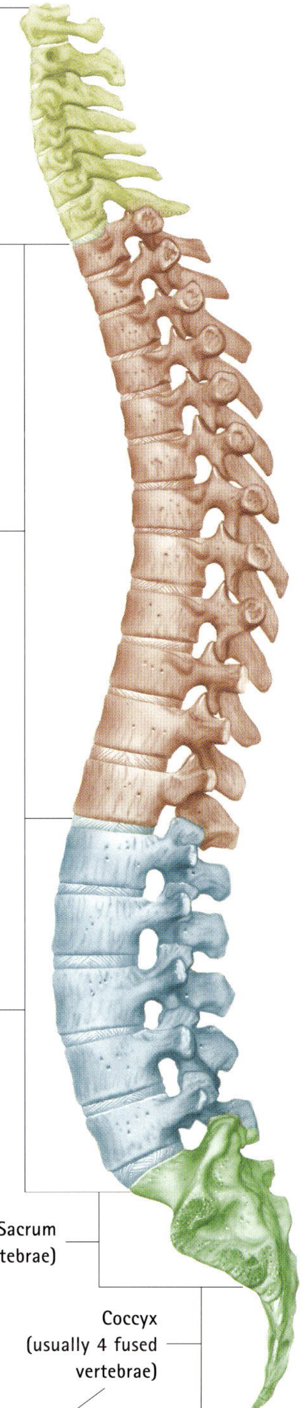

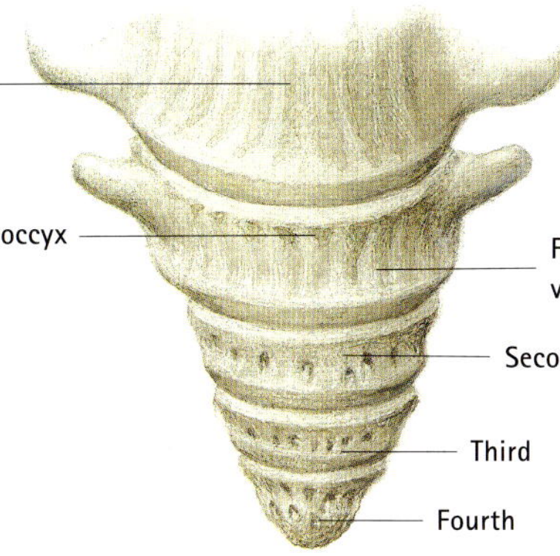

Coccyx

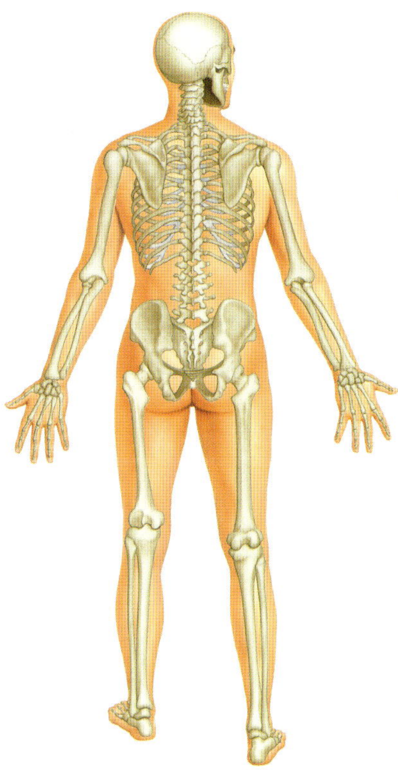

them. It also hold the nucleus in position. The nucleus pulposus does not sit exactly in the center of the disk. It is located more toward the back, so the annulus is thinner behind than in front.

Intervertebral disks in young people have a high water content (80–90 percent). This gradually reduces during the day because water is squeezed out of each disk as it bears weight during standing and movement. The lost water is however reabsorbed into the disk during sleep when lying down. The average young adult is around ¾ inch (2 centimeters) taller upon waking than at the end of the day.

As a person ages, the ability to replace this water gradually reduces and the disks become drier and thinner. This partly explains why people lose height, or appear to shrink, as they get older.

Sacrum

The sacrum and coccyx form the lower end of the spine. The sacrum is formed by fusion of the five sacral vertebrae. The sacrum forms a part of the bony pelvis, being joined to the hip bones at the sacroiliac joints. Above, it articulates with the fifth lumbar vertebra, and below with the coccyx. Passing through the sacrum from top to bottom is the sacral canal, which is a continuation of the spinal canal of the rest of the vertebral column. The spinal roots (cauda equina) of the spinal cord pass through the sacral canal, giving off sacral spinal nerves which leave through pelvic and dorsal passages in the bone (sacral foramina).

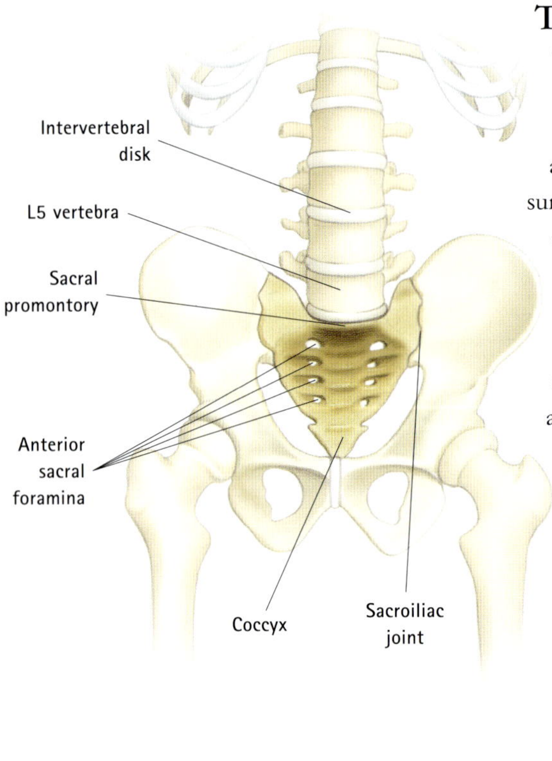

Sacrum

The sacrum is located at the base of the spine above the coccyx, and consists of five fused vertebrae.

Coccyx

The coccyx is the lowest bone of the spine. Situated just above the anus, it is formed from three to five rudimentary vertebrae, which are fused and joined to the sacrum above. The coccyx is also called the tailbone. The word "coccygeal" derives from the Greek word kokkyx meaning "cuckoo bird" as the coccyx was thought to look like a cuckoo's bill.

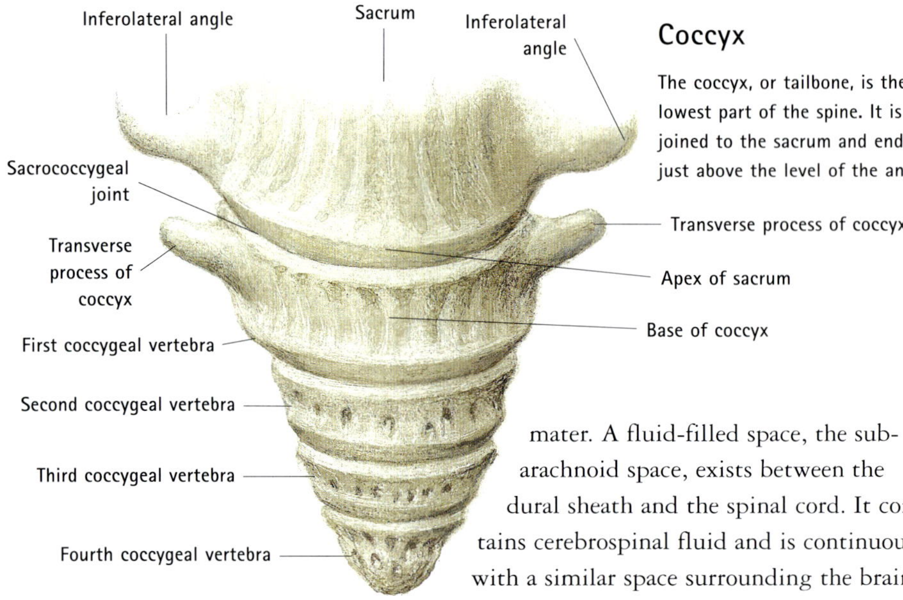

Coccyx

The coccyx, or tailbone, is the lowest part of the spine. It is joined to the sacrum and ends just above the level of the anus.

The vertebral canal

The vertebral canal is a hollow cavity that extends the length of the vertebral column and contains the spinal cord and the roots of the spinal nerves. It is surrounded at the front by the vertebral body and at the sides and back by the vertebral arch. The vertebral canal is lined by a membrane known as the dural sheath.

The dural sheath consists of an outer fibrous layer called the dura mater and a thin, inner layer called the arachnoid mater. A fluid-filled space, the subarachnoid space, exists between the dural sheath and the spinal cord. It contains cerebrospinal fluid and is continuous with a similar space surrounding the brain. The spinal cord usually ends at the level of the first or second lumbar vertebrae and below this level the canal is occupied only by spinal nerve roots, passing along its sides. It is therefore safe to place a needle between the spines of the fourth and fifth lumbar vertebrae to take samples of cerebrospinal fluid for neurological examination. This procedure is known as a spinal tap or lumbar puncture.

The dural sheath is separated from the bone surrounding the canal by a narrow epidural space, which is filled with fat and some veins. The sacral spinal nerves, which supply the organs of the pelvis, can be anesthetized by placing anesthetic into the epidural space of the sacrum. This type of anesthesia, known as epidural anesthesia, is commonly used when babies are born by cesarean section, allowing the mother to remain conscious during the birth.

Sacroiliac joint

The sacroiliac joint is located where the ilium of the hip bone and the sacrum at the base of the spine meet. Because the weight of the trunk passes through this joint, powerful interosseous ligaments are required to unite the bones. At the front, the joint is synovial and at the back it is fibrous. The synovial part of the joint has an L-shaped surface, which is usually related to the first three fused segments of the sacrum in the male and the first two segments in the female. The sacral and iliac articular surfaces are reciprocally ridged and furrowed to increase stability.

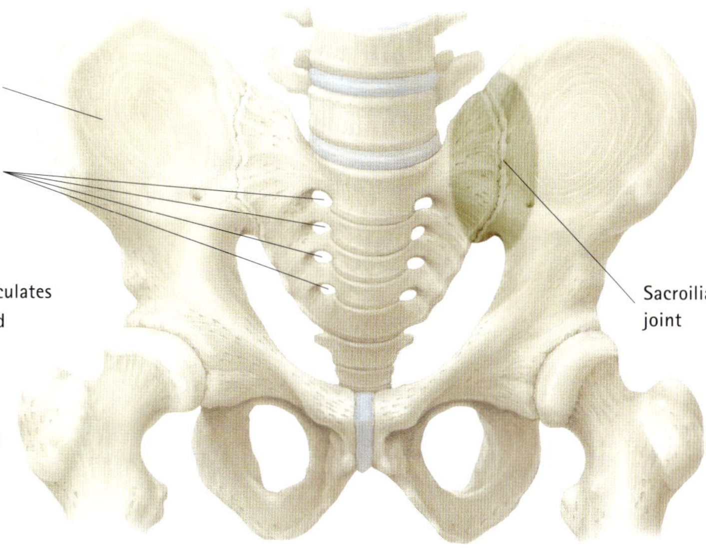

Sacroiliac joint

The sacroiliac joint articulates between the sacrum and ilium bone of the hip. It transfers body weight to the pelvis. The range of movement is relatively small.

Herniated disk

Intervertebral disks comprise a soft nucleus and a fibrous outer casing (annulus) that holds the disk in place. If the annulus is weakened by age or injury, the disk may bulge into the vertebral canal compressing nearby spinal nerves, referred to as a "slipped" or herniated disk.

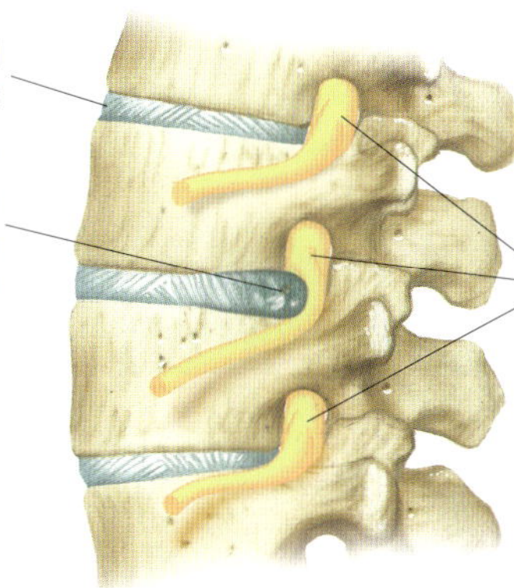

The interosseous sacroiliac ligament is strong and unites the roughened areas of bone behind the synovial part of the joint. The weight of the body tends to drive the upper end of the sacrum downward, tightening this ligament and drawing the joint surfaces together. This same force tends to tilt the lower end of the sacrum upward, which is resisted by the powerful sacrotuberous and sacrospinous ligaments.

The sacroiliac joint is capable of only a small amount of movement and is subject to great stress, from the downward pressure of the body's weight and the upward thrust of the legs and pelvis. It must also be able to cope with the movements of the body, for example as it turns, twists, pulls and pushes. Movement of the sacroiliac joint increases during pregnancy. An excessive motion can cause a strain on the joint. With increasing age, the joint cavity may become partially or completely obliterated by fibrous tissue or fibrocartilage, and may even show bony fusion in the very old.

Spinal canal

The vertebral column provides a protective casing for the spinal cord, a length of nervous tissue that runs from the brain to the level of the first or second lumbar vertebrae. Cerebrospinal fluid and a layer of membrane fill the space between the vertebral canal and cord to provide additional cushioning.

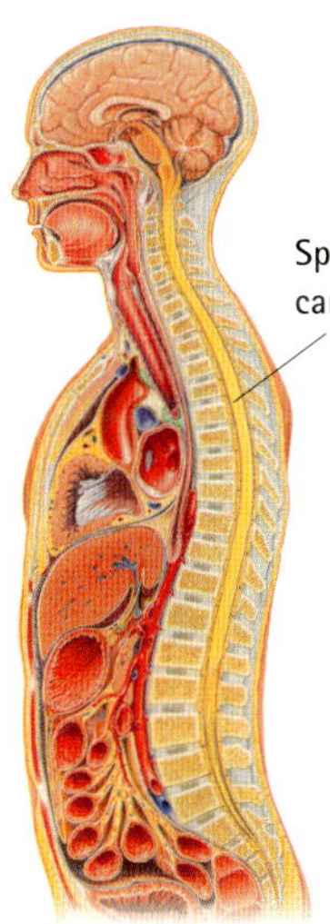

Disorders of the spine

The predominant symptom of spinal disorders is back pain. In the USA, back pain is the most common cause of limitation of activity for people under the age of 45. Approximately 85 percent of the population will experience back pain in their lives.

See also *Bone marrow biopsy on page 435; Epidural anesthesia on page 454; Lumbar puncture on page 436; Traction on page 455; Treating the musculoskeletal system on page 444; X-ray on page 431*

Lumbago

Although lumbago means simply "backache," it is usually an ache in the lower back associated with spinal disk damage. When humans first stood upright, the spine became a weight-bearing organ, something for which it was not designed. As a consequence, awkward weight bearing and unusual strains can disrupt the disks between the vertebrae. When damaged disk tissue projects into the spinal canal, nerves are frequently compressed, producing neuralgia (such as sciatica). An operation to remove the displaced piece of disk or injections of cortisone into the disk to make it shrink back into position may be required. Surgery is imperative if muscle weakness occurs.

Diagnosis depends largely upon the result of CAT scan, x-rays or MRI (magnetic resonance imaging). The patient may need advice on back-strengthening exercises.

Lumbago may also be due to referred pain from pelvic organs, such as prostate inflammation and cancer, or associated with menstrual pains or uterine or ovarian disease.

Prolapsed intervertebral disk

A prolapsed intervertebral disk (also known as slipped disk or herniated nucleus pulposus) occurs when the soft center of a disk (nucleus pulposus) ruptures through a tear or fracture in the fibrous tissue of the disk. This can happen in the lumbar (lower back) or cervical (neck) regions, putting pressure on spinal nerves (radiculopathy).

Lumbar radiculopathy (sciatica) is characterized by severe lower back pain that radiates to the buttocks, legs and feet, and may be combined with tingling in the legs, numbness, muscle weakness, muscle spasms and groin pain. Cervical radiculopathy brings pain in the sides and back of the neck, down the shoulders and arms and sometimes the hands and fingers.

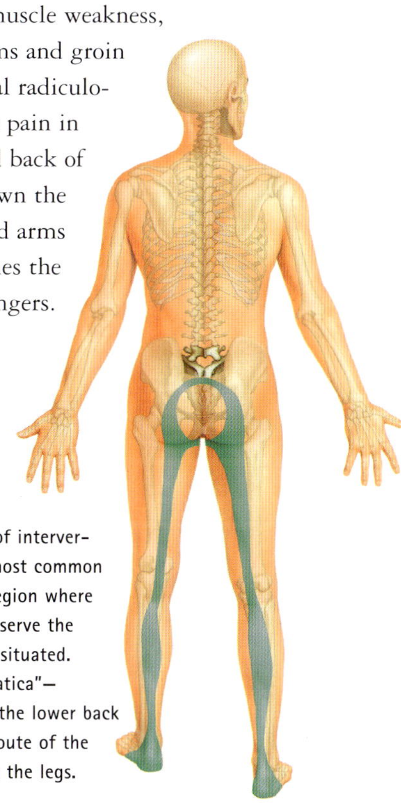

Damaged lumbar disks

The herniation of intervertebral disks is most common in the lumbar region where the nerves that serve the lower limbs are situated. This causes "sciatica"—a sharp pain in the lower back and along the route of the sciatic nerves in the legs.

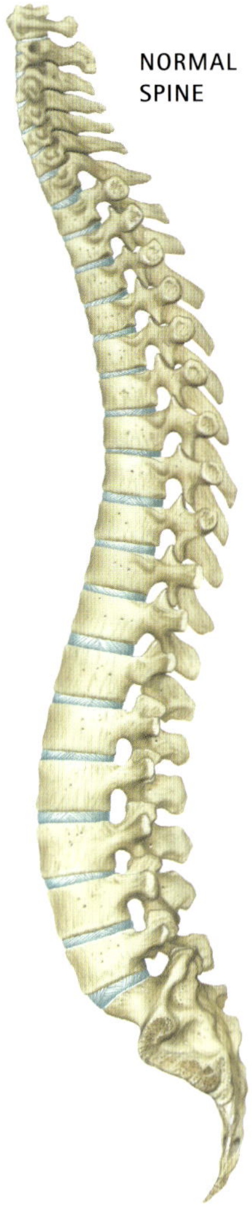

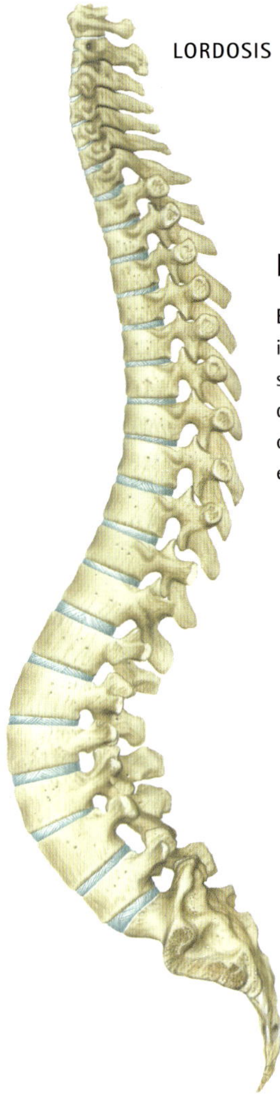

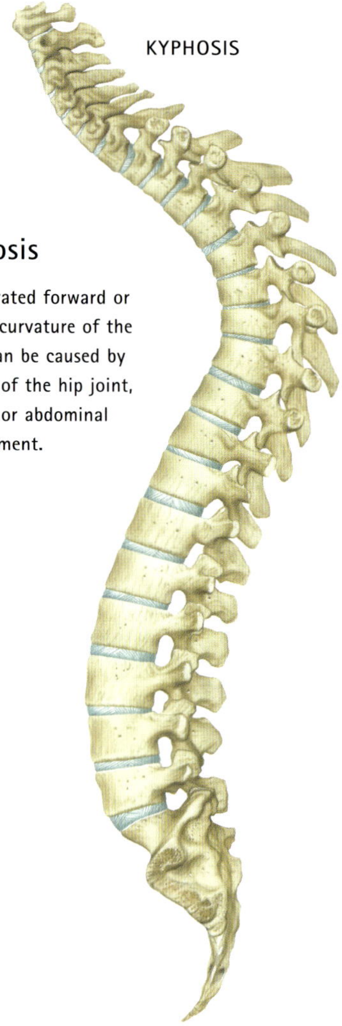

SPINAL CURVATURE

The spinal column is normally curved, but abnormal or excessive curvature (as in kyphosis or lordosis) may be caused by disease, injury or congenital disease.

Lordosis

Exaggerated forward or inward curvature of the spine can be caused by disease of the hip joint, obesity or abdominal enlargement.

Kyphosis

Those with kyphosis of the spine have a "hunchbacked" appearance. Kyphosis can be caused by a number of different reasons, such as injury or a congenital abnormality.

Scoliosis

Scoliosis is lateral curvature of the spine. The first visible sign is likely to be unevenness of the level of the hips or shoulders.

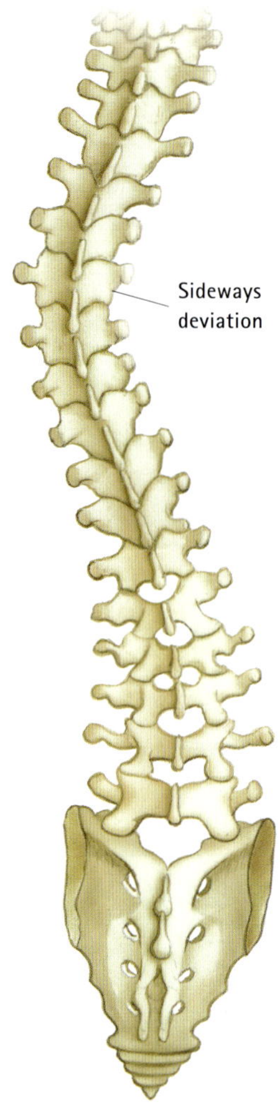

NORMAL SPINE

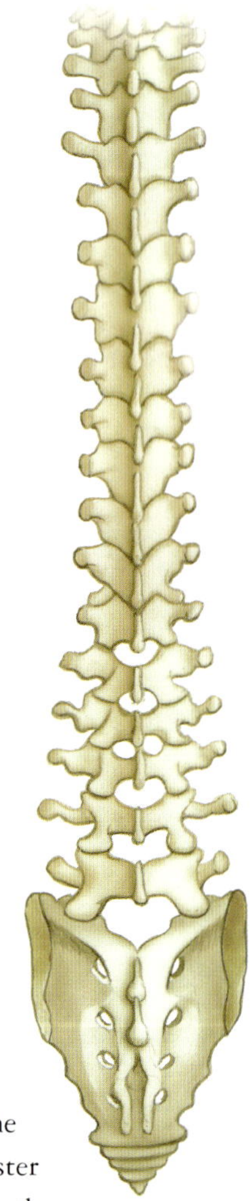

The natural ageing and degeneration of the spine cause most prolapsed disks. Middle-aged and older men are most at risk, especially those who undertake strenuous physical activity. Bed rest on a firm mattress, medication and perhaps physical therapy are usually prescribed to reduce inflammation; surgery may be needed in some cases.

Lordosis

Lordosis, or swayback, is one of a group of spinal deviations. It is usually an exaggeration of the normal forward or inward curvature of the lumbar (middle to lower) spine. It can occur by diseases of the hip joint, obesity or abdominal enlargement. Physical therapy and exercises may help.

Kyphosis

Kyphosis is an abnormal curvature of the spine in an anterior or forward direction. It usually occurs in the upper part of the spine, causing a hunched back and shoulders, with back pain and stiffness. It may be a congenital condition, appearing in children or a dolescents, but more commonly it develops later in life (in older women it is sometimes known as "dowager's hump").

Kyphosis is caused by compression fractures of the spine, due to osteoporosis, ankylosing spondylitis, infection, endocrine diseases, arthritis, Paget's disease, cancer or tuberculosis of the vertebrae. The curvature, and any degenerative changes in the vertebrae, is seen in an x-ray of the spine.

Treatment aims to reverse the underlying cause; back pain and stiffness may be helped by exercises, a firm mattress for sleeping, and a back brace. Bed rest and sometimes traction are recommended for severe pain. Surgery can be used to correct the defect, but is not usually feasible if the condition is due to osteoporosis or other degenerative disease.

Scoliosis

Scoliosis is an abnormal sideways deviation of the vertebral column (spine). It affects the muscles and ligaments connected to the spine and if left untreated can lead to deformities of the rib cage, which in turn can result in heart and lung problems. Treatment may involve wearing a corrective back brace, the use of traction, exercise, a plaster cast or, in more severe cases, orthopedic surgery, in which a metal rod is inserted to support the spine.

Coccygodynia

Coccygodynia (also called tailbone pain) is inflammation of the bony area (tailbone or coccyx) located at the lowest part of the spine. It may occur for no apparent reason, but usually follows an injury to the coccyx—for example when someone falls heavily backward in a sitting position. The pain is worse when sitting or passing feces, but goes when the person stands.

Treatment with analgesic and anti-inflammatory drugs can relieve the symptoms. People with coccygodynia should avoid long periods of sitting, or use a padded cushion or seat when they sit. Persistent coccygodynia can be treated with cortisone injected into the area. In severe cases that don't respond to other treatments, the coccyx can be surgically removed.

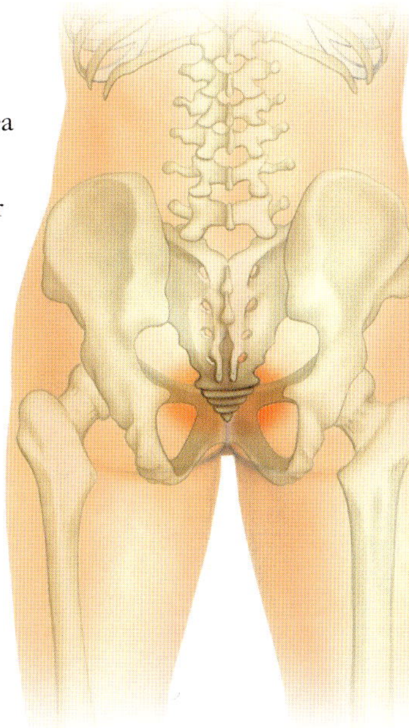

Coccygodynia

Coccygodynia is a painful inflammation of the coccyx (tailbone) that usually occurs following a direct fall onto the area. Treatment with anti-inflammatory drugs may be necessary to relieve the symptoms.

Ankylosing spondylitis

Ankylosing spondylitis is a form of arthritis affecting young men between 15 and 40 years of age. It involves the spine, the sacroiliac joints in the pelvis, the hip and the shoulder. Over the years these joints gradually become inflamed and eventually stiff and immovable. The cause is unknown, though the disease tends to be inherited and to run in families.

The first sign of the illness is often low back pain and stiffness, which is worse in the morning, gets better during the day with exercise, but occurs again at night, often waking the sufferer from sleep. As the disease progresses, back pain and stiffness eventually affect the upper part of the spine and sometimes the neck. The vertebrae in the spine may become fused, creating an abnormal curve in the upper spine. Hips and shoulders are affected in a third of cases. A quarter of cases develop uveitis, or inflammation of the front part of the eye.

There is no cure for ankylosing spondylitis, but some measures can lessen the effects of the disease. Breathing exercises, and exercises to maintain posture, help with the curvature of the upper spine. Analgesics and anti-inflammatory drugs are commonly prescribed. Surgery, for example hip replacement, may be required in severe cases.

Spinal cord

The spinal cord is a cylindrical nervous structure which occupies the vertebral canal, a cavity extending the length of the vertebral column. A series of rootlets (made up of nerve cell fibers) attaches in a line along the front and back of each side of the cord. Those attaching to the front (ventral rootlets) and to the back (dorsal rootlets) group together to form 31 pairs of spinal nerves, giving the cord a segmented appearance. These nerves supply skin, bone, muscles and joints of the limbs and trunk.

The spinal cord is usually 16½–17¾ inches (42–45 centimeters) long and does not extend the full length of the vertebral column. In the adult its lower end (conus medullaris) is usually located at the level of the L1 or L2 vertebra. Below this, the vertebral canal is occupied largely by the elongated rootlets of the lumbar and sacral spinal nerves (cauda equina, which is Latin for "horse's tail") traveling down to lower levels before they exit from the canal.

The spinal cord is made primarily of nerve cell bodies and their associated fibers that function to process and transmit sensory information to the brain and motor information from the brain. The cell bodies group together in the center of the cord to form a column of gray matter, which is H-shaped in cross-section, the H being formed by pairs of dorsal "horns" at the back and ventral horns at the front, separated by an intermediate zone.
The dorsal horns are specialized to process sensory information (for example, touch, pain, temperature, joint sensation) and to relay this information up to the brain.
The ventral horns contain motor neurons, which transmit messages out to the muscles via spinal nerves. The intermediate zone contains many interneurons involved

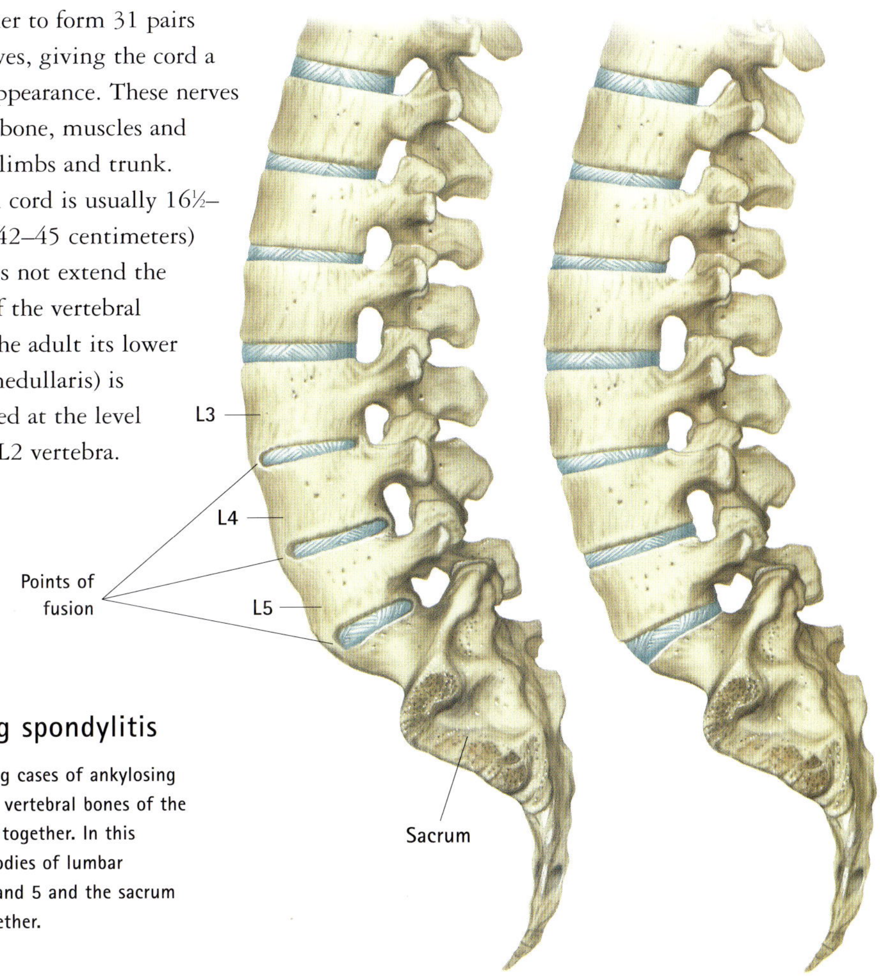

Ankylosing spondylitis

In long-standing cases of ankylosing spondylitis, the vertebral bones of the spine may fuse together. In this example, the bodies of lumbar vertebrae 3, 4 and 5 and the sacrum have fused together.

in linking incoming sensory neurons with outgoing motor neurons to bring about automated (reflex) responses which do not involve the brain at all. Simple spinal reflexes include the stretch reflex such as the knee jerk, in which tapping the patellar ligament below the kneecap brings about contraction of the quadriceps muscle of the thigh; and the withdrawal reflex, in which a pain stimulus to the skin (such as touching a hot iron) causes a reflex withdrawal from the stimulus.

The gray matter of the spinal cord is surrounded by white matter (nerve fibers), transmitting information to and from the brain. Fibers carrying similar types of information tend to group together into bundles as tracts.

SEE ALSO *Nervous system on page 64*

Spinal nerves

Spinal nerves emerge from the sides of the vertebral column and function to transmit information in both directions between the spinal cord and the peripheral structures of the body. Each nerve is made up of sensory fibers (transmitting information from skin, muscles, bones and joints to the spinal cord) and motor fibers (transmitting messages away from the spinal cord toward the skeletal muscles). Some spinal nerves also carry sympathetic (autonomic) fibers (transmitting messages to sweat glands and blood vessels) and parasympathetic fibers to mucous membranes and smooth muscles in the pelvis.

Each nerve is formed within the vertebral canal by the union of its dorsal and ventral roots. The dorsal roots, made up of sensory fibers, emerge in a line along each side of the back of the spinal cord. The ventral roots, made up mainly of motor fibers, emerge in a line along each side of the front of the cord. Once the two roots unite, the spinal nerve leaves the vertebral canal through a space on each side between each pair of adjacent vertebrae. Soon after it leaves these vertebrae, the spinal nerve divides into a small branch, which supplies structures in the back, and a large branch, which supplies the limbs and the rest of the trunk.

The 31 pairs of spinal nerves are named according to the level that they exit from the vertebral column. There are eight cervical spinal nerves (C1–C8), twelve thoracic (T1–T12), five lumbar (L1–L5), five sacral (S1–S5) and one coccygeal nerve. Each nerve supplies a circumscribed area of skin (dermatome) and a specific group of muscles. The thoracic nerves supply skin and muscles of most of the trunk wall, whereas the ventral rami of cervical, lumbar and sacral nerves form complex networks, known as plexuses, from which branches emerge to supply the motor and sensory needs of the limbs.

One common disease which affects spinal nerves is shingles, in which the varicella-zoster virus that causes chickenpox infects the cells of the sensory roots of one or more spinal (or cranial) nerves.

SEE ALSO *Nervous system on page 64*

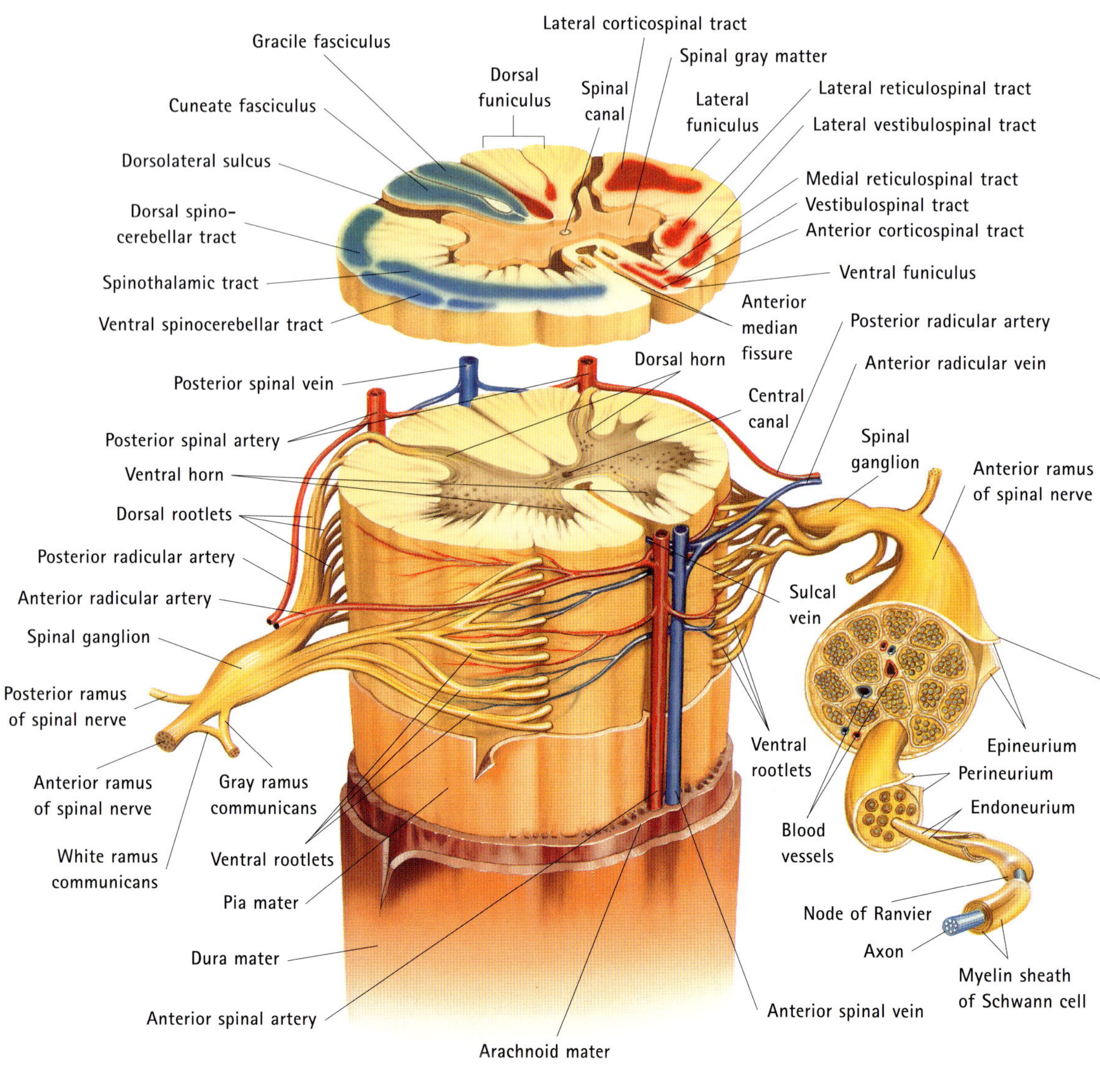

Spinal cord—cross-section

The spinal cord is a column of nerve tissue that runs along the vertebral canal. It comprises nerve cells that process sensory information and transmit it to the brain. It also contains motor neurons which control the actions of muscles in the limbs and trunk. Spinal nerves attached to the cord pass these messages to the body's skin, bones, muscles and joints via the peripheral nerves.

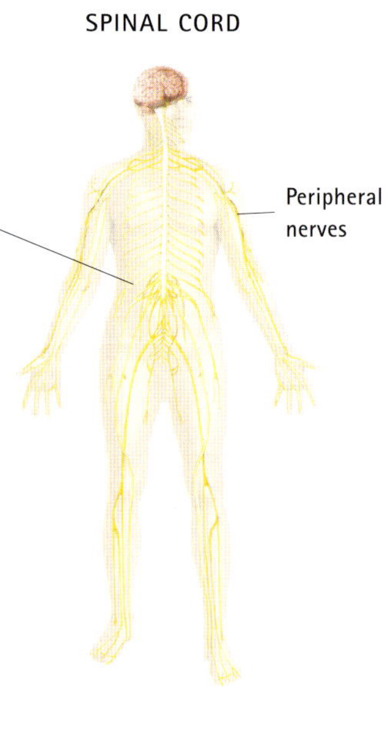

Disorders of the spinal cord and spinal nerves

The most common disorder affecting the spinal cord is spinal cord injury occurring as a result of motor vehicle accidents, falls, and sporting injuries. Spinal cord injuries impose a dramatic change in a patient's life, and patients require extensive social support and rehabilitation.

Neural tube defect

Neural tube defects are disorders resulting in the abnormal development of the brain and spinal cord, and/or the membranes (meninges) surrounding them, in the embryo. In week 3 or 4 of pregnancy, a narrow sheath normally folds and closes to form what is known as the neural tube. This is the first stage of development of the central nervous system (brain and spinal cord); if the tube fails to close perfectly such defects as spina bifida, anencephaly and hydrocephalus can occur. Anencephaly is a condition in which a large part of the brain, skull and scalp are absent due to the incomplete closure of the end of the neural tube nearest to the head.

Research has shown that ensuring a daily intake of 0.4 milligrams of folic acid before and during pregnancy can halve the risk of neural tube defects. Folic acid occurs naturally in foods such as dark green leafy vegetables; supplements may be taken in tablet form. Neural tube defects can be detected by prenatal tests.

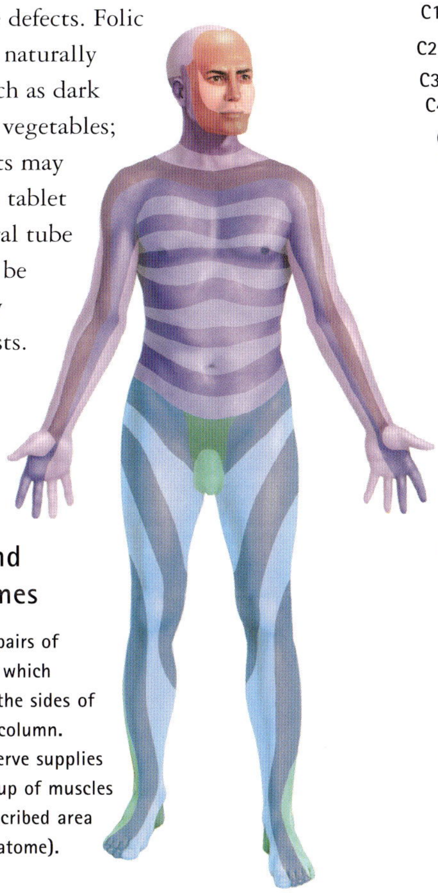

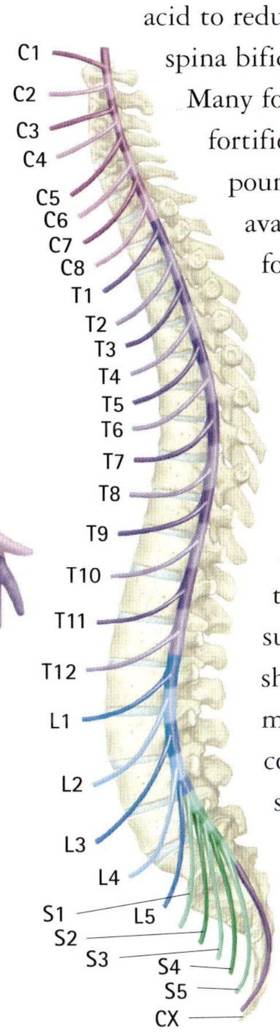

Spinal nerves and dermatomes

There are 31 pairs of spinal nerves, which emerge from the sides of the vertebral column. Each spinal nerve supplies a specific group of muscles and a circumscribed area of skin (dermatome).

Spina bifida

Spina bifida is a congenital defect of the spinal column resulting from the abnormal formation of the neural tube very early in embryonic development (usually during weeks 3–4 of fetal life). It can appear in several forms. In its most benign form, spina bifida occulta, neither the spinal cord nor the meninges (the covering of the spinal cord) protrudes through the opening left in the vertebrae.

In its most severe form, meningomyelocele, the spinal cord and nerves are exposed. Weakness in the feet (sometimes paralysis), problems with reflexes, and spinal defects indicate spina bifida. A soft fatty deposit on the skin covering the defect, or a cyst protruding over the spine, may also be present. Babies born with meningomyelocele usually undergo surgery soon after birth. Early surgical correction of the defect is important to minimize the risk of meningitis and further neurological damage. Although surgery will improve quality of life for the child, disorders such as limb paralysis and bladder and bowel problems may still occur.

During pregnancy it is important that the mother's diet contains adequate folic acid to reduce the risk of spina bifida in the baby. Many foods are now fortified with this compound and it is also available in tablet form if prescribed. Women of childbearing age should aim to ingest 400 micrograms of folic acid a day. It is recommended that folic acid supplementation should begin 3–4 months prior to conception and should continue for the first 3 months of the pregnancy. Spina bifida can be screened for in pregnancy with the alpha-fetoprotein (AFP) test, ultrasound and amniocentesis.

Spina bifida

Spina bifida occurs when the bony column that surrounds the spinal cord does not fuse properly during embryonic development. The spinal cord and its covering (the meninges) may then protrude through the opening between the vertebrae, creating a cystic swelling filled with cerebrospinal fluid.

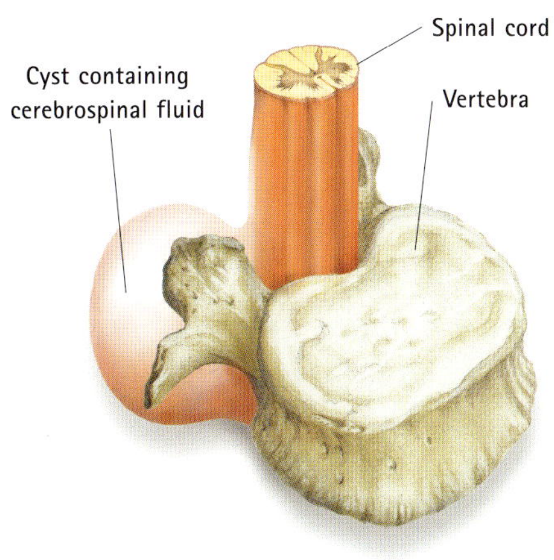

Paraplegia and quadriplegia

Paraplegia is partial or full paralysis of the body below the chest or waist, involving the trunk and lower limbs. Quadriplegia (also referred to as tetraplegia) is partial or full paralysis and loss of sensation below the neck, involving all four limbs and the trunk of the body.

In paraplegia, all body parts below the level at which the spinal cord is damaged are affected.

Quadriplegia is caused by damage to the spinal cord in the neck, usually at the level of the fourth and fifth cervical vertebrae (if injuries occur above this level the diaphragm is often paralyzed and the victim may die). Injury is usually from trauma, most commonly road accidents and sporting injuries. Males between 15 and 35 years old make up the group that is most likely to be affected.

Less frequently, quadriplegia may be caused by spinal-cord tumors or birth defects. As well as muscle paralysis, paraplegia and quadriplegia lead to dysfunction of other organs and body systems whose nerve supply from the spinal cord has been disrupted. There may be loss of urinary and bowel control, impaired sexual function, loss of normal blood pressure, loss of body-temperature control, poor healing of tissues, and constipation.

Treatment as soon as possible after the injury gives the injured person the greatest chance of minimizing the extent of the damage and of recovering some function. Surgery may be performed to remove fluid or tissue that is pressing on the spinal cord (decompression laminectomy); to remove bone fragments or foreign objects; or to stabilize fractured vertebrae by fusion of the bones or insertion of hardware. Bed rest and spinal traction (which immobilizes the spine and reduces dislocation) promotes healing.

If movement or sensation return within a week after the injury, then most function will eventually be recovered (although it may take 6 months or more). Losses of function that remain after 6 months are likely to be permanent. Approximately one-third of sufferers will be permanently wheelchair-bound. Thus, rehabilitation remains the mainstay of treatment. Physical therapy can help joint stiffness, and passive exercises will help prevent contractures.

Prolonged immobility can give rise to complications such as constipation, pressure sores and ulcers, so frequent position changes and good skin care are important. Other complications, such as bladder and lung infections and kidney stones, will also need to be treated. Psychotherapy or counseling may relieve depression and sexual problems.

With rehabilitation, some lost functions can be restored or compensated for. However, paraplegics and quadriplegics will have permanently reduced mobility and will require help with many facets of their lives, including accommodation, employment, transport, and access to buildings.

The condition can often be prevented by following common-sense safety precautions such as the wearing of seat belts in cars and of protective headgear for contact sports. When swimming, one should not dive into shallow pools or into water of unknown depth.

SEE ALSO *Nervous system on page 64; Spinal cord on page 217*

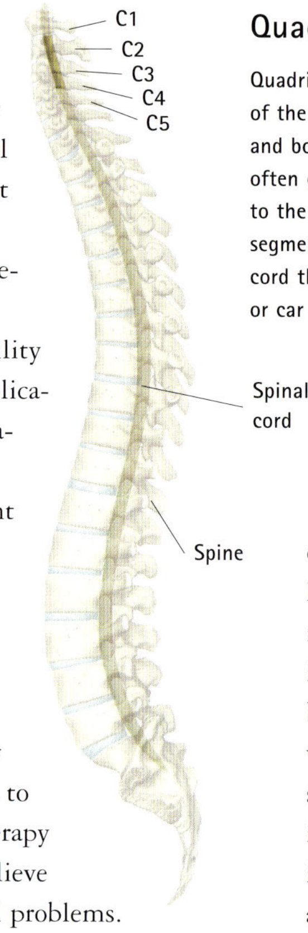

Quadriplegia

Quadriplegia is the paralysis of the torso, both arms and both legs. It is most often caused by injury to the fourth or fifth segment in the spinal cord through sporting or car accidents.

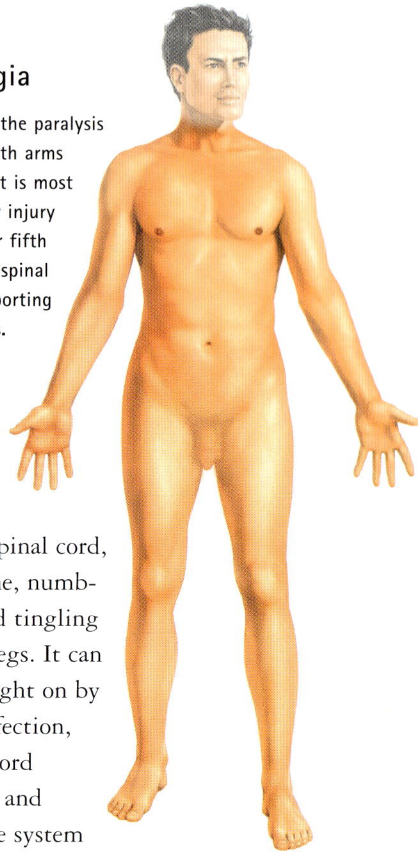

Paralysis—paraplegia

Paraplegia is the result of injury or disease to the spinal cord between the T1 (thoracic) and L2 (lumbar) segments. It spares the arms but depending on the nerves damaged may involve the legs, pelvic organs and trunk.

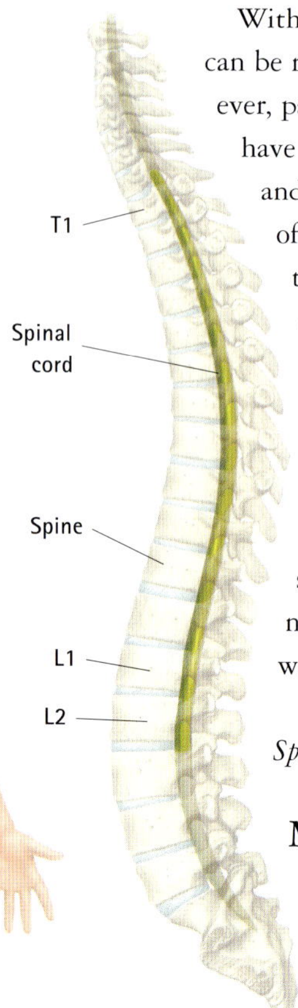

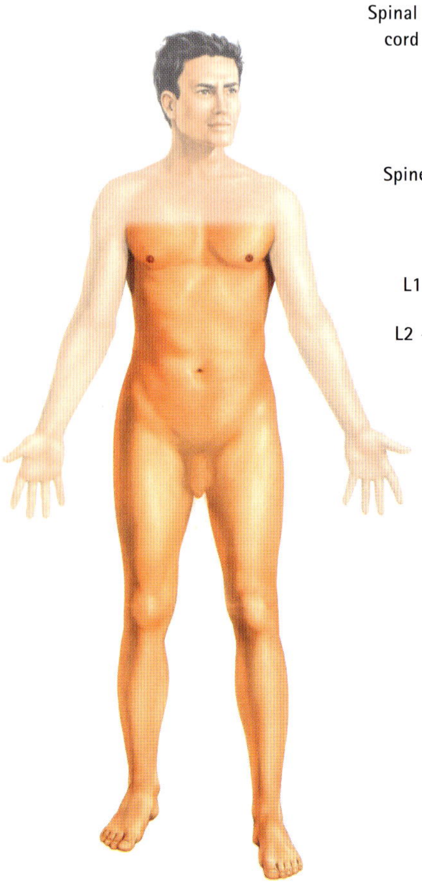

Myelitis

Myelitis is a general term for inflammation of the spinal cord. It involves the loss of fatty tissue (myelin) around the nerves. One of the most common types is acute transverse myelitis, a neurological syndrome that involves inflammation through one level of the spinal cord. This rapidly developing condition obstructs the path of motor nerve fibers, causing low back pain, muscle spasms, dysfunction of the spinal cord, headache, numbness and tingling in the legs. It can be brought on by viral infection, spinal cord injuries and immune system abnormalities in which the spinal cord is attacked.

Chest wall

The chest wall consists of the sternum, rib cage, thoracic vertebrae and surrounding musculature. The upper portion of the rib cage is covered by the collar bones. The chest wall plays an important role in breathing, as well as protecting the underlying heart and lungs from injury.

SEE ALSO *Muscular system on page 48; Skeletal system on page 30*

Trapezius muscle

The trapezius muscle is a flat, triangular muscle, lying under the skin of the back of the neck and upper part of the back of the chest. The triangular muscles on each side of the back meet in the midline, forming a trapezoidal (four-sided) shape.

The trapezius muscle arises, directly or through ligaments, from the spines of the vertebrae of the thorax and neck, and from the occipital bone of the cranium. The muscle fibers converge upon the shoulder, where they are attached to the collar bone (clavicle) and shoulder blade (scapula). The trapezius muscle acts mainly in steadying the shoulder during arm movements; it also assists in movements of the scapula.

Trapezius muscle

The trapezius muscle acts to draw the shoulder back and assists some movements of the arm. This muscle is flat and is situated at the back of the neck and upper chest.

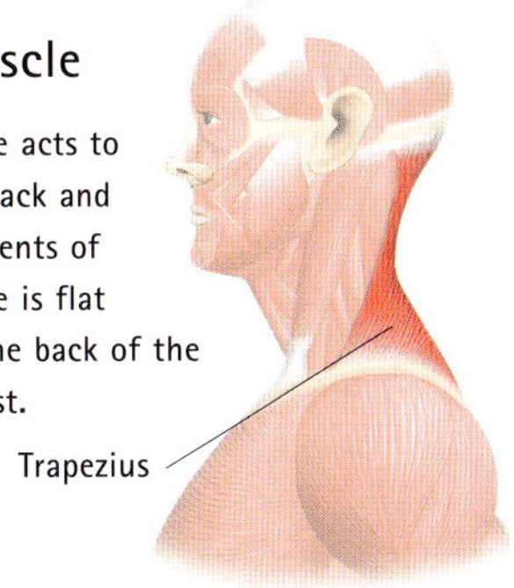

Ribs

The rib cage helps shield the heart and lungs from injury. Movement of the rib cage also assists the diaphragm in controlling the intake and expulsion of air during breathing.

Typically, the human skeleton has 24 ribs, arranged in 12 pairs. At the back, they join the thoracic vertebrae in the spine. At the front, each of the upper seven pairs connects to the sternum (or breastbone) directly by a costal cartilage. Pairs 8, 9 and 10 also connect to costal cartilages but these join with each other and then join the cartilage of the seventh rib. As they are not directly attached to the sternum, they are termed false ribs. The top seven pairs are directly attached to the sternum, and so are called true ribs. Pairs 11 and 12 are not attached at all at the front and are called floating ribs.

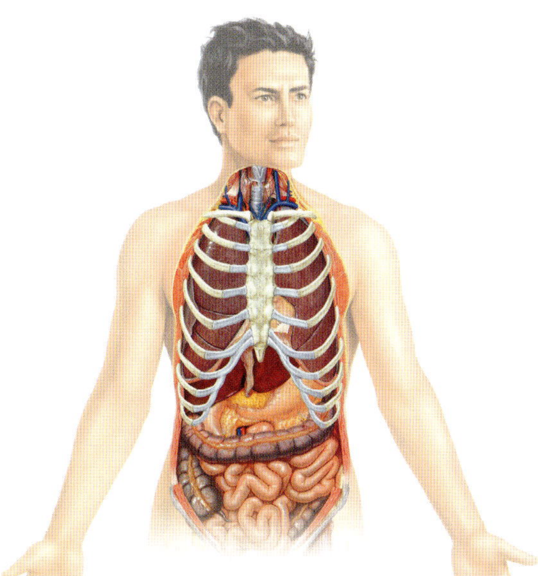

Rib cage

The rib cage is designed to protect the organs within the chest, particularly the heart and lungs. Damage to the rib cage may lead to serious problems, such as a punctured lung.

Collar bone

The collar bones, or clavicles, are a pair of short horizontal bones above the rib cage. They are attached to the breast bone (sternum), and the two shoulder blades (scapulas) on either side. The function of the collar bones is to stabilize the shoulders.

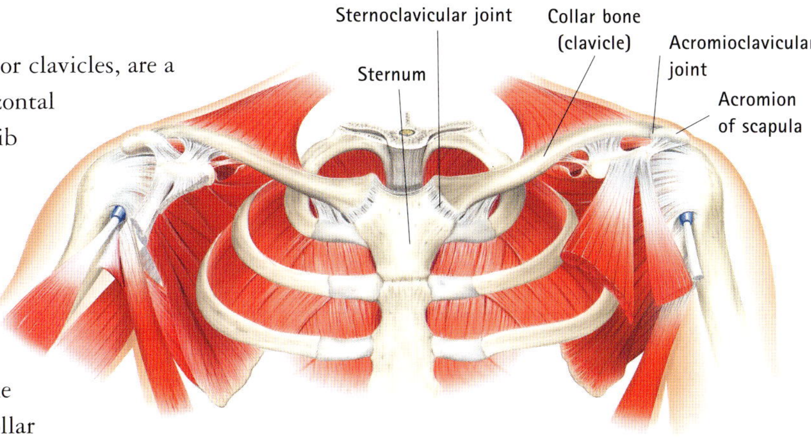

Sternum

The sternum (or breastbone) forms part of the skeleton of the thorax. It is situated in the front wall of the chest, in the midline. It consists of three parts, from top to bottom: the manubrium; the body of the sternum; and the xiphoid process (pronounced "ziffoid"). The manubrium articulates with the clavicles (collar bones), thereby assisting in stabilizing the shoulders. The ribs are connected to the sides of the manubrium and body of the sternum by their costal cartilages. The pectoralis major muscle is attached in part to the sternum. The sternum contains a marrow cavity, which is usually a convenient site for bone marrow biopsy.

Ribs

The 24 ribs are arranged in 12 pairs. The first 7 pairs are referred to as "the true ribs" because they are directly attached to the breastbone (sternum). Pairs 8–10 are known as "false ribs" because they are not attached to the sternum. The lowest two pairs (11 and 12) are not attached at the front and are called "floating ribs."

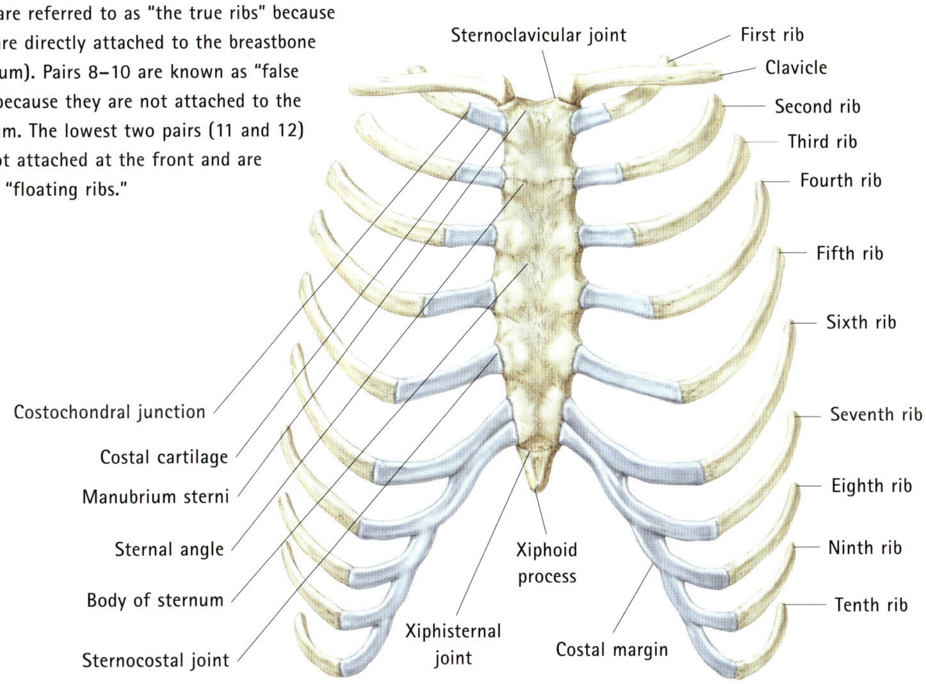

Collar bone

The collar bone, or clavicle, helps to stabilize the shoulder joint. It is attached to the scapula (shoulder blade) at one end and the sternum (breastbone) at the other.

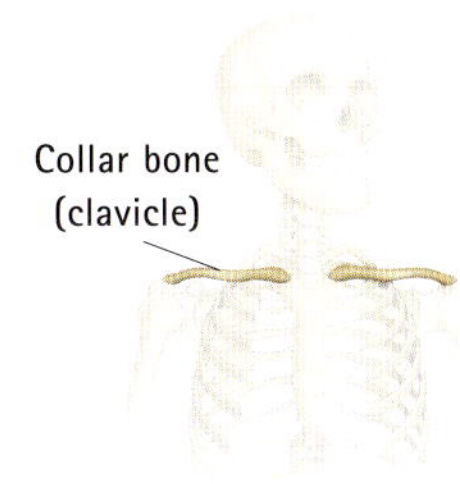

Pectoral muscles

The pectoral muscles lie in the front of the chest, under the breasts. There are two on each side: the pectoralis major and pectoralis minor. They arise mainly from the collar bone (clavicle), sternum and rib cage, and are attached to the humerus in the upper arm and the coracoid process of the shoulder blade (scapula). The pectoralis major can be felt under the breast when the muscle is tensed by attempting to pull a fixed object sideways toward one's midline.

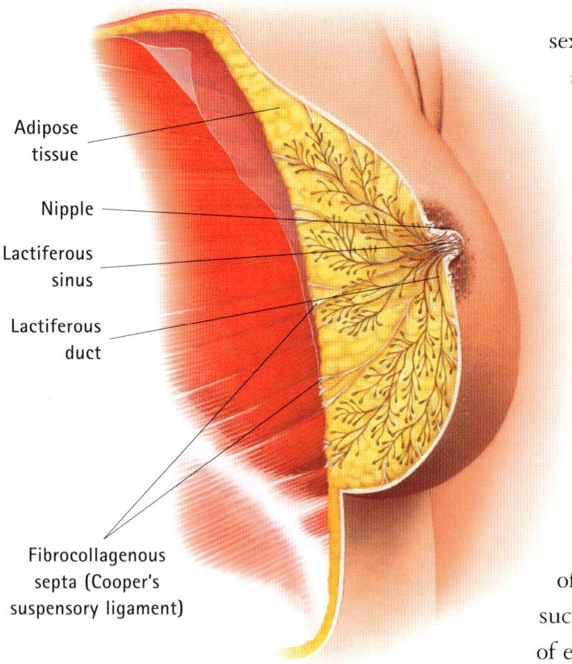

Normal breast

The development of breast tissue is governed by hormones, in particular the sex hormone estrogen. Toward the end of each menstrual cycle the breasts tend to swell and may become painful, but after menstruation, return to their normal size. After the menopause, the breasts usually shrink in size.

Breasts

Mammary glands, or breasts, are modified sweat glands and are present in all mammals. They develop in the embryo along two narrow elongated regions (milk lines) extending from the armpit (axilla) to the groin. In humans a single pair of mammary glands develops under the skin of the upper chest although it is not uncommon for rudimentary extra mammary glands to develop elsewhere in the milk line. Normally only functional in the female, they are composed mainly of fat cells (cells capable of storing fat) interspersed with sac-like structures called lobules. These lobules are glands that can produce milk in females when stimulated by certain hormones such as prolactin. The lobules empty into a network of ducts (channels) that transport milk from the lobules to the nipple.

Differences in breast size and shape are largely due to inherited factors; also, being overweight increases the amount of fatty tissue in the breasts and hence their size.

Female mammary glands undergo considerable changes during the individual's life. At birth the mammary glands of both sexes are alike, consisting of a limited amount of glandular tissue leading into 15–20 openings in the nipple (some domestic animals have only one opening into the nipple). There is little fat or fibrous tissue. Some temporary secretory activity, traditionally known as witch's milk, may occur briefly just after birth under the influence of placental hormones. After puberty, development involves elevation and pigmentation of the nipple and a considerable increase in fat and fibrous tissue. There is also some proliferation of the glandular tissue. Transient changes such as swelling may occur toward the end of each menstrual cycle.

During pregnancy, under the influence of placental hormones, there is a decrease in the amount of fat and fibrous tissue and an enormous increase in the amount and complexity of the glandular components. During the third trimester a liquid called colostrum is secreted. Colostrum contains milk fat and immunoglobulins but few nutrients. It continues to be secreted for the first few days after birth and then is followed by the production of true milk, induced by a lactation-stimulating hormone, prolactin.

Release of milk from the mammary glands is controlled by another pituitary hormone, oxytocin. Breast milk is rich in lactose, protein, calcium and fat, and also contains vitamins and immunoglobulins. Lactation can continue for several years if suckling is maintained. After weaning, most of the glandular tissue breaks down and is replaced by fat and fibrous tissue.

The glandular and fibrous components of mammary glands gradually decrease with age up to and after menopause, and there is an increased amount of fat. Fat is more translucent to x-rays which facilitates mammograms in older women.

The mammary glands in males have a similar structure throughout life to that of prepubescent females. Male mammary tissue is, however, sensitive to hormonal influences with temporary enlargement sometimes occurring at puberty and more permanently in ageing males. Considerable development may also accompany certain pituitary tumors which secrete excessive amounts of prolactin. Mammary enlargement in males or prepubescent girls is called gynecomastia. In older men, it may be due to treatment with estrogens or steroids.

SEE ALSO *Adolescence on page 394; Breast feeding on page 414; Hormones on page 112*

Nipple

The nipple is the raised area in the center of a breast. It is surrounded by a disk-shaped pigmented area called the areola. The milk ducts of the breast empty into the nipple. The nipples, which contain some erectile tissue, can be an erogenous zone, common to both men and women.

The nipples need extra care when a woman is breast feeding. Good breast hygiene and the use of breast pads will maintain dryness between feedings and relieve symptoms of tenderness and irritation. Breast creams may be used to help keep the nipple area lubricated and supple.

Disorders of the Breast

Breast cancer is the most serious disorder affecting the breast. One in every 11 American women will develop breast cancer during her lifetime. In recent years mass screening programs, utilising mammography, have helped identify breast cancer at an earlier and more treatable stage.

SEE ALSO *Biopsy on page 435; Mammography on page 432; Mastectomy on page 465*

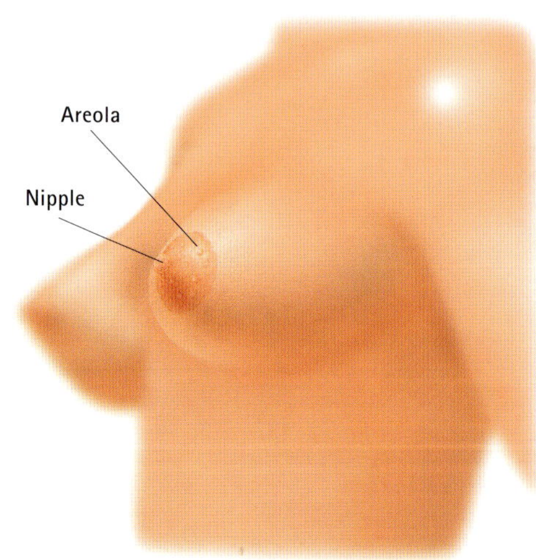

Nipple

The nipple is a conical projection of each breast. The areola is the reddish brown region of skin around each nipple.

Fibrocystic disease

Fibrocystic disease is a very common condition of women in their reproductive years. It is characterized by a number of small cysts (fluid filled sacs surrounded by fibrous tissue) that give the breast a dense, irregular and lumpy consistency. Usually both breasts are involved.

The breasts become enlarged and tender just before the menstrual period, with symptoms improving after the period finishes. The condition improves after menopause and often after commencing the contraceptive pill. No treatment is required, though any single lump that stands out should be investigated by a physician.

Mastitis

Mastitis is inflammation of the breast. Acute mastitis is caused by an infection that enters the breast through a cracked nipple, most commonly while breast feeding. The breast becomes hot and swollen and there may be a fever. Mastitis is treated with antibiotic drugs. Breast feeding should be continued if possible.

If an abscess forms, breast feeding should cease and the abscess must be incised and drained. Chronic mastitis is another name for fibrocystic disease of the breast.

Breast cancer

Cancer of the breast is the most important malignant tumor in women. It is by far the most common cancer in women in the economically developed nations of the world, where approximately one woman in ten will develop the disease. (Cancer of the cervix is almost as common in many developing nations.) Breast cancer is also one of the leading causes of cancer death in women (the other is lung cancer). Men can develop breast cancer as well, but this is approximately 100 times less common than in women.

CAUSES

Except during pregnancy and breast feeding, there are no functional milk-producing glands in the breast. There are, however, branching tubes called ducts which radiate out from the nipple and subdivide into smaller and smaller ducts. The layer of cells lining these small ducts gives rise to almost all breast cancers. As in most cancers, not enough is understood about how breast cancers develop, but two influences do seem to be important. Firstly, the growth and division of the lining cells of the ducts is controlled by hormones (especially estrogen), and stimulation of cell growth by a relative excess of estrogen (without a counterbalancing effect of progesterone) may create an environment in which breast cancer is more likely to develop. This could explain why breast cancer is often associated with other non-cancerous lumps in the breast. It could also account for the greater risk of breast cancer later in life, in association with early onset of menstruation and late menopause, in women who have not had children and in women whose first child was born relatively late.

Secondly, some breast cancers arise because of genetic factors. Genes associated with a familial risk of developing breast cancer have been identified. Despite this, environmental influences are probably much more important in causing breast cancer, but we have very little knowledge about exactly which factors play a role or how they affect an individual's risk of developing the tumor.

HOW BREAST CANCER DEVELOPS

Most breast cancers develop in women aged over 30 years and arise either in the upper outer part of the breast (which includes tissue that extends into the armpit) or in tissue just under the nipple. At first, the tumor cells grow entirely within the ducts of the breast, but later they invade through the walls of the ducts and into the surrounding fat and connective tissue.

Fibrocystic breast

The breast has benign (non-cancerous) lumps which often become tender in the days before a period starts.

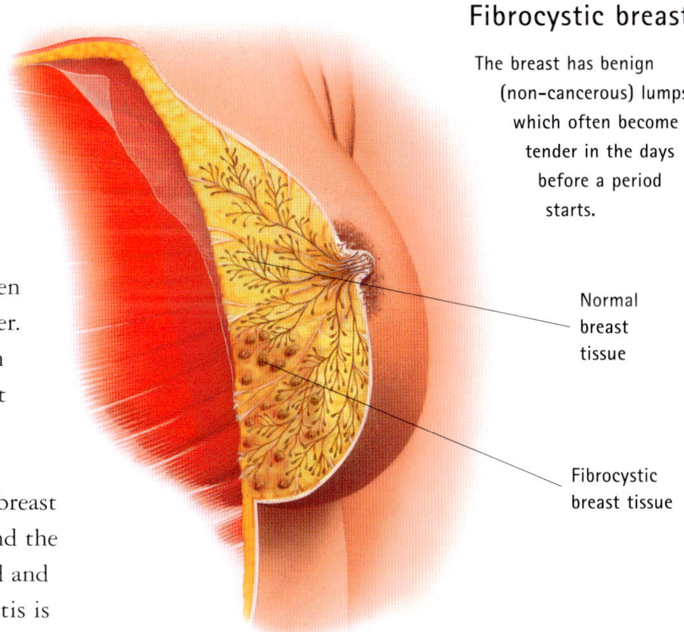

Breast abscess

A red, swollen, painful lump in the breast is usually a sign of a breast abscess. It often occurs if the nipples are cracked or inflamed, as bacteria can more easily enter the breast tissue.

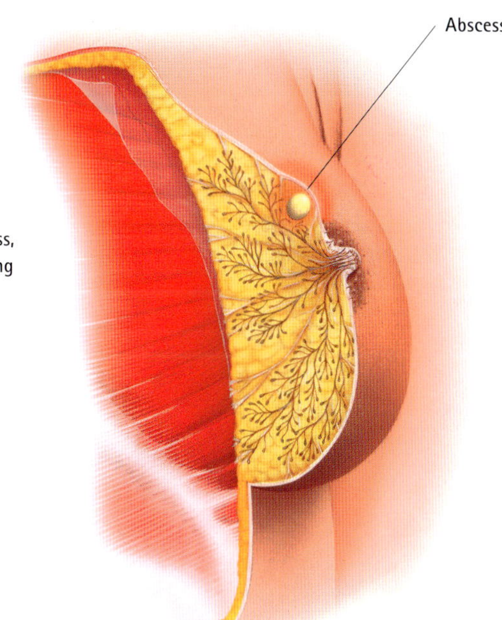

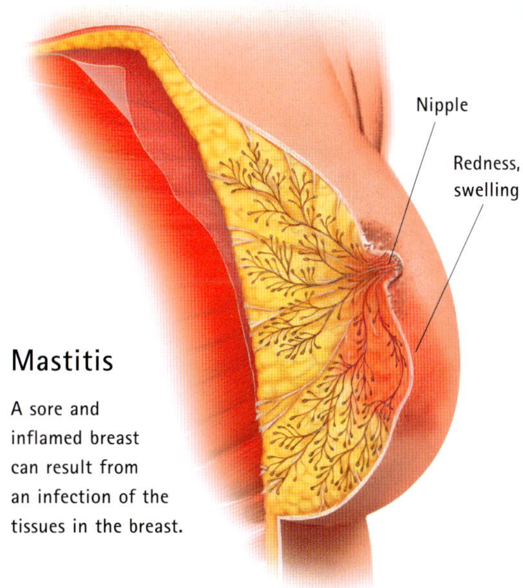

Mastitis

A sore and inflamed breast can result from an infection of the tissues in the breast.

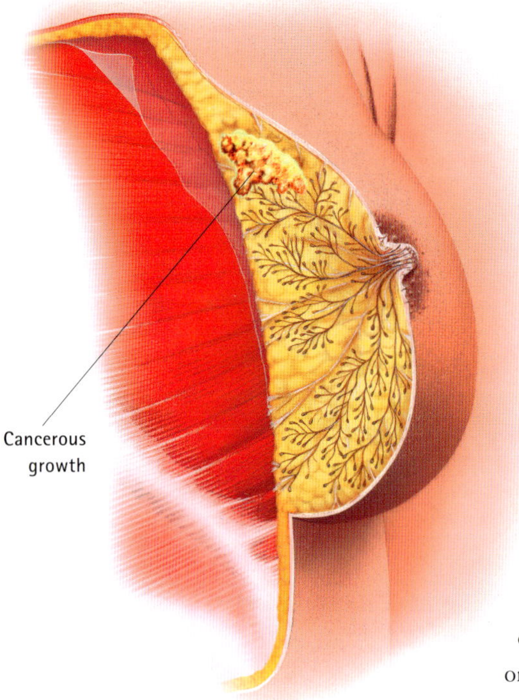

Breast cancer

The most common indication of a breast cancer is a hard lump in the breast that is often immovable. The skin over the lump may look dimpled, the nipple may turn inward and there may be lumps under the arm.

Still later, breast cancer cells enter the lymphatic vessels and spread to the nearest groups of lymph nodes, which include nodes in the connective tissue of the armpit and nodes above the collar bone. At about the same time, the cells may enter blood vessels and thus spread to the lungs, or throughout the body. Early diagnosis is critical for breast cancer. Some of these tumors are more rapidly growing than others and more likely to spread.

However, the stage at which a breast cancer is diagnosed is the single most important factor that determines the course of the disease and the patient's chances of survival. Tumors diagnosed at the stage where the cancer cells are still confined to the walls of the ducts can be treated very effectively. In such cases, the likelihood of the patient surviving at least five years from the time of diagnosis is over 90 percent. Once the cancer cells have spread to nearby lymph nodes, the survival figure drops considerably. When tumors are diagnosed at the stage when secondary cancer is widespread, the five-year survival rate is less than 15 percent. Unfortunately, breast cancers may not produce obvious symptoms until relatively late, by which time they may have spread too far for effective treatment. Therefore, an active effort to identify them at an early stage must be made.

Two useful approaches to achieving early identification are self-examination for the presence of a lump, and screening by mammography.

To make a reliable diagnosis, any mass that is detected must be sampled for microscopic examination. This may involve removal of cells using a fine needle, removal of a core of tissue using a biopsy needle, or complete removal of the lump. The pathologist can provide information about the likely behavior of an individual cancer based on the appearance of the cells.

TREATMENT

Current treatment for breast cancer involves much less removal of tissue by surgery than was the case 20 or 30 years ago. Radiation therapy, chemotherapy and hormonal therapy all have a place in modern management. Newer treatments being tested include drugs and antibodies that can block signals to cancer cells. Treatments may include lumpectomy, mastectomy (partial, total or radical, i.e. extending to the armpit) and radiation therapy. Chemotherapy and hormonal therapy with antiestrogen drugs such as tamoxifen may also be used. After the procedure, except in the case of radical mastectomy, a breast implant or prosthesis may be inserted.

Fibroadenoma

A fibroadenoma is a benign tumor formed from glandular tissue. It occurs most commonly in the female breast, most often among women going through their reproductive years. Many times, fibroadenomas are not tender. However, they generally need to be surgically removed to confirm they are benign, and because they can continue to grow and cause an unsightly lump. A fibroadenoma can usually be removed under local anesthesia.

PELVIS

The term pelvis (from the Latin meaning "basin") encompasses a number of structures. It can refer to the bony pelvis, a ring of bone between the trunk and thigh. It can refer to the lesser or true pelvis, the part of the bony pelvis below the pelvic inlet, or may refer to the pelvic cavity, a funnel-shaped region within the lesser or true pelvis that contains pelvic organs, including the bladder, rectum and internal genitalia. Usually the term refers to the lesser pelvis.

SEE ALSO *Skeletal system on page 30*

Bony pelvis

The bony pelvis forms the skeletal framework for the pelvis and is mostly covered by muscles. It functions to transfer weight from the vertebral column to the lower limbs as well as to provide protection for the pelvic and lower abdominal organs. It comprises the hip bones, sacrum and coccyx. Each hip bone is made up of three bones (ilium, ischium and pubis) that are separate in a child, but later fuse. Each hip bone unites in front at the pubic symphysis and joins the sacrum behind at the sacroiliac joints. The coccyx (tail bone) forms a joint with the lower end of the sacrum. The pelvic inlet demarcates a region called the greater or false pelvis above (part of the abdominal cavity) and the lesser or true pelvis below. The lower borders of the bony pelvis form the pelvic outlet. In women, the pelvis is constructed to accommodate the fetus during childbirth. Therefore, the pelvic inlet and outlet are larger than in the male pelvis, the length of the canal is shorter and its walls are more parallel than those of the male.

Lesser pelvis

The lesser pelvis (or true pelvis) is bounded at the back and above by the sacrum and coccyx. Muscles of the wall include the piriformis muscle toward the back, and the obturator internus muscle on the side wall. The pubic bones and pubic symphysis lie in front and below the lesser pelvis, and the floor is formed by the pelvic diaphragm; below this is the urogenital diaphragm. The lesser pelvis is open above to the abdominal cavity.

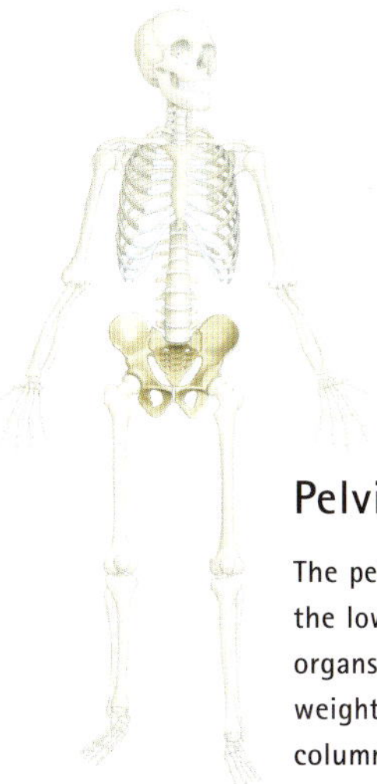

Pelvis

The pelvis helps protect the lower abdominal organs and transfers weight from the vertebral column to the lower limbs.

Pelvis—female

Unlike its male counterpart, the female pelvis is designed to support the fetus during pregnancy. The inlet and the outlet are larger than the male, while the canal is shorter.

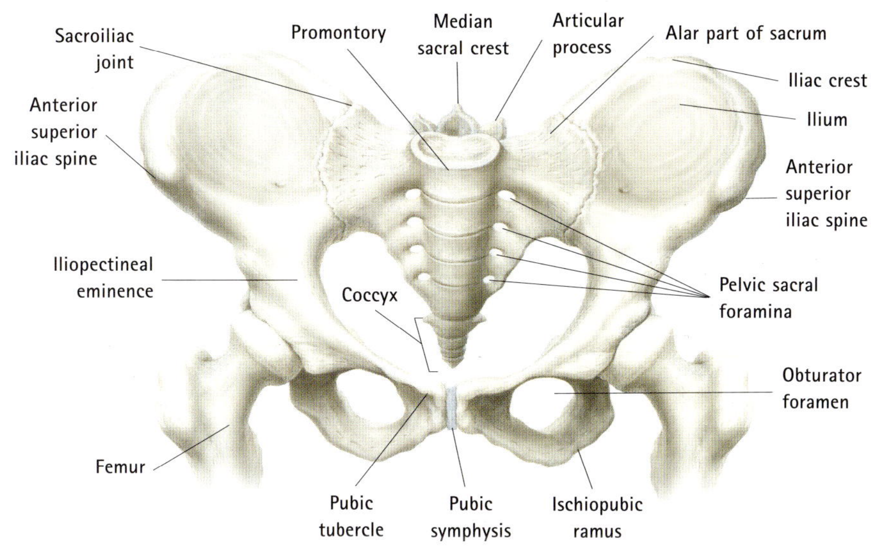

Pelvic cavity

Within the lesser pelvis is a funnel-shaped cavity or region, known as the pelvic cavity. It is defined as the area between the pelvic inlet above and the thin, sheet-like muscle of the pelvic diaphragm below. It contains and protects the pelvic organs: the bladder and rectum in both sexes; the uterus and vagina in the female; and the prostate and seminal vesicles in the male.

Pelvic diaphragm or floor

The pelvic diaphragm (or pelvic floor) is important in supporting the pelvic organs. It is formed by the coccygeus and the levator ani muscles, and has a sphincteric (constrictive) action on the rectum and vagina, and assists in increasing intra-abdominal pressure. The puborectalis part of the pelvic diaphragm is important in fecal and urinary continence.

The pelvic floor muscles form a "floor" or diaphragm across the pelvis, running from the back to the front and in from the sides. Strung like a hammock between the sacrum at the back and the hip bones at the front and sides, these muscles support the bladder and bowel and, in women, the uterus.

Pelvic injuries

A direct blow or compression injury may cause the pelvic bones to fracture and/or the joints to dislocate. Falls on the feet may fracture the part of the hip bone associated with the hip joint—this is called fracture of the acetabulum.

Soft tissue injury must be considered in pelvic fractures, as there is potential for damage to bladder, urethra, rectum, blood vessels and nerves.

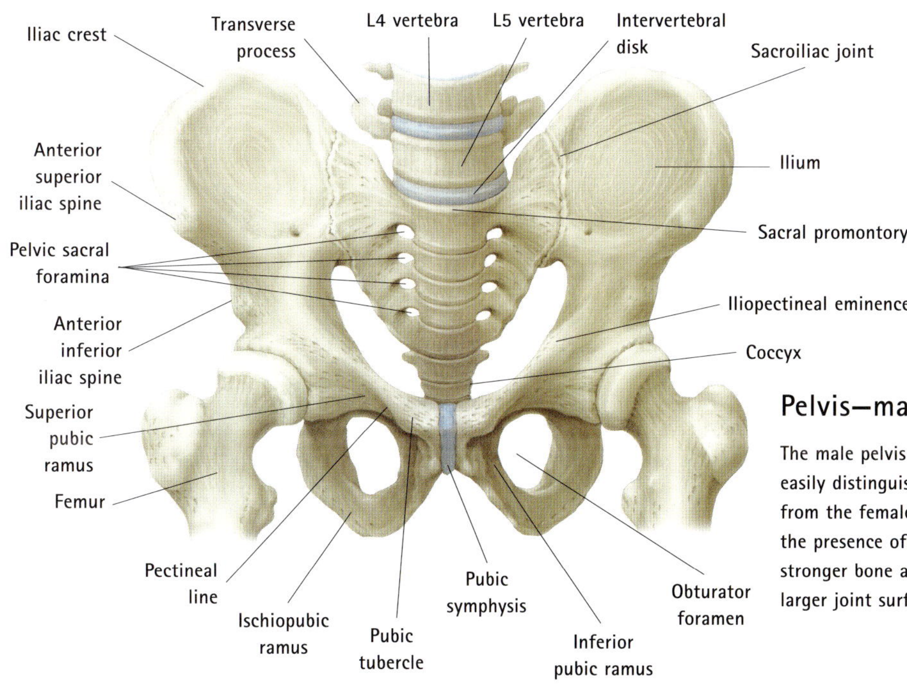

Pelvis—male

The male pelvis is easily distinguished from the female by the presence of stronger bone and larger joint surfaces.

Pelvic floor muscles

Stretching from the sacrum at the back to the hip bones at the front, the pelvic floor muscles form a muscular floor across the pelvis.

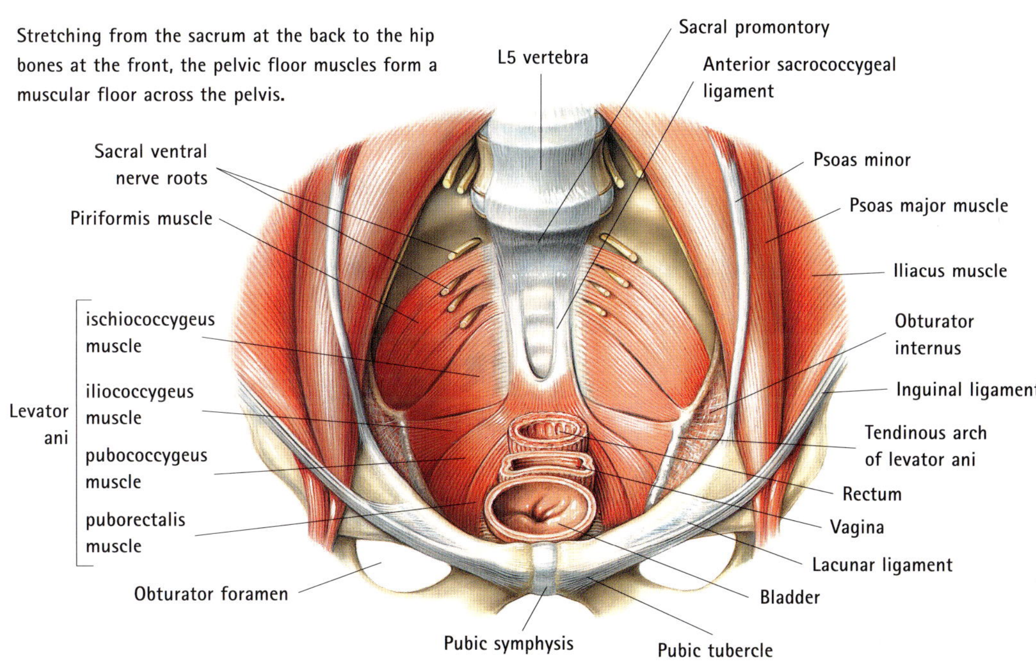

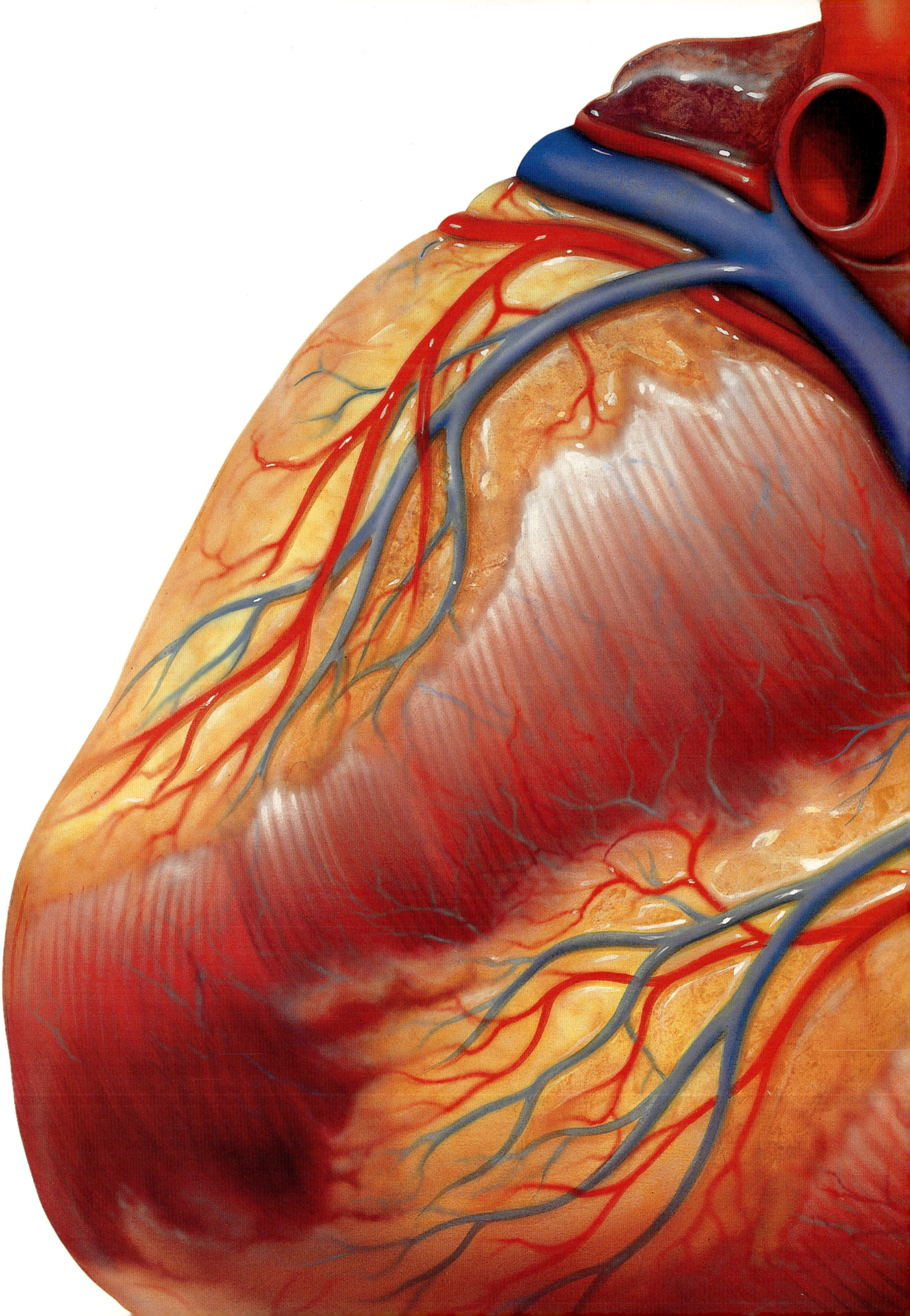

The Chest Cavity

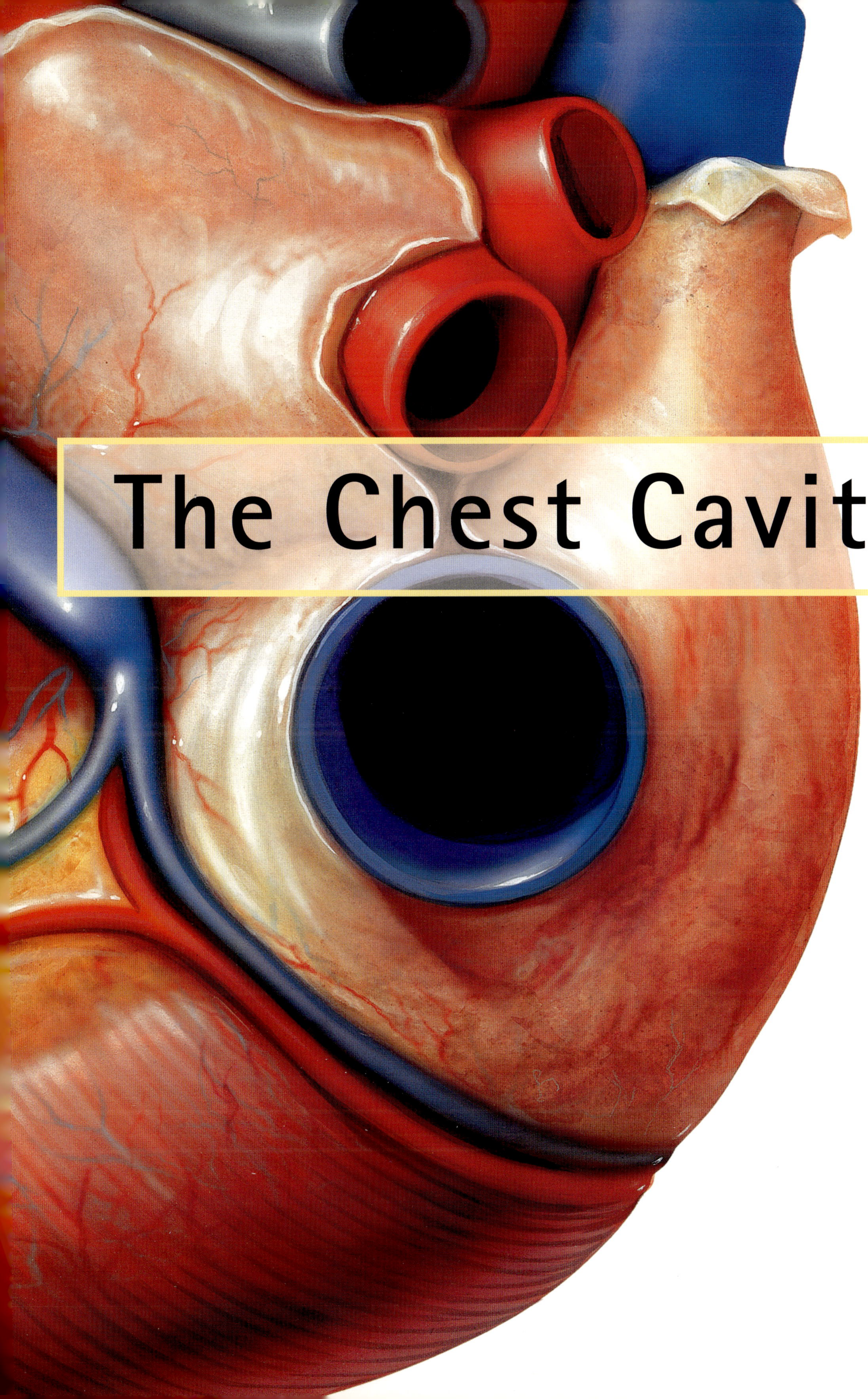

THE HEART

Heart—front

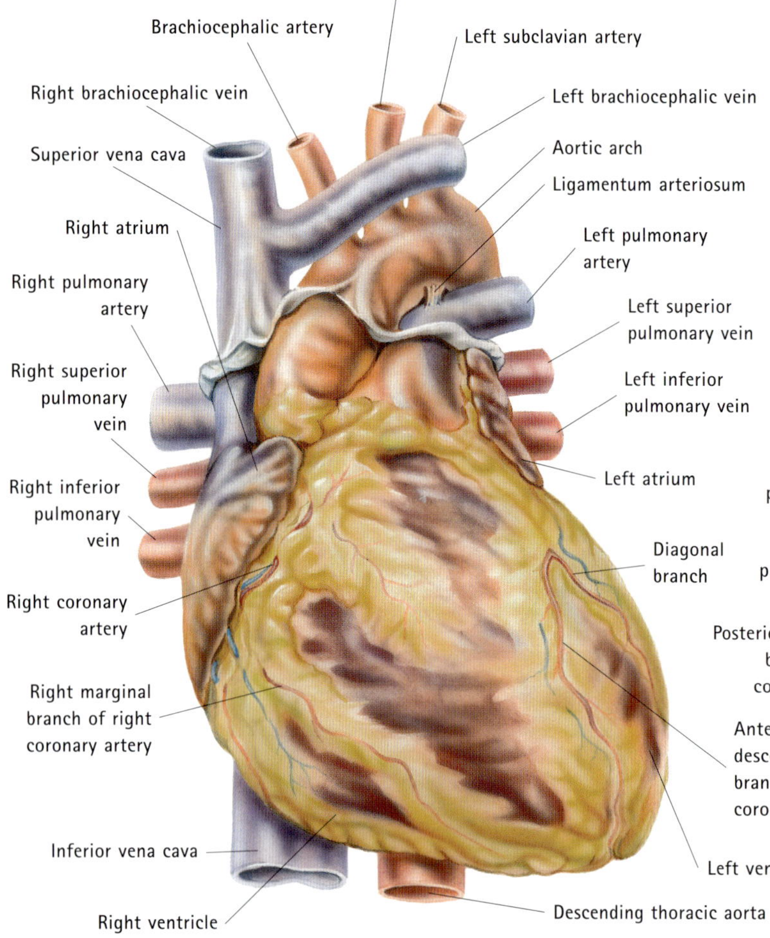

Heart—back

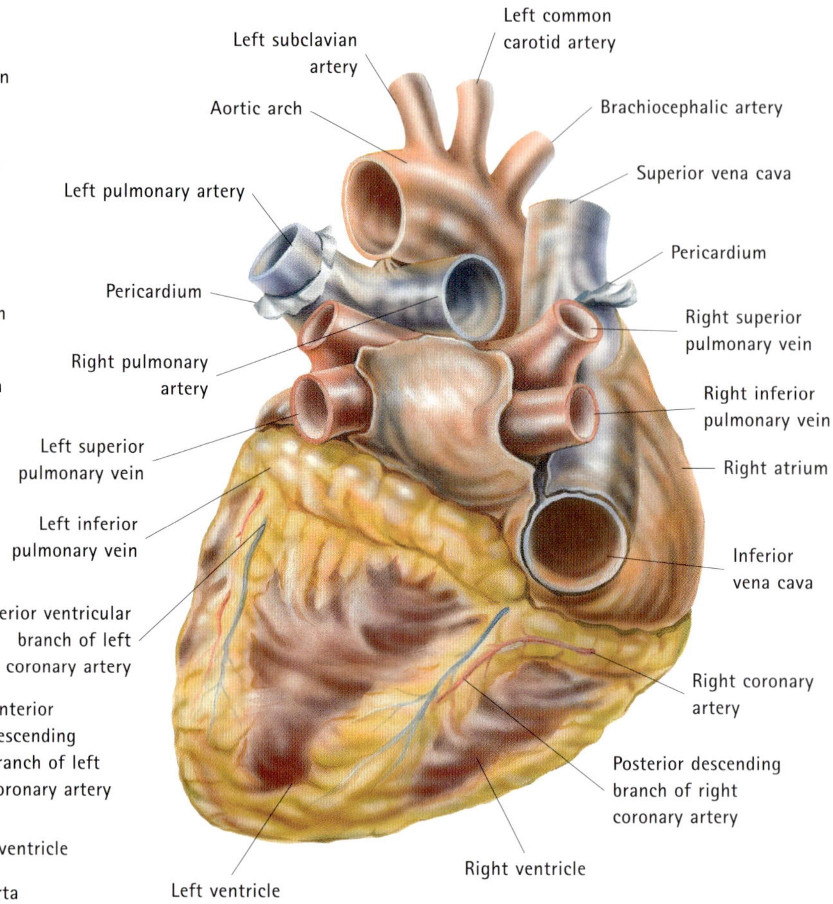

Heart—section

SHOWING ALL FOUR CHAMBERS

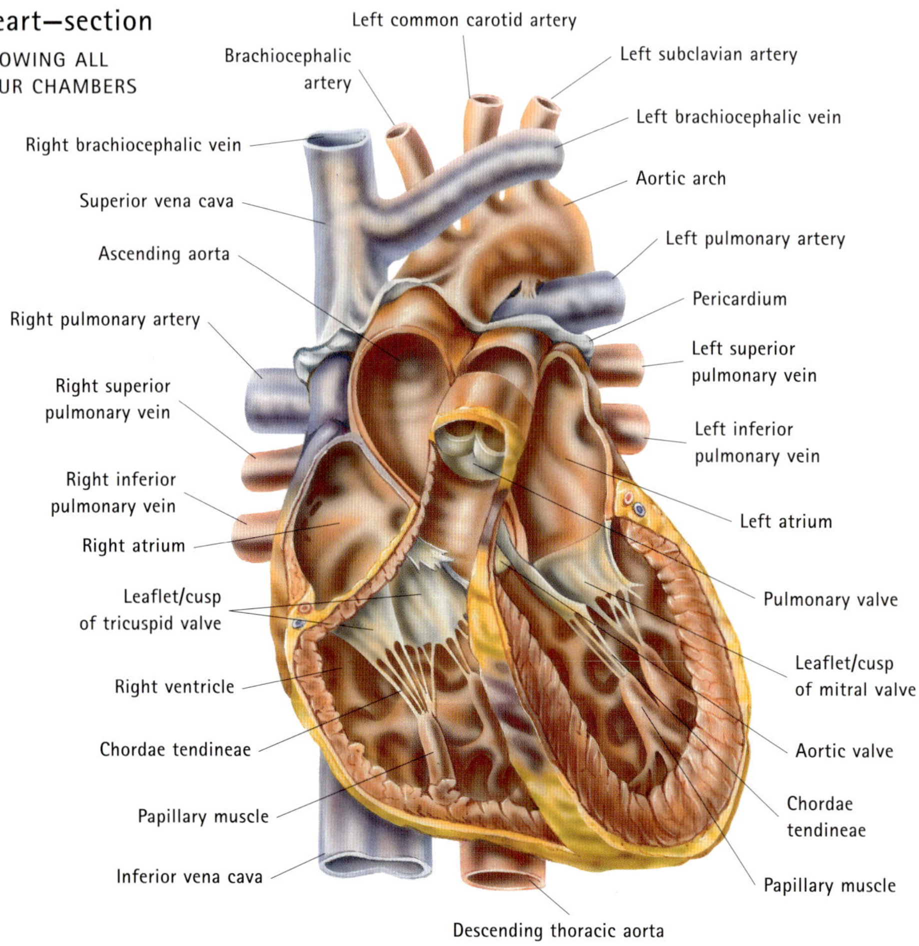

Heart

The heart is essentially a muscular pump, which is responsible for moving blood around the vessels of the body. It is made up of two halves, separated by the septum, a thick muscular wall. Each half is again divided into an upper and lower chamber.

The two chambers on the left side of the heart (left atrium and ventricle) are responsible for receiving oxygen-rich blood from the lungs and pumping it out to the body along its largest artery, the aorta. The two chambers on the right side of the heart (right atrium and ventricle) are responsible for receiving relatively deoxygenated blood from the body and distributing it via the pulmonary trunk to the lungs for gas exchange to occur. These two circuits of blood flow are called systemic and pulmonary circulations respectively. The muscle of the heart receives a rich blood supply from the coronary arteries, which branch from the aorta. Cardiac blood is returned to the right atrium by the cardiac veins.

SEE ALSO *Circulatory system on page 78; Disorders of the heart on page 230*

Heart valves

Within the heart there are four valves whose function is to ensure that blood flows in one direction: from atrium to ventricle and out through its appropriate artery. The two atrioventricular valves are located between the atria and ventricles on each side of the heart. Between the right atrium and right ventricle lies the tricuspid valve, while the mitral valve lies between the left atrium and left ventricle. The function of atrioventricular valves is to prevent backflow of blood from the ventricles to the atria during ventricular contraction.

The other two cardiac valves (semilunar valves) are located at the outlets from the ventricles: the pulmonary valve lies at the point where the right ventricle expels blood to the pulmonary trunk, while the aortic valve lies at the outlet of the left ventricle. The role of semilunar valves is to prevent backflow of blood from the aorta and pulmonary trunk into their respective ventricles when the ventricles relax.

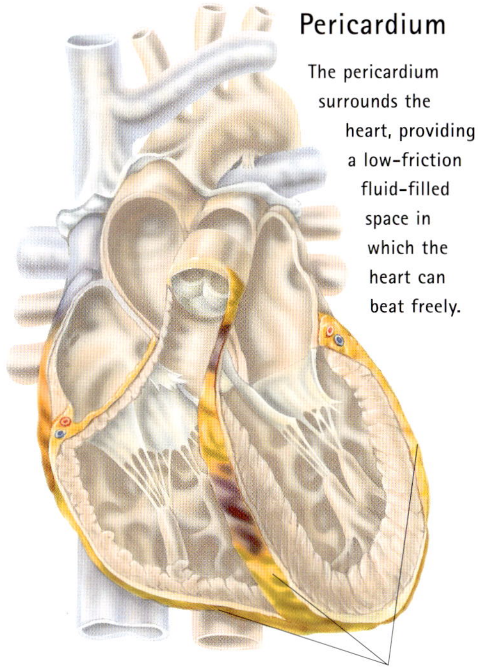

Pericardium

The pericardium surrounds the heart, providing a low-friction fluid-filled space in which the heart can beat freely.

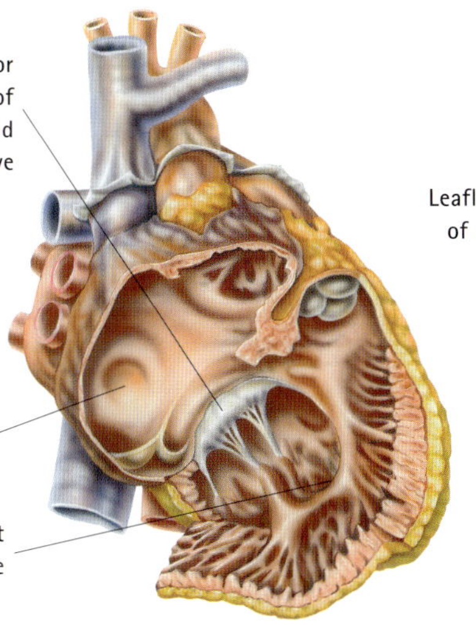

Tricuspid valve

Comprising three triangular flaps of tissue, called cusps or leaflets, the tricuspid valve helps to control the direction of blood flow between the right atrium and right ventricle of the heart.

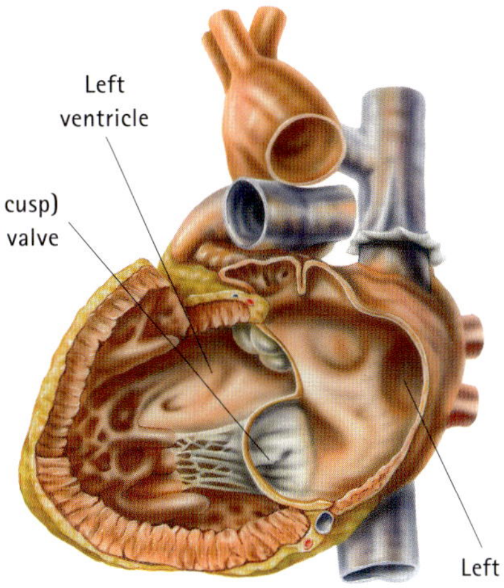

Mitral valve

Located between the left atrium and left ventricle, the mitral valve, also known as the bicuspid valve, prevents the backflow of blood from the ventricle to the atrium during systolic contraction of the heart.

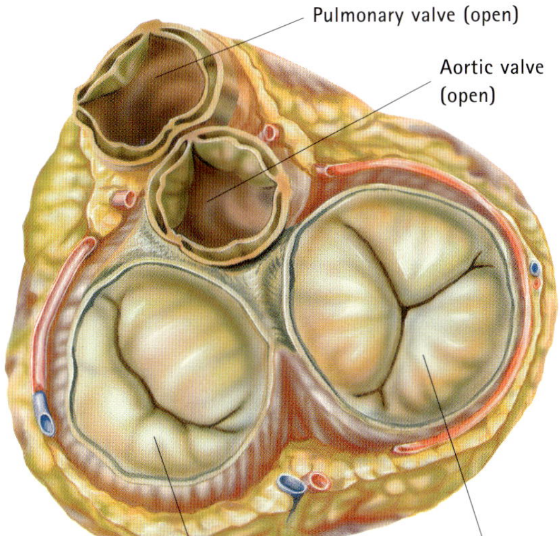

Heart in ventricular systole

The ventricles of the heart contract, pushing oxygenated blood into the aorta (for circulation around the body) and deoxygenated blood into the pulmonary artery (to be sent to the lungs).

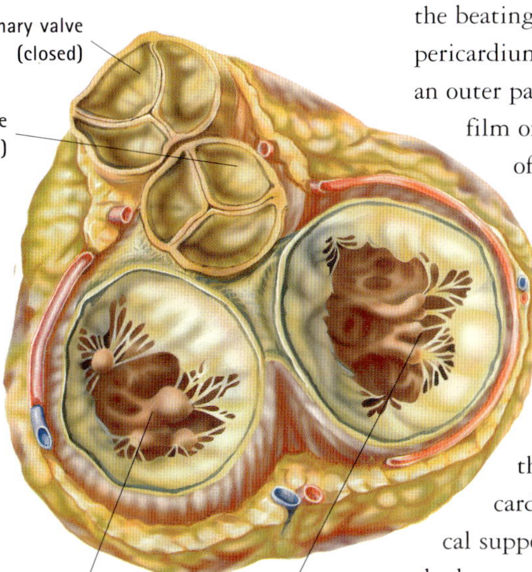

Heart in ventricular diastole

At the end of each contraction the mitral and tricuspid valves open, allowing blood to fill the left and right ventricles of the heart.

Coronary arteries

Coronary arteries are the arteries that supply the heart with oxygenated blood. The right and left coronary arteries arise from the aorta and branch out into smaller arteries supplying the right and left sides of the heart.

Pericardium

The pericardium is a series of sacs which enclose the heart. The inner set of sacs is called the serous pericardium and provides a low-friction, fluid-filled space to permit the beating heart to move freely. The serous pericardium has an inner visceral layer and an outer parietal layer, separated by a thin film of fluid. Outside the parietal layer of the serous pericardium lies the fibrous pericardium. This is a tough layer of connective tissue that is attached at the top around the great vessels entering and leaving the heart. This tissue is fused below with the central part of the diaphragm. The fibrous pericardium provides a strong mechanical support to maintain the position of the heart within the center of the chest.

Pumping action of the heart

The heart has a system of specialized cells, which either set the rhythm of cardiac contraction or allow for rapid spread of electrical impulses through the heart.

The sinoatrial node, in the right atrium, is a specialized tissue that acts as the heart's pacemaker; it controls the frequency of the heart's rhythmic contractions. Electrical impulses are then transmitted through the atria to the atrioventricular node, which then passes the impulse down the atrioventricular bundle, resulting in coordinated contractions of the ventricles.

Although divided into a left and right side with distinct functions, the heart is organized in such a way that its pumping action serves both sides at once. In the relaxation phase (diastole) blood pours from the left and right atrium into its corresponding ventricle. In the next contraction phase (systole) the blood is forced from the left and right ventricles into the aorta and the pulmonary artery, respectively. The valves control the direction of blood flow. At the start of each contraction, the atrioventricular valves close and the pulmonary and aortic valves open. At the end of each contraction the aortic and pulmonary valves close and the atrioventricular valves open.

Heartbeat

The beating of the heart against the chest wall is known as the apex beat. It results from the contraction of the ventricles of the heart and can be felt on the lower left side of the chest, immediately below the left nipple.

The rate and rhythm of contraction of the heart chambers is normally determined by the sinoatrial node, located near the entrance of the superior vena cava into the right atrium. The sinoatrial node sends out regular electrical impulses to produce the orderly contraction of the heart chambers. It, in turn, is under the influence of sympathetic and parasympathetic nerves, which increase or reduce the heart rate, respectively, and is also influenced by circulating hormones like epinephrine (adrenaline).

Disorders of the Heart

Heart disease can be congenital (existing from birth), or can develop later in life.

See also *Imaging techniques on page 431; Angioplasty on page 452; Defibrillation on page 453; Electrocardiogram on page 430; Electromyography on page 431; Holter monitor on page 430; Pacemaker on page 453; Radioisotope scan on page 433; Treating the cardiovascular system on page 439*

Coronary artery disease

Disease of the coronary arteries usually leads to the narrowing and reduction of blood supply to the cardiac muscle.

ATHEROSCLEROSIS

The most common type of coronary artery disease is due to atherosclerosis, which involves the accumulation of fats (atheroma) within the vessel wall leading to the formation of fatty/fibrous plaques and deposition of calcium. This disease may also occur in arteries of the neck, legs and abdomen, but is particularly dangerous in the brain and coronary arteries. Lifestyle factors, such as smoking and a diet rich in saturated fats, increase the risk of atherosclerosis. High blood pressure is also an important contributing factor to this disease.

The interior of a vessel may be blocked by fatty plaques, which cause thickening

HEART CYCLE

In the cardiac cycle, the chambers of the heart pass through a relaxation phase (diastole) and a contraction phase (systole).

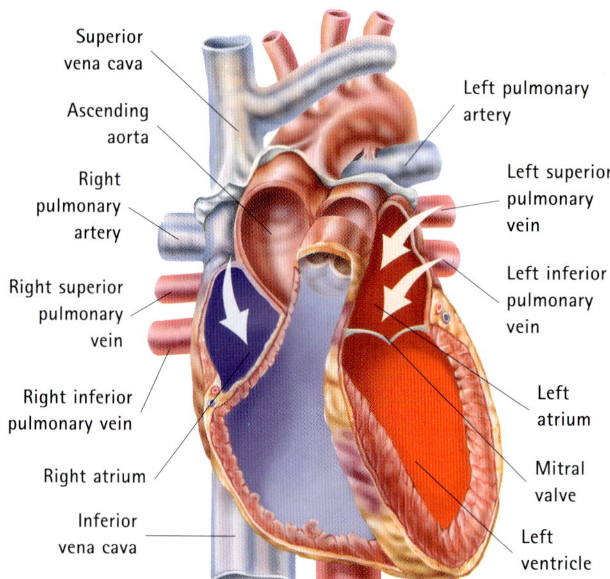

Heart cycle 1

In atrial diastole (at the beginning of ventricular diastole), deoxygenated blood from the systemic circulation and oxygenated blood from the lungs enter the left and right atria (upper chambers) of the heart.

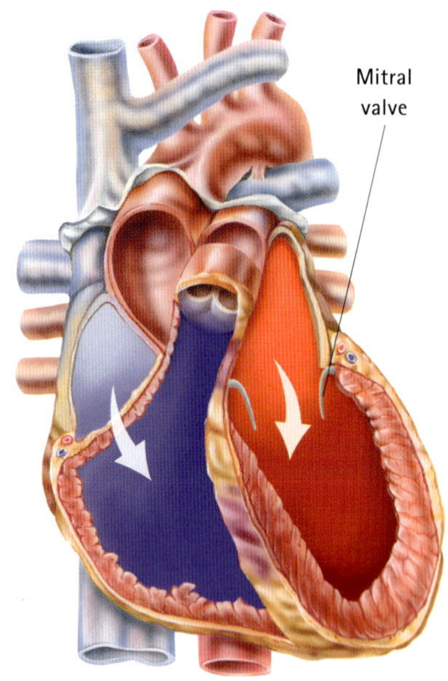

Heart cycle 2

Toward the end of ventricular diastole, the atria contract (atrial systole) and pump blood into the left and right ventricles.

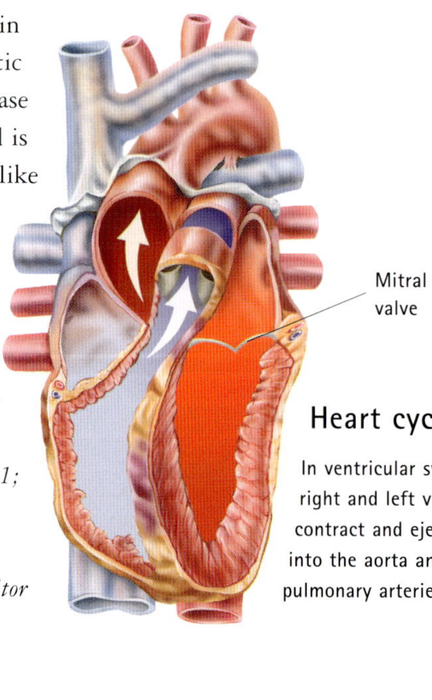

Heart cycle 3

In ventricular systole, the right and left ventricles contract and eject blood into the aorta and pulmonary arteries.

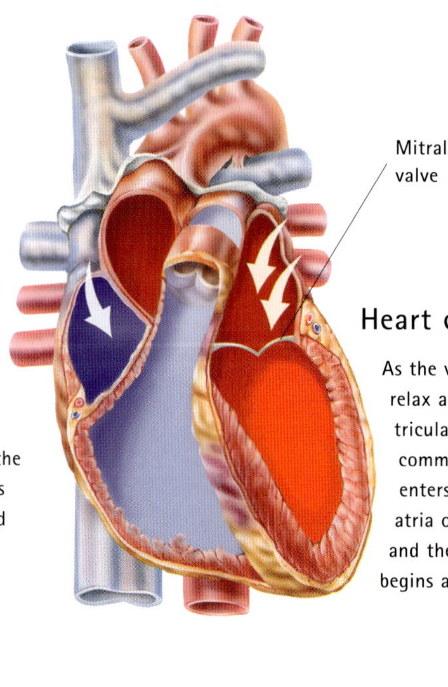

Heart cycle 4

As the ventricles relax and ventricular diastole commences, blood enters the two atria once again and the cycle begins again.

Heartbeat

The sinoatrial node, which is known as the pacemaker of the heart, determines both the heart's rhythm and rate.

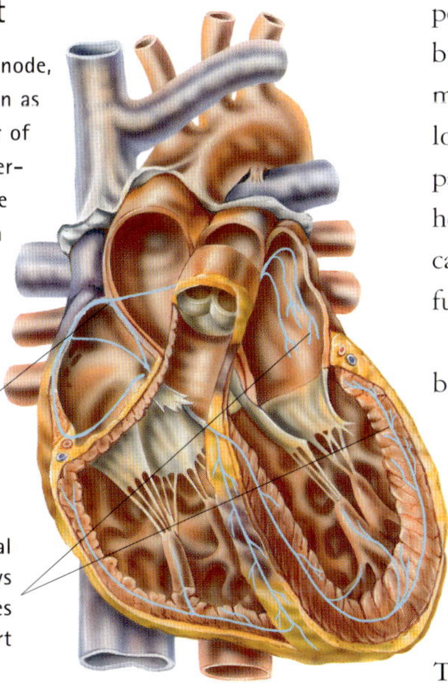

and loss of elasticity in the wall. Plaques may lose their surface, with the result that blood coagulates on them (thrombus formation) which also contributes to obstruction of coronary arteries (coronary heart disease).

Partial obstruction of the coronary artery may severely limit the patient's ability to perform physical activity. Such patients will complain of chest pain that has a crushing feeling and is usually located in the center of the chest (angina pectoris).

MYOCARDIAL INFARCTION

The myocardium is the name for the muscle of the heart, while infarction is the death of tissue due to interference with its blood supply. Myocardial infarction is the death of heart muscle due to loss of arterial blood supply. This generally occurs as a result of blockage of the coronary arteries which supply the heart with blood, usually due to atherosclerosis, which leads to the accumulation of fatty and fibrous tissue in the walls of arteries in discrete patches called plaques.

If plaques lose their surface layer, blood may coagulate on the rough surface of the vessel wall, causing sudden obstruction of the entire vessel. A patient with a myocardial infarction will experience a crushing pain centered mainly in the middle of the chest, but possibly spreading to the left arm, neck or upper abdomen.

The pain will not be relieved by rest nor by prescription medication, and patients should immediately seek medical help. With very early intervention it may be possible to dissolve the coagulated blood, but in most cases treatment is aimed at minimizing the amount of heart muscle lost, supporting the heart and providing pain relief. Patients may die if the loss of heart muscle is so great that heart function cannot be maintained, or if the electrical function of the heart is impaired.

Obstruction of the coronary arteries may be treated before too much damage has occurred by several techniques. Partial or mild obstructions may be treated by dilating the blocked segment with a balloon, removing the obstructing matter (plaque) with laser pulses, or inserting a "stent" to hold the artery open. These are all methods of angioplasty. More severe obstructions may be bypassed surgically by grafting a piece of leg vein or chest wall artery alongside the obstructed segment (coronary artery bypass).

CONGESTIVE CARDIAC FAILURE

If the death of cardiac muscle occurs slowly, the patient may develop congestive cardiac failure. In such cases, the heart is unable to move blood around the body effectively, so that blood accumulates in the veins of the legs, abdomen or lungs, causing swelling. Pooling of fluid in the lungs will cause shortness of breath. Treatment of congestive cardiac failure is usually aimed at improving the strength of the heart and removing excess fluid from the body by increasing the output of urine.

ANGINA PECTORIS

Angina pectoris is the name given to chest pain caused by lack of oxygen to the heart, due to arteriosclerosis in the coronary (heart) arteries. It is usually brought on by exercise. In severe arteriosclerosis, the pain may be felt at rest, when it is then known as unstable angina. The pain is felt across the chest and may also be felt in the neck, between the shoulder blades and in the arms.

In severe cases sufferers describe the pain as "crushing," and may also feel cold, sweaty and anxious at the same time, and have difficulty in breathing.

Angina pectoris can be confirmed by an electrocardiogram (EKG), which shows certain changes while the pain is present. If pain is not present, changes can be demonstrated while the patient is undergoing an exercise (stress test). Other tests include cardiac angiography and perfusion scanning.

The condition is commonly treated with drugs that dilate the coronary arteries and increase the blood flow to the heart muscle. These include nitroglycerin tablets, which can be taken under the tongue (they usually relieve pain within seconds and confirm the diagnosis of angina).

Nitrates can also be given as tablets or skin patches. Beta-blockers and calcium antagonists (channel blockers) are also used to treat angina. Balloon angioplasty and bypass surgery are other options.

Angina can be treated at home if it responds to drug treatment. If not, or there is unstable angina (pain at rest), hospitalization may be necessary, as the angina may then be a sign of myocardial infarction (heart attack).

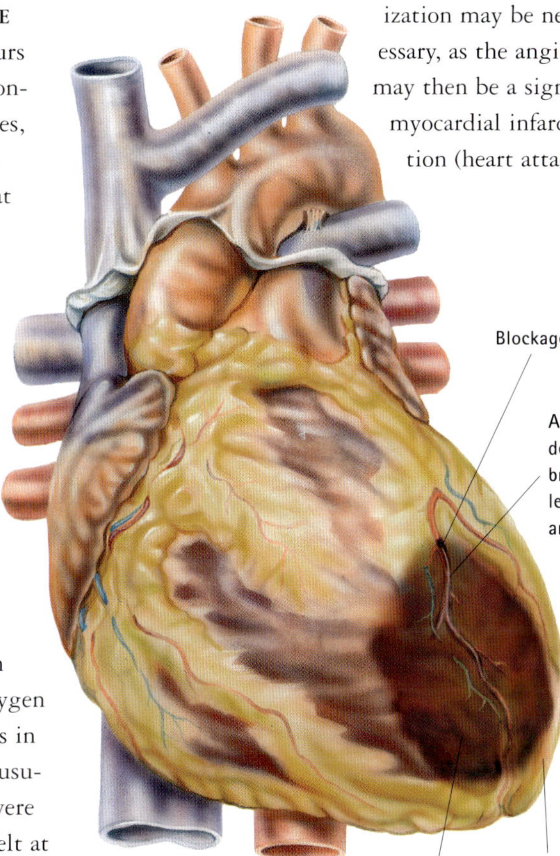

Coronary artery disease—damage to heart tissue

Heart attack (myocardial infarction) occurs when some of the heart's blood supply is suddenly severely restricted or cut off. This causes part of the heart muscle to die from lack of oxygen.

Electrical abnormalities

Problems with the conducting and pacemaking tissues of the heart may lead to abnormalities of the cardiac rhythm (cardiac arrhythmias) or delayed and ineffective contraction of cardiac muscle (bundle branch blocks or heart block). Cardiac arrhythmias may affect the atria (atrial tachycardia, atrial fibrillation) or the ventricle (ventricular tachycardia, ventricular fibrillation). Many electrical abnormalities are linked with problems involving the coronary arteries, although some can be caused by rheumatic heart disease, fever, excess thyroid hormone, excessive coffee and alcohol.

ARRHYTHMIA

An arrhythmia (also known as a dysrhythmia) is a variation in heartbeat from the normal rhythm. An arrhythmia may be a normal variation in the heartbeat, or it may be due to disease that has damaged the heart muscle and caused irregularities in nerve conduction and muscle contraction. Certain drugs, including caffeine, cocaine, psychotropics, and sympathomimetics, can also cause arrhythmias.

Not all arrhythmias are dangerous. In sinus arrhythmia—a normal occurrence in children—the pulse rate increases or decreases with breathing. Sinus arrhythmia usually does not require treatment.

Some other arrhythmias (varying in their degrees of danger) include atrial fibrillation, paroxysmal supraventricular tachycardia, sinus tachycardia, sinus bradycardia, bradycardia associated with heart block, sick sinus syndrome and ectopic heartbeat.

A common type of minor arrhythmia is the ventricular ectopic beat (ventricular extrasystole). These occur when an abnormal focus of electrical activity in the ventricles causes a premature contraction. This may occur in healthy people, especially if they are tired or have consumed a lot of caffeine. The patient may feel nothing or may notice the occasional heavy heartbeat.

Certain types of arrhythmia can be life-threatening. Ventricular fibrillation and ventricular tachycardia, for example, can cause cardiac arrest and severe decrease in blood flow to tissues and organs.

Symptoms of arrhythmia include noticeable changes to the rhythm or pattern of the pulse, a sensation of awareness of the heartbeat (palpitations), chest pain, shortness of breath, lightheadedness, dizziness or unconsciousness. In ventricular fibrillation, the first symptom may be loss of heartbeat; sudden death can occur in this situation.

To confirm and diagnose an arrhythmia, tests such as an electrocardiogram (EKG) and 24-hour Holter monitoring are often performed. An echocardiogram and coronary angiography could show heart disease that may be producing an arrhythmia.

Arrhythmias are usually treated with either traditional antiarrhythmic drugs such as digitalis and quinidine, or modern antiarrhythmics such as beta-blockers, calcium antagonists and disopyramide. Some arrhythmias are treated by electrical destruction of diseased tissue, and sometimes a surgical implant is inserted.

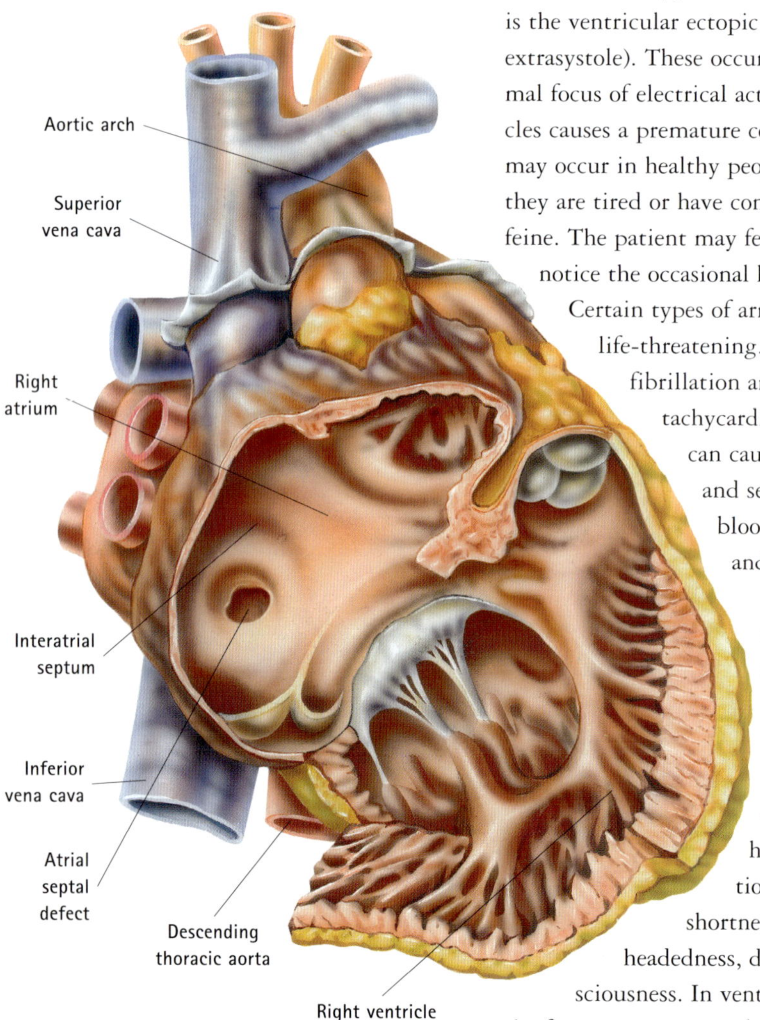

Atrial septal defect

The normal fetus has a hole between the two upper chambers of the heart, which allows blood to bypass the lungs. It usually closes at birth, but in some people it remains open and may need to be closed by surgery.

HEART BLOCK

Another type of disorder involving electrical conduction in the heart is heart block. In this condition the conduction of electrical impulses from the atria to the ventricles is blocked (AV block) so that the atria beat at a faster rate than the ventricles. Conduction blocks may also occur within the conduction pathways of the ventricle.

Stokes-Adams syndrome (also known as Adams-Stokes syndrome) is a form of heart block. The condition is most common in the elderly and is associated with coronary heart disease. Stokes-Adams syndrome is best treated with an artificial pacemaker. Alternatively, drug treatment may be used.

TACHYCARDIA, PAROXYSMAL

Paroxysmal tachycardia refers to episodes of rapid heart rate (140–220 beats per minute). It may arise from problems with either the atria or ventricles in the heart. In both types the patient complains of palpitations, the sensation of a rapidly beating heart fluttering in the chest. The ventricular type is more serious, and is often a complication of severe heart disease.

Ventricular tachycardia often causes chest pain and breathlessness: these symptoms are less common in atrial tachycardia. The two types can be distinguished by an electrocardiogram (EKG) and treated with appropriate antiarrhythmic medication.

FIBRILLATION

Fibrillation is irregular electrical activity of the heart muscle, with the result that the heart muscle cannot contract effectively. It may occur in either the atria or ventricles of the heart. Atrial fibrillation may be due to disease of the mitral valve, myocardial infarction, infections or thyroid disease. The great danger of atrial fibrillation is that stagnant blood in parts of the atrium may coagulate. Small pieces of coagulated blood may then break off and become lodged in the arteries of the brain or kidneys. Ventricular fibrillation is even more serious, because the ventricles are unable to pump blood while fibrillating.

Patients with ventricular fibrillation will have no pulse and will die within minutes unless resuscitation and electrical defibrillation are used immediately.

Congenital heart diseases

Congenital heart diseases arise before birth, and affect about 1 in every 100 babies. About half of all babies born with a congenital abnormality of the heart die during the first year of life. Often there is no known cause, but some may be due to abnormal genes, viruses (rubella) or drugs (thalidomide). Congenital heart disease may also accompany other congenital abnormalities.

Congenital heart disease is usually due to one (or a combination) of the following defects: abnormal openings between the two sides of the heart (atrial or ventricular septal defect), blockage of large arteries or valves (coarctation of the aorta, pulmonary stenosis), or an abnormal positioning of heart chambers or vessels.

ATRIAL SEPTAL DEFECT

Atrial septal defect (ASD) is a congenital "hole in the heart" between the two upper chambers of the heart (the atria). In the fetus, this opening is normal, and allows blood to bypass the lungs (which are not in use). At birth, this opening usually closes to allow blood to be pumped into the lungs once breathing starts. In persons with atrial septal defect, the hole doesn't close.

Most people with atrial septal defect do not have symptoms, but if the opening is large enough there may be symptoms of shortness of breath and irregular heartbeats (palpitations). These symptoms often do not develop until adulthood.

Through the stethoscope, a physician may hear abnormal heart sounds in a person with ASD. An ultrasound of the heart or an echocardiogram will show the hole. An EKG (electrocardiogram) may also be abnormal, as the hole interferes with the normal conduction of the heartbeat through the heart.

No treatment is necessary if there are no symptoms; if there are symptoms or the hole is large, surgical closure of the defect may be needed. People affected with ASD will need to take antibiotics during dental and surgical procedures, as they are at increased risk of developing bacterial endocarditis.

VENTRICULAR SEPTAL DEFECT

A ventricular septal defect (VSD) is a hole in the wall separating the two ventricles, or lower chambers, of the heart. Because of pressure differences between the two ventricles, blood passes through the hole from left to right. Blood in the left ventricle then leaks into the right ventricle with the result that more blood than normal is pumped to the lungs. This often produces chest infections and breathing difficulties. A large opening in a very small child can also cause symptoms of heart failure. Although medium-sized and small ventricular septal defects often close by themselves, surgery is often needed to repair larger holes.

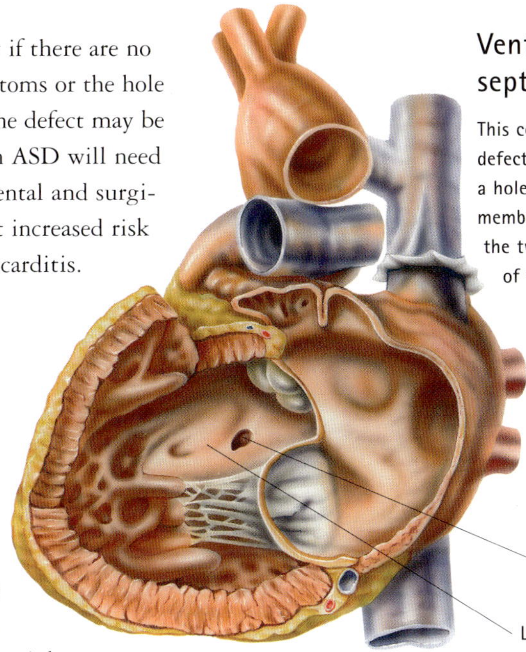

Ventricular septal defect

This congenital heart defect occurs when a hole forms in the membrane dividing the two ventricles of the heart.

PATENT DUCTUS ARTERIOSUS

Normally the fetus receives oxygen from the placenta via the umbilical cord, so the lungs are not needed. Hence, there is a channel between the pulmonary artery and the aorta called the ductus arteriosus. The function of this duct is to allow the blood flowing through the fetal heart to bypass the lungs and be pumped straight into the systemic circulation and around the body.

Shortly after birth the duct normally closes, and the heart commences pumping blood into the lungs as well as the rest of the body. In about 60 out of 100,000 infants, the duct doesn't close properly—this condition is known as patent ductus arteriosus. It is similar to atrial septal defect except that in that condition, the hole that fails to close is between the two upper chambers of the heart (the atria).

The cause is unknown, but it is more common in premature infants. Patent ductus arteriosus causes mild shortness of breath and failure to thrive, which may progress over time to heart failure.

The condition is diagnosed by a physician using a stethoscope (usually a heart murmur is audible) and an echocardiogram (ultrasound of the heart). Surgical ligation of the patent ductus corrects the condition and is usually done between 6 months and 3 years of age, or earlier if heart failure develops.

TETRALOGY OF FALLOT

Tetralogy of Fallot is a congenital disease of the heart; in other words, the basic disorder is present before the time of birth. The word tetralogy implies that there are four main abnormalities comprising the disease.

The two most important disorders are pulmonary stenosis and ventricular septal defect.

In addition, there is a shift in the position of the aorta (the main artery passing from the heart to the rest of the body) so that it lies over the ventricular septal defect. The wall of the right ventricle thickens as a consequence of the narrowed pulmonary artery. The result of these disorders is that insufficiently oxygenated blood is pumped from the heart to the body. Blood flow to the lungs is decreased, which compounds the cyanotic effects.

Children with this disease usually develop a blue tinge to the lips and mouth (cyanosis) within the first year of life. Breathlessness and fainting will become increasingly more severe and frequent. These episodes of fainting are referred to as "Tet spells."

It is estimated that about 50 in 100,000 infants develop tetralogy of Fallot, with a

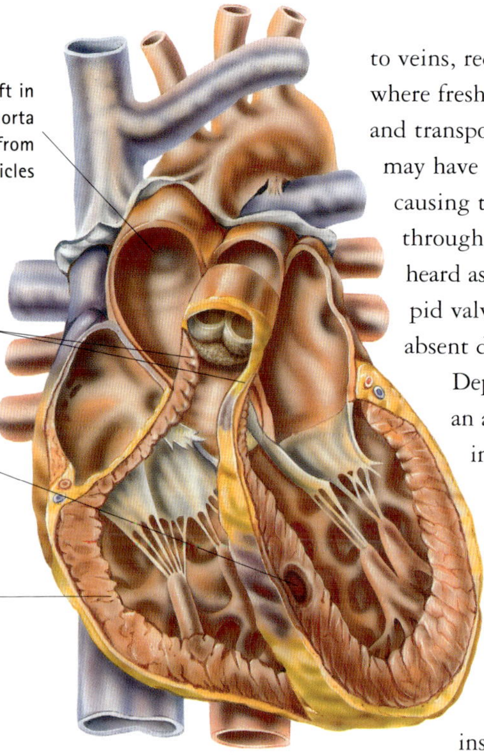

Tetralogy of Fallot

This congenital condition is a combination of four heart defects causing low oxygen levels in the blood.

higher than normal incidence in children with Down syndrome. As with most congenital heart defects, the cause of this group of disorders is unknown, but prenatal occurrences associated with such conditions also include viral illnesses, alcohol abuse, insufficient nutrition and diabetes. Treatment is by corrective surgery.

Diseases of the heart valves

The valves of the heart may be involved in disease. The valve openings of the heart may become narrowed (valvular stenosis) with the result that blood does not readily flow through; or be unable to prevent the backward flow or regurgitation of blood (valvular incompetence or insufficiency).

Many valve diseases give rise to murmurs—sounds produced by the turbulent flow of blood across roughened valve surfaces, or the excessive movement of blood through normal valves. These may be heard with the aid of a stethoscope, or if loud enough, the vibrations may be felt on the chest wall.

Often a heart murmur may not indicate serious disease, but should be investigated.

TRICUSPID VALVE DISEASE

A disease-damaged tricuspid valve may cause regurgitation from ventricle to atrium to veins, reducing blood flow to the lungs where fresh supplies of oxygen are gathered and transported back to the heart. The valve may have become too narrow (stenotic), causing turbulence as blood passes through the constricted passage. This is heard as a rumbling murmur. The tricuspid valve may also be obstructed or even absent due to congenital heart disease.

Depending on the heart condition, an artificial valve may be surgically inserted to replace a damaged or abnormal valve.

MITRAL VALVE DISEASE

Mitral valve disease usually is divided into two types: mitral stenosis, where the valve opening is too narrow, and mitral insufficiency (also called incompetence) where the mitral valve is unable to prevent backflow of blood from the left ventricle to the left atrium.

Most mitral valve disease is generally caused by rheumatic heart disease. However, mitral insufficiency may also usually be caused by chordae tendineae or papillary muscles that may have been ruptured or are defective.

Patients with mitral stenosis usually present to their doctor with symptoms of failure of the left side of the heart, including shortness of breath, breathlessness while lying down, and coughing up blood. They may develop palpitations and chest pain.

Patients may also develop atrial fibrillation, which can have serious consequences if blood coagulates in the left atrium and throws off tiny fragments to block arteries in the brain, limbs or kidneys. There will be a murmur (a sound produced by abnormal blood flow through the heart) heard best on the lower left side of the chest.

Medical treatment for mitral stenosis is aimed at treating arrhythmias (abnormal rhythm) and usually includes digoxin, beta-adrenergic blocking drugs and maybe quinidine in various combinations, but would depend on the clinical circumstances. Anticoagulants like warfarin prevent formation of thrombosis in the atrium that can dislodge and block other arteries.

Antibiotics are used to treat infections on valve surfaces. Surgical treatment involves separating the valve cusps (valvotomy), which are often stuck along their edges. In some cases, the entire valve may need to be replaced by an artificial valve (prosthesis), or a valve graft from the heart of a pig.

Patients with mitral insufficiency often complain to their physician of fatigue, shortness of breath and episodic breathlessness at night. Symptoms may start suddenly if the condition is due to sudden rupture of the chordae tendineae. A murmur is also present, but the timing of the sound differs from that heard in mitral stenosis. Medical treatment may be similar to that used in mitral stenosis; surgical treatment involves replacing the damaged mitral valve with a prosthetic valve or graft.

Replacement of the mitral valve in either mitral stenosis or insufficiency will require open-heart surgery, in which the patient's body is cooled and the function of the heart is temporarily taken over by a heart–lung machine. This procedure will allow the heart to be stopped and opened, so that the surgeon may reach the damaged valve.

PULMONARY AND AORTIC VALVE DISEASE

The pulmonary valve may be affected in disease to produce either pulmonary stenosis or pulmonary insufficiency (incompetence).

Pulmonary stenosis is usually a congenital disease in which the cusps of the pulmonary valve are fused in the form of a membrane or diaphragm with a narrow central opening, rather than the normal three cusp shape.

Pulmonary insufficiency means that the pulmonary valve is unable to prevent backflow of blood from the pulmonary trunk into the right ventricle when the ventricle stops contracting. It most often arises in patients who have had rheumatic heart disease, in which the valve cusps become contracted and fused with the wall of the artery, or in cases of increased blood pressure in the lung circulation.

Aortic stenosis is narrowing of the aortic valve, caused by valvular heart disease such as rheumatic fever. The narrowing of the valve causes an additional strain on the heart that may lead to angina and heart failure.

Valve replacement with an artificial valve through surgery is usually the most preferred treatment option.

Rheumatic heart disease

In some children between the ages of 5 and 15, streptococcal infection of the tonsils or throat may be followed 2 weeks later by an immune reaction called rheumatic heart disease or rheumatic fever. It manifests as fever and arthritis; the joints become tender, swollen and red. It can also damage the tissues of the heart, central nervous system and skin.

With bed rest, treatment with aspirin (and sometimes corticosteroid drugs), the symptoms resolve in a few weeks. However, the endocardium (the internal lining of the heart) may be permanently scarred, causing heart murmurs and valvular heart disease.

People with rheumatic heart disease should take antibiotics before undergoing medical and dental procedures to prevent bacterial endocarditis (infection of the internal lining of the heart).

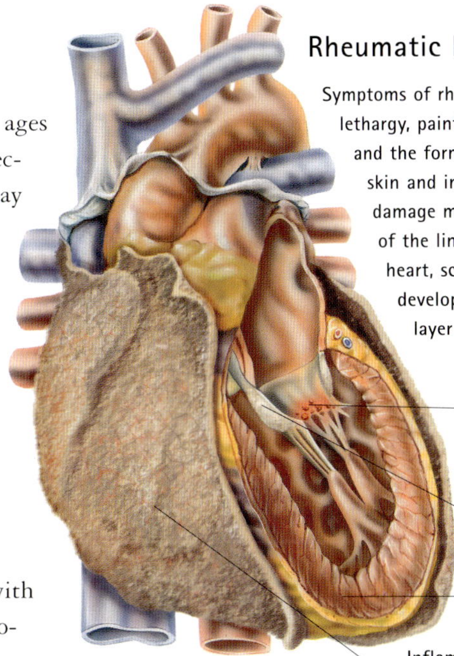

Rheumatic heart disease

Symptoms of rheumatic fever include fever, lethargy, painful swelling of the joints, and the formation of nodules under the skin and in the heart. Permanent heart damage may result from inflammation of the lining, valves and muscles of the heart, scarring of the valves and the development of a fibrous, thickened layer around the pericardium.

Diseases of the myocardium (heart muscle)

Most disease of the heart muscle is usually caused by blockage of the coronary arteries, but there are also some rare but serious heart diseases that primarily affect the heart muscle itself.

MYOCARDITIS

Myocarditis is inflammation of the heart muscle, which usually accompanies a generalized disease of the body (such as diphtheria or toxoplasmosis) or a viral infection (such as poliomyelitis). Patients often experience rapid heart rate in mild cases.

In more severe cases, as the heart begins to fail, there may be shortness of breath, an enlarged heart and and heart murmurs.

CARDIOMYOPATHY

Cardiomyopathy is a general term for diseases of the myocardium. As cardiomyopathy progresses, the heart becomes weakened, enlarged and beats irregularly. In the terminal stages, the heart fails and the person may require a heart transplant. The condition may be caused by chronic diseases such as coronary atherosclerosis, excessive alcohol intake, infection due to viruses, or beriberi and other vitamin B deficiency disorders. In less common cases it is usually caused by the inflammation of heart muscle due to rheumatic fever or other immune disorders.

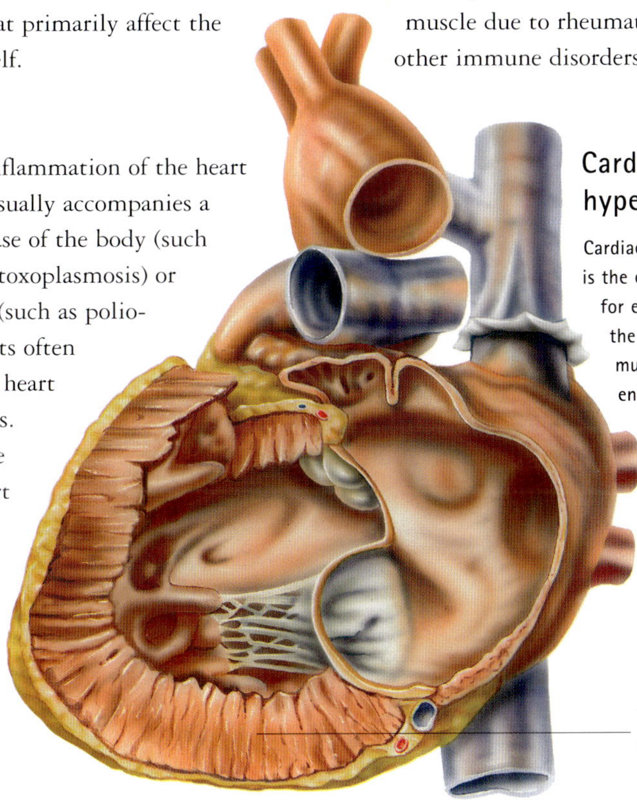

Cardiac hypertrophy

Cardiac hypertrophy is the clinical term for enlargement of the heart. The heart muscle cells become enlarged due to an increased demand for the heart to work harder.

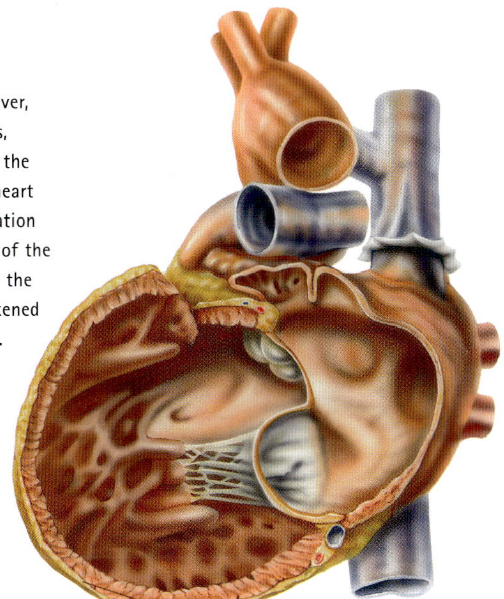

Cardiomyopathy

Cardiomyopathy is a disease that alters the structure or function of muscular wall of the heart ventricles (lower chambers). The diseased heart becomes enlarged, weak, and has trouble pumping blood effectively. The heart rhythm may also become abnormal.

To begin with, the person notices a shortness of breath (dyspnea) and a decreasing ability to tolerate physical exertion. Chest pain, fainting (syncope), and palpitations may be present. When the heart fails, swelling of the ankles and abdomen may occur.

Tests such as an x-ray of the chest, coronary angiography, echocardiogram, chest CAT or MRI scan will show the extent of the condition. Blood tests and a biopsy of the heart muscle may be required to find the cause. Treatment is aimed at correcting the underlying cause.

Medication will relieve the workload of the heart and stabilize the patient's condition. Rest and oxygen (given by mask) will reduce the workload of the damaged heart muscles. Hospitalization is advised if symptoms of severe heart failure are present.

A heart transplant may be necessary if the patient is suitable and a donor is available.

CARDIAC HYPERTROPHY

Cardiac hypertrophy is enlargement of the heart. It is caused by enlargement of heart muscle cells in response to a need for the heart to pump harder over months or years, in conditions such as hypertension or restricted flow of blood from the heart to the aorta (aortic stenosis). Because the heart must work harder, it requires more oxygen than normal; therefore, someone with cardiac hypertrophy is more vulnerable to angina and myocardial infarction. It is necessary to treat the underlying cause.

Hypertrophic cardiomyopathy is a group of diseases in which the cardiac muscle enlarges for no apparent reason. About half the cases are inherited and the disease affects mainly young adults. Symptoms are breathlessness, tiredness, fainting and chest pain.

The condition is treated with beta adrenergic and calcium channel blocking drugs. In severe cases, an operation to cut or excise the muscle may be required.

Diseases of the endocardium

The endocardium is the smooth inner lining of the heart.

ENDOCARDITIS

Endocarditis is inflammation of the endocardium caused by bacteria or fungi that enter the blood and infect and damage the endocardium of the heart and the valves and the heart muscle itself. Endocarditis is more likely if there is pre-existing damage to the heart. It may follow the delivery of a baby, surgery or dental work, or the use of intravenous drugs, all of which may introduce organisms into the bloodstream.

Symptoms begin gradually with fatigue, chills, aching joints and intermittent fever. Blood culture of the bacterium or fungus confirms the diagnosis.

Cardiac tamponade

Cardiac tamponade is a condition in which blood or fluid accumulates in the area surrounding the heart (the pericardium). The fluid puts pressure on the heart and interferes with its ability to pump blood efficiently. This is a serious condition requiring urgent treatment.

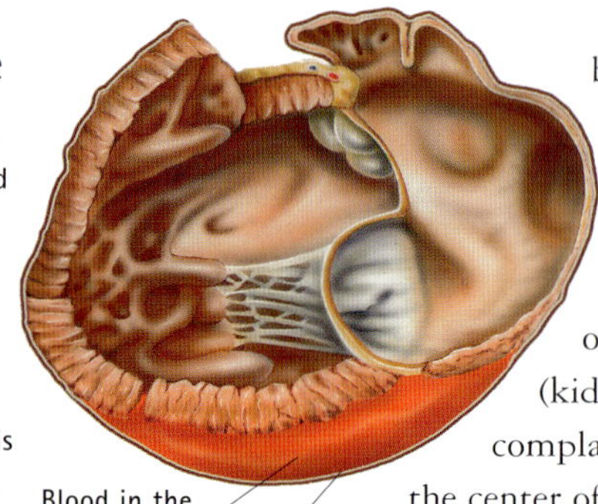

Treatment is with antibiotics which may be given intravenously. If untreated, death from heart failure is usual, but, with early treatment, most patients survive. Persons who have pre-existing heart valve disease should take antibiotic drugs before any dental or surgical procedure.

Diseases of the pericardium

The pericardium is the tissue sac that surrounds the heart. The sac consists of an inner double-layered part and an outer part which is tough and fibrous.

PERICARDITIS

Pericarditis is literally inflammation of the pericardium and may be the result of infection (viruses, tuberculosis or pus-forming bacteria), invasion by cancer cells (leukemic infiltration), connective tissue diseases (rheumatoid arthritis, systemic lupus erythematosus) or changes in blood chemistry (kidney failure and gout). Patients complain of sharp, strong pain in the center of the chest. Unlike myocardial infarction, the pain is made worse by breathing in, moving and lying flat on the back. Patients also commonly have fever and an audible friction rub, heard when a stethoscope is applied to the chest; they may develop failure of the right ventricle.

CARDIAC TAMPONADE

Cardiac tamponade (also called pericardial tamponade) is a potentially life-threatening condition in which blood or fluid accumulates in the pericardium (the sac which encloses the heart), causing pressure on the heart. This prevents the ventricles from expanding fully, so they cannot adequately fill or pump out the blood.

The condition may result from a wound which ruptures blood vessels in the heart muscle, from bacterial or viral inflammation of the pericardium (pericarditis), or from cancer invading the pericardial sac. It may also be caused by radiation therapy to the chest, hypothyroidism, or systemic lupus erythematosus (SLE).

Symptoms involve chest pain radiating to the neck, shoulder, back or abdomen, difficulty in breathing, a weak or absent pulse, and low blood pressure. A chest x-ray, echocardiogram or CAT or MRI scans of the chest may show blood or fluid in the pericardium if there is any.

Cardiac tamponade is a medical emergency. Treatment is aimed at stabilizing the patient, removing the fluid by means of a needle with a suction syringe inserted through a surgical puncture, and correcting the underlying cause.

Heart failure

Heart failure may be a complication of a myocardial infarction (heart attack), myocarditis (disease of the heart muscle),

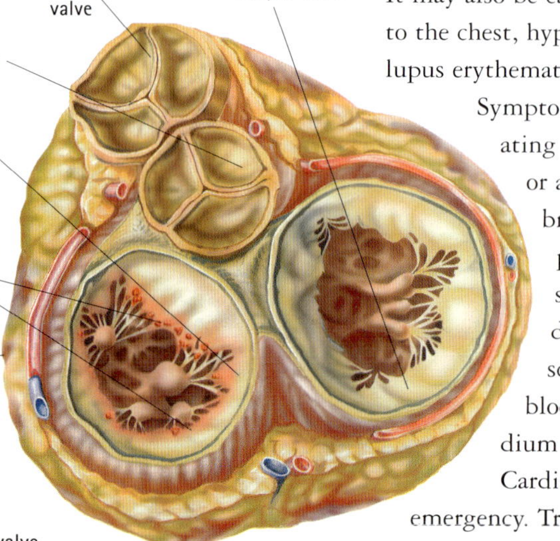

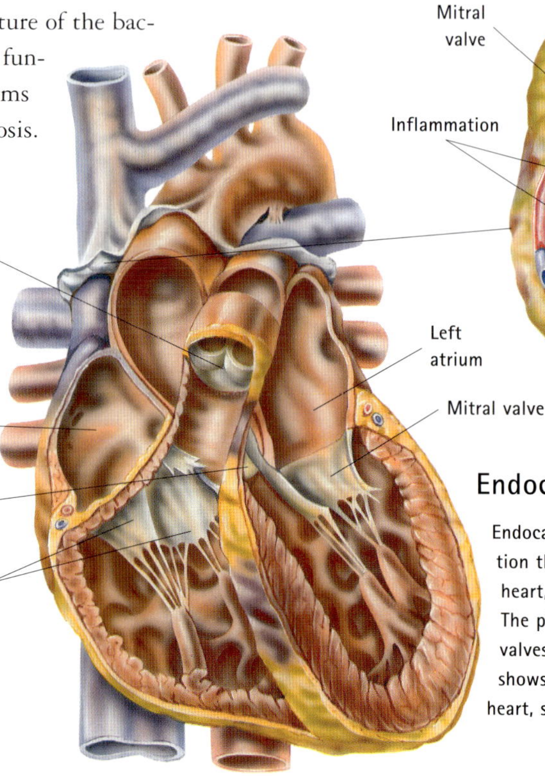

Endocarditis

Endocarditis is a bacterial or fungal infection that affects the inner lining of the heart, as well as the valves and muscle. The picture above shows diseased heart valves seen from above; the one at left shows the location of the valves in the heart, seen from the front.

mitral or aortic valve disease, or other heart disorders. Heart failure may reflect impairment of the left or right ventricle and may develop gradually or suddenly (as in the case of pulmonary edema).

PULMONARY EDEMA

Pulmonary edema is a build-up of fluid in the lungs, most often due to left ventricular failure. The symptoms of pulmonary edema are shortness of breath, occurring initially on exertion and later even while at rest, especially when lying flat.

Treatment requires urgent admission to hospital for intravenous diuretic medication to remove fluid from the lungs via the kidneys. The patient is given oxygen, and is treated for any underlying condition that may have brought on the heart failure.

COR PULMONALE

Cor pulmonale is a condition in which the right side of the heart enlarges, weakens and may fail. It is caused by high blood pressure in the pulmonary circulation, which forces the right ventricle (which pumps blood into the lungs) to overwork. Usually the high blood pressure in the pulmonary circulation is caused by lung diseases such as chronic bronchitis or emphysema.

Cor pulmonale is usually chronic and incurable, though affected people may live 10 or 15 years with the condition. In the meantime, symptoms can be relieved or controlled with medications such as diuretics, digitalis, antibiotics, and vasodilators. Plenty of rest and a low-salt diet are recommended. Oxygen (inhaled using a mask) may be needed in later stages of development. Ultimately, only a lung transplant will cure the condition.

LUNGS

The lungs are paired organs in the chest that are responsible for gas exchange between the atmosphere and the blood. Inhaled oxygen is supplied to the blood, and carbon dioxide is exhaled. The lungs are enclosed within pleural sacs, which provide a smooth, low friction surface so that the lungs can move

Lungs

The lungs are divided into lobes. The left lung has only two lobes and is smaller than the right lung as the heart and its vessels take up more space in the left side of the chest. The trachea carries inhaled air down into the bronchial tree within the lungs, where oxygen and carbon dioxide are exchanged.

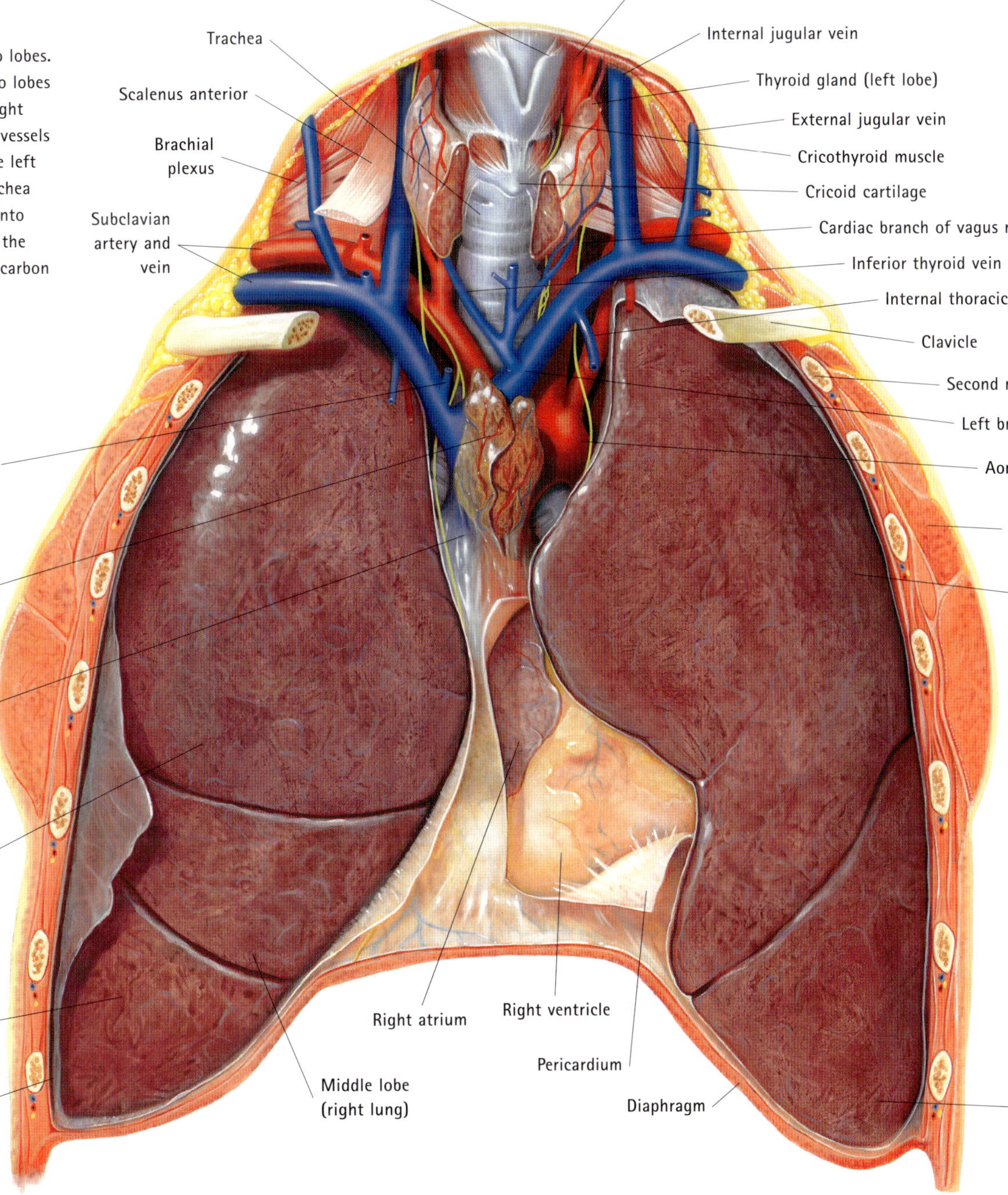

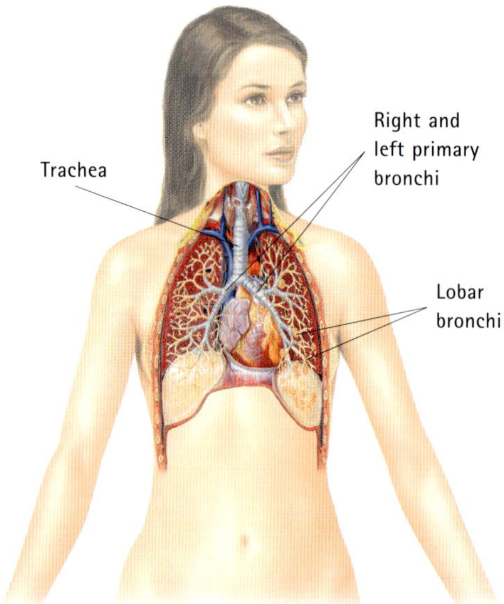

Lungs

The lungs are the two main organs of the respiratory system, lying on either side of the heart within the chest cavity. The trachea and a network of tubes (the bronchial tree) supply the lungs with air. Alveoli, tiny sacs at the end of the bronchioles, transfer oxygen from the air into the bloodstream.

NB: In this illustration the top two-thirds of the lungs and pleura have been cut away to show the heart and the bronchial tree.

freely inside the chest. Between the two pleural sacs, with their enclosed lungs, lies the mediastinum. The mediastinum contains the heart, esophagus, trachea and major vessels and nerves.

Each lung has a roughly conical or pyramidal shape, with a base sitting on top of the diaphragm; sides in contact with the rib cage (costal surface), the mediastinum (mediastinal surface), and the backbone (vertebral surface); and an apex. The lung apex is encircled by the first rib and actually lies above the first rib in the hollow at the angle between the neck and the shoulder.

On the mediastinal surface of each lung is a region known as the lung hilum, where the large airways (the left and right main bronchi) enter the lung and the major lung vessels (pulmonary arteries and veins) enter and leave, respectively. The bronchi subdivide into the bronchioles—smaller tubes which in turn subdivide into the alveoli, which are tiny clusters of air sacs.

The lungs are divided into lobes, usually three in the right lung and two in the left lung, by a series of clefts or fissures.

SEE ALSO *Disorders of the lung on page 241; Respiratory system on page 95*

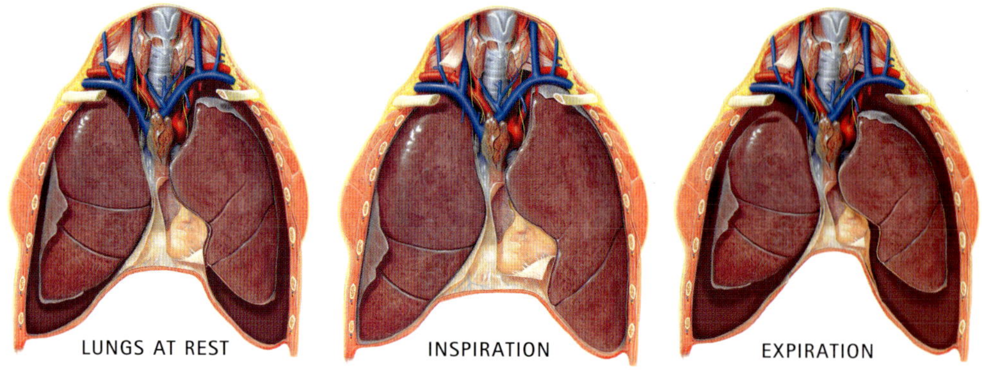

Inflation and deflation of the lungs

Most of the expansion of the lungs with each intake of breath, or inspiration, is due to the contraction of the diaphragm, the muscle separating the chest and abdominal cavities. Breathing out air from the lungs depends mainly on the passive recoil of elastic tension built up in the lungs and chest during inspiration. Some lung expansion is produced by the intercostal muscles, which lie in the space between the ribs and can actively raise or lower the ribs. As the lungs compress and expand, the pressure within also rises and falls in relation to the outside atmospheric pressure. The pleural sacs are at a pressure below that of the outside atmosphere. This ensures that when the chest expands or the diaphragm muscle descends, the lungs are also expanded as air flows in to equalize the pressure.

Breathing

When we breathe in, the intercostal muscles move the ribs upward and outward and the diaphragm pushes downward. This draws air into the expanded lungs.

To reach the lungs, air must flow through the mouth or nose, pharynx, larynx, trachea and main bronchi. The walls of the windpipe (trachea), the main, lobar and smaller bronchi are strengthened by the presence of

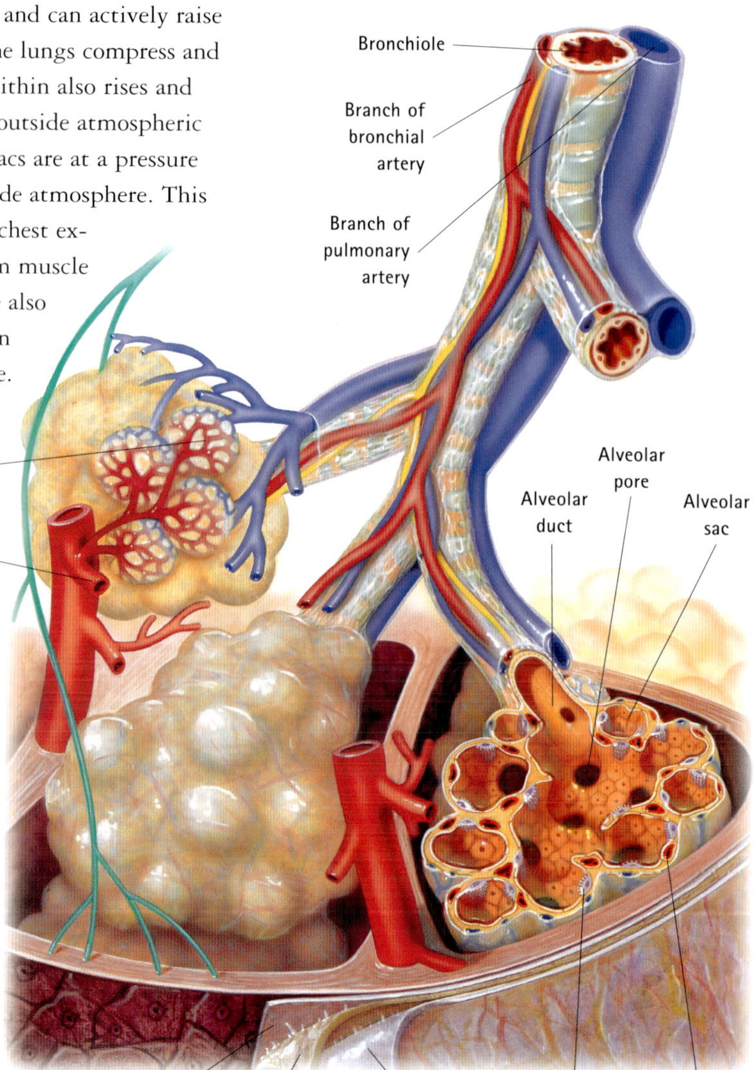

Lung function—gas exchange

The lungs contain millions of tiny air sacs (alveoli) which are located at the ends of the branches of the bronchial tree. The alveolar walls are extremely thin and coated with capillaries. This allows oxygen to pass into the blood from inhaled air and carbon dioxide to pass from the blood to the alveoli where it can be exhaled.

cartilage either as incomplete rings around the airway (as in the trachea and main bronchi) or as large plates (as in the finer divisions of the bronchi). Air reaches the lungs via the two main bronchi entering at the hilum of each lung. There the bronchi divide repeatedly, as many as 23 times, into finer and finer divisions (bronchioles), until the tiny air sacs known as alveoli are reached.

The inner surfaces of the larger airways are lined with special cell types. Some of these cells produce mucus to trap inhaled debris and bacteria, while others have fine hairs, called cilia, on their surfaces, which beat rhythmically toward the larynx, thus wafting the debris toward the throat, where it may be swallowed or coughed up as sputum. In the air passages, particularly concentrated toward the alveoli, are scavenger cells called macrophages, which clean up debris and defend against invasion by bacteria. The inner surfaces of the alveoli are covered with a thin film of fluid, which contains a special chemical known as pulmonary surfactant. The surfactant helps to reduce the surface tension in the alveoli and thus prevent the tiny air sacs from collapsing when air is breathed out.

Gas exchange

The alveoli are the sites where gas exchange between blood and inhaled air occurs. The walls of the alveoli are extremely thin—less than a few ten-thousandths of an inch (hundredth of a millimeter) thick, and are richly supplied with thin-walled capillaries, filled with blood. Venous blood from the right side of the heart, which has relatively high levels of carbon dioxide and low levels of oxygen, flows through the capillaries of the lung. As this takes place, carbon dioxide in the blood diffuses into the air spaces of the alveoli, and oxygen diffuses from the air spaces into the blood.

Oxygen entering the blood is bound to a protein called hemoglobin in the red blood cells. Hemoglobin contains another type of chemical called the heme group. It also contains several bound iron atoms, each within a heme group, which assist with the transport of oxygen. Hemoglobin is the chemical that gives blood its red color.

Gas exchange in the lungs is most effective at sea level and is usually driven by the pressure difference for that particular gas between the alveolar air and the blood. At higher altitudes the partial pressure of oxygen in the air is lower, so that oxygen loading of hemoglobin may not be complete.

Bronchus

The trachea, or windpipe, branches at its lower end into two large (primary) air tubes—the right primary bronchus leading into the right lung, and the left primary bronchus serving the left lung. The route from the trachea through the right primary bronchus is more vertical, wider and shorter than it is to the left, and so inhaled objects are more likely to end up in the right lung.

The outer wall of each bronchus is supported by cartilage. The interior is lined with mucous membrane and many microscopic, mobile, hair-like projections called cilia that shift mucus and trapped particles upward.

At the lung's entrance, the primary bronchus branches into smaller (secondary) bronchi. These are also called lobar bronchi because one of them serves each lung lobe, of which there are usually three in the right lung and two in the left.

Each secondary bronchus divides into smaller (tertiary) bronchi. These split into bronchioles that separate into hair-like terminal bronchioles (of which there are about 64,000) ending at microscopic bubbles known as alveoli, where the exchange of carbon dioxide for oxygen occurs.

Together, all these branching air passages are referred to as the bronchial tree.

Alveolus

An alveolus, also known as a pulmonary alveolus, is one of the tiny air spaces in the lung where blood exchanges carbon dioxide for oxygen.

There are about 300 million alveoli in each human lung, grouped in grape-like clusters (alveolar sacs). Alveoli give the lungs their huge surface area, estimated to be 1,000 square feet (93 square meters), and the thin-walled blood vessels in each provide the means for gases—carbon dioxide and oxygen—to be exchanged by diffusion.

Bronchus

A bronchus is a tube that conducts air from the trachea to the lung tissue. The two main bronchi divide several times into smaller branches until they give rise to thin, delicate airways called bronchioles. These connect to small air sacs known as alveoli.

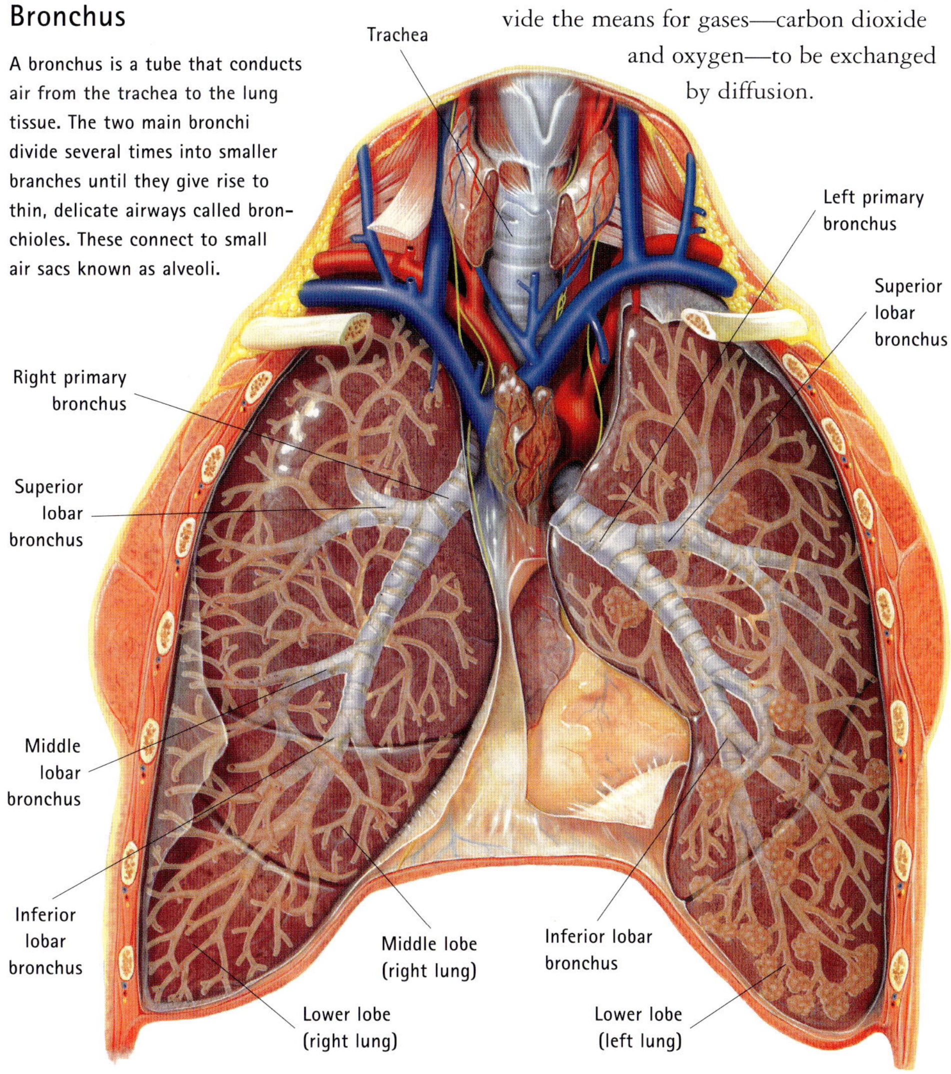

Pulmonary circulation

In the human body, blood passes through the heart twice during each complete passage of the circulatory system. In this so-called "double circulation" system, found in all mammals, the blood follows two complementary but separate routes from the heart. Systemic circulation takes the blood around the body and back; pulmonary circulation transports the blood to the lungs and back. The purpose of pulmonary circulation is to pass the blood close to the thin-walled air sacs (alveoli) of the lungs so that the blood can exchange carbon dioxide collected from the body for oxygen.

Before being sent into the pulmonary blood vessels, the blood returns, deoxygenated and dark red, from the head, limbs and internal organs, to the heart. The oxygen-depleted blood enters the right atrium then passes, via the tricuspid valve, to the right ventricle located immediately below. From there the blood is pumped out along the route of the pulmonary circulation.

The right ventricle sends the blood through the pulmonary valve and on to the lungs via the pulmonary arteries, the only arteries that carry deoxygenated blood after birth. The right pulmonary artery runs to the right lung and the left goes to the left lung. Each pulmonary artery divides repeatedly to culminate in a network of tiny, thin-walled capillaries around the air sacs of a lung. Here, carbon dioxide diffuses out of the blood into the lungs, from where it is exhaled. At the same time, oxygen contained in air breathed into the lungs diffuses into the blood.

From each lung the oxygenated blood (now bright red) flows from the capillaries to venules and eventually into the two pulmonary veins. Most veins after birth carry deoxygenated blood but the pulmonary veins, transporting oxygenated blood from the lungs back to the heart, are the exception. The pulmonary circulation is completed when blood is delivered back to the heart at the left atrium. From there the oxygenated blood passes through the mitral valve to the muscular left ventricle and then on to the rest of the body via the systemic circulation.

Pulmonary circulation comes "on-line" in humans only after birth. Before that, the lungs (along with the gastrointestinal tract) do not function. Instead, the fetal blood collects oxygen from and dumps carbon dioxide into the mother's circulation through the placenta.

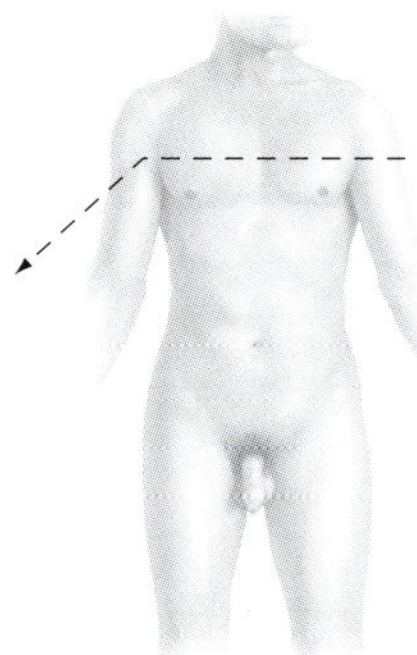

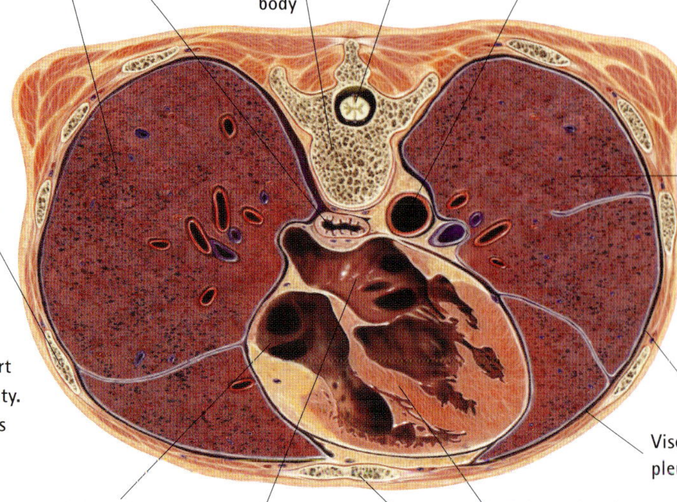

Thorax cross-section

Protected by the rib cage, the heart and lungs dominate the chest cavity. The synergy of the heart and lungs ensures a constant supply of oxygenated blood circulating through the body.

Breathing

Breathing—also known as respiration or ventilation—is the inspiration and expiration of air into and out of the lungs, by the contraction and relaxation of the diaphragm, the chest wall and abdominal wall. Breathing is the means by which a person absorbs oxygen from the air through the lungs into the bloodstream, and exhales carbon dioxide from the bloodstream through the lungs into the air.

At rest, the average person (at sea level) breathes 10 to 15 times a minute. In times of stress, or during exercise or other physical activity, the demand for oxygen by tissues is greater, and breathing is deeper and faster. Respiration is more labored at higher altitudes (unless the person is acclimatized) and during the later stages of pregnancy.

If breathing stops, oxygenation of blood ceases, and irreversible brain damage or even death may occur in four to six minutes.

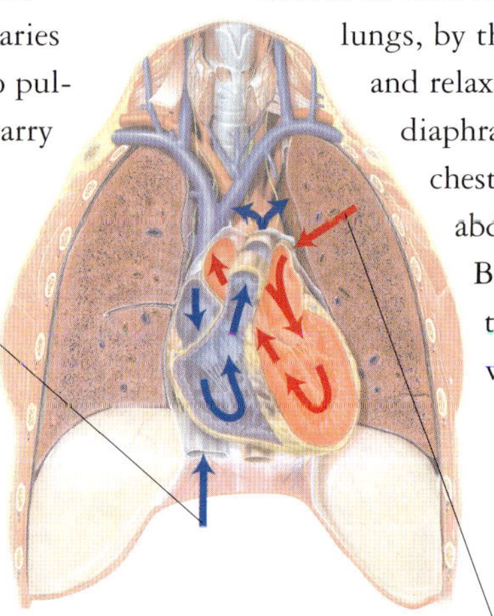

Oxygen-depleted blood enters the right ventricle of the heart and is pumped into the lungs to be oxygenated by the alveoli.

Oxygenated blood flows out of the lungs to the left side of the heart and is pumped out into the body for systemic circulation.

Pulmonary circulation

The primary purpose of pulmonary circulation is to offload carbon dioxide, which the blood has collected from the body, and pick up oxygen through the thin-walled air sacs of the lungs.

NB: In this illustration the top two-thirds of the lungs and pleura have been cut away to show the heart.

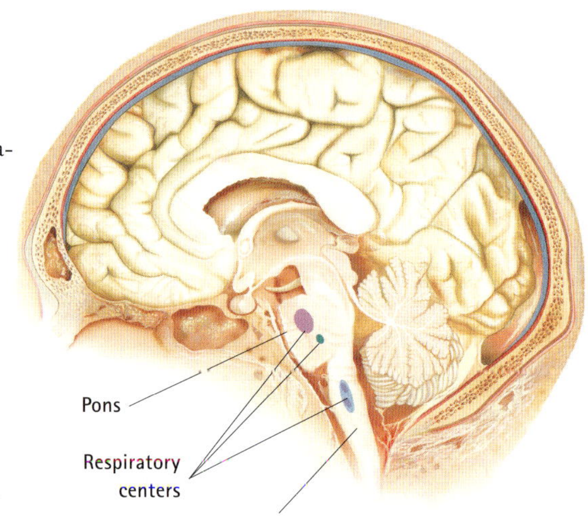

Respiratory centers in the brain

Specialized areas in the pons and the medulla receive information about blood carbon dioxide and blood oxygen levels from monitors elsewhere in the body. These respiratory centers send messages to the respiratory muscles to increase or decrease the rate and depth of breathing.

Disorders of the lungs

Some lung disorders are inherited. The best-known is cystic fibrosis, a genetic disorder affecting one in every 2,500 live births.

Some lung problems arise because of disease of other organs. One example is pulmonary edema, which is the accumulation of fluid in the lower lungs due to failure of the heart to pump blood effectively.

SEE ALSO *Endoscopy on page 433; Imaging techniques on page 431; Pulmonary function tests on page 431; Sputum tests on page 431; Treating the respiratory system on page 441*

Asthma

Asthma is a disease of the bronchi (or air passages) of the lung, characterized by periodic and repeated attacks of wheezing alternating with periods of normal breathing. The muscles of the bronchi contract, reducing airflow and producing shortness of breath, coughing and wheezing. Mucus production increases and the lining of the bronchi becomes inflamed.

In many industrial countries, asthma currently affects one in twenty people, and the number of sufferers is rising. Asthma is more prevalent in children than adults, affecting one in ten children; by adulthood, many children seem to have outgrown it. The disorder is more common if there is family history of asthma, eczema or allergies. In some cases asthma does not develop until adulthood. Commonly an adult asthmatic is short of breath early in the morning. Cough is less frequent than in childhood asthma. Adult asthma is often associated with cigarette smoking or chronic lung diseases such as chronic bronchitis.

Asthma attacks typically come and go. There may be intervals of days, months or years between attacks, or there may be attacks every day. Symptoms can occur spontaneously or can be set off by trigger factors. The most common form of asthma (allergic bronchial asthma) is caused by an allergic reaction. Many pollens, molds, dusts (especially house dust and wood dust), cigarette smoke, feathers and animal hair can cause allergic-type asthma attacks. Asthma is sometimes associated with hay fever.

Respiratory infections, exposure to cold, industrial fumes and certain emotional and psychological states can trigger an asthma attack. Drugs such as aspirin, beta-blockers and anti-inflammatory drugs are other precipitating factors. Asthma from these causes may occur in people who have no history of allergic reactions, as well as in those who do. Exercise may trigger asthma; symptoms of exercise-induced asthma (coughing, wheezing, chest tightness lasting several minutes to an hour or more) are different from the deep and rapid breathing that quickly returns to normal after exercise.

In a mild attack, wheezing may be barely audible and the wheezing occurs only during exhalation. As the attack worsens, the wheezing becomes louder and may also be present with inhalation; the sufferer breathes rapidly, gasps for breath, and becomes agitated. In a severe asthma attack, if the bronchioles become totally blocked, airflow may diminish and wheezing may stop. In children especially, this is a sign of serious trouble. An attack of asthma may be prolonged and may not respond to treatment. This condition is called status asthmaticus.

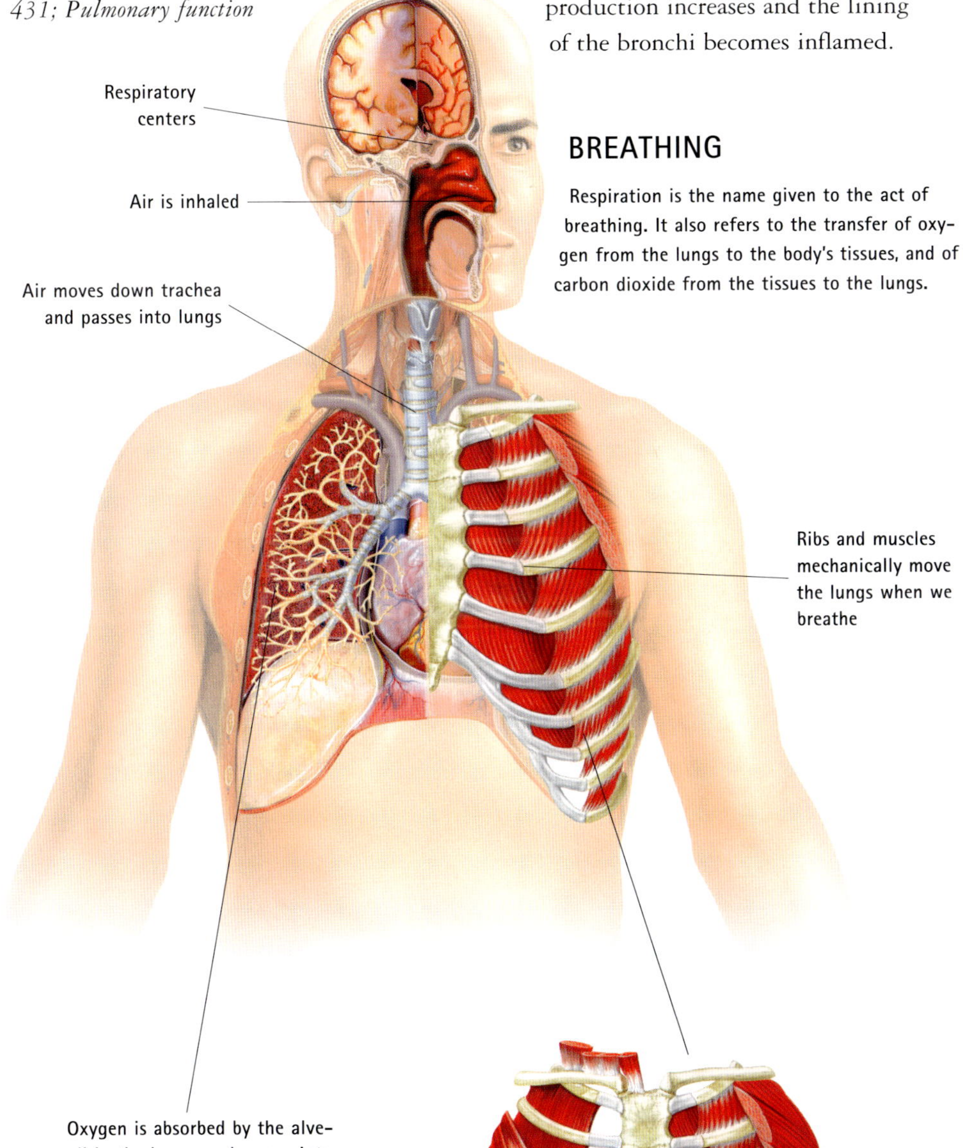

BREATHING

Respiration is the name given to the act of breathing. It also refers to the transfer of oxygen from the lungs to the body's tissues, and of carbon dioxide from the tissues to the lungs.

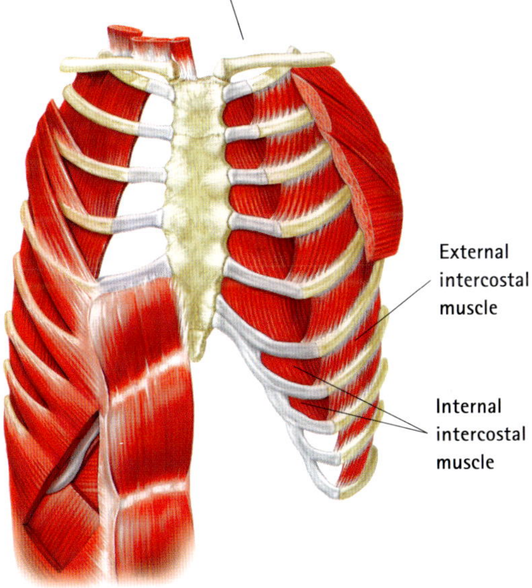

Breathing muscles

Breathing is caused by the actions of the intercostal muscles (the muscles between the ribs) and the diaphragm, a muscular dome that separates the abdomen from the chest. The external intercostal muscles assist by elevating the ribs, resulting in an increased volume of air in the lungs.

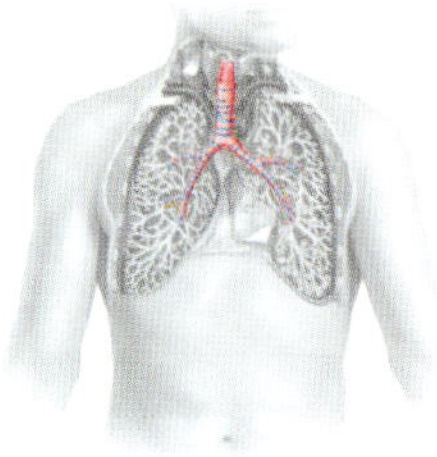

Asthma

In asthma, the lining of the bronchi become inflamed and swollen. The smooth muscles of the bronchial walls go into spasm, and the bronchi produce excessive mucus.

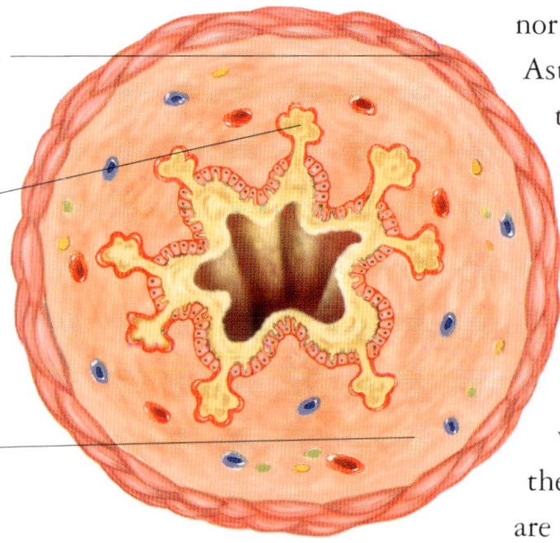

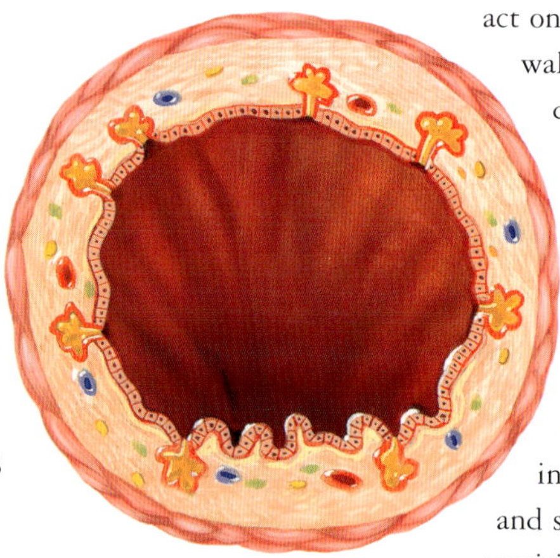

NORMAL BRONCHUS

An attack of status asthmaticus requires hospitalization and urgent treatment.

It is usually not difficult for a physician to diagnose an asthma attack, especially if the sufferer is known to be asthmatic. Wheezing sounds, audible through a stethoscope, and rapid improvement of symptoms after treatment confirm the diagnosis. Breathing tests such as spirometry show reduced airflow across the airways during attacks, which improves after treatment. Chest x-rays are usually normal. In some asthmatic people, eosinophils (white cells associated with allergies) are found in the blood and sputum.

Management of asthma

Unlike other lung diseases, asthma is usually a reversible condition; the narrowing of the airways caused by bronchospasm may improve spontaneously or in response to one or more of the medications available.

Several simple measures can reduce the risk of attack. Trigger factors such as pollens, animal furs and foods known to cause asthma should be avoided. A person with allergic asthma should sleep in a room without carpets or rugs. Blankets and pillows of synthetic fiber reduce the risk of house dust and mites. Asthmatics should not smoke, nor should others smoke in the same house. Asthma medication should be taken prior to events known to trigger an episode—before exercise, for example.

Various medications are available for treating acute attacks, and for long-term prevention of asthma. Often they are used in combination.

Bronchodilators dilate the bronchial wall, allowing air through and relieving the symptoms of asthma. Some of these are "beta agonists," so called because they act on the beta adrenoreceptor of bronchial wall muscle, which results in bronchodilation. Others, known as anticholinergic agents, also act on the bronchi, though using a different mechanism to relax and open the airway passages. Both types of bronchodilator may be used together.

Side effects of bronchodilators include nervousness, restlessness, insomnia and headache. Elderly patients and some young children may be more sensitive to the effects of bronchodilators.

When symptoms of asthma are frequent and difficult to control with bronchodilators, preventive medications are used. These medications do not cure an attack of asthma once it has begun. Instead, they prevent attacks from recurring. Physicians believe that long-term damage to the bronchi is minimized by using preventive medications.

Commonly, bronchodilators and preventive medications are used together—the bronchodilators to treat the symptoms of an attack and the preventive medication to stop attacks from occurring. Corticosteroids are the most commonly used preventive medications. They greatly help in reducing bronchial inflammation and airway obstruction and improve lung function.

Sodium cromoglycate is also a preventive medication; it stops the release of chemicals such as histamine into the bronchi, which can cause asthma. Sodium cromoglycate is useful for asthma that is triggered by exercise, cold air and allergies, such as to cat fur.

In the treatment of asthma, inhaled medications are generally preferred over tablet or liquid medicines. Inhaled medications act directly on the surface of the airways; absorption into the rest of the body is minimal so side effects are fewer compared with oral medications. Inhaled medications are given via a metered dose inhaler, or a puffer. In children who have difficulty with inhalers or puffers, or in adults with a more severe attack, asthma medications can be given by nebulizer, which administers the medication in the form of a fine mist inhaled through a mouth mask.

Oral corticosteroids, such as prednisone, methylprednisone and hydrocortisone, may be used as preventive medications. However, they have more side effects than inhaled corticosteroids, so long-term use is not recommended, except when other treatments have failed to restore normal lung function and the risks of uncontrolled asthma are greater than the side effects of the steroids. Corticosteroids may also be given intravenously in severe attacks requiring hospitalization, for example in status asthmaticus. Intravenous epinephrine (adrenaline) and oxygen may also be needed during a life-threatening attack.

As the severity of an asthma attack is variable from person to person, it is helpful for an asthmatic to have an individual management plan in case of an attack. The plan should be made available to carers, teachers, nurses, parents and anyone else who has responsibility for that person.

Acute respiratory distress syndrome

Acute respiratory distress syndrome (ARDS), sometimes called adult respiratory distress syndrome, is a name given to several conditions in which fluid leaks out of the blood vessels and into the alveoli of the lung, causing pulmonary edema, respiratory failure, and in some cases death.

The syndrome, also known as pump lung, stiff lung or white lung, may be caused by a number of conditions, including shock, fluid overload, narcotic overdose, disseminated intravascular coagulation and massive trauma or burns.

Clinical signs include breathlessness, rapid heartbeat, cyanosis (a bluish discoloration) and hypoxia (lack of oxygen). A lung x-ray shows evidence of interstitial and alveolar edema.

The condition is treated with mechanical ventilation. Diuretics such as frusemide may be used to decrease the amount of fluid in the lung, and antibiotics may be needed to combat lung infection. If the underlying condition is treatable the prognosis is good.

Atelectasis

A partial or complete collapse of a lung is known medically as atelectasis. It is most often caused by a blockage in a bronchus or bronchioles. Obstructions can include tumors, inhaled objects, or thick mucus from infections or a disease such as cystic fibrosis. A form of atelectasis can also occur as a postoperative complication caused by the effects of surgery.

Acute atelectasis can lead to a sudden and major collapse accompanied by chest pain, rapid and uncomfortable breathing, dizziness and shock. Alternatively, the signs of atelectasis may include a less dramatic, more gradual collapse, cough, shortness of breath and a fever.

Atelectasis is not normally fatal and the affected lung usually reinflates once any obstruction has been extracted or dislodged. A procedure called a bronchoscopy may be required to remove a blockage. Pneumonia and permanent scarring of lung tissue are potential complications of atelectasis.

Chronic obstructive pulmonary disease

Another important group of lung diseases in Western countries is chronic obstructive pulmonary disease (COPD). There are two serious diseases within this classification: chronic bronchitis and emphysema, which can occur together.

CHRONIC BRONCHITIS

Chronic bronchitis is usually characterized by cough and sputum production, occurring on most days during at least three consecutive months for more than two successive years. The sputum may contain just mucus or may include some pus in it if additional infection is present.

Chronic bronchitis results from inhalation of airborne irritants over a long period. By far the most common cause of this disease process is cigarette smoking, although it can also be triggered by repeated exposure to high levels of dusts, irritant gases or other pollutants. In the early stages of chronic bronchitis (simple chronic bronchitis) the main change in the large airways is not inflammation, but rather an increase in mucus secretion by cells lining the airways and glands within the airway walls.

This leads to cough and the production of white or clear sputum (mucus secretion). This is often accompanied by bacterial infection and inflammation, which may cause coughing up of yellowish or greenish sputum that is a mixture of mucus and pus (mucopurulent bronchitis).

In some people, continuing exposure to the inhaled irritant leads to progressive involvement of smaller airways. This is an important complication because it causes widespread inflammation and scarring of the walls of these airways, leading to narrowing which limits the flow of air (chronic obstructive bronchitis). The person becomes increasingly breathless. The irritants that trigger chronic bronchitis usually also damage and destroy the alveoli of the lung in parallel, leading to the development of emphysema, which causes even more breathlessness. The combination of chronic bronchitis and emphysema may also overlap with asthma (asthmatic bronchitis).

Eventually, these forms of chronic respiratory disease may be complicated by development of heart failure. They are major causes of chronic disability and death.

Because damage to the small airways and the lungs is difficult or impossible to reverse, stopping smoking or avoiding exposure to other inhaled irritants can ensure that severe disease does not develop. In the early stages of chronic bronchitis, stopping smoking can lead to complete disappearance of symptoms. In the late stages, treatment with antibiotics, anti-inflammatory agents, bronchodilators and oxygen may be required.

EMPHYSEMA

Emphysema often accompanies chronic bronchitis. Whereas chronic bronchitis mainly involves the larger airways, emphysema involves the destruction of very fine airways and alveoli. This is caused by a combination of repeated infection, tissue degeneration and over-distension of alveoli due to obstruction of the larger airways in chronic bronchitis.

Bronchitis

In bronchitis, the walls of the airways (the bronchi) become inflamed. The bronchial glands then secrete excessive amounts of mucus, which may be coughed up as sputum.

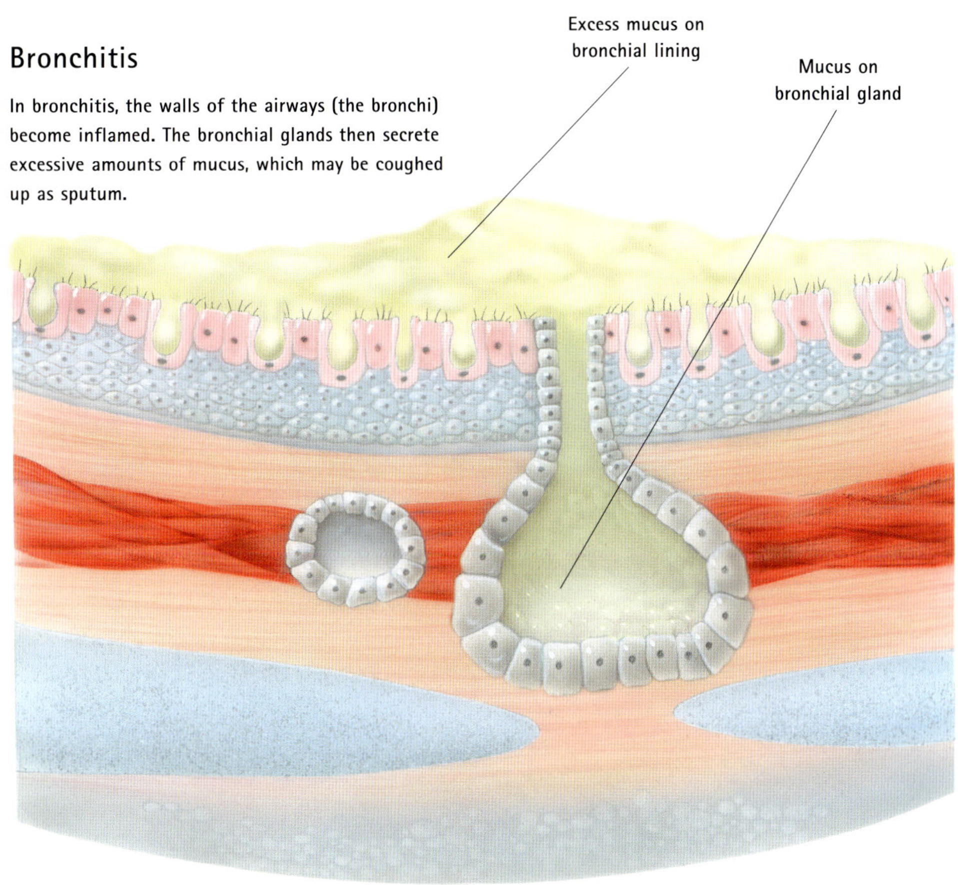

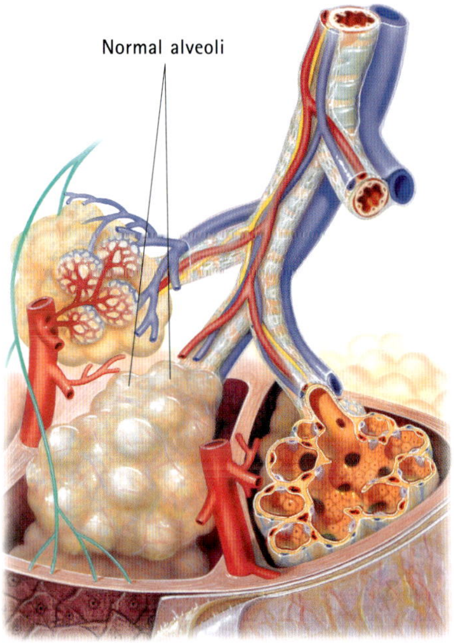

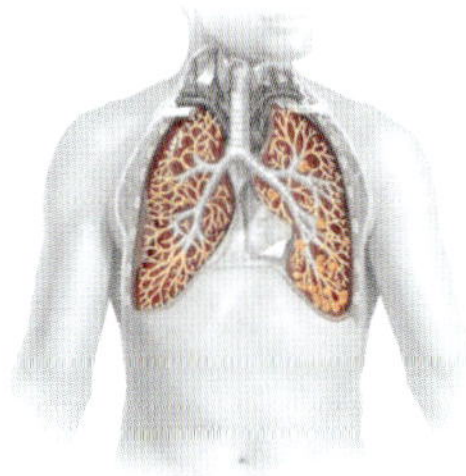

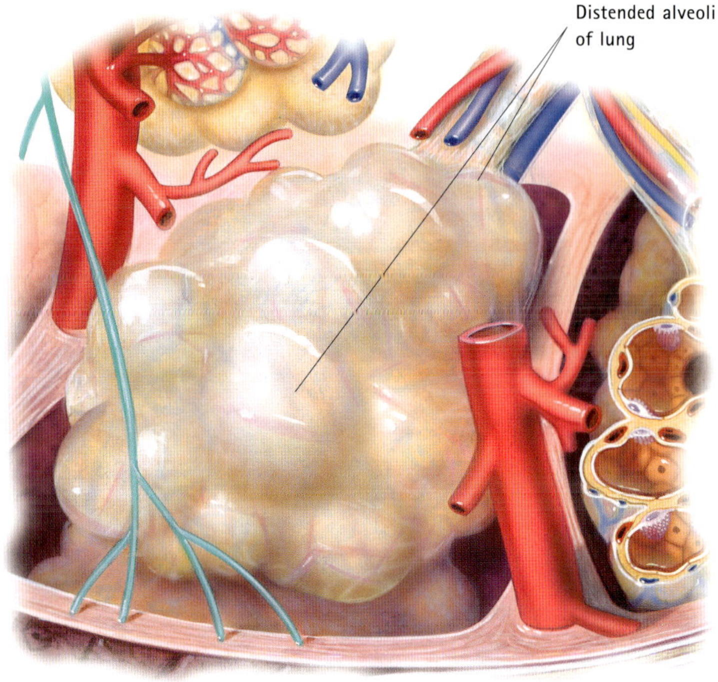

Emphysema

In emphysema, the air sacs in the lungs (the alveoli) become damaged and distended. As a result, the lungs are less able to supply the oxygen that the body requires.

Emphysema or pulmonary emphysema is a chronic obstructive lung disease marked by wheezing, breathlessness and increasing loss of lung function. The disease occurs most commonly in smokers and people exposed to polluted air and airborne dust or similar irritants, but can also affect children who suffer asthma or bronchitis.

In the early stages of emphysema the lining of the lung's airways, the bronchi and bronchioles, is stimulated by irritation of smoke or other pollutants to produce abnormal quantities of mucus. Over time, the greater quantity of mucus leads to persistent coughing, "smoker's cough," and a greater susceptibility to colds, which can lead to chest infections.

The early symptoms are similar to those of bronchitis: coughing and bringing up mucus combined with asthmatic wheezing as the airways narrow. In emphysema, there is the added complication that the alveoli decay. As the alveoli become less efficient, air is trapped in the lungs and the lung tissue decays. The trapped air decreases the volume of fresh air that can be inhaled and the lungs may expand permanently in the effort to counter this, resulting in a permanently expanded chest and a "barrel-chested" appearance that is characteristic of the disease. The lungs are unable to supply the oxygen the body needs, and the smallest exertion may produce severe breathlessness.

In addition to breathlessness, the symptoms of emphysema include a bluish tinge to the skin, a build-up of carbon dioxide caused by inefficient reoxygenation of the blood, loss of weight, swelling in hands and feet, tightness in the chest and increased respiratory distress in cold or smoky air.

Reduced oxygen intake causes the heart to pump faster, and the strain can lead to heart failure. Bullous emphysema is a condition in which the alveoli distend and form cysts on the lungs that may rupture, causing the lungs to collapse.

The exact cause of emphysema is not known, but it is commonly associated with cigarette smoking and long-term exposure to air pollutants in mining and industrial processes. A deficiency in a protein in the liver (alpha1-antitrypsin) has been found in some sufferers. Antitrypsin counteracts trypsin, an enzyme produced by many types of bacteria that decays tissue.

When emphysema is diagnosed it is essential to stop further damage by giving up smoking and avoiding exposure to pollutants and dust. No treatment can reverse the lung damage but existing lung function can be improved with the use of drugs such as bronchodilators, which relax muscles that restrict the airways, and by supplying oxygen to the sufferer for inhalation at home. Steroids can reduce lung inflammation and antibiotics are used to clear up infections. A more recent development is the surgical removal of damaged lung tissue to allow the functioning tissue room to expand and work better. Lung transplantation is also an option.

Infections of the lung

Infections of the lung are of many different types and are named according to the sites of the lung which are affected.

Bronchitis is infection of the large airways by bacteria and may follow viral infections of the upper respiratory tract. Bronchiolitis is infection of the finer airways (bronchioles) in infants and is usually due to respiratory synctial virus. Bronchiectasis is the dilation of the large airways with the accumulation of secretions and chronic infection. Pulmonary tuberculosis is due to infection of the lung by mycobacteria. Tuberculosis is now much rarer than in the nineteenth century, but is still a serious public health problem.

ACUTE BRONCHITIS

Acute bronchitis is an inflammation of the major airways, following upper respiratory infection by a virus such as influenza, causing cough and production of sputum. It generally clears up on its own, unless a bacterial infection also occurs, in which case the sputum usually becomes thick and yellow. Because acute bacterial bronchitis can lead to the development of pneumonia, especially in elderly patients, treatment with antibiotics is advisable.

PNEUMONIA

Pneumonia is a serious lung disease in which inflammation, caused by bacteria and/or viruses, results in the accumulation of fluid and cellular debris in the air spaces of the lungs. This prevents the exchange of carbon dioxide and oxygen in the lungs. Pneumonia is very common, affecting about 1 percent of the population every year. It is most common in winter and spring because of sudden drops in air temperature, overcrowding in poorly ventilated rooms and the prevalence of bacteria and viruses. Smoking, alcoholism, a poor immune system and air pollution are associated with an increased risk of pneumonia. As many as 5–10 percent of those developing pneumonia may die from the disease, usually the very young or the elderly. Bedridden patients with dementia, hip fractures or chronic heart disease are particularly prone to developing pneumonia, which is often the ultimate cause of death.

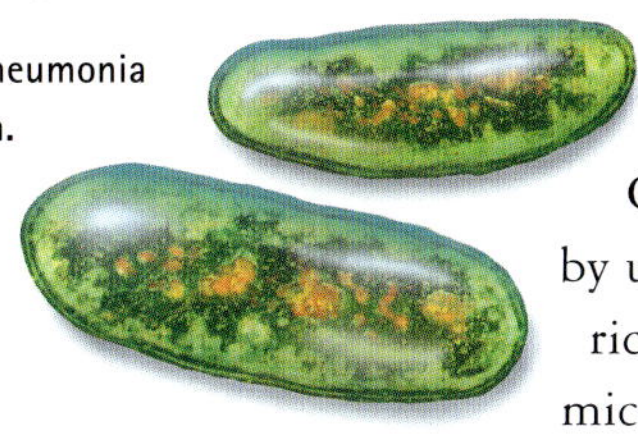

Bacterial pneumonia

Culture of the sputum of pneumonia sufferers will reveal bacteria. These need to be tested for sensitivity to antibiotics to help guide treatment.

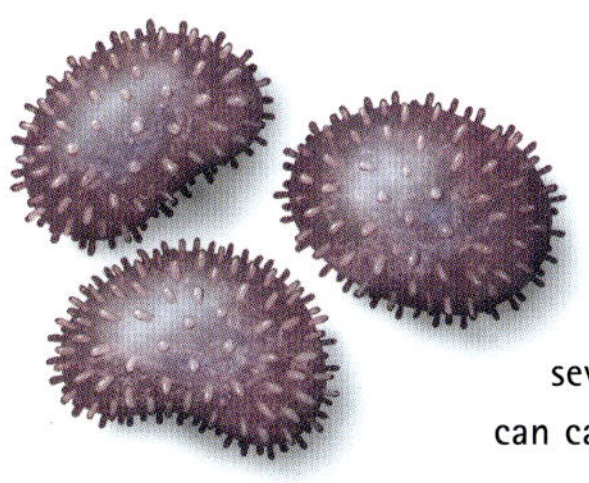

Viral pneumonia

The influenza virus is one of several viruses that can cause pneumonia.

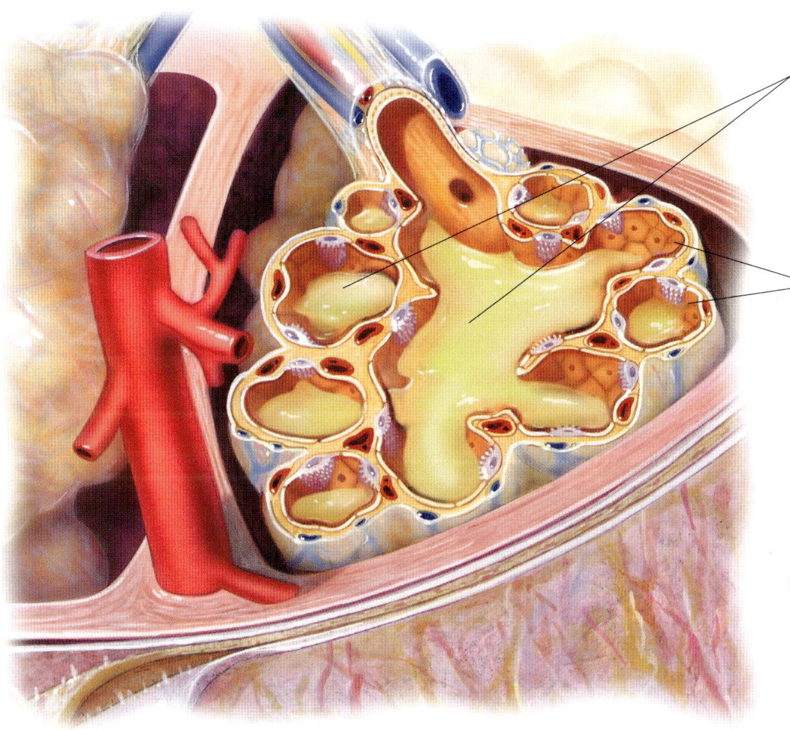

PNEUMONIA

In pneumonia, air spaces in the lungs fill with fluid and cellular debris preventing gas exchange. This can result in respiratory or cardiac failure.

Depending on the site, different types of pneumonia may be described: lobar pneumonia affects an entire lobe of the lung; lobular pneumonia affects lung tissue around the major airway branches; bronchopneumonia is lobular pneumonia affecting both lungs.

The microorganisms responsible for pneumonia include bacteria, such as *Streptococcus pneumoniae*, *Staphylococcus aureus*, *Klebsiella pneumoniae* and *Haemophilus influenzae*, and viruses, such as influenza, adenovirus and respiratory syncytial virus. Occasionally pneumonia may be caused by unusual microorganisms such as fungi, rickettsiae and mycoplasma. Unusual micro-organisms such as mycoplasma and fungi and are found to be responsible when the patient's immune system is functioning poorly, as in those with AIDS.

Patients with pneumonia may initially complain of cough and sputum (coughed-up lung fluid), breathlessness, sharp chest pain aggravated by coughing or by taking a deep breath, fever and a general feeling of being unwell. The sputum is usually yellow or green and may occasionally contain blood, particularly if vessels in the lung have been ruptured by vigorous coughing. A chest x-ray will show loss of the normal air-filled spaces, so that the lung fields appear white, either uniformly, within a lobe of the lung (lobar pneumonia) or in patches (bronchopneumonia). Culture of the sputum will reveal bacteria, which may be the actual causative organism or present as secondary bacterial infection of a pneumonia caused by a virus. The bacteria need to be tested for sensitivity to antibiotics, because in recent years many bacteria have developed resistance to some available antibiotics. The concentration of oxygen in the arterial blood may be reduced, and in the severely ill the concentration of carbon dioxide in the blood may rise.

Treatment of pneumonia involves prescribing the appropriate antibiotic for the causative microorganism. The patient should be supported by inhaled oxygen, receive physical therapy to clear the airways and analgesics to relieve pain. Complications that may arise include respiratory failure, cardiac failure, abscesses (pus-filled cavities) in the lung and fluid in the pleural sacs around the lung (pleural effusion). Occasionally pus may collect in the pleural sacs around the lung (pleural empyema).

Lung cancer

Lung cancer is a common cancer, ranked second in frequency in both men (after prostate cancer) and women (after breast cancer). It is usually more aggressive than other cancers and is thus the most common cause of cancer death in most countries.

Tobacco smoking plays a very important role in causing lung cancer. A variety of pre-cancerous changes can be seen in the inner layer of cells (the epithelium) lining the major airways of smokers.

Cigarette smoking is particularly incriminated, although pipe and cigar smoking also increase the risk of developing lung cancer. Long-term heavy smokers have at least a 20-fold higher risk of developing lung cancer than non-smokers.

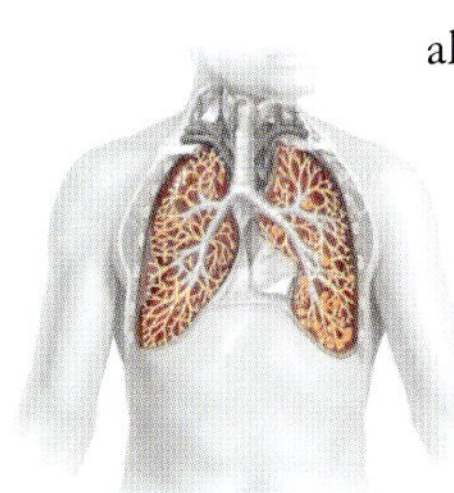

The risk decreases for those who stop smoking, approaching that of non-smokers after 15 years, with a progressive disappearance of the abnormalities in the epithelium. Therefore, lung cancer can be regarded as a preventable disease.

Lung cancer

Malignant growths of the lung are among the most common types of cancer. Most lung cancers arise in the cells that line the major airways, but the disease encompasses a number of different tumor types. The cancer usually progresses through the walls of the airway and may then spread through the blood to other organs.

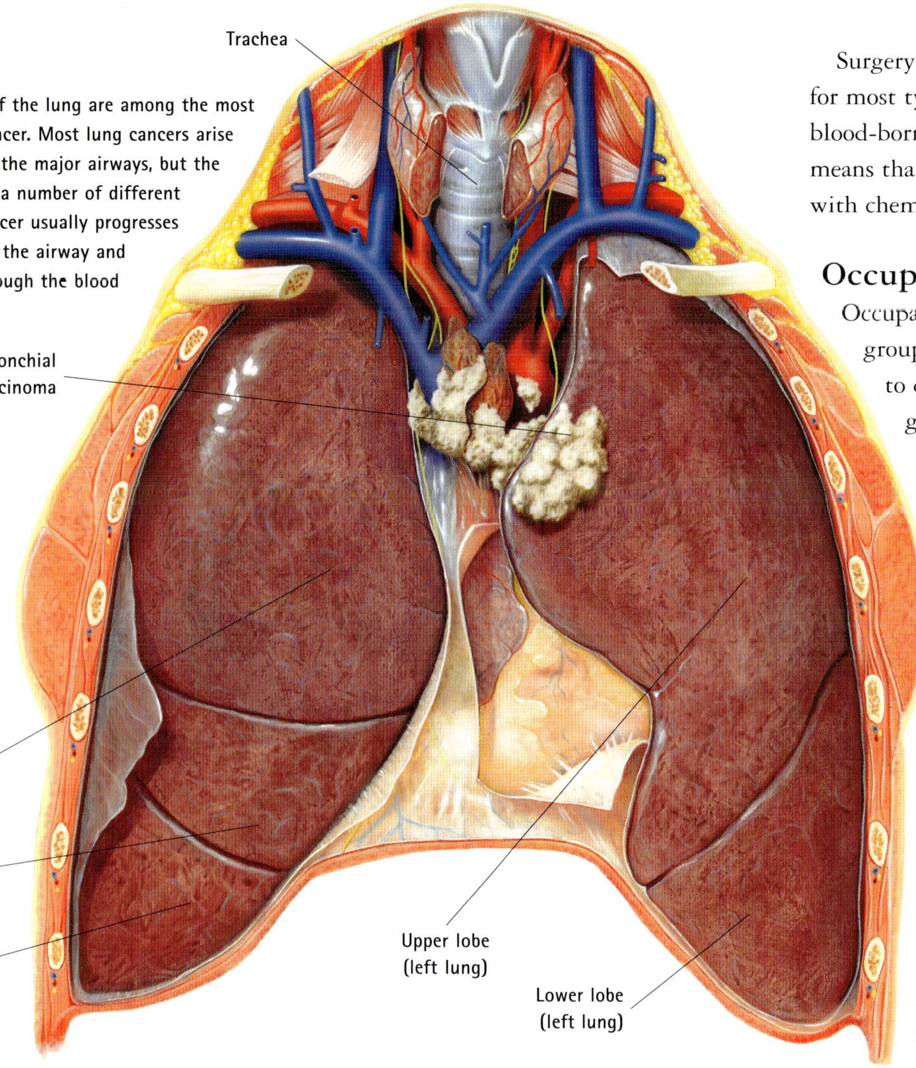

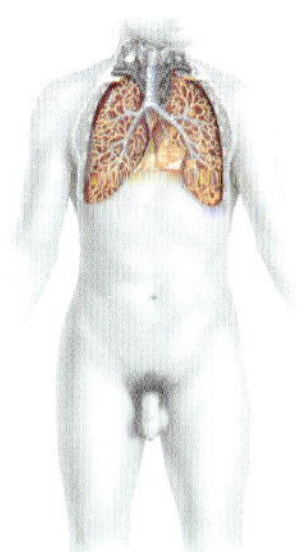

Given that there is also a strong association between smoking and vascular disease, and increased risk of a number of other cancers in smokers, the implementation of public health strategies is only partially successful. Other risk factors for the development of lung cancer include exposure to asbestos, a variety of metal dusts, ionizing radiation and certain industrial chemicals.

Although most lung cancers develop in the epithelium of the airways, lung cancer is not a single type of tumor. Several different varieties can be recognized microscopically, of which one known as small cell cancer is of particular importance.

As they grow, lung cancers usually invade through the airway wall, causing cough and sometimes the coughing up of blood. The cancer may also grow into the airway tube, which tends to cause obstruction which leads to wheezing, repeated chest infections and/or breathlessness.

In a proportion of patients, the first evidence of lung cancer is the appearance of secondary growths, especially in the bones (causing pain), in the liver (causing enlargement) and in the brain (causing seizures and various other complications). This pattern is especially likely to occur in small cell lung cancer, which is spread quickly through the body via the blood. Unusual symptoms can also develop as as a result of hormone secretion by the tumor cells.

Early diagnosis is the key to effective treatment of cancers. In the case of lung cancers, however, few symptoms develop early in the course of the disease. There is as yet no available diagnostic test that is sensitive, easy to perform and inexpensive. While x-rays, examination for cancer cells in the sputum, biopsy via bronchoscope and other techniques allow a specific diagnosis to be made, these techniques are unsuitable for population screening. Thus, by the time symptoms become obvious and a diagnosis is made, the tumor is already quite large and has typically invaded surrounding structures, spread via the lymphatics and the blood, or both. This is why the success rate for the treatment of lung cancer is still very poor.

Surgery and radiation therapy are used for most types of lung cancer, but early blood-borne spread of small cell cancers means that these tumors can only be treated with chemotherapy.

Occupational lung diseases

Occupational lung diseases are a varied group of disorders caused by exposure to dusts, gases or fumes. The largest group are the pneumoconioses, which are caused by prolonged inhalation of inorganic dusts such as coal dust, silica and asbestos.

Asbestosis is a particularly serious form, because asbestos fibers can cause fibrosis of the lung and cancer of the pleural sac (mesothelioma) and lung (bronchial carcinoma).

Another important lung disease is extrinsic allergic alveolitis, caused by inhalation of organic dust which produces an allergic reaction in the lung (for example, "farmer's lung," due to moldy hay; "bird fancier's lung," due to pigeon and budgerigar droppings; and "malt worker's lung," due to moldy barley and malt).

Other occupational respiratory conditions include occupational asthma, in which dusts and fumes constrict the airways within a few minutes to hours of exposure and toxic reactions to irritant gases like chlorine.

ASBESTOSIS

Asbestosis is a disease in which the lungs become inflamed through the inhalation of asbestos particles. It can develop into asbestos cancer (mesothelioma), which is a malignant tumor of the lining of the lung, chest or abdominal cavity.

Asbestos is an extremely poor conductor of heat and so was used industrially as an insulating material for many years. It is fibrous, and the fibers are easily shed and inhaled with air. The fibers can then accumulate in tissues of the airways and lungs and inhibit the exchange of oxygen between inhaled air and blood. Asbestos is no longer used because of its harmful effects; removing it from old installations requires special breathing apparatus and protective suits.

SILICOSIS

Also known as miner's phthisis, grinder's disease and potter's asthma, silicosis is a respiratory disorder caused by inhaling silica, mainly in industries. It involves close contact with stone, sand or ceramics such as mining, quarrying, tunneling and the manufacture of ceramics and pottery. The most serious effect of the inhaled silica dust is impaired gas exchange associated with the development of small hard nodules and fibrosis in the lungs.

The severity of the disease depends on the level and length of time of exposure. Chronic or simple silicosis develops after more than 10 years of low-level exposure. Higher levels over 5–10 years cause accelerated or complicated silicosis. Acute silicosis results from highly concentrated exposure over a short period of time.

Symptoms range from shortness of breath and a dry cough in mild forms to lethargy, restless sleep, appetite loss, chest pain, a cough that produces blood, and hypoxia (lack of oxygen in the blood) which shows as blueness of the lips and skin. There is no cure for silicosis and it commonly reduces life expectancy and quality of life. Records of the disease date back two thousand years. It was prevalent in the nineteenth and early twentieth centuries but is now becoming rarer due to improved occupational health and safety standards.

Diseases of the pleura

The pleura is a thin, two-layered membrane that lines the lung and chest cavity and enables the lungs to move about smoothly against the wall of the chest during breathing.

PLEURISY

Pleurisy is inflammation of the pleura and is a consequence of other lung conditions rather than a disease in itself. Conditions that may lead to the development of pleurisy include bronchitis, pneumonia, a blood clot in the lung, lung cancer, collagen vascular diseases (such as systemic lupus erythematosus or rheumatoid arthritis), congestive heart failure or kidney and liver disorders.

The most common symptoms of pleurisy are sudden chest pain that worsens with breathing, and coughing. Other symptoms of the underlying disease may also be present; for example, chest pain, cough and fever.

Treatment is directed toward the underlying cause and may include painkillers, anti-inflammatory drugs and antibiotics.

PNEUMOTHORAX

Pneumothorax literally means "air in the chest." It occurs when air enters one or both of the pleural sacs which enclose the lungs. Pneumothorax may occur spontaneously, following rupture of an air space in the outer part of the lung, or as a result of injury, which leads to perforation of the chest wall. Traumatic pneumothorax may follow penetrating chest wall injuries, blast injuries and diving accidents.

A particularly dangerous type of pneumothorax is tension pneumothorax, which occurs when a flap-valve effect in the damaged pleural membrane allows air to enter the pleural sac, but not escape. Pressure in the pleural sac rises rapidly, causing complete collapse of the lung on the affected side and a movement of the organs in the center of the chest to the opposite side of the chest, thus compressing the other lung. Both heart and lung function may be seriously affected, and rapid reduction of the air pressure in the pleural sac may be required.

Pneumothorax is treated by insertion of a tube into the affected side of the chest. This tube is connected to a drain and water valve mechanism, which allows air to

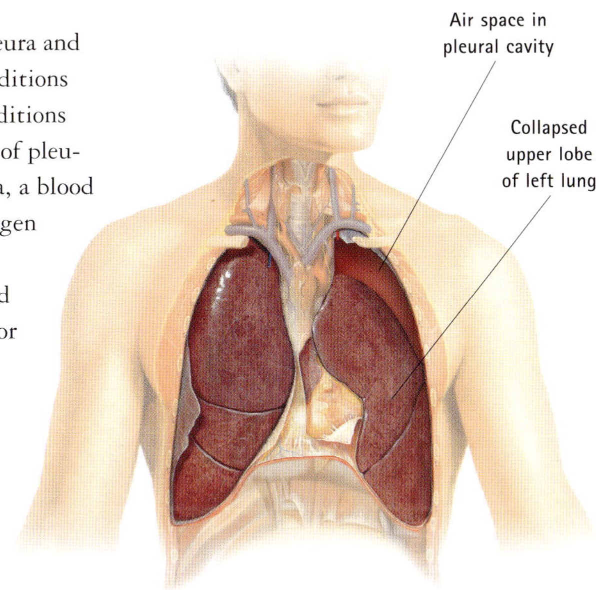

Pneumothorax

In this condition, air enters one or both of the pleural sacs which enclose the lungs. This may occur if the chest wall is pierced. When air is trapped in the chest cavity, increased pressure can result in the collapse of the lungs.

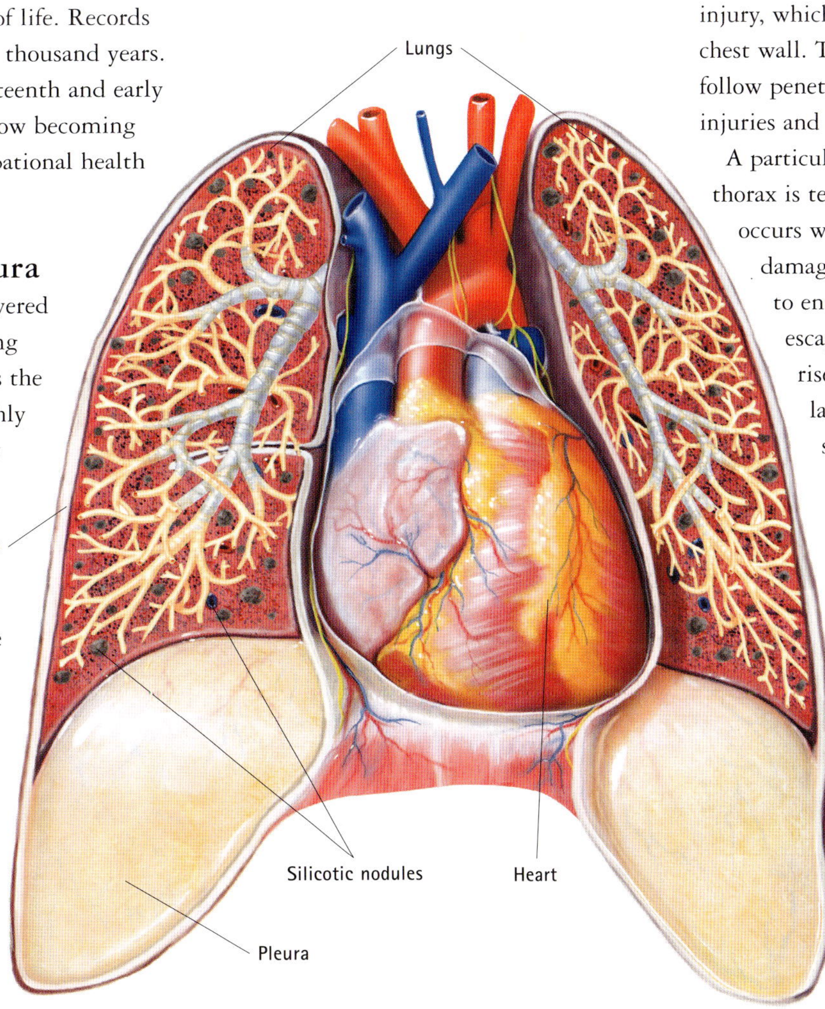

Silicosis

Inhalation of silica dust causes the development of hard nodules in the lungs. The nodules impede air exchange, depriving body tissues of oxygen and causing coughing and breathing difficulties.

NB: In this illustration the upper two-thirds of the lungs have been cut and the pleura has been peeled back to reveal the bronchial tree and heart.

leave the pleural sac but prevents re-entry. Often pneumothorax is accompanied by bleeding into the pleural sac (hemothorax), which may require drainage if a large amount of bleeding has occurred.

Diseases of the pulmonary circulation

Pulmonary hypertension and pulmonary embolism are two of the most commonly encountered serious disorders of the pulmonary circulation. Pulmonary hypertension is abnormally high blood pressure in the lungs. It is not necessarily related to high blood pressure in the systemic circulation and can interfere with oxygen levels in the blood. Possible causes include tissue damage in the lungs due to disease or the inflammation of blood vessels in the lungs.

In pulmonary embolism, a condition requiring urgent medical attention, a pulmonary artery becomes blocked by fat, an air bubble, tumor or blood clot (which will usually have arisen in the deep leg veins). Symptoms include breathing discomfort, sudden chest pain, shock and a blue discoloration to the skin caused by an excess of deoxygenated blood in the body. Very large pulmonary embolisms are often fatal.

Symptoms of lung disorders

The most common symptoms of lung disease include cough, breathlessness and wheezing.

COUGH

A cough is one of the most common symptoms of lung irritation or disease; almost any lung disease or infection will cause a cough. A cough is a reflex action resulting in contraction and a sharp expiration of the diaphragm to expel something that is causing irritation to the lining of the bronchi (airways).

Coughing may be provoked by irritants such as smoke, allergens, viral or bacterial infection or other inflammation, foreign bodies or a growth in the bronchi.

In children and adults, a cough is commonly caused by bronchitis (inflammation of the airways). This commonly follows an upper respiratory tract infection such as the common cold or a viral throat infection. A bacterial bronchitis may result, especially in those with asthma or other lung disease. In adults, smoker's cough may be due to emphysema and chronic bronchitis, conditions caused by damage to the airways and connective tissue of the lung as the result of smoking. A cough may also be a symptom of lung cancer, or pulmonary fibrosis, and seen in a variety of lung disorders and connective tissue diseases.

If a cough lasts longer than a few weeks, the physician may wish to investigate further, with x-rays, CAT scans and MRI scans of the lung. Sputum can be collected and sent to the laboratory for microbial analysis.

Cough

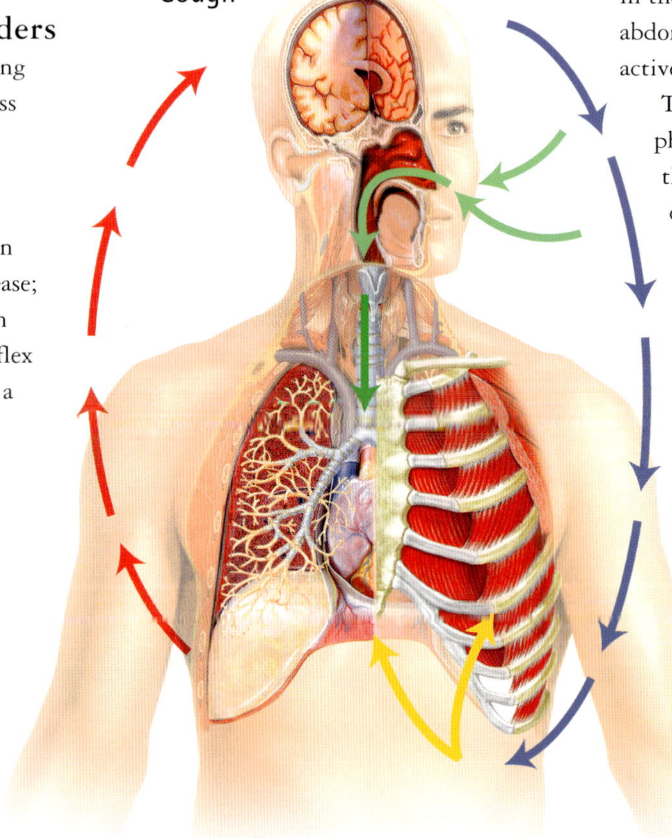

A Irritants are inhaled and stimulate nerve receptors in larynx, trachea and bronchi

B Nerve receptors in larynx, trachea and bronchi send signals to brain stem via vagus nerve

C Brain stem triggers coughing reflex via phrenic nerve

D Diaphragm rises and chest muscles contract, forcing air out of lungs as a cough

To cure a cough, the underlying cause has to be treated. Most cough symptoms can usually be suppressed by medications such as codeine. Often a cough suppressant will be combined with anti-flu preparations for greater effectiveness.

DIAPHRAGM

The diaphragm is a muscular layer which separates the chest cavity from the abdominal cavity. It is attached at the back to the vertebral column (spine), to the ribs along the side of the chest, and to the sternum at the front of the chest. The diaphragm is essential for life because it is the main muscle used for breathing.

When the diaphragm is at rest (i.e. not contracting) it forms a high dome; when the diaphragm contracts, this dome descends, thus increasing the height of the chest cavity. Increasing the height of the chest cavity in turn draws air into the lungs within the chest. This means that the diaphragm is the main muscle for inspiration (drawing air into the lungs).

Expiration, or the passage of air out of the chest cavity, is usually passive, i.e. it occurs because of the relaxation of tension in the soft and hard tissues of the chest and abdomen. The diaphragm does not play any active part in expiration.

The phrenic nerves controlling the diaphragm come from the upper parts of the spinal cord in the neck. If the nerve cells which control the diaphragm are separated from the brain stem control centers for breathing (e.g. by a high spinal cord injury), then the patient will be unable to breathe without assistance from a ventilator.

The diaphragm is pierced by several structures which pass between the chest and abdominal cavities. The three largest of these are the esophagus, carrying food to the stomach; the aorta, carrying oxygenated blood from the heart to the lower body; and the inferior vena cava, a large vein which carries deoxygenated blood from the lower body back to the heart. Several other nerves and lymphatic channels also pass through the diaphragm. The central part of the diaphragm is known as the central tendon. It is fibrous rather than muscular and has the pericardial sac, which surrounds the heart, firmly attached to its upper surface. The central tendon descends with every inspiration.

SEE ALSO *Respiratory system on page 95*

Hiccups

Hiccups are spasmodic involuntary contractions of the diaphragm that cause a disturbance in the rhythm of breathing. They develop when some stimulus, such as rapid eating, triggers a sudden spasm of the diaphragm. The irregular rhythm causes a sudden closure of the vocal cords, resulting in the characteristic hiccup noise.

Hiccups are most likely to occur after a meal when the stomach is stretched. They usually begin without warning and stop in the same way. There are many folk remedies but most do not work and the hiccups disappear of their own accord. Remedies which do work involve increasing the carbon dioxide in the blood, and include holding the breath, and breathing into a paper bag (though this is not advised as it is dangerous). Persistent hiccups may require medication.

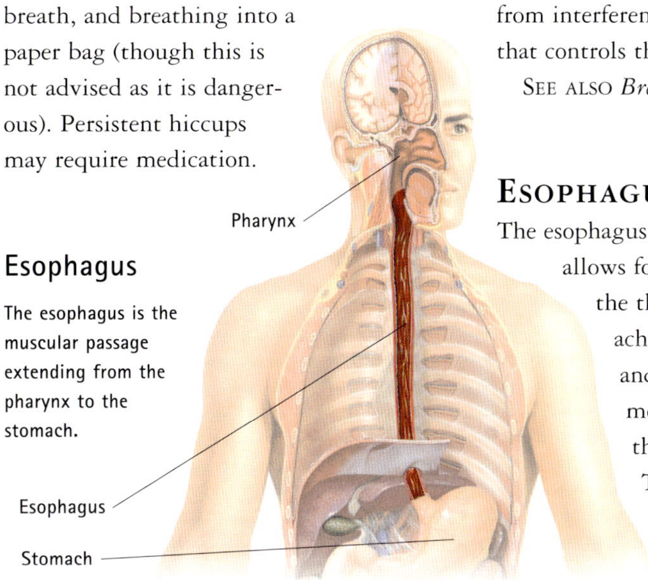

Esophagus

The esophagus is the muscular passage extending from the pharynx to the stomach.

Far less commonly, hiccups may also be caused by irritation of the diaphragm during pneumonia, or after surgery; or from harmful substances in the blood, such as those which result from kidney failure; or from interference with the part of the brain that controls the breathing, due to a stroke.

SEE ALSO *Breathing on page 240*

Esophagus

The esophagus is a muscular tube that allows food to be transported from the throat (pharynx) to the stomach. It passes through the neck and chest and into the abdomen, with the portion through the chest being the longest. The esophagus must pass via the diaphragm on its way to the stomach.

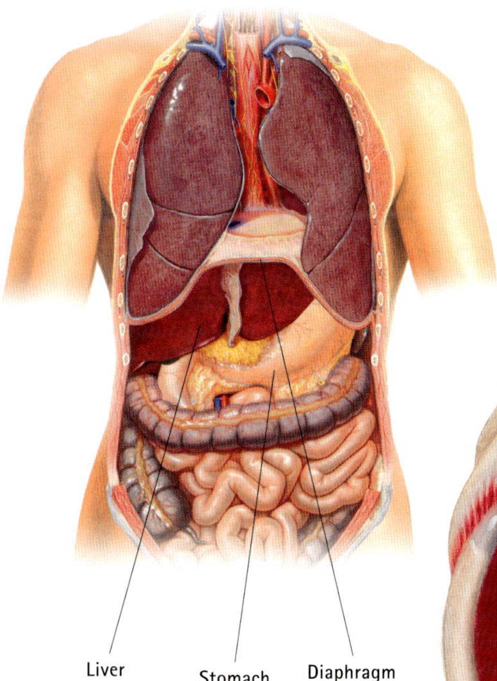

Diaphragm

The diaphragm is the muscular layer that separates the chest cavity from the abdominal cavity. It is the main muscle used for breathing. The heart and the lungs rest on the upper convex surface of this muscle. The lower concave surface forms the roof of the abdominal cavity, lying over the stomach on the left and the liver on the right. The illustration at left shows the diaphragm as seen from below.

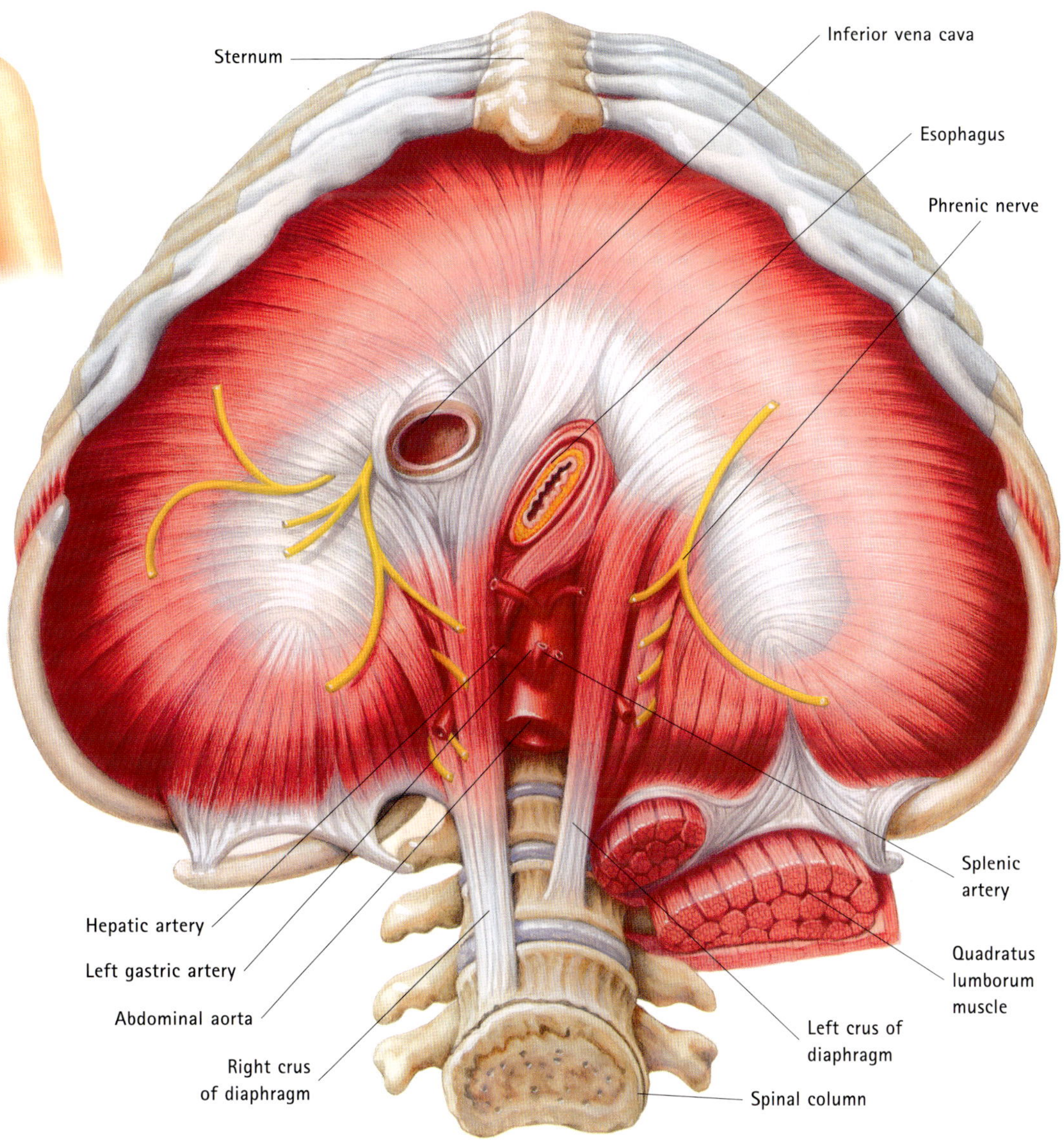

It can expand and contract at its upper and lower ends using circular muscles called sphincters. The upper sphincter relaxes to accept food from the pharynx; the food is moved by muscular contractions to the lower sphincter, which relaxes to let food into the stomach. It then closes to prevent gastric reflux (return flow).

SEE ALSO *Digestive system on page 101*

Disorders of the esophagus

Disorders of the esophagus most commonly involve an obstruction of food or the regurgitation of food and stomach juices.

Esophageal atresia

Sometimes the esophagus does not form properly during development and may be obstructed; this condition is known as esophageal atresia. The infant has great difficulty in swallowing milk; the regurgitated milk may be inhaled, causing pneumonia.

Achalasia

Achalasia of the esophagus is a condition in which nervous control of the esophageal muscle is impaired, due to degeneration of nerve cells in the esophageal wall. This condition results in difficulty in swallowing (dysphagia), retention of food in the esophagus and dilation of the esophagus.

Esophageal diverticulum

A diverticulum of the esophagus may occur when the esophagus "blows out" between strands of encircling muscle. This usually occurs in the upper part of the esophagus.

Esophageal varices

In cirrhosis of the liver, caused by chronic alcoholism or viral hepatitis, small veins in the lower end of the esophagus may become enlarged (esophageal varices). These varices protrude through the inner surface lining of the esophagus and may rupture, discharging large amounts of blood into the lower esophagus. This blood is either vomited or passed out through the anus as a dark, black, tarry stool. This is a serious condition, because most patients who have massive bleeding from esophageal varices die as a result of the initial bleed.

Injury to the esophagus

Three areas of the esophagus are relatively narrow and may be sites where swallowed corrosive substances (such as caustic soda or sulfuric acid) are slowed up and may cause major damage. These sites of narrowing are at the beginning of the esophagus, where the esophagus crosses the arch of the aorta just above the heart, and where the esophagus passes through the diaphragm. As already mentioned, the esophagus may be damaged by swallowed corrosive substances such as strong acids or caustic soda (corrosive esophagitis). These substances cause chemical burns to the lining of the esophagus, which may cause perforation of the esophagus with spilling of food into the chest cavity, bleeding from esophageal ulcers and constriction (esophageal strictures) when scar tissue forms several weeks after injury.

Heartburn

Heartburn is a feeling of burning pain in the chest behind the breastbone combined with a sour or bitter taste at the back of the throat. These symptoms are caused by the regurgitation (reflux) of the contents of the stomach, a mixture of food and digestive acids which flows back up the esophagus causing pain and irritation.

Chronic cases of heartburn are termed GERD (gastroesophageal reflux disease) and may be associated with intestinal disorders. Chronic exposure of the esophegeal lining to stomach acids can lead to a number of complications such as ulceration of the esophagus, scarring and narrowing (strictures), and changes in the lining (epithelium) that can lead to cancer.

Relaxation of the lower esophageal sphincter, a muscle which closes after allowing swallowed food to pass to the stomach, is the cause of reflux. The sphincter may become weakened, or may react to stimulus from certain foods, alcohol or tobacco, or may also open under pressure from the stomach that results from overeating or from sitting, lying or taking up activity too soon after eating.

Heartburn is quite common in the overweight, can be caused by hiatus (hiatal) hernia, and is also common in pregnancy and the hormonal changes that reduce the tone of all muscles, including the stomach and esophageal sphincters.

The symptoms are usually treated with antacids or, for more severe cases, anti-ulcer drugs. It is also important to reduce or avoid coffee, very spicy foods, give up smoking, lose weight, eat smaller and less-rich meals, reduce alcohol, not lie down or bend over until food is digested. Certain drugs such as aspirin, anti-inflammatory or anti-arthritic drugs should be avoided. Raising the head of the bed when you sleep may help avoid heartburn at night.

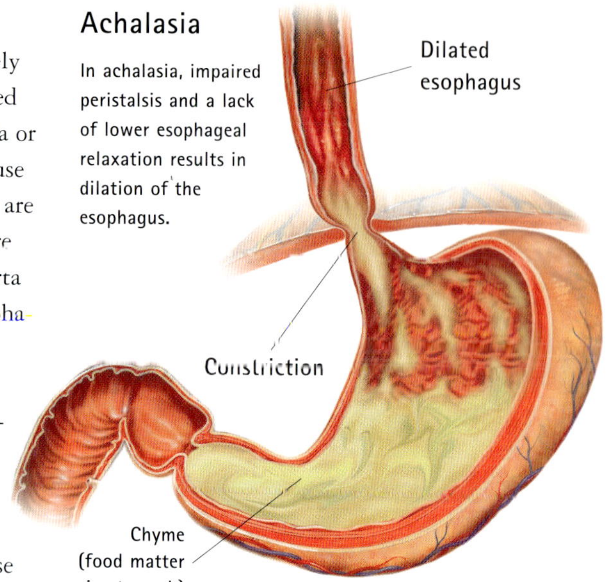

Achalasia

In achalasia, impaired peristalsis and a lack of lower esophageal relaxation results in dilation of the esophagus.

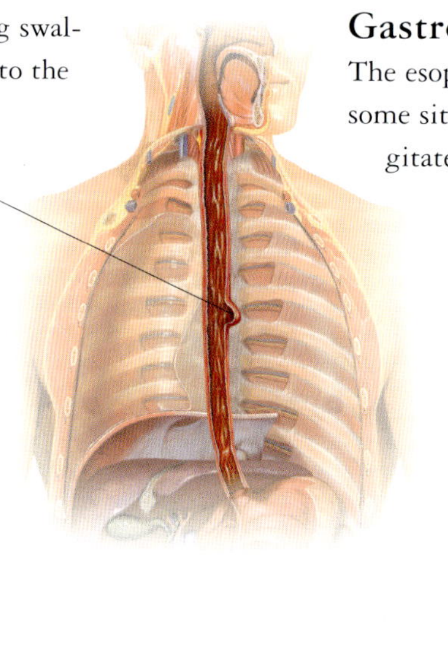

Esophageal diverticulum

When the esophagus "blows out" between strands of encircling muscle, a diverticulum may occur.

Gastroesophageal reflux

The esophagus leads into the stomach. In some situations, stomach juices may regurgitate from the stomach into the lower esophagus (gastroesophageal reflux). This usually occurs when part of the stomach slides through the diaphragm into the chest cavity (sliding hiatus hernia), thus interfering with the ability of the diaphragm to prevent reflux of stomach juices back into the esophagus.

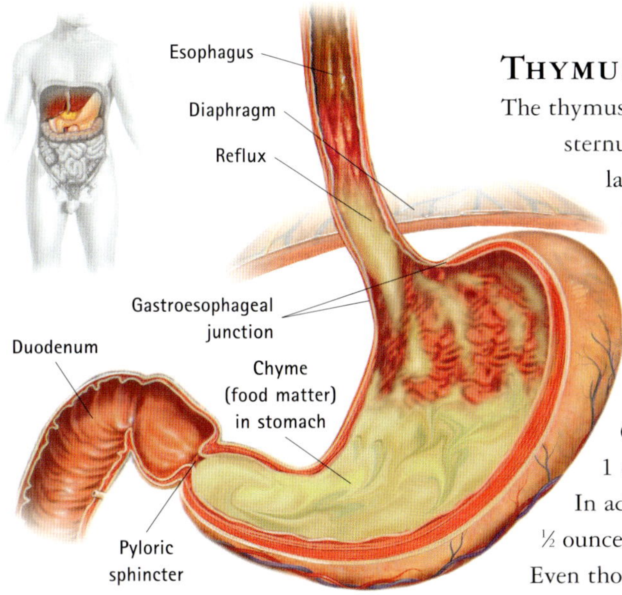

Gastroesophageal reflux

The backflow of acid fluid from the stomach (gastroesophageal reflux) may damage the lining of the lower esophagus.

Esophagitis

Esophagitis literally means inflammation of the esophagus, the tube connecting the back of the throat (pharynx) with the stomach. An inflammation of the esophagus usually damages the inner, or epithelial, lining most of all. It may be due to the accidental swallowing of corrosive chemicals such as caustic soda or concentrated acid (corrosive esophagitis), or due to reflux of acidic stomach juices into the lower esophagus (reflux esophagitis). The latter may occur in sliding hiatus hernia when the mechanism which keeps stomach juices in the stomach is impaired and cannot function properly.

Reflux esophagitis is a serious disease and may be accompanied by heartburn and regurgitation. In severe cases, surgery to treat the sliding hiatus hernia may be necessary.

Esophageal cancer

Cancer of the esophagus is a relatively rare, but serious, condition, usually involving the malignant growth of the cells lining the esophagus. The patient complains of progressive difficulty in swallowing, beginning with solids and eventually involving liquids. There can be progressive weight loss and, occasionally, bleeding.

Complications of the disease include fatal bleeding, pneumonia, obstruction of the trachea and problems with heart rhythm.

Thymus gland

The thymus gland is found beneath the sternum, at about the level of the large vessels leaving the heart. It is much more obvious in children and adolescents, when it is active, than in adults, when it has shriveled to a fatty, fibrous remnant. At birth the thymus weighs about ½ ounce (14 grams) and reaches about 1 ounce (28 grams) by puberty. In adults the gland reduces to only ½ ounce (14 grams).

Even though it is called a gland, the thymus is actually a lymph organ, manufacturing T lymphocytes (a type of white blood cell) for distribution to the rest of the body. T lymphocytes are involved in the defense of the body against viruses and cancer cells, in delayed-type hypersensitivity reactions and in graft rejection.

If the thymus is removed from a newborn rat, other lymphoid organs in the body fail to develop and there is a decrease in the number of T lymphocytes in the blood. The animal is unable to make an adequate defense against viruses or to reject transplanted foreign tissue. Animals subsequently become weak, lose weight and die about 3–4 months after birth, due to widespread infections occurring throughout the body. Removal of the thymus in adults (sometimes advised for the treatment of the disease myasthenia gravis) does not have such serious effects because T lymphocytes have already been distributed throughout the body. In humans there are some diseases where T lymphocytes fail to develop. Children so affected will die soon after birth if the disease is left untreated.

SEE ALSO *Lymphatic Immune system on page 55*

Thymus

Part of the endocrine system, the thymus is a ductless gland found just under the sternum in the chest. It is a lymph organ and uses the hormones it produces to manufacture mature T lymphocytes, which play an important role in the body's immune defenses.

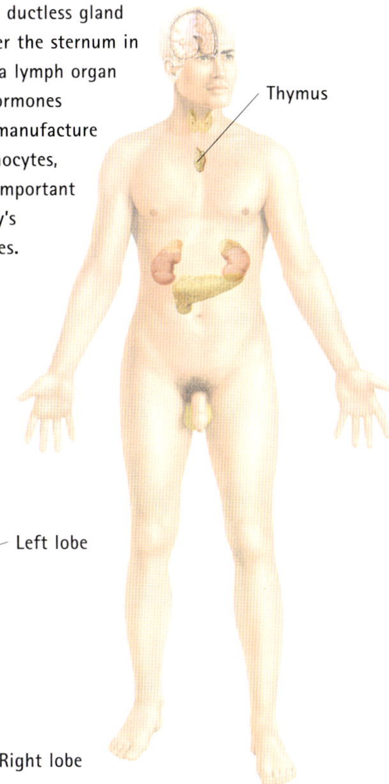

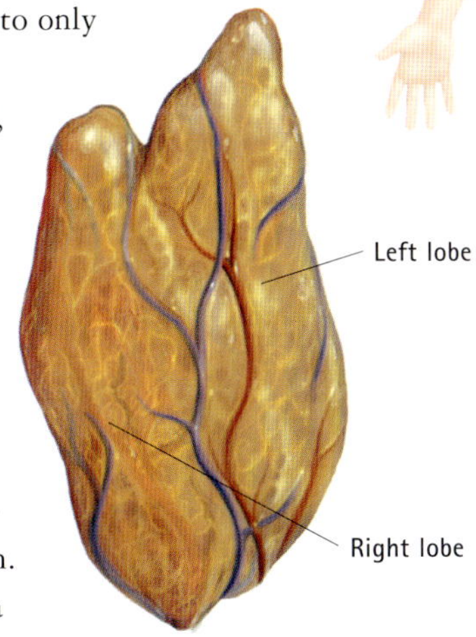

Thymus microstructure

T lymphocytes (a type of white blood cell) divide in the cortex of the thymus, moving into the medulla when mature. These T cells leave the thymus after about 3 weeks, entering the body's circulation via the blood vessels. Degenerated epithelial cells in the medulla form distinctive structures known as Hassall's corpuscles.

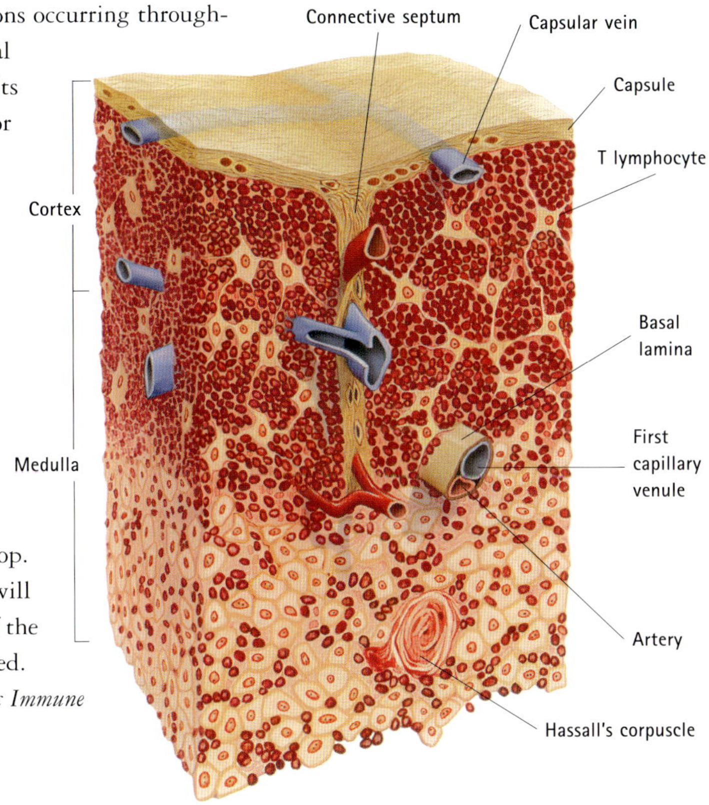

The Abdominal Cavity

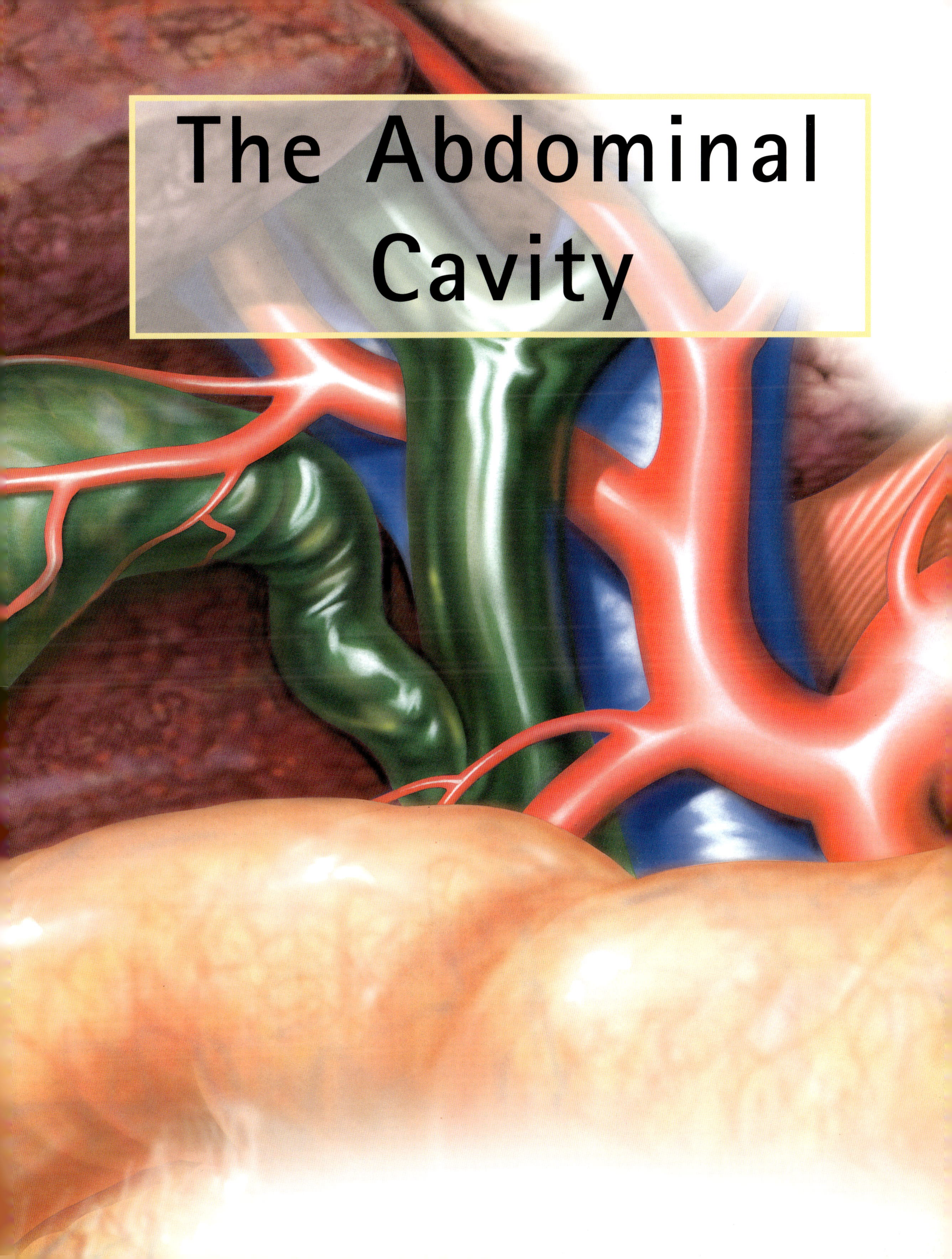

Abdomen

Situated between the thorax and the pelvis, the abdomen is the larger of the body's two major cavities (the other is the thorax, or chest). The abdomen contains some of the body's most important organs—primarily those organs associated with digestion (the stomach, intestines, liver, gallbladder and pancreas). The organs of elimination (kidneys, ureters and bladder) and reproduction (ovaries and uterus) are also contained in the abdominal cavity but have been allocated their own chapter in this book.

SEE ALSO *Digestive system on page 101; Female reproductive system on page 104; Urinary and reproductive organs on page 288; Urinary system on page 98*

Parts of the abdomen

The sides (the lateral walls) and the front (the anterior wall) of the abdomen are made up of layers of muscle covered by fat and skin. When they contract, the muscles raise the pressure in the abdomen to aid breathing and passing of feces. The back (or posterior wall) of the abdominal cavity is formed by the vertebral bones of the spinal column and by the muscles that run up and down them. The roof of the abdominal cavity is formed by the diaphragm, a dome-shaped sheet of muscle separating the abdomen from the thorax. When the diaphragm contracts, it raises the pressure in the abdominal cavity and assists breathing.

The pelvic cavity is usually also considered to be part of the abdomen (though some consider it separately, while others refer to both the pelvic cavity and the abdomino-pelvic cavity). The pelvic cavity contains the bladder, rectum and, in females, the reproductive organs of the uterus and the ovaries. The floor of the pelvic cavity is formed by muscles and bones.

Lining the abdominal cavity, and also extending out into the cavity to cover the organs within, is a thin lubricating membrane, the peritoneum. Folds of the peritoneum attach the organs to the back of the cavity, while allowing the intestines to move relatively freely, in order to aid movement of food down the alimentary canal. Other folds of the peritoneum, called mesenteries and omenta, supply the organs with nerves, blood vessels and lymph channels.

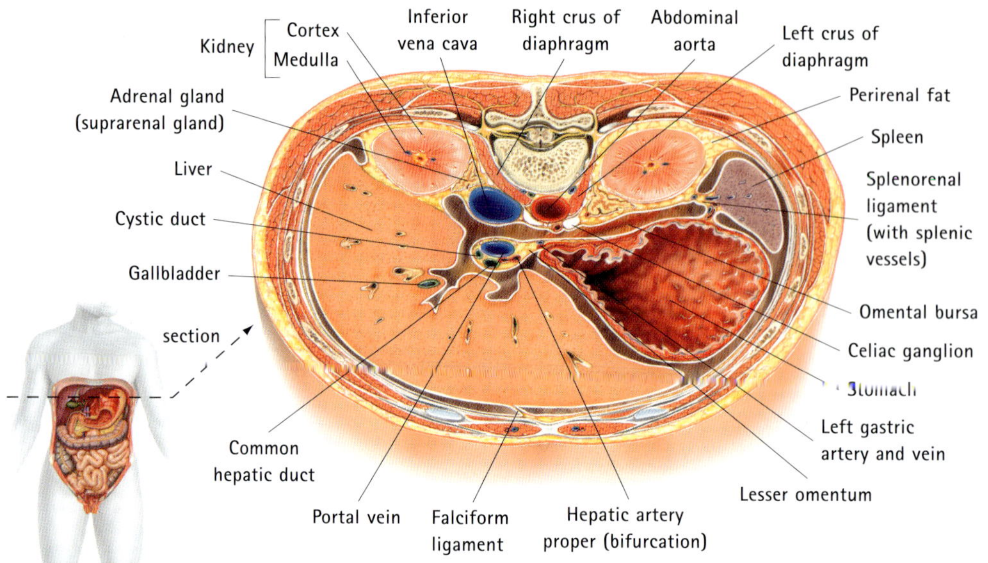

Cross-section of the abdomen

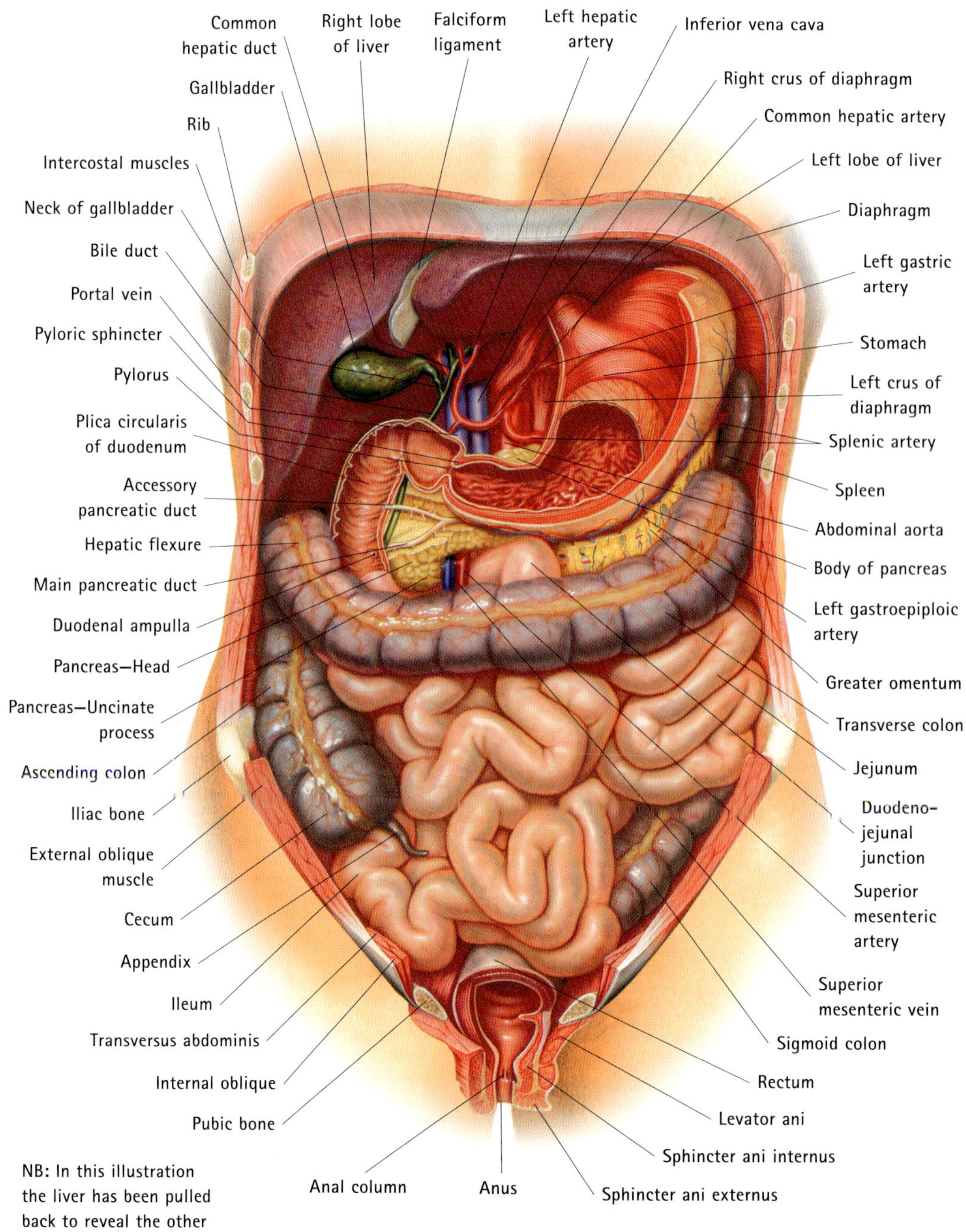

NB: In this illustration the liver has been pulled back to reveal the other abdominal organs.

Organs of the abdomen

To make it easier to identify internal organs and to help locate exactly where abdominal pain is coming from, health professionals divide the abdomen into four quadrants: the right upper, right lower, left upper and left lower.

Peritoneum

The peritoneum is a thin membranous layer that lines the abdominal cavity. Mesenteries and omenta—folds of the peritoneum—supply blood vessels and nerves to the abdominal organs.

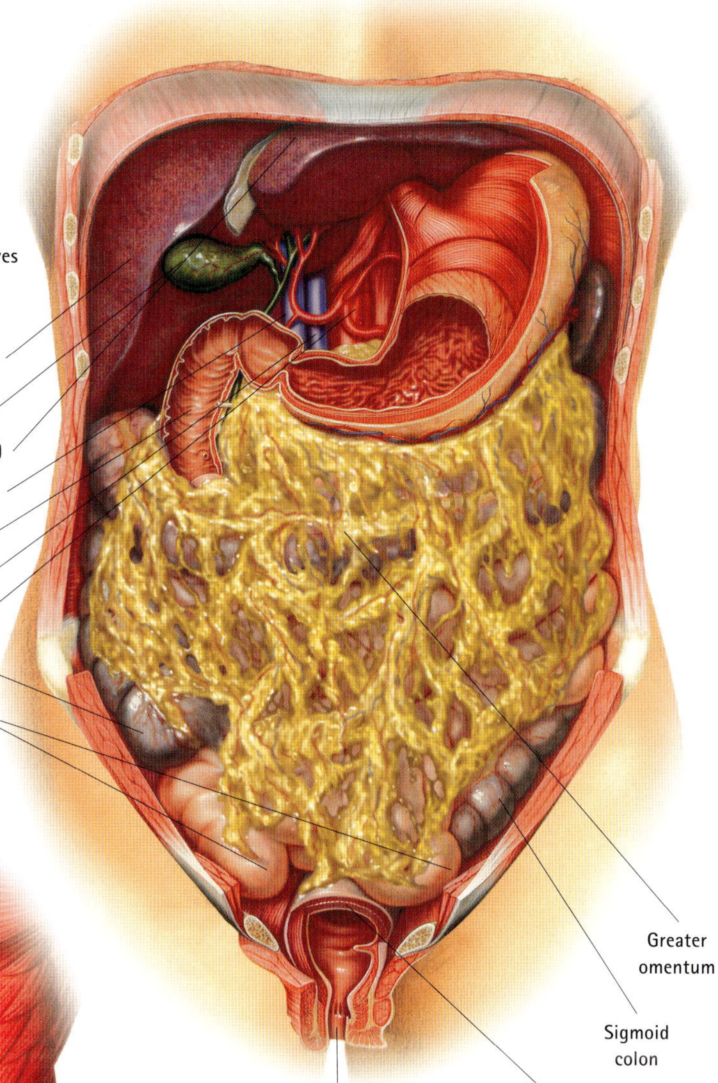

Abdominal disorders

Because the abdomen contains a large number of important organs, it is not surprising that abdominal disorders are common. Symptoms of abdominal disease include heartburn, pain, cramps, constipation, diarrhea, vomiting and nausea. Indigestion (known as dyspepsia), food poisoning and gastroenteritis are common maladies that can be treated at home or by a primary care physician (general practitioner).

More serious conditions, which will require some sort of treatment by a specialist physician or surgeon, include a gastric or duodenal ulcer, hepatitis (inflammation of the liver), colitis (inflammation of the large bowel), as well as cancers of the stomach, liver and large bowel.

An abdominal illness can be an emergency. An acute abdominal infection such as peritonitis, appendicitis, or cholecystitis (inflammation of the gallbladder) can be serious and life-threatening. A perforated gastric or duodenal ulcer and injury from trauma are other common abdominal emergencies. "Acute abdomen" is the term health professionals generally use for an abdominal emergency.

SEE ALSO *Digestive system on page 101; Imaging techniques on page 431; Paracentesis on page 454; Treating the digestive and urinary systems on page 444*

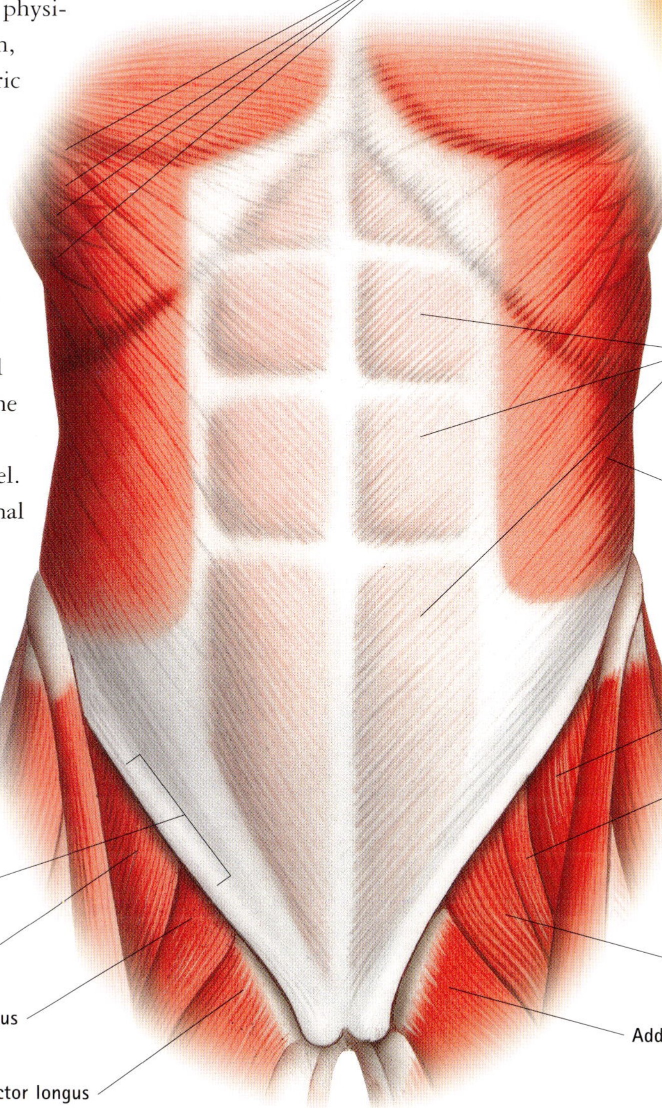

Muscles of the abdomen

Ascites

Ascites is an abnormal accumulation of fluid around the tissues and organs of the abdominal cavity. It can indicate a range of disorders, but cirrhosis of the liver, cancer or heart failure are among the most common causes. If enough fluid is present, ascites can cause the abdomen to become distended and painful.

Ultrasonography is one of the best ways to confirm the condition, particularly in overweight patients. The cause of ascites is indicated by its appearance and chemical content. Collection of the fluid for analysis is done using a medical procedure called paracentesis. This may also be used to drain fluid to relieve discomfort. Treatment of ascites usually targets the underlying cause.

Peritonitis

Inflammation of the peritoneum (the membrane that lines the wall of the abdomen and covers the organs) is known as peritonitis. It can result from infection (such as an abdominal abscess), injury or occasionally other diseases. Symptoms usually include abdominal pain, distension and tenderness, fever, nausea and vomiting.

The cause must be quickly identified and treated, for example with intravenous antibiotics in the case of bacterial peritonitis. Untreated, the mortality rate is high.

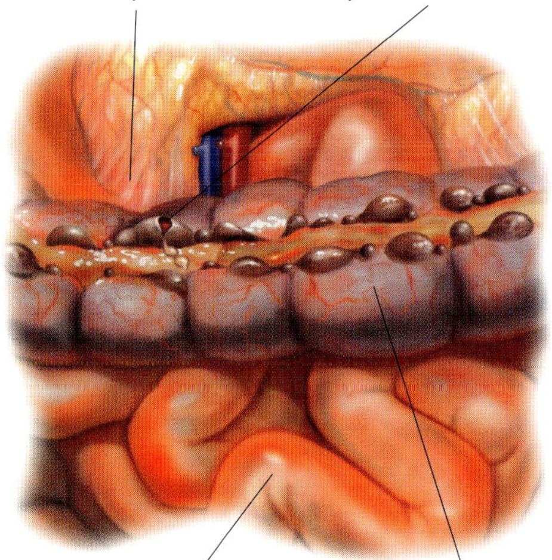

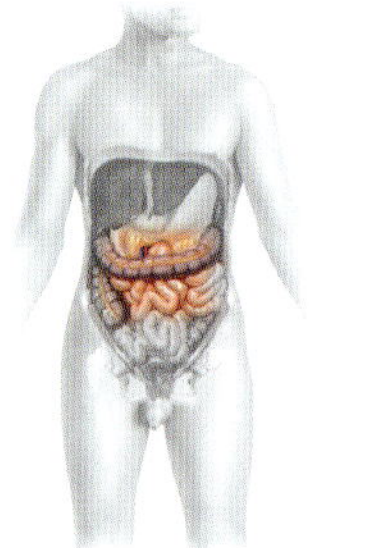

Peritonitis

An infection in the abdomen, if untreated, can lead to inflammation of the peritoneum—peritonitis—as in this example of a ruptured diverticulum.

Alimentary canal

The alimentary canal (otherwise known as the digestive tract) is a long muscular tube (about 30 feet, or 9 meters) extending from the mouth to the anus. It is made up of layers of circular muscles that contract and move the food in waves along the tract: this process is known as peristalsis. Its function is to break food down into smaller and smaller particles, absorb the nutrients along the way and expel the rest as waste.

Different parts of the tract have different functions. Food enters through the mouth, or oral cavity. Jaws, tongue and teeth work to mash and mix the food into smaller pieces. Saliva, secreted from the salivary glands placed around the oral cavity, moistens the food to make it easier to transport, and contains enzymes that begin the process of breaking down the food.

From the mouth, food passes down a muscular tube called the esophagus into an acid-rich pouch—the stomach.

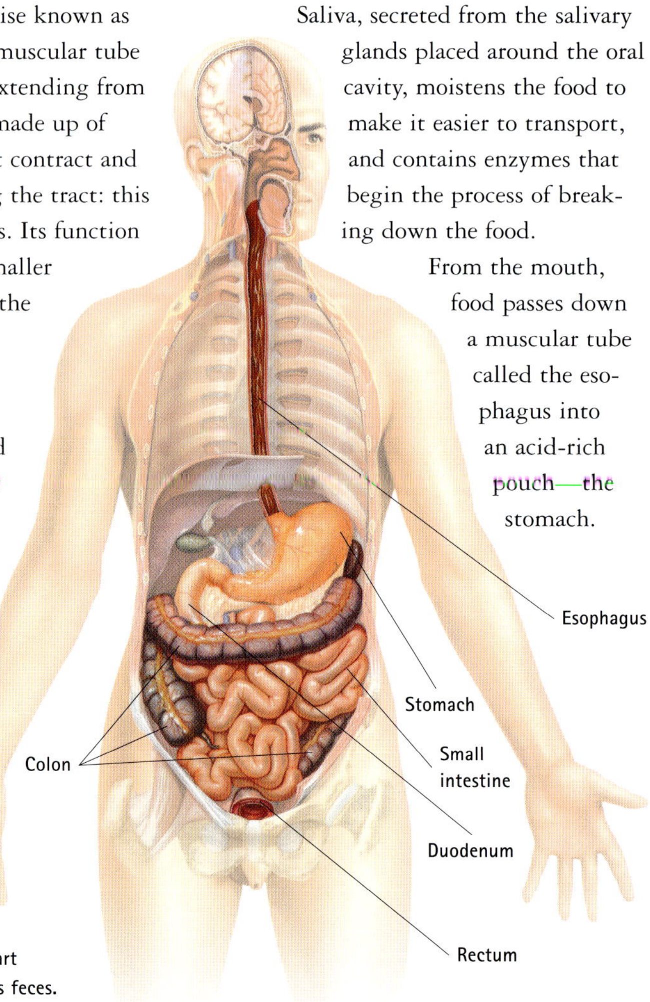

Alimentary canal

The alimentary canal is the body's "food processor." Along the first part of its length, food is broken down into fats, carbohydrates and proteins. Further along the canal these are absorbed into the blood, to provide energy and to build and repair body tissues. Water is absorbed along the final part and anything left over is excreted as feces.

In the stomach, the food is broken down by pepsin, an enzyme, into a semi-liquid stew of food, acid and digestive juices.

Next, the mixture passes into the duodenum, which is the first part of the small intestine. More digestive juices are added to the mix: bile from the liver and digestive enzymes from the pancreas. As the food particles move further down the intestine, they are reduced to smaller and smaller constituents—carbohydrates, proteins, fats, vitamins and water. These are then absorbed through the lining of the small intestine and pass into the bloodstream. They are stored in the liver to be later used in the body's metabolic processes.

Two other organs are essential in this process. One is the liver, which, as well as storing nutrients, also produces bile. Via the gallbladder and bile ducts, bile travels into the small intestine where it aids in the absorption of fats and fat-soluble vitamins. Then there is the pancreas, which makes the amylase, protease and lipase enzymes needed to break down carbohydrates, proteins and fats in the small intestine.

Once food has reached the end of the small intestine, what is left of it is essentially waste material. The waste enters the large intestine, or colon—a tube that is wider but shorter than the small intestine. As it travels along the colon, water is reabsorbed, the waste hardens and becomes feces. The fecal material is then stored in the last part of the large intestine, the rectum, before being expelled through the anus.

SEE ALSO *Digestive system on page 101; Endoscopy on page 433; Imaging techniques on page 431*

Stomach

The stomach receives food from the esophagus and continues the process of digestion. It acts as a reservoir, permitting the intake of large amounts of food every few hours. When empty of food, the stomach contains only about one-twelfth of a pint (50 milliliters) of liquid, but can expand to accommodate up to 2 pints (1,200 milliliters) after a large meal is eaten.

The stomach mixes the food with the acidic gastric juices, which digest protein and carbohydrates, and delivers semi-

STOMACH FUNCTION

Arrival in stomach

The arrival of food from the esophagus stimulates the stomach lining to produce hormones and gastric juices (acids and enzymes) needed for digestion.

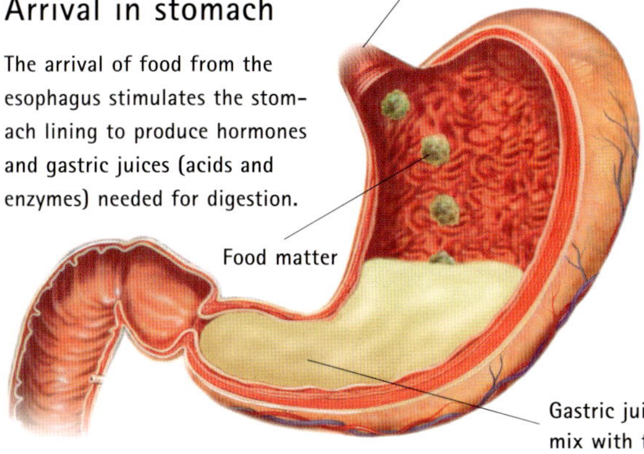

Digestion

As food begins to be broken down, the muscles in the stomach wall contract, mixing the food and gastric juice into a thick substance called chyme.

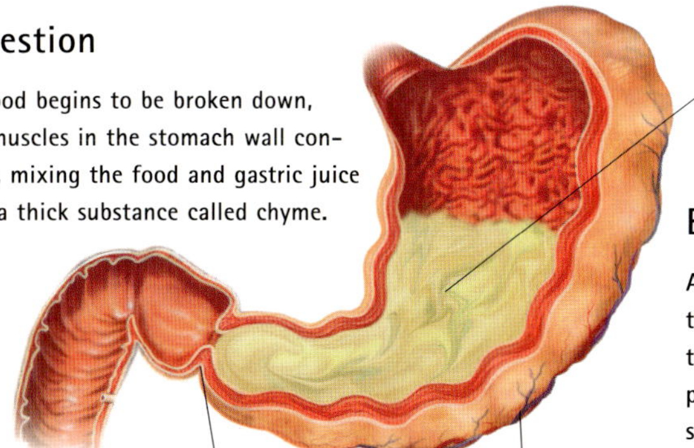

Exiting the stomach

After a few hours of processing, the waves slow. With each contraction, chyme stimulates the pyloric sphincter to open, and small amounts of chyme pass from the stomach into the duodenum.

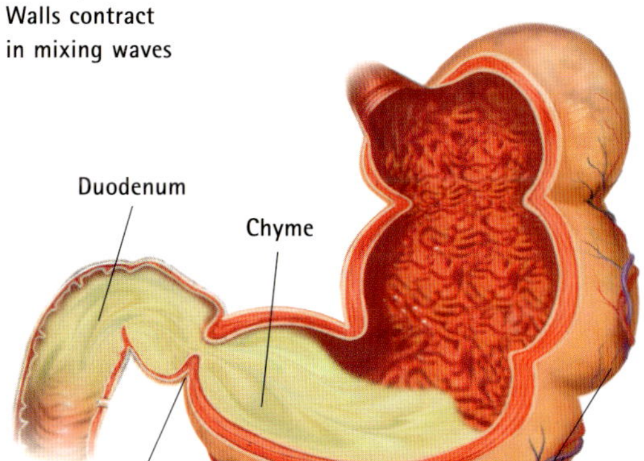

digested food to the next part of the gut, the duodenum. The stomach is also the site of absorption of some drugs such as aspirin.

SEE ALSO *Digestive system on page 101*

Structure and function

The stomach has many parts. The cardia is located at the entrance of the esophagus into the stomach (cardioesophageal junction). The fundus lies to the left of the cardia, while the body of the stomach is the large central part which extends from the fundus to the pylorus. The pylorus is the final part of the stomach and is made up of a pyloric antrum, which leads to the pyloric canal. This canal is encircled by a ring of muscle known as the pyloric sphincter, which controls the passage of stomach juices and food into the duodenum.

The inner lining of the stomach has many types of specialized cells. Some, such as the mucous cells, produce a thick layer of mucus to protect the stomach from its own acid juices. Another group, the parietal or oxyntic cells, produces hydrochloric acid to keep the stomach interior at the optimal acidity level for protein digestion, while yet another group, the zymogen cells, produces the enzymes to digest protein and fat. There are also several cell types in the stomach that produce hormones for controlling acid secretion and regulating the levels of nutrients in the blood.

The stomach is a very muscular organ and can churn the partially digested food it receives to break it up into more easily digested fragments. Waves of muscular contraction, called peristalsis, move down the stomach from the body to the pylorus. When the food is ready to be moved on down the digestive tract, these waves propel the food into the pyloric canal and on to the duodenum.

The secretion of stomach acid is controlled in several phases or stages. The sight, smell, taste and even the thought of food act upon the brain to stimulate acid secretion. The nerve impulses which control this are carried from the brain stem by the vagus nerve. When the food reaches the stomach, local mechanical and chemical stimulation leads to the release of gastrin, a hormone that controls gastric juice secretion.

DISEASES AND DISORDERS OF THE STOMACH

There are many problems, diseases and disorders that can affect the stomach including gastroenteritis, ulcers and stomach cancer.

SEE ALSO *Gastrectomy on page 461; Imaging techniques on page 431*

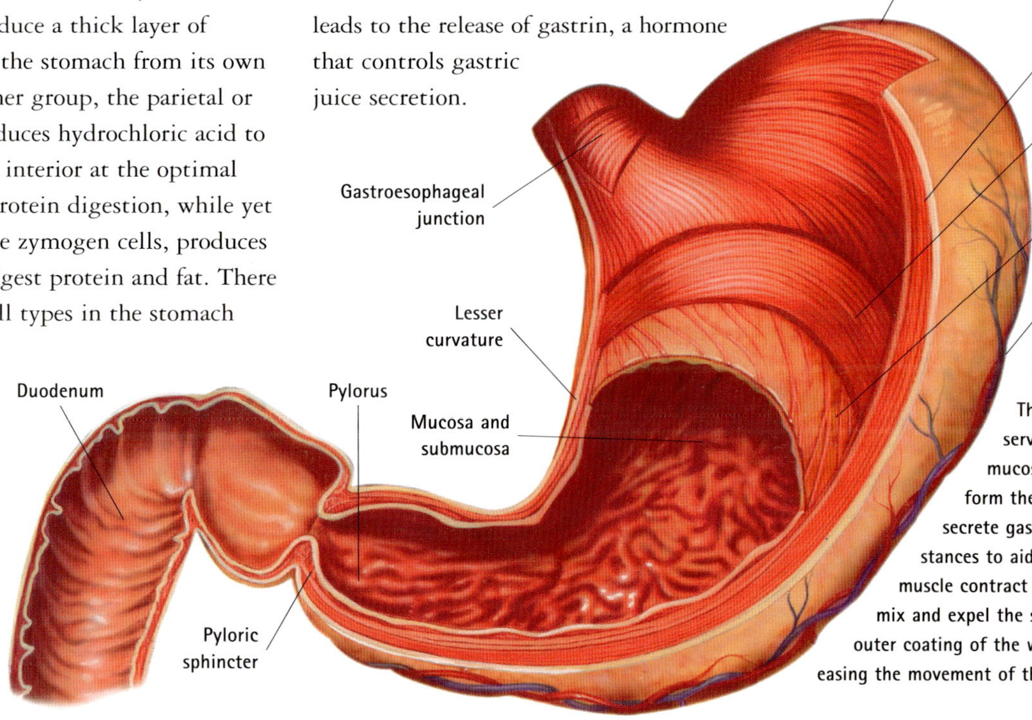

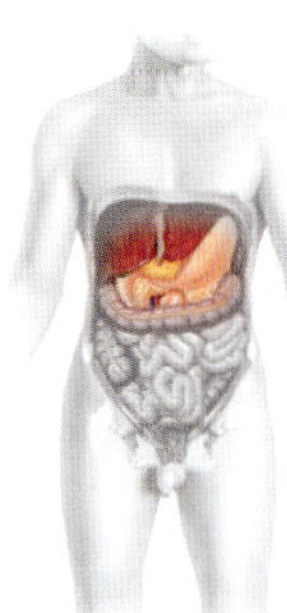

Stomach

The walls of the stomach serve several purposes. The mucosa and submucosa, which form the stomach's inner lining, secrete gastric juices and other substances to aid digestion. The layers of muscle contract and expand in order to mix and expel the stomach contents. The outer coating of the wall is smooth and slippery, easing the movement of the stomach.

Gastroenteritis

Infection of the stomach and intestines may occur as a result of ingesting disease-causing viruses and bacteria (food poisoning). This causes an inflammatory condition called gastroenteritis. The sufferer usually experiences nausea, vomiting and upper abdominal discomfort if the infection involves the upper gastrointestinal tract.

Stomach cancer

Stomach cancer is fortunately becoming rarer in Western countries. The incidence of stomach cancer is higher in Japan and in eastern and central European countries, compared to the USA. Cancer of the stomach, rare in people aged under 40, is about twice as common in men as in women.

There are several different types of stomach cancers. About one-quarter of cancers produce ulcers and can be mistaken for peptic ulcers. A further quarter produce bulky bulbous growths in the interior of the stomach, while about 15 percent spread superficially through the surface lining of the stomach. Unfortunately, many stomach cancers are advanced at the time of diagnosis and are found to be both partly within and partly outside the stomach. Most tumors arise in the pyloric region, the bulk of the remainder developing in the stomach body.

Most patients who have stomach cancer will initially experience a vague feeling of heaviness after meals. Weight loss will follow and may be accompanied by vomiting of a dark brown, almost black, material. A palpable lump may be present in the upper abdomen and the liver may be enlarged if the tumor has already spread to that organ.

The diagnosis may be made by barium meal x-ray and examination with a fiber-optic instrument (gastroscopy). The only curative treatment is the surgical removal of the tumor (gastrectomy), any local lymph nodes and even portions of nearby organs.

The long-term prospects for patients with this disease are poor. Of all patients with stomach cancer, only about 12–15 percent survive 5 years; in patients where the tumor is localized at the time of diagnosis, 40–50 percent may survive for 5 years.

Peptic ulcer

Peptic ulcers result from the damaging action of acidic stomach juices on the vulnerable mucosal lining of the esophagus, stomach or duodenum. Peptic ulcers affect men three times more often than women. Among patients under 50 years of age, duodenal ulcers are ten times more common than stomach (gastric) ulcers.

A patient with a peptic ulcer will experience pain in the upper abdomen, which may be relieved by food or antacid preparations. The ulcer may erode through a blood vessel, leading to the vomiting of large amounts of blood. If left untreated, the ulcer may perforate the gut wall, spilling stomach or duodenal juices into the abdominal cavity and causing painful chemical peritonitis. Swelling and scarring of the gut wall, which may cause obstruction of the gut, may arise with ulcers in the lower esophagus and pyloric sphincter.

At one time it was thought that peptic ulcers arose from excessive acid secretion as a result of psychological stress. It is now recognized that infection of the upper gut with acid-resistant bacteria and use of non-steroidal anti-inflammatory drugs play a major role in their initiation. Treatment of peptic ulcers makes use of antacid preparations, antibiotics and drugs to control acid secretion (triple therapy).

Gastric ulcer

Gastric ulcers involve the loss of the inner lining (mucous layer) of the stomach to produce a lesion (wound). Most patients with gastric ulcers are aged 40 to 60 years, about 10 years older than patients who develop duodenal ulcers.

Most gastric ulcers develop on the inner surface of the right edge of the stomach, usually within a few inches (5–6 centimeters) of the last part of the stomach (pylorus). Some gastric ulcers are associated with duodenal ulcers, but most appear separately. Some gastric ulcers develop in conjunction with gastric cancers and it is of the utmost importance that these are identified early to improve the patient's chances of survival.

Causes vary, but environmental and genetic factors play important roles. Gastric ulcers can be related to a number of drugs (aspirin, non-steroidal anti-inflammatory drugs, and steroids) and dietary and personal factors. It is now recognized that infection of the stomach with acid-resistant bacteria *(Helicobacter pylori)* is an important contributing factor in the development of gastric ulcers. Also, people with a family history of gastric ulcers are at greater risk.

Ulceration of the lining of the stomach may also occur as a result of stressful illnesses such as acute blood loss, serious infection, burns, brain injury and brain

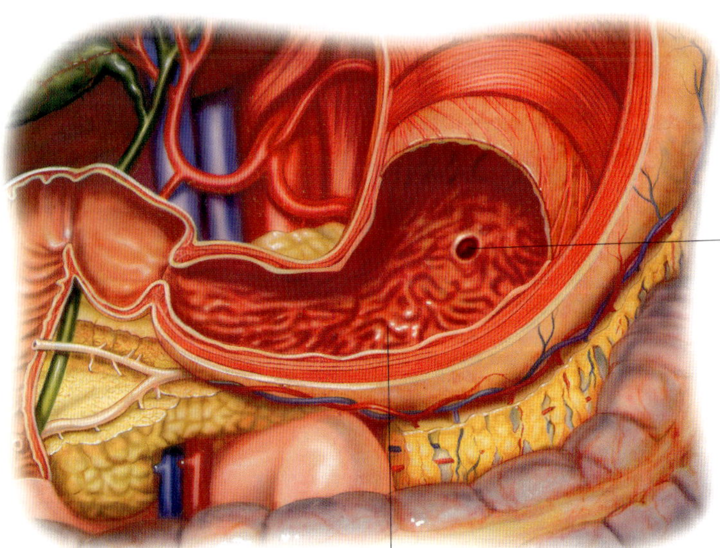

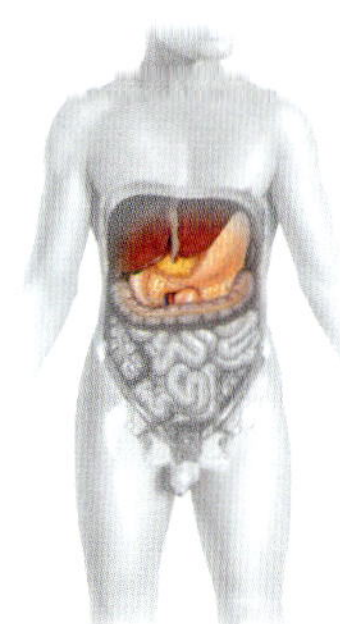

Peptic ulcer

Damage to the mucous membrane can result in the formation of a peptic ulcer. Ulcers can form in the lining of the stomach, esophagus and duodenum.

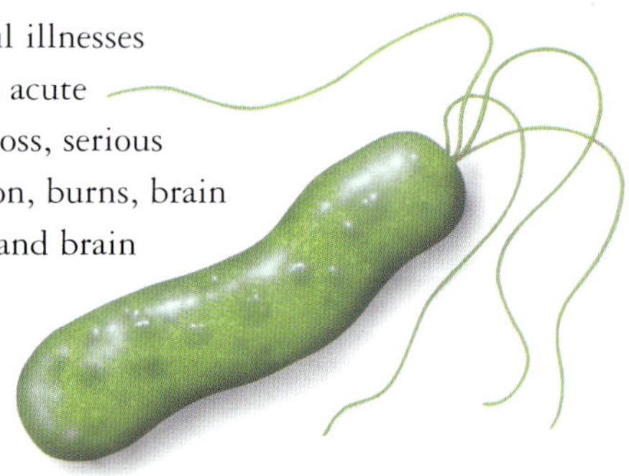

Ulcer-causing bacteria

The presence of an acid-resistant bacteria (Helicobacter pylori) in the upper intestine plays a major role in the formation of peptic ulcers.

tumors. Patients with stress ulcers from burns or infection typically develop bleeding from the stomach and duodenum, and some patients may have perforation of the stomach or duodenum wall. Ulcers developing in patients with brain injury or tumors are usually due to increased gastrin production and excessive stomach acid secretion.

Patients with gastric ulcers usually complain of experiencing pain in the upper abdomen within 30 minutes of eating a meal. With duodenal ulcers, the pain usually comes on more than one hour after meals.

The severity and duration of attacks of pain from the gastric ulcer are usually greater than from duodenal ulcers, but in both cases food and antacids often temporarily relieve the pain. Some patients with gastric ulcers experience vomiting, loss of interest in (or aversion to) eating, and pain aggravated by meals. Treatment is by antibiotics, antacids or surgery.

Hiatus hernia

Parts of the stomach may protrude (herniate) through the diaphragm in a condition called hiatus (hiatal) hernia. Hiatus hernia can be of two types: paraesophageal (known as a rolling hiatal hernia) or sliding hiatus hernia. The paraesophageal or rolling hiatal hernia involves the herniation of the stomach fundus through the esophageal opening. Sliding hiatus hernia is more common than the rolling type and involves the movement of the cardioesophageal junction through the diaphragm into the chest cavity. Rolling hiatus hernia may not be accompanied by any symptoms, while patients with sliding hiatus hernia usually experience heartburn, particularly on lying down, and also a sensation of regurgitation.

Sliding hiatus hernia is the most common (about 90 percent) and involves the sliding of the upper stomach (cardiac part and cardioesophageal junction) through the esophageal opening in the diaphragm.

Patients complain of heartburn, a burning pain in the lower front of the chest and the upper abdomen, particularly after meals and when lying down. They may also experience regurgitation of bitter or sour-tasting fluid (waterbrash) as far as the throat and mouth, particularly at night. Difficulty in swallowing may also occur due to swelling of the lining of the lower esophagus. Continued regurgitation of stomach juices into the esophagus can result in reflux esophagitis.

Sufferers are often advised to take antacids, eat smaller meals more frequently, elevate the head of the bed and avoid lying down after meals. Surgical treatment of sliding hernias is aimed at anchoring the gastroesophageal junction in the abdomen and tightening the esophageal opening through the diaphragm. Most patients will experience a good outcome from surgical treatment.

Rolling hiatus hernia is also known as paraesophageal hiatus hernia. The cardioesophageal junction, where the esophagus and stomach join, stays within the abdominal cavity; it is usually the fundus of the stomach which protrudes into the thorax. Patients complain of gaseous eructations (burping), a sense of pressure in the lower chest and, occasionally, irregular heartbeat.

Rolling hiatus hernia is usually treated by the use of surgery which aims to fix the part of the stomach that herniates to the back of the anterior abdominal wall.

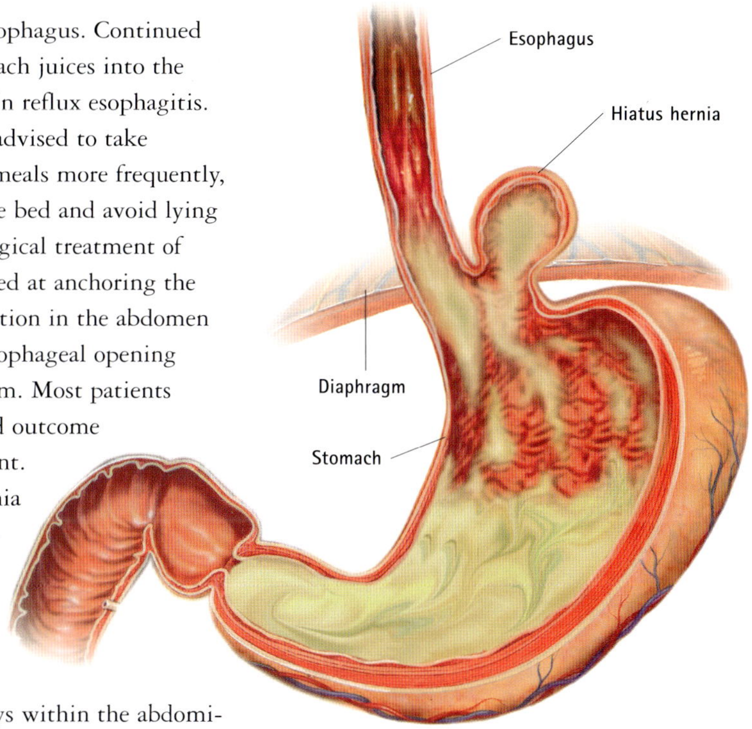

Hiatus hernia

The stomach bulges out of the weakest part of the diaphragm, through which the esophagus passes. Sufferers can experience heartburn and indigestion.

Gastritis

Gastritis is an inflammation of the stomach lining. It tends to become more prevalent with age and can appear as a short-lived acute (sudden onset) condition or may be ongoing and chronic (long-term).

Depending on the cause and type of gastritis, symptoms can range from mild stomach aches and cramps, burping, diarrhea, a swollen abdomen, chest pain, loss of appetite and an unpleasant (often acidic) taste in the mouth due to vomiting. Although gastritis is a common and usually minor ailment, signs that may indicate the need for medical attention include black tarry feces (melena) due to bleeding in the stomach, vomit containing blood, a fever and severe pain.

Causes of acute gastritis include alcohol, caffeine, nicotine, overeating, medications that irritate the stomach lining (such as aspirin and non-steroidal anti-inflammatory drugs), physiological stress brought on by another illness or major surgery, food allergies and bacterial or viral infections.

Treatment depends on the cause, but the symptoms of acute gastritis usually disappear within a few days with rest and avoidance of drugs or food that irritate the stomach.

Over-the-counter antacid medications usually provide short-term relief of

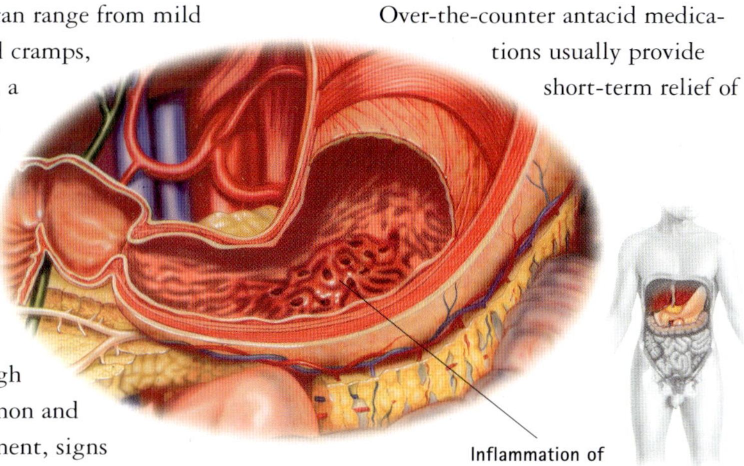

Gastritis

Inflammation of the stomach (gastritis) is a common disorder. It often manifests as a sudden pain resulting from the irritation of the stomach wall.

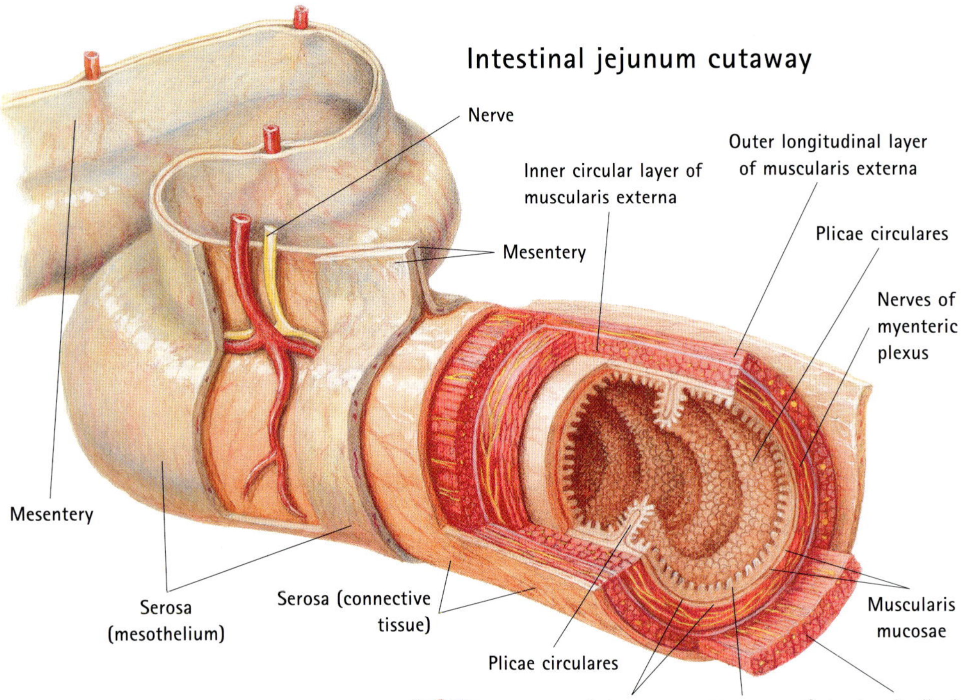

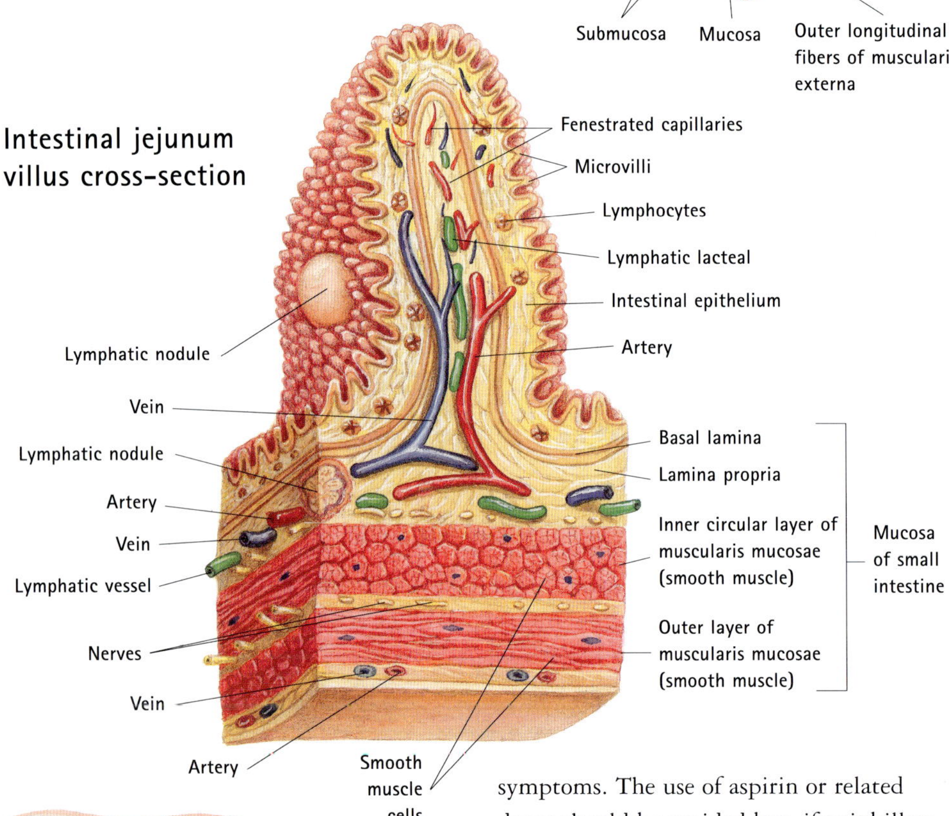

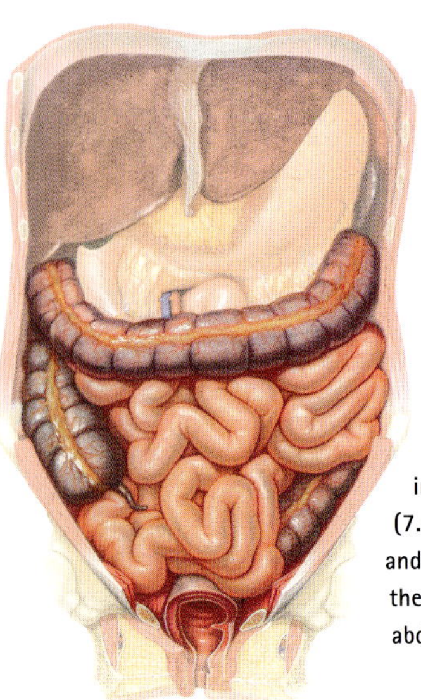

INTESTINES

The small and large intestines together measure approximately 25 feet (7.5 meters) in length, and lie folded up in the lower part of the abdominal cavity.

symptoms. The use of aspirin or related drugs should be avoided but, if painkillers are required, acetaminophen (paracetamol) may help. The only definitive way to identify the cause (and sometimes, to confirm the presence) of chronic gastritis is through tests such as endoscopy and biopsy.

One of the most common causes of chronic gastritis is a bacterium called *Helicobacter pylori*, first described in the medical literature in the 1980s by two Western Australian medical researchers. Not all people infected with these bacteria develop symptoms. Many people become infected in childhood and live normal lives without ever developing any complications. However, *H. pylori* infections (which are readily treated with antibiotics) are a major cause of stomach ulcers and cancers.

Chronic gastritis can also be due to an autoimmune disorder in which the cells in the stomach lining are attacked by the body's own immune system. This causes a loss of stomach cells and a reduction in a person's ability to absorb vitamin B_{12} which leads to a condition known as pernicious anemia. The gastritis itself may not cause any symptoms but may be identified when causes of the anemia are investigated.

Another type of chronic gastritis is hypertrophic gastritis, in which the folds of the stomach wall become enlarged and inflamed. Ménétrier's disease, a rare form of this condition, is most often seen in elderly patients; protein loss is a common complication. Drug treatments are available but, if they fail, surgery may be required.

Pyloric stenosis

Pyloric stenosis is a disease that usually affects only newborn infants. It is more common in boys than girls and results from the thickening of the muscle (pyloric sphincter) surrounding the exit from the stomach to the duodenum. Infants with pyloric stenosis are usually born at full term and feed and grow well for the first two weeks of life. After this time they begin to regurgitate milk with increasing force over the next few days until the vomiting becomes projectile. Unless the problem is corrected, the infant may become dehydrated, lose weight and die.

Treatment may be by antispasmodic drugs and rehydration with intravenous fluids, but usually surgery to sever the fibers of the enlarged pyloric sphincter will be required.

Intestines

The intestines consist of two parts: the small intestine (duodenum, jejunum and ileum) and the large intestine (colon and rectum). The intestines occupy the lower two-thirds of the abdominal cavity. The small intestine (or small bowel) leads on from the stomach, and is mainly concerned with absorption of nutrients from food after digestion (simple sugars, amino acids from protein, fatty acids and glycerol from fats).

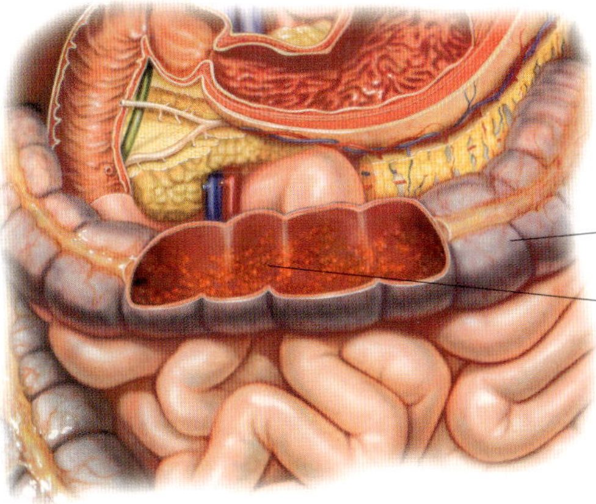

Ulcerative colitis

This condition is characterized by patches of ulceration on the mucous membranes.

Some vitamins are also absorbed here (for example, vitamin B_{12} is absorbed from the end of the ileum).

The large intestine leads on from the small intestine and is concerned with absorption of water and salts (electrolytes), thereby forming the stool or fecal mass by the time the rectum is reached. It is arranged like a picture frame around the margins of the abdomen and is composed of an initial part called the cecum, with the small vermiform appendix attached to it. It also includes the ascending colon, transverse colon, descending colon, sigmoid colon and the rectum.

The small intestine occupies the central part of the frame provided by the large bowel. It contains around 3½ fluid ounces (100 milliliters) of gas, while the large intestine has considerably more. The average person produces up to approximately 1 quart (about 1 liter) of gas per day as flatus. This gas is mainly nitrogen with some carbon dioxide, oxygen, hydrogen and methane. The last four are produced by bacteria in the large bowel, while nitrogen may enter the bowel from the bloodstream.

SEE ALSO *Digestive system on page 101*

DISORDERS OF THE INTESTINES

Many disorders affect both the small and large intestine, disturbing both the absorption and processing of nutrients and the excretion of waste from the body.

SEE ALSO *Endoscopy on page 433; Colostomy on page 461; Imaging techniques on page 431; Treating the digestive and urinary systems on page 444*

Ulcerative colitis

Ulcerative colitis is a disease of the large bowel, but it may also involve the end of the small bowel. The disease usually arises between 15 and 20 years of age, although some patients develop the disease in their sixties. The disease involves loss of the surface lining of the bowel (ulceration) with bleeding. Patients complain of rectal bleeding and diarrhea, with frequent discharges of watery stool mixed with blood, pus and mucus. Many patients have abdominal cramps, together with fever, vomiting, weight loss and dehydration.

The complications of ulcerative colitis are very serious. About 30–40 percent of patients will develop cancer in the affected bowel within 20 years of onset.

Other complications include a possibly life-threatening dilation of the colon (toxic megacolon), which occurs in about 3 percent of patients, and perforation of the bowel in a further 3 percent.

Treatment may involve the use of anti-inflammatory drugs, or surgery, particularly if chronic symptoms or cancer develop. Due to better treatment, the mortality rate is much lower now than 30 years ago.

Crohn's disease

Crohn's disease is a long-term, progressive disease that may involve both the small and large intestine. Most patients develop the disease in their twenties or early thirties. The disease involves swelling of the lymphoid tissue in the wall of the gut. This leads to ulcers on the inner lining of the bowel and cracks or fissures in the wall itself. Patients complain of diarrhea, lower abdominal pain, weakness, weight loss and fever. Many patients develop anemia (defined as reduced number, size and/or hemoglobin content of red blood cells in the blood) due to iron or vitamin B_{12} deficiency. Complications include bowel obstruction, abscesses in the bowel wall and bleeding from the bowel.

Intussusception

Intussusception is a condition in which one part of the bowel telescopes inside an immediately adjacent segment (forming a tube within a tube). It is the most common cause of bowel obstruction in children under two and is more common in boys than girls. The disease can have serious consequences if left untreated, because gangrene may arise in the telescoped part of the bowel due to the loss of blood supply.

Intussusception appears to be associated with previous viral infection, which causes

Intussusception

Intussusception is an abdominal disorder in which the intestine folds back inside itself, causing a painful obstruction.

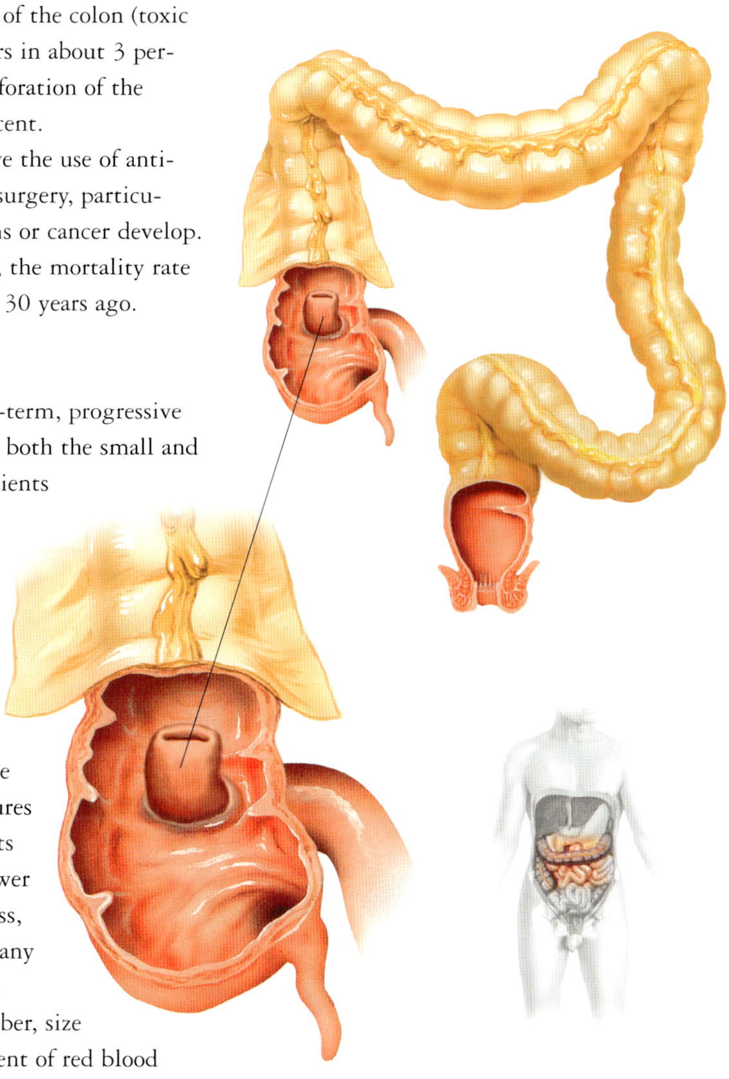

enlargement of clumps of immune system cells (Peyer's patches) in the bowel. Often the cause is not known, and the condition may become apparent when an otherwise healthy infant experiences sudden, severe abdominal pain.

Treatment of the disease involves reversal of dehydration, therapeutic barium enema to distend the intussusception, or surgery may be needed if bowel perforation and/or peritonitis occur.

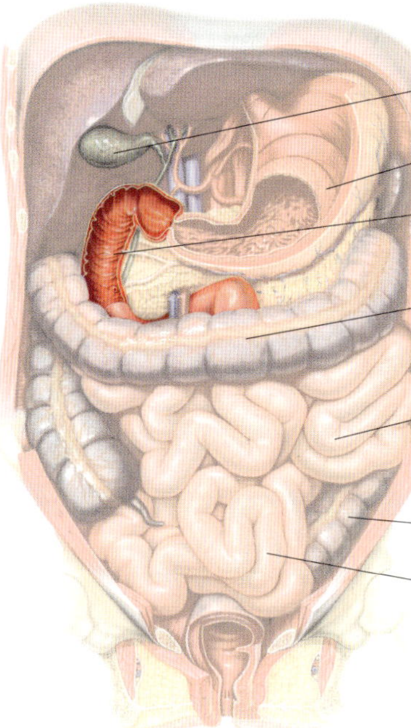

Duodenum

The first 10 inches (25 centimeters) of the small intestine is known as the duodenum. It receives the contents of the stomach, including digestive enzymes and acid, and is the site where bile is brought into contact with food.

Small intestine

The small intestine is made up of the duodenum, the jejunum and the ileum. The primary function of the small intestine is the absorption of nutrients.

SEE ALSO *Digestive system on page 101*

Duodenum

The duodenum is the part of the gut directly beyond the stomach and is the first part of the small intestine. It has a "C" shape, with the curvature of the "C" encircling the head of the pancreas. The duodenum receives the contents of the stomach once the stomach's digestive enzymes and acid have been added to the food and the digestion of protein has begun. This means that the duodenum must be able to withstand periodic exposure to the highly acid stomach contents.

The duodenum is also the place where bile from the liver and gallbladder, and pancreatic enzymes from the pancreas are brought into contact with food.

The interior of the duodenum has a folded surface, which increases the available surface area for absorption of sugars, fats and amino acids. While some of the absorption of these nutrients will begin in the duodenum, the other parts of the small intestine, the jejunum and the ileum, continue the process.

Jejunum

Located in the left upper region of the abdomen, the jejunum is the middle part of the small intestine, extending from the duodenum to the ileum. The jejunum is about 8 feet (2.4 meters) long, and together with the duodenum, makes up about two-fifths of the length of the small intestine.

The jejunum, like the rest of the small intestine, is covered by smooth muscle with an inner, circular layer that is thicker than the outer, longitudinal layer. Food passes from the duodenum into the jejunum by a series of muscular contractions and relaxations known as peristalsis.

In the jejunum, as in the duodenum, food is digested and absorbed, passing though the lining of the walls of the jejunum into the lymphatic vessels and the hepatic portal vein to the liver.

Like the duodenum, the jejunum is suspended from the back of the abdominal wall by a fold of membrane called the mesentery. A number of blood vessels, nerves and lymphatic vessels travel through the mesentery and supply the jejunum.

Ileum

The ileum forms the last part of the small intestine. It continues from the jejunum and ends into the cecum at the ileocecal junction. The ileum is about 12 feet (3.5 meters) long and, like the jejunum, is suspended from the back abdominal wall by a thin membrane called the mesentery.

While most nutrients (fats, sugars and amino acids) are absorbed from the duodenum and jejunum, there are important substances absorbed primarily from the ileum. These include bile acids, which are produced by the liver and secreted into the duodenum to break up fat globules.

Most of the bile acids are then recycled by being reabsorbed from the ileum and returned to the liver. The ileum is also important as the absorption site for vitamin B_{12} (cyanocobalamin).

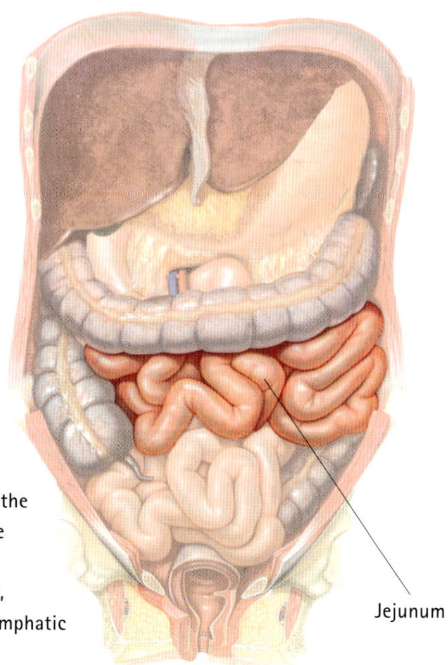

Jejunum

The jejunum is the part of the small intestine between the duodenum and the ileum. It digests and absorbs food, sending nutrients to the lymphatic vessels and liver.

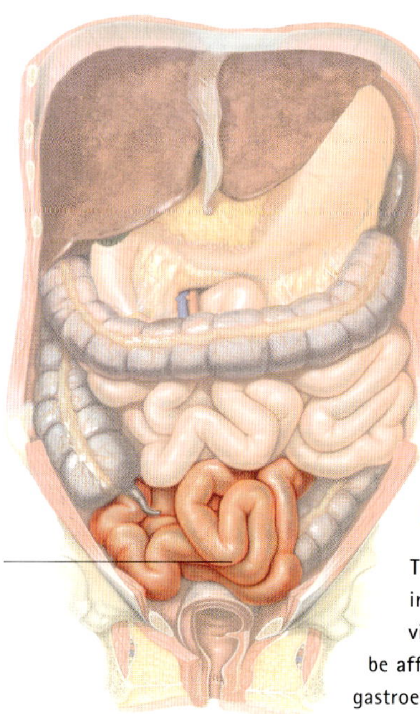

Ileum

The ileum is the last part of the small intestine, which absorbs bile acids and vitamin B_{12}. The working of the ileum can be affected by a number of illnesses, such as gastroenteritis and Crohn's disease.

Diseases and disorders of the small intestine

The duodenum is a common place for ulcers to occur because of its periodic exposure to acidic stomach juices. It is rarely the primary site of malignant diseases (cancer).

The jejunum can be affected by several diseases, including bowel obstruction, gastroenteritis, celiac disease and Crohn's disease. Peptic ulcers, though occurring commonly in the duodenum, usually do not occur in the jejunum. Tumors are also rarely formed in the jejunum.

The jejunal button and jejunostomy tube (the J-tube) are surgically installed feeding devices that deliver nutrients directly into the jejunum. They are used to assist patients with nutritional inadequacies or those who are unable to eat due to illness.

The ileum can be affected by bowel obstruction, Crohn's disease and gastroenteritis. Crohn's disease is most common in the terminal ileum, with patients usually complaining of diarrhea, recurrent abdominal pain, weakness, weight loss and fever. Obstruction of blood vessels supplying the ileum may cause gangrene of the bowel, requiring immediate surgery.

Tumors are rare in the small intestine, comprising only 1–5 percent of all tumors of the gastrointestinal tract.

SEE ALSO *Treating the digestive and urinary systems on page 444*

Celiac disease

Celiac disease, also known as non-tropical sprue, is a true allergic reaction to gluten in the diet. This hypersensitivity results in the destruction of the food-absorbing surface of the jejunum in the small intestine. Gluten is a protein found in many grains, especially wheat, rye, barley and oats.

Normally the immune system cells, known as T lymphocytes, protect the body against invaders, but in celiac disease they attack gluten, thus inflaming the small intestine lining and resulting in inefficient absorption of nutrients.

Celiac disease is thought to be inherited, though the actual cause is not known. It is rare in non-Caucasians but affects 1 in 1,500 to 2,000 Caucasians, being particularly common in people of western Irish descent, of whom it affects around 1 in 300.

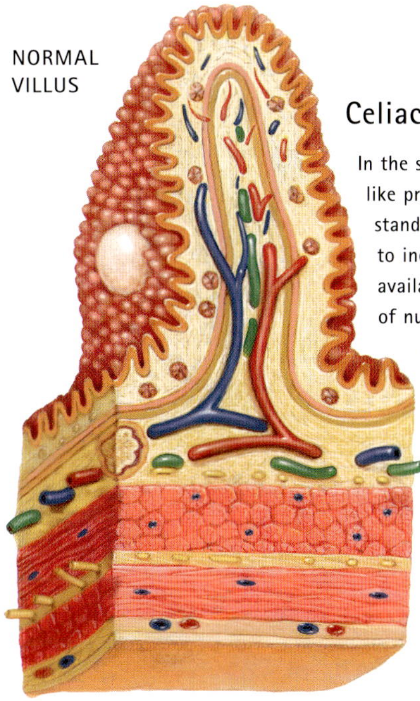

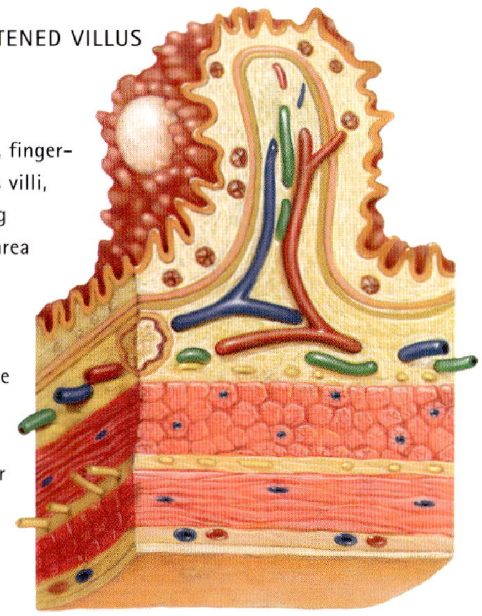

Celiac disease

In the small intestine, tiny, finger-like projections, known as villi, stand out from the lining to increase the surface area available for absorption of nutrients. In celiac disease, the villi in the jejunum in the small intestine are flattened, making it harder for the nutrients in food to be absorbed.

About half of those suffering from the disease are diagnosed during childhood. It usually manifests itself between the ages of 6 and 21 months, often following an infection. Symptoms in children are diarrhea, failure to thrive (stunted growth) and weight loss. Teenagers can suffer delayed puberty and loss of some hair (alopecia areata). In adults, symptoms include flatulence, chronic diarrhea (which does not respond to treatment), weight loss and chronic fatigue. Vitamin deficiency can also result, and produce symptoms such as scaly skin, bruising, blood in the urine, tingling and numbness, muscle spasms and bone pain. It is usually diagnosed through screening blood tests together with a biopsy from the jejunum.

Osteoporosis is a serious illness which generally occurs in people with celiac disease, and symptoms can include night-time bone pain. Five percent of adults with celiac disease will also have anemia, and lactose intolerance is a common feature of celiac disease at all ages.

Once the disease is diagnosed, gluten must be totally eliminated from the diet. The diet needs to be supervised by a specialist who will suggest certain strategies. There are support groups offering valuable resources, and cookbooks that feature gluten-free recipes. In young babies, breastfeeding and postponing the introduction of foods containing gluten can offer some protection.

Celiac disease can go into spontaneous remission in children at about the age of 5 years, and in others, strict adherence to the diet can heal the intestine. However, for most people it will be necessary to stay on a gluten-free diet for the rest of their lives.

Lactose intolerance

Lactase, the enzyme that helps break down lactose, the sugar in milk, is found in the brush border cells of the small intestine. If the enzyme is missing, the lactose cannot be digested and will therefore produce increased gas, colic and sometimes violent, watery and acid diarrhea.

There are several forms of lactose intolerance. Premature infants of under 30 weeks are often unable to cope even with breast milk because the lining of the intestine is immature. There is also a form of lactose intolerance that runs in families and can be extremely severe. It is usually associated with intractable vomiting from infancy.

Temporary damage to the lining of the small intestine often follows an episode of gastroenteritis in children, and results in temporary lactose intolerance. The lining heals in a week or so; until then, the child should avoid milk and dairy products.

The commonest form of lactose intolerance is genetic: up to 90 percent of Asians, 70 percent of African-Americans and up to 25 percent of Caucasians have lactase deficiency in varying degrees.

Diagnosis is made by the lactose tolerance test, when blood glucose does not rise after ingesting lactose, but the patient becomes flatulent and develops diarrhea.

Treatment means avoiding milk and milk products; soy milk is often advised instead.

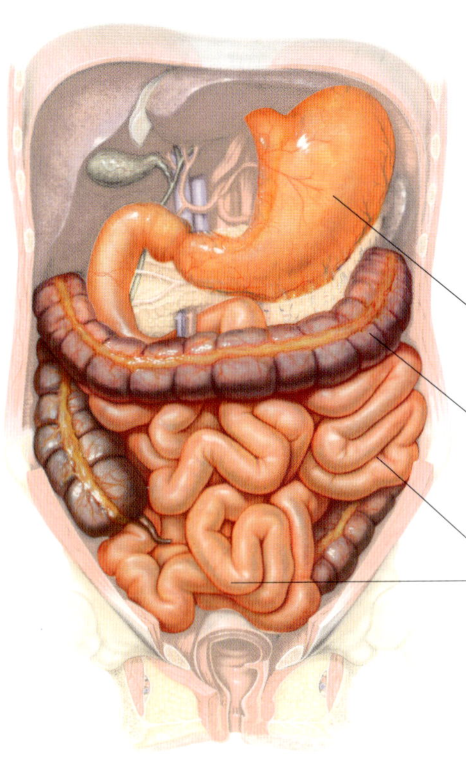

Enteritis

The small intestine, stomach and colon can all be affected by a form of enteritis, an inflammatory condition usually caused by bacterial or viral infection.

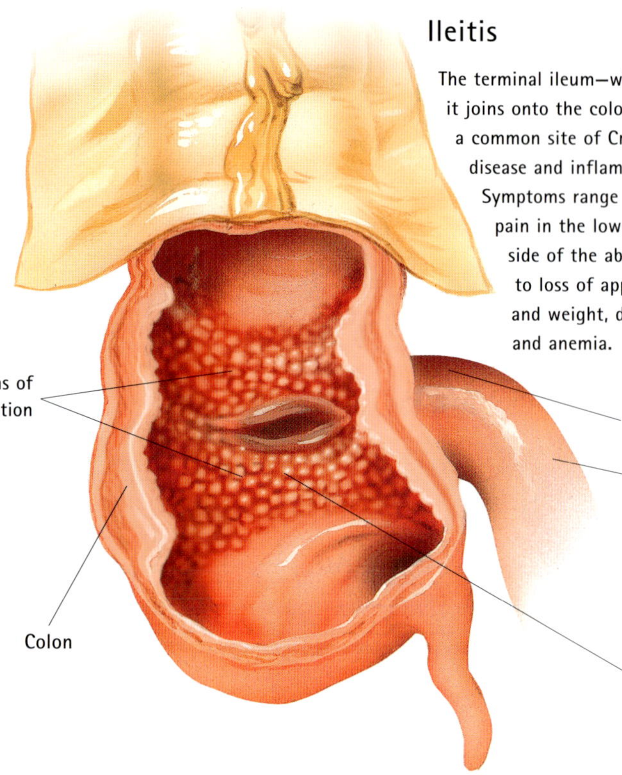

Ileitis

The terminal ileum—where it joins onto the colon—is a common site of Crohn's disease and inflammation. Symptoms range from pain in the lower right side of the abdomen to loss of appetite and weight, diarrhea and anemia.

Enteritis

Enteritis is inflammation of the small intestine, caused by a bacterial or viral contamination of ingested food or liquids. The stomach is often also involved; the condition is then known as gastroenteritis. If the colon is involved, it is called enterocolitis. Symptoms of abdominal pain, diarrhea, fever and dehydration can begin as soon as 4 hours or as late as 72 hours after exposure.

Mild cases usually need no treatment and clear up in 1 to 3 days. Consuming clear fluids such as apple juice or broth is recommended, and milk should be avoided. Antidiarrheal medications may be useful in relieving symptoms. Diarrhea can cause rapid and extreme dehydration in infants and in such cases medical advice should be sought immediately.

Ileitis

Ileitis is inflammation of the ileum, which is the furthest end of the small intestine. Ileitis can occur as the result of infection with viruses or bacteria, and includes common conditions such as viral gastroenteritis and rarer conditions such as tuberculosis of the intestine, which affects the ileum more than other parts of the bowel.

Ileitis may also occur following radiation therapy of the abdominal organs. A person suffering from ileitis may experience pain in the abdomen and loss of appetite and weight. Treatment involves medication to remove the source of infection.

Ileus

Ileus refers to paralysis of bowel movement (a normal function known as peristalsis). It may arise in several different situations: after major abdominal surgery, infection in the abdomen, low output of blood from the heart, and even due to pneumonia. The abdomen is distended and bowel sounds are absent, although occasional faint or irregular bursts of bowel movement may be heard. An x-ray may reveal the loops of bowel to be distended.

In newborn infants with cystic fibrosis, ileus may occur due to the presence of a thick plug of meconium (meconium ileus). In normal newborn infants, the meconium, which is a paste filling the bowel before birth, is cleared from the bowel within a few days. Surgical intervention may be required to remove the blockage.

Adhesions

Adhesions are thin bands of scar tissue that form in a body cavity after an operation or a severe infection. They occur most commonly in the abdomen. Though adhesions are painless, they may restrict movement and function of surrounding organs. For example, the intestines may become pinched or entangled by them. When this happens, the normal movement of food through the intestines may be prevented—abdominal obstruction, vomiting, abdominal swelling and pain can occur. If untreated a bowel obstruction can lead to death.

Treatment is by surgery. While the patient is under general anesthetic, the surgeon opens the abdomen, locates the adhesion(s) and cuts them, releasing the trapped bowel. Unfortunately they usually grow back after the operation. Some sufferers undergo multiple operations.

Colic

Colic is pain resulting from the distension of a hollow internal organ, usually the intestine. Babies and children are especially prone, but adults can suffer too. Typically colic in babies begins at 2–4 weeks and lasts until around 3 months. Symptoms are excessive crying (3 hours or more a day, 3 days a week is excessive) and curling up the legs. Older children complain of pain. One common cause is wind that has passed into the stomach or intestine rather than being released through a burp. Colic can also be caused by a groin hernia.

There is no safe medicine for babies with colic. Aromatherapy and osteopathy may help. If the baby is being breastfed, this should be continued, as ceasing will not solve the problem. If the baby's crying is causing tension in the mother and it affects her breastfeeding, then relaxation and reassurance that the colic will pass may help. Making sure the baby has a routine and learning what the baby's cries mean can also help. It is important that the mother of a colicky baby has regular breaks and is offered support.

Large intestine

The large intestine consists of the colon and rectum. While the small intestine's primary function is the absorption of nutrients, the large intestine's primary function is the reabsorption of water and the movement of waste material toward the anus.

SEE ALSO *Digestive system on page 101*

Colon

The colon forms the majority of the large intestine and moves waste material to the anus, absorbing salt and water. It is made up of the cecum, ascending colon, transverse colon, descending colon and sigmoid colon. The appendix is attached to the cecum.

Appendix

The appendix is a thin, worm-shaped pouch, 3½ inches (9 centimeters) long, and is attached to the first part of the colon (large intestine). It has no function and appears to be a relic of evolution.

Rectum

The rectum is the second last part of the digestive tract and leads into the last part, the anus. The function of the rectum is to receive fecal material from the sigmoid colon and store it for a short time until it is convenient to expel the stool. It also receives gas, which is passed as flatus.

The upper rectum has a series of folds in its wall called the rectal valves. At the lower end of the rectum are longitudinally running folds called anal (or rectal) columns. The lower ends of these columns are joined together by anal valves to form the pectinate line. Immediately above each valve lies an anal sinus, into which open the anal glands. These glands may also become sites of infection to form anorectal abscesses. Below the pectinate line is the anal canal, which leads to the external environment.

Anus

The anus is a short tube approximately 1½ inches (3–4 centimeters) long leading from the rectum through the anal sphincter to the anal orifice, through which feces are expelled.

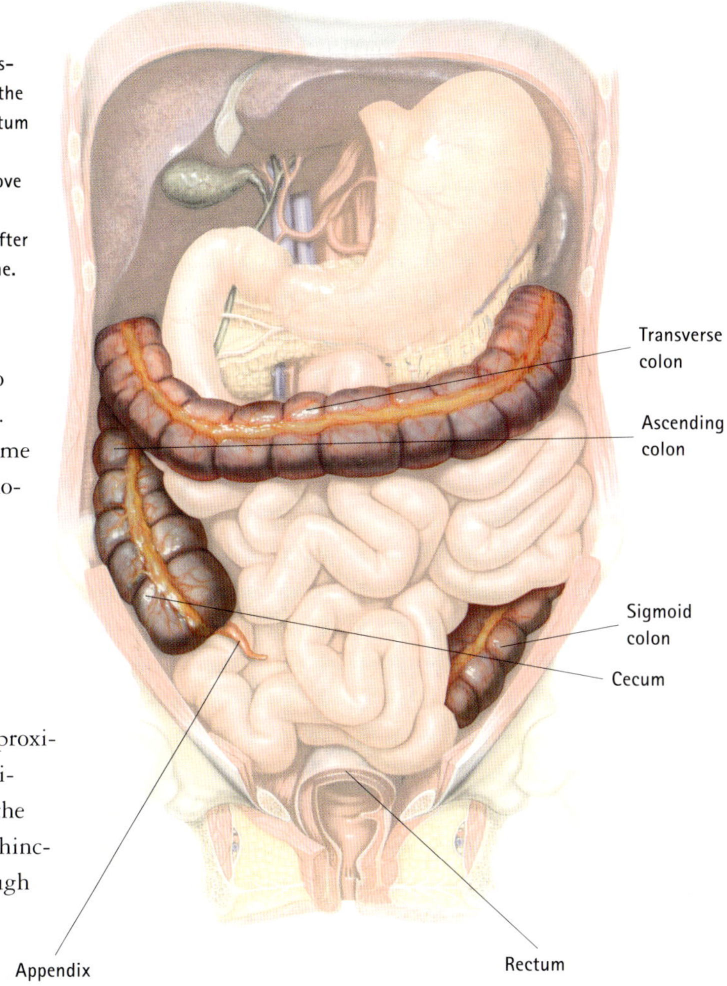

Colon

The colon is part of the large intestine. It stretches from the end of the small intestine through to the rectum and is made up of several parts. The function of the colon is to move solid material to the anus and to absorb salt and water remaining after passage through the small intestine.

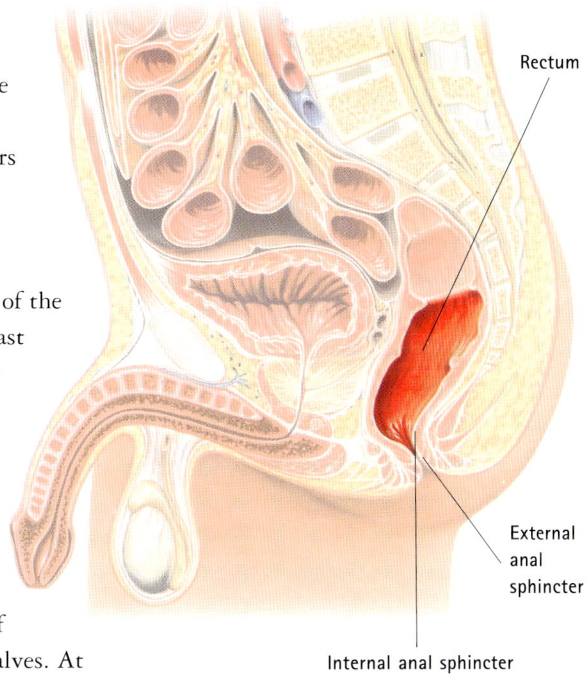

Rectum

The rectum and the anal canal measure 6–8 inches (15–20 centimeters). Rings of muscle called the anal sphincters keep the anus closed when fecal matter is not being expelled.

Diseases and disorders of the large intestine

Many disorders can affect the large intestine. Common investigations of the colon include barium enema and colonoscopy.

SEE ALSO *Colostomy on page 461; Endoscopy on page 433; Imaging techniques on page 431*

Bowel cancer

Bowel cancer invariably arises from the layer of cells lining the inside of the large intestine (which includes the colon and rectum); it is also known as colorectal cancer. In the economically developed nations of the world, it is one of the three most common cancers (the others being lung cancer, and prostate cancer in men and breast cancer in women) and is one of the major causes of death from cancer.

In fact, each year in the USA, more than 100,000 new cases of bowel cancer are diagnosed and about half that number of people die from the disease each year. The causes may be genetic, with some families being particularly at risk, and/or dietary: diets

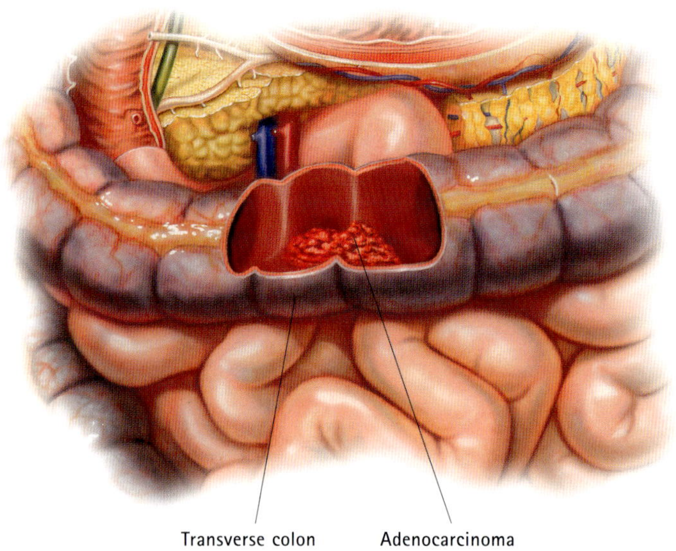

Bowel cancer

This bowel cancer is growing out from the wall of the transverse colon. It will need to be surgically removed, along with some surrounding bowel and nearby lymph nodes. If the cancer has not spread beyond the walls of the bowel, then the prospects for long-term survival are very good.

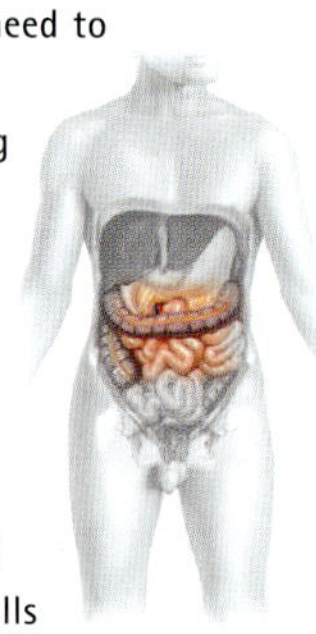

high in protein and fat are shown to be contributing factors. A large consumption of meat may change the type of bacteria in the bowel, favoring bacterial species that turn bile acids and other chemicals into cancer-causing agents (carcinogens).

There is good evidence that for most bowel cancers a sequence of abnormal changes in the bowel wall precedes the emergence of a cancer. One of these is the development of pre-cancerous polyps (outgrowths of the lining of the bowel), although not all polyps are pre-cancerous. Several inherited disorders that lead to the formation of multiple colorectal polyps have been identified. There are other inherited conditions associated with bowel cancer that do not exhibit polyp formation. Also, there is an increased risk of bowel cancer in patients with long-term inflammatory bowel diseases, such as ulcerative colitis or Crohn's disease.

Cancer of the large intestine occurs in persons over 50. It is more common in men. The symptoms are variable and often vague.

About half of all large bowel cancers develop in the lower third of the colon or the rectum. Bleeding from these tumors may be visible, but often goes unnoticed. Such cancers may produce crampy abdominal pain, constipation, or alternating constipation and diarrhea.

Bowel cancers grow relatively slowly, so early recognition of symptoms by the patient and prompt diagnosis of the disease by the doctor has enormous potential to save lives. Blood screening of feces can also help recognize the disease early. Diagnosis is by barium enema (an x-ray investigation), colonoscopy (using a flexible fiberoptic instrument) and sampling of blood for a chemical released by cancer cells—the carcinoembryonic antigen (CEA). Many bowel cancers are, however, not diagnosed until the tumor has spread.

The spread of bowel cancer is initially through the wall of the bowel, then by lymphatics to the lymph nodes draining nearby and later via the blood to the bones, liver, lungs and other organs. As with many cancers, the extent of spread of the cancer at the time of diagnosis is the most important factor that determines the course of the disease and the patient's chances of survival.

Treatment of bowel cancer is by surgery, removing the tumor and surrounding bowel and lymph nodes, and sometimes radiation therapy for those patients for whom major surgery may be too risky. Early detection and diagnosis is the key to surgical cure. Colonoscopy with biopsy is the most effective method of early diagnosis currently available.

Intestinal polyps

A polyp is a growth or tumor protruding from a mucous membrane. It may be shaped like a grape on a stalk or may be a small lump. While polyps are usually benign, they may lead to complications.

Intestinal polyps are found most often in the rectum and sigmoid colon. They usually cause no symptoms, though in some cases there may be bleeding from the rectum. They should be removed, as they may become malignant. Surgery to remove a polyp is usually done via a proctoscope or sigmoidoscope inserted via the rectum; the polyps are snipped off or destroyed by electric cauterization. Anyone with a personal or family history of polyps should have regular sigmoidoscopic examinations.

Polyposis of the colon

Polyposis of the colon is a rare genetic condition. Patients develop multiple polyps (mushroom-shaped lumps which usually have a stalk, by which they are attached to the bowel lining) in the large intestine. Polyps develop in other conditions, but not as many as seen in polyposis. The danger is that each polyp has a chance of cancer developing within it; the more polyps there are, the greater the chance of cancer. This means that patients with polyposis are almost certain to develop cancer within their lifetime. The only cure is to completely remove the colon.

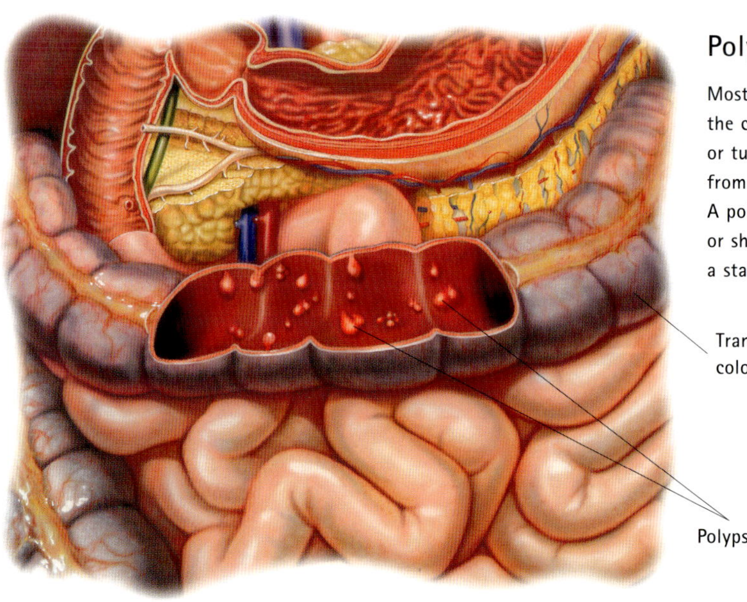

Polyps

Most commonly found in the colon, polyps are growths or tumors which protrude from the mucous membrane. A polyp may be a small lump, or shaped like a grape with a stalk.

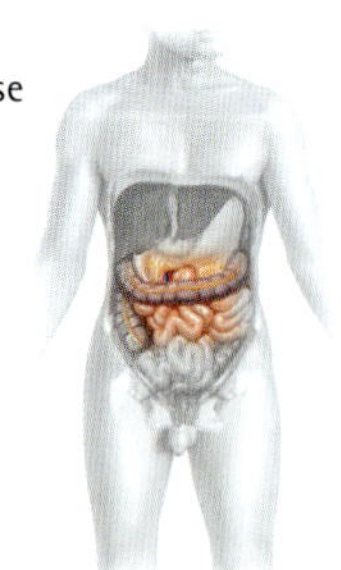

Irritable bowel syndrome

Irritable bowel syndrome (IBS) is a common condition, affecting 15–25 percent of people in Western societies. Symptoms include abdominal pain (especially cramps and spasms), bloating and abnormal function of the bowel. Some patients may have hard stools and difficulty in passing motions; others may have loose stools and a feeling of urgency to pass motions. Patients with IBS appear to have a greater degree of pain sensitivity in the lower bowel. The cause remains obscure, but appears to involve increased sensitivity of the bowel to the sensations of bowel function.

The diagnosis is usually made in the absence of other physical evidence of gut disease: the condition is thought to be a functional, rather than a structural or biochemical, change in the bowel.

Treatments include antidepressants and drugs which inhibit the function of a neurotransmitter known as serotonin. Sufferers should avoid irritants such as tea, coffee and alcohol, and check for possible allergies, particularly to dairy or wheat.

Diverticulosis

In diverticulosis, the wall of the bowel "blows out," forming outpockets which protrude between the muscle bands of the bowel wall. A diet which is low in dietary fiber and high in fat is thought to contribute to this condition of the colon.

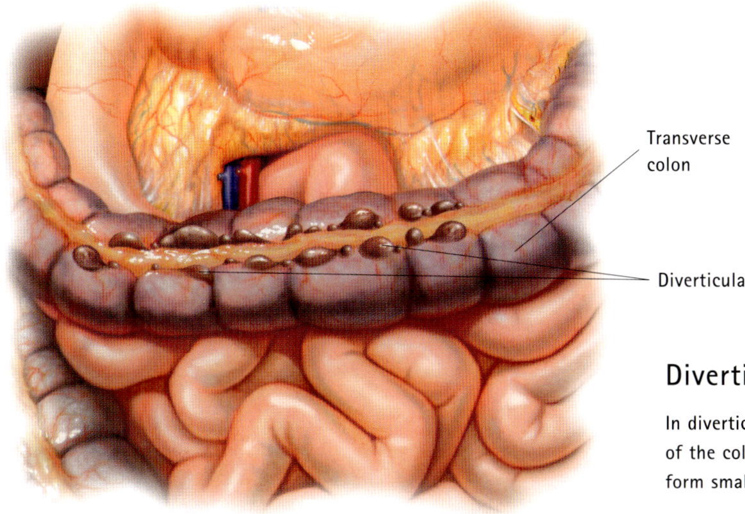

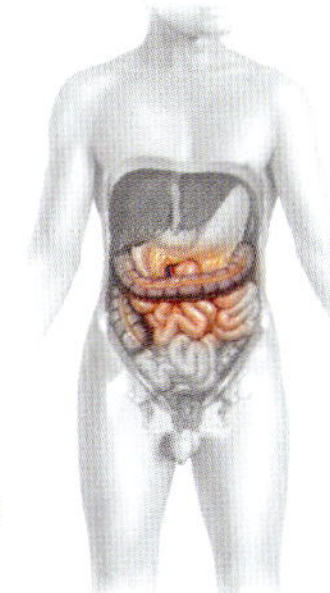

Diverticulosis

In diverticulosis, inner layers of the colon wall bulge out to form small pockets (diverticula).

Diverticulitis

Diverticulitis is a disease of the large bowel or colon, which arises when small outpockets (protrusions) in the wall become inflamed; these outpockets are called diverticula, thereby giving the disease its name.

The patient experiences pain usually in the lower left part of the abdomen, which is mild to severe in intensity and may be either persistent or cramping in nature. Patients may also have mild fever, tenderness of the lower abdomen, swelling of the lower abdomen, constipation and, sometimes, even blood in their feces.

Occasionally diverticula rupture, and the contents of the bowel leak into the tissues of the bowel wall and, sometimes, into the cavity of the abdomen. The bacteria from the bowel multiply within these regions.

Complications of diverticulitis include abscesses in the bowel wall and nearby abdominal cavity, fistulas, peritonitis and obstruction of the bowel. About one-quarter of patients with acute diverticulitis require surgery to either drain or remove abscesses, relieve obstruction or close fistulas.

Volvulus

A volvulus is a twisting of the intestine around itself, most commonly toward the lower end of the digestive tract—in the ileum, cecum or sigmoid colon of the bowel. It blocks the intestine and constricts the blood vessels serving the intestine. Abdominal pain and swelling, vomiting and constipation result. Surgery to untwist the volvulus is usually necessary. If gangrene has set in, the affected portion of the intestine must be surgically removed.

A volvulus in a very small child may relate to an intestinal malformation during

Irritable bowel syndrome

Irritable bowel syndrome is characterized by cramping abdominal pain and spasms plus unusual diarrhea or constipation, but does not have the inflammation associated with other intestinal tract diseases.

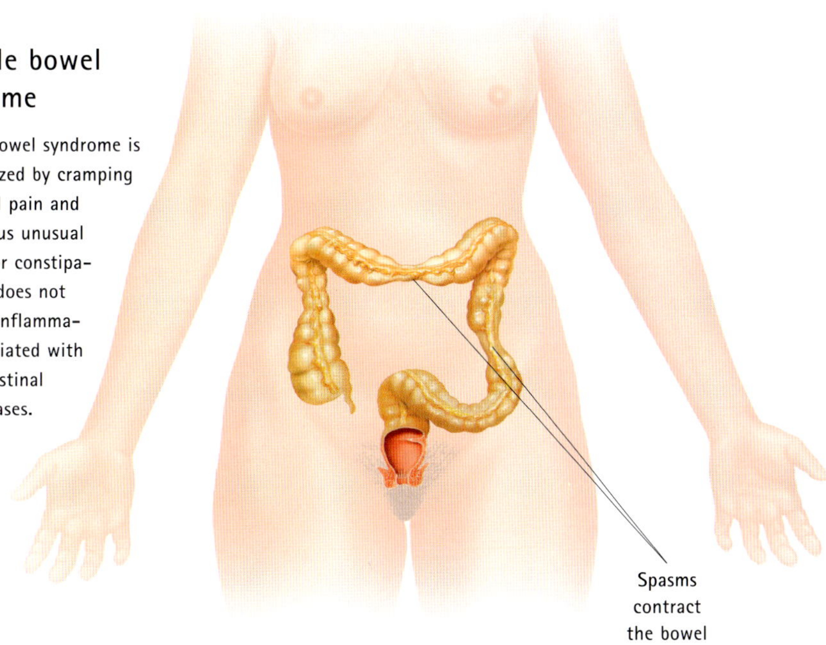

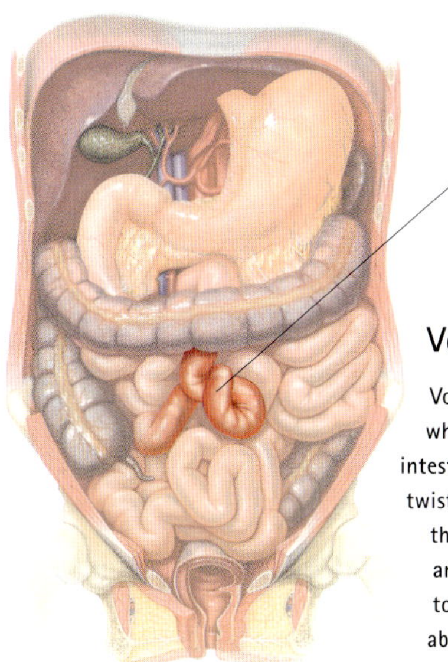

Volvulus

Volvulus occurs when a loop of intestine becomes twisted, blocking the intestine and leading to severe abdominal pain.

NORMAL APPENDIX

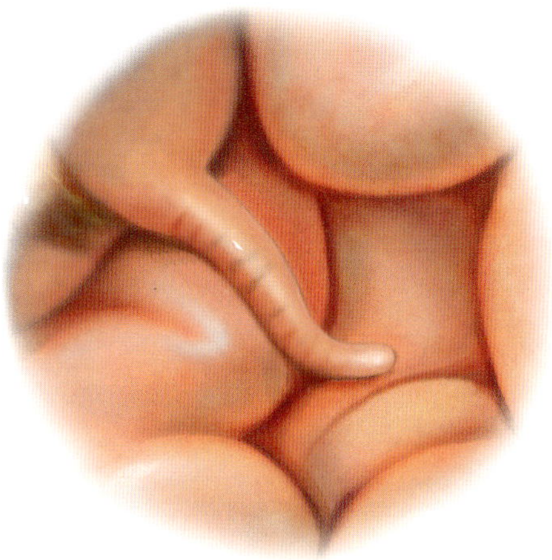

INFLAMED APPENDIX

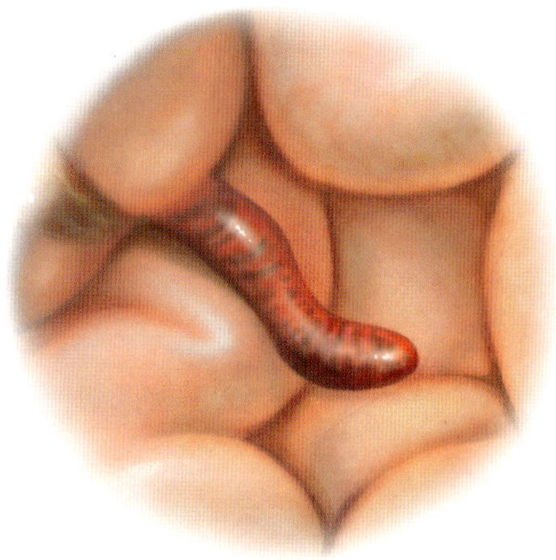

Appendix

The appendix is a redundant organ that is attached to a part of the large bowel. In children especially, it may become inflamed and if so, will need to be surgically removed in hospital. The term for an inflamed appendix is appendicitis.

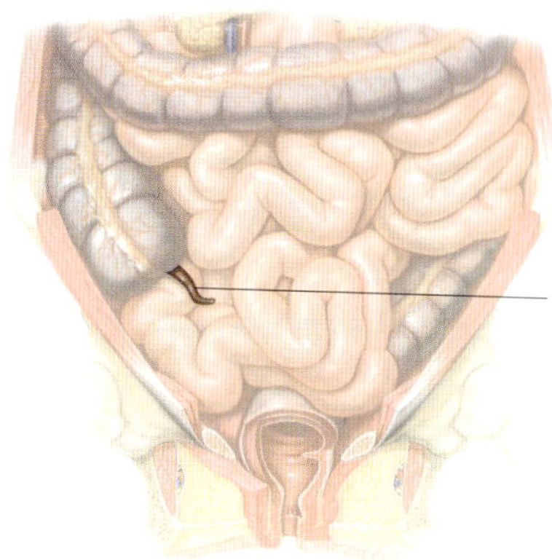

fetal development. In older people, the twisting may be due to scar tissue caused by surgery or infection.

Hirschsprung's disease

Hirschsprung's disease is a congenital condition (present at or before birth). The basic abnormality is the absence of nerve cells in the wall of the lower large bowel.

Without these nerve cells, the bowel is unable to make the peristaltic movements which propel the feces toward the anus. Consequently, the infant will experience constipation, abdominal swelling, reluctance to consume food, and vomiting.

Temporary relief may be obtained by washing the colon with a saline solution. In the long term, however, the child may require a colostomy to allow for normal evacuation of the large bowel.

Appendicitis

Appendicitis is thought to be caused by an obstruction of the opening into the appendix, possibly by feces or lymph tissue; inflammation of the appendix follows. Pain develops in the lower right side of the abdomen, which is spasmodic at first, but then becomes constant. Other symptoms include fever, nausea and vomiting. In some cases, the inflammation can turn into an abscess which can burst, causing peritonitis.

Appendicitis is usually treated by urgent surgery. The surgeon removes the appendix through an incision in the lower right side of the abdomen. Recovery from appendectomy takes about a week, and usually there are no further problems.

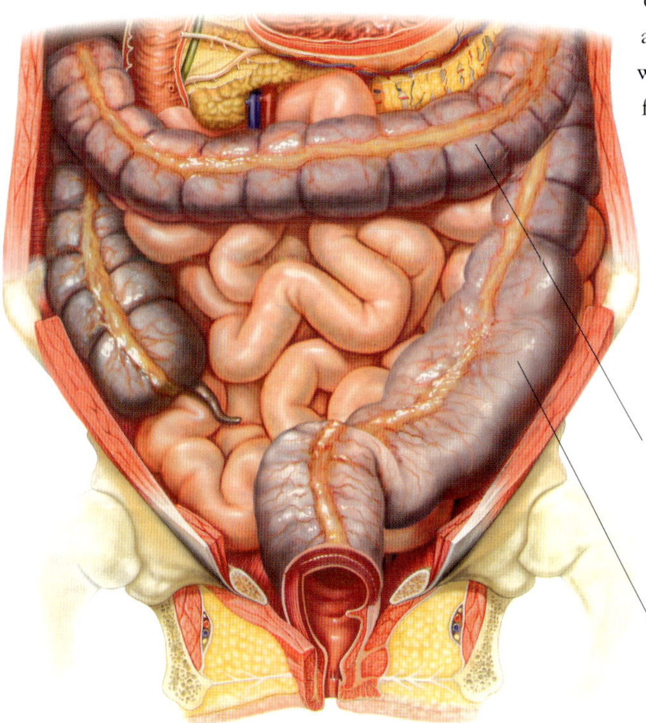

Hirschsprung's disease

In this condition, the nerve cells in the wall of the colon are absent, resulting in enlargement, constipation and obstruction.

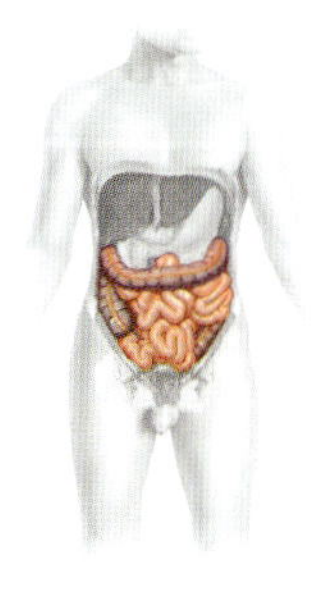

Disorders of the rectum

The exterior anus and rectum are common sites of disease and can be involved in hemorrhoids (dilated veins of the anorectal junction), cancer, as well as inflammatory disease (also known as proctitis).

In rectal prolapse, the rectum may be turned inside out through the anal canal. This may occur as a congenital problem in babies or in adults whose pelvic muscles are weakened, injured or paralyzed.

SEE ALSO *Endoscopy on page 433; Treating the digestive and urinary systems on page 444*

Proctitis

Proctitis is inflammation of the rectum. It can occur as part of ulcerative colitis. It may also be due to bacterial infection, as is sometimes seen in male homosexuals with gonorrhea of the rectum from anal intercourse. These patients will complain of rectal irritation, itching, pain and pus-containing (purulent) discharge. Gonorrhea of the rectum is treated with the appropriate antibiotic (penicillin and/or tetracyclines).

Occasionally proctitis may be due to physical factors, as seen in radiation proctitis, a side effect of radiation therapy of tumors in the uterine cervix, bladder, uterus or prostate. In those patients the inflammation may be reduced with steroids.

Prolapsed rectum

A prolapsed rectum is an abnormal movement of the internal mucous membranes of the rectum, the end section of the large intestine, down to or through the anus. It appears as a red mass up to several inches (about 10 centimeters) long, which may bleed. Prolapsed rectums are most common in children under the age of 6 years. The condition usually corrects itself in young children—a physician may gently push the protruding mass back inside.

Surgery is often needed when the condition occurs in adults and may involve attaching the rectum to surrounding muscle for support. Alternatively, the anus may be tightened via the insertion of a circle of wire or nylon. Prolapsed rectum may be associated with conditions such as cystic fibrosis, constipation, malnutrition and infestation with pinworm or whipworm.

Disorders of the anus

Common anal disorders include anal fissures, hemorrhoids and anal itching.

SEE ALSO *Endoscopy on page 433*

Anal fissure

An anal fissure is a slit or tear in the mucosa (or lining) of the anus that produces a tearing pain when feces are passed. Most common in infants and children, the condition is very painful, and may be accompanied by bleeding while passing feces. The best treatment is a high-fiber diet and anti-inflammatory ointments. A chronic fissure may result in anal spasm or stenosis (narrowing), requiring minor surgery.

Anal itching (pruritus ani)

Anal itching is usually caused by irritation of the skin in or around the anus. The itching can be caused by diarrhea, infections (especially yeast infections), skin diseases, or other problems such as hemorrhoids and/or tearing of the anal skin. Treatment is to clean and dry the area thoroughly. Cortisone cream applied to the anal skin will often relieve the symptoms.

Hemorrhoids

Commonly known as piles, hemorrhoids are enlarged veins in the lower portion of the rectum and anus, which become swollen due to straining when passing feces. They occur frequently in people who suffer constipation and those who usually keep sitting down for prolonged periods.

Hemorrhoids are common, affecting 1 in every 500 persons. They are especially common during pregnancy, after childbirth and in hepatic hypertension, caused by tumors or cirrhosis. Though not dangerous, they can cause irritation by bleeding or passing through the anus.

Symptoms include bright red blood in the stool, anal itching and pain during bowel movements. Painful thrombosis (clotting of blood in the veins) sometimes occurs.

The diagnosis is usually made by a primary care physician (general practitioner). The physician may use a small tube called an proctoscope for anoscope to visualize the hemorrhoids along the anal canal. More serious conditions, such as colonic or rectal polyps and cancer, which can also cause bleeding, must be ruled out.

Hemorrhoids are treated with a high-fiber diet, topical steroid ointment, and surgery in troublesome cases. Surgical techniques include rubber band ligation, cryosurgery and hemorrhoidectomy.

These treatments are usually effective, but the condition may recur unless preventive measures—treatment of constipation and a more active lifestyle—are adopted.

Anorectal fistula

Anorectal fistula (or "fistula in ano") is a condition in which there is an abnormal passage between the inner surface of the anus and the surface of the skin around the anus. It is due to infection on the anus, usually by pus-causing bacteria that results in

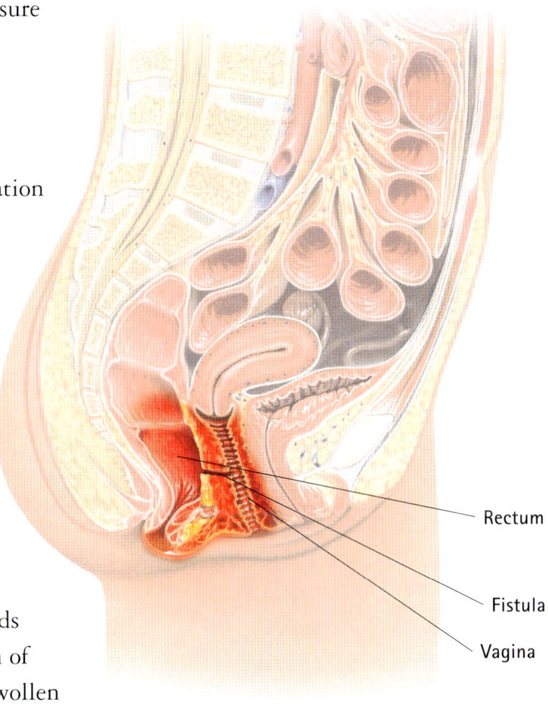

Fistula

For a woman, a problematic labor may lead to the formation of a fistula between the rectum and the vagina, which can lead to feces passing into the vagina.

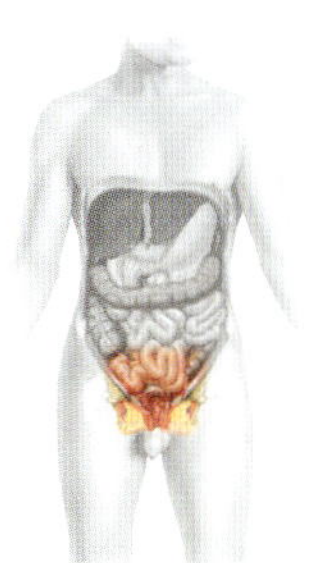

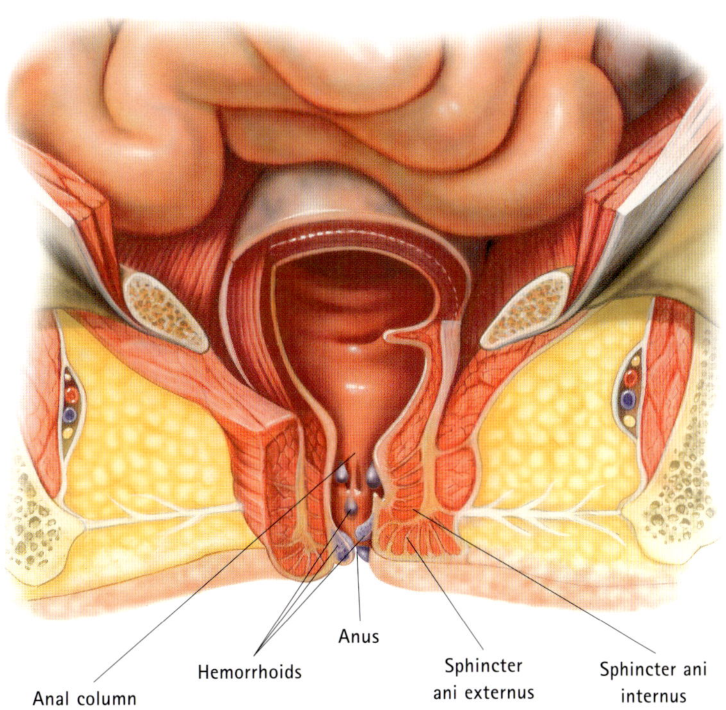

Hemorrhoids

Hemorrhoids are swollen veins in the lower part of the rectum or anus, often due to constipation or prolonged periods of sitting.

a break in the wall of the anal canal or rectum. Other cases may be due to cancer of the anus and rectum, Crohn's disease or diverticulitis. An untreated anorectal fistula may lead to bacterial infection throughout the body, and some may become the sites of origin of cancers. Small anorectal fistulas may repair themselves, but for larger ones the site of infection should be removed and the fistula opened to allow pus to drain from the site.

Liver

The liver is an organ associated with the digestive tract. It is the heaviest single organ in the body, weighing about 3½ pounds (1.5 kilograms) in an adult, and makes up about one-fiftieth of total body weight. The liver is normally reddish brown in color, and lies under the cover and protection of the lower ribs on the right side of the upper abdominal cavity.

The liver has an upper (diaphragmatic) surface, which is in contact with the diaphragm, and a lower (visceral) surface, which is in contact with organs in the abdominal cavity. These two surfaces are separated at the front by a sharp inferior border, which may sometimes be felt when the liver becomes enlarged and protrudes below the line of the ribs. The visceral surface of the liver is in contact with the gallbladder (which is usually attached to the liver by connective tissue), with the kidney, part of the duodenum, the esophagus, the stomach and a part of the large bowel.

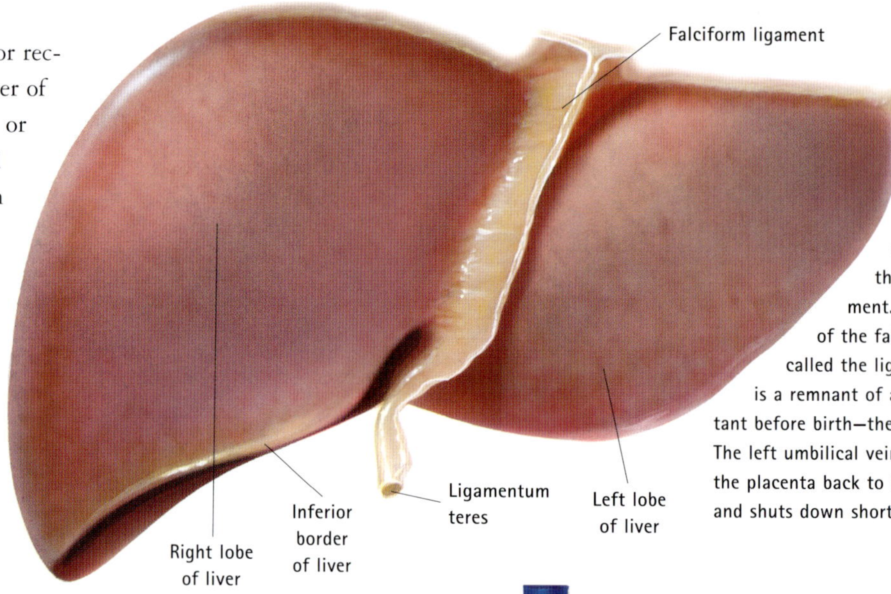

Liver

The liver as seen from the front is divided into two lobes (left and right) by a fold of peritoneum called the falciform ligament. The lower edge of the falciform ligament is called the ligamentum teres and is a remnant of a structure important before birth—the left umbilical vein. The left umbilical vein carries blood from the placenta back to the developing fetus and shuts down shortly after birth.

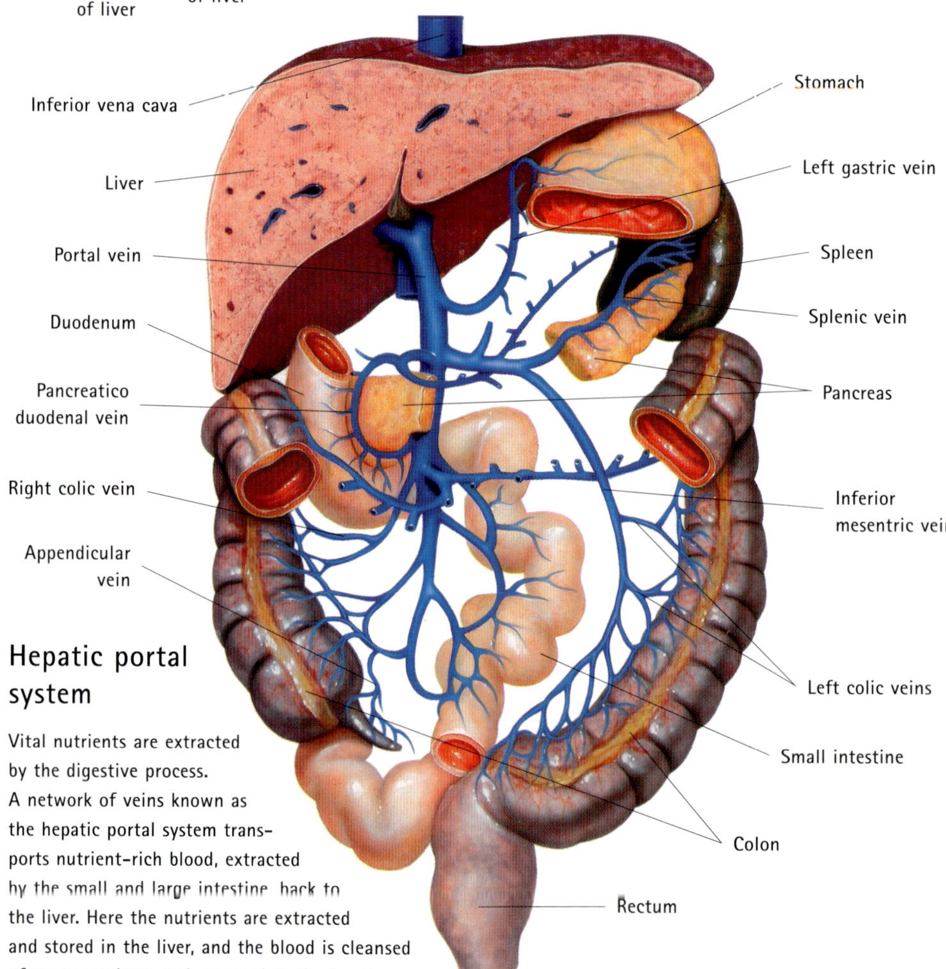

Hepatic portal system

Vital nutrients are extracted by the digestive process. A network of veins known as the hepatic portal system transports nutrient-rich blood, extracted by the small and large intestine, back to the liver. Here the nutrients are extracted and stored in the liver, and the blood is cleansed of waste products and returned to the heart.

Liver

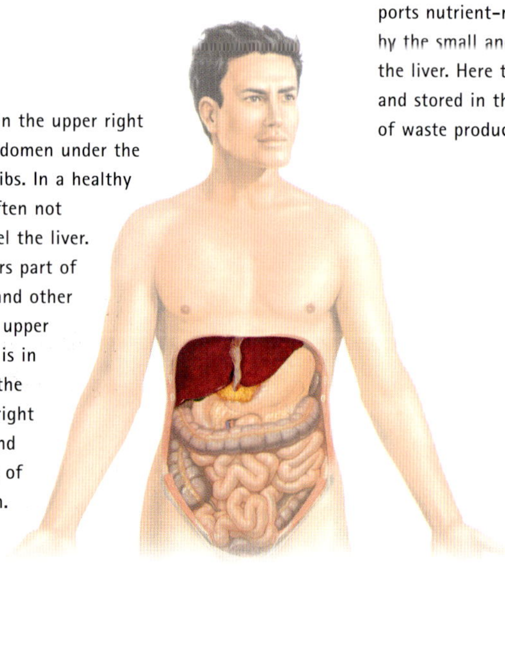

The liver lies in the upper right side of the abdomen under the cover of the ribs. In a healthy person it is often not possible to feel the liver. The liver covers part of the stomach and other organs of the upper abdomen and is in contact with the right kidney, right large bowel and the beginning of the duodenum.

Hepatic artery and portal vein

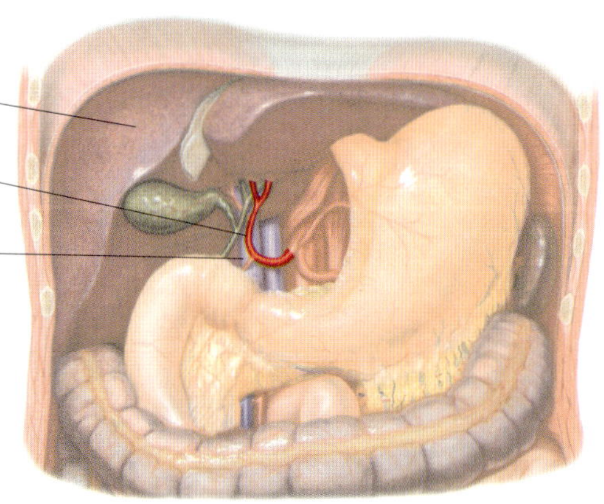

NB: In the illustration at right, the liver has been peeled back to show the position of the hepatic artery and the portal vein.

The liver is attached to the diaphragm—the muscle which separates the chest and abdominal cavities—by a series of folds of membrane called the falciform, triangular and coronary ligaments.

The liver is also connected to the stomach and duodenum by folds of membrane called the gastrohepatic and hepatoduodenal ligaments respectively.

On the visceral surface of the liver lies a region known as the porta hepatis. This is the site at which vessels and ducts enter and leave the liver. Within the porta hepatis are the portal vein, which carries blood from the gut to the liver, the hepatic artery, which carries blood from the aorta to the liver, and the common hepatic, cystic and bile ducts, all of which are part of the biliary system of ducts that store bile and deliver it to the duodenum.

SEE ALSO *Digestive system on page 101*

Microscopic structure

Microscopically, the liver contains sheets of cells (hepatocytes) arranged in hexagonal prism-shaped lobules. Each hepatocyte is about one-thousandth of an inch (25 thousandths of a millimeter) across, and they are piled up in sheets one cell thick, much like bricks in a wall. Hepatocytes contain a large amount of glycogen, which is an energy storage chemical made from glucose.

The space between the sheets of hepatocytes is filled with small blood vessels called liver sinusoids. In the walls of the liver sinusoids are special cells called macrophages (Kupffer cells) that are capable of engulfing debris. A system of bile ductules runs between the hepatocytes. These ductules carry bile, which is produced by the hepatocytes. The ductules eventually join together to form hepatic ducts, which in turn unite to form the bile duct.

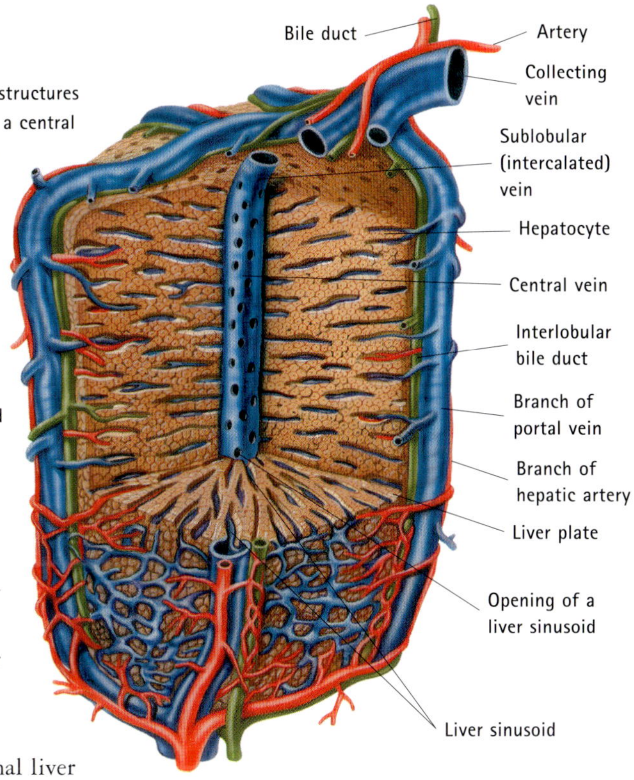

Liver lobule

The liver is made up of many hexagonal structures called liver lobules. Each liver lobule has a central vein at its core, which drains blood from the lobule to the hepatic veins that lead out of the liver. Blood arrives in the liver from the gut via the portal vein, which has branches at the corners of the liver lobule. More blood is brought into the liver by the hepatic artery and its branches. The bile produced by liver cells passes into the branches of the bile duct, which eventually drains toward the duodenum.

At the corners of each hexagonal liver lobule lie branches of the portal vein, hepatic artery and hepatic ducts, while the center of each lobule is occupied by a central vein.

Venous blood from the gut flows past the sheets of liver cells on its way to the central vein. Nutrients, bile salts and toxic and waste substances are removed from the portal blood by hepatocytes and processed as necessary. The central veins of all the lobules join together and contribute blood to the hepatic veins. These drain into the inferior vena cava, which transports blood back to the heart.

Metabolic functions

The liver serves many metabolic functions. It receives all the blood returning from the gastrointestinal tract, which is laden with glucose derived from the breakdown of digested food. The liver converts much of this glucose to a storage molecule called glycogen. Glycogen can be converted back to glucose for release into the bloodstream whenever sugar is required.

This means that the liver plays a key role in maintaining a relatively constant concentration of glucose in the blood, regardless of the time of day or the energy demands of the body. Two hormones released from the pancreas, insulin and glucagon, are important in the control of this function.

The liver also plays a key role in the metabolism of other sugars, as well as fats and proteins. Liver cells have much of the cellular machinery (granular endoplasmic reticulum) associated with making proteins

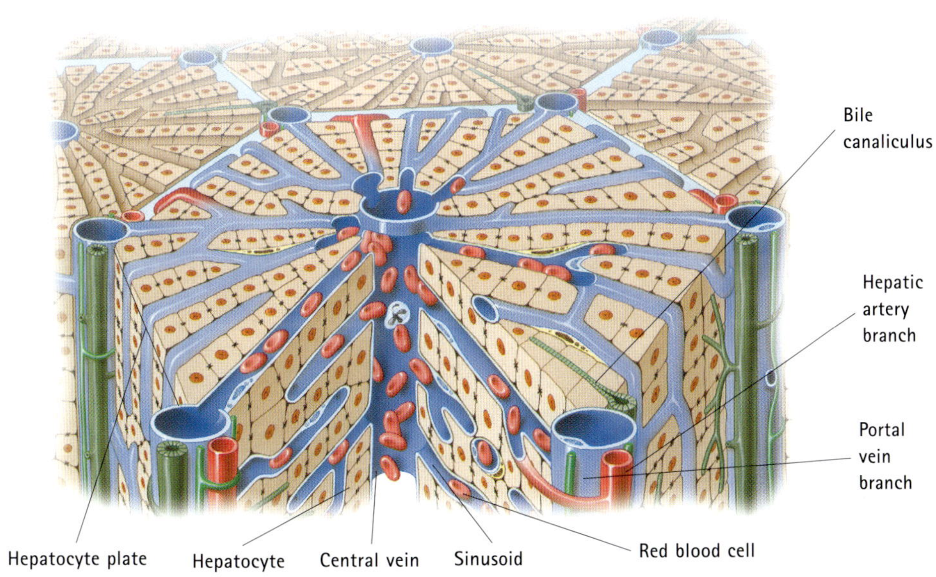

Liver—microstructure

Each liver lobule consists of radially arranged sheets of specialized epithelial cells interpenetrated and ensheathed by supporting connective tissue. Small blood vessels called liver sinusoids run past the sheets of liver cells. Branches of both the portal vein and hepatic artery feed into these sinusoids. Nutrients and toxic substances are drawn from the passing blood by liver cells and processed appropriately. Other fine tubular structures between the liver cells, called bile canaliculi, collect bile and carry this fluid toward the branches of the bile duct located at the corners of the liver lobule.

and glycogen. Under the electron microscope, granules of glycogen with the appearance of berries are often visible in liver cells. Liver cells also contain many mitochondria—tiny chemical powerhouses whose presence indicates very high levels of metabolic activity in the liver. The position of the liver in the flow of blood from the gut also allows it to remove and destroy any toxic substances that may be ingested along with food and water. These toxic substances include alcohol, drugs and the chemicals produced by microorganisms. Bacteria entering the body from the gut must also get past the defensive scavenger cells (known as Kupffer cells) of the small vessels of the liver before they can reach the body at large.

The liver stores vitamin A, which is essential for the well-being of the surface-lining tissues of the body, and stores iron, used in the production of hemoglobin in the blood. Bile, which is used to aid the digestion of fats, is made in the liver and released into the duodenum through the biliary system of ducts. The liver also produces albumin, which is an important plasma protein in the blood that helps to control fluid movement between the inside of blood vessels and the spaces between the cells throughout the body.

Finally, the liver makes several important substances involved in the control of blood clotting, including the clotting factors prothrombin and fibrinogen.

Bile is a yellow-orange fluid produced and excreted in the liver. It consists of water, bile salts and a chemical called bilirubin. Bile is produced by the hepatocytes and flows along tiny channels (bile canaliculi) toward the bile duct branches at the corners of the lobules.

Bile is also continually recycled, because it is reabsorbed from the gut once the digestion of fats has taken place. About 90 percent of the bile secreted by the liver is actually recycled from the gut. The bile flows back to the liver in the portal vein blood, where it is transferred by the hepatocytes to the bile duct branches for delivery back to the gut for further fat digestion.

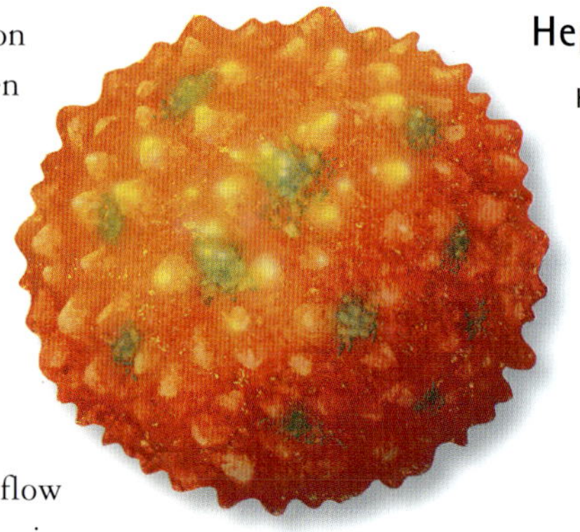

Hepatitis virus

Hepatitis is inflammation of the liver. When caused by a virus, it is called viral or infectious hepatitis. Some types of viral hepatitis can be transmitted in drinking water.

Disorders of the liver

The most common disorders of the liver are hepatitis and cirrhosis. Primary cancer of the liver is relatively rare.

Investigation of liver disease often involves taking blood specimens to test for liver enzymes, which are released from damaged liver cells. Removal of a small specimen of liver for examination (liver biopsy) may be performed to investigate jaundice, liver enlargement or suspected cirrhosis. Patients are given medication to calm nervousness, and a special biopsy needle is inserted between ribs 8 and 9 on the right side of the body below the nipple. The tissue is examined by a pathologist to assist in making a diagnosis. Other diagnostic tools to determine liver disease include ultrasonography, computerized tomography, magnetic resonance imaging techniques and isotope scans to detect regions of high blood flow and metabolic activity.

SEE ALSO *Imaging techniques on page 431; Liver function blood tests on page 437; Viral hepatitis on page 375*

Congenital diseases

Several congenital diseases may affect the liver. These include congenital bile duct atresia, in which the bile ducts fail to develop, and problems with the enzymes responsible for bile metabolism. Children with these disorders develop jaundice, or yellowing of the skin, during the first few days to weeks of postnatal life. This should be distinguished from the normal mild jaundice that occurs to some extent in all infants shortly after birth, and which usually disappears in a few days.

Hepatitis

Hepatitis is an inflammation of the liver that reduces its ability to function. It can be

DAMAGED LIVER MICROSTRUCTURE

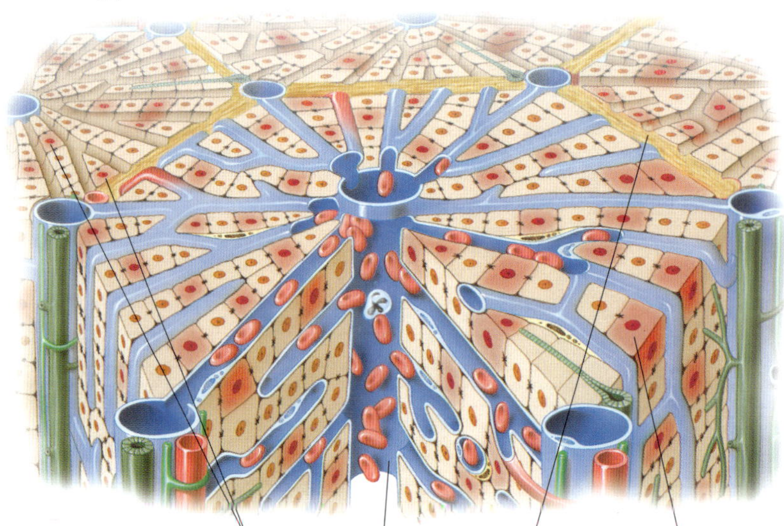

NORMAL LIVER MICROSTRUCTURE

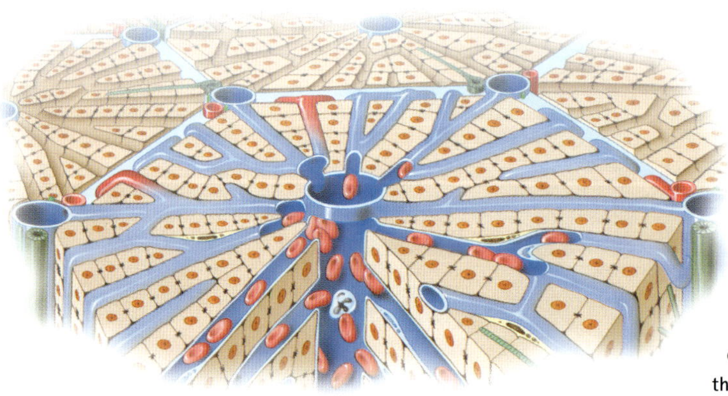

Scarring of the liver microstructure

Chronic hepatitis causes cirrhosis, a disease in which the normal microscopic lobular architecture of the liver is destroyed. Scarring and distortion of the hepatocytes (liver cells) and connective tissue that form each hexagonal lobule can disrupt the flow of blood through the liver.

CIRRHOTIC LIVER

Cirrhosis

If it is not treated, chronic hepatitis can lead to cirrhosis of the liver. When this occurs, nodular, fibrous tissue replaces damaged liver cells and connective tissue in the liver, distorting its smooth surface and internal microstructure.

NORMAL LIVER

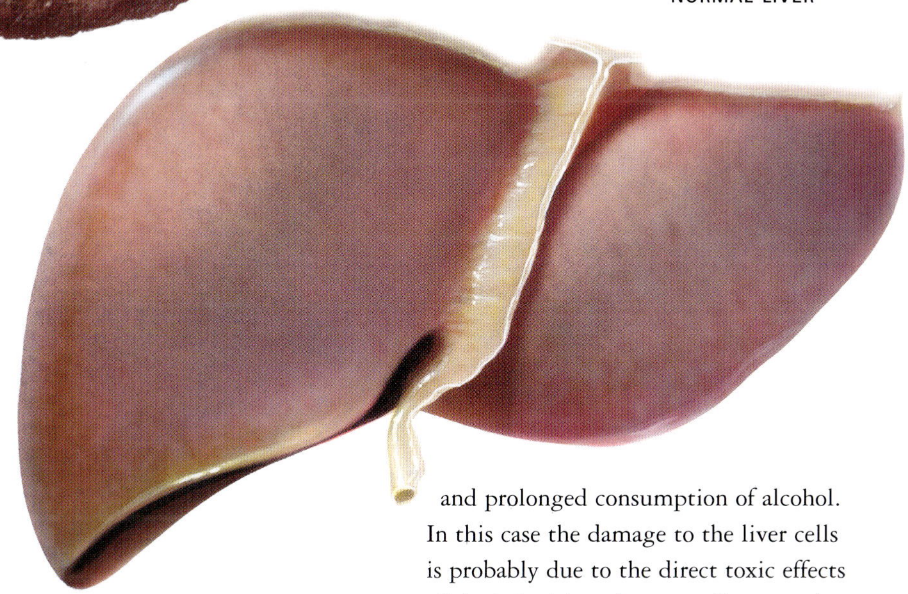

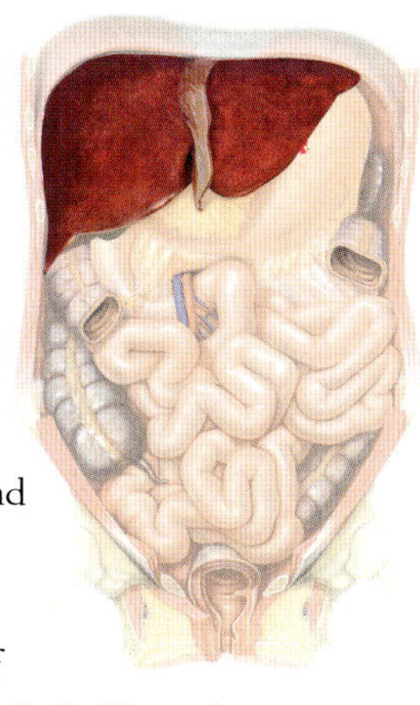

infectious, when caused by viruses and parasites, or non-infectious, when caused by alcohol, certain drugs, and toxic agents such as chemicals found in aerosol sprays and paint thinners. Alcoholic hepatitis is the result of sustained consumption of excessive quantities of alcohol. If caught early enough it can be reversed; if not, it leads to alcoholic cirrhosis.

Hepatitis can also be caused by an autoimmune disorder when the body mistakenly fights its own healthy tissue with its own cells, and can be associated with some illnesses such as Wilson's disease. Most cases are caused by a viral infection.

Some cases of hepatitis are difficult to recognize, but when symptoms are present there may be general weariness, loss of appetite, fever, vomiting, abdominal pain and jaundice (a yellowing of the skin and eyes). This yellowing is a result of the damage the virus inflicts on the liver cells. Viral hepatitis in its acute phase usually lasts from a few days to several weeks. Some types of viral hepatitis (including hepatitis B and hepatitis C) may subside into chronic hepatitis, which may continue for years. Other types of hepatitis (e.g. hepatitis A) only cause acute infection.

The different types of viral hepatitis are transmitted in different ways: for example, hepatitis A by the gut through contaminated food or water, and hepatitis B and C by blood, intravenous drug use with shared needles, or sexual intercourse. The incubation period for hepatitis A is 2–6 weeks, while type B and C have longer incubation periods of about 6 weeks to 6 months. Patients experience a period of feeling generally unwell, followed by nausea, vomiting and disinterest in eating. They may also develop fever, upper abdominal pain and yellowing of the skin and eyes (jaundice) due to high levels of a chemical called bilirubin in the blood.

Sometimes, widespread destruction of liver cells occurs, with a progression to complete liver failure, coma and even death (though fortunately this is rare).

Cirrhosis of the liver

Cirrhosis of the liver is a disease in which there is death of liver cells followed by production of fibrous tissue and regeneration of liver cells in lumps or nodules. The nodules distort the normal structure of the liver and prevent the easy flow of blood through the liver. Cirrhosis may be caused by excessive and prolonged consumption of alcohol. In this case the damage to the liver cells is probably due to the direct toxic effects of alcohol, although a poor diet may also contribute. Chronic hepatitis may also lead to cirrhosis of the liver.

Patients with alcoholic cirrhosis may also show Dupuytren's contracture, a contraction of connective tissue in the hand, deterioration of the brain, heart problems and enlargement of the parotid salivary gland in the cheek. Men may develop breast enlargement (gynecomastia) and wasting of the testes. Other causes include problems with the storage of iron (hemochromatosis) and an inherited disease to do with the metabolization of copper (Wilson's disease).

Patients with cirrhosis often commonly develop jaundice. They may also experience swelling of the abdomen and ankles, mental confusion, disorientation and coma. The disordered regrowth of liver cells interferes with blood flow through the cirrhotic liver. This raises the pressure of blood in the portal vein, leading to a condition known as portal hypertension, which can have serious consequences because small veins in the lower esophagus swell (esophageal varices) and may rupture. Patients with cirrhosis can die from blood loss due to bleeding from esophageal varices, or they may develop low blood pressure, high levels of ammonia in the blood, coma and ultimately kidney failure. About 20 percent of patients with cirrhosis will eventually develop liver cancer.

Treatment of cirrhosis will depend on the cause of liver damage. Alcoholics must abstain completely from alcohol. Hemochromatosis is treated by removal of blood, while Wilson's disease is treated by drugs which serve to bind the excess copper.

Portal hypertension

Portal hypertension is an increase in the pressure of the blood in the portal venous system. This is the system of veins that drains blood from the intestines to the liver.

The most common cause of this condition is chronic liver disease leading to cirrhosis, which obstructs blood flow through the liver. This causes life-threatening bleeding into the lower esophagus and stomach. The patient may vomit blood (hematemesis) or pass black, tar-like feces containing altered blood (melena). The associated swelling of the abdomen is due to the accumulation of fluid in the abdominal cavity (ascites). Fluid collects in the abdomen and lower legs because the diseased liver is unable to produce the protein albumin that helps to control fluid movement between the tissues of the body and the bloodstream.

Hepatic encephalopathy

Hepatic encephalopathy is an acute complication of liver disorders in which nitrogen wastes such as urea and other toxins build up in the body and eventually affect the brain and nervous systems.

When liver cells are damaged due to conditions such as alcoholic cirrhosis or hepatitis, they can no longer do their job of cleansing the body of toxins effectively. This usually causes metabolic abnormalities marked by symptoms that may range from confusion and memory loss to muscular tremors and speech impairment.

Hepatic encephalopathy may be chronic, leading to dementia, coma and death. It is due to liver failure as a result of diseases like hepatitis, alcoholic cirrhosis, and cancer, some medications, or eating excess protein when you have such diseases.

Treatment involves addressing contributing causes and may include the removal of toxins from the intestinal tract, preventing ammonia absorption from the intestines, adopting a reduced-protein diet and avoiding medications that are normally metabolized by the liver. Those who are suffering from hepatic encephalopathy may also require hospitalization with respiratory and cardiovascular support.

Cancer of the liver

Cancer in the liver may have spread from other parts of the body, including the large bowel, breast, lung, pancreas, stomach, kidney and uterus, or arise in the liver itself (primary cancer of the liver). In Western countries, cancer which has spread from other sites to the liver is about 20 times as common as primary liver cancer. Cancer may be spread to the liver by blood from the gut, along the lymphatic vessels, or through the arterial blood supply.

Treatment may include surgery, if only an isolated nodule of tumor is present, or chemotherapy delivered directly to the liver. Life expectancy once a tumor has spread to the liver is very low.

Liver cancer usually occurs in patients with previous cirrhosis of the liver. In developing countries, infestation with liver flukes or ingestion of a fungal toxin (aflatoxin) present in grains and peanuts can play important roles in its initiation. A certain type of liver cancer (angiosarcoma of the liver) is caused by industrial exposure to vinyl chloride. With primary liver cancer, removal of part of the liver, or liver transplant if the whole liver must be removed, is the only hope of a total cure. Chemotherapy may delay the advance of the tumor.

HEPATOMA

Hepatoma, or hepatocellular carcinoma, is a malignant tumor of the liver. It is a primary tumor, that is, originating in the liver rather than spreading to the liver from another site. It is usually associated with an underlying liver disease such as alcoholic cirrhosis or hepatitis B or C infection, and it is especially common in South Africa and Southeast Asia.

The symptoms of hepatoma are a hard mass in the right upper quadrant of the abdomen, unexplained weight loss and appetite loss, abdominal aches and pains, and sometimes jaundice with yellow skin.

Blood tests for liver function, an abdominal CAT scan and a liver biopsy are generally carried out by the physician to confirm a diagnosis of hepatoma.

Treatment is not usually successful: in only about 25 percent of cases can the tumor be fully removed and it tends to spread to other organs such as lungs and bones. It is usually considered incurable.

Jaundice

Jaundice is a condition in which there is yellowing of the skin and eyes due to an increased concentration of bilirubin in the blood. The condition may be due to excessive breakdown of blood cells (hemolytic jaundice), excessive production of pigments (pigment overload), problems with liver cells (hepatocellular jaundice), or obstruction of the ducts leading from the liver to the gut (obstructive jaundice). Many cases of jaundice often present a combination of all these types.

All normal newborn infants have a small degree of jaundice, which is due to immature enzyme systems in the liver. This "physiological" jaundice will usually disappear over a few days, but may become a serious problem in infants who are born prematurely or who have increased destruction of red blood cells due to other diseases. If the concentration of bilirubin in the blood rises too high, the brain may be damaged in a condition called kernicterus.

Jaundice may also be caused by drugs including phenacetin, paracetamol, sex steroids, antipsychotic drugs, some anesthetics, selected antibiotics and anti-tuberculosis drugs. Some toxic chemicals like carbon tetrachloride can also damage the liver and cause jaundice.

GALLBLADDER

The gallbladder is a sac-shaped organ that stores and concentrates bile prior to its release into the small intestine. It is part of the biliary tree, a series of ducts that conveys and stores bile. The gallbladder is usually firmly attached to the lower surface of the liver and lies on the right side of the abdomen, just below the ribs, at the front.

Bile is a body fluid which contains pigments, lecithin and bile salts. The pigments are made from cholesterol and bilirubin, and give the bile fluids a yellow to orange color during life. Bile is very important because, when it is released into the small intestine, it serves to break down relatively large globules of fat into smaller droplets, thus increasing the available surface area of the fat particles so as to improve their digestion and absorption.

Bile is produced by liver cells, passes along the bile ducts, and is stored and concentrated in the gallbladder before being released into the initial part of the small intestine, known as the duodenum.

Once digestion and absorption of fats has taken place, the bile is reabsorbed at the end of the small intestine, carried back to the liver by a group of veins called the portal system, and there re-excreted into the bile ducts to begin the process again. This process of digestion and reabsorption is called enterohepatic biliary circulation.

The gallbladder is joined to the bile duct by the cystic duct. The bile duct passes from the junction with the cystic duct down through the head of the pancreas to drain bile into the duodenum. Just before it enters the duodenum, the bile duct is joined by the main duct of the pancreas. The passage of the bile duct via the pancreas means that a tumor (cancer or, more correctly, carcinoma) in this pancreatic head can obstruct the lower bile duct and cause a build-up of bile in the biliary tree. This back-up finally reaches as far up the biliary tree as the gallbladder and liver, causing an enlarged gallbladder leading to jaundice when the bile salts reach the bloodstream.

SEE ALSO *Digestive system on page 101*

DISORDERS OF THE GALLBLADDER

A common disorder of the gallbladder is gallstones, which may in turn lead to acute or chronic cholecystitis (inflammation of the gallbladder). Cancer of the gallbladder is quite rare and is associated with gallstones in 70 percent of cases.

Certain x-ray examinations help to diagnose disorders of the biliary tree. During these examinations, the patient swallows a chemical similar in structure to normal bile constituents, and one that is also opaque to x-rays. The chemical is absorbed from the small intestine and transported to the liver by the portal system. It is subsequently excreted into the biliary tree and concentrated in the gallbladder. X-rays at this stage will reveal the ducts of the biliary tree (cholangiography) and gallbladder (cholecystography) standing out against the darker upper abdominal contents.

Further investigations include ultrasound and retrograde cholangiography, which involves backfilling the biliary tree from

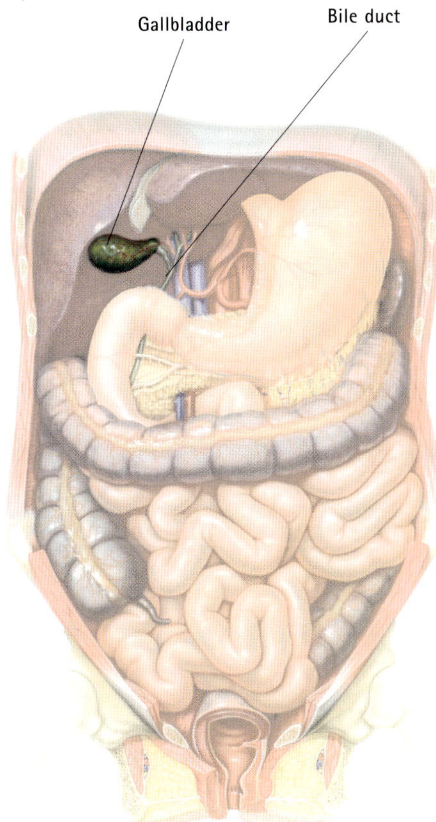

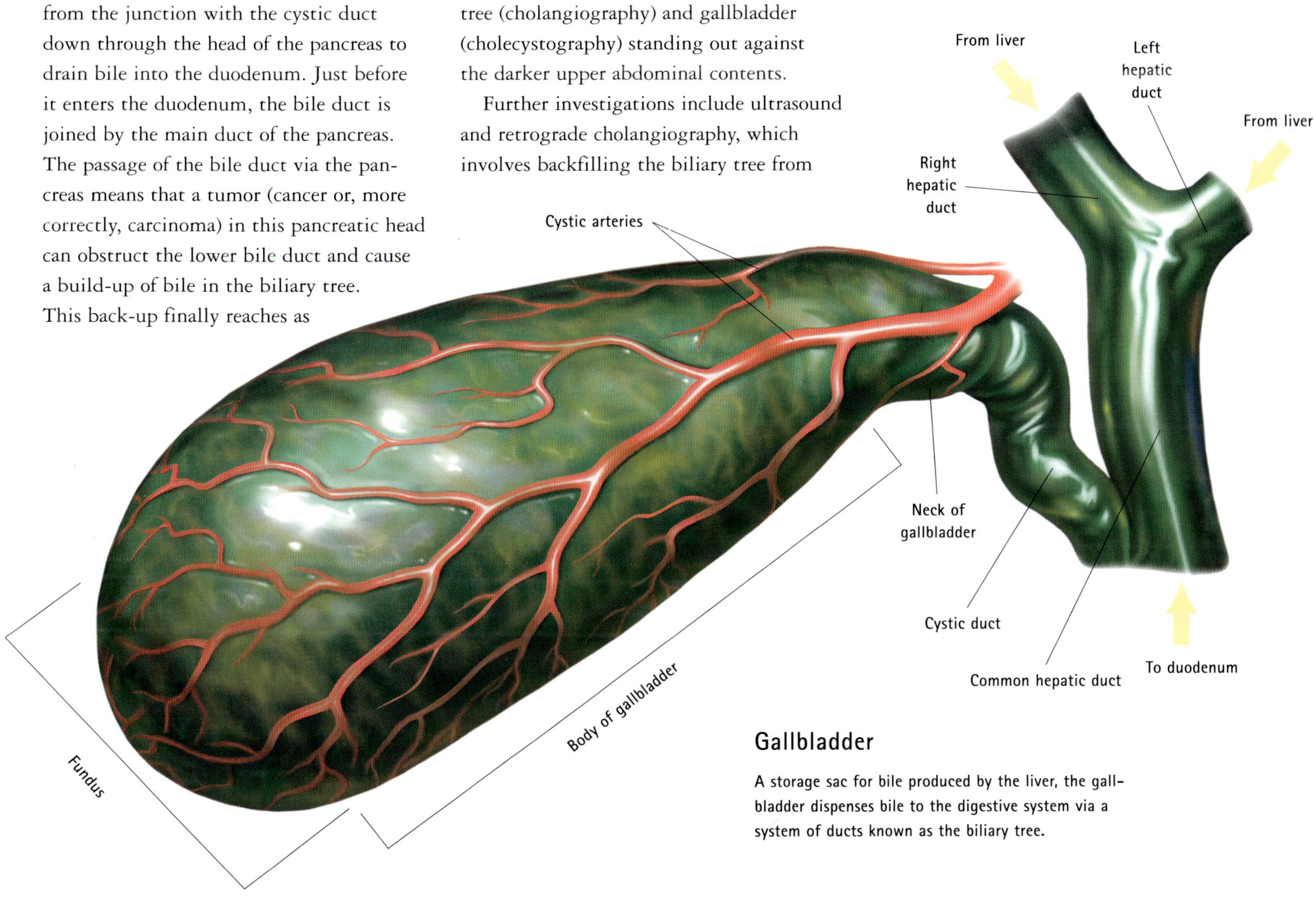

Gallbladder

A storage sac for bile produced by the liver, the gallbladder dispenses bile to the digestive system via a system of ducts known as the biliary tree.

the duodenum with a chemical opaque to x-rays. This test requires the insertion of an endoscope, a device for viewing the interior section of the gut.

The endoscope is inserted through the stomach and into the duodenum, in order to place a catheter into the lower bile duct.

Removal of the gallbladder is known as cholecystectomy—this may be done by open surgery or by laparoscopy, a technique whereby instruments are inserted into the abdomen through small punctures in its wall. Gallbladder disease can be prevented or minimized by losing weight and reducing one's intake of fatty foods.

SEE ALSO *Cholecystectomy on page 460; Imaging techniques on page 431; Lithotripsy on page 452*

Gallstones

These are very common—as many as 7 percent of adults in the USA have gallstones (cholelithiasis) in their gallbladders. Each year in the USA approximately 350,000 people have operations for gallstones, and as many as 6,000 people die from associated complications. Gallstones are more common in women in the 40–65 years age group than in any other group in the community, suggesting that there is a hormonal link. In fact, taking the contraceptive pill contributes to the formation of gallstones in susceptible women.

About 75 percent of gallstones are made of cholesterol, while the remainder are composed of bile pigments. Problems arise when cholesterol or bile pigments come out of solution in the bile fluids concentrated in the gallbladder. This means that gallstones are usually encountered in the gallbladder itself, where they may cause chronic inflammation of the gallbladder lining (chronic cholecystitis). Occasionally the gallstones may leave the gallbladder and become lodged in the bile duct, where they can cause obstruction to bile flow. Jaundice results as bile builds up in the liver and eventually the bloodstream.

Actually, 70 percent of people with gallstones never require surgery, but several problems can arise. Chronic cholecystitis is the most common form of gallbladder disease. Patients experience episodes of abdominal pain (biliary colic) whenever the gallstones cause transient obstruction of the cystic duct, which leads out of the gallbladder. This pain is usually felt in the right upper quadrant of the abdomen and may be accompanied by nausea and vomiting. Patients may not be able to tolerate fatty foods, and may also complain of indigestion, heartburn and flatulence.

If the impaction of the gallstone in the cystic duct leads to inflammation, the abdominal pain may become more severe and persistent. The pain may be accompanied by fever and increased numbers of white blood cells in the blood. This is known as acute cholecystitis. Resulting complications include gangrene and formation of pus in the gallbladder (empyema, or suppurative cholecystitis), perforation of the gallbladder with spilling of infected contents into the abdominal cavity, or the formation of abscesses near the gallbladder.

Gallstones may pass through the cystic duct into the bile duct. In this situation, known as choledocholithiasis, patients experience biliary colic accompanied by moderate to severe jaundice, chills and fever. The condition may result in the pancreas becoming inflamed (pancreatitis).

Gallstone disease is detected in a number of ways. During an oral cholecystogram, the patient is given a chemical that is opaque to x-rays and is excreted by the liver into the gallbladder and bile duct. Ultrasound is a technique whereby sound waves are used to form an image of the gallbladder and the stones within. Retrograde cholangiography is an investigation in which an endoscope is used to insert a catheter into the lower bile duct and backfill the biliary tree from the duodenum.

The symptoms of gallstones can be prevented or minimized by losing weight and reducing one's intake of fatty foods. However, if symptoms persist, then the gallstones (along with the gallbladder)

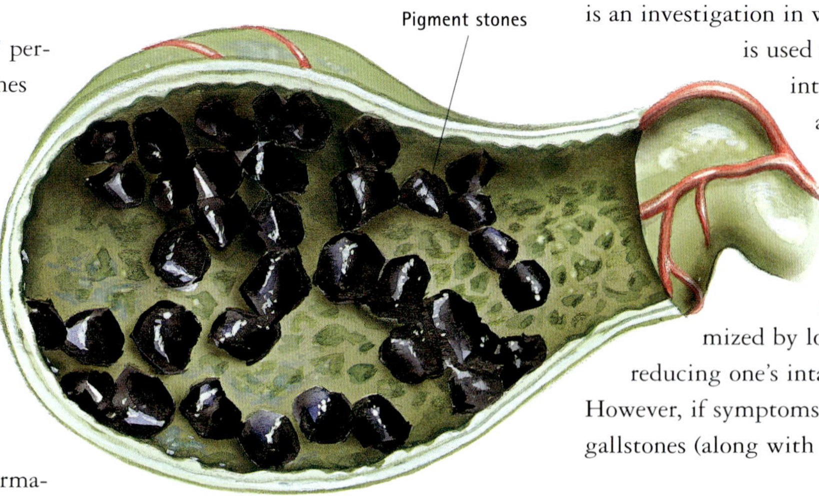

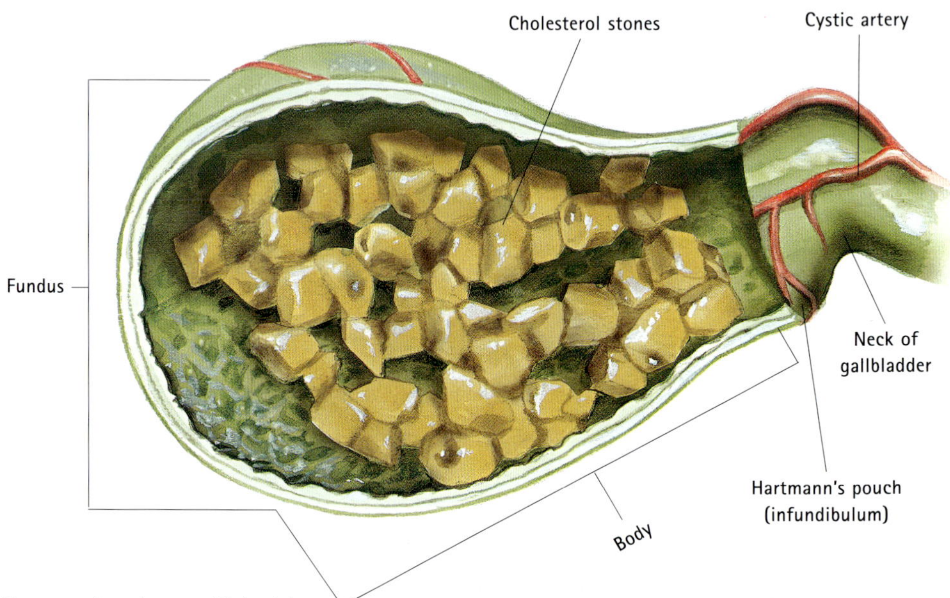

Stones in the gallbladder

Gallstones are composed of bile (pigment gallstones) or cholesterol that have settled in the gallbladder. In advanced cases the entire gallbladder may be filled with gallstones.

should be surgically removed. This is done either by open abdominal operation or through a laparoscope. In general, laparoscopic surgery is the preferred method for most people as it results in a shorter hospital stay and a much faster recovery.

If there is bacterial infection of the gallbladder (for example, cholecystitis) urgent treatment in hospital with intravenous antibiotics is necessary; after the inflammation has subsided, the gallbladder can be removed at a convenient time in the subsequent weeks or months.

Cholecystitis

Cholecystitis is an inflammation of the gallbladder. It is usually caused by gallstones obstructing the outlet of the gallbladder into the cystic duct, which empties into the bile duct system. It may be chronic (long term) or acute (sudden), when there is often accompanying bacterial infection of the gallbladder. Cholecystitis occurs in middle-aged people, especially overweight women who are on the contraceptive pill.

The condition causes severe pain in the right upper quadrant of the abdomen, with nausea, chills, vomiting and high fever. An ultrasound shows gallstones in the gallbladder and cystic duct.

Treatment is with intravenous fluids and antibiotics in hospital. Surgical removal of the gallbladder (cholecystectomy) is required if the inflammation does not settle down with treatment.

Gallstones

Gallstones have different shapes and sizes. They can be multifaceted or round, and vary in color from yellow to brown to black. Small black gallstones are made of calcium bilirubinate. Yellow-brownish gallstones are made from cholesterol and can range from about ½–1½ inches (1–4 centimeters) in size. A single gallstone can grow large enough to fill the gallbladder.

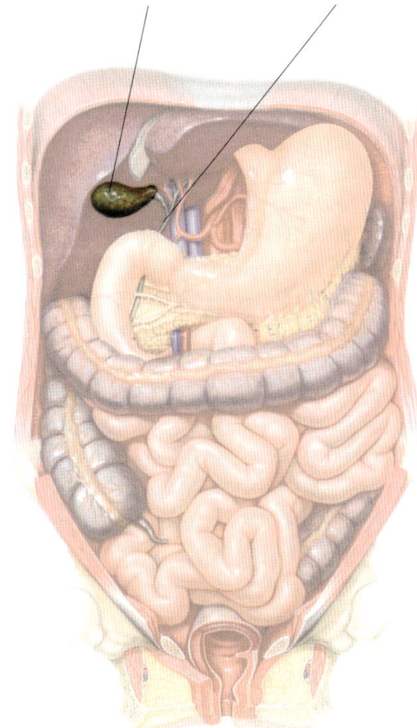

Cholecystitis

Cholecystitis is inflammation of the gallbladder, resulting in attacks of severe, sharp abdominal pain. It is usually caused by gallstones, which will be visible on ultrasound examination.

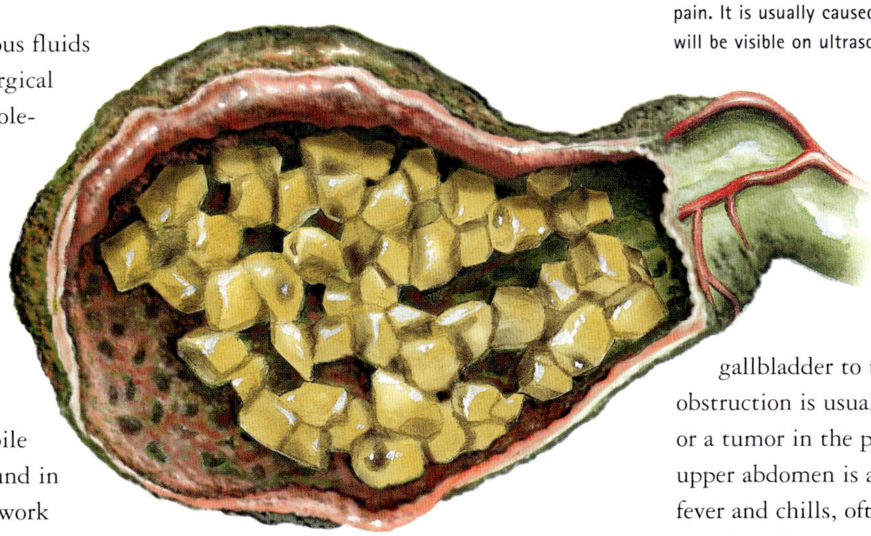

Bile ducts

The bile ducts (collectively called the biliary tree) are the narrow tubes through which bile flows in the liver. They are found in various parts of the body, but work together as a system.

The hepatic ducts transport bile from the liver (where it is produced) to the rest of the biliary system. The hepatic ducts join up to form the common bile duct, whose function it is to transport bile to the duodenum. The cystic duct transports bile from the gallbladder (where bile produced by the liver is stored) to the common bile duct.

SEE ALSO *Digestive system on page 101*

Disorders of the bile ducts

Bile ducts can be blocked by gallstones or cancers, or may become inflamed.

Cholangitis

Cholangitis is inflammation of the bile ducts. It is usually caused by an obstruction of the duct that transports bile from the gallbladder to the small intestine. The obstruction is usually caused by gallstones or a tumor in the pancreas. Pain in the upper abdomen is accompanied by a high fever and chills, often with vomiting and jaundice. The urine may be dark and the feces pale. Treatment with antibiotics may cure the condition, but a severe obstruction may require surgical removal of the gallstones. Sclerosing cholangitis is a chronic disorder of the liver in which the bile ducts become inflamed, thickened, scarred (sclerotic) and obstructed, leading to cirrhosis of the liver. It is fatal unless a liver transplant is carried out.

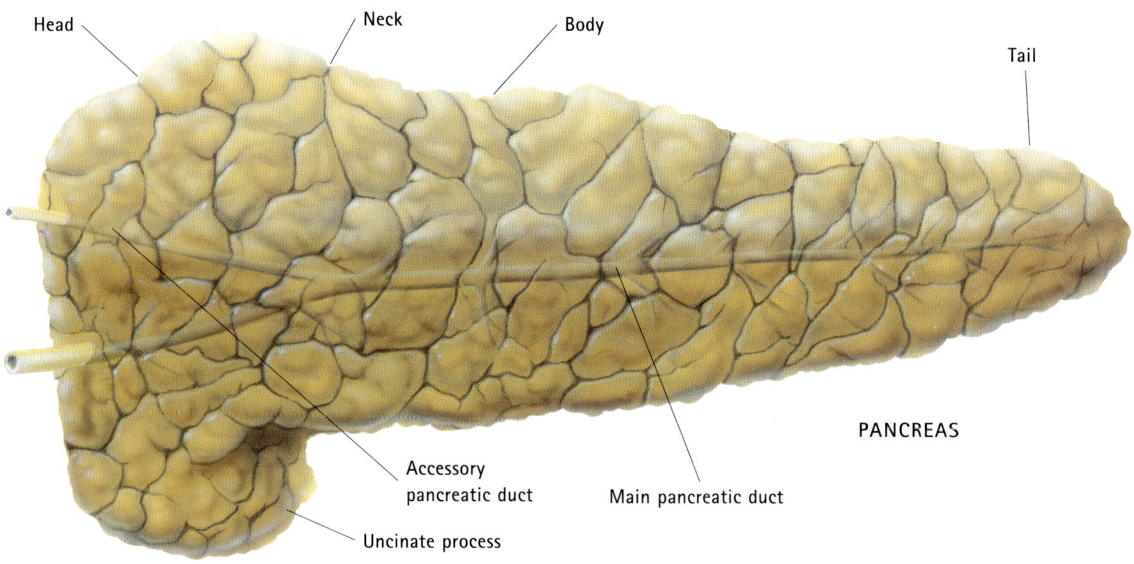

PANCREAS

The pancreas plays two different roles in the body. Most of its cells (the exocrine acinar cells) secrete enzymes into the gut and are part of the digestive system. Other cells (in the islets of Langerhans) produce hormones which enter the bloodstream and are part of the endocrine system.

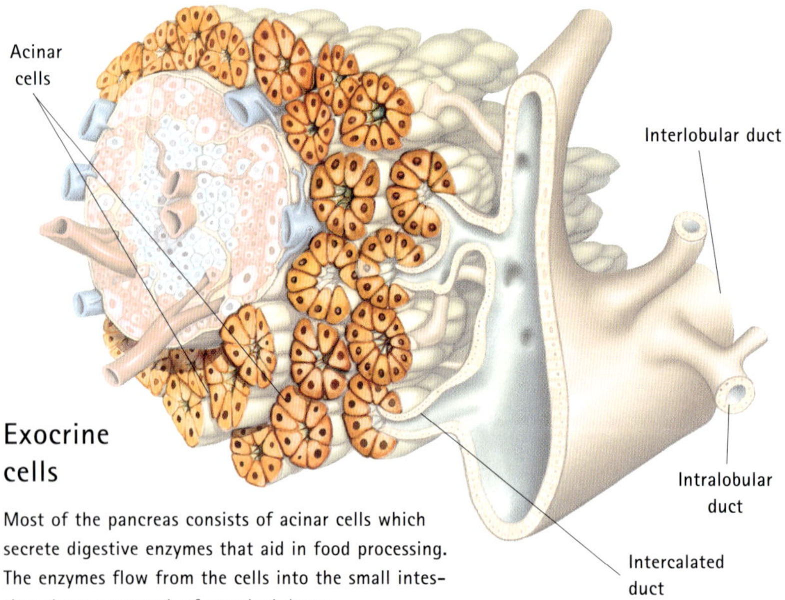

Exocrine cells

Most of the pancreas consists of acinar cells which secrete digestive enzymes that aid in food processing. The enzymes flow from the cells into the small intestine along a network of attached ducts.

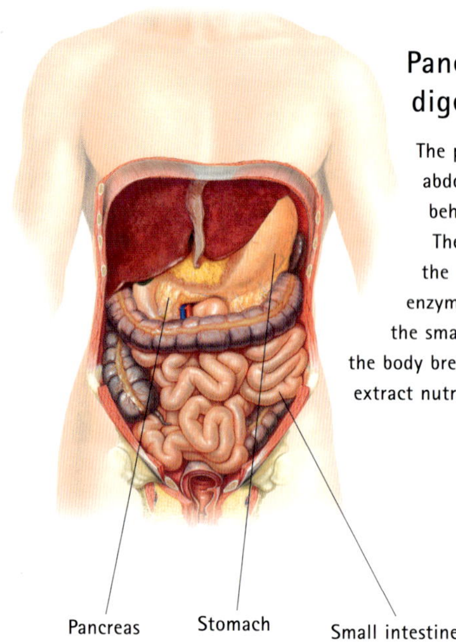

Pancreas—digestive function

The pancreas lies in the abdominal cavity just behind the stomach. The exocrine part of the pancreas secretes enzymes which flow into the small intestine and help the body break down food and extract nutrients.

PANCREAS

The pancreas is unusual in that it is a mixed gland, comprising some cells that secrete enzymes into the gut (exocrine pancreas) and other cells that produce hormones which enter the bloodstream (endocrine pancreas). In other words, the pancreas is part of both the digestive and endocrine systems of the body.

The pancreas lies within the abdominal cavity, behind the stomach and in front of the large artery and vein which pass down the center of the abdomen (the aorta and inferior vena cava respectively).

The pancreas has a head region, which is encircled by the four parts of the duodenum. Leading off to the left from the head region are the neck, body and tail of the pancreas. The pancreatic tail meets the spleen on the left of the abdomen.

The pancreas has a series of ducts which allow digestive enzymes to flow into the interior of the duodenum. The point at which the larger of these ducts enters the duodenum is called the ampulla of Vater.

The exocrine secretions of the pancreas are slightly alkaline to neutralize the acid juices coming into the duodenum from the stomach. The function of enzymes in the pancreatic juices is to help digest proteins, fats and starches in the food.

Under normal circumstances, enzymes of the pancreas are prevented from digesting the pancreas itself by three mechanisms. Firstly, the enzymes are stored within cells of the pancreas in compartments separate from the other cell proteins. Secondly, the enzymes are secreted in an inactive form. Thirdly, there are chemical inhibitors of the enzymes, which are present within the pancreatic ducts and tissue.

The endocrine function of the pancreas is concerned with both storage of foodstuff after meals and their release during fasting. The two pancreatic hormones responsible for these functions are respectively insulin and glucagon, which are produced in special cell types within a number of tiny spherical clumps of pancreatic tissue—these are also known as the pancreatic islets, or the islets of Langerhans.

SEE ALSO *Endocrine system on page 110; Digestive system on page 101; Hormones on page 112; Insulin on page 113*

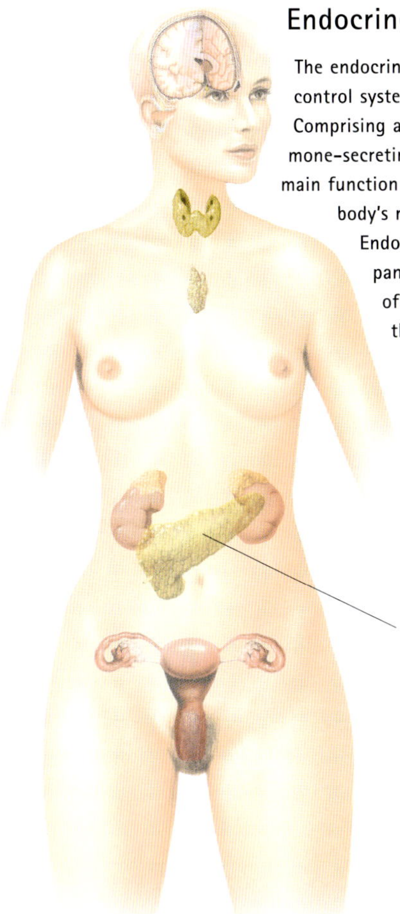

Endocrine function

The endocrine system is a major control system in the body. Comprising a number of hormone-secreting glands, its main function is regulating the body's metabolic activities. Endocrine cells in the pancreas—the islets of Langerhans—produce the hormones insulin and glucagon, which control sugar levels in the body.

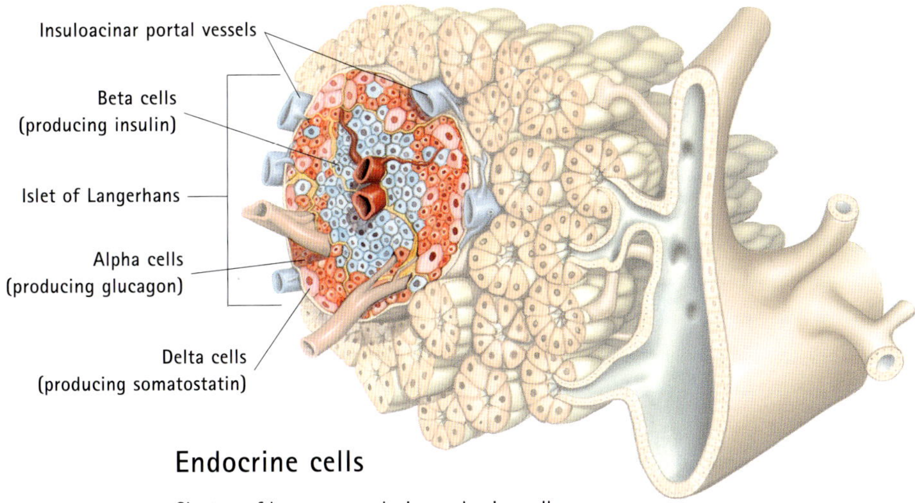

Endocrine cells

Clusters of hormone-producing endocrine cells—the islets of Langerhans—are scattered throughout the pancreas. Alpha cells secrete glucagon, which elevates blood sugar. Beta cells secrete insulin, which affects the metabolism of fats, proteins and carbohydrates. Delta cells secrete somatostatin, which can inhibit the release of both glucagon and insulin.

Islets of Langerhans

The pancreatic islets are a type of endocrine gland, that is, a gland that secretes products directly into the bloodstream rather than onto the surface of the gut or skin. The islets are made up of several different types of cell. The first two types (alpha and beta cells) are involved in the control of blood sugar concentration and are responsive to changes in that concentration. The alpha cell type produces a hormone known as glucagon, which acts to increase blood sugar level. The beta cell type produces the hormone insulin, which has the effect of reducing blood sugar level. A third type, the delta cell, secretes somatostatin, which inhibits the release of glucagon and insulin.

Insulin release is stimulated by rising blood levels of glucose and amino acids. The hormone stimulates the uptake of sugars, protein and nucleic acids by the body's cells. Glucagon is released in response to low blood glucose and amino acid concentration. It then stimulates release of glucose, amino acids and fats from body stores.

Disorders of the Pancreas

The most common disorders of the pancreas are pancreatitis (inflammation of the pancreas, usually caused by alcohol or gallstones) and cancer. Damage to the endocrine part of the pancreas by viruses or unknown agents may be one cause of diabetes mellitus, a condition of inadequate control of blood sugar level. In fact, some patients with chronic pancreatitis may develop diabetes mellitus as a complication.

SEE ALSO *Blood sugar tests on page 437; Diabetes on page 117*

Congenital disorders

Some congenital conditions involve the pancreas. One of these is annular pancreas, in which a ring of pancreatic tissue surrounds the descending duodenum, causing upper gut obstruction in infants and adults. These patients may vomit after meals and x-rays often show a dilated stomach and upper duodenum. Surgery to bypass the blocked segment will correct the problem.

Cancer of the pancreas

Cancer of the pancreas is a very serious disease that is a significant cause of death among men aged between 35 and 60 years. It appears to be more frequent in cigarette smokers and diabetics. It is particularly serious because early spread of the disease to nearby structures, lymph nodes and the liver is common, thus making complete surgical removal of the cancer impossible. Cancer in the head of the pancreas may often obstruct the duct system of the pancreas, causing weight loss and jaundice from the build-up of bile salts.

Other pancreatic cancers located in the tail of the pancreas, away from the pancreatic duct system, will produce weight loss and abdominal pain as the initial symptoms. The prognosis for pancreatic cancer is poor, with most patients dying within a year of diagnosis. Only about 10 percent of patients will survive 5 years, and complete cures are extremely rare.

Tumors of the islet cells

Occasionally tumors may arise from the islet cells of the pancreas. Insulinomas, for example, arise from the beta cells of the islets. They produce insulin in excess amounts, which gives rise to symptoms of low blood-sugar level. Patients show bizarre behavior, memory lapses, palpitations, sweating and unconsciousness. Some may even be mistakenly treated for psychiatric illness. Symptoms are relieved by food, and patients usually gain weight due to overeating. Treatment may be with drugs, which suppress release of insulin from the tumor, or by surgical removal if the tumor can be located.

Very rarely, gastrin-producing tumors may arise in the pancreas or duodenum. Gastrin is a hormone which controls the amount of stomach acid produced, and it is usually released in response to distension of the stomach by food. These gastrin-producing tumors are often cancers of pancreatic islet cells. They cause a condition known as Zollinger-Ellison syndrome.

Pancreatitis

Pancreatitis is a non-bacterial inflammation of the pancreas. It is caused by the digestion of the tissue of the pancreas by its own enzymes. Pancreatitis may be acute, with a sudden onset for a relatively short duration, or chronic, lasting for weeks and up to months with frequent relapses.

Individuals suffering acute pancreatitis experience a sudden onset of pain in the upper abdomen, nausea and vomiting, and have increased concentrations of the digestive enzyme amylase in their blood. Acute pancreatitis may be caused by gallstones (about 40 percent of cases), excessive alcohol intake (a further 40 percent of cases), increased levels of calcium or fats in the blood, surgery or drugs, such as corticosteroids, diuretics and oral contraceptives.

Complications of acute pancreatitis can include the formation in the pancreas of abscesses, or pus-filled cavities, and a pancreatic pseudocyst (a cavity filled with fluid that is rich in digestive enzymes).

Treatment does not usually involve surgery unless complications are experienced. Medical treatment can also include fluid replacement, pain relief, gastric suction and the control of blood calcium levels.

Chronic pancreatitis is often caused by alcoholism. Patients with chronic pancreatitis suffer recurrent bouts of abdominal pain and problems with absorbing food, and frequently develop diabetes mellitus.

Treatment may be medical, as described for acute pancreatitis, or may involve surgery to relieve chronic pain.

Adrenal glands

The 2 adrenal (or suprarenal) glands lie one on top of each kidney at the back of the abdomen. Each adrenal gland is 1–2 inches (3–5 centimeters) long, somewhat triangular in shape and yellowish brown in color. Each gland has 2 parts: an outer region, the adrenal cortex, and a core, the medulla. Both produce and secrete hormones, but differ in structure, function and development.

Cortical hormones

The adrenal cortical (cortex) hormones are produced from cholesterol and are called steroids. They are divided into glucocorticoids such as cortisol, mineralocorticoids such as aldosterone, and androgens, which are similar to sex hormones.

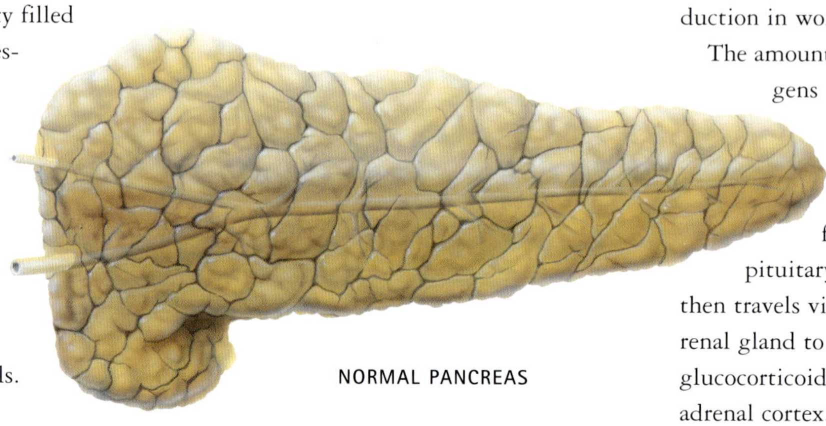

NORMAL PANCREAS

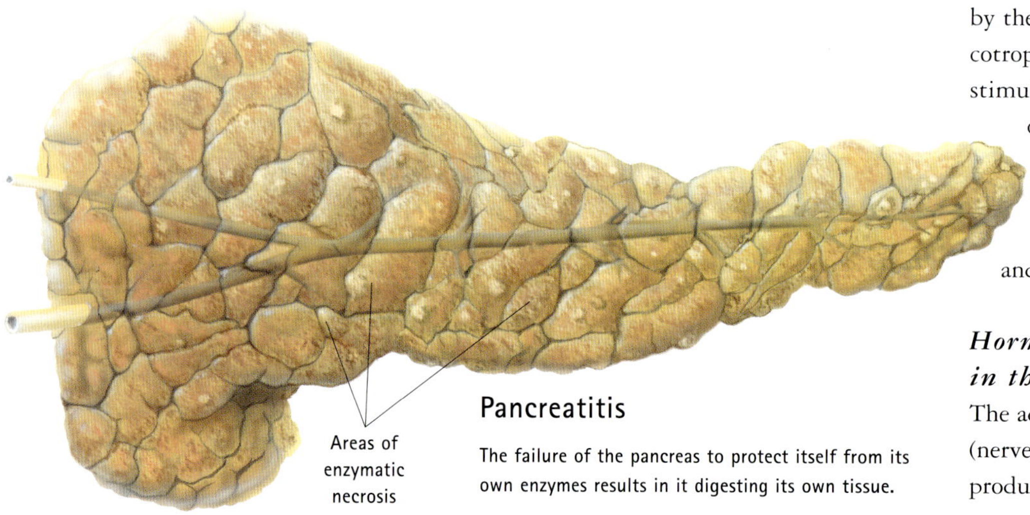

Pancreatitis

The failure of the pancreas to protect itself from its own enzymes results in it digesting its own tissue.

Glucocorticoids are involved in the metabolism of glucose and the response of the body to injury. One of the effects of glucocorticoids is to reduce the body's immune response. This has led to their use in treating tissue rejection after organ transplants, and in reducing allergic responses.

Mineralocorticoids are another class of hormones produced by the adrenal gland. They maintain adequate fluid volume, blood pressure and heart output in the body. A deficiency of mineralocorticoids, such as in Addison's disease, can cause reduced cardiac output and fatal shock.

Aldosterone is the most important of the mineralocorticoids. It increases sodium retention in the kidneys and other tissues such as the sweat and salivary glands; it also causes secretion of potassium. The effect of aldosterone on sodium retention is important in controlling blood volume. If there is a decrease in blood flow or blood volume in the kidney, the kidney releases renin and acts on a second hormone, angiotensin, to increase aldosterone levels. This helps conserve sodium and restore blood volume.

Androgens contribute to the development of male sexual characteristics. Excess production in women causes masculinization.

The amount of glucocorticoids and androgens secreted by the adrenal glands is controlled by adrenocorticotropic hormone (ACTH), which is secreted by the front (anterior) lobe of the pituitary (the hypophysis). ACTH then travels via the bloodstream to the adrenal gland to stimulate the release of the glucocorticoids and sex steroids from the adrenal cortex. ACTH secretion is regulated by the hypothalamus, which secretes corticotrophin-releasing hormone (CRH), which stimulates ACTH secretion. Cortisol acts to control both CRH and ACTH, by a mechanism known as a negative feedback loop, in which an increase in cortisol leads to a decrease in CRH and ACTH and vice versa.

Hormones produced in the adrenal medulla

The adrenal medulla is derived from neural (nerve) tissue and is concerned with the production and secretion of epinephrine (adrenaline) and norepinephrine (noradrenaline). These hormones can cause increased heart rate, widening of airways and breakdown of glycogen to glucose for energy. All of these enable the body to be better equipped to handle emergency situations.

The hormone epinephrine (adrenaline) is released by the central part (medulla) of the adrenal glands in response to stress. Epinephrine increases heart rate, blood pressure, and flow of blood to the muscles. It causes the liver to release glucose into the blood. These changes enable the body to respond to stress and danger. Epinephrine can be produced chemically and is used as a drug to treat shock, allergy attacks, anaphylaxis and asthma. It is also used in surgery to decrease bleeding or prolong the effect of local anesthetics, and can be given as a heart stimulant during cardiac arrest.

SEE ALSO *Electrolytes on page 114; Endocrine system on page 101; Hormones on page 112*

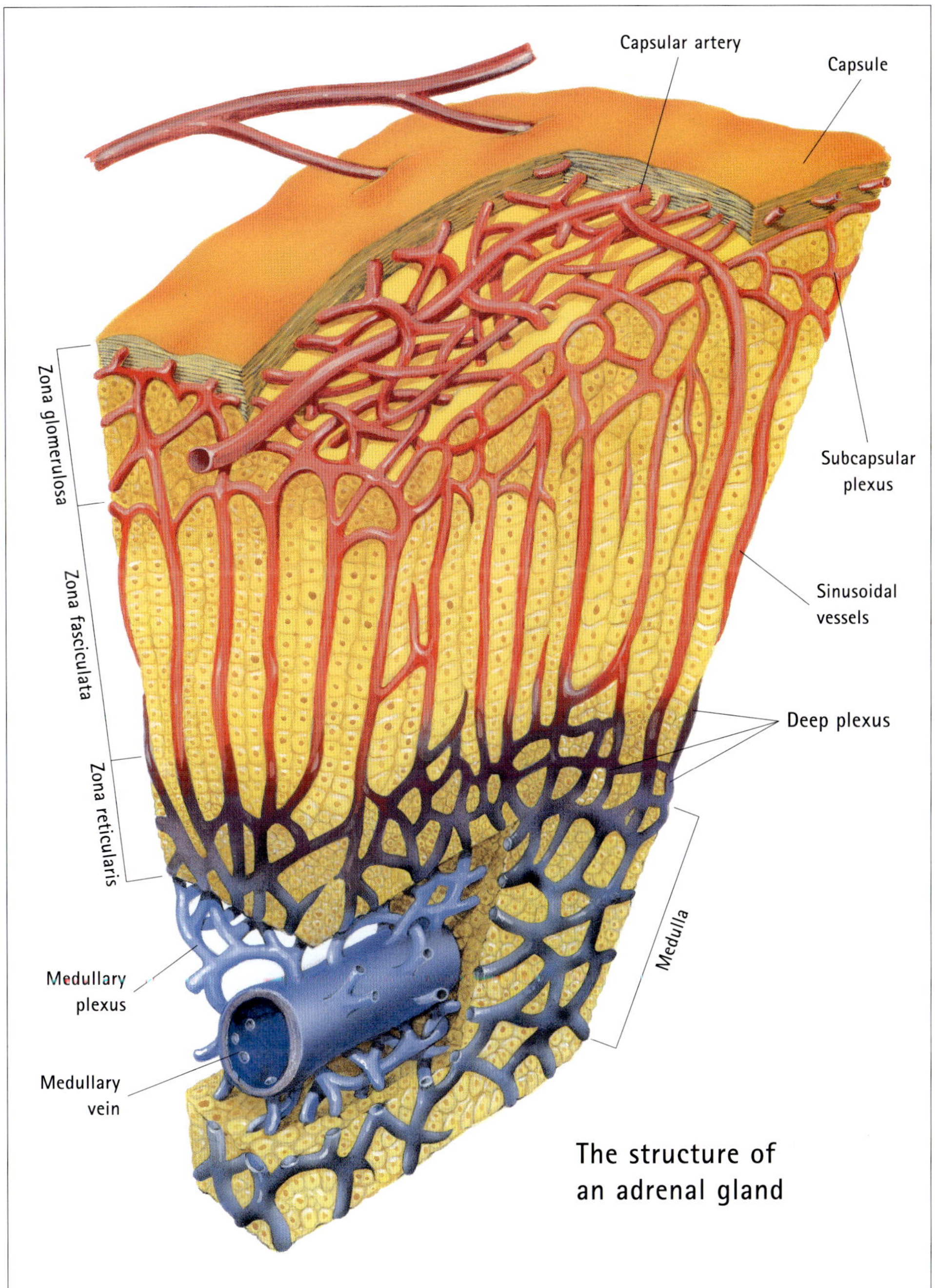

The structure of an adrenal gland

ADRENAL GLANDS

These tiny glands are just as essential to our health as the much larger organs all around them. They secrete steroids, which deal with glucose, help the body respond to injuries, keep the volume of blood at the right level, and, in males, contribute to the development of sexual characteristics.

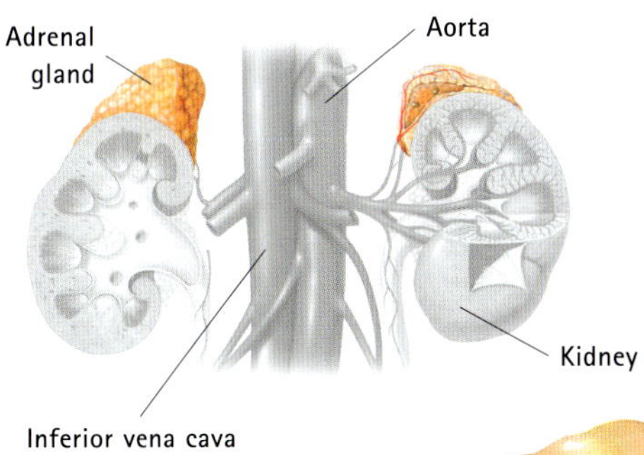

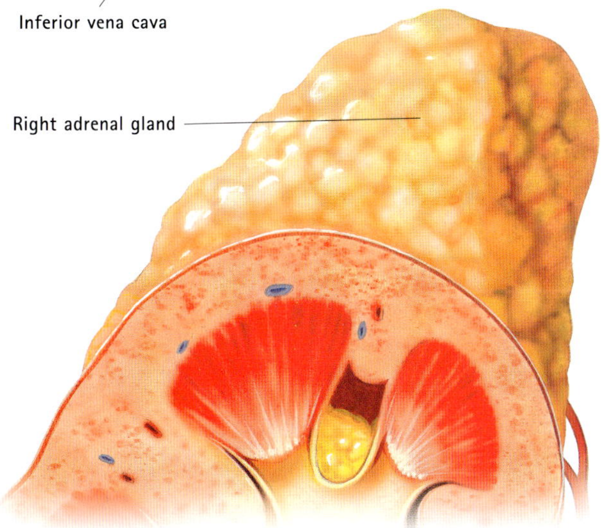

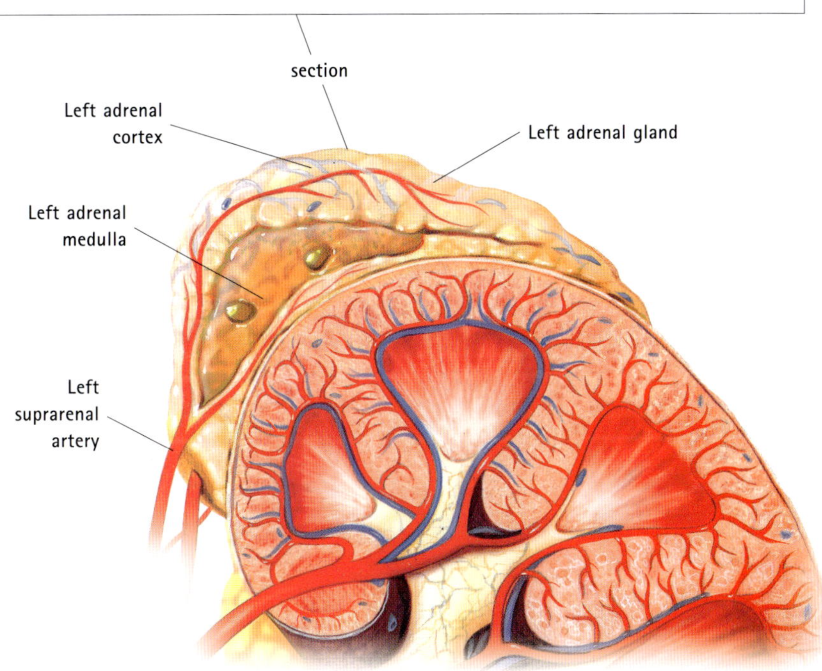

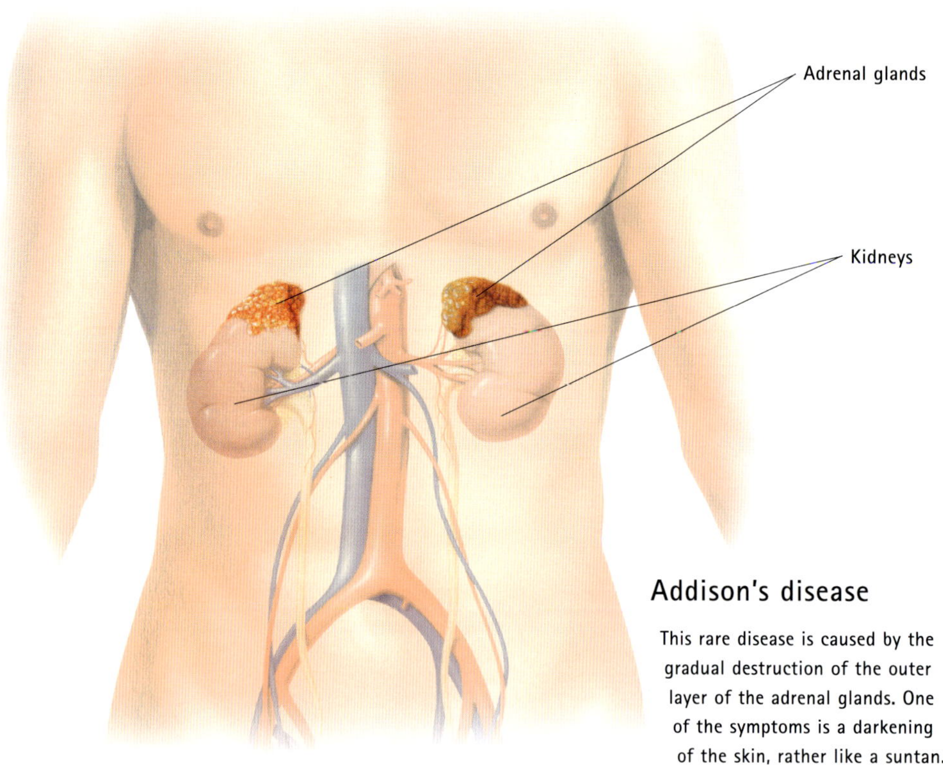

Addison's disease

This rare disease is caused by the gradual destruction of the outer layer of the adrenal glands. One of the symptoms is a darkening of the skin, rather like a suntan.

Neuroblastoma

Neuroblastoma is a cancer that affects fetal nerve cells, causing a tumor in early childhood. It often begins in the adrenal gland in the abdomen and may spread quickly to parts of the body such as the bone marrow and lymph nodes. Treatment is surgical removal of the tumor, radiation therapy, chemotherapy and bone marrow transplant.

Cushing's syndrome

Cushing's syndrome occurs when there is an excess, or prolonged use, of cortisol. Symptoms of Cushing's syndrome include weight gain, weakness of the muscles, high blood pressure, a "moon" face and facial hair.

Diagnosis is usually based on symptoms, with blood and urine tests performed to confirm increased cortisol levels and magnetic resonance imaging (MRI) or CAT scan to reveal any tumors.

Adrenal disorders

If the levels of the hormones produced by the adrenal glands fall below or increase above normal, various disorders will result.

SEE ALSO *Imaging techniques on page 431*

Adrenal hyperplasia

Adrenal hyperplasia is an enlargement of the adrenal gland and affects the outer region, the cortex. The condition can be caused by overproduction of ACTH from the pituitary, often occurring with pituitary tumors. Elevated levels of ACTH lead to excess stimulation of the adrenals, resulting in enlargement (hyperplasia) and excess production of cortisol and androgens. Adrenal hyperplasia may also be caused by a defect in cortisol production.

High levels of ACTH lead to masculinization in women, including deepening of the voice, excess facial hair, and absence of menstrual periods.

If the cortisol defect is present in a fetus, the baby can be born with congenital adrenal hyperplasia (this condition is known as adrenogenital syndrome).

Addison's disease

This is a rare disease that most commonly affects people between 30 and 50 years of age. It is caused by the gradual destruction of certain portions of the adrenal glands.

In Addison's disease, it is the outer layer, also known as the adrenal cortex, which is destroyed so that the adrenal glands are unable to produce enough hydrocortisone for the body's needs. The most common cause of the destruction is an autoimmune process in which the body destroys its own cells. Diagnosis follows a blood test showing low hydrocortisone levels.

The early symptoms of Addison's disease are loss of appetite and weight, and a feeling of increased tiredness and weakness. There may also be abdominal symptoms, such as diarrhea or constipation, and mild indigestion with nausea and vomiting. In most cases, the skin becomes darker. If left untreated, the condition leads to acute adrenal failure and coma, requiring emergency treatment in a hospital.

The treatment for Addison's disease is replacement hydrocortisone hormones in the form of tablets, which will need to be taken daily and continued for life. Affected people who take their tablets regularly can lead a normal healthy life.

Cushing's syndrome

Symptoms produced by an excessive production of cortisol include fatty swellings of the face and trunk, and general weakness. It is usually a result of a tumor in the pituitary gland, which overstimulates the adrenal glands. The tumor may have to be removed surgically to cure the condition.

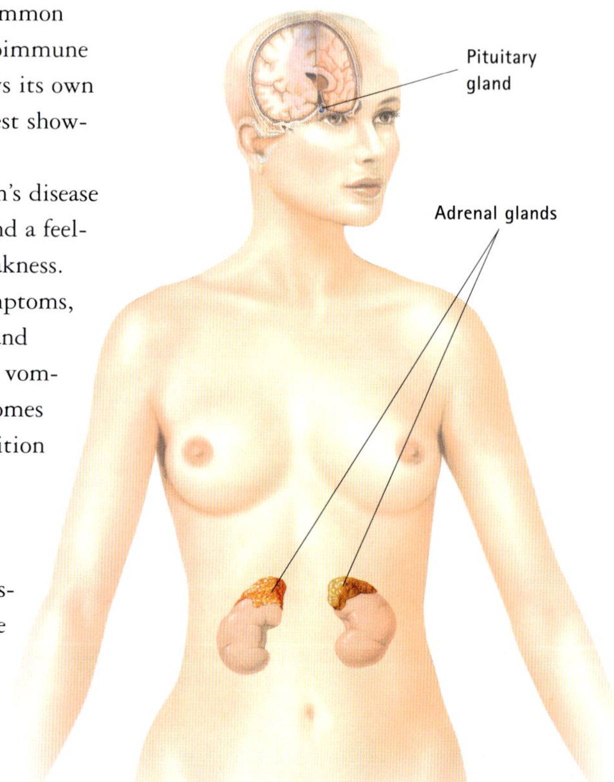

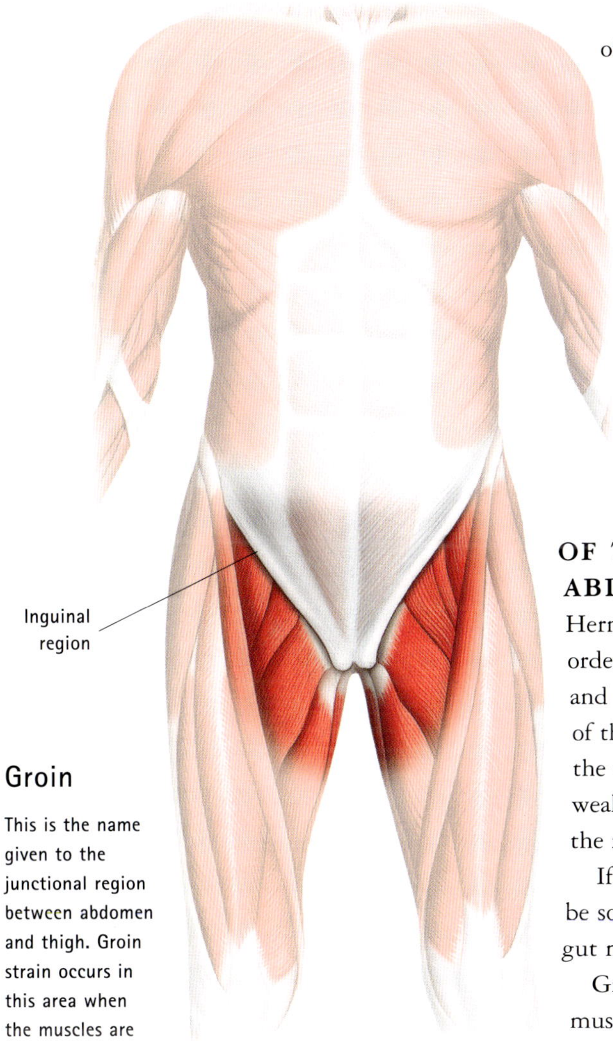

Groin

This is the name given to the junctional region between abdomen and thigh. Groin strain occurs in this area when the muscles are overextended.

Pheochromocytoma

Pheochromocytoma is a tumor involving the medulla of the adrenal gland or associated tissues. It causes excess production of epinephrine (adrenaline) and norepinephrine (noradrenaline). These tumors are most common in young and mid-adult life, and are generally not malignant. Symptoms include severe headache, rapid heart rate and palpitations, sweating, abdominal pain, nervousness, irritability, increased appetite and loss of weight. Patients may have high blood pressure, and diagnosis is based on raised hormone levels in the blood or urine. Treatment is by medication to block the hormones, or maybe removal of the tumor.

Groin and abdominal wall

The groin (inguinal region) is where the abdomen joins the front of the thigh. The abdominal wall is made up of three layers of muscle covered by skin and fat. The external oblique muscle forms the outer layer, the internal oblique muscle forms a middle layer, while the innermost layer of the abdominal wall is formed by the transverse abdominis muscle.

SEE ALSO *Lymphatic/ Immune system on page 55; Muscular system on page 48*

Disorders of the groin and abdominal wall

Hernias are the most common disorders affecting the abdominal wall and groin. Hernias occur when part of the gut is forced into the front of the thigh under pressure through a weakness in the abdominal wall in the region of the inguinal canal.

If painful, medical advice should be sought immediately, because the gut may become damaged.

Groin strain generally occurs when muscles in this region are pulled or strained. The groin also contains lymph nodes (also known as lymph glands), and these can swell up if inflamed, for instance with infections in the leg. Tinea cruris (jock itch) is a common fungal infection that occurs in the groin region, and is more common in men.

SEE ALSO *Ultrasound on page 432*

Hernia

A hernia is a protrusion of tissue or an organ through an abnormal opening. A hernia may be acquired or congenital, and may occur in various parts of the body, though most hernias involve the abdomen. Hernias occur most often when pressure in the abdomen (for example, during coughing or lifting a heavy weight) forces the soft inner abdominal tissue through a weakness in the muscles of the abdominal wall.

The most common types of hernia are inguinal and femoral hernia. Both involve protrusion of the small intestine from the abdomen into the groin. Inguinal hernias are more common than femoral, and more common in men than women, as the hernia can enter the inguinal canal more easily. Femoral hernias, on the other hand, occur below the inguinal canal at the top of the thigh.

An incisional hernia occurs when tissue has protruded through the site of a previous surgical operation. An umbilical hernia may be seen in newborn infants and involves protrusion of tissues through the navel.

Some hernias are reducible—they can be pushed back into the abdomen. A hernia which cannot be pushed back is called irreducible; such a hernia may become pinched ("strangulated"). If the supply of blood is cut off, the hernia may become gangrenous and cause death. If there is a section of intestine caught in the hernia, intestinal obstruction, infection and gangrene may follow. This requires emergency surgery. Non-urgent treatment of a hernia depends on where it is located. In an infant, an umbilical hernia will usually disappear by itself by the age of about four; if not, it can be surgically corrected. An inguinal, femoral or incisional hernia is treated with an operation called a hernia repair, in which a surgeon will close the weakness in the abdominal wall under general or local anesthesia.

Inguinal hernia

When a hernia occurs, the intestine protrudes through a weakened muscle in the abdominal wall, causing a lump under the skin of the groin.

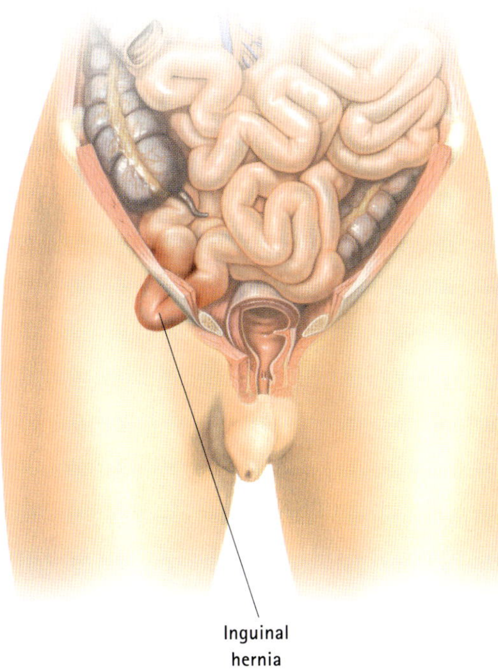

Hernias appearing in the midline of the abdomen usually do not require treatment. If a person with an inguinal hernia is too old or too unwell for an operation, wearing a truss will sometimes help.

Spleen

The spleen is an organ about the size of a fist, which is situated high in the left side of the abdomen, beneath the diaphragm. It is an important blood-forming organ during fetal life but is not essential to life in the adult. The spleen receives a disproportionately large blood supply for its size, which it filters through channels called sinuses. Red blood cells squeeze through narrow pores in the sinuses and older, more rigid cells are destroyed there. The part of the spleen that forms blood cells in fetal life and filters the blood through the sinuses is called the red pulp. The other portion, called the white pulp, functions as part of the body's immune system.

The spleen is part of the immune system and contains large numbers of lymphocytes. It assists the body in fighting certain infections, especially pneumonia and meningitis. If the spleen is removed (splenectomy), vaccinations are given to boost the body's immunity against such infections. Even so, a small risk remains of severe infection and people who have undergone splenectomy are advised to carry a medical alert card or necklace to alert medical staff of the fact in the event of a sudden febrile illness.

The spleen is a soft organ and is easily ruptured by injuries to the upper abdomen, which often occur in car collisions or during contact sports, or with certain infections, such as malaria. If the spleen is enlarged, for example by glandular fever, it is even more susceptible to rupture. Rupture results in severe pain and internal bleeding. This is a medical emergency and immediate medical attention should be sought.

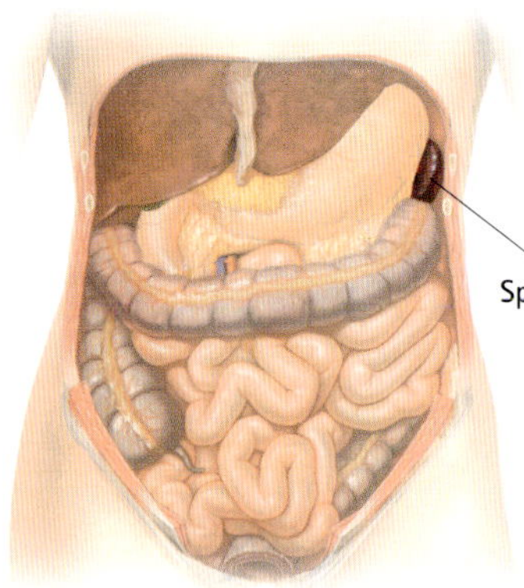

SPLEEN

The spleen is designed to filter blood and also functions as part of the body's immune system. Its red color and pulpy texture are due to its high blood content. The organs that surround the spleen (the stomach, colon and kidney) leave impressions on its soft surface. An enlarged spleen may indicate a disease or disorder elsewhere in the body.

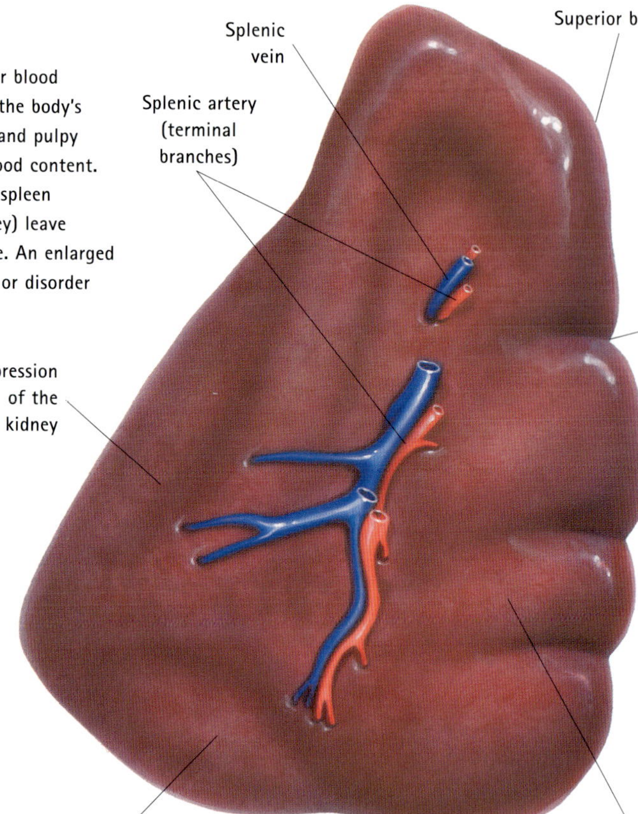

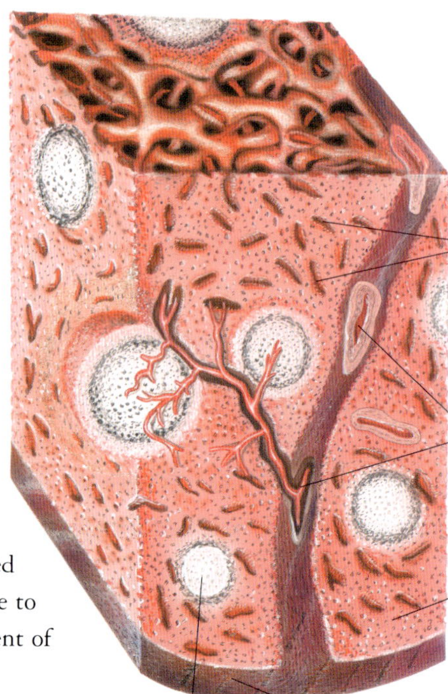

Spleen—microstructure

Red blood cells are filtered through channels, called sinuses, in the spleen which remove old and abnormal cells. The capillaries in the spleen are surrounded by lymphatic tissue.

Spleen—immune function

The spleen is the largest unit of lymphatic tissue in the body. It plays a vital role in the body's immune system by producing and storing lymphocytes, a type of white blood cell. These cells attack invading bacteria and viruses and make antibodies against them.

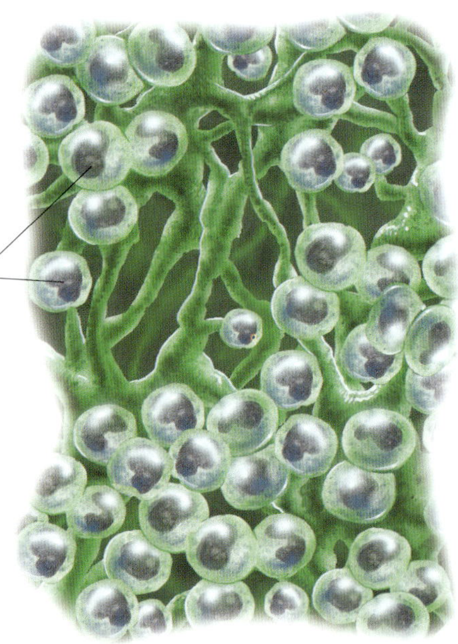

Splenectomy may need to be performed as a life-saving procedure. Sometimes it is possible for the surgeon to repair the damage and preserve part of the spleen, so that it can continue to function.

SEE ALSO *Lymphatic/Immune system on page 55*

Enlargement of the spleen

There are many problems that can affect the spleen, causing enlargement.

The spleen can enlarge due to a build-up of pressure in the splenic vein—the result of portal hypertension complicating cirrhosis of the liver. Other causes of splenomegaly include autoimmune diseases such as rheumatoid arthritis and systemic lupus erythematosus (SLE), and storage diseases such as amyloidosis and Gaucher's disease.

Whenever the spleen is markedly enlarged, a reduction in circulating red blood cell counts may occur because of the destruction or pooling of blood cells in the spleen. This condition is called hypersplenism and can result in conditions such as anemia, leukopenia and thrombocytopenia.

If the spleen is enlarged because of a blood disorder it may cause pain, or hypersplenism needing treatment.

In some cases the spleen can be shrunk using medication or chemotherapy. Sometimes a small dose of radiation therapy can be directed over the spleen to relieve the patient's symptoms. Otherwise splenectomy may be recommended as treatment.

Hemolysis

In conditions in which there is increased destruction of red blood cells, called hemolysis, the spleen becomes enlarged (a condition known as splenomegaly) and can be felt below the rib cage.

Hemolysis can result from a variety of disorders—some inherited, some developing during adult life. The treatment for these conditions may include splenectomy.

The most common cause of hemolysis is malaria, in which there is an infection of the red blood cells by parasites transmitted by *Anopheles* mosquitoes. This infection results in high fevers, shaking, chills, anemia and jaundice. The damaged red cells are removed by the spleen, which becomes enlarged and painful.

In places where malaria is endemic, the spleen may eventually become very large in people exposed to repeated infections.

Lymphomas

Due to its link with the lymphatic system, the spleen may be affected by the same sorts of diseases that affect the lymph nodes. For example, lymphomas (or tumors of the lymph nodes) may cause enlargement of the spleen. Generally this does not cause pain but is detected during abdominal examination. The physician may then confirm that the spleen is enlarged by an ultrasound study, isotope scan or CAT scan.

The spleen may be affected by Hodgkin's or non-Hodgkin's lymphomas. Splenectomy is only usually performed when the diagnosis of non-Hodgkin's lymphomas indicates that the spleen alone is affected.

Leukemia

Leukemias can also cause enlargement of the spleen, which may be massive in the case of chronic myeloid leukemia. Other disorders of the blood cells that cause enlargement need to be distinguished from leukemias. In particular, a myeloproliferative disease—myelofibrosis (also known as agnogenic myeloid metaplasia), can also cause massive enlargement of the spleen.

In this disease, while the adult spleen regains the capacity for blood cell formation, its massive enlargement results in a net decrease in circulating red blood cells.

SEE ALSO *Blood count on page 437; Ultrasound on page 432*

Solar plexus

The solar plexus is a dense network of nerve cells on the abdominal aorta behind the stomach. The solar plexus is known anatomically as the celiac plexus due to the fact that it is situated around the celiac artery just below the diaphragm.

The solar plexus is part of the autonomic nervous system and nerve fibers branch out from it to all the abdominal viscera. Through these, the solar plexus helps to regulate vital bodily functions such as intestinal contraction and adrenal secretion, as well as controlling the kidneys, spleen, liver and pancreas. The region of the solar plexus is often referred to as the "pit of the stomach." A blow to the solar plexus can cause severe pain and difficulty in breathing. This is because there is temporary paralysis of the diaphragm and of the chest and abdominal muscles.

A celiac plexus block (known also as a neurolytic celiac plexus block, or NCPB) is a form of long-term pain relief used occasionally in the palliative care of patients terminally ill with certain forms of abdominal cancer, particularly those affecting the pancreas. The block involves a procedure during which a chemical is injected into the solar plexus, leaving it paralyzed so that the transmission of pain signals from the abdomen to the brain is impeded. This treatment is usually used in conjunction with other methods of alleviating pain and discomfort in the patient.

SEE ALSO *Autonomic nervous system on page 75*

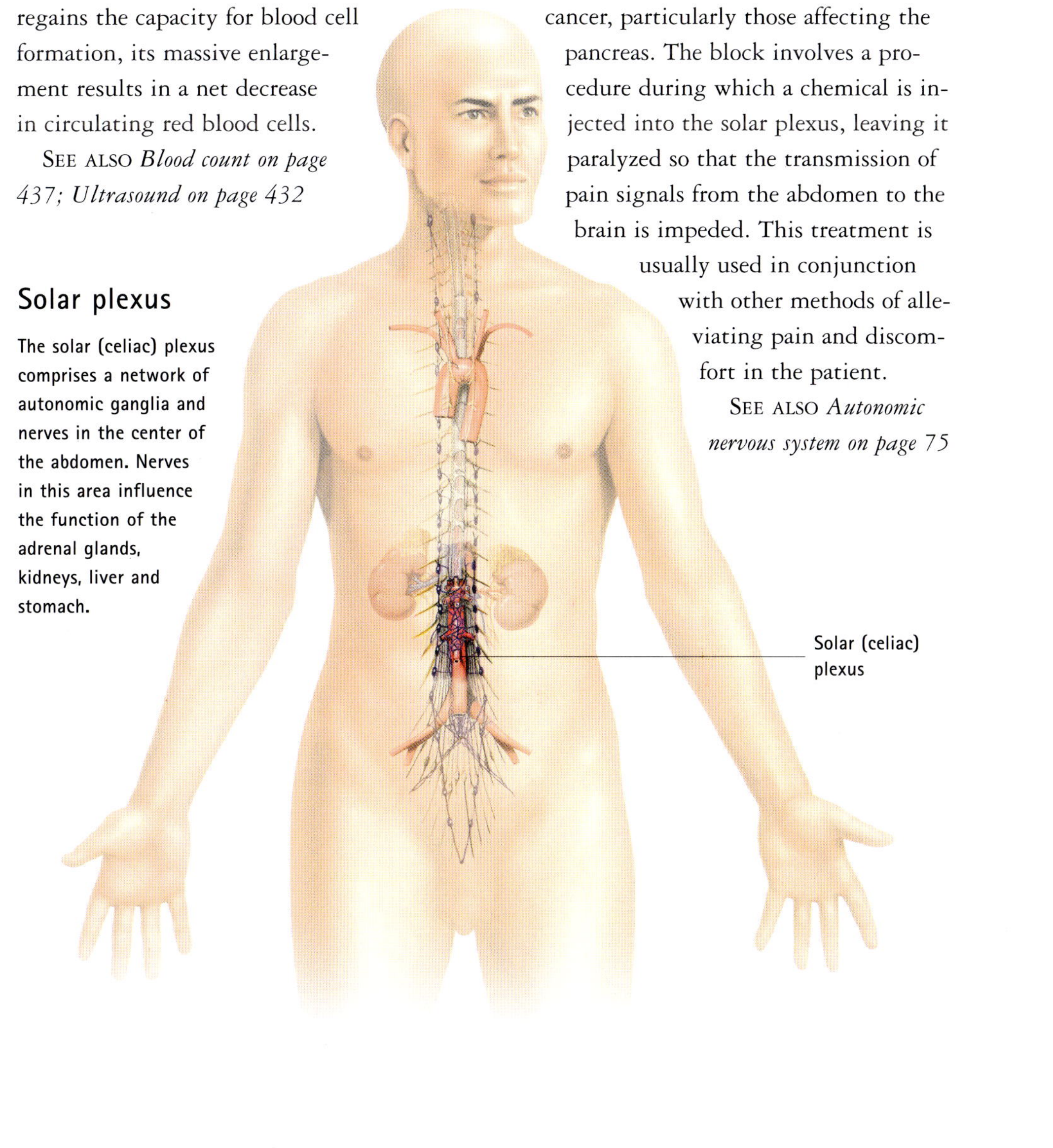

Solar plexus

The solar (celiac) plexus comprises a network of autonomic ganglia and nerves in the center of the abdomen. Nerves in this area influence the function of the adrenal glands, kidneys, liver and stomach.

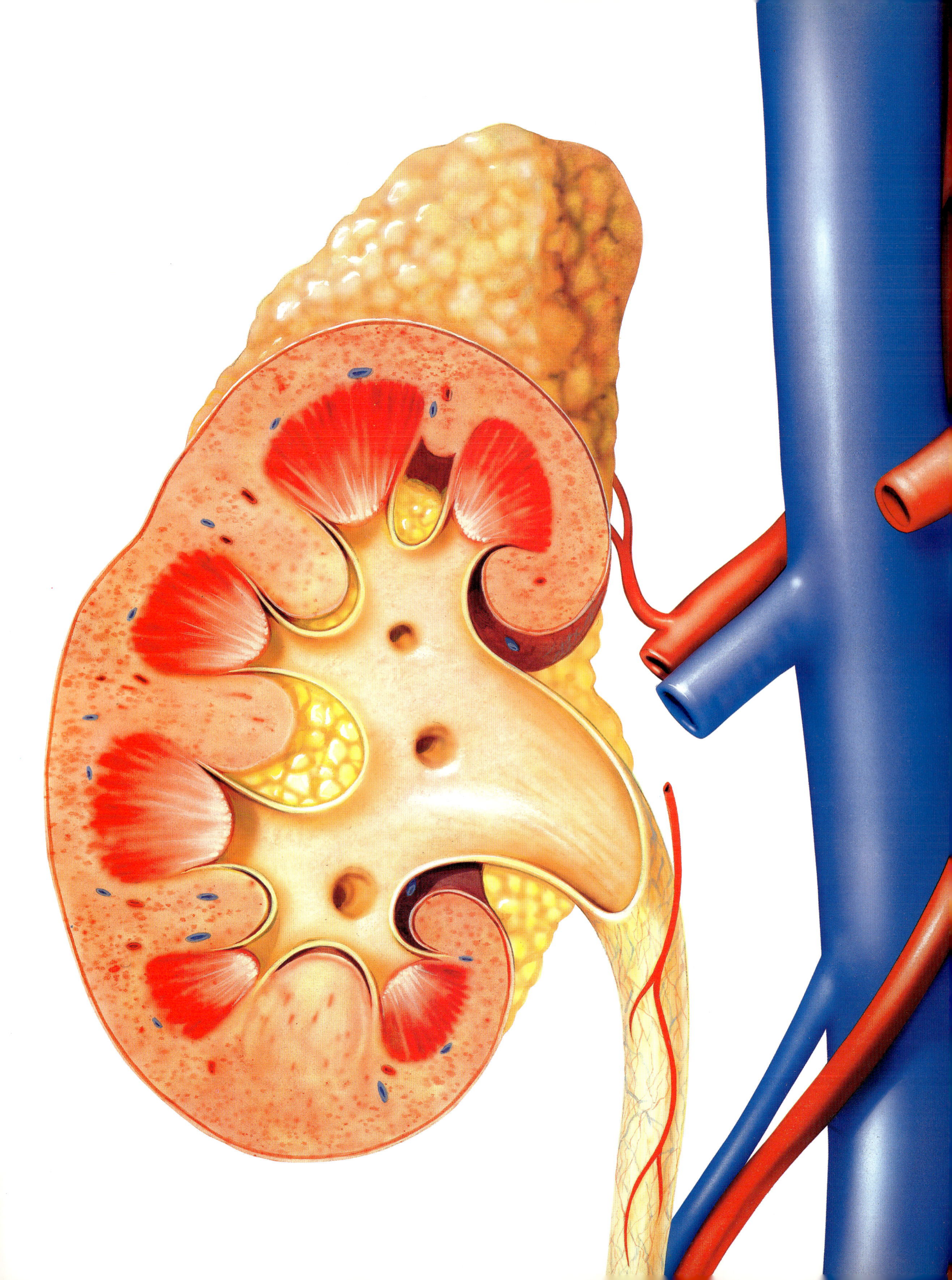

Urinary and Reproductive Organs

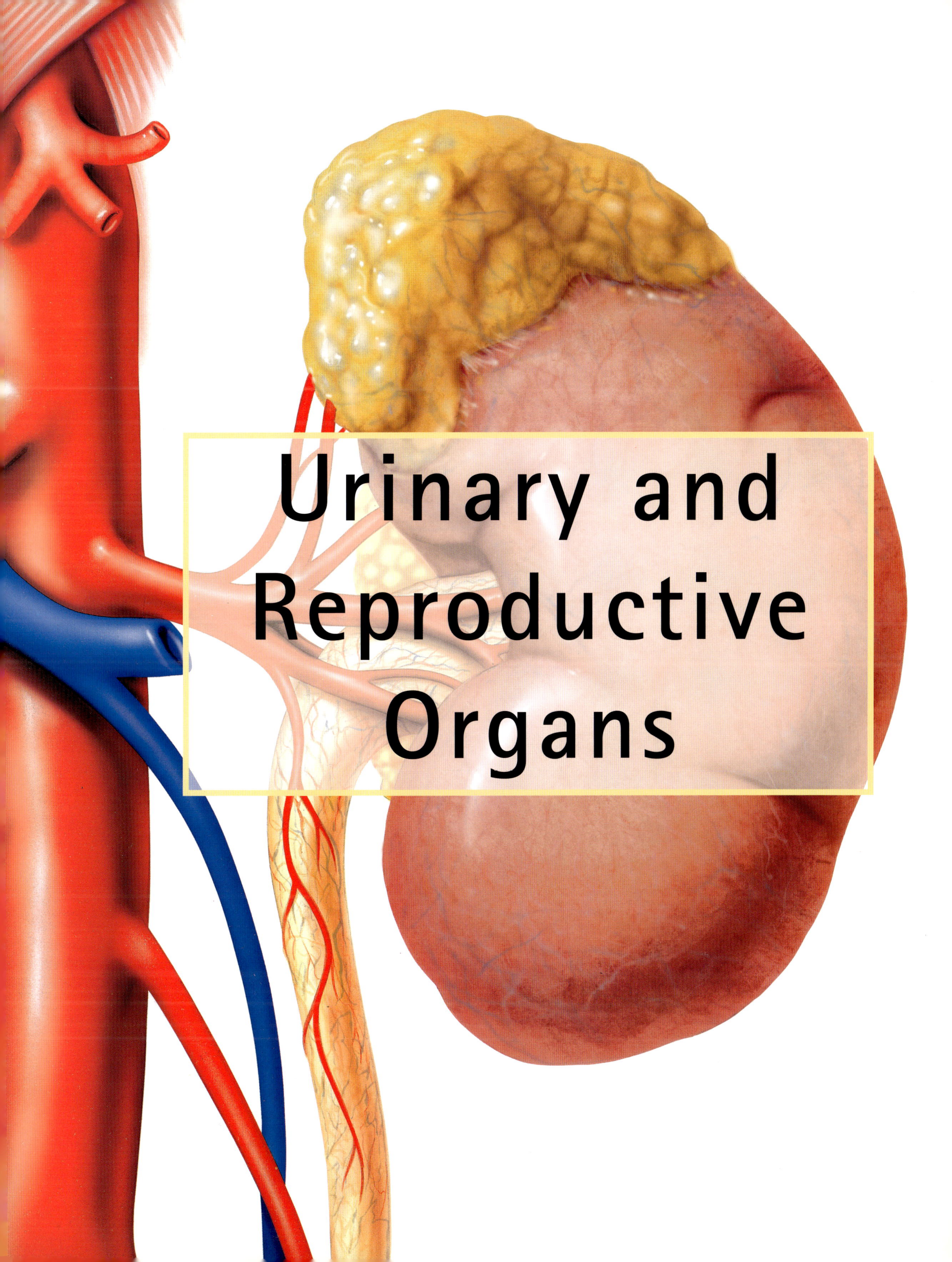

Urinary tract

The urinary tract is a specialized filtering and recycling system that excretes certain fluid wastes produced by the body in the process of metabolism, such as urea, creatinine and ammonia. These wastes, along with excess water, are removed from the blood as urine by the two kidneys. Urine then passes through tubes called ureters into the bladder. From the bladder the urine passes out of the body through another tube called the urethra.

The urinary tract is subject to various disorders, such as bacterial infection. Urinary tract blockage occurs when there is a blockage in the urethra, bladder or ureters.

SEE ALSO *Urinary system on page 98*

Kidney

The kidneys are a pair of bean-shaped, red-brown organs whose function is to dispose of the waste matter produced by the normal functioning of the body, and to keep the salts and water of the body in correct balance. They do this by filtering excess water and chemicals from the blood and excreting them as waste in the form of urine.

The kidneys are located at the back of the abdomen, one on each side of the spine, at the level of the lowest ribs. Because of the position of the liver, the right kidney in most people is located slightly lower than the left. Each kidney is about 4 inches (10 centimeters) long and 1 inch (2.5 centimeters) thick and weighs approximately 5 ounces (140 grams). Each has an outer layer (the cortex), an inner layer (the medulla), and a pelvis, a hollow inner structure that joins with the ureters, the tubes that conduct urine to the bladder.

At the center on one side of each kidney is an indentation known as the renal hilus, the exit point for the ureter and the location where nerves, blood and lymphatic vessels enter and exit. Enclosing each kidney is a protective membrane, the renal capsule. Surrounding each capsule is a cushion of fatty tissue and a layer of connective tissue which attaches the kidneys to the back wall of the abdomen. An adrenal gland sits on top of each kidney.

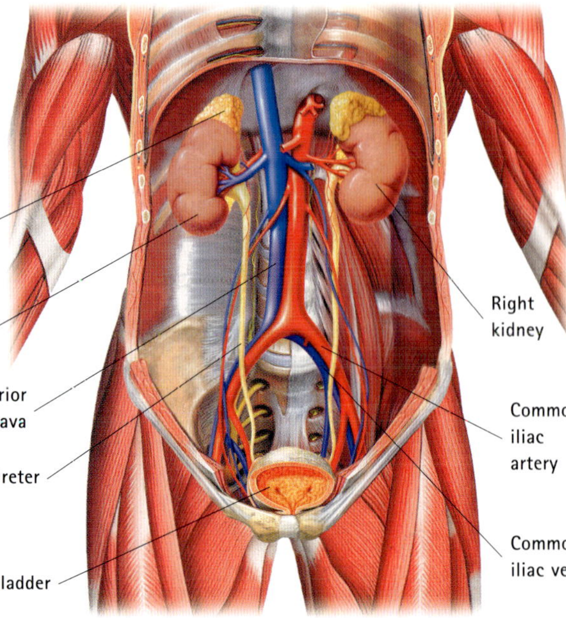

Urinary tract

The two kidneys filter the blood and send the waste products (as urine) to the bladder via the ureters. This waste is then expelled via the urethra.

The renal medulla contains between 8 and 18 renal pyramids, triangular shaped as their name implies and with a striped appearance. The pyramids are positioned with their tips, the renal papillae, facing

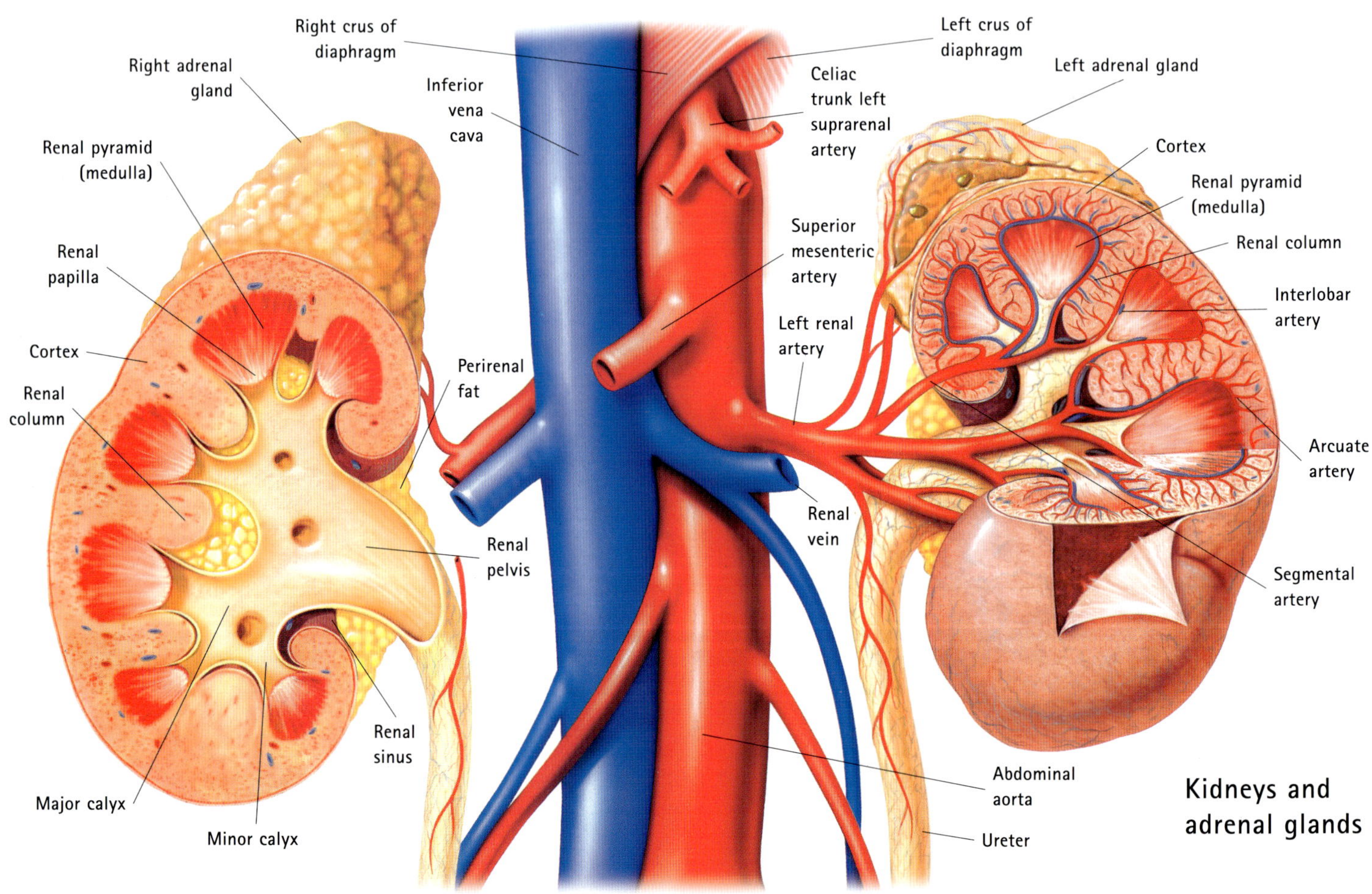

Kidneys and adrenal glands

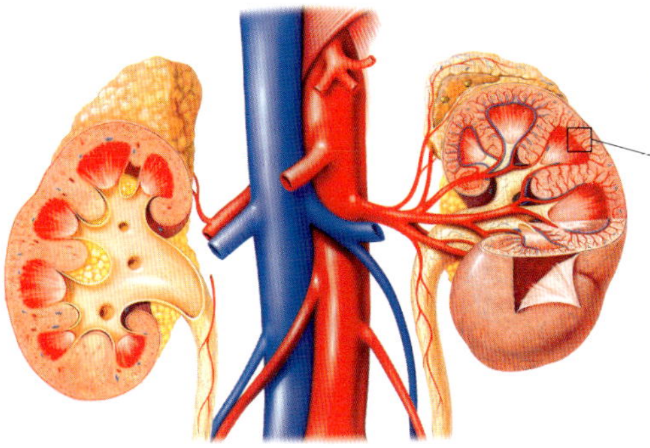

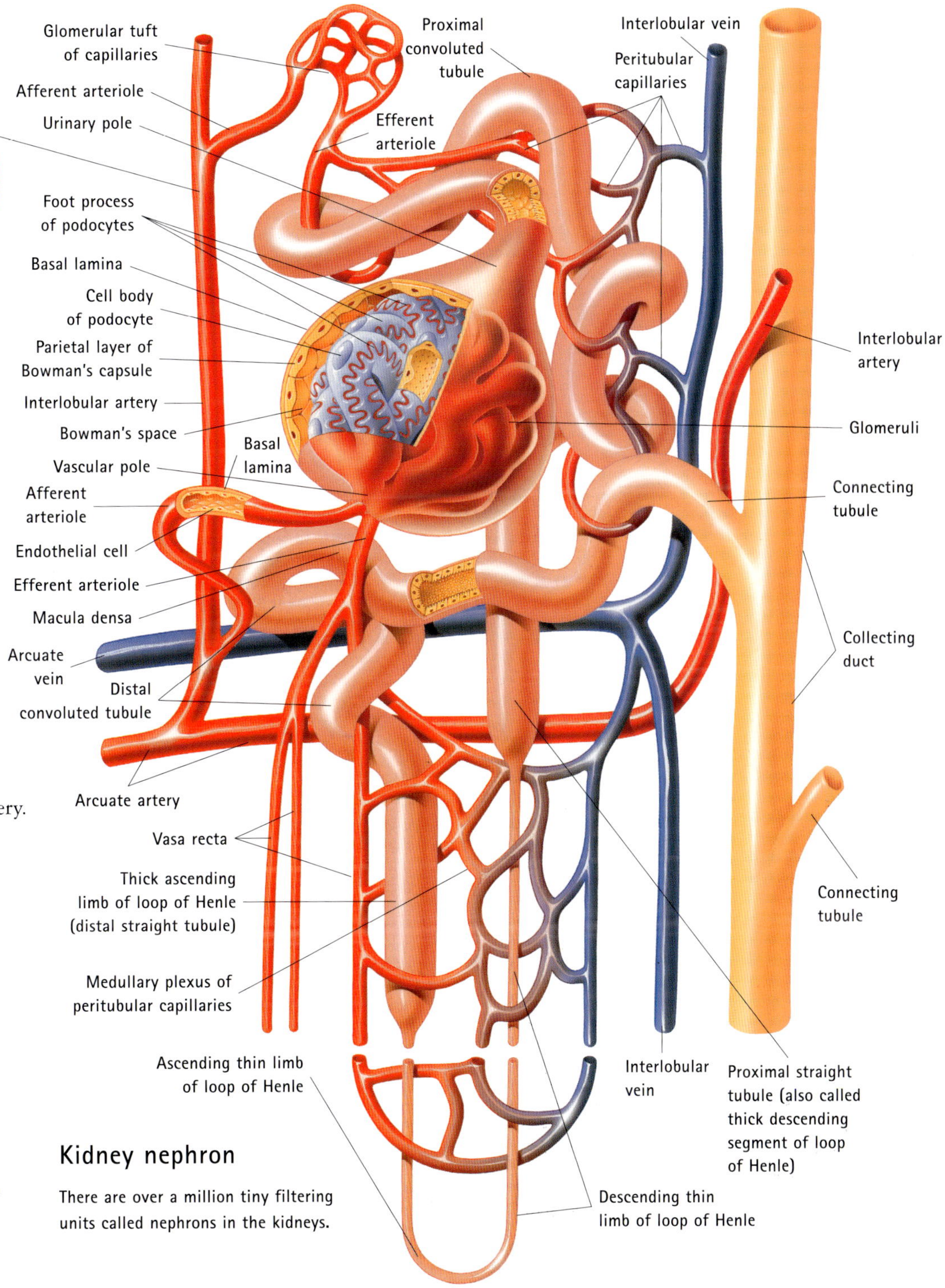

Kidney nephron

There are over a million tiny filtering units called nephrons in the kidneys.

toward the renal hilus and their bases aligned with the edge of the renal cortex. The cortex continues in between each pyramid creating areas known as renal columns.

The renal arteries are two large blood vessels that branch off either side of the abdominal aorta to supply the two kidneys. They pass through the hilum of the kidney, where each artery gives off small branches to the adrenal gland and ureter and then divides into two large branches, which are called the anterior and posterior divisions of the artery.

Each branch divides into smaller and smaller branches, eventually forming the capillaries which supply oxygen to the kidney tissue and take part in kidney filtration via their role in the nephrons.

SEE ALSO *Urinary system on page 98*

Nephron

The functional units of the kidneys are microscopic structures called nephrons, of which there are estimated to be about 1.2 million in each kidney. Each nephron has a renal corpuscle, which lies in the renal cortex, and a renal tubule which runs through a renal pyramid. The renal corpuscle is comprised of an extensive ball-shaped capillary network called the glomerulus surrounded by a double-walled cup of epithelial tissue also known as the glomerular or Bowman's capsule. Together, all these structures filter the blood, producing a liquid (the filtrate) that contains minerals, wastes and water.

The purified blood is returned to the body while the filtrate passes into the renal tubule, which comprises the proximal convoluted tubule, the descending limb of the loop of Henle, the ascending limb of the loop of Henle and the distal convoluted tubule. As the filtrate passes along the renal tubule, a network of tiny blood vessels called the peritubular capillaries reabsorbs useful substances from it and secretes additional wastes into it. About 99 percent of the filtrate is reabsorbed in this way and returned to the general circulation. The rest—1 percent, or about 1–1½ quarts or 1–1¼ liters a day—collects in the pelvis and is transported to the bladder as urine.

If the body needs to conserve water (or needs to dilute salt in the blood), the kidneys return more water to the capillaries. If the body has more water than it needs, more is excreted in the urine. In this way, the precise balance of salts and water in the body is maintained. Toxins, such as urea, are not reabsorbed but are excreted in the urine and in this way the body rids itself of the unwanted products of metabolism.

Disorders of the Kidney

Kidney disease may be caused by many factors, such as injury, infection, cancer, or be part of a more generalized disease affecting other parts of the body. In some cases, there may be no obvious cause of kidney disease.

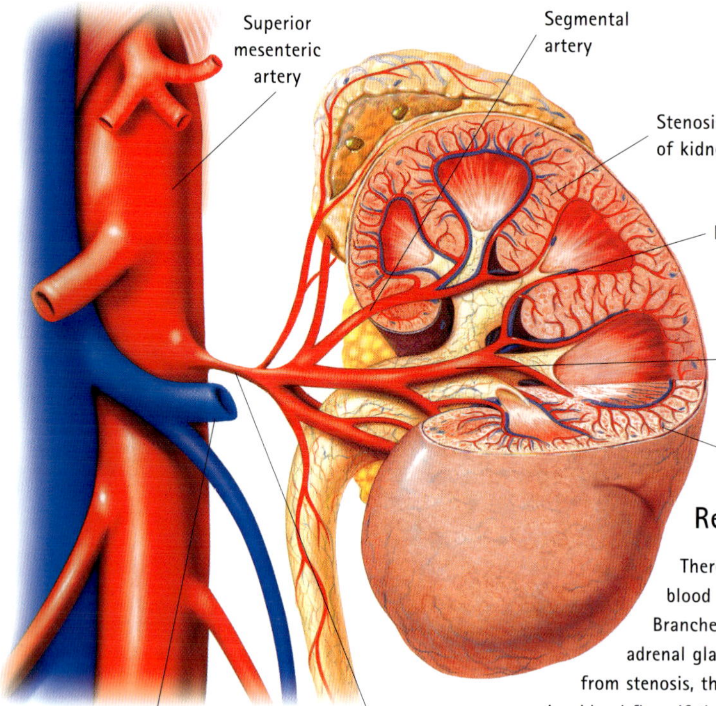

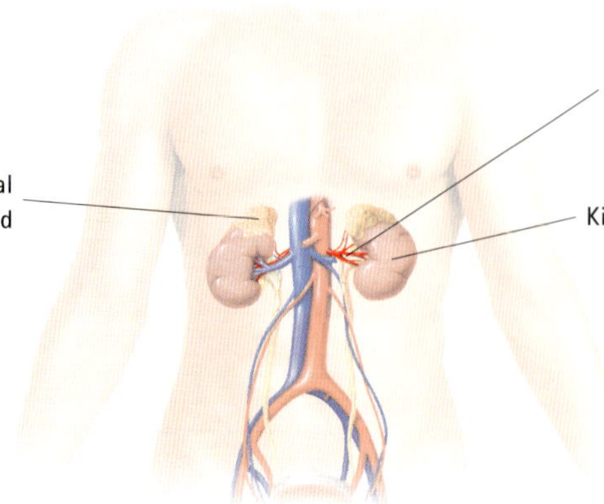

Renal artery

There are two renal arteries, supplying blood and oxygen to the kidneys. Branches of the renal artery supply the adrenal gland and ureter. In people suffering from stenosis, the arteries are narrowed, restricting blood flow. If the blood supply is completely disrupted, death of kidney tissue will occur and the kidneys will cease to function.

The symptoms of kidney disease often vary and may be vague until the disease is well advanced. Treatment depends on the cause and may involve the care of a nephrologist, a specialist in kidney diseases.

The renal arteries can be affected by arteriosclerosis, which may cause clots (thrombi), or a narrowing (stenosis). Blockage of the arteries can disrupt the blood supply to the kidneys, causing death of kidney tissue and loss of function.

Treatment is by angioplasty (e.g. balloon angioplasty) or bypass surgery. Less commonly, arteriosclerosis may sometimes cause an aneurysm in the artery, which may rupture, causing abdominal pain and sometimes even death. Treatment is by surgical repair of the aneurysm.

SEE ALSO *Angioplasty on page 452; Dialysis on page 454; Lithotripsy on page 452; Metabolic imbalances and disorders of homeostasis on page 117; Radioisotope scans on page 433; Transplant surgery on page 459; Treating the digestive and urinary systems on page 444; Ultrasound on page 432; Urinalysis on page 431*

Uremia

Uremia is the medical term for retention in the blood of urinary substances, normally caused by severe kidney failure. It occurs when the diseased kidney is no longer able to clear the blood of waste products such as creatinine, urea and ammonia. The term uremia has been used for over a century. It was originally used because it was presumed that the symptoms of renal failure were due to retention of abnormal amounts of urea in the blood. It is now clear that the symptoms of renal failure are attributable not so much to the accumulation of urea but to disturbances of water and electrolyte balance and the accumulation of many other endproducts of metabolism.

Uremia can cause a wide range of clinical symptoms. Symptoms vary from patient to patient, depending on the degree of reduction in renal function and the rapidity with which renal function is lost. At an early stage patients will often have no symptoms, but as renal function deteriorates, symptoms develop. Loss of appetite, nausea, hiccups and vomiting are the common initial symptoms of uremia. Stomach ulcers occur in one-quarter of uremic patients. The patient's skin often develops a sallow complexion (a yellow-brown appearance). This is due to the combined effect of impaired excretion of urinary pigments (urochromes) combined with anemia. High blood pressure (hypertension) is the most common complication of uremia. Fluid retention may result in pulmonary edema (fluid in the lungs) and/or congestive heart failure. Pericarditis (inflammation of the lining of the heart) may occur. The patient may experience disturbances of the nervous system that can include loss of concentration, memory loss and minor behavioral changes.

A peripheral sensory neuropathy is common and usually results in loss of sensation, typically in the legs. Other symptoms include muscle cramps and muscle twitching. In terminal uremia patients become drowsy and finally sink into a coma.

Diabetes and high blood pressure are the commonest causes of uremia. Other causes include glomerulonephritis (inflammation of the kidney), analgesic nephropathy (due to long-term ingestion of large quantities of analgesics), systemic vascular disease (for example, systemic lupus erythematosus), obstruction of the urinary tract (by congenital defect, kidney stones or tumors) and polycystic kidneys (a congenital abnormality).

TREATMENT

The treatment is to correct the underlying kidney disease if possible. A diet low in protein, salts and water will help minimize the production of wastes by the body and alleviate symptoms. Eventually, kidney dialysis or a kidney transplant may be necessary.

Early stage kidney failure can be controlled through the restriction in the diet of salt, fluid and protein. However, if the kidneys are not able to fulfil more than 10 percent of their normal function, this is considered end-stage kidney disease and dialysis is necessary. Dialysis is a method of removing toxic substances (impurities or wastes) from the blood when the kidneys cannot do so.

There are two types of dialysis: hemodialysis and peritoneal dialysis. Hemodialysis involves filtering the blood gradually through an artificial kidney machine called a dialyzer which removes specific soluble materials from the blood. The purified blood is then fed back into one of the patient's veins. Peritoneal dialysis uses the person's abdominal peritoneal membrane to act as the dialyzer. It involves filling the abdominal cavity via a catheter with a special solution that absorb toxins.

Usually, dialysis needs to be performed three times a week for periods ranging for four to six hours. Dialysis may take place in a hospital, at a special dialysis center, or at the patient's home. In chronic renal failure, dialysis will need to be performed for the rest of the person's life or until a kidney transplant is performed.

Renal transplant is the surgical implantation of a healthy kidney into a patient with kidney disease or kidney failure. It allows a patient suffering from kidney disease to live a life without dialysis. Transplantation is usually preceded by a period of dialysis while a donor can be found. The donor may be living (usually a blood relative) or someone recently deceased. As with other transplants, the main problem with kidney transplantation is rejection of the new kidney by the recipient's immune system. Hence, the recipient needs life-long treatment with medications that suppress the immune response. Transplants from a blood-related living donor are slightly less likely to be rejected. Other problems include finding a donor and the high expense.

Nephritis

Nephritis is a term for inflammation of one or both kidneys and may be acute or chronic. It may involve the glomeruli (glomerulonephritis), the main tissue of the kidney and pelvis (pyelonephritis), or the spaces within the kidney (interstitial nephritis).

GLOMERULONEPHRITIS

Glomerulonephritis is inflammation of the glomeruli, the clusters of tiny blood vessels in the kidney that filter waste products from the bloodstream, forming urine. Damaged glomeruli cannot filter these waste products, leading to serious kidney complications. The disease may be caused by specific problems with the body's immune system, but the precise cause of most cases is unknown. There are two types; acute and chronic.

Acute (sudden) glomerulonephritis sometimes follows a sore throat which caused by a streptococcal infection and is more common in children than in adults. A few weeks after the onset of infection, the affected person notices smoky or slightly red urine, puffy eyes and ankles, a general ill feeling, drowsiness, nausea or vomiting, and headaches. Blood tests for kidney function reveal biochemical abnormalities in the blood, and the urine is found to have blood and protein in it. Mild cases are easily quite treated with bed rest, and by restricting salt and fluid intake. Sometimes, and especially in adults, kidney dialysis is necessary until the kidneys recover.

Chronic glomerulonephritis develops slowly, and may not be detected until the kidneys fail, which may take 20–30 years. Because symptoms develop gradually, the disorder may only be discovered during a medical examination for some other problem. Or it may be discovered as an unexplained cause of hypertension. In other cases, there are symptoms, such as blood in the urine, or unexplained weight loss, nausea, vomiting, a general ill feeling, fatigue, headache, muscle cramps, seizures, increased skin pigmentation, bruising, confusion, or delirium. There may be signs of chronic renal failure such as edema and fluid overload. The diagnosis can be confirmed with blood tests that show reduced kidney function, and a urine test that may show blood and proteins in the urine. Abdominal ultrasound, CAT scan, or intravenous pyelography may show scarred, shrunken kidneys. A kidney biopsy will reveal inflammation of the glomeruli.

In some cases of glomerulonephritis there is a spontaneous remission. In other cases, treatment with corticosteroid or immunosuppressive drugs will often bring about an improvement. However, many cases cannot be cured so the goal will be to manage symptoms for as long as possible.

Antihypertensive medications can control high blood pressure along with dietary restrictions on salt, fluids and protein. In the end stages, regular kidney dialysis or kidney transplantation may be necessary.

PYELITIS

Pyelitis is generally an inflammation of the pelvis of the kidney. It is often caused by bacteria that make their way to the kidney via the blood or the bladder.

Pyelitis is reasonably common, particularly in small children. It can be easily treated but requires prompt attention to ensure that it does not lead to the development of pyelonephritis.

PYELONEPHRITIS

Pyelonephritis is an infection of the kidney and the renal pelvis. Acute (sudden onset) pyelonephritis is usually the result of the

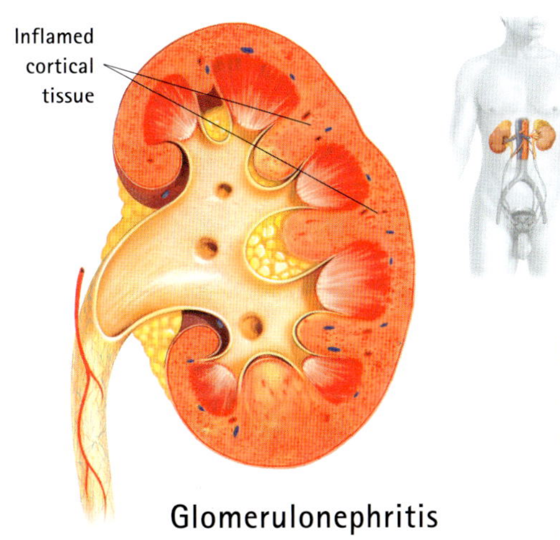

Glomerulonephritis

Due to the inflammation of the capillary loops in the glomeruli, the kidneys are unable to filter waste products from the blood. This may lead to fluid retention in the legs, face and arms.

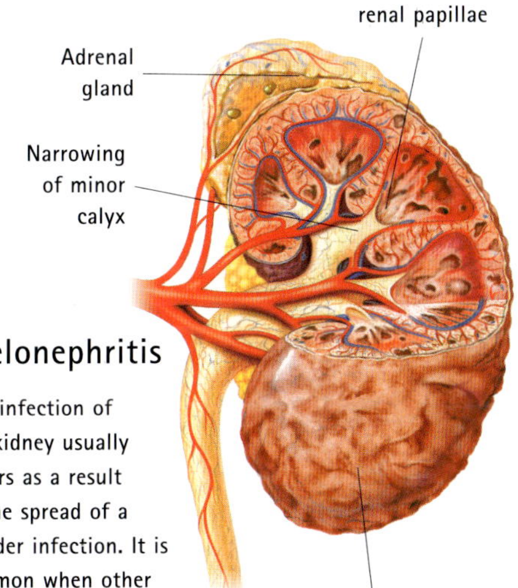

Pyelonephritis

This infection of the kidney usually occurs as a result of the spread of a bladder infection. It is common when other urinary tract diseases are present, such as bladder and kidney stones.

upstream spread of a bladder infection (cystitis). It is more common if there is pre-existing urinary tract disease, such as kidney or bladder stones. It is characterized by the sudden onset of pain in the lower back, fever with chills, nausea, and vomiting, pain passing urine (dysuria) and frequent urination. In children, the symptoms may be milder and less obvious.

A urine test shows white blood cells and bacteria in the urine, while a culture of the urine will show the specific bacteria that is causing the infection (usually *E. coli*), and its antibiotic sensitivity. Acute pyelonephritis generally responds to treatment with antibiotics given intravenously in hospital. The infection usually clears in 10–14 days.

Chronic pyelonephritis is caused by destruction and scarring of the kidney tissue as a result of recurrent or untreated bacterial infection. There is often an associated abnormality of the urinary tract, which leads to repeated infection. Treatment is surgical correction of any abnormality and a prolonged course of antibiotics. Eventually there may be renal failure requiring dialysis or transplantation.

INTERSTITIAL NEPHRITIS

Interstitial nephritis is inflammation of the spaces between the renal tubules and (sometimes) of the tubules themselves. It is usually a temporary, reversible condition, occurring as a side effect of certain drugs such as analgesics or antibiotics. Interstitial nephritis causes varying degrees of impaired kidney function; if severe, dialysis may be needed temporarily. Corticosteroids or anti-inflammatory medications may be of benefit in some cases. Occurring more often in the elderly, it may be chronic and progressive, and eventually, long-term dialysis or renal transplantation may be necessary.

Cancer of the kidney

This may occur in the renal pelvis, or in the body of the kidney itself. Hypernephroma, or Grawitz's tumor, is the most common form of kidney cancer and occurs mainly in adults. Less common is nephroblastoma (also called Wilms' tumor) which usually occurs in children under the age of seven.

Treatment of both types of malignant cancer is nephrectomy (surgical removal of the affected kidney) plus radiation treatment and anti-cancer drugs.

Kidney stones

Stones may form in the kidneys (nephrolithiasis) due to an underlying metabolic disorder, or may form for no obvious reason. Stones may form in the tissue of the kidney causing damage, or form in the kidney and pass into the ureter, causing severe colicky pain. Sometimes they may be painless, but they silently obstruct the ureter, raising the pressure in the renal pelvis and produce distension of the pelvis (hydronephrosis) and progressive loss of kidney function.

Sometimes kidney stones may be passed spontaneously. If not, they need to be surgically removed. This may be done via cystoscopy. In lithotripsy, a machine called a lithotriptor sends sound waves into the body to break up stones which then pass out in the urine.

Nephrosis

Nephrosis (nephrotic syndrome) is any abnormal kidney condition in which damage to the glomeruli leads to high loss of protein in the urine. Other symptoms include edema (swelling of the body tissues), fatigue, weakness and loss of appetite. Nephrosis occurs at any age, but is more common in children.

The leakage of protein, especially albumin, out of the glomeruli and into the urine lowers the concentration of protein in the blood, causing a shift in osmotic pressure. Because of this, fluid then leaks out of the blood and into the tissues, causing puffy eyes and ankles, a swollen abdomen and other signs of edema.

Nephrosis can be caused by damage to the glomerulus in conditions such as glomerulonephritis, diabetes, systemic lupus erythematosus and other autoimmune disorders. It is likely that several pathologic processes are involved, all of which affect the glomeruli. Nephrosis may also be caused by any condition that causes an increase in the pressure of blood in the vessels of the glomerulus, such as heart failure or following a thrombosis in the renal vein. Elevated pressure in the blood vessels of the kidney also forces protein out into the urine. In some cases there is no obvious cause.

Nephrosis is incurable but can be controlled with immunosuppressive drugs or cortisone, which reduce kidney inflammation; diuretics, which reduce fluid retention and edema; and a low salt diet. Severe cases may need dialysis or kidney transplant.

Hydronephrosis

Hydronephrosis is a condition that occurs when the kidney's pelvis and calyces become distended because urine is unable to drain into the bladder. This distension occurs because the ureter is obstructed, for example by a tumor or a stone in the ureter or bladder or by enlargement of the prostate gland. If the condition progresses unchecked, this distention destroys the tubules in the cortex of the kidney, which may then become atrophied and scarred.

Urinary tract infection is a common accompaniment, leading to further kidney damage. If both kidneys are affected, kidney failure may be the end result. The symptoms are recurrent back pain, pain on passing urine, fever and chills. The urine may be cloudy and may contain blood. Sometimes

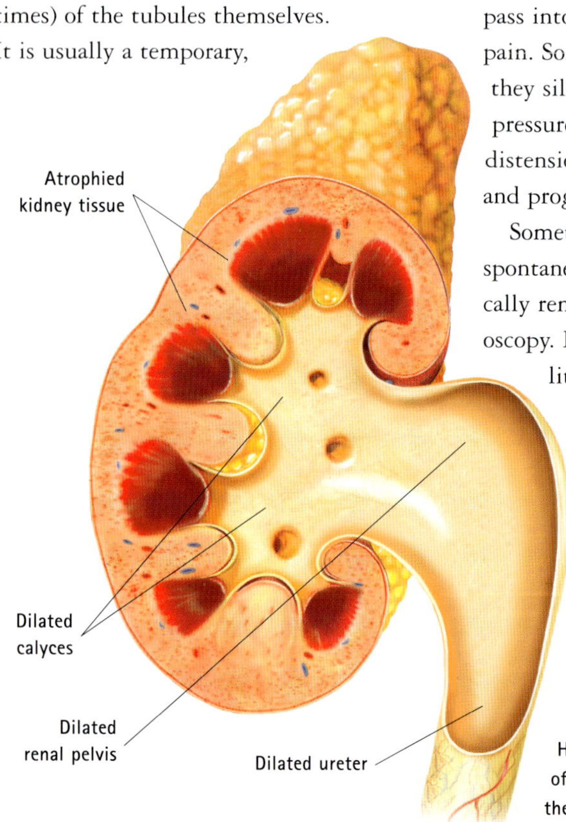

Hydronephrosis

Hydronephrosis is caused by an obstruction of the urinary tract resulting in distension of the renal pelvis and calyces.

the condition presents no symptoms and is discovered only during investigations for some other disease or infection.

Diagnosis can be confirmed with a pyelogram (an x-ray investigation in which the kidney is outlined with the use of dye), ultrasound, CAT scan of the kidneys or abdomen or abdominal magnetic resonance imaging. These tests will also help in revealing the underlying cause of the condition.

The treatment is to treat the underlying cause. It may be necessary to surgically drain the dilated renal pelvis and remove the cause of the blockage. If there is a urinary tract infection has occurred, it will have to be treated with antibiotics.

Ureter

The ureters are two thin, muscular tubes that pass urine from the kidneys to the bladder. The walls of the ureters contain smooth muscles, which contract and propel urine in waves to the bladder. The ureter may become blocked by a stone or a tumor. It may become inflamed, along with the rest of the urinary tract, in pyelonephritis.

Disorders of the ureter can be identified by a pyelogram, an x-ray taken after dye has been injected, showing the outline of the ureter and other organs in the urinary tract.

SEE ALSO *Urinary system on page 98*

Urethra

The urethra is the tube through which urine travels from the bladder to the outside of the body. In women, the urethra is short; it opens between the vagina and clitoris.

The urethra is longer in men and also serves as the passage for semen.

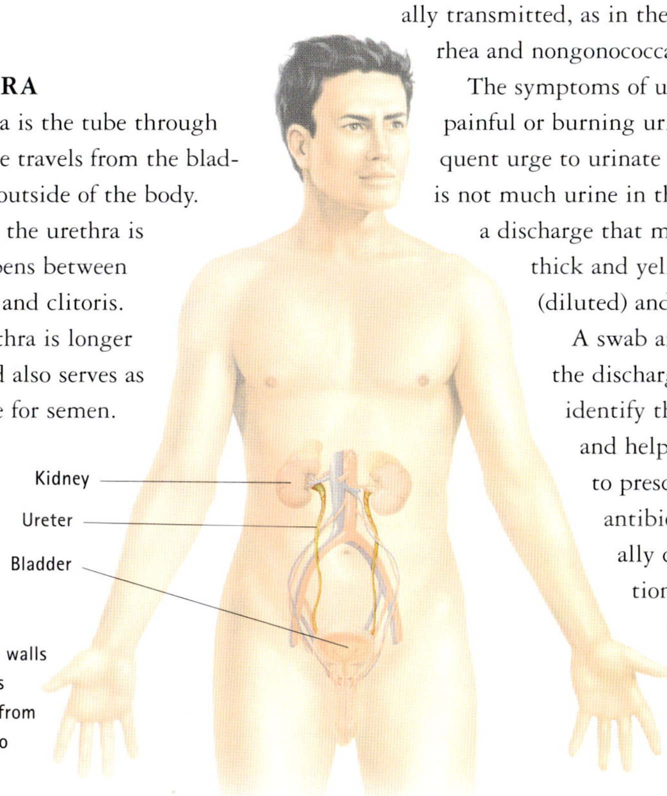

Ureter

The muscular walls of the ureters propel urine from the kidneys to the bladder.

It passes through the prostate gland, where it is joined by the sperm ducts, and opens at the tip of the penis.

SEE ALSO *Urinary system on page 98*

Disorders of the urethra

The most common condition affecting the urethra is urethritis (inflammation of the urethra), which is usually sexually transmitted. If left untreated, urethritis may, in men, result in urethral stricture.

Congenital defects

Epispadias is a congenital defect in which the urethra opens abnormally on the top of the penis, either in the head of, or along the entire length of, the penis. In hypospadias, the urethral opening appears on the underside of the penis. Both conditions cause urinary incontinence and sexual dysfunction in adulthood. Reconstructive surgery usually cures the defects.

Urethritis

Urethritis is inflammation of the urethra—the tube through which urine travels from the bladder to the outside of the body. It often occurs with cystitis (inflammation of the bladder), prostatitis (inflammation of the prostate) or epididymitis (inflammation of the epididymis). Urethritis is often sexually transmitted, as in the case of gonorrhea and nongonococcal urethritis.

The symptoms of urethritis are painful or burning urination, a frequent urge to urinate even when there is not much urine in the bladder, and a discharge that may either be thick and yellow, or watery (diluted) and white.

A swab and culture of the discharge will usually identify the organism and help the physician to prescribe the correct antibiotic which usually cures the condition. Urethritis may also result in scarring and narrowing of the urethra in men.

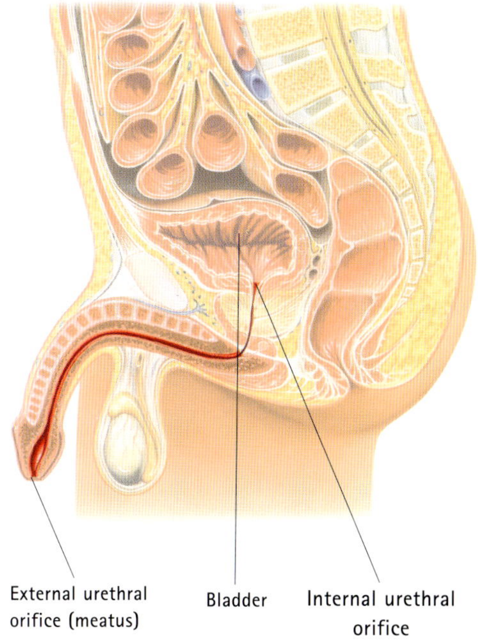

Male urethra

In men, the long urethral tube (usually 8 inches or 20 centimeters long) passes from the bladder through the prostate gland and penis. It transports urine and semen out of the body.

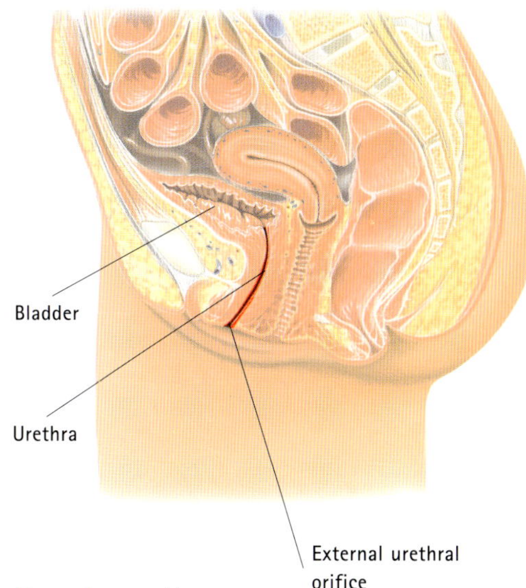

Female urethra

The female urethra is relatively short. As a result, invading bacteria can pass more easily into the body, leading to frequent bladder infections such as cystitis.

Urethrocele

A urethrocele is a bulge of the urethra along the front wall of the vagina. It is caused by a weakness in the vaginal wall, usually after childbirth. A minor urethrocele does not produce symptoms; a larger one causes a lump inside the vaginal opening and may cause urinary incontinence. Surgery through the vagina will repair the weakness in the vaginal wall and the bulging urethra.

Bladder

Found in the pelvic cavity, the bladder is the part of the urinary tract that collects and stores urine via the ureters from the kidneys. When the bladder is full it empties, expelling urine from the body via the urethra. In most adults, the bladder holds about one pint (475 milliliters) of urine when full. It passes from 24–68 fluid ounces (700–2000 milliliters) of urine a day.

SEE ALSO *Urinary system on page 98*

Male bladder

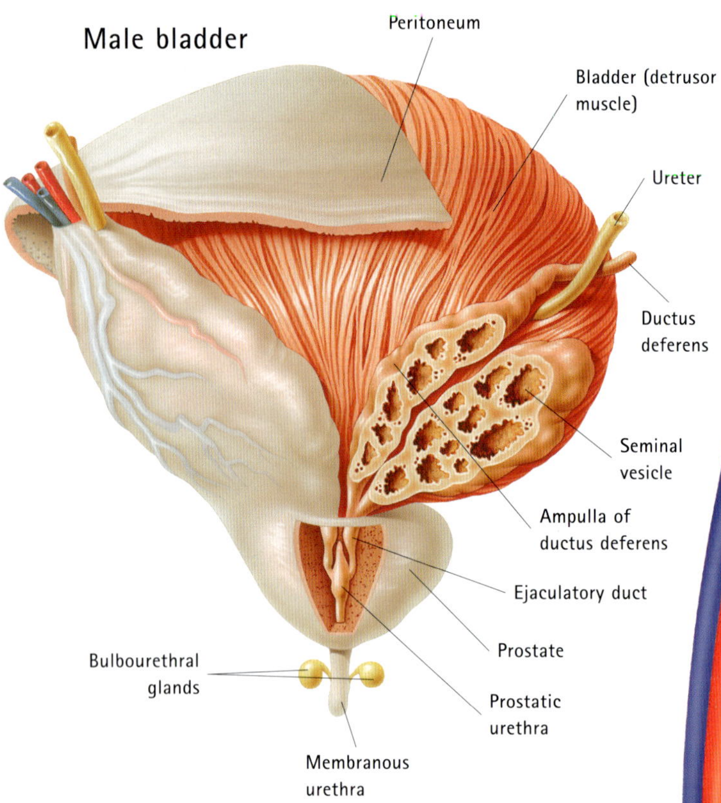

Male urinary system

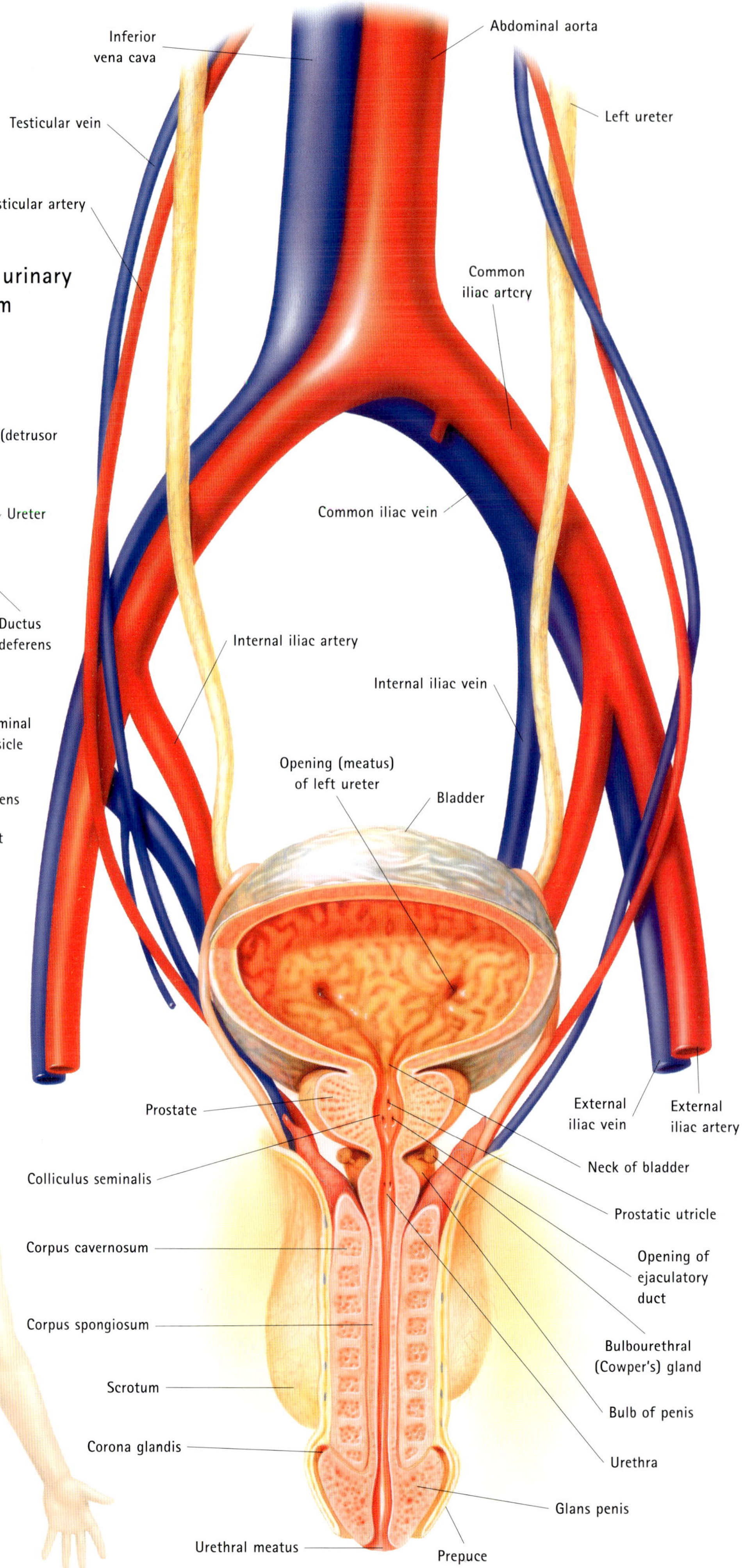

Male bladder

Urine is stored in the bladder and expelled through the urethra, which passes through the prostate gland and travels the length of the penis. In old age an enlarged prostate can obstruct the flow of urine at the outlet of the bladder and surgery may be required.

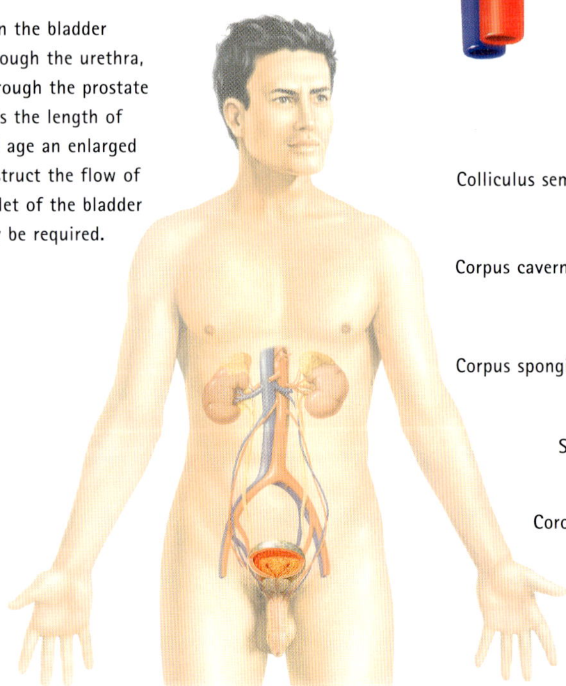

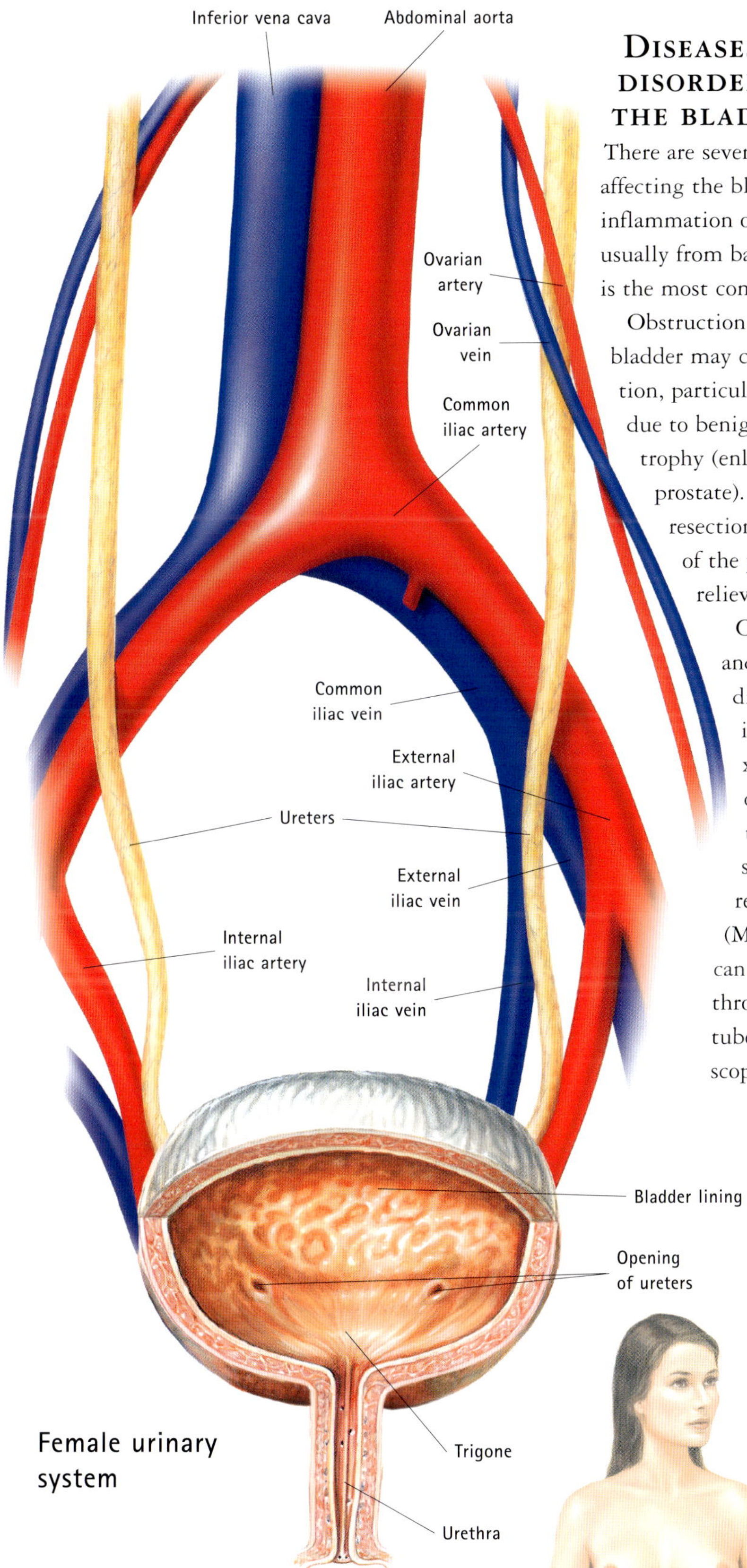

Female urinary system

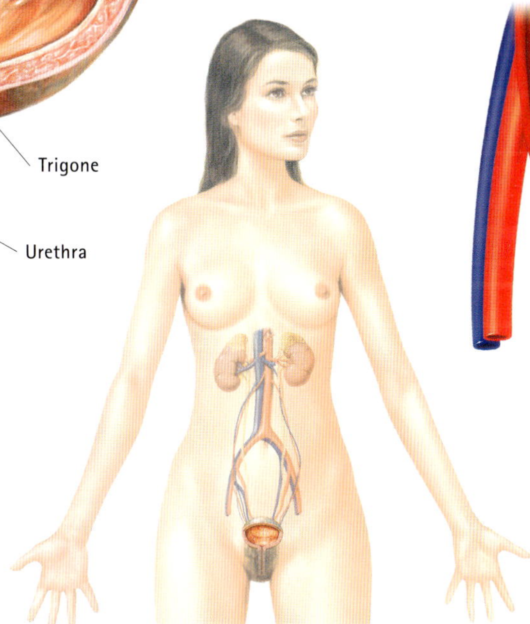

Female bladder

In the female, the urethra is much shorter than in the male. The shorter urethra makes it easier for bacteria from outside the body to enter the bladder and cause cystitis (inflammation of the bladder). As a result, cystitis is more common in women than in men.

Diseases and Disorders of the Bladder

There are several conditions affecting the bladder. Cystitis or inflammation of the bladder, usually from bacterial infection, is the most common.

Obstruction of the outlet of the bladder may cause urinary retention, particularly in older men, due to benign prostatic hypertrophy (enlargement of the prostate). Trans-urethral resection (TUR) of a part of the prostate usually relieves the obstruction.

Calculi, cancers and other bladder disorders can be identified using x-ray, ultrasound, computerized axial tomography (CAT) scan or magnetic resonance imaging (MRI). The urologist can view the bladder through a fiberoptic tube known as a cystoscope, which can also be used to take a biopsy. A cystogram is an x-ray taken of the bladder after dye that is opaque to x-rays is injected into a vein. Filtered out by the kidneys, the dye passes into the bladder, showing stones, papillomas, cancers and other conditions.

SEE ALSO *Lithotripsy on page 452; Ultrasound on page 432; Urinalysis on page 431*

Bladder stones

Stones, or calculi, may form in the bladder. Composed mainly of calcium or uric acid, calculi may grow in the bladder slowly without causing symptoms. When they grow to a certain size, they may block the outlet to the urethra and cause urinary obstruction. Cystitis often accompanies calculi. Treatment is usually surgical. This may mean lithotomy—an operation to remove a stone through a surgical incision. Alternatively, the stone may be destroyed by lithotripsy, a procedure in which an instrument called a lithotrite is passed through the urethra into the bladder and used to crush the stone or shatter it with an electrical spark. Ultrasonic lithotripsy is similar except that high-frequency sound waves are used to destroy the stone without inserting any instruments into the body. The crushed fragments of the stone can then be passed out of the body.

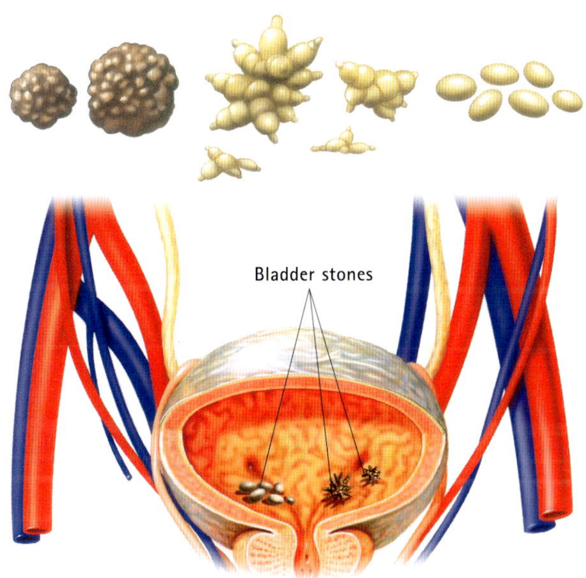

Bladder stones

Bladder stones grow slowly and often cause no symptoms until they are large. They may block the outlet of the bladder, or cause bladder infections. These (from left, mulberries, jackstones and gravel) are shown at their actual size.

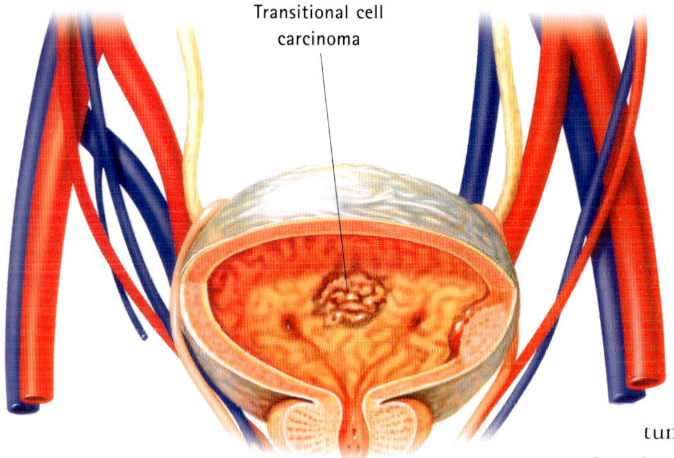

Bladder transitional cell carcinoma

Bladder cancers such as this one are often caused by cancer-producing chemicals such as cigarette smoke, dyes, paint and rubber. Most can be easily cured by removal or cauterization (burning) with an instrument that is passed through the urethra and into the bladder.

Cystitis

Cystitis is inflammation of the bladder, usually from bacterial infection. Bladder infections cause the sufferer to urinate more frequently than normal. Urination is accompanied by a burning or stinging sensation and there may be blood in the urine. If the infection spreads upstream to the kidneys, inflammation of the kidneys (pyelonephritis) may occur, with fever and back pain.

Cystitis is usually caused by bacteria gaining entry into the bladder via the urethra. Because the urethra is shorter in women than in men, cystitis is more common in females. It sometimes occurs in women after sexual intercourse. Any abnormality in the bladder makes cystitis more likely. A tumor or stones in the bladder, an enlarged prostate, or a distended uterus during pregnancy may obstruct the normal flow of urine and cause the disorder.

Cystitis is confirmed by a urine test which isolates the bacteria responsible and determines which antibiotic(s) will be the most effective. A short course of the correct antibiotic usually clears up the infection. The patient should drink copious amounts of water to help flush the bacteria out of the urinary tract.

Should another attack of cystitis occur, it may be because of an abnormality, in which case the bladder should be more fully investigated. This may involve blood tests, an intravenous pyelogram (IVP) and cystoscopy.

Cancer of the bladder

Bladder cancers usually arise from cells lining the bladder. The most common form of bladder cancer is a papilloma—a slow-growing wart-like growth, or tumor, attached to a stalk. Tumors other than papillomas are less common but have a poorer prognosis, spreading by penetrating the bladder muscle, infiltrating surrounding fat and tissue, and eventually invading the bloodstream and lymphatic system. In its early stages, cancer of the bladder may not have obvious symptoms. However, later symptoms may include blood in the urine, frequent urinary tract infections, frequent and painful urination, abdominal or back pain, persistent low-grade fever and anemia.

Bladder cancer has no known cause. However, it has been linked to exposure to cancer-promoting chemicals (carcinogens). Cigarette smokers, painters, truckers, metalworkers, leatherworkers, machinists, rubber and textile workers, and people exposed to industrial dyes are at increased risk. It is more common over the age of 40. If detected early, papillomas can usually be treated successfully by transurethral resection (TUR). In this procedure the urosurgeon inserts a small tube into the bladder and removes or cauterizes the tumor.

It may be combined with chemotherapy or radiation therapy. Larger and more invasive cancers require radical cystectomy (removal of the bladder) and construction of an artificial storage organ.

Cystitis

Bacteria gain entry to the bladder, usually via the urethra, leading to inflammation of the bladder—this is known as cystitis.

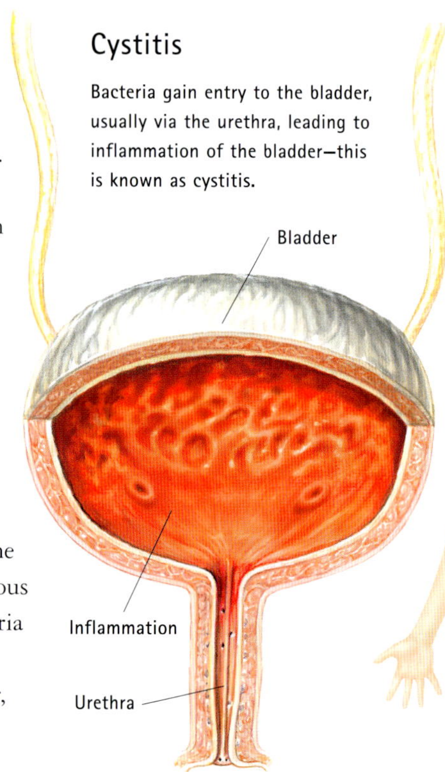

Incontinence

In an infant, daytime control of the bladder is achieved around the age of two, and night-time control some years later. Lack of voluntary control of the bladder beyond this age is called incontinence. It is a problem most common in the elderly.

There are several causes of incontinence in adults. Shortening of the urethra and loss of the normal muscular support for the bladder and floor of the pelvis, may cause incontinence. This occurs during pregnancy, after childbirth (especially after multiple pregnancies) and as a consequence of ageing.

Any neurological disorder affecting the bladder such as spina bifida, multiple sclerosis, and nerve degeneration that occurs with conditions like diabetes mellitus can cause incontinence.

Stress incontinence is the involuntary leakage of urine during exercise, coughing, sneezing, laughing, lifting heavy objects, or during other body movements that put pressure on the bladder. Urge incontinence is the inability to hold urine back long enough to reach a toilet. It is a major complaint of patients with urinary tract infections. Overflow incontinence is the leakage of small amounts of urine from a bladder that is always full. In older men, this can occur when the flow of urine from the bladder is blocked, e.g. in enlargement of the prostate gland (prostatomegaly) or following surgery or cancer.

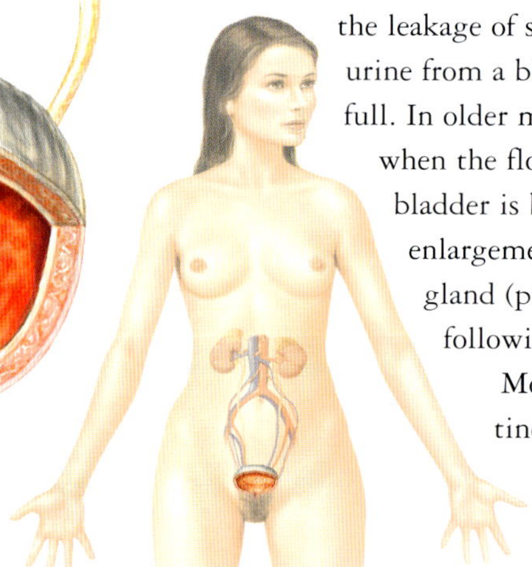

Most people with incontinence can be cured. For mild incontinence, the use of a portable urinal or bedside commode, and wearing

INCONTINENCE

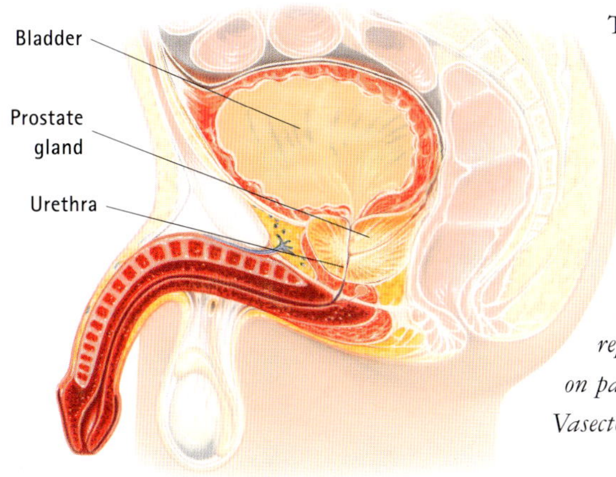

Male incontinence

Incontinence in men is often caused by an enlarged prostate gland that compresses the urethra, obstructing the flow of urine. Urine then collects in the bladder and eventually leaks out.

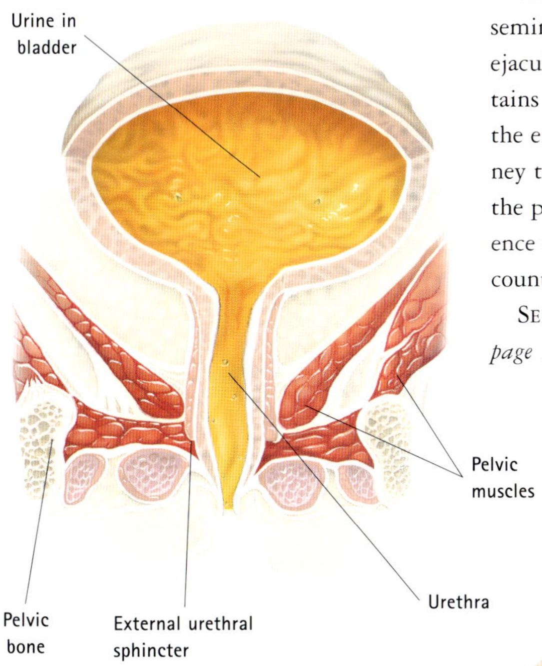

Female incontinence

Women may suffer incontinence during pregnancy and after childbirth due to a weakening of the muscles that support the bladder and the floor of the pelvis.

sanitary pads or panty liners may be enough to manage the disorder. Other treatments include strengthening of bladder muscles and pelvic floor muscles, and surgery to tighten relaxed or damaged bladder muscles, or to remove a blockage due to an enlarged prostate.

MALE REPRODUCTIVE ORGANS

The male reproductive organs consist of the testes, the epididymis, the prostate gland and the penis.

SEE ALSO *Contraception on page 448; Development of the fetus on page 398; Endocrine system on page 110; Fertility on page 396; Fertilization on page 397; Infertility on page 397; Male reproductive system on page 105; Puberty on page 421; Sexual behavior on page 107; Vasectomy on page 461*

PROSTATE GLAND

This is a part of the male sex organs. About the size of a walnut, it surrounds the neck of the bladder and the urethra, and is composed of both glandular and muscle tissue.

Secretions from the prostate and the seminal vesicles make up the seminal fluid ejaculated during sexual orgasm that contains the glucose and enzymes that provide the energy spermatozoa need for their journey toward the ovum. Also, secretions from the prostate can be tested to reveal the presence of any infections that may affect the count and quality of sperm.

SEE ALSO *Male reproductive system on page 105; Urinary system on page 98*

DISORDERS OF THE PROSTATE

Benign prostatic hypertrophy and prostate cancer is a significant cause of health problems in the elderly male. However, prostatitis is mainly a disease of young men.

Benign prostatic hypertrophy

By the age of 50, it is common for the prostate to show some signs of enlargement. It may increase in weight from less than an ounce (20 grams) to five or six times this. Known as benign prostatic hypertrophy, this enlargement may eventually obstruct flow of urine from the bladder, and if left uncorrected, can eventually cause bladder and kidney damage.

Symptoms of urethral obstruction include nocturia (the need to empty the bladder often at night); dysuria (pain or difficulty with urination); and the sudden urgency for urination. Usually, the bladder will not empty completely and urine may stagnate, causing infection or the formation of stones. Straining to pass urine may make things worse and can cause bleeding and abdominal pain. A catheter may be inserted to effectively drain the bladder.

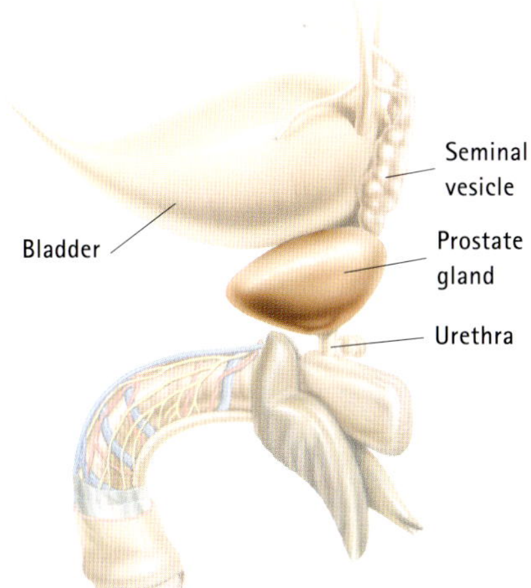

Prostate gland

The prostate gland is a structure about the size of a walnut. It secretes some of the fluid present in semen.

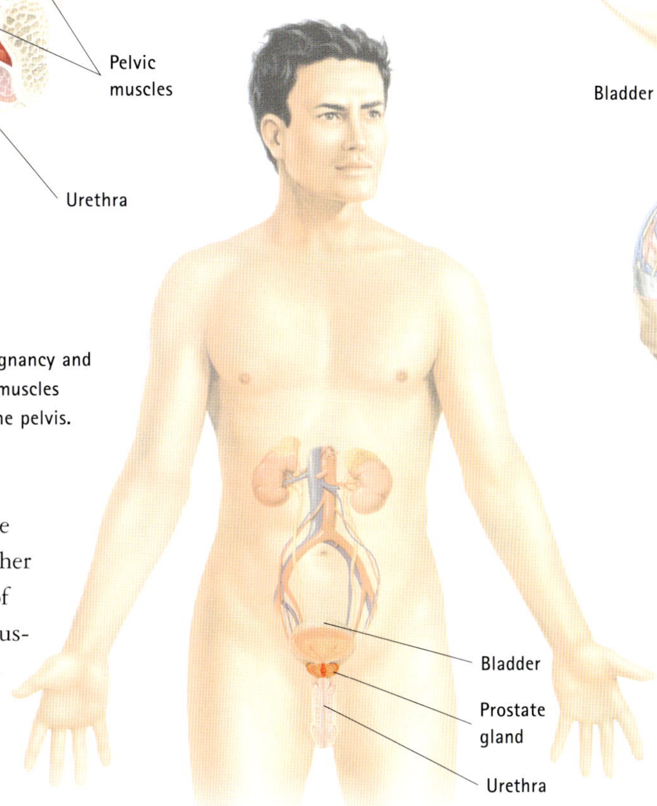

Prostate–urinary link

The prostate gland surrounds the neck of the urethra. Because of its location, enlargement of the prostate (which may occur with age or disease) causes difficulty in urination.

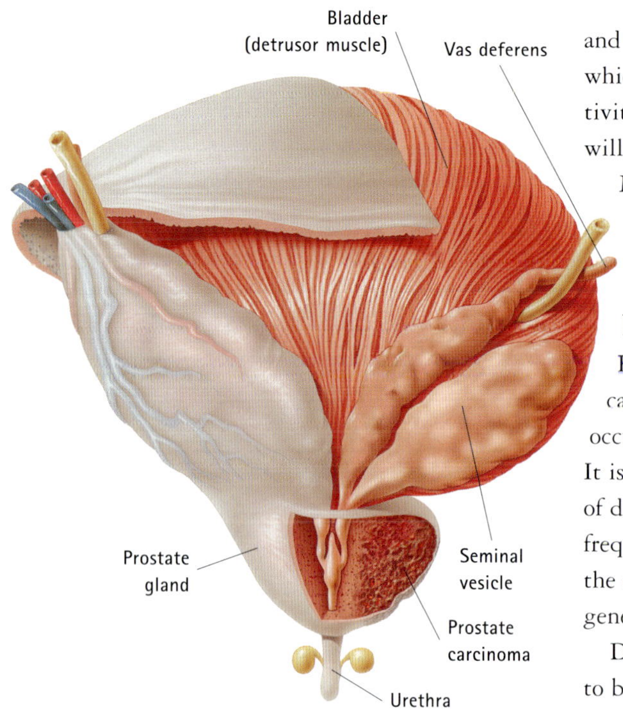

Prostate cancer

The most common cancer in men, prostate cancer usually arises in the outer section of the prostate and is seldom diagnosed until it has spread to other organs.

Diagnostic procedures to confirm an abnormality include excretory urography, which is an x-ray of the urinary tract, and cystourethroscopy, an internal visual examination of the bladder.

Prostatitis

Prostatitis is inflammation of the prostate gland, the walnut-sized organ that encircles the urethra at the base of the bladder. The condition is caused by bacteria such as *E. coli* (transmitted from the bowel) or gonorrhea, chlamydia or trichomoniasis (transmitted through sexual contact with an infected partner). Prostatitis commonly occurs along with urethritis (inflammation of the urethra), epididymitis (inflammation of the epididymis), and/or orchitis (inflammation of the testis).

The symptoms of prostatitis are a burning pain on urination, a diminished urine stream and pain on ejaculating. There may be fever, chills and low back pain. Blood may be found in the urine or the semen, and there may be a frequent desire to urinate urgently. The testes may be painful.

A physician will diagnose the condition after finding an enlarged and tender prostate on rectal examination. A urethral swab and a urine sample are taken to determine which bacteria are involved, and their sensitivity to antibiotics. Oral antibiotic therapy will need to be continued for 6–8 weeks.

Most cases are successfully treated, but recurrence is common and in some men the condition becomes chronic.

Prostate cancer

Prostate cancer is the most common cancer typically found in men, mostly occurring in individuals aged over 60 years. It is also one of the most common causes of death from cancer in men. Despite its frequency, almost nothing is known about the causes of prostate cancer, although genetic factors may play a role.

Development of prostate cancer appears to be age-related and latent tumors (tumors which are demonstrable by microscopic examination of the prostate but do not produce any clinical effects) are present in more than 50 percent of men aged over 80 years.

The growth of prostate cancer cells often depends at least in part on stimulation by testosterone, but the precise role of hormones in the development of this cancer remains poorly understood.

Prostate cancer commonly produces discomfort during urination or symptoms suggesting obstruction of urine flow. Because the cancer usually arises in the outer part of the prostate gland, whereas the urethra runs through the central portion, symptoms relating to urinary outflow may not develop until quite late in the course of the disease. In some patients, prostate cancer first becomes apparent as a result of its spread to other organs. In this case, the outcome is less favorable.

A major site of spread of prostate cancer is into bone, especially the spine and pelvis, producing pain. Unlike most other tumors, prostate cancer that has spread to bone is typically associated with new bone formation around the secondary deposits, making these easily visible on x-rays.

Benign (non-cancerous) enlargement of the prostate is very common among men in the same age group (much more common than cancer of the prostate) and causes similar urinary obstruction. This makes it quite difficult to distinguish the two conditions, a situation that is complicated by the fact that they often co-exist. However, much as is the case for other cancers, early diagnosis is critically important in achieving a good response to treatment.

One way of demonstrating the possible presence of a cancer of the prostate is by digital rectal examination (DRE), because many cancers can be felt directly by the medical practitioner by examination through the rectum. Specific diagnosis usually requires needle biopsy of the gland under ultrasound guidance.

Currently there is no effective screening test that allows an early diagnosis of prostate cancer to be made easily and cheaply. A blood test for prostate-specific antigen (PSA), an antigen produced by prostate cells, is available but is not considered to be sufficiently sensitive or specific for use as a screening test for early prostate cancer. The PSA test is still valuable in following the progress of prostatic cancer after a person has had treatment for the cancer.

The extent of spread during diagnosis has an important bearing on the approach to treatment of prostate cancer. Surgical removal or radiation therapy may be the initial treatment. Because tumor cell growth is partly controlled by testosterone, treatments that reduce the levels of testosterone or block its activity are also valuable. Patients with early cancers often have an excellent response to therapy but those with late-stage disease have a poor outcome.

Testes

The testes, or testicles, are the major organs of reproduction in the male. They are two ovoid organs contained in the scrotum, a sac which lies directly behind the penis. In this location they are kept cooler by about 3.5°F (2°C) than within the body, functioning better at this lower temperature. The testes produce male sex hormones (primarily testosterone) and manufacture sperm, the microscopic cells which carry the man's genetic material to combine with that of the woman after fertilization of her ovum.

At puberty the reproductive organs mature to become fully functional. This change is triggered by hormones released from the pituitary gland which enable the testes to start producing sperm and testosterone. The testes produce sperm continually from puberty to old age.

SEE ALSO *Development of the fetus on page 398; Endocrine system on page 110; Infertility on page 397; Male reproductive system on page 105; Puberty on page 421*

Testosterone

Testosterone is the hormone responsible for the development of secondary sexual characteristics in the male. Secreted by the testes, it stimulates growth of facial and pubic hair, enlargement of the larynx and deepening of the voice, enlargement of the penis and testes, and an increase in muscle strength. Earlier in life, testosterone plays an important part in the development of external genitalia in the male fetus.

Testosterone production is controlled by the follicle stimulating hormone (FSH) and luteinizing hormone (LH). Both of these hormones are secreted by the front lobe of the pituitary gland. Small amounts of testosterone are also synthesized from cholesterol in the adrenal glands, from the ovaries and from the placenta.

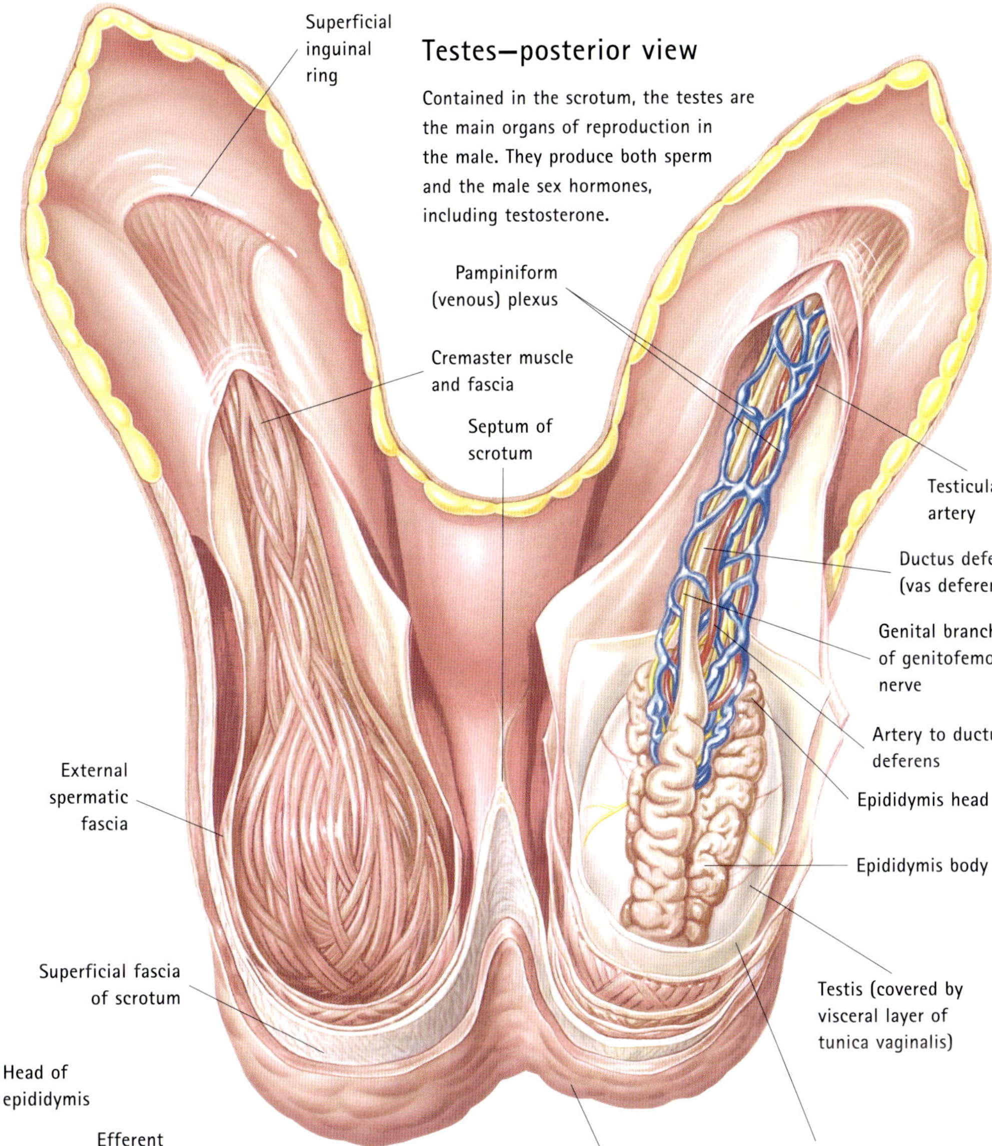

Testes—posterior view

Contained in the scrotum, the testes are the main organs of reproduction in the male. They produce both sperm and the male sex hormones, including testosterone.

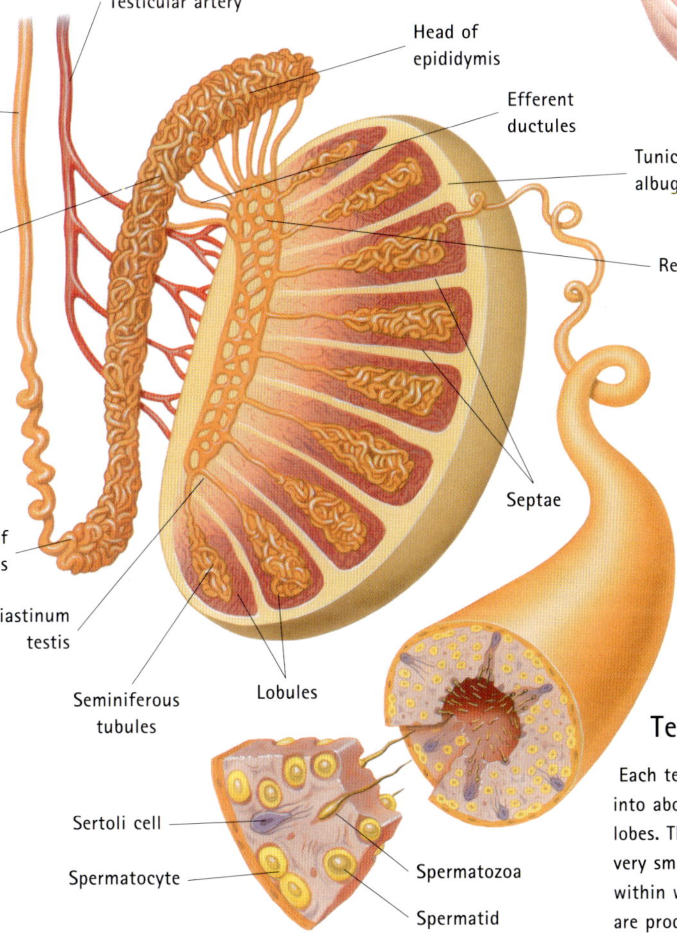

Testes—structure

Each testis is divided internally into about 250 compartments or lobes. These contain one to three very small convoluted tubules within which spermatozoa (sperm) are produced.

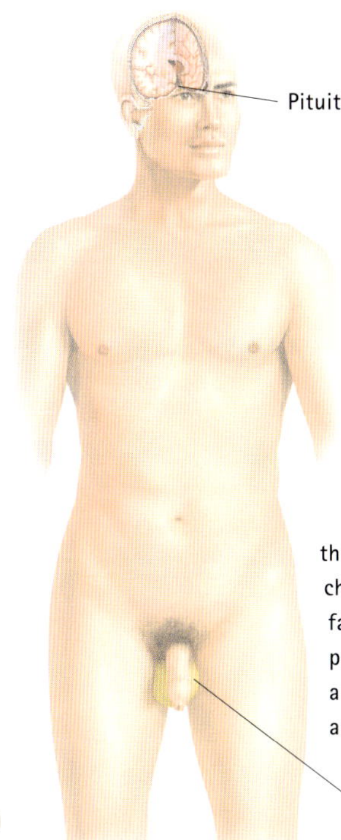

Testosterone

The hormone testosterone is produced by the testes under the control of the pituitary gland. Testosterone controls the development of male characteristics, such as facial hair and sperm production, and also plays a role in muscle growth and the body's metabolism.

Drugs based on synthetic testosterone have clinical uses such as the suppression of milk supply in lactating women and the treatment of female frigidity, breast cancer and testicular disorders.

Sperm

Sperm are the mature male reproductive cells that combine with the ovum, the female reproductive cell, to begin the process leading to pregnancy and the development of a baby. Sperm are produced at the rate of about 50,000 per minute every hour from puberty until late in life. They are made in the testes (or testicles) which hang outside the body in a sac of skin—the cooler scrotum (sperm do not develop properly at full body temperature).

Newly produced sperm pass into the epididymis, a tube behind each testis, where they mature and wait to be ejaculated. During ejaculation, muscle contractions squeeze the sperm along the sperm duct and into the urethra. Along the way, the seminal vesicles and prostate gland add seminal fluids, which mix with the sperm and mobilize them. On average, between 80 million and 300 million sperm are ejaculated each time a man has an orgasm.

Although sperm can live up to 72 hours in the female reproductive tract, their death rate is high; if ejaculation occurs when the woman is fertile, generally only one sperm will penetrate the ovum. Sperm which are not ejaculated are reabsorbed into the man's lymph glands over a period of time.

Sperm carry the man's genetic potential. One half of the chromosomes which carry the genetic makeup of a new human being are to be found in the head of the sperm, the other half coming from the mother's ovum. The sperm decides the sex of the baby; one half of sperm cells contain the Y chromosome that produces a male child, the other half contain the X chromosome that produces a female child.

The microscopic sperm, which is like a tadpole in shape, has a tail which enables it to "swim" from the vagina, through the cervix and uterus and into the fallopian tube to meet the ovum. In each sperm is a cap (acrosome) containing enzymes that can penetrate the coating around the ovum. Where subfertility (infertility) is a problem, tests are carried out on both partners to determine the reason. A man may fail to manufacture sperm (azoospermia); sperm may be weak or few in number (oligospermia: fewer than 20 million per milliliter) or there may be damage to the ductus deferens.

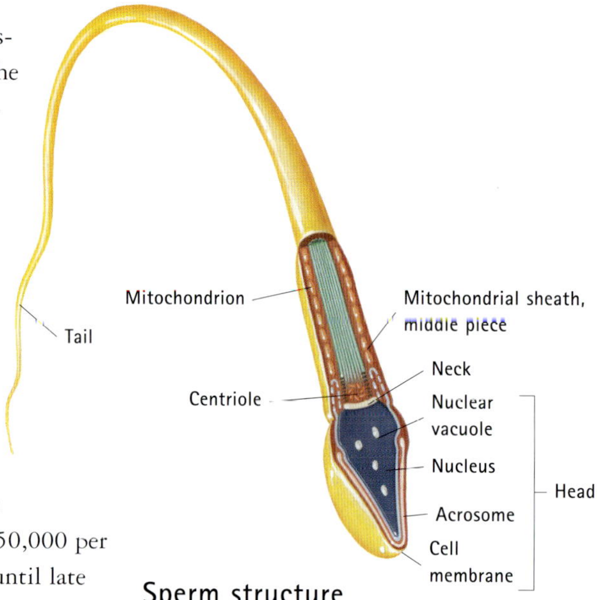

Sperm structure

The head of each sperm has a nucleus, containing chromosomes, and an acrosomal membrane which holds enzymes needed for fertilization. The tail of the sperm helps it move in a corkscrew action on its journey from the testes to the female reproductive organs.

Semen

Also called seminal fluid, semen is a liquid that is emitted from the male reproductive tract. It comprises sperm cells and the fluids that nourish and support them. The fluids are produced and secreted by the various tubules and glands of the reproductive system, including the prostate gland and seminal vesicles. These ensure that the semen contains the correct concentration of nutrients and electrolytes needed to keep the sperm healthy. These nutrients include sugars, amino acids, phosphorus, potassium, and prostaglandin hormones. A small amount of mucus is also secreted into the semen by the bulbourethral and urethral glands. The secretions of the testes and other glands are controlled by the male hormone testosterone.

Semen contains millions of sperm cells, needed to fertilize female ova (eggs). Sperm cells are produced by the testes, stored in the epididymis, then transported down a muscular tube called the ductus deferens (or vas deferens).

During ejaculation, muscles around the epididymis and ductus deferens contract, forcing semen into the urethra. The fluid is then expelled out of the body by spasmodic contractions of the bulbocavernosus muscle in the penis.

When a man ejaculates, at the height of sexual excitation, he will normally discharge about $\frac{1}{20}$–$\frac{1}{5}$ ounce (1.5–5 milliliters) of semen from his urethra. Each ejaculation normally contains between 200 and 300 million sperm, which comprise only about 2–5 percent of the ejaculate.

Semen analysis is usually carried out to assess a man's fertility—about one-third of infertility problems seen in couples are due to male problems. The semen must be tested within 2 hours of collection for a valid result. Low sperm counts or insufficient semen production may be the cause of infertility.

Disorders of the Testes

The most serious disorder of the testes is testicular cancer, which most often affects young men. Early detection aids in treatment, so all young men should undertake regular testicular self-examination.

SEE ALSO *Infertility on page 397*

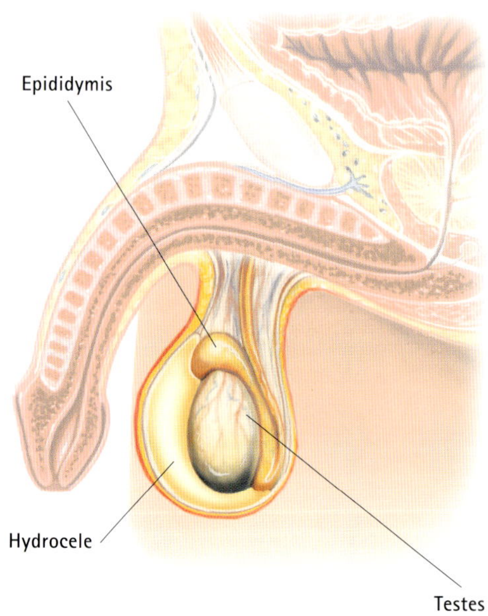

Hydrocele

A hydrocele is a collection of clear amber fluid in the sac that covers the testes. Surgery is the most effective long-term treatment.

Hydrocele

Hydrocele is an accumulation of fluid in the sac which covers the testes, the tunica vaginalis. It is a common cause of swelling in the scrotum and may be due to injury or inflammation in the testes, or obstruction of drainage. The fluid may reappear even after removal. Hydroceles may be painless, but, if causing discomfort, can be treated surgically.

Varicocele

A varicocele is a mass of enlarged and distended veins, like varicose veins, arising from the spermatic cord, which feels like a "mass of worms" on the testicle. The mass is more obvious when standing; it may disappear on lying down, as blood pressure to it decreases. A varicocele more commonly develops on the left side and can result from defective valves in the testicular vein or by problems with the renal or kidney vein. A varicocele may produce an ache but is often painless. If it prevents proper drainage of blood from the testes it may raise their temperature and reduce sperm production.

Orchitis

Orchitis is an inflammation of one or both of the testicles. Often caused by viruses (especially mumps), bacteria and sexually transmitted organisms such as those that cause gonorrhea or chlamydia, orchitis occurs with infection of the prostate (prostatitis), infection of the urethra (urethritis) and infection of the epididymis (epididymitis).

Symptoms of orchitis are pain in the groin and testicle, pain with urination (dysuria), pain with intercourse or ejaculation and, at times, a discharge from the penis. Lymph nodes in the groin may be tender and enlarged and, on rectal examination, the prostate gland may be tender and enlarged.

Urinalysis and blood tests will confirm the diagnosis and isolate the organism responsible. The condition is treated with analgesics and antibiotics (if caused by bacteria). If the condition is sexually transmitted, sexual partners must also be treated. A full recovery is usual, though sterility may follow mumps orchitis.

Epididymitis

Epididymitis is inflammation of the epididymis, an oblong structure attached to the upper part of each testis. The inflammation causes the epididymis to become swollen and painful. The condition may be caused by a bladder infection (cystitis), or it may be a complication of a sexually transmitted disease. The affected person experiences pain when urinating and increased frequency of urination. The scrotum may be painful, enlarged and tender. Treatment with antibiotic drugs and painkillers (analgesics) is usually effective.

Testicular torsion

Testicular torsion, the twisting of one testicle and its spermatic cord, can be a congenital defect or result from injury or exertion; symptoms are pain and swelling, nausea and vomiting. Ultrasound may be used for diagnosis and the condition must be resolved within 24 hours as blood flow to the testicle is cut off, causing tissue death.

Testicular cancer

Testicular cancer is rare but is most often found in men between 15 and 35 years of age. It is not caused by injury but is often found on inspection after trauma. Generally painless, it begins as a noticeable lump that can often be detected by self-examination. Treatment is to remove the testicle (orchiectomy) and repplacing it by a cosmetic prosthesis if required; the other testicle will produce a normal amount of sperm and hormones. In prostate cancer, a complete orchiectomy may be performed to stop the production of testosterone, which some tumors need for growth.

Self-checking monthly by feeling each testicle between the thumb and forefinger can reveal early changes. Normal testicles are slightly soft but firm and smooth to the touch. Any hardness or lumpiness, or marked differences between testicles should be a signal for a medical check-up.

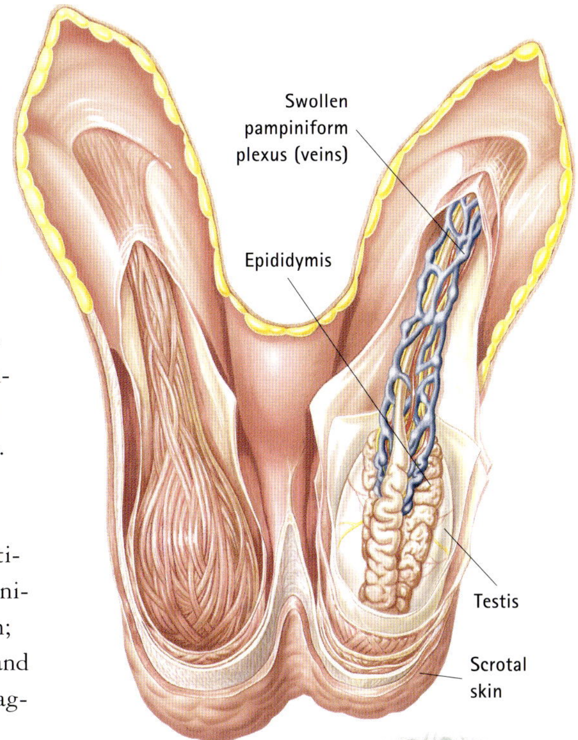

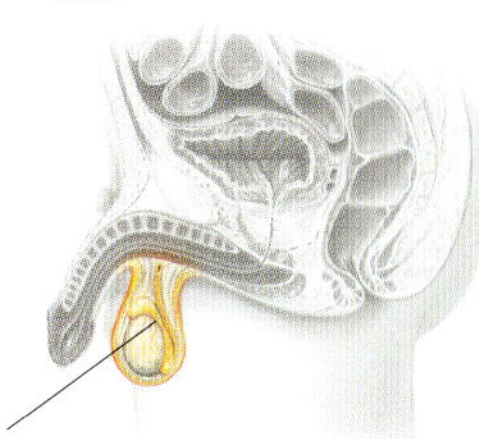

Varicocele

In this condition, the veins of the spermatic cord distend like varicose veins, leading to a swelling of the scrotum.

Scrotum

The scrotum is the bag of skin and soft tissue attached to the perineum in the male, between the thighs. It contains the testes and the lower parts of the spermatic cords. Under the skin of the scrotum is a thin layer of muscle (dartos muscle), whose contractions make the skin of the scrotum wrinkle. The testes normally descend before or just after birth from the abdominal cavity, along the inguinal canal, and into the scrotum. In rare cases, the testes fail to descend into the scrotum.

SEE ALSO *Male reproductive system on page 105*

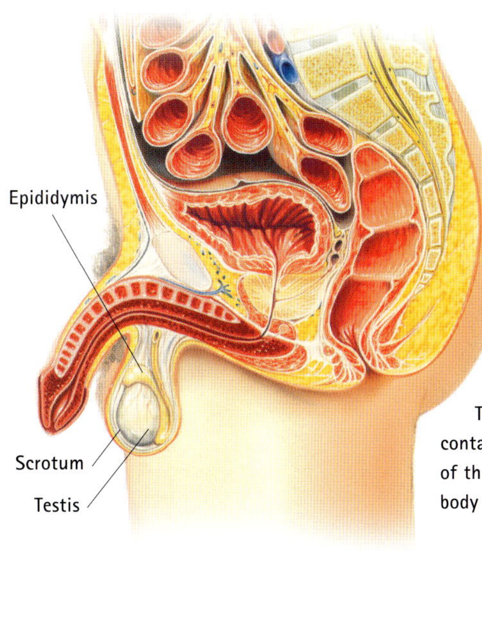

Scrotum

The scrotum is a bag of skin hanging just behind the penis, containing the testes. The cremaster muscle adjusts the tension of the scrotal skin, moving the testes toward or away from the body in response to changes in temperature.

Disorders of the scrotum

An inguinal hernia, particularly the indirect type, will often descend into the scrotum along the route taken by the testis and will need to be surgically corrected.

Other conditions that can cause a lump in the scrotum. They include epididymitis, tumors such as carcinoma of the testes, varicocele, hematocele (a collection of blood within the scrotum) and a spermatocele (a cyst-like mass within the scrotum containing fluid and dead sperm cells).

A physician will often diagnose the lump with the help of an ultrasound of the scrotum, or a biopsy performed during surgery.

Treatment of disorders of the scrotum depends on the cause; most conditions can be easily treated and some such as hematoceles and spermatoceles do not need treatment. A scrotal support (jockstrap) may be worn to help relieve pain and discomfort.

It is a good idea for males to regularly self-examine each testis and the scrotum and consult a physician if a lump is found.

Penis

The penis is the external male reproductive and urinary organ, through which semen and urine leave the body. It is attached at its base to the pelvic bone by connective tissues. It is comprised primarily of two cylinders of sponge-like vascular tissue (corpora cavernosa). A third cylinder contains the urethra, a tube that carries the urine and the ejaculate. The urethra ends in an external swelling at the tip of the penis, the glans. The glans is particularly sensitive and, in an uncircumcised penis, is covered by a protective foreskin (prepuce).

After physical or psychological sexual stimulation, the two spongy cylinders become engorged with blood and the penis becomes erect and hard. This enables the male to insert the erect penis into the female's vagina during sexual intercourse. The blood is unable to drain out through the veins in the penis because they are temporarily closed by pressure from arterial blood in the corpora cavernosa. An erection ceases when the veins open, allowing the blood to flow back into the body's circulation. Sildenafil (Viagra) works by relaxing smooth muscle in the corpora cavernosa and allowing the inflow of blood.

Circumcision is the surgical removal of all or part of the foreskin of the penis. In infancy, it is usually performed (often without anesthetic) for social or cultural reasons, as there are no known medical reasons for routine circumcision of newborn boys. In adults, it is generally performed for various medical reasons, for example, for phimosis or paraphimosis.

SEE ALSO *Male reproductive system on page 105; Sexual behavior on page 107; Urinary system on page 98*

Disorders of the penis

The most common disorders affecting the penis are impotence and sexually transmitted disease. However, diseases related to the foreskin are also common, and are seen most frequently in uncircumcised males.

SEE ALSO *AIDS on page 377; Gonorrhea on page 383; HIV on page 377; Sexual dysfunction on page 109; Sexually transmitted diseases on page 365; Syphilis on page 384*

Balanitis

Balanitis is inflammation of the glans of the penis. Usually caused by a yeast such as candida (thrush), it is more common in males who have not been circumcised, and who have not been keeping the glans clean. Symptoms are a red, shiny glans with itchiness and a slight discharge. It is treated with antifungal creams or ointments and by keeping the glans washed and clean.

Sexually transmitted diseases

The penis is often affected by venereal (sexually transmitted) diseases. Syphilis or chancroid form ulcers (chancres), on the skin of the penis; other venereal diseases, such as gonorrhea, cause infections of the urethra (urethritis). Treatment is with an appropriate antibiotic.

Cancer of the penis

Cancer of the penis occurs most commonly in elderly, uncircumcized males who have had chronic balanitis. It manifests as a small ulcer that bleeds easily and does not heal. It is treated by amputating the end of the penis or by radiation therapy.

Phimosis

Phimosis is a condition in which the foreskin of the penis is so tightly constricted that it cannot be easily pulled back over the glans. Balanitis may accompany it. Paraphimosis is a condition in which the foreskin of the penis is retracted and "stuck" and cannot be returned to its normal position, making

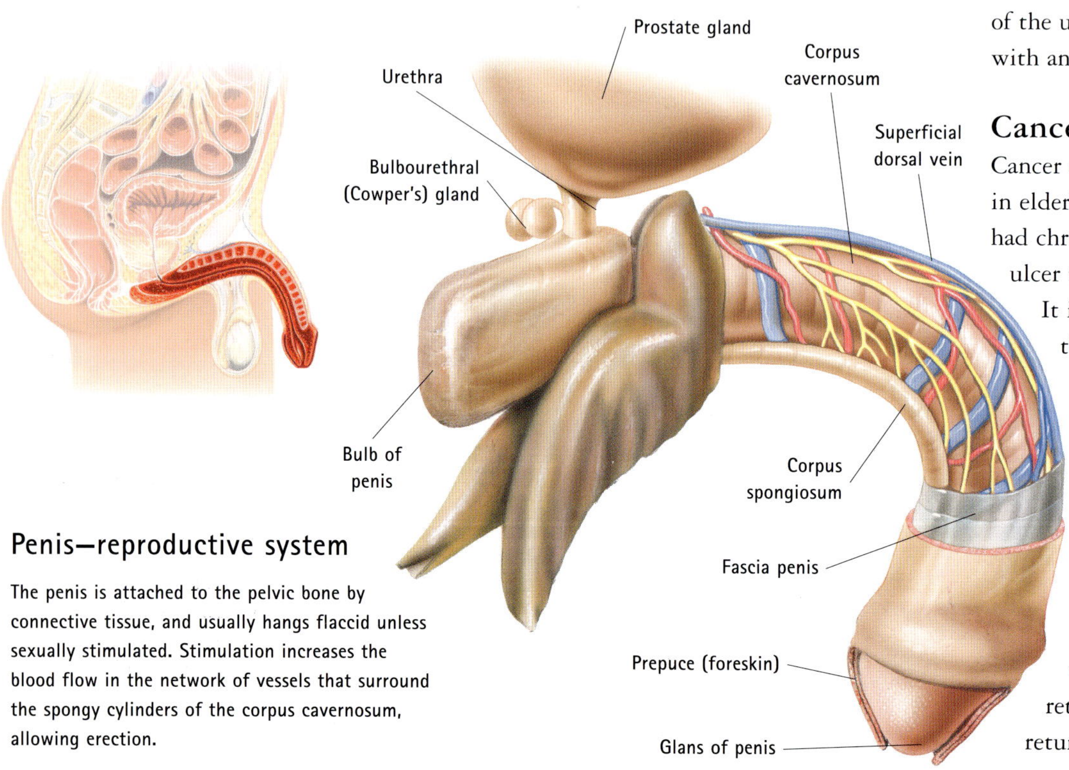

Penis—reproductive system

The penis is attached to the pelvic bone by connective tissue, and usually hangs flaccid unless sexually stimulated. Stimulation increases the blood flow in the network of vessels that surround the spongy cylinders of the corpus cavernosum, allowing erection.

the glans swollen and painful. Circumcision is the usual treatment for these conditions.

Priapism

This is a condition in which the penis remains persistently and painfully erect without any sexual arousal or desire. It is caused by a blockage of the veins that carry blood from the penis, trapping blood in the penis and causing erection. It may be brought on by excessive sexual stimulation or by certain drugs, including corticosteroids, anticoagulants and antihypertensives.

Treatment options include surgery, the injection of anesthesia into the spinal cord, and the draining (aspiration) of blood from the penis. Without treatment the penis may be permanently damaged, making normal erections impossible.

FEMALE REPRODUCTIVE ORGANS

The female reproduction organs consist of the ovaries, the fallopian tubes, the uterus and the vagina. Disorders of these organs are common and often require specialist gynecological management.

SEE ALSO *Childbirth on page 408; Contraception on page 448; Development of the fetus on page 398; Endocrine system on page 110; Female procedures on page 462; Female reproductive system on page 104; Fertility on page 396; Fertilization on page 397; Infertility on page 397; Menopause on page 424; Menstruation on page 422; Pregnancy on page 404; Puberty on page 421; Sexual behavior on page 107*

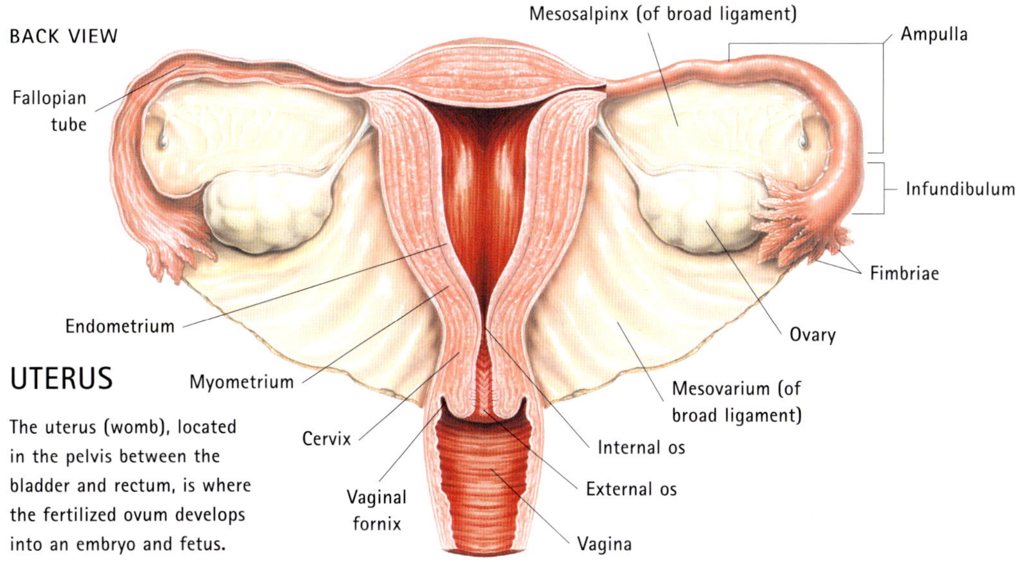

UTERUS

The uterus (womb), located in the pelvis between the bladder and rectum, is where the fertilized ovum develops into an embryo and fetus.

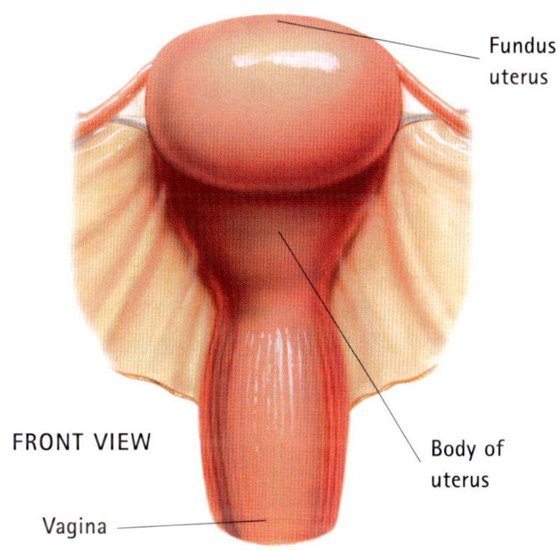

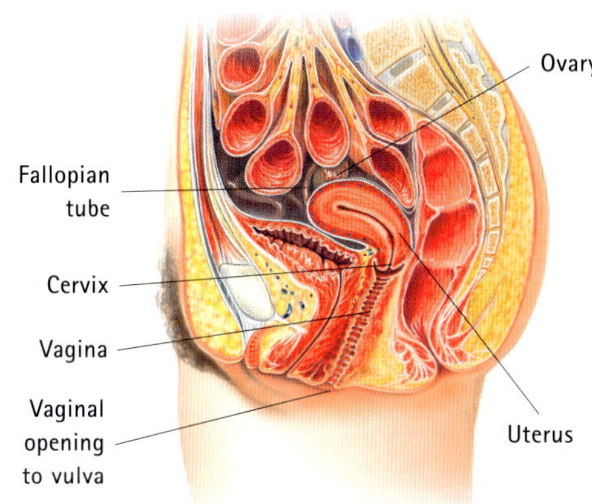

UTERUS

The uterus (womb), the organ of gestation, is located in the pelvis, between the bladder and the rectum. It undergoes changes during the menstrual cycle. The non-pregnant uterus is pear-shaped and flattened from front to back. It communicates with the vagina below. The two fallopian tubes open into its upper part, one on each side. In about 80 percent of women the uterus is anteverted (tilted forward); in the remaining 20 percent it is retroverted (tilted backward). The uterus should be somewhat mobile and can be displaced vertically by a distended bladder. This displacement allows a clearer view of a non-pregnant uterus during ultrasound examination.

The uterus consists of the fundus (top), body and cervix. The wall of the body consists of an endometrium or inner lining adjacent to the central cavity; a thick muscular myometrium; and an outer layer of peritoneum or perimetrium. The endometrium contains large numbers of glands and blood vessels. It undergoes proliferation and secretion during much of the menstrual cycle, followed by sloughing of all but the deepest parts at menstruation. The remaining tissue forms the basis for regeneration of a new endometrium during the next cycle. The myometrium undergoes mild contractions during menstruation.

The uterus enlarges enormously in pregnancy and its hollow center is obliterated during the third month. It reaches the top of the pubic bone by about 12 weeks, the level of the umbilicus at about 20 weeks and the diaphragm at 36 weeks. The pregnant uterus also displaces or compresses adjacent abdominal organs.

The uterus is attached to the lateral walls of the pelvis by the broad ligament. Further support is provided by ligaments attached to the cervix, which cross the pelvic floor and attach to the walls of the pelvis. Pelvic floor muscles are also important in maintaining the uterus in its correct position.

The cervix has a thick muscular wall and a mucus-secreting lining which is not sloughed at menstruation. The mucus is usually thick and fills the cervical canal to form a protective plug and a barrier to sperm penetration. The mucus becomes thinner around the time of ovulation. Some of the mucus glands may become blocked and enlarged to form cysts. The layer of cells adjacent to the cervical canal, which connects the vagina to the uterus, is thin but becomes thicker just above the opening of the cervix with the vagina.

Hysterectomy is surgical removal of the uterus. This is described as subtotal when only the uterus is excised and total if both the uterus and cervix are removed. A radical hysterectomy involves the removal of uterus, cervix, fallopian tubes, ovaries, the upper third of the vagina and at least some of the adjacent lymph nodes. Hysterectomy may be carried out via an abdominal incision or via an incision around the cervical opening (vaginal hysterectomy).

SEE ALSO *Abortion on page 464; Contraception on page 448; Dilation and curettage on page 463; Female reproductive system on page 104; Fertility on page 396; Menstruation on page 422; Pregnancy on page 404*

Endometrium

The endometrium is the inner glandular layer of the uterus. It consists of simple, tubular-shaped glands overlying a layer of connective tissue called the lamina propria. In a woman of reproductive age (approximately 13 to 50 years), the endometrium undergoes cyclical changes in its thickness and composition as part of the normal 28-day menstrual cycle.

During the first part of the cycle (the proliferative phase—lasting from 5 to 14 days after the appearance of menstrual blood), estrogen from the ovaries makes the endometrium thicker by inducing the multiplication of gland cells.

The second part of the cycle (the secretory or luteal phase, from 15–28 days after the appearance of menstrual blood) starts after the release of the egg from the ovary and the formation of the corpus luteum, which secretes progesterone. Progesterone makes the glands of the endometrium secrete and blood vessels of the endometrium be coiled.

The final stage of the cycle is the menstrual phase; it begins at the end of the secretory phase and lasts 4 days. It ends when the proliferative phase of the next cycle begins. The menstrual phase occurs only if the egg is not fertilized and no embryo embeds in the endometrium.

During the menstrual phase, hormonal support for the endometrium is lost, leading to the endometrium breaking down with bleeding. Blood and the discarded endometrium is expelled from the cervix into the vagina as menstrual fluid.

The endometrium is also where the embryo will implant to continue its development into a fetus during pregnancy. It is the tissue of the endometrium that contributes blood vessels to the maternal part of the placenta, supplying the fetus with nutrient-rich blood.

After menstrual cycles have stopped (usually at about 50 years of age—a stage of a woman's life called menopause), the hormonal support of the endometrium is lost and the endometrium becomes thinner, less vascular and not prone to cyclical changes.

Disorders of the Uterus

Many disorders of the uterus present with heavy menstrual bleeding. In the past, hysterectomy was often seen as a panacea for these conditions but the modern trend has been to reserve hysterectomy only for cases in which it is absolutely necessary.

SEE ALSO *Endometrial biopsy on page 436; Female procedures on page 462; Hysteroscopy on page 434; Infertility on page 397; Laparoscopy on page 435; Pap smear on page 436; Ultrasound on page 432*

Developmental disorders

Complete absence of the uterus is rare but deformities resulting in a bicornate (double) uterus or a single uterus divided by an internal septum can occur. These reflect the embryological origin of the uterus, which starts as two separate tubes that fuse during fetal life. A double or septate vagina may accompany bicornate uterus.

Retroversion of the uterus

The uterus is usually tilted forward in the pelvis (anteverted). Retroversion, where the uterus is tilted backward, exists in about 20 percent of women. The condition may be present from puberty but, providing the uterus remains mobile and is able to expand during pregnancy, no symptoms or complications are likely to occur.

In some instances retroversion may occur later in life because the pressure of either a tumor or adhesions fixes the uterus in an abnormal position. In such cases symptoms such as backache and pelvic pain may be experienced and there may be difficulties in conception and pregnancy.

Pelvic inflammatory disease

Pelvic inflammatory disease involves infection of the female reproductive organs above the level of the cervix. This type of infection is serious for several reasons. Firstly, infection may spread throughout the abdominal cavity, giving rise to peritonitis. Secondly, infection may spread to nearby pelvic organs to create an abscess. Thirdly, it is a common cause of infertility because the fallopian tubes may become blocked by scar tissue during healing.

Most pelvic infections are usually caused by bacteria. They may occur after childbirth or termination of pregnancy, in which case invasion of the tissues by bacteria (*E. coli*, *Clostridium*, *Streptococcus*) is rapid and the patient may be at great risk of dying from septicemia. Where the infection is due to the bacteria *Neisseria gonocccus* (gonorrhea), the progress is usually slower, and can cause vaginal discharge, fever, burning when passing urine, and pelvic pain.

Long-term pelvic infection may cause abnormal periods, menstrual pain, painful intercourse and infertility. Treatment for

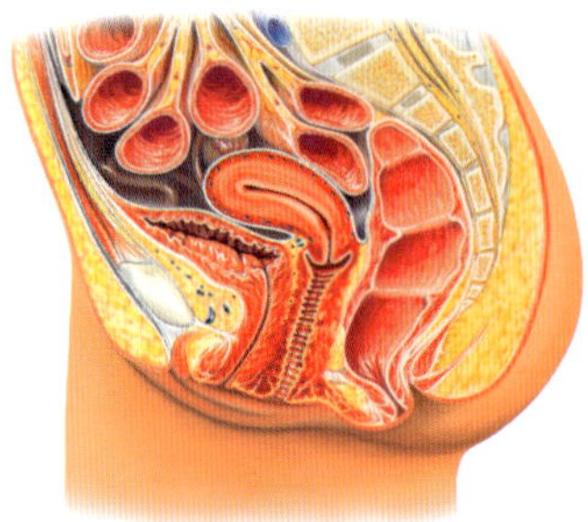

NORMAL UTERUS

A normal uterus is tilted forward in the pelvis.

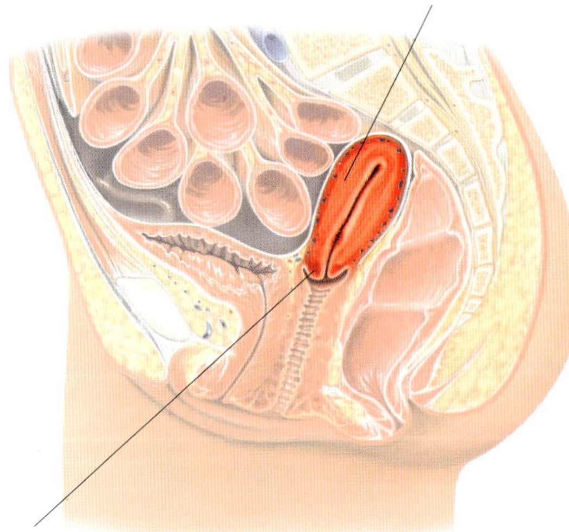

Retroverted uterus

One-fifth of all women have a retroverted uterus, in which the uterus is tilted backward in the pelvis.

pelvic inflammatory disease includes the appropriate antibiotic and pain relief, although surgery may be required for complications of chronic infection.

Fibroids

Leiomyomata uteri, known colloquially as fibroids, are benign (non-malignant) tumors of the uterus which are believed to arise from smooth muscle cells of the wall of the uterus. They are extremely common, occurring in 20 percent of Caucasian women and 50 percent of African-American women. Where symptoms do occur, they include abdominal distension (swelling), abdominal discomfort, constipation, heavy periods (hypermenorrhea), increased frequency of urination, and bleeding between periods.

Women with fibroids may get pregnant, but with complications such as miscarriage, premature or prolonged labor and excess bleeding after birth (postpartum hemorrhage). Patients with fibroids may suffer from anemia due to excessive blood loss following heavy menstrual periods.

Treatment of fibroids is not necessary where there are no symptoms, but such tumors should be carefully watched to determine the rate of growth. If the woman is experiencing symptoms, the uterus may be removed (hysterectomy) or the individual lump cut out (myomectomy).

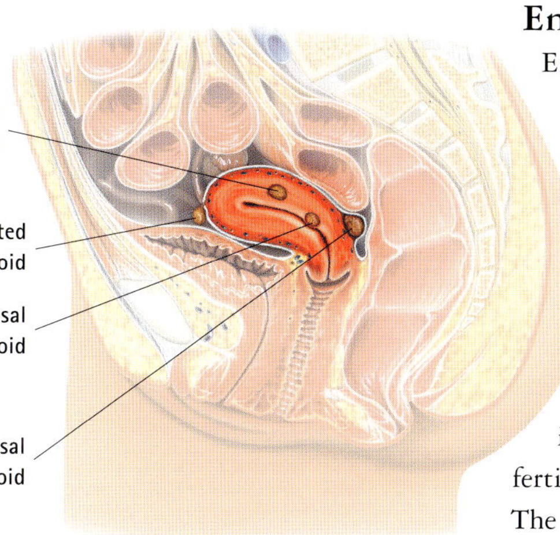

Fibroids

Fibroids are non-malignant tumors of the uterus. Intramural fibroids grow in the muscular tissue in the uterus wall. Submucosal fibroids grow in the endometrial layer. Subserosal fibroids occur just under the surface of the uterine wall. Pedunculated fibroids develop a stalk which can protude from the uterus or cervix.

Myomectomy is used in younger women who wish to retain the ability to have children. Unfortunately, myomectomy is usually less effective than hysterectomy and fibroids may recur.

Prolapsed uterus

A prolapsed uterus is the displacement of the uterus into the vaginal canal as a result of weakness in the supporting ligaments. This weakness may be due to the normal ageing process or the result of stretching during childbirth.

Hormonal changes during menopause are also thought to be a cause and, rarely, a pelvic tumor. Obesity, chronic bronchitis, asthma, excessive coughing, and chronic constipation may increase the risk of developing the condition. Sufferers may experience feelings of pressure or pulling in the pelvis, pain in the anus or lower abdomen, urinary tract infections, excessive vaginal discharge and difficulty having sexual intercourse. They may also urinate when coughing, laughing, or straining to lift heavy objects. In severe cases the neck of the uterus (cervix) protrudes from the vagina.

A pelvic examination determines the severity of the prolapse. Treatment may be via vaginal pessary, a ring-shaped object inserted into the vagina, or by surgery.

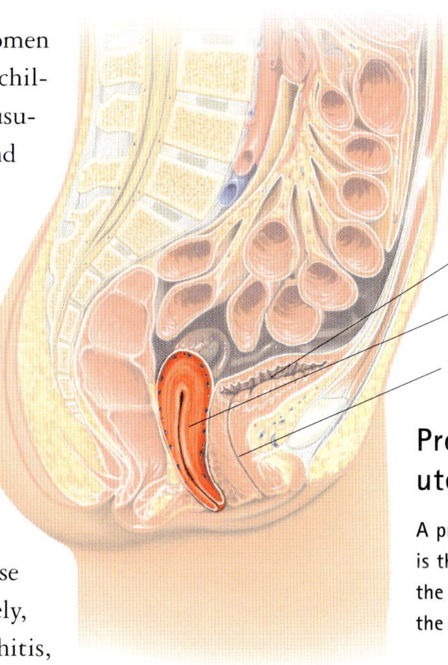

Prolapsed uterus

A prolapsed uterus is the descent of the uterus into the vaginal canal.

Endometriosis

Endometriosis is a condition in which tissue normally found on the lining of the uterus grows in other parts of the body such as the pelvic cavity, ovaries, bowel, bladder and rectum. This tissue (called the endometrium) acts as it would in the uterus, swelling before each period as if to prepare for nourishing a fertilized egg and then bleeding. The result is scarring and adhesions (clusters of endometrial cells), which may implant in the ovaries and fallopian tubes and obstruct the passage of the ovum.

Symptoms include increasingly painful and abnormal periods, lower back pain, pelvic cramps, pain during intercourse, abdominal pain before or during menstruation, blood in the urine, fatigue, bloating, diarrhea and constipation. Some women may not experience pain. Side effects are blood cysts in the ovaries and infertility.

The cause of endometriosis is not known, though women whose mother or sisters have had the condition have a greater chance of suffering from it. Mild cases are treated with painkillers. Symptoms may be reduced by creating a state of pseudo-pregnancy or menopause using hormonal drugs or oral contraceptives. Where pregnancy is not desired, options include removal of the ovaries or a hysterectomy. Scar tissue and

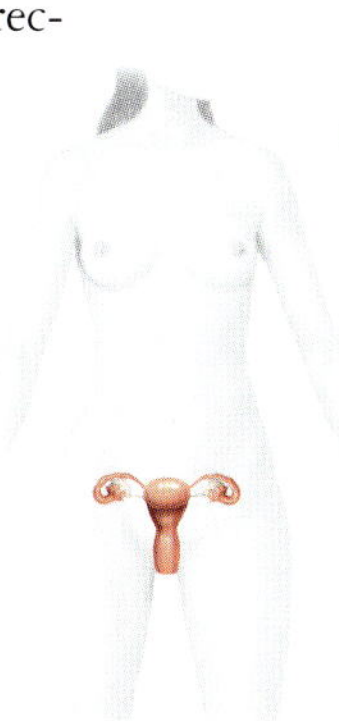

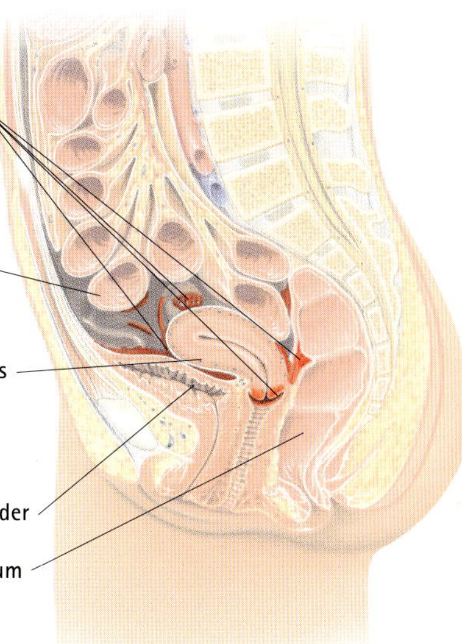

Endometriosis

In this condition, endometrial tissue is found outside of the uterus in other parts of the body, for example the fallopian tubes, cervix, ovaries, rectum, bladder and bowel.

adhesions may be surgically removed in some cases. Pregnancy has been known to cure the condition.

Cancer of the endometrium

Cancer of the endometrium is usually seen only in post-menopausal women. The most common symptom is bleeding after menopause. In fact, about 40 percent of women who have vaginal bleeding after menopause have a cancer of the reproductive tract. Pain is not a common symptom, but mild uterine cramping may occur.

Diagnosis is made by taking a scraping of the uterus wall (curetting), and examining the tissue under a microscope. Treatment consists of removal of the uterus (hysterectomy), which can be curative if the cancer is confined to the uterus, or radiation therapy along with with hormone therapy and surgery, if some spread has occurred to areas located outside the uterus.

Menstrual problems

The days preceding menstruation are a time of discomfort for some women; this is known as premenstrual tension or premenstrual syndrome and should be differentiated from dysmenorrhea, which is discomfort or pain during menstruation. Dysmenorrhea is usually most severe for the first 12 hours or so of the period. It is most prevalent in younger women and usually appears 2–4 years after the onset of menstruation and is uncommon after the thirtieth year. It is widely believed that childbirth will cure dysmenorrhea but this is not always the case.

The pain characterizing dysmenorrhea ranges from a dull ache to a severe ache with colic. It is usually located in the lower abdomen but may also be referred to the inside and front of the thighs. Nausea, vomiting, diarrhea and migraines sometimes accompany the pain.

It is now generally accepted that the pain is the result of the actions of a group of substances produced in the endometrium called prostaglandins. These are involved in both endometrial breakdown and control of excessive bleeding during menstruation. They also stimulate contractions of the uterine and cervical muscle during menstruation which may become painful (prostaglandins are involved in uterine

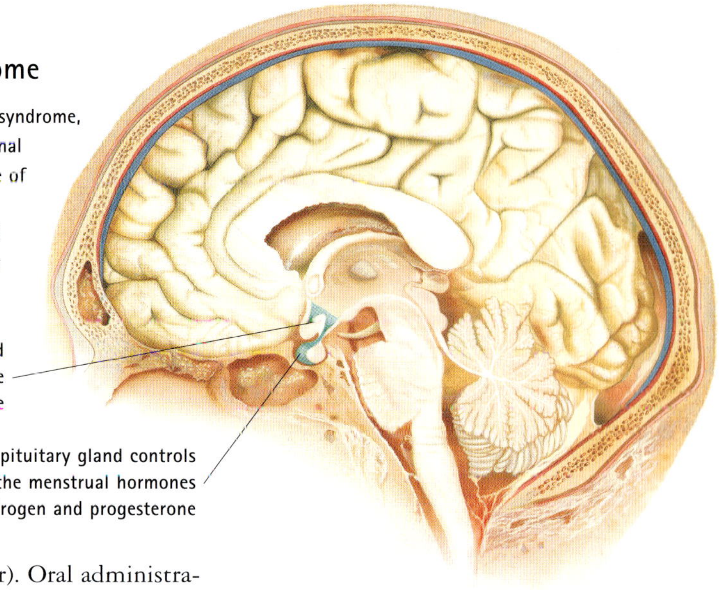

Premenstrual syndrome

In some woman, premenstrual syndrome, which may be linked to hormonal fluctuations, can cause a range of physical and psychological symptoms. These include mood changes, depression, irritability and lack of concentration.

contractions during labor). Oral administration of prostaglandin inhibitors for the first few days of the period help to alleviate the dysmenorrhea. Prostaglandin production does not occur without previous stimulation of the uterus by progesterone, so an alternative therapy is to use oral contraceptives which suppress ovulation.

Dysmenorrhea mostly occurs in the absence of any recognizable abnormality, when it is called primary dysmenorrhea. It may also be a symptom of various gynecological disorders such as endometriosis, when it is called secondary dysmenorrhea.

Amenorrhea is absence of menstruation, either primarily in women who have never menstruated or secondarily, when menstruation ceases after having occurred previously. Amenorrhea of either sort may be the result of pituitary malfunction due either to injury or to a tumor or, rarely, to the congenital inability of the pituitary to produce gonadotropins. Stress, excessive or heavy exercise or low body weight are also implicated. Primary amenorrhea may be the result of chromosomal or genital abnormalities or ovaries that are insensitive to gonadotropins.

PREMENSTRUAL SYNDROME

Premenstrual syndrome (PMS) describes a range of physical and psychological symptoms which usually occur for a few days prior to menstruation. About 80–90 percent of women experience PMS at some time during their reproductive lives, although only about 5 percent report severe symptoms. In some cases PMS may extend as far back as ovulation; a few women may experience PMS-like symptoms only around the time of ovulation and not later in the cycle. PMS is most common in women over 30, especially those who have had years of natural menstrual cycles uninterrupted by pregnancies or oral contraception.

It has been suggested that PMS is less common in non-Western societies but this is probably related to more frequent pregnancy and lactation than to any other social factors. Behavior characteristic of PMS has also been reported in non-human primates (rhesus monkeys and baboons) which also have menstrual cycles.

Physical disturbances associated with the syndrome may include weight gain, fluid retention, a sensation of pelvic heaviness or bloating, breast tenderness or enlargement, skin blemishes, headaches, constipation, and frequency of urination. Pre-existing conditions such as varicose veins, migraine and acne may worsen. The thyroid gland may be enlarged and blood sugar slightly elevated.

Psychological or behavioral changes may include an inability to concentrate, tiredness, mood swings, irritability and depression. PMS has been taken as a mitigating circumstance in certain criminal trials but this also undermines female responsibility in other spheres. Suicide or suicidal thoughts and accidental death have been reported as being more common in the latter part of the menstrual cycle. Symptoms should be classified as due to PMS only where they occur between ovulation and menstruation and are resolved within a few hours of the onset of menstruation.

Many theories have been put forward to explain PMS. Possible psychological causes include negative social or cultural attitudes towards menstruation and the unconscious association of menstruation with preexisting psychological or psychiatric problems. It has been hypothesized that endocrine imbalances involving prolactin, insulin, cortisone and androgens may be the cause.

Many practitioners believe that a simple explanation can be found in progesterone deficiency or in altered progesterone-to-estrogen ratios. However, studies of hormone levels of women suffering from PMS have failed to conclusively confirm these theories, although the occurrence of PMS at times of hormonal fluctuations, such as at puberty or after pregnancy, indicates at least some involvement of endocrine factors.

The wide range of symptoms and the uncertainty regarding their cause makes effective treatment of PMS difficult. Keeping a daily chart of physical and psychological symptoms may help in deciding which symptoms should be considered for treatment. Vitamin B_6, fluid tablets, prostaglandin inhibitors, antidepressants and low doses of progesterone have all been tried but, although they give relief in some women, are not universally successful. Altering the hormonal pattern of a normal menstrual cycle with oral contraceptives may alleviate PMS in some women but others report aggravation of symptoms. Some success has also been reported with dietary changes, including reduction in caffeine or salt intake, and with increased exercise.

Surgical removal of the ovaries (oophorectomy, also known as ovariectomy) has been used as a final resort in severe cases. There is little evidence that pregnancy or tubal ligation can have any effect on PMS.

Uterine bleeding

Uterine bleeding, similar to normal menstruation, may occur even if ovulation has not taken place. The hormonal environment prior to endometrial breakdown in an an ovulatory cycle will be different from that in a normal cycle. If ovulation does not occur, there is no corpus luteum to produce progesterone and so the bleeding is not the result of progesterone withdrawal. Instead the follicle that should have ovulated continues to secrete estrogen for a variable length of time till it degenerates and blood estrogen levels drop. The endometrial response is to undergo breakdown in a similar way to that which occurs under progesterone withdrawal, although often to a lesser extent, so that loss of blood and tissue debris is lighter than with a normal period. Withdrawal bleeding when an oral contraceptive is stopped for a few days occurs in a similar way.

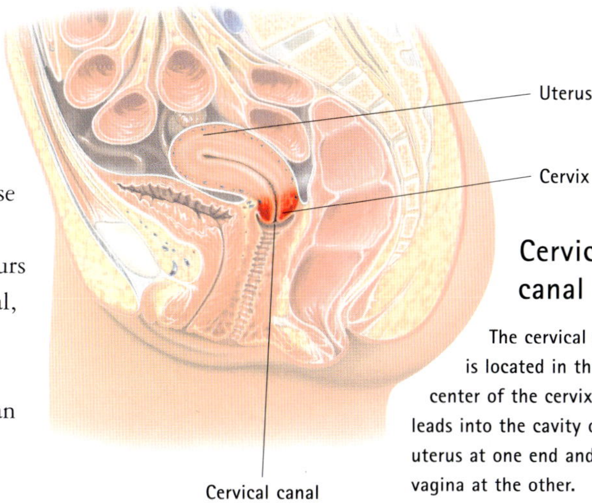

Cervical canal

The cervical canal is located in the center of the cervix, and leads into the cavity of the uterus at one end and the vagina at the other.

Anovulatory cycles are normal for the few months after menarche and just before menopause. Anovulation may be established by biopsy of the uterine lining and looking for changes in the tissue which indicate the influence of progesterone.

Uterine bleeding is considered to be abnormal if menstruation is heavy, of normal intensity but occurring too frequently, or if bleeding or spotting occurs consistently between normal periods. This can be due to disturbances to the normal production of pituitary gonadotropins, ovarian hormones or prostaglandins.

Other possible factors include cancer of the uterus, cervix, ovary or fallopian tube, retained fragments of placenta, an infection such as pelvic inflammatory disease, ectopic pregnancy, polycystic ovary syndrome and, occasionally, coagulation disorders that prevent normal blood clotting.

Irregular abnormal bleeding may be associated with the use of an intrauterine device for contraceptive purposes, or emotional disturbances. Bleeding may also be associated with ovulation. Benign tumors in the uterine muscle called fibroids or leiomyomas are also a cause of excessively heavy menstruation ("flooding") in older women.

Irregular or heavy bleeding is frequently associated with menopause. In the absence of any abnormality this is a symptom of the alteration in the pattern of female and pituitary hormones at this time. Bleeding occurring after menopause should be investigated thoroughly as it may be a symptom of a malignant disease of the uterus.

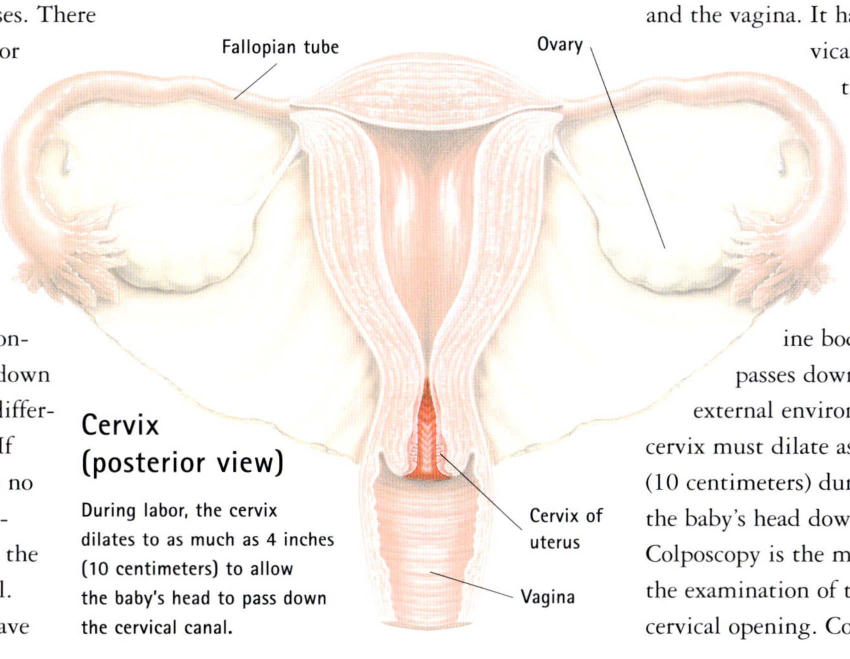

Cervix (posterior view)

During labor, the cervix dilates to as much as 4 inches (10 centimeters) to allow the baby's head to pass down the cervical canal.

Cervix

The cervix is the lower part of the uterus. It is situated between the body of the uterus and the vagina. It has a centrally placed cervical canal, which leads into the cavity of the uterine body at one end and opens to the vagina at the other.

Sperm must pass through the cervix on their way to the uterine body, and menstrual blood passes down the cervix to the external environment. In labor, the cervix must dilate as much as 4 inches (10 centimeters) during the passage of the baby's head down the birth canal. Colposcopy is the medical term used for the examination of the upper vagina and cervical opening. Colposcopy is used

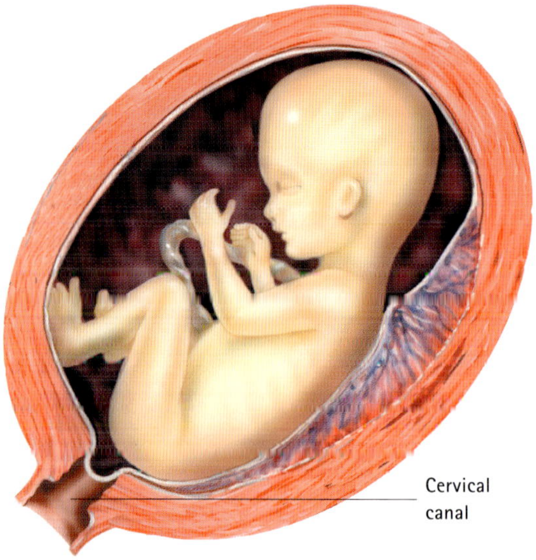

Incompetent cervix

This condition can cause miscarriage, as the weakened cervix prematurely dilates allowing the amniotic sac to rupture.

principally for diagnosis, whereby samples of cervical tissue and fluid are taken for further examination.

SEE ALSO *Artificial insemination on page 464; Childbirth on page 408; Contraception on page 448; Dilation and curettage on page 463; Female reproductive system on page 104; Fertility on page 396; Fertilization on page 397; Infertility on page 397; Menstruation on page 422*

Disorders of the Cervix

The most serious disorder affecting the cervix is cervical cancer. Fortunately, the incidence of cervical cancer has decreased significantly in recent years through the widespread use of the Pap smear, which makes early detection possible.

SEE ALSO *Hysteroscopy on page 434; Laparoscopy on page 435; Pap smear on page 436*

Viral infections

The sensitive surface tissue of the cervix is in contact with the external environment, particularly during sexual intercourse when the penis may introduce viruses and bacteria into the vagina. Two viruses of particular concern are the human papilloma virus (HPV) and the herpes simplex type II virus. HPV may cause warts to grow on the vaginal and cervical tissue. Human papilloma virus is believed to cause cervical cancer (carcinoma), by inducing changes in the cells of the cervix.

Cancer of the cervix

Cervical cancer is the second most common cancer affecting women, accounting for almost 10 percent of cancers in females in developed countries. The incidence of the disease is more common in those women who have frequent intercourse with many different partners and is also more common in women who have given birth.

During the early stages the cancer of the cervix may be confined to the cervical tissue and can be easily removed by surgically cutting out a cone-shaped block of tissue. It is for this reason that early detection, by regular use of the Papanicolaou ("Pap") smear test, is very important.

Cervical polyps

Cervical polyps are small tear-shaped structures that protrude through the vaginal opening of the cervical canal. They are composed of tissue derived from the inner lining of the cervical canal. They can cause bleeding and may be removed surgically.

Incompetent cervix

During pregnancy, the cervix is sealed with a plug of mucus and stays tightly shut, keeping the fetus safely within the uterus, until labor begins. However, in some cases, especially where the cervix has been damaged by surgical procedure such as a cone biopsy, the cervix can be weakened and may open prematurely during the third or fourth month of pregnancy. When this happens, the amniotic sac may pass through the cervix into the vagina and may rupture, causing loss of amniotic fluid and miscarriage.

The condition is not usually diagnosed until a miscarriage has occurred; however, subsequent miscarriages can be prevented by a procedure in which the cervix is stitched closed with strong thread. This procedure tightens the cervix and keeps it from dilating prematurely. The stitches are removed at approximately 37 weeks to allow delivery of the fetus to take place.

Ovaries

Situated close to the side walls of the pelvis and supported by the broad ligament of the uterus, the ovaries are elliptical organs roughly 1½ inches (3 centimeters) long and ½ inch (1 centimeter) wide.

Ovaries contain thousands of small undeveloped follicles, each of which consists of an ovum surrounded by specialized secretory cells that produce female hormones. Ovaries in women of reproductive age also contain a few follicles undergoing enlargement and development prior to each ovulation. Most of these developing follicles will degenerate so that only a single ovum from one of the ovaries will finally be ovulated.

After ovulation, a corpus luteum is formed from the remains of the ovulated follicle. About two days before the start of the next menstrual cycle it will begin to degenerate. Degenerating follicles and corpora lutea persist for a number of cycles and in some women continue to produce small amounts of estrogen and progesterone.

Changes in the normal function and structure of the ovaries occur with age. Several years prior to menopause the number of follicles (both undeveloped and developing) begins to decrease, there is a decline in the overall amount of estrogen in circulation and loss of other ovarian tissue, resulting in an overall decrease in size of the ovaries.

SEE ALSO *Contraception on page 448; Endocrine system on page 110; Female reproductive system on page 104; Fertility on page*

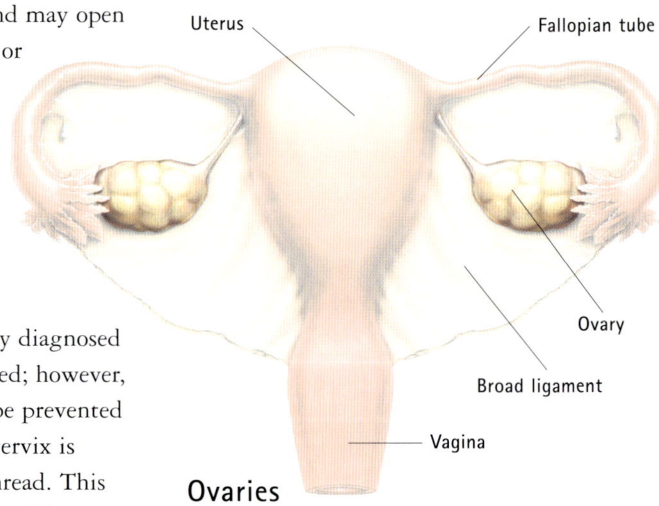

Ovaries

The two ovaries are the female gonads in which the ova (eggs) are formed. They resemble large almonds in both shape and size and are situated on either side of the uterus, supported by the broad ligament.

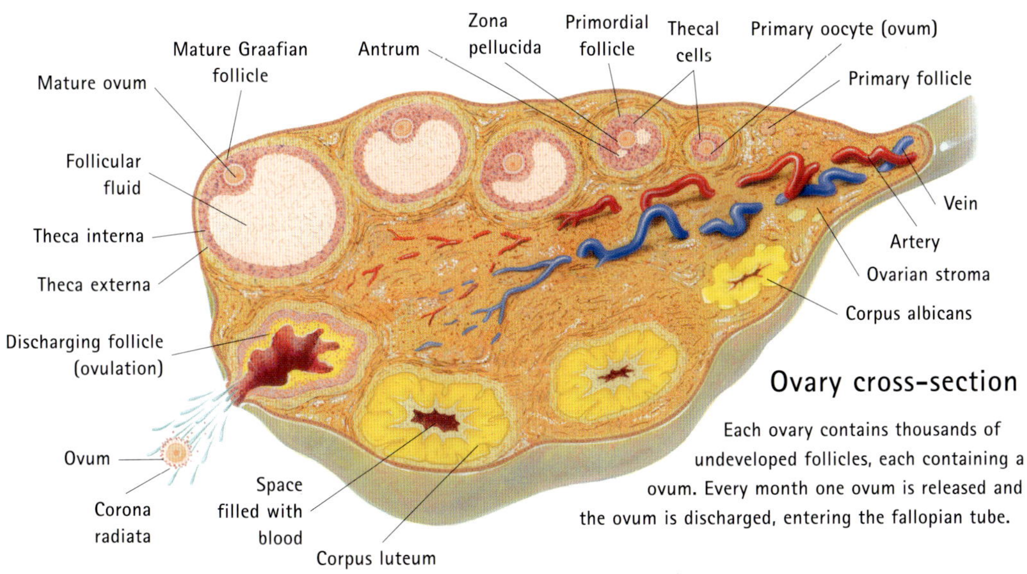

Ovary cross-section

Each ovary contains thousands of undeveloped follicles, each containing a ovum. Every month one ovum is released and the ovum is discharged, entering the fallopian tube.

396; Fertilization on page 397; Hormone replacement therapy on page 450; Infertility on page 397; Menopause on page 424; Menstruation on page 422; Puberty on page 421

Graafian follicle

At puberty the average woman has about 300,000 egg cells (ova) lying dormant within her ovaries. When she becomes sexually reproductive, some 20 ova (each contained within its own sac-like structure known as a follicle) begin to ripen at the beginning of each menstrual cycle.

Usually just one follicle becomes dominant and continues development while the others, and the eggs they contain, shrivel up. This remaining follicle ultimately matures into a Graafian follicle.

The Graafian follicle has a diameter of about ½ inch (12 millimeters), is composed of an outer wall three to four cells thick surrounding follicular fluid, and a mature ovum; it produces the hormone estrogen, which prepares the uterus for pregnancy. During ovulation it ruptures, releasing the ovum, which is then swept into the fallopian tubes where it may be fertilized by a sperm. The remaining follicle collapses on itself, its cells enlarge and the structure develops into the corpus luteum, which produces the hormones progesterone and estradiol.

If conception fails to take place, the corpus luteum ceases development after about 14 days and degenerates. If, however, the ovum is fertilized, the corpus luteum continues development and plays an important role during pregnancy.

Ovarian hormones

The two main hormones produced by the ovaries are estrogen and progesterone, which play a unique role in ovulation, pregnancy and the maturing female's development of secondary sexual characteristics.

ESTROGENS

Estrogens are a group of steroid hormones produced in the ovaries and, in lesser amounts, by other organs. Steroid hormones are fatty substances derived from cholesterol and are taken up from the blood. The conversion of cholesterol to estrogens takes place via a series of steps, each of which is controlled by an enzyme. Intermediate substances in the pathway include progesterone and testosterone; if one of the enzymes is deficient or inactive, testosterone may be secreted instead of estrogen.

The estrogens produced in the ovary are estradiol and a related substance, estrone.

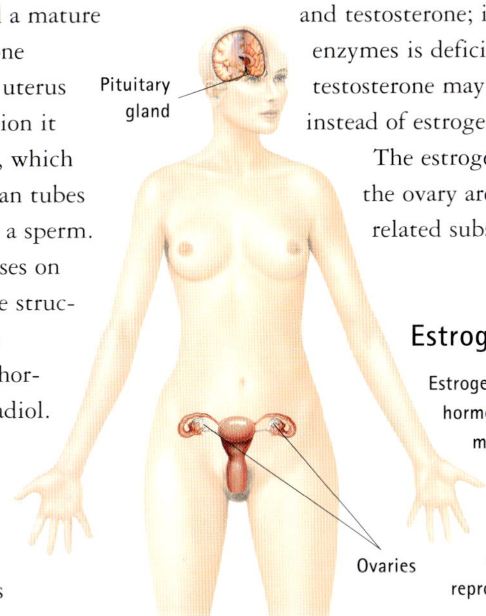

Estrogens

Estrogens are female sex hormones produced mainly by the ovaries. They control the development of the female sex characteristics and reproductive system.

These are both released into the circulation and are broken down in the liver to form estriol, found in urine. The ability of target organs, such as the uterus, to respond to estrogens and other hormones is determined by whether the cells of that organ have receptors for the hormone on their surfaces.

Estrogens are secreted in high quantities in pregnancy. The placenta and the fetal and maternal adrenal glands are all involved. Estrogen levels fall a few days after giving birth. Estrogen production declines before and during menopause, and its loss may cause thinning of the lining of the vagina, vulvitis, osteoporosis and increased risk of cardiovascular disease. Hormone replacement therapy, using a combination of estrogen and progesterone reduces or eliminates these effects.

PROGESTERONE

Progesterone is a steroid sex hormone produced in the corpus luteum (ruptured follicle) in the ovaries during the second half of the menstrual cycle. Progesterone is one of the main hormones of pregnancy. It stimulates the endometrium to secrete a fluid, which protects and nourishes the fertilized ovum in the uterus before implantation. It also fosters placental growth.

If pregnancy does not occur, then the corpus luteum only functions until about day 26 of an average cycle, after which

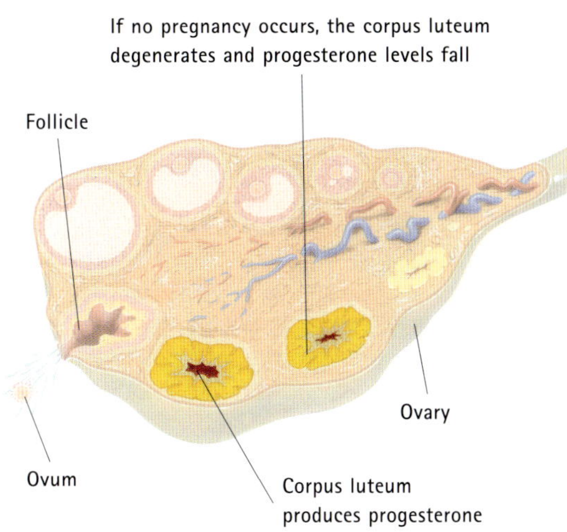

Progesterone

After releasing its ovum (egg), an ovarian follicle turns into a gland-like structure, the corpus luteum. This produces progesterone, which prepares the uterus for pregnancy.

progesterone production decreases rapidly; this causes changes in the lining of the uterus that lead to menstruation. If the woman becomes pregnant, the placenta secretes human chorionic gonadotropin (HCG), which prolongs the production of ovarian progesterone until the end of the first trimester, after which progesterone is secreted by the placenta.

Progesterone drops to very low levels a few hours after birth. During lactation the pituitary gland produces large amounts of prolactin, and this often disrupts development of follicles in the ovary and suppresses ovulation. Even when menstruation is restored, which can be as early as three months after birth, the first few cycles are often anovulatory. Anovulatory cycles do not produce a corpus luteum.

Diminished secretions of progesterone can lead to spontaneous abortion (miscarriage) in pregnant women.

Synthetic progesterone is used as an oral contraceptive, most commonly in combination with estrogen but sometimes on its own if use of estrogen is inadvisable.

Ovulation

Ovulation is the release of an ovum from the ovary on about day 14 of an average menstrual cycle. It can take place from either ovary; the selection of left or right ovary seems to be a random process.

Before ovulation, during the first half of each menstrual cycle, a few follicles, each consisting of an ovum and the secretory cells surrounding it, start to develop. Most of these follicles will degenerate at various stages of maturity until usually just one is left to undergo ovulation. Development of the follicles is stimulated by the follicle stimulating hormone (FSH), a gonadotropin, released by the pituitary gland.

As the follicles develop, the secretory cells produce increasing quantities of estrogen, which is released into the blood. When blood estrogen reaches a particular concentration, the pituitary is stimulated to produce a surge of another gonadotropin called luteinizing hormone (LH). This induces ovulation through a temporary break in the surface of the ovary.

The open end of the fallopian tube adjacent to the ovary draws the ovum inside the tube. The ovum is either fertilized here or degenerates within the next few days. Transport into the uterus occurs only if fertilization has taken place.

Occasionally the fallopian tube may fail to collect the ovum, which then falls into the peritoneal cavity. If fertilization then occurs, ectopic implantation may result on one of the pelvic or peritoneal organs.

After ovulation, the secretory cells in the ovary transform into a corpus luteum which will then produce progesterone and small amounts of estrogen until day 26 of a non-pregnant cycle. The corpus luteum degenerates at this time with a rapid decline in concentrations of progesterone in the blood resulting in menstruation.

Determining whether and when ovulation has taken place is important in establishing the cause of infertility and also in "natural" methods of birth control and in timing intercourse to maximize the possibility of conception.

Tests should be made over a number of cycles in order to establish a pattern of ovulation. The simplest method is to take daily body temperature readings and to record these on a temperature chart. Temperatures should be lower during the first half of the cycle and there may be a further drop at the time of ovulation. Alternatively, cervical mucus may be sampled throughout the cycle. This dries in a characteristic fern-like pattern around the time of ovulation; it is not seen during the rest of the cycle.

Other techniques rely on the fact that a functional corpus luteum and progesterone production can only occur if ovulation has taken place. These can be established either by measuring blood progesterone levels during the second half of the cycle or by biopsy of the uterine lining in order to see if changes consistent with progesterone production have occurred. Monitoring of the developing follicles with ultrasound is also used in some instances.

In some women pelvic pain ("mittelschmerz"), possibly accompanied by bleeding, may be associated with ovulation.

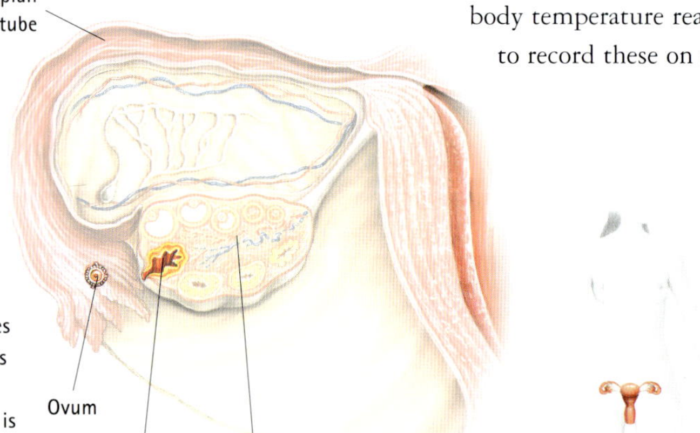

Ovulation

The release of an ovum (egg) from the ovary is known as ovulation and takes place on about day 14 of the menstrual cycle. A surge of hormone encourages the follicle that carries the mature ovum to burst open. The ovum is discharged and is swept up in the fallopian tube.

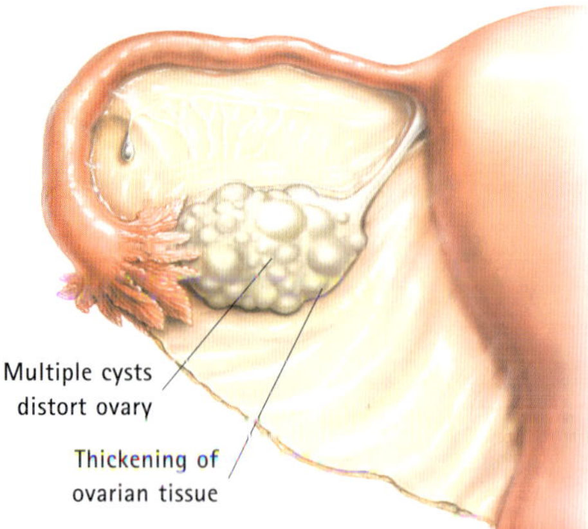

Ovarian cysts

In polycystic ovary syndrome multiple fluid-filled cysts form in the ovaries from follicles that fail to rupture and release eggs.

Disorders of the ovaries

Ovarian abnormalities are likely to cause infertility, for example if the ovary is unable to respond normally to the stimulating gonad hormones (gonadotropins) from the pituitary. This occurs in polycystic ovary syndrome (Stein-Leventhal syndrome). Decreased estrogen secretion is involved in the failure to ovulate and menstruate normally. Other types of ovarian cysts may also affect ovulation.

SEE ALSO *Female procedures on page 462; Infertility on page 397; Laparoscopy on page 435; Laparotomy on page 435*

Ovarian cysts

Ovarian cysts are relatively common and can occur in females of all ages, including both newborns (neonates) and those who have already been through menopause. They are hollow, fluid-filled structures which may be derived from the abnormal development of ovarian follicles or from a corpus luteum (an ovarian follicle that has matured and released its egg).

Ovarian cysts are multiple and affect both ovaries. They contain a watery or bloody fluid and are quite small although some may be as large as 2 inches (5 centimeters) or more in diameter. They are often symptomless and ovulation may occur in their presence; however, complications are not uncommon. These include pain, disturbances to the menstrual cycle and to ovulation with resulting infertility, excessive hair growth, rupture the cysts and degeneration of adjacent normal ovarian tissue. Larger cysts may damage the ovarian blood supply with possible loss of the entire ovary.

The cause of ovarian cysts of this type is complex but has been linked to imbalances in the production of pituitary hormones, collectively called gonadotropins, which are involved in the control of ovarian function.

In women of childbearing age some of the complications of ovarian cysts may be reduced with weight loss. Estrogen and/or progesterone or fertility drugs are other possible treatments. In some cases, drainage or surgical removal of the larger cysts or even of the whole ovary may be necessary.

Certain ovarian cysts are frequently associated with endometriosis (a condition in which a tissue-like uterine lining is found outside the uterus).

These cysts are lined with fragments of displaced uterine lining (endometrium) and are known as chocolate or tarry cysts because they contain a viscous brown material, arising from bleeding into cavities of the cysts during menstruation. Chocolate cysts are commonly multiple, large and affect both ovaries. These can be treated by giving hormones for the endometriosis, removing the cysts surgically.

Several forms of cystic tumors also exist and, unlike other types of ovarian cyst, may become malignant. One of the most common is multilocular cystadenoma. Usually only one ovary is affected, with a mass of tiny, mucus-filled cysts surrounding a single very large cyst. This can grow to enormous size, 44 pounds (20 kilograms) or more, which will distend the abdomen. The cyst may either be drained, or surgically removed. Sometimes the whole ovary will need to be removed. Other types of cystic tumor do not reach the considerable size of an untreated multilocular cystadenoma.

FOLLICULAR CYST

Following degeneration of an ovum (egg), the unruptured Graafian follicle enclosing the ovum begins to secrete fluid from its lining cells. This leads to the formation of a cyst in the ovary, which may grow to up to 2 inches (5 centimeters) in diameter. When this happens, the menstrual cycles sometimes become longer and excessive menstrual bleeding can ensue. If the cyst is larger than 2 inches (5 centimeters) in diameter, it needs to be surgically removed.

STEIN-LEVENTHAL SYNDROME

Stein-Leventhal syndrome, also known as polycystic ovarian syndrome, is a disorder of the follicles in the ovaries characterized by multiple cysts, the absence of menstruation, obesity, infertility and increased hair growth on the face and body.

Instead of releasing ova, the swollen follicles fill with fluid and turn into cysts. The ovaries can consequently become two to five times larger than normal. Abnormal hormone levels are thought to cause the condition, which commonly occurs just after puberty and may have a hereditary link.

Oral contraceptives are one of the medications used to treat the disorder. Pregnancy may be possible with medical help.

Cancer of the ovaries

Ovarian malignancies are the most common cause of death from gynecological cancers. This can occur at any age, although they are rare in childhood.

Ovarian cancers are often diagnosed late because they are relatively asymptomatic until well developed or have spread to other organs.

It is common for ovarian cancer to be described as if it were a single entity but there are many different types with different rates of spread and response to therapy. There is even a particular type of tumor called a teratoma, usually benign but sometimes malignant, in which fragments of skin, hair, cartilage, thyroid-like tissue and even teeth develop.

Some types of cancer are more common after menopause and they can cause postmenopausal bleeding.

Ovaries are also a fairly common site for secondary cancers arising from other organs.

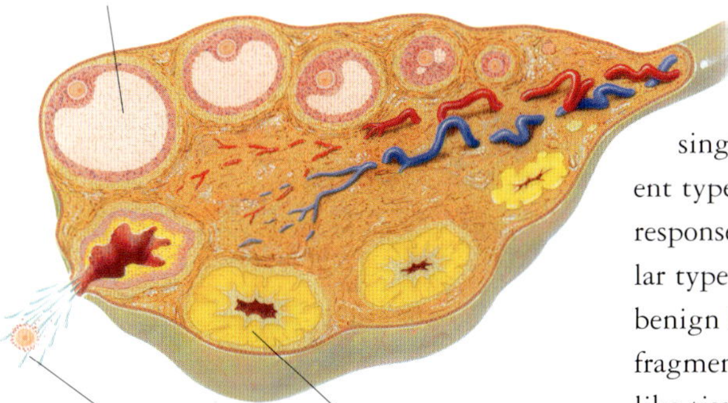

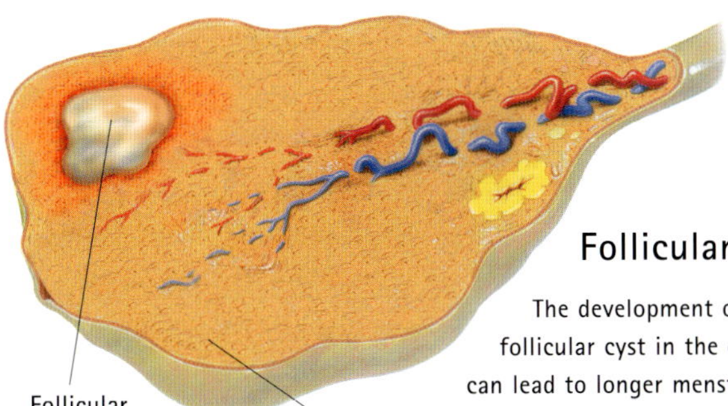

Follicular cyst

The development of a follicular cyst in the ovary can lead to longer menstrual cycles and excessive bleeding.

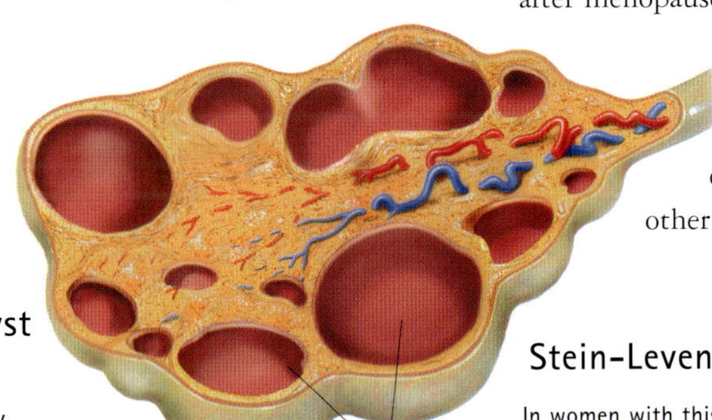

Stein-Leventhal syndrome

In women with this disorder, follicles in the ovaries do not release an ovum at ovulation but swell and turn into fluid-filled cysts.

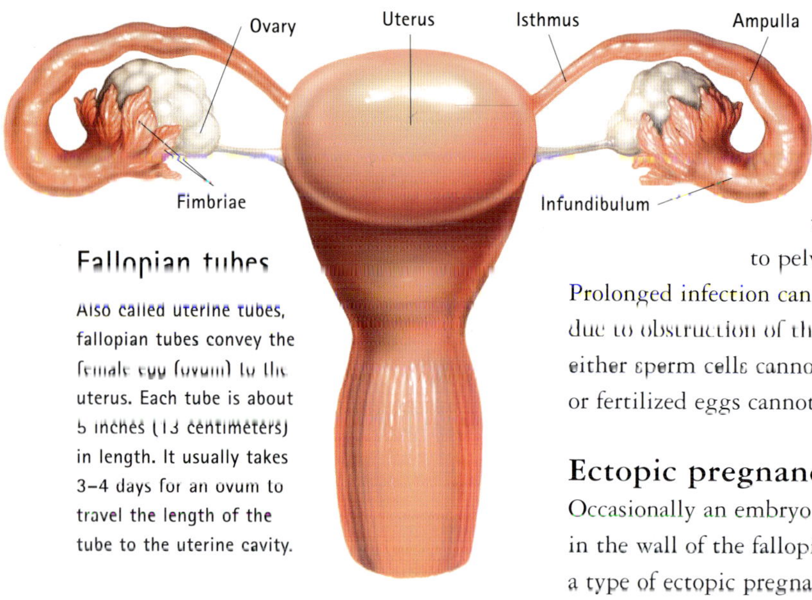

Fallopian tubes

Also called uterine tubes, fallopian tubes convey the female egg (ovum) to the uterus. Each tube is about 5 inches (13 centimeters) in length. It usually takes 3–4 days for an ovum to travel the length of the tube to the uterine cavity.

Fallopian tubes

Each of the two fallopian (uterine) tubes leads from its corresponding ovary, which is located on the lateral wall of the pelvis, to the body of the uterus.

They are shaped a little like an alpine or medieval trumpet, with a narrow end joined to the uterus and a broad, flared end next to the ovary. The fallopian tubes convey the woman's egg (ovum) to the uterus. Also, shortly after sexual intercourse, sperm cells may reach the fallopian tubes, traveling in the opposite direction. In fact, the fallopian tubes are a common site for fertilization (the junction of the egg and sperm) to occur. The fertilized egg becomes an embryo, and moves down the tube and implants itself in the wall of the uterus.

SEE ALSO *Contraception on page 448; Female reproductive system on page 104; Fertility on page 396; Fertilization on page 397; Pregnancy on page 404; Sterilization on page 461; Tubal sterilization on page 465*

Salpingitis

Salpingitis is inflammation of the fallopian tubes. The condition is usually due to pelvic infection. Prolonged infection can lead to infertility due to obstruction of the tubes so that either sperm cells cannot reach the eggs, or fertilized eggs cannot reach the uterus.

Ectopic pregnancy

Occasionally an embryo may be implanted in the wall of the fallopian tube, known as a type of ectopic pregnancy.

Ectopic pregnancies are unable to proceed to full term, because the embryo quickly outgrows the available blood supply and may rupture the tube. The mother's life may also be at risk when this happens, because profuse bleeding into the mother's abdominal cavity may occur, leading to potentially fatal blood loss. In such cases surgery to remove the damaged tube may be needed. However, this procedure normally leaves the other tube intact and capable of performing its reproductive function.

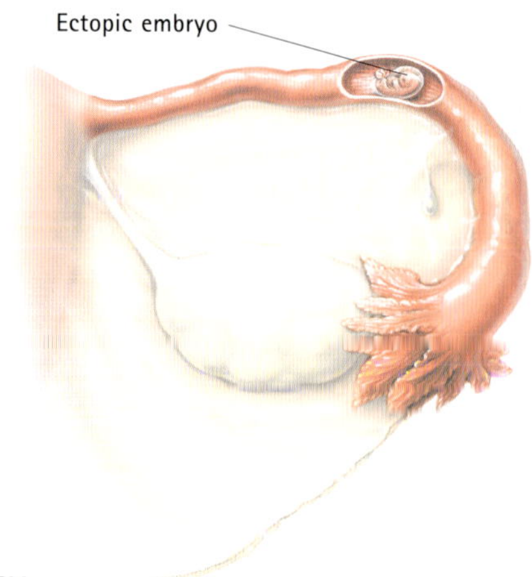

Ectopic pregnancy

An ectopic pregnancy occurs when a fertilized egg implants in the wall of the fallopian tube instead of in the wall of the uterus. At first it feels like a normal pregnancy until severe pelvic pain and vaginal bleeding occur. If not quickly treated, the condition can turn into a life-threatening medical emergency.

Disorders of the fallopian tubes

The most common disorder of the fallopian tubes is salpingitis due to pelvic infection. Salpingitis accounts for approximately 20 percent of the cases of infertility.

SEE ALSO *Female procedures on page 462; Infertility on page 397; Laparoscopy on page 435*

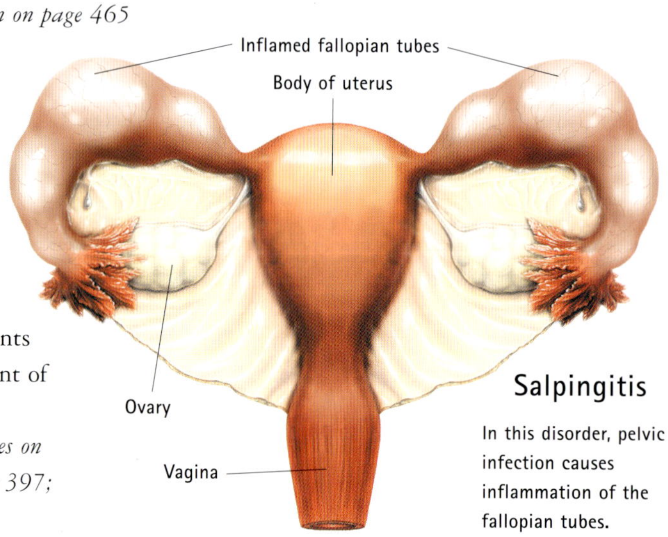

Salpingitis

In this disorder, pelvic infection causes inflammation of the fallopian tubes.

Vagina

The vagina is the passage connecting the uterus to the outside of the body. Situated in the lower pelvis and pelvic floor, it is located between the bladder and the rectum. The lower opening in virgin women is partly closed by a thin membrane called the hymen. The vaginal wall consists of a thick external muscle, an inner mucosa, and a cavity with the inner surfaces in direct contact. The cavity tilts slightly backward and is at about 90° to the cervical canal, which connects the vagina with the cavity of the uterus. The vagina contains fluid consisting of cervical mucus and plasma from capillaries in the mucosa. The external opening of the cervix protrudes into the upper part of the vagina and is surrounded by a groove or fornix. The vagina is supported by the cervical ligaments and pelvic floor muscles.

The vaginal mucosa is lined with a thick layer of cells containing glycogen. The superficial cells are sloughed into the vaginal fluid and replaced by cell divisions deeper in the lining. The sloughed cells are broken down by bacteria *(Lactobacillus)*, which are normal vaginal flora. The lactic acid thus produced discourages invasion by pathogens.

SEE ALSO *Artificial insemination on page 464; Childbirth on page 408; Contraception on page 448; Female reproductive system on page 104; Fertility on page 396; Menstruation on page 422; Sexual behavior on page 107*

Hymen

In a female infant the vaginal opening is closed by a thin membrane called the hymen. The hymen usually ruptures before puberty to allow menstrual blood to escape. An intact hymen used to be considered evidence of virginity but, in fact, the hymen usually ruptures, at least in some part, during physical exercise. Further rupture usually occurs during the first sexual intercourse.

After childbirth there is little of the hymen left in the mother. In rare cases the hymen may have failed to rupture before the attainment of puberty, causing menstrual blood to accumulate in the vagina. The hymen then must be cut surgically.

Disorders of the vagina

Vaginitis is a common disorder affecting the vagina, and its treatment depends on the underlying cause. Vaginal tumors are rare and occur most commonly in postmenopausal women.

SEE ALSO *Infertility on page 397; Pap smear on page 436*

Vaginitis

Vaginitis is a common gynecological complaint characterized by a discolored discharge and vaginal irritation and redness. Vaginitis is specific if it can be attributed to a particular irritative agent or bacterium. This is usually the case with vaginitis in women of reproductive age. Pre- and postmenopausal women have low levels of estrogen and the vaginal lining is very susceptible to a wide range of noxious agents. It is often impossible to pinpoint the cause of this type of vaginitis and it is described as non-specific.

One of the most common causes of vaginitis in young women is thrush resulting from infection with candida (yeast). Candida infections are not usually transmitted sexually and are due, in some cases, to loss of *Lactobacillus* as a result of antibiotic administration. Candida infections are especially prevalent in diabetes or pregnancy. Treatment is with fungicides. Vaginitis can also be the result of bacterial and protozoal infections, spread chiefly by sexual contact.

The vaginal mucosa atrophies after menopause, with little glycogen produced to maintain an acid environment. This renders the vagina prone to infection or irritation leading to post-menopausal or atrophic vaginitis, which is characterized by miniscule ulcers, often with a blood-stained discharge accompanied by itching or soreness. The condition is treated with estrogen pessaries or oral estrogens.

Tumors of the vagina

Benign vaginal tumors can arise in the fibrous tissue of the vaginal wall or from remnants of embryonic tissue. These may be removed surgically if they cause bleeding, discharge or local irritation. Vaginal malignancies are often the result of secondary spread from growths elsewhere. Primary vaginal carcinomas are more common in post-menopausal women. Treatment is by irradiation, or by removal of the carcinoma or the entire vagina and possibly adjacent organs. Primary vaginal adenocarcinoma may occur in women aged under 25 who were exposed in utero to diethylstilboestrol (a non-steroidal synthetic estrogen) administered to their mothers early in pregnancy.

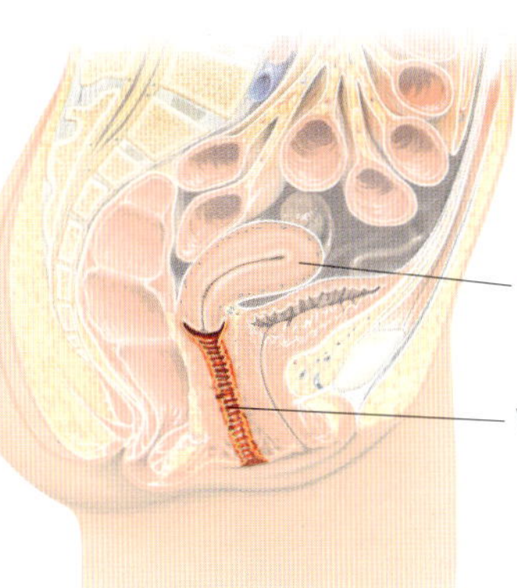

Vagina

The vagina is the passage that links the uterus to the exterior of the body. Its muscular walls are flexible enough to allow the passage of the baby during childbirth.

Vulva

The female external genitalia are collectively known as the vulva. They are made up of paired folds, called the labia majora, which are covered by skin and pubic hair and have a moist internal lining. The labia minora are fleshy folds within the labia majora which lie on either side of the vestibule containing the vaginal and urinary openings and mucus-secreting glands. The clitoris is erectile tissue like the penis and has erotic functions. The upper ends of the labia minora join around the clitoris. Bartholin's glands are lubricating glands that are situated on either side of the vaginal opening at the innermost part of the labia.

The clitoris lies about ½–1 inch (1–2 centimeters) in front of the external opening of the urethra, which carries urine from the bladder. The shaft of the clitoris is about ½ inch (about 1 centimeter) in length. The tip of the clitoris has a glans or head, usually hidden within a fold of thin skin known as the prepuce. The prepuce of the clitoris is a forward extension of the labia minora.

SEE ALSO *Female reproductive system on page 104; Sexual behavior on page 107*

Disorders of the vulva

Inflammation and irritation of the vulval skin (known as vulvitis) can be the direct result of local irritants or be secondary to vaginal or urinary tract infection. Carcinoma of the vulva accounts for about 5 percent of all genital cancers and is usually treated with irradiation or surgery.

Perineum

The perineum encloses the base of the pelvis in both men and women. It is made up of a sheet of fibrous tissue and muscle and provides support for the pelvic floor muscles immediately above. It also contains the urethral and anal sphincters.

Vulva

The vulva is the collective term for the external parts of a woman's genitals, including the labia majora, labia minora, mons pubis and clitoris.

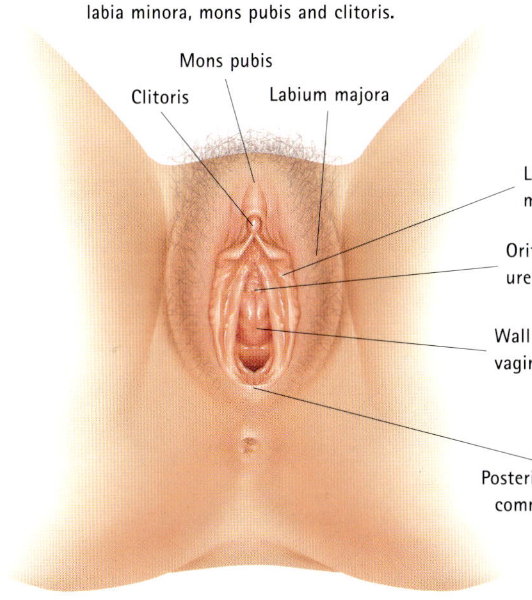

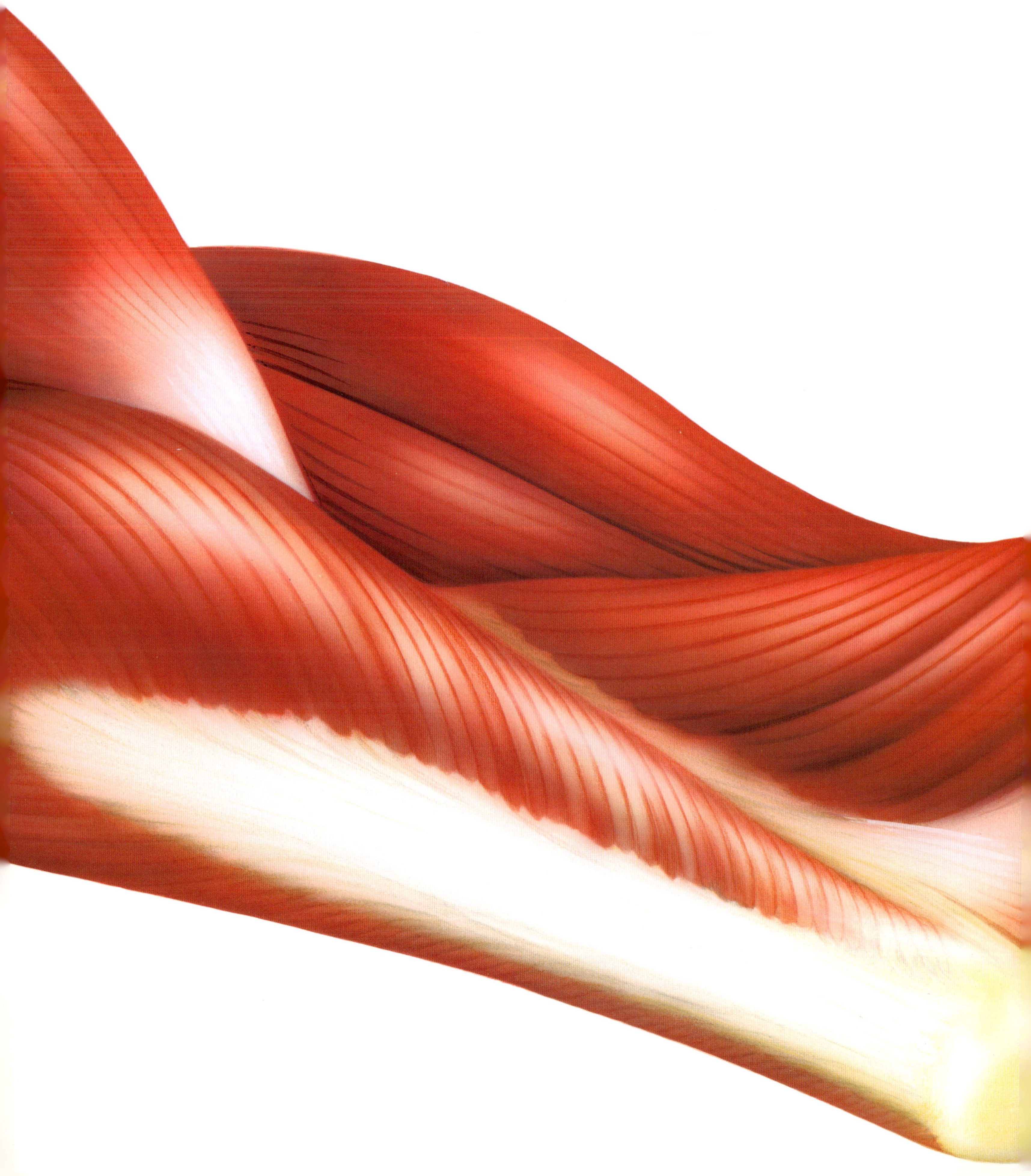

The Shoulders, Arms and Hands

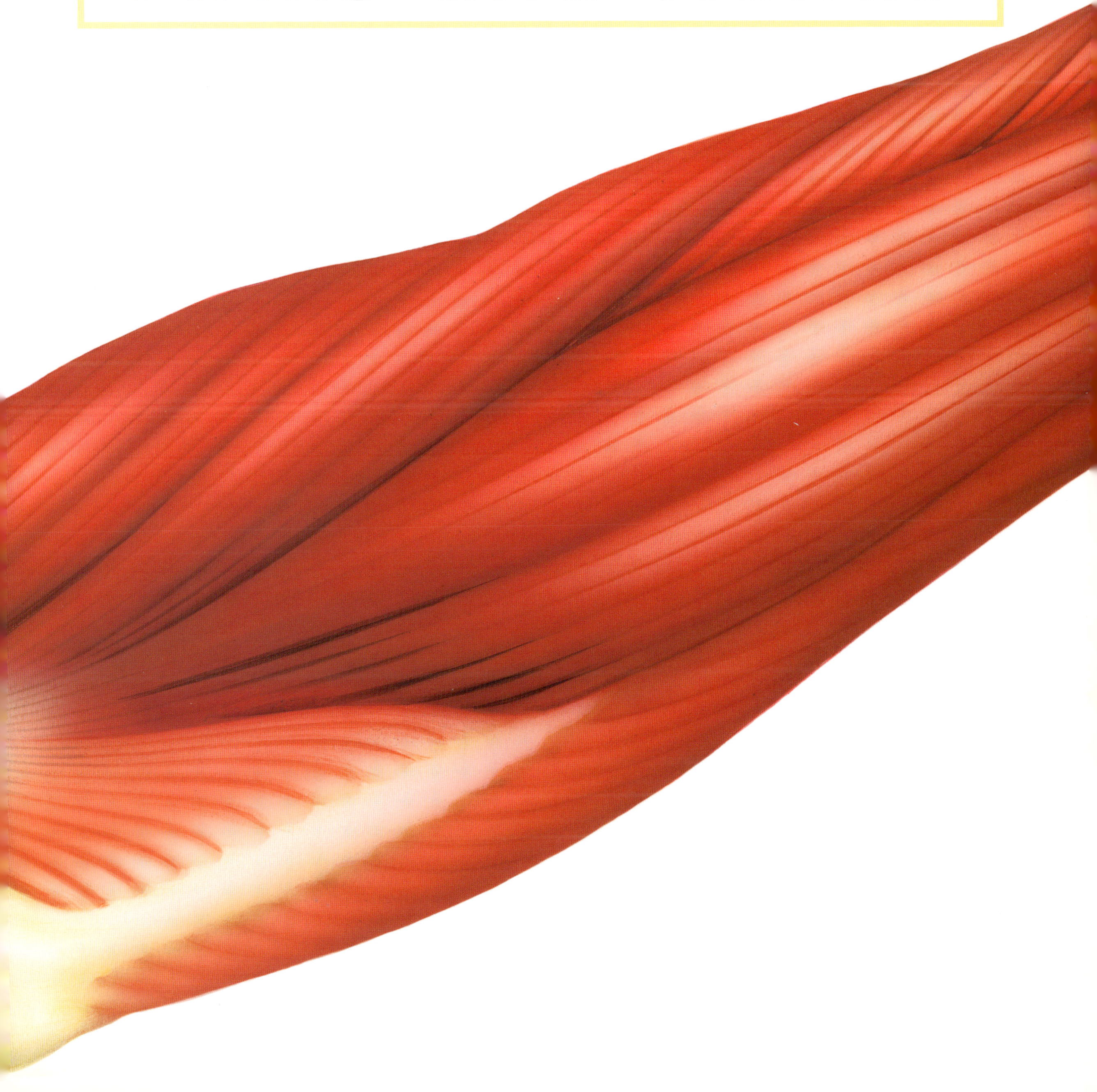

Arm

The arm, or upper limb, extends from the shoulder, where it is attached to the trunk; to the wrist, where it joins the hand. It has two parts: the upper arm, the section between the shoulder and elbow, and the forearm, which extends from the elbow to the wrist. The whole limb is designed for mobility. It gives the hand such a range of motion that it can reach most regions of the body and manipulate external objects.

SEE ALSO *Muscular system on page 48, Nervous system on page 64, Skeletal system on page 30*

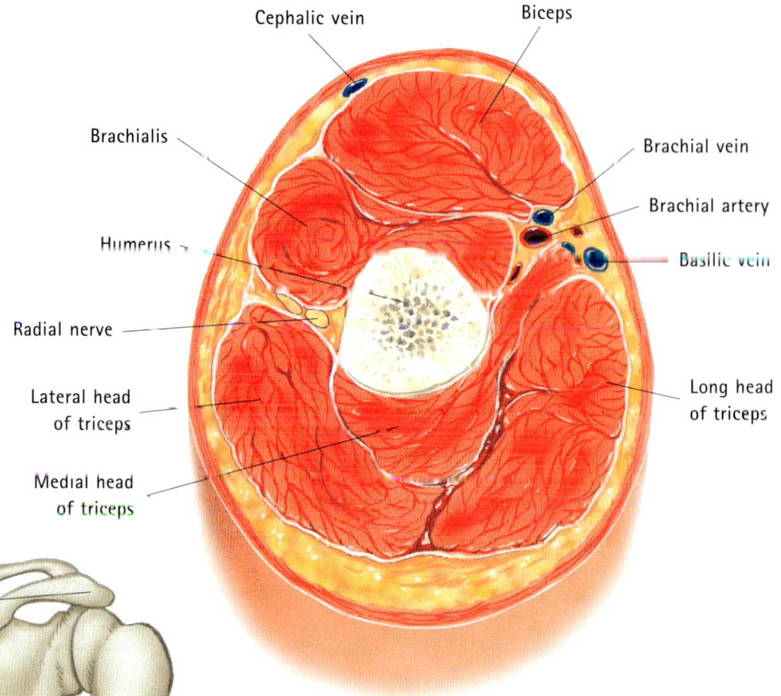

Cross-section of the arm

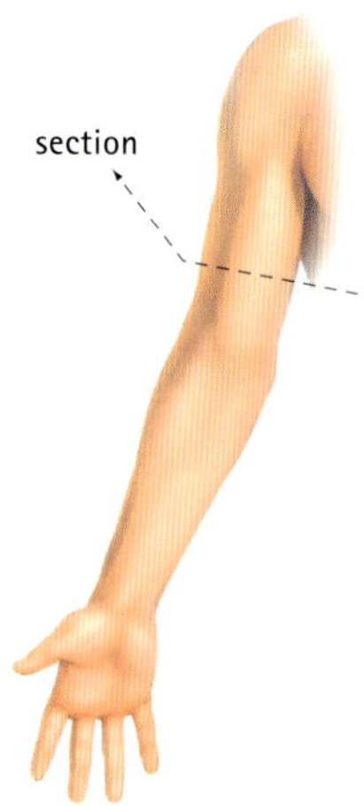

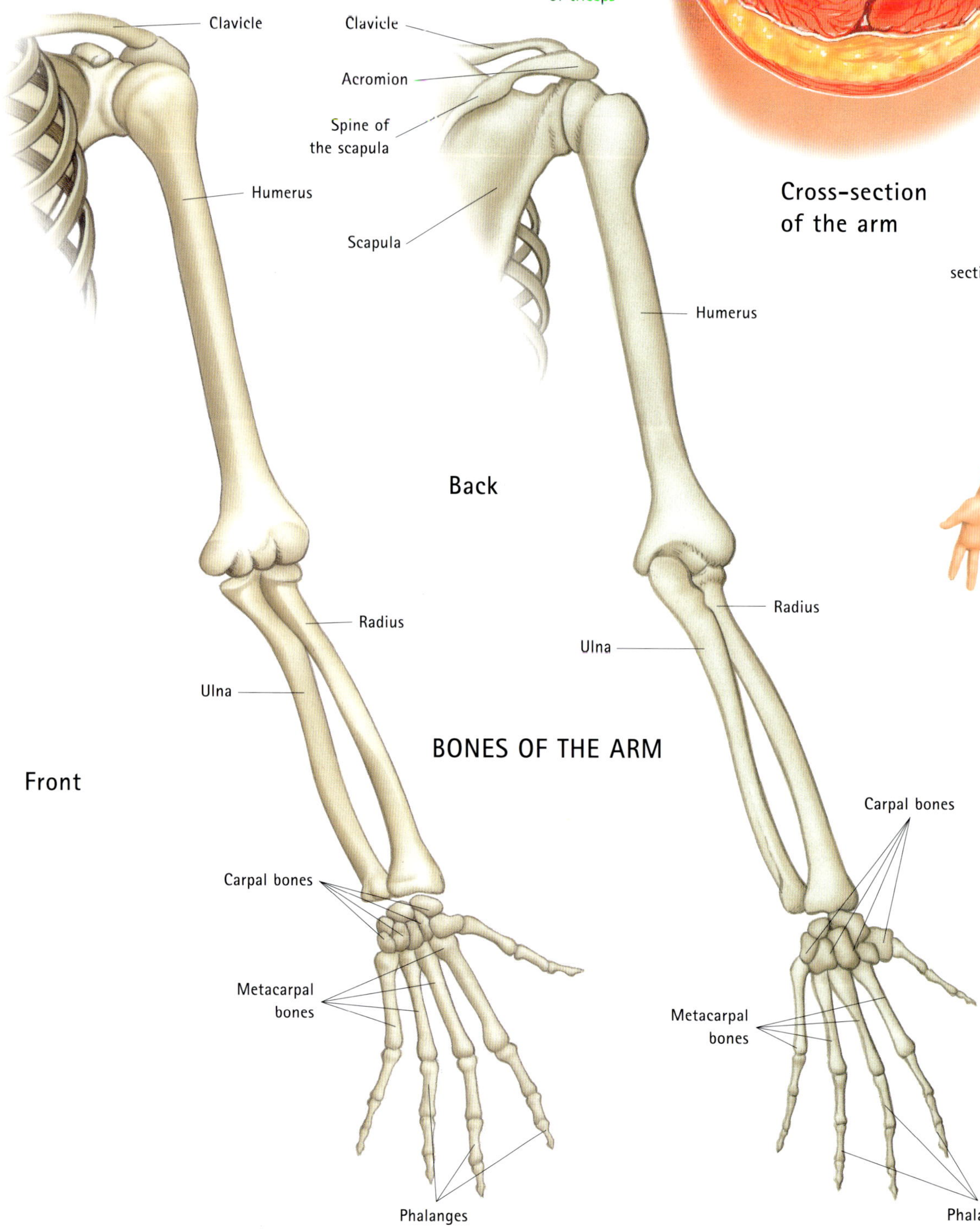

BONES OF THE ARM

SURFACE MUSCLES OF THE ARM

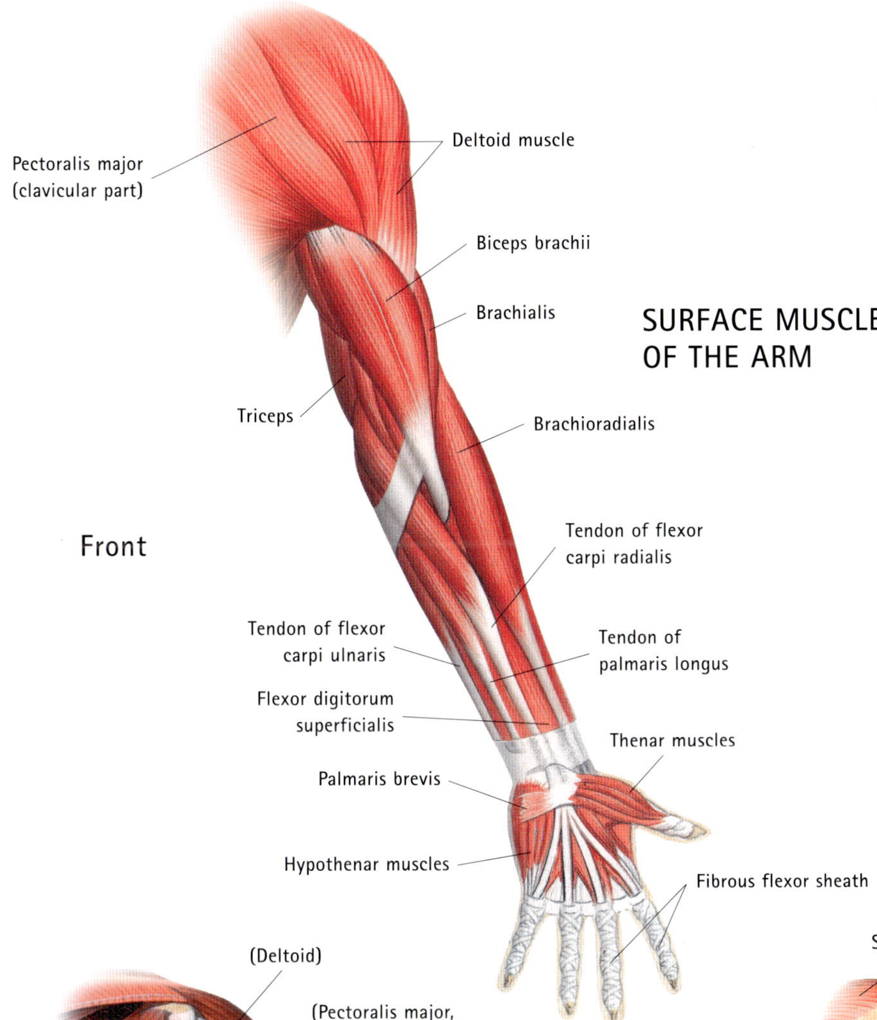

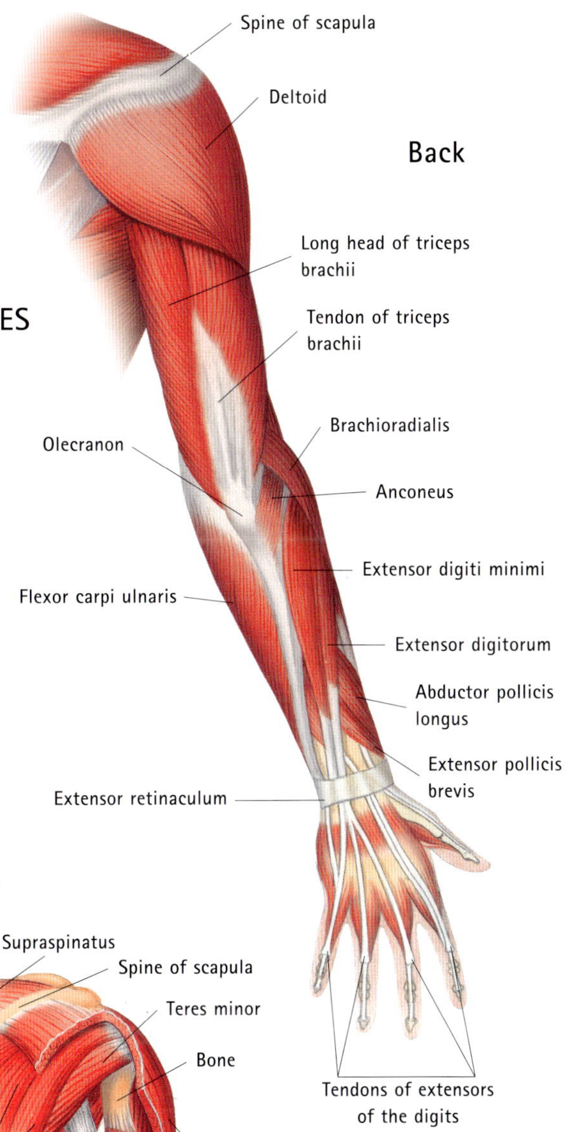

DEEP MUSCLES OF THE ARM

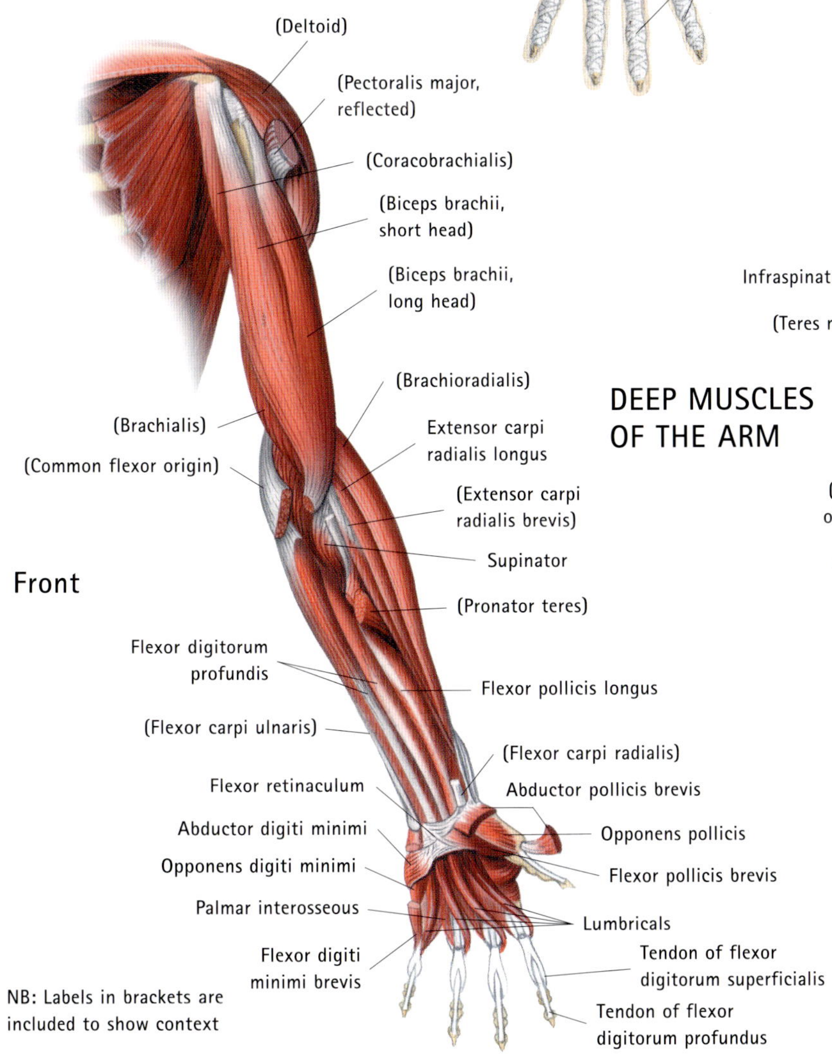

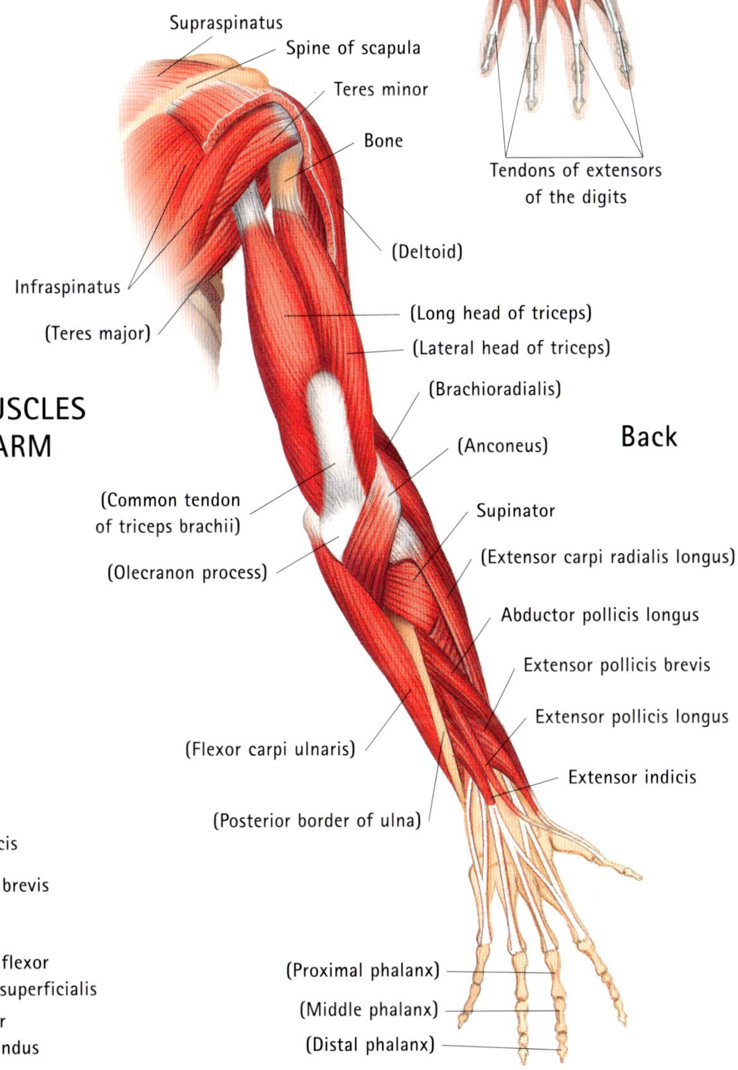

NB: Labels in brackets are included to show context

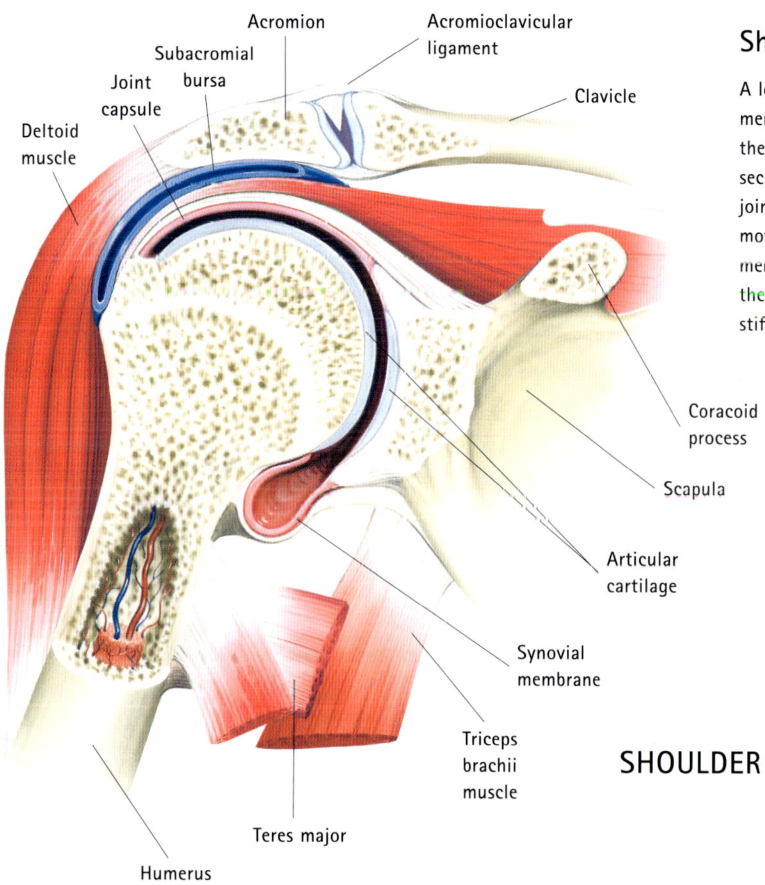

Shoulder cross-section

A loose capsule lined with a synovial membrane joins the humerus bone to the scapula. The synovial membrane secretes a fluid which lubricates the joint, reducing friction and easing movement. Damage to the synovial membrane or cartilage that cushions the joint may cause pain and stiffness in the shoulder.

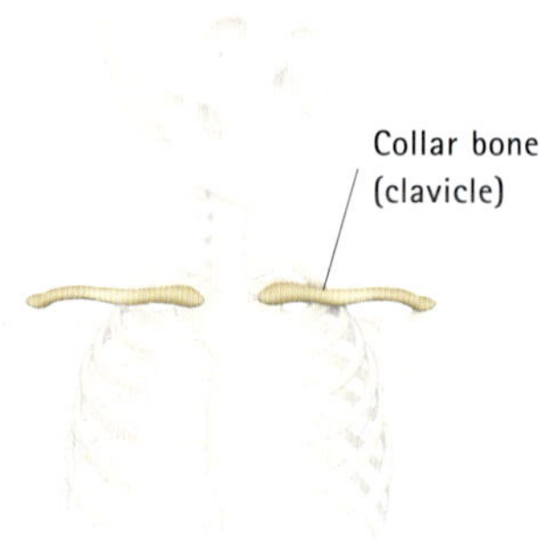

Collar bone

The collar bone, or clavicle, helps to stabilize the shoulder joint. It is attached to the scapula (shoulder blade) at one end and the sternum (breastbone) at the other.

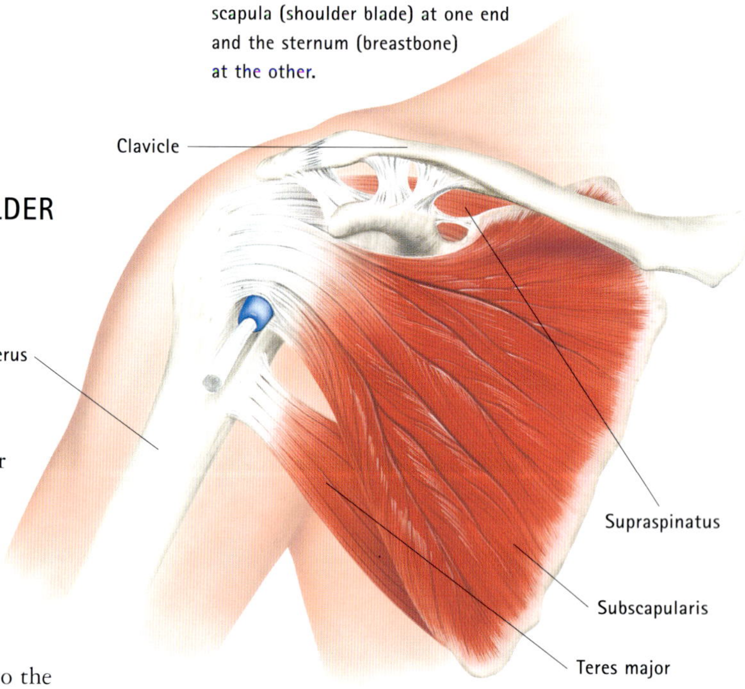

Shoulder

The shoulder is made up of three bones—the clavicle (collar bone), the scapula (shoulder blade) and the humerus—and their associated joints and muscles, as well as major nerves and blood vessels passing to and from the arm.

The arm is attached to the trunk by the pectoral girdle, which consists of the clavicle and the scapula. The clavicle acts as a strut to hold the arm away from the center of the body. It forms joints with the sternum (breastbone) at one end and with the scapula at the other end. The scapula is a flat triangular-shaped bone which covers part of the upper back. It is largely enclosed by muscle but has a prominent spine, which can be felt extending across the back toward the shoulder, where it expands to form the acromion.

Major bones of the shoulder and arm

The shoulder is formed by the interconnection of the collar bone (clavicle), the shoulder blade (scapula) and the humerus.

See also *Skeletal system on page 30*

Collar bone

The collar bones, or clavicles, are a pair of short horizontal bones above the rib cage. They are attached to the breastbone (sternum), and the two shoulder blades (scapulas) on either side. The function of the collar bones is to stabilize the shoulders.

Scapula

The scapula (shoulder blade) forms part of the shoulder at the back. It is a triangular, flattened bone, with several projections. The scapula is attached to the outer end of the collar bone (clavicle) at the acromioclavicular joint. Its outer end provides a socket for the head of the upper arm bone (humerus), forming the bony articulation of the shoulder joint. The scapula is held in place by strong muscles, which can move it in relation to the chest wall. It provides attachment for many of the muscles of the shoulder and upper arm, including the biceps.

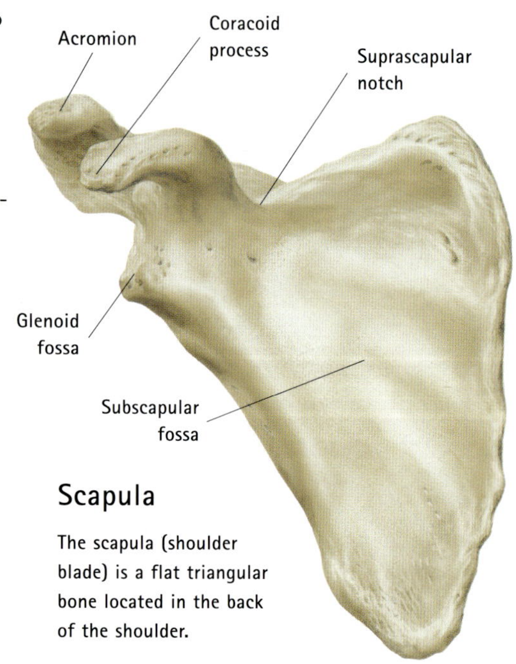

Scapula

The scapula (shoulder blade) is a flat triangular bone located in the back of the shoulder.

Scapula—back view

Each scapula is connected to the humerus bone and clavicle (collar bone).

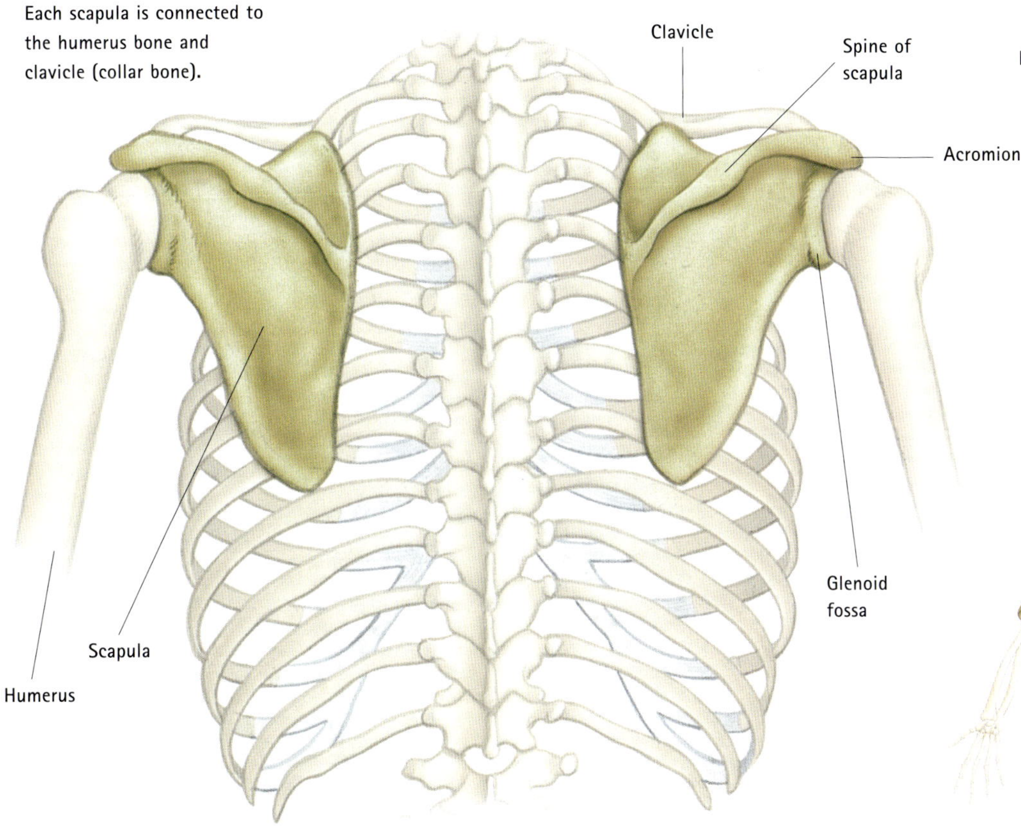

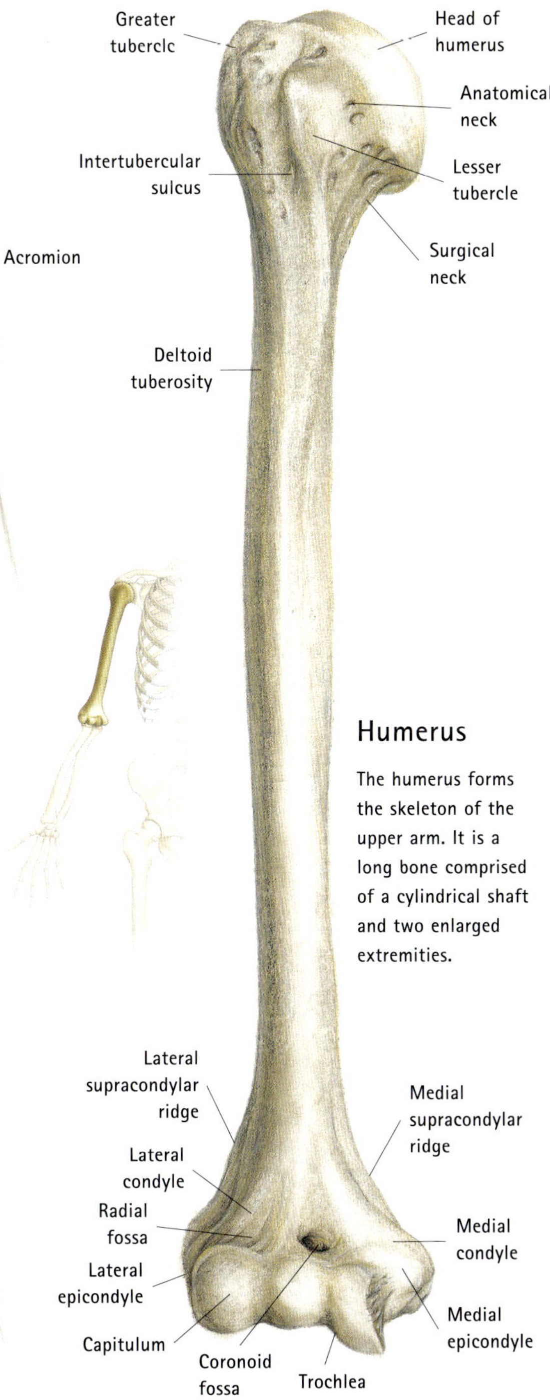

Humerus

The humerus forms the skeleton of the upper arm. It is a long bone comprised of a cylindrical shaft and two enlarged extremities.

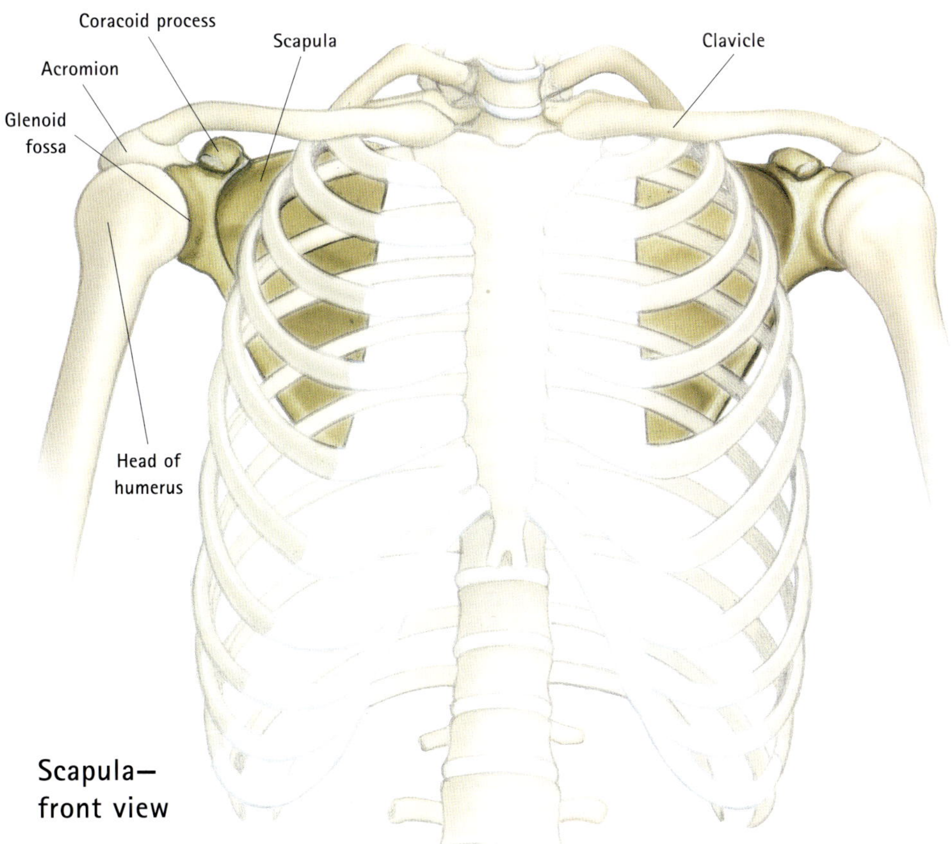

Scapula—front view

The connection between the scapula and the head of the humerus forms the shoulder joint.

Humerus

The humerus is a long bone that forms the skeleton of the arm. It consists of a cylindrical shaft, a rounded head (upper end), which articulates with the scapula to form the shoulder joint, and paired condyles (lower end), which articulate with the radius and ulna to form the elbow joint.

Several major nerves lie in contact with the humerus. The familiar tingling sensation which we experience in our hand when we "hit our funny bone" is actually caused by hitting the ulnar nerve as it passes behind the humerus on its way to the hand.

Radius

The radius is one of the two bones of the forearm. Located on the thumb side, it lies parallel to and rotates around the other forearm bone, the ulna. Near the uppermost end of the radius, which forms part of the elbow, is a raised and roughened area called the radial tuberosity. This is an attachment point for the biceps brachii (commonly called the biceps), the long muscle of the upper arm. At its other, larger end, the radius forms part of the wrist. The rotational movement of the radius around the ulna allows the wrist to move from side to side.

Ulna

The ulna lies on the medial (inner) side of the forearm, extending from the elbow to the wrist. The ulna is a long bone of irregular cross-section, thickest at the elbow end and tapering toward the wrist. It projects above and behind the elbow. At the elbow, the ulna forms a hinge joint with the humerus. The radius articulates with the ulna at both ends, and is firmly bound to it along most of its length by a fibrous membrane which permits the radius to swing around the ulna, carrying the wrist with it. The ulna is not involved in the wrist joint.

Acromioclavicular joint

The inner surface of the acromion meets the clavicle to form the acromioclavicular joint, which can be felt at the top of the shoulder. This joint allows a small amount of gliding movement to occur between the two bones in conjunction with movements of the arm. The acromioclavicular joint is supported by a capsule and several ligaments, the most important of which is the coracoclavicular ligament. These ligaments also help to support the weight of the arm.

Shoulder joint

The outer (lateral) angle of the triangular scapula is flattened to form a shallow cavity, in which the head of the humerus (the long bone of the upper arm) sits, forming the glenohumeral or shoulder joint. The shoulder joint is a ball and socket joint, allowing movement of the arm to occur in almost any direction. The socket, which is formed by the glenoid cavity of the scapula, is very shallow and has a small contact area, relative to the head of the humerus, which forms the ball.

Only a small part of the head of the humerus is in contact with the glenoid cavity at any time, making the joint extremely mobile but also making it relatively unstable (easy to dislocate). A ring of fibrocartilage, the glenoid labrum, which encircles the edge of the glenoid cavity, deepens the socket slightly, thereby increasing the contact area.

The joint surfaces are covered by smooth, glassy cartilage and the two bones are held together by a relatively loose capsule. The inside of the capsule is lined by a synovial membrane, which produces an oily synovial fluid that is released into the joint cavity to lubricate the cartilage surface and reduce friction. The capsule is reinforced on the top, front and back by a group of muscles known as the rotator cuff muscles, whose tendons blend with the capsule as they pass over it. In addition to enabling certain movements at the joint, these muscles are said to act as "dynamic ligaments," holding the head of the humerus in the socket during movement. The rotator cuff muscles, subscapularis covering the front, infraspinatus and teres minor covering the back and supraspinatus covering the top of the capsule, are the most important factors preventing dislocation of the head of the humerus from the socket when force is placed on the joint.

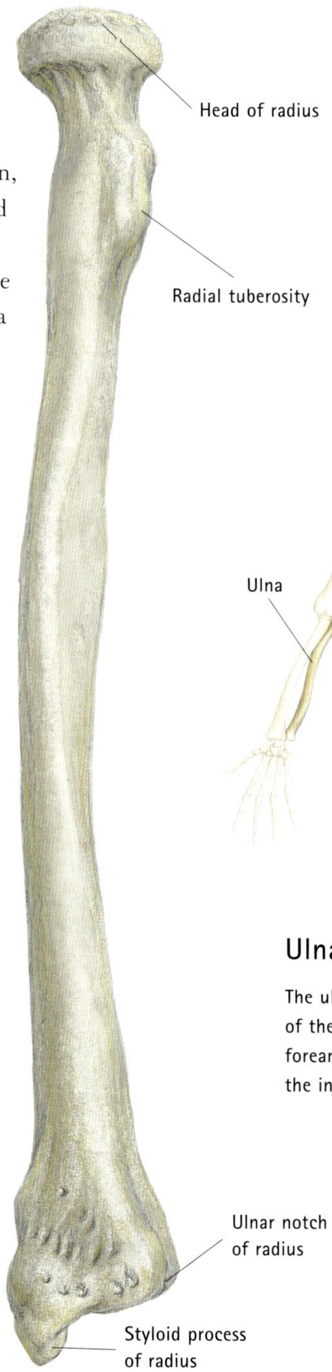

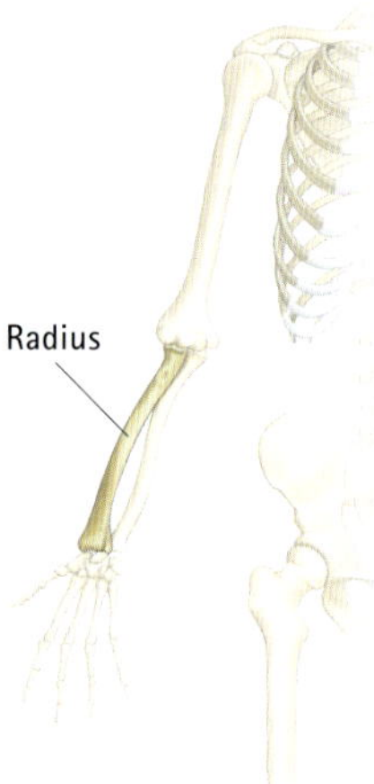

Radius

The radius is one of the two bones of the forearm. The forearm can rotate around its axis, because the head of the radius forms a pivot joint at the elbow.

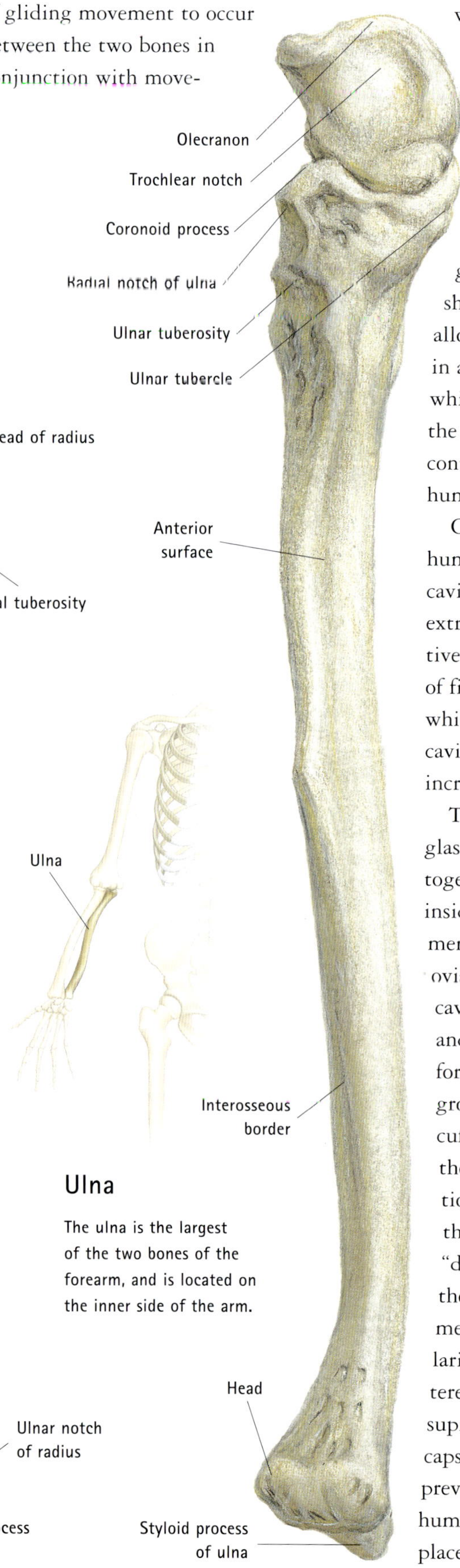

Ulna

The ulna is the largest of the two bones of the forearm, and is located on the inner side of the arm.

The shoulder joint is bridged and protected above by a structure known as the coracoacromial arch, which consists of the acromion behind, the coracoid process in front and the coracoacromial ligament passing between them. A cushion of synovial fluid enclosed by a synovial membrane, and known as the subacromial bursa, lies between the supraspinatus tendon (which blends with the joint capsule) below and the coracoacromial arch above. It functions to reduce friction between the greater tubercle of the humerus and the arch when the arm is elevated.

Shoulder joint

The shoulder is a ball and socket joint which allows the arm to move in almost any direction. Only a small part of the head of the humerus (the ball) makes contact with the glenoid cavity (the socket) at any time, which provides maximum mobility but increases the risk of dislocation.

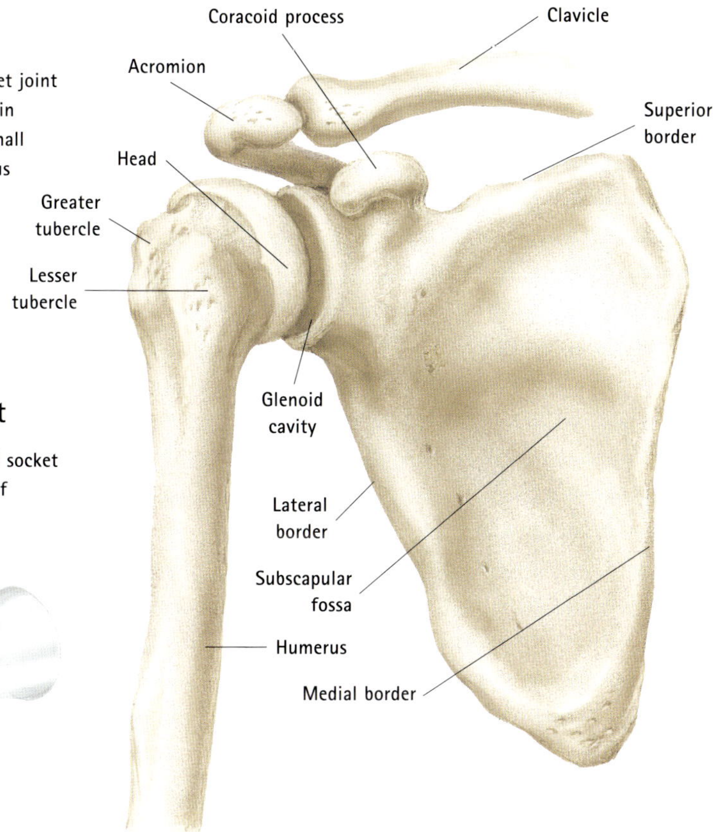

Ball and socket joint

The shoulder joint is a ball and socket joint. It has the widest range of movement of all the joints.

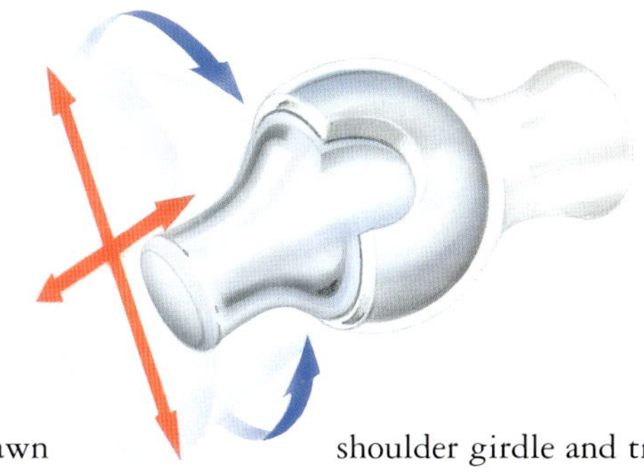

Movements of the shoulder and arm

The shoulder is a multi-axial ball and socket joint, which allows movement in almost any direction. The arm (humerus) can be drawn forward (flexed), drawn backward (extended), elevated (abducted), drawn downward (adducted) and rotated (around its own axis). Movements of the shoulder joint are always accompanied by movements of the pectoral girdle (clavicle and scapula), which increase the range of movement.

SEE ALSO *Joints on page 212*

Muscles of the shoulder

The muscles associated with the shoulder and pectoral girdle fall into two groups, those attaching the humerus to the shoulder girdle and trunk wall, and those attaching the shoulder girdle to the trunk.

The first group includes the rotator cuff muscles: the deltoid (covering the shoulder and giveing it its rounded contour); the pectoralis major (covering the front of the chest); the latissimus dorsi (large flat muscle covering the lower back and converging on a tendon which attaches to the humerus); and the teres major (a small bulky muscle passing from the scapula to the humerus).

The second group of muscles includes the rhomboids and levator scapulae, which pass from the inner (medial) side of the scapula to the vertebral column; the trapezius (a large triangular muscle extending from the skull and vertebral column across to the spine of the scapula and the clavicle); the latissimus dorsi (covering the lower back and converging on a tendon which attaches to the humerus); and the serratus anterior and pectoralis minor (both extending from the scapula to the front of the chest wall).

Pectoral girdle

The bones and muscles of the pectoral girdle provide a support network for the shoulder joint. One set of muscles attaches the humerus bone of the arm to the shoulder girdle, another set attaches the shoulder girdle to the trunk of the body.

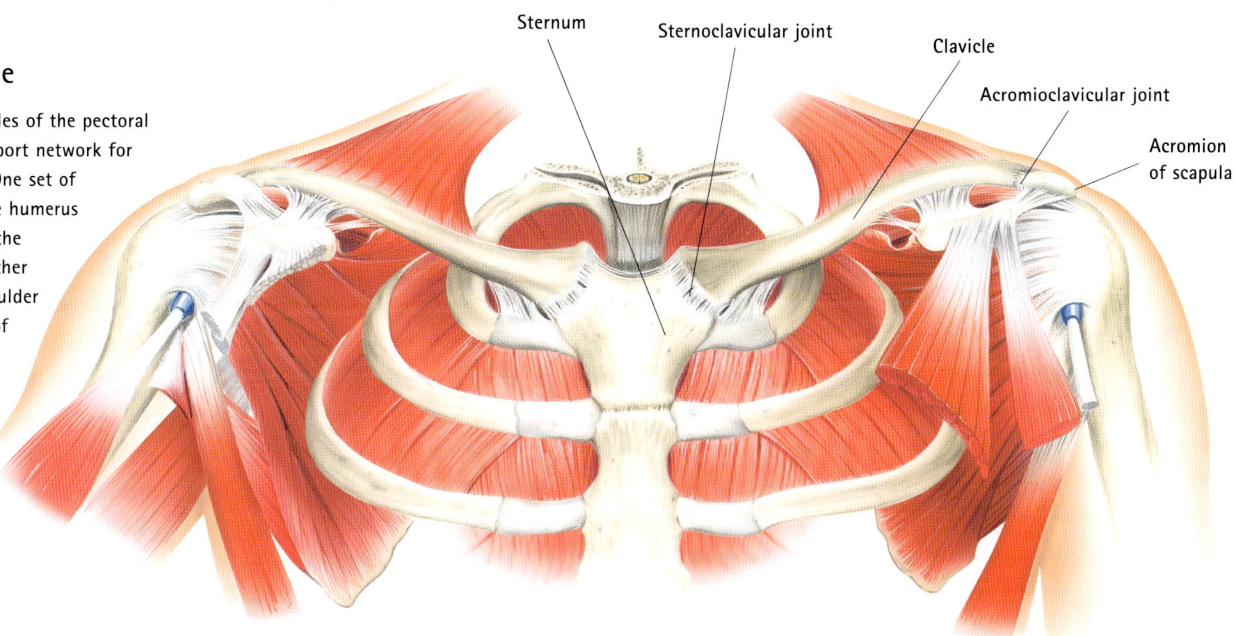

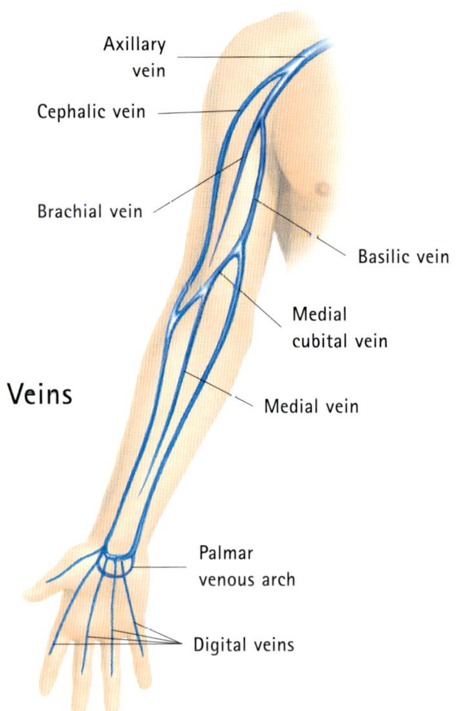

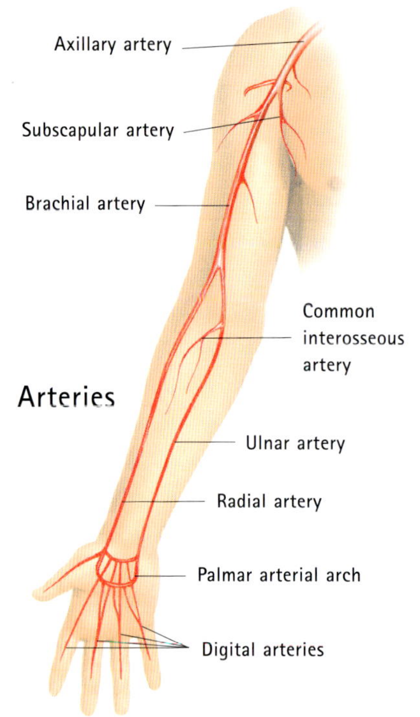

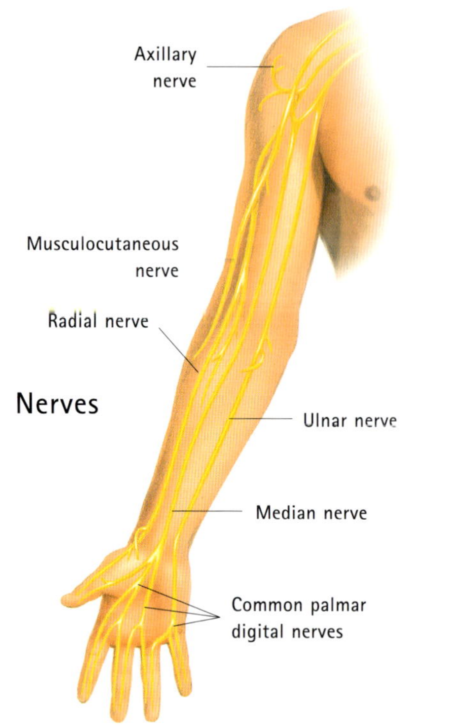

As a general rule, muscles which pass in front of the shoulder joint act to flex or rotate the humerus, those passing behind the joint extend and/or rotate the humerus, and those passing above abduct the humerus. Because the large deltoid muscle passes over three sides of the joint (front, back and top), it is involved in most shoulder movements. Muscles which are most important for flexion of the shoulder joint are the pectoralis major and deltoid muscles. Muscles which are important for extension include the teres major, latissimus dorsi and deltoid muscles.

Abduction is brought about by the deltoid assisted by supraspinatus; medial rotation by subscapularis, latissimus dorsi, teres major, pectoralis major and part of the deltoid muscle. Lateral rotation is brought about by infraspinatus and teres minor; and adduction mainly by latissimus dorsi at the back and pectoralis major at the front and teres major; although teres major only adducts when the movement is resisted.

Rotator cuff muscles

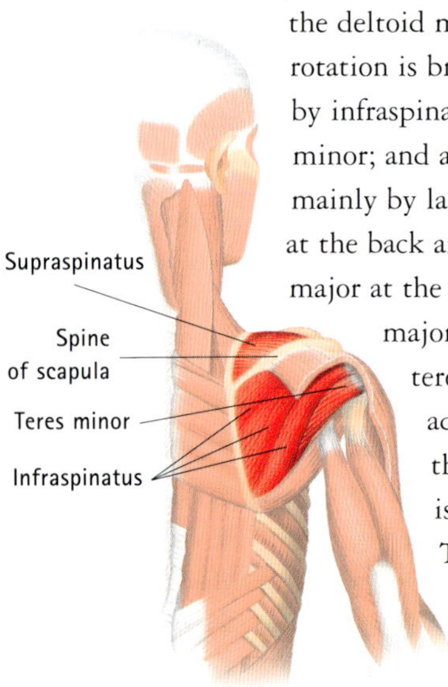

The trapezius muscle has fibers passing up to the skull and across to the vertebral column and is important in shrugging the shoulders and in rotating the scapula upward (when lifting the arm above the head). The serratus anterior is also essential for scapular rotation.

Weakness or paralysis of any of these muscles will impair shoulder movement.

SEE ALSO *Electromyography on 431; Muscular system on page 48; Treating the musculoskeletal system on page 444*

NERVES AND BLOOD VESSELS OF THE SHOULDER AND ARM

Most of the nerves and blood vessels supplying the arm pass through the armpit (axilla), just below the front of the shoulder joint as they travel to and from their target structures in the shoulder region and arm. In this area these structures are held together by loose connective tissue (axillary sheath).

Blood vessels of the shoulder, arms and hands include the large axillary artery and vein, the brachial, radial and ulnar arteries, and the axillary, radial, musculocutaneous, median and ulnar nerves. The major blood vessel of the arm is the brachial artery, which starts at the top of the arm and passes down the inside of the arm to the front of the elbow. The radial artery provides blood to muscles of the forearm and, along with the ulnar artery, contributes to the palmar arches in the hand, from which the digital arteries arise to supply blood to the fingers.

The subclavian vein is a continuation of the axillary vein and is a major vein draining the arm. It joins with the external and internal jugular veins, draining the superficial and deep tissues of the head and neck to form the brachiocephalic vein. Prior to its termination is the last valve through which blood passes before reaching the heart.

The nerves arise from a complex known as the brachial plexus, located on the side of the neck, where it can be felt as cords passing toward the arm, just above the clavicle. The axillary nerve is of particular importance to the shoulder because it supplies the deltoid muscle from its deep surface by encircling the surgical neck of the humerus.

The ulnar nerve arises from the brachial plexus in the shoulder. It runs down the inner side of the upper arm, initially accompanying the brachial artery, winds around the inner side of the elbow joint to enter the forearm, and runs along the inner side of the forearm to enter the hand on the little finger side. In the forearm, its position can be mapped by a line from the medial epicondyle of the humerus to the inner edge of the pisiform bone.

Branches of the ulnar nerve activate the flexor carpi ulnaris muscle and half of the flexor digitorum profundus muscle, and small muscles of the hand. The ulnar nerve provides sensation for the skin on the little finger side of the hand. At the elbow, the ulnar nerve lies in a groove on the upper surface of the medial epicondyle, where it is liable to injury.

The radial nerve starts in the brachial plexus. It supplies extensor muscles in the back of the arm, such as the triceps and extensors for the wrist, fingers and thumb. It also supplies skin over the back of the arm and hand, on the thumb side.

SEE ALSO *Circulatory system on page 78; Nervous system on page 64*

DISORDERS OF THE SHOULDER

Disorders of the shoulder are common. The most common problem affecting the shoulder is inflammation of the supraspinatus tendon, a condition known as "painful arc syndrome."

SEE ALSO *Arthroscopy on page 434; CAT scan on page 432; Joint diseases and disorders on page 42; Magnetic resonance imaging on page 433; Treating the musculoskeletal system on page 444; X-ray on page 431*

Dislocation of the shoulder

Dislocation of the shoulder is a relatively common occurrence and usually occurs as a result of a sudden force being transmitted along the arm when it is in the elevated (abducted) position. In this position, the force from the head of the humerus is transmitted to the lower part of the capsule where it is weakest and not reinforced by tendons, forcing the head to pop out of the lower part of the socket. Although the humerus can be placed back into the socket with the assistance of local anesthetic and muscle relaxants, the damage that is usually sustained by the capsule and labrum in the dislocation makes the joint vulnerable to dislocation in the future, and surgery may be required to overcome the problem.

Painful arc syndrome

Stress on the supraspinatus tendon in some people may cause degeneration of the tendon with age, resulting in the formation of crystalline calcium deposits in the tendon (tendinitis), which cause friction and consequent swelling of the bursa. When this occurs the person experiences pain on elevating the arm because of pressure on the swollen bursa as the humerus impinges on the acromion, a condition known as "painful arc syndrome." The syndrome is usually treated with heat, physical therapy and anti-inflammatory medication.

Frozen shoulder

Frozen shoulder (adhesive capsulitis) is caused by inflammation of all the rotator cuff tendons causing generalized thickening of the shoulder capsule, which may adhere (stick) to the humerus. It may or may not be preceded by trauma to the joint. Frozen shoulder is characterized by increasing pain and stiffness of the shoulder. Over time the pain subsides but the stiffness continues to increase. The stiffness usually outlasts the pain by a few months before movement gradually returns to normal. The course of the disease may take 1–2 years, but a good recovery can usually be expected.

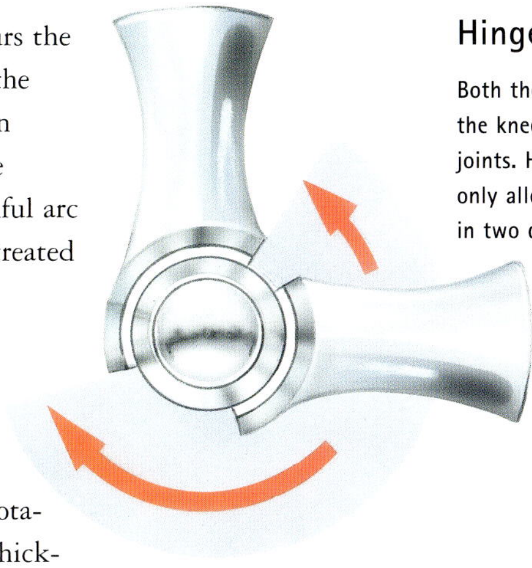

Hinge joint

Both the elbow and the knee are hinge joints. Hinge joints only allow movement in two directions.

ELBOW

The elbow is the joint between the expanded lower end of the bone of the upper arm (the humerus) and the two bones of the forearm (the radius and ulna).

The radius and ulna articulate with the expanded end of the humerus. There are bony swellings on the humerus, either side of the joint, called the lateral and medial epicondyles. These bony protrusions can be felt through the skin on either side of the elbow.

The brachial artery is the main artery of the upper arm, which runs down the medial side of the humerus and across the inside surface of the elbow, below which it divides into radial and ulnar arteries.

DISORDERS OF THE ELBOW

Elbows are prone to certain types of sports injuries, commonly called "tennis elbow" and "golfer's elbow."

Elbow, articular

Cross-sectional view of the elbow joint.

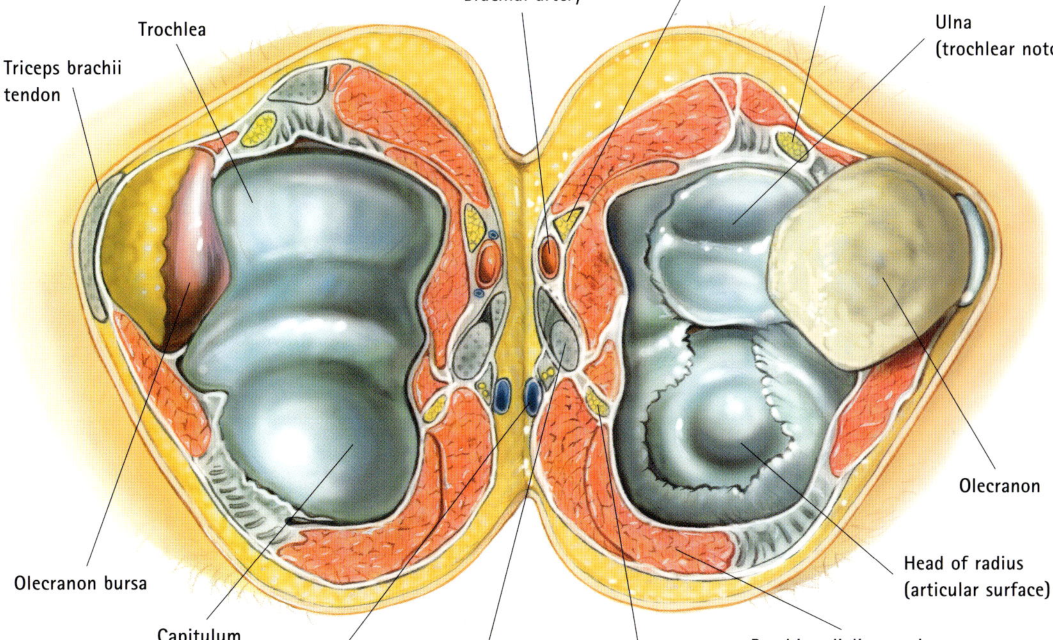

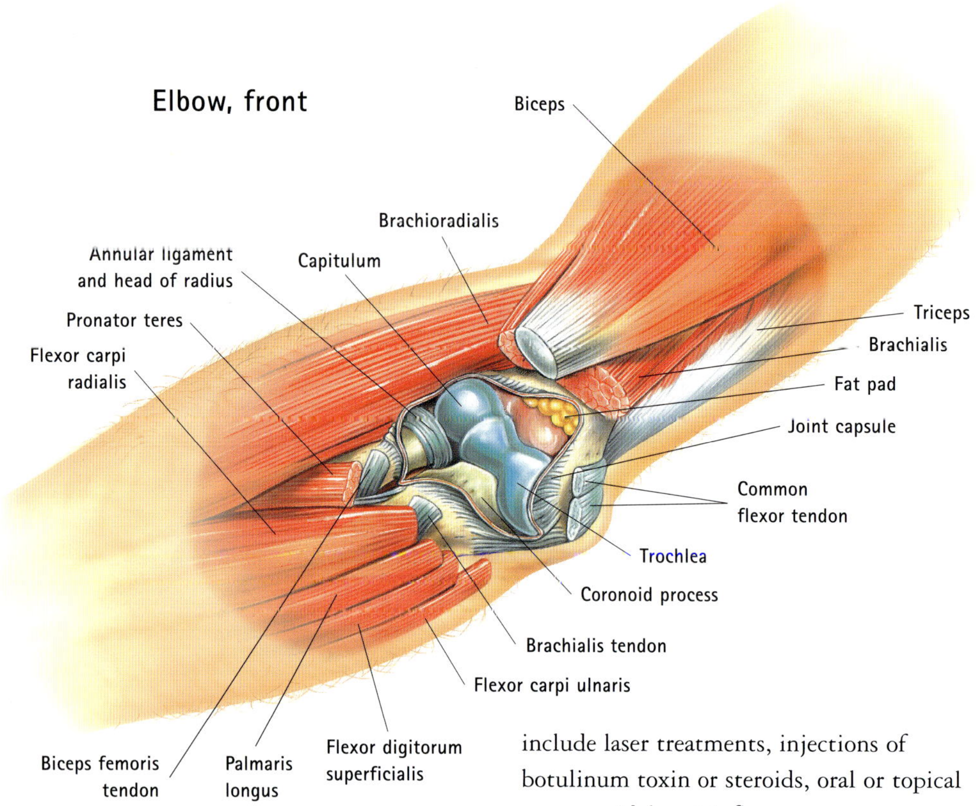

Tennis elbow is a painful sensation in the vicinity of the lateral epicondyle (lateral epicondylitis). The pain may also radiate down the forearm. Some of the muscles which extend (straighten) the wrist attach to the lateral epicondyle and it is thought that repetitive wrist movements cause strain or damage to this attachment.

Golfer's elbow is pain in the vicinity of the medial epicondyle where some of the muscles which flex (bend) the wrist attach. Repetitive movements of the wrist are believed to place stress on this part of the elbow. A similar condition can arise from excessive throwing using poor technique.

Despite the common names for the two conditions, tennis and golf can affect either epicondyle, and epicondylitis can also result from other sports requiring repetitive wrist or forearm movements, including bowling, gymnastics, baseball, fencing, swimming and karate.

Epicondylitis does not usually resolve without some sort of treatment. Therapies include laser treatments, injections of botulinum toxin or steroids, oral or topical non-steroidal anti-inflammatory agents, elbow support bands, remedial exercise and acupuncture. Some of these remedies seem to give relief in many cases.

There is no universal agreement as to which method of treatment is best, nor do there seem to have been any clinical trials which satisfy the majority of practitioners treating these conditions.

SEE ALSO *Treating the musculoskeletal system on page 444*

WRIST

The wrist (carpus) contains a group of eight bones, known as the carpal bones, which join the radius and ulna to the hand. This complex joint allows a wide range of movement, in which the hand can be bent forward (flexion) or backward (extension) or moved from side to side. The scaphoid bone lies on the thumb side of the wrist and forms the floor of a region known as the "snuff box." The adjacent bone is the lunate. The shape of the carpal bones forms a concavity or U-shape. The roof of this concavity is closed over by a dense band of connective tissue, forming the carpal tunnel. Long flexor tendons to the fingers and thumb pass through this tunnel, as well as the median nerve.

As well as the long flexor tendons that cross in front of the wrist through the carpal tunnel, long tendons associated with extension of the fingers and thumb pass across the back of the wrist. All these tendons are surrounded by synovial sheaths that lubricate them.

Three nerves cross the wrist to supply the skin and muscles of the hand—the median, ulnar and radial nerves. The ulnar nerve is particularly important for fine

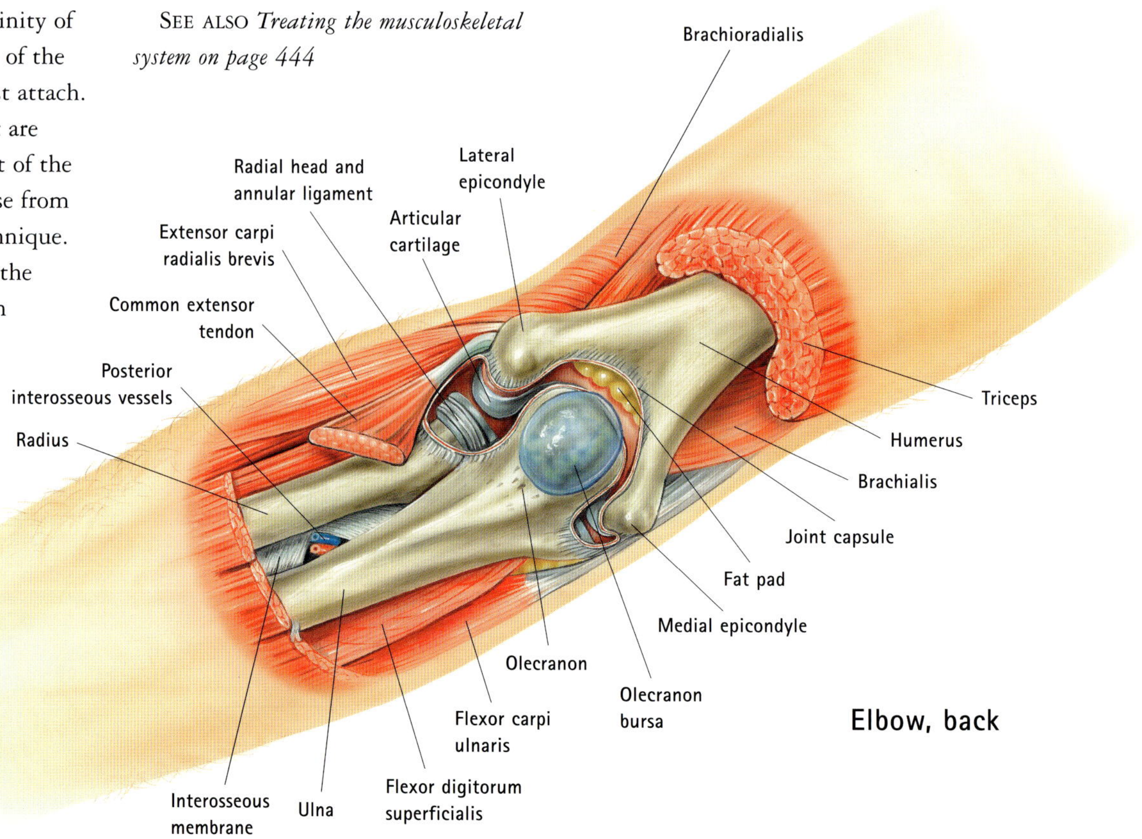

finger movements. The wrist is also crossed by the ulnar and radial arteries.

SEE ALSO *Circulatory system on page 78; Nervous system on page 64; Skeletal system on page 30*

Disorders of the wrist

Disorders of the wrist may also involve the wrist joint itself or the overlying tendons or nerves.

SEE ALSO *Imaging techniques on page 431, Treating the musculoskeletal system on page 444*

De Quervain's disease

Inflammation of two tendons to the thumb, a condition known as de Quervain's disease, causes pain which is exacerbated by gripping and twisting movements, as in wringing clothes. Resting the tendons by avoiding the painful movements usually helps, and anti-inflammatory medication may be recommended. The tendon sheaths can also form a small swelling or ganglion, usually on the back of the wrist. The ganglion may not be painful, but if necessary it can be removed surgically.

Arthritis

The wrist is one of the joints commonly involved in rheumatoid arthritis. There is pain, joint stiffness and reduced movement, with joint deformity and swelling which may lead to carpal tunnel syndrome.

Osteoarthritis can occur in the wrist, especially if there has been a previous injury. Osteoarthritis is also common in the joint between the wrist and first metacarpal bone, which is involved in thumb movements (carpometacarpal joint). Arthritis in this joint, causing pain at the base of the thumb, can be the result of a common fracture of the metacarpal bone called Bennett's fracture. Treatment includes resting the joint, anti-inflammatory medication and steroid injections.

Carpal tunnel syndrome

This is a syndrome resulting from compression of the median nerve in the carpal tunnel in the wrist. The carpal tunnel is a gap formed by the wrist bones (called the carpal bones) and the tough ligament that forms the roof of the tunnel (the flexor retinaculum). The passageway is rigid, so swelling of any of the tissues in this area can cause compression of the nerve, causing a numbness or pain in the wrist, hand, and fingers (except the little finger). The symptoms are usually worse at night.

Carpal tunnel syndrome is most commonly found in middle-aged women. It may occur during pregnancy, before the menstrual period, or during menopause. The condition is also found in diseases such as rheumatoid arthritis, acromegaly and hypothyroidism, or following injury or trauma to the area.

Treatment consists of splinting the wrist to immobilize it for several weeks. Anti-inflammatory drugs may help to improve the condition. If these fail, a physician may inject a corticosteroid drug into the ligament. In severe cases, surgical excising (resection) of the flexor retinaculum may be needed to relieve the pressure on the nerve.

Ellipsoidal joint

The radius and the scaphoid bone of the hand meet to form an ellipsoidal joint. Ellipsoidal joints allow flexion and extension and movement from side to side, but rotation is limited.

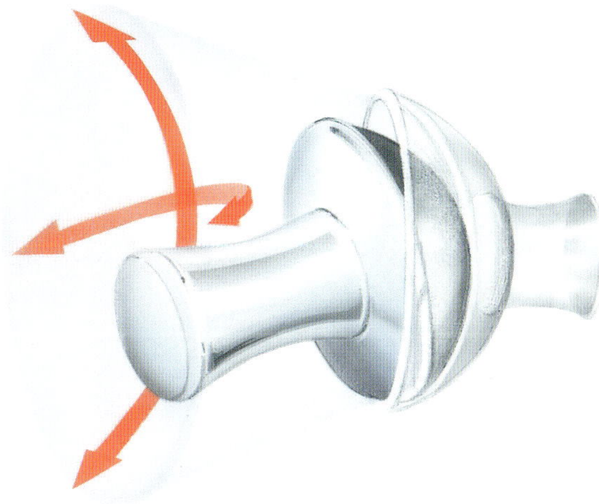

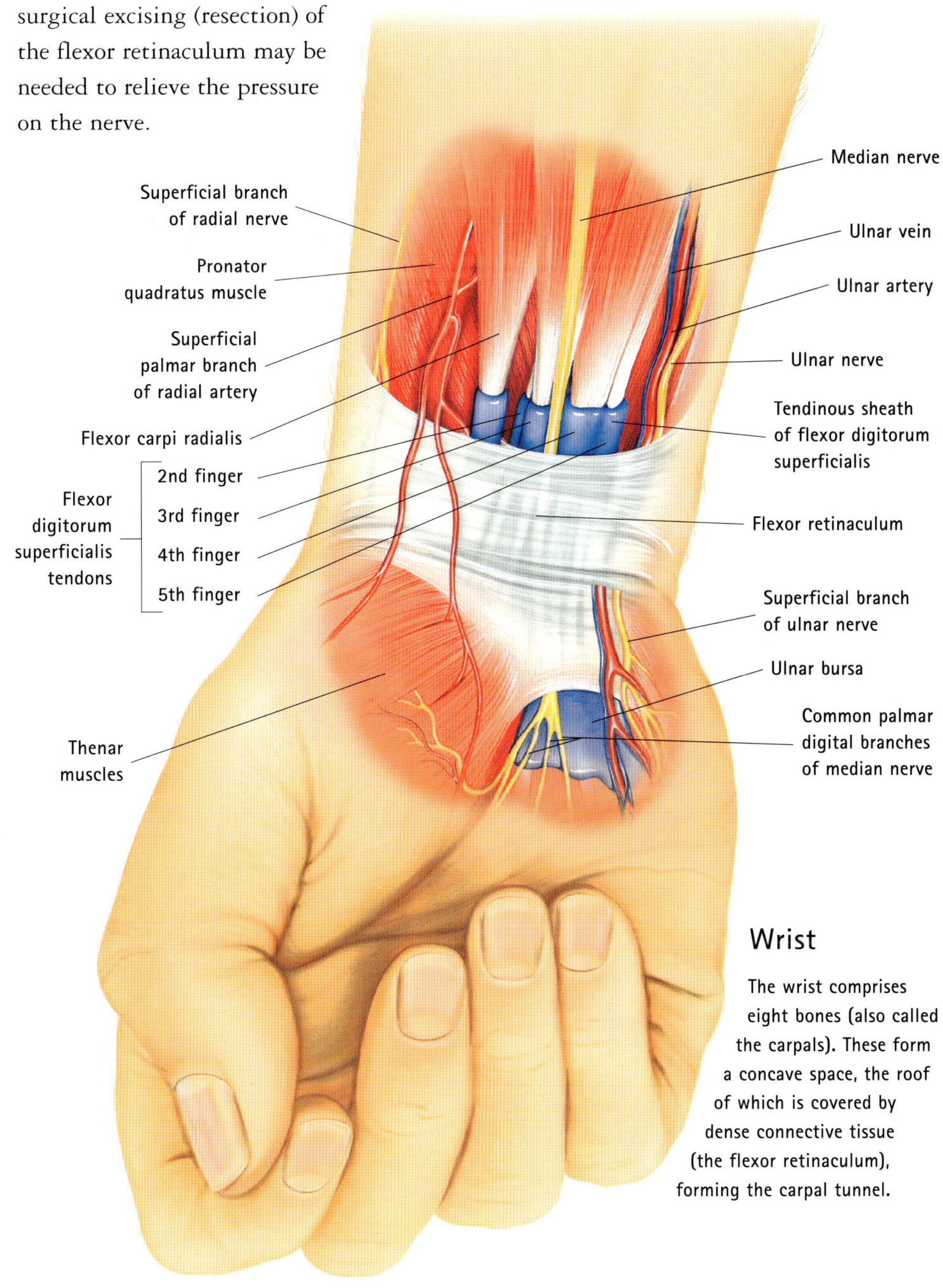

Wrist

The wrist comprises eight bones (also called the carpals). These form a concave space, the roof of which is covered by dense connective tissue (the flexor retinaculum), forming the carpal tunnel.

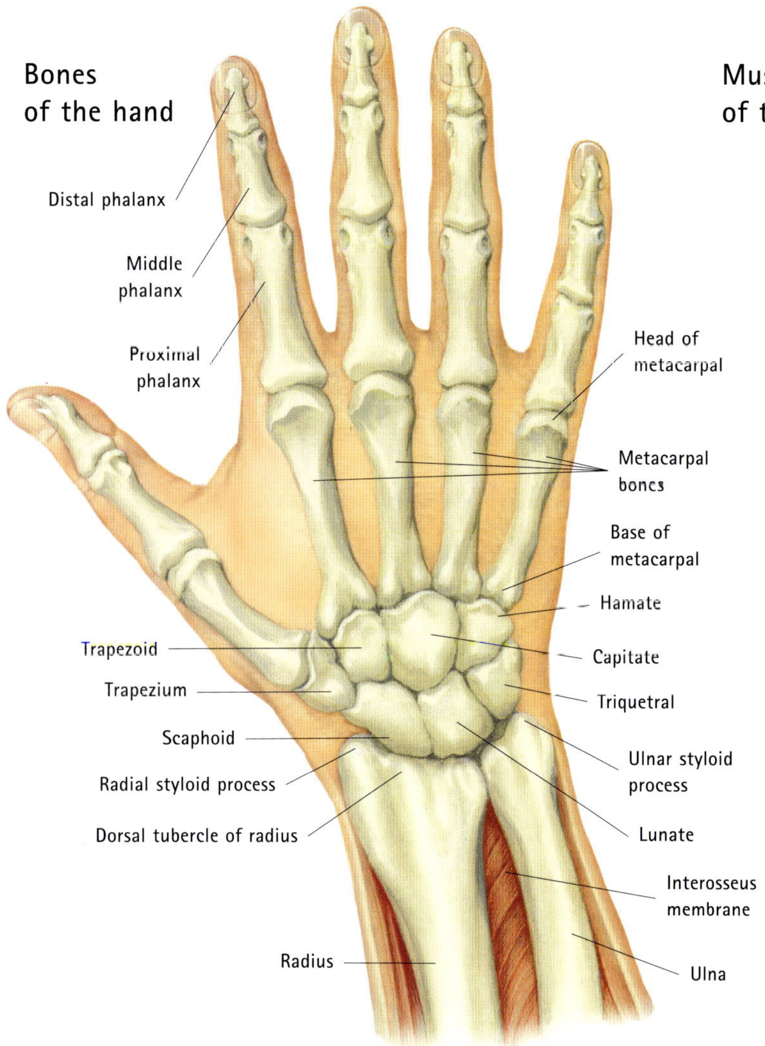

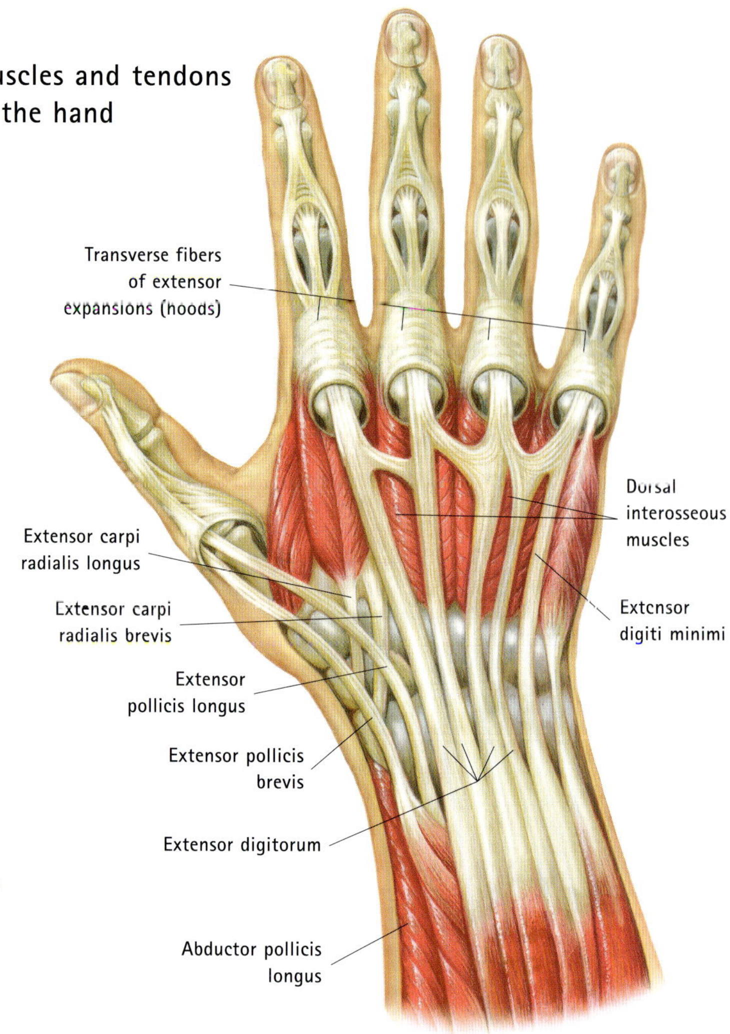

HAND

The hand is designed to grasp and manipulate objects. It consists of the palm, or front of the hand, the dorsum or back, and the thumb and fingers.

SEE ALSO *Muscular system on page 48; Skeletal system on page 30*

Bones of the hand

Between the wrist bones (carpals) and finger bones (phalanges) are five metacarpal bones. The metacarpals join the bones of the fingers, the phalanges ("phalanx" is the term for a single one) in the metacarpophalangeal (MCP) joints.

The metacarpal bones can be felt through the skin over the back of the hand. The first metacarpal, between the wrist and thumb, is particularly mobile, allowing the thumb to perform a wide range of movements. The fifth metacarpal can be fractured relatively easily. Between the metacarpals and the dorsal skin of the back of the hand are long tendons which pull the fingers and thumb backward (extension) and a network of veins draining blood from the fingers and hand.

The palm

The palm has a slightly hollowed surface, that helps with gripping objects, and it is covered by thick skin which is tightly bound to the tissue below.

The creases on the skin of the palm are the result of flexing the thumb and fingers and their joints. The palm also contains tendons, which bend the fingers forward (flexion), and two soft bulges due to the thenar and hypothenar muscles.

Muscles and tendons of the hand

The thenar muscles form the fleshy prominence between the wrist and the thumb and contribute to thumb movements. These include the important movement of opposition by which the thumb can be touched to the tips of the fingers.

The hypothenar group lies along the side of the palm, between the wrist and little finger. The interosseus muscles lie between each of the metacarpal bones and move the fingers apart (abduction) and back together (adduction), as well as helping with flexion of the metacarpophalangeal (MCP) joints and extension of the fingers. The thumb has a separate adductor muscle to move it toward the palm.

Besides these muscle groups there are four thin, worm-like muscles, the lumbricals, which connect between the long flexor tendons and the fingers and also assist with MCP flexion and finger extension.

Sheets of connective tissue, known as septa, separate the palm into compartments: the thenar and hypothenar compartments; a central compartment containing the long flexor tendons, blood vessels and nerves; and an adductor compartment for the thumb adductor.

The long flexor tendons are surrounded by synovial sheaths containing fluid which lubricates the tendons.

The four fingers are referred to as the index, middle, ring and little finger. They each contain three bones or phalanges, with hinge joints between them.

The thumb has only two bones, the proximal and distal phalanx, again with a hinge joint between. The fingers contain no muscles, and are moved by tendons conected to muscles in the palm or forearm. Their tips are protected by nails on the dorsal (back) surface.

The fingertips are covered in a unique pattern of skin ridges, and these ridges give us our fingerprints.

Nerves and blood vessels of the hand

The hand is supplied by two arteries, the radial artery on the same side as the thumb, and the ulnar artery on the other side. These two arteries join to form two arches in the palm. The digital arteries branch off from these arches and run down each side of the fingers.

Three large nerves—the ulnar, median and radial—supply the muscles and skin of the hand. The ulnar nerve supplies the hypothenar, interosseus muscles, the thumb adductor and two of the lumbricals, as well as the skin over the little finger and the adjacent side of the ring finger. The median nerve supplies the thenar muscles and the remaining two lumbricals, together with the skin of the palm, thumb, index, middle and adjacent side of the ring finger.

The sensory nerve supply (innervation) of the fingertips is particularly rich, providing for sensitive, delicate tactile discrimination. The radial nerve does not supply any hand muscles, but innervates the skin over most of the back (dorsum) of the hand and the back of the thumb.

Movements of the hand

The hand is involved in holding objects in two rather different ways, referred to as the power grip and precision grip.

In the power grip, used for carrying heavy bags or for holding tightly onto a support, objects are grasped in the palm, with much of the muscle power coming from flexor muscles in the forearm. Long flexor tendons extend from the forearm to the fingers and thumb, so that the digits can be held tightly around the object. Muscles within the hand may also be active but the large forearm flexors are particularly important.

The precision grip is used for delicate manipulation of an object, for example, when writing, sewing or drawing. The thumb is opposed to one or more of the fingertips, involving the thenar and adductor muscles. Precision grip particularly involves muscles within the hand, most of which are controlled by the ulnar nerve, except for the thenar muscles of the thumb, controlled by the median nerve.

Disorders of the hand

The hand is the most commonly injured part of the body. A disorder of the hand can involve the nerves, bones, connective tissues or tendons, and can result in either temporary or permanent loss of usage.

SEE ALSO *Nerve conduction tests on page 430; Treating the musculoskeletal system on page 444*

Nerve damage

Injury to any of the hand nerves causes a unique group of problems. Damage to the ulnar nerve causes loss of abduction and adduction movements of the fingers, and loss of sensation in the little finger. The hand develops a deformity called clawhand in which the fourth and fifth metacarpophalangeal (MCP) joints are extended and the fingers flexed. The median nerve can be damaged by cuts to the wrist or by compression, such as in carpal tunnel syndrome. This causes weakness or loss of opposition of the thumb, and loss of sensation in the thumb and fingertips.

Radial nerve injury in the hand only affects sensation over the back of the hand. However, radial nerve injury in the arm leads to inability to extend the hand, and the hand flexes at the wrist (wrist drop). Since it is not possible to have a firm power grip when the wrist is flexed, wrist drop makes it difficult to hold implements (knife, fork, hairbrush, etc). This can be helped by a brace which holds the wrist in a slightly extended position.

Trigger finger

"Trigger finger" is a finger locked in the flexed or bent position, often a result of tenosynovitis, where the finger tendon is inflamed and swollen. As the finger is bent, the swollen tendon moves out of its sheath but does not slide back in to allow the finger to straighten. However, the finger can be straightened with additional effort or with pressure from the other hand, often with a snap (hence "trigger finger").

Rest or anti-inflammatory drugs are the usual methods of treatment.

Ganglion

A ganglion is a cyst in or around a tendon or joint, especially in the hands, wrists and feet. Ganglion cysts can be treated with ice packs applied to the affected area, with oral medication for pain.

A ganglion cyst may be removed by a ganglionectomy or cyst aspiration.

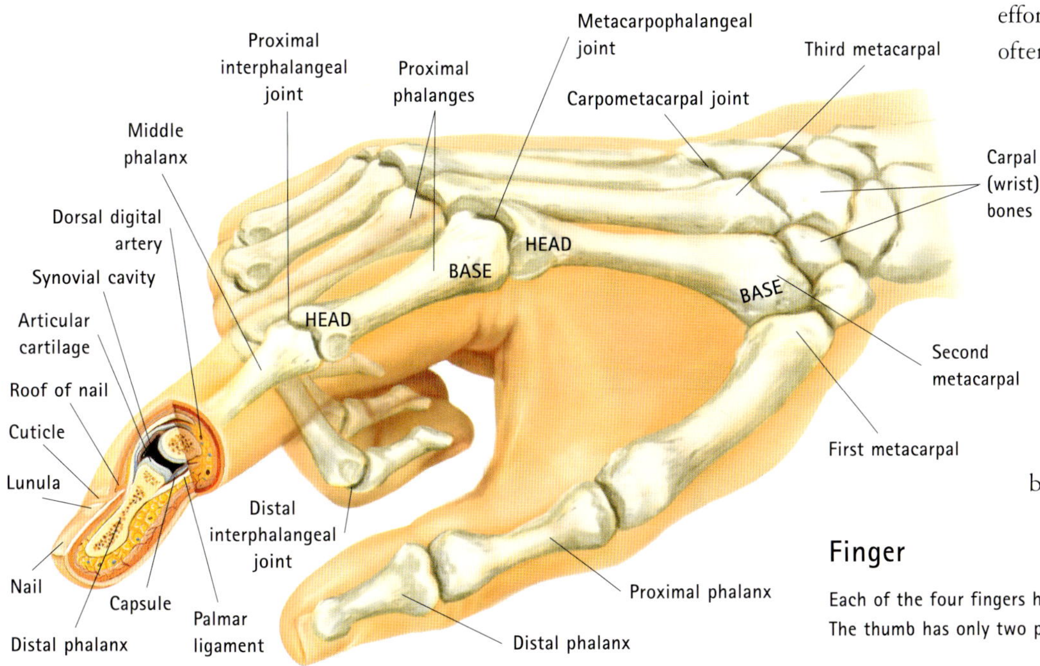

Finger

Each of the four fingers has three slender bones known as phalanges. The thumb has only two phalanges, but has a wider range of movement.

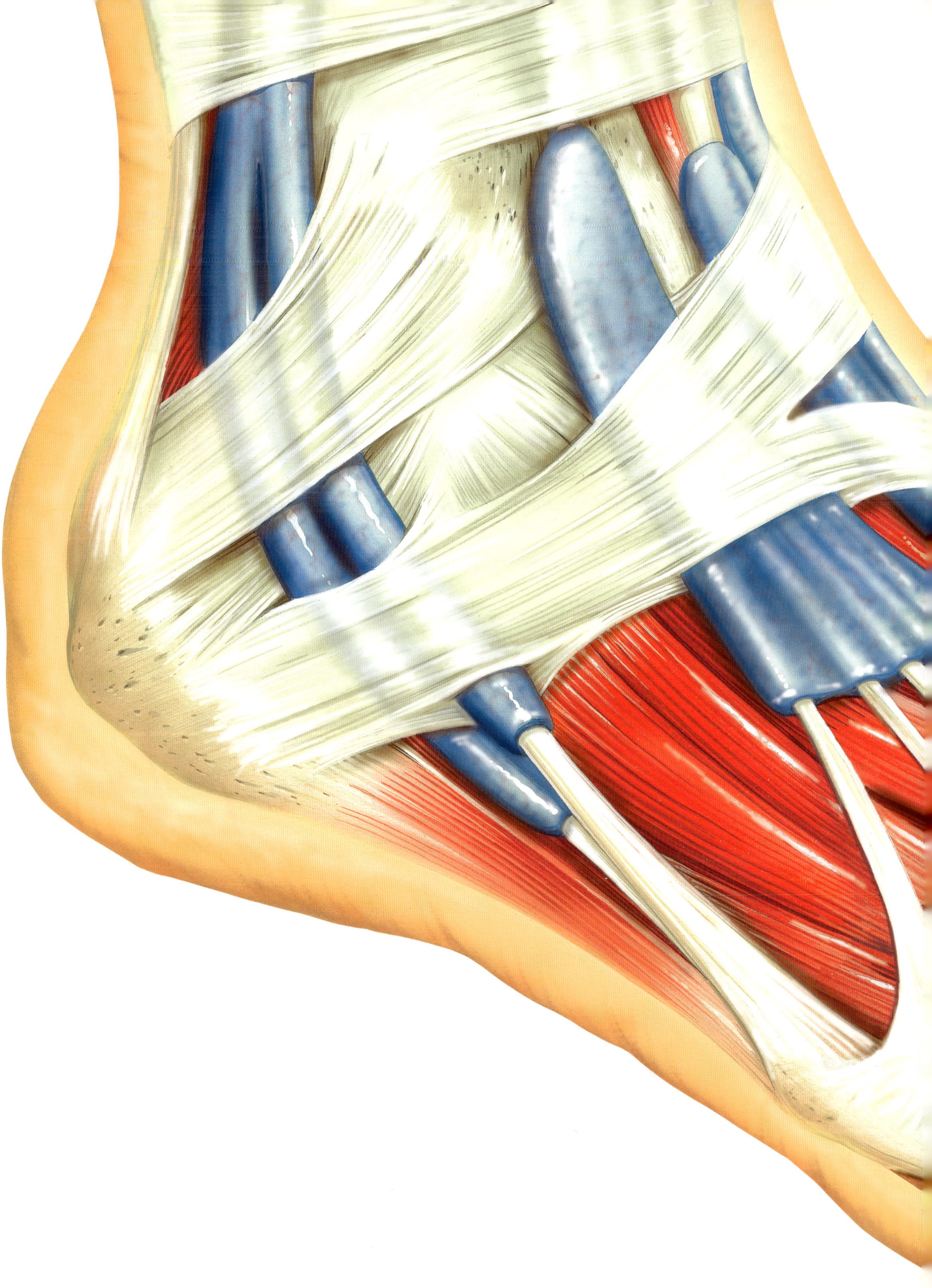

The Hips, Legs and Feet

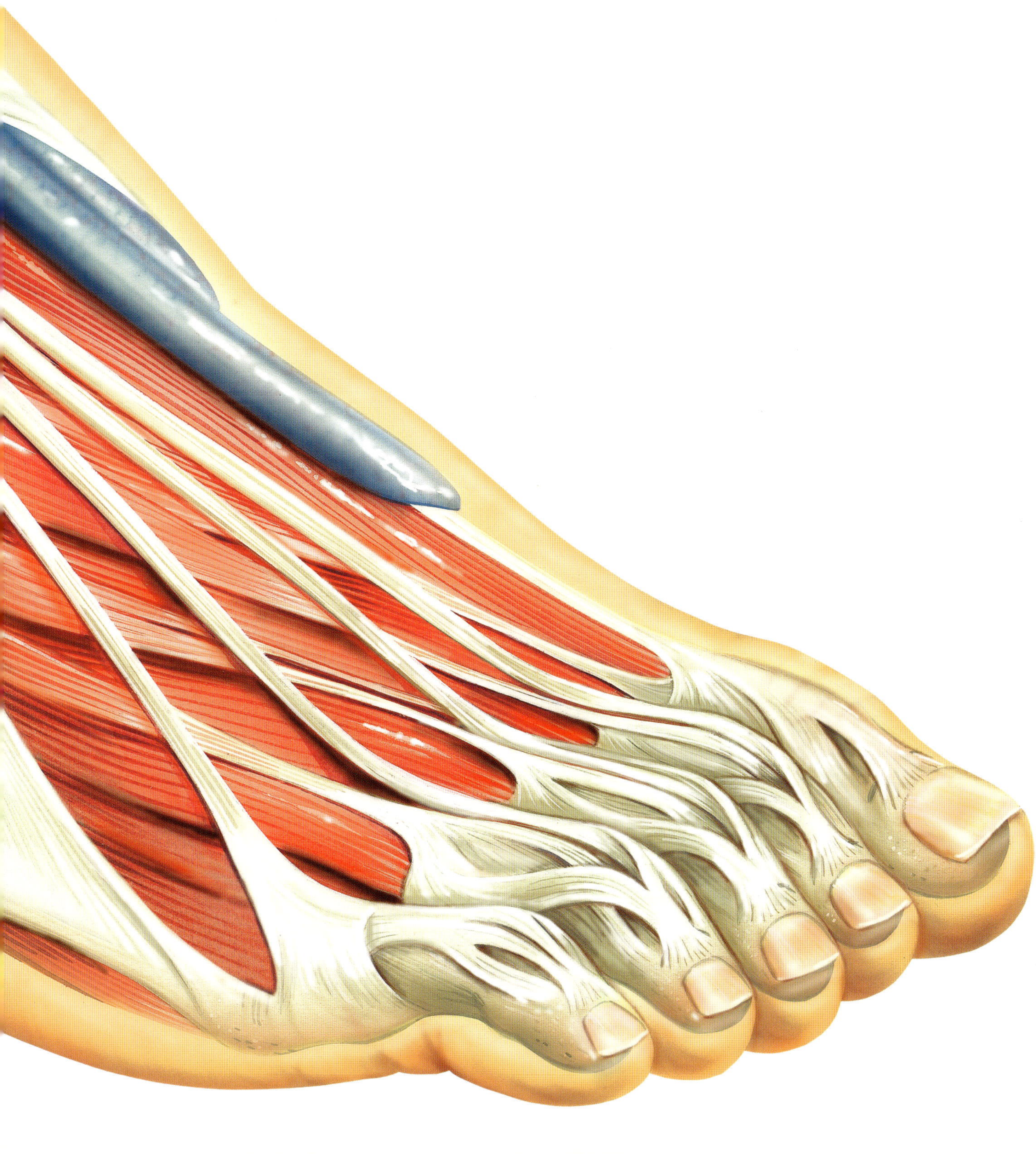

Leg

In general usage "leg" refers to the thigh, knee, calf and ankle, excluding the foot. The leg is designed on a similar structural plan to the arm, but in humans is adapted to bipedal locomotion, that is, walking upright on two legs. The bones are long to increase the stride, the joints are large and bound by strong ligaments, and the muscles controlling locomotion are powerful.

SEE ALSO *Circulatory system on page 78; Muscular system on page 48; Nervous system on page 64; Skeletal system on page 30*

Bones and Joints of the Leg

The two hip bones (the pelvic girdle) connect the lower limb to the vertebral column and join together in front at the pubic symphysis. Together with the sacrum, they are known as the bony pelvis. Body weight is transferred from the vertebral column, through the relatively immobile sacroiliac joints to the hip bones, and from the socket of each hip bone to the femur.

The hip joint is a ball-and-socket joint, and is one of the most mobile joints in the body.

The femur is the longest bone in the body, and it articulates with the hip bones above and with the tibia (shin bone) and patella (kneecap) below.

The knee joint, where the femur meets the tibia and patella, is a hinge joint that allows bending and straightening movements and also has a degree of gliding and rotational movement.

The tibia is the second longest bone in the body. The tibia and fibula comprise the bones of the lower part of the leg. The fibula is thin, and bears little weight. It serves as a structure for muscle attachment, and stabilizes the outside of the ankle joint.

The tibia and fibula articulate with the talus bone at the ankle joint. It is a mortice-like joint, with the tibia and fibula projecting down on either side of the talus to prevent any sideways movement.

SEE ALSO *Skeletal system on page 30*

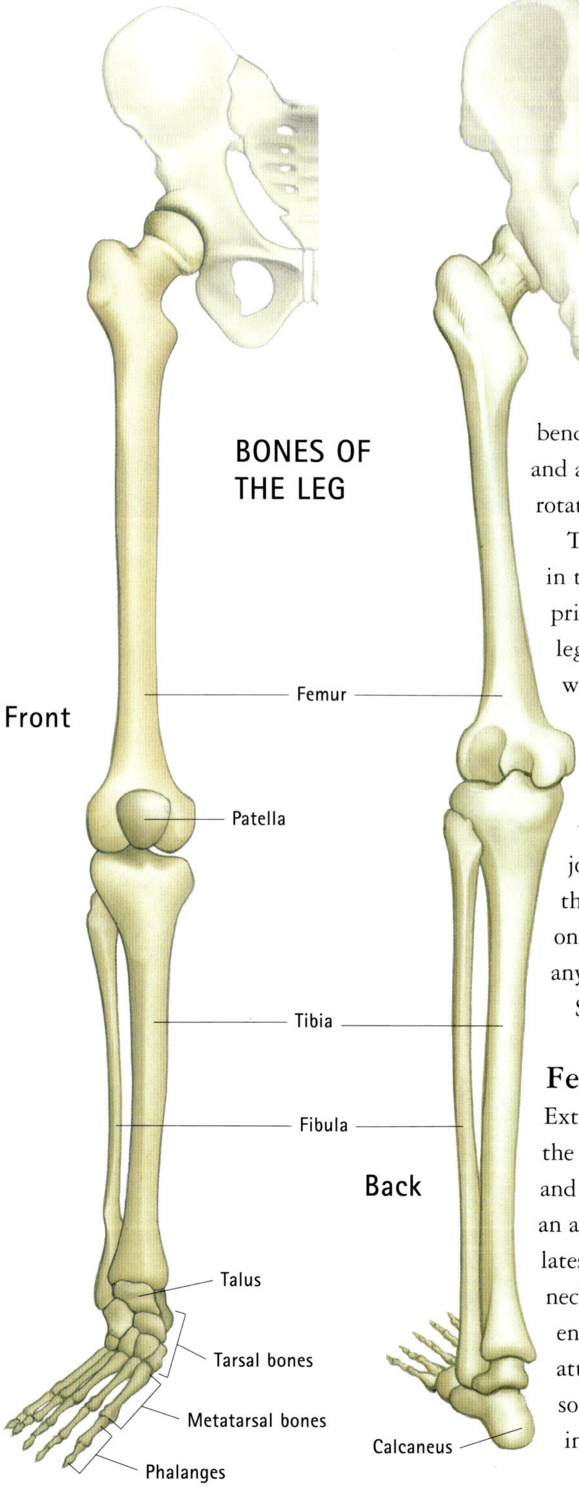

Femur

Extending from the hip to the knee, the femur (thighbone) is the longest and strongest bone in the body. It has an almost spherical head, which articulates with the hip bones, and a long neck, at the end of which are two enlargements (trochanters) for muscle attachment. The shaft angles inward, so that the knees come to lie together in the midline below the trunk. This arrangement reduces lateral body sway, and hence reduces energy expenditure in walking. The two femoral condyles at the lower end take part in the knee joint. The femur

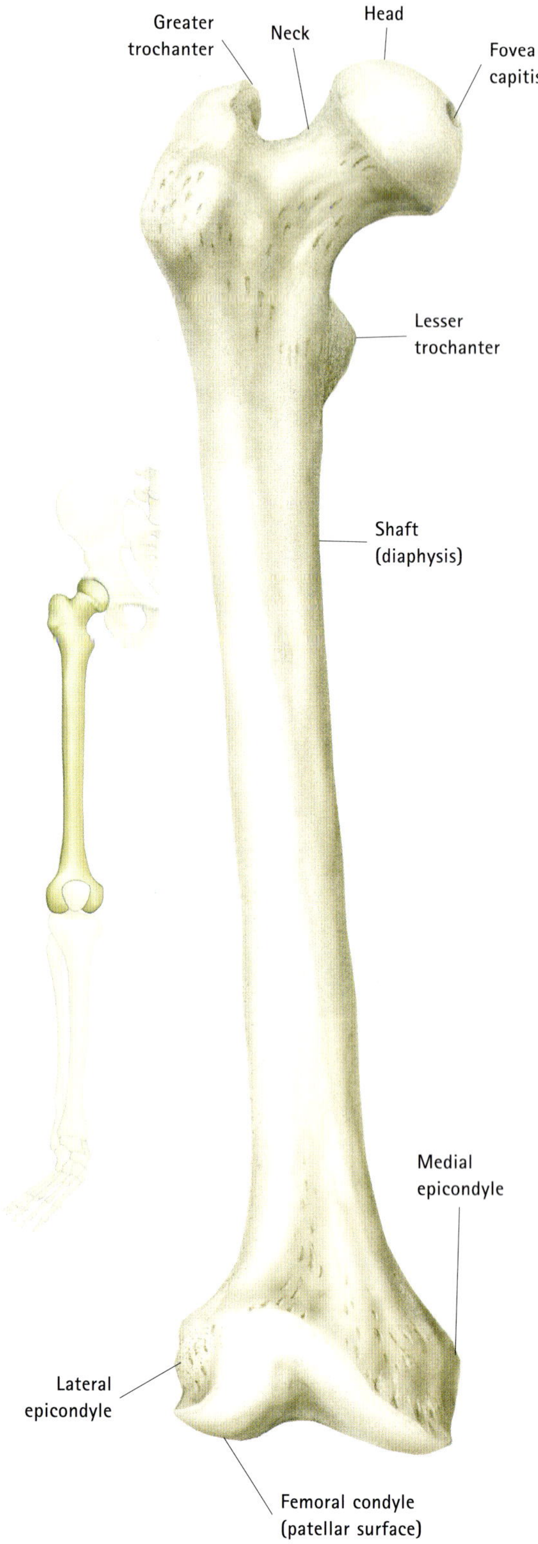

Femur

The femur, the strongest and longest bone in the body, must support the weight of the torso and provide an attachment for the powerful muscles of the lower limb.

articulates with the kneecap (patella), a bone within the tendon of the quadriceps muscle. It also articulates with, and transfers weight to, the tibia. The tendons of the powerful muscles that move the leg are attached to the femur.

Tibia

The tibia, or shin bone, is the inner and thicker of the two bones of the lower leg, the other being the fibula. It is also the second longest bone in the body, after the thighbone (femur). At its upper end, the tibia meets the femur to form the knee joint. At its lower end it meets with the fibula and a small bone called the talus to form the ankle joint. The small bump felt protruding on the inside of the ankle is part of the tibia that articulates with the talus. It is known as the medial malleolus.

Fibula

The fibula is the long, slender bone on the outside of the lower leg. It extends from just below the knee to the ankle, where its lower end forms the outer side of the ankle joint. The fibula does not bear weight like the shinbone (tibia), but instead serves as an attachment for some of the leg muscles.

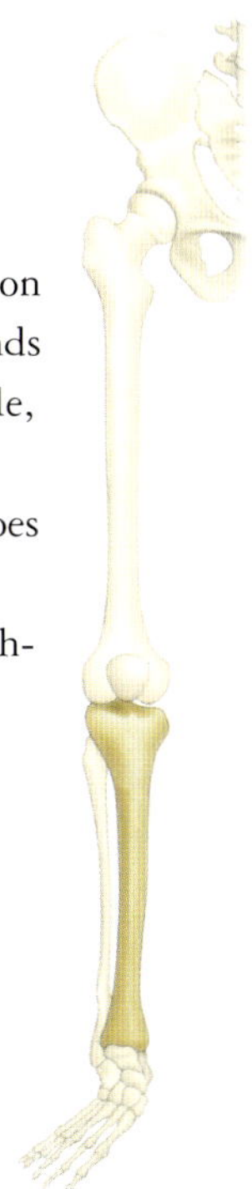

Disorders of the Bones and Joints of the Leg

The bones and joints of the leg may be damaged through injury (such as fractures) and disease (including rickets). They may also be affected by many inherited disorders.

SEE ALSO *Imaging techniques on page 431; Treating the musculoskeletal system on page 444; Osteoarthritis, Osteomalacia, Paget's disease, Rickets and other individual disorders in Index*

Bow legs

In bow legs, the lower part of the leg angles outward toward the knee. It may occur as a consequence of injury (such as a fracture) or disease (such as rickets, osteomalacia, Paget's disease, or osteomyelitis affecting growth plates in children).

Perthes' disease

Perthes' disease is a chronic disorder that affects children, in which the head of the femur (the ball part of the ball-and-socket hip joint) becomes inflamed and flattened, due to an interruption in its blood supply. The cause of the condition is unknown. Movement of the affected joint becomes limited, resulting in a limp, with pain in the thigh and groin. The condition occurs most frequently in boys aged 4–10 years and tends to run in families. In most cases,

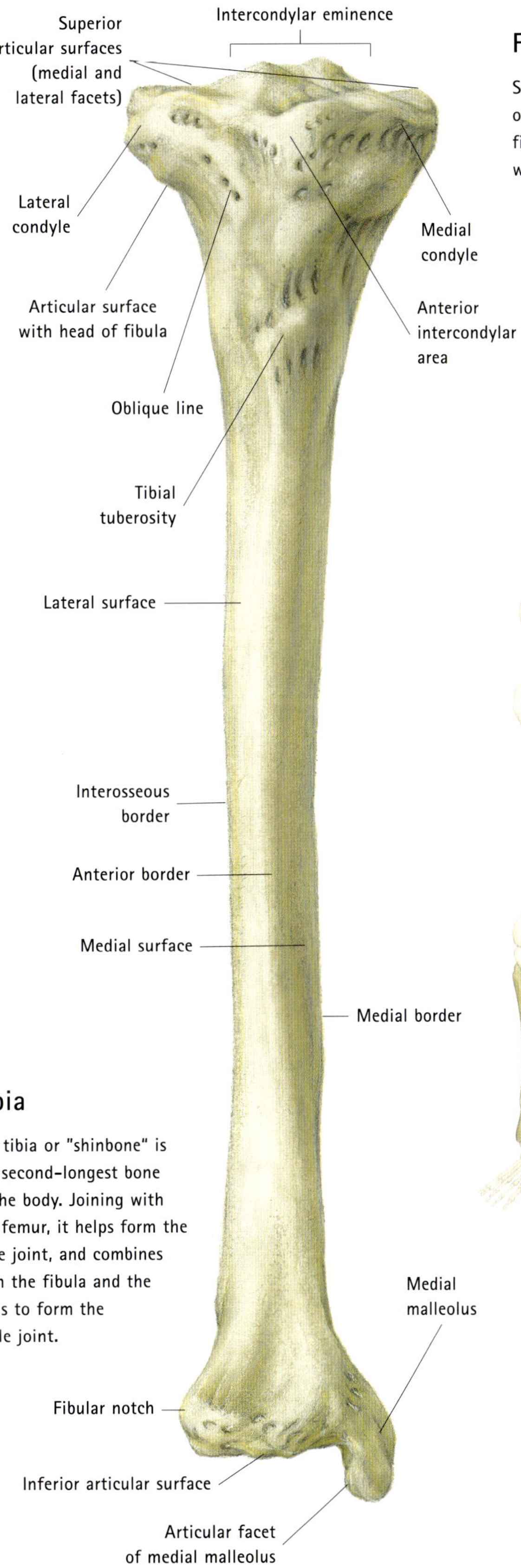

Tibia

The tibia or "shinbone" is the second-longest bone in the body. Joining with the femur, it helps form the knee joint, and combines with the fibula and the talus to form the ankle joint.

Fibula

Serving as a point of attachment for some of the musculature of the lower leg, the fibula is a slender bone that articulates with the talus bone at the ankle joint.

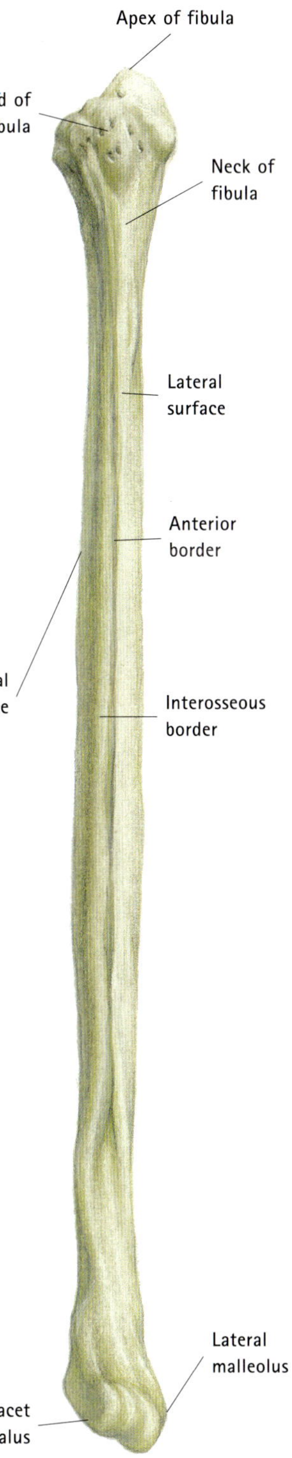

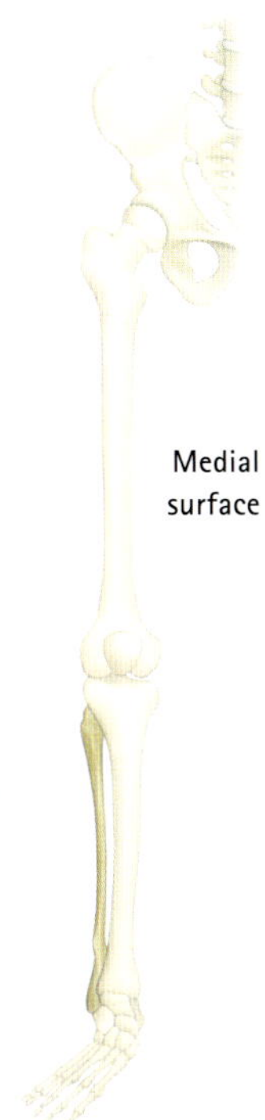

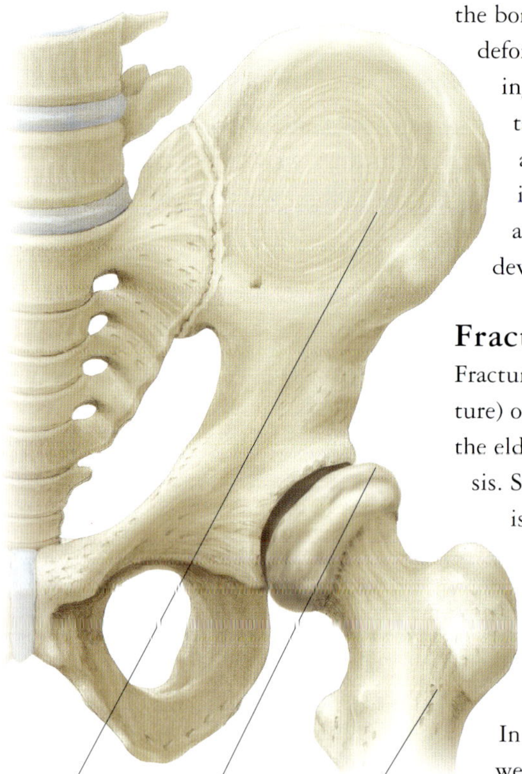

Perthes' disease

In this condition, the ball part of the femur degenerates due to an interruption to the blood flow, resulting in pain in the groin and thigh. Osteoarthritis may occur in affected adults.

the bone heals itself without any resulting deformity. Treatment is aimed at protecting the bone and joint while healing takes place. Bed rest and the using appliances like a brace, cast or splint is usually recommended. Osteoarthritis of the affected hip may develop in adulthood.

Fractures

Fracture of the neck of the femur (hip fracture) often follows a fall, and is common in the elderly, especially those with osteoporosis. Surgery followed by physical therapy is the recommended treatment.

A fracture of the tibia is a common injury, especially in childhood. It is usually caused by a direct blow to the child's leg, most often during a contact sport. In adulthood, a fracture may occur after weakening of the bone during jogging, running or walking.

After the injury, there is severe pain in the leg at the fracture site, swelling of the tissue surrounding the fracture and, if the fracture is complete (that is, broken all the way through), the leg is deformed.

The fracture must be set under general anesthesia in hospital. X-rays of the tibia will confirm that the ends of the bone have been correctly aligned. The bone is held in place with a cast which extends above the knee and below the ankle. Healing usually takes 6–8 weeks. Then physical therapy is needed to restore muscle strength and eliminate stiffness in the ankle and knee.

Fracture of the fibula is common in children, often occurring in contact sports such as football or hockey. It can occur with an ankle sprain. In most cases fractures are not serious, and setting is usually unnecessary; however, surgery is occasionally necessary.

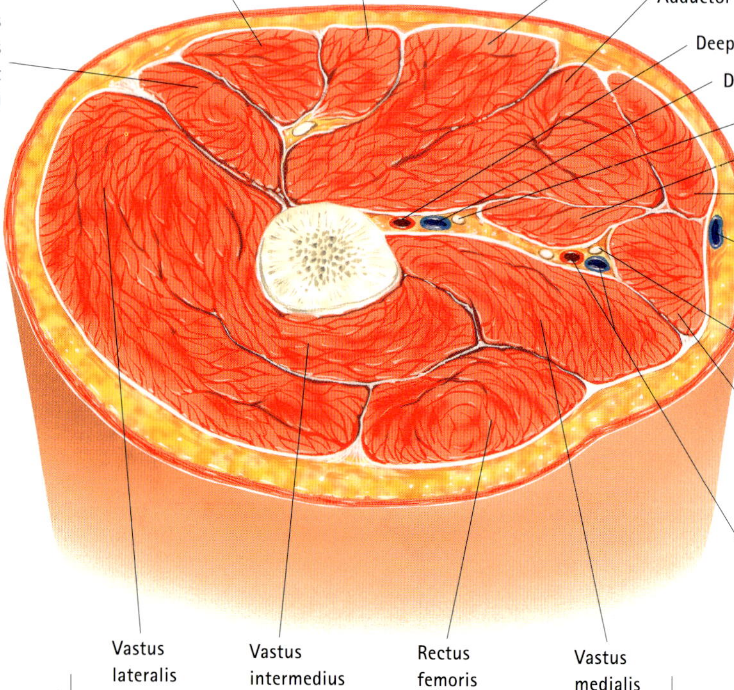

Muscles of the leg

Powerful muscles surround and stabilize the hip region. The characteristic shape of the buttocks in humans is a result of the gluteus maximus muscle. This is the largest muscle in the body and powerfully extends the thigh when running or climbing. Gluteus medius and minimus muscles are laterally placed and are important in keeping the pelvis level, and swinging the opposite side forward, during walking.

The thigh has two distinct muscle compartments, separated by connective tissue (deep fascia). The quadriceps (Latin for "four heads") muscle lies in the anterior (front) compartment and extends the knee joint. The quadriceps muscle forms the major muscle mass of the front and outer side of the thigh, covering most of the front and sides of the thigh bone (femur). The muscle is known as the quadriceps femoris muscle because it has four parts. These are the rectus femoris, vastus lateralis, vastus medialis and vastus intermedius.

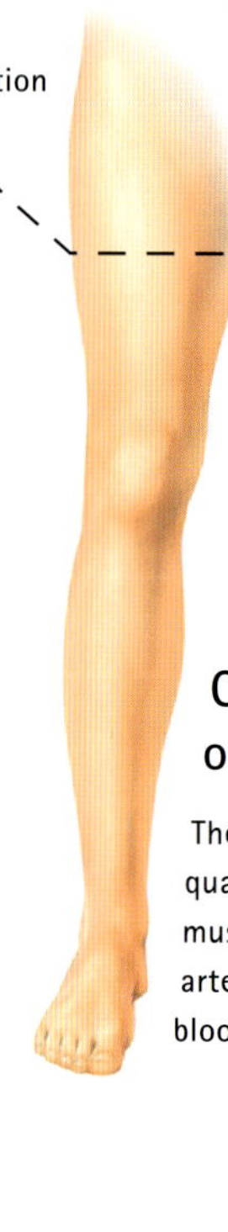

Cross-section of the thigh

The thigh contains the quadriceps and hamstring muscles as well as the femoral artery, the principal supplier of blood to the leg.

SURFACE MUSCLES OF THE LEG

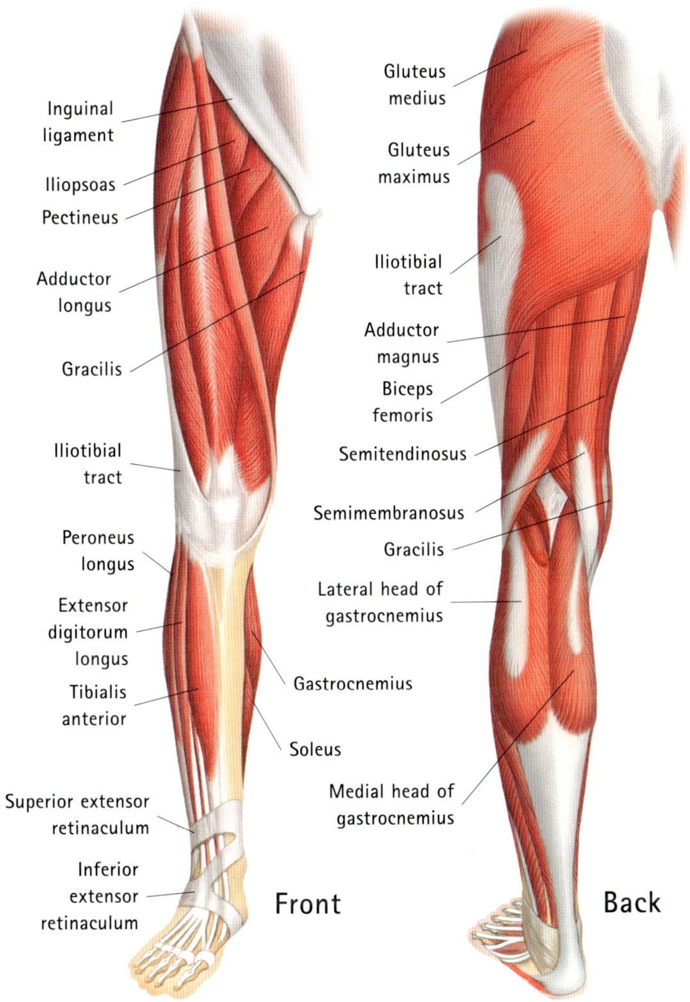

DEEP MUSCLES OF THE LEG

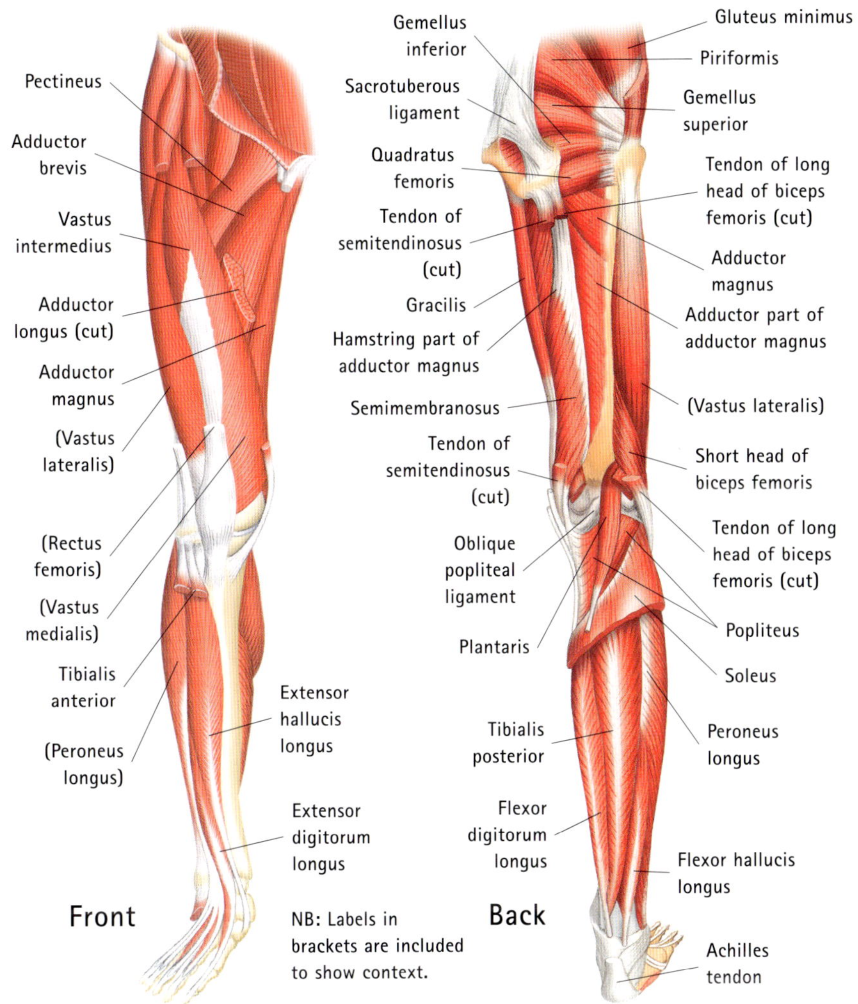

The quadriceps muscle arises from the upper two-thirds of the thigh bone and from the bony pelvis. At its lower end, the tendons of the component muscles blend to form a single tendon, which is attached to the upper surface of the kneecap (patella). The tendon extends below the patella as the "patellar ligament," which is attached to the tubercle of the shin bone (tibia) and is spread on both sides to enforce the capsule of the knee joint.

The hamstring muscles are located in the posterior compartment (at the back). The hamstring group consists of the semimembranosus, the semitendinosus and the biceps femoris muscles, which are attached above to the ischium (part of the pelvis). In the middle of the thigh they separate so the biceps femoris tendon passes behind the outside of the knee to reach the fibula, and the semimembranosus and semitendinosus tendons pass behind the inside of the knee to reach the tibia. They extend the hip joint and flex the knee.

The posterior compartment also contains the adductor muscles, which pull the leg toward the midline. They are important in counterbalancing the action of the gluteus medius and minimus muscles (adductors) when walking.

The lower part of the leg is divided into three compartments by deep fascia. The anterior (front) contains muscles that move the foot upward (dorsiflex the ankle), an action that is important in allowing the toes to clear the ground when swinging the leg forward in walking.

The lateral compartment (on the outside of the lower leg) contains only two muscles, that turn the sole of the foot outward (eversion). The calf region (posterior compartment) contains the greatest number of muscles, and is divided into two groups—those closest to the skin (superficial) and the deeper group. The superficial group contains the powerful gastrocnemius and soleus muscles, both critical to pushing off from the ground (plantarflexion of ankle joint—in which the foot moves downward) during walking, running and jumping, and when standing on one's toes.

The deeper group of muscles pass behind the ankle joint and attach to bones of the foot. The largest of these, flexor hallucis longus muscle, is critical to pushing off from the big toe during walking.

SEE ALSO *Muscular system on page 48*

DISORDERS OF THE MUSCLES AND SOFT TISSUES OF THE LEG

The muscles and soft tissues of the leg can easily become inflamed and swollen through injury or overuse.

SEE ALSO *Electromyography on page 431*

Compartment syndrome

Due to the inelastic nature of the fascia bounding the compartments of the lower part of the leg, any swelling in a compartment results in pressure build-up, which

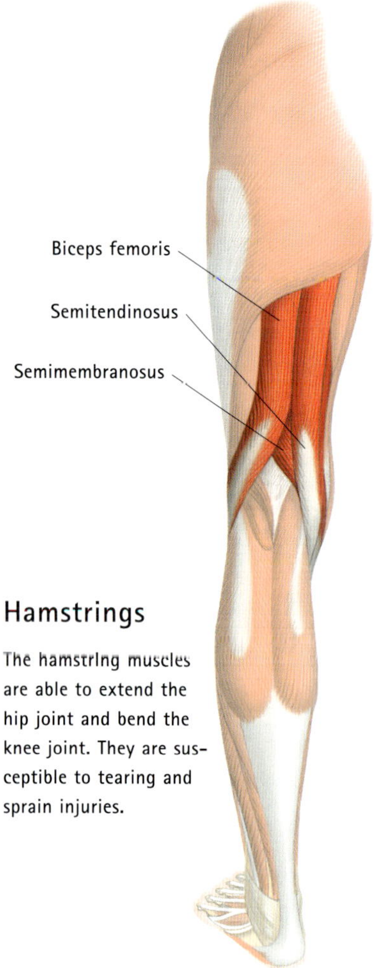

Hamstrings

The hamstring muscles are able to extend the hip joint and bend the knee joint. They are susceptible to tearing and sprain injuries.

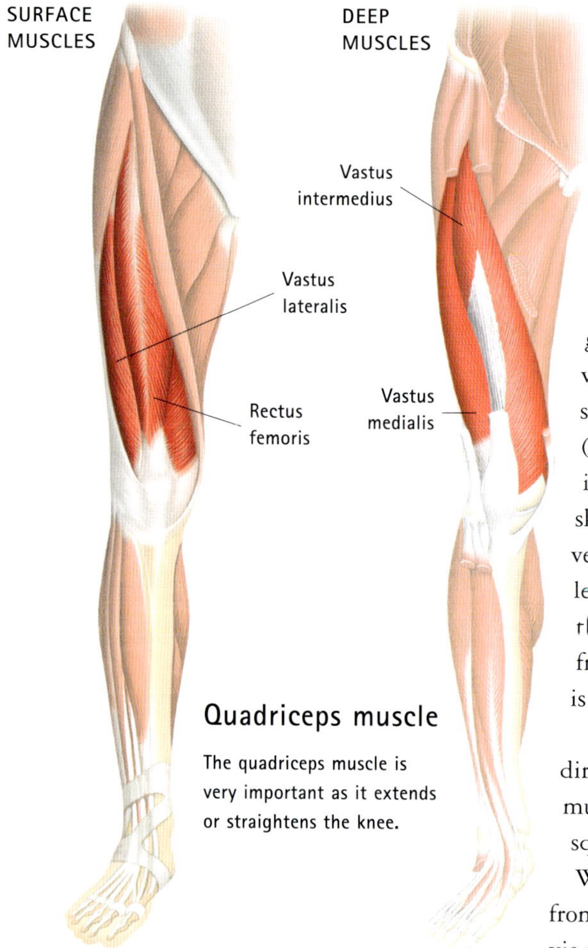

Quadriceps muscle

The quadriceps muscle is very important as it extends or straightens the knee.

can impair the blood supply. This is called compartment syndrome and may result from inflammation of muscles caused by excessive exercise in an unfit person, or trauma, such as fracture. With muscle overuse, it commonly involves the anterior, or deep posterior, compartments of the lower part of the leg and is characterized by shin pain (commonly known as shin splints) that increases during exercise and reduces at rest. Compartment syndrome may also be associated with pins and needles sensations and muscle weakness.

Nerves and Blood Vessels of the Leg

The femoral, obturator and sciatic nerves are the principal nerves of the leg, supplying the muscles and much of the skin. The sciatic nerve, which is the major nerve of the back of the thigh arises from the lumbar and sacral plexus at the base of the spine. It leaves the pelvis through the greater sciatic foramen, passes under the gluteus maximus muscle and runs down the back of the thigh. The sciatic nerve sends branches to supply several muscles, namely the biceps femoris, semitendinosus, semimembranosus, and the ischial head of the adductor magnus. The sciatic nerve also provides sensation for the hip joint. In the lower third of the thigh, it divides into two branches, the tibial and common peroneal nerves, which supply all the muscles of the lower part of the leg and foot.

The femoral artery is the principal artery of supply to the leg. A major branch of the femoral artery, called the profunda femoris artery, supplies blood to much of the thigh. The femoral artery descends in the front of the thigh. Two-thirds of the way down, it passes backward behind the knee and is renamed the popliteal artery. It then divides into anterior and posterior tibial branches that descend in the anterior and posterior compartments of the lower part of the leg.

Veins are divided into two groups: the deep group and the superficial group. The deep group of lower limb veins travel with the arteries and are similarly named. The superficial veins (e.g. great and small saphenous) travel in the superficial tissue just below the skin. The function of the great saphenous vein is to return blood from the foot and leg to the femoral vein below the groin; the small saphenous vein returns blood from the calf to the popliteal vein, which is located behind the knee.

Lower limb veins have numerous valves directing blood toward the heart. When muscles in the leg contract, the veins are squeezed and blood is forced upward. When the muscles relax, blood can flow from the superficial veins into deep veins via perforating or communicating veins. Faulty valves in the communicating veins can cause the superficial veins to become elongated, tortuous and enlarged (varicose veins).

SEE ALSO *Circulatory system on page 78; Nervous system on page 64*

Disorders of the Nerves and Blood Vessels of the Leg

Disorders such as sciatica can lead to pain, which may be relieved by anti-inflammatory drugs. However, a thrombosis or blood clot in the leg is a more serious matter as it can travel to the lungs or brain and result in death.

SEE ALSO *Imaging techniques on page 431; Nerve conduction tests on page 430; Ultrasound on page 432*

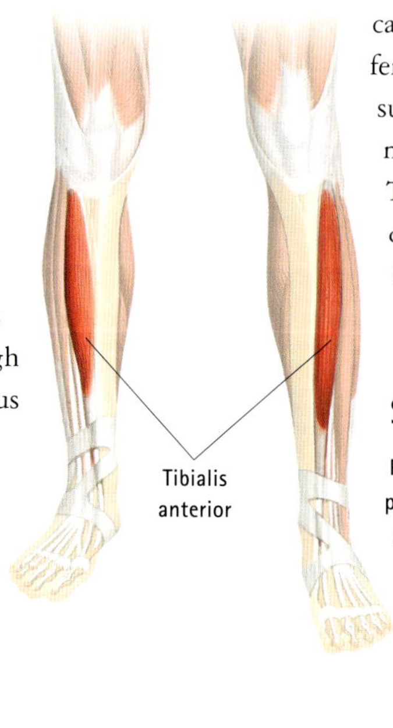

Shin splints

Running on hard surfaces such as bitumen or concrete for extended periods can cause inflammation of the tissues that join the muscles and bones of the lower leg. The soft tissues linked to the tibia (such as the tibialis anterior muscle) are most commonly affected.

NERVES AND BLOOD VESSELS OF THE LEG

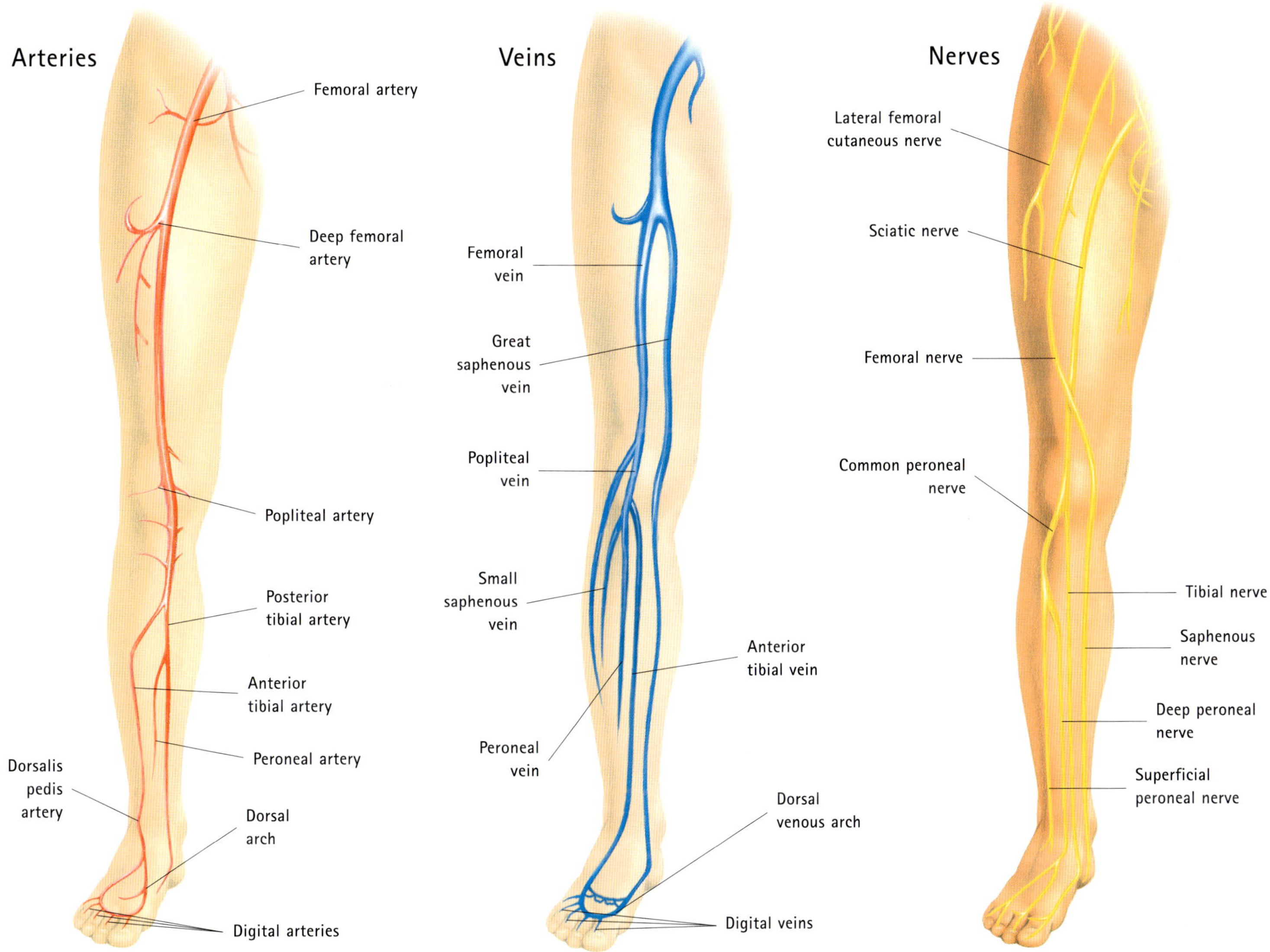

Thrombosis

Stasis of blood in veins may result in the formation of clots (thrombosis). When formed in the deep veins of the lower part of the leg it is called deep vein thrombosis (DVT). Increased risk is associated with advanced age, bed rest, immobilization and oral contraceptives. The greatest risk of thrombosis is that a piece of clot may dislodge and become lodged in the lung; this is called a pulmonary embolus, and is a major cause of death in the USA. Painful white leg ("milk leg") from venous thrombosis can also occur in the last three months of pregnancy or in the postdelivery period. Thrombophlebitis is the tenderness associated with a blood clot in a vein.

Varicose veins

Varicose veins are dilated, elongated, tortuous veins. They arise in the superficial veins of the leg—the great and small saphenous veins and their tributaries. Varicose veins are defined as primary when they arise as a result of problems in the superficial veins themselves (many sufferers have a family history). Alternatively they may be secondary to the formation of clots (venous thrombosis) in the deep leg veins (deep vein thrombosis, DVT), or secondary to faulty valves in the veins that normally direct blood from the superficial veins into the deep leg veins. In the latter two cases, pressure in the superficial veins increases, either because blood cannot flow into the blocked deep veins, or because faulty valves in the communicating veins result in backflow into superficial veins when the leg muscles contract.

Valves in the veins may become incompetent when veins are overstretched by excessive pressure for long periods (weeks to months), for example when standing for prolonged periods or in pregnancy. The pressure causes the veins to become dilated but the valves do not increase in size, therefore affecting the efficiency of the valves.

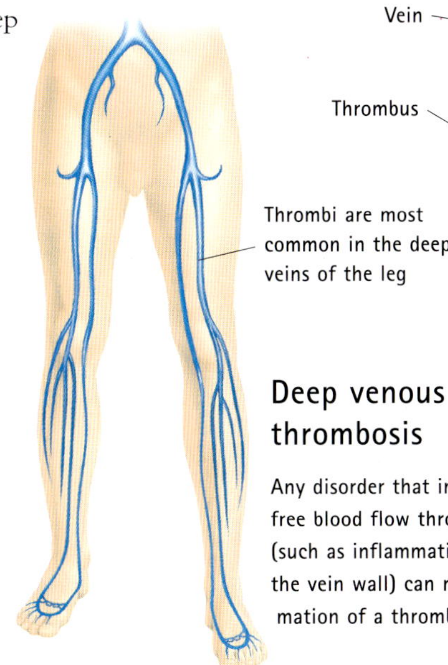

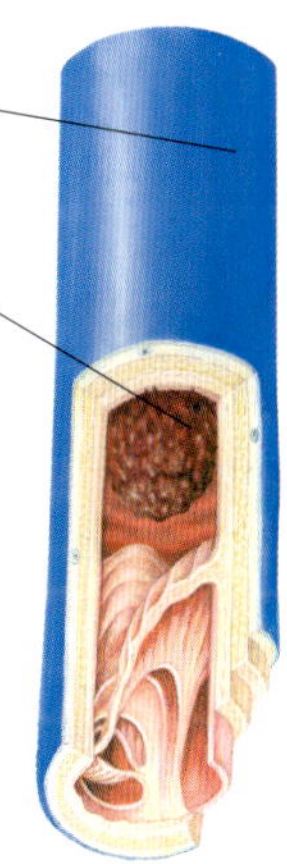

Deep venous thrombosis

Any disorder that interferes with free blood flow through the veins (such as inflammation or injury to the vein wall) can result in the formation of a thrombus (blood clot).

Varicose veins

Faults in the walls or valves of the veins, or the formation of blood clots, can disrupt blood flow in the veins of the legs causing swollen and knotted (varicose) veins. Faulty valves allow blood to pool, swelling and distending the veins.

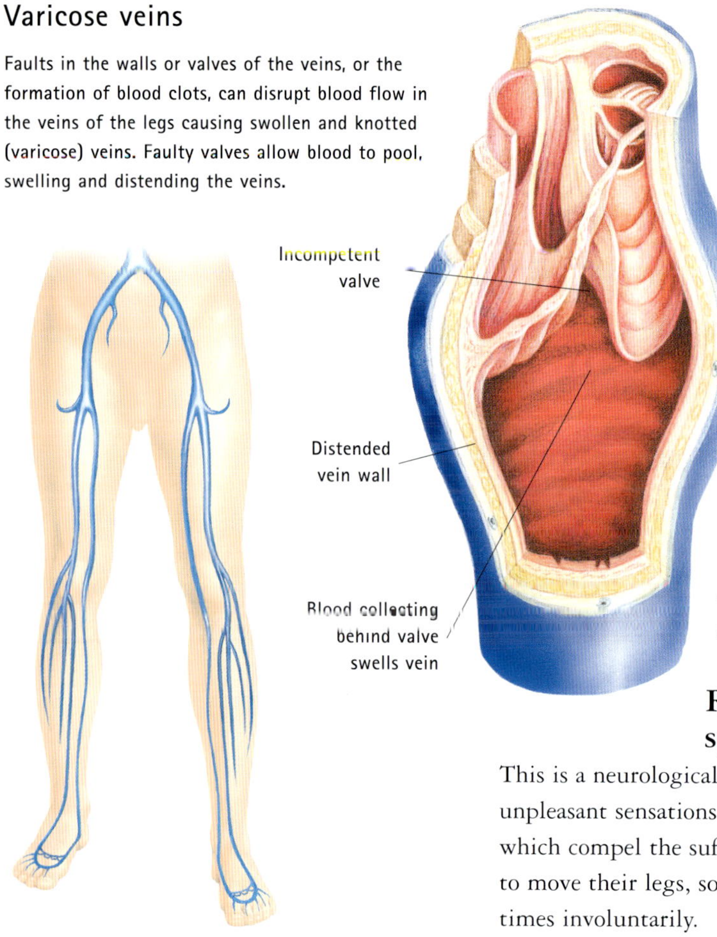

External compression with elastic support stockings and elevating the legs can relieve varicose veins and relieve the swelling or edema in the lower leg that accompanies them. Sluggish flow (stasis) of blood in varicose veins can affect the nourishment of the skin in the area, and even minor trauma can lead to the formation of a varicose ulcer. This is especially common over the subcutaneous surface of the tibia.

Sciatica

Sciatica is usually caused by stress on the nerve around the point where it emerges from the vertebral column.

Sciatica occurs most commonly when the nerve becomes pinched by a herniated disk between vertebrae or is placed under pressure around the base of the spine due to arthritis or pregnancy.

Occasionally, sciatica may be due to more serious conditions such as a tumor, blood clot or abscess in the spine. The severity and type of pain varies, from pins and needles in the toes to excruciating pains that run the length of the limb. Weakness in the lower leg muscles can result; in severe cases, it may be difficult to bend the knee or even move the foot, making standing near-impossible.

Sciatica usually gets better with simple treatments, often within a few days, although very occasionally surgery may be necessary.

In most cases, over-the-counter pain killers and anti-inflammatory drugs ease the pain.

Restless legs syndrome

This is a neurological disorder marked by unpleasant sensations in the lower legs, which compel the sufferer to move their legs, sometimes involuntarily. The exact cause is not known but conditions such as iron-deficiency anemia, arthritis and pregnancy may play a part. There is no specific treatment but treatment of any suspected underlying cause may alleviate discomfort.

Hip

The hip (coxal or innominate) bone is made up of three bones: the ilium, ischium and pubis. The three bones fuse with each other at the acetabulum (part of the hip joint) at 14–16 years of age.

The crest of the ilium is found at the waist, laterally, below the ribs. The tuberosities of the two ischia are the bony knobs that we sit on. The pubic bones are at the front at the lower limit of the soft anterior abdominal wall. The left and right hip bones and the sacrum form the bony pelvis. The hip joint is formed by the acetabulum, a cup-shaped socket and the rounded head of the femur.

The head of the femur is covered with cartilage and lubricated by synovial fluid. The whole joint is enclosed in a fibrous capsule which is loose enough to permit free movement yet strong enough to hold the femoral head securely in place. The muscles of the hip joint are large and powerful to hold the joint firm and to move the thigh for walking and running. They include the gluteus maximus and the gluteus medius at the back, and the rectus femoris at the front.

More than half the round head of the femur is held within the acetabulum and surrounding cartilage, making the hip joint

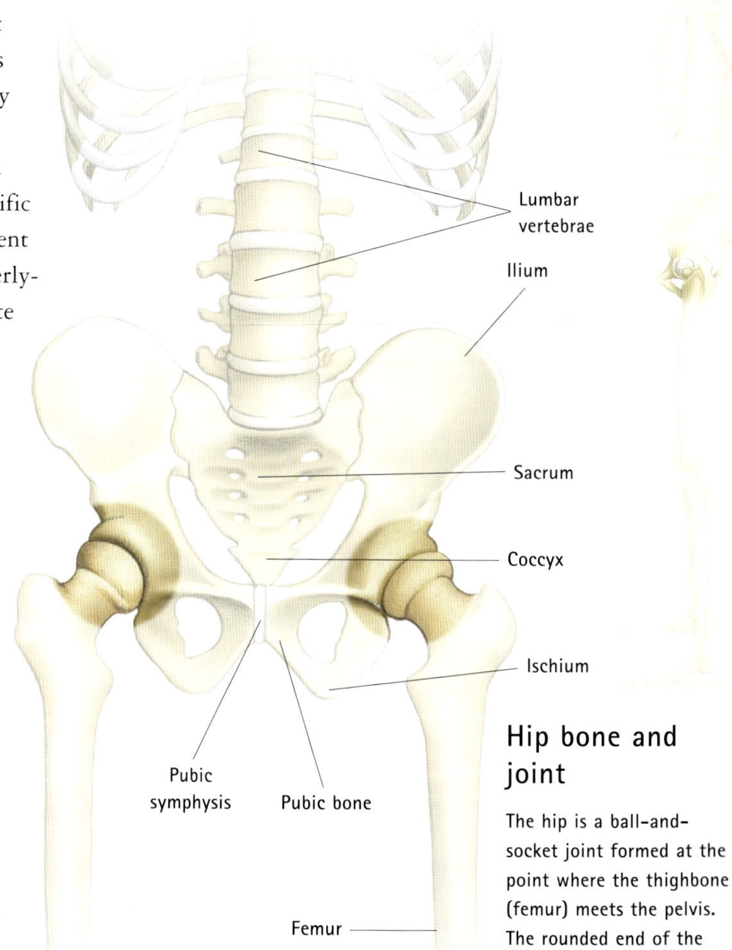

Hip bone and joint

The hip is a ball-and-socket joint formed at the point where the thighbone (femur) meets the pelvis. The rounded end of the thighbone fits into the cup-shaped socket.

extremely stable and strong, with capability of rotation second only to the shoulder joint and limited only by the flexibility of its supporting ligaments. The hip transmits the entire weight of the upper body to the head and neck of each femur and is at its most solid during weight-bearing activities.

The strength and flexibility of the pelvis and hip allows the femur to descend vertically while bearing the full weight of the body, an ability vital for the development of standing, running and jumping.

Blood supply to the hip joint is from the femoral arteries. Several nerves supply the hip joint and pain in the hip can be misleading as it may be generated from the spinal column. Loss of blood supply due to accident or dislocation can result in pain which radiates to the knee.

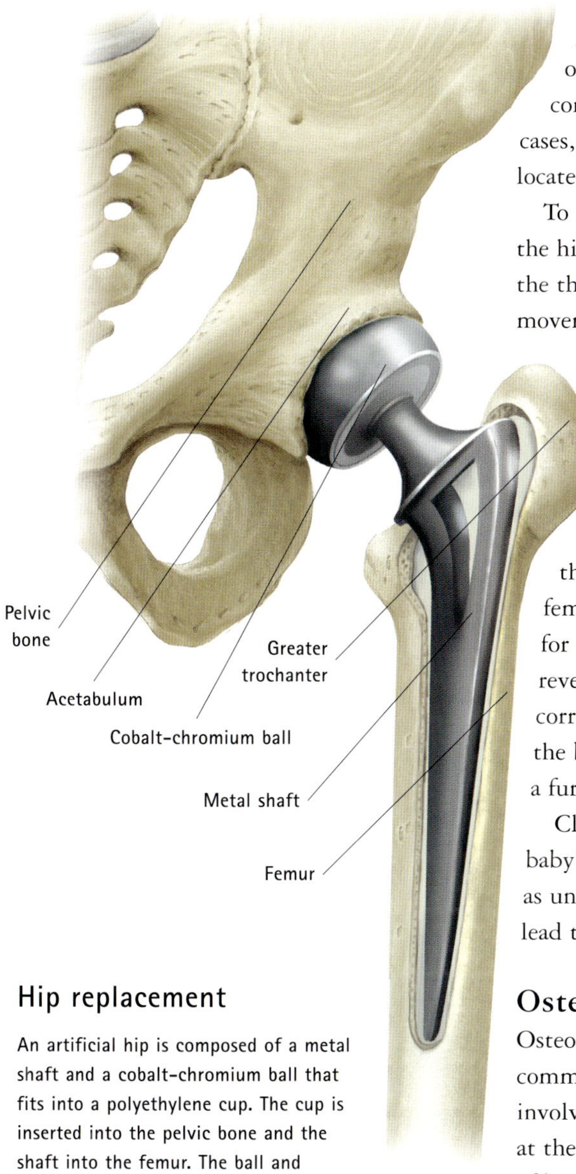

Hip replacement

An artificial hip is composed of a metal shaft and a cobalt-chromium ball that fits into a polyethylene cup. The cup is inserted into the pelvic bone and the shaft into the femur. The ball and cup form a replacement joint where they meet.

Disorders of the hip

Joint abnormalities, deformities or malalignments are revealed by radiography or, in newborns, by observation. Disorders affecting growth of bones during childhood, such as rickets, can affect the formation of the head of the femur and the joint. Slipped epiphysis of the femur may occur in late childhood and adolescence and is a dislocation of the growing head (epiphysis) of the femur from the femoral neck. Perthes' disease also occurs around this age. Symptoms are increasing discomfort at the hip with pain referred to the knee.

SEE ALSO *Imaging techniques on page 431*

Congenital dislocation of the hip

Congenital dislocation of the hip is found in approximately 15 babies per 10,000 births and may affect one or both joints. It is eight times as common in girls as in boys. In such cases, the end of the femur is not properly located in the socket (acetabulum).

To test for congenital dislocations of the hip, the baby is laid face upward and the thighs are moved sideways to see if movement produces a clunk as the head of the femur enters the pelvic socket. Instability of the joint at this stage can correct itself but if present after five days, treatment will usually be recommended.

The pelvic harness, which holds the thighs so that the head of the femur is securely in its socket, is worn for a few months, until radiography reveals that the abnormality has been corrected. If further treatment is needed, the baby may wear abduction splints for a further few months.

Clicks or noises heard on moving a baby's hips should always be investigated, as untreated abnormalities of the hip can lead to limping and permanent deformity.

Osteoarthritis

Osteoarthritis is a degenerative disorder common in adults of middle age, which involves the abrasion and loss of cartilage at the surfaces of joints with outgrowths of bony ridges at abraded surfaces. Osteoarthritis of the hip joint may be secondary to a structural abnormality, or of a primary nature, involving no underlying abnormality. One underlying cause is congenital dysplasia, dislocation or subluxation of the hip, where the head of the femur and socket fit badly. This can arise in infancy due to genetic factors or swaddling, which leaves the thighs extended. This delays development of the hip joint, which can become deformed once the child begins to walk.

Osteoarthritis of the hip joint in people under age 40 often follows from disease or disorder and may require surgery to correct it. This may involve osteotomy, reshaping the end of the femur, or removal of diseased tissue and replacement with an artificial hip. As age advances, degenerative diseases and hip replacements become more common.

Hip injuries

Trauma or violent stresses can result in dislocation or fracture of the hip bone or fracture of the neck of the femur. Such injuries commonly result from falls onto hard surfaces in sports such as ice-skating or athletics. Automobile accidents are also a common cause because of the high velocity of secondary impacts, such as when the knee strikes the dashboard. This type of stress can dislodge the head of the femur from its socket, causing dislocation with injury to surrounding tissues.

Persons over 60 are at risk of serious fracture from even trivial falls, because their bones and supporting tissue structures are likely to be weaker than those of younger persons. This applies especially if the person has osteoporosis.

Hip replacement

This is an increasingly common and successful option in instances where a joint has deteriorated through rheumatoid arthritis, osteoarthritis or osteoporosis.

The materials being used are constantly improving, extending the useful life of replacement joints and reducing the possibility of an adverse response from the body's tissues. The stainless steel that was previously used in early artificial joints has been replaced by titanium alloys, which are usually lighter and more stable. The artificial hip usually consists of a highly polished metal (cobalt-chromium) ball, which replaces the head of the femur, and a cup made of extremely tough polyethylene, to replace the socket (the acetabulum). It is important to avoid any abrasion between the two surfaces, as small particles of debris can cause inflammation.

The materials used in the artificial joint must also simulate the mechanical properties of bone. Otherwise uneven stresses may occur, which would weaken bones and induce reshaping.

Knee

The knee joint, between the thigh and the lower leg, is one of the most important and complicated joints in the body. Because its adjacent surfaces do not fit closely together, it relies mainly on ligaments and muscles for stability. As a mobile but weight-bearing joint, the knee is under a great deal of strain and is vulnerable to injury when excessive force is put on it.

The knee is a hinge type of joint with its main movements being flexion (bending) and extension (straightening), but some backward and forward gliding movements also occur in association with the hinge movements of flexion and extension. A little rotation also occurs at the end of extension (to "lock" the knee) and at the beginning of flexion (to "unlock" the knee).

The knee joint is formed by three bones: the femur above, the patella (kneecap) in front and the tibia below. The lower end of the femur has a concave surface at the front, into which the back of the patella fits, and two rounded bulges at the bottom called condyles. The upper surface of the tibia has two rounded slightly concave areas (also called condyles) separated by an intercondylar space. Each of the rounded condyles of the femur fits into the shallow sockets formed by the corresponding condyle of the tibia. The adjacent (contacting) surfaces are lined by cartilage.

Knee joint capsule

The three bones of the knee joint are united by a fibrous capsule (the knee joint capsule) which encloses a single, large joint cavity between the bones. The inside of the capsule is lined by a membrane, which produces an oily fluid called synovial fluid that acts to lubricate the joint surfaces to keep them friction-free. In some areas, pouches of synovial membrane extend beyond the confines of the joint capsule forming sacs known as bursae, and making the synovial membrane of the knee joint the most extensive of any joint in the body.

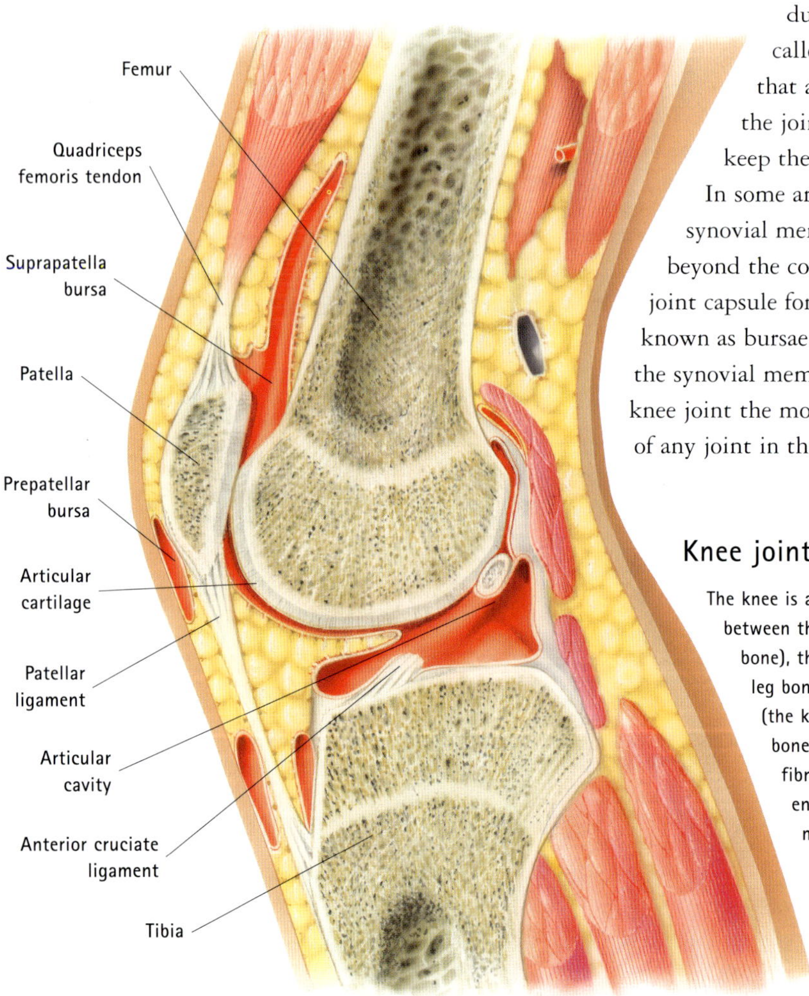

Knee joint

The knee is a complex hinge joint between the femur (upper leg bone), the tibia (the lower leg bone) and the patella (the kneecap). The three bones are united by a fibrous capsule which encloses a large membrane-filled cavity. This membrane produces synovial fluid to lubricate the joint.

Consequently, when the knee is traumatized, the synovial membrane responds by producing more fluid to protect the joint, and the swelling that occurs can be quite considerable.

The knee joint capsule is continued in front by the quadriceps tendon above, the kneecap and the patellar ligament below (attaching the kneecap to the tibia). The kneecap is incorporated into the joint, fitting into a concavity on the front of the femur. It glides up and down on the femur during contraction and relaxation of the quadriceps muscle and may be fractured in falls onto the front of the knee.

The knee joint cavity contains two circular fibrocartilaginous disks, or menisci, which are attached to the top of each tibial condyle. These are the "cartilages" often referred to in descriptions of injuries. Each disk is wedge-shaped in cross-section, with the thicker surface on the outside, so they act to deepen the sockets on top of the tibia and to allow a small amount of rotation to occur between the tibia and femur. The menisci may be torn if they get caught between the tibia and femur when the knee is forcibly rotated in the flexed (bent) or semiflexed position, as may occur when someone is tackled or tripped around the ankles or legs, and their thighs and body twist forward.

In osteoarthritis the cartilage that protects the bone surfaces from rubbing against each other degenerates and the joint may become so damaged and painful that it needs to be replaced with a prosthesis. A knee replacement may also be needed after damage from rheumatoid arthritis or injury. Although it will greatly reduce the pain and restore full mobility to the joint, a replacement joint is not as strong as a natural one. Extensive physical therapy will help rebuild muscle.

Ligaments of the knee

Because the sockets on the tibia are so shallow, the ligaments are important in strengthening the knee joint and limiting excessive or unwanted movements. The inner (medial) side is reinforced by the medial collateral ligament, which is a long flat band of fibers, about ½ inch (1 centimeter) wide and 4½ inches (10 centimeters) long, extending from the sides of the femoral condyle down onto the shaft of the tibia. As it passes the joint it is fused to the capsule and the medial meniscus. The medial ligament can be stretched and torn when the knee is struck forcefully from the outer (lateral) side. In serious injuries the medial meniscus, which is fused to the ligament, may also be damaged.

The outer (lateral) side is reinforced by the lateral collateral ligament, which is a narrow cord-like band about 1–1½ inches (2–3 centimeters) long, which extends from the lateral femoral condyle down to the head of the fibula. Damage to this ligament is much less common than to the medial ligament but can occur when the knee or leg is struck from the inner side.

Two important ligaments exist internal to the joint capsule, in the space between the medial and lateral pairs of condyles. Because the two internal ligaments cross over each other in the form of an X they are known as the cruciate ligaments. The anterior cruciate ligament passes from the front of the tibia up and backward to the back of the femur. The posterior cruciate ligament passes from the back of the tibia upward and forward to the front of the femur. The cruciate ligaments guide the tibia in its movement around the end of the femur and prevent excessive forward and backward gliding during flexion and extension of the knee.

The anterior cruciate ligament may be damaged when the knee is hit from the front and overextended or in twisting injuries of the knee.

The anterior cruciate ligament is only half as thick as the posterior cruciate ligament and so is more frequently damaged. If the ligament is completely severed, a "knee reconstruction" may be performed, in which parts of tendons surrounding the knee are used to replace the damaged ligament.

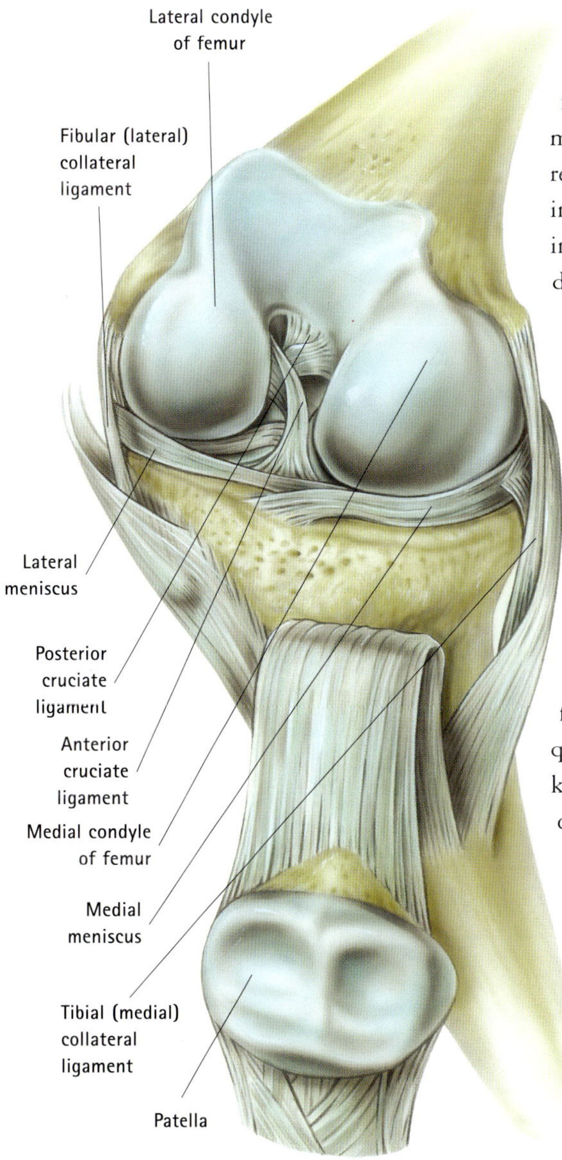

Muscles of the knee

The muscles around the knee joint maintain its stability. The tendons of the hamstrings, which pass over the inner and outer sides of the joint, help to reinforce it, but the most important muscle is the quadriceps femoris muscle on the front of the thigh, whose tendon reinforces the joint and ensures correct movement of the patella on the femur. Weakness in the inner side of the quadriceps muscle, which occurs when the knee has been inactive after previous injury, can cause the patella to dislocate.

Ligaments and bones of the knee

Ligaments and muscles keep the knee strong and stable, but excessive force on the joint, like that experienced during sport, can damage it. In the illustration to the left, the patella has been separated to show the inner ligaments and bones of the knee joint.

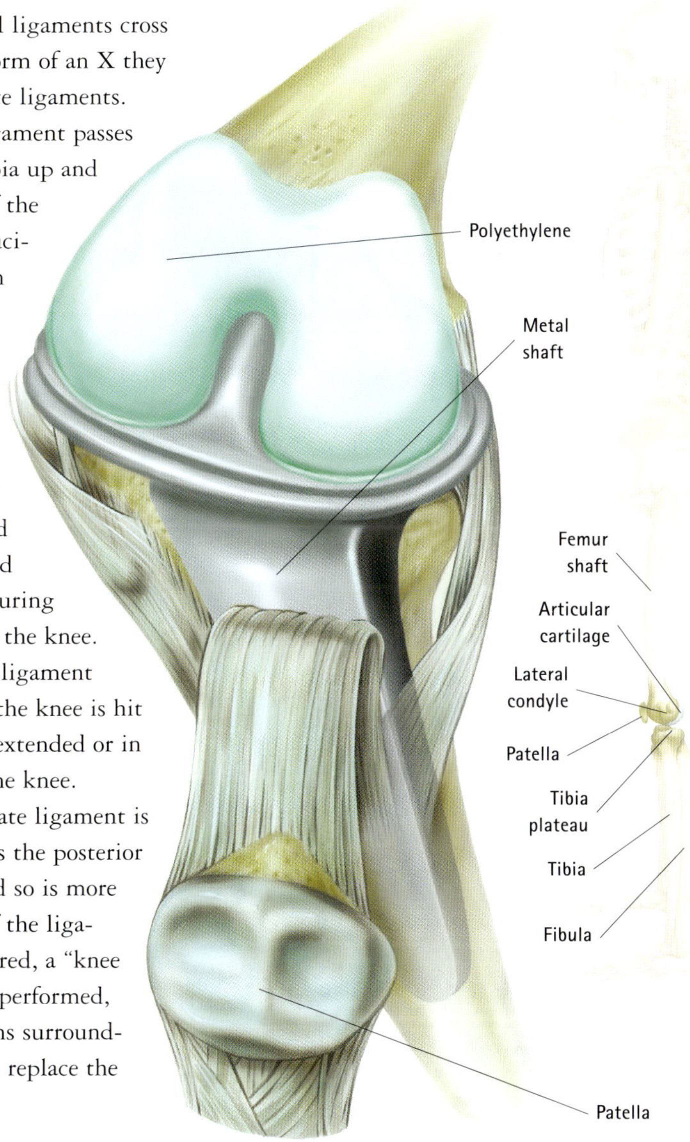

Knee replacement

The knee joint is particularly vulnerable to stress injuries, and reconstruction or replacement may be necessary if the ligaments have been badly torn or severed. In knee replacement the damaged joint is repaired by inserting metal shafts into the tibia and femur. A strong polyethylene coating covers the end of the shafts (in place of cartilage) and the knee ligaments are re-attached to hold the joint together.

DISORDERS OF THE KNEE

The ligaments and muscles which stabilize the knee and allow it to move can become weak, inflamed or swollen due to developmental problems, injury or overuse.

SEE ALSO *Imaging techniques on page 431*

Knock-knee

Knock-knee (genu valgum) is a condition in which there is a significant space between the ankles when the knees are touching in the normal stance position; it is commonly

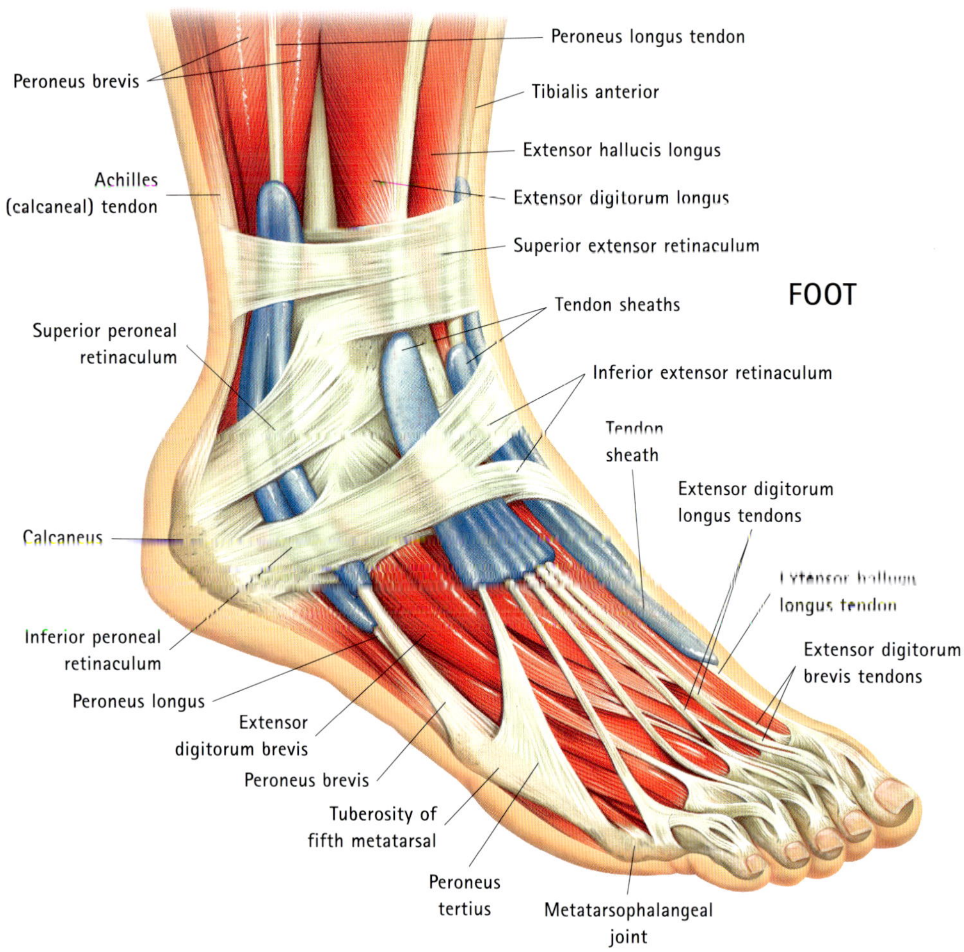

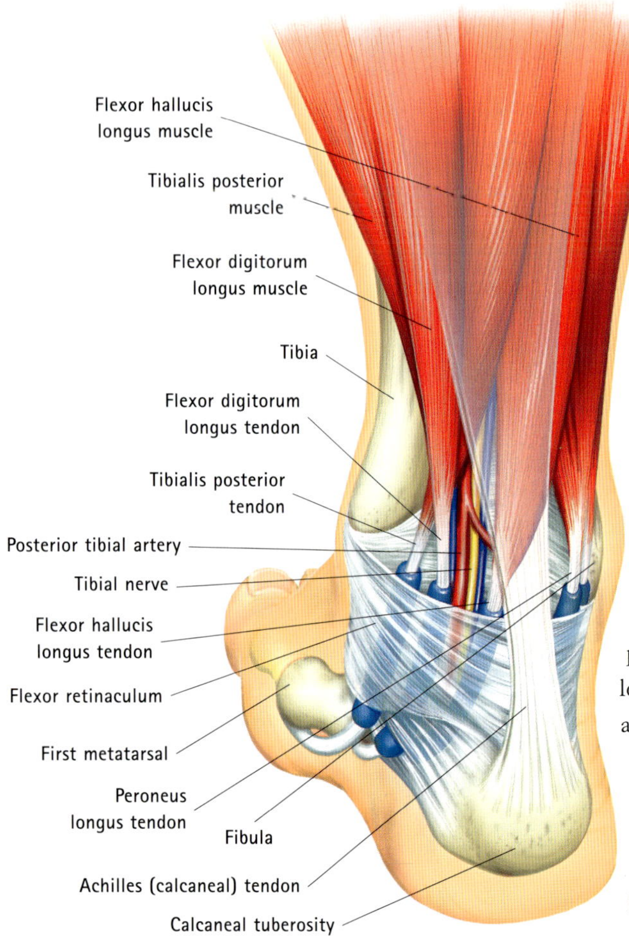

seen in childhood. This outward angulation of the lower legs causes the knees to knock together when walking.

Knock-knee often appears at about three years of age as a normal part of development and usually corrects itself by the time a child reaches puberty. The condition may also be symptomatic of disease, irregular bone growth or weak ligaments, however. It is rare that knock-knees cause difficulty walking, although they may lead to foot and back pain, and damage to the knees. Most problems are associated with low self-esteem due to the appearance of the legs.

While knock-knee is usually not treated, strengthening exercises or braces may be used to encourage the correction of bone alignment. Surgery may be considered in severe cases to prevent or correct excessive abnormal wear on the cartilage of the knee joint.

A procedure known as osteotomy may be performed, where a wedge of bone is taken from the inside of the upper leg bone just above the knee. The bone is then fixed with plates and screws.

Housemaid's knee

Housemaid's knee is one example of bursitis, inflammation of a bursa. A bursa is a fluid-filled sac that protects ligaments, tendons, skin or muscle where they rub across bone. Here, the prepatellar bursa in front of the patella becomes inflamed. Bursitis may result from injury, pressure or overuse. The condition is painful and limits movement. It may be relieved by rest or anti-inflammatory drugs.

Osgood-Schlatter disease

Osgood-Schlatter disease is a common, minor ailment of late childhood and early adolescence. Characterized by pain, tenderness and often swelling at the attachment of the quadriceps tendon just below the knee on the front of the shin (the tibial tuberosity), it usually affects both legs. It is more common in boys than girls and occurs in highly active or athletic children. It is caused by excessive use of the quadriceps muscles and inflammation in the quadriceps tendon where it inserts into the front of the tibia. Avoidance of strenuous exercise, particularly jumping, usually corrects the problem although, in severe cases, the affected limb may need to be immobilized in a cast for several weeks.

Ankle

The ankle is the region where the lower leg joins the foot. The ankle attaches the bones of the lower leg, the tibia and fibula, to the talus, one of the bones of the foot. The ankle joint is what's known as a synovial hinge joint and is fairly stable. It allows the heel to be raised from the ground, as in pointing one's toes (plantarflexion) and for the upper surface of the foot to be brought closer to the front of the leg (dorsiflexion).

Prominent features of the ankle joint include the medial malleolus, a bony

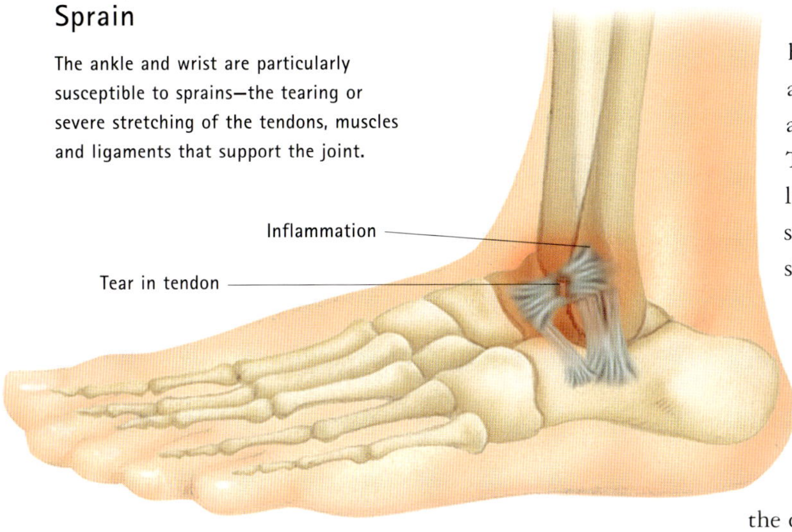

Sprain

The ankle and wrist are particularly susceptible to sprains—the tearing or severe stretching of the tendons, muscles and ligaments that support the joint.

protrusion on the end of the tibia, and the lateral malleolus, a similar bony landmark at the lower end of the fibula. The two malleoli, together with a part of the tibia, form a socket in which the talus can move. Because the talus is wider at the front, the joint is most stable when the heel is raised, where the fit between the talus and socket is tightest.

Several ligaments help to make the ankle joint stronger and more stable. The deltoid or medial ligament is a broad, strong, triangular-shaped ligament which connects the medial malleolus to three of the tarsal bones (talus, navicular and calcaneus). On the outside of the ankle there are three cord-like ligaments that attach the lateral malleolus to the talus and calcaneus. These cords are not as strong as the deltoid ligament; they can be easily damaged. Ligaments between the tibia and fibula also help ankle joint stability by strengthening the socket.

The main muscles producing dorsiflexion of the ankle lie at the front of the lower leg while the main muscles producing plantarflexion are at the back of the lower leg (calf muscles). The Achilles tendon is a band of fibrous tissue that connects the calf muscles to the back of the heel bone.

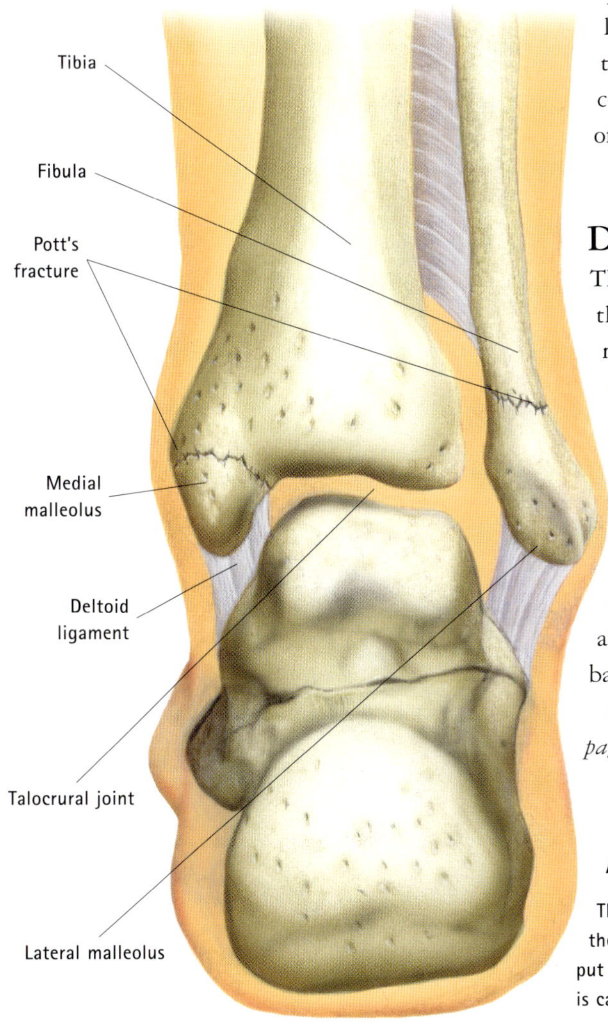

Ankle fracture

The ankle bones are prone to fracture because the ankle is a weight-bearing joint and is easily put under stress. The ankle fracture shown here is called a Pott's fracture.

Disorders of the Ankle

The ankle joint, between the talus and the bones of the leg, is the most commonly injured joint in the body, with injury usually resulting from a fall which forces the ankle into a position of excessive inversion. This causes damage to ligaments which reinforce the outside of the ankle (ankle sprain) and swelling around the joint. More severe sprains may also involve fractures of the fibula or base of the fifth metatarsal.

SEE ALSO *Imaging techniques on page 431*

Sprain

A sprain is an over-stretching or tearing of ligaments; sprains are particularly common around the ankle joint. A frequent injury, referred to as "twisting the ankle," occurs when the foot is forcefully turned inward. This usually occurs as a result of a fall on an uneven surface and leads to tearing of the ligaments on the outside of the ankle. This will cause pain and swelling in front and below the lateral malleolus and can make the ankle joint quite unstable. As for most sprain injuries, ice packs, rest and elevation will reduce swelling and aid healing.

Fractures of the ankle

Fractures of the ankle are fairly common, and include breaking of the lateral malleolus and an injury known as Pott's fracture. Pott's fracture occurs when the foot is forcefully turned outward. The medial malleolus is broken and there is also an associated fracture of the fibula, either at the lateral malleolus or higher, on the shaft of the bone.

Rupture of Achilles tendon

The Achilles tendon is a weak spot and if subjected to sudden stress—running, jumping, pushing forward—it may snap. This injury most commonly occurs in sportspeople and middle-aged men. When it happens, there is a snapping sensation and immediate pain is felt in the heel or the back of the leg. Flexing the foot downward becomes difficult.

To treat an Achilles rupture, the doctor usually puts the back of the patient's foot, ankle and the lower part of the leg into a plaster cast, with the toes pointing downward, so that the two ends of the tendon are close and can heal. The patient uses crutches, keeping weight off the foot, for six weeks. After removal of the plaster, physical therapy is needed to restore the function of the calf muscles and ankle joint. It usually takes about six months to return to normal.

Another option is surgery. Under general anesthetic, the two ends of the ruptured tendon are sewn together, and the foot goes into plaster for six weeks, followed by months of physical therapy.

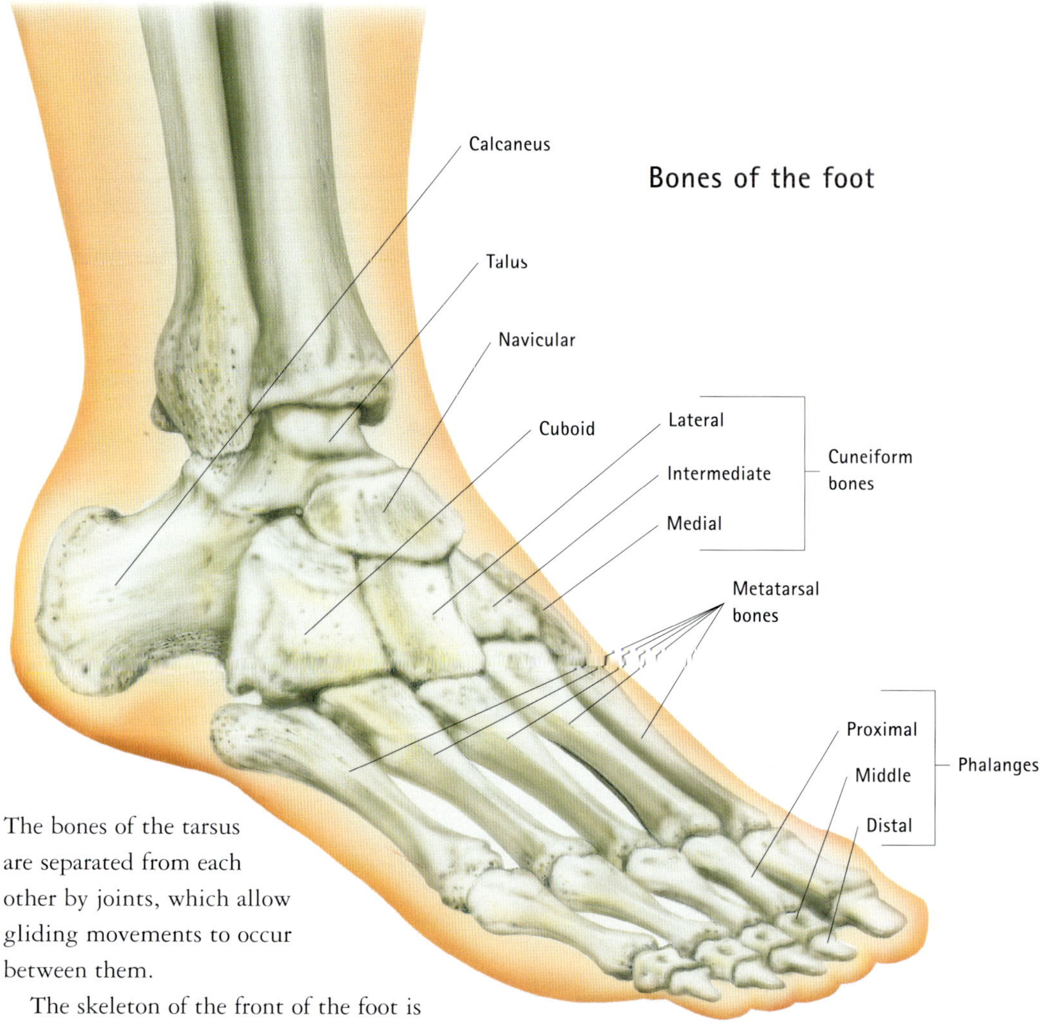

Achilles tendinitis

Achilles tendinitis—inflammation of the Achilles tendon—can develop if the tendon suffers too much wear and tear. It happens most often after strenuous activity, especially by those not used to exercise. Wearing shoes with high heels or with worn heels, both of which place abnormal stresses on the tendon, may also cause inflammation.

Sufferers feel pain in the affected tendon, which becomes worse with activity. The Achilles tendon may become thicker than normal and can be painful or tender when it is touched. An inflamed Achilles tendon is susceptible to rupture.

The treatment for Achilles tendinitis is rest, analgesics and anti-inflammatory drugs, which relieve the symptoms but do not cure the condition, which tends to recur. In rare cases, the tendon can be removed under anesthetic by an orthopedic surgeon and replaced with artificial or natural tissues.

The best treatment, though, is prevention. Warming up before exercise, muscle stretching before and after running, and the wearing of proper shoes during exercise will help prevent the condition from developing.

Foot

The foot is designed to support the body weight and to act as a lever to propel the body forward during walking. Rather than being just one rigid bone, its skeleton is segmented, allowing it to adapt to the shape of any surface and to enhance its propulsive effect during running.

Movements of the foot can be defined as dorsiflexion (feet point upward), plantarflexion (feet point down), inversion or supination (sole faces inward) and eversion or pronation (sole faces outward).

Bones and tissues of the foot

The back half of the foot (the tarsus) is formed by seven irregularly shaped tarsal bones. One of these, the talus, fits into a socket formed by the bones of the leg to form the hinge-like ankle joint. The talus sits on top of the largest tarsal bone, the calcaneus, which forms the heel and has the Achilles tendon attached to it.

The other tarsal bones are the navicular, the cuboid and the three cuneiform bones. The bones of the tarsus are separated from each other by joints, which allow gliding movements to occur between them.

The skeleton of the front of the foot is formed by the metatarsals and phalanges. The metatarsals are long bones that form joints behind with the tarsus, and in front with the phalanges, or toe bones. The big toe has only two phalanges, the other toes have three phalanges each.

The upper surface of the foot is formed by thin skin, which overlies tendons extending from the front of the leg to the tarsus and toes. In contrast, the sole of the foot is covered by thick skin which is separated from the underlying tissues by fat pads. Fibrous tissue extends longitudinally along the sole, deep into the fat pads, from the calcaneus to the toes. It is important in binding the overlying skin and fat firmly to the deeper bones and ligaments of the foot, so that it does not slip around during walking. The plantar fascia, which also acts like a bow-string to maintain the longitudinal arches of the foot, becomes stretched when the foot is flattened after the heel hits the ground.

The sole of the foot contains long tendons and a number of small muscles which are arranged in four layers. Unlike on the hand, the big toe is prevented from coming into contact with the other toes. However, the toes do have the potential to be used for grasping; this can be seen in people born without arms, who can learn to write, draw and manipulate objects with their feet.

Disorders of the foot

Arthritis or tight-fitting shoes can deform the metatarsophalangeal joint of the big toe. This causes the toe to be forced against the other toes and the head of the metatarsal to jut out onto the medial side of the foot, forming a lump known as a bunion.

Ill-fitting footwear can also cause bunions on the outside of the foot, as well as deformities of the toe joints, or thickening of the skin (corns). Deformities of the toe joints, usually caused by shoes which are too short, include hammer toe and mallet toe.

SEE ALSO *Imaging techniques on page 431*

Clubfoot

Clubfoot is a relatively common condition in which the foot is turned so that the sole faces inward, the toes point downward (inverted and plantarflexed) and the foot cannot be straightened. It usually occurs in newborn babies as a result of genetic factors or, more commonly, from the developing foot being forced into an abnormal position in the uterus. If treated early, clubfoot is curable by gradually straightening the foot over time using a series of plaster casts.

Bunion

A bunion is a solid growth that forms a lump at the base of the big toe, or hallux. The lump results from friction and distortion of the first metatarsal bone, plus fluid and bony growths, where the bone meets the base of the big toe.

Although susceptibility to bunions is often inherited, there is no precise cause. They usually develop in middle age. Bunions can be tolerated and are not usually painful unless they are cramped by badly fitting footwear. Over-tight shoes will cause pain but have not been proven to actually cause bunions. Treatment is by surgery and is usually successful but recovery can be painful.

Plantar warts

Plantar warts occur on the soles of the feet. Constant pressure and friction on the soles prevents the normal outward expansion of warts at this site, and instead plantar warts grow inward toward the dermis. This "iceberg" configuration of plantar warts can make their complete removal difficult.

Plantar warts can be treated with liquid nitrogen, topical salicylic acid or surgical removal under local anesthetic.

Heel spur

A painful condition, heel spur syndrome is often most severe upon standing after a period of rest. It is typically associated with development of a spur-like projection from the bone of the heel, visible on x-ray examination. The spur itself is not the cause of the pain—it is usually secondary to injury or inflammation of the plantar fascia (a ligament that extends from the heel bone across the sole of the foot) which becomes calcified to form the visible spur. Other causes of heel pain include stress fractures and bursitis. Treatment of the heel spur syndrome may involve providing suitable physical support to the heel, anti-inflammatory drugs or injections, or surgery.

Hammer toe

A hammer toe is a deformity in which the toe is bent in the shape of a hammer. Although any toe may be affected, it usually affects the second toe.

The condition may be congenital or acquired, occurring most often in children who have outgrown their shoes. In older people this condition may often be caused by pressure from a bunion.

Hammer toe is best prevented by ensuring footwear fits correctly. A physician can treat mild cases by manipulating the toe and then splinting it.

In severe cases, surgery may be needed to straighten the joint.

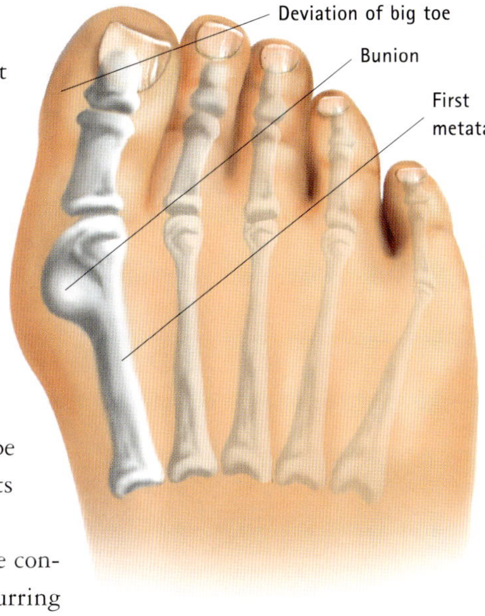

Bunion

A bunion is a lump at the base of the big toe. If painful, it can be treated by wearing properly fitted shoes. If that does not ease the pain, surgery may be required.

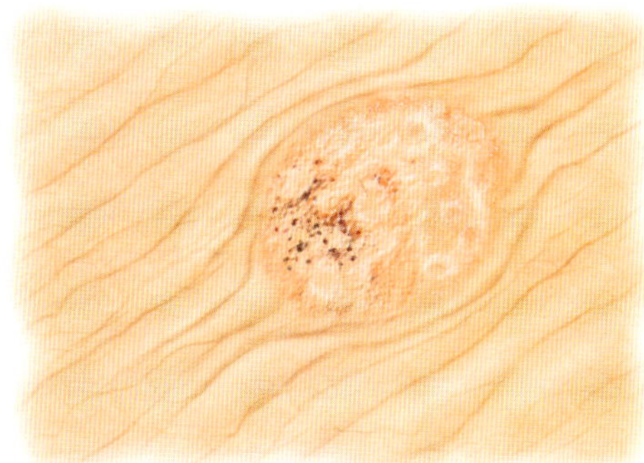

Plantar wart

Plantar warts occur on the soles of the feet and are transmitted by direct contact. People who use communal showering facilities are particularly at risk.

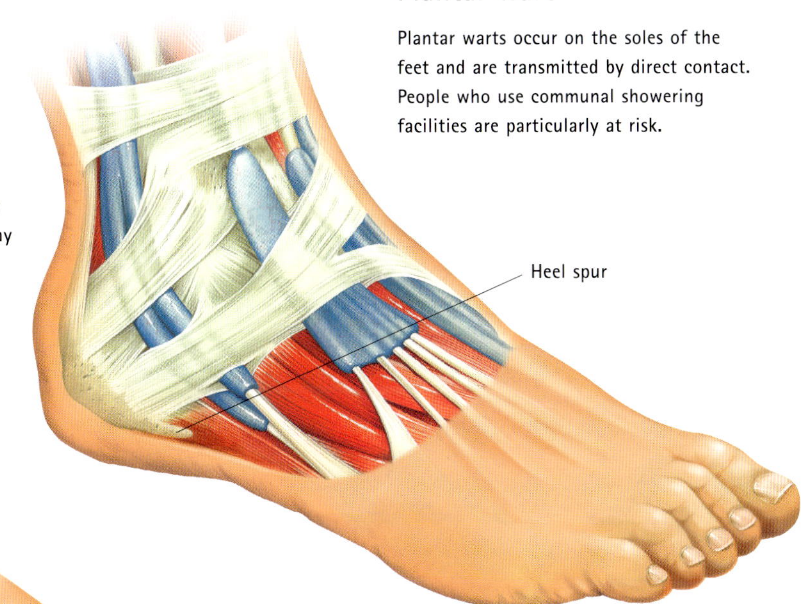

Heel spur

Heel spur syndrome is characterized by a bony outgrowth that develops along the undersurface of the heel bone, causing inflammation.

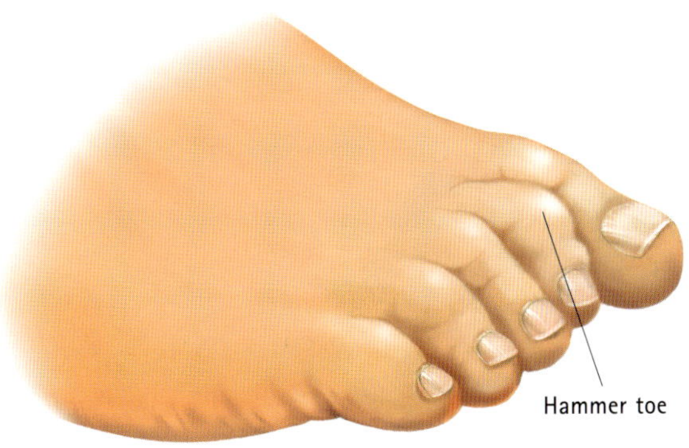

Hammer toe

Hammer toe refers to a fixed flexion deformity of the interphalangeal joint of the toe. The second toe is usually affected. The skin over the top of the flexed joint becomes hardened from pressure against the shoe. Treatment may be by protective padding or surgery.

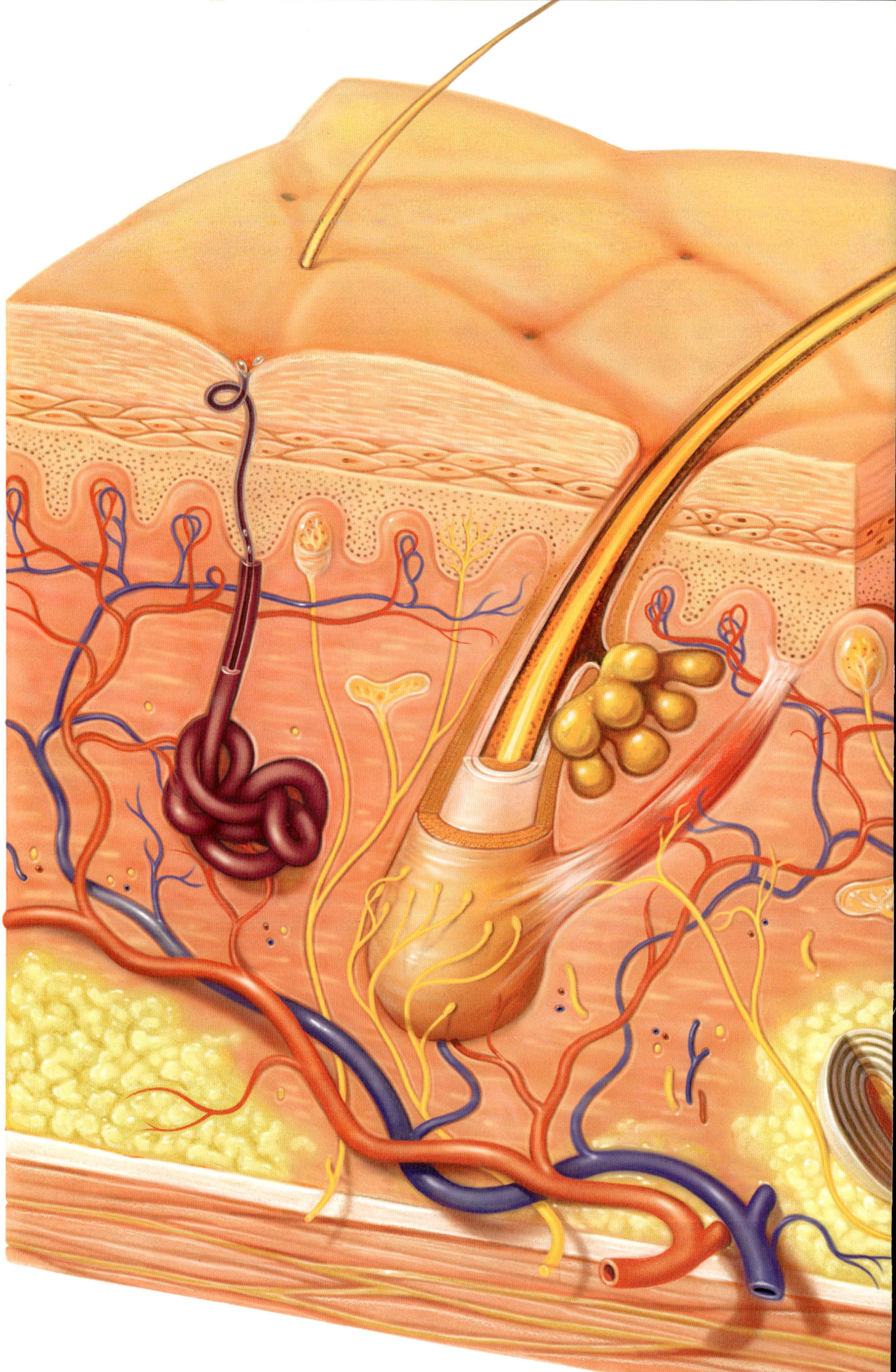

The Skin

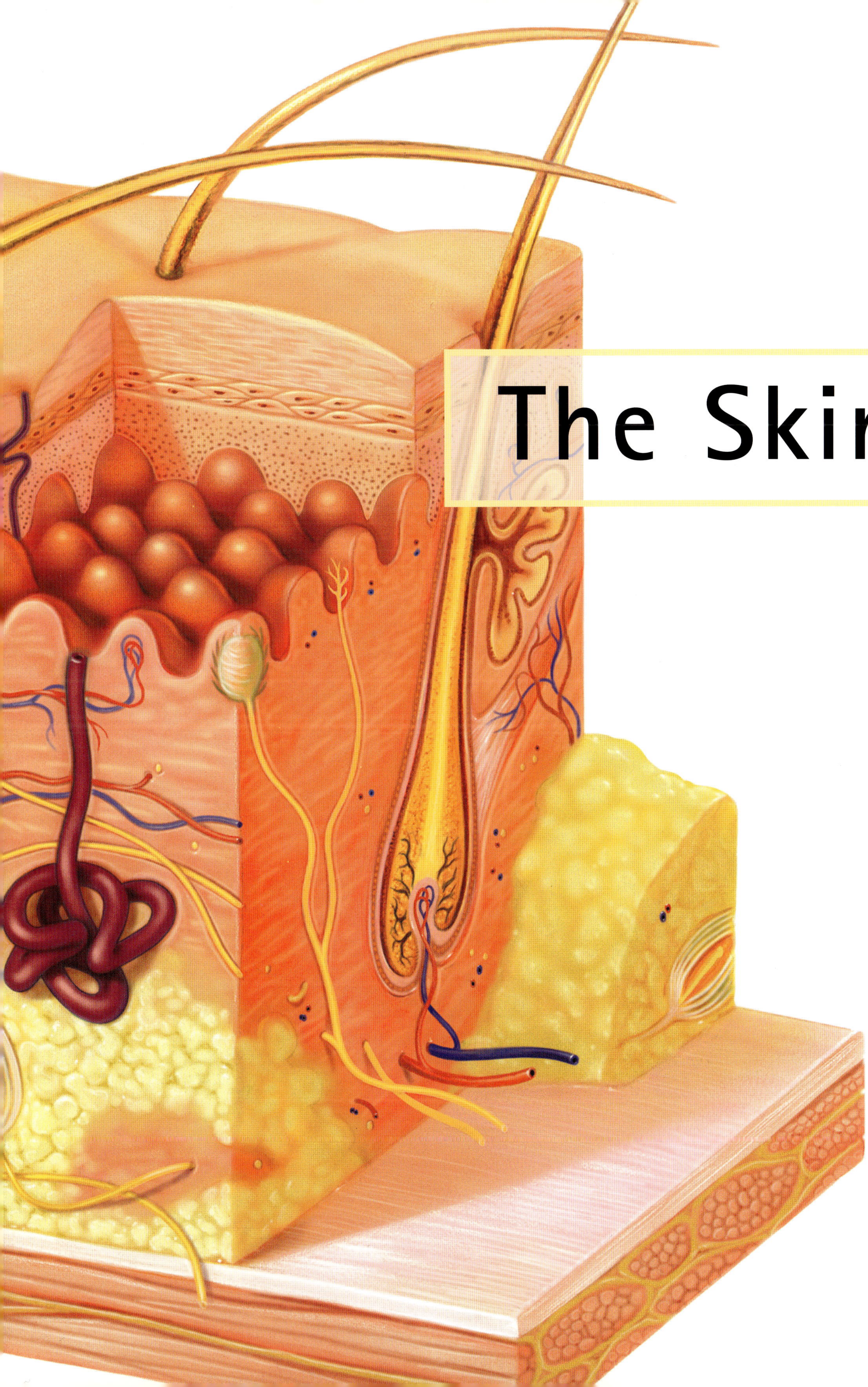

The skin

The skin is a protective organ that covers the body, merging with mucous membranes at the openings of the body such as the mouth and anus. It is loosely attached to underlying tissues, and varies in thickness from 0.5 millimeter on the eyelids to ⅙ inch (4 millimeters) or more on the palms and soles of the feet.

The skin protects the body from external agents, extremes of temperature and invading organisms such as viruses, bacteria, fungi and parasites. Specialized nerve receptors in the skin allow the body to sense pain, hot and cold, touch and pressure. Skin plays an important role in temperature regulation, protection from ultraviolet light, in the manufacture of vitamin D as well as in attracting the opposite sex.

As the body ages, the features of skin change. In infancy the skin is soft, dry and free of wrinkles. At adolescence, pigmentation increases and body hair becomes longer and thicker, especially in the male. Exposure to the elements, along with physiological changes, result in ageing skin becoming dry, wrinkled and flaccid.

Skin may be destroyed by injury, burns or disease. In most cases regeneration takes place from normal skin at the margins of the injured area. If the injury is severe, scar tissue forms, over which skin cannot regrow. The damaged area can be covered in skin taken from another site in the body. This is known as skin grafting.

SEE ALSO *Cells on page 16, Diseases and disorders of cells and tissues on page 25; Healing on page 24; Nervous system on page 64; Senses on page 140; Temperature regulation on page 348; Tissues on page 20*

Structure of the skin

The skin has three layers of tissue: epidermis, the outermost layer; dermis in the middle; and subcutaneous tissue, the innermost layer.

SEE ALSO *Cells on page 16; Tissues on page 20*

Epidermis

The epidermis is made up of five sublayers, each with its own function. As cells in the top layer die and are sloughed off, they are continually replaced by cells from the lower layers. It takes approximately four weeks for cells formed in the bottom layer to reach and replace those in the top layer.

The bottom layer of the epidermis is the basal layer (stratum basale or stratum germinativum). This layer is three to five cells thick; cell mitosis, or cell reproduction, occurs here, and it is also the layer in which pigment-producing cells called melanocytes are found. They produce the pigment melanin, which absorbs dangerous ultraviolet light and gives skin its dark appearance in dark-skinned races. Sunlight also stimulates the production of melanin, which results in a tanned appearance after prolonged exposure to sunlight.

Above the basal layer lies the spiny layer (stratum spinosum), which is eight to ten cells thick and consists of flattened cells with spiny projections. Above this is the grainy layer (stratum granulosum). This layer is three to five cells thick; the cells in this layer produce keratin, a waxy substance that forms a tough surface layer in the outermost epithelial cells.

Next is the clear layer (stratum lucidum), a thin, clear, keratin-rich layer three to four cells thick, which is thickest on areas subjected to heavy wear and tear, such as the palms of the hands and the soles of the feet. The outermost layer is the horny layer (stratum corneum). Most of the cells in this layer (25–30 cells thick) are dead; it takes about two weeks for cells to reach the surface, where they are sloughed off. Thus, the skin is in a constant process of renewal.

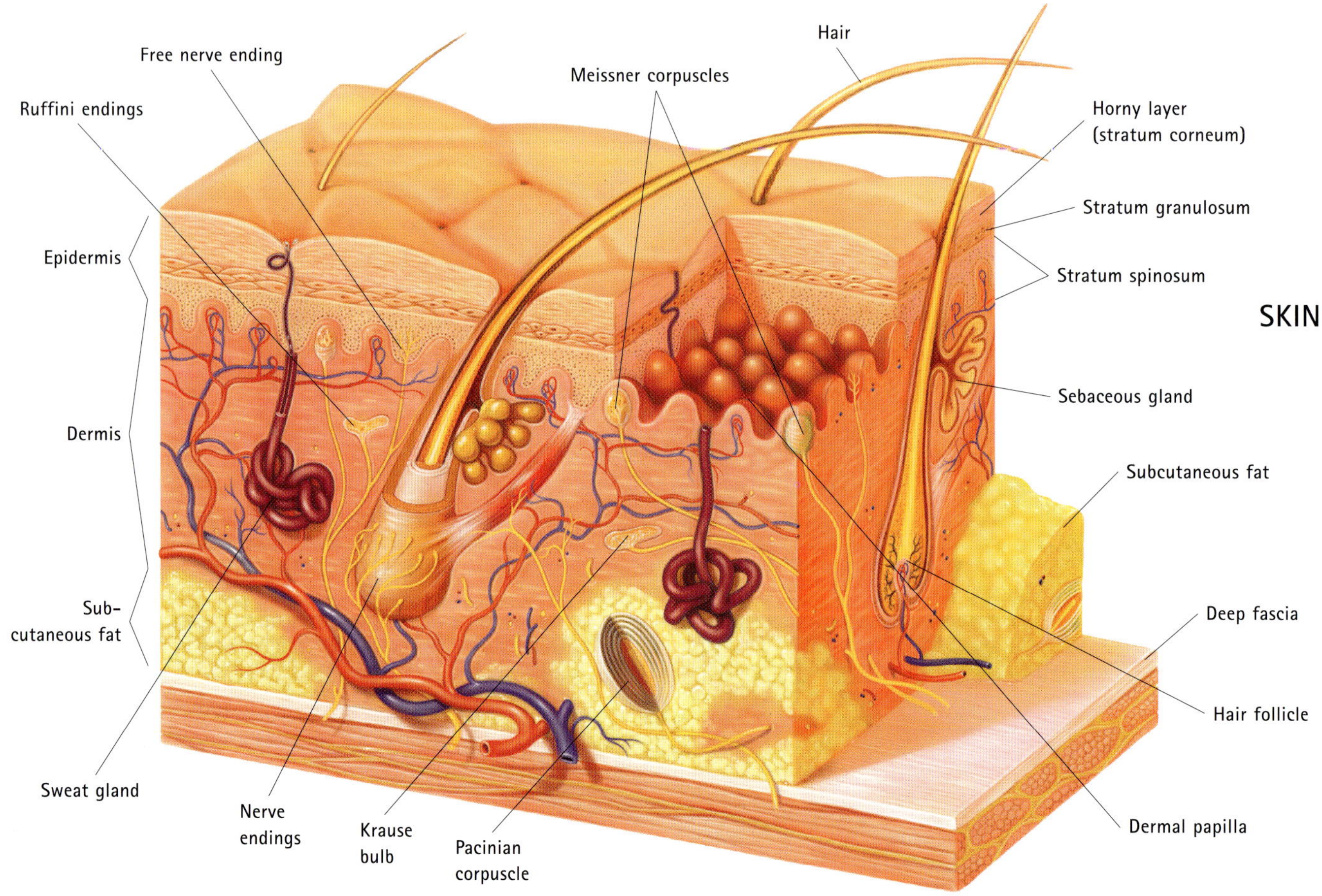

SKIN

The stratum corneum is the main barrier to skin infection; if a bacterium, virus or fungus manages to penetrate this layer, dermatitis (inflammation of the skin) may follow.

Dermis

The inner layer of the skin, the dermis, is composed of a network of fibers made from the proteins collagen and elastin, which provide strength and support to the skin. Among them are networks of blood vessels, nerves and fat lobules. The junction of the dermis and epidermis is irregular, with finger-like projections of dermis called papillae running up into the epidermis and causing elevations in the surface in the palms of the hands and the soles of the feet. In the fingers they create fingerprints.

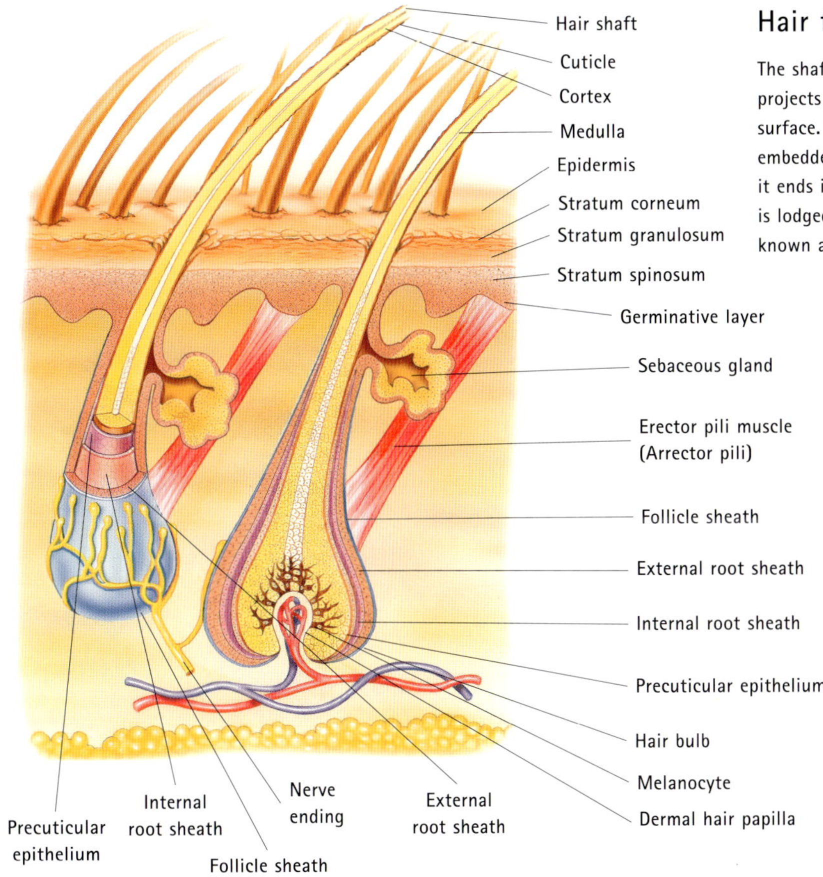

Hair follicle

The shaft of the hair projects from the skin's surface. The root is embedded in the skin—it ends in a bulb, which is lodged in a pit known as the follicle.

Specialized structures

Among the layers of the epidermis and the dermis are the specialized structures—the hair follicles, the sweat glands and the sebaceous glands. These develop from the epidermis and extend into the dermis. Beneath the dermis is a layer of fat cells arranged in lobules, which insulates the body against extremes of temperature and stores fat. The thickness of this layer varies.

SEE ALSO *Glands on page 22; Temperature regulation on page 348; Virilization on page 421*

Hair

Hair is a fine, thread-like structure, made of a tough protein called keratin. It is produced only by mammals and is an identifying character of that group of animals. Hair consists of a root embedded in the skin, and a shaft, projecting from the skin surface. The root ends in a soft, whitish enlargement, the hair bulb, which is lodged in an elongated pit in the skin, called the follicle.

The hair grows upward from the base of the follicle at the rate of about ⅓ millimeter a day. Blood vessels arranged in a small protrusion known as a papilla extend up into the follicle and the root of the hair to nourish it. Attached to each hair follicle is a tiny muscle called the erector pili. This muscle is controlled by the autonomic nervous system. Under certain conditions, such as in the cold, it contracts to make the hair stand on end.

Each hair is composed of dead epithelial cells containing keratin, arranged in columns around a central core. The cells also contain the dark pigment melanin, which is responsible for hair color. With age, less pigment is deposited into these cells, so hairs tend to become white.

Hair has a protective function. Hair around the eyes, ears and nose serves to prevent dust, insects and other matter from entering these organs. Eyebrows decrease the amount of light which reaches the eyes.

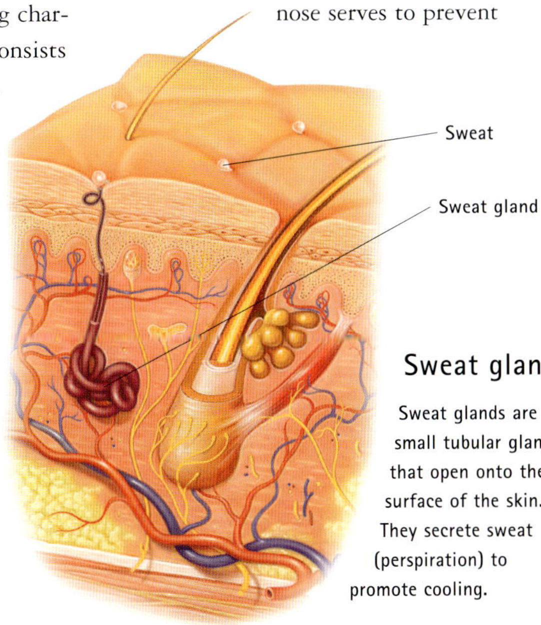

Sweat gland

Sweat glands are small tubular glands that open onto the surface of the skin. They secrete sweat (perspiration) to promote cooling.

Hair color is genetically determined. Dark hair color usually dominates over light hair color. For example, if a child has one parent with black or brown hair and one with red or blond hair, the child's hair is likely to be dark.

Sweat glands

Sweat glands are of two types: eccrine and apocrine. Both secrete a watery fluid—perspiration—onto the surface of the skin.

The eccrine sweat glands are distributed over the body, except on the lips and some parts of the genital regions. They are small tubular glands, opening at pores onto the surface of the skin. They can secrete large quantities of sweat, which cools the body by evaporation. Sweat glands are activated when the body becomes overheated (due to the environment or exercise), and occasionally by emotions such as fear ("cold sweat").

The apocrine sweat glands are special sweat glands found in the axillary and pubic regions and in the areolae of the breasts. These tubular glands have a particularly wide lumen (internal cavity), which opens into hair follicles rather than directly onto the skin surface. They secrete an odoriferous secretion which probably acts as a pheromone for sexual attraction.

Body odor is determined primarily by sweat from both types of glands, and by interactions between bacteria and sweat, particularly the sweat produced by the apocrine glands, and to control the spread of bacteria and fungi. Sebaceous glands are under hormonal control (testosterone in men, ovarian and adrenal androgens in women); this means that production of sebum increases after puberty.

Sebaceous glands

Sebaceous glands are skin glands that produce a fatty liquid called sebum. They are found in the lower layer of skin (the dermis) throughout the body surface, except on some hairless areas such as the soles of the feet and palms of hands, usually associated with hair follicles. They often release their sebum along the hair shaft. In some areas of the body (for example, the lips, glans penis and clitoris), sebaceous glands release their oils directly to the skin surface. Sebum contains fatty acids and glycerides, and helps to minimize loss of water from the skin.

Sebaceous glands

Found in the skin, usually attached to hair follicles, sebaceous glands secrete a fatty liquid called sebum, which lubricates the hair and skin and controls the spread of bacteria.

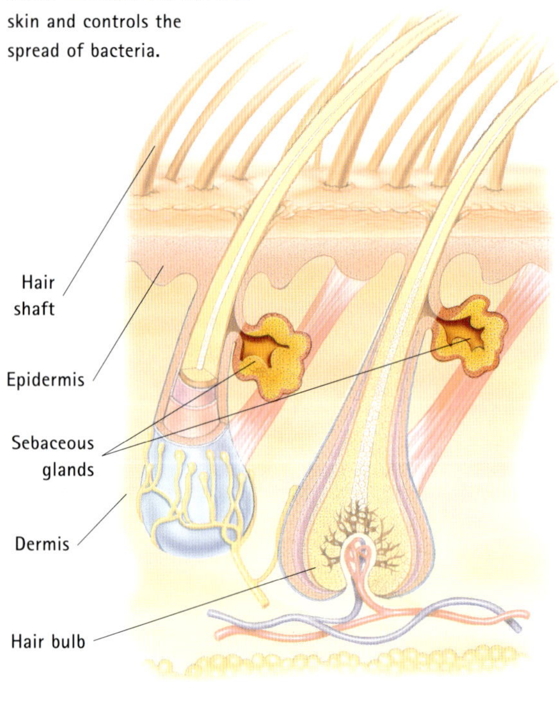

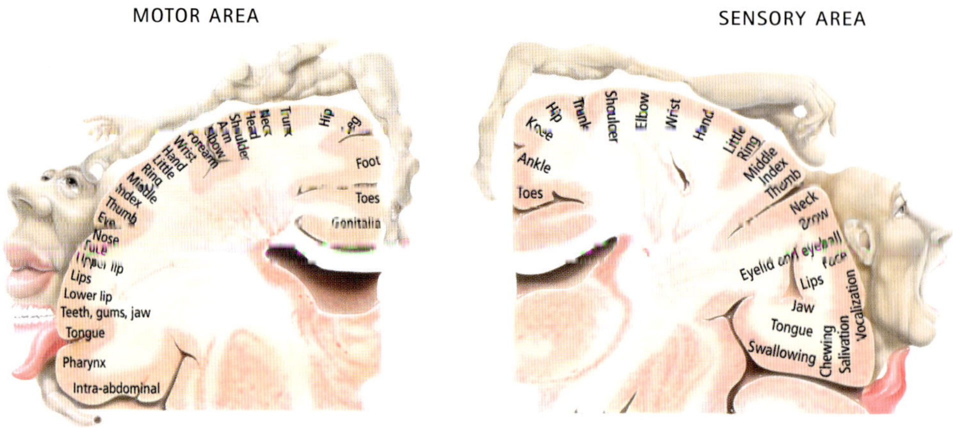

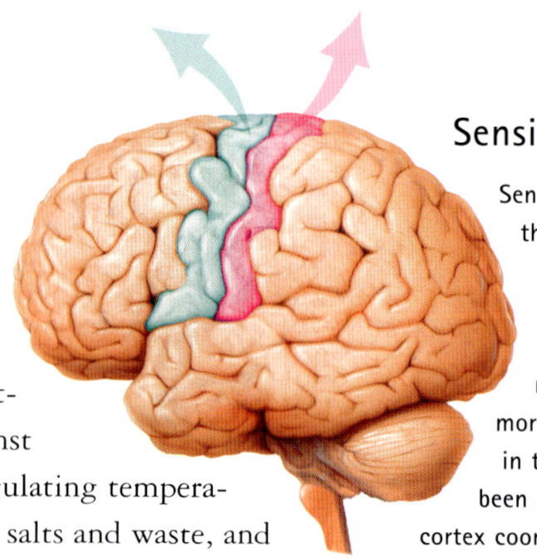

Sensitivity

Sensitivity to touch varies greatly over the body depending on the number of nerve endings in different body parts. The fingertips, lips, and tongue are highly sensitive, containing many receptors. This allows them to send more information to the sensory cortex in the brain. Once this information has been processed, the motor part of the cortex coordinates an appropriate response.

Specialized functions

The skin has numerous specialized functions. These include acting as a sensory organ, protecting body tissues against injury and attack, regulating temperature, excreting water, salts and waste, and producing vitamin D.

SEE ALSO *Glands on page 22; Nervous system on page 64; Senses on page 140; Temperature regulation on page 348*

Temperature regulation

The skin plays a major role in temperature regulation. Tiny glands in the skin secrete sweat, a salty, watery fluid. As the sweat evaporates it cools the body and prevents overheating. Tiny hairs embedded in the skin become erect in cold conditions, forming a fine blanket that insulates the skin.

The skin has blood vessels that dilate in hot conditions, allowing heat to escape. This also happens when body temperature is abnormally high, as in a fever. When outside conditions are cold, the blood vessels in the skin contract, diverting blood flow away from the skin to minimize the loss of heat.

Melanin and vitamin D

The skin contains a dark pigment called melanin, which protects tissues from harmful ultraviolet light by absorbing it. The amount of melanin in the skin varies according to a person's racial origins and is passed on as a genetic trait. The amount of melanin in the skin can be temporarily increased following periods of exposure to sunlight (suntanning). Sunlight also leads to the production of vitamin D in the skin—essential for the absorption of calcium from the gut and the maintenance of bone density.

Sebum production

Sebaceous glands in the skin open into hair follicles and secrete sebum, which is responsible for the waxy feel of skin. It lubricates, softens and protects the skin from damage by water, chemicals and microorganisms.

Touch

The sense of touch (the perception of the skin coming in contact with an object) is called the tactile sense. The surface of the skin has thousands of sensory nerve endings known as cutaneous receptors, which detect levels of pain, pressure and vibration, and temperatures ranging up to 113°F (45°C) and down to 50°F (10°C). The degree of sensitivity to touch varies greatly over the body, as nerve endings are concentrated in particular parts of the body, such as the

TOUCH

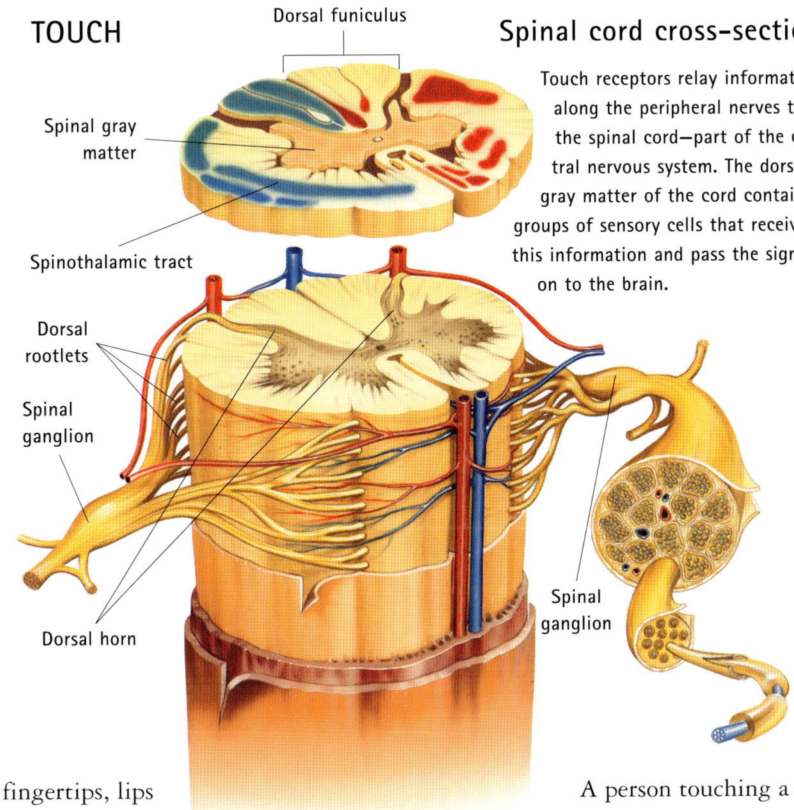

Spinal cord cross-section

Touch receptors relay information along the peripheral nerves to the spinal cord—part of the central nervous system. The dorsal gray matter of the cord contains groups of sensory cells that receive this information and pass the signals on to the brain.

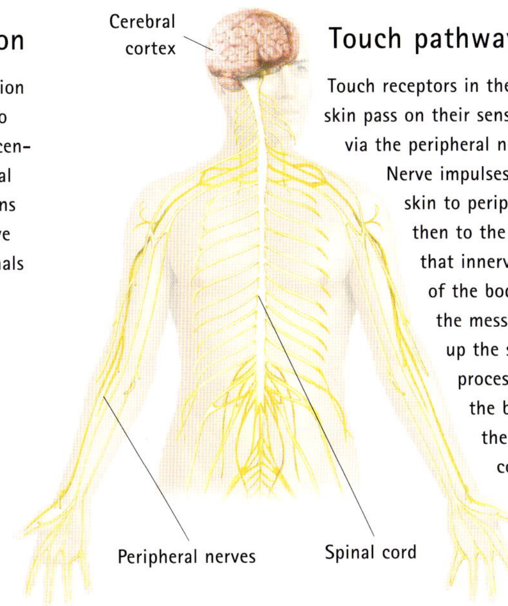

Touch pathways

Touch receptors in the surface of the skin pass on their sensory information via the peripheral nervous system. Nerve impulses pass from the skin to peripheral nerves, then to the spinal nerve that innervates that region of the body. From here, the message is relayed up the spinal cord to processing centers in the brain stem and then to the cerebral cortex in the brain.

fingertips, lips and tongue. The fingertips can distinguish between objects which are barely ½ inch (2 millimeters) apart, whereas on the back of the hand, which has fewer receptors, the objects must be 2 inches (50 millimeters) apart before they can be defined as more than one object. Fingertips can also detect vibration of as little as one 10-millionth of a yard (meter). The area of skin served by one nerve fiber is called the receptive field of that fiber. The tongue, lips and fingertips are particularly sensitive to touch. The tongue has about twice the concentration of pressure spots as a fingertip.

Touch is important to a newborn baby because it provides comfort and reassurance. The first skin to skin contact between a mother and her new baby is known to help in bonding their relationship. It is recognized that even babies in intensive care need to be touched in order to achieve optimal development. The sense of touch is also vital to the development and learning experiences of a baby.

A baby uses touch to understand the environment around it. The experience of different textures helps the baby to relate visual perceptions to touch. Touch continues to be an important part of communication, both good and bad, between humans throughout their lives.

A person touching a sharp spike or a hot surface will pull the hand away in a reflex action. In a newborn baby, reflex actions include: the grasp reflex, which can also be found in the baby's foot; the Moro or startle reflex, which is used by doctors to test muscle tone; and the galant reflex, which is tested by gently stroking a finger along one side of a baby's back while supporting the baby under the abdomen. The baby's body will bend like a bow, pulling the pelvis toward the side stroked. This reflex indicates the state of development of the spinal nerves.

Certain medicines, injuries to the nervous system and illnesses can damage the peripheral nerves and impair the sense of touch. A lessening sense of pain, vibration, cold, heat, pressure and touch occurs with ageing.

The Skin and Sun Exposure

Exposure to the sun is essential to human growth and development, but it can also have harmful effects. Exposure to sunlight in infancy and childhood aids production of vitamin D, vital for bone formation and growth, and prevents the development of rickets. The harmful effects of over-exposure to sunlight include sunburn, premature ageing of the skin and skin cancers. Severe or frequent sunburn in childhood increases the risks of developing skin cancers (especially melanoma). Outdoor workers and others with high sun exposure develop squamous or basal cell carcinomas, most of which are treated by removal. Fair-skinned people or those with a history of severe childhood sunburn are at considerably higher risk of melanoma, an aggressive cancer which can spread very quickly through the body.

Sunstroke is a form of heatstroke by sun exposure and caused by the inability to sweat sufficiently to reduce internal body temperature. Developing quickly, the condition causes heart rate and breathing to speed up; internal temperature may rise markedly, bringing disorientation and loss of consciousness and a risk of death.

SEE ALSO *Cancer on page 27; Skin cancer on page 350*

Photosensitivity

Photosensitivity is a reaction to sunlight or ultraviolet light. It can cause skin conditions such as solar urticaria where itching wheals appear. Photosensitivity can cause blistering in sufferers of the enzyme deficiency disease, porphyria cutanea tarda. It can also cause rosacea, a type of acne related to long-term exposure to sunlight. Certain chemicals or drugs, such as some antibiotics, can produce photosensitivity.

Disorders of the Skin

Skin can be affected by various diseases and conditions. Skin disorders are frequently treated by a primary care physician; more difficult conditions are usually treated by a specialist dermatologist.

SEE ALSO *Allergies on page 60; Biopsy on page 435; Cyanosis on page 86; Diseases and disorders of cells and tissues on page 25; Healing on page 24; Raynaud's disease on page 86; Skin tests on page 438; Treating cancer on page 446; Treating infections on page 445; Chickenpox, Jaundice, Measles, Ulcers and other individual disorders in Index*

Skin cancer

There are several types of skin cancers. They are among the most common cancers, especially in fair-skinned people. Exposure to the sun is the highest risk factor for development of skin cancer. In dark-skinned individuals, the skin pigment melanin provides some protection against damage by ultraviolet rays. Persons who lack this protection and/or those whose exposure to the sun is excessive may develop premature ageing of the skin, pre-cancerous skin changes such as scaly spots, and eventually cancers.

Exposure to certain industrial chemicals, exposure to x-rays, and certain hereditary disorders may also lead to skin cancer.

The three major varieties of skin cancer, known as basal cell carcinoma, squamous cell carcinoma and melanoma, arise from the surface layer of cells (the epidermis). For all skin cancers the best outcomes are achieved if the diagnosis is made at an early clinical stage and treatment is begun promptly. Since development of skin cancers is related to sun exposure, minimizing ultraviolet-induced injury is the best way to reduce the risk of these tumors. Wearing protective clothing and using suitable sunscreens are two methods of limiting such injury to the skin. A better but more difficult way to reduce risk is to change lifestyle and behavior.

SEE ALSO *Cancer on page 27; Treating cancer on page 446*

BASAL CELL CARCINOMA

Basal cell carcinoma (BCC) is a form of skin cancer. It usually appears on the face, and is caused by excessive exposure to sunlight. It begins as a small, pearl-like nodule, which slowly grows and ulcerates in the center to form a small scab. Basal cell carcinoma is treated with cautery (application of heat or electric current), liquid nitrogen, or low-dose radiation therapy. Larger tumors need to be removed surgically. Early treatment achieves a cure rate of over 95 percent.

Ongoing checkups should be carried out, as once a basal cell carcinoma has occurred, the chance of getting further basal cell carcinomas is increased.

SQUAMOUS CELL CARCINOMA

Squamous cell carcinoma is another common skin cancer and may develop on sun-exposed skin, including the face, neck and back of the hands. It usually appears as a scaly patch that may be red or ulcerated. Unlike basal cell carcinoma, squamous cell carcinoma of the skin eventually spreads to nearby lymph nodes or other body organs.

MELANOMA

This is a dangerous form of cancer, which arises from pigment-producing cells (melanocytes), normally located in the deepest part of the surface layer of skin cells (the epithelium). Melanomas usually occur in the skin, but they can also develop at other sites. These tumors can spread through the lymphatic system and the blood, with a fatal outcome. However, melanoma can be cured if it is diagnosed early and surgically removed. Therefore it is vital for a person who develops a pigmented area or lump on the skin to seek early medical attention. Since melanomas arise from pigment-producing cells, they may have a dark color, but this is by no means necessary.

Development of a melanoma of the skin is related to exposure to sunlight. This is not the only predisposing factor, as a tumor may arise in a pre-existing mole (nevus) or in skin that has not been subjected to repeated or prolonged sun exposure. Heredity plays a role in a minority of cases.

SKIN CANCER

Basal cell carcinoma is the most common form of skin cancer. It usually forms a localized tumor and very rarely spreads.

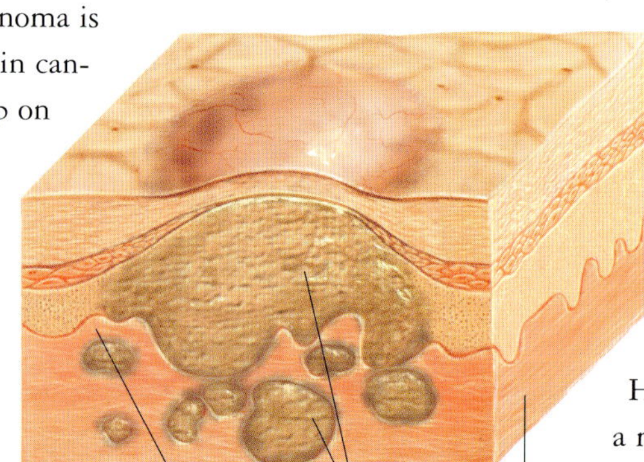

Melanoma

Melanomas are dark, cancerous growths found on the skin. They can begin as ordinary moles which become malignant.

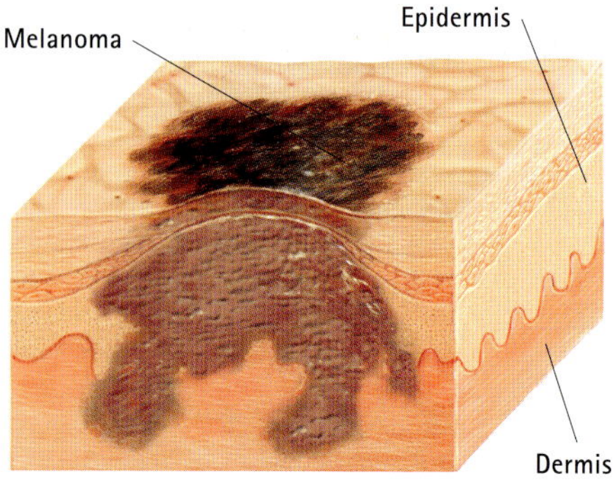

Basal cell carcinoma

A basal cell carcinoma, as it appears on the skin and in cross section. The best treatment is prevention; anyone who spends a lot of time outdoors, and in particular children, should wear sunscreen and a hat and avoid exposure to direct sunlight.

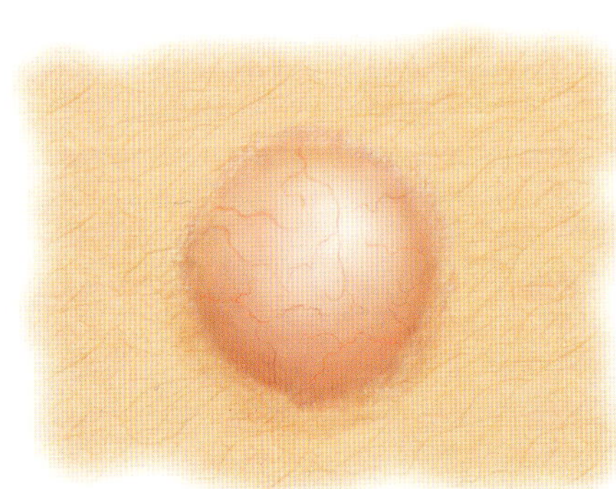

Squamous cell carcinoma

Malignant changes in the squamous cells that form the skin's surface layer can result in the formation of a cancerous tumor called a squamous cell carcinoma.

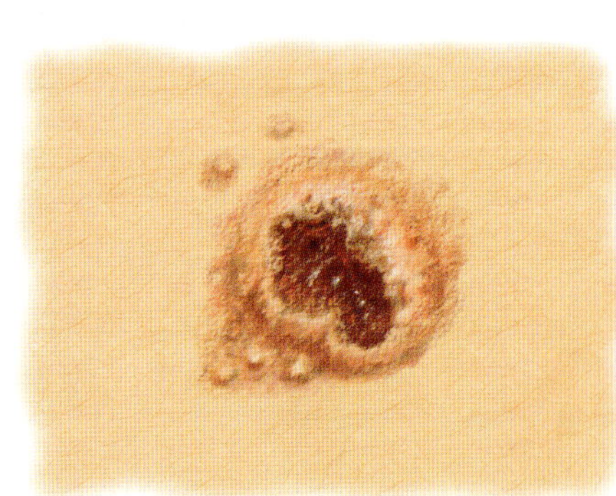

Distinguishing a melanoma from a mole may not be easy, especially in children, in whom one variety of nevus used to be referred to as juvenile melanoma. Most adults can recognize an early melanoma through self-examination.

Development of an enlarging new mole in an adult, increase in size of an existing mole, changes in the surface or color, development of irregularity of the borders or additional pigmented areas around the edges, itching or pain, and bleeding are all important warning signs. The diagnosis can only be established by microscopic examination of a biopsy sample. In the early stages of a melanoma, the tumor spreads radially, growing outward from its origin to form a larger but still superficially located collection of tumor cells. Later, it shows vertical growth, invading deeper layers of the skin.

If a melanoma is found microscopically to still be in the phase of radial growth, the most likely outcome is much better than if it shows vertical growth. The deeper the invasion into tissues, the higher the likelihood of spread to lymph nodes (lymph glands) or distant sites and the poorer the response to treatment.

Mycosis fungoides

This can be mistaken initially for a simple fungal rash. It is, however, a rare and frequently fatal form of cancer of the immune system that often takes years to progress. The first outward indications are raised, scaly, red or brown, itchy skin lesions that tend to form open sores. As the disease advances, the patient develops swollen lymph glands and then tumors on the skin. During the late stages, the skin reddens, peels and becomes scaly and extremely itchy. Eventually, cancer cells spread to the lymph glands and internal organs. If treated early enough, mycosis fungoides may be cured. Remission by various cancer treatments is also possible in later stages.

Kaposi's sarcoma

A malignant skin tumor often seen in late-stage AIDS, Kaposi's sarcoma is possibly caused by an unknown virus. Flat, reddish brown or purple patches appear first on the toes or feet and then slowly spread over the skin and in the mouth, developing into plaques or nodules. They may also occur in the digestive tract or lungs. Treatment is with radiation therapy and anti-cancer drugs.

Benign tumors of the skin

Damage, blockage or abnormal growth of skin cells may lead to benign tumors. Any abnormal growth on the skin should be assessed by a doctor to ensure that the tumor is benign and to rule out malignancy.

SEE ALSO *Biopsy on page 435; Tumors on page 26*

Papilloma

A papilloma is a non-cancerous growth or tumor on the skin or a mucous membrane. Warts are among the best known papillomas and are caused by viruses (human papilloma virus, or HPV). Corns, which develop with repeated rubbing of the skin, are also regarded as papillomas.

Hemangioma

A hemangioma is a benign, congenital tumor consisting of a cluster of blood vessels. There are two main types: capillary and cavernous. Capillary (strawberry) hemangiomas are raised red lumps resembling strawberries and are caused by dilated blood vessels. They may disappear between the ages of 5 and 10 and can also be removed by laser treatment.

Cavernous hemangiomas are raised purple or red patches on the body. Treatment may by surgery or, for children, corticosteroids.

Keratosis

A keratosis is a thickened lesion of the outermost skin layers. Small lumps on the upper arms, and less commonly the thighs, characterize a harmless condition known as keratosis pilaris. It tends to run in families and can be controlled but not cured.

Seborrheic keratoses (also called "barnacles of ageing") are harmless, slightly raised, dark spots that appear with age. Actinic or solar keratoses are precancerous lesions caused by sun exposure. They develop into skin cancer and should be removed.

Lentigo

Lentigines ("lentigo" is the singular form) are flat, round, light-brown spots that appear mostly on sun-affected areas such as the face and hands, and are frequently found on Caucasians and Asiatics. Often confused with freckles, they do not fade in winter. Liver spots (solar lentigines) on the back of the hands of people who spend a lot of time outdoors rarely become malignant. Lentigines are occasionally up to an inch (2.5 centimeters) across, irregular in shape and staining. If they change in size or shape, indicating possible melanomas, they need immediate attention from a doctor.

Mycosis fungoides

Mycosis fungoides generally develops over three stages commencing with itchy skin lesions, followed by large raised skin lesions and finally mushroom-like tumors that frequently ulcerate.

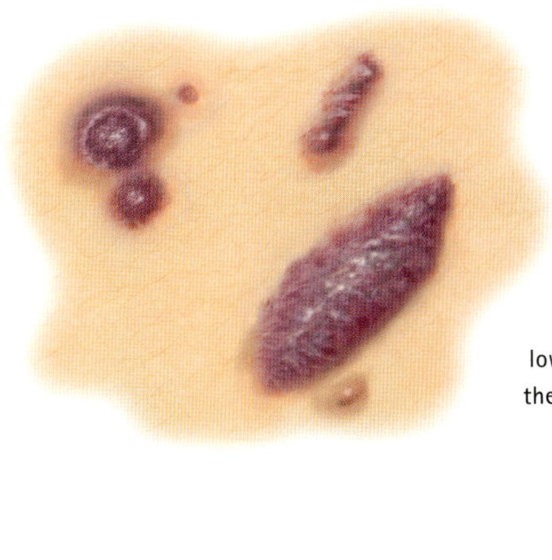

Kaposi's sarcoma

These cancerous growths on the skin first appear on the lower extremities and then spread up the body.

Papilloma

Warts and corns are types of papilloma, a non-cancerous growth on the skin or a mucous membrane.

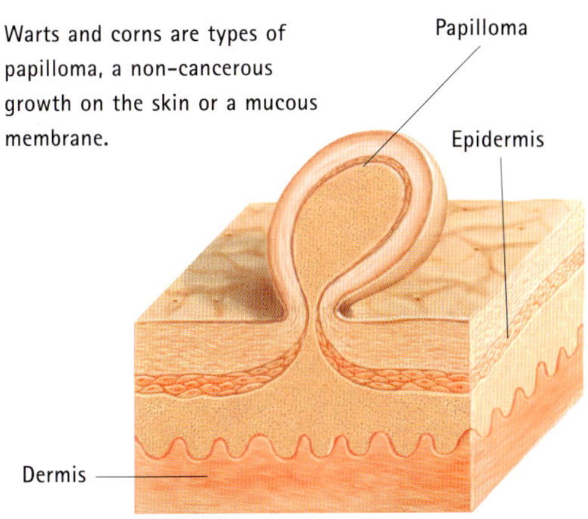

Xanthoma

A xanthoma is a small, firm, yellow to red-brown raised lesion that develops beneath the skin and appears on the skin's surface as a nodule, papule, plaque or benign tumor. It is a painless deposit from the bloodstream of excess fats (lipids) such as cholesterol and triglycerides. Xanthomas usually tend to be seen on the elbows, knees, hands, feet, buttocks, joints and tendons. Although they can be unsightly, xanthomas cause few problems. They may be indicators of underlying metabolic disorders characterized by undesirable elevated blood lipid levels.

Clustered around the joints in a condition known as xanthoma tuberosum, these nodules can suggest cirrhosis of the liver or thyroid disorders. In xanthoma tendinosum, xanthomas occur in clumps on the tendons and indicate a hereditary lipid storage disease. When they occur on the eyelids, xanthomas indicate high cholesterol levels. The sudden appearance of large clusters on the legs, arms, buttocks or trunk indicate critical triglyceride levels in the blood. Most xanthomas disappear eventually, if the underlying cause is treated successfully.

Dermatitis

Dermatitis means inflammation of the skin. Though it is not contagious, and not life threatening, it can be debilitating as it tends to recur and become chronic. There are many different types of dermatitis.

SEE ALSO *Allergies on page 60; Inflammation on page 24; Skin tests on page 438*

Eczema

Also known as atopic dermatitis, eczema is a chronic allergic skin disorder, occurring most commonly in infants, beginning between the ages of 1 month and 1 year. Most infants will outgrow it by the time they are 2 to 3 years old, but it may recur. In adults also, it is a recurring condition.

The condition is a hypersensitivity reaction similar to an allergy, causing chronic inflammation of the hands, scalp, face, back of the neck or skin creases of elbows and knees. The inflammation causes the skin to become itchy and scaly. Repeated irritation and scratching can cause the skin to thicken and become leathery. There may be blisters with oozing and crusting, and the skin may get infected by scratching.

Eczema may occur for no known reason, or as an allergy to things including foods, woolen clothing, skin lotions and ointments, soaps, detergents, cleansers, plants, tanning agents used for shoe leather, dyes, topical medications, moisture, overheating, common house dust, dog or cat dander, cigarette smoke and stress. There is often a family history of asthma, hay fever, eczema or other allergy-related disorders.

The first step in treating eczema is to identify whatever is causing it and if possible remove or avoid it. Dry skin makes the condition worse, so to keep the skin healthy, reduce the frequency of bathing to once or twice a week in lukewarm water, and use a small amount of very mild soap (or no soap at all). Apply a moisturizing lotion to the affected areas twice a day and after bathing; this will keep the skin moist and protect it from any other irritants.

If this does not improve the condition, a physician may prescribe mild cortisone cream or ointment, or antihistamines to reduce itching. Keeping nails short and wearing soft gloves at night will also minimize scratching. Sometimes exposure to sunlight helps heal the rash.

Chronically thickened areas of skin may be treated with medium- to very high-potency steroid creams. In extremely severe cases, corticosteroids ingested by mouth may be needed.

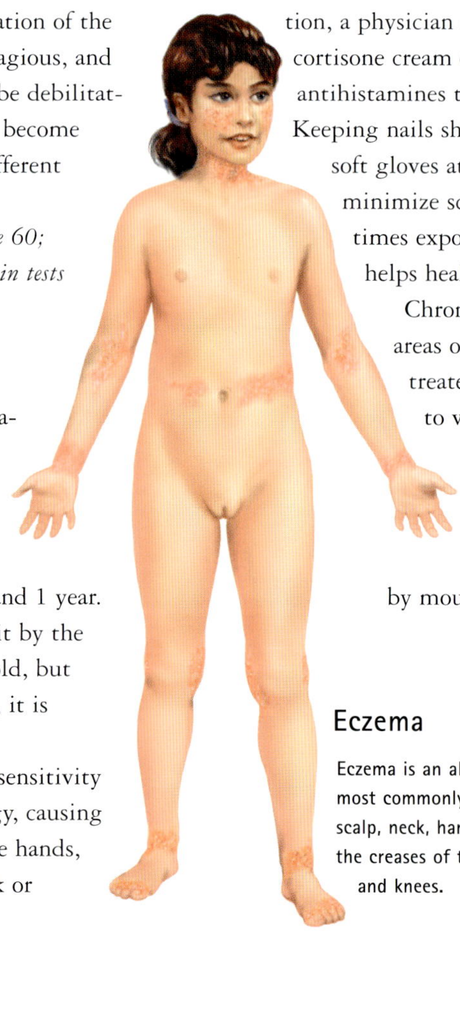

Eczema

Eczema is an allergic condition that most commonly affects the face, scalp, neck, hands and feet, and the creases of the trunk, elbows and knees.

Seborrheic dermatitis

Seborrheic dermatitis is a patchy inflammation of the skin, characterized by greasy, oily, reddish areas of skin with white or yellowish flaking scales, which may appear on the eyebrows, nose, forehead, or ears. It is painless but may be mildly itchy.

Seborrheic dermatitis is a chronic and heredity condition enhanced by stress, fatigue, and cold weather; it may improve in the summer months.

Contact dermatitis

Contact dermatitis is caused by contact with an irritating substance that damages the skin, causing itching, redness, cracks and fissures and in severe cases, bright red, weeping areas. The hands, feet and groin are often affected.

Many substances can cause contact dermatitis, including topical drugs, cosmetics, chemicals, soaps and detergents. Treatment is to avoid contact with the irritant, wear protective gloves and protective clothing, bathe in lukewarm water, and apply topical creams, ointments or lotions. These may include lubricants to preserve moisture, or steroid preparations to reduce inflammation. Contact dermatitis usually clears up within two or three weeks.

Stasis dermatitis

Stasis dermatitis is the result of fluid build-up under the skin. It is caused by varicose veins, poor circulation, and conditions that cause swelling of the extremities, especially the feet and ankles (peripheral edema). This causes surrounding tissue to become fragile;

Atopic eczema

Eczema is characterized by itching, redness, blisters, oozing and crusting. Scratching can cause further irritation.

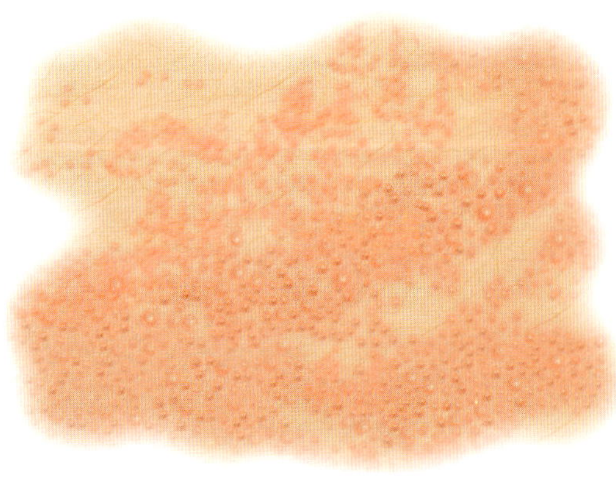

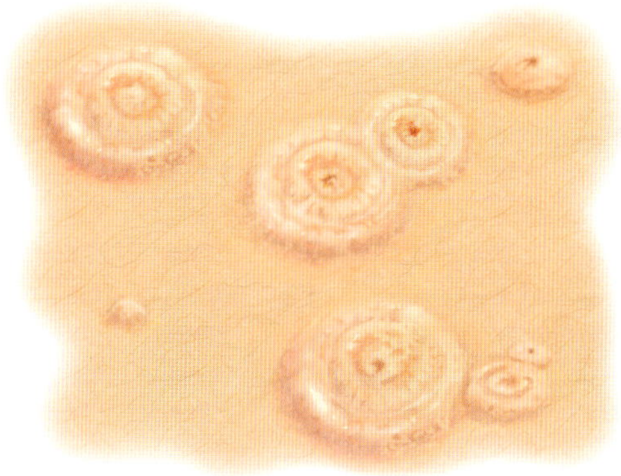

Erythema multiforme

Erythema multiforme is characterized by pimples, spots or blisters that usually appear on the face, hands and legs.

the skin darkens and becomes thin and inflamed. Ulcers may form and heal very slowly. Itching and scratching of the area may cause the skin to thicken. Elevation of the affected limbs, wearing elastic stockings and gentle exercise will relieve the swelling. The underlying condition must be controlled by surgical correction of varicose veins, and diuretics to remove excess fluid.

Erythema multiforme

Erythema multiforme is a type of hypersensitivity (allergic) reaction that occurs in children and young adults. It is caused by infection (especially with the herpes simplex virus), drug sensitivity or other allergic reactions. Spots, pimples or vesicles appear on the face, hands and legs, caused by damage to the blood vessels of the skin and to underlying skin tissues.

The condition may subside in 2–3 weeks with oral corticosteroid treatment and skin dressings. In one variant, Stevens-Johnson syndrome, a rare and potentially fatal skin condition, lesions are extensive, involving multiple body areas and requiring treatment in a burns unit in hospital. In Stevens-Johnson syndrome, there is a high risk of death. Complications may include shock due to the loss of body fluids, systemic infection and scarring.

Erythema nodosum

In erythema nodosum, painful, red, oval nodules appear on the skin, usually on the shins. The nodules turn purple, then brown, and disappear after several weeks. Fever and malaise are accompanying symptoms.

The condition usually follows a streptococcal throat infection, but may be associated with diseases such as tuberculosis or ulcerative colitis. Drugs such as penicillin, salicylates and birth control pills may also cause the condition. Treatment is with oral hormones (corticosteroids), bed rest and aspirin; topical creams do not help.

Chilblain

Chilblain is an inflammation of the skin, usually occurring on the ears, face, toes or fingers. The inflammation is due to cold, damp weather that can damage small blood vessels and nerves in the skin. The condition causes pain, itching, swelling, redness, and sometimes blistering and ulceration of the skin. People with poor circulation are more susceptible to chilblains. Treatment is to gently warm the affected areas and provide warm, protective clothing.

Acne

Acne (cystic acne, acne vulgaris) is a skin condition that occurs in adolescence. It can begin at puberty and continue in adulthood.

The hormones active at puberty stimulate the sebaceous glands, and the excess oil produced makes a fertile breeding ground for bacteria. Bacterial infection creates debris and waste matter (pus) that blocks the gland. The area becomes red and inflamed and a whitehead appears, developing into a pimple: this may become a blackhead, or develop into a painful cyst or boil. Affected areas are those that are naturally greasy, including the forehead, face, nose, chin, chest and back.

Pimples form where skin pores are clogged by dried sebum (the oil from the sebaceous glands), flaked skin and bacteria. Bacterial growth in the blocked pore irritates the skin and a white head of pus forms from debris, largely white blood cells. A blackhead may form from dead skin cells if there is no bacterial infection in the pore. Drugs such as anabolic steroids, corticosteroids, iodides, bromides and phenytoin can cause acne and eruptions of pimples.

Acne affects women as well. Polycystic ovaries can cause continuing acne for women. An excess of the hormone androgen in adulthood can be another cause. Acne is not due to poor hygiene but may be caused by particular foods. Acne is a non-infectious skin disease, and taking good care of the skin can help to control the symptoms and lessen the chance of scarring.

Severe acne, acne vulgaris, will require medical treatment. Medication may be prescribed to fight the acne bacillus, usually antibiotics such as tetracycline or retinoic acid, which is derived from vitamin A. Isotretinoin is used to treat recurrent acne. It is similar in structure to vitamin A but acts differently, reducing the oil content of the skin. It also inhibits the growth of

Seborrheic dermatitis

Seborrheic dermatitis mainly appears on the head and trunk. Skin becomes greasy or dry and red patches appear topped by white scales.

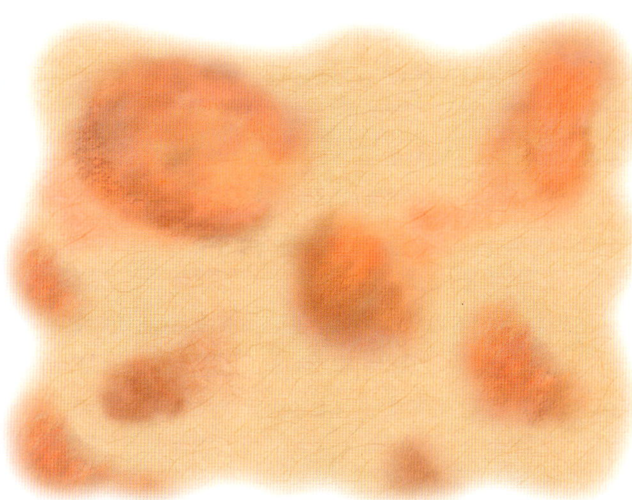

Erythema nodosum

Erythema nodosum is a painful condition in which red nodules appear, usually on the shins, fading to bruise-like patches that disappear after several weeks.

bacteria in the skin that are the cause of inflammation and the pus associated with acne. Significant side effects may include dry skin and liver upsets. Isotretinoin is harmful to a fetus during pregnancy.

There are several skin cleansers available for acne. Some may contain benzoyl peroxide, which induces peeling in the skin's surface layer; they may also contain retinoic acid, which dissolves blackheads—but this is not to be used during pregnancy.

Acne rosacea

Acne rosacea, also called rosacea, is an inflammation of facial skin. It is more common in males, especially alcoholics.

Unsightly red, thickened excess skin tissue forms on the nose, cheeks and forehead, associated with changes in the hair follicles, sebaceous glands and surrounding connective tissue. The exact cause of acne rosacea is unknown, though it may result from an overuse of steroid creams on the face.

Antibiotics are taken orally or applied in a cream or lotion on the affected area to treat this condition. The excess tissue may be removed with a scalpel or laser.

Lichen planus

Lichen planus is a rare and recurrent skin inflammation characterized by small and slightly raised bumps that itch. It usually starts at the wrists or on the legs and may spread to the trunk. Lesions may also appear in the mouth, vulva or penis, and the nails may be affected as well.

The exact cause of the condition is unknown, but is thought to be a result of an allergic or immune reaction from potential allergens such as certain medications, dyes and other chemical substances.

This disease mostly affects the middle-aged and elderly. The dark red or purple lesions may be almost 1½ inches (4 centimeters) long, with distinct borders and a shiny or scaly appearance. They may occur singly or in clusters. Symptoms are a dry mouth with a metallic taste, ridges on the nails and hair loss. Creams or lotions containing corticosteroid drugs control the condition, which may subside after a few months. However, symptoms may keep recurring for years.

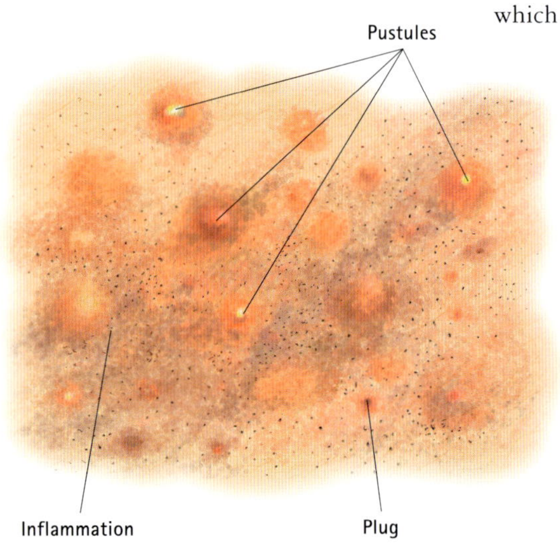

Acne

Right: Acne affects the skin, with inflammation, pustules (pimples) and plugs (blackheads). Here some of the pustules have broken through the skin.

Below: Over-production of sebum may block the outlet of a sebaceous gland and its hair follicle, forming a blackhead. Later, the area becomes inflamed and pus may form under the skin, producing a pustule.

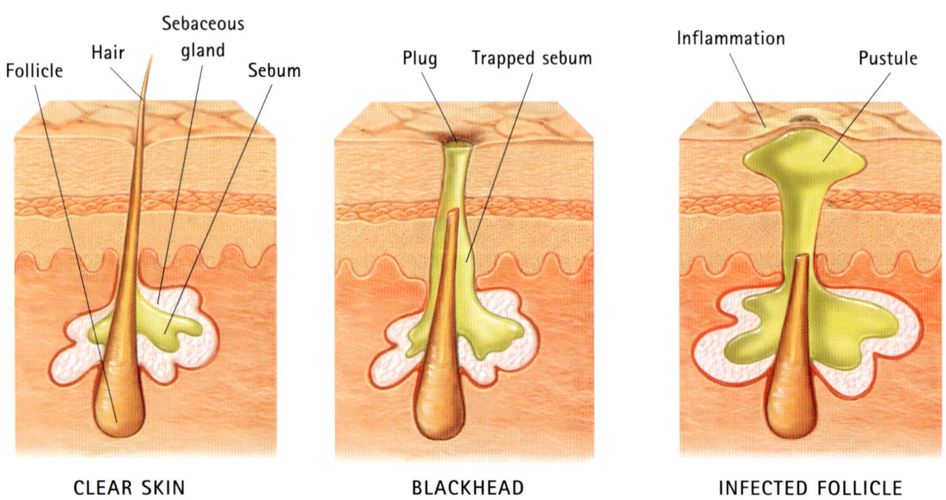

Lichen planus

Small slightly raised bumps with an accompanying itch characterize this inflammation of the skin.

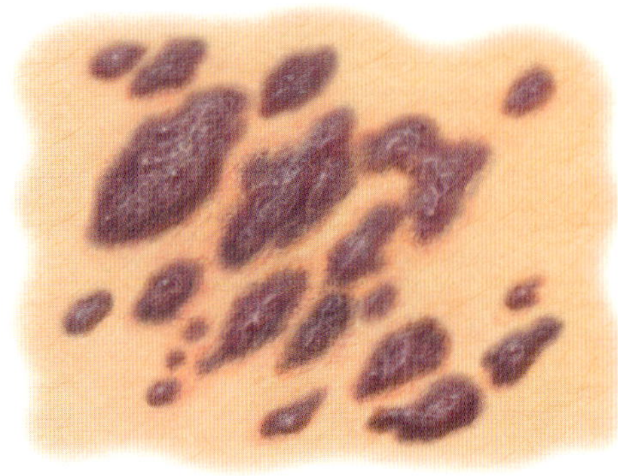

Sometimes lichen planus may be a precursor to squamous cell cancer. Mouth ulcers may develop into oral cancer.

Psoriasis

The exact cause of this mild but frequent skin disorder is uncertain, although it is known to involve an abnormal reaction by the body's immune system against the skin.

A person with a father with psoriasis has about a 30 percent chance of developing the disease; the child of a mother with psoriasis has a 20 percent chance. The disease is rarely seen during childhood, often making its first appearance at puberty.

The impact of the disease varies widely. Typically, it causes a rash of sensitive red patches covered by silvery-white scales on the elbows, knees, lower back and scalp. The outer layer of skin at the sites of these flaking sores grows approximately seven times faster and thicker than normal. These areas can range in diameter from a few millimeters to several centimeters. In severe cases they may join to form very large "plaques" that can cover large areas of skin, such as a person's entire back.

Psoriasis does not normally affect the face, is not contagious, does not cause scars and rarely leads to hair loss. It often affects the nails, causing pitting, discoloration and separation from the nail bed. About 10 percent of sufferers develop a form of arthritis that causes swollen and painful joints and can become debilitating.

A rare but potentially fatal, form of the disease is erythrodermic psoriasis, in which the skin is inflamed and sufferers have problems regulating body temperature.

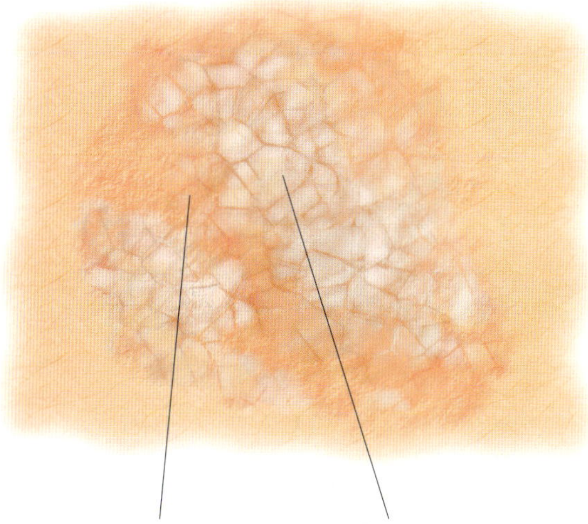

Psoriasis

In this condition, sensitive red patches covered by white scales can appear on the elbows, knees, lower back and scalp.

Psoriasis is not an allergic reaction and avoidance of certain foodstuffs will not control the disease. It is, however, aggravated by excessive alcohol intake. It also has a tendency to repeatedly recur, with possible triggers including stress, streptococcal throat infections, certain medications and wounds.

Although the condition is incurable and difficult to treat, there are many therapies and medications to control symptoms. Complete remission is rare. Sufferers usually need to experiment to find the treatment that works best for them. A variety of creams and ointments, intended to reduce the scaling, is available. Oral medications for psoriasis are normally prescribed in only the most serious cases because they often have dangerous or unpleasant side effects.

Blistering diseases

Fluid accumulating between skin cells or skin layers causes blisters. In adults there is a small group of diseases in which blisters appear early and are the predominant feature. Diagnosis of these diseases is usually made by skin biopsy.

SEE ALSO *Biopsy on page 435*

Pemphigus

Previously, the rare and incurable autoimmune disease pemphigus was invariably fatal. These days, treatment with steroids and immunosuppressive drugs make it possible for sufferers to lead a near-normal life. Characterized by large blisters on the skin and mucous membranes, pemphigus may initially appear similar to the disease pemphigoid but it is not related. The blisters are extremely painful and tend to appear mostly on the scalp, face, nose, mouth, throat, chest, armpits, groin, navel area and, in women, the vagina.

People of all ages and both sexes can develop pemphigus although the condition is rarely seen in children. The cause of the disease is unknown.

Pemphigoid

In the autoimmune disease pemphigoid (or bullous pemphigoid), large blisters appear on the skin. Uncomfortable, but rarely fatal, the condition can be controlled by strong drug treatments. It is incurable, but sometimes disappears after several years. Pemphigoid usually occurs in the over-50s, and more often in women than in men. The cause is unknown but it is not contagious.

Infections of the skin

The skin is a common site of infection because of its defensive role against invading bacteria, viruses and other microorganisms.

SEE ALSO *Infectious diseases on page 364; Inflammation on page 24; Treating infections page 445*

Boil

A boil, also known as carbuncle or furuncle, is a bacterial infection of a hair follicle. It appears as a painful, swollen red lump, usually on the face, neck, armpit, buttocks or thigh. A boil contains pus, a thick yellow fluid that is a byproduct of inflammation, which consists of white blood cells, dead tissue matter in combination with bacteria.

Boils may heal spontaneously, or they may continue to grow in size, burst, drain and then heal on their own. Most boils usually need to be drained surgically by a primary care physician (general practitioner).

Erysipelas

Erysipelas is an acute inflammation of the tissues below the skin, usually on the face, caused by infection by *Streptococcus* bacteria. The bacteria enter via a break in the skin, often following a respiratory infection such as a cold. A bright red spot appears on the nose or cheeks, enlarges and becomes hot and painful, accompanied by fever, chills, muscle pains and malaise. Infection of the blood (septicemia) may follow if the condition is not treated. Erysipelas can also occur in the legs, where veins have been cut.

The inflammation subsides in a week or 10 days with antibiotics. Painkillers and anti-febrile medications (antipyretics) help relieve the symptoms.

Impetigo

Also known as "school sores," impetigo is a highly contagious bacterial skin infection that occurs mainly in children. It usually affects the mouth and nose area, but can also occur under diapers in babies, or anywhere else on the body. It is not painful, but can be itchy and lead to scratching that will spread the infection. The surface of the skin is infected with bacteria, either *Streptococcus pyogenes* or *Staphylococcus aureus*, which feed on a wound such as a cut or insect bite, or a skin condition such as eczema.

Appearing first as red blisters, the spots become filled with pus and form a scab. Impetigo is treated with antibiotics and/or antibiotic ointment.

If the sore is surrounded by red skin or the person is unwell, then further medical

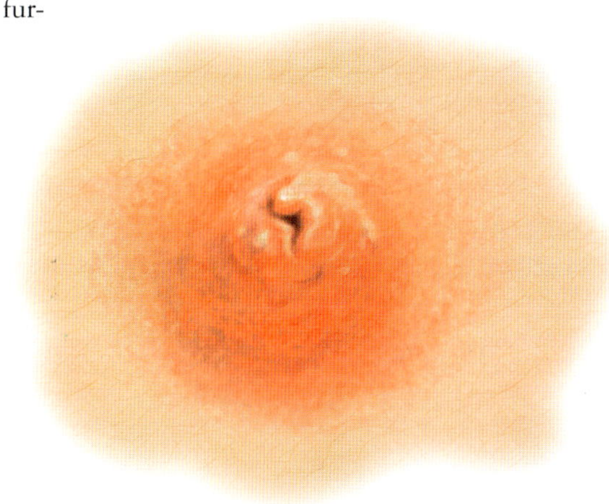

Boil

A boil is a bacterial infection of a hair follicle. This boil has burst through the skin and is discharging pus.

attention should be sought. As impetigo is highly contagious, the sufferer needs to bathe or shower daily, keep linen and towels separate, keep fingernails and hands clean, and avoid contact with people until the sores have healed.

Cellulitis

Cellulitis is a bacterial (usually streptococcal) infection of the skin and underlying tissues. Cellulitis can arise after dermatitis, ulcers, injury or animal bites. The symptoms are redness, swelling and tenderness of the skin, swollen lymph nodes which may be accompanied by fever, chills and rapid heartbeat. Medical aid should be sought.

Folliculitis

Folliculitis is a bacterial or fungal infection of the hair follicles. It can occur anywhere on the body and often arises when follicles are damaged by shaving or wearing tight clothing. The condition is marked by itching, reddened skin, a rash and pustules around the follicles that may dry out and crust over. It is treated with antiseptic creams and antibiotics.

Whitlow

A pus-filled inflammation on the end of a finger or toe may be termed a whitlow. A herpetic whitlow is a swollen painful fingertip caused by the herpes simplex virus entering through a wound after exposure to infected oral or respiratory secretions; it may also occur as the result of nail-biting. Treatment may include the release of pus.

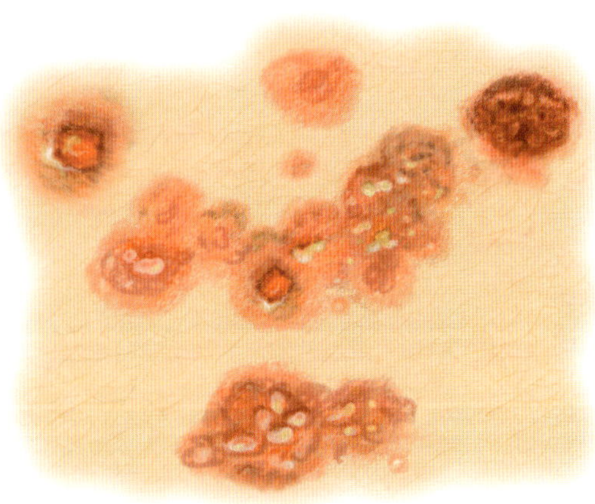

Impetigo

This contagious bacterial infection of the skin involves blister-like spots that fill with pus, rupture and become itchy yellow crusts.

Ringworm

Ringworm is a skin infection caused by various fungi, which results in a characteristic red, ring-shaped rash. The fungi dermatophytes invade the top layer of skin and affect the tissue below. Ringworm is common among children and can be treated with non-prescription anti-fungal creams and powders. Oral medications and stronger topical creams may be used for more severe and persistent cases. Preventive measures include attention to personal hygiene and keeping the skin clean and dry.

Since the infection is highly contagious, hairbrushes, clothing and personal items should be cleaned after use. The complications of ringworm include secondary skin infections and spread of tinea to other parts of the body. Tinea is ringworm of the feet (athlete's foot), scalp, groin and nails.

Warts

Warts are benign tumors that occur in the outer layers of the skin. They appear as a raised, rough, round or oval lump that may be skin-colored, or lighter or darker than surrounding skin. Caused by the papillomavirus, they are mildly contagious and usually occur on the hands, feet, and face.

Warts are often named for where they appear. Plantar warts occur on the soles of the feet, genital warts in the skin on or around the genitalia. If they occur on the hands, arms or legs they are called common warts, or verrucae vulgaris. Multiple pinhead size warts occurring in children are called verrucae planae juveniles.

Warts usually cause no symptoms (though plantar warts may be painful) and disappear spontaneously within 2–3 years. If unsightly or painful they may be treated with an over-the-counter paint containing a mildly corrosive agent such as salicylic acid and/or lactic acid, which is applied daily for several weeks. They can be surgically removed with cryotherapy (freezing), electrocautery (burning), or laser treatment. Recurrence is common.

Shingles

Shingles is the common term for herpes zoster infection, a reactivation of the

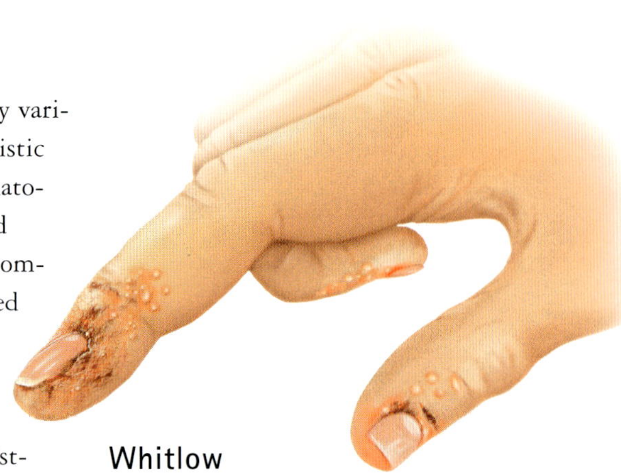

Whitlow

Small fluid-filled blisters are a feature of herpetic whitlows, caused by the highly contagious herpes simplex virus.

varicella-zoster virus that causes chickenpox. Following chickenpox infection the virus remains dormant in the body and reactivation can occur years later, particularly if the body's immunity decreases. This occurs most commonly with age.

While dormant, the virus is in spinal ganglia beside the spinal cord; on activation it migrates along a sensory nerve to the area of skin it supplies (its dermatome). It is here that the infection appears as a rash composed of small blisters with surrounding inflammation. The rash is therefore localized, and usually confined to one side of the body. It may be associated with severe burning or sharp nerve pain. Pain may precede the rash and may persist afterward.

This is often the most distressing aspect of the infection, since it is otherwise generally benign. More severe disseminated infection and neurological complications can occur in immunocompromised patients,

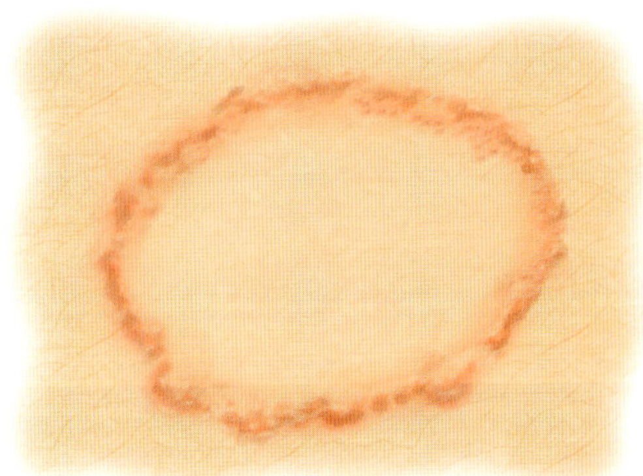

Ringworm

Also called tinea, this disease is caused by a fungus, not a worm as the name suggests.

such as those with lymphoma or following bone marrow transplantation. Antiviral medications can now treat shingles, including oral medications, and can reduce the duration of infection and the pain during and after it.

Scabies

Tiny mites *(Sarcoptes scabiei)* cause the highly contagious skin condition scabies. It spreads by close contact with an infected person, by sharing their bed or wearing their clothes. Female mites burrow under the skin—favoring hands, toes, groin and bends of elbows and knees—to lay eggs. It takes approximately 10 days for the mites to mature and continue the cycle. Within weeks, the skin develops an allergic reaction and itches intensely. An eczema-like rash often develops and the mite burrows may become visible.

Treatment includes chemical washes, but the itching may persist for weeks.

Molluscum contagiosum

Characterized by multiple little blisters or blebs ⅒ inch (2–3 millimeters) across with a central dimple, molluscum contagiosum is an acutely infectious disease transmitted by skin to skin contact.

Due to a pox-type virus, it is common to find the infection on children where it is probably spread by skin contact. In adults, it can spread by sexual contact and is commonly found in the genital area in this case.

People with AIDS may develop florid molluscum all over the face, with the blebs being up to ⅕ inch (5 millimeters) across. The incubation period is relatively long, ranging from a week to several months.

Treatment includes freezing with liquid nitrogen, using electric high-frequency currents (diathermy), or applying various chemicals injected into each spot with a sharpened sterile toothpick.

Congenital skin disorders

Congenital disorders of the skin range from the common birthmark to rare inherited disorders, such as pachyonychia and ichthyosis.

Warts

Caused by a virus, warts are small benign tumors that most commonly occur on the skin of the hands, feet and face.

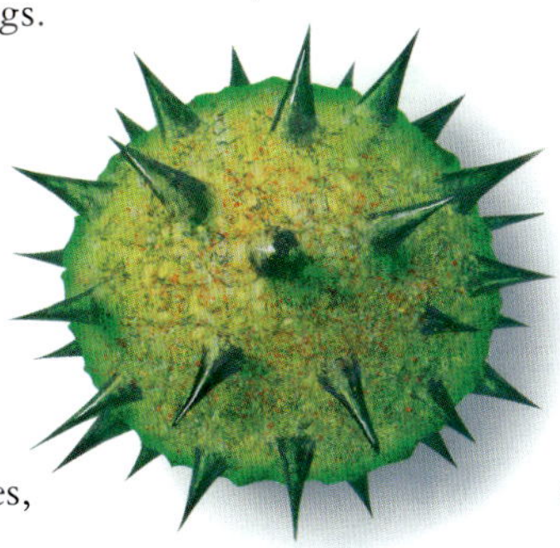

Wart virus

The papillomavirus is contagious and can be spread by scratching or rubbing infected skin.

Pachyonychia

Pachyonychia ("elephant nails," also called nail-bed hypertrophy) is a rare condition involving the overgrowth and excessive thickening of the nails. It is often a congenital disorder and may occur with other disorders of the skin and mucous membrane.

Ichthyosis

Ichthyosis is an inherited disorder in which the skin becomes thick, forming cracks and fissures, giving it a fish-scale appearance.

Legs, arms, hands and trunk are the areas most affected. The condition typically begins before the age of 4 and improves during adulthood; however, it may recur when a person becomes elderly. There are various types of ichthyosis, but most are inherited genetic traits. Some forms may be acquired or develop in association with other diseases.

There is no cure for ichthyosis, but the use of mild soaps, moisturizing creams and ointments will help the condition. Ointments that contain catalytic agents, such as lactic acid and salicylic acid, are useful, as they help the skin to shed.

Birthmarks

Babies are sometimes born with marks, spots or patches on their skin, known as birthmarks. These marks remain, unlike bruises that are caused by the trauma of birth. The most common birthmarks include Mongolian blue spots, port-wine stains, salmon patches, cafe au lait spots and congenital pigmented nevi.

Mongolian blue spots are blue-colored "bruises" or nevi, common on dark-skinned and Asian babies. These spots are harmless and will disappear during childhood.

Salmon patches are red marks that are found on the eyelids, nose or on the back of the head. Formerly called "stork bites," they are very common and will usually disappear over time. They are a type of hemangioma.

Port-wine stains are also a type of hemangioma. They are reddish purple-brown lesions that do not fade away but can be treated with surgery.

Shingles

Caused by the herpes zoster virus, shingles is characterized by painful blisters on the skin.

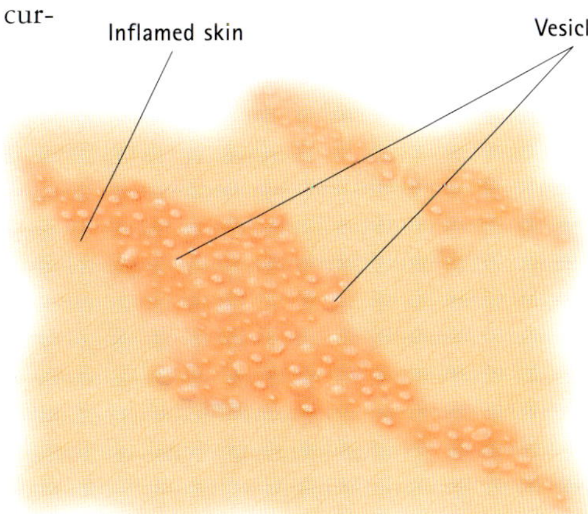

Scabies

Scabies is caused by female parasitic mites burrowing under the skin to lay their eggs. An allergic reaction to the mites' feces causes an itchy rash to appear.

Cafe au lait spots appear as coffee-colored patches on the skin. They are quite common and will not disappear.

Congenital pigmented nevi are moles that vary in color from light brown to black and which may have hairs. Congenital nevi (those present at birth) have an increased risk of developing into malignant melanoma, a form of skin cancer. Those with a mixture of colors, those that are irregular in shape, and those that are large have the highest potential for malignancy. All congenital nevi should be examined periodically; if there are changes in the size, color, or surface texture, or sudden ulceration, bleeding, or itching in the birthmark, a physician should be consulted and the mole should be removed.

Diagnosis of the type of mole is made by a physician based on the appearance of the mole, when it appeared, and any symptoms or pre-cancer type features. Usually the mole doesn't need to be treated; however, if it looks ungainly and is causing concern, it can be covered with cosmetics. Alternatively, it can be surgically removed.

If a mole has an increased risk of cancer, surgical removal is desirable. Usually, moles can be removed under local anesthesia in a physician's office. Alternatively a mole can be removed using liquid nitrogen, electrotherapy or radiation therapy.

Rash

Rash is a general term used to describe the temporary occurrence of raised or differently colored spots or patches on the skin. Rashes have a variety of causes and will vary in appearance depending on how severely the person is affected.

SEE ALSO *Allergies on page 60; Antihistamines on page 439; Infectious diseases on page 364*

Urticaria

Urticaria (known as hives or nettle rash) is a skin rash in the form of wheals that are red, itchy and raised. It appears suddenly and may disappear just as quickly, leaving no trace or permanent damage. It can affect either sex and appear at any age. Urticaria of pregnancy may appear in the last 2–3 weeks and disappears after delivery.

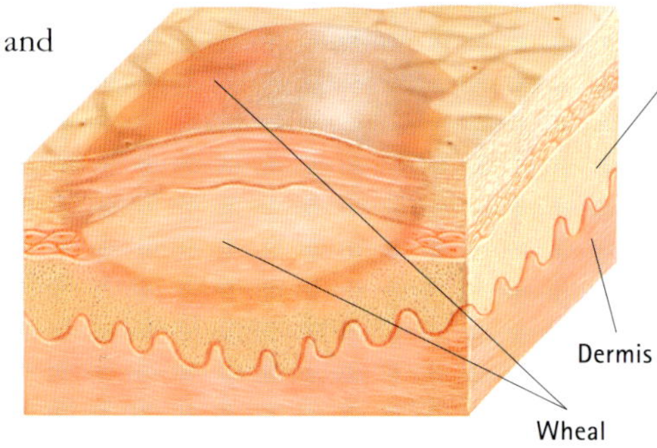

Urticaria

More commonly known as hives, urticaria is a rash of itchy red wheals that can be caused by an allergic reaction to food or plants, illness or emotional stress.

Urticaria can occur after illness, as a result of skin infection, or as the result of an allergic response when the chemical histamine is produced within the body.

Occasionally urticaria may cause swellings in the throat and restrict breathing. If the rash appears to be below the skin surface and there is burning or pain rather than itching, a medical examination may be needed to establish whether it is urticaria or angioedema, a more serious condition.

Visible symptoms and the torment of itching can be relieved by taking a lukewarm bath with soothing additives, or by applying cool compresses to the rash.

Medical advice should be sought in all cases where urticaria is persistent or is causing distress or discomfort. Drugs used in treatment include corticosteroid creams, antihistamines and, where there are breathing difficulties, epinephrine (adrenaline).

Heat rash

Also known as miliaria or prickly heat, heat rash consists of tiny blisters at the site of sweat pores. When the sweat ducts become blocked, the sweat escapes into other levels of the skin and the blisters result. Extremely hot weather is usually the cause. In newborn babies the sweat glands are not fully developed and will become blocked if the baby is overheated. Babies may also suffer heat rash when they have a fever.

Treatment for babies is a tepid bath, fresh air and lightweight clothing; sometimes the application of calamine lotion may also be necessary. The rash usually disappears in a couple of days; if it is persistent it is advisable to check with a health professional. Other types of heat rash include miliaria crystallina, seen in people who have a fever or are suffering from sunburn.

These blisters are tiny and shaped like dewdrop. Miliaria rubra, commonly called prickly heat, is the most common form of heat rash and produces dense, itchy red papules (solid cone-shaped lumps), which generally appear on the trunk. Miliaria pustules are blisters filled with pus, and miliaria profunda are firm papules found in the vascular layer of the skin.

Treatment for heat rash concentrates on reducing body heat, minimizing sweating, and avoiding irritants such as tight clothing.

Diaper rash

Diaper rash (nappy rash) is a common inflammation of the skin in babies mainly due to wetness and heat beneath the diaper, resulting in red blotches, spots or lesions. It can also be caused by contact friction, chemical allergies and the blockage of sweat glands and may be exacerbated by the interaction of the skin with feces and the ammonia in urine. The rash tends to be worse in the creases of the skin.

Diaper rash can be treated or prevented by keeping the skin as dry

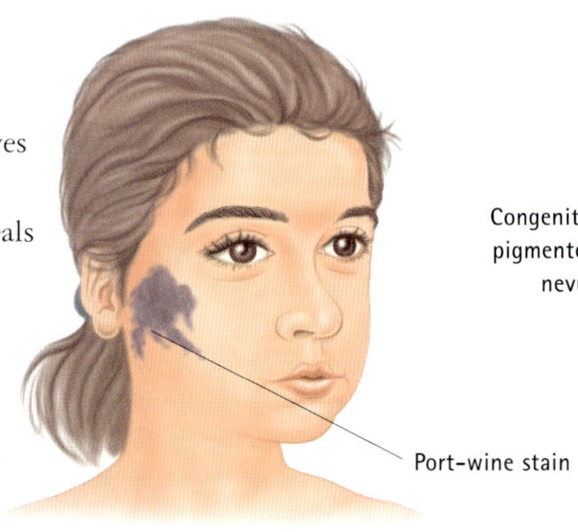

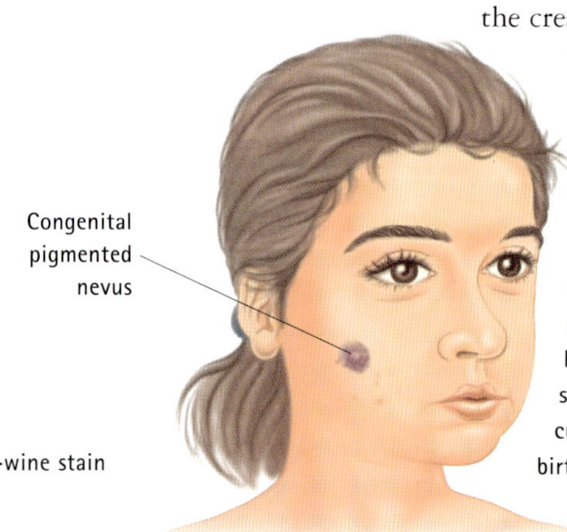

Birthmarks

Although most birthmarks are harmless, some may be unsightly. They may be hidden by cosmetics or treated by plastic surgery, laser surgery, cautery (electric current) or cryosurgery (freezing). Some birthmarks disappear of their own accord. Their cause is unknown.

as possible, changing diapers frequently, air drying between changes and leaving the diaper off for as long as is practical each day. Plastic pants used to cover diapers can make the condition worse, and wipes containing potential irritants such as alcohol should be avoided. Simply cleanse with mild soap, water and a soft cloth.

Ointments containing zinc oxide are helpful for reducing any friction between the diaper and the baby's skin. Secondary infections can be caused by fungi (such as *Candida*) or bacteria (such as *Staphylococcus* or *Streptococcus*) which are normally found in the skin and which thrive in damaged areas. These infections can be treated with topical antibiotics.

Discoloration and damage

The skin may be damaged by too much exposure to the sun, natural ageing, or trauma. In disorders such as vitiligo and melasma, however, discoloration is caused by abnormalities in melanocytes—cells in the epidermal layer of the skin that synthesize the pigment melanin.

SEE ALSO *Cyanosis on page 86; Fibrosis on page 24; Inflammation on page 24; Raynaud's disease on page 86; Scar on page 25; The skin and sun exposure on page 349*

Vitiligo

Up to 2 percent of the world's population, or some 50 million people, suffer from vitiligo. This disorder involves destruction of pigment-producing cells (melanocytes), which occur in the skin, linings of the nose, mouth, genital and rectal areas, and in the retinas. The result is white (pigment-free) patches on the skin. Hair growing at affected sites also turns white.

The cause of vitiligo is unknown, although it is most prevalent in people with certain autoimmune disorders. One of the most beneficial treatments is psoralen photochemotherapy, which involves careful, controlled and long-term use of certain drugs combined with ultraviolet light exposure.

Vitiligo

In this condition white patches develop on various parts of the body, such as the skin, as a result of damage to pigment-producing cells.

Keloid

Overproduction of collagen in scar tissue at the site of a wound creates a keloid, or keloid scar. People with darkly pigmented skin have a greater tendency than others to develop keloids. Firm, raised, hard and slightly pink, keloids may arise following surgery or after a burn. Keloids can also sometimes arise from a minor scratch on the skin.

Keloids may appear anywhere on the skin, but commonly form on the breastbone, upper back and shoulder. They may itch, cause pain or be tender to the touch; although they are otherwise harmless. In some cases the keloid scar continues to grow and may develop claw-like projections into the surrounding skin.

Keloids can be treated by surgical excision, cryosurgery, or injection of corticosteroids into the keloid, but they may recur.

Stretch marks

Stretch marks are streaks or lines that appear on the skin as a result of rapid growth, or as reactions to certain diseases and topical medications. They are commonly seen on the abdomen and breasts of pregnant women, but can also occur on the buttocks, hips, thighs and sides, wherever the elastic fibers in the skin stretch and rupture.

Hormonal changes and the rapid growth of puberty can cause stretch marks in both males and females. They begin to appear as soft, red-purple glossy streaks and generally can't be prevented or cured. However, they may often fade or disappear with time, particularly when the initial cause of the skin stretching has passed.

Bruise

Bruising is a discoloration of the skin that takes place after an injury. If bruising occurs frequently, for no apparent reason, it may indicate an underlying bleeding disorder.

Bruise

A bruise (also called contusion) is a discoloration of the skin. It occurs when blood vessels are damaged or broken as the result of a blow to the skin, such as bumping against something. The discoloration is caused by blood leaking out from damaged vessels into the skin. It first appears reddish, then, one or two days later, blue or purple. By day six, the color changes to green and after a week or so, the bruise will appear yellowish-brown. The skin color will return to normal in 2 to 3 weeks.

Bruising is usually more extensive in older persons, because of the greater fragility of the blood vessels. Some medications, especially those that cause bleeding such as aspirin and anticoagulants, make individuals more prone to bruising. However, cortisone medications such as prednisone, clotting disorders like hemophilia, or liver diseases can also cause serious bruising and bleeding in some people.

To prevent or minimize bruising after an injury, apply ice to the affected area. Apply pressure by hand or with a bandage, but do not use a tourniquet.

Scalp

The scalp is the skin and connective tissue covering the skull. The hair that grows from it protects against heat loss, minor abrasions and ultraviolet light. The adult human scalp contains around 100,000 hair follicles.

SEE ALSO *Head on page 124*

Disorders of the scalp

Hair may disappear because the follicles are damaged, either permanently or temporarily; because the person is going bald for heredity reasons or hormonal imbalance; or

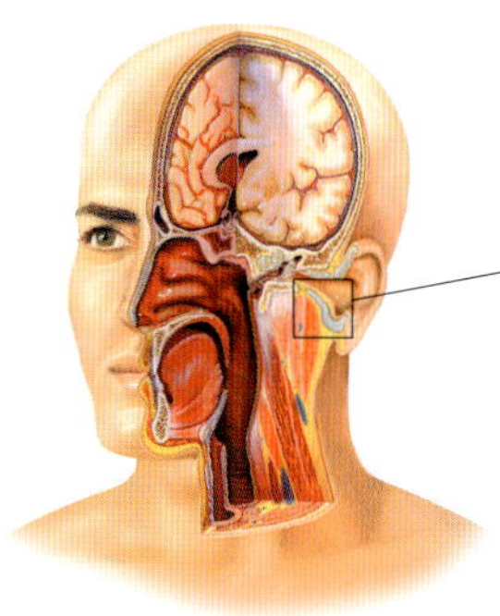

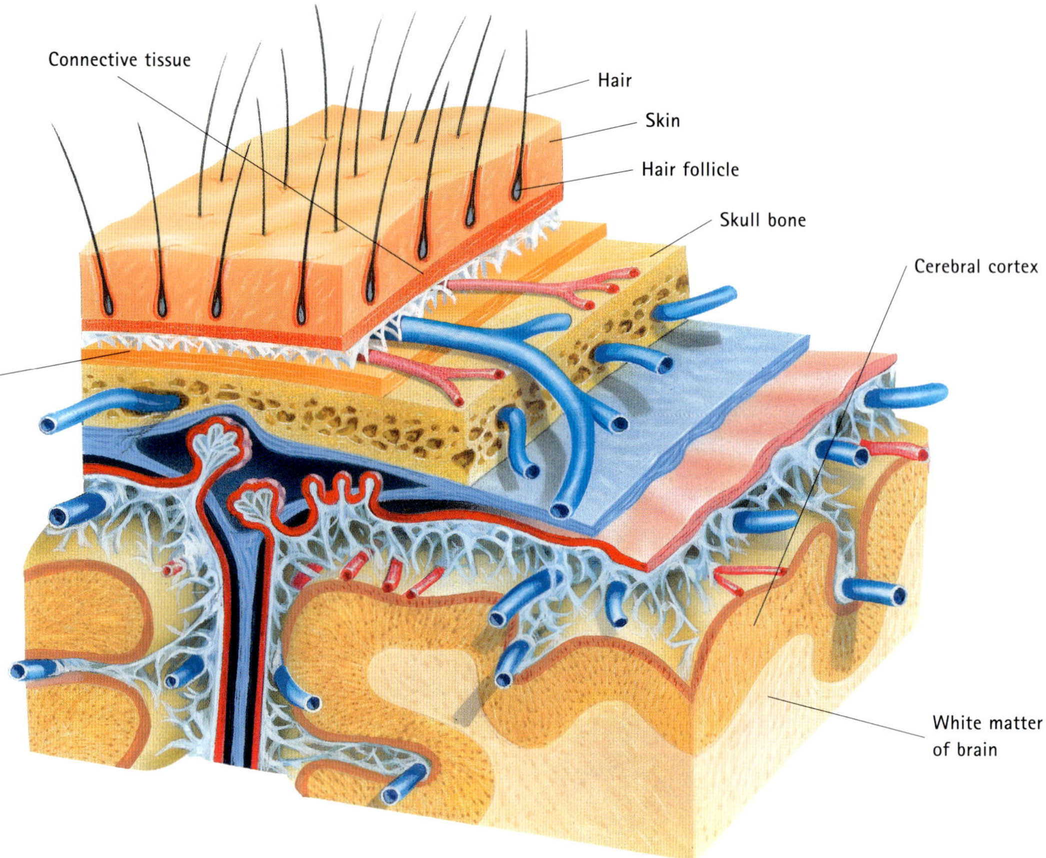

Scalp

The scalp is the layer of skin and connective tissue that covers the skull, and contains the follicles for the hair on the head.

because it has been voluntarily removed. Conditions that cause itching in the scalp include dandruff, scalp psoriasis and lice.

SEE ALSO *Lice on page 391*

Cradle cap

Cradle cap, a skin condition that can occur on a baby's scalp. It is yellow and crusty. However the disorder is not infectious. It is a form of seborrheic dermatitis that occurs when a baby's head is not cleaned properly, usually because the parents are concerned that they may damage the soft spot (fontanelle) on the baby's head. Sebum, which is secreted from the sebaceous glands, forms in layers on the scalp, creating patches or a "cap" on the head.

Treatment is simple and usually effective if carried out when the crusts first appear: the affected areas should be rubbed with a little olive oil or paraffin in the evening. The next day the area should be washed gently with soap and warm water and the scales should lift off easily. This should be repeated until the crusts disappear.

Careful shampooing will prevent cradle cap from returning. Persistent cases may require the application of a special preparation available from a pharmacist or prescribed by a doctor.

Dandruff

Dandruff is the everyday term for mild seborrheic dermatitis of the scalp. It is related to cradle cap in infants, and in adults, manifests as greasy or dry, white scales that are shed from the scalp. Other areas may also be affected, including eyebrows, forehead, and behind the ears.

Dandruff is sometimes associated with excessive production of grease or oil from the sebaceous glands, and sufferers tend to have oily skin. The cause is still unknown, but it tends to be worse in hot, humid weather or cold, dry weather and during periods of stress and fatigue.

Dandruff is a chronic condition with periods of improvement followed by deterioration. Although it is not dangerous and rarely needs medical treatment, it can often be annoying and socially embarrassing.

Shampooing should be done frequently (daily), prolonged (at least 5 minutes) and vigorous enough so that the scales are loosened with the fingernails.

In addition, a non-prescription dandruff shampoo containing selenium sulfide or zinc pyrithione should be used at least once a week. If the condition worsens, a doctor should be consulted, as corticosteroid creams or ointments may be needed.

Baldness

The most common cause of loss of hair (alopecia) is male hormones. Androgenetic alopecia (male pattern baldness) occurs in both sexes and is an inherited condition associated with sexual development. It is characterized by a receding hairline, eventually leaving only a peripheral rim of hair.

Another type is alopecia areata, a non-scarring, inflammatory hair loss disease that gives the hair a patchy, moth-eaten look.

Baldness is common among people of the European and Australian races. In Australian Aborigines alopecia is often accompanied by balding of the calves of the legs; it is less common in native Americans, Africans and Asians. It affects between 50–80 percent of Caucasian men and between 20–40 percent of Caucasian women. In men, baldness is age related, affecting about 30 percent of

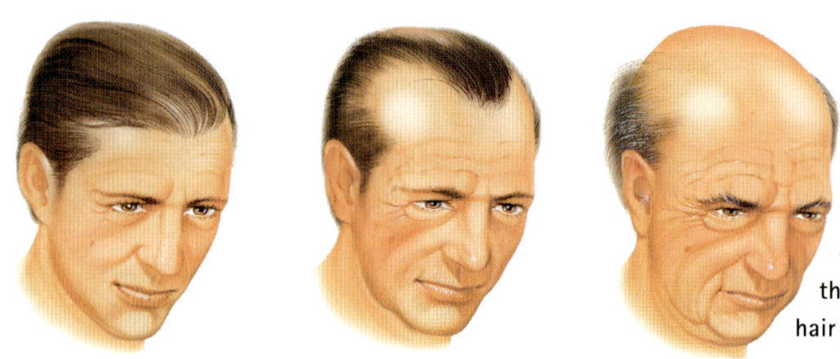

Male pattern baldness

Hair loss follows a typical sequence in male pattern baldness. It begins at the temples, continues at the top and back of the head until finally, in severe cases, hair grows only at the sides of the head.

men in their thirties, 40 percent in their forties, and so on. However, it is not related to age in women.

Temporary hair loss can be caused by illnesses that are accompanied by high fever, but can also be caused by pregnancy, chemotherapy, x-rays, ingestion of metals, malnutrition, some skin diseases, endocrine disorders, chronic wasting diseases and trauma such as chemical damage to the hair.

Not all hair loss requires treatment. If it is caused by illness, trauma or pregnancy, regrowth will take place on its own within three to four months.

There are three treatments: hair transplantation from an area where hair is growing; the drug minoxidil, which stimulates regrowth and is applied to the scalp; and the drug finasteride, which may promote growth. If hair loss is not hereditary, tests for thyroid disease, autoimmune conditions and anemia should be done.

Head lice

Minute wingless insects that live on the scalp and suck blood, head lice are spread through direct contact and by sharing hats and combs. Though anyone can have them, young children, particularly those who attend preschool, school or child care, are more likely to suffer because of the contact they have with other children.

Lice are attracted to clean rather than dirty hair, and are very common. Head lice cause itching. Their presence can be confirmed by a fine black powder of louse feces or pale flecks on the pillow, by the lice themselves, and by their eggs (nits—white specks stuck to hairs, near their roots).

Head lice can be eradicated by the use of special shampoos, which everyone in the household should use. Wash the hair with the shampoo, then have the sufferer sit in the sunlight. Check behind the ears and the back of the neck, then comb the hair carefully using a special lice comb, rinsing or wiping the comb between strokes. Dry the hair with a hair dryer, as the heat will also help to eradicate the pest.

Daily combing is important once it is certain that the lice have been eradicated. Hair brushes, combs, pillow cases, hair accessories and hats should also be washed.

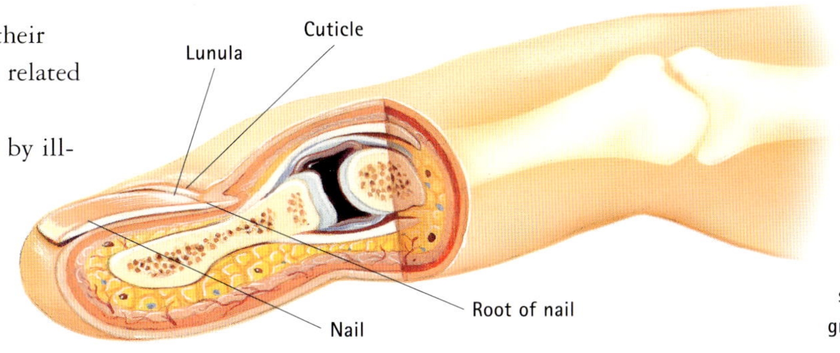

Nails

Nails provide protection for the sensitive tips of the fingers and toes. The cuticle is the fold of skin where the nail meets the skin. The top layer is visible, while the unseen bottom layer seals off the site of new nail growth to prevent infection.

Nails

The primary function of our nails is to provide protection for the sensitive tips of our fingers and toes. They are made mainly of the tough protein called keratin and most of what we can see of them is actually dead—attributes that make them reasonably resilient to everyday knocks.

The only living part of a nail is the root, where growth occurs. This is located under the cuticle at the nail's base. Beneath the nail is the nail bed, visibly pink through the nail in healthy circumstances due to a rich blood supply. If a nail becomes detached from the bed it is inevitably replaced. As fingernails grow at an average rate of about 2 inches (5 centimeters) a year, it takes about 6 months for a nail to regrow fully. Toenails take 12–18 months to grow back. Nails can also be excellent indicators of a person's more general state of health. A change in color from normally healthy pink can indicate a problem with a major organ.

Disorders of the Nails

Although most nail afflictions are minor and easily remedied, neglect or incorrect treatment can turn them into painful, lingering and unsightly problems.

Paronychia

Paronychia is an infection of the skin around a nail. It is caused by bacteria, fungi (especially candida) or both. Usually the infected skin is already damaged, from biting, picking or trimming nails or from immersing hands in water for long periods. The skin around the nail becomes red and swollen, and in the case of bacterial paronychia, there may be tiny abscesses. The nail is often infected as well, becoming discolored and misshapen. Treatment is with antibiotic or antifungal creams or ointments. Fungal paronychia may take some months to clear. Diabetics are especially prone to paronychia and need to take special care.

Onychomycosis

A fungal infection in a nail (onychomycosis) is usually very persistent. This is most likely to occur on a toe, rather than a finger, and is often indicated by a discolored, brittle or peeling nail. Without treatment, onychomycosis can cause numbness or pain in the affected area. The nail may separate from the nail bed and eventually be destroyed. Treatment often includes taking an oral antifungal medication over several months.

Ingrown nails

Another normally minor problem that is far more likely to affect toes than fingers, particularly the big toe, is an ingrown nail. This occurs mostly when a nail is trimmed overly short or shoes are too tight, forcing the sharp corner of a growing nail to push into the skin. The result is a painful swelling and sometimes infection.

Regular massage of the skin around the ingrown corner often corrects the problem. Gently taping the skin back from the nail edge may also help, as can going barefoot for a while. If the problem persists, minor surgery can help.

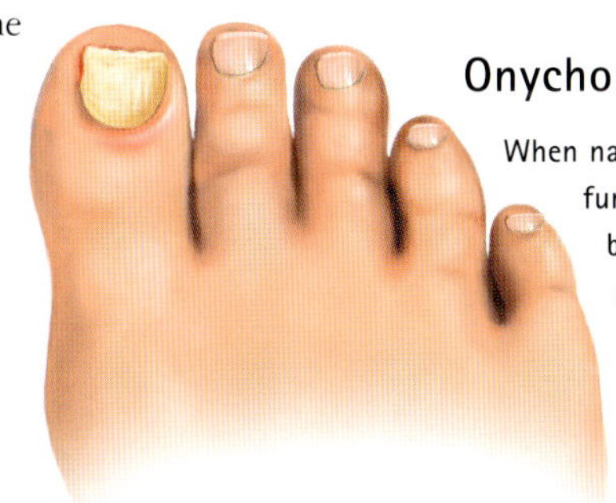

Onychomycosis

When nails are affected by fungal infections they become discolored and brittle. Toenails or fingernails may be affected.

Infectious Diseases

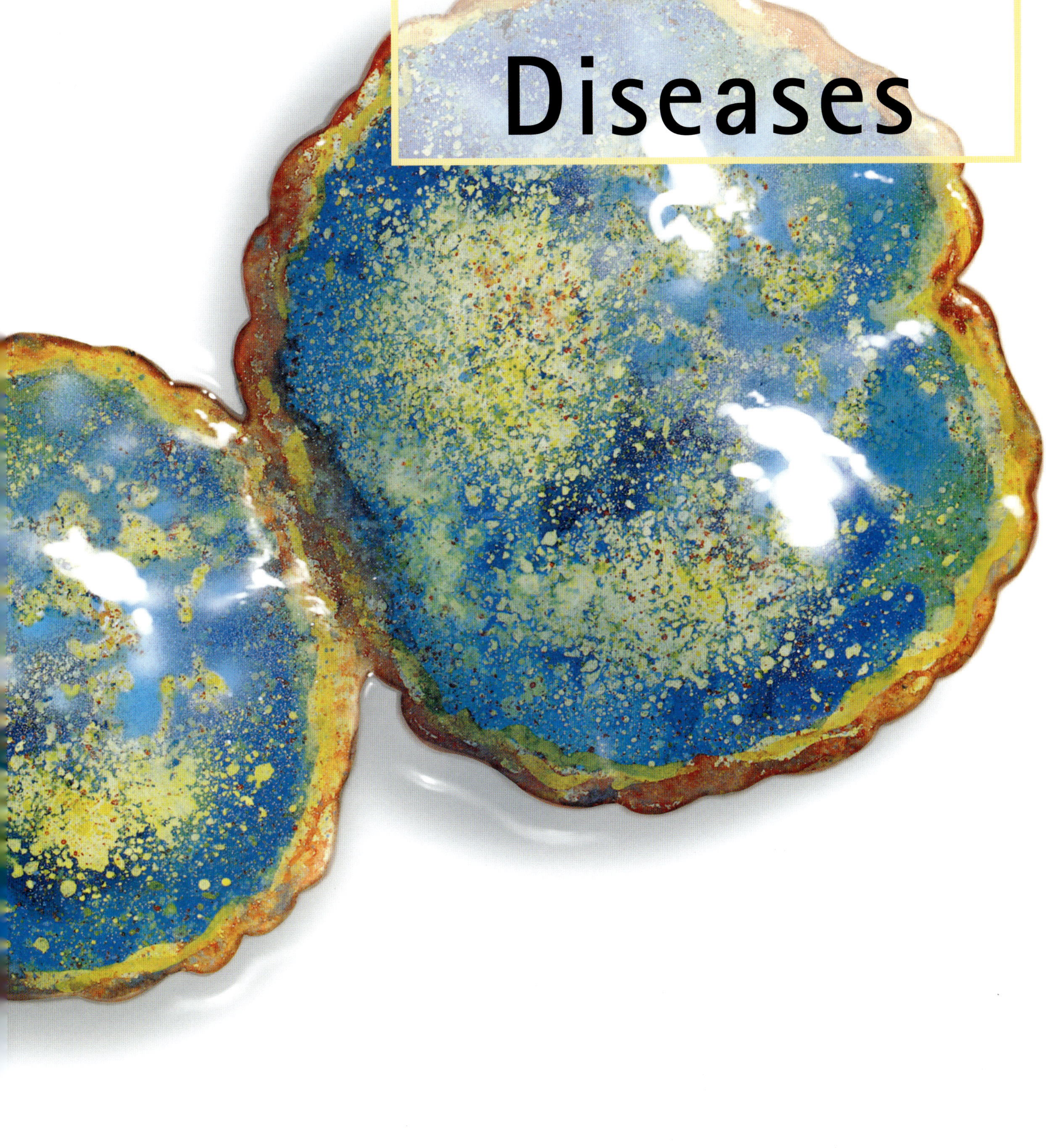

INFECTIOUS DISEASES

An infectious disease is a disease caused by the invasion of and multiplication in the body's tissues of microorganisms that have been passed from one person (or animal) to another. These microorganisms include bacteria, viruses, fungi, protozoa or rickettsiae. Some diseases may be caused by the toxins (poisons) these microorganisms produce.

For infection to occur, a number of factors must be present: an infectious agent; an environment in which it can reproduce, such as contaminated food; a mode of transmission; and a susceptible host.

Transmission can occur in many ways. Infectious agents can be spread in droplets through the air when infected persons sneeze or cough. Whoever inhales the droplets can then become infected. Some diseases can be passed through contaminated eating or drinking utensils. Similarly, ingesting contaminated food or liquids exposes a healthy person to the disease. Other diseases can be spread through sexual activity. Common entry routes into a host are the skin (especially if it has been injured) and the mucosal surfaces of other body openings. A pregnant mother may also transmit infections from her blood supply to that of the fetus. Occasionally, infections can be spread in the course of medical or surgical treatment, or through using dirty injection equipment, as occurs with some frequent drug users.

Every infectious disease has an incubation period. This is the length of time between the entry of the infectious agent into the body and the appearance of the first symptoms of the disease. It may be as short as a few days, as with the common cold, or it may be months or years, as in some slow viral diseases that affect the brain.

Not everyone who has an infectious agent contracts the disease. The virulence of the microorganism may be insufficient for the exposed person to develop the disease. However, this person (known as an asymptomatic carrier) may be harboring the disease and may spread it to another. Carriers may also include someone who is incubating the disease and is yet to develop it, or someone who may have recovered from the disease but is still capable of transmitting it to others.

A person whose immune system has been damaged by diseases of the immune system, leukemia and cancer is more likely to prove susceptible to disease. A disease that a normal person would fight off easily but which causes often serious illness in someone with a damaged immune system is called an opportunistic infection.

An infection may be local and confined to one area or it may be generalized. If it is localized it often produces characteristic symptoms of pain, swelling and reddening at the site of infection. This response is the body's way of fighting the infection and is known as inflammation. If the infection is generalized it will spread through the blood and affect the whole body, causing fever, chills, and a rapid pulse.

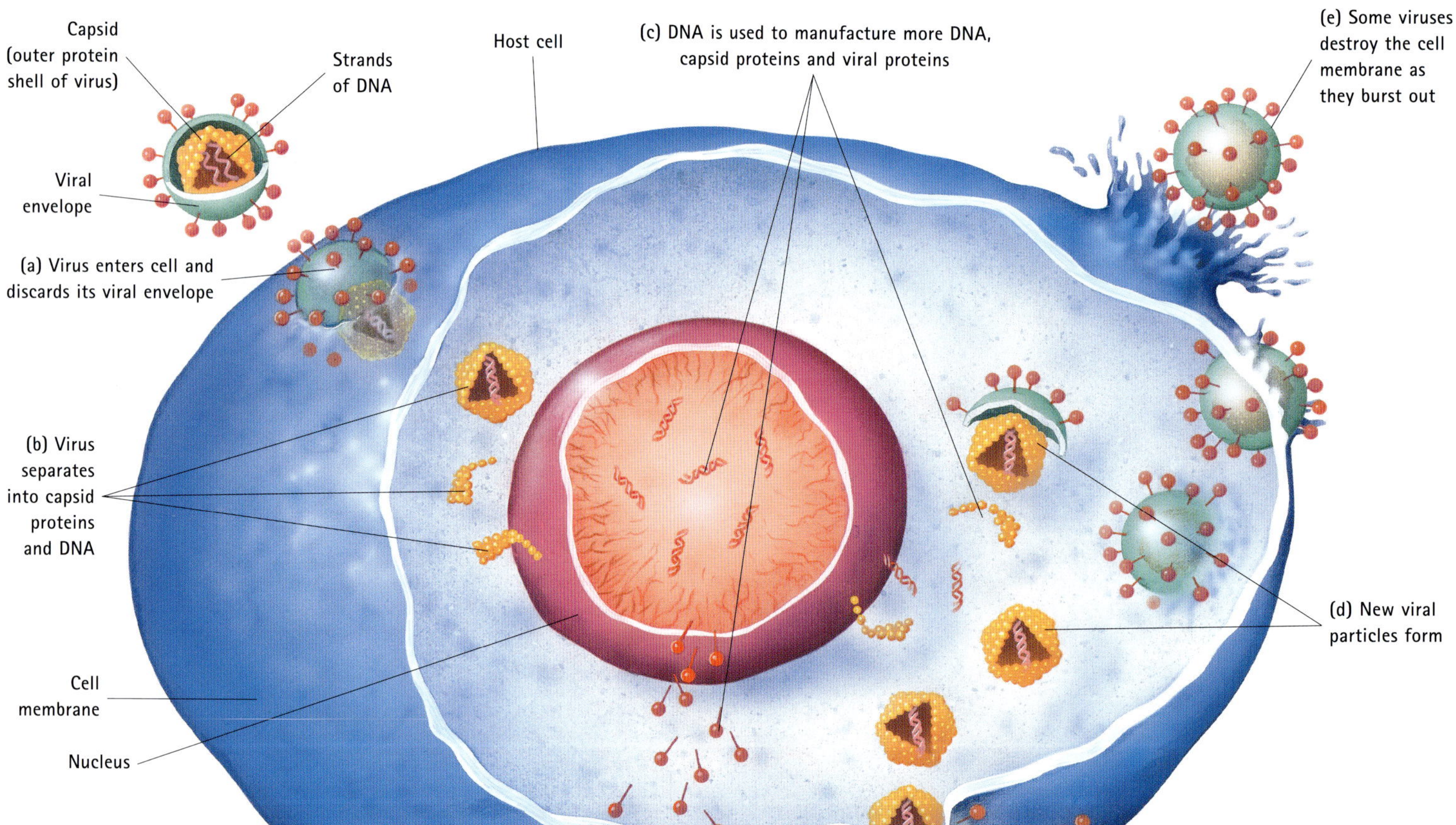

Infection: viral replication

A virus reproduces by entering a cell and using the elements of the cell to clone itself. These "copies" of the virus then invade the rest of the body.

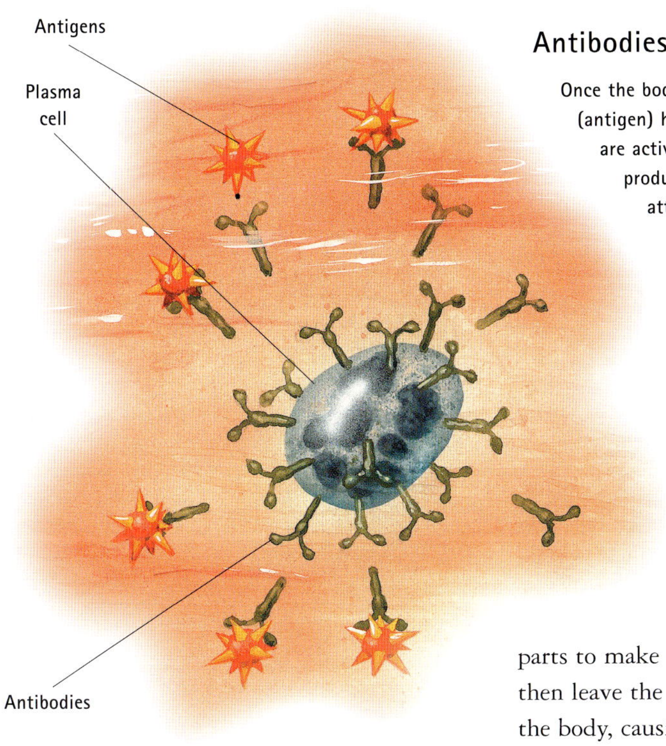

Antibodies

Once the body recognizes a foreign substance (antigen) has entered the body, B lymphocytes are activated, become plasma cells and begin producing antibodies. The antibodies attach to the antigens, which are eventually neutralized.

Different types of infectious agents (pathogens) cause different types of disease.

SEE ALSO *Blood on page 82; Blood tests on page 436; Cells on page 16; Inflammation on page 24; Immunization on page 447; Lymphatic Immune system on page 55; Skin tests on page 438; Treating infections on page 445*

Bacteria

Bacteria are tiny, single-celled organisms with a cell wall but no nucleus. They may need oxygen to live (aerobic bacteria) or they may be able to live without oxygen (anaerobic bacteria). Some bacteria are spherical (cocci), corkscrew-shaped (spirilla or spirochetes) or rod-shaped (bacilli). They are also classified as gram-negative or gram-positive, according to whether their cell wall holds a special laboratory stain called Gram's stain.

Most bacteria can be treated by antibiotics, which kill bacteria by interfering with their metabolism. Antibiotics can be injected or taken orally. Certain antibiotics work only against a particular type of bacteria, for example, they may work against gram-positive but not gram-negative bacteria. Bacteria gradually tend to develop resistance to antibiotics over time.

Viruses

Viruses are between 20 and 100 times smaller than bacteria. They contain either deoxyribonucleic acid (DNA) or ribonucleic acid (RNA). They are not considered to be alive, since they cannot reproduce outside a living cell. However, a virus can reproduce by entering a cell and using the cell's parts to make more copies of itself, which then leave the cell and spread elsewhere in the body, causing disease. These are called intracellular infections.

Viruses have also been implicated in causing some types of cancers. Antibiotics are ineffective against viruses, although antiviral drugs are available for some viral diseases. However, many viral diseases can be vaccinated against, including poliovirus, influenza, rabies, rubella, yellow fever, measles, mumps and chickenpox.

Fungal infections

Fungal infections are diseases caused by the growth of fungi in or on the body. Fungal infections are usually mild, normally involving only the skin, hair, nails or other superficial sites, and they usually clear up spontaneously. Such infections include athlete's foot and ringworm. In someone with a damaged immune system, fungi may invade the internal organs of the body and may cause serious disease.

Fungal disease can usually be treated with antifungal drugs, which are administered intravenously, orally,or applied to the skin.

Antibodies

An antibody forms part of the body's defense against infection. When an invader (antigen) such as a virus or bacterium enters the body during an infection, specialized white blood cells known as lymphocytes react by making proteins called antibodies. These combine with the invader and neutralize it. The presence of antibodies indicates past exposure to a disease. Many blood tests for diseases work by identifying antibodies in the blood.

Antibodies can be created artificially in the body by immunization. This involves exposing the body to a weakened or killed form of virus or other invader. It causes the body to manufacture antibodies, so that if later exposed to the real disease, it can launch a prompt and effective immune response.

Opportunistic infection

Infections that take advantage of a weakness in the immune defenses are called "opportunistic." They are common in HIV/AIDS, cancers, blood disease such as leukemia, bone marrow disease and aplastic anemia.

Examples of opportunistic infections include candidiasis (thrush)—a fungal infection of the mouth, throat, or vagina; cytomegalovirus, a viral infection that causes eye disease that can lead to blindness; herpes simplex, viruses which can cause oral herpes (cold sores) or genital herpes; *Pneumocystis carinii*, which can cause a fatal pneumonia; and toxoplasmosis, a brain infection.

Treatment involves correcting (if possible) the underlying condition and treating the invading organism with antibiotic, antifungal or antiviral medications, usually by intravenous injection in an isolation ward in hospital. In the absence of a normal immune system, opportunistic infections can often prove to be fatal.

Sexually transmitted diseases

A sexually transmitted disease (STD) is any disease transmitted from one person to another through sexual contact. Such diseases may also be transferred from mother to child before, during, or immediately after birth, or through kissing, tainted blood transfusions or the use of unsterilized hypodermic syringes. The diseases may be bacterial, viral or parasitic in origin. STDs have been around for all of recorded history and up until the end of the twentieth century they were known as venereal diseases.

The annual rates of reported new cases of STDs in the USA are the highest of any country in the industrialized world, higher than in some developing countries. Chlamydia is the most frequently reported infectious disease in the country.

Food poisoning

Food poisoning is an acute gastroenteritis caused by eating food which is contaminated or poisonous. Though not common in the Western world thanks to health regulations governing food vendors, it still occurs when food preparation is poor or when food is reheated or partly refrigerated. Food poisoning can be caused by bacteria (*Salmonella*, *Shigella* or *E. coli*, for example) which survive in poorly cooked or unrefrigerated meats.

The bacteria are swallowed and produce toxins which affect the gut. *Staphylococcus* bacteria, for example, can be transmitted from someone who has a boil, abscess or any other infection. These bacteria produce a toxin which may usually survive the entire process of cooking food.

One to eight hours after ingesting the contaminated food, the patient experiences nausea, followed by vomiting, abdominal cramping, and sometimes diarrhea, along with general symptoms such as fever and chills, weakness and headache. Other people who ate the same food may be similarly affected. Children, the elderly, and those with poor immune systems (for example, with HIV infection) are worst affected.

Fortunately, food poisoning is rarely fatal (with the exception of botulism) and recovery usually takes place after about 6–24 hours. Affected people should avoid dehydration by drinking electrolyte solutions to replace fluids lost by diarrhea. Those unable to take oral fluids due to nausea, and young children who can dehydrate very rapidly, may need intravenous fluids administered in hospital.

Food poisoning is best prevented in the first place by storing and preparing food carefully and cooking it thoroughly.

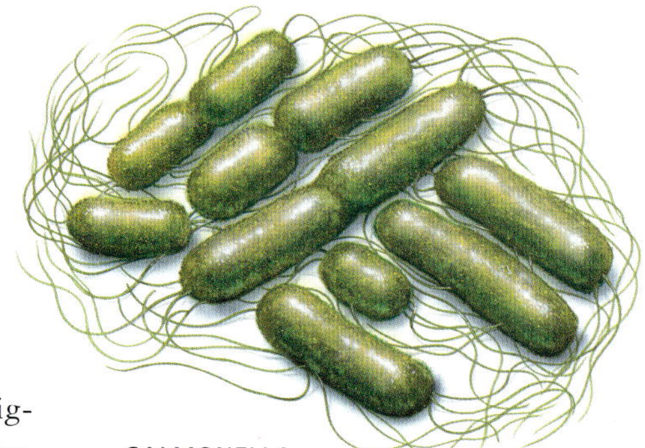
SALMONELLA

SHIGELLA

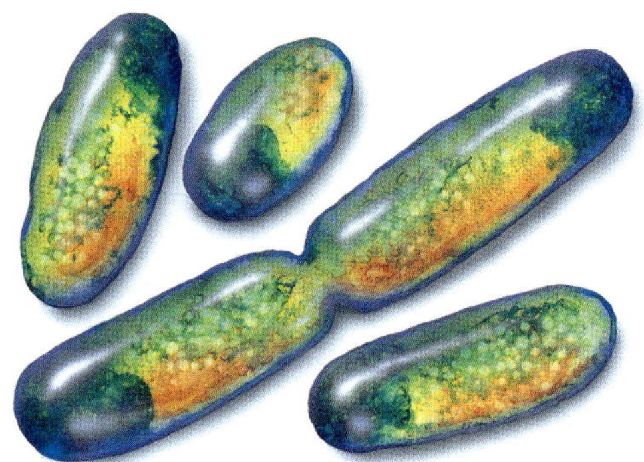
E. COLI

Food poisoning

Bacteria such as these can cause food poisoning, producing vomiting, abdominal pain, headache and diarrhea.

Dysentery

Dysentery is an inflammatory disease of the large bowel, common in tropical areas where living conditions are crowded and sanitation is poor. In areas of Africa, Latin America, Southeast Asia and India, it is endemic.

In humans there are two main forms: bacterial dysentery (caused by the bacterium *Shigella sonnei*), and amebic dysentery (caused by an ameba, *Entamoeba histolytica*). Both forms are transmitted by fecally contaminated drinking water or food, or hand to hand contact. Amebic dysentery is also spread by flies and cockroaches.

Both forms of dysentery cause severe diarrhea often with blood and mucus in the stools, abdominal pain and sometimes contracting spasms of the anus with a persistent desire to empty the bowels. The infection may spread through the blood to the liver, lungs, brain or other organs.

Shigella infections are mild and usually curable in a week or so with antibiotics such as ampicillin or norfloxacin. However, in a severe attack, excessive dehydration can be fatal (especially in infants and young children); serious cases require hospital care and intravenous fluid supplements.

Amebic dysentery causes attacks that come and go for months before the diagnosis is made, and which may be complicated by abscesses, particularly liver abscesses. Drugs such as metronidazole or idoquinol are usually successful in treating the condition. In tropical areas where food or water may be contaminated, one should avoid eating uncooked foods and ensure that foods are hygienically prepared. Drinking water should be boiled and foods covered to prevent flies from contaminating them.

Septicemia

Septicemia (blood poisoning) is an infection of the bloodstream, which can occur directly or as a complication of infection at another site. It is a serious illness resulting in high fever and often violent shaking, called rigors. Infective organisms are more likely to gain access to the blood in people with decreased immunity or in hospitalized patients with intravenous catheters or undergoing invasive procedures. The effects of septicemia are due to the combination of bacterial toxins and the body's immune response to the infection. The elderly may fail to mount a high fever.

Treatment is with prompt introduction of intravenous antibiotics and supportive therapy in hospital.

Viral diseases

Viruses are a group of infectious organisms so minute that they can only be visible through electron microscopes. Just about all viruses contain enough genetic material to duplicate themselves. They are much smaller than bacteria and cannot provide their own energy, nor can they replicate themselves outside living cells. All viruses rely on another invading organism to survive, taking over its cellular machinery and using it to reproduce.

Viruses infect all body tissues, but individual viruses show a preference for particular parts; for example, the poliomyelitis virus only infects part of the nervous system, the herpes virus infects the skin.

Some viruses cause acute disease lasting for only a short duration and others cause recurring or chronic disease, while others

Viral attack at cell level

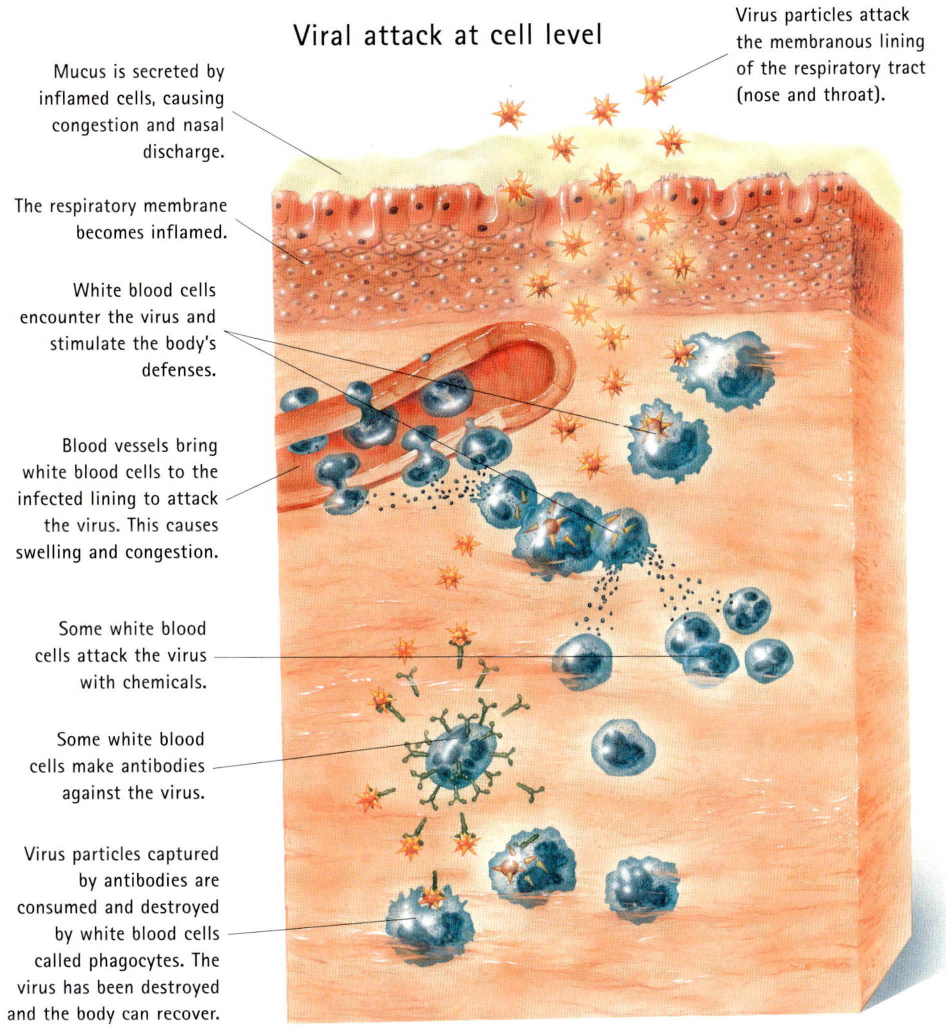

VIRUSES

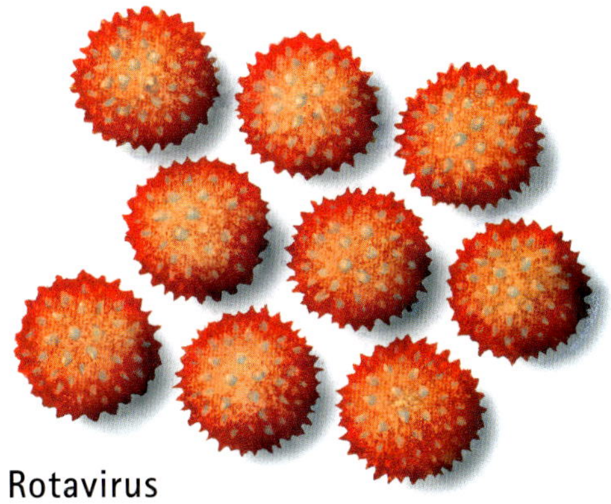
Rotavirus

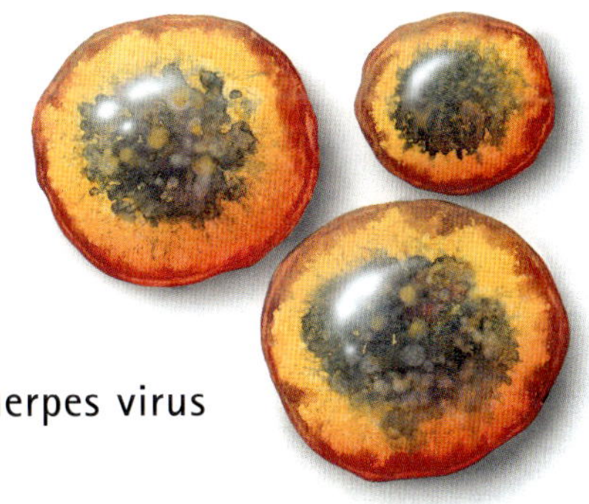
Herpes virus

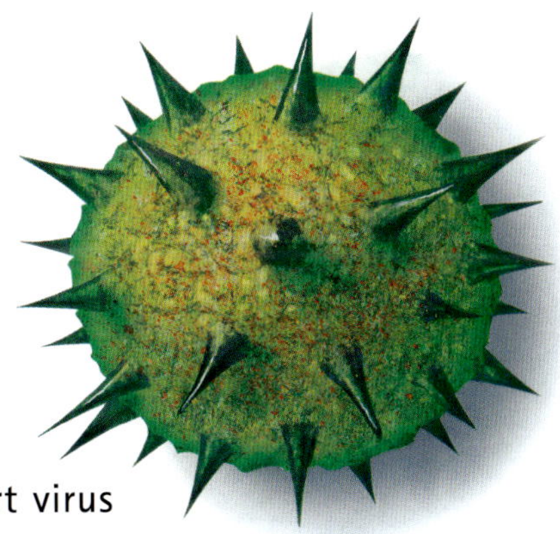
Wart virus

Polio virus

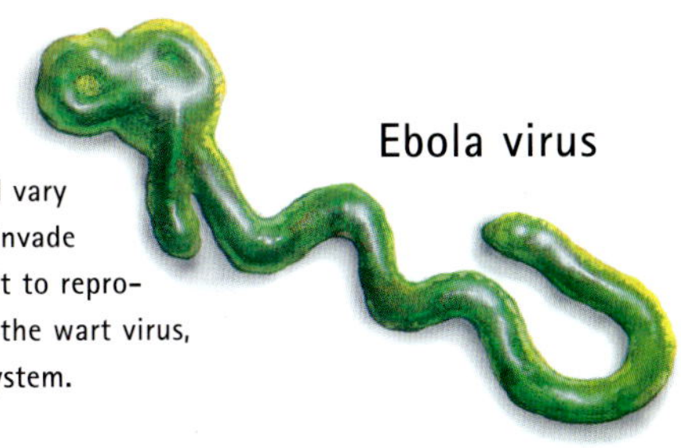
Ebola virus

do not cause any disease. The acute viral infections are of two types, local and systemic, as the result of the effect of the invading virus on the host.

Local infections occur at the site of the viral infection, such as the common cold, which infects the area around the nose, or enteritis, which causes bowel inflammation. Many viruses enter the body via the nose or mouth and begin their cycle of infection in the nose and throat. They then enter the bloodstream where they are spread to other parts of the body as in, for example, measles, mumps and chickenpox.

Other viral diseases are transmitted by the bites of insects, ticks and mites. These diseases begin in the skin or lymph nodes and spread rapidly into the bloodstream.

Some viruses remain in the tissues after the initial infection even though there are specific antibodies for them circulating in the blood and tissues. It is thought that these viruses reside inside cells where they are protected from antibodies, which are unable to penetrate the cell membrane. Measles and herpes viruses fit into this category. Other viruses can remain in the body for years before producing any symptoms.

Once a virus has entered the body it will find little resistance, apart from the presence of lymphocytes, a type of white blood cell that produces antibodies, and a small amount of interferon, which also helps to destroy viruses. After a few days the body begins to produce antibodies and greater amounts of interferon.

Because viruses are so intimately involved in the vital processes of cells, they are difficult to eradicate with medication without damaging the cells, even though there are a few antiviral drugs available. Antibiotics

Viruses

These tiny infectious organisms are much smaller than bacteria and vary considerably in shape and structure. In order to survive they must invade another cell, taking over their host's cellular machinery and using it to reproduce. Each virus has a preference for a different part of the body—the wart virus, for example, infects the skin; the polio virus attacks the nervous system.

are not effective against viruses because they work on the elements which are found in bacteria and not in viruses.

Many viral diseases can be prevented by good hygiene. This means maintaining efficient sanitation and waste disposal combined with personal cleanliness and the use of clean water. Immunization by vaccine can prevent epidemics caused by certain acutely infectious viruses and has been particularly effective against viruses such as smallpox and poliomyelitis.

SEE ALSO *Antivirals on page 446; Blood on page 82; Blood tests on page 436; Cells on page 16; Immunization on page 447; Inflammation on page 24; Lymphatic/Immune system on page 55; Skin tests on page 438*

Common cold

The common cold can be caused by one of five viral families that, between them, encompass a couple of hundred unique viral strains. Most typical of these are the rhinoviruses and coronaviruses, which affect the upper respiratory tract. Secondary infections may occur in the eye or middle ear, particularly in children. Adults may also suffer from inflamed sinuses. The main difference between the common cold and other respiratory infections, including the flu, is the absence of fever (except in children), as well as the general mildness of the symptoms.

Because the viral strains are sufficiently different from one another it is possible to catch one and later be infected by another. The cold is spread by contact between people, which is thought to be the reason why colds are more prevalent in winter when people spend more time indoors and in contact with each other. Colds are transmitted by droplets breathed, coughed or sneezed onto another person. The incubation period is short—between one and four days.

First symptoms can be a sore throat, feeling tired, nasal discharge and/or aching muscles followed by sneezing, coughing, headaches, a chill and nasal discharge. Cold symptoms will vary from person to person, but will usually take from seven to ten days from start to finish.

Treatment consists of easing the symptoms; plenty of fluids and acetaminophen (paracetamol) or ibuprofen may help. Over-the-counter preparations are plentiful for treatment of cold symptoms. Children should never be given aspirin because of the possibility of Reye's syndrome, which can prove to be fatal.

Antibiotics are of no value against a virus, though they may be prescribed for an infectious complication.

Influenza

This viral disease is quite distinct from the common cold and other upper respiratory infections that are often incorrectly referred to as "the flu," although the symptoms of influenza are varied and can resemble a severe common cold.

Influenza is transmitted by airborne droplet infection and occurs more commonly in winter. The virus is remarkable because of the frequency with which its outer coat proteins change. Since immunity to viral infections depends on the binding of antibodies to such proteins, the immune system of a previously infected person cannot recognize the influenza virus with a new outer coat and thus infection can recur.

Relatively minor variations in the recognizable surface proteins (antigens) of the influenza virus occur almost yearly, producing new infective strains. Major antigenic shifts occur less frequently, but effectively produce completely new viruses, leading to large-scale epidemics of influenza that can spread worldwide (known as pandemics).

Influenza typically causes acute onset of fever with chills, headaches, aching muscles and extreme tiredness. There may also be a dry cough, a sore throat and loss of appetite. Fever usually reaches 100–104°F (38–40°C) and persists for 3 or 4 days. Other symptoms last for 1 to 2 weeks. Although influenza is usually self-limiting, the patient usually feels most unwell and the illness is responsible for considerable time lost from work and school.

What makes influenza potentially dangerous are the major complications that are associated with this infection. Persons with chronic lung disease may suffer exacerbations of their condition when they develop influenza. The virus can affect the lungs, producing a pneumonia with severe breathlessness, which may develop in patients of any age.

A complicating bacterial infection, leading to typical pneumonia with cough and sputum production, is usually more common in older persons. Pneumonia associated with influenza can be fatal, especially during major outbreaks.

It is important for persons at high risk of developing influenza to be immunized. High-risk groups include individuals aged 65 or more, persons with chronic heart or lung disease, diabetics and immunosuppressed persons. Immunization is also recommended for persons in nursing homes because of the likelihood of transmission of infection in such an environment. It is essential that individuals be reimmunized each year with the vaccine developed against the most recent infective strains.

Avian Influenza

Avian influenza is a contagious viral infection that can affect all species of birds. On rare occasions, however, it has spread to human populations. It is also called bird flu. There are 15 types of avian influenza virus. The virus spreads through bird feces and contaminated water or dust.

The strain that causes the greatest number of deaths is called Highly Pathogenic Avian Influenza (HPAI). This strain was first recognized in Hong Kong in 1997 after 18 people were infected, with 6 cases resulting in death. During this outbreak millions of poultry were slaughtered —this containment measure effectively confined the virus to Hong Kong.

The symptoms of avian influenza in humans resemble other forms of influenza, including fever, sore throat, cough, headache and muscle aches and pains.

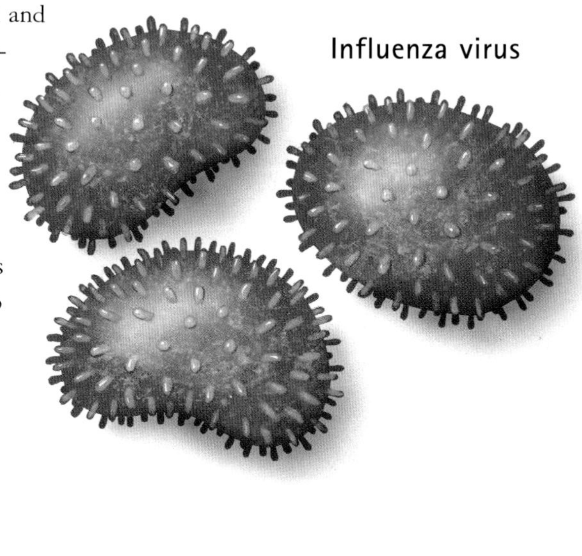

Influenza virus

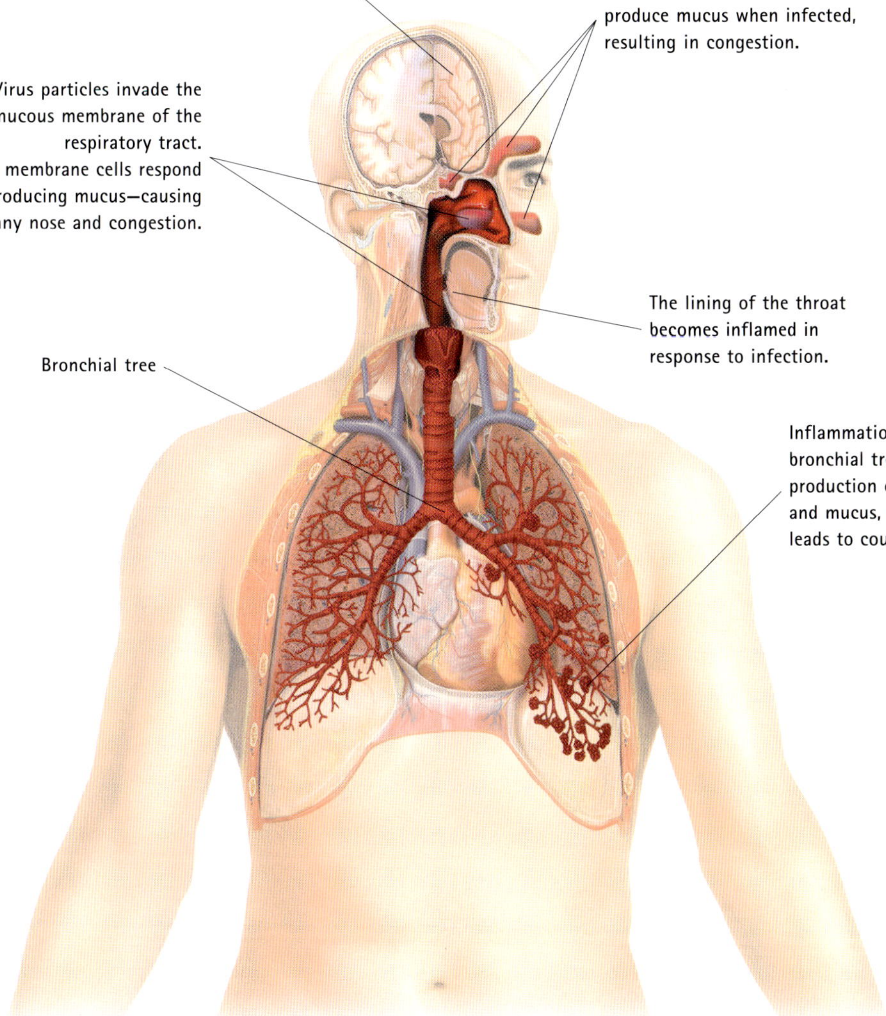

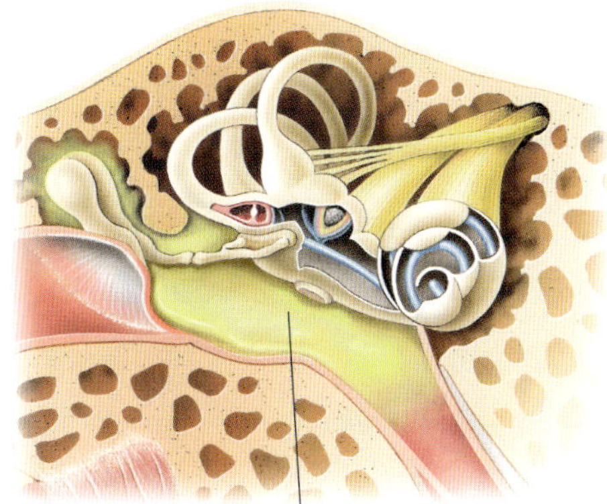

Ear infection

Virus particles traveling into the middle ear cause infection, leading to swelling and accumulation of fluid. This causes earache.

COLD

The common cold is caused by one of many viruses. Millions of cold viruses are easily transmitted via infected droplets that are coughed or sneezed into the air. When a droplet is inhaled, the virus attacks the lining of the upper respiratory tract, causing cold symptoms to develop.

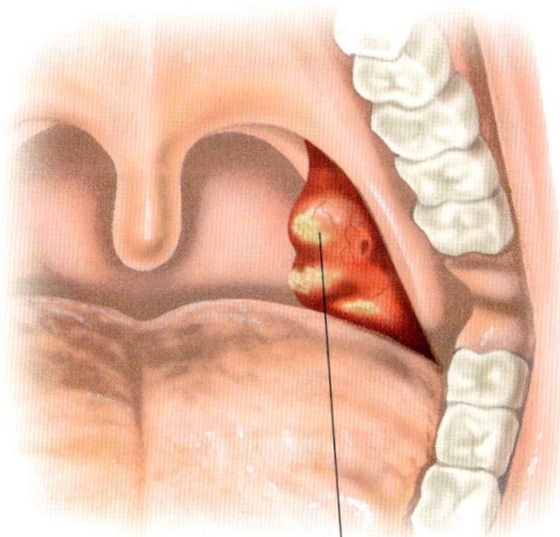

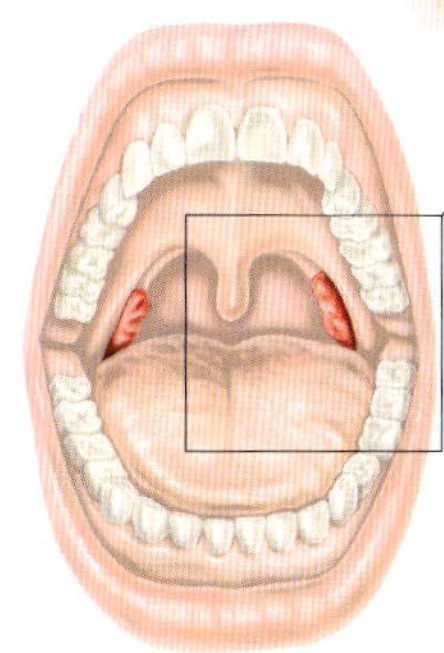

THROAT AND MOUTH

Tonsillitis

The tonsils protect the membrane of the mouth and throat from invading cold viruses. They become swollen and inflamed as part of the body's defense system to stop infection moving from the exterior to the interior of the body.

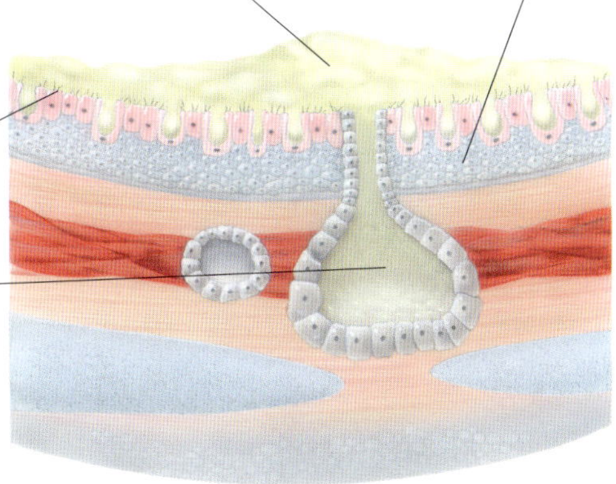

Lungs and bronchial tree

The cold virus attacks the tiny hairs (cilia) and cells on the lining of the bronchial tree in severe cases of cold. The tissue swells and glands produce mucus, resulting in coughing.

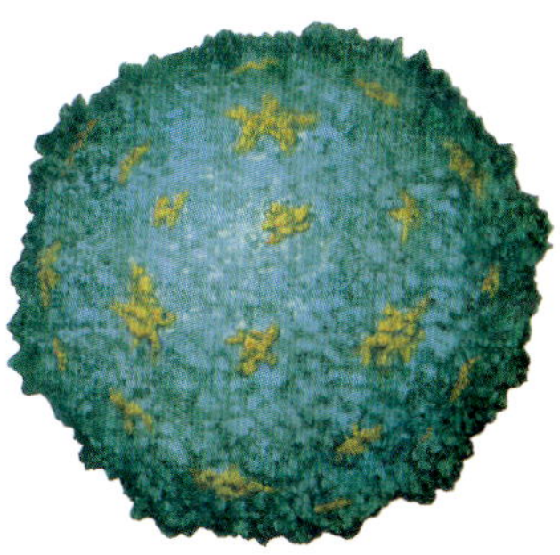

Cold virus

No vaccine is currently available for avian influenza, although researchers are currently working to develop one.

Croup

Difficulty inhaling, combined with making a noise like a barking seal (known as stridor), is the most obvious symptom of croup. It is usually caused by viral infections and in the early stages is contagious. If the child is very distressed, sits bolt upright and has a high fever, emergency medical attention should be sought. Usually associated with a cold, croup is normally worse at night.

If the child has difficulty breathing and is distressed, then medical attention should be sought. The previously recommended treatment of a session in a steamy room, such as a bathroom, has recently been shown to be of no benefit. The introduction of oral corticosteroids to treat croup has been a major change in management in recent years and is the recommended treatment in moderate to severe croup.

Severe acute respiratory syndrome

In early 2003 the new infectious disease severe acute respiratory syndrome (SARS) swept across the world. The syndrome, which arose in Guangdong Province in southern China in late 2002, spread to 29 countries. Within only a few months of the outbreak, the novel coronavirus responsible for SARS was identified and tests are now available to confirm the infection. By July 2003, when the epidemic was eventually brought under control, more than 8,000 people had been infected and over 800 people had died. Health screening of outgoing passengers from SARS affected countries and incoming passengers from SARS affected countries greatly assisted in containing the epidemic.

The SARS virus is extremely contagious and is thought to be spread either by direct contact or droplet spread. The clinical symptoms of SARS are non-specific and are similar to that of most serious atypical respiratory infections.

Symptoms include fever, lethargy, dry cough, headache, sore throat, myalgia (muscle pain) and progressive breathlessness.

The current treatment for SARS is supportive. More than 20 percent of patients develop progressive respiratory failure requiring ventilatory support. No vaccine is currently available.

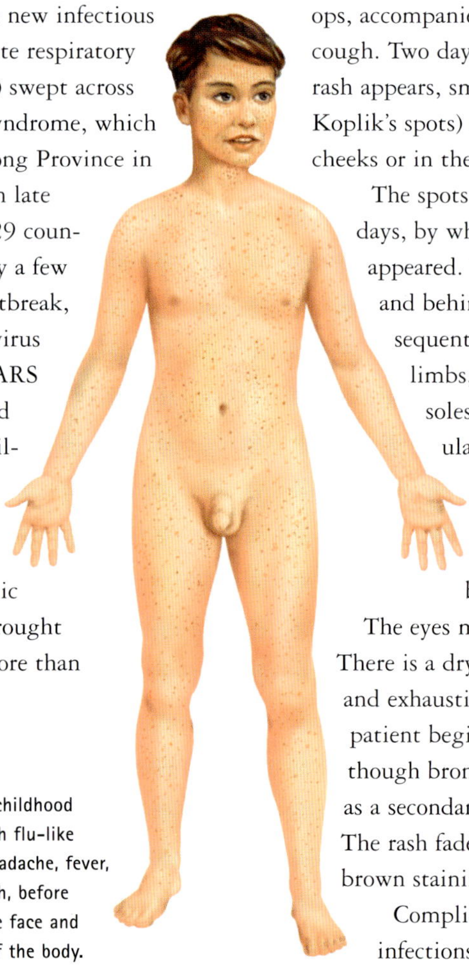

Measles

Measles is usually a childhood disease. It begins with flu-like symptoms such as headache, fever, runny nose and cough, before a rash appears on the face and spreads to the rest of the body.

Measles

An extremely infectious viral disease, measles is spread by droplet infection. It is normally a childhood illness, but adults can also contract it. The incubation period is 10–14 days. Initially a high fever develops, accompanied by a runny nose and dry cough. Two days before the characteristic rash appears, small white spots (called Koplik's spots) may be seen inside the cheeks or in the region of the back teeth.

The spots fade within two or three days, by which time the rash will have appeared. The rash begins on the face and behind the ears, then spreads sequentially onto the body and limbs, including the palms and soles. The rash tends to be irregular and in patches.

The lymph glands enlarge generally, while the eyes become inflamed, bleary and sensitive to light. The eyes may discharge secretions. There is a dry cough, unproductive and exhausting. Within a week the patient begins to improve rapidly, though bronchitis, frequently appearing as a secondary infection, may persist. The rash fades but can leave temporary brown staining and may flake slightly.

Complications include middle ear infections, bronchitis, pneumonia and, most seriously, encephalitis or meningitis. If encephalitis occurs, usually three to seven days after the rash begins, the patient may sink into coma, have convulsions and vomit. One in five people with measles encephalitis can die. Those who survive may remain epileptic or become retarded.

Beyond general nursing measures and controlling extreme fevers, little can be done to treat measles. Antibiotics may be of value if a middle ear infection or pneumonia occurs (as happens in 15 percent of cases). Vitamin A may be beneficial.

Since the introduction of immunization with an attenuated live measles virus in infancy (12–15 months), the periodic epidemics seen earlier are now rare. A booster dose is recommended at the age of 10–12, while people over 20 traveling into developing countries are advised to have a further booster. When immunized the child may develop a mild fever, slight cough and even a transient rash. However, this is not infectious. The measles vaccine is now combined with mumps and rubella (German measles) vaccine (MMR). The vaccine, because it contains a live virus, should not be given in pregnancy or immune deficiency.

Rubella

Rubella (German measles) is caused by a virus which is transmitted by droplet inhalation. The incubation period for rubella is 2–3 weeks. Highly infectious, rubella is usually a mild illness lasting days only, but may be occasionally complicated by arthritis. If an expectant mother develops rubella during early pregnancy there is a high risk of congenital abnormalities developing in the fetus.

In childhood the condition usually begins with slight irritability and minor but tender enlargement of lymph glands in the back of the neck and head. A transient spotty rash will usually appear initially on the face and behind the ears, and then spread rapidly over the body. Occasionally the rash is noted when a child is being bathed in warm water, yet will have faded by the time the child is seen by the doctor and not recur. A few spots may be seen on the palate at the same time. There may be mild fever but the child is usually only minimally ill, with a runny nose. Occasionally it is noted that

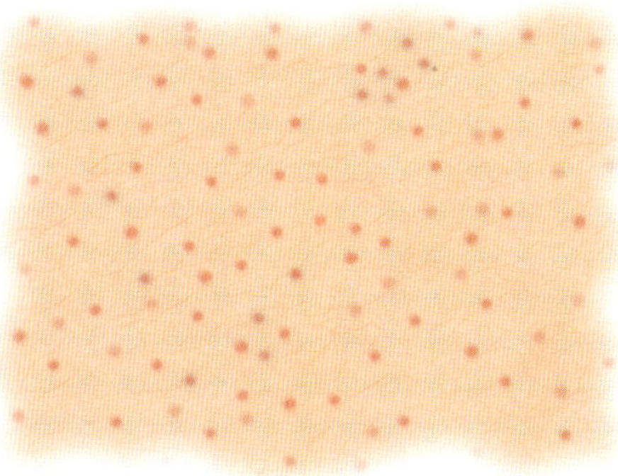

Rubella

Also known as German measles, rubella is similar to but less contagious than measles. The rash begins on the face and scalp, and then quickly spreads to the body and arms.

a child develops bruising due to a transient shortage of platelet cells in the blood.

Adults often, and children rarely, develop various painful joints, particularly in the fingers. A very low percentage of patients develop an encephalopathy where there is inflammation and damage to the brain. Mortality in that instance may be 20 percent but recovery otherwise is usually complete.

If a woman is planning a pregnancy she is advised to have her rubella antibody level checked well beforehand. If there are adequate antibodies present, then there is no problem; if antibodies are absent, she is advised to be vaccinated at least three months before starting on a pregnancy.

If the embryo is infected before 14 weeks, a variety of different problems can occur, ranging from miscarriage to congenital heart defects, cataracts, glaucoma and deafness. The brain may not develop adequately and retardation can occur, even if the developing infant is past 14 weeks.

Rubella vaccination, usually combined with measles and mumps vaccines, is given initially at the age of 12–15 months. A second vaccination may be given in the early teens. Arthritis is an occasional complication and may occur with older children and women but otherwise there are few complications following immunization.

Mumps in childhood

Mumps is usually a childhood disease and symptoms are often more severe if it is contracted by adults. Immunization against mumps, measles and rubella is generally given at the age 12-15 months.

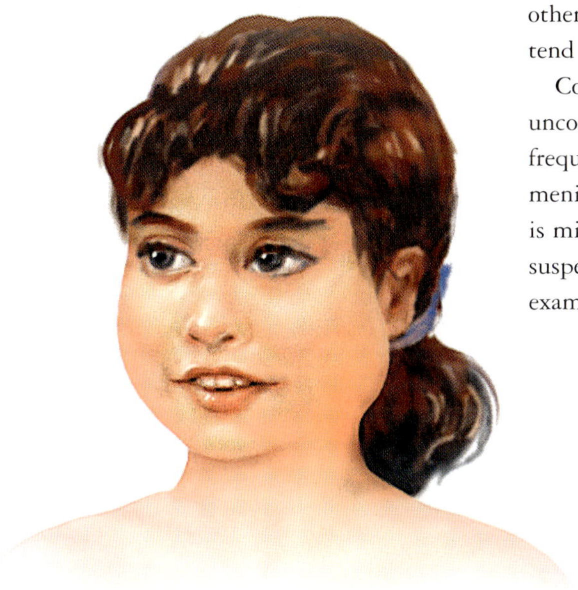

Mumps

In its simplest form, mumps is a viral illness of childhood due to a paramyxovirus which produces a mild febrile condition lasting a few days and characterized by painful swelling of the salivary glands that lie under and in front of the ears. After a 2–3 week incubation period, the patient becomes infectious a day or so before the swelling occurs and for 3 or 4 days thereafter. The illness is spread by droplets transmitted by coughing or breathing. Occasionally one side of the neck only will swell. The other side may swell up several days later. Adults usually suffer more than children. Occasionally the other salivary glands will also tend to enlarge.

Complications are not uncommon, the most frequent being aseptic meningitis, which is mild and often not suspected unless an examination of the fluid around the brain and spinal cord (cerebrospinal fluid) is carried out. Twenty-five percent of adult males can develop swelling of one or both testes, occasionally leading to sterility. Similar painful swelling of the ovaries occurs in adult women. Inflammation of the pancreas in the abdomen can produce severe upper abdominal pain, sometimes with nausea and vomiting.

Far less common complications include encephalitis (often with high fever and disorientation), inflammation of the thyroid gland, inflammation of the heart muscle (myocarditis), arthritis, kidney inflammation (nephritis) and thrombocytopenia, a condition in which the platelet cells in the blood decrease in number, leading to otherwise unexplained bruising. Deafness can also be a complication.

As far as it is known, the fetus is not affected. Mortality is minimal, most deaths resulting from encephalitis.

Treatment for mumps is largely simple nursing in bed. Testicular swelling (orchitis) may need surgical intervention, although high doses of hydrocortisone may help. The pain is often lessened if the scrotum is suspended in a scrotal support.

Live virus vaccine is easily available and is usually given in conjunction with live measles and rubella vaccines, at the age of 12–15 months. Complications of immunization are rare but can include mild fever and minor swelling of one or both salivary glands. Even so the child is not infectious.

A booster dose is recommended when the child is about 14 years of age. The vaccine should not be given to people who lack immunity (such as HIV infected people or cancer patients undergoing treatment), or to women who are pregnant.

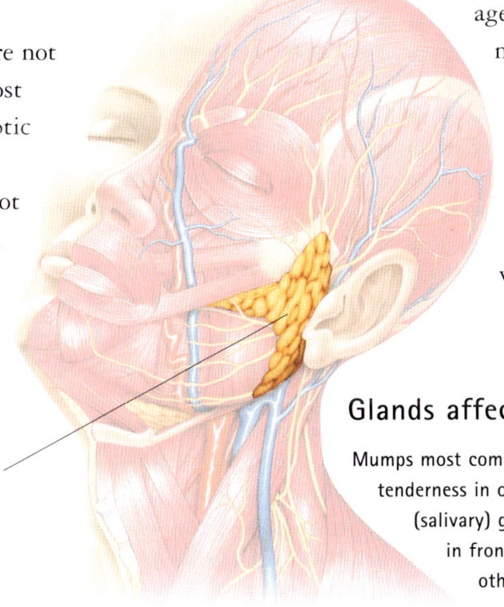

Glands affected by mumps

Mumps most commonly causes swelling and tenderness in one or both of the parotid (salivary) glands that lie just under and in front of the ears. Occasionally the other salivary glands, testes and ovaries may also be affected.

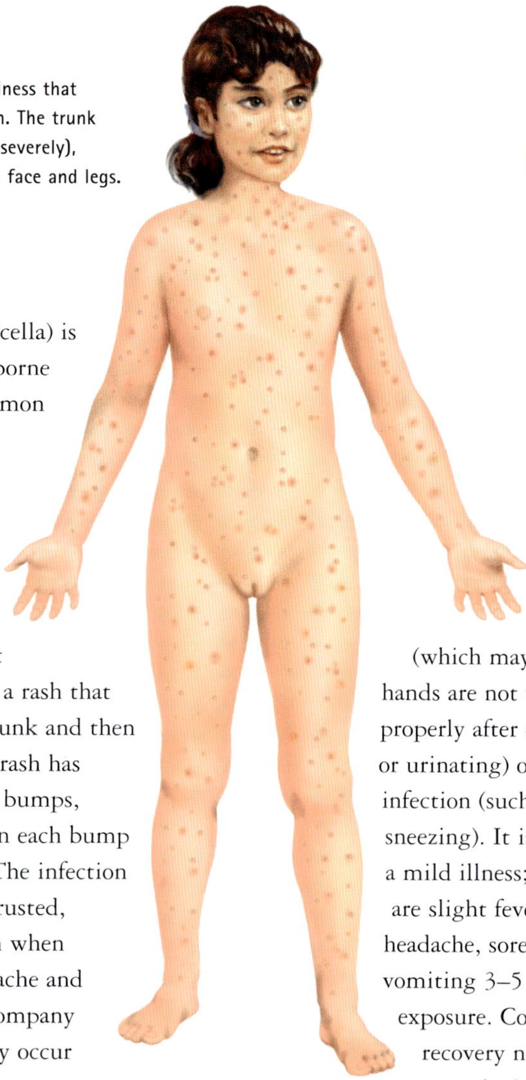

Chickenpox areas

Chickenpox is a contagious viral illness that produces a characteristic itchy rash. The trunk is usually affected first (and most severely), and spots then spread to the arms, face and legs.

Chickenpox

Chickenpox (also called varicella) is a highly contagious and airborne viral disease. It is most common among school-age children, though it can occur at any age, and one attack usually protects a person for life, unless it is very mild.

The main symptom, which usually appears about 13–17 days after contact, is a rash that is at first apparent on the trunk and then spreads over the body. This rash has three stages: little red itchy bumps, followed by a clear blister on each bump and finally crusts or scabs. The infection lasts until all bumps have crusted, which takes 7–10 days from when the rash first appears. Headache and cold symptoms can also accompany chickenpox but these usually occur before the rash appears.

Treatment revolves around relieving the itching, as scratching of the scabs can lead to life-long scarring. A lukewarm bath with cornflour or an oatmeal preparation added can provide relief. Calamine lotion can also be applied and the fingernails should be kept very short. Some children may need antihistamines, but these should only be taken on medical advice. Acetaminophen (paracetamol) may reduce fever, but aspirin must never be given as it can lead to the complication of Reye's syndrome.

Anyone who is pregnant, has a chronic illness or weak immune system, should seek medical advice if they have been in contact with someone with chickenpox. After chickenpox clears, the virus which causes the disease may lie dormant and emerge later to cause shingles.

Poliomyelitis

Poliomyelitis (commonly known as polio) is a viral illness, usually affecting children. It is spread from the infected individual to others by fecal-oral infection (which may occur if hands are not washed properly after defecating or urinating) or droplet infection (such as with sneezing). It is usually a mild illness; symptoms are slight fever, malaise, headache, sore throat, and vomiting 3–5 days after exposure. Complete recovery normally occurs in 24–72 hours. In about 10 percent of cases poliomyelitis causes inflammation of the spinal cord and of the brain stem. Symptoms are fever, severe headache, stiff neck and muscle pain. In some of these cases, this may progress to weakness or paralysis of muscle groups, causing difficulty swallowing and breathing, and paralysis of the muscles of the legs and lower torso.

The condition is diagnosed by identifying the virus in the cerebrospinal fluid, in the throat or in feces. Treatment involves physical therapy to aid muscle function. In the most severe cases, a tracheostomy (cutting an opening in the windpipe to insert a breathing tube) and an artificial respirator may be necessary.

The disease can be prevented by immunization. Polio vaccine may be given by injection, or by mouth, at the ages of 2 months, 4 months, 6–18 months and 4–6 years. The development of polio vaccines has almost eliminated the disease in industrialized countries.

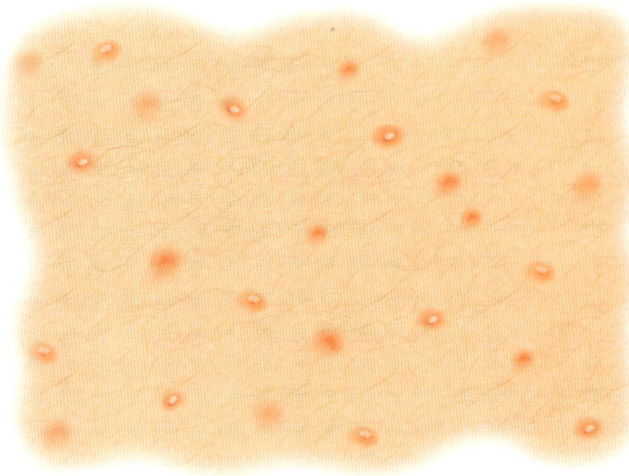

Chickenpox

The chickenpox rash usually begins as small flat spots, which then become raised and form fluid-filled blisters. These then crust over to form scabs. The virus is contagious until all spots have crusted over.

Herpes simplex

The family of herpes viruses includes the herpes simplex viruses (two closely related types) as well as the viruses responsible for chickenpox and for infectious mononucleosis (glandular fever). All these viruses may produce symptoms at the time of initial

Poliomyelitis

The success of vaccines against the polio virus has greatly reduced the incidence of poliomyelitis. Jonas Salk developed the first effective vaccine, which has been largely superseded by an oral vaccine developed by Albert Sabin.

infection, but in 50 percent of people they persist indefinitely in a dormant form in the sensory nerve cells of an infected person, with later reactivation and recurrent disease.

Herpes simplex virus types 1 and 2 (HSV-1 and HSV-2) infect the skin and mucous membranes. Transmission requires close personal contact, but the initial infection is often inapparent. Both HSV-1 and HSV-2 infect the cells of the nerves that supply the infected area and persist in these cells. Reactivation of herpes simplex virus infections may be triggered by stress, menstruation, exposure to sunlight or by other illnesses. Severe disease may develop if reactivation follows suppression of the immune response.

Most initial infection by HSV-1 occurs in children and is symptomless, although it may also be associated with fever, a sore throat, or ulceration of the mouth. Because infection with HSV-1 usually involves the skin and mucous membrane of the mouth, recurrences are manifested as "cold sores" that typically develop on the lip margins.

On the other hand, HSV-2 is mainly spread by sexual transmission and thus produces genital herpes, although some 10–15 percent of cases of genital herpes are due to HSV-1. The initial infection usually produces small blisters which burst and turn into sores that are often painful or itchy. These involve the genital area, buttocks and thighs, and can be accompanied by fever, headache and an illness similar to flu. Recurrences are associated with similar skin and mucous membrane changes and may be preceded by flu-like symptoms.

Diagnosis of the common forms of herpes virus infection is largely based on medical history and physical examination; in addition, various laboratory tests are available. While there is no treatment available that will eliminate the infection, it can be quite effectively controlled using different anti-viral drugs, which need to be administered as early as possible in the course of the primary infection or episode of recurrence.

Severe forms of infection are uncommon but include inflammation/ulceration of the cornea of the eye (mostly HSV-1); encephalitis (brain inflammation and tissue destruction) (also mostly HSV-1); and disseminated infection in the newborn (mostly HSV-2) and in individuals with HIV (AIDS).

Herpes virus

Once the herpes simplex virus has entered the system, it stays with the person for life.

Cold sores

A cold sore, also known as herpes simplex type 1 or HSV-1, results from a viral infection. It attacks the skin and nervous system, producing small, sometimes painful, fluid-filled blisters around the mouth and nose. After a first infection the virus will continue to live in the nerve system in a dormant state in which it can be reactivated by a trigger. These triggers may include sunlight, physical or emotional stress, hormonal changes, certain foods or drugs. The trigger can also be unknown.

An attack begins with a tingling sensation at the spot where the sore will erupt, followed by a rash, then blisters or spots. These can come in clusters, fill with fluid, rupture and form crusts. Cold sores can take up to three weeks to disappear and are highly contagious until healed. Anyone with a cold sore must be diligent about washing hands and scrubbing fingernails, and avoid kissing and other oral contact. It is also important not to touch the eye after touching a sore as this can cause an infection or corneal ulceration.

While cold sores are unpleasant to look at, they are not a serious risk to general health. Most cold sores will clear up without treatment. In most cases, however, acyclovir, an anti-viral drug, can be prescribed. It is most effective if used at the first signs of a sore. Aromatherapists recommend neat lavender oil.

Once the herpes virus has entered the system it stays with the person for life, though attacks usually diminish and often disappear over a period of time. From 10 to 15 percent of cases of genital herpes are caused by the cold sore virus (herpes simplex type 1) and it is possible to sexually transmit cold sores to the genitals.

SMALLPOX VIRUS

Finding a cure

Homeopathy uses the principle of "like cures like"—the idea that the same substance can both cause and cure an illness. Edward Jenner's development of a vaccine against smallpox in the nineteenth century was based on a similar principle.

Smallpox

Smallpox was formerly an acute and contagious viral infection that was eradicated worldwide by a vaccination campaign launched by the World Health Organization. The program began in 1967, when smallpox caused about 2 million deaths. The last known naturally occurring case of the disease was reported in 1977.

Smallpox was spread by contact, the virus being exhaled or expelled in saliva by the infected person. The infection could cause death before the characteristic skin pustules appeared, or be so minor that symptoms went unnoticed; in those cases the virus could continue to be spread. The virus could also live in bedding, clothing or dust for up to 18 months, although it would not replicate outside the human body. The vaccination program was so successful that it has now generally been discontinued.

Rabies

Rabies is an acute viral disease that affects the nervous system of animals and is transmitted to humans via saliva, commonly after being bitten or licked on broken skin by a rabid animal, such as a bat, dog or cat. The average incubation time is 1–2 months, though it may occasionally be more than a year. It starts with fever, depression, nausea and vomiting. In "furious" rabies the victim becomes highly agitated, with uncontrollable behavior, spasms of the throat muscles, excessive saliva and frothing at the mouth which makes them unable to drink water.

For this reason, rabies was once known as hydrophobia, which means "fear of water." "Dumb" rabies is characterized by sluggishness, weakness and paralysis. Once symptoms appear, death is inevitable, and usually occurs within a week.

Vaccination soon after exposure to infection offers protection. An injection of rabies immune globulin is followed by a course of rabies vaccine over 28 days.

Vaccination does not cancel out the risk of infection but reduces the intensity of the disease in those infected. Possible side effects of vaccination include headache, dizziness, nausea, muscle aches and, less commonly, neurological disorders and paralysis. Family pets should also be vaccinated.

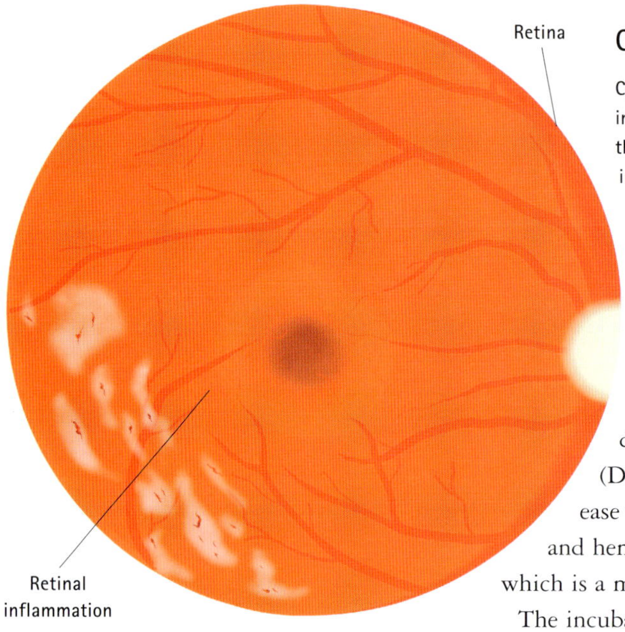

Cytomegalovirus

Cytomegalovirus causes many illnesses, including retinitis, an inflammation of the retina in the eye, which can result in loss of vision if untreated.

Yellow fever

The bite of the mosquito *Aedes aegypti* can transmit the virus infection yellow fever to humans. Common in tropical climates, particularly Africa and South America, yellow fever is an acute infectious disease. There are two different patterns of transmission of the disease—either from person to person via the mosquito or from a mammalian host, often a monkey, to a forest mosquito and from there to a human.

Symptoms appear abruptly. Shivering, a high fever, severe headache, bone pains, dizziness, backache, nausea and vomiting strike suddenly, within three to six days of being infected. The virus destroys liver cells, and jaundice (yellowing of the skin) is common. While the majority of sufferers recover completely, others may become delirious and go into a coma, with death often the result. Those who recover have lifelong immunity. There is no cure and the only treatment is administration of intravenous fluids, anti-nausea medication, kidney dialysis and skilled medical care.

Immunization with an extremely mild yellow fever virus will give protection for 10 years. This vaccine cannot be given to children under the age of one.

Adenovirus

Adenoviruses are a group of DNA viruses that mainly cause diseases of the upper respiratory tract. They most commonly affect infants and children, particularly between autumn and spring, when they cause an acute upper respiratory tract infection, with sore throat, fever and swollen lymph glands in the neck, sometimes with bronchiolitis and pneumonia. They may also cause a condition called pharyngoconjunctivitis (sore throat and fever with inflammation of the conjunctiva) often seen in children on summer camps. Adenoviruses can cause an acute diarrheal illness in children, and in immunosuppressed people, such as those with AIDS, may cause severe pneumonia.

There are no effective treatments for adenoviruses. However the course of the infection is usually mild and the child makes a good recovery. Vaccines for some adenoviruses have been developed but their use is currently limited.

Dengue

Dengue fever is an acute viral infection of the body, transmitted by mosquitoes. It is similar to malaria in that it is a very common cause for hospitalization of children in tropical countries. It is spread to humans by mosquito bites. There are two types of this disease—ordinary dengue fever (DF), which is usually a mild disease with no serious complications, and hemorrhagic dengue fever (HDF), which is a much more virulent form.

The incubation period from the mosquito bite to the appearance of the first symptoms is about 5 to 6 days. In DF, the symptoms are headache, fever, vomiting, muscle pain, joint pain, and enlarged lymph nodes. The fever rises and falls in cycles of 1 to 2 days, and the illness usually lasts about 10 days.

HDF is mostly common in Southeast Asia. This form causes the capillaries in the skin and body organs to rupture, which results in a petechial rash (hemorrhaging in the skin), bleeding from the nose, bowels and kidneys, and can lead to serious and even fatal complications. Treatment for DF is symptomatic—analgesic for the fever, and rest and fluid replacement. There is no vaccine and the most important advice for travelers is to take measures to prevent being bitten by mosquitoes.

Cytomegalovirus

The human cytomegalovirus (CMV), a member of the herpes family of viruses, causes cytomegalic inclusion disease. This is an extremely common virus and some 90 percent of people in the over-70s population have antibodies to CMV in their blood.

Infants commonly acquire the virus from their mother in the uterus, during birth or through breast feeding. It can be spread by close contact later in life and can be reactivated in adults after a period of dormancy.

Most healthy people do not develop any significant symptoms to CMV, though it may a cause a flu-like illness lasting a few weeks. However, people with suppressed immune systems (such as those on immunosuppressive drugs or with AIDS) who become infected with CMV may develop serious diseases such as pneumonia, hepatitis, encephalitis, colitis, and retinitis. They may need to be hospitalized and receive treatment with antiviral drugs.

Infectious mononucleosis

Infectious mononucleosis (also known as glandular fever) is an illness commonly resulting in swollen lymph nodes (also known as lymph glands), fatigue, fever and a sore throat. Often referred to as "the kissing disease," infectious mononucleosis is thought to spread through saliva, through nasal secretions, sexual contact, blood transfusions and respiratory droplets. The disease is called mononucleosis as the blood of sufferers contains unusually large numbers of the white blood cells known as mononuclear leukocytes or monocytes, which are formed in the spleen and bone marrow.

The illness is caused by the Epstein-Barr herpes virus (EBV) which often produces no symptoms in young children and can stay in the body without effect for a long time before being activated by, for instance, a weakening of the immune system due to disease. EBV commonly results in mononucleosis in people aged 15–35 years.

Lethargy is often the first sign of illness. Other symptoms include muscular aches, loss of appetite, enlarged spleen and, occasionally, a faint pink rash. After about ten days acute symptoms subside, but fatigue, a general feeling of discomfort and sometimes depression may continue for up to about three months. Bed rest may cure the illness within six weeks, while painkillers and other non-prescription medications may be used to treat symptoms.

Viral hepatitis

Hepatitis is an inflammation of the liver, which affects the liver's capacity to function effectively.

Most cases are caused by a viral infection, and the symptoms manifested by the disease include general weariness, loss of appetite, fever, vomiting, abdominal pain and jaundice. Some cases of hepatitis are difficult to recognize, but when symptoms are present there may be general weariness, loss of appetite, fever, vomiting, abdominal pain and jaundice (a yellowing of the skin and eyes). Acute viral hepatitis can last from a few days to several weeks, and in some instances it will usually develop into chronic hepatitis. Other hepatitis viruses only present as an acute illness and are not associated with chronic infection.

There are seven known hepatitis viruses, labeled A, B, C, D, E, F and G. Hepatitis A (HAV), the most common form worldwide, is spread through oral-fecal transmission. This virus can also be transmitted in drinking water, or water in contact with food which is infected with raw sewage. Time between exposure and developing symptoms is around 28 days. Treatment is usually bed rest and adequate intake of fluids and most patients recover completely. Hepatitis A does not cause chronic infection. A vaccine against the disease (made from inactivated hepatitis A virus) is available to those considered at risk. This generally provides at least ten years protection.

Hepatitis B (HBV), though not as common as HAV, is becoming increasingly frequent. Spread through blood transfusions, intravenous drug use or unprotected sexual intercourse, this virus remains in the body for many years. Babies born to a mother with HBV have a 95 percent chance of being infected. A more serious disease than hepatitis A, it can become chronic and can lead to permanent liver damage such as cirrhosis or liver cancer. Symptoms appear between 40 days to six months after exposure, and include fatigue, abdominal pain, loss of appetite, nausea, diarrhea, dark urine and jaundice. Treatment is bed rest for acute infection. Treatment for chronic infection includes interferon and the antiviral agent, lamuvidine. A vaccine is available that provides long-lasting protection.

Hepatitis C (HCV) was first identified in the 1980s and initially was known as non-A or non-B. It is spread through blood, most commonly through sharing needles among intravenous drug users, occasionally during sexual acts or from an infected mother to her baby. It can also spread through sharing toothbrushes or razors. Symptoms are rarely acute and may be similar to HBV. Of 100 people infected with HCV about 20 will find the virus has cleared up of its own accord after about six weeks, the other 80 will suffer chronically. Treatment is via interferon formulations in combination with the antiviral agent ribavirin, which is effective in about 30 percent of patients. In many countries blood and sperm donors are screened for HCV. There is no vaccine.

Hepatitis D (HDV), also known as delta agent, is a parasite of HBV using the B virus to survive and can occur only at the same time as HBV. It is transmitted only through infected blood, and has similar symptoms to HBV. Between 70 and 80 percent of those infected will develop cirrhosis. It can be prevented with the same vaccine

Infectious mononucleosis

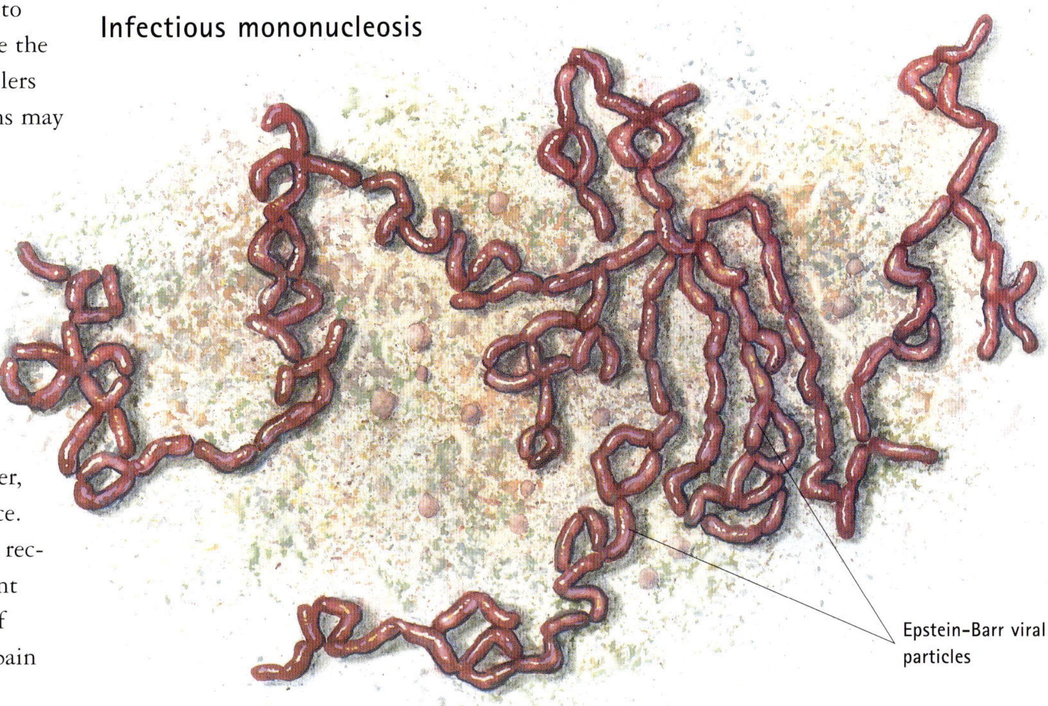

Cirrhotic liver

If it is not treated, chronic hepatitis can lead to cirrhosis of the liver. When this occurs, nodular, fibrous tissue replaces damaged liver cells and connective tissue in the liver, distorting its smooth surface and internal microstructure.

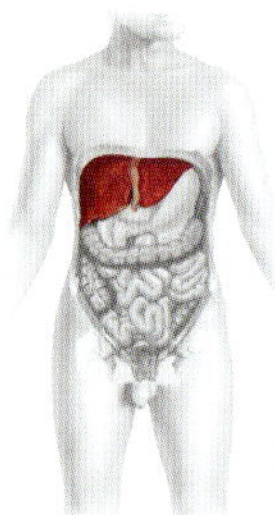

Hepatitis virus

Hepatitis is inflammation of the liver. When caused by a virus, it is called viral or infectious hepatitis. Some types of viral hepatitis can be transmitted in drinking water.

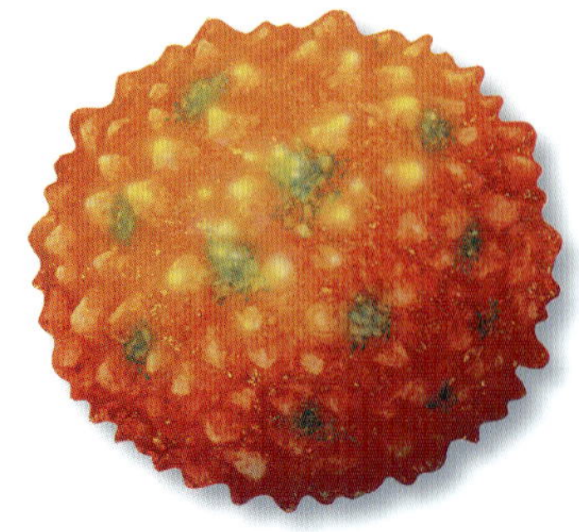

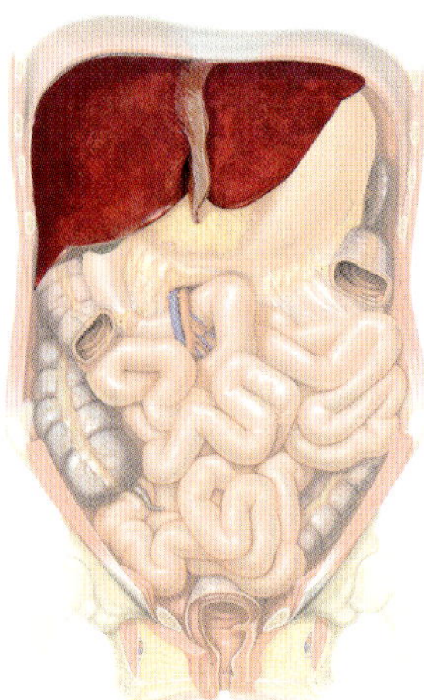

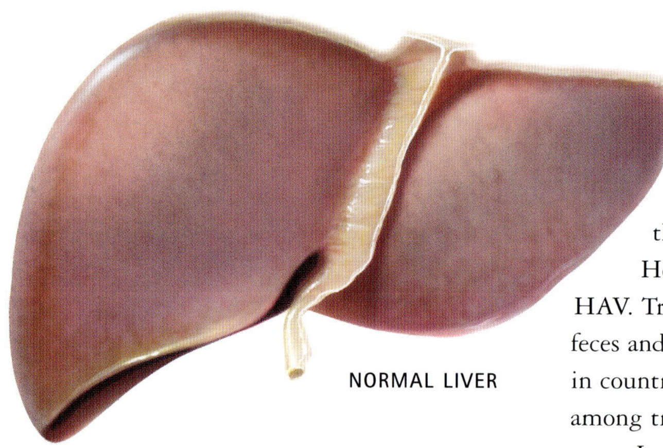

NORMAL LIVER

DAMAGED LIVER MICROSTRUCTURE

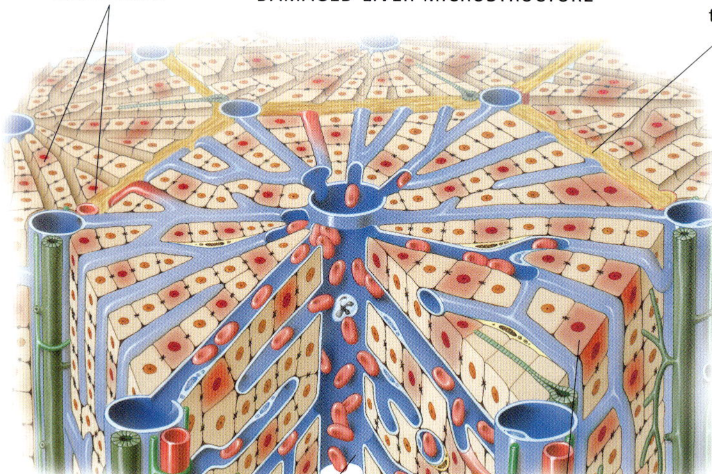

Scarring of the liver microstructure

Chronic hepatitis causes cirrhosis, a disease in which the normal microscopic lobular architecture of the liver is destroyed. Scarring and distortion of the hepatocytes (liver cells) and connective tissue that form each hexagonal lobule can disrupt the flow of blood through the liver.

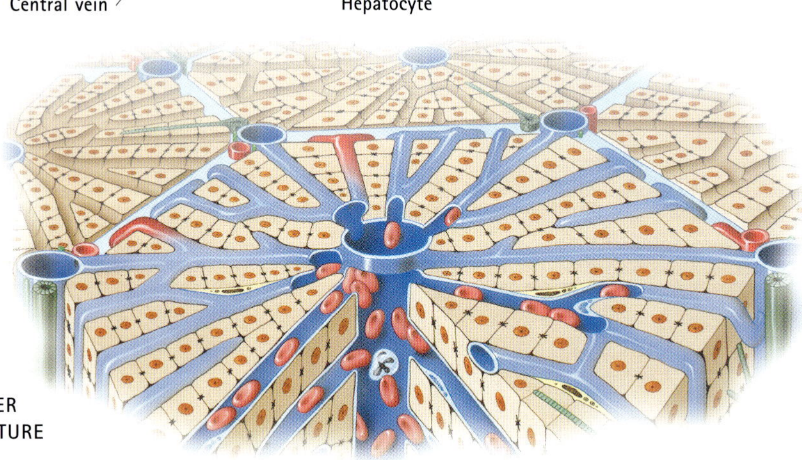

NORMAL LIVER MICROSTRUCTURE

used against HBV and treatment is with alpha interferon, though this is not always effective.

Hepatitis E (HEV) is similar to HAV. Transmitted in the same way, via feces and oral ingestion, it is found mostly in countries where sanitation is poor or among travelers returning from high-risk areas. It is particularly dangerous for pregnant women, in whom infection can prove fatal. The symptoms of fatigue, abdominal pain, loss of appetite, nausea, diarrhea, dark urine and jaundice are similar to HAV. There are few chronic cases and treatment is rest for about a fortnight.

Hepatitis F (HFV), was first reported in 1994 and is spread in the same way as HAV and HEV. Hepatitis G (HGV) is thought to be the cause of large numbers of sexually transmitted and blood-borne cases of hepatitis. Symptoms are not yet fully determined though it does cause both acute and chronic forms of the disease and can infect a person already infected with HCV. There is no vaccine and treatment is rest.

When symptoms associated with hepatitis are present, tests will be conducted on liver function to determine whether the illness is hepatitis or some other problem, such as gallstones, or even cancer. Laboratory tests, including a biopsy, may be necessary.

Anyone traveling to areas where hepatitis is a risk needs to consider vaccination and be aware of the need for good hygiene. People working in high-risk professions (physicians, nurses and dentists) also need to consider vaccination. Immunization may also advised for newborns and adolescents.

HIV

HIV (human immunodeficiency virus) is a retrovirus, one of a unique family of viruses consisting of RNA surrounded by a protein envelope. It attacks a type of white blood cell critical to the immune system known as helper T lymphocytes, or T4 helper cells. This may eventually cripple the immune system and leave the body vulnerable to a variety of life-threatening illnesses that are ordinarily harmless otherwise.

TRANSMISSION

The HIV virus is transmitted, among other ways, through sexual contact (including oral, vaginal and anal sex). It is also transmitted via blood through transfusions, needle sharing or accidental needlestick injury. It is possible for a pregnant woman to pass the virus to the fetus, and a nursing mother can infect her baby through her milk. The infection is not spread by touching and hugging, or by contact with inanimate objects.

High-risk behaviors include promiscuity, especially when involving anal intercourse, and intravenous drug use with shared needles. Others at high-risk include infants born to mothers with HIV, the sexual partners of those exhibiting high-risk behavior and people who received blood transfusions before screening for the virus was introduced (around the mid-1980s).

Contrary to some popular perceptions, AIDS is not a "homosexual disease;" in Africa and other developing regions, in particular, transmission is predominantly through heterosexual contact.

A person may be HIV-positive for many years before developing illnesses that indicate a serious deterioration of the immune system. At that stage, a person is said to have acquired immunodeficiency syndrome (AIDS). There is, at present, no cure for AIDS, but drugs have been developed that suppress replication of HIV virus in the body, and so effectively arrest or stop the progress of the disease.

AIDS

AIDS stands for acquired immune deficiency syndrome. It is caused by the HIV (human immmunodeficiency virus). AIDS is the final and most serious stage of HIV disease. The illness is characterized by severe immune deficiency, leaving the body vulnerable to life-threatening illnesses.

The HIV attacks and destroys certain types of white blood cells called T4 lymphocytes (also known as CD4 or T helper lymphocytes), which are responsible for patrolling the body and destroying foreign invaders. Because the HIV destroys these cells, they are no longer available to fight common bacteria, yeast and viruses which normally would not cause disease.

The body's lowered defenses, then, leave it susceptible to these invaders which cause opportunistic infections.

It is these infections, not the HIV, that eventually cause the death of AIDS sufferers. First recognized in the USA in 1981, the disease has grown rapidly to become one of the world's major health problems. The World Health Organization estimates there are more than 20 million cases of HIV infection worldwide and most of these are in the developing countries of Africa and Asia.

DEVELOPMENT OF THE ILLNESS

The initial illness resembles a mild flu, with fever, headache, fatigue, loss of appetite, swollen lymph nodes and skin rashes. The symptoms appear within two to four weeks of exposure. The illness can then lie dormant for as long as ten years. During this time there may be no symptoms at all.

The sufferer may develop low-grade fever, chronic tiredness and weakness, appetite loss and loss of weight, with swollen lymph nodes especially in the neck, jaw, groin and armpits. Diarrhea, malnutrition and minor infections such as oral thrush are common. This stage is sometimes known as AIDS-related complex, or ARC.

In a small percentage (between 1 and 10 percent) of those infected with HIV, the illness doesn't progress further. But in a majority, immunity levels eventually fall off to below critical levels and infections become more serious and life-threatening. These include *Pneumocystis carinii* pneumonia, toxoplasmosis, tuberculosis and a range of viruses, including cytomegalovirus, herpes simplex virus, varicella zoster and Epstein-Barr viruses.

This stage represents full-blown AIDS; people with full-blown AIDS may die within two years if not treated.

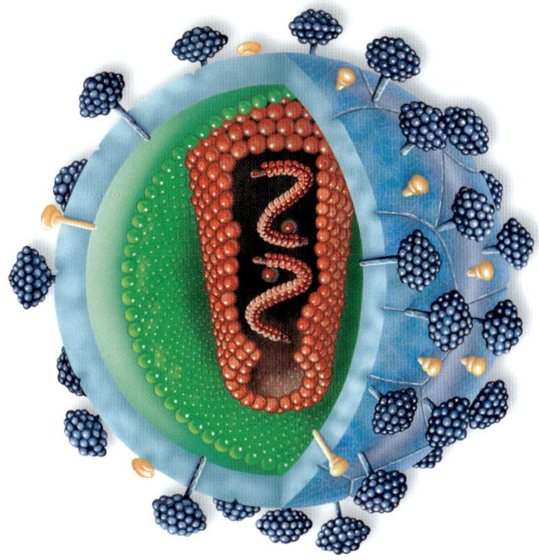

Human immunodeficiency virus (HIV)

DIAGNOSIS

HIV infection is confirmed by an HIV antibody test, which looks for HIV antibodies in the blood formed in response to infection with HIV. If the test is positive, a follow-up is always performed to confirm it. The test becomes positive within three months of exposure. Someone who has been recently exposed, yet has a negative result, should be tested again three months after exposure. Progress of the disease can be monitored by regularly measuring the T4 cell count in the blood. The lower the count, the further the disease has progressed. Serious infections are likely to develop at counts of below 200 cells per cubic millimeter.

TREATMENT

AIDS and HIV are treated by a primary care physician (general practitioner) together with a specialist physician in an outpatient department or hospital clinic. In the later stages of the illness, when serious infections develop, the condition must be treated in hospital.

No cure exists for HIV infection itself. Until recently, the only treatment for HIV was to treat the opportunistic infections with antibiotic and antiviral drugs. However, over the past few years, antiviral drugs such as AZT and the newer protease inhibitors have been developed to try and slow down the body's loss of immune function and susceptibility to disease.

These drugs suppress the HIV virus replicating itself in the body and effectively arrest the progress of the disease. They are

Candidiasis

Fungal infections are common in HIV infection. This picture shows flat white patches of Candida albicans in the soft palate of the oral cavity.

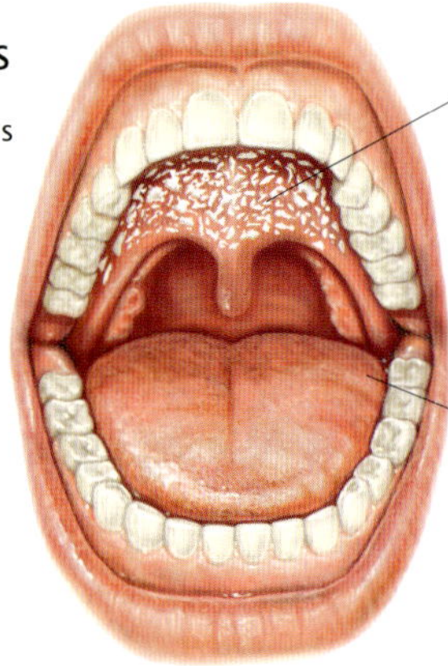

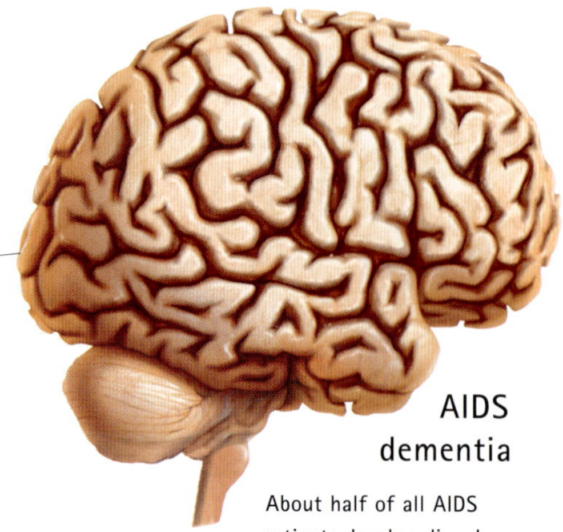

AIDS dementia

About half of all AIDS patients develop disorders of the brain. Opportunistic infections from viruses and other organisms are common. So is dementia, a condition in which concentration and memory fail. This brain shows the atrophy (shrinking in size) that occurs in AIDS dementia.

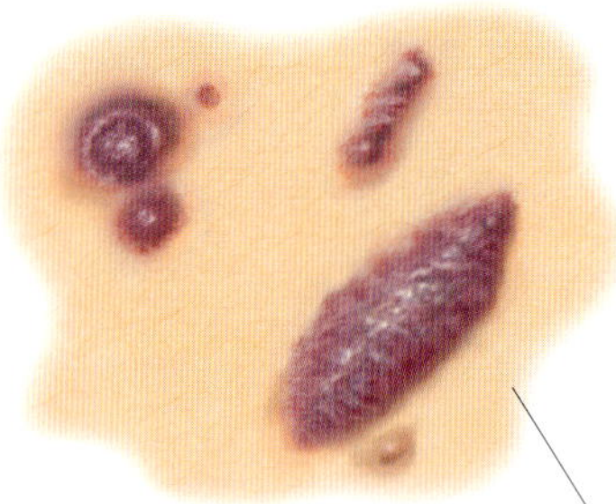

Kaposi's sarcoma

The most frequently occurring cancer in AIDS sufferers, Kaposi's sarcoma produces raised, purple-brown skin lesions. In late stage of the disease it may also affect the lungs and other internal organs.

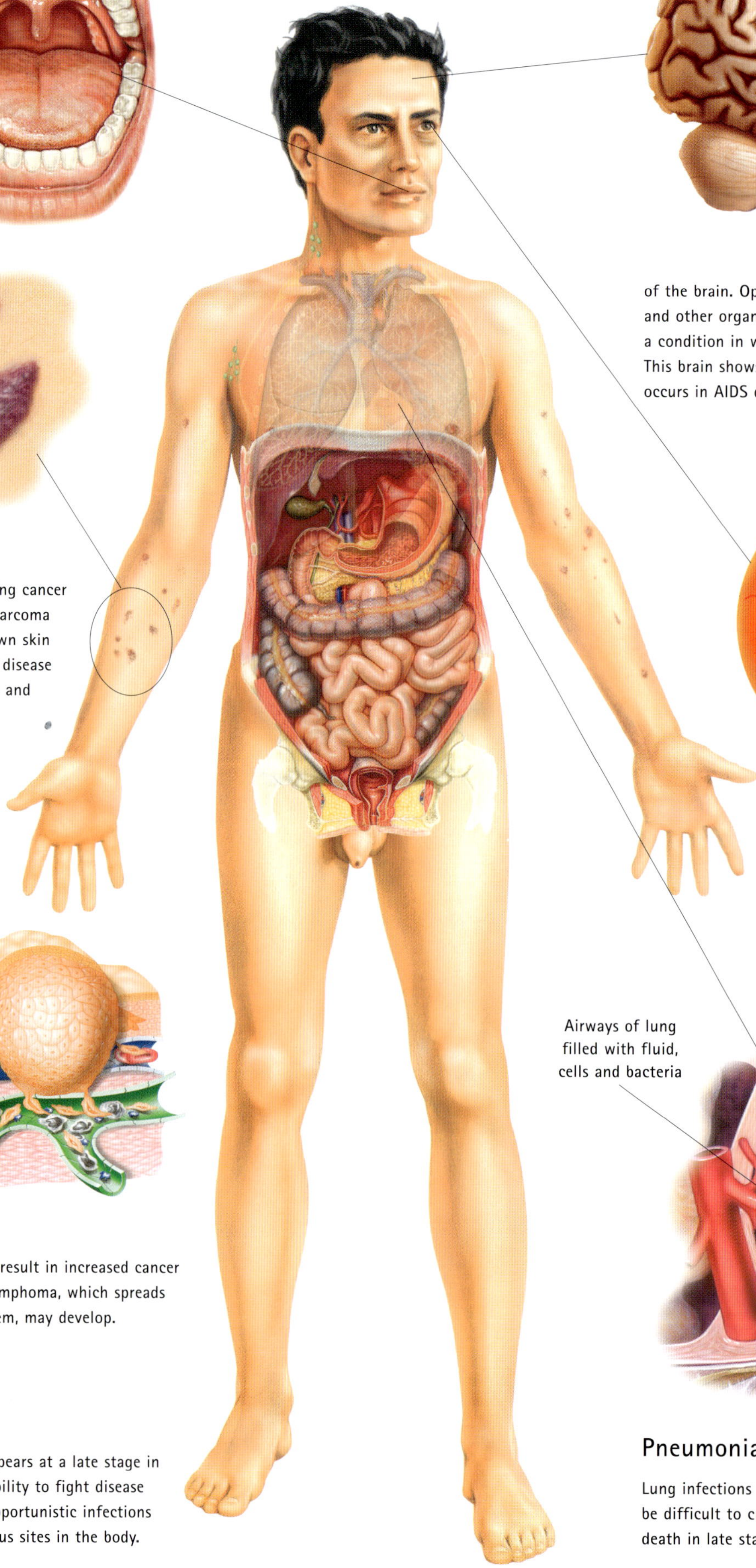

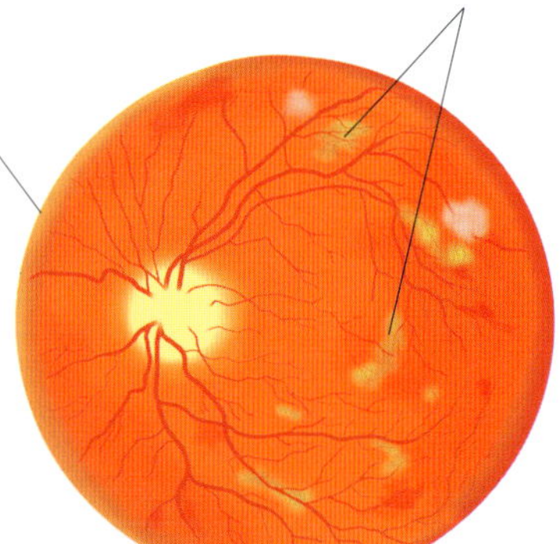

Retinopathy

AIDS can cause eye disorders such as retinopathy, which can result in loss of vision.

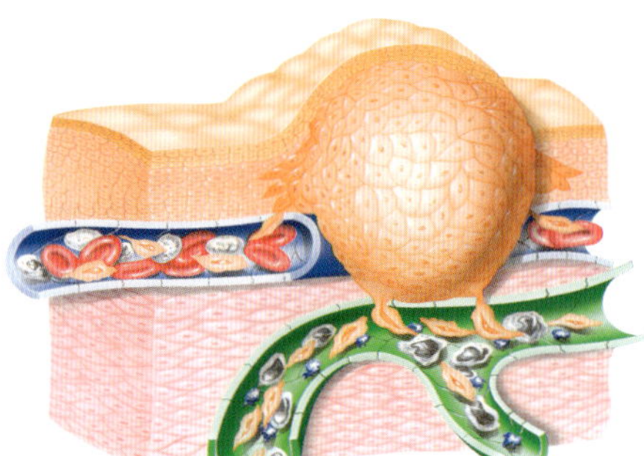

Lymphoma

Low T4 cell counts in AIDS result in increased cancer incidence. Non-Hodgkins lymphoma, which spreads through the lymphatic system, may develop.

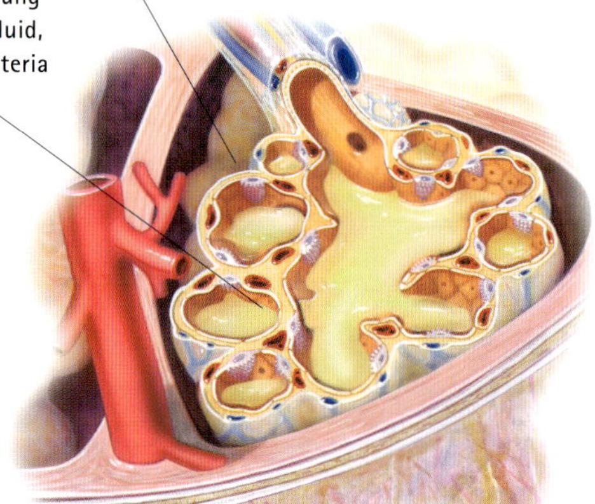

AIDS

AIDS is a syndrome that appears at a late stage in HIV infection. The body's ability to fight disease lessens progressively and opportunistic infections and cancers appear at various sites in the body.

Pneumonia

Lung infections may progress to pneumonia which may be difficult to cure and is often the final cause of death in late stage AIDS.

used in combination. They are expensive, and they may not be well tolerated by the sufferer. Nevertheless, when used with conventional treatment of opportunistic infections, they can prolong life indefinitely. Prior to combination antiviral therapy, the mortality from full-blown AIDS was generally thought to be 100 percent.

Still, prevention remains the major tool in combating HIV and AIDS. Practising safe sex and using condoms (which have the added advantage of protecting against other sexually transmitted diseases such as chlamydia and gonorrhea) are the most effective means of stopping HIV transmission. Intravenous drug users should never share needles. Before entering into a sexual relationship with anyone who is at risk of having or contracting HIV, it is wise to find out about their HIV status first.

Thanks to AIDS awareness and safe sex campaigns, rates of new HIV infections are falling, at least in the developed world. Unfortunately, however, because of the cost of combination therapy drugs, widespread use of them in developing countries is impractical. Management of AIDS and HIV in these countries depends on public awareness campaigns and on the hope of finding a cheap and effective method of immunization.

Bacterial infection

Bacteria are simple organisms of microscopic size and were one of the early forms of life to evolve. Many are beneficial and live in harmony with humans—in the digestive system aiding the breakdown and absorption of food, and in soil and water breaking down dead matter and animal wastes, a process which maintains conditions for life on our planet. Some are harmful and can cause and spread infections such as cholera, pneumonia, tuberculosis and whooping cough, or release deadly toxins that cause illness or death. Botulism, a serious type of food poisoning, results from bacterial growth in unsterilized stored food.

Immunization is a way of stimulating the body to make antibodies to a specific disease without first catching the infection. Health authorities in most countries recommend immunization for all citizens against the most common diseases of childhood. Widespread immunization has been very successful in preventing the spread of many once devastating bacterial (as well as viral) diseases. Antibiotics are drugs commonly prescribed to fight bacterial infections.

Immunization and the use of antibiotics have controlled many serious bacterial diseases within human society and eradicated smallpox. Unfortunately, antibiotics may eliminate beneficial as well as harmful bacteria; this can be a cause of digestive disorders and secondary infections. Overuse of antibiotics may also be the reason that some bacteria, notably *Staphylococcus aureus*, are resisting treatment.

Shape is a major feature used in the classification of bacteria. There are four main forms: spheres (cocci), rods (bacilli), coils (spirochetes) and commas (vibrios). Within these groups, there is a lot of variation. Rods, for example, can be either thick or thin, long or short, and have pointed or rounded ends. Bacteria may also occur as single cells or in groups such as chains, pairs or clusters.

SEE ALSO *Antibiotics on page 445; Blood on page 82; Blood tests on page 436; Immunization on page 447; Inflammation on page 24; Lymphatic/Immune system on page 55; Penicillins on page 446; Skin tests on page 438*

Scarlet fever

Scarlet fever is caused by bacteria known as Group A betahemolytic streptococci. Thanks largely to the advent of antibiotics, which offer quick and effective treatment, this once common disease is no longer the deeply feared scourge it was during the 1800s and early 1900s.

As immunity usually develops after one bout of scarlet fever, the disease tends to occur mainly in children. It begins suddenly with a fever and sore throat within two days of contact with an infected person. Shivering, headaches and vomiting may follow. Within two days of the first symptoms, a small bright red rash begins appearing, first on the neck and chest but eventually spreading to the rest of the body. The rash, which feels like sandpaper, is a reaction to a toxin released by the bacteria. It may last up to a week, after which the skin peels as if it has been sunburnt. During the early stages of scarlet fever, the tongue is coated in white and the taste buds are red and swollen—known as a "strawberry tongue." Later, the white coating disappears to leave a "raspberry tongue," which is red all over.

Before antibiotics were developed, scarlet fever was often followed by meningitis or rheumatic fever. These are now rare as complications of streptococcal infection.

Diphtheria

Diphtheria is an acute infectious disease of the larynx, tonsils, and throat, caused by the bacterium *Corynebacterium diphtheriae*. As a result of widespread immunization against diphtheria over many years, this disease is now rare and usually seen in nonimmunized children, under ten.

Diphtheria is transmitted from person to person by airborne droplets from infected persons or asymptomatic carriers. The illness develops after a period of one to four days after exposure to the bacterium. The child feels weak and unwell with fever and a sore throat, and may find the lymph nodes in the neck will become swollen.

A toxin produced by the bacteria damages the lining of the throat, causing a tough, fibrous gray or greenish yellow membrane to form at the back of the throat, which may obstruct breathing. The toxin may enter the bloodstream causing damage to the heart, kidneys and nervous system. Treatment involves bed rest in hospital and use of diphtheria antitoxin and antibiotics such as penicillin or erythromycin.

Diphtheria immunization is generally carried out in the first year of life. It is usually combined with pertussis and tetanus immunization (DPT). Falling childhood immunization rates in some Western countries have led to a resurgence of the disease.

Streptococcal infections

Streptococcal infections are caused by perhaps the most common human bacterial pathogen, *Streptococcus*. *Streptococci* are classified into several groups and cause a spectrum of diseases, ranging from minor to life-threatening ones. The same groups of streptococci that cause minor infections can occasionally cause severe disease.

The commonest streptococcal infection is streptococcal pharyngitis, or "strep throat." This is generally due to infection by Group

Streptococcus

Meningococcus

Gonococcus

BACTERIA

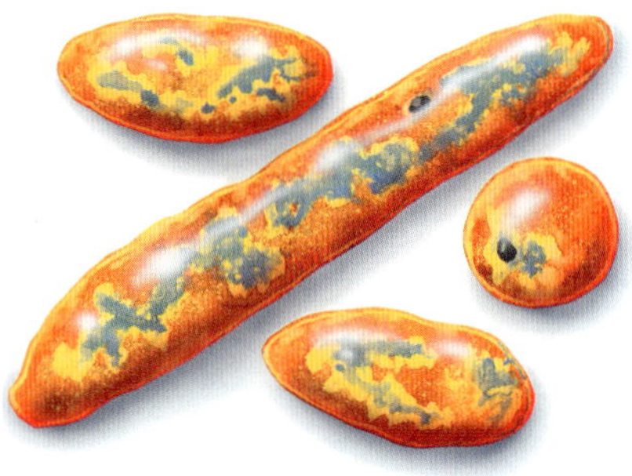
Legionella bacillus

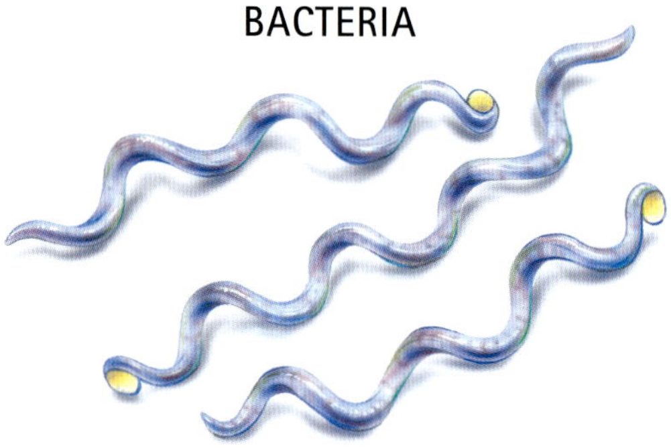
Syphilis spirochete

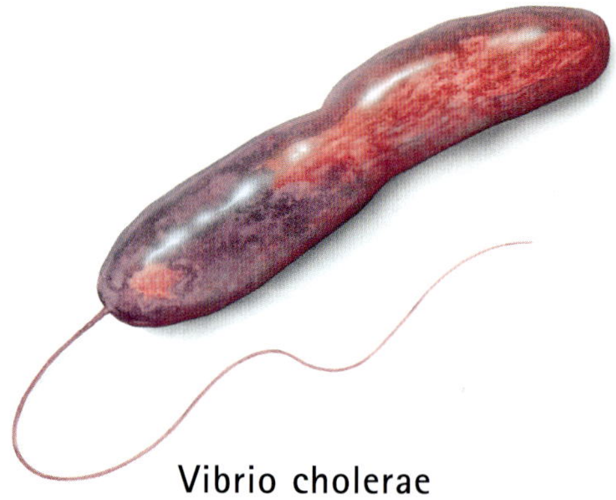
Vibrio cholerae

A streptococci, which most often causes sore throat and fever in childhood. Although most cases of throat infection are viral and associated with the common cold, throat swabs should be taken if streptococcal infection is suspected, particularly in children aged 5–15 years. Antibiotic treatment can prevent the development of possible serious long-term complications.

Apart from otitis media, sinusitis and abscess formation, there may be occasional immune-mediated complications. These are acute rheumatic fever and post-streptococcal glomerulonephritis. Acute rheumatic fever can be prevented by a full 10-day course of penicillin, avoiding possible later damage to the heart valves.

Another common form of superficial Group A streptococcal infection is impetigo. Occasionally Group A streptococci cause severe invasive infections, such as streptococcal toxic shock syndrome or necrotizing fasciitis ("flesh-eating bacteria").

Group B streptococci are the commonest cause of infection of the uterus after childbirth and neonatal infection. Group D streptococci are major causes of pneumonia and endocarditis (infection of the heart valves).

Staphylococcal infection

Staphylococci are bacteria responsible for a wide range of human infections. Several species exist including *Staphylococcus epidermidis*, which is part of the normal bacterial flora of the skin, and *S. aureus*, commonly known as "golden staph," which has a greater potential to cause infection.

Staphylococcal infection most commonly involves the skin, in the form of abscesses, but the bacteria may spread to deeper tissues to cause deep abscesses, osteomyelitis or septicemia. Some strains of *S. aureus* produce a toxin that results in toxic shock syndrome—high fever, a red rash, low blood pressure and organ dysfunctions. *Staphylococci* are very common causes of infections in hospitals: *S. epidermidis* causes infection of intravenous cannulas and *S. aureus* can cause severe infections in diabetics, patients with renal (kidney) failure and intravenous drug users. Treatment of staphylococcal infection is with antibiotics and surgical drainage of abscesses when required.

The widespread use of antibiotics in hospitals has led to the emergence of multiresistant staphylococci, which are important pathogens in surgical and intensive care

Staphylococcus

units. The prevention of cross-infection by handwashing is important for reducing staphylococcal infections.

Tetanus

Tetanus is an acute infectious disease, which occurs in both humans and animals, produced by the bacillus *Clostridium tetani* entering the body through a dirty wound, particularly a puncture wound.

Symptoms include muscle stiffness and cramps, which appear first around the mouth and jaw (hence the disease's previous name, lockjaw), a sore throat and difficulty breathing and swallowing, leading to severe muscle spasms and convulsions.

Tetanus can incubate for between 2 days and 2 weeks, but sometimes as long as 3 months; the longer the incubation period the milder the disease.

The severity of the symptoms is related to the amount of toxin produced by the bacterial infection and the resistance of the person to the disease. Of those who contract tetanus, 30–40 percent will die if they are not treated. Tetanus is more prevalent in older people and agricultural workers who regularly come into contact with animal manure. In nearly half of tetanus cases no puncture wound is evident; any wound can serve as the entry point for tetanus germs, even a superficial abrasion. The spores of *Clostridium tetani* are most commonly found in topsoil and are spread by animal feces. They may live on anything on the ground.

Wounds should be thoroughly cleaned and any dead tissue removed. Recovery from a tetanus attack does not guarantee immunity from the disease. Complications include hypertension, fractures of the spine or long bones, abnormal heartbeat, coma, general infection, blood clots in the lungs, pneumonia and eventually death.

Immunization against tetanus is available as part of immunization programs in most industrialized countries, starting with babies from 6 weeks of age. It consists of a series of injections, the number depending on which type of tetanus toxoid is used; it is important that the immunization be repeated every 10 years. Redness and a hard lump are the most common side effects of the vaccine; if other side effects are noticed a medical check-up is a wise precaution. Accident victims are usually routinely administered with the vaccine. Treatment of tetanus includes antibiotics, sedatives and muscle relaxants.

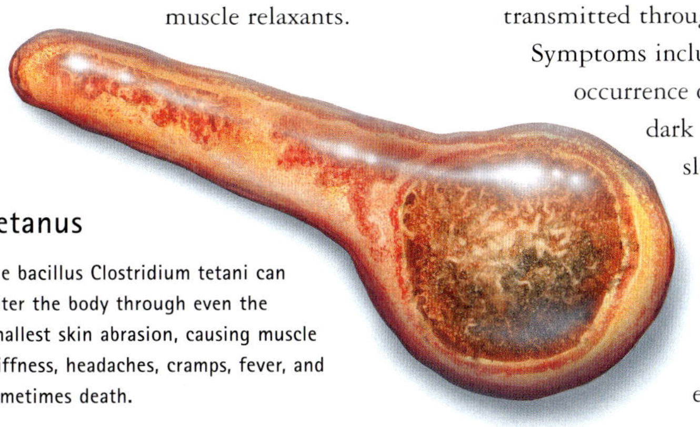

Tetanus

The bacillus Clostridium tetani can enter the body through even the smallest skin abrasion, causing muscle stiffness, headaches, cramps, fever, and sometimes death.

Anthrax

Anthrax is an infectious disease transmitted from farm animals. In this case it has caused an ulcer on the skin of a finger. The dark area is dead skin and tissue.

Meningococcal disease

Meningococcal disease is a rapid, potentially fatal form of bacterial infection, due to *Neisseria meningitidis*. It is most commonly seen in children under 5 years of age, and its incidence has increased in recent years. As a general rule, meningococcal disease is a combination of meningitis and septicemia. The illness develops rapidly, with a flu-like infection, headache, confusion, and the appearance of a blotchy, purplish rash. Early treatment with intravenous antibiotics can be lifesaving.

The infection occurs mainly in winter, is spread through respiratory secretions and can occur in epidemics. People who are in close contact with the patient require preventive antibiotics. Vaccination against meningococcal disease is advised for travelers to epidemic areas and those with reduced immunity.

Anthrax

Anthrax is an infectious disease caused by the bacterium *Bacillus anthracis*, rarely seen in the Western world today, but which still exists in Africa, Asia and the Middle East. The infection is transmitted to humans most commonly by farm animals such as sheep, cattle, horses, goats and swine, and is transmitted through a break in the skin.

Symptoms include nausea, fever and the occurrence of a skin boil that forms a dark scab. The boil forms slowly and may spread to form other boils.

In another rarer form, anthrax spores are inhaled and cause a rapidly fatal pneumonia—hence experimentation by some governments with anthrax as a biological weapon. A few years ago in the USA a number of people became infected after inhaling anthrax spores that had been sent to them via mail.

Immediate treatment with penicillin or tetracycline is usually effective in treating the skin form of anthrax. Ciprofloxacin may also be used. A vaccine is available for travelers at risk of exposure to animals or animal products in affected areas.

Whooping cough

Whooping cough, also known as pertussis, is caused by the organism *Bordetella pertussis*. It is a serious, common and highly infectious illness in young children, particularly children under two years. It is spread by coughing, sneezing and close personal contact. Whooping cough incubates for between one and two weeks and can last for as long as three months. The affected person remains infectious for up to a month after the onset of the cough.

Beginning like a cold, whooping cough turns into exhausting coughing bouts with a characteristic whooping sound. Pneumonia is the most common complication and middle ear infections, nosebleeds, hemorrhages inside the eye, loss of appetite and dehydration are other complications.

Adults and older children who contract the disease may not suffer as severely, but can still spread the disease. For those under two years, hospitalization is often necessary. Full immunization is the best precaution.

Typhoid

Also known as enteric fever, typhoid is a debilitating intestinal disease caused by infection with the bacterium *Salmonella typhi*. It is rare in the industralized countries. The disease incubates for between 1 and 2 weeks. Symptoms include headache, loss of appetite, fatigue and constipation, followed by abdominal pain and rosy spots on the abdomen and chest which last for 3–4 days, and diarrhea, which is the main problem; pneumonia may be a complication in severe cases. A blood test will determine if typhoid has been contracted. The disease can be treated with antibiotics.

Typhoid vaccines give partial protection and the risk of the disease can be reduced by proper sanitation, good hygiene, by boiling or purifying all water meant for drinking, pasteurizing milk and washing fruit and vegetables. People who have had typhoid may become carriers.

Paratyphoid

Paratyphoid is a gastrointestinal disease caused by certain forms of *Salmonella* bacteria. It occurs throughout the world but, because it is spread via food or water contaminated by feces or urine from an infected person, outbreaks occur mainly where sewerage and sanitation systems are inadequate. Symptoms, similar to but usually less severe than those of typhoid, appear between 1 and 10 days after consuming a contaminated product. They include headaches, watery diarrhea, a rosy chest and abdominal rash, and dry cough. In severe cases, the disease can cause intestinal bleeding, mental fogginess and minor deafness. Death is rare when medical attention is provided. Treatment commonly involves antibiotics.

SEE ALSO *Gastroenteritis on page 258*

Cholera

Cholera is a bacterial infection of the intestines. It is spread in contaminated water supplies, in contaminated food and, rarely, by person-to-person contact. The disease occurs 1–5 days after ingesting the *Vibrio cholerae* bacteria. The sufferer passes large volumes of pale watery diarrhea, and this can quickly lead to dehydration and to death if not treated. Cholera is particularly dangerous for young children whose relatively small body mass results in a more rapid onset of dehydration.

Treatment is oral rehydration through salt solutions by mouth, or drip, which may be combined with antibiotics.

SEE ALSO *Dehydration on page 120; Infectious diseases on page 364; Intestines on page 260*

Leprosy

Leprosy is caused by a rod-shaped bacterium, *Mycobacterium leprae*, a relative of the tuberculosis bacillus. It is prevalent in Central and South America, in East Asia, and in the tropical countries of Asia and Africa. There are two main forms: tuberculoid leprosy and lepromatous leprosy.

Tuberculoid leprosy is an infection in the deep skin layers, which destroys the hair follicles, sweat glands and nerve endings at the site of infection. The skin above the site becomes dry and discolored and loses the ability to sense touch, heat and cold, and pain. Fingers and toes are easily injured and may become mutilated and fall off.

In lepromatous leprosy, the organism multiplies freely in the skin. Large, soft bumps, or nodules, appear over the body and face. The mucous membranes of the eyes, nose, and throat may be invaded. In extreme cases the voice may change drastically, blindness may occur, or the nose may be destroyed completely.

Both types of leprosy are only mildly contagious (via the respiratory tract) and the infection is very slow to develop, ranging from 6 months to 10 years. Typical early signs of the disease include one to three slightly raised patches on the skin which are non-itching, may be reddish in color and on which there is sensory loss. The diagnosis can be confirmed through a biopsy of the edge of an affected skin area or nerve.

Leprosy most commonly strikes people aged 10–20 years, and is seen more in men than women. It is thought to be transmitted through the inhalation of dust particles which are laden with bacilli.

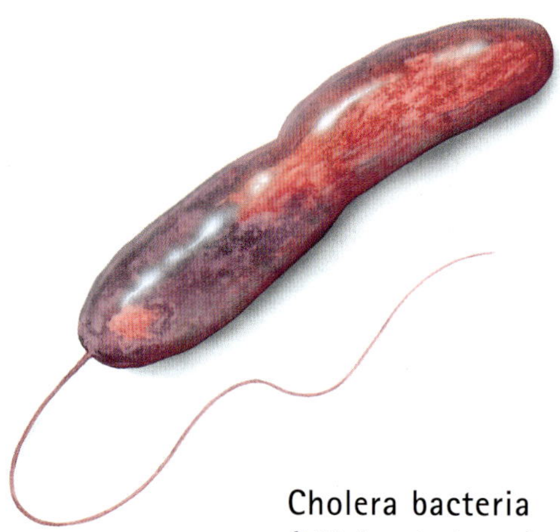

Cholera bacteria (Vibrio cholerae)

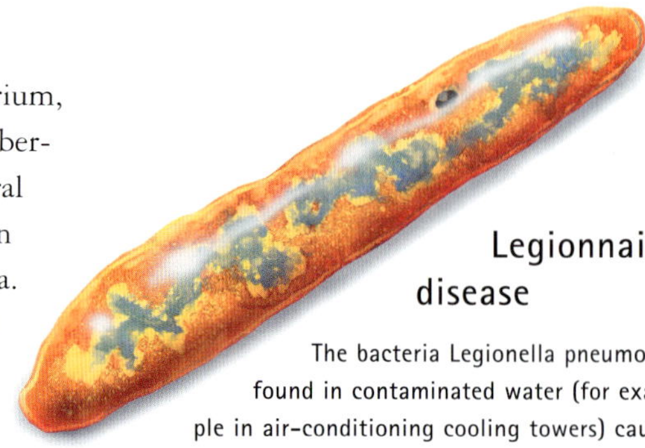

Legionnaires' disease

The bacteria Legionella pneumophila found in contaminated water (for example in air-conditioning cooling towers) causes Legionnaires' disease.

Leprosy has existed for thousands of years and has a huge stigma attached to it because of the dreadful deformities that can result from infection. For this reason it is now often referred to by another name—Hansen's disease—in an effort to avoid fear and hysteria that may disadvantage efforts to treat the disease and reduce its incidence.

Early treatment is important in preventing deformities and other physical handicaps. Drugs such as dapsone, rifampin and clofazimine are used in combination to prevent drug resistance and may cure the disease within a year. Treatment usually needs to be continued for several years after the disease becomes inactive.

Legionnaires' disease

Named after an occasion when several members of the American Legion (ex-servicemen) became ill at a reunion in Philadelphia, USA, in 1976, this disease has symptoms that are similar to pneumonia. More severe in heavy smokers with lung disease and in those who lack immunity, the disease is caused by a bacterium of the *Legionella* species. The strain identified in the original outbreak was *L. pneumophila*, but other strains can also cause the disease. It has been attributed to contaminated water, particularly in air-conditioning units and is contracted by breathing in fine water droplets or aerosols that contain bacteria. It cannot be acquired by drinking contaminated water and is not passed from person to person. *Legionella* bacteria may also thrive in spa pools, humidifiers, garden potting mix and reticulated water systems where water temperature is kept between 20 and 45°.

The flu-like symptoms of Legionnaires' disease usually appear 2 to 10 days after infection. These include headache, loss of appetite, muscle aches, a dry cough progressing to grey or blood-stained sputum, disorientation, fever, and at times stomach cramps and diarrhea. Diagnosis is confirmed by a sputum culture and many antibiotics appear to be effective. Untreated, the condition can usually be fatal.

Tuberculosis

Tuberculosis is an infectious, inflammatory disease caused by bacteria belonging to the *Mycobacterium* group (usually *M. tuberculosis*). Tuberculosis is primarily a disease of the lungs, but it may spread to other parts of the body, particularly in patients whose immune systems have been weakened and are in the last stages of disease. About 10–20 new cases per 100,000 population are diagnosed each year in industrialized nations, so it is still a very important public health problem.

Initial infection with tuberculosis usually occurs in childhood. The bacteria are inhaled and cause a small patch of pneumonia in the middle or lower areas of the lung. The initial site develops a tubercle, which is a clump of immune system and other cells surrounding a cavity filled with cheese-like material derived from dead lung tissue. The infection also spreads to lymph nodes in the center of the chest, but may not spread further at this stage. Children infected in this manner may have no symptoms at all, or complain only of a mild fever, cough and feeling unwell.

If the initial infection occurs in adults, the tubercle may develop in the upper parts of the lung. In most cases this initial infection is halted and the bacteria are walled-off inside the tubercle. The bacillus can then lie dormant for some years, with the possibility of becoming active again later.

The serious complications of tuberculosis arise when the bacteria escape from the initial site of infection and spread through the lung or the rest of the body. Widespread infection of the lung may cause the collapse of lung lobes and further infection of the pleural sacs around the lung.

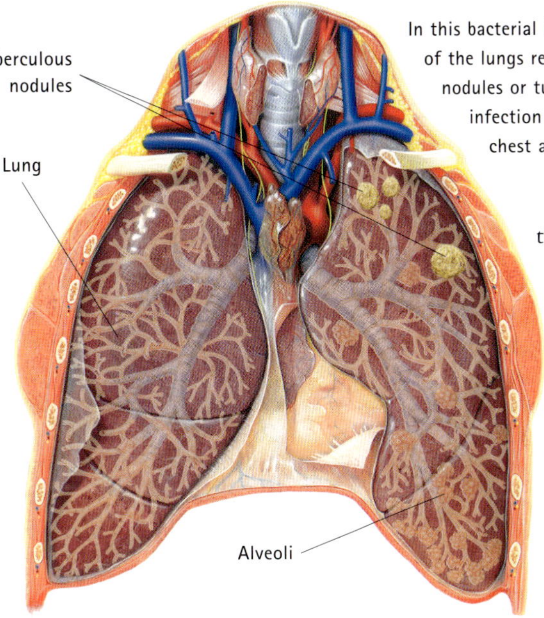

Tuberculosis

In this bacterial infection, inflammation of the passageways of the lungs results in the development of lesions and nodules or tubercles in the lungs. From here the infection may spread to the lymph nodes in the chest and to other organs in the body.

The bacteria may also spread via the bloodstream to the spleen, liver, kidneys, fallopian tubes and brain membranes, and testes, with consequences ranging from sterility to death from extreme infection.

Treatment of tuberculosis is by the appropriate combination of antibiotics. The tuberculosis bacteria are likely to develop drug resistance if only a single drug is used, so a combination of three drugs is usually given. Treatment must also continue for long periods to avoid recurrence.

Pott's disease is a form of tuberculosis of the spine that affects the vertebrae and may also damage the intervertebral disks.

TUBERCULOSIS BACTERIA

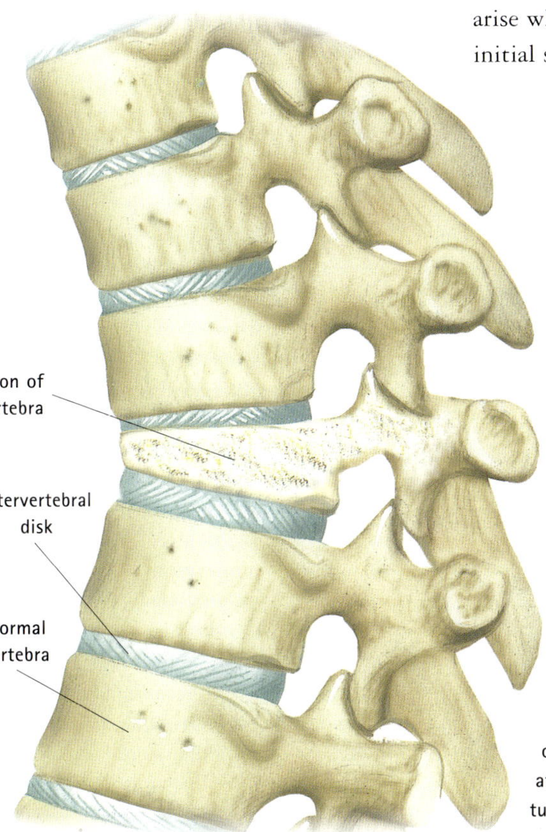

Pott's disease

Caused by the tuberculosis bacteria, Mycobacterium tuberculosis, this disease of the spine first affects the vertebrae, then attacks the intervertebral disks. Spinal curvature may result if the disease is left untreated.

Listeriosis

Listeriosis is an infection caused by *Listeria monocytogenes*, a bacterium found in nature and in some foods. Infection is not common and there are usually few or no symptoms in healthy people though it can cause a flu-like illness. In pregnancy, however, listeriosis is dangerous to the fetus and can cause miscarriage, stillbirth or premature birth.

About half of the babies infected at or near birth will die. Signs that a newborn may be infected include a red skin rash, whitish nodules on mucous membranes, respiratory distress, shock, vomiting, lethargy and jaundice. The condition often manifests as meningitis in babies aged 2 weeks or older.

In adults, the infection can take many forms depending on the body part affected. It may manifest as meningitis, septicemia, pneumonia, endocarditis or, in less severe cases, skin lesions, conjunctivitis or abscesses.

The organism can be found in such foods as soft cheese, cold cooked chicken, cold meats and paté, raw or smoked seafood and pre-prepared salads. Observing good hygiene and avoiding these foods during pregnancy minimizes the infection risk.

Lyme disease

A multisystemic recurrent inflammatory disease, Lyme disease (also known as Lyme borreliosis or Lyme arthritis) is transmitted by tick bite, with symptoms ranging from skin lesions to chronic arthritis. It is caused by a corkscrew-shaped bacterium or spiro-chete called *Borrelia burgdorferi* which ticks, especially the *Ixodes* variety, collect from the bodies of white-footed field mice and other rodents. The ticks are dispersed by deer and migratory birds.

Lyme disease is difficult to diagnose due to the variety of symptoms, which are similar to those seen in many other disorders, and to the fact that most sufferers do not recall the tick bite. The most recognizable symptom is a rash which may have concentric red rings and is accompanied by chills and fatigue. Days to weeks afterward, there may be joint pain and problems with the nervous system or heart, and up to a year later skin disorders, arthritis and neurological problems such as facial palsy may appear. Treatment with antibiotics in the early stages may prevent these later symptoms. However, if the disease has progressed, long-term treatment (including several weeks of intravenous antibiotics) may be necessary.

People living, traveling or working in tick-infested areas should wear light-colored long-sleeved shirts, long pants, socks, closed shoes, and a hat; and carry tick repellent.

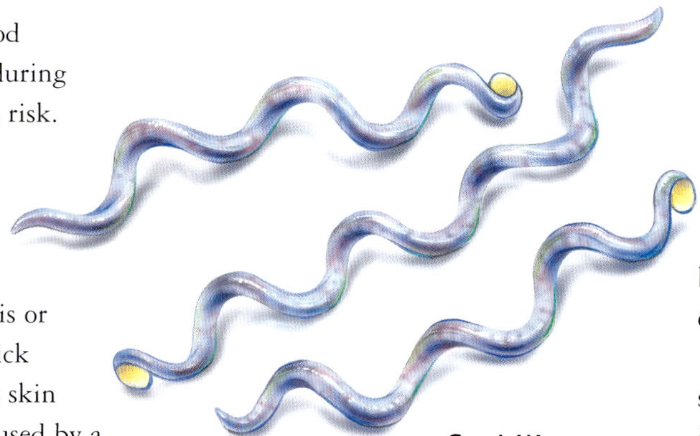

Syphilis

A serious sexually transmitted disease caused by the bacteria Treponema pallidum, syphilis may take years to develop fully. In its final stages, the disease affects the nervous system.

Syphilis

Syphilis is a serious, sexually transmitted disease which is caused by the organism called *Treponema pallidum*. Clinically it can resemble many other diseases, including gonorrhea. A disease that progresses through three stages, each separated by months and even years, cases of infection are on the rise in many parts of the world, including the USA.

Syphilis begins with a lesion, a painless, circular chancre that may appear on the lips, mouth, tongue, nipple, rectum or genitalia; nearby lymph nodes may enlarge but are not painful. The chancre heals and weeks, or even months later, the secondary symptoms appear when microbes spread to organs and tissues in the body. Non-painful skin rashes keep appearing and disappearing, sometimes in association with fever, headache and hair loss. Tertiary lesions, which can appear years later, may destroy normal skin, bone and joints by ulceration.

Tertiary syphilis also affects the nervous system. It can take three forms: cardiovascular syphilis, which affects the heart severely; neurosyphilis, which affects the brain and the nervous system; and benign late syphilis.

Difficult to diagnose (a series of blood tests is often necessary), syphilis can be treated with penicillin, or an alternative for those allergic to this drug. Anyone who has sex with a person known to have syphilis, or someone known to have another sexually transmitted disease, should be tested for syphilis. It can pass into the fetus during any stage of pregnancy and cause growth deformities in the baby or stillbirth; this can be prevented by treatment early in the pregnancy.

Tabes dorsalis is one of the later effects of syphilis, occurring 20–30 years after the initial infection. It involves damage to the part of the spinal cord involved in sensory inputs—the posterior (dorsal) roots and columns. The damage causes sudden sharp pains, usually starting in the legs. The sense of limb position is affected so that there is difficulty walking, especially in the dark. Tendon reflexes are also lost and there may be bladder and bowel incontinence, and impotence. There is no cure available, but medication may slow the course of the damage and can provide pain relief.

Gonorrhea

Gonorrhea is a common sexually transmitted disease caused by a bacterial infection *(Neisseria gonorrhoeae)*. It is frequently transmitted during sexual intercourse, including both oral and anal sex. Gonorrhea is most common in people aged 15–30 years. Risk of infection increases with multiple partners, partners with a history of infection, and unprotected intercourse.

In women the infection usually involves the cervix, although it may also spread to the vulva and vagina, urethra and fallopian tubes. Signs of infection include a vaginal

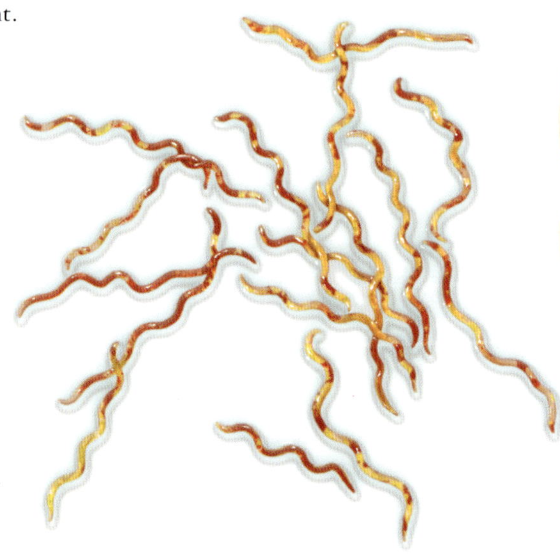

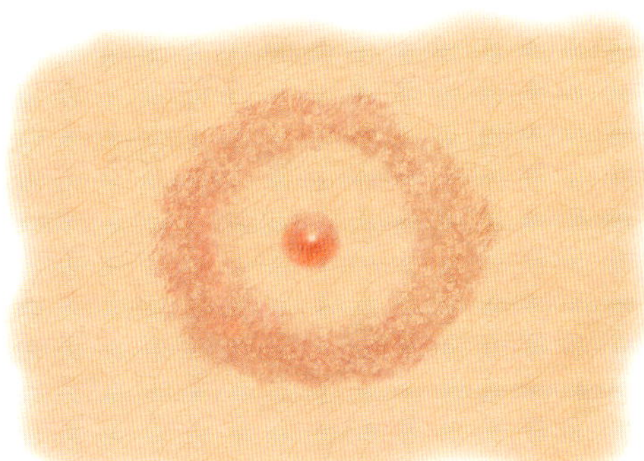

Lyme disease

Lyme disease is caused by a bacterial infection spread by tick bites. The most obvious symptom of this disease is a rash which may look like small "bull's eyes."

Gonorrhea bacteria

discharge and pain on urinating although, in about half the cases, no symptoms may be noted. In some cases the rectum may become infected, causing discomfort in the anal region. Throat infections can occur following oral sex. Symptoms start 1–3 weeks after infection.

If the bacteria spread to the fallopian tubes (as happens in 10–15 percent of untreated cases) the condition is termed pelvic inflammatory disease (PID). PID can cause abdominal pain and may lead to blocked fallopian tubes and infertility.

Approximately 50 percent of women will be unaware they have the disease and therefore may pass it on to unsuspecting sexual partners. However, on rare occasions the disease may be transmitted from mother to baby during childbirth, and may result in infection of the baby's eyes.

In males, gonorrhea usually affects the urethra and is associated with a discharge and pain on urinating. In homosexuals, infections of the throat, anus and rectum are common. In men the infection can spread to other regions of the reproductive tract, such as the epididymis and the prostate. In both sexes gonorrhea can occasionally (in about 1 percent of cases) lead to widespread infection of the peritoneum, joints and blood, with abdominal pain, arthritis and fever.

Diagnosis of gonorrhea is made by identifying the bacteria in the discharge, for example from the cervix, urethra or rectum. As many strains of gonorrhea have become resistant to common antibiotics, specific antibiotic courses are used. If the disease is treated early the prognosis is good. Gonorrhea is often associated with other sexually transmitted infections, including HIV, which should also be tested for. Patients should be advised to abstain from any sexual contact until treatment for gonorrhea has been successfully completed. Also, all sexual contacts should be traced and tested for infection.

Chancroid

Chancroid, or soft chancre, is a sexually transmitted disease (STD) most common in tropical regions. Shallow painful ulcers appear in the infected areas, usually the genitals, 3–5 days after exposure through sexual contact. They are caused by the bacterium *Haemophilus ducreyi* and can easily be mistaken for the first symptoms of syphilis, in which similar but hard and painless chancres appear. In chancroid, localized swelling and inflammation of lymph glands follow the appearance of ulcers. The disease is often associated with HIV transmission.

Chancroid can be treated successfully with sulfonamides, or the antibiotics azithromycin and erythromycin, although the ulcers may leave scars.

Diseases caused by Chlamydia

Chlamydia is a genus of three bacterial parasites: *Chlamydia psittaci*, which causes psittacosis; *C. trachomatis*, which causes trachoma, conjunctivitis and a variety of sexually transmitted diseases; and *C. pneumoniae*, which causes infections in the respiratory tract.

It is estimated that up to 30 percent of sexually active women have had a sexually transmitted chlamydial infection, which is six times as common as genital herpes and thirty times as common as syphilis. There may be no symptoms, or mild ones: women may suffer urethritis, slight menstrual-like discomfort and vaginal discharge; men may notice a frequent urge to urinate, a whitish yellow discharge and redness at the tip of the penis. The relative absence of symptoms increases the possibility of unknowingly passing the disease to others.

Chlamydia has often been linked with pelvic inflammatory disease, ectopic and premature birth, conjunctivitis in babies or pneumonia. A test for chlamydia in those planning a baby is wise.

Chlamydia pneumoniae, which was identified as a separate species in the 1980s, causes a mild atypical pneumonia with fever, cough and sore throat. Fortunately, most forms of chlamydia infection are easily treated with antibiotics.

Psittacosis

Psittacosis is a common infectious disease of many bird species that is sometimes contracted by humans. It is caused by the bacterium *Chlamydia psittaci*. Psittacosis can be acquired by inhaling dust from dry and infected bird droppings or from handling birds already affected by the disease. Symptoms usually appear within 10 days of contact and include fever, headache, fatigue, chills, muscle aches, chest pains and cough.

A severe form of pneumonia can develop so much so that, if left untreated, may be fatal, particularly among the elderly. The disease is, however, treatable with tetracycline antibiotics. Psittacosis is also known as ornithosis and parrot fever.

Chlamydia bacteria

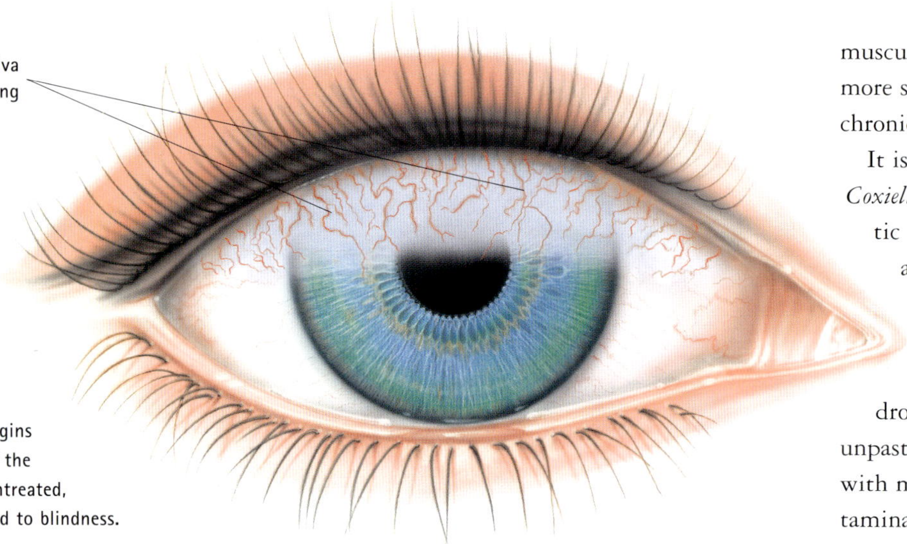

Trachoma

This condition begins with infection of the conjunctiva. If untreated, trachoma can lead to blindness.

Lymphogranuloma venereum

Lymphogranuloma venereum (also called LGV, lymphogranuloma inguinale, climatic bubo or Nicolas-Favre disease) is a sexually transmitted disease common in tropics and spread by unprotected sexual intercourse. It is caused by the bacterium *Chlamydia trachomatis* and develops between 3–12 days after contact with the infection.

First symptoms of LGV are a small painless blister on the penis or in the vagina, which may become an ulcer and heal without being noticed. Lymph glands then become swollen and tender and may develop sinuses—openings to the skin surface which discharge fluid. Fever, headaches and joint pains may develop without treatment, which is normally the antibiotic tetracycline or, in pregnant women, erythromycin.

Trachoma

Also called granular conjunctivitis, trachoma is a disease of the eye caused by infection with the organism *Chlamydia trachomatis*. It is the world's most common cause of blindness, with 500 million people affected around the world. The people affected are mainly in developing countries (an inadequate supply of running water often being a factor). The disease is spread by direct contact from person to person.

The condition begins slowly as a mild conjunctivitis which develops into a severe infection with copious amount of eye discharge. Erosions form in the cornea of the eye, which becomes infiltrated by blood vessels and scarred, causing blindness. Trachoma is easily treated with antibiotics such as oral erythromycin. If treated early, the eye will recover completely. However, once scarring and blindness have occurred, vision usually cannot be restored.

Rickettsial diseases

Rickettsial diseases are most commonly caused by the microorganisms *Rickettsia* and *Coxiella*. The most notorious rickettsial disease is typhus. In the early twentieth century epidemic typhus was a leading cause of suffering and death. However, the subsequent development of methods for the prevention and treatment of rickettsial disease have greatly decreased the incidence in many countries.

Typhus

Typhus is a general term for any of several related diseases caused by species of the microorganism *Rickettsia*, which are transmitted by a louse or a flea from infected rats or mice. They include epidemic typhus, endemic typhus and scrub typhus. Symptoms are fever, headache and pink spots on all parts of the body except the face, hands and feet. Vomiting and delirium may occur. Antibiotics usually eradicate the infection.

Vaccines are available against epidemic and endemic typhus, but not scrub typhus. Insecticides, mite-repellent creams and clothing covering arms and legs are recommended in places where typhus is common.

Q fever

This is an infectious disease acquired from animals; it causes high fever, chills and muscular pains and may be accompanied by more serious illnesses such as pneumonia, chronic hepatitis and encephalitis.

It is caused by the microorganism *Coxiella burnetii*, which is found in domestic animals such as cattle, sheep, goats and cats, as well as in wild animals and ticks. The disease may be passed to humans when they inhale contaminated dust or droplets, consume contaminated food or unpasteurized milk or come into contact with materials, such as soil, which are contaminated with infected blood or feces.

The incubation period is 9–28 days after which a fever suddenly occurs, accompanied by symptoms resembling influenza, such as severe headaches, shivering, muscle pain and sometimes chest pain. After a week a dry cough may develop and the fever may continue for up to 3 weeks. This early form of the disease is known as Q fever (early) and may include complications such as pneumonia and hepatitis. Q fever (late) is a rare relapse of the illness that may cause problems with the aortic heart valve.

People who work with animals, such as farmers and veterinarians, are most at risk. Q fever rarely causes death and is treated with antibiotics. It is widely found in Europe, North America and parts of Africa.

SEE ALSO *Encephalitis on page 144; Hepatitis on page 272; Pneumonia on page 244*

Diseases caused by protozoa

Protozoa are the simplest organisms of the animal kingdom. One of the most feared protozoal infections is malaria, which is endemic throughout most of the tropics and affects over 100 million people annually.

Malaria

Malaria is a tropical febrile illness caused by the protozoan parasite *Plasmodium*, transmitted to animals and humans by the *Anopheles* mosquito. There are four varieties of the parasite, the most common being *Plasmodium vivax*, followed by *P. falciparum* (which causes malignant or cerebral malaria), and then *P. malariae* and *P. ovale*. Usually two weeks after a bite by an infected mosquito, the patient develops violent chills and shivering, high fever and drenching sweats. Headache, muscle pains, cough and diarrhea may all occur. *P. falciparum* may also produce "blackwater" (dark urine) from the massive breakdown of red blood cells and the excretion in the urine of the blood pigment (hemoglobin).

Malaria causes over 1 million deaths yearly worldwide, especially from cerebral malaria due to *P. falciparum*. This type may progress rapidly, with confusion, convulsions and coma possibly leading to death within a day. Diagnosis of malaria is made by the examination of blood films taken over three days. As malaria may not appear for between four weeks and several months after infection, tourists are advised to treat any unexplained fever on their return as potential malaria (or other possible infections like dengue fever).

Many countries where malaria is common regularly spray insecticides to eliminate mosquitoes. As the mosquito needs water and animals to breed, malaria is not likely to be present in major cities and most holiday resorts. Even so, travelers should take precautions against being exposed to malaria and other diseases caused by mosquitoes. They should use mosquito repellent, especially at dawn and dusk, wear long sleeves and pants, wear light-colored clothing, avoid perfumes, perfumed soaps and deodorants, have mosquito nets on the beds, and use an insecticide coil in the room.

There are now kits for self-diagnosis and treatment. Travelers into malarial areas are advised to take prophylactic tablets such as doxycycline, chloroquine, primaquine or mefloquine. Unfortunately, resistance to these drugs is developing rapidly. For the treatment of an acute attack, quinine and artemisin-type drugs may be used together with combinations of the prophylactic drugs.

Malaria, especially the *P. vivax* type, can become a chronic illness and keep recurring.

Giardiasis

Caused by the flagellate protozoa *(Giardia lamblia)* that is found in contaminated water, giardiasis is a common infection of the small intestine. It is characterized by stomach ache and large, bad-smelling, frothy stools containing mucus. It is infectious and is usually transmitted as cysts through oral-fecal contact or by ingesting food or water contaminated by feces. *Giardia* is one of the most common intestinal parasites; it is estimated that up to 20 percent of the world's population is infected with it at any one time. It is common in tropical regions and in developing countries with poor sanitary conditions, inadequate water quality control and overcrowding.

Children are more likely to be affected than adults, and families with young children who attend preschool or day care centers, as well as homosexual men and women, and anyone who drinks untreated water from creeks or rivers, are at most risk.

The period from infection to the onset of acute symptoms ranges from several days up to two weeks, distinguishing it from food poisoning. Without treatment, the disease can go on for months, with recurrent mild symptoms, such as digestive troubles, intermittent diarrhea, and weight loss. Diagnosed by laboratory testing of fecal matter, giardiasis is treated with drugs.

Leishmaniasis

Leishmaniasis is a parasitic disease caused by various protozoan parasites of the genus *Leishmania*, which live on dogs and rodents in many parts of world, especially tropical and subtropical countries. The parasites are transmitted from infected animals or people to new hosts by sandfly bites.

There are two main types. One is visceral leishmaniasis, also called kala-azar, which attacks the internal organs, and causes fever, enlargement of the spleen, anemia and skin darkening. Symptoms include cough, fever, weight loss, diarrhea, general abdominal discomfort, thinning hair and scaly, ashen skin. Not all symptoms appear at the same time, however, and as many of them are associated with other diseases it increases the difficulty of diagnosing leishmaniasis. After a sandfly bite the bone marrow, spleen and lymph nodes are invaded by parasites. In children there may be a sudden onset of vomiting, diarrhea, fever and cough, while adults may suffer fever for up to 2 months, as well as general fatigue, loss of appetite and weakness. As the disease progresses the immune system is damaged and death may occur within 2 years from complications such as infection. It can be fatal if untreated.

The other type is cutaneous leishmaniasis, also known as Delhi boil or oriental sore, which attacks the skin, causing skin lesions and ulcers. It can attack the mucous membranes, causing nasal congestion, nose bleeds, mouth and nose ulcers, and difficulty breathing and swallowing. The characteristic skin lesions may look like those of cancer, tuberculosis or leprosy. They can cause disfigurement requiring plastic surgery.

Treatment of both types is with antimony compounds or pentamidine. There is a good chance of a cure if the disease is diagnosed before the immune system is damaged. Some cases of cutaneous leishmaniasis heal spontaneously and do not require treatment.

Leishmaniasis has been reported in all continents except Australia. Travelers to endemic areas should avoid sandfly bites by using insect repellent, wearing appropriate clothing, and ensuring that windows and beds are screened with fine netting.

Toxoplasmosis

Toxoplasmosis is an infection caused by the parasite *Toxoplasma gondii*. It is usually acquired after eating raw or undercooked meat, raw eggs or unpasteurized milk containing infective cysts. The organism spreads from the intestines throughout the body, even crossing the placenta and infecting a fetus if a woman is pregnant.

Handling cat feces or soil contaminated with cat feces can also cause an infection. Most people have no symptoms, although sometimes there may be flu-like symptoms or a lymph gland may enlarge.

The main concern is infection during pregnancy, as the disease can seriously harm the developing fetus; people with weakened immune systems are also at greater risk of suffering serious illness. A month-long course of medication is the only effective cure for the condition.

Cryptosporidiosis

Cryptosporidium is a parasite of domestic and wild animals. In people with normal immune systems, it causes no illness or only mild symptoms. In people with deficient immune systems, for example those with AIDS or those on long-term corticosteroid or immunosuppressive medication, it may cause severe diarrhea, abdominal pain, malnutrition, dehydration, weight loss, and, occasionally, death. The organism is transmitted by the oral-fecal route, through ingestion of contaminated water or through person-to-person contact. Treatment for people suffering from the condition includes rehydration and antidiarrheal medications. There is no completely effective drug treatment for cryptosporidiosis.

Diseases caused by worms

A worm is an invertebrate animal with a soft slender body and no limbs. Four main groups cause parasitic infections in humans: flatworms (Platyhelminthes) such as the tapeworm, roundworms (Nematoda), ribbon worms (Nemertea) and segmented worms (Annelida). Infection is caused by ingestion of eggs or penetration of the skin by larvae.

Tapeworms

Parasitic tapeworms, members of a group of parasitic flatworms, can infect the intestinal tract. The condition is not contagious. Most sufferers have no symptoms; some people, however, may experience pain in the upper abdomen, diarrhea, unexplained weight loss or symptoms of anemia. Bowel movements will contain worms and worm eggs.

Tapeworm infestation results from eating infected or improperly cooked beef, pork or fish containing encysted larvae. It is most common in Africa, the Middle East, Eastern Europe and South America. A drug may be prescribed to kill the parasite. Hygiene before eating is important, as well as avoiding food which could be infected. Proper cooking provides the most certain protection.

Hydatid disease

Hydatid disease is an infection, usually of the liver, that is common in southern South America, the Mediterranean, the Middle East, central Asia and Africa. It is caused by the larvae of a type of tapeworm *(Echinococcus granulosus)*, which can infest dogs, foxes, wolves, cattle and sheep. Humans become infected when they eat food that is contaminated with tapeworm eggs. The larvae lodge in the liver, where they form cysts which grow slowly for 10–20 years before producing symptoms of a lump and a dull ache on the right side of the abdomen. Occasionally, organs such as the lung, brain and bones may also become affected. X-rays and CAT scans will highlight the cyst(s), which must be surgically removed.

Tapeworms

After they enter the body via infested meat, parasitic tapeworms lodge themselves in the wall of the intestines, and can cause abdominal cramps, diarrhea, nausea and flatulence. In some cases, tapeworms migrate through the circulatory system to affect the liver, lungs and brain.

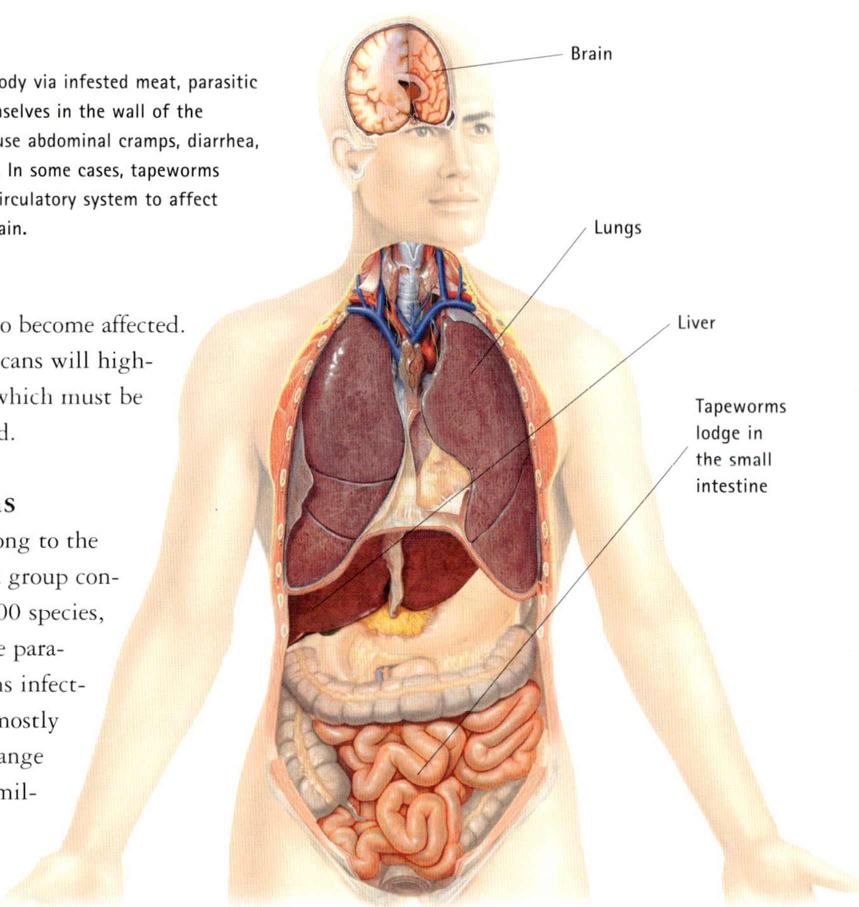

Roundworms

Roundworms belong to the class Nematoda, a group containing over 10,000 species, many of which are parasites. Roundworms infecting humans live mostly in the intestine, range in length from a millimeter to many centimeters, and include hookworms, pinworms and whipworms.

The most common roundworm is a long pale yellow worm with the scientific name of *Ascaris lumbricoides*, which infects the human gastrointestinal tract and lungs.

Ascariasis infection affects an estimated 1 billion people across the world, especially in developing countries where untreated human sewage is used as a fertilizer for food crops. It develops when people consume food, water or soil traces contaminated with roundworm eggs.

In heavy infestations, adult worms can cause intestinal obstructions or malnutrition, and larvae can irritate the lungs. Oral medications are usually effective in eradicating infections although surgery may be needed to clear blockages.

Ancylostomiasis is an intestinal infection caused by the roundworms *Ancylostoma duodenale* and *Necator americanus*.

Pinworms

Pinworms *(Enterobius vermicularis)* are common gut parasites which usually cause only mild symptoms such as itching around the anus, and minor diarrhea. They infect people when their eggs are swallowed and hatch inside the colon. Adult pinworms, which look like white threads up to 1/2 inch (1 centimeter) long, lay their eggs in the rectum. Poor personal hygiene, particularly among children, is responsible for their spread—typically, an infected person scratches their anus, picks up the eggs on their fingers and reinfects themselves or passes the parasite on.

Effective medications are available but, even without treatment, infections usually disappear within weeks if strict personal hygiene measures are taken. Other names include seatworm and threadworm.

Hookworm

The hookworm *(Ancyclostoma duodenale)* is a common parasitic roundworm found in tropical and subtropical climates, and flourishing in unsanitary conditions. The eggs are found in infected human feces, and hatch into infective larvae which can then infect another human being, either by direct contact, usually through bare feet, or by swallowing contaminated soil—a common source of infection for children in areas that have poor sanitation. Where the hookworm enters the skin it creates an itchy patch, hence its other name, "ground itch." Upon entering the body, it travels to the intestines where it attaches to the intestinal wall and sucks blood from it for nourishment. In severe cases, symptoms can include abdominal pain, diarrhea, loss of appetite and weight loss.

A stool sample will usually determine if the infection is severe enough to cause anemia or protein deficiency as a result of the blood loss. This can retard growth as well as mental development in children, and hookworm infection can prove to be fatal in babies. Treatment is usually drugs over a period of one to three days, sometimes along with an iron supplement.

Strongyloidiasis

Strongyloidiasis is a widespread tropical infection of humans by worms. Microscopic worm larvae penetrate the skin and migrate through the lungs, where they may cause cough and shortness of breath. On reaching the intestine they multiply and are excreted in the stools. Diagnosis is suspected by blood eosinophilia and confirmed by microscopic examination of the stool.

Trichuriasis

Trichuriasis is an intestinal infection with the roundworm *Trichuris trichiura*, caused by eating the worm's eggs in contaminated food. The infection is common in tropical areas with poor sanitation, where soil is contaminated with feces. The larvae mature in the intestine and migrate to the colon where they multiply. It may produce no symptoms, but with heavy infestations can cause bloody diarrhea, weight loss, stomach pain and nausea. Extreme cases can result in dehydration and anemia. Treatment is by antiparasitic medication and prevention can be achieved by improving sanitary and waste disposal systems.

Trichinosis

Trichinosis is a roundworm infection in which the invading parasite migrates through the body causing cyst-like growths in the muscle fibers. The infection commonly occurs due to eating undercooked pork infected with larvae of the *Trichinella spiralis* roundworm. Pigs pick up the parasite by eating uncooked garbage. High temperatures kill the larvae; undercooking pork allows them to survive. The larvae mature and reproduce in the intestine, sending new larvae through the circulatory system to the muscles. In most cases there is little or no pain, but severe infection may result in muscular rheumatism.

Treatment is by antiparasitic medication. Infection can be prevented by cooking meat thoroughly at 150°F (65°C) or more, or freezing it for 2 days at –16°F (–26°C).

Toxocariasis

Toxocariasis is an infection by nematode worms of the genus *Toxocara*, parasites found in the gut of the dog and the cat. Infection by *T. canis* (found in dogs) is more common than infection by *T. cati* (found in cats). Eggs from the feces of dogs and cats, found in the soil, can be transferred directly to the mouth via the hands. This infection may have no obvious symptoms, or quite mild symptoms, and can only be detected by a serological blood test to detect antibodies to *Toxocaris* or a liver biopsy. However, the infection may disappear without treatment after 6–18 months.

If symptoms such as skin rash, enlarged spleen, recurring pneumonia or an eye lesion are present, treatment will be necessary, together with the worming of animals and the adoption of hygienic practices, such as washing hands more frequently.

Filariasis

Characterized by irregular swinging fevers, and swelling and inflammation of lymph nodes and vessels, this tropical disease is due to infection by the larvae of the nematode *Wuchereria bancrofti*.

Monkeys, cats and human beings may be infected by the bites of mosquitoes, after which the larvae develop into adult worms, which may grow up to 3 inches (8 centimeters). Over time the adult worms invade the lymph nodes and vessels causing both inflammation and obstruction. Massive swelling of legs, arms, breasts and scrotum can occur (known as elephantiasis).

In the early stages, antibiotics may be required for secondary infection. While drugs can suppress, if not always cure, the disease, surgery may be required when swelling or elephantiasis occurs.

Schistosomiasis

Schistosomiasis is an infection caused by the *Schistosoma* genus of trematode worms, and is second only to malaria in prevalence worldwide. The worms develop in snails and are released into fresh water where their larvae penetrate the skin to infect humans. The most common form of the disease occurs in South America, Africa and the Middle East, while another form occurs in the Far East.

Local inhabitants may not develop symptoms of initial infection, whereas visitors may develop intense itching (swimmer's itch). Weeks later a febrile illness occurs with cough and diarrhea, due to an immune reaction to the developing worms and their eggs in the intestines. Prolonged exposure to the worm infection can result in scarring of the liver (cirrhosis) and in the enlargement of the liver and spleen.

Infection is diagnosed by increased blood eosinophils, antibody tests and microscopic stool examination for eggs. Drugs are available which are active against the infection.

Fungal infections

Fungal infections commonly present as skin disease or as systemic infections. Many fungal infections (for example, Histoplasmosis) cause no problems in healthy individuals, but in people with compromised immune systems (e.g. AIDS sufferers) they may cause serious systemic infection.

SEE ALSO *Antifungals on page 446*

Candidiasis

Candidiasis, also called moniliasis, is an infection caused by the candida fungus (usually *Candida albicans*). This organism normally grows harmlessly in the intestinal tract, mouth and the vagina, but under certain conditions it can proliferate and cause infection. These conditions include damaged skin, moisture and warmth. A person is also more susceptible to candidiasis during pregnancy, when taking antibiotics or corticosteroids, or when the immune system is weakened as in AIDS.

In the mouth and the vagina, candidiasis appears as small, white patches on a red, inflamed background. In the skin, candida forms moist, bright red flat patches with

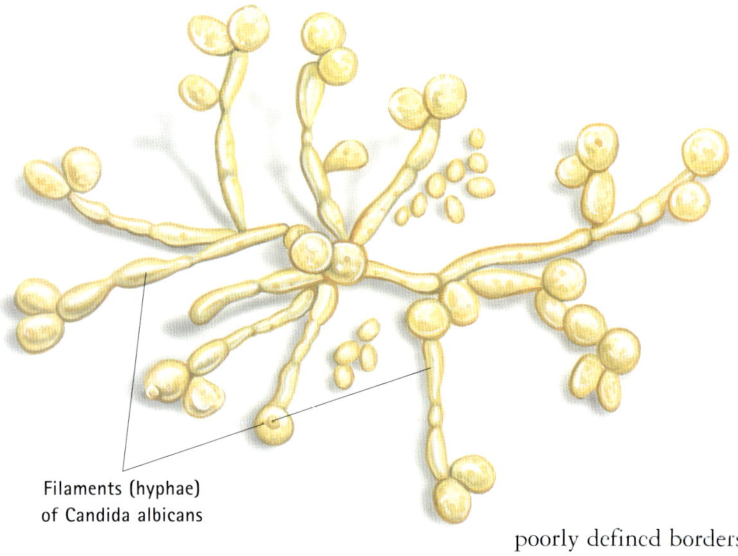

Candidiasis

Candidiasis is a fungal infection that affects the mucous membranes of the body. Infection of the mouth (thrush) causes red, inflamed areas and creamy, white, painful patches to form inside the mouth. It can be treated by applying antifungal drugs directly to the area or by taking medications orally.

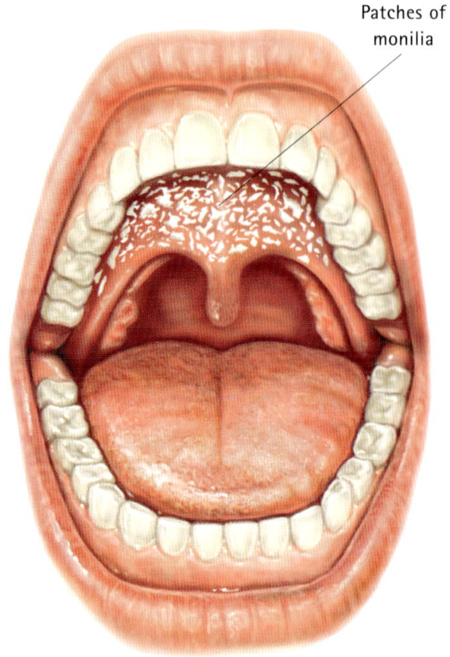

poorly defined borders. These occur usually in the groin, around the anus, beneath the breasts (particularly in overweight women with pendulous breasts), in folds of skin of obese people, and in the armpits. In babies, they may occur as diaper rash. It is important to keep the affected areas cool, dry and exposed to sunlight where possible. Eating yoghurt, buttermilk or sour cream, or taking acidophilus helps prevent candidiasis.

Antifungal topical medications such as nystatin, haloprigin, miconazole, or clotrimazole may help, and are available as vaginal pessaries, creams, and as suppositories.

Tinea

Tinea is a fungal infection caused by various species of *Trichophyton* or *Microsporum*, which grow on the skin, in hair or the nails, with signs such as dry, scaly, red or cracked skin, but no permanent damage. Ringworm, an infection of the skin causing reddish circular marks, is caused by a tinea fungus—not a worm.

Tinea has become increasingly common in recent times because the spores can survive for long periods in flaked-off skin cells, which may be picked up on the bare feet from floor coverings. This is an easy method of transporting infection between users of hotel rooms and sports clubs, and between residents in private homes.

Tinea grows fast in warm moist areas of the body, particularly in skin folds. The skin rash is quite characteristic, appearing as a slightly raised patch with sharp borders, which, as it expands and spreads, leaves a dry or scaly clearing in the center. Between the toes it may cause painful itchy cracks in the skin and a characteristic unpleasant smell. Effects can vary from redness to blisters, which can become sites of secondary bacterial infection, particularly if blood circulation is poor.

Tinea infections are named after the part of the body where they are found. *Tinea capitis* affects the scalp, creating itchy red areas and hair loss; *tinea corporis* is tinea found on any part of the trunk, arms or legs, particularly in skin folds, where it will create red spots that spread as the fungus grows; *tinea cruris*, also called jock itch, is tinea in the groin. There are other forms of the disease as well, such as *tinea pedis*, or athlete's foot, probably the most common form of this fungal infection. It usually presents as a patch of white scaly skin under the little toe—it may not be evident elsewhere; tinea unguium infects toenails and

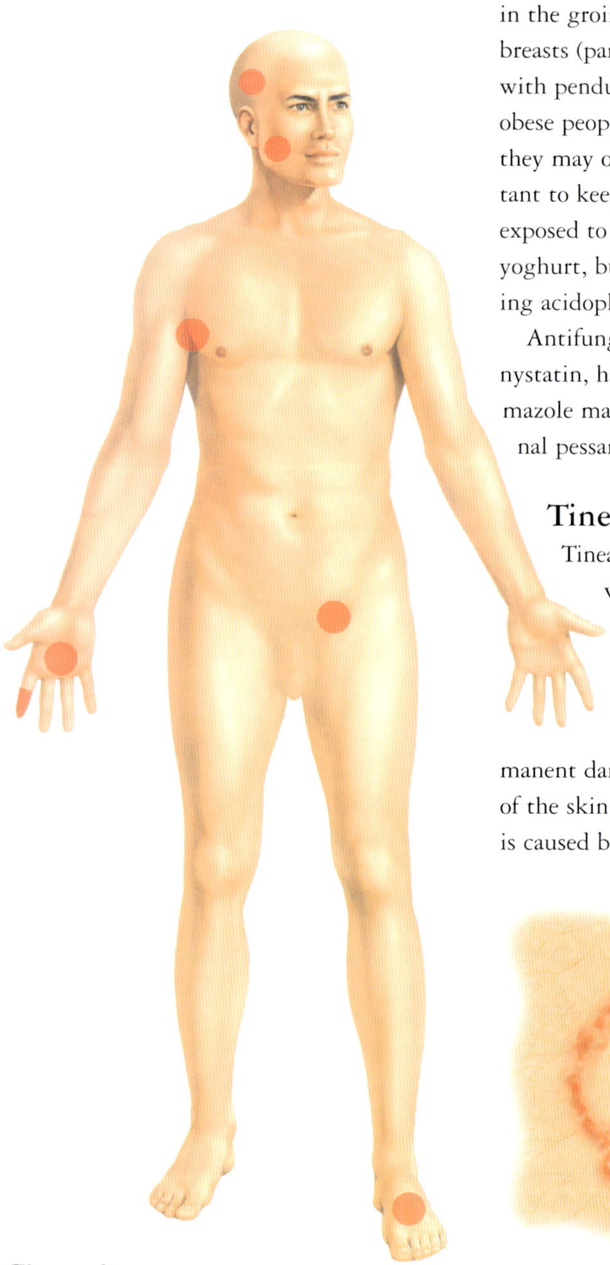

Tinea sites

Tinea is a fungal infection of the skin, especially in warm moist areas of the body, such as between the toes and in the groin area.

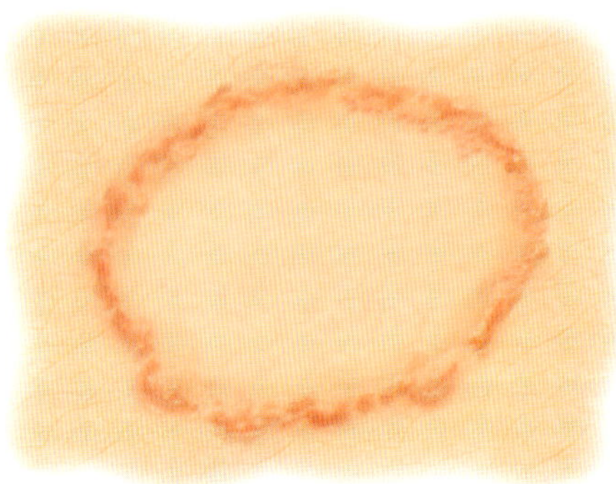

Ringworm

A popular name for tinea, ringworm is a fungal infection of the skin and is not caused by a worm.

Scabies

Scabies is caused by female parasitic mites burrowing under the skin to lay their eggs. An allergic reaction to the mites' feces causes an itchy rash to appear.

fingernails (onychomycosis), appearing as white or powdery patches, thickening the nail or even making it fall off; tinea versicolor, an infection caused by the yeast *Pityrosporum orbiculare*, changes the color of skin and prevents tanning.

Tinea is easily recognizable by the physician. If there is doubt, a skin lesion biopsy can be performed followed by microscopic examination or culture, which shows the fungus responsible.

Most tinea infections are mild and can be treated with antifungal creams or powders, such as those that contain miconazole or clotrimazole. In some cases, topical corticosteroids may be added as well. In cases of severe or chronic infection, oral antifungal medications containing grizeofulvin or terbinafine may be needed.

Antibiotics may be needed to treat secondary bacterial infections, which may take approximately 4–6 weeks to clear.

As the fungal spores are so hardy, it is important to eradicate them from shoes, socks and floor surfaces to avoid continual reinfection. Preventive measures are to ensure feet and toes are completely dry before dressing, to go without shoes and socks when indoors, to avoid skin contact with areas in common use at public places such as in pools or gyms.

Other preventive measure include keeping home floor coverings clean; it may also be wise to to throw away shoes that have been worn for some time.

Parasitic infections

Parasites includes lice, mites and ticks. They can live, eat and breed on the human body, and cause symptoms like itching and rash as well as disorders such as scabies.

Lice

Lice are tiny parasites, about the size of sesame seeds, which live on the skin of the human body where they suck blood for food.

HEAD LICE

Head lice *(Pediculus humanus capitis)* live in human hair. They hatch out from tiny eggs known as nits, which are attached with a glue to the human hair, and can affect anyone. Contrary to popular belief, they do not thrive on dirty hair in unsanitary conditions; they prefer clean hair and are no respecters of social or economic status. Once laid the nits will usually stay attached to the hair shaft, unless dislodged, for approximately 10 days, when they will hatch and mature in about 2 weeks. A female louse can live for up to 30 days and lay about 6 eggs a day.

Lice are passed from one person to another by direct contact, which is why children in constant contact with each other are most likely to spread them. They can be transmitted on combs, brushes, hats, pillowcases and towels. Scratching and occasionally small white specks are the signs.

More easily seen in sunlight and on dark hair than light, nits and lice can be difficult to eradicate. They are mostly found on the scalp behind the ears and near the neckline at the back. Eradication involves the use of a fine "nit" comb, insecticidal shampoo, a warm to hot hairdryer and disinfecting everything the head may have come in contact with—by washing in hot water and drying in hot air or strong sunlight.

Head lice and nits on the eyelids need to be physically removed. Bed linen and stuffed toys that cannot be washed or dry cleaned should be placed in sealed plastic bags for 2 weeks. Carpet and furniture should be thoroughly vacuumed and if possible, disinfected. It is also important that other household members ensure they have not been infected.

PUBIC LICE

Pubic lice or crabs *(Phthirus pubis)* live in pubic hair and thigh area. They are contracted during sexual activity but can also contaminate bedding and clothing. Lice on toilet seats are usually injured and not likely to be looking for a new host. They can live in wet towels and be passed on in a gym or household.

Pubic lice can be easily visible attached to or moving in the pubic hairs, and look like a crab under magnification. They feed on blood and can live up to 30 days, mating frequently in that time. The most common symptom is an itch in the genital area; a rash, or tiny blue spots, may also be evident.

Treatment is with an insecticidal lotion or shampoo (often permethrin). The itching continues for some days after the lice have been killed and all bed linen and clothing must be disinfected in hot water and dried in hot conditions. Bathrooms, towels and linen need to be disinfected by washing with hot, soapy water and hot drying. Sexual contact should be avoided until the infection has been treated successfully so that the lice are not passed on to partners.

BODY LICE

Larger than head and pubic lice, body lice *(Pediculus humanis corpons)* usually live on people in unhygienic, crowded conditions.

These are the lice responsible for carrying diseases such as typhus, trench fever and relapsing fever (or tick fever). They too can be eradicated by hot washing of everything that has come into contact with the body and by the use of an insecticidal shampoo.

Scabies

Tiny mites *(Sarcoptes scabiei)* cause the highly contagious skin condition scabies, spread by close contact with an infected person, by sharing their bed or clothes. Female mites burrow under the skin—favoring hands, toes, groin and bends of elbows and knees—to lay eggs. It takes about 10 days for the mites to mature and continue the cycle. Within weeks, the skin develops an allergic reaction and itches intensely. An eczema-like rash often develops and the mite burrows may be visible. Treatment includes chemical washes, but the itching may persist for weeks.

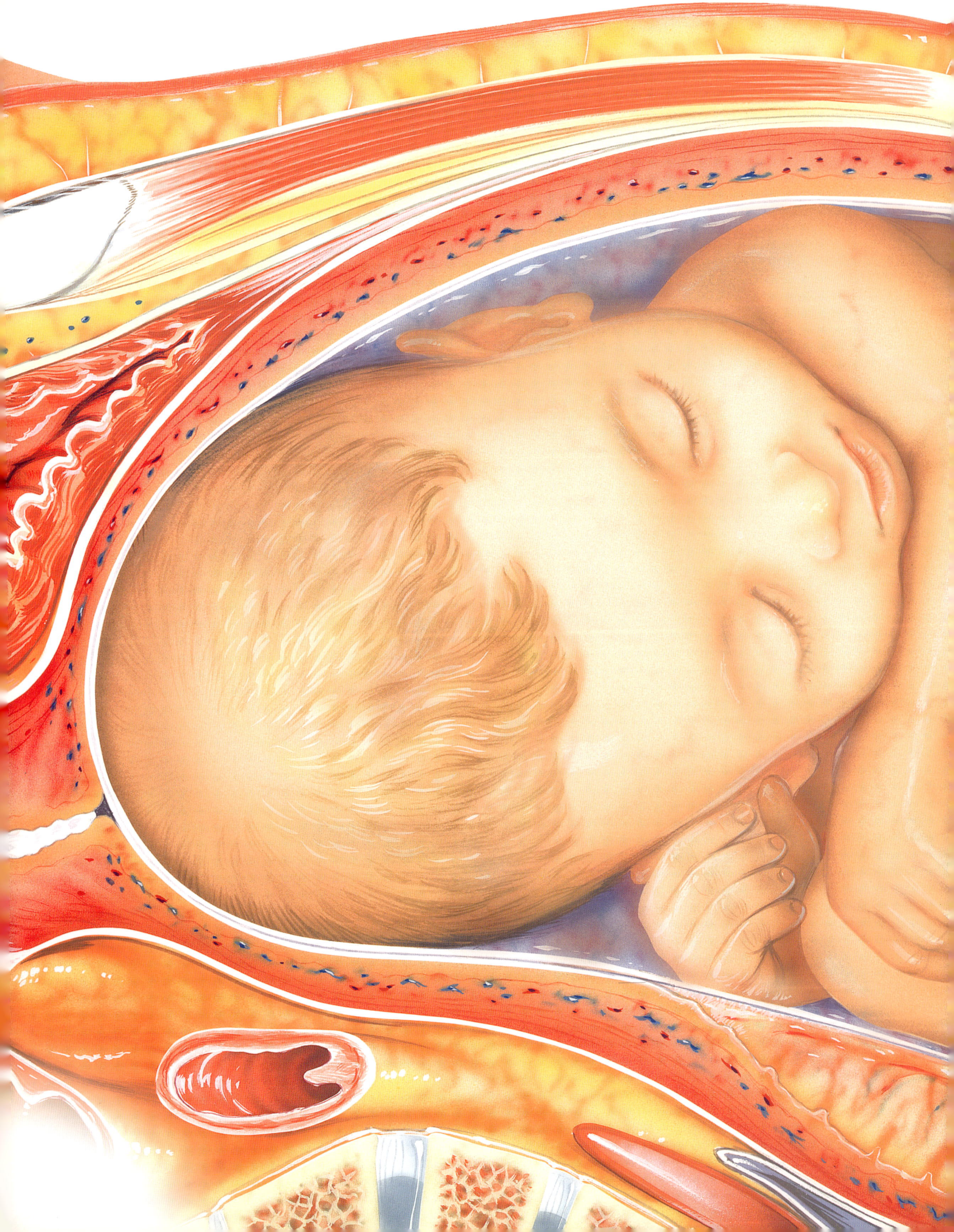

The Human Life Cycle

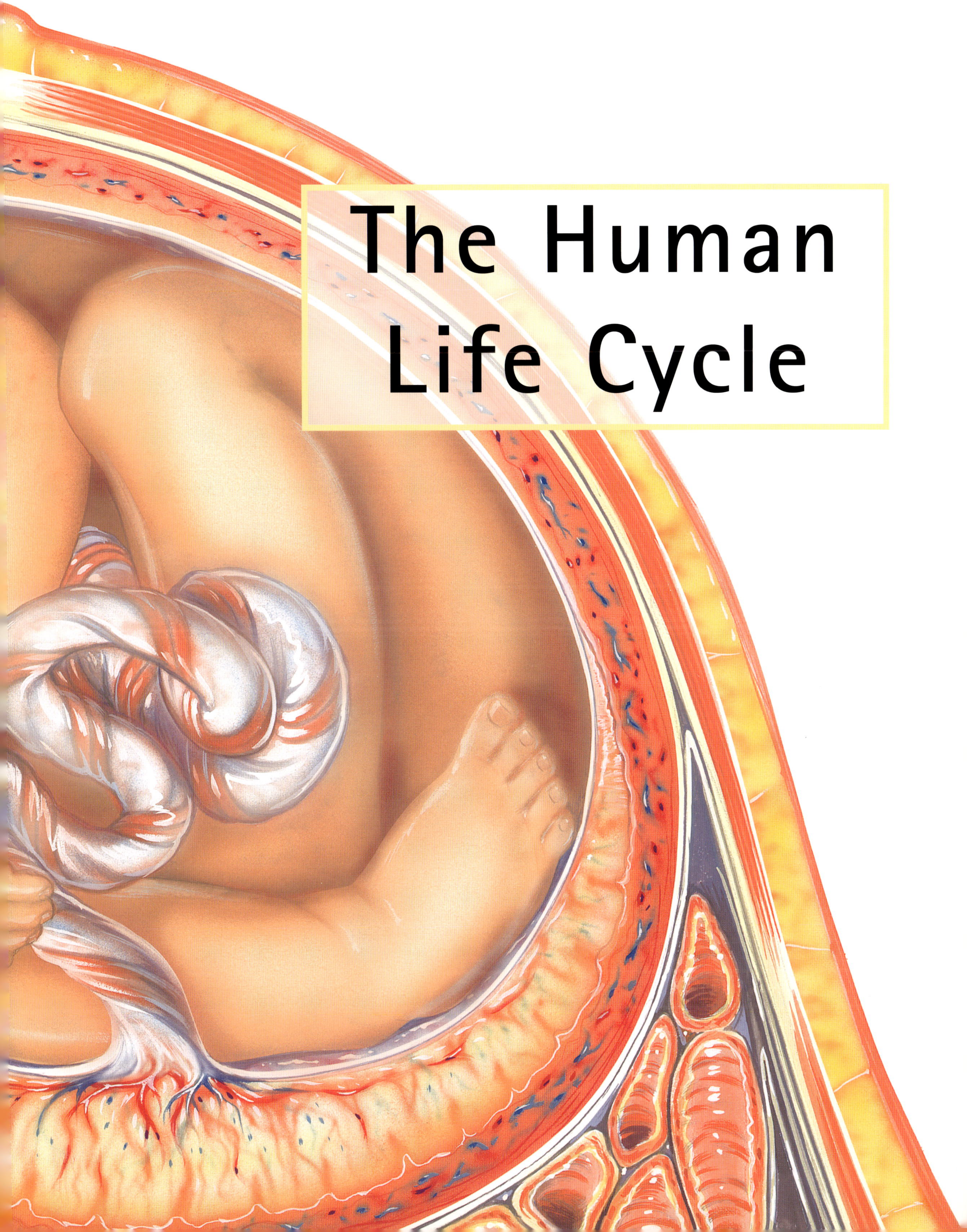

Ageing

A human develops through the stages of infancy, childhood and adolescence to become an adult. After physical development peaks by mid-adulthood, the ageing process begins and continues until death. There is no fixed or universal time frame, but the periods in the human life span are characterized by changes in outward structure and in the function of body systems.

See also *Cells and tissues on page 14; Chromosomes on page 18; DNA on page 17; Genes on page 18; Homeostasis and metabolism on page 112; Hormone replacement therapy on page 450*

Childhood

Childhood begins after the first year of infancy and ends with the changes of puberty. A child is not an adult in miniature. A baby's head at birth makes up about a quarter of the baby's total length; however the eyes are relatively large because they have almost reached their adult size.

Fat cells increase in both size and number in the first two years of life, then change very little after that, except in children who are severely obese.

Children's bones of the arms and legs still have cartilage at the ends. Growth of the cartilage plates causes bone to lengthen. During the first twelve years of life, ossification (bone-forming) centers appear in the heads of long bones (epiphyses) and bone gradually replaces the cartilage.

The first set of teeth appears in the first two years, then from the sixth year they are shed and replaced by permanent teeth.

The brain grows rapidly in childhood: from half the adult size at two years of age to about 90 percent of adult size at age six. By a process called myelination, nerve fibers are progressively coated in myelin, a substance that improves conduction of nerve signals. The cortex of the brain is almost fully myelinated by age seven, when children can use language effectively and can think logically about concrete things that they experience in everyday life.

Adolescence

Adolescence is the stage characterized by structural, physiological and psychological changes that occur when a boy or girl undergoes puberty. The most noticeable structural change is the growth spurt that begins at about the age of 10 or 11 years in girls and about two years later in boys. Height typically increases by 3 inches (8 centimeters) or more each year, and the body proportion becomes more like an adult's. The growth spurt of muscles results in a dramatic increase in physical strength along with endurance.

The appearance of pubic hair is followed by underarm hair one to two years later, and facial hair a year after that. However, growth of fat under the skin layer slows down and the lymphatic system, especially the thymus, even regresses. The sebaceous and sweat glands of the skin are more active, making acne a common problem of puberty.

The key event of puberty is the maturation of the reproductive system. In girls, the onset of menstruation (menarche) usually occurs somewhere between 11 and 14 years of age. The uterus develops its adult shape and the first period occurs. Increased secretion of sex hormones by the ovaries leads to development of breasts and female pattern of pubic hair. The external genitalia also mature, with a thickening of the fat pad over the pubic bone and enlargement of the labia majora and clitoris.

In boys, the testes grow rapidly and begin to secrete testosterone and produce sperm. The first ejaculation usually occurs around age 14 or 15 but may not contain sperm. Testosterone is responsible for secondary sex characteristics such as facial and pubic hair. The scrotum grows, and its skin becomes darker and wrinkled. The penis increases in length and girth, and develops a glans at its tip. Compared to girls, boys have broader shoulders, narrower hips and smaller buttocks; also the cartilage of the larynx (voice box) is more prominent

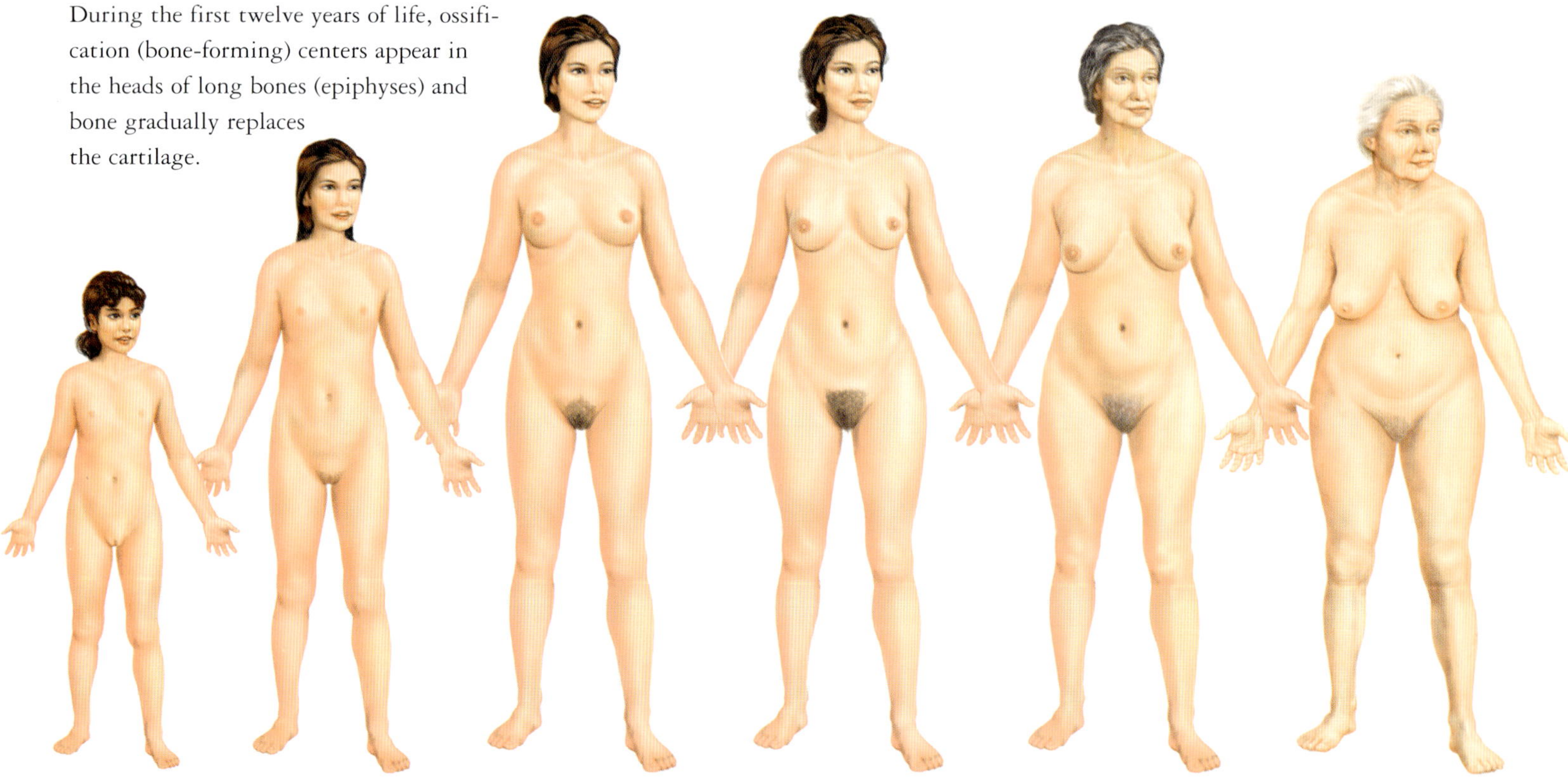

AGE 5 AGE 10 AGE 18 AGE 30 AGE 55 AGE 70

(making what is known as the "Adam's apple") and the voice becomes deeper as the vocal cords are longer. The voice begins to break at about age 14 and this process is complete after about a year.

Puberty also brings about cognitive changes. Adolescents progress beyond the concrete thinking of childhood to develop more abstract and conceptual thinking. But they often see only their own point of view, and have difficulty appreciating the views of others. This transition from childhood to adulthood is a critical time and carries risks of emotional problems, as adolescents have to develop a personal identity, of which sexual identity is an important part.

Adulthood and senescence

There is no definite age for the beginning of adulthood, although in most societies it is legally set at 18 or 21 years of age. Adulthood is understood as the stage when the individual has reached full anatomical, physiological and sexual maturity.

The bones have ceased to grow because the cartilage growth plates have now been entirely replaced by bone. All the body systems are now fully functioning. However, the body continues to change in response to environmental changes, and to physical as well as emotional trauma. The bones, for example, heal after fractures and are continually remodeled to adapt to pressure applied to them. The processes of adaptation and repair are vital for survival of the individual. However, adaptation and repair are not always completely successful. The genetic coding which controls them is altered by external influences such as radiation, or by errors introduced by cell division over the years. This decline in function is called the ageing process. Although old age (senescence) is said to begin at age 65, the ageing process has already begun much earlier.

The structural changes of ageing are well known. As an individual's ages, the skin becomes thin, fragile and wrinkled. On the face, the skin sags and forms bags under the eyes and a double chin. Dark "age spots" appear and begin to increase in number. Baldness begins as hair loss exceeds replacement rate. Poor function of melanocytes (cells that produce the dark pigment melanin) causes hair to turn gray.

Muscle mass is reduced and muscle tone declines. Bones become more porous and weak because of loss of bone substance and minerals, especially calcium, and osteoporosis may develop. Spongy bones such as vertebrae partially collapse, reducing height. Hip and wrist fractures are common. Wear and tear of joints leads to osteoarthritis.

All body systems deteriorate. For many, by 50 years of age, hearing is reduced and the lens of the eye loses some elasticity, making small print difficult to read. For some the lens may also become clouded by a cataract. The immune response is reduced because of the decline in the activation of both T and B lymphocytes, which fight infection. Thus people over the age of 60 should be immunized against common infections such as influenza in winter. The arteries become thicker and less elastic, predisposing the elderly to hypertension. The heart is less efficient as a pump. The lungs begin to lose their elasticity, making breathing more laborious and oxygen exchange less efficient. Digestion becomes slightly less efficient because of reduced secretion of digestive enzymes. Absorption of some important molecules extracted from food, such as calcium and vitamin B_{12}, is impaired. Vitamin B_{12} deficiency leads to impaired nerve function and anemia.

The kidneys shrink and their filtering function declines, resulting in the build-up of waste products like urea. The bladder which collects urine from the kidney and contracts to empty its contents now does not expand and contract as effectively, resulting in more frequent trips to the bathroom. In males, the prostate, which is located under the neck of the bladder and encircles the urethra (the tube conducting the urine out from the bladder), frequently becomes enlarged, constricting the urethra. The reproductive system undergoes dramatic changes.

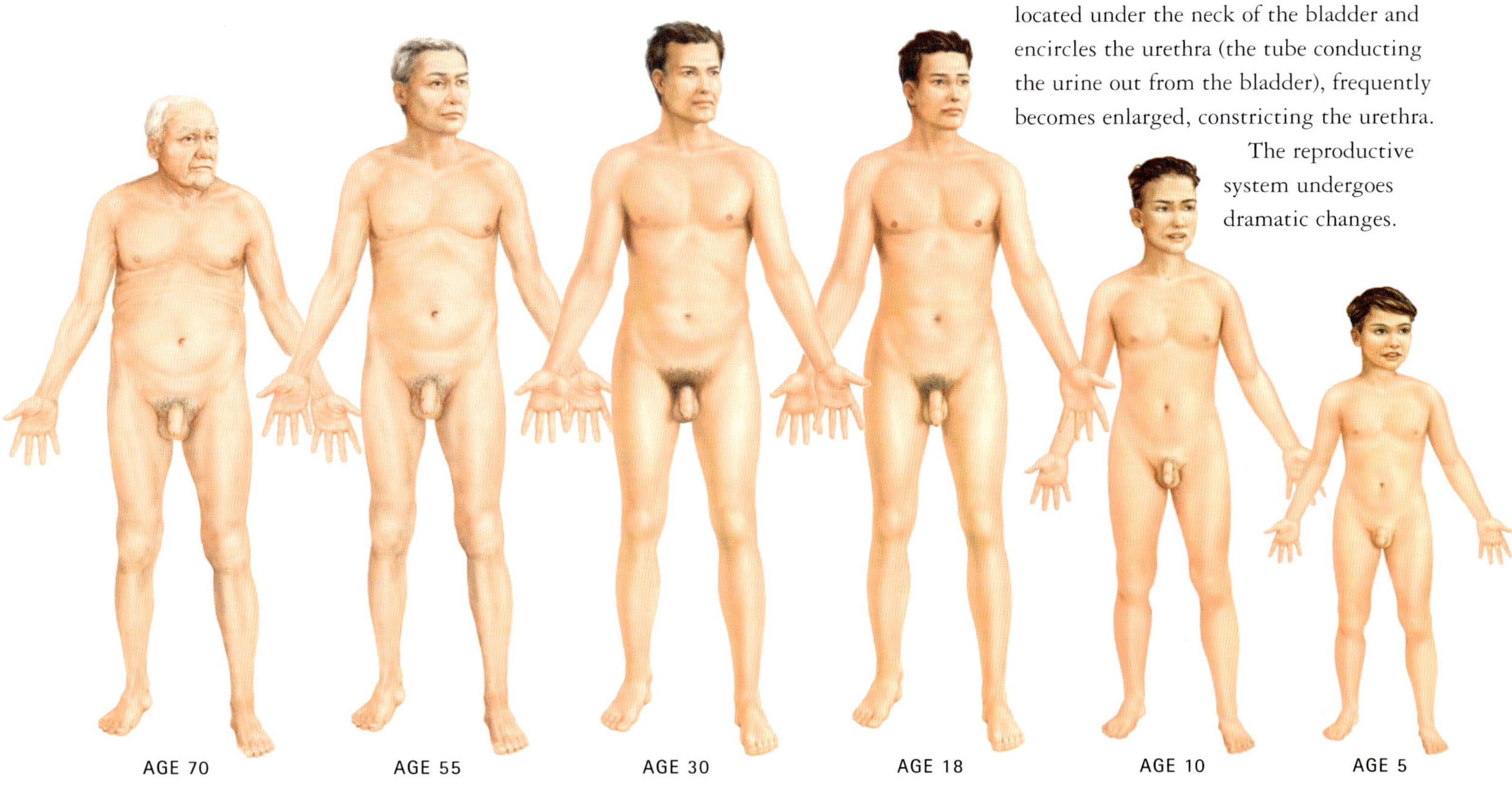

At the climacteric (the male equivalent of menopause), the production of testosterone and sperm in the testes are reduced. The sperm count may be low, but is still enough in some instances to produce offspring until a very old age. Semen volume is also reduced. Erection is more difficult to obtain and to maintain, even in the absence of problems in the blood vessels of the penis. Research is progressing in this field.

In females, the production of eggs and sex hormones by the ovaries is reduced. From the mid-thirties, the risk of chromosomal abnormalities such as Down syndrome increases sharply because the eggs may be more genetically defective. The complex balance of sex hormones begins to be disturbed by the fourth or fifth decade, resulting in irregular menstruation, hot flashes and mood swings. Without the stimulation of the hormone estrogen, the breasts lose their firmness, distribution of fat in the body changes, and pubic hair becomes sparse. The external genitalia become wrinkled because of loss of pubic fat, the vagina is less elastic and its lining thins, making sexual intercourse painful. Loss of the protective effects of estrogen results in a higher risk of cardiovascular diseases. Hormone replacement relieves many of these changes but, recent studies have shown an increased risk of breast cancer. Isoflavones extracted from plants such as soy beans may be a safer alternative.

As the nervous system ages, cognitive function begins to decline. The brain shows increasing atrophy with age, and loss of nerve cells is more localized in some special areas. Attention is not affected by age, but loss of neurons (nerve cells) in the hippocampus results in impaired memory function. Older people take longer to learn new information or a new skill, but can retain the knowledge for as long as younger people. The ability to form new concepts also declines with age, as does performance in intelligence tests, although this only drops significantly after 70 years of age. Dementia is also a problem in the ageing population; Alzheimer's disease and incidents causing lack of oxygen in the brain.

Fertility

Fertility is the ability to conceive a child; it depends on several factors in both male and female reproductive function.

Males must have an adequate number of vigorous sperm cells. Normal human semen has a volume of approximately 0.1 fluid ounces (3 milliliters) per ejaculate, with 60 to 100 million sperm per milliliter. At least 50 percent of sperm should be moving four hours after ejaculation and at least 60 percent of sperm should be normal in shape.

In women, the vagina must be hospitable to the sperm and the mucus of the cervix must be thin enough for the sperm cells to penetrate. The interior of the uterus and uterine tube must allow movement of sperm to the egg. Egg production depends on many factors, including the normal cyclical production of sex hormones by the hypothalamus of the brain and normal production of eggs by the ovary. Once the egg has been fertilized, there must be a favorable environment for implantation of the early embryo into the wall of the uterus; infections and tumors in the uterus may prevent the early embryo developing.

Fertility tests include an examination of a sample of the man's ejaculate, looking at the percentage of sperm with normal shape and movement as well as the absolute number; a hysterosalpingogram and/or a laparoscopic examination of the woman to ensure that the fallopian tubes will allow the passage of sperm cells and fertilized egg; a cervical mucus sperm penetration

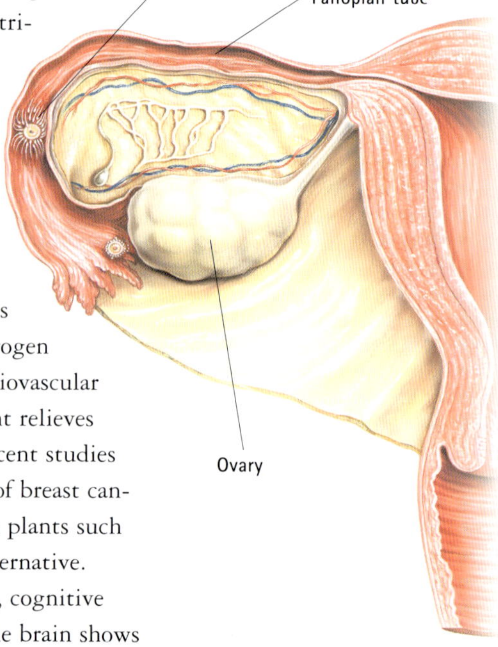

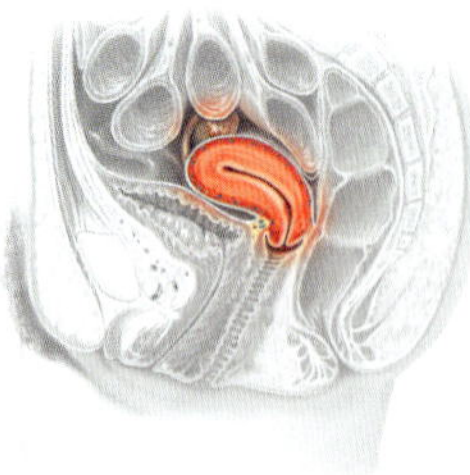

Female fertility

Egg production and a favorable uterine environment are the key factors in female fertility. Once an egg has been produced successfully in the ovaries it moves down the fallopian tubes where it may be fertilized.

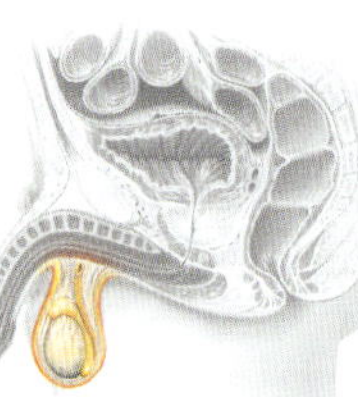

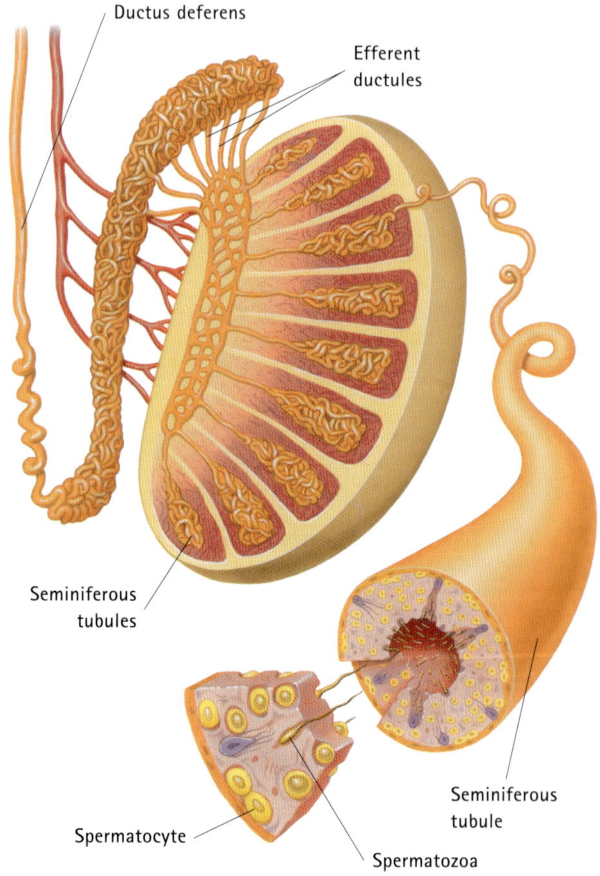

Male fertility

Male fertility is dependent on the production of an adequate number of vigorous sperm cells. Sperm are produced in the seminiferous tubules of the testes.

test to determine if the cervical mucus is hostile to the sperm cells; and analysis of blood levels of the woman's pituitary hormones and progesterone to check for normal hormonal control of egg production and evidence of ovulation. It is also thought to be advisable for a woman to prepare a body temperature chart to help identify times of egg release.

SEE ALSO *Female reproductive organs on page 303; Female reproductive system on page 104; Male reproductive organs on page 297; Male reproductive system on page 105; Sexual behavior on page 107*

INFERTILITY

In a fertile couple having unprotected sexual intercourse two or more times a week, a pregnancy will result within 12 months in 90 percent of cases. Only after this time will the couple be considered for infertility treatment, unless there are unusual circumstances. The term used today is subfertility rather than infertility and it is used to describe failure to conceive after 12 months of regular sex without contraception.

Couples who have never had a baby are said to suffer from primary subfertility; those who have had one or more children are said to have secondary subfertility. After treatment, around half of these couples will achieve a pregnancy and in most cases this takes about 12 months—though it can take up to 10 years. In about 35 percent of cases it is the man who has a fertility problem, in 35 percent it is the woman, while in the remaining 30 percent both have problems.

Subfertility rates in industrialized countries are rising alarmingly, mostly because women are delaying childbearing. Discovering that conception will need medical assistance and may even be impossible has a profound effect on the lives of couples who have planned a child. Counseling as well as medical assistance is important.

SEE ALSO *Female reproductive organs on page 303; Female reproductive system on page 104; Imaging techniques on page 431; Laparoscopy on page 435; Male reproductive organs on page 297; Male reproductive system on page 105; Sexual behavior on page 107*

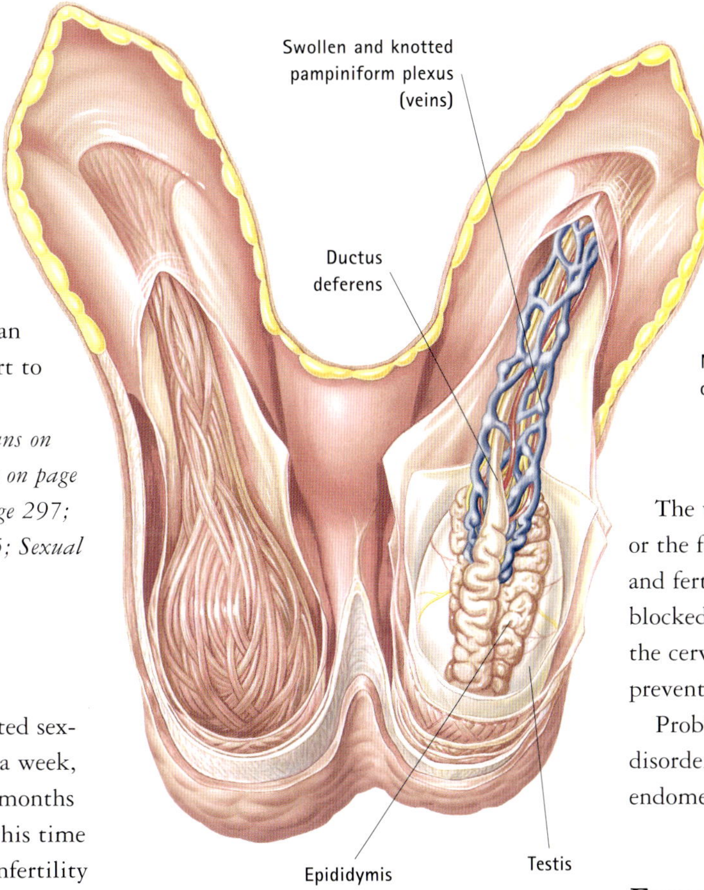

Male infertility—testicular disease

Any damage to, or disease of, the testes and the vas or ductus deferens (which link the testes to the seminal vesicles) can cause infertility. In a varicocele, veins in the testes become swollen and knotted, disrupting blood flow and affecting the production and movement of sperm.

NB: Illustration shows a posterior view of the testes.

Causes of infertility

One of the causes of subfertility can be recurrent miscarriage or recurrent spontaneous abortion (RSA). The two principal reasons for these are, firstly, a chromosomal abnormality that prevents normal development of the embryo or, secondly, something wrong with the woman's reproductive system. Hormonal and anatomical problems, such as a structural abnormality of the uterus, can be responsible for infertility. For some, miscarriage may be "unexplained," which may occur due to immunological factors or could also be due to environmental factors such as exposure to pollution or chemicals.

Male causes may be the failure to manufacture sperm (azoospermia); the sperm may be weak or few in number (oligospermia: fewer than 20 million per milliliter), or the tubes (vas deferens or ductus deferens) that link the testicles to the seminal vesicle where the sperm is stored may be damaged. Problems may be caused by testicular disease; a varicocele (a lump of varicose veins in the scrotum); occupational factors such as working with certain chemicals; general health disorders; mumps; treatment such as radiation therapy; excessive smoking or too much alcohol intake.

The woman may not produce an ovum; or the fallopian tube along which the sperm and fertilized ovum must travel may be blocked. In some cases it is thought that the cervical mucus is hostile to the sperm, preventing them from reaching the ovum.

Problems with sexual intercourse and disorders with the lining of the uterus (the endometrium) can also prevent conception.

FERTILIZATION

Fertilization is the fusion of ovum and sperm to form a zygote, which is a one-celled embryo. This occurs in the ampulla (middle region) of the fallopian tube.

The ovum is released from the ovary along with two important coverings: a thin non-cellular membrane called the zona pellucida, which forms around the ovum while it is still in the ovary and, outside this, some of the cells from the outer wall of the ovarian follicle. The unfertilized ovum can remain in the ampulla for only two days, after which it degenerates.

At ejaculation, 60 to 100 million sperm are deposited in the vagina, but only a few of these travel through the female reproductive tract, with just 200 or so sperms reaching the ampulla. A major barrier to sperm travel for most of the menstrual cycle is the cervical mucus, but this becomes watery and readily penetrable around the time of ovulation. If this does not happen, or if the original number of sperm in the ejaculate is abnormally low, fertilization is unlikely to occur. Sperm motility is also an important factor, as transport to the ampulla is dependent on sperm movement, as well as on contractions of the muscular walls of the female reproductive tract.

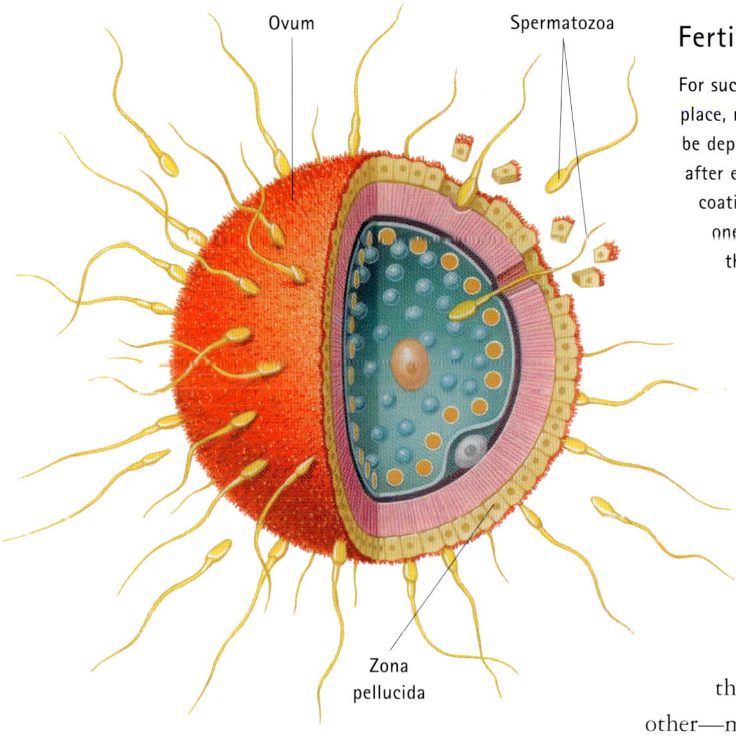

Fertilization

For successful fertilization to take place, millions of tiny sperm must be deposited in the female vagina after ejaculation. They dissolve the coating of the ovum allowing only one sperm to finally penetrate the outer surface, fertilize the ovum and join with it to produce a zygote.

Sperm undergo some maturation in the male reproductive tract and a process called capacitation in the female reproductive tract completes maturation. Capacitation takes about seven hours and is induced by secretions produced in the uterus and fallopian ovum, and the chromosomes of ovum and sperm blend to form the nucleus of a new cell, the zygote. This starts dividing almost immediately and begins to move down the fallopian tube toward the uterus.

Penetration of the ovum by two sperm occasionally happens. The resulting embryo has 69 chromosomes in each of its body cells instead of the normal 46 and has severe development problems which in almost all cases lead to early spontaneous abortion (miscarriage). Some infants with this problem are born alive, but die soon after birth.

Fertilization with the combination of genetic material from two different individuals has the advantage of providing genetic variability in the embryo. This is enhanced by the rearrangement of genes on both maternal and paternal chromosomes during meiosis. It also allows for both male as well as female individuals.

SEE ALSO *Artificial insemination on page 464; Chromosomes on page 18; Female reproductive organs on page 303; Female reproductive system on page 104; Genes on page 18; Male reproductive system on page 105; Male reproductive organs on page 297; Sexual behavior on page 107*

Sex

Sex is the sum of anatomical and physiological features, which divides the members of a species into two groups that complement each other—male and female. In humans the two sexes, sexuality and sexual intercourse are all interwoven with the propagation and survival of the species, although each can also stand apart.

Sex is determined genetically at conception when the X chromosome from the ovum is paired with an X or Y chromosome from the sperm. An XX pairing leads to a genetic female, an XY pairing to a genetic male. Other genetic elements, as yet poorly understood, may modify the expression of the basic genetic coding.

Fetal development

Fetal development refers to the process of the development of the individual in the uterus, from fertilization to the formation of the embryo and the fetus, to birth.

The term "embryo" covers the time from fertilization to the end of week 8 of gestation. After this, the term "fetus" is used. During the first eight weeks the embryo grows from a single cell to a complex multicellular organism. The placenta and fetal membranes develop and all the major organ systems are acquired, although not necessarily in functional form. External features are sufficiently developed by the end of week 8 so that the fetus has a recognizably human form, including the head, arms and legs.

SEE ALSO *Amniocentesis on page 436; Cells and tissues on page 14; Chorionic villus sampling on page 436; DNA on page 17; Spina bifida on page 219*

Fertilization and early development

Fetal development begins with the fertilization of the egg, when the head of a sperm penetrates a mature ovum high in a fallopian tube. Both sperm and ovum have 23 chromosomes each, but after fertilization, the resulting cell, which is called the zygote, has the full complement of 46 chromosomes.

The zygote grows by dividing its cells, a process called mitosis. This process continues as the zygote travels down the fallopian tube, brushed along by fine hairs (cilia). After division has occurred several times, the solid cluster of cells is called the morula, which, after several more days of dividing, becomes a hollow sphere called the blastocyst. Eventually, after about three to five days, the blastocyst reaches the uterus.

Further division sees the blastocyst separate into a cluster of inner cells in one part of a fluid-filled sac with an outer layer known as the trophoblast. The cluster of inner cells continues to divide until three separate layers of cells have formed.

These are the so-called germ layers from which organs will form. The outermost layer is called the ectoderm; the innermost, the endoderm. Between these two layers develops a third layer, the mesoderm.

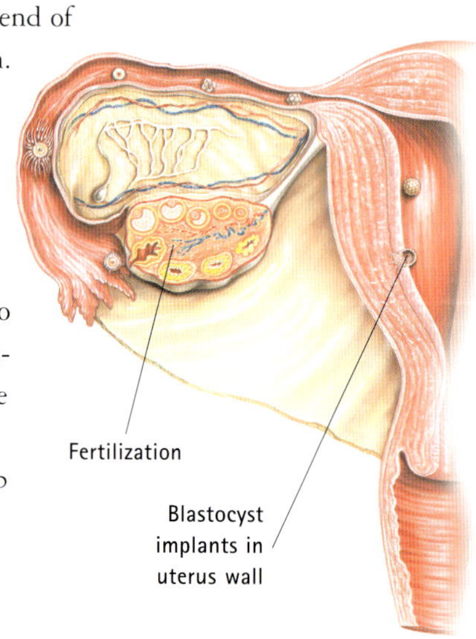

Pregnancy—early stages

Once fertilized, an ovum (egg) is called a zygote and begins to divide immediately. This developing mass moves along the fallopian tube and reaches the cavity of the uterus 5–6 days after fertilization. The mass of cells (now called a blastocyst) implants in the wall of the uterus and begins to develop into an embryo.

Now the blastocyst implants itself in the lining of the uterus (endometrium) until, about ten days after fertilization, it is completely buried. By this time, from the germ cells, unique tissues and organs begin to take shape. Bone, muscle, heart, connective tissue, blood cells and vessels begin to grow from the mesoderm. From the endoderm, digestive and respiratory tracts form, and from the ectoderm, skin, hair, and the tissues of the central nervous system.

Formation of the embryo and amnion

The inner mass of cells undergoes rearrangement to form a flat elongated sheet covered by a hollow balloon-like structure (rather like a gondola under a hot air balloon). The hollow structure will form the amnion, a fluid-filled sac in which the embryo will eventually float, connected to the placenta by the umbilical cord. The flat sheet of cells will become the embryo. The amnion does not form part of the embryo's body but its normal functioning is important in development. If the amniotic sac is punctured prematurely, loss of amniotic fluid may lead to miscarriage or infection.

Damage to the amniotic membrane may cause the formation of strands of tissue (amniotic bands) within the sac. If one of these becomes wrapped around a developing body part it may constrict its blood supply and cause distortion or loss of the part. Amniotic bands can usually be detected by ultrasound in some cases.

If there is insufficient amniotic fluid the loss of its cushioning effect can also result in pressure damage to the developing embryo; this mostly results in spontaneous abortion (miscarriage). A sample of amniotic fluid (obtained by amniocentesis, a common test for fetal abnormalities) contains molecules and cells shed from the embryo that can be used to detect some abnormalities such as Down syndrome and spina bifida. It is difficult prior to week 14 to safely remove enough amniotic fluid for diagnosis, so this is not done during the embryonic period.

FORMATION OF THE PLACENTA

About two weeks after fertilization, blood vessels begin to develop within the embryo. At the same time, tiny extensions from the outer trophoblast layer of the blastocyst reach out into the endometrium and become blood vessels, joining with the mother's circulation. They develop into an essential organ called the placenta.

Across the placenta, separated by only a few layers of cells, the mother's blood circulation and that of the embryo come into close proximity. Across this barrier, oxygen, nutrients and antibodies (infection-fighting proteins) are able to pass from mother to embryo and waste products can pass from embryo to mother. The placenta also secretes hormones that maintains the endometrium.

Organ development

Considerable rearrangement is required for a flat sheet of cells to become a recognizably human embryo. Folding and growth during week 3 results in a rolling under of the sides, head and rear of the embryo so that it becomes a hollow tube with the gut suspended inside it.

This coincides with the start of formation of many of the organ systems. The diaphragm and the compartments in the chest cavity which contain the lungs and heart begin to develop just after folding. These are important for normal postnatal function and in embryonic development.

HEART AND BLOOD VESSELS

The heart and blood vessels are the first organs to function. This is important as even a small embryo is so functionally active that it requires a good blood supply.

During week 3, groups of blood vessels form throughout the embryo and, as they grow, join and form a network. This network communicates with the heart which develops separately. The heart then starts to beat. Heartbeat can be detected by ultrasound during week 5 of gestation. The heart is initially just a simple tube; the partitions which will convert it into a 4-chambered structure that will develop later.

Fetal skull development

In the skull, as in the rest of the skeleton, bone gradually spreads out to replace cartilage.

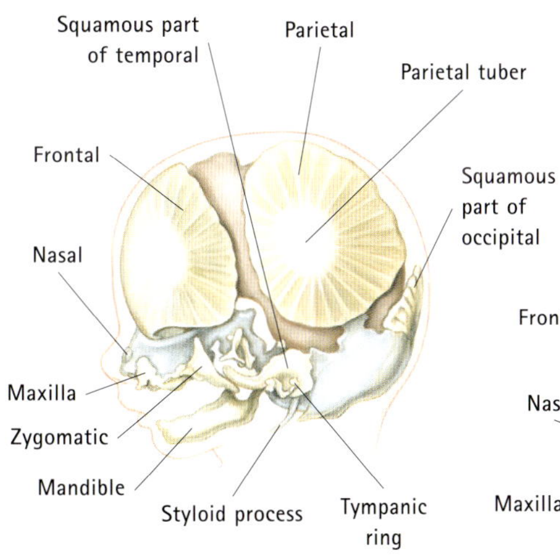

12 WEEKS
Cartilaginous material has been laid down, and has become a base for the skull.

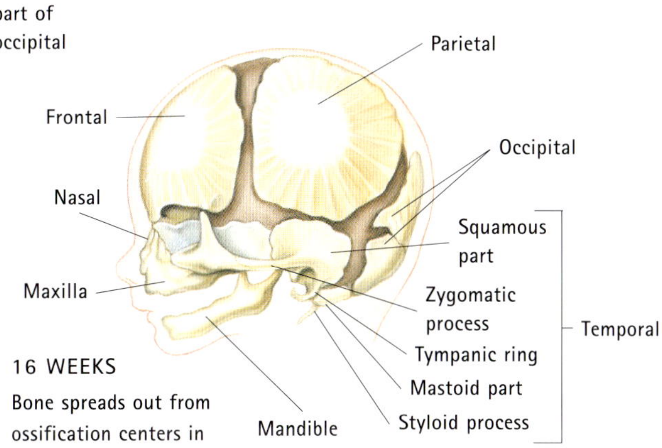

16 WEEKS
Bone spreads out from ossification centers in the cartilage.

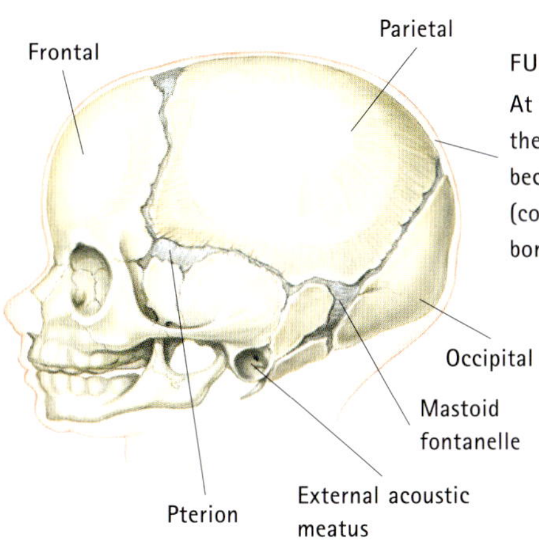

FULL TERM
At full term, the skull has become ossified (converted to bone).

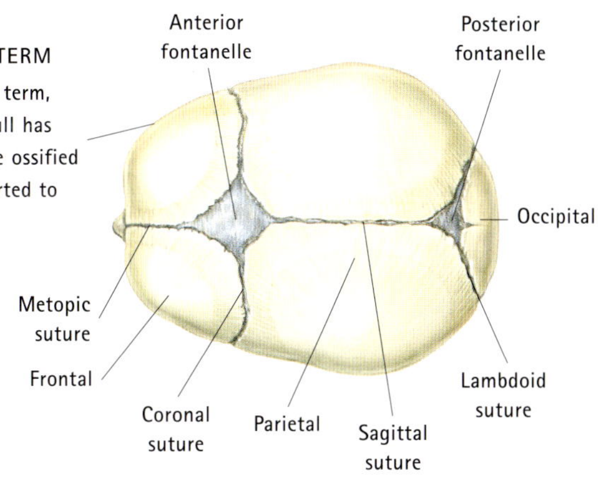

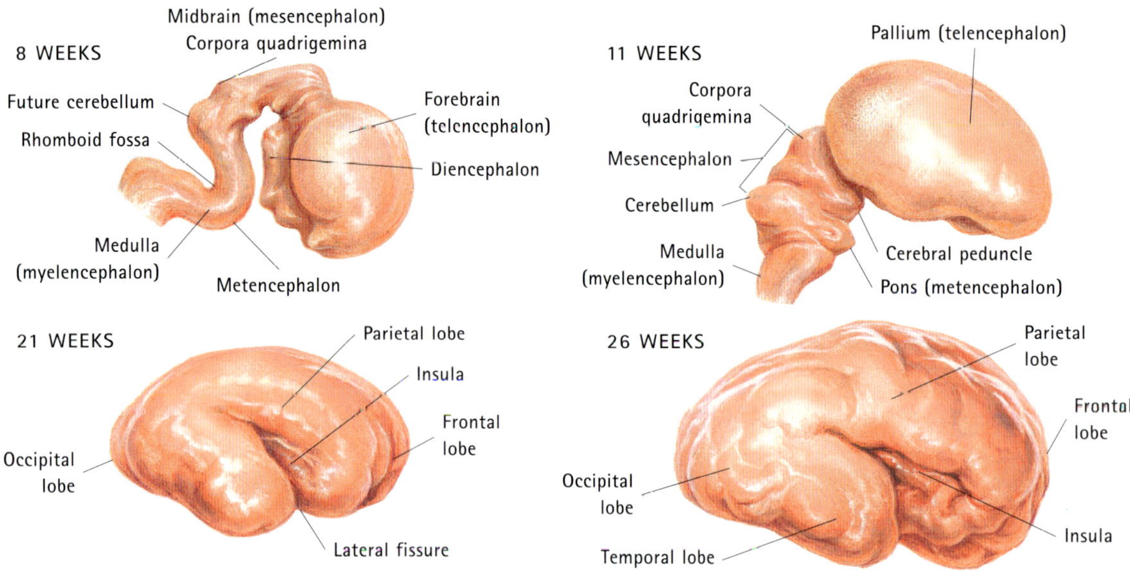

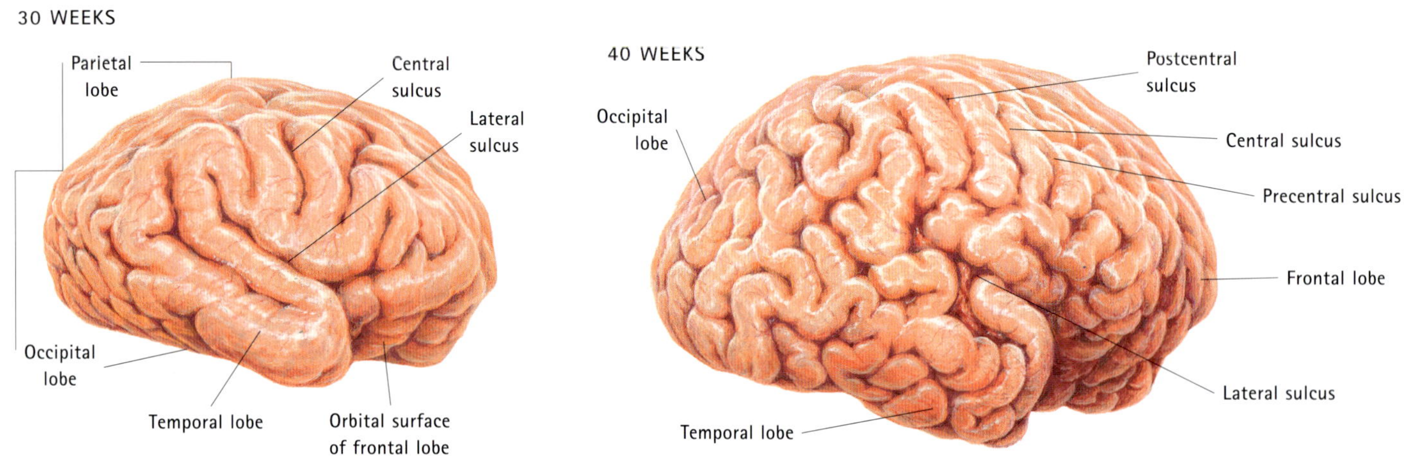

Fetal brain development

Over a period of 9 months, the primitive neural tube forms into the prosencephalon, the mesencephalon and the rhombencephalon, which in turn develop into the various sections of the mature brain. By 4 weeks the prosencephalon has become the telencephalon and the diencephalon, and the rhombencephalon has become the myelencephalon and the metencephalon. The telencephalon develops into the cerebral hemispheres. At full term, all the surface features of the adult brain are already present.

BRAIN AND SPINAL CORD

Development of the brain and spinal cord (the central nervous system or CNS) is initiated during the third week when a shallow groove forms along the back of the embryonic disk. This groove becomes deeper and the edges fuse together to form a tube. The fusion starts in the region of the neck and extends forward and backward to fully enclose the tube by the end of week 4 of gestation. Sometimes an abnormal opening remains at one or, rarely, both ends.

After closure of the tube, there is extensive further growth and development of the CNS which continues during the first two years of postnatal life. It is important to realize that the CNS is prone to abnormal development in the embryonic period due to extensive cell division. A common cause of CNS malformations is heat stress, which can be caused by infections or strenuous exercise in early pregnancy. The hollow center of the neural tube remains and becomes filled with cerebrospinal fluid. The vertebrae and skull form around the developing CNS.

Continuing growth of the CNS is possible after bone formation because of remodeling, that is, resorption of old bone from the inside and laying down of new bone on the outside, which results in a larger structure of the same shape. In the case of the developing brain, the fontanelles and the flat bones of the skull also play a role in permitting growth.

Some nerves grow out from the CNS to innervate muscles and organs of the body. Other nerves, including those which carry sensory information, develop separately and later make communication with the appropriate parts of the CNS.

DIGESTIVE SYSTEM

The digestive system starts as a simple tube, formed when the underside of the flat embryo is rolled inward during folding. The upper part becomes the esophagus and this joins to the mouth when it develops. The part of the tube below the diaphragm expands to form the stomach. Outgrowths from this region will form the liver, gallbladder and pancreas. The rest undergoes enormous change to form the intestines.

The growth of the intestines is so considerable that they temporarily herniate into the body cavity in the umbilical cord during weeks 6 to 10. After this the abdominal cavity has enlarged sufficiently to permit return of the intestines to their normal position. This procedure occasionally fails and the infant is born with a type of umbilical hernia.

RESPIRATORY SYSTEM

The respiratory system arises at the end of week 4 as a tubular outgrowth from the front of the upper digestive tract. The first part of the respiratory system remains unbranched and forms the larynx and trachea. Development of the other respiratory structures involves a process of repeated branching and later differentiation. The first branches are the main lower airways (bronchi). Later branching forms the smaller airways and the alveoli where, in postnatal life, oxygen is supplied to the blood.

The lungs are immature until around 32 weeks of gestation; an infant born before this has a reduced chance of survival, though with sophisticated treatment some do survive from as early as 23 weeks. The lungs develop slowly because the embryo lives in a fluid rather than a gaseous environment, and the lungs are non-functional before birth, oxygen being obtained from the placenta via the umbilical arteries.

REPRODUCTIVE SYSTEM

The gonads develop at about the same time in the same region of the posterior body wall as the mesonephros. By week 7 it is possible to distinguish the ovaries from the testes. Most of the development of the gonads and other reproductive organs takes place during fetal life, including relocation of the ovaries into the pelvis in the female and the testes into the scrotum in the male.

FACE AND JAWS

The face and jaws start to develop during week 4, arising from a series of thickenings on either side of the developing head, which fuse in the midline and rearrange to form small but recognizable jaws, mouth and nose by the end of the embryonic period.

EYES AND EARS

The retina of each eye starts out as an outgrowth of the brain, which then elongates toward the surface of the head. As the retina goes through the process of development, it assumes a shape rather like a cup on the end of a stalk. The cup becomes the retina and part of the iris, and the stalk becomes the optic nerve. The presence of the developing retina induces superficial structures of the head to form lens, cornea and eyelid. Final maturation of the retina takes place after birth as a result of exposure to light.

The inner, middle and external ears all develop from surface structures on the side of the developing head. The precursor to the inner ear (which contains the organs of hearing and balance) starts as a hollow vesicle which sinks into the underlying tissues, undergoes profound development and connects to nerves from the brain during weeks 5 to 7 of gestation. The middle ear is filled with air and contains three tiny bones (the ossicles), which transmit sound from the external ear to the organ of hearing. These bones develop near a tubular outgrowth from the pharynx, which will become the eustachian tube. The external ear grows inward to make contact with part of the middle ear to form the eardrum.

Many noxious influences, such as rubella (German measles) or heat stress, can disrupt the normal development of these complex organs. Rubella or heat stress during the embryonic period can cause congenital cataract (lens opacity) and deafness.

LIMBS

Limbs start as small protrusions from the surface of the embryo during week 4. The ends of the limb buds become paddle-shaped and during week 6 develop thickenings where the fingers and toes will develop.

Controlled (or programmed) cell death is an important mechanism in many parts of the embryo for disposing of tissue that is no longer required. The skeletal support of both the limbs and digits is initially cartilage, which is then replaced by bone from week 7. It is not known what actually determines the development of an arm rather than a leg or why particular fingers or toes form where they do.

Limb abnormalities are common and most are caused by genetic factors. Drugs such as thalidomide can be involved as well. The most critical time is during weeks 5 to 7, when the basic pattern of the limbs is structured.

KIDNEYS

Three successive sets of kidneys develop in intrauterine life. The first set is rudimentary and the second, the mesonephros, functions from weeks 4 to 9. It is then gradually replaced by the final pair of kidneys, called the metanephrons. Renal function is important in intrauterine life as the urine helps to maintain the correct volume of the amniotic fluid.

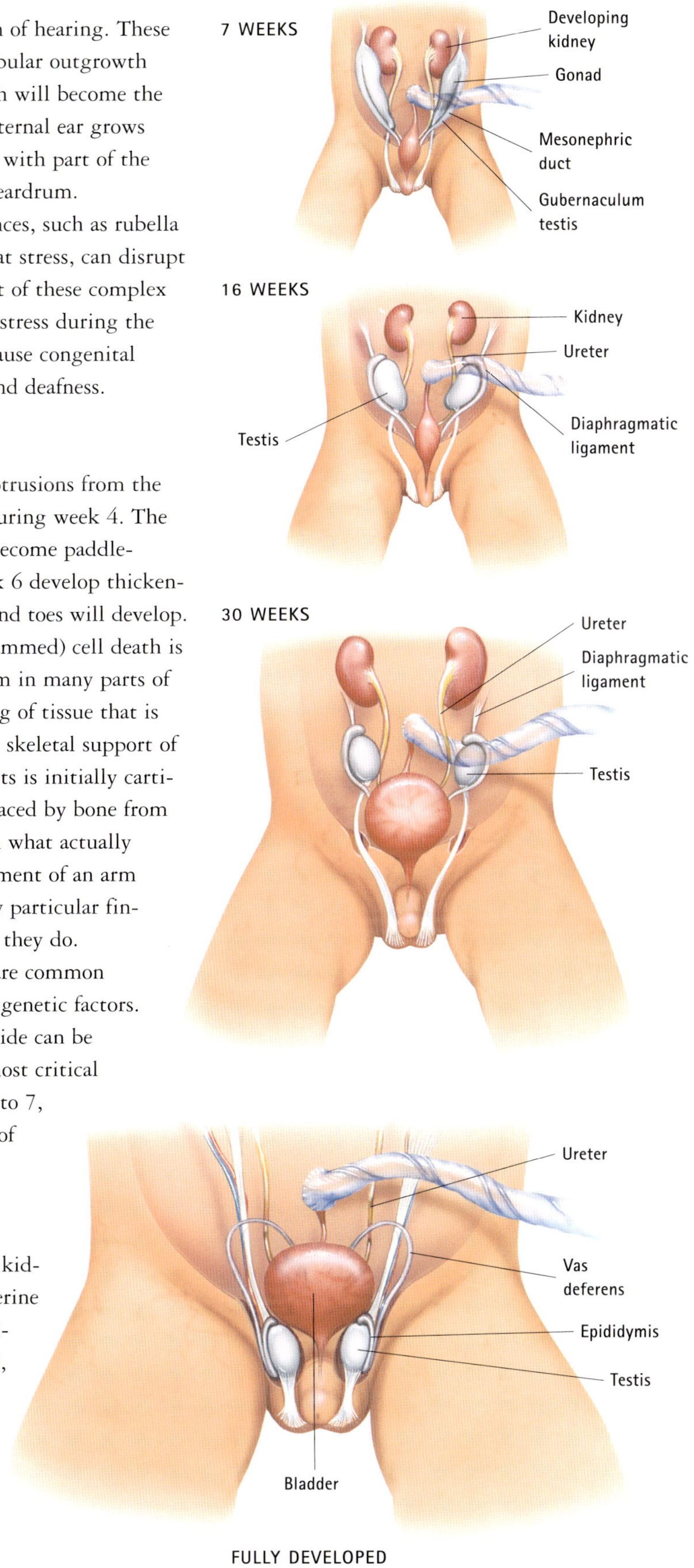

Descent of testes

The testes (male gonads) form in the embryo from a piece of tissue at the back of the abdomen, near the kidneys. When they are fully developed at about 30 weeks, the testes make their way down the inguinal canal, reaching the scrotum as the time of birth approaches.

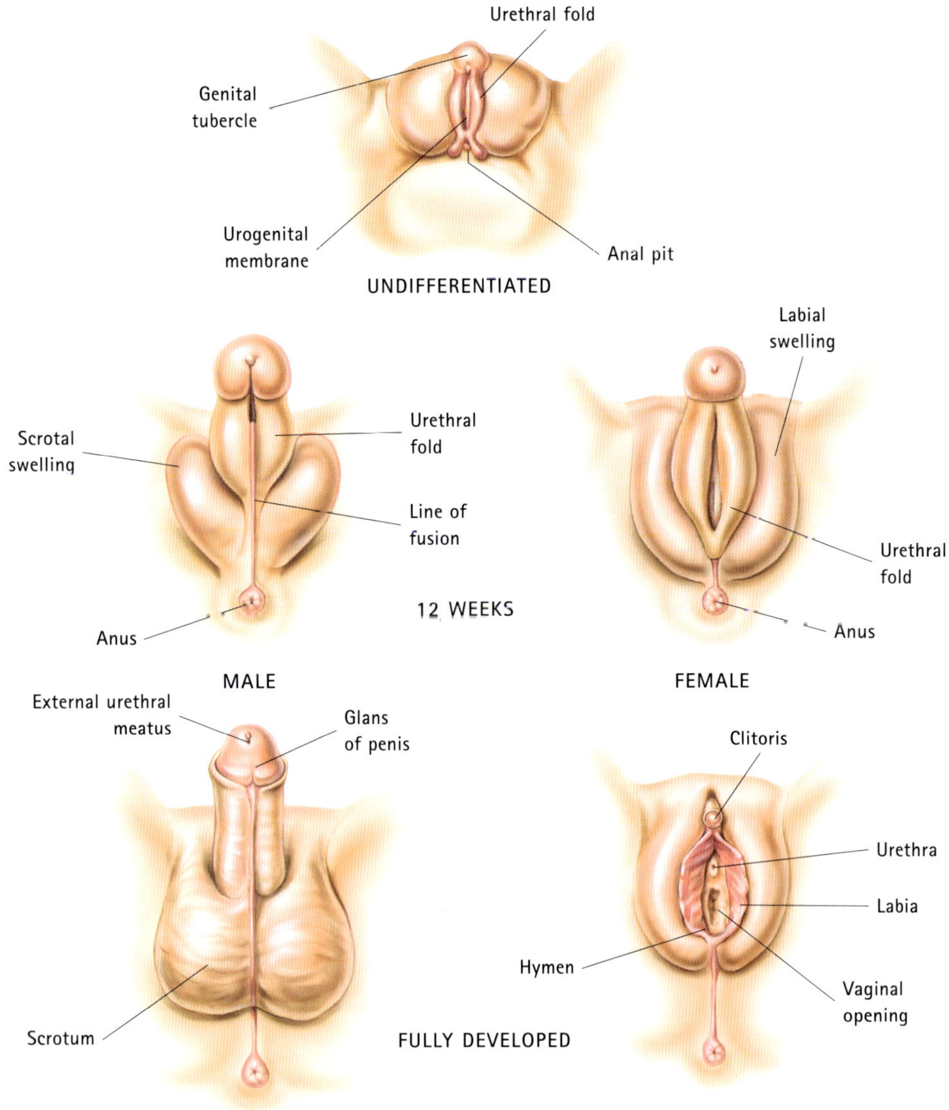

Fetal sex differentiation

Up to 12 weeks, there is no difference between the male and female external genital organs, which consist of a genital tubercle, a urogenital membrane, a pair of urogenital folds and a genital swelling. After 12 weeks, these differentiate into the penis and scrotum in the male, and the clitoris and vulva in the female.

Growth of the fetus

From eight weeks, the embryo is called a fetus. It is only tiny, at 1 inch (2.5 centimeters) long, and weighing about 1 gram, but all the human features are present. Limbs, eyes, ears, nose, mouth can all be seen, and the internal organs such as the heart, kidneys, liver, lungs, brain and digestive tract have formed. The limbs have begun to move (though the movements are not discernible by the mother until the twentieth week). From now on the main changes will be in size. Growth becomes more rapid; from now on it grows at about $\frac{1}{20}$ inch (1.5 millimeters) each day. During this embryonic stage, material (either cartilage or membrane derived from mesenchyme) is laid down. This becomes a template for the skull. The base of the skull develops from cartilage, which gradually becomes ossified (bony) as bone spreads out from ossification centers in the cartilage.

The rest of the skeleton develops in a similar way. Gradually, bone spreads out and replaces the cartilage until, by birth, much of the skeleton is fully ossified.

The genital organs begin to develop in the second month, but there is no difference in appearance between the sexes until about the seventh week. After this the sex glands develop differently, becoming the testes in the male and ovaries in the female. Both the testes and the ovaries gradually move to lower positions in the body, with the testes coming to lie in the scrotum by the end of the eighth month. In the male, a pair of tubules form that join up to the testes, and then open into the urethra; pouches in the ducts become the seminal vesicles. In the female, a pair of ducts also develops; one end of these ducts comes to lie next to the ovaries; the other end fuses into a common tube that becomes the uterus and vagina.

In both sexes, the external genital organs develop from a genital tubercle, along with a pair of urogenital folds with genital swellings on either side of the fold. At twelve weeks these differentiate into external male or female genitalia; in the male the tubercle and the united urogenital folds become the penis and the genital swellings fuse together and become the scrotum, while in the female, the tubercle becomes the clitoris and the urogenital folds and genital swellings become the lesser and greater lips of the vulva.

By twelve weeks, the fetus has a definite face with discernible facial expressions, though the head is disproportionately large because of development of the brain. In the eleventh or twelfth week the external genitals become evident, and by the fourth month the fetus is clearly recognizable as human. By twelve weeks, tiny nails are growing on its fingers and toes. The external ears, the eyelids (which will remain fused until the sixth month or so), and 32 permanent teeth buds have formed.

During the fourth month, simple reflexes have developed, and the mother first becomes aware of the movements of the fetus. The fetus will now respond to stimuli, such as a loud noise or a change in the mother's position, by moving vigorously. During the fifth and sixth months a downy covering (lanugo) develops on the body, and the body becomes increasingly larger in proportion to the head.

Weight gain begins during the seventh month, when fat is deposited under the skin all over the body. In the last weeks before birth a special brown fat will also be deposited in the upper part of the body. During the seventh month the skin, which is red and wrinkled, is covered with a creamy white substance known as vernix, produced by glands under the skin. This keeps the fetus waterproof. At 40 weeks, the fetus weighs 6–9 pounds (2.5–4 kilograms), is about 20 inches (50 centimeters) long, and is mature and ready for birth.

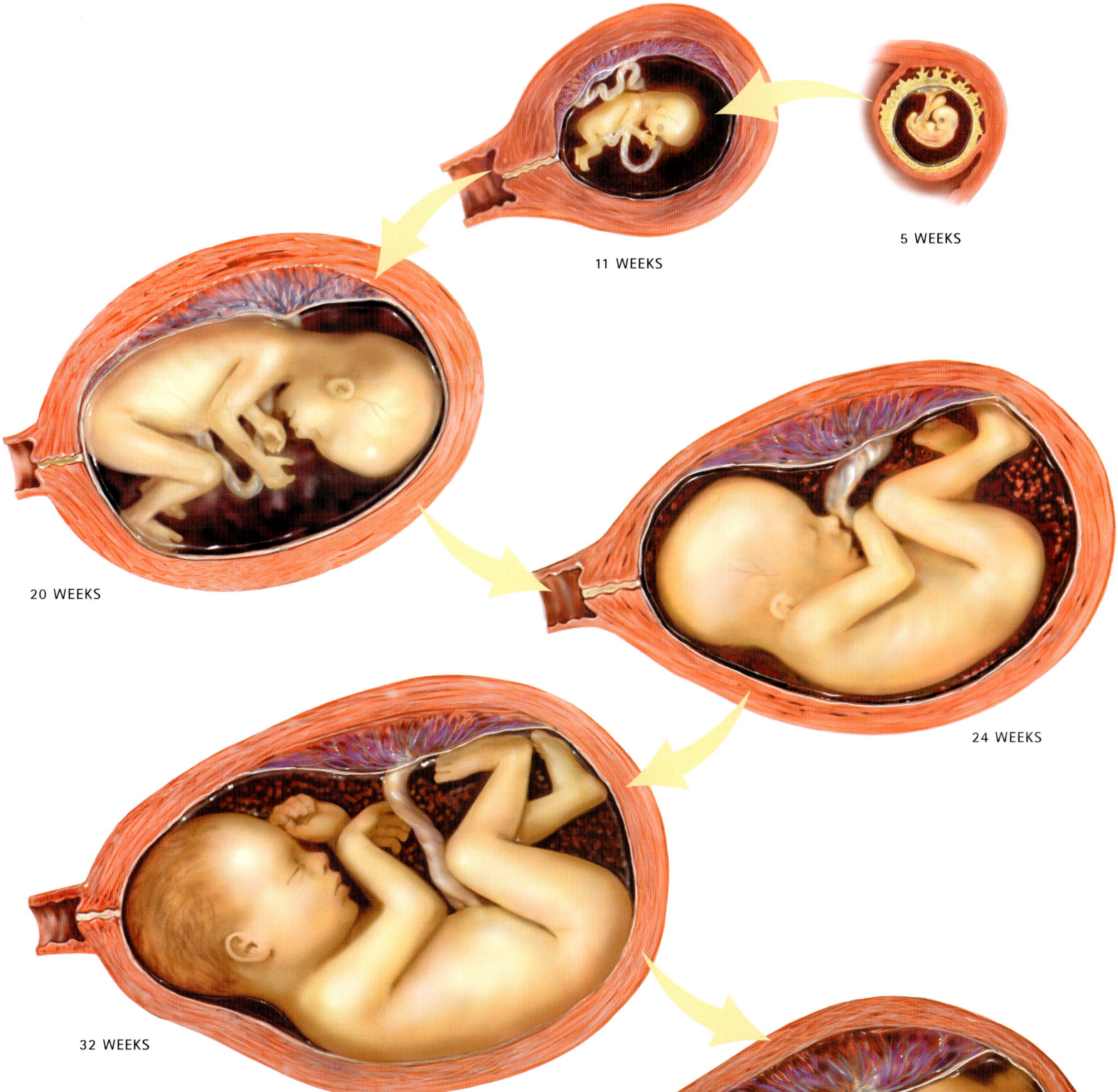

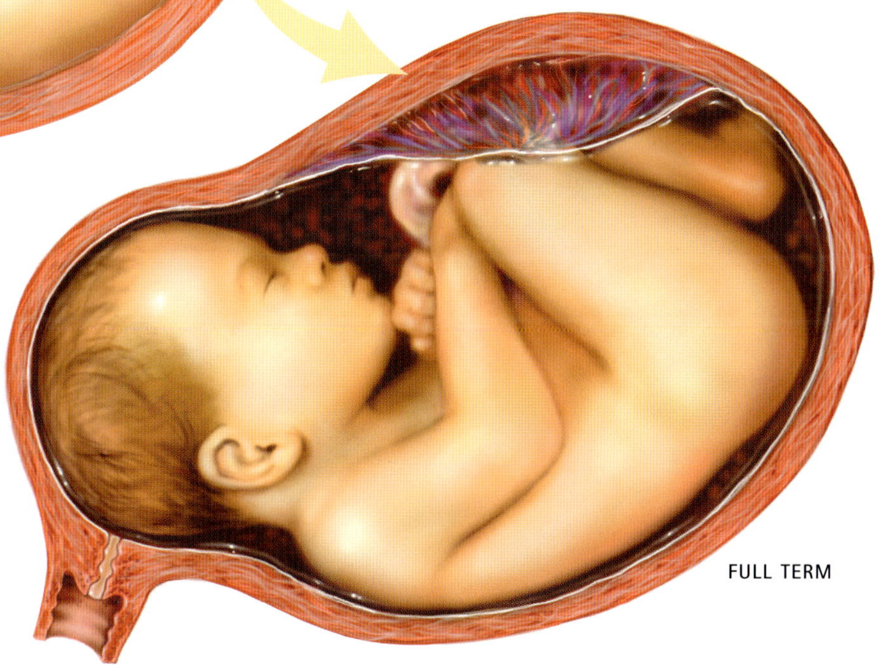

FETAL DEVELOPMENT CYCLE

Fetal development is a gradual process, but certain events take place at particular times. At 5 weeks, facial features have formed. At 11 weeks, the organs have developed. By week 20 the fetus is active, and movements are easily felt. By 24 weeks most of the organ systems become ready to function normally. At 32 weeks the growth rate has slowed down but weight has markedly increased. At 40 weeks, the fetus weighs 6–9 pounds (2.5–4 kilograms) and is ready for birth.

The physician or the obstetrician will monitor fetal development in various ways—through measurements of the mother's weight and abdominal girth, by listening to the fetal heartbeat through a special stethoscope, and by ultrasound, which is performed routinely during pregnancy.

Fetal abnormalities

Several factors affecting the mother's health can cause delayed growth of an otherwise normal fetus. These factors include poor nutrition, heart disease or high blood pressure, smoking, drug dependence, multiple pregnancies, heart disease, preeclampsia or eclampsia, and high altitude.

Babies born to these mothers may be below normal weight. The general outlook for the development of these infants is poorer than for normal weight children.

Sometimes the fetus may develop abnormally. A disorder of development while the fetus is growing in the uterus is known as a congenital abnormality. It may be caused by a genetic disorder, so that the abnormality is inherited, or it may be acquired while the fetus is growing, due to exposure to an agent that is teratogenic, that is, causes abnormal development.

Though they are generally rare, the range of possible congenital abnormalities is wide. Limbs or organs may be absent, duplicated or malformed. Organs may fail to move to the correct place, as in undescended testes (cryptorchidism); they may fail to open properly, as in imperforate anus, or fail to close at the correct time, as in patent ductus arteriosus (when the channel bypassing the lungs does not close).

Congenital abnormalities may be obvious at birth, or they may take months or years to become evident; as, for example, in the case of Huntington's disease (a degenerative brain disease), which does not manifest until the person is in middle age.

Inherited congenital anomalies are caused by abnormal genes. There are more than 2,000 known inherited congenital anomalies; some are common, for example sickle cell disease, Down syndrome, and cleft palate, while others such as phenylketonuria, an inherited metabolic disorder, are rare.

Sometimes genetic abnormalities may occur if there is a spontaneous mutation of the parental genes, which may happen during sperm production or egg formation. Severe mutations are not compatible with life and result in miscarriage. Others allow the fetus to survive to birth but it may not live long. Other abnormalities, such as Down syndrome, are compatible with life well into adulthood, but the affected person may have disabilities.

Acquired congenital abnormalities can be caused by a variety of agents that affect the fetus in the uterus, including drugs, infections and toxins. Infection in the mother is a common cause of acquired congenital abnormality. Rubella (German measles) contracted in the first three months of pregnancy may cause deafness, cataracts, heart disease, jaundice or other abnormalities in the child. Cytomegalovirus (CMV) and *Toxoplasma* are two organisms that cause congenital anomalies.

Some drugs taken during pregnancy are responsible for abnormalities in the child. Corticosteroids, anticonvulsants, anti-cancer drugs, narcotics, sedatives, tranquilizers, antidepressants, antibacterials (especially tetracycline), anticoagulants, and drugs prescribed to treat cardiac conditions and hypertension can all cause congenital defects in a few cases. A pregnant woman should avoid taking medication without first obtaining medical advice. Environmental toxins, x-rays, or injury to the fetus may cause an abnormality in the fetus. The age of the mother may also be a factor. Down syndrome (trisomy 21), for example, occurs more frequently when conception takes place after about 35 years of age.

If there is a known or suspected risk of a congenital abnormality developing, it can often be detected during pregnancy by screening procedures. The most reliable procedure is to examine a sample of fluid obtained by amniocentesis from the amniotic sac; this is done between the fifteenth and eighteenth week of pregnancy. Chorionic villus sampling (CVS) is a technique used in the first trimester of pregnancy in which a tissue sample is extracted from the placenta and analyzed for evidence of genetic defects in the fetus. Ultrasound can also reveal some genetic abnormalities.

Termination of pregnancy (abortion) may be considered if fetal disorders are found early in a pregnancy. The decision to abort rests with the parents and should be taken after they are made fully aware by their physician of the nature of the disorder and the consequences of abortion.

Genetic counseling should be considered by anyone who has a history of chromosomal abnormalities, and by all pregnant women over 35 years of age. There may be a higher risk than normal of congenital abnormality in a second child if the first is born with an abnormality; parents in this situation should undergo genetic counseling if they wish to have another child.

Congenital abnormalities cannot be reversed, but they can often be successfully treated with surgery, hormone treatment, diet and physical therapy, depending on the condition and its severity. In the future, gene therapy may provide a treatment.

FETAL ALCOHOL SYNDROME

Fetal alcohol syndrome (FAS) is the name given to a group of symptoms characterized by physical and mental abnormalities in an infant, and linked to alcohol consumption by the mother during pregnancy. Opinions differ as to how much alcohol is too much, and many doctors recommend abstinence for this reason. It is known that binge drinking, particularly in the first trimester (12 weeks) of the pregnancy, is potentially dangerous and the baby may be born with varying degrees of fetal alcohol syndrome.

A baby born with the most severe form of FAS will suffer from intellectual disability, growth deficiencies (most babies with FAS are shorter and weigh less than normal babies), and facial abnormalities, such as narrow eyes, low nasal bridges and a thin upper lip. Permanent heart and joint abnormalities are also likely.

Pregnancy

Pregnancy is the state of having a developing fetus in the uterus, which, in a human female and other mammals, extends from conception (union of an ovum and spermatozoon) to labor (parturition).

In the human female, a pregnancy takes about 283 days (10 lunar months) from the first day of the last menstrual period or approximately 267 days from conception.

Fetus

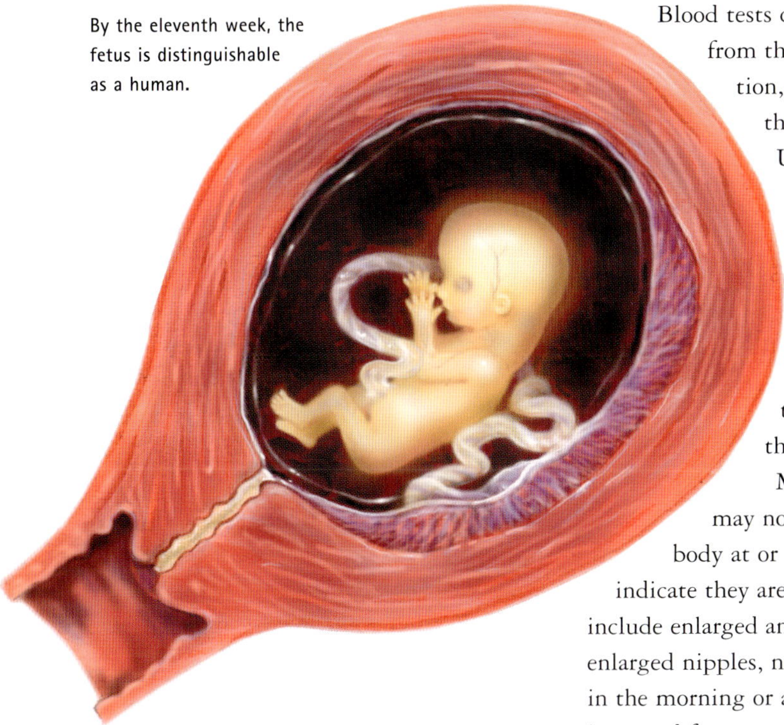

By the eleventh week, the fetus is distinguishable as a human.

Only 5 percent of babies whose mothers go into labor and give birth naturally are born on their estimated date of delivery—in the majority of cases, the baby will born within the 10 days either side of this date.

There are about 24 hours in every menstrual cycle when a woman is fertile, and sperm can survive for up to 5 days in the woman's reproductive system; this means that for about 5 days in every 28 sexual activity could result in pregnancy.

Approximately 1 in 4 women who are planning pregnancy will conceive in the first 3 months. When a couple is fertile and have unprotected sexual intercourse at least 2 or 3 times a week, a pregnancy will result within 12 months in 90 percent of cases. If fertilization does occur, within 5–6 days the fertilized egg will have implanted itself into the womb's lining (the endometrium) and the placenta will have begun to develop and to produce hormones.

A pregnancy that ends prematurely with the loss of the embryo or fetus either spontaneously or by artificial induction is an abortion. A spontaneous abortion is also known as a miscarriage.

SEE ALSO *Abortion on page 463; Amniocentesis on page 436; Chorionic villus sampling on page 436; Embryoscopy on page 434; Female reproductive organs on page 303; Fetoscopy on page 434; Ultrasound on page 432*

TESTS AND SIGNS OF PREGNANCY

Blood tests can confirm a pregnancy from the ninth day after ovulation, or 4 or 5 days before the monthly period is due. Urine tests can confirm a pregnancy from the eleventh day following ovulation. Tests are generally conducted whenever the period is missed; this is often the first indication of the pregnancy.

Many women, however, may notice changes in their body at or before this time that indicate they are pregnant. These can include enlarged and tender breasts, enlarged nipples, nausea (which can occur in the morning or any other time), an increased frequency of urination, tiredness, dizziness, moodiness and skin changes—acne can either break out or clear up as a result of the pregnancy hormones.

In medical terminology, a pregnancy is divided into three trimesters (periods of 3 months) and many of the symptoms described above have usually passed by the second trimester. During this trimester, the woman generally feels fit and well, and it is often now that it becomes more obvious to others that she is pregnant. Sometime between weeks18 and 22 she will feel the fetus move for the first time, and a woman may have an ultrasound to determine whether the fetus is developing according to the dates calculated.

In the last two months of the pregnancy new symptoms of discomfort may appear or worsen. These include practise contractions of the uterus (known as Braxton Hicks contractions), which can be quite painful; constipation, which can be alleviated by increasing the intake of fiber and water in the diet; hemorrhoids, which may need a cream prescribed; varicose veins, which, like hemorrhoids, are due to the relaxing or softening of the vein walls by the pregnancy hormones and are best relieved with rest and support hosiery; edema or swelling of the ankles, which also responds to rest; increased frequency of urination; leg cramps, particularly at night; itchy skin; stretch marks due to breaks in tissue deep below the skin (not removable by massage or creams); backache, which can be the result of poor posture; skin changes, including the linea nigra (a brown line down the middle of the abdomen); and chloasma, or pregnancy mask, which is a brown patch on the face.

Amniocentesis

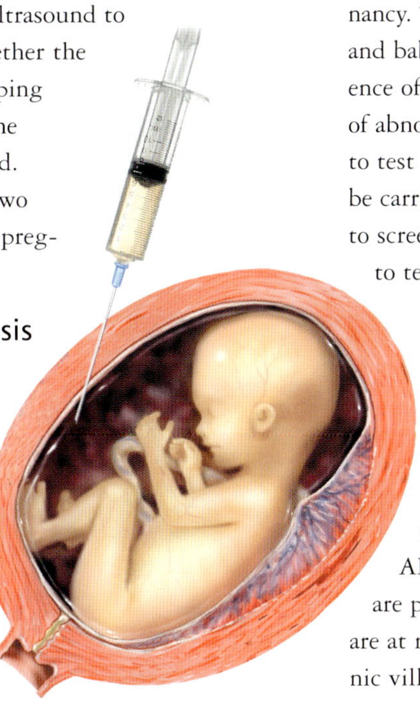

Amniocentesis is a procedure for testing a sample of amniotic fluid, which surrounds the fetus. The test method can detect many abnormalities.

ANTENATAL CARE

This includes regular checkups by qualified health care providers and information about a healthy lifestyle, and has reduced many of the problems which women previously encountered during pregnancy. With modern obstetrics the risks to both mother and baby have been greatly reduced.

MEDICAL CHECKS

Many tests can be conducted during a pregnancy. Tests to monitor the health of mother and baby include urine tests for the presence of protein, to exclude the possibility of abnormal kidney function, and for sugar, to test for diabetes. Blood tests may also be carried out to determine blood group, to screen for blood group antibodies, and to test for anemia and rubella.

Tests which are available but not routinely performed are the alphafetoprotein test to detect a multiple pregnancy and spina bifida, and screening for syphilis, a disease which can cause malformations in the fetus. An HIV antibody test for AIDS can also be done. Other tests that are performed on women whose fetuses are at risk of abnormalities include chorionic villus sampling and amniocentesis.

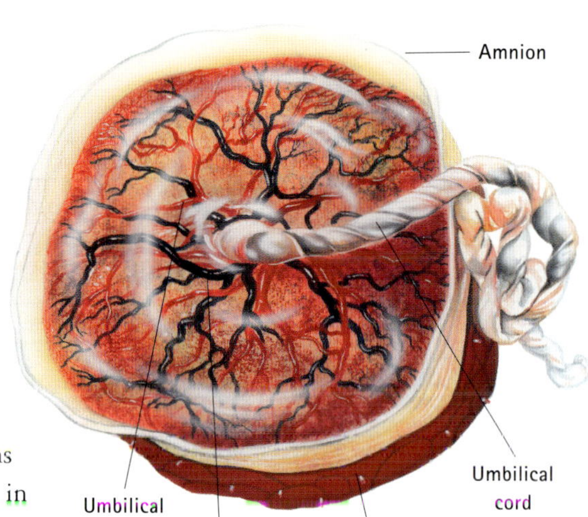

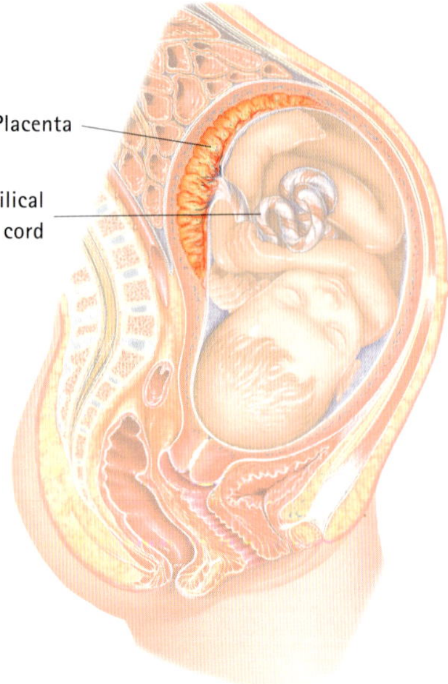

Placenta

The placenta connects the baby to the mother via the umbilical cord. It keeps the baby in position and ensures the baby receives adequate nutrients and oxygen. The placenta also acts as an endocrine organ, producing hormones such as placental prolactin, estrogen, and human chorionic gonadotropin during the pregnancy.

Placenta

An organ of pregnancy, the placenta joins the mother with her offspring and helps in providing selective exchange of substances through its membrane.

FORMATION

At conception, the ovum and sperm combine to form a single-celled embryo, the zygote. This divides rapidly and forms a hollow structure called a blastocyst upon reaching the uterus on day 4 of gestation. Some of the cells inside the blastocyst will become the embryo and fetal membranes, the outer cells will form the placenta.

On about day 6 the blastocyst adheres to the lining of the uterus (the endometrium). The outer cell mass undergoes changes which make it highly invasive and the entire blastocyst burrows its way into the endometrium which heals over it. That part of the outer cell mass that will form the placenta becomes elaborately folded and, as further erosion into the endometrium occurs, maternal blood vessels are broken down and blood seeps into spaces between the folds. Gradually a circulation is set up with blood flowing in from the arteries and out via the veins. Fetal blood vessels develop within the folds of the placenta and these connect with the blood vessels in the body of the embryo via the umbilical cord.

The parts of the placenta overlying the fetal capillaries become very thin in order to allow diffusion of nutrients and oxygen from the maternal blood into the fetal blood and for fetal wastes to pass back to the mother's blood. Despite their close proximity, the two bloodstreams never actually have direct contact. As pregnancy proceeds, the fetus and the amnion enlarge so much that the cavity of the uterus is totally filled.

The blastocyst can implant anywhere in the uterus, usually in the upper posterior wall. Implantation near the opening of the cervix results in a placenta previa which partly or wholly covers the cervical opening. This forms a physical barrier at birth between the fetus and the birth canal; a placenta previa may also prematurely separate from the uterine wall. This can cause severe or fatal hemorrhaging of the mother's blood, either before or just after birth.

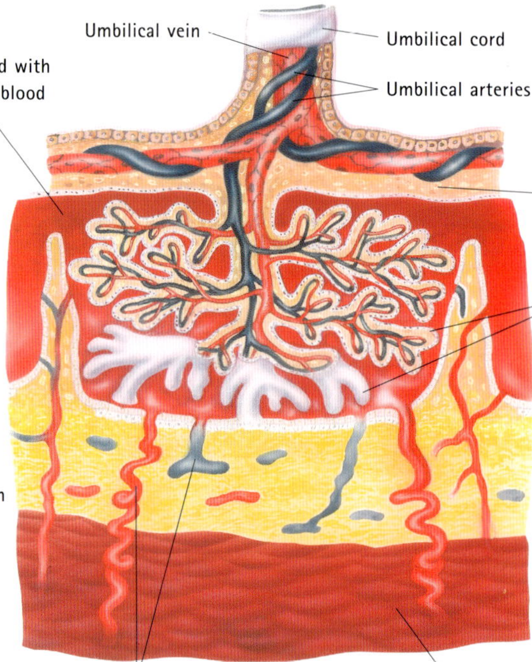

Placenta cross-section

The placenta contains tissue from both the mother and baby, allowing for the diffusion of nutrients and oxygen and the removal of fetal waste.

OTHER FUNCTIONS

The placenta produces many hormones including human chorionic gonadotropin (HCG), estrogen, progesterone, growth hormone and placental prolactin. HCG is the earliest placental hormone to be produced and is first secreted on day 6 of gestation. HCG maintains the corpus luteum (the ovarian follicle from which the ovum burst) and ensures that it continues to create progesterone and estrogen until the placenta can produce adequate amounts of both, usually by the third month of gestation, when HCG levels decline.

The placenta secretes progesterone in increasing amounts in the last two trimesters and is necessary for the maintenance of pregnancy. HCG crosses the placenta into the maternal blood and is the basis for many of the tests for pregnancy. The placenta forms a protective barrier between the maternal and fetal blood although certain noxious agents can travel across it and infect the fetus. These include many (though not all) viruses, notably rubella and HIV, anti-Rhesus factor antibodies, alcohol, pesticides, drugs such as thalidomide, and hormones such as diethylstilbestrol. Microbes much larger than viruses usually cannot cross the placenta but to this exceptions include the bacterium of syphilis and the protozoan parasite which causes toxoplasmosis.

Umbilical cord

The umbilical cord joins the unborn baby (fetus) to the afterbirth (placenta), by which the fetus obtains nourishment from its mother. The fully developed cord is ½–1 inch (1–2 centimeters) in diameter and about 20 inches (50 centimeters) long, but its length is very variable: 8–48 inches (20–120 centimeters). It contains two umbilical arteries and one umbilical vein, embedded in a mucoid connective tissue known as Wharton's jelly. It is covered by a thin layer of epithelium.

Immediately after birth, blood from the placenta and umbilical cord passes into the baby. After delivery, the umbilical cord is tied in two places and cut between the ties (the ties prevent loss of blood). The remnants of the cord initially remain attached to the navel (umbilicus), then shrivel up and fall off after 3–4 days.

Twins

The chance of a natural pregnancy being a multiple pregnancy varies according to heredity, age, race and the number of children a woman has already conceived.

Twins occur in approximately one out of every 90 natural pregnancies; the rate is higher in pregnancies resulting from infertility treatments. Seven out of ten pairs of twins are the result of two eggs being released by the woman's ovaries and fertilized at the same time, producing fraternal (dizygotic or non-identical) twins. Fraternal twins may be of different sex and develop with separate placentas (occasionally these may fuse into one).

Identical (monozygotic) twins are produced after, rather than at the time of, conception, often after the fertilized egg has implanted itself into the uterine lining, when it splits into two.

In this instance, the twins will share the same placenta and be of the same sex. Multiple births involving three or more babies, which are less common, occur through similar processes.

Twins, or a greater number of babies, fill the available space in the uterus more quickly than a singleton; the mother is at greater risk of hemorrhoids, heartburn, backache and premature (preterm) labor. The babies are also at an increased risk of prematurity, poor fetal growth, perinatal death and, in the long term, cerebral palsy.

Multiple pregnancies require additional support, although there is no medical evidence for the common belief that hospitalization for bed rest is beneficial.

Siamese (conjoined) twins are identical twins formed from a single fertilized ovum, joined together by a part, or parts, of their bodies at birth. Most conjoined twins will usually be joined at the hip, head or chest, and some share limbs and internal organs such as the liver and heart. Most are delivered by cesarean section and less than half survive after birth.

Disorders of pregnancy

Disorders of pregnancy can range from a simple feeling of nausea during the first trimester (morning sickness) to the life-threatening condition known as eclampsia.

SEE ALSO *Dilation and curettage on page 463; Rhesus (Rh) factor on page 91*

Miscarriage

A miscarriage or spontaneous abortion generally occurs before week 12 of a pregnancy. Miscarriage occurs in one in every five pregnancies. Before week 20 of the pregnancy the loss of a fetus is a miscarriage; after that date it is a stillbirth. Around 85 percent of miscarriages are related to fetal abnormalities. Of the remaining 15 percent, two-thirds are due to problems in the mother and one-third have an unknown cause. Miscarriage can follow a severe fever, particularly if it is caused by a virus. It can

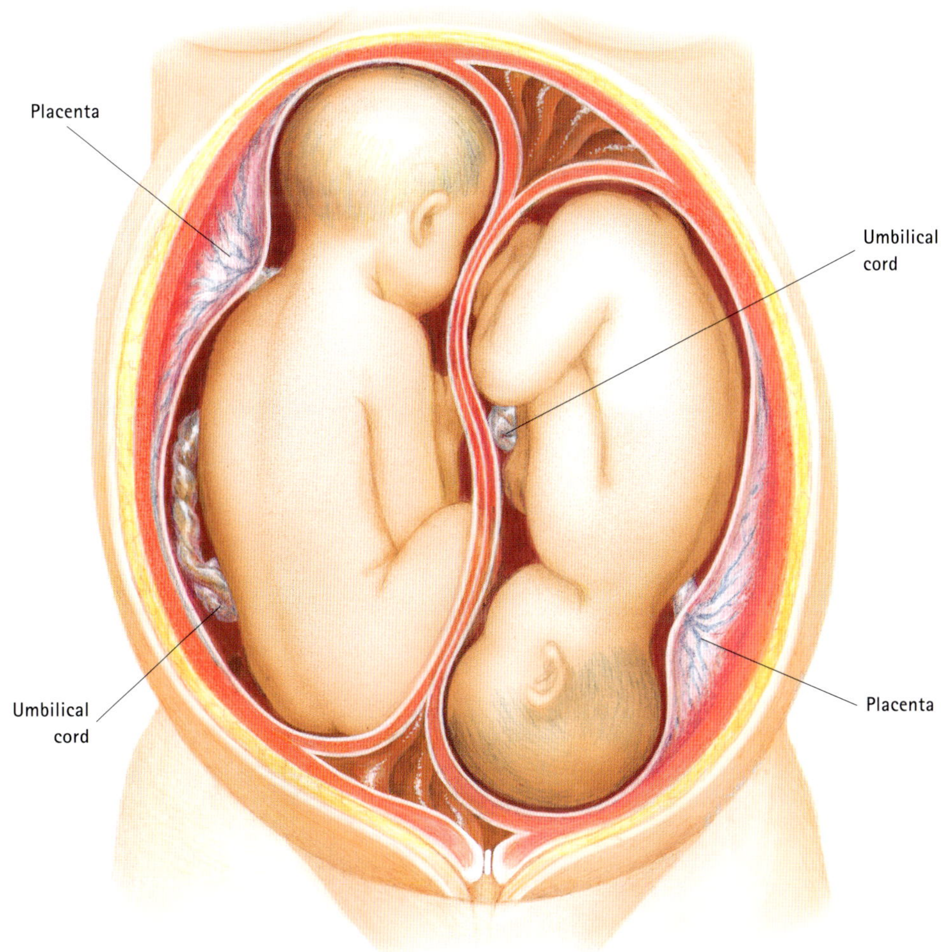

Twins

Twins can be fraternal or identical; fraternal twins result from the fertilization of two eggs, while identical twins occur when a single egg splits into two.

be due to some abnormality of the uterus. It is less common in women under 25, and more likely after the age of 35.

Bleeding, first spotting then a heavier discharge, is usually the first sign of a miscarriage, though some minor bleeding can also occur as part of a normal pregnancy. Cramping of the uterus follows. If an ultrasound detects a fetal heartbeat, then there is around a 90 percent chance that the pregnancy will proceed. Although some doctors still recommend bed rest there is no evidence that this is necessary.

When the fetus is not longer viable any longer, cramping will continue and pieces of tissue may be expelled. It is important to seek medical advice if miscarriage seems possible. Sometimes a curette may be suggested to scrape out any remaining contents of the uterus; however, ultrasound will usually indicate that this is not necessary.

Late miscarriage (between weeks 12 and 20) can be caused by a weak cervix (in approximately 20 percent of women). If this is found to be the cause, a stitch (cervical cerclage) may be used. After 20 weeks a miscarriage can be the result of placental insufficiency. Recurrent miscarriage (after a woman has had three or more miscarriages in succession) will need investigation by a specialist. The next pregnancy will need careful monitoring.

Preeclampsia

A condition that may develop during the second half of pregnancy, preeclampsia is characterized by edema, high blood pressure, and protein in the urine. It occurs in about 5 percent of pregnancies and the exact cause is still unknown, although poor nutrition is possibly involved.

Particularly at risk are first-time mothers, those carrying multiple babies, women in their teens or over 40 years of age, those with high blood pressure or chronic nephritis (kidney inflammation or infection), and those who have had preeclampsia before.

There are no symptoms in the early stages so it is important that women in these risk categories have regular antenatal check-ups to screen for the condition.

Initial signs include a sudden increase in edema (a little swelling is normal in pregnancy), sudden weight gain, nausea and dizziness. Then there may be abdominal pain and vomiting, severe headaches and disturbed vision. If left untreated it may develop into eclampsia, a condition that can trigger life-threatening seizures.

Preeclampsia is potentially fatal for the developing fetus because it restricts the supply of blood to the placenta, causing the baby to grow more slowly. The only cure is the birth of the baby. If the pregnancy is not far enough advanced for the baby to be delivered, bed rest or hospitalization may be recommended. Medication for lowering blood pressure lowering may be required. Preeclampsia is also known as toxemia or pregnancy-induced hypertension.

Eclampsia

Eclampsia is a serious condition that can occur in pregnant women anytime between the fifth month of pregnancy and the end of the first week after delivery. It is characterized by headaches, high blood pressure, visual disturbances, irritability, abdominal pain and convulsions.

In the most severe cases eclampsia causes coma and death. It is one of the most dangerous complications of pregnancy, and the best treatment is prevention. Regular measuring of blood pressure and the testing of urine during pregnancy is used to detect pregnancy-induced hypertension (PIH).

If detected and treated early, complications can be prevented. When eclampsia does occur, expert hospital care is absolutely essential as the sufferer will need heavy sedation. Any woman who suffers from preeclampsia is at risk of eclampsia.

Hydatidiform mole

A hydatidiform mole is a benign (or noncancerous) growth that develops within the uterus from a degenerating embryo. The abnormal growth produces multiple cysts, which often resembling a bunch of grapes. The tumor usually causes bleeding similar to that from a threatened miscarriage. It may be spontaneously expelled from the uterus; if not, a dilation and curettage (scraping) is performed to surgically remove it. In some cases, a hydatidiform mole will progress to form a malignant tumor called a choriocarcinoma. Choriocarcinomas are treated with chemotherapy (alone or with radiation therapy) and prognosis is generally good. Hysterectomy may not be needed.

Childbirth

Childbirth is the act of giving birth to a baby, and can be done with varying degrees of assistance and intervention. Most babies born in industrialized societies are born in hospitals, under the direction of the medical profession. This has been associated with huge drops in maternal death rates. The rate of maternal death is 5 per 100,000 births for normal vaginal delivery; for cesarean section it is approximately 40 per 100,000 births.

In many countries, home birth is also an option. Home birth has a very low rate of medical intervention (a priority for many women) and a low rate of complications and hospital transfers. There are, however, medical reasons why a home birth may not be possible—or sensible—and these include certain complications such as toxemia or sudden unpredictable hemorrhage, medical problems such as heart disease, and fetal problems such as the baby presenting in transverse (across the uterus) position.

Some mothers favor a water birth, which is often only possible in the home environment. For those women who want to go through labor and give birth in water, some birth centers and hospitals will provide a large enough bath. As soaking in a warm bath can relieve pain, many institutions have a protocol which allows laboring in water, but not birthing. However, around half the women who labor in water will also give birth in water. When this occurs, the baby is lifted straight from the birth canal, out of the water and into a warmed room where both mother and baby are kept warm.

SEE ALSO *Anesthesia on page 454; Female reproductive organs on page 303*

Labor

Labor—the process of giving birth—is described in three stages. The first stage is actual labor, during which the cervix is thinned and dilated, and is followed by a period known as transition; the second stage is the birth of the baby; and the third stage is the expulsion of the placenta, which nourished the baby during the pregnancy.

During the first stage of labor, the cervix must dilate fully to allow the baby to move into the birth canal. The endocrine signal that the baby sends to the placenta stimulates the production of estrogen. The uterus then produces prostaglandins and contracts more strongly and more frequently as it works to shorten and soften (efface) the cervix by pressing the baby's head against it. This effacement and stretching of the cervix then triggers the posterior pituitary gland to produce oxytocin; the contractions become stronger, working the uterus to stretch and dilate the cervix.

Once the cervix is fully dilated the baby can move into the birth canal. This stage of labor can be very short or as long as 36 or more hours. In many hospitals, labor is augmented and accelerated after 12 hours.

The second stage of labor begins when the mother feels an urgent need to bear down and push the baby into the birth canal. This stage is much shorter than the first, with an average three to five "pushes" before the baby's head presents at the opening to the vagina. This is known as the "crowning." The baby's head then passes beneath the pelvic bone, with the face toward the mother's spine. The shoulders will quickly follow the head and the baby will soon be born.

The third stage of labor is the expulsion of the placenta. Once the baby is born the uterus continues to contract and there is a rush of oxytocin that prompts the placenta to separate from the uterus in a peeling action. The umbilical cord will still be joined to the placenta and to the baby at this stage. Once the placenta separates from the uterus it will slide down the birth canal.

Labor, though it can be avoided with a cesarean birth, is beneficial for the well-being of the baby. Stress hormones, known as catecholamines, surge through the baby's system during a vaginal delivery in response to the contractions of the uterus and the baby's head being squeezed. This surge protects the baby from asphyxia and prepares the baby for the environment outside the uterus. It clears the baby's lungs and prepares them to breathe, and sends a rich supply of blood to the baby's heart and brain.

OXYTOCIN

Oxytocin is a hormone secreted by the pituitary gland and by the ovary and placenta. Secretion of this hormone results from

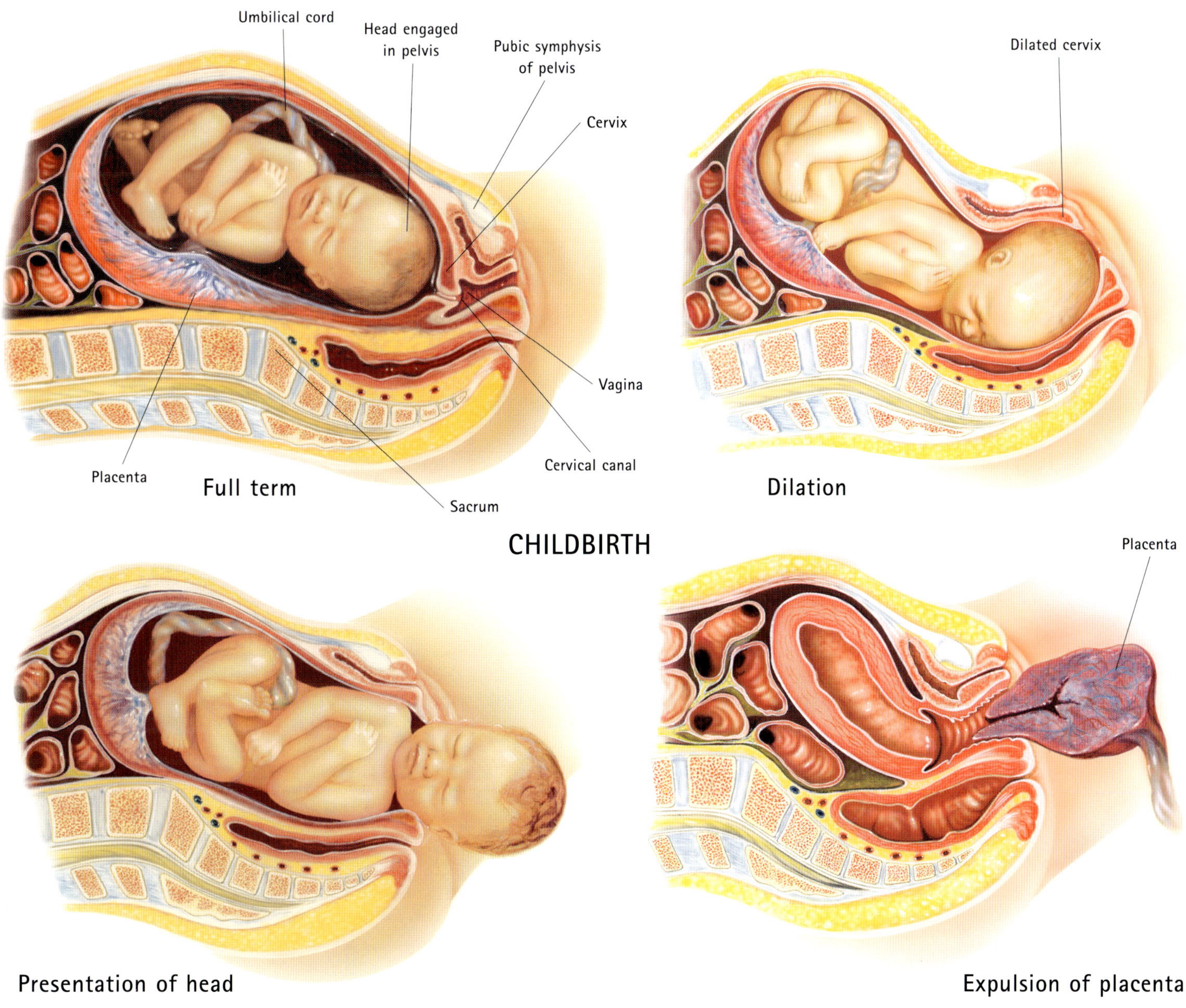

CHILDBIRTH

stimulation of nerves in the nipple during suckling. Oxytocin is responsible for the release of milk during breast feeding (lactation), whereas the manufacture of milk by the glandular tissue is under the control of a different pituitary hormone, prolactin.

It is also thought that oxytocin is important in maximizing uterine contractions during the later stages of labor. It may also help to reduce bleeding after delivery by causing mild uterine contractions which compress damaged blood vessels. However, oxytocin does not seem to be involved in the initiation of labor.

SIGNS OF LABOR

Signs of labor include: the show, which is the expulsion of a mucous plug at the mouth of the uterus (the cervix) and which can appear as many as two or three weeks before actual labor; contractions, which can be regular or irregular but which indicate labor when they become stronger and more painful as they progress; and the waters breaking, which means the amniotic sac which is holding the baby has broken.

PAIN CONTROL

Labor is a painful experience and many women in industrialized societies who have little or no experience of pain use medication to help them cope with it. Epidural anesthesia is a very common form of pain relief, as are nitrous oxide gas and pethidine (a narcotic). Less invasive methods include acupuncture, aromatherapy, hypnosis, massage, relaxation, transcutaneous electrical nerve stimulation, nipple stimulation, and warm baths and showers.

ASSISTED BIRTHS

Some obstetricians believe in active management of labor which involves breaking the waters (amniotomy), the use of synthetic oxytocin to speed up labor and continuous monitoring of the labor's progress. This can involve electronic fetal monitoring, an epidural anesthetic, episiotomy (an incision in the perineum to facilitate the delivery of the baby's head and shoulders) and instrument delivery either with the help of forceps or by vacuum extraction.

In an episiotomy the woman's perineum is cut under local anesthetic to give the baby an easier passage and to prevent tearing. The World Health Organization says the systematic use of episiotomy is unjustified and advises protecting the perineum in other ways. When labor is allowed to progress naturally and the woman gives birth in an upright position (squatting, sitting or standing), a tear is less likely and gravity helps the birth of the baby.

Premature delivery

The birth of a live baby or babies at less than 37 weeks gestation is described as a premature (or preterm) birth; prematurity is not defined by birth weight. Around 5 percent of babies are born prematurely and usually the cause is unknown. Nearly half of all multiple births will occur prematurely.

Other causes of prematurity are: maternal illness; placental problems such as placenta previa; unexplained premature rupture of the membranes; cervical problems such as cervical incompetence; health problems in the mother such as diabetes, smoking, anxiety, distress or poor nutrition; health problems in the baby; maternal age (for those over 40 and under 17 years, the risk is greater) and a previous low birth weight baby. About 6 percent of all babies are born underweight, the majority being premature—the rest are termed small-for-dates.

A pregnancy that ends prematurely with the loss of the embryo or fetus either spontaneously or by artificial induction is an abortion. A spontaneous abortion is also known as a miscarriage.

Little can be done to prevent premature delivery. Drinking plenty of fluids, particularly water (about 8 glasses a day), may help prevent premature labor in some instances. A woman in an industrialized country and who goes into premature labor will be closely monitored in a hospital environment. Attempts will be made to continue pregnancy and if that is not possible, then the baby will be cared for after birth in a neonatal intensive care unit.

Problems and complications

The most common position (85 percent of births) for a baby to be born in is the occipito anterior position (head down, facing the mother's spine). However, babies can also be born in various positions, including: occipito posterior (with the spine against the mother's, a position which causes long backache during labor); breech (bottom first); or transverse (lying across the uterus). Breech babies can be buttocks first or have a foot presenting; this is known as footling breech. Breech babies are sometimes turned in the uterus by the obstetrician or midwife or by using natural therapies prior to the birth; they are then born either by cesarean section or born naturally.

Even a baby in the most natural position for birth can present problems that require medical help. Cesareans are life-saving in an emergency and are often used with breech babies, though this is not always justified. Other methods, such as forceps and vacuum extraction, can assist the delivery of a baby.

Forceps (two instruments shaped like large flat spoons that fit over the baby's head) are often used when the mother's blood pressure rises and the baby needs to be delivered, or when there are signs of fetal distress, or when the baby is in an unusual position. Vacuum extraction is used in similar situations. A suction instrument is used to guide the head down the canal.

Despite all the best intentions and most sophisticated calculations, babies are still born unexpectedly. When this happens it is important to make the environment as warm as possible, to support the laboring

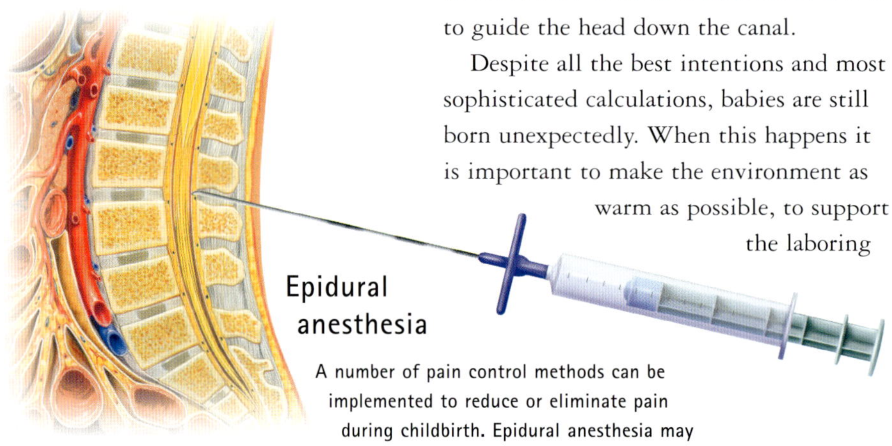

Epidural anesthesia

A number of pain control methods can be implemented to reduce or eliminate pain during childbirth. Epidural anesthesia may be used to numb the lower half of the body.

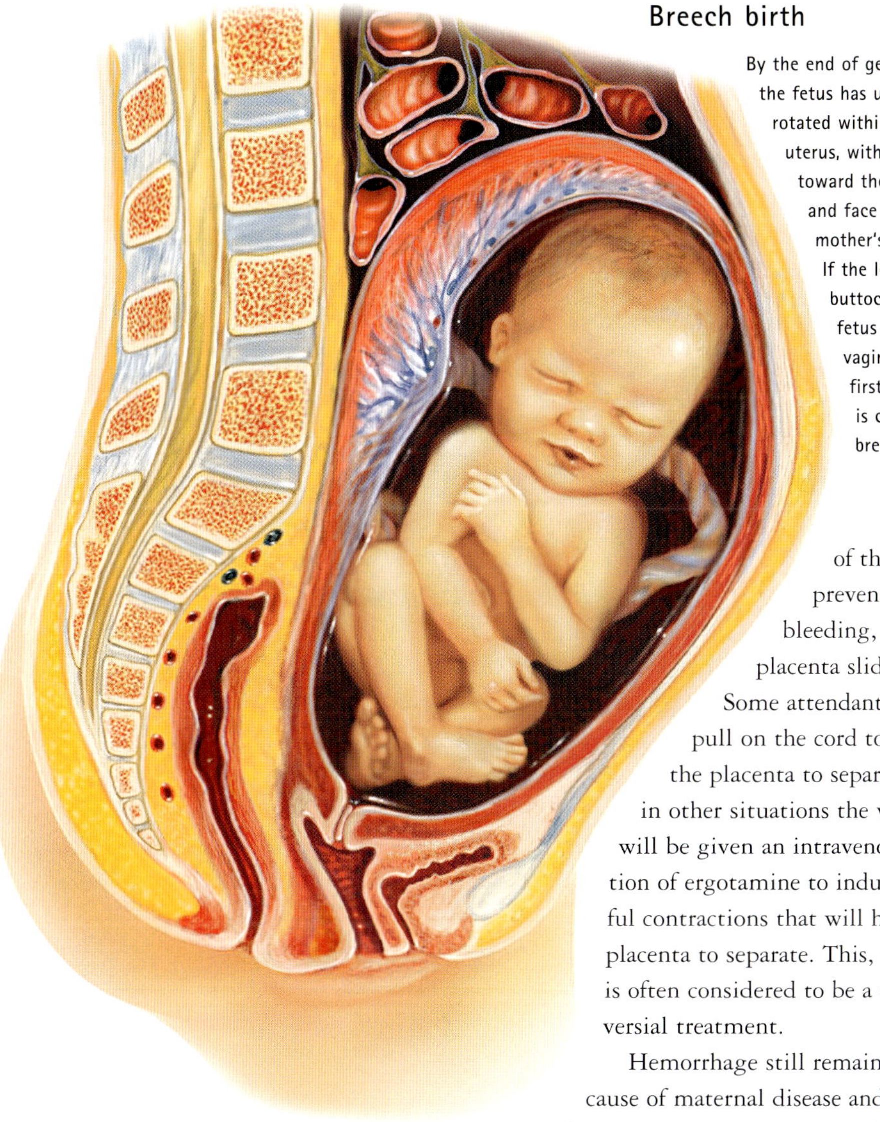

Breech birth

By the end of gestation, the fetus has usually rotated within the uterus, with its head toward the birth canal and face toward the mother's sacrum. If the legs and buttocks of the fetus enter the vaginal canal first, then this is called a breech birth.

woman both physically and emotionally, and to be ready to assist if needed. This can involve gently unlooping the cord from around the newborn baby's neck and making certain there is no membrane over the baby's face preventing breathing. It may also involve keeping both the mother and baby warm and comfortable until assistance arrives. It is not vital to clamp and cut the cord which attaches the baby to the placenta; as long as the placenta detaches from the wall of the uterus during the third stage of labor without problems. The placenta can remain attached to the baby for some time. But, if the placenta does not detach then hemorrhage can result.

In the normal progress of labor, the expulsion of the placenta occurs in the third stage, often passing almost unnoticed. The uterus continues to contract, the placenta separates from the lining, the contractions of the uterus prevent excess bleeding, and the placenta slides out. Some attendants will pull on the cord to help the placenta to separate and in other situations the woman will be given an intravenous injection of ergotamine to induce powerful contractions that will help the placenta to separate. This, however, is often considered to be a controversial treatment.

Hemorrhage still remains a major cause of maternal disease and death, and a retained placenta may need manual removal. Clamping the cord early has not been found to have any effect on blood loss or on hemorrhage.

Cesarean section

Cesarean section, the delivery of the baby through an incision made across the abdomen, in a line just below the pubic hair line, is the surgical alternative to vaginal birth.

In many modern countries the cesarean rate has increased due to fear of litigation (about deaths of babies born vaginally). The World Health Organization has found that cesarean rates should not be higher than 10–15 percent.

In many industrialized countries, apart from the Scandinavian countries, the rate is climbing through the 20s, and this climb is attributed to clinical practice.

Cesarean birth is a major operation. The potential benefits are great when done appropriately, but there are also substantial risks for both mother and baby. Cesareans are often used for the delivery of multiple pregnancy though the indications have not yet been properly established.

Overdue births

It is unreasonable to expect all babies to spend 280 days *in utero*; just as some are premature, some will naturally be overdue. Only around 5 percent of babies are born on the due date of delivery or estimated date of confinement. With ultrasound, pregnancies can be closely monitored and a pregnancy can be allowed to go two or more weeks past the due date through the use of this technology, which can monitor the baby and the uterine environment. Babies born overdue are termed postmature.

After giving birth

During the labor and birth the woman's hormones help her to cope physically and emotionally with the whole experience. Once the baby is born, if there have been no drugs, she is usually in a state of euphoria. This euphoria can continue for some time and in some women it buoys them along through the early days and weeks of motherhood. For others, there is an emotional drop in mood sometime in the first week, which is often called the "baby blues," and for a few there is the possible trauma of postnatal depression.

A hospital environment is usually not conducive to maternal–infant bonding and often restrictive hospital practices will not allow the mother to spend precious time with her newborn, thus interfering with the establishment of breast feeding and the forming of this bond.

This, along with the hormonal changes and a new mother's lack of experience with babies, can lead to postnatal depression. Postnatal psychosis is a rare mental illness that needs medical treatment.

The postpartum period is the first days after the delivery of a baby. In this time a woman's body will change again as it returns to the non-pregnant state. Excess fluid will be excreted and muscle tone begins to return. The body also adapts to its new role of feeding the baby. Women who are supported by friends and others find that coping with being a new mother is easier.

Newborn

A newborn is a baby less than 6 weeks old. Healthy newborn babies begin to breathe almost as soon as they reach the outside world and before the umbilical cord has stopped pulsating. Usually the cord is left intact until it has stopped pulsating, then it is clamped, or in some countries it is bound.

The head of a newborn is disproportionately large in relation to the body; the hair, if present, is wet; the skin is peeling, wrinkly or furry, or all three; the complexion may be red, pink or blue. The baby may be alert or crying, quiet or sleepy.

From the third month of pregnancy, the unborn child (fetus) develops hair follicles which initially produce very delicate hairs all over the body, mainly on the forehead, cheeks, back and shoulders. These hairs are called lanugo (from the Latin *lana*, meaning wool). The lanugo is shed before birth except on the eyebrows, eyelids and scalp, where it persists until it is replaced by stronger hairs a few months after birth. No more hair follicles are formed after birth.

A newborn baby has more individual bones than an adult. (Many bones ossify and fuse together with growth.) A newborn baby's skull is made up of a number of individual bones held together by fontanelles, the most obvious being the soft spot (often called the fontanelle) in the top of its skull. A newborn has a low sloping forehead, a receding chin and often a slightly misshapen head which is from pressure and only temporary when the baby is born vaginally. The newborn's skin may be covered in a creamy substance known as vernix. The molding of the skull may be high and pointed if the baby was delivered in the posterior position (with its spine against the mother's); the face may be swollen and bruised if the baby was born face first. If the contractions of the uterus pushed the baby's head against a cervix that was not fully dilated, the baby may have a large bump on the head like a blister (called a caput).

Enlarged genitals, a slight menstrual-like bleeding in some girls, and "milk" in the breasts of both sexes are caused by the presence of the mother's hormones in the baby's bloodstream, and soon disappear.

Immediately after birth the Apgar test (named after Virginia Apgar, the US anesthesiologist who devised it) is performed on newborns in many industrialized countries. The Apgar test scores an aggregate of observations about the baby's vitality based on heart rate, breathing, skin color, muscle tone and reflex response. It is repeated again 5 minutes after the first test. Another test, the Neonatal Behavioral Assessment Scale, is now used in hospitals around the world. Devised by American pediatrician T. Berry Brazelton, it assesses the baby's behavioral response to human and non-human stimuli.

Other checkups done on a newborn include measuring the diameter of the head and the length from head to foot, recording the weight, checking hips and jaws for dislocation and the mouth for cleft palate. Babies' reflexes are also tested.

A pH test may be done to test the blood in the umbilical cord artery for acidity. A low pH indicates that blood and tissues contain too many metabolic products and that the newborn's oxygen supply is lacking.

Vitamin K may be administered, either intravenously or orally, to prevent vitamin K deficiency bleeding (hemorrhagic disease of the newborn). Newborns have only minimal stores of vitamin K, which is essential to the clotting of blood. Bleeding from the umbilicus, intracranially, or from a circumcision, can occur up until week 26 after birth, and is most likely to occur between weeks 4 and 6 after birth in babies who are deficient in vitamin K.

SEE ALSO *Rhesus (Rh) factor on page 91*

Infancy

An infant, or baby, is a child of 12 months of age or less (although some authorities define this stage as 2 years or less). It is an important period, during which the child gains in weight and height, begins to walk and talk, begins teething, and develops sensory discrimination.

A baby grows and develops more rapidly in the first 12 months than at any other time. At birth, an average newborn infant weighs 7½ pounds (3.4 kg) and is about 20 inches (51 cm) long and gains weight at an average of 6–7 ounces (170 to 200 grams) per week for the first 3 months. After this the rate declines to an average of 2 ounces (60 grams) per week by 12 months.

A baby's skeleton includes a greater percentage of cartilage than an adult skeleton in that some parts are still made of cartilage and have not yet ossified. Some cartilaginous parts do not become bone until the child is almost an adult.

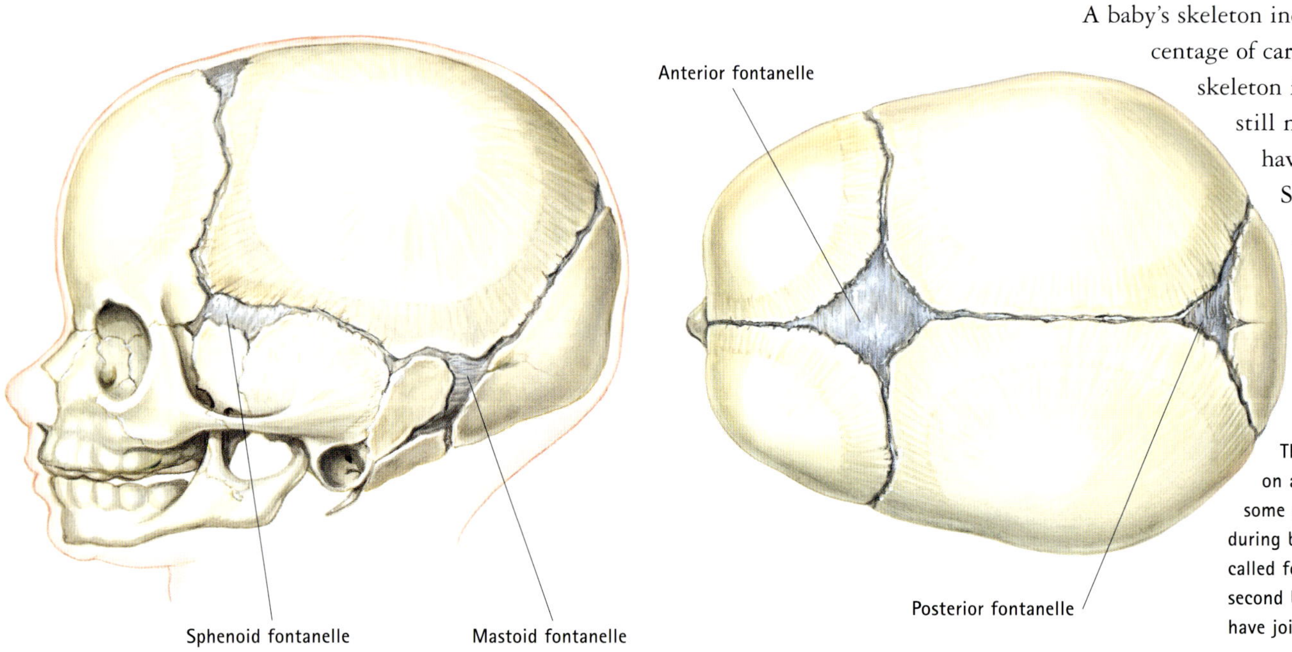

Newborn's skull

There are six "soft spots" on a baby's skull, which allow some compression of the skull during birth. These regions are called fontanelles. By a baby's second birthday these bones have joined together.

Infancy

Infants develop at their own rate, but there is a specific order of developmental changes. For example, babies must learn the muscular skill of head control before they learn to sit up.

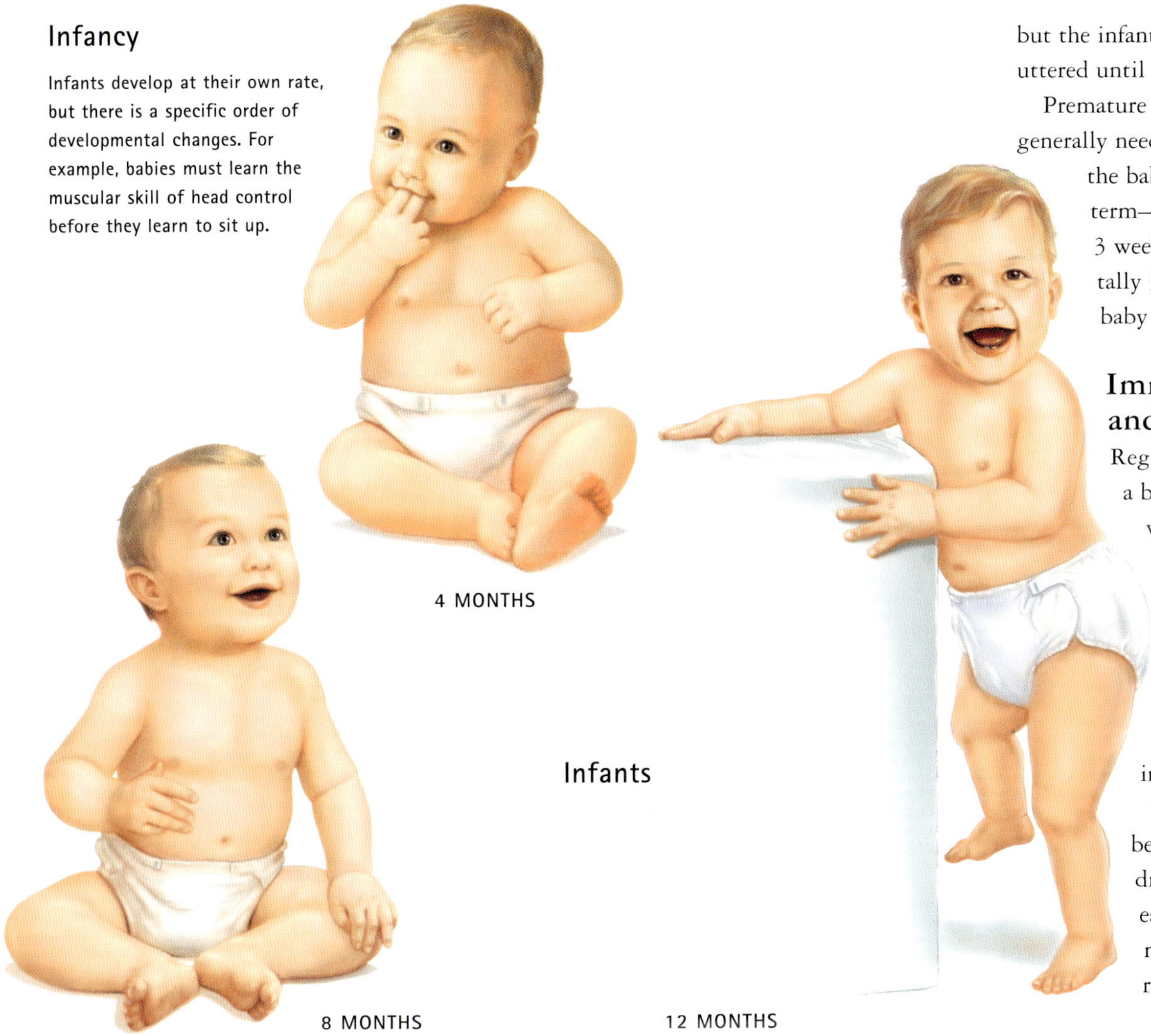

Infants

A baby's brain grows at a remarkable rate in the first year, doubling in weight as the number of brain cells (glia) increases along with the number of connections between cells and parts of the brain.

SEE ALSO *Cradle cap on page 360; Immunization on page 447; Teething on page 415*

Developmental stages

Babies develop individually. Some develop physically more quickly than others, others develop mental or social skills sooner than their peers. Skills develop in a particular order because of the way the body and the nervous system mature, but not every child will go through every stage.

The infant's means of exploring and understanding the world begins with reflex movements—sucking, grasping, throwing and kicking. By 4 months of age the infant can reach out and grasp an object, and can hold a small object between thumb and forefinger by the tenth month.

At 4 months old most babies are able to sit up for a short span of time without any support, and can do so without support for 10 minutes or more by the age of 9 months. Crawling usually begins between 7 and 10 months, and by 12 months most infants can stand up alone. With help, most babies can walk by 12 months and can walk by themselves at 14 months.

The senses also develop rapidly during the first 3 months of life. Newborns are able to distinguish between sounds and objects that are close to their face; within 3 months the infant can distinguish color and form. Mimicking of sounds begins soon after birth, but the infant's first full words will not be uttered until between 12–18 months.

Premature babies' development stages generally need to be corrected to the age the baby would be if born full term—thus a baby which is born 3 weeks early will be developmentally 3 weeks behind a full-term baby born at the same time.

Immunization and illnesses

Regular checkups to ensure that a baby is developing normally will allow any problems to be detected and treated quickly. Checkups are also important to ensure that the baby is protected against a range of childhood diseases which can have serious repercussions, including death.

Immunization programs have been designed to protect children against a number of diseases, including diphtheria, measles, mumps, poliomyelitis, rubella (German measles),

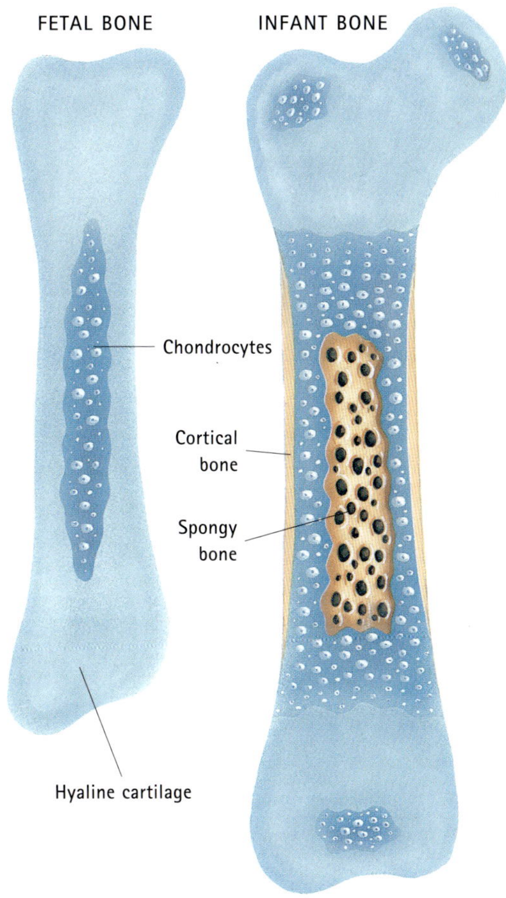

Bone development

Bone growth begins very early in fetal development. At 6 weeks, chondrocytes in the center of the cartilage begin to enlarge— the first stage in bone development. By the time the baby is born, bone has been laid down in the center of the shaft and in a collar around the middle of the shaft.

tetanus (which can cause lockjaw), whooping cough (also known as pertussis), *Hemophilus influenzae* type B (which can result in meningitis) and hepatitis B.

Other common illnesses which babies can suffer in the first year include bronchiolitis; bronchitis; conjunctivitis (also known as pink eye); croup; ear infections; encephalitis; gastrointestinal upset; hand, foot and mouth disease; herpes simplex; hepatitis; influenza; pneumonia and roseola infantum.

Feeding

Until they are around 6 months old, babies receive all the nutrients they need from breast milk (the preferred choice) or specially prepared formula. After this time, small amounts of soft, nutritious food may be added to the baby's diet.

BREAST FEEDING

Breast feeding means feeding a baby with milk directly from the mother's breast. Within each breast are about 15 to 20 milk glands, or lobules, in clusters. It is here that the milk is produced. From the alveoli, canals called ductules lead into larger canals called milk ducts and the milk flows via these to pools which lie under the areola, the brown circle around the nipple. The baby takes the whole of the nipple into its mouth and milks the breast by pressing and pumping the milk from these pools.

One of the first actions of the healthy baby after birth is to seek its mother's nipple. The first "milk" is actually a nutrient and antibody, a rich yellowish substance known as colostrum. Only after two or three days will real milk start to appear. This will provide all the nourishment, both food and drink, which the baby will need for the next four to six months. Ideally, babies should be breast fed for six to twelve months, but even just a few weeks of breast feeding will get them off to a good start.

Breast feeding is a learned skill for both mother and baby, and without some family support and knowledgeable advice many mothers encounter difficulties. However, many more mothers have no problems and find breast feeding a very pleasurable and rewarding experience. Breast milk is a unique, constantly changing substance containing antibodies, hormones, enzymes, growth factors and immunoglobulins, as well as vital nutrients.

The keys to successful breast feeding are correct positioning of the baby—chest to

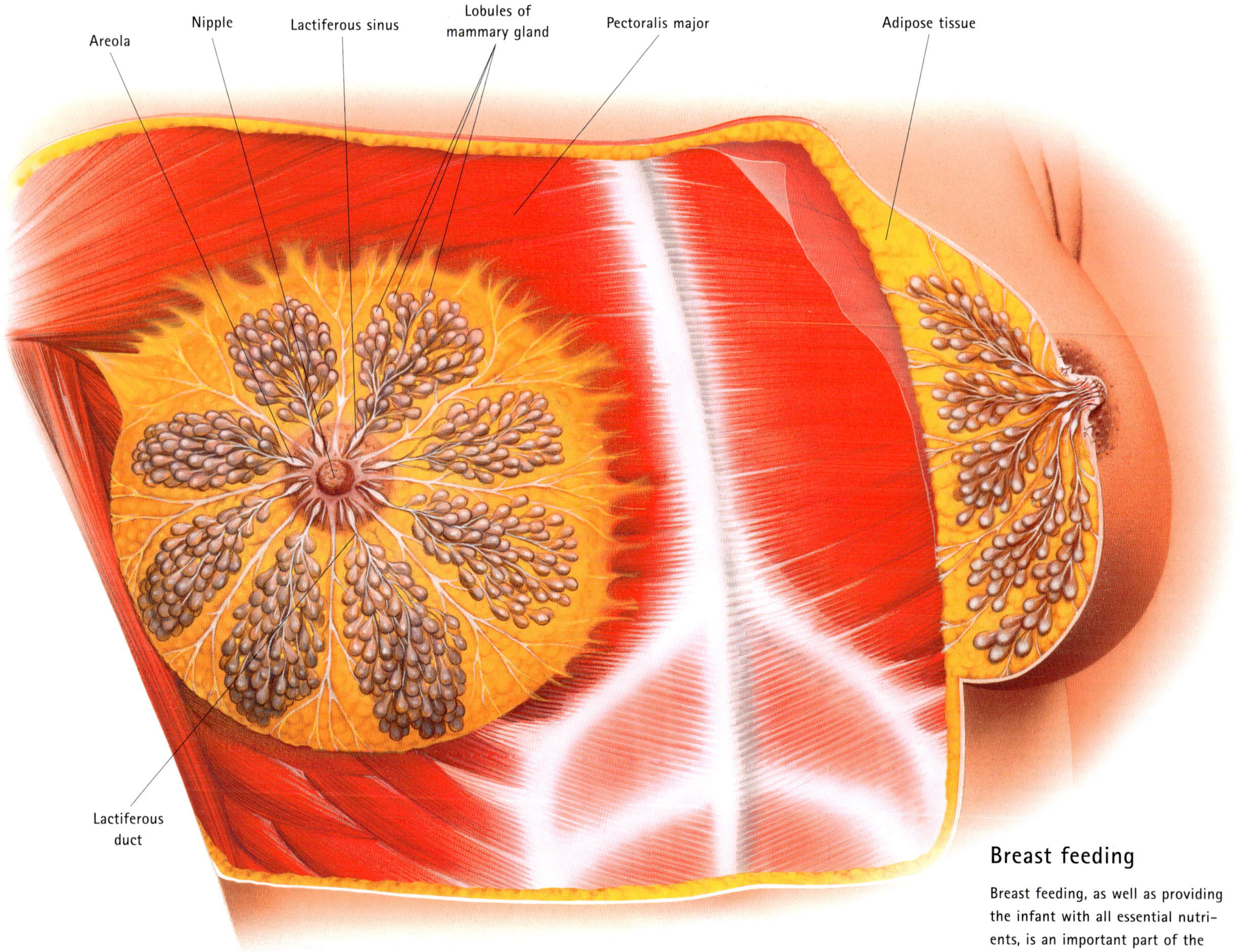

Breast feeding

Breast feeding, as well as providing the infant with all essential nutrients, is an important part of the bonding process.

chest and chin to breast—in a relaxed and comfortable environment, allowing the baby to feed for as long and often as necessary, understanding how the human body produces milk and how it is released, and support and advice of someone who understands how to breast feed. The mother knows her baby is getting enough milk if the baby wets six to eight diapers a day, has pale yellow urine, is bright-eyed, grows steadily and feeds well at the breast.

One of the most common reasons women stop breast feeding is reduced milk supply: very few women are actually unable to breast feed. The only way to increase the amount of breast milk available is to breast feed. Supplementing a baby's feeds with artificial baby milk will reduce the demand for breast milk and result in a subsequent drop in supply. To keep up her milk supply, it is important that the mother drinks plenty of water, gets as much rest as possible, and eats healthy, nutritious food. Women also stop breast feeding because of sore, cracked nipples or breast infection (mastitis). However, these problems can be easily remedied with medical treatment.

Breast milk can be "expressed" (pumped) by hand or by using a hand-operated or electric pump. This milk, if stored correctly, can then be fed to the baby in a bottle, or for sick or premature (preterm) babies via a tube. Expressing milk allows the mother some time away from her baby, which is important if she must return to the workforce. Many companies in industrialized countries now make provisions for women to continue to breast feed their babies after they return to work.

Artificial baby milks, if mixed according to directions, can be bottle fed to babies if breast feeding is unsuccessful or not desirable. They are nutritionally complete but they do not contain all the properties of natural breast milk.

The woman who breast feeds has a lower risk of developing breast and ovarian cancer, osteoporosis and heart disease. Breast feeding is convenient, costs little or nothing and also helps the mother's body to return to its pre-pregnant shape. Both mother and baby benefit from the intimacy of the close physical contact, which is an important part of the bonding process.

Sleeping

Understanding how a baby sleeps (which is very different from the way adults sleep) and how much sleep is needed, is the second greatest problem faced by parents and care givers. Many sleep problems are really problems of adults' perceptions of a baby's sleep needs. A baby sleeps better in the presence of others and with some noise around; baby sleep is much more restless than adult sleep; a baby's brain is more alert during sleep than an adult's and in the first 3 months a baby will generally wake after an REM (rapid-eye-movement) stage of sleep. Not until the end of month 3 can a baby fall into a deep sleep. By 6 months a baby has the ability to sleep for longer periods, but will still need adult comfort and attention when awake.

Bedtime routines, including story reading, bathing and soft singing, work for some families; other families sleep together or in the same room and find that this gives everyone the required night sleep. There are many theories and methods to help parents come to terms with, and develop strategies for, getting the night's sleep they want.

Teething

Teething is eruption of the first teeth in a baby. A baby's first tooth will generally erupt between the fifth and twelfth month of the first year of life; a few babies will get their teeth earlier or later. A very small number, about 1 in 2,000, will be born with one or more teeth. The last tooth usually erupts around the age of 3 years. There are 20 primary teeth (also called deciduous or milk teeth), which are smaller and usually whiter than the permanent set of teeth.

The lower central teeth (incisors) arrive first, followed by the central upper incisors, the lateral incisors (lower and upper), the first molars, the canines and the second molars, though not all children get their teeth in this order. These first teeth guide the permanent teeth into place and aid in the growth of the jawbone. It is important that they are kept healthy and clean—they are not dispensable.

A baby's first teeth can be cleaned with a clean wet cloth, sitting the baby on an adult's lap and gently rubbing the teeth. Once the child is accustomed to this routine twice a day, a very soft toothbrush

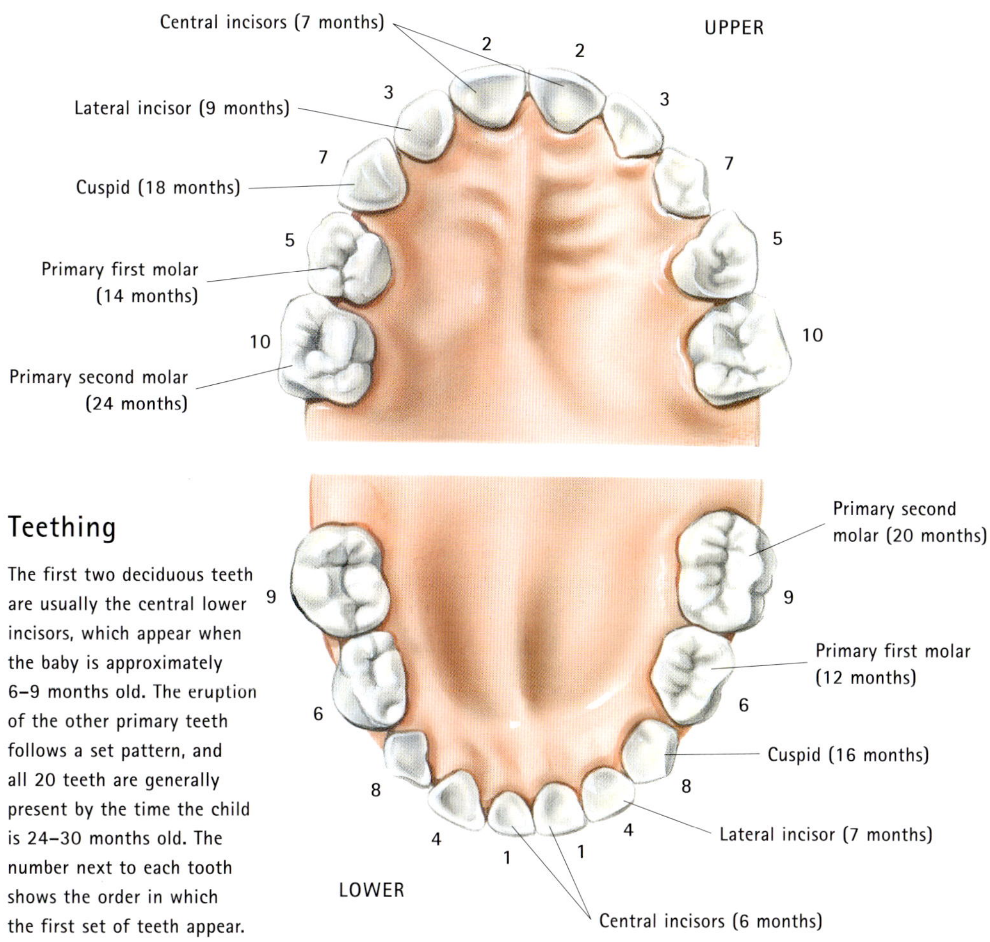

Teething

The first two deciduous teeth are usually the central lower incisors, which appear when the baby is approximately 6–9 months old. The eruption of the other primary teeth follows a set pattern, and all 20 teeth are generally present by the time the child is 24–30 months old. The number next to each tooth shows the order in which the first set of teeth appear.

can be used. Children will need assistance cleaning their teeth until middle childhood.

Signs that a tooth is about to erupt can include dribbling and a need to chew on anything and everything. A baby who is teething may also be unhappy and pull on the ear, because the ear canal is connected to the same nerve as the lower jaw. There is much folklore surrounding teething, inaccurately blaming it for fever, diarrhea, constipation, loss of appetite, diaper rash and convulsions. All these symptoms require medical attention—they are not symptoms of teething and indicate other conditions. The canines and the molars usually tend to cause the most distress.

The best treatment for sore gums due to teething is a cool (not cold), hard, clean teething ring or similar. The dribbling that accompanies teething results in some fluid loss and babies are generally more thirsty when they are teething; they wake up at night and need to drink.

The arrival of teeth is not a reason to stop breast feeding as babies do not usually bite the nipple, except occasionally in an attempt to relieve teething pains. The baby's tongue normally protrudes over the bottom teeth while sucking, thus protecting the nipple.

Teething gels that contain anesthetics can cause allergies and must be treated with caution; acetaminophen (paracetamol) may be used to relieve discomfort.

Communicating

Babies are born with the ability to communicate with the adults responsible for their care. From day one, babies can gaze into the eyes of a person holding them at chest height (if drugs are used during labour it may take a little longer). They can imitate their carers' facial expressions and one day, some time in the first 6 weeks, will smile. Babies need the love and attention of those who care for them; this ability to smile and gaze at parents or carers is vital in establishing emotional bonds between baby and carer.

Crying is a babies other method of communication and the most important way they communicate in the first 12 months. Many adults have little experience of a baby's cries until they have responsibility for a baby for the first time. Some babies cry more than others and need more attention. This is normal, but crying can sometimes become a problem in itself and it is important that parents learn strategies to help them cope with their babies' needs without being overwhelmed.

Babies cry for one or more of the following reasons: they are hungry, they are too hot or too cold, they have a pain in the stomach, they need their diapers changed, they are feeling insecure or lonely, they are bored, or they have become over-excited and need to be calmed. Unfortunately, determining which of these problems it is can sometimes be difficult. Over the age of 6 months one of the most common reasons for crying is boredom. A stimulating environment with changing patterns, shapes, sounds and colors will amuse a baby, as will the presence of an adult or older child who spends time interacting with the baby.

PARENTAL CONCERNS

Sudden infant death syndrome (SIDS) is an issue which frightens many parents. Understanding and taking heed of the risk factors are the best safeguards parents can have to protect their child.

Teething is often blamed for illnesses such as diarrhea, stomach upsets and infections, but these are not symptoms of this normal part of development. Teething can cause considerable discomfort and unhappiness but not an illness.

Failure to thrive is another issue that often confronts parents. A baby who has not gained weight in accordance with standardized baby growth charts may be described as "failing to thrive." In many instances it can be explained that parents' expectations are unrealistic, but in some infants there can be feeding or malabsorption problems which are affecting growth rate.

CHILDHOOD

Childhood is regarded as the period between infancy and puberty. It is the beginning of the journey toward full independence, a period of fast physical and mental growth during which education and experience build a foundation for adulthood. Childhood ends in the teenage years when growing sexual maturity brings a marked change in appearance, attitude and interests.

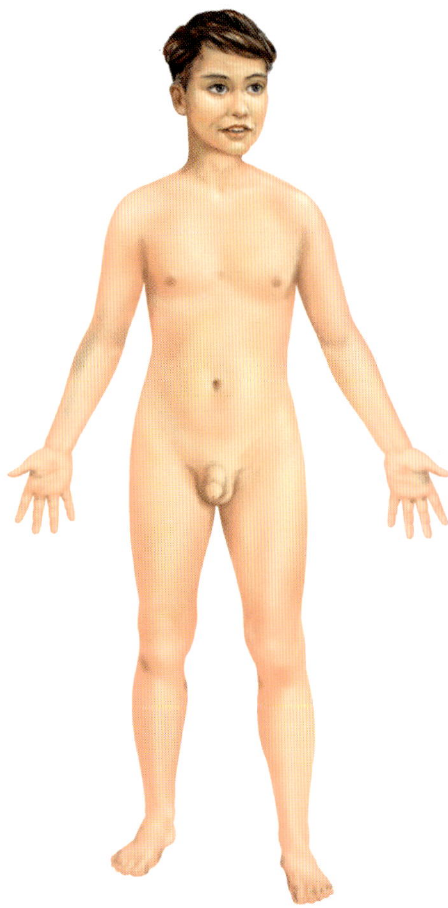

5-year-old boy

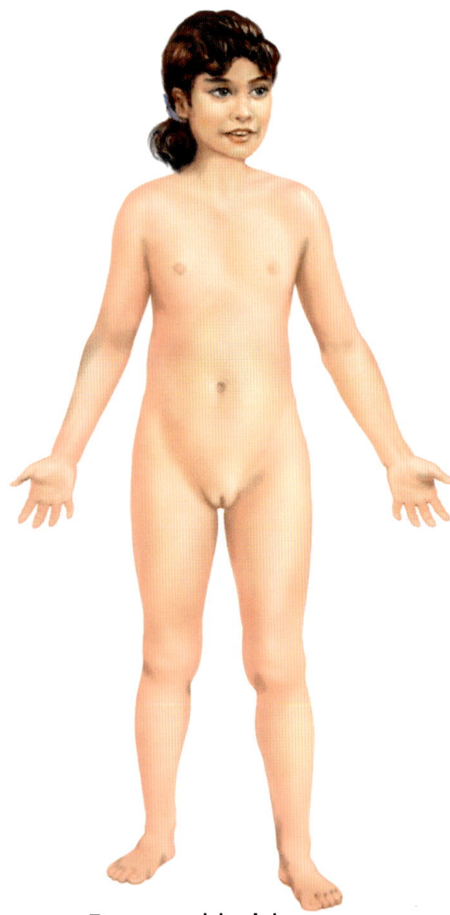

5-year-old girl

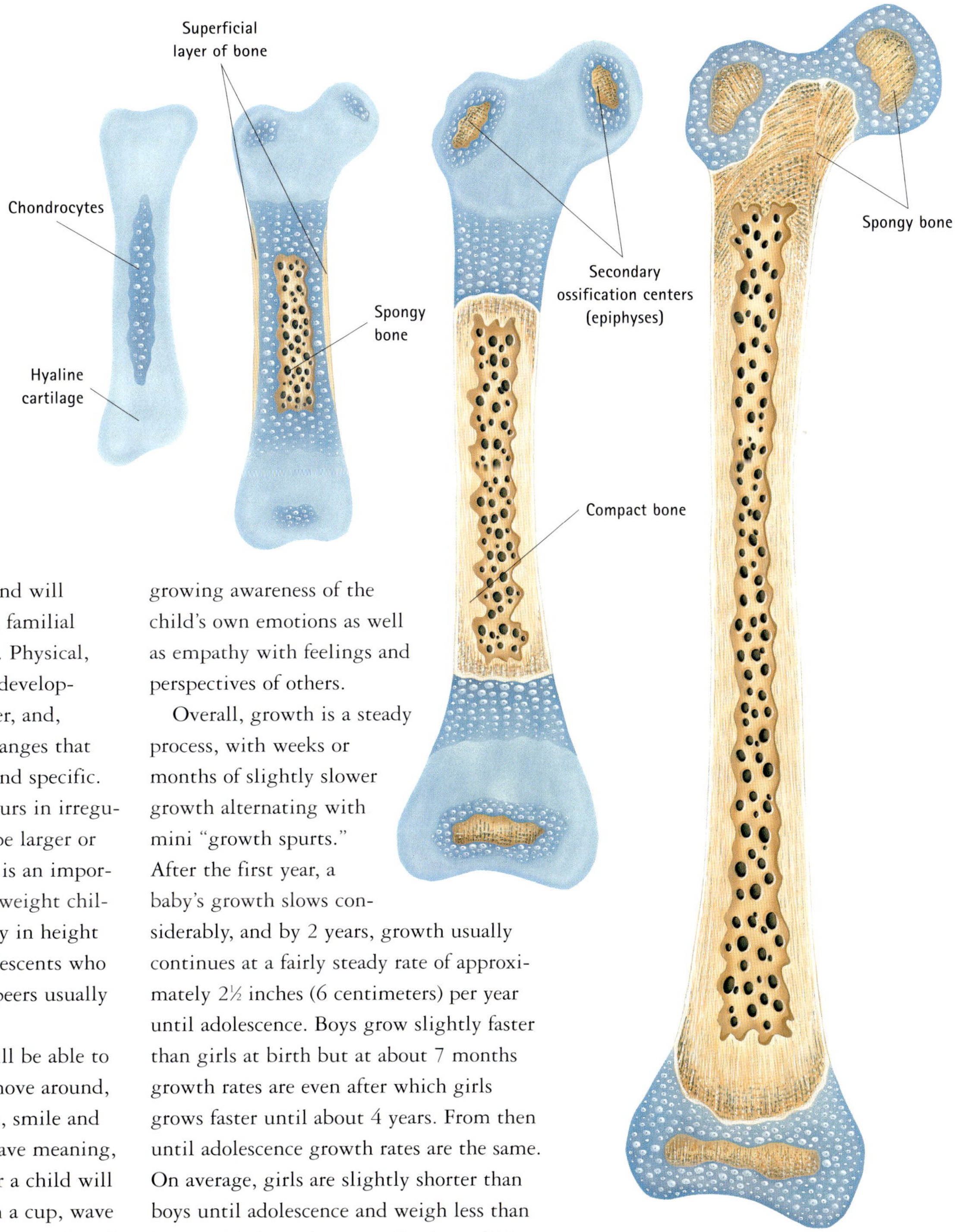

Bone formation during childhood

Long bones begin as cartilage in the embryo. By birth, ossification (development of bone) has reached almost to the ends of the cartilage models. New centers for bone growth then develop at either end of the bone. A plate of cartilage (growth plate) develops between these two areas of bone, and is where the increase in bone length occurs. The growth plate moves steadily away from the center of the bone toward the ends until all cartilage has ossified. Growth in bone length is then complete. Long bones are modeled to be wider at the ends than the middle, providing extra support at the joints where it is needed.

Each child is an individual and will develop by genetic inheritance, familial traits and environmental input. Physical, emotional, social and spiritual developments will impact on each other, and, though highly complex, the changes that a child undergoes are ordered and specific.

In most children growth occurs in irregular spurts. Some children will be larger or smaller than their peers. Build is an important factor in growth, and overweight children may appear to grow slowly in height compared to their weight; adolescents who reach puberty later than their peers usually catch up eventually.

At 12 months old a child will be able to sit without support, crawl or move around, come up to a standing position, smile and babble, and say words which have meaning, or appear to. In the second year a child will learn to walk alone, drink from a cup, wave goodbye, understand simple commands and questions, and combine words.

By around the age of 3 or 4 years, most children will have mastered the major physical tasks, including walking, jumping, running, climbing stairs, grasping and manipulating objects. By the third year, the child will able to use sentences containing 5 or 6 words, and can understand grammar and meaning by 6 years old. At the age of 3 or 4, the child will have a vocabulary of several hundred words. Between 2 and 6 years of age the child develops cognitive skills such as knowing how to perceive, think, recognize, and remember. There is growing awareness of the child's own emotions as well as empathy with feelings and perspectives of others.

Overall, growth is a steady process, with weeks or months of slightly slower growth alternating with mini "growth spurts." After the first year, a baby's growth slows considerably, and by 2 years, growth usually continues at a fairly steady rate of approximately 2½ inches (6 centimeters) per year until adolescence. Boys grow slightly faster than girls at birth but at about 7 months growth rates are even after which girls grows faster until about 4 years. From then until adolescence growth rates are the same. On average, girls are slightly shorter than boys until adolescence and weigh less than boys until about the age of 8, after which they are heavier until about 14 years of age.

By the time a child is 6 years of age all the primary teeth will have appeared and the secondary teeth are starting to emerge. The secondary teeth erupt in a pattern, gradually replacing the primary teeth.

During childhood the long bones, such as those in the arm and leg, start as cartilage "models" which are then slowly converted into bone (ossified). The centers of ossification are the shaft (diaphysis) and regions near the ends of the bone (epiphyses). Between the diaphysis and epiphyses is a growth plate where increase in length of the bone takes place. By the age of about 20, ossification reaches and includes the growth plate, at which time growth stops.

Similarly the skull has certain bones which develop directly from the soft connective tissue membrane through the process of ossification. Other parts of the skull are derived from pre-existing cartilage. As the brain grows, the flat bones of the skull enlarge by expansion at the margins. The brain (along with the skull, the eyes and the ears) develops earlier than other parts of the body. At birth the brain is already 25 percent of its adult weight, at age 5 about 90 percent, and at age 10, about 95 percent. Other internal organs such as the liver, the spleen, and the kidneys grow in line with the child's general increase in height.

The different bones of the wrist ossify at known times during development, so that x-rays of a child's wrist are used as a measure of a child's growth compared to chronological age. These can also be used to estimate the expected height of a child during investigation of height problems.

Normal growth depends on a sufficient intake of nutrients and vitamins, exercise (which helps prevent obesity) and rest. Sleep patterns vary by age and individual child, but most children need an average of 10 to 12 hours of sleep per night. Also important is the production of various hormones by the body, especially the growth hormone. There is a great deal of variation among children. Some will develop intellectually ahead of their peers, while others will develop physically or socially ahead of their peers. Milestones are guides only, and failure to match them does not necessarily mean that there are any problems.

However, parents who suspect that their child may not be developing normally should seek advice. During regular visits to the doctor, the child's height and weight can be recorded on a growth chart. This enables the doctor and the child's parents to compare the child's height and weight compared with that of other children the same age, and determine whether the child is growing at an appropriate rate. Most children who are short or delayed in development are healthy and normal.

Childhood problems

Problems that occur during childhood may be related to physical, emotional or learning disorders; most can be successfully treated.

SEE ALSO *Chickenpox on page 372; Croup on page 370; Leukemia on page 93; Measles on page 370; Mumps on page 371; Rubella on page 370; Scarlet fever on page 379; Whooping cough on page 381*

Behavioral problems

Although behavioral problems cause parents to worry they are not usually serious, and unless there are severe underlying emotional disorders, the children will usually grow out of them. However, counseling may be needed in some cases.

AGGRESSION

When children play together some aggressive behavior is normal until they learn from the adults in whose care they are how to get on with others, when to stand up for themselves, when to wait and how to share. Children who grow up in a violent family environment often grow up into violent adults who are unable to solve their problems, except in an aggressive manner.

Adults who are never able to learn these skills usually continue with their aggressive physical behavior of hitting, fighting or being unable to share. By the time children reach school age they should have learnt to be able to control aggressive behavior; if it is still a problem then it is wise to seek skilled advice from a counselor.

BULLYING

Being bullied is not just a problem among children. The deliberate act of hurting other people—frightening them with words or deeds—can be practised by adults, teenagers and children. Bullying is a problem for both the victim and the perpetrator. People who bully will choose a victim who seems easy to hurt, and being bullied can destroy someone's self-esteem.

At the same time, research has discovered that it is people who already have low self-esteem who are the targets of bullies. Strategies which can help to avoid being bullied include learning empowerment. Adults should be involved in the case of a child who is or has been bullied.

Bullies have also been found to have low self-esteem as well as low self-confidence and likewise need strategies—usually from professional counselors—to help them find ways of being accepted.

TANTRUMS

These are not just the province of toddlers. Tantrums are thrown by adults, teenagers as well as children, who have no other outlet for their frustration or stress. The frustration can be mixed with other emotions such as jealousy or feeling unwanted and these can add to the fury. Tantrums in young children can be dealt with by an adult who stays in control, is ready to come to the child's assistance, and who does not give in to the child's demands. With older children, the causes and consequences of their behavior should be discussed. Tantrums in adults generally are the sign of deeper problems and may need counseling.

WHINING

On the whole, it is usually unhappy children who whine. Causes of whining can be tiredness, ill health, boredom, frustration or a feeling of insecurity. Reacting swiftly to deal with the cause of the problem and not giving in to the demands of the person who exhibits this behavior, together with providing a secure emotional environment, will help to avoid any whining.

Bed-wetting

Also called enuresis, bed-wetting commonly refers to the involuntary discharge of urine, usually during sleep at night. Children can only be said to be bed-wetting if they are still unable to control urination during sleep at the age of five. Even at this age, however, it is still quite common. In extreme cases, enuresis can continue into adulthood unless treatment is sought.

Bladder control develops in the following stages. When the child is up to one year old the bladder empties automatically when it is full. At one to two years of age an awareness of when the bladder is full develops. At three, the ability to hold urine grows as the bladder is able to contain larger quantities. At four, the ability to stop urine at will is fully developed. Most children are dry at night by this age. By the age of six, the ability to pass urine at will is possible.

There are three main factors which affect bed-wetting. The first factor is heredity—if both parents were bed-wetters then there is a three-out-of-four chance their child will experience similar problems. The second factor is the ability, or inability, of the bladder to send a signal to the brain strong enough to wake up the sleeping person. The third factor is over-productive kidneys; in some cases the kidneys usually produce the same quantity of urine at night as they do during the daytime.

It is usually recommended that parents wait until a child is seven or eight years old before seeking assistance; however, if the problem is causing a great deal of distress to the child or family, or if the child suddenly starts to wet the bed after being dry, then help should be sought. The first step should be a medical checkup to make sure that the cause of the bed-wetting is not organic, such as developmental or physical problems or infections of the urinary tract.

The usual forms of treatment are counseling with behavior therapy, use of alarms, and drug therapy. Bladder training programs also have a good record of success.

Learning disorders

The areas of the brain associated with learning activities include Broca's area, the parietal lobe, cerebellar cortex and auditory lobe. Problems in the nervous system that affect information processing by these areas can result in learning disorders. Learning difficulties may also arise if the brain is damaged by disease or accident. The inability of the brain to link or interpret information received or stored results in difficulties retaining or understanding language, calculations, or reason.

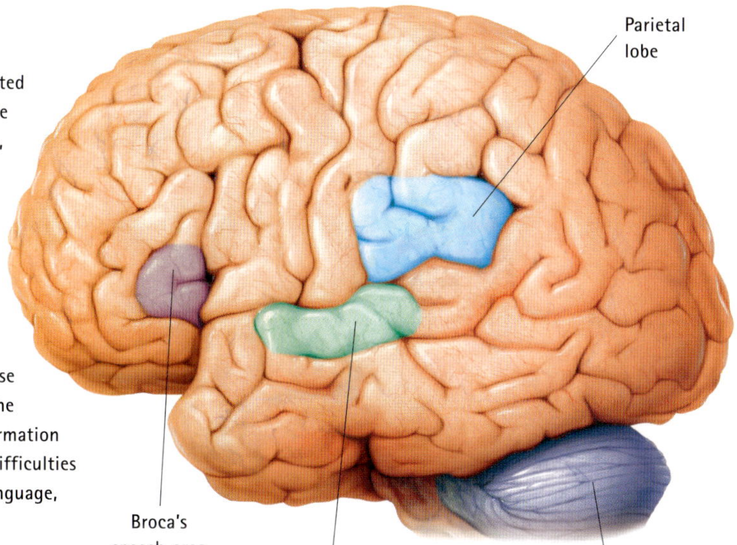

Toilet training

There is no specific age when a child is ready to use a toilet. Up until the age of 15 months a child does not have the muscle control necessary to hold in the contents of the bladder or bowel until an appropriate moment. Emotional readiness is also an important element of toilet training; this sometimes does not develop until the end of the third year. Once children are aware of their bodily functions and of what society expects of them, by observing others, they will usually begin to use a toilet without much "training." Emotional stress or distress, anxiety over performance and fear of failure can all hinder the process.

It is advisable for parents to let children take their time in learning to use a toilet and to allow the child to proceed gradually at their own pace. In some societies a "potty" (chamberpot) is introduced before the toilet; children will usually progress naturally from this to a toilet before they reach school age. It is very important that children are also taught the importance of hygiene when they empty their bladder or bowels.

Thumbsucking

Many babies suck their thumbs while still in the uterus and can be born with blisters on the top lip, commonly referred to as sucking blisters. Although frowned upon in some societies, sucking the thumb (the fingers or some other part of the hand) is a comfort mechanism used by many babies and children; it is emotionally comforting and physically harmless. Children who continue to suck their thumbs into the school years, however, may be in need of emotional support or counseling, as such long-term sucking is often a replacement activity for some need. By this age the action of sucking may also damage the alignment of the teeth, a problem which can usually be rectified quite easily with orthodontic treatment if it does not right itself with time.

The other result may often be calluses on thumbs or fingers, but these too will disappear when the habit ceases. In most cases no treatment is needed.

Learning disorders

A learning disorder affects a person's ability either to link the information within the brain or to interpret what they see and hear, resulting in the person learning differently from someone without a malfunction. This type of disorder is usually concerned with spoken and written language; memory and reasoning; coordination; social competence; or emotional maturation, including self-control or attention span.

Learning disorders affect around 15 percent of school children—more often boys than girls. The disorders may be caused by a problem in the nervous system that affects the receiving, communication and processing of information. Some children with learning disabilities also have a short attention span or are hyperactive. Children with reading problems are sometimes described as dyslexic although many professionals no longer use this term.

Some children may be developmentally delayed, intellectually disabled or suffering from a physical problem such as cerebral palsy. Others who have been, or are, chronically ill will be at risk of learning problems as they miss or drop behind in schoolwork.

Signs that a child has a learning disability are: difficulty following and understanding simple instructions; an inability to master reading, writing and/or mathematical skills (these children may have difficulty processing words as they hear them, or they may have difficulty with word-finding); a difficulty differentiating left from right; a lack of coordination that indicates difficulties with motor functions—using pens and scissors may pose a problem; an inability to understand the passage of time (these children have a visual memory problem which makes the understanding of sequential organization a difficulty); difficulty remembering what they have been told (these children may have suffered recurrent ear infections which can mean their auditory memory is not intact and they find it difficult to work or sound out new words).

The longer a learning disability is left untreated, the worse it will become. Once it has been determined that a child has a learning disability and after any underlying health problems have been treated, treatment by a remedial teacher or a psychiatrist trained in working with children suffering from learning disabilities may help the child to overcome the disability.

Dyslexia

Dyslexia is a disability which makes learning to read difficult. It can affect both

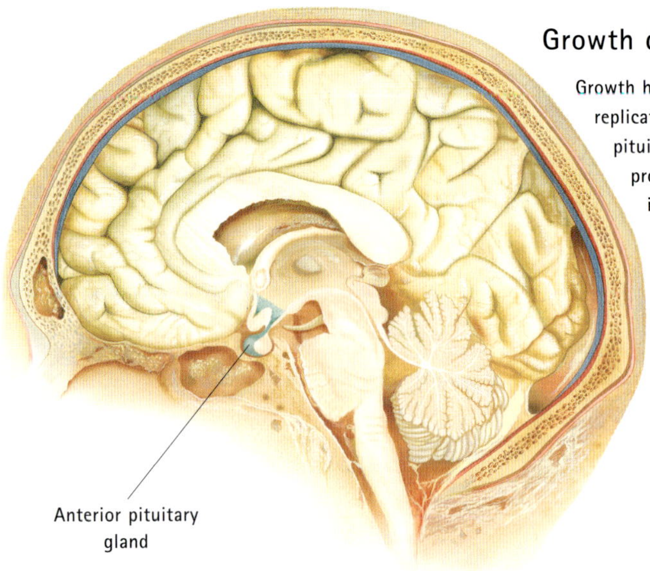

Growth disorders

Growth hormone controls cell growth and replication and is produced by the anterior pituitary gland. A deficiency in the production of growth hormone inhibits development, causing disorders such as dwarfism. The production of too much growth hormone stimulates abnormal growth of bones and muscles as in gigantism and acromegaly.

children and adults despite a normal intellect and satisfactory education. Originally a catchword which covered various disorders, the term dyslexia came to refer to all types of reading, writing and spelling problems. There is currently no internationally agreed definition of dyslexia and for that reason not all educators and psychologists will use the term. Dyslexia, as it is now generally understood, refers to its key feature: a substantial difficulty in gaining effective reading skills.

Dyslexia was originally believed to be a disability produced by poor sight, but it is now seen as result of abnormal brain function. There is no proven genetic basis for the condition, but current research into families with many affected members suggests that it may be inherited. The degree of intellectual ability plays no part, as dyslexia sufferers often score above average in non-language based intelligence tests. Unlike the ability to speak, which is innate in humans, reading skills are learned.

Children with normal vision learn to read after a gradual acquisition of pre-reading skills—the ability to follow a sequence of characters, the development of a vocabulary of language, the identification of sounds, and the recognition that sounds can be represented by letters.

This last process is one which dyslexic children are unable to develop. They are unable to decode speech into the individual sound components (phonemes) which are used to build words. There are 40 or more phonemes in the English language. The word pit, for example, is a combination of three sounds: "puh," "ih" and "tuh." It is essential to understand this and be able to associate the individual letters P, I and T with the sounds they represent before one can recognize the written word. Where the brain cannot hear the individual sounds it is not able to make the association between sounds and letters, and reading skills will not properly develop.

Dyslexia is suspected when reading skills fail to match a child's intellectual level, with the key indicator being the inability to decode phonemes. This may affect one child in five severely enough to persist into adulthood, and at this level it may not improve with time or instruction, causing academic standards to fall gradually behind.

Treatment should always be attempted. It will usually take the form of special instruction aimed at developing awareness of phonemes before moving on to improving word recognition, pronunciation and reading comprehension. With early identification and special instruction it is possible to make dramatic improvements in reading ability in most affected children. Without help, problems may become extremely difficult to fix, with devastating effects on academic achievement, adult life and future employment. Specialist remedial help can minimize the possibility of problems extending into adulthood, when they may be considerably more difficult to alleviate.

Growth disorders

Human growth is controlled by hormones, especially growth hormone. Excessive production of growth hormone by the anterior part of the pituitary gland in adulthood causes acromegaly, a disorder in which the bones in the arms and legs, hands, feet, jaw, and skull get thicker and longer. Facial features become coarser, and the voice may become deeper. Before puberty the condition results in gigantism, causing excessive size and stature. Treatment is surgery or radiation therapy on the pituitary gland.

A deficiency of growth hormone causes dwarfism. The affected person has normal body proportions but is smaller than normal. Administrating growth hormones to patients with dwarfism may induce skeletal growth. Dwarfism may also be caused by malnutrition and chronic illness.

Puberty

Puberty is the 2–6 year period between childhood and adolescence when normal hormonal changes cause a rapid increase in body size, changes in the shape and composition of the body, and rapid development of the reproductive organs and the secondary sexual characteristics.

In females, this process includes the development of the breasts, widening of the hips, rapid growth of the uterus, and the appearance of hair on the underarms and around the vulva. Menstruation, a monthly discharge of blood and uterine tissue, begins about 2 years after the onset of puberty and continues to be irregular for 2 years or so before becoming more regular. The changes are caused by the actions of the sex hormones, estrogen and progesterone, released by the ovaries under the control of hormones from the pituitary gland.

In males, secondary sexual characteristics include rapid growth in the size of the testes and the penis; an increase in the size of the larynx, which deepens the voice and gives the "Adam's apple" look to the throat; and the appearance of facial, underarm, pubic and body hair. Height increases rapidly and the first ejaculation occurs, usually about a year or so after the penis begins enlarging. These changes are governed by the male sex hormone testosterone.

In both females and males, there is also increased glandular activity. The apocrine glands, located in the underarms, the anus,

PUBERTY

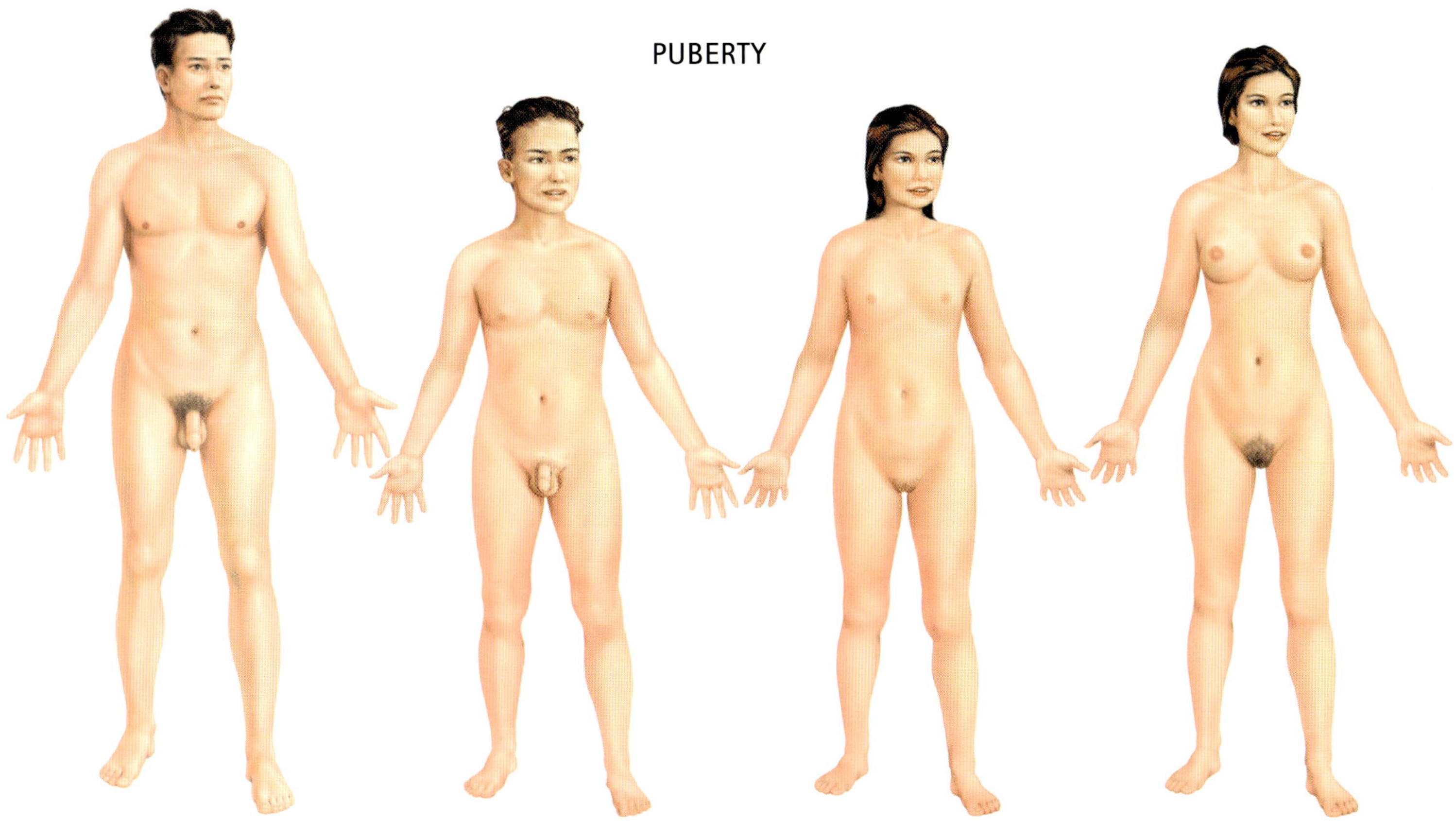

17-year-old male

Appearance of facial hair along with body and pubic hair, and increase in the size of the testes and penis are the predominant signs of male puberty. The larynx also increases in size creating the "Adam's apple" look at the front of the neck.

12-year-old male

Puberty in males occurs about 2 years later than in females. At age 12, a boy has little or no body hair, a small penis and testes, and still appears childlike.

10-year-old female

There is little or no breast development, no pubic hair and an almost boyish appearance at this age.

17-year-old female

Fully developed breasts, hair growth around the vulva and widening of the hips are the outward signs of female puberty.

the genitals and the breasts, become active at this time, giving off their characteristic odors. The sebaceous glands increase their production of sebum—an oily substance that lubricates the skin. This may lead to the familiar problem of acne in adolescence.

Girls undergo greater physical change during puberty than boys; they reach puberty earlier and take less time to reach maturity. Puberty in girls begins around age 11 and continues until about age 16. In boys, it begins about age 13 and continues until about age 18. Differences in the times and rates of growth and sexual maturity are caused by a combination of inherited tendencies, nutrition, and the environment.

Emotional and behavioral changes accompany these physical changes. Hormonal changes awaken sexual feelings, and many adolescents have some sexual experience. Dating normally begins during puberty.

In some cases puberty occurs later than usual or does not occur at all. Sexual development can be retarded by many different factors, including metabolic defects, hereditary conditions, hormonal disorders along with poor nutrition.

SEE ALSO *Acne on page 354; Endocrine system on page 110; Female reproductive organs on page 303; Female reproductive system on page 104; Hormones on page 102; Male reproductive organs on page 297; Male reproductive system on page 105*

Virilization

Normal virilization or masculinization is the appearance of male secondary sexual characteristics in the male. Virilization occurring in adult females or children usually indicates a serious underlying condition. It may be associated with hirsutism (abnormal male pattern hair growth) which is also a signal of an underlying disorder.

In boys, the changes of puberty may appear early due to abnormal production of testosterone, the male sex hormone. This abnormal production of testosterone first accelerates bone growth and then stops it earlier than normal, resulting in lower final height. Treatment with synthetic hormones can achieve normal growth.

Virilization can occur in females (also called virilism) when ovarian tumors make the hormones testosterone and progesterone. If unchecked, these hormones affect body shape, causing increasing muscularity and growth of facial and body hair. Also, the breasts may atrophy or fail to develop, the voice deepens and the clitoris enlarges. Menstruation and ovulation both cease. Disorders of the adrenal gland in females, causing production of excessive quantities of androgens (male hormones), will create the same symptoms of virilization.

Girls who are deficient in adrenal hormones will appear normal at birth but may fail to undergo puberty or to menstruate. Urine tests can detect abnormal hormone levels and aid diagnosis of tumors or adrenal problems. Medication with dexamethasone can prevent adrenal production of androgens. Surgical removal of tumors or of the adrenal gland is sometimes necessary. The exposure of a female fetus to high levels of androgens (sex hormones) early in pregnancy, through drugs prescribed to prevent a miscarriage or the presence of a hormone-producing tumor in the mother, may result in genital abnormalities, with the external genitals developing a male appearance. This may include a greatly enlarged clitoris and absence of a vaginal opening, while the internal reproductive organs develop normally. This condition is known as female pseudohermaphroditism. Surgery may be needed to correct the genital abnormalities.

Feminization

Feminization, the normal sexual development of females, occurs during puberty. It is a process which is governed by the sex hormones estrogen and progesterone. Under the influence of the sex hormones, the breasts develop, the hips widen, hair appears on the underarms and around the vulva, the uterus grows, and menstruation begins. Feminization can be delayed in some circumstances, or be absent if the female is suffering from certain diseases, such as Turner's syndrome and hypopituitarism.

Testicular feminization syndrome is the development of female sex characteristics in a male, giving rise to a genetic male with the appearance of a female. It is caused by lack of response of tissues to male hormones produced by the testes.

Menstruation

Menstruation is the cyclic vaginal discharge of blood and tissue debris derived from breakdown of the uterine lining. The menstrual cycle is considered to start with the onset of menstruation, which occupies approximately 5 days out of the full cycle of about 28 days (both these figures can vary considerably from the average). Apart from menstruation, the cycle can be divided into two phases, the days before (proliferative phase) and the days after ovulation (secretory phase). The menstrual cycle is an unusual reproductive cycle found only in humans and a few other primates.

The most obvious changes during the menstrual cycle take place in the uterine lining (endometrium). Menstruation involves the degeneration and loss of a large amount of endometrium along with about 1–2 fluid ounces (35–50 milliliters) of blood from broken endometrial vessels.

Toward the end, a small amount of the lining (about 1 millimeter thick) remains, which is essential for regeneration. This regeneration occurs under the influence of estrogen from follicles developing in the ovary during the proliferative phase, and proceeds until the lining is about 1/10 inch (3 millimeters) thick. The fully formed endometrium is a richly vascular tissue that also contains glands.

Progesterone derived from the corpus luteum is the predominant hormone from just after ovulation to about 2 days before menstruation (approximately day 26). The main function of progesterone is to stimulate the endometrial glands to produce a secretion capable of nourishing the embryo for the initial days of pregnancy, before a functional placenta has formed. The secretion occurs regardless of whether pregnancy has taken place or not.

If the woman is not pregnant, the corpus luteum lasts until about day 26 of the cycle when it starts to degenerate and progesterone production decreases rapidly. The blood vessels in the uterine lining are sensitive to levels of circulating progesterone and respond to reduced levels by contracting and cutting off the blood supply to all but the deepest parts of the endometrium.

This results in tissue breakdown and bleeding from the damaged blood vessels, which leads to the onset of the next menstruation. Menstrual flow is facilitated by muscular contractions and by substances produced in the endometrium which prevent clotting of the menstrual blood.

Menstruation does not take place if the woman has become pregnant because the corpus luteum survives and continues to secrete progesterone under the influence of hormones derived from the early placenta. This is essential to maintain the pregnancy until the placenta is able to produce its own progesterone toward the end of the first trimester. It is only then does the corpus luteum gradually cease to function.

SEE ALSO *Endocrine system on page 110; Female reproductive organs on page 303; Female reproductive system on page 104*

Menarche

The first menstruation (also known as menarche) normally occurs between 10 and 16 years. Puberty is a more general term describing both the onset of menstruation and the development of secondary sexual characteristics, which usually precedes menarche. This includes the growth of pubic hair, development of the breast and a spurt in body growth.

The age of menarche appears to be related to many social factors. Menstruation tends to occur earlier in urban girls of higher socioeconomic status and relates to heredity and better nutrition. It was widely believed in the past that menarche took place earlier in girls living in tropical climates, but this seems to be a racial rather than a climatic effect. The age of puberty has dropped by several years over the last century but this trend now seems to have slowed down.

Menopause

Menopause literally means "pause in menstruation." Menopause does not accurately describe what is happening to a woman's body at the time when ovarian function and hormonal production decline and the body adjusts itself. Climacteric is a more accurate term because it describes a time of life, just as puberty describes the time after the menarche, the first menstruation.

During the climacteric period, which can last several years, the woman's body goes through three phases. Premenopause is before the menstrual cycle ceases; this term can also refer to a woman's fertile years, but more commonly to the years when the menstrual periods are irregular and heavy. The next phase, the perimenopause, are the years on either of the two sides of menopause when a woman may experience some of the symptoms commonly associated with menopause, including hot flashes (also known as hot flushes), depression and loss of interest in sexual intercourse. The menopause starts after the final menstrual period, which can only be determined when 12 months have passed without a menstrual bleed. After this is the postmenopause, extending to the end of the woman's life.

There are three important hormones produced by the ovaries: estrogen, progesterone

Ovulation

During the proliferative phase of the cycle (days 8–10), follicle cells in the ovary divide and increase in size, creating a sac that contains an oocyte (female sex cell). By days 10–14, a mature follicle holding a secondary oocyte (mature egg) has formed. A surge of hormone from the pituitary gland triggers ovulation—the follicle releases its ovum into the fallopian tube. The ruptured follicle then transforms into the corpus luteum, which secretes progesterone and estrogen. If the egg is not fertilized the corpus luteum breaks down, hormone levels fall and menstruation occurs.

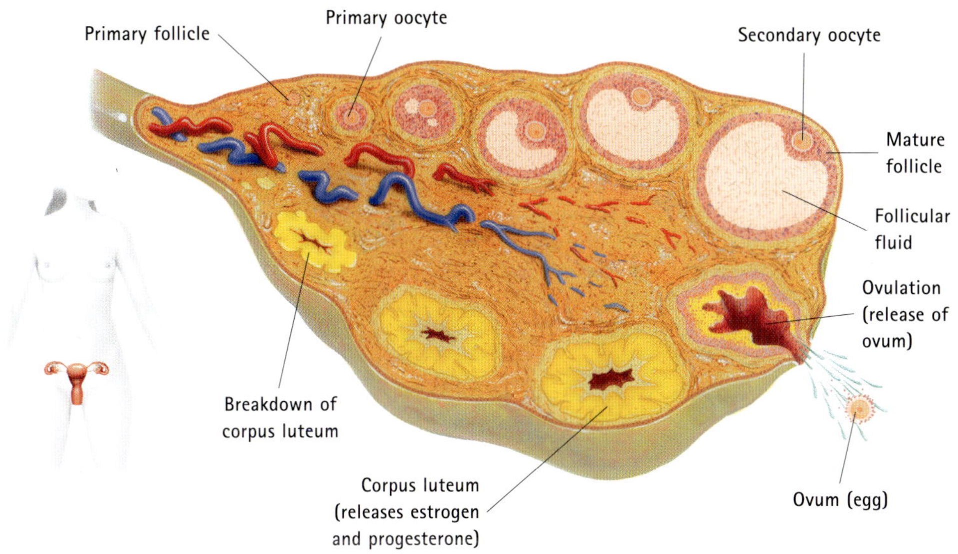

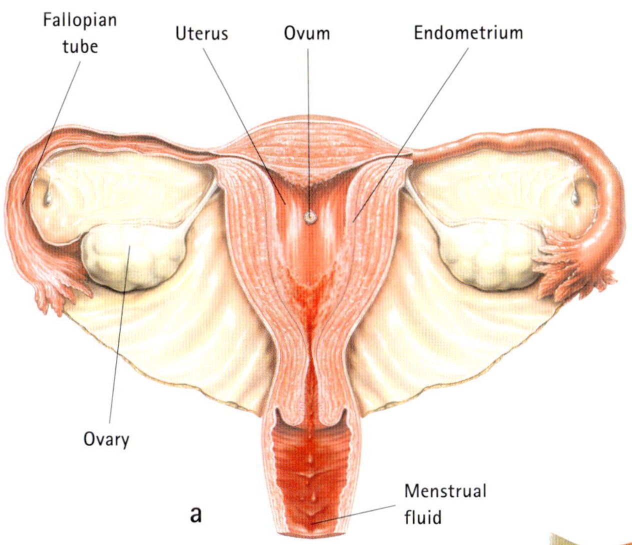

DAYS 1-6
MENSTRUATION

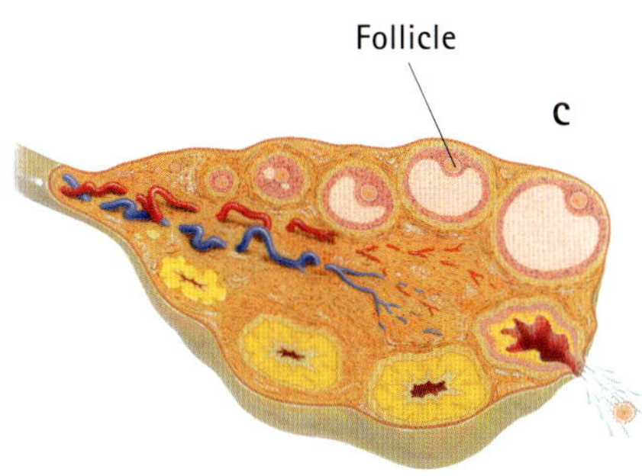

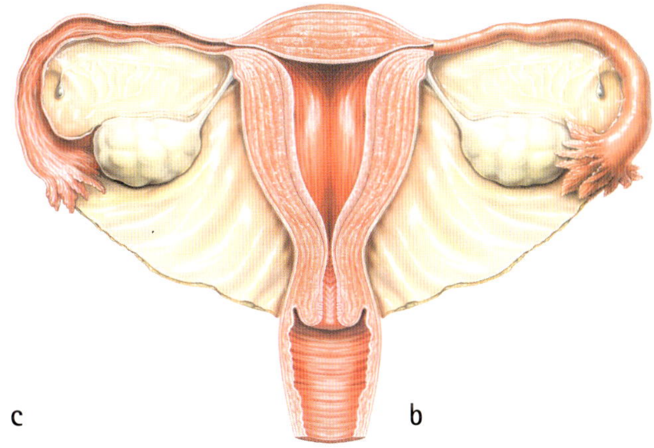

DAYS 7–13
PROLIFERATIVE PHASE

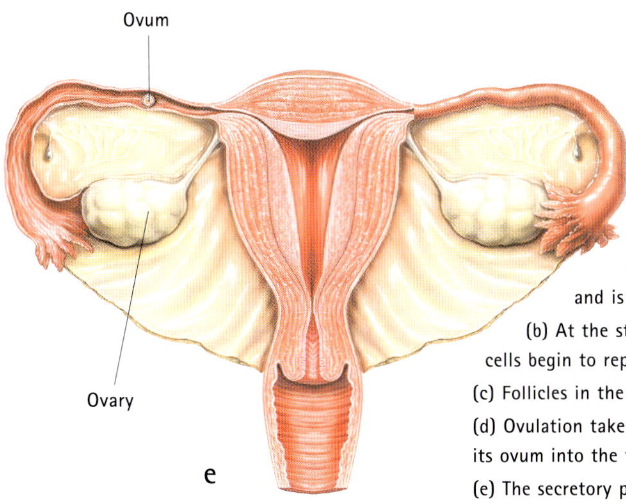

DAYS 15–28
SECRETORY PHASE

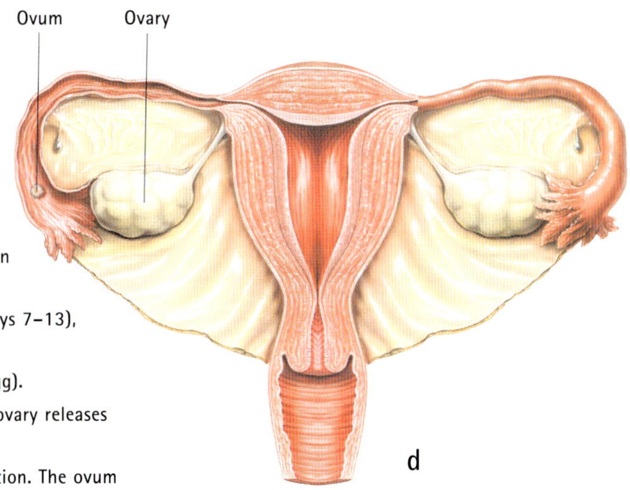

DAY 14
OVULATION

Menstruation

(a) During menstruation (days 1–6) the lining of the uterus (endometrium) breaks down and is discharged.

(b) At the start of the proliferative phase (days 7–13), cells begin to repair the endometrium and

(c) Follicles in the ovary begin to form an ovum (egg).

(d) Ovulation takes place (around day 14) and the ovary releases its ovum into the fallopian tube.

(e) The secretory phase (days 15–28) follows ovulation. The ovum travels up the fallopian tube toward the uterus and hormones thicken the endometrium in readiness to support a fertilized ovum. If the ovum is not fertilized, hormone levels fall and blood vessels in the uterus constrict. This causes the unused endometrial blood and tissue to be discharged from the body and the menstrual cycle begins again.

and relaxin. Once a woman has passed the menopause, estrogen and progesterone will no longer be produced by the ovaries, though in some obese women estrogen may still be produced by fat tissue, thus reducing some of the symptoms.

Estrogen is needed to maintain healthy body tissues and long term estrogen deficiency can result in stroke, dowager's hump, angina, genital degeneration and fractures, commonly of the hip, due to osteoporosis.

SEE ALSO *Bone diseases and disorders on page 34; Female reproductive organs on page 303; Hormone replacement therapy on page 450; Female reproductive system on page 104; Hormones on page 112*

PHYSICAL EFFECTS

On reaching menopause a woman may experience some or all of the following symptoms: hot flashes, night sweats, palpitations and sleep problems (due to changes in the normal working of the blood vessels), anxiety and depression, vaginal dryness, and decreased or a complete lack of interest in sexual intercourse.

A number of physiological changes accompany the climacteric and many women are unaware of most of them. They include changes to the urogenital tissue; changes to the reproductive organs; poor bladder control; changes to the menstrual cycle; dry skin that is more prone to wrinkles; changes in hair growth patterns; osteoporosis; loss of muscle strength and mobility; increased chance of heart disease; and breasts flattening and losing their fullness.

The menopause can produce different responses in a woman. Some of the above symptoms, together with a decline in the male hormones, can be responsible for a loss of interest in a sexual relationship.

For some women, male hormone replacement may benefit. However, if a woman maintains a healthy lifestyle and seeks help for any symptoms she encounters, she often finds the menopause brings a new lease of life including a liberated sex life, because as compared to earlier, there are no more contraceptive worries involved.

The climacteric is a natural life event and the state of hormone deficiency that generally accompanies it may, or may not, require any medication.

GERIATRIC MEDICINE

Geriatric medicine, or geriatrics, is an area of medical specialization that deals with health care of the elderly, and the disorders and diseases that arise with age. It is not to be confused with gerontology, which is the study of the ageing process and the social effects associated with it.

Industrialized countries have undergone a change in their population structure throughout the last century. As standards of nutrition and public health have improved, life expectancy for men and women has also improved, and the proportion of aged persons has increased. People over 70 years make up an increasing percentage of the population in Western countries.

Although people in the industrialized nations are living longer, they still experience the declines associated with ageing, and many diseases requiring full-time or nursing home care are becoming more common. In the UK, one-third of all hospital beds are occupied by people over age 65, about half of those being psychiatric patients.

Aged patients often have more than one disorder, and often suffer from chronic or incurable conditions, meaning they require proportionately more specialized attention. Women tend to live longer than men and constitute over half of the aged population. Geriatric medicine is increasingly concerned with disorders affecting women.

Complicating the provision of health care to elderly people are factors such as unreported illness, multiple disorders, loneliness and potential loss of independence. As infirmity grows, people may be reluctant to visit their doctor with what seem only minor symptoms. These may become major problems by the time they are reported and may be complicated by other disorders that have arisen in the meantime.

If mobility or normal dexterity is affected, elderly persons may not be able to care of themselves. In severe cases, they may not even be able to leave the home alone. This can result in depression, which may require treatment as well as the underlying causes. Care may therefore involve a team of health professionals as well as the family doctor.

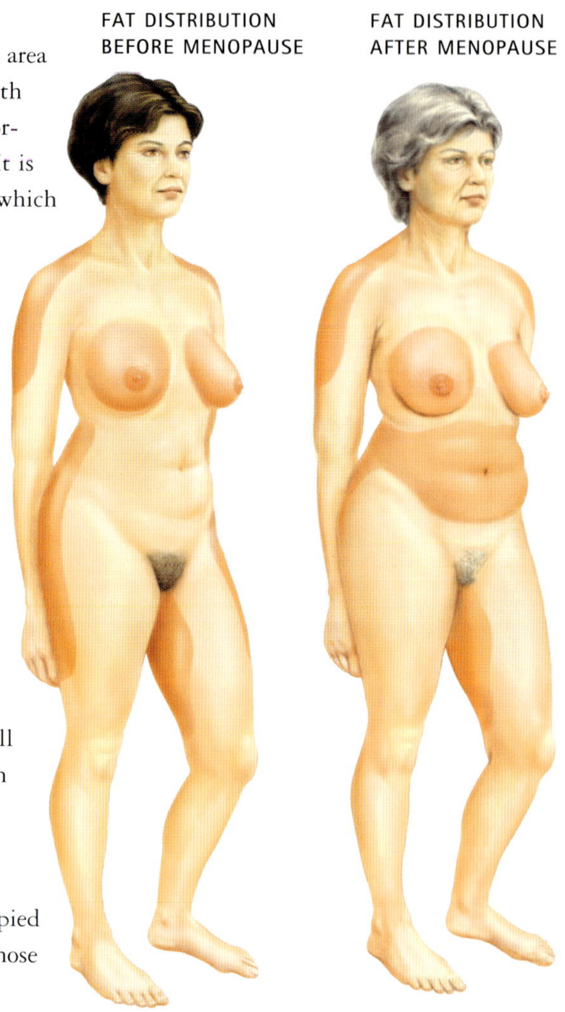

Menopausal fat distribution

Falling estrogen and progesterone levels during menopause affect fat distribution in women. Before menopause, most women have some fat deposits around the thighs, hips, breasts and upper arms. After menopause, fat tends to collect around the breasts, abdomen and waist.

Geriatric medicine developed in response to the special health needs of the elderly and infirm. Fast help in emergencies, a thorough assessment of each patient's needs, comprehensive treatment aimed to reduce disability (or minimize its effects) and assistance in keeping patients in their homes while addressing the needs of their caregivers are all elements of this medical specialty. The practice of geriatric medicine is based on acceptance that medical disorders in elderly people are not necessarily curable.

Treatments prescribed may be different than in general medicine. The emphasis is on retaining quality of life for patients, rather than solely achieving a cure. The aims and expected outcomes of any treat-

Common fracture sites

A reduction in the level of estrogen in the body during menopause can result in a loss of calcium and lead to brittle bones (osteoporosis). There is an increased risk of fractures during this period of life; the most common include fractures of the collar bone, upper and lower arm, radius (Colles' fracture), femur, and compression fractures of the spine.

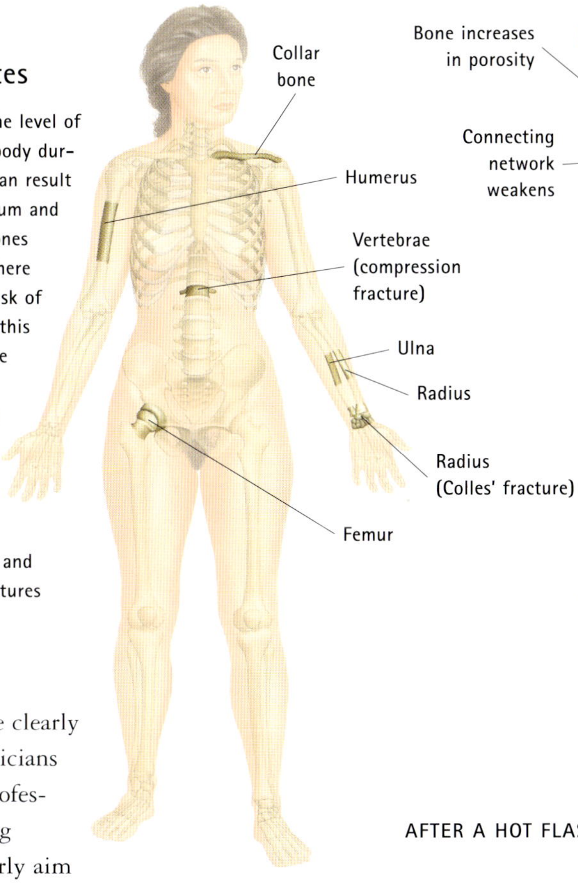

Osteoporosis

Women entering menopause are encouraged to take calcium and vitamin D supplements and to maintain exercise levels, to avoid bones becoming porous and weak.

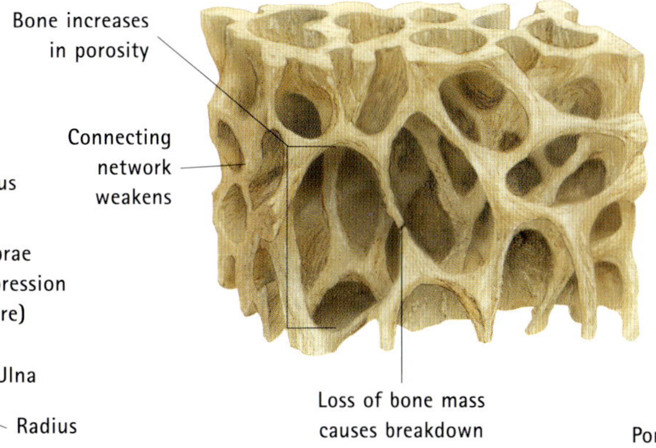

DURING A HOT FLASH

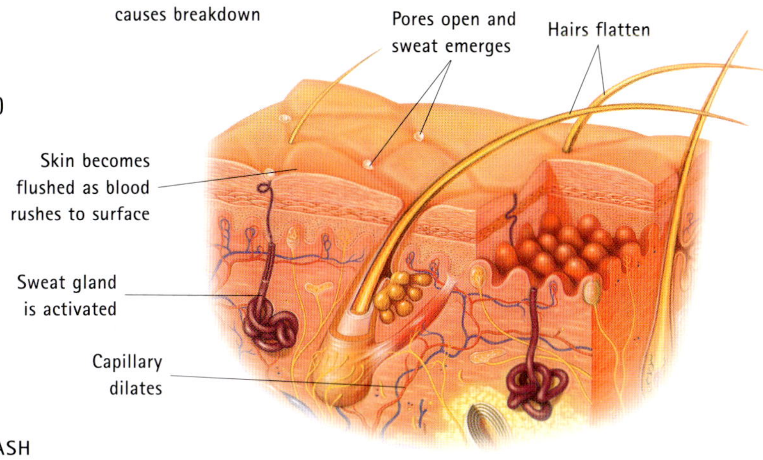

AFTER A HOT FLASH

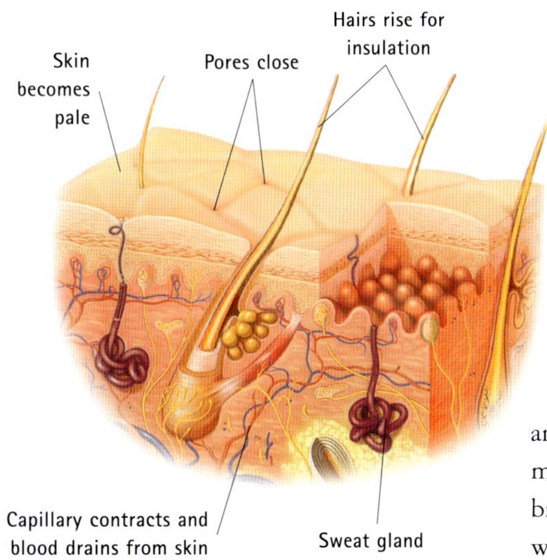

Hot flash

Hormone imbalances during menopause cause the capillaries of the skin to dilate, sending a sudden flood of warm blood to the surface. After the flash has passed, the capillaries rapidly constrict, the skin becomes pale and cold as blood drains away and hairs rise to provide insulation.

ment must be clearly defined. Physicians and health professionals dealing with the elderly aim to relieve discomfort and treat symptoms, no matter what the predicted outcome or the nearness or certainty of death.

Disorders or their effects are kept under control in order to allow patients to live as well as possible and maintain physical mobility and independence for as long as possible. It is important to maintain access to all medical services to fulfil these aims.

Central to geriatric medicine is the provision of care in the home. This relieves the pressure on hospital beds and in many cases retains the quality of life that comes from familiar surroundings and maintenance of domestic routines. Home care includes nursing, many forms of therapy, counseling, providing special equipment, rehabilitation, personal care services, preparation or delivery of meals, and even making modifications to the home.

Palliative care for the terminally ill and bereavement counseling for relatives are also part of a web of services that support health workers in geriatric medicine.

Ageing is an inescapable fact of life, and a process that begins at conception. Cells mature, die and are subsequently replaced. Life expectancy and general health are products of our genetic heritage, from parents, grandparents and their parents, and also of our environment, particularly nutritional standards in childhood years, and health care. As age advances, soft tissues become less flexible and the internal organs lose their efficiency. Sharpness of the senses is lost or gradually declines. Eyes lose their ability to focus on close objects and reading glasses become necessary. Hearing high-pitched tones becomes more difficult, taking some of the impact and enjoyment out of listening to music and affecting how speech is heard and understood.

Although the effects of age can be reduced, delayed and deferred, they cannot be eliminated. As the body ages, circulatory and respiratory ailments become more common. Blood flow to the liver, kidneys and brain is reduced, affecting the clearance of waste products from the bloodstream.

Lung capacity decreases, reducing the effective reoxygenation of blood and leaving increasing amounts of unexpired air in the lungs. Intellectual impairment is a common problem in the elderly and is a serious problem, affecting many aspects of an elderly person's life, including the ability to live independently, manage financial affairs, or drive even a vehicle.

About 10 percent of persons over the age of 65 years have some degree of mental impairment. It may occur due to irrversible conditions, such as Alzheimer's disease or multiple small strokes, or it may be treatable, as in diseases of the thyroid gland, sleep disorders, or depression.

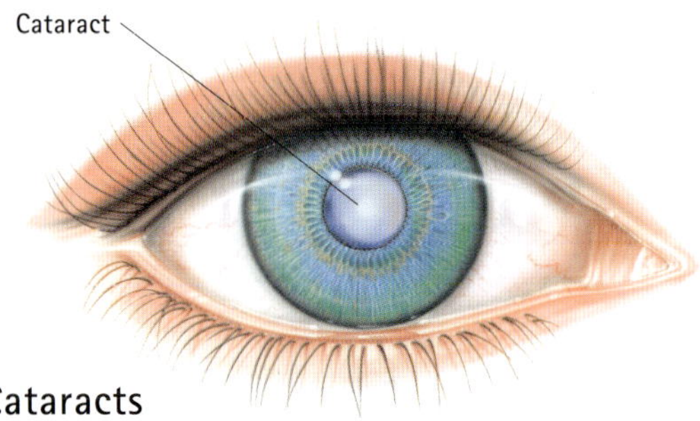

Cataracts

About one in five people over the age of 60 will develop cloudy spots in the lens of the eye. Cataracts are more common in older people because, as we age, the lens may deteriorate and become less transparent.

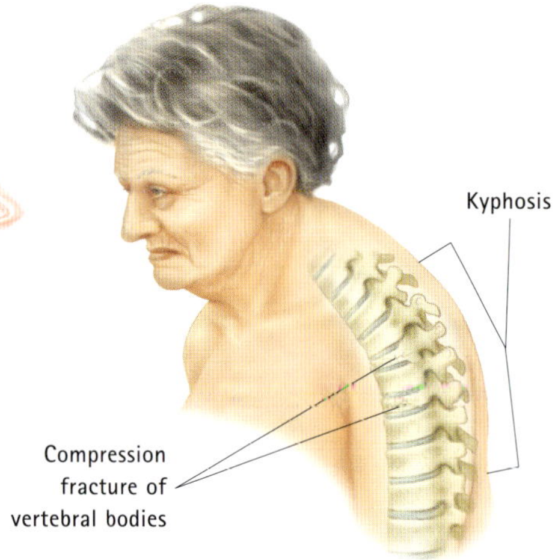

Dowager's hump

Older women with severe osteoporosis may develop dowager's hump. The vertebrae in the spine compress and become distorted, leading to a forward curvature of the thoracic spine which may worsen progressively.

ELDERLY WOMAN

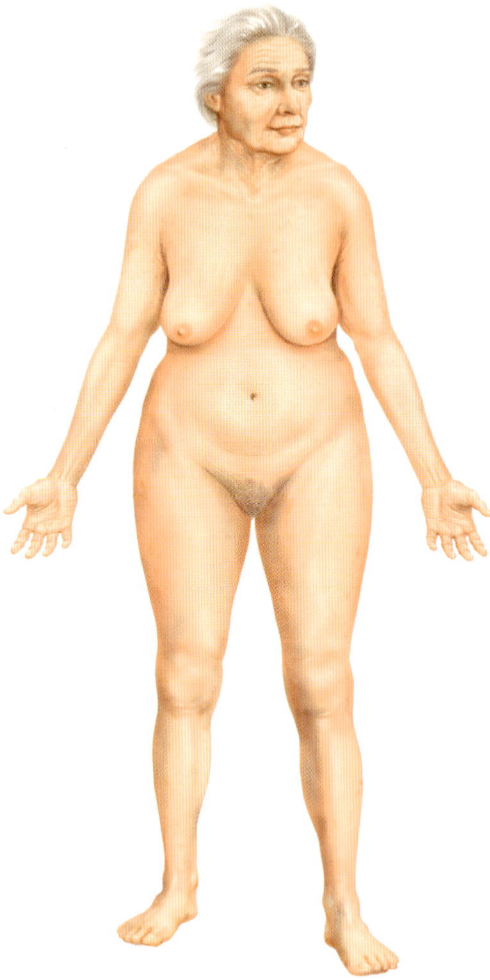

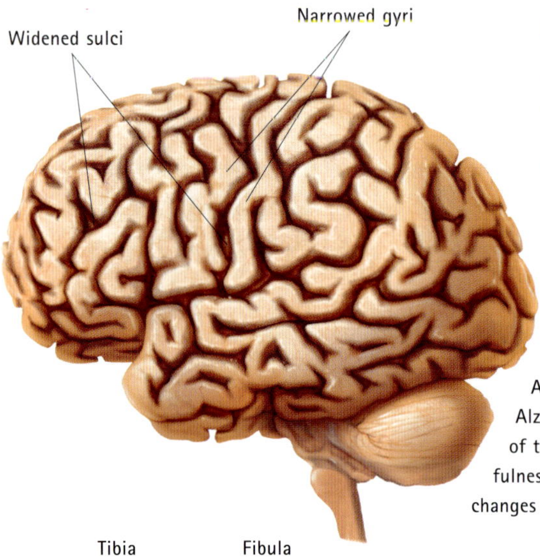

Alzheimer's disease

After the age of 65, the risk of developing Alzheimer's disease doubles every 5 years. Signs of this form of dementia include increasing forgetfulness, diminished intellectual capacity, personality changes and loss of motor skills and coordination.

ANKLE FRACTURE

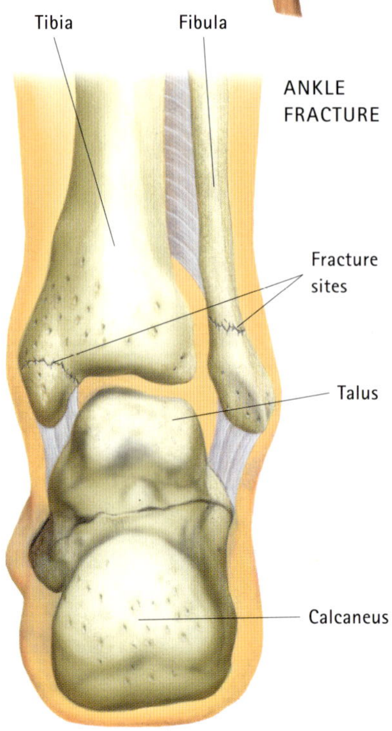

Fractures

In older age groups, especially women after menopause and people suffering from osteoporosis, bones lose calcium and phosphate and become less dense. Consequently they become weak and prone to fracture. Wrist, humerus, ankle, hip and vertebral fractures as a result of falling are common.

Decreasing mobility is also a feature of old age. Decreased balance, poor gait, muscle weakness, poor coordination and arthritis in the joints may also restrict movement. As elderly people become more sedentary, medical problems may develop; those who are bed- or chair-bound may develop edema, contractures, incontinence or pressure sores.

As a person ages, the senses deteriorate. Most older adults have at least some degenerative disease of the eye; 16 percent of those aged 75 to 84, and 27 percent of those older than 85 are blind in both eyes. Visual impairment reduces the ability to drive, read, shop, and even walk.

Hearing loss affects one-third of 65 year olds, two-thirds of those over age 70, and three-quarters of those 80 years of age and older. Typically, elderly people have difficulty hearing sounds in the higher frequencies. Visual and hearing losses increase their sense of alienation and loneliness and are often a contributing factor in falls and injuries.

Maintaining adequate nutrition is essential in the elderly, who are often malnourished because of poverty, social isolation, depression, dementia, pain or immobility. As a consequence of malnourishment, they tend to suffer pressure sores and take longer to recover from illnesses.

As a rule, elderly people tend to take too many medications. It has been estimated that an elderly person over the age of 65 takes on average 13 different medications in a year. These may interact with toxic effects. To make matters worse, the function of the kidneys deteriorates in old age, so toxic levels of the drugs may build up.

The immune system functions more poorly in old age, making elderly people more susceptible to cancers and infections. Progressive diseases also tend to become more severe in old age. Other diseases that are common in the elderly include fracture (hip, ankle); heart attack; incontinence; kidney disorders; Parkinson's disease; osteoarthritis; osteoporosis; shingles; prostate problems; and stroke.

Geriatric medicine can now ensure a comfortable and functional independent

ELDERLY MAN

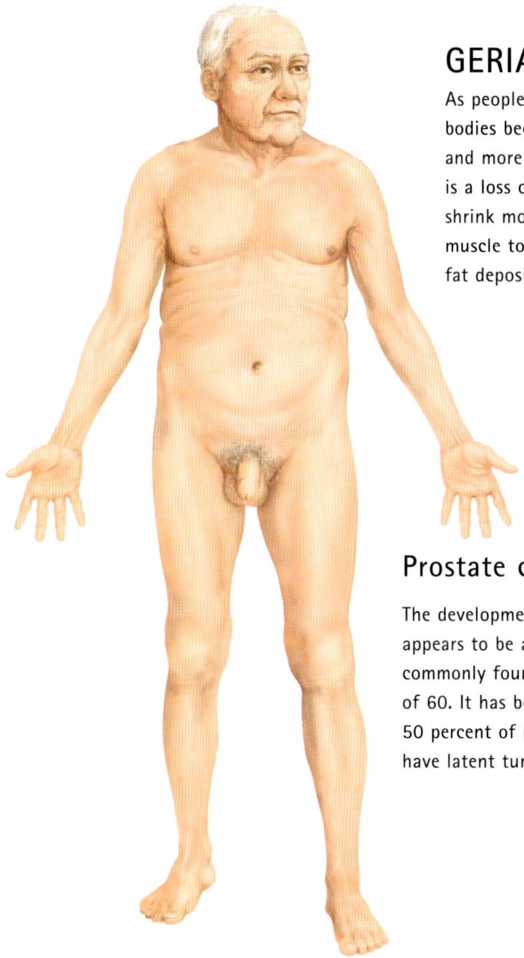

GERIATRICS

As people grow older their bodies become less flexible and more inefficient. There is a loss of height—women shrink more than men—and muscle tone is reduced and fat deposits increase.

Osteoporosis

Both elderly men and women can suffer from osteoporosis. As we age, new bone growth no longer replaces old bone as quickly as it used to, so the bones become porous and brittle.

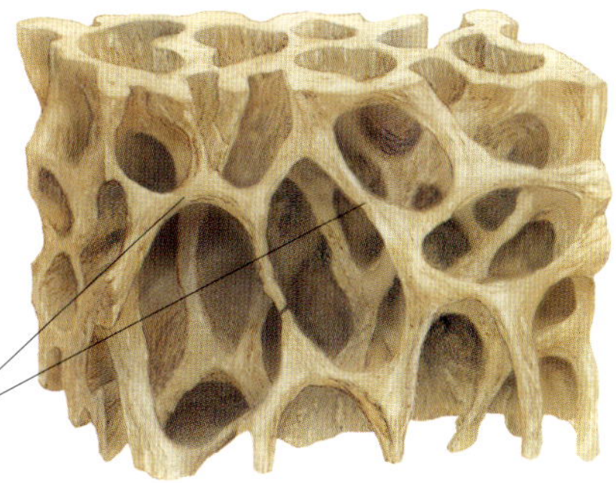

Prostate cancer

The development of prostate cancer appears to be age-related, as it is most commonly found in those over the age of 60. It has been found that more than 50 percent of men aged over 80 years have latent tumors of the prostate.

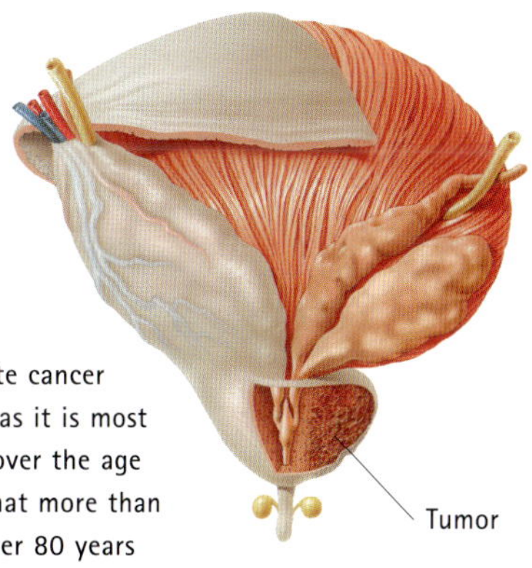

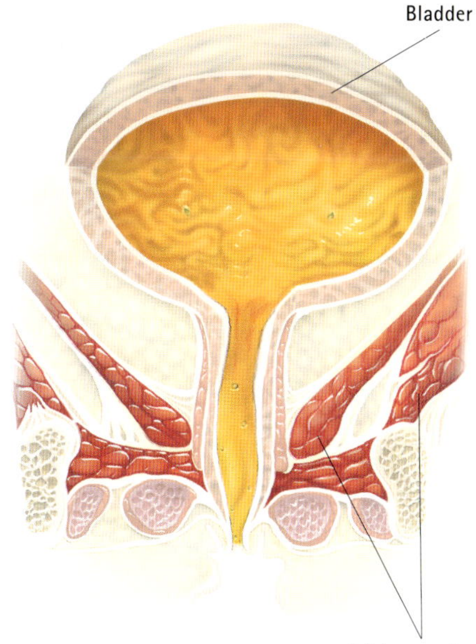

Incontinence

In the elderly, incontinence can be caused by a number of different factors. The muscles that support the bladder and floor of the pelvis can weaken with age, or incontinence may be associated with disorders such as senility, stroke or diabetes mellitus.

life for sufferers of some chronic conditions that previously resulted in death or disability.

SEE ALSO *Alzheimer's disease on page 148; Bone diseases and disorders on page 34; Cataracts on page 165; Diabetes on page 117; Disorders of the prostate on page 297; Incontinence on page 296; Joint diseases and disorders on page 42; Osteoarthritis on page 43; Parkinson's disease on page 152; Stroke on page 142*

Palliative care

Palliative care is defined by the World Health Organization as "the active total care of a person whose disease is not responsive to curative treatment." The aim of palliative care is to meet the needs of the whole person—their physical, spiritual and psychological needs—as well as of their family to give them the best quality and quantity of life possible under the circumstances. Palliative treatments relieve but do not cure.

Many treatments and therapies for terminal illnesses can prolong the life of the patient for many years, which makes palliative care a complex mix of managing symptoms and therapies, and providing emotional support to prepare the patient and their family for the inevitability of death.

To be effective, palliative care must provide for the needs of the patient coping with the rigors of therapy as well as maintain a level of physical, mental and social functioning that is satisfactory to the patient. Palliative care has become a subspeciality of medicine in many countries.

When a person is dying, the most important goal is comfort. Forms of comfort include: pain medication that is tailored to the needs of the individual patient; drugs, oxygen therapy and emotional support to relieve shortness of breath; the management of incontinence with drugs, catheters or disposable products; products to relieve dry mouth, such as fluids, humidified air, ice to suck on, or medication; the support and care needed in order to sleep as much as necessary; and an understanding that a patient's appetite may alter or fluctuate.

Emotional and spiritual support is also very important. People with religious or spiritual beliefs may find that they receive the support and comfort they need from priests or counsellors.

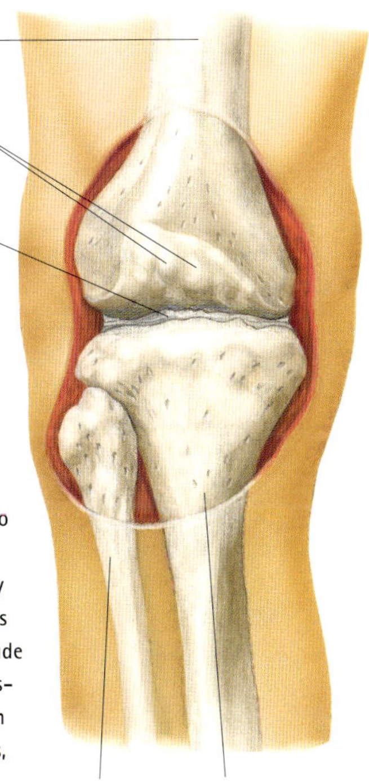

Osteoarthritis

As part of the ageing process, the cartilage that covers the ends of the bones begins to disintegrate, and the bones themselves may wear down. Symptoms of osteoarthritis include pain, stiffness and discomfort in joints such as fingers, knees, hips, toes and the spine.

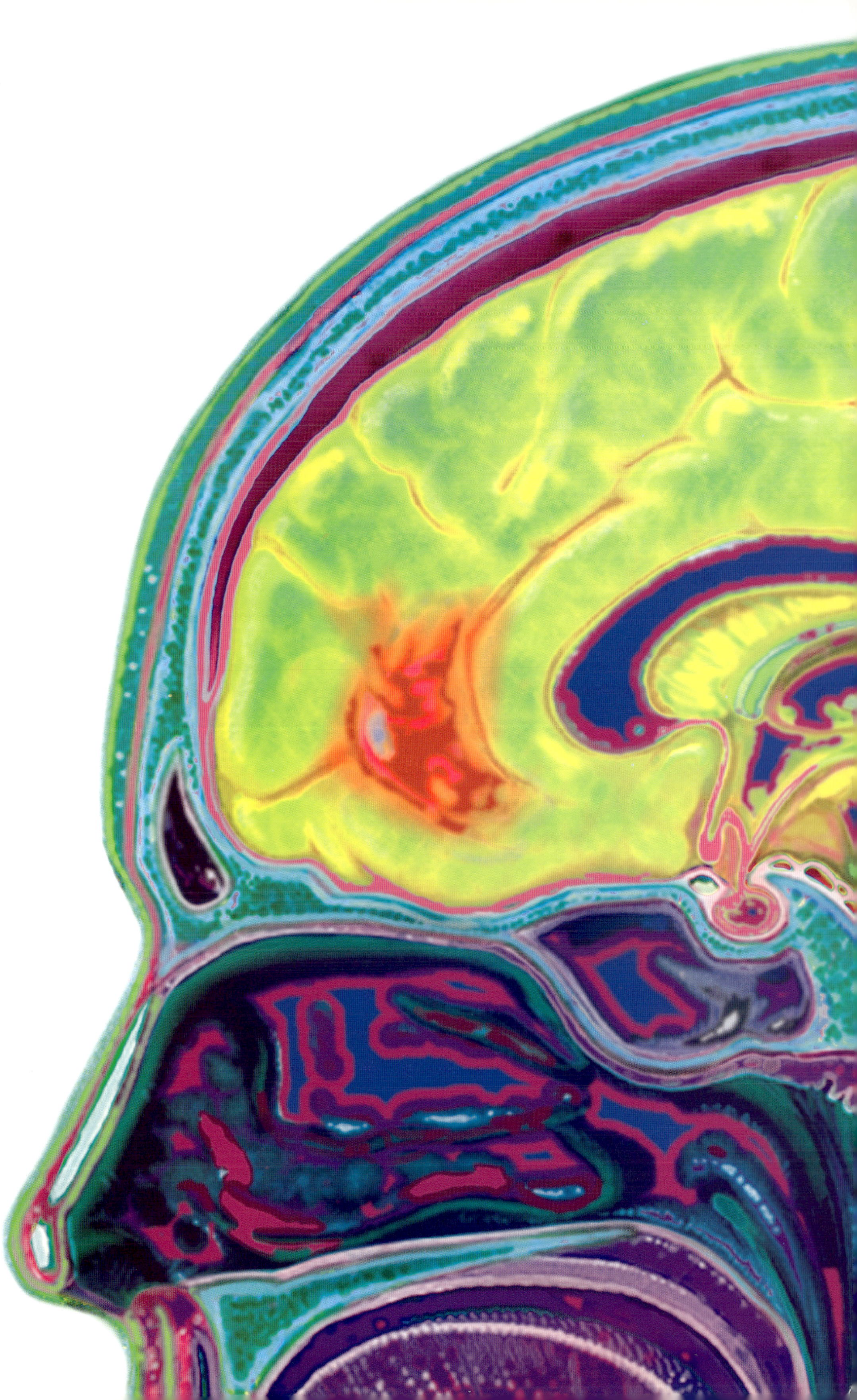

Diagnosis and Treatment

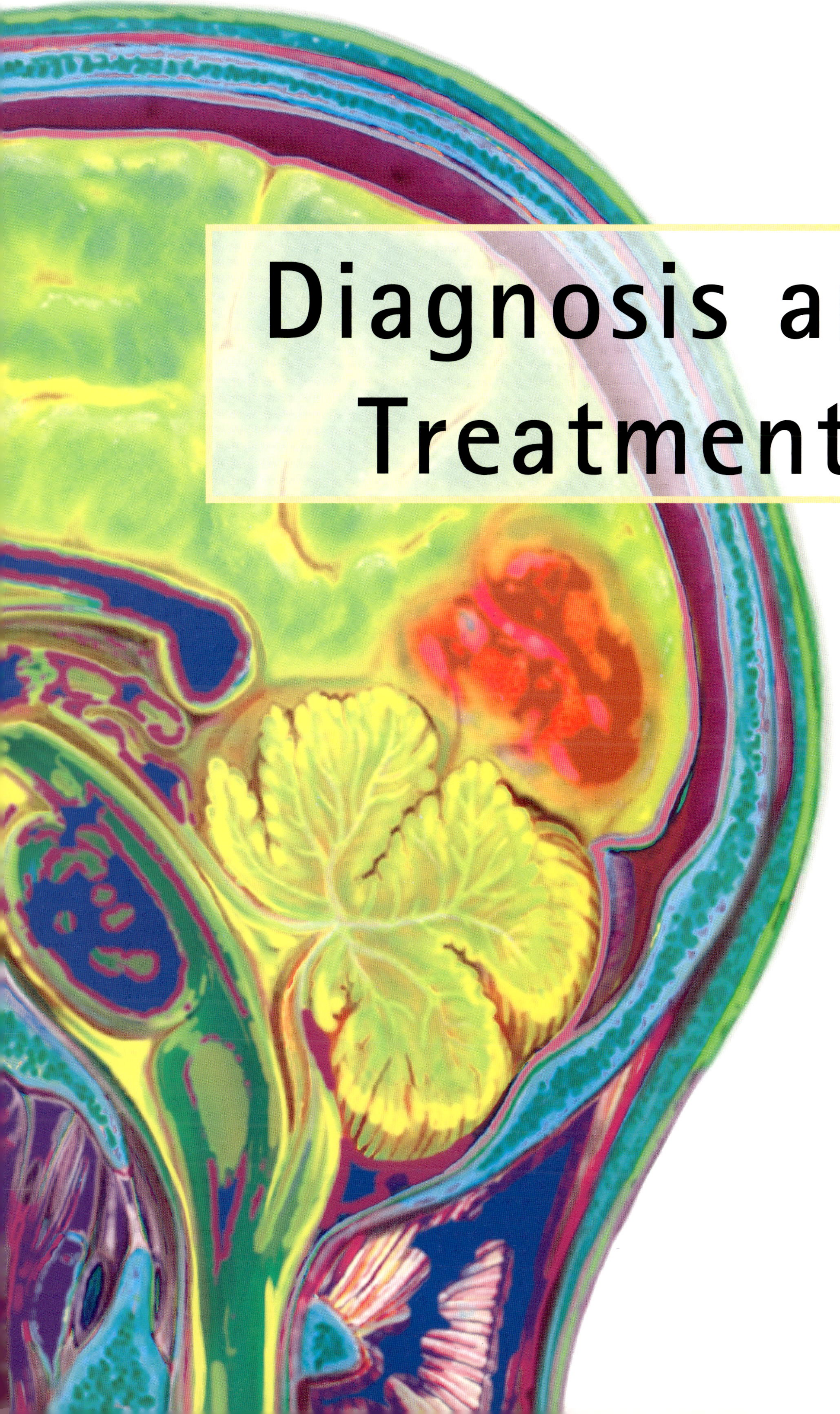

Non-invasive diagnostic techniques

Non-invasive methods for diagnosis range from simply listening to a patient's symptoms to the use of advanced imaging technology. Sound waves, x-rays and electrical impulses can all be used to provide detailed data on the body. Imaging techniques are particularly useful for assessing the function of organs such as the brain and heart, where invasive procedures can be hard to perform.

Audiometry

Audiometry is a very precise method of measuring a person's ability to hear. In normal hearing, sound reaches the eardrum and inner ear through the air and through the mastoid bone behind each ear. An instrument called an audiometer electronically produces pure tones at a specific pitch and volume; it is used to test each ear, with different tests for measuring air conduction and bone conduction hearing. In both tests the volume of each tone is reduced to the point at which the person can no longer hear it. These sound thresholds are charted or graphed to produce an audiogram.

Audiometry can also measure the speech threshold to show how words are heard and understood at specific volumes. This involves listening to a series of words and repeating each one as it is heard. A separate speech test measures the ability to discriminate between similar-sounding words.

Tympanometry is a test used to establish the cause of hearing loss, particularly in children. It can differentiate between problems caused by a blocked eustachian tube, fluid in the middle ear and malfunction of the bones of the middle ear.

Ophthalmoscopy

Ophthalmoscopy is examination of the interior and the retina of the eye with an ophthalmoscope, an instrument about the size of a flashlight, which projects a beam of light through the pupil. The instrument allows the physician to view the interior and back portion of the eye, including the retina, optic disc and blood vessels, and to detect and evaluate symptoms of eye disease such as glaucoma, or the effect of other diseases such as diabetes, atherosclerosis or hypertension on the eye.

Tonometry

Tonometry is the measurement of intraocular pressure (pressure within the eye), which is typically elevated in people with glaucoma. Various techniques are available, but applanation tonometry is considered to be the most accurate. This procedure involves bringing the measuring instrument into contact with the anesthetized cornea.

Electroencephalogram

An electroencephalogram (EEG) is a record of the overall activity occurring in the cerebral cortex of the brain. It is obtained by recording electrical signals from electrodes placed at various points on the surface of the head and appears as a series of spikes on a graph. These spikes, known as brain waves, can be classified according to their frequency into four groups known as alpha, beta, theta and delta waves. Their pattern varies during different types of brain activity such as rest, sleep and mental concentration.

In certain brain conditions the EEG reading is found to be abnormal. An EEG can be used in the diagnosis of epilepsy (the test will help determine the type of epilepsy), brain injuries, abscesses, meningitis, encephalitis and brain tumors. The procedure is usually carried out in a hospital clinic or a doctor's office. A sleep encephalogram, performed on a person after they have been kept awake the previous night, can help evaluate some types of sleep disorders.

Absence of electrical activity in an EEG is sometimes used as legal proof of brain death in a person who is in a coma and is being kept alive only by artificial means.

Nerve conduction tests

Nerve conduction tests evaluate the health of a nerve by recording how fast an electrical impulse travels through it. Two electrodes are placed on the skin at different points along the path of a nerve which is being tested. One electrode initiates a nerve impulse which travels along the nerve and the other records the impulse. The time between the stimulus and response is recorded to determine how quickly and thoroughly the impulse travels along the nerve.

Different nerve diseases will affect conduction in different ways, and the test can be used as a diagnostic aid. The person being tested feels a small electric shock, but most people tolerate the test well.

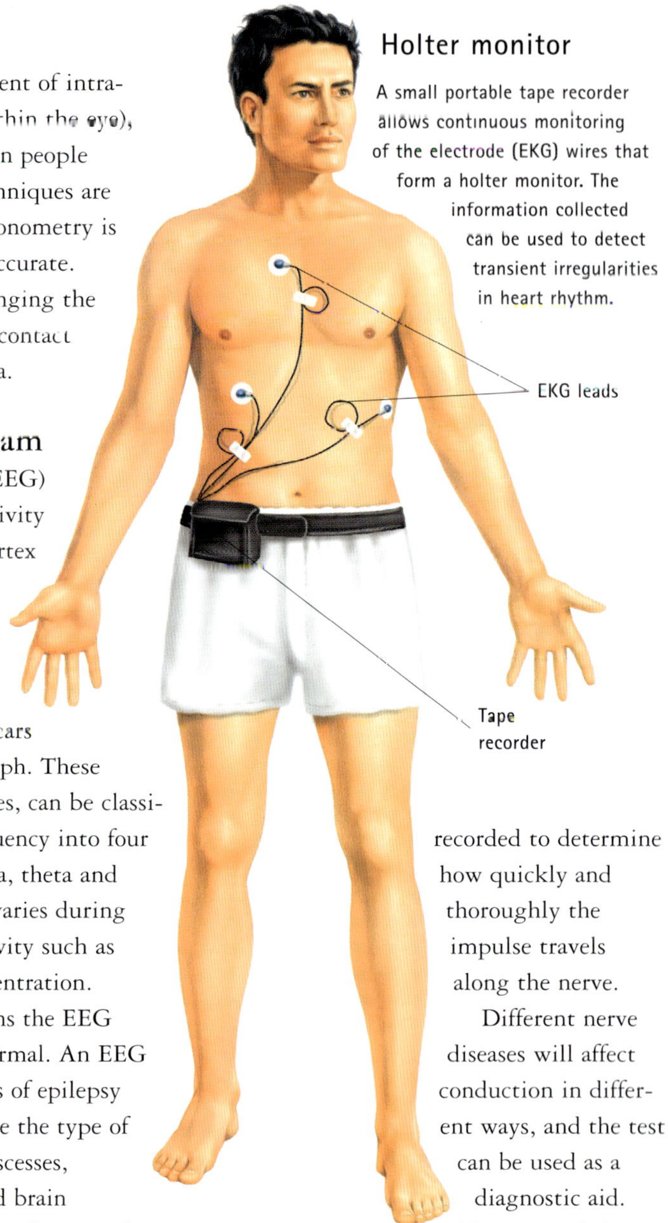

Holter monitor

A small portable tape recorder allows continuous monitoring of the electrode (EKG) wires that form a holter monitor. The information collected can be used to detect transient irregularities in heart rhythm.

Electrocardiogram

The activity of the heart can be measured through minute electrical impulses from the cardiac muscle that reach the skin surface. This electrical activity can be detected by taping electrodes to the skin in positions dictated by the information required. Electrical signals are amplified and recorded on an electrocardiogram, a chart scribed on a moving strip of paper.

Results show the efficiency of the heart's ability to transmit the electrical impulses that control the cardiac cycle, and its rate and rhythm. In a person with a normal heart, the EKG tracing has a characteristic pattern, showing upward and downward deflections. The first upward deflection, called a P wave, is known as the atrial complex and is due to the electrical activity associated with the contraction of the upper chambers (atria) of the heart. A further series of deflections, the Q, R, S and T waves, are the ventricular complexes and are due to the electrical activity associated with the contraction of the lower chambers (ventricles). Delayed transmission of electrical impulses, abnormally-shaped complexes and abnormal rhythms in the EKG tracing may indicate heart disease.

Some EKG changes only become apparent when the subject is exercising. An EKG carried out in a physician's office when a person is exercising vigorously is called a stress test. Other abnormal electrical activity in the heart may occur intermittently and may not be present when an EKG is performed. A Holter monitor is a device that makes a 24-hour EKG recording which can be used to detect sporadic arrhythmias.

Electromyography

Electromyography is a way of diagnosing the health of muscle tissue by measuring its electrical activity. At rest, muscles are electrically silent, producing no electrical impulses, but when stimulated or voluntarily contracted they generate an electrical current. These impulses are recorded on a cathode ray tube by continuous tracing in the form of a wave and monitored as a sound through a loudspeaker.

The visual tracing or electromyogram records the electrical impulses, and these can be interpreted to evaluate the health of the muscle, to reveal any weakness or wasting indicating impairment of nerve functions or muscle disease.

Pulmonary function tests

Pulmonary function tests (also known as lung capacity tests) measure the effectiveness of ventilation. They are used in lung diseases such as asthma, bronchitis, emphysema and fibrosis of the lung to determine the severity of the disease.

Spirometry is a test using an instrument called spirometer, which measures how well the lungs take in air, how much air they hold, and how well the lungs exhale air. A peak flow meter is a smaller device that measures how quickly the lungs expel air.

Sputum test

Sputum is the fluid that patients cough up from their lungs, usually during infective or allergic lung conditions, or after inhaling dust. Sputum may contain a lot of clear thick mucus, as seen in asthma or chronic bronchitis; pus, as in pneumonia; frothy fluid tinged with blood, as in heart failure with pulmonary edema; or black material, commonly found in coal-miner's lungs and heavy air pollution. Sputum may also be analyzed for the presence of bacteria, which can be cultured and tested for antibiotic resistance; pus cells, indicating infection; malignant cells, which indicate cancer; or asbestos bodies, seen in asbestosis.

Urinalysis

Urinalysis is a test performed to analyze cells, proteins and other chemicals in the urine. It is used to detect disease, especially of the kidney and urinary tract. For example, the presence of blood in the urine (hematuria) may be due to a urinary tract infection, a stone, a polyp or cancer.

Pus cells and bacteria may indicate urinary infection, while protein in the urine (proteinuria) may indicate nephritis or myeloma. Bilirubin in the urine may possibly be a sign of jaundice, and glucose in the urine is usually caused by diabetes mellitus.

The sample of urine may be cultured for 24–48 hours. If bacterial colonies form in the culture it may be a sign of infection in the urinary tract.

Imaging Techniques

These are techniques used in the production of diagnostic images. Since the discovery of x-rays (electromagnetic radiation) in 1895, imaging techniques have played an important role in diagnosis.

X-ray

X-rays are a form of electromagnetic radiation, similar to light but of a much shorter wavelength. They can penetrate soft tissue such as skin and muscle rather easily, but are absorbed by bones and other objects containing heavy atoms.

X-rays, or their penetration, are recorded or observed by use of photographic film or fluorescent screens, so they can therefore be used to study the internal structure of the body, particularly bones. Also observable are the presence and location of foreign objects, such as swallowed coins, and surgically-inserted needles and metal plates.

Cavities, such as those of the gut, bladder, heart and blood vessels, can be studied by filling them with substances which are either radiopaque (for example, barium) or

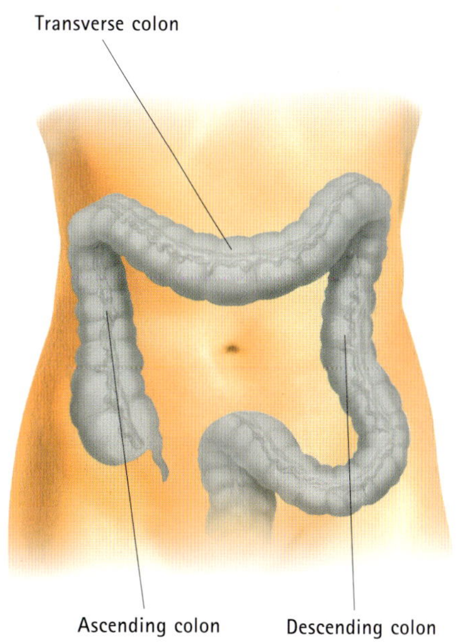

Barium enema

In a barium enema, a white-colored liquid called barium is introduced into the rectum and moved around the colon (shown at left) by a radiologist. Bowel movements may be white or lighter colored in the days following the test.

Barium meal

Also called a barium swallow, a barium meal gives the doctor an idea of the condition of the esophagus and stomach (pictured), and how well foods and liquids travel through them.

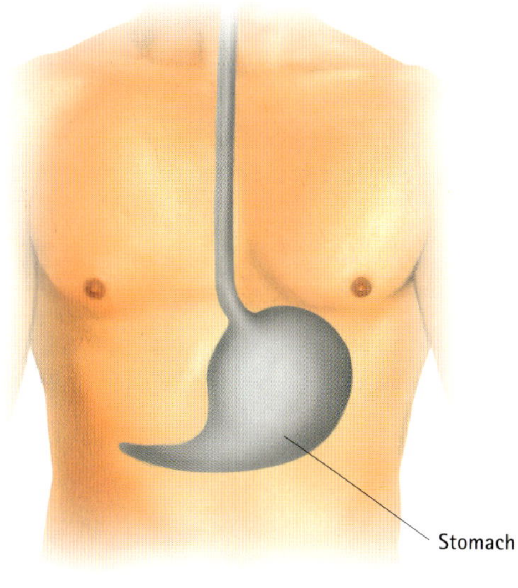

Angiography

Cardiac angiography is performed by introducing a catheter (a thin hollow tube) through the right or left femoral artery. The catheter is advanced into the aorta and then into the left ventricle of the heart. The tip of the catheter is placed in the left or right coronary artery and dye is injected. X-rays taken of the heart after the injection of the dye will reveal any narrowing or blockages in the coronary arteries.

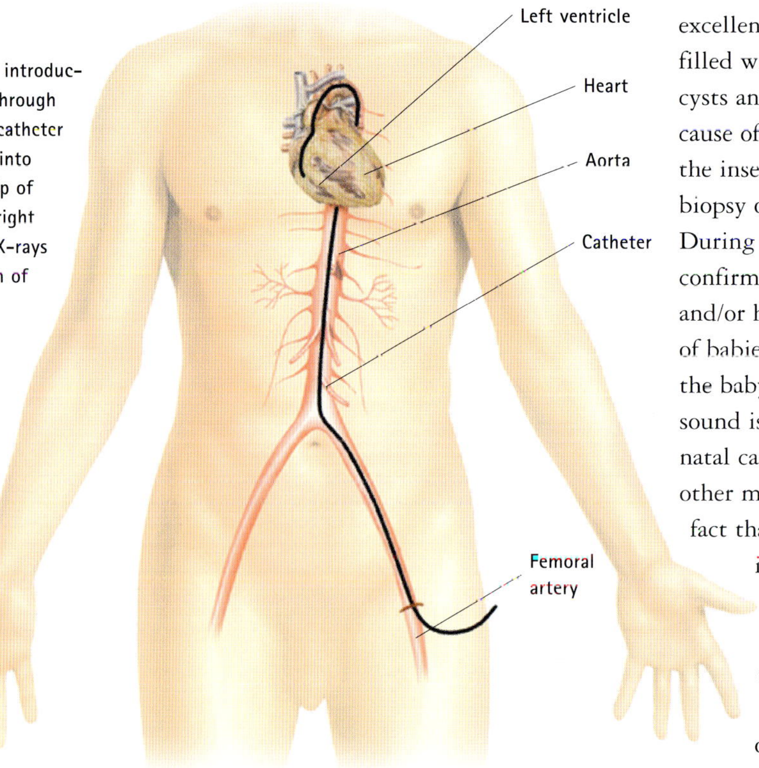

radiolucent (for example, carbon dioxide). The radiopaque substances offer resistance to x-rays, and appear as light areas on exposed film; the radiolucent substances permit their passage and appear as dark areas on exposed film.

Radiopaque dyes are most commonly used in angiography (heart), pyelography (kidney and urinary tract), cystography (bladder), myelography (spine), hysterosalpingography (uterus), cholangiography (gallbladder) and venography (veins).

Because x-rays are a form of ionizing radiation, they are potentially damaging to tissue. Excessive exposure to x-rays must be avoided. Exposure is minimized by a narrow beam of x-rays, covering sensitive parts of the body with lead-containing rubber shields, and using sensitive recording devices that require only low x-ray dosage. The damaging effects of x-rays can be of therapeutic value in the treatment of cancer and other tumors.

Cross-sectional pictures of body structures are obtained by computerized axial tomography (CAT scanning), in which a series of x-ray images focused on different planes are analyzed by computer and presented as sections. CAT scans reveal the soft tissue in more detail than plain x-rays.

MAMMOGRAPHY

A mammogram is an x-ray picture of the breasts. It is used to detect tumors and cysts and help differentiate malignant (cancerous) tumors from benign (non-cancerous) ones. Any woman with breast symptoms such as a lump, nipple discharge, breast pain, dimpling of the skin on the breast, or a recent retraction of the nipple, should have a mammogram. Approximately 90–95 percent of breast cancers are detected with mammography. The test is also used to screen for breast cancer in women with no symptoms. The American Cancer Society recommends a screening mammogram around the age of 40; annually or every two years between the ages of 40 and 49; and every year thereafter.

Ultrasound

Also known as ultrasonography or sonography, ultrasound is a diagnostic procedure that uses sound waves above 20,000 cycles per second to create an image. It produces excellent images of soft organs or organs filled with fluid; it is used to detect both cysts and solid tumors, to investigate the cause of abdominal pain and also to guide the insertion of a needle during a needle biopsy or a test such as amniocentesis. During pregnancy ultrasound is used to confirm the age of the fetus, the position and/or health of the placenta, the number of babies and the physical development of the baby. In industrialized societies ultra sound is treated as an expected part of antenatal care. It has numerous advantages over other methods of imaging, including the fact that the transmission of sound makes it safer than x-rays, which use ionizing radiation.

Ultrasound is usually performed as an outpatient procedure. A small amount of gel is applied to the skin over the area to be scanned, which improves the transmission of sound waves. The ultrasound transducer is moved over the area, transmitting the reflections of the sound waves to a computer, which converts them to an image on a screen.

Doppler ultrasonography is a specialized type of ultrasound used to investigate the flow of blood through blood vessels, and to

Ultrasound

This procedure uses sound waves to form a two-dimensional image of internal organs. It is commonly used in antenatal care to determine the health and development of the fetus. The transducer transmits reflections of the sound waves to a computer, which converts them into an image on a screen.

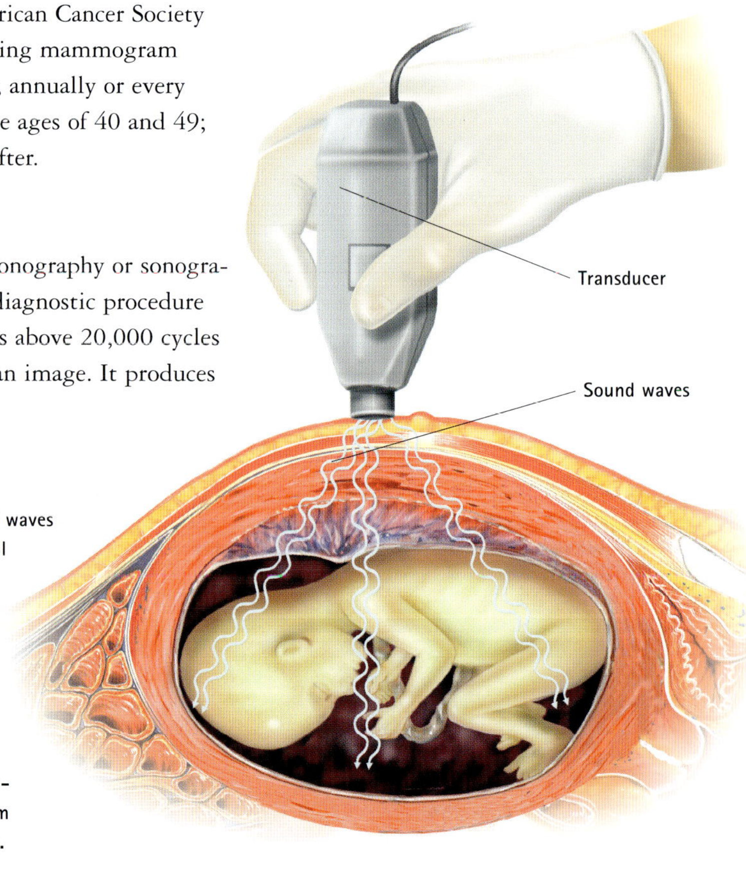

look at the movement of parts of organs such as heart valves. Very high energy ultrasound produces a heating effect that is used in sports physical therapy to treat soft tissue injury. Ultrasound is also used in one form of lithotripsy, where the sound waves are used to shatter kidney stones.

Echocardiogram is a diagnostic method of assessing cardiac health using ultrasound. High frequency sound waves are directed at the heart wall and valves and the resultant echo is used to detect an infection, damage or tumors and to measure the heart wall and chamber. The test shows how efficiently the heart pumps blood and whether valves are working properly or not.

Magnetic resonance imaging

Magnetic resonance imaging (MRI) is a non-invasive procedure used for imaging tissues that have high water and fat content. MRI uses an external magnetic field created by a series of powerful electromagnets in a scanner to excite hydrogen atoms in the body, which give off radio signals to the scanner. The signals are read by a computer and converted into a detailed image. Because it uses a magnetic field instead of x-rays, MRI scanning does not expose people to radiation.

During an MRI, the patient lies on a table that slides into a cylindrically shaped machine, and must remain motionless for up to 90 minutes. The test is painless but noisy, and can cause claustrophobia in some people; a mild sedative can help.

Positron emission tomography

Commonly referred to as a PET scan, positron emission tomography is an imaging technique for monitoring body processes such as blood flow and metabolism. Chemical compounds with radioactive isotopes are injected into an organ, and the body is scanned using a special machine. Positrons (positively charged electrons) emitted by the isotopes collide with electrons within the body, creating gamma radiation that is detected by the scanner. This information is converted to computer images to give a cross-sectional view of the organ being studied. This technique is used to assess such things as muscle damage after a heart attack and the effects of chemotherapy drugs on body tissue.

Radioisotope scan

Also known as a nuclear scan, a radioisotope scan uses radioactive isotopes in the diagnosis of disorders of the bones, the heart, the lungs, thyroid gland and kidney. It can measure the size, shape, function and position of an organ, and can also detect any abnormalities such as cysts, abscesses, tumors or other diseases in these organs.

During the test, the radioactive isotope is introduced into the body in one of two ways: either it is swallowed and concentrated by the organ being examined, or it is introduced into the organ via a catheter that is inserted into a vein or artery and guided to that organ. A camera, which can detect the gamma rays emitted by the isotope, then takes an image of the organ, and a physician interprets this picture. The patient is exposed to a small amount of radiation, but this is outweighed by the benefits of the scan; the dose is not dangerous. The test takes a few hours and takes place in a doctor's office or hospital radiology clinic.

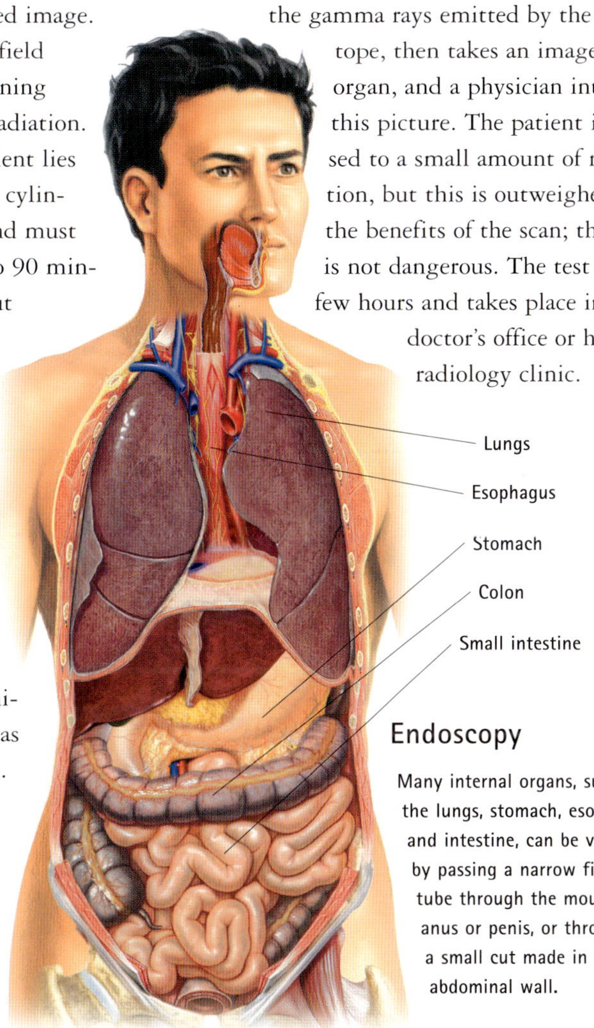

Endoscopy

Many internal organs, such as the lungs, stomach, esophagus and intestine, can be viewed by passing a narrow fiberoptic tube through the mouth, anus or penis, or through a small cut made in the abdominal wall.

Thermography

Thermography is a technique involving measuring the temperature in different parts of the body by scanning with a heat-sensitive infrared camera. It is especially known for its use in the diagnosis of breast cancer by measuring skin temperature—a tumor is marginally hotter than surrounding tissue. Before the scan takes place, the skin is exposed to the air for 10 minutes to stabilize its temperature and increase the accuracy of the test.

The scanning procedure results in an infrared photo of the body's surface temperature, called a thermogram, which may show up such things as disease-causing plaque on the arteries of the heart, various cancers and infection. The procedure has also been used in the diagnosis and treatment of pain, such as in the back and in the wrists where carpal tunnel syndrome is suspected. Pain shows up as cool colors on the thermogram as it causes blood vessels in the skin to constrict, which reduces skin temperature. It is estimated that about a quarter of thermographic tests bring false positive responses, making the technique unreliable for the screening of serious disease such as cancer.

Invasive diagnostic techniques

The use of invasive diagnostic techniques is often necessary to obtain information on the internal health of organs such as the colon and the stomach. Minor surgical procedures such as biopsy and arthroscopy, allow investigation and analysis of tissue and bone for disease.

Endoscopy

An endoscopy involves looking inside the body through an endoscope to investigate suspected abnormalities. Endoscopes are narrow tubes containing optical fibers and lights, and are extremely flexible. They are also capable of removing tissue samples or destroying abnormal tissue.

Endoscopy is also used to examine the esophagus (esophagoscopy), the stomach (gastroscopy), the lungs (bronchoscopy), the small intestine (upper gastrointestinal endoscopy), the lower portion of the large

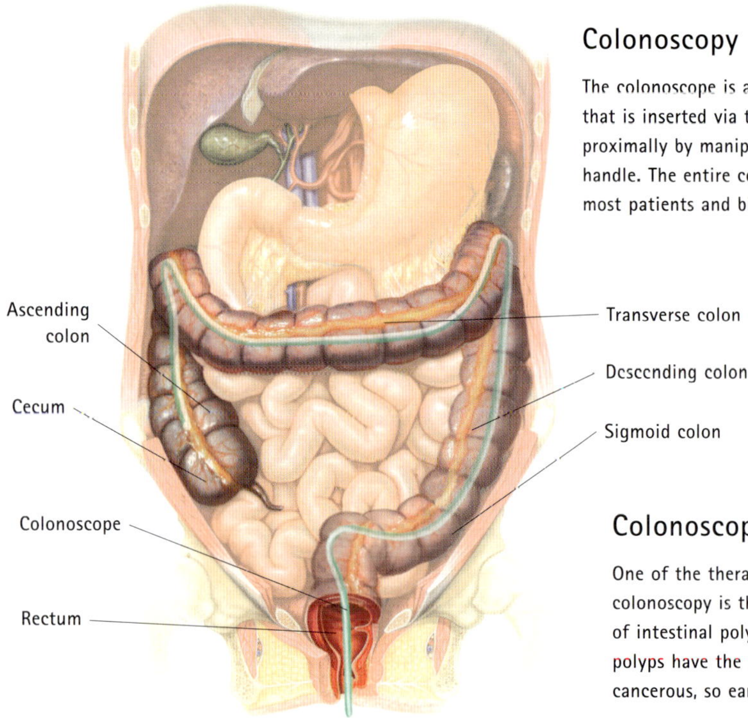

Colonoscopy

The colonoscope is a flexible instrument that is inserted via the rectum and advanced proximally by manipulating controls on the handle. The entire colon can be examined in most patients and biopsies can be taken.

intestine (sigmoidoscopy), all the large intestine (colonoscopy), the anal canal and rectum (proctoscopy) and bladder (cytoscopy). Some conditions can also be treated during an endoscopy.

The use of the endoscope has simplified diagnosis and treatment of a number of problems which in the past required barium meals and enemas, though these are still used. Any complications from an endoscopy are rare.

Laryngoscopy

Laryngoscopy is an examination of the interior of the larynx or voice box. Indirect laryngoscopy is the simplest type and involves holding a small mirror against the back of the palate, with the mirror angled down towards the larynx. This procedure may not be suitable for those with a strong gag reflex. Direct laryngoscopy is another option and, as the names suggests, is a way of looking directly at the larynx. It is generally performed under general anesthetic because of problems with the gag reflex, and involves the insertion of an instrument called a laryngoscope in the mouth. During this procedure a microscope can be used for a magnified view of the vocal cords and other parts of the larynx. Tissue biopsies can also be taken through this method.

Other optical instruments can also be used to examine the larynx. A flexible

Colonoscopic ligation

One of the therapeutic uses of colonoscopy is the ligation and removal of intestinal polyps. Many intestinal polyps have the potential to become cancerous, so early removal is important.

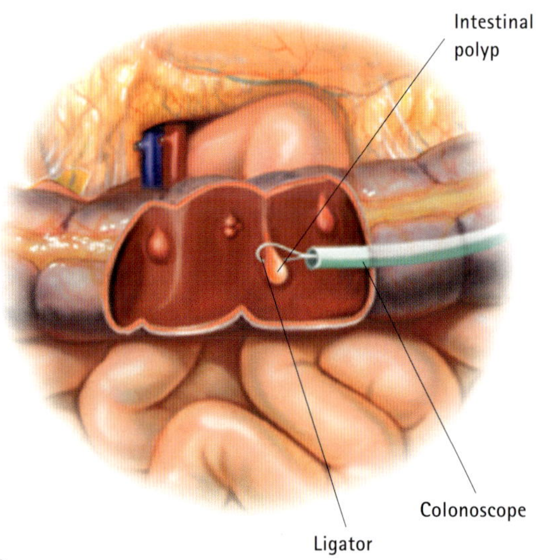

nasopharyngoscope can be inserted through the nose to look at the vocal cords during normal speech. This may cause a little gagging and the image obtained is not as clear as when looking at the larynx directly. A rigid instrument called a 90 degree telescope may be placed at the back of the throat for a clear, magnified image of the vocal cords. A camera may be attached to these instruments to record information during the examination.

Arthroscopy

Arthroscopy is a surgical procedure in which a small telescope known as an arthroscope is inserted into the cavity of a joint through a small incision to allow examination of joints for damage or disease. Surgical instruments have been devised for use in arthroscopy, and operations conducted in this way generally heal more quickly because of the small incision made.

Hysteroscopy

Hysteroscopy is the use of a hysteroscope, a uterine speculum with reflector, to remove by excision a fibroid tumor that is bulging into the uterine cavity. It is also used in visual examination of the canal of the uterine cervix and the uterus.

Fetoscopy

Fetoscopy is the use of high-resolution fiberoptic equipment in surgery on a pregnant woman. It is used when performing operations on the growing baby (fetus) or its environment, in the first and early second trimester of the pregnancy. The surgery is performed through a small incision in the abdomen and uterus.

Embryoscopy

Using the same high-resolution fiberoptic equipment as employed in fetoscopy, embryoscopy allows surgeons to see the embryo and fetus (two stages of a growing

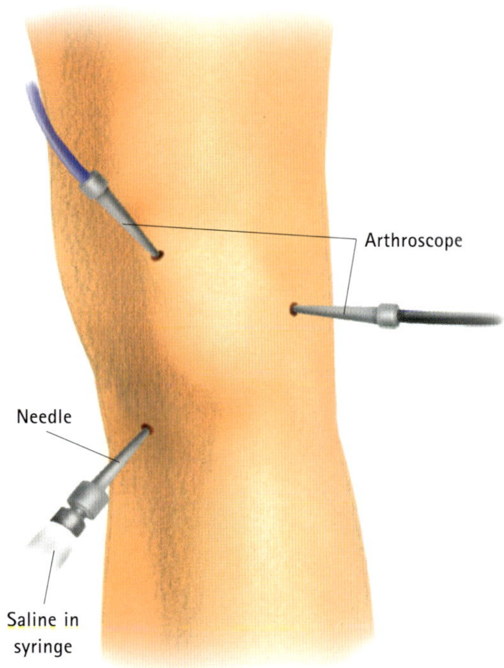

Arthroscopy

In arthroscopy, the joint cavity is filled with saline through a hollow needle and the arthroscope (a small telescope) is introduced through a small incision. The interior of the joint can be viewed by maneuvering the arthroscope and, if necessary, by using alternative sites of entry.

baby) in early pregnancy in order to identify abnormalities. Fetal blood sampling can also be performed using this technique, known as embryofetoscopy.

Laparoscopy

Laparoscopy is a procedure for examining the abdomen internally, which is sometimes referred to as "keyhole" surgery.

Two rods inserted through small incisions in the abdominal wall below the navel, one for viewing and one with lighting, allow intricate operations and diagnostic surveillance to be undertaken. After the incision is made, carbon dioxide gas is pumped into the abdomen to elevate the abdominal wall. This procedure makes it easier for the surgeon to see and manipulate organs in order to collect samples of tissue. When the female reproductive organs are being examined during laparoscopy, a dye may be injected through the cervical canal, so that the fallopian tubes are easier to see.

Through the same incision, surgical procedures can also be carried out. These include gallbladder removal, hernia repair, removal of ectopic pregnancies, sterilizing procedures and even removal of uterus, ovaries and fibroids. In vitro fertilization is also undertaken by this method.

The laparoscopy procedure is less invasive than normal surgery and, depending on the severity of the medical problem involved, may be performed on an out-patient basis. The greater risk of perforation of organs than in conventional surgery is counterbalanced by far shorter convalescence.

Laparotomy

Laparotomy is the exploration of the abdominal cavity by surgical means and is used in unexplained illness where a diagnosis remains obscure despite investigation with ultrasound and CAT scans. Laparotomy may also be used following trauma such as motor vehicle accidents, or puncture wounds from implements such as knives and bullets, to check for bowel perforation.

Laparotomy can detect conditions such as an inflamed appendix, various infections and cancer of the liver, ovary, colon and pancreas. It can be used to correct hernias in the abdominal wall and to remove diseased organs and abnormal tissue. During the procedure the surgeon may take samples of fluid in the abdominal cavity for laboratory examination. Where internal bleeding is suspected in ectopic pregnancies, laparatomy may be carried out.

Laparotomy has now largely been superseded by laparoscopy.

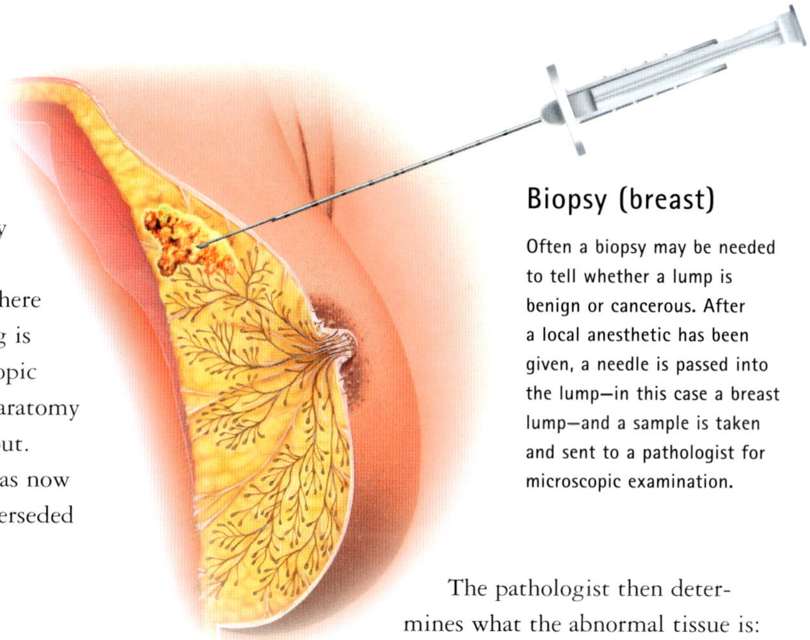

Biopsy (breast)

Often a biopsy may be needed to tell whether a lump is benign or cancerous. After a local anesthetic has been given, a needle is passed into the lump—in this case a breast lump—and a sample is taken and sent to a pathologist for microscopic examination.

Biopsy

Biopsy is a medical procedure used to obtain live tissue for examination. Unwanted growths in the body (tumors) must sometimes be examined in this way to see whether they are benign (harmless) or malignant (cancerous). Under anesthetic a small piece of tissue is surgically removed and sent to a laboratory for examination.

Biopsy can aid diagnosis of various diseases. A small sample of tissue or cells from lymph nodes, internal organs such as the liver, or from bones, can be obtained simply for intensive examination and the correct diagnosis made without unnecessary surgery.

NEEDLE BIOPSY

A needle biopsy is a procedure to sample an abnormal lump or mass in the body. It may also be used to determine whether disease is present in normal tissue, for example in the lung or liver. A needle biopsy is usually performed by a radiologist.

During the procedure, the radiologist inserts a small needle into the abnormal area and removes a sample of the tissue, which is then sent to a pathologist.

The pathologist then determines what the abnormal tissue is: cancer, non-cancerous tumor, infection or scar. The procedure is quicker and causes much less damage to normal tissue than an open biopsy, which leaves a surgical scar.

BONE MARROW BIOPSY

This is a procedure in which a needle is inserted into a bone to obtain a sample of the cells of the bone marrow. This is known as aspiration. Bone marrow biopsy can also involve the removal of a piece of bone tissue, including the marrow; this is correctly termed biopsy. Bone marrow samples are taken from the sternum (breastbone) or iliac crest (hip bone), under local anesthetic.

Bone marrow biopsy is valuable for establishing a specific diagnosis for several groups of diseases of the blood and bone marrow such as anemia, leukemia and lymphoma.

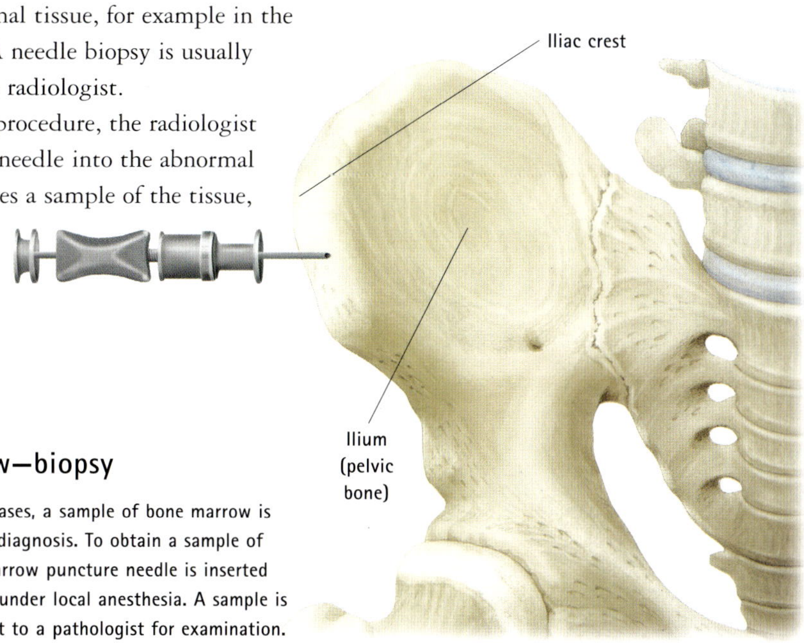

Bone marrow—biopsy

In some blood diseases, a sample of bone marrow is needed to make a diagnosis. To obtain a sample of bone marrow, a marrow puncture needle is inserted into a pelvic bone under local anesthesia. A sample is drawn out and sent to a pathologist for examination.

ENDOMETRIAL BIOPSY

During an endometrial biopsy, a sample of tissue is removed from the uterine wall to be tested in a laboratory. This is done to determine the cause of abnormal bleeding. A small tube is passed through the cervix and into the uterus and the sample taken from the uterine lining, the endometrium.

If the purpose of the test is to exclude the likelihood of endometrial cancer, a curette may be preferred. This is a different procedure requiring anesthetic, usually general, and involves removing tissue by scraping the wall with a curette. A biopsy can be performed with no anesthetic.

Chorionic villus sampling

Chorionic villus sampling (CVS) is a prenatal test used to diagnose genetic defects in the fetus. Prenatal testing is often recommended for mothers over 35, as they are at an increased risk of having a child with a chromosomal abnormality such as Down syndrome. CVS has an advantage over amniocentesis; it can detect problems much earlier in the pregnancy, but has a slightly higher incidence of miscarriages due to the procedure (about 1.5 percent) and has been associated with limb deformities.

This involves inserting a catheter via the cervix and into the uterus and sampling the chorion, a membrane that forms around the embryo in early stages of pregnancy before the placenta forms. The catheter is guided by abdominal ultrasound.

Amniocentesis

Amniocentesis involves a sample of amniotic fluid being taken from the amniotic sac. A hollow needle is inserted into the mother's uterus through the abdominal wall, using ultrasound to determine the needle's position. The fluid is cultured in a laboratory, a process that can take up to four weeks. Women found to be carrying a baby with an abnormality are usually offered the option of an abortion.

Amniocentesis is performed between the 14th and 18th week of a pregnancy on women deemed to be at high risk of having a baby with physical defects of the central nervous system (such as spina bifida or anencephaly) or genetic abnormalities (such as Down syndrome). Those at greater risk are women whose family history indicates there may be a problem, or women over 35 years of age.

Counseling should be offered to both the woman and her partner before the procedure takes place, and after the procedure if an abnormality has been detected.

Amniocentesis is sometimes performed late in the pregnancy where there is a risk of placenta previa (in which the placenta obstructs birth) or premature birth.

There is also a risk of miscarriage with amniocentesis of between 1 and 2 percent. There is also a 0.5 percent risk that the baby will be born with a very low birth weight.

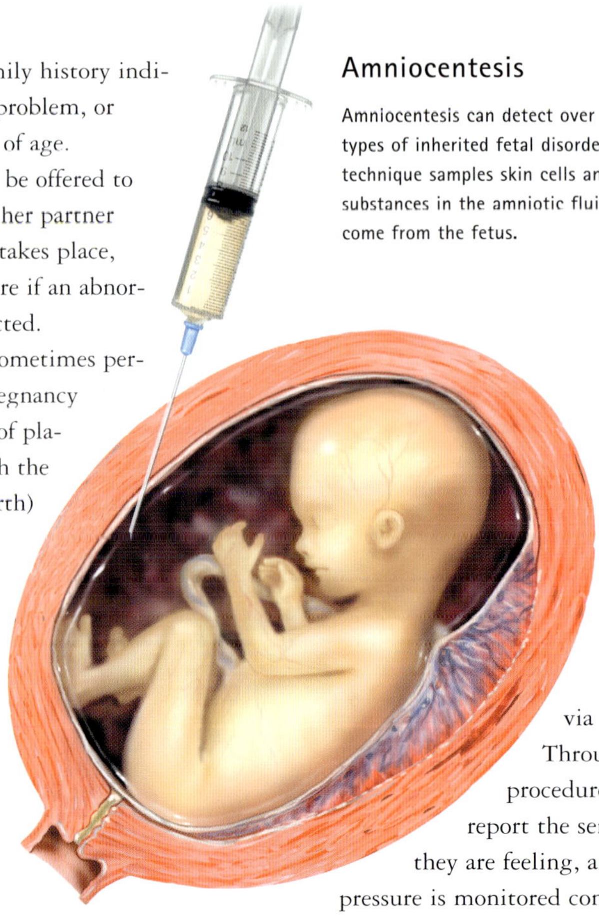

Amniocentesis

Amniocentesis can detect over 40 different types of inherited fetal disorders. The technique samples skin cells and biochemical substances in the amniotic fluid that have come from the fetus.

Pap smear

The Pap smear (or Papanicolaou smear) is a cervical cell sample prepared for viewing under a microscope. It can often reveal the presence of precancerous and cancerous cells in the cervix and is regarded as a crucial screening test that must be performed regularly throughout a woman's life, starting from the time she becomes sexually active.

Since its introduction in the 1940s, the Pap smear has contributed to a massive worldwide reduction in deaths due to cervical cancer. Taking the smear is a simple and painless procedure that can be performed in a doctor's office. Medical advice should be sought on the frequency with which Pap smears need to be taken because it may vary according to a patient's personal history.

Cystometography

Cystometography is a medical procedure used to assess the function of the bladder in people who are experiencing problems with urination. It is also commonly known as filling cystometography.

The patient is made to lie down and remains as relaxed as possible while the bladder is filled with sterile water. Abdominal pressure is also applied to the bladder via the rectum. Throughout the procedure, patients report the sensations that they are feeling, and bladder pressure is monitored continuously.

Lumbar puncture

Lumbar puncture is a procedure involving the insertion of a needle into the spinal cavity (usually below the third lumbar vertebra) to take a sample of cerebrospinal fluid or to administer drugs or an anesthetic. The five lumbar vertebrae are situated in what is known as the "small" of the back. The procedure is also known as a spinal tap.

BLOOD TESTS

Analysis of the different substances in blood can provide important information on the status of the body's immune system and on the presence or progress of disease.

SEE ALSO *Blood on page 82*

Blood groups

Blood is classified into groups, or types, based on antigens (substances which produce immune responses) found on the red blood cells. The main blood groups are A, B, AB and O. In any of these groups individuals can be either Rh positive or, more rarely, Rh negative.

During pregnancy, Rh incompatibility will cause the mother to generate antibodies against the fetus' red blood cells, and in

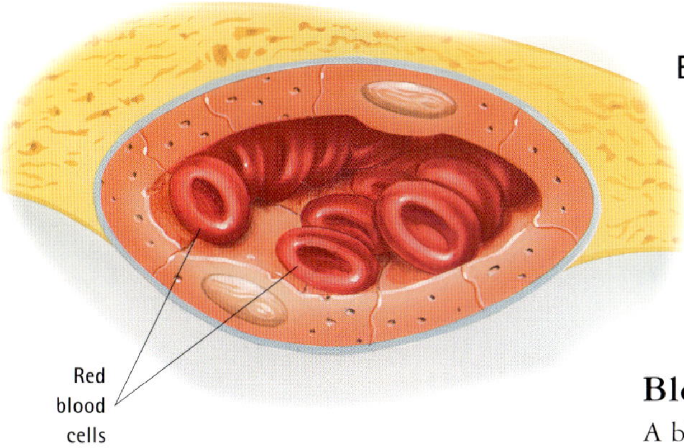

Erythrocyte sedimentation rate

Erythrocyte sedimentation rate (ESR) measures the rate at which red blood cells (erythrocytes) settle in a column of blood standing in a tube. It is a test used to help diagnose many illnesses.

later pregnancies a fetus may suffer anemia or jaundice, or not be able to survive.

Blood typing is vital for safe blood transfusions. Mixing incompatible blood types will cause the red cells to clump together, blocking blood vessels and even leading to death. Blood typing, as well as a test called cross-matching, performed prior to blood transfusion, ensures that blood of donor and host are thoroughly compatible.

Blood count

A blood count is a test to determine the number of red cells, white cells, platelets and hemoglobin in the blood. A blood sample is taken and then the number of blood cells counted electronically, or visually using a microscope.

Other laboratory tests establish the hemoglobin level and the percentage of red and white cells in the sample. Some infections can be suspected if the white cell count is too high, while too few white cells could indicate damage to bone marrow.

Blood

Blood is composed of red blood cells, various types of white blood cells (leukocytes) and platelets in a solution of water, electrolytes and proteins called plasma. About 40 percent (by volume) of blood is red blood cells. This illustration shows all the different types of blood cell—it does not accurately represent the proportions present in the blood.

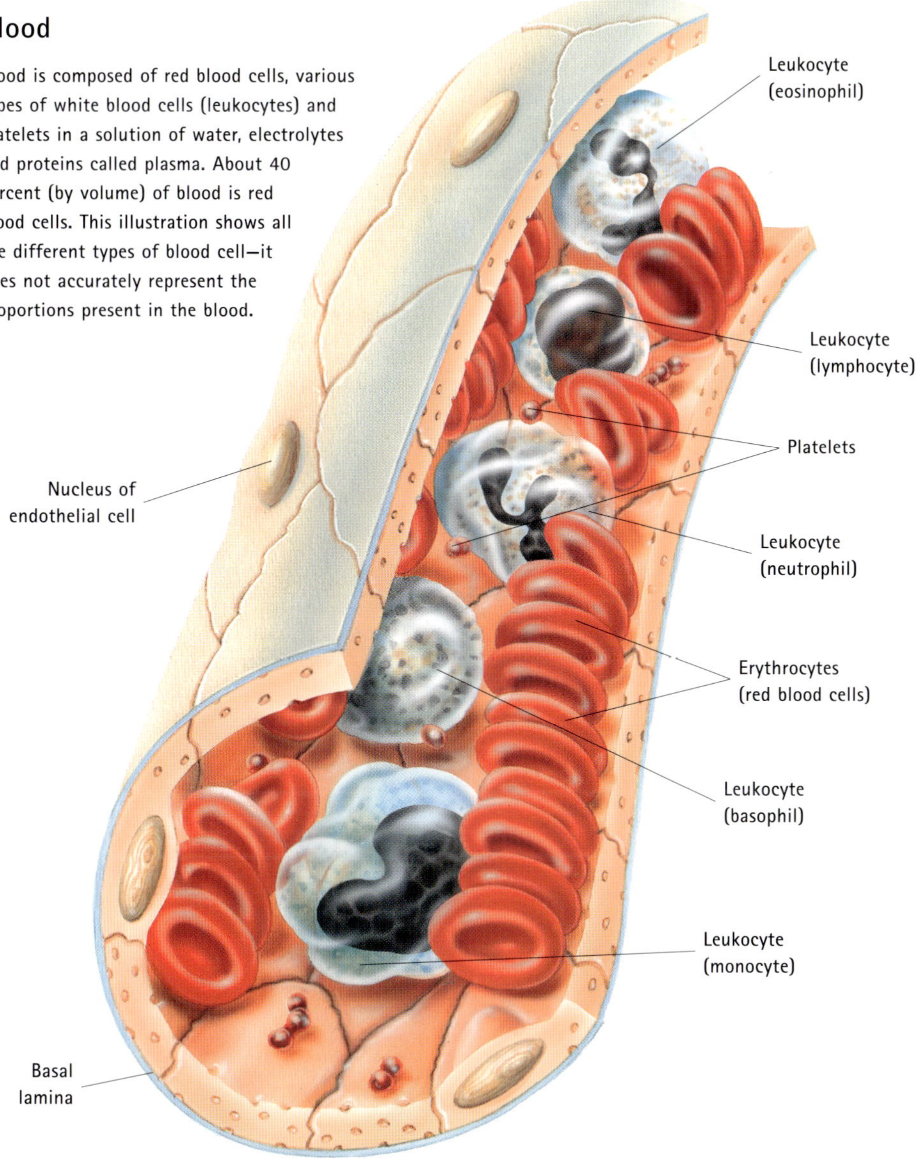

Blood sugar tests

Blood sugar tests determine the level of glucose in the blood. Glucose is a simple sugar obtained from digesting carbohydrates in food, and is needed to nourish cells throughout the body. It circulates in the blood. If sugar is present in the urine this will indicate a high level in the blood, and a blood sugar test may then be done in order to confirm this.

A person with a low blood sugar level (hypoglycemia) will feel hungry, weak and tired, may have a headache, be sweating and feel faint, and may even lapse into a coma. A high blood sugar level (hyperglycemia) means the body is not properly controlling the absorption of glucose; this can be a symptom of diabetes.

Blood gas analysis

Blood gas analysis is a test conducted on arterial and venous blood to measure oxygen and carbon dioxide levels and hydrogen ion concentration (pH).

The sample can be taken from an appropriate artery or vein. The levels will show how well carbon dioxide is being removed from the blood and how efficiently the lungs are working to re-oxygenate it. Gas analysis of arterial blood also gives a correct measure of the blood's acidity.

Erythrocyte sedimentation rate

A common pathology test, erythrocyte sedimentation rate (ESR) is a measure of the rate at which red cells settle in a column of blood standing in a thin tube.

Infections, anemia, certain rheumatic diseases and other autoimmune disorders are all associated with a rapid rate compared to normal ESR (1–20 millimeters/hour). Certain cancers, for example multiple myeloma can also, though not invariably, produce a high ESR reading.

The efficacy of treatment of infections can be monitored over a period until cure—judged by a normal ESR—has occurred. Therapy for tuberculosis and bone infections is checked in this way.

Liver function blood tests

Liver function blood tests are tests done on the blood which provide information about the health of the liver. The liver performs many essential functions. It makes many important proteins, such as albumin and clotting factors. Albumin is the major protein in the blood, and is responsible for keeping fluid in the blood vessels and for transporting chemicals.

Liver function blood tests reveal levels of albumin in the blood. In severe liver disease, the albumin level decreases and fluid leaks into the limbs (edema) and abdominal cavity (ascites). A reduction in clotting factors also occurs, resulting in easy bruising and a tendency to bleed.

Other abnormalities such as abnormal enzyme levels may also be detected. Liver cells contain enzymes which are released into the blood when they are damaged. An alcoholic binge will produce this, as well as damage by viruses or chemicals.

Nucleic acid testing

Nucleic acids are present in body cells as well as in viruses, bacteria and other agents of infection. Tests exist that can detect the presence of foreign nucleic acids in the body, and are a very useful diagnostic tool.

Enzyme-linked immunosorbent assay

Enzyme-linked immunosorbent assay (ELISA) is a family of versatile immunologic techniques which rely on the binding of an antibody or antibodies to their corresponding antigen. An enzyme linked to one of the antibodies (directly or indirectly) is used to generate a reaction product, which is usually colored. Measurement of this product allows the amount of bound antibody to be determined.

Uses of ELISA techniques include detection of antibodies in blood serum, which is valuable in the diagnosis of infections, and measurement of minute amounts of circulating proteins or drugs.

Skin tests

The body's sensitivity to different allergens and microorganisms can be tested by applying these substances to the skin or injecting them into the skin layers.

SEE ALSO *Allergies on page 60; Immunity on page 58; Infectious diseases on page 364*

Mantoux test

The Mantoux test is a test for tuberculosis in which a solution made from dead tuberculosis bacteria is injected between the skin layers, usually on the forearm. If a minor swelling or redness develops at the site within two to three days, this indicates the presence of infection-fighting antibodies which in turn indicate a past or present infection with tuberculosis. Infections less than two weeks old may not show up with this test. Named after French physician Charles Mantoux, who developed the test in 1908, the Mantoux test is also known as a purified protein derivative (PPD) test.

Antigen tests

The immune system responds to molecules that are recognized as foreign: these are called antigens and are carried by bacteria (e.g. salmonella), viruses (e.g. hepatitis B), toxins and other substances such as allergens. Antigen tests are used to detect which foreign substance is causing illness or other reactions in the affected person.

It is possible to test for the development of different types of immune responses to an antigen in a number of ways, but an important group of antigen tests is that used to detect responses to allergens in the environment. These allergens include, for example, cat and dog fur; feathers; house dust mites; cockroaches; various molds; and the pollens of grasses, weeds or trees. Testing for an allergic response involves introducing small amounts of antigen into the superficial layers of the skin by a prick or scratch, then measuring the resulting zone of redness and wheal formation 30 to 60 minutes later. Multiple antigens are usually tested at once. People predisposed to developing allergic responses often react to several antigens.

Medication

Medicinal drugs are normally chemicals designed to alter the processes of the body in order to prevent, treat or manage disorders and to relieve symptoms or pain.

Drug interactions

Drugs can interact with other chemicals in the body, and these usually include other drugs. Chemical interactions between drugs may change the make-up of those drugs and render them ineffective, less (or more) effective, or toxic (poisonous).

The risk of adverse drug interactions is higher in older people, because their vital

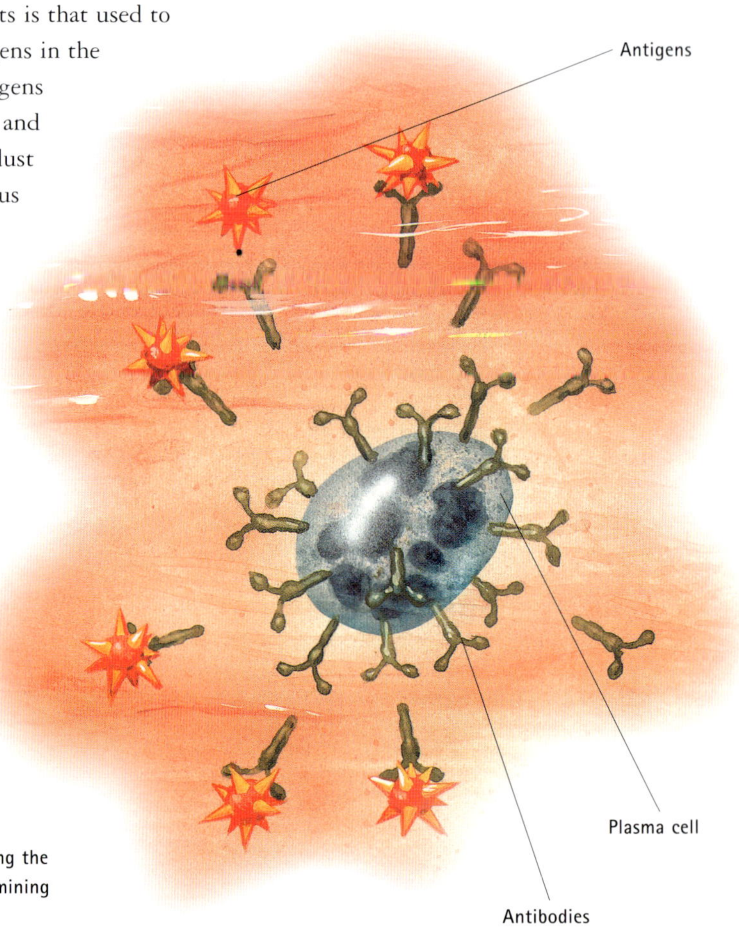

Detecting antibodies

Once the body recognizes a foreign substance (antigen) has entered the body, B lymphocytes are activated, become plasma cells and begin producing antibodies. The antibodies attach to the antigens, which are eventually neutralized. Immunologic tests, such as the enzyme-linked immunorsorbent assay, can detect antibodies circulating in the blood. This is useful for diagnosing infection, monitoring the progress of diseases and determining drug efficacy.

organs process drugs less easily and because they tend to have more diseases and take more drugs than younger persons.

The following types of drugs commonly interact with other medications: antibiotics, anticoagulants, anticonvulsants, antidepressants, antihypertensives, decongestants, and sedatives. Antibiotics (such as ampicillin, amoxycillin, neomycin and tetracycline) and anticonvulsants can interfere with the effect of oral contraceptives (birth control pills), and increase the risk of pregnancy.

Special care should be taken with heart drugs. The combination of beta-blockers and certain types of calcium channel blockers, for example, may slow the heart rate excessively or cause heart failure. Beta-blockers can aggravate asthma and prevent diabetic patients from recognizing the signs of low blood sugar due to an insulin reaction.

The combination of calcium channel blockers and digoxin can also slow the heart rate. Angiotensin-converting enzyme (ACE) inhibitors can often cause dangerously high potassium levels in people who often use potassium supplements. Diuretics combined with digoxin can cause dangerously low potassium or magnesium levels, which can cause heart arrhythmias.

Certain foods or beverages can interfere with drugs as well. Foods most likely to interfere with medications include dairy products, alcohol, caffeine, salt, and fruit juices. For instance, some antidepressants (those known as MAO inhibitors) are dangerous when consumed with anything containing tyramine (such as red wine, cheese and beer). It is advisable not to take over the counter drugs with other medications without first seeking medical advice. Many common medications contain the same antihistamines as anti-cold preparations; you can get an unexpected double dose by taking an allergy drug along with some cold remedies.

Harmful interactions may occur between herbal products and drugs. Warfarin taken with ginseng, garlic, ginkgo, ginger and feverfew may result in bleeding. Zinc and echinacea may negate the effects of cyclosporine. Ginseng may also interfere with digoxin and may cause headaches, trembling, or manic episodes if taken with phenelzine sulfate. Saint John's wort and saw palmetto may inhibit anemia drugs.

Drugs can interact with alcohol and tobacco; smoking reduces the effectiveness of many medications. Ask a physician or pharmacist if the drug you have been prescribed can interact with foods, other medications, or herbal remedies you may be taking.

Antihistamines

Antihistamines are the main group of medicines used in treatment and control of allergies. They work by blocking the action of histamine, a chemical produced in the body as part of an allergic reaction. They can also fight nausea and so can be used to counter travel sickness. The side effect of most antihistamines is drowsiness (though some newer types are less sedating), so they should not be taken before driving or operating machinery, and it is not safe to drink alcohol while on the medication. They are often used at night, and their sedative effect can be useful for treating insomnia.

Treating the Cardiovascular System

Medications which act on the blood, blood vessels and heart muscle can be used to treat cardiovascular disorders. These drugs can alter the rate and pressure of blood flow, the pumping action of the heart, and the clotting action and flow of blood.

SEE ALSO *Circulatory system on page 78; Disorders of the heart on page 230*

Antihypertensives

Antihypertensives are drugs used to treat high blood pressure (hypertension). They may be given orally or injected for rapid effect, and used alone or in combination.

Beta-blockers decrease heart rate and cause relaxation of small peripheral blood vessels, lowering blood pressure. They are not suitable for asthmatics as they may cause an asthma attack. Thiazide diuretics relax small blood vessels and also cause the kidneys to increase the amount of salt and water eliminated in the urine. Potassium levels may drop, so doctors may need to prescribe a potassium supplement.

Calcium channel blockers influence the movement of calcium ions into the cells of the heart and blood vessels. CCBs work by obstructing the flow of calcium ions into vascular and heart muscle cells. This causes the muscle cells to relax. As a result the arteries dilate, which improves blood flow to the heart and leads to a drop in blood pressure. The heart rate is also slowed by some CCBs. There can be side effects associated with certain CCBs. These vary according to the drug used but may include swelling in the legs and headache.

Alpha-adrenergic blockers as well as angiotensin-converting enzyme (ACE) inhibitors act by dilating blood vessels. There is evidence that antihypertensive treatment reduces vascular and organ complications (such as to eyes and kidneys).

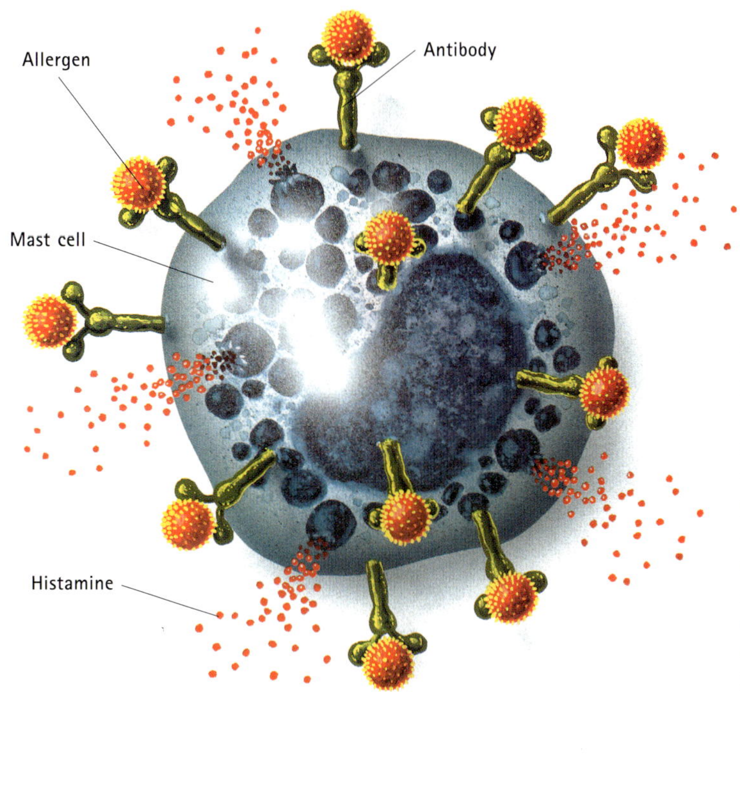

Antihistamines

When mast cells are activated by invading allergens they release a chemical called histamine. This is responsible for the unpleasant effects of allergies like hay fever or hives. Antihistamines block the release of histamine.

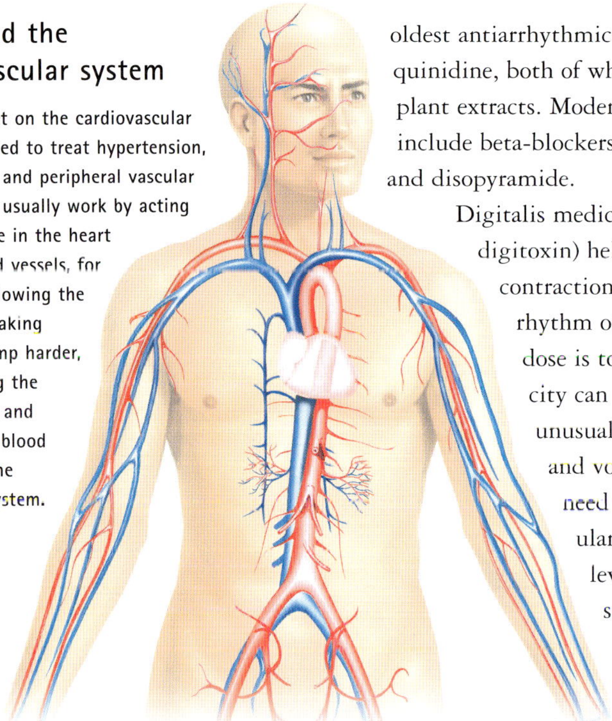

Drugs and the cardiovascular system

Drugs that act on the cardiovascular system are used to treat hypertension, heart disease and peripheral vascular disease. They usually work by acting on the muscle in the heart and the blood vessels, for example by slowing the heart rate, making the heart pump harder, or by relaxing the blood vessels and lowering the blood pressure in the circulatory system.

Because there are often no symptoms, people with high blood pressure may feel normal whether or not they take their medication. But medication should be taken according to the doctor's instructions; treatment usually needs to be continued for life. Non-compliance with treatment is a significant problem in the management of high blood pressure. If the patient experiences annoying side effects, alternative blood pressure medication should be used.

Antiarrhythmic drugs

Antiarrhythmic drugs control irregularities of the heartbeat. The oldest antiarrhythmics are digitalis and quinidine, both of which were originally plant extracts. Modern antiarrhythmics include beta-blockers, calcium antagonists and disopyramide.

Digitalis medications (digoxin and digitoxin) help control the rate of contractions and regulate the rhythm of the heart. If the dose is too high, digitalis toxicity can occur, producing unusual visual effects, nausea and vomiting; blood tests need to be conducted regularly to monitor blood levels. Beta-blockers such as propanalol, atenolol and pindolol decrease heart rate by affecting conduction in the heart. Beta-blockers can cause asthma attacks, so are unsuitable for people who are asthmatics.

Vasodilators

Vasodilators are drugs that dilate small blood vessels, such as arterioles (small arteries) and venules (small veins). They increase blood flow to the tissues, lower the blood pressure of the circulation, and make the work load of the heart easier.

Some, like nitroprusside and nitroglycerine, are used in ischemic heart disease and congestive heart failure, while others like hydralazine (a smooth muscle relaxant), diltiazem (a calcium channel blocker) and enalapril maleate (an ACE inhibitor) are used in the treatment of hypertension, often in combination with other hypertensive drugs such as diuretics and beta-blockers.

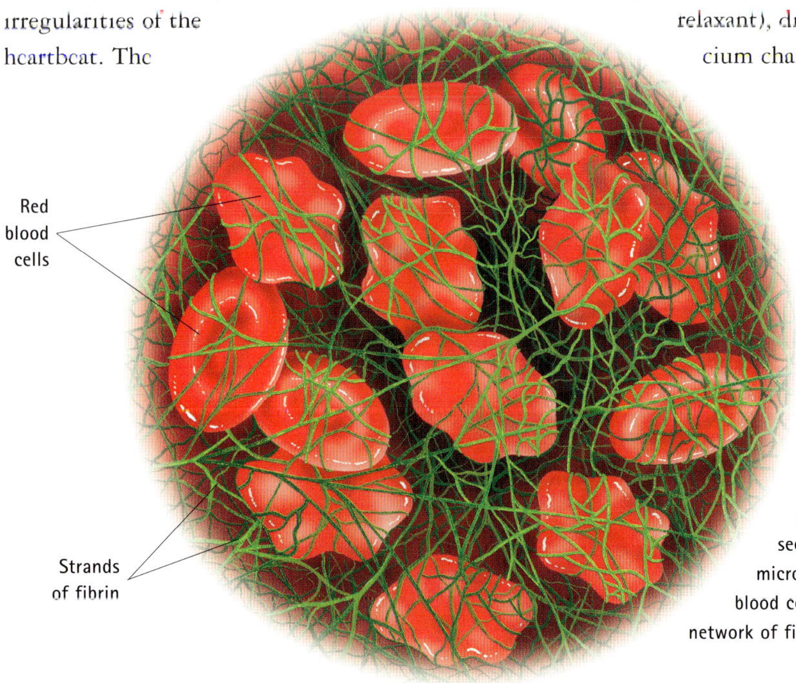

Blood clot

This picture of a blood clot as seen by an electron microscope shows red blood cells trapped in a network of fibrin fibers.

Anticoagulants

Anticoagulants are drugs that affect the normal clotting of blood. They are used in myocardial infarction (heart attack), strokes, embolism and during surgery to prevent blood clotting (thrombosis). Warfarin is an oral preparation suitable for long-term therapy. Heparin is a quick-acting intravenous or subcutaneous preparation used over short periods in a hospital setting. Low-dose aspirin is taken by many people as a low-grade anticoagulant.

As too much anticoagulant can cause bleeding, people taking them require regular blood tests to ensure correct dosage. They should also carry a card or wear a bracelet identifying the drug and dose in case of accident. Some medications affect the action of anticoagulants.

THROMBOLYTIC DRUGS

Thrombolytic therapy is treatment used to dissolve, or lyse, a clot. The drugs employed to dissolve clots act by enhancing the body's own anti-clotting mechanism, the fibrinolytic system. This is a series of proteins whose actions result in the breakdown of fibrin, the hard clot formed when the coagulation system is activated.

Thrombolytic therapy has been used mainly to treat coronary artery thrombosis which can cause heart attack. Studies have shown that thrombolytic therapy can signif icantly reduce the resulting damage to heart muscle and the risk of death from heart attack. However, the treatment must be given within hours of onset to be effective.

TREATING THE RESPIRATORY SYSTEM

Respiratory disorders often hinder the passage of air and the effective exchange of oxygen and carbon dioxide by the body. Medications that reduce swelling and inflammation of the membranes of the respiratory tract act on muscles surrounding the airways and those that counter allergies and infections can often be used to relieve the symptoms and disorders that affect the lungs, nasal passages and bronchi.

SEE ALSO *Respiratory system on page 95; Disorders of the lungs on page 241*

Decongestants

Decongestants are drugs that shrink swollen membranes in the nose and make it easier to breathe. They are useful in relieving the symptoms of a stuffy or runny nose in cold and flu infections, sinusitis and nasal allergies. Decongestants can be taken orally and also through nasal spray.

Oral medications usually contain pseudoephedrine and phenylpropanolamine; they are effective, but may cause insomnia. People who suffer from high blood pressure (hypertension) should seek advice from a medical practitioner before taking oral decongestants, as they may result in an increase in blood pressure.

Many decongestant nasal sprays have a rebound effect if used over a long time, resulting in irritation and inflammation of the nasal membranes when the treatment is stopped. A doctor's advice should be sought if treatment with a nasal decongestant is to be continued for more than five days.

Expectorants

Expectorants are compounds used in cough and cold products to help loosen phlegm and make coughing more productive. They can be bought over-the-counter without a doctor's prescription. Guaifenesin is the most widely used expectorant and is found in a range of over-the-counter preparations.

Expectorants cause mucus and other substances blocking the airways to become thinner, so they are easier to cough up. Coughing becomes easier and less irritating, and chest congestion is relieved. For example, someone with pneumonia may find an expectorant helps clear the airway of mucus, making breathing easier.

Some medical professionals question whether the use of an expectorant speeds recovery or really does relieve symptoms of respiratory illness. Some compounds sold as expectorants do not in fact have expectorant properties at all; these include ammonium chloride, a bitter plant extract called horehound, pine tar and spirits of turpentine.

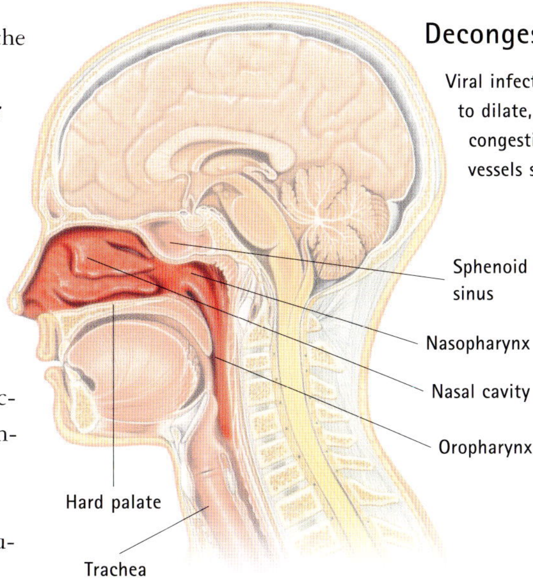

Decongestants

Viral infections and allergies affecting the nose cause blood vessels to dilate, leading to swelling of the membranes inside the nose and congestion. Decongestants are drugs that constrict the blood vessels so that the membranes shrink, making it easier to breathe.

If the cough does not produce mucus, then an expectorant is of no use. In this case, a cough suppressant may be more appropriate because it allows the sufferer to cough less, feel more comfortable and sleep better.

It is important to follow instructions on the labels of cough suppressants to avoid overuse and possible side effects. If a cough persists, medical advice should be sought.

Bronchodilators

These are drugs for treating asthma. After inhalation, they provide rapid relief from breathlessness by relaxing the smooth muscle that surrounds the airways. Most bronchodilators are short-acting and are used for temporary relief of acute symptoms—more than one dose may be needed. The drugs do not treat the airway inflammation (the basis of asthma) and so should be used in combination with anti-inflammatory agents.

Long-acting bronchodilators are also used but they are not rapidly effective, and therefore are not useful for treating an acute attack of asthma. They can, however, be used in combination with anti-inflammatory drugs.

Treating the respiratory system

The respiratory system is susceptible to infection because of its role in filtering out airborne bacteria, infectious agents and other microorganisms before they damage the body. Allergies and disorders such as asthma also affect breathing and the efficiency of air exchange. A range of medication and therapies are available to target infection and inflammation, unblock the airways and improve respiratory function.

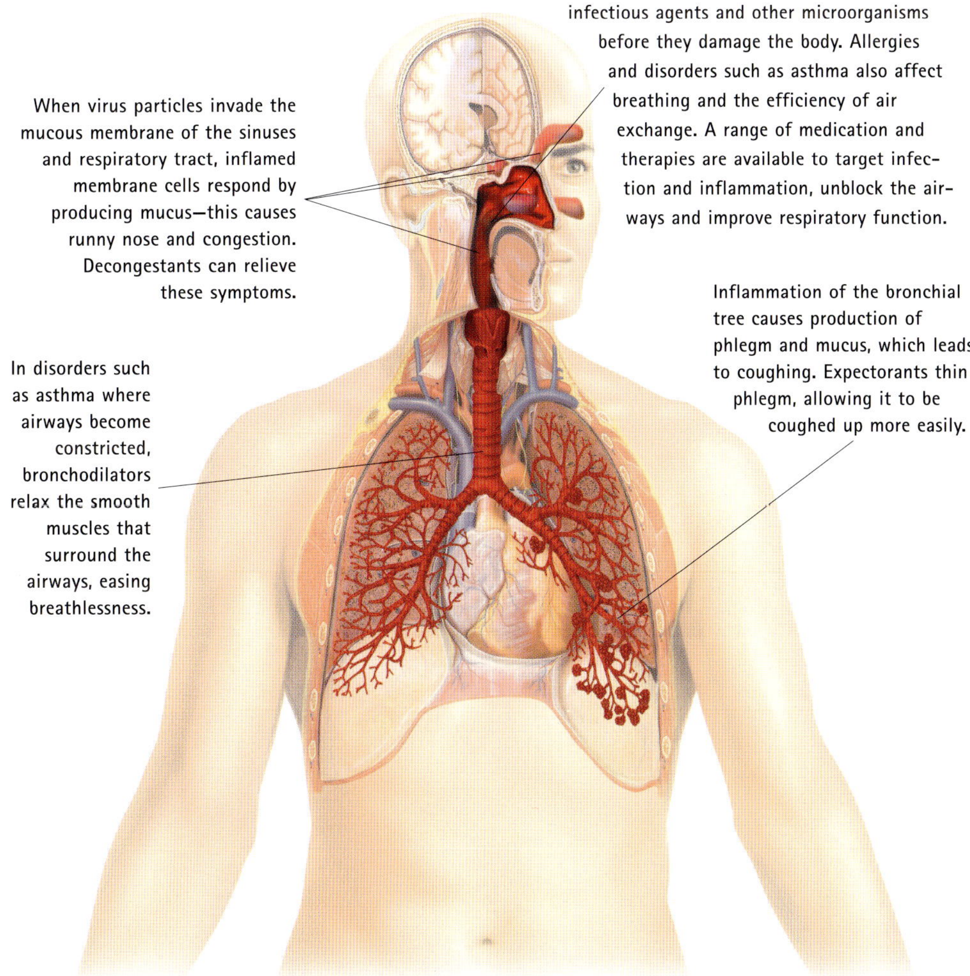

Treating the Central Nervous System

Drugs that restore or adjust the balance of chemicals and electrical impulses in the brain can be used to modify or control brain activity and also to treat disorders of the central nervous system.

SEE ALSO *Nervous system on page 64*

Tranquilizers

Tranquilizers, drugs which calm nervous activity, are of two types—antipsychotics and antianxiety drugs. The most commonly prescribed types of antianxiety drugs are the benzodiazepines, which are usually prescribed for anxiety states, panic or sleep problems, to control epileptic fits or the symptoms of alcohol withdrawal. Tranquilizers are absorbed into the bloodstream and affect the central nervous system, slowing down physical, mental and emotional responses. They can affect judgment, memory and the ability to concentrate, and cause drowsiness, dizziness, confusion and mood swings.

In the long term they can be responsible for nausea, loss of libido, increased appetite and weight gain and lethargy. Taken at the same time as alcohol, painkillers or antihistamines (even cold remedies) they can cause unconsciousness and failure to breathe. It is important to take tranquilizers only in the manner prescribed. Tranquilizers can easily become a drug of dependence even within 4–6 weeks of regular usage. Withdrawal creates its own symptoms—sleep problems, tension, muscle pain, panic attacks and depression—and must be undertaken carefully with medical supervision.

Sedatives

Sedatives are drugs that calm anxiety or mental disturbance, induce sleep or drowsiness and reduce the body's functional activity. Sedatives may also be prescribed for insomnia and epilepsy. They can be taken as tablets or by injection. Anxiety is often treated with sedative drugs because they relax muscles, thus reducing tension and allowing sleep to provide a temporary relief from symptoms. Travelers with a fear of flying or who suffer claustrophobia may use sedatives to reduce anxiety before and during a flight.

Side effects include drowsiness, impaired judgement, lack of coordination, lowered heart rate and blood pressure, nausea and diarrhea, and dependence. More rarely sedatives produce memory defects, hallucinations, constipation, vomiting and headaches, and loss of consciousness. Sedatives should be taken only on prescription and under medical supervision, used only intermittently, and not taken in pregnancy.

Long-term use can lead to dependence, and stopping use should be a gradual process under medical supervision. All sedatives may interact with other similar drugs, such as anti-epilepsy, antidepressant and anti-psychotic drugs, and especially with alcohol, which often increases the risk of side effects and overdose.

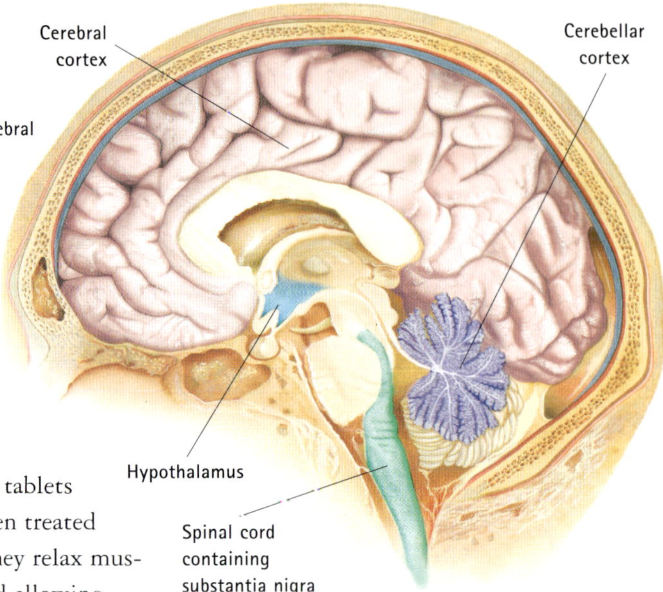

Hypnotics

Hypnotics are central nervous system depressants, acting mainly on the cerebral cortex area of the brain and on the parts that control alertness and consciousness (the substantia nigra and reticular activating system).

Barbiturates

Barbiturates are sedative and sleep-inducing drugs derived from barbituric acid. They act by depressing the central nervous system. In larger doses, they lower the blood pressure and slow down breathing. Once commonly prescribed for insomnia and anxiety, they have fallen from favor in recent years because they are addictive, and if taken with alcohol or other drugs, can result in a fatal overdose.

Hypnotics

Hypnotics are a group of drugs used to ease anxiety or to produce sleep. They function by depressing the central nervous system. Hypnotics are sometimes called sedatives or tranquilizers, though neither of these induces sleep, as hypnotics do.

In the past, barbiturates and chloral hydrate were often used as hypnotics, but they have now been replaced with benzodiazepines, which are safer and have a lesser likelihood of overdose.

Hypnotics can be useful in cases where sleep is important to recovery. Nevertheless, they are not without dangers.

While overdose is very rare with benzodiazepines, it may occur where benzodiazepines are taken in combination with alcohol or other drugs.

Because of possible side effects such as drowsiness, loss of coordination, and the loss of judgement, these drugs should not be used with alcohol and other similar types of drug. The dose should be kept low when prescribed for the elderly. Warnings should also be given about driving and working with machinery.

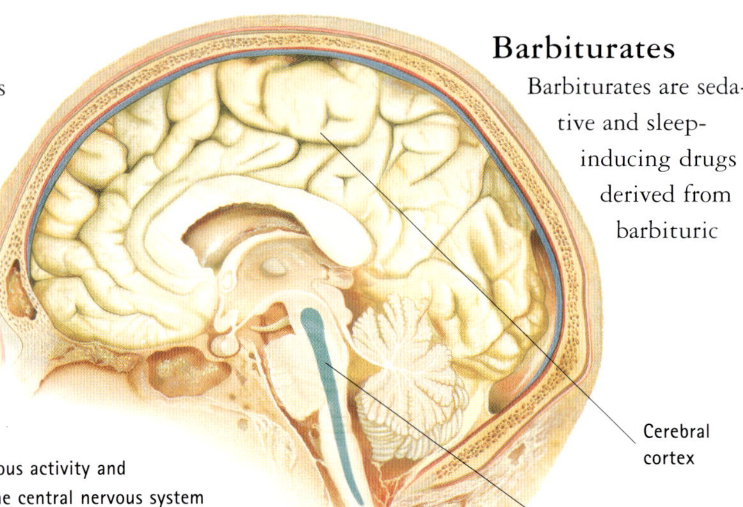

Tranquilizers

These drugs calm nervous activity and affect wide areas of the central nervous system including the cerebral cortex and the reticular formation of the brain.

Psychotropic drugs

Medications which affect the mind and behavior are called psychotropic drugs. They include antidepressants, neuroleptics, mood stabilizers, and benzodiazepines. These drugs may be prescribed by a psychiatrist to treat conditions such as depression, anxiety, phobias and insomnia. They may be used in conjunction with psychotherapy.

Determining the right medication(s) and doses that work best for a particular person is usually a matter of trial and error. The drugs do not always work, but when they do the effect may be dramatic, allowing a sufferer to function in day-to-day living. However, in many cases people may become dependent on them, or use them after they are no longer needed. Also, withdrawal from psychotropic drugs may cause unpleasant symptoms. (On the other hand, some drugs must be continued indefinitely.) It is important to seek medical advice before coming off these drugs.

Medication by itself cannot solve a psychological problem and will not substitute for learning the coping skills necessary to adapt to and enjoy life.

Antidepressants

Antidepressants are drugs that alleviate depression. They work by correcting the biochemical imbalance in the brain that is thought to be the cause of depression.

The oldest are the tricyclic antidepressants such as amitriptyline and imipramine, which elevate the levels of the neurotransmitters serotonin and norepinephrine (noradrenaline) in the brain. Side effects may include dry mouth, blurred vision, sweating, constipation, urinary problems and impotence (in males).

The monamine-oxidase (MAO) inhibitors are less popular as they interact dangerously with foods such as cheese, some meats, alcohol and yeast extracts. They are usually used only if other antidepressants fail.

The newest category of antidepressants are the serotonin-specific reuptake inhibitors (SSRI). As these work quickly and have few side effects, they are now the most commonly prescribed. Examples include fluoxetine (Prozac) and paroxetine.

It takes from 2–6 weeks for an antidepressant to begin to work. The initial dose is usually kept low to minimize side effects and is increased over time until the desired result is reached. Most side effects disappear in a few days or within a few weeks.

Anticonvulsants

Anticonvulsant drugs prevent epileptic attacks by depressing the activity of the brain. Some are more suited to particular types of seizures than other types. They are used in combination. Anticonvulsants need to be taken for years, or even for life. However, once the sufferer has been free of seizures for several years, the dose can be reduced or stopped.

Commonly used anticonvulsants include carbamazepine, phenytoin, lamotrigine, gabapentin, topiramate and valproate for generalized ("grand mal") seizures; and valproate or ethosuximide for smaller ("petit mal") seizures. Possible side effects include drowsiness, rashes, dizziness, headache, nausea and indigestion.

Antiemetic drugs

Antiemetic drugs are used to treat the symptoms of nausea and vomiting. Phenothiazines, a type of major tranquilizer, are the most potent antiemetics, but they also have the greatest number of side effects, including drowsiness, hypotension (low blood pressure) and movement disorders.

Antihistamines (used to treat allergies) are also useful in treating nausea, especially when it is associated with motion sickness. Antihistamines may sometimes produce drowsiness, so people who take them should avoid drinking alcohol, driving automobiles or operating machinery.

Anticholinergics, which slow the actions of the smooth muscle in the bowel, are used in the relief of nausea and vomiting associated with vertigo and motion sickness. Side effects of these drugs can include blurred vision, dry mouth and tachycardia (rapid heart rate). They should not be used by people with glaucoma and urinary retention as they worsen these conditions.

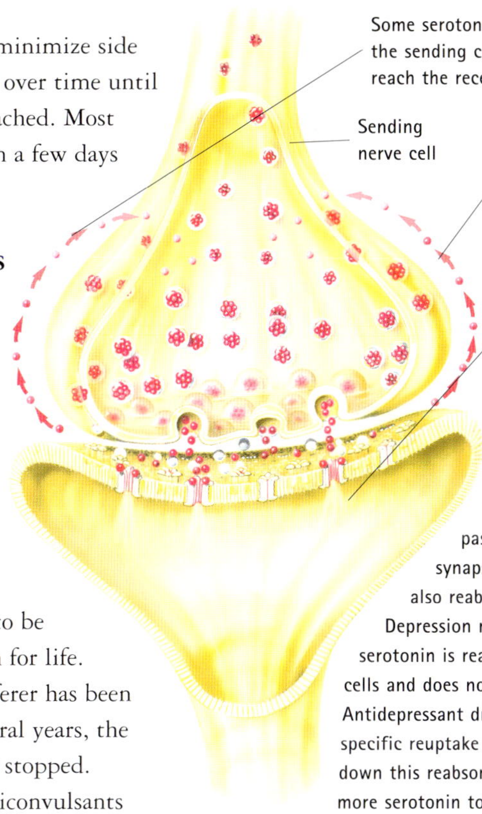

Antidepressants

Serotonin is a neurotransmitter that activates nerve cells in the brain. Sending cells pass serotonin across the synapse to receiving cells but also reabsorb a certain amount. Depression may occur when too much serotonin is reabsorbed by the sending cells and does not reach the receiving cells. Antidepressant drugs called serotonin-specific reuptake inhibitors work by slowing down this reabsorption process, allowing more serotonin to reach receiving cells.

Sympathomimetic drugs

Sympathomimetic drugs are medications whose actions are similar to those of the hormone epinephrine (adrenaline). They work by mimicking the effects or stimulating the release of epinephrine (adrenaline) or norepinephrine (noradrenaline). They are often used in the treatment of conditions such as asthma, shock and cardiac arrest. They are used in over-the-counter preparations to relieve nasal congestion and allergic disorders, and to suppress appetite.

Side effects include rapid heart rate (tachycardia), high blood pressure, increased body temperature, agitation, cardiac arrhythmias and seizures. Amphetamines and ephedrine are both sympathomimetic drugs. An overdose can be fatal.

Endorphins

Endorphins are a group of naturally occurring opiates that have pain-relieving properties found in the brain. They are related to pain-killing opiates such as opium, morphine and heroin; the word "endorphin" comes from the word "endogenous" (meaning "produced within the body") and the word "morphine".

Endorphins were discovered in 1973, following the discovery of receptors in the brain that morphine binds to. This led to the belief that the body must contain its own naturally occurring opiates; since then, several related molecules called enkephalins have also been discovered.

Endorphins and enkephalins are released into the circulation during vigorous physical exercise, such as running. They are believed to account for the painkilling (analgesic) and euphoric effect that exercise can produce in most people (the "runner's high"), especially if they are mildly depressed or anxious. (Acupuncture is also thought to stimulate the production of endorphins.)

Dependence on and tolerance to morphine and other narcotic analgesics is thought to be caused by suppression of the body's normal production of endorphins by these opiates. When the effects of morphine wear off, withdrawal symptoms may occur because the body lacks endorphins.

Treating the musculoskeletal system

Drug treatments for muscle and skeletal disorders are used to control the inflammation, pain and swelling that usually accompany musculoskeletal injuries and disorders. Most do not provide a cure but can help relieve many of the symptoms.

SEE ALSO *Muscular system on page 48; Skeletal system on page 30*

Anti-inflammatory drugs

The function of anti-inflammatory drugs is to reduce inflammation, and are used in conditions such as rheumatoid arthritis, osteoarthritis and connective tissue disorders. They do not cure these disorders, just control the symptoms.

Aspirin is a popular anti-inflammatory drug. Common side effects following its usage include irritation of, and bleeding from, the lining of the stomach. The drug should not be given to children, as it may cause Reye's syndrome.

Nonsteroidal anti-inflammatory drugs (NSAIDs) cause fewer digestive problems than aspirin and are often used to treat headache and menstrual cramps. Ibuprofen and naproxen are two common NSAIDs.

For some time there has been considerable concern regarding the adverse gastrointestinal side effects of the NSAIDS, which include gastric ulcers, bleeding and perforation. The newer COX-2 inhibitors (celecoxib and rofecoxib) have been shown to have significantly less gastrointestinal side effects and are now widely used.

To control more serious inflammatory disorders, corticosteroids such as hydrocortisone (cortisol) are often used. However, because of potential side effects such as bruising, osteoporosis, infections, diabetes and high blood pressure, their use is limited to short courses when other therapies fail.

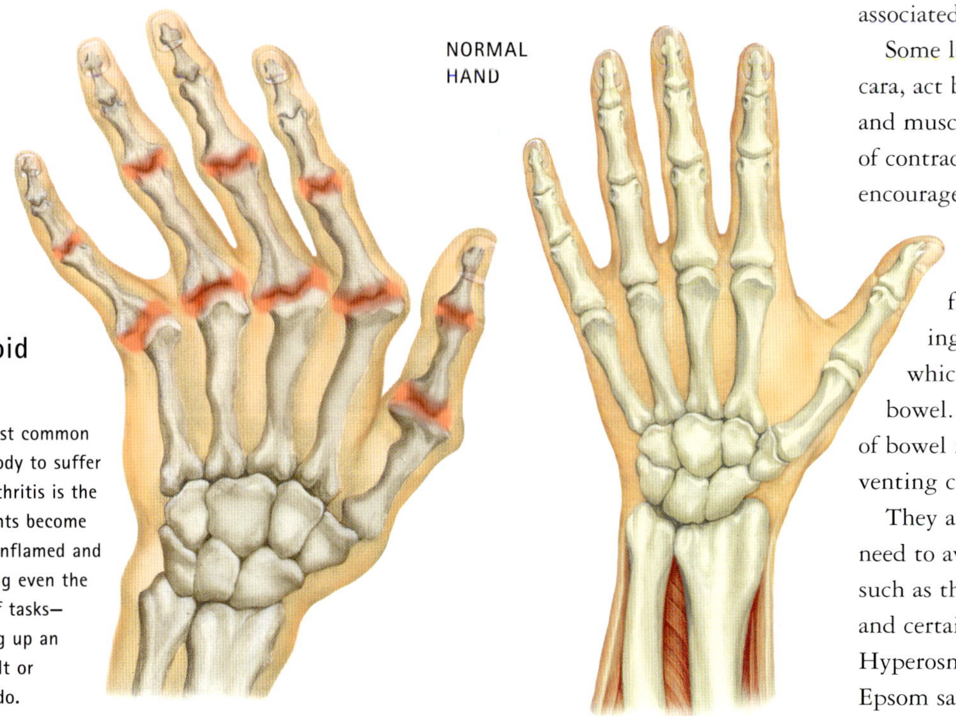

Rheumatoid arthritis

One of the most common areas of the body to suffer rheumatoid arthritis is the hands. The joints become stiff, painful, inflamed and swollen, making even the most simple of tasks—such as picking up an object—difficult or impossible to do.

Treating the digestive and urinary systems

Medications that stimulate or inhibit the passage of food, liquid and nutrients via the body are used to treat a variety of digestive and urinary disorders. Some treatments also act on the contents of the digestive tract and kidneys, attracting water into the bowel for example, for therapeutic effect.

SEE ALSO *Abdominal cavity on page 254; Digestive system on page 101; Urinary system on page 98; Urinary organs on page 288*

Enema

Enema is the injection of fluid into the rectum through the anus to expel the rectum's contents. Enemas were once given routinely to pregnant women but are now used to treat cases of constipation or colonic inertia that cause fecal impaction—the compaction of hard stools in the rectum which can cause painful cramps and block normal defecation. Enema fluids may be a saline solution or contain an oil such as olive oil. A barium enema is given to outline the lower portion of the large intestine in x-ray procedures. It is expelled by taking a mild laxative.

Laxatives

A laxative is a drug that speeds the passage of stool or feces through the intestinal tract and causes a bowel movement. Laxatives are used to treat or prevent constipation (infrequent bowel movements), thereby relieving associated abdominal discomfort.

Some laxatives, such as senna and cascara, act by directly stimulating the nerves and muscle of the bowel. This starts a series of contractions known as peristalsis which encourage a bowel movement. Others, called hyperosmotics, work by attracting water into the bowel from surrounding tissues, increasing the bulk and volume of the stool, which speeds its passage through the bowel. These do not increase the number of bowel movements and are more for preventing constipation than treating it.

They are recommended for people who need to avoid straining while defecating, such as those recovering from childbirth and certain types of abdominal surgery. Hyperosmotics include milk of magnesia, Epsom salts (magnesium sulfate) and

Glauber's salts (sodium sulfate). Bulk laxatives such as bran, psyllium and vegetable fiber, and general roughage absorb water in the intestinal tract, swelling, and forming soft, bulky stool. They are the safe and effective forms of laxatives and are available in powder form for mixing in a drink, or as wafers, granules or tablets containing bran.

To be safe and effective, these must be taken with a least one glass of water, so are not suitable for people suffering from other conditions that demand a restricted fluid intake such as kidney failure. Using natural measures, such as increasing the amount of fiber and fluid in the diet, is generally better than taking laxatives.

Laxatives should generally be a short-term treatment for constipation only. Some harsh stimulant varieties that may be habit-forming and large doses can cause side effects such as cramps, dehydration and malnutrition. If taken for too long, abnormalities of the bowel wall may occur.

Laxatives should not be taken during constipation that occurs with abdominal pain or fever as these symptoms may indicate a bowel obstruction.

Antidiarrheals

Antidiarrheals are drugs used for the relief of symptoms of diarrhea. Some are simple absorbent substances such as kaolin, chalk or charcoal. They absorb water and help harden the feces. Others, such as diphenoxylate (often used in combination with atropine), slow down the contractions of the bowel muscle so that the contents are propelled more slowly and more water is absorbed by the bowel. Overtreatment with anti-diarrhea drugs can cause constipation and abdominal cramps, so excessive or extended use is not recommended. Codeine, an analgesic, also relieves diarrhea, though prolonged use may cause addiction.

Diuretics

Diuretic drugs act on the kidneys to increase urine output. They cause the kidneys to increase the amount of salt and water eliminated from the body in the urine, and so are often called "water pills." Alcohol, tea and coffee also have a mild diuretic effect.

Diuretics are used to treat excessive build-up of fluid in the body (edema) or the lungs (pulmonary edema). Edema can be caused by disorders of the heart (such as congestive heart failure or heart attack), liver dysfunction (by not producing protein) or kidney dysfunction (by leaking protein). Diuretics are used to treat hypertension (high blood pressure), congestive heart failure and high fluid pressure within the eye (glaucoma).

Some diuretics, for example thiazide diuretics, can sometimes cause the kidneys to lose potassium, which must be replaced by adding foods rich in it to the diet, or by taking potassium supplements.

Narcotics

Narcotics (also known as narcotic analgesics) are drugs that produce relief from pain, a state of stupor or sleep and often addiction or physical dependence.

Opium, produced from the opium poppy, has been in use for thousands of years. In 1803 the alkaloid compound morphine was discovered and in 1898, heroin (diacetylmorphine) was discovered and used as a treatment of morphine addiction (though it was later found to be even more addictive and dangerous than morphine).

Modern narcotics are either opium derivatives, such as morphine and codeine, or synthetic drugs known as meperidine and methadone. They are used to control severe pain, such as pain from kidney stones, cancer pain and post-operative pain.

Codeine is sometimes used in cough mixtures because it suppresses coughing and sometimes also as a treatment for diarrhea. Methadone is sometimes used in the treatment of heroin addiction, though it is itself addictive. Side effects of narcotics can be problematic; they include constipation, nausea and vomiting and urinary retention.

Morphine

Morphine is a bitter-tasting crystalline alkaloid derived from the opium poppy and related to codeine and meperidine. It was discovered in 1803 and called morphine after Morpheus, the Greek god of dreams. A powerful narcotic painkiller (analgesic), it is usually prescribed for people in severe pain, such as the pain associated with terminal cancer. It is also used in the treatment of acute heart failure and shock.

Morphine may be taken orally, but because of its slow and poor absorption by the intestine, it may also be injected or given via an infusion pump. If used for longer than a week or a fortnight, withdrawal symptoms may be experienced on cessation of the treatment.

Morphine depresses respiration, and it should be used with care in people with lung disease such as chronic bronchitis, and in the aged or the very young (who are particularly sensitive to the drug's respiratory depressant effects). Constipation is another common side effect.

Methadone

Methadone is a synthetic narcotic painkiller (analgesic) similar to morphine. It blocks the effects of heroin withdrawal, and so is used primarily in the treatment of heroin users, though it may sometimes be used in the management of cancer pain. It can be given once a day as an oral preparation. Side effects are similar to those of other narcotics and include nausea, constipation and urinary retention. Methadone is itself addictive but it is thought to be easier to withdraw from than heroin, though still not easy.

Though not a cure for drug dependence in itself, methadone gives people the opportunity to manage their lives and reduces their dependence on crime and the illegal drugs market, and therefore minimizes the social and legal consequences and the risks of transmission of infections such as HIV (AIDS), hepatitis B and C.

Treating infections

Infections can be treated by administrating drugs that either kill invading bacteria, viruses, parasites, worms or fungi or that prevent the multiplication of the invading microorganisms and allow the body to mount its own immune response.

SEE ALSO *Infectious diseases on page 364; Lymphatic/Immune system on page 55*

Antibiotics

Antibiotics are drugs that fight bacterial infection. They work either by preventing an infection from growing (bacteriostatic antibiotics) or by destroying an existing infection (bactericidal antibiotics). Some are

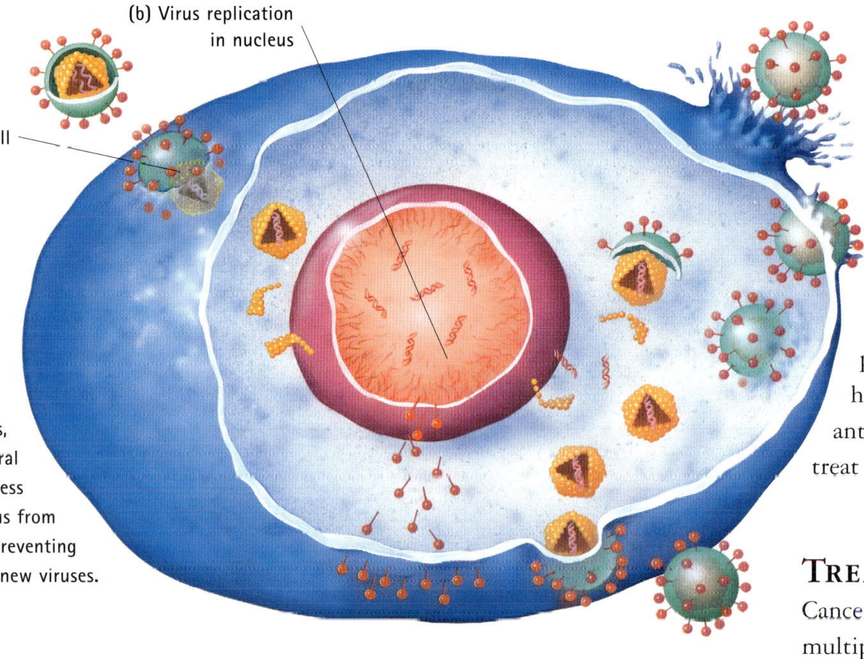

Antivirals

A virus reproduces by entering a cell and tricking the cell's genetic code into making copies of the virus in large numbers, as shown here. Antiviral drugs inhibit the process by preventing the virus from entering the cell, or preventing the cell from making new viruses.

effective against a broad range of bacteria (broad-spectrum antibiotics); others are only effective against certain types of bacteria.

Antibiotics are produced either from a mold or a fungus, or are produced synthetically. Common forms include aminoglycosides, macrolides, penicillins, tetracyclines and cephalosporins. They may be given orally, or via intravenous or intramuscular routes in more serious infections.

The modern trend is against using antibiotics indiscriminately or for the wrong reasons—for example, in treating viral illnesses (on which they have no effect).

A course of antibiotics must be finished to prevent bacteria re-establishing infection and developing immunity against the drug.

Side effects may include rashes, nausea, diarrhea and secondary infections such as thrush. In rare cases, anaphylaxis may occur. Tetracyclines should not be given during pregnancy, or to young children as they may discolor developing teeth. Aminoglycosides, for example gentamicin, amikacin and tobramycin, can also cause damage to the auditory nerves and to the kidney.

Penicillins

First discovered in 1938 by British physician Alexander Fleming, penicillins are a class of antibiotics used to kill bacteria. They were originally extracted from molds of the genus *Penicillium*, but are now synthesized. They work by preventing bacterial cells from forming a cell wall, killing them in the process. Penicillins can be administered orally via tablet, syrup or capsule, or by injection. In some people, they may cause allergic reactions such as skin rashes, swelling of the joints, and in rare instances, anaphylaxis, which may be fatal.

Excessive use of penicillins has resulted in the development of bacterial strains that are resistant to them. Newer penicillins such as amoxycillin, methicillin, oxacillin and dicloxacillin have been developed that are effective against many resistant strains.

Antifungals

Antifungals are used to treat infections such as ringworm, candidiasis and athlete's foot (tinea pedis). Those commonly used include clotrimazole, miconazole and ketoconazole.

Since fungi are more resistant to treatment than other microorganisms, a lengthy duration of treatment is usually required to cure a fungal infection.

Antifungals may be taken as tablets or applied as creams, ointments or pessaries. Antifungal injections can treat serious internal fungal infections such as actinomycosis, blastomycosis and histoplasmosis.

Antivirals

Antivirals are drugs used to treat viral infections or to prevent them from developing. Usually they work for only one kind of virus infection. Many viruses, including the common cold, do not respond yet to antiviral drugs. Those that do respond include herpes simplex (cold sores), herpes zoster (shingles) and influenza.

Treatment with antiviral drugs must be started early if they are to work. Acyclovir is used to treat the symptoms of chickenpox, shingles and herpes virus infections of the genitals, skin, brain, lips and mouth. It is also used to prevent recurrent genital herpes infections. Although it doesn't cure herpes, it helps relieve pain and discomfort and promotes healing of the sores. Idoxuridine is an antiviral drug that is commonly used to treat viral infections of the eye.

TREATING CANCER

Cancer treatments aim to kill or prevent the multiplication or development of cancer cells. This may be achieved by chemotherapy, which targets cancer cells, or by hormonal therapy, which prevents cancer growth by altering the body's internal environment.

SEE ALSO *Cancer on page 27*

Chemotherapy

Chemotherapy is the use of chemicals to kill cancer cells. There are many different anticancer drugs; some are given by mouth and some are injected. There is no drug yet available that is 100 percent effective in killing cancer cells, nor is there one that does not potentially harm healthy cells.

Research on chemotherapy is constant, and treatment methods are improving to minimize its harmful effects. Chemotherapy may be given over a brief period, or may extend over several years; it can be given in hospital, outpatient clinics, at a doctor's rooms or even at home. There are many different courses of treatment and all are being improved through sharing scientific knowledge and by international trials of new drugs and procedures.

Drugs used in chemotherapy are basically poisons, some of which developed out of research into chemical warfare. Standard combinations, or cocktails, of drugs have been set for treating most forms of cancer. After starting treatment and observing the patient's response, the combinations and doses are adjusted. This has made chemotherapy less unpleasant and more effective.

Hormone therapy can be used to restrict the growth of certain cancers. Immunotherapy uses substances such as interferon to attack specific tumors. Combination therapy uses surgery or radiation as well as chemotherapy to fight cancer: surgery or radiation to treat the cancer directly, chemotherapy to wipe out cancer cells that have spread to other parts of the body.

Anti-cancer drugs attack all fast-dividing cells, so that normal, healthy fast-dividing cells are killed along with the cancer cells. Most of the side effects are due to the killing of healthy cells. The major side effects of chemotherapy include: nausea and vomiting; hair loss; low blood cell count (cytopenia); damage to liver and kidneys; mouth ulcers and digestive disorders; and infertility and fetal abnormalities.

Chemotherapy today involves the careful monitoring of patients, including regular blood tests. As a result, problems are quickly detected and drug doses can be adjusted and other treatments given to minimize any side effects and discomfort. Many people now have chemotherapy as outpatients, and do not require a stay in hospital.

Cytotoxic drugs

Cytotoxic drugs destroy cells or prevent their multiplication. Penicillins or other antibiotics used in the treatment of bacterial infections are cytotoxic, but the term usually refers to drugs that are used in the treatment of cancer, such as methotrexate and cyclophosphamide. They can sometimes also be used to treat other disorders such as rheumatoid arthritis.

Cytotoxic drugs affect not only cancer cells but other cells as well, especially those that grow rapidly, such as cells of the gastrointestinal tract skin and bone marrow. They may cause side effects such as serious blood disorders, hair loss (alopecia) and reduced resistance to infection. Dosages and side effects of cytotoxic drugs need to be carefully monitored by the physician.

Hormonal therapy

Hormones are natural chemicals produced by the body to regulate various processes such as blood sugar metabolism, bone growth or milk production. Some cancers will only grow in the presence of certain hormones; for example, certain types of breast cancers need the female hormones estrogen and progesterone to grow. Hormonal therapy seeks to prevent these cancers from growing by altering the hormonal environment around them. It is usually used in conjunction with other cancer treatments such as surgery and radiation therapy.

Cancers that are stimulated by hormones have certain areas on their surface called receptor sites. By using a drug that blocks these sites, the growth of the cancer can be slowed or even stopped. Many breast cancers have estrogen and progesterone receptors, and are stimulated by these hormones. This means they may respond to treatment with a drug such as Tamoxifen that blocks the effects of estrogen on breast cancer.

Treatment with Tamoxifen reduces the recurrence of cancer and is used after surgery or radiation therapy. Not all breast cancers respond to the drug; before commencing treatment with Tamoxifen a section of breast cancer must be tested in a pathology laboratory to see whether it will respond to the drug.

In males, prostate cancer grows more quickly when exposed to the male hormone testosterone. By reducing the amount of testosterone in the environment around the prostate, hormone therapy can be used to reduce the growth and spread of these cancers. This is done by surgically removing the testicles or treating the patient with drugs that block the action of testosterone on the prostate. Alternatively, estrogens or luteinizing-hormone-releasing hormones may be administered.

The treatment is effective in slowing down the progress of prostate cancer metastases (secondaries). It is generally used in conjunction with surgery, radiation therapy, or chemotherapy.

Side effects associated with hormonal therapies include loss of libido, weight gain, diarrhea, tiredness, hot flashes, bone loss, and, in women, irregular menstrual periods and vaginal dryness or bleeding.

Procedures and Therapies

The range of available medical procedures and therapies is continually being developed and improved to help us manage our bodies better in both health and disease.

Immunization

Immunization uses a killed virus or bacterium, a weakened strain ("attenuated"), a deactivated toxin ("toxoid"), or sometimes a synthetically or genetically engineered vaccine to help the body to build up its immunity to certain diseases. After being immunized, the body's immune system manufactures antibodies which are special proteins that can recognize and help destroy viruses and bacteria or foreign toxins when they invade the body. In addition, other parts of the immune system are activated to combat an infection.

A single dose of some vaccines will give immunity for life, however others need to be given according to a specific schedule and need boosters or additional reinforcing doses to maintain immunity. When children are immunized against diphtheria, tetanus and pertussis (whooping cough) they are given a vaccine known as DTP. This has recently been replaced with a new vaccine called DTaP, which contains a different form of pertussis vaccine called "acellular pertussis" vaccine and has fewer side effects than the older form.

When children are immunized against measles, mumps and rubella, they are given a vaccine called MMR. A vaccine is available in some parts of the world for varicella (chickenpox), and a vaccine that against all four of these, called MMRV, is being developed. Vaccinations against *Haemophilus influenzae* type B and poliomyelitis are also part of the usual childhood immunization schedule in most countries.

It is important that when children are being immunized, they should receive the full schedule of initial and booster vaccinations so that the immunization process is effective. The timing is also important in immunization. Schedules have been established by health authorities to achieve the most effective results.

There are other vaccines which are available or in the process of being developed, and many of these are also intended for travelers and people at high risk of infection such as the elderly or people with underlying diseases, who may have weakened immune systems. Some of these vaccines include hepatitis A, Japanese encephalitis, yellow fever, tuberculosis (BCG vaccine),

Contraception

An IUD consists of a ring, coil, spiral or loop. Once inserted in the uterine cavity, it can be left in place for as long as a year. Its presence renders implantation of a fertilized egg almost impossible.

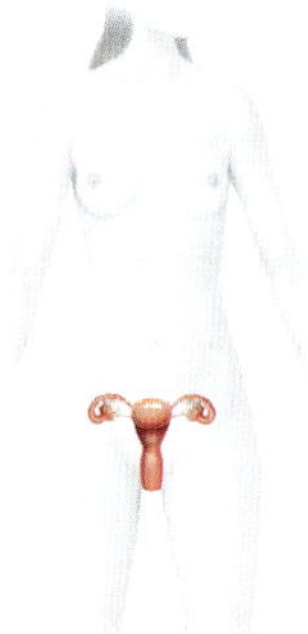

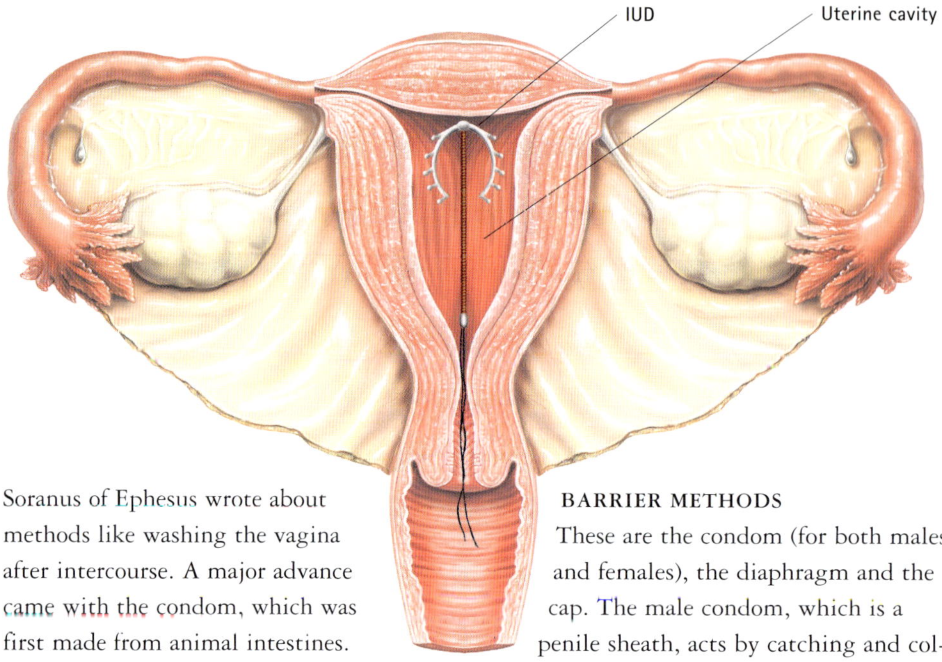

pneumococcal infection, meningococcal infection, influenza, varicella-zoster (chickenpox and shingles), plague and typhoid. Some of these vaccines are not available in all countries, and there are other vaccines currently under development. Even today, there are no effective vaccines for dengue fever, malaria or HIV (AIDS).

It is advisable for travelers to be aware of the schedule for certain immunizations when they are traveling to countries where vaccine-preventable diseases are a risk.

The vaccines used in immunization have been tested thoroughly and are safe and effective. Though minimal risks have been found with some vaccines, serious reactions are rare. The complications associated with the disease far outweigh the risks of complications from the vaccine.

Immunization against infectious diseases of all types has probably saved more lives than any other public health measure, apart from the provision of sanitation and clean water. Research into new vaccines is ongoing with a vaccine for infants against meningitis and blood poisoning being among the newest to become available.

Contraception

Contraception (birth control) is, quite simply, any action taken to avoid conception. The only sure guarantee against pregnancy is to avoid vaginal intercourse. For conception to occur, the female's egg (ovum) must be met by sperm-loaded semen from the male. In natural conception the semen is deposited in the vagina. It then travels through the cervix and uterus to fertilize the egg. This egg must then implant itself into the lining of the uterus (endometrium).

Written details of birth control—a term first used by the reformer Margaret Sanger in 1915—date back to 1550 BC in Egypt. Classical writers such as Pliny the Elder and Soranus of Ephesus wrote about methods like washing the vagina after intercourse. A major advance came with the condom, which was first made from animal intestines. Vaginal barriers such as diaphragms and caps were mentioned in 1823 by the German physician F. A. Wilde. Vasectomy was used in the nineteenth century and the first documented female sterilization was performed in 1881. An Australian couple, the Billings, first used the monitoring of changes in cervical mucus to determine fertile times in the 1960s, around the same time as the first contraceptive pills became available. Contraception should be discussed with qualified medical professionals to determine which method should be used, according to the partners' current lifestyle, medical history and future plans.

There are numerous contraceptive options ranging from the oral contraceptive pill that need to be taken daily. There is even a hormonal implant available which is inserted under the skin and gives contraceptive protection for up to three years.

INTRAUTERINE DEVICE

This is a small device inserted into the uterus to prevent conception. It is still not sure how it works but it seems that almost any foreign body in the uterus will prevent conception. Made of plastic, metal or other material, and inserted under sterile conditions by a trained professional, intrauterine devices (IUDs) have a failure rate of between 1 and 6 pregnancies per 100 women. The introduction of the progestogen releasing IUD has broadened the indication for IUDs from just being contraceptive to being therapeutic for several gynecological conditions.

BARRIER METHODS

These are the condom (for both males and females), the diaphragm and the cap. The male condom, which is a penile sheath, acts by catching and collecting the semen so it does not get into the vagina. The female condom is a thin silicone membrane that partly or completely covers the outside of the female genitals, acting as a barrier for sperm.

Diaphragms and caps are placed inside the vagina to cover the entrance to the uterus and stop sperm getting into the womb. They come in a range of sizes and must be fitted to the user by a trained health professional. Many family planning clinics and doctors recommend the use of a spermicide (which kills the sperm) at the same time. When used carefully, barrier methods have about the same effectiveness as IUDs.

FERTILITY AWARENESS METHODS

Not having sexual intercourse at a fertile time is the basis of the methods variously described as fertility awareness, periodic abstinence, the Billings method, rhythm method and the temperature method.

Changes in a woman's body during the menstrual cycle can be interpreted to indicate fertility. The cervical mucus changes under the influence of estrogen and at the time when a woman is most fertile, it is plentiful, clear and sticky.

The woman's basal body temperature, taken with a special thermometer, also rises slightly for around three days. Most women have cycles which can be documented so that they can work out when they are most likely to be fertile. There is a wide range of pregnancy rates depending on the method used and the couple involved.

ORAL CONTRACEPTIVES

The most common oral contraceptive in current use is the low-dose combination pill containing both estrogen and progesterone. Progesterone-only pills are also available but tend to be used only when the combined pill is contraindicated. Some combination pills contain the same concentrations of estrogen and progesterone throughout the dose cycle, in others the relative amounts of the hormones are varied in pills to be taken at different times of the cycle. The rationale of these latter preparations is that variable hormone levels resemble the natural situation more closely.

Estrogen and progesterone in the oral contraceptive have to be in a form which can be taken by mouth and still remain active, which precludes use of natural hormones. If natural hormones are taken through the mouth, they are broken down in the liver before they enter the general blood circulation of the body.

Synthetic hormones resemble the natural ones closely but have a slight modification to their molecular structure, which usually overcomes this problem. These estrogens can be either ethinyl estradiol or mestranol. The most commonly used synthetic progesterones are levonorgestrel, norgestrel as well as norethindrone.

Oral contraceptives are usually taken for 21 days and then stopped for 7 days, during which time withdrawal bleeding occurs. This mimics but is not the same as a period. The regimen may be varied if it is desired to avoid withdrawal bleeding at a particularly inconvenient time. Anecdotal evidence also suggests that at least some women elect to take the pill continuously rather than cyclically and thus avoid the nuisance of withdrawal bleeding altogether.

During a normal menstrual cycle, a few follicles develop in the ovaries during the first half of the cycle (the follicular phase) under the influence of follicle stimulating hormone (FSH) from the pituitary. These follicles produce increasing amounts of estrogen and this in turn stimulates a surge of luteinizing hormone (LH) from the pituitary which induces ovulation. The amount of estrogen and progesterone in the pill is approximately the same as that present at the start of the follicular phase.

Estrogen levels in a normal cycle rise by about six- to ten-fold during the natural follicular phase. However, this does not occur with the contraceptive pill. Low levels of estrogen and progesterone result in decreased production of FSH and LH by the pituitary, which in turn suppresses ovarian function and prevents ovulation.

Oral contraceptives affect other parts of the reproductive tract besides the ovary. Progesterone makes the cervical mucus thick and impermeable to sperm, and both estrogen and progesterone affect the motility of the fallopian tube and thus may reduce gamete or zygote transport. Changes occur in the endometrium that render it hostile to implantation and it also becomes thin after prolonged contraceptive use. These changes are reversible when the pill is stopped and regular ovulations are restored in the vast majority of women within 3 to 6 months.

Progesterone-only pills often do not inhibit ovulation but instead function by causing changes in cervical mucus and the endometrium as described above. It is also thought that they may interfere with capacitation of sperm in the female tract.

The most obvious advantage of the pill is its almost 100 percent reliability and it is thought that the occasional failures are more often due to non-compliance than to method failure. The pill also offers many non-contraceptive benefits, including relief from a wide range of menstrual disorders and reduction in the incidence of ovarian cysts, rheumatoid arthritis, ectopic pregnancies and pelvic inflammatory disease. There is also decreased risk of endometrial and ovarian cancer.

The most serious of the reported disadvantages are disorders of the cardiovascular system in some individuals. These include abnormal blood clots (thromboembolism), stroke and some elevation of blood pressure. Risks usually increase slightly with age but are reduced considerably in non-smokers and with low-dose pills.

Many of the other problems reported in the past have been shown to be due to other causes or of minimal significance with the low-dose pill. Some drugs, including certain antibiotics and anticonvulsants, may reduce the effectiveness of oral contraceptives to some extent.

The emergency pill (also known as the morning-after pill) is a short course of hormones which must be started within 72 hours of unprotected intercourse. Hormones from the pill delay ovulation and change the lining of the uterus. Fewer than 5 in 100 will conceive when this method is used correctly. Emergency IUD insertion or copper IUD insertion, if done within five days of unprotected sex, will prevent conception and also provide on-going contraception.

SPERMICIDE

The main function of spermicide is to kill or immobilize sperm. The spermicide should be inserted into the vagina at least 15 minutes before intercourse and can be a foam, cream, jelly, film or suppository. It can also be a sponge which carries spermicide. Spermicides can be used with a barrier method or on their own when they have about a 30 percent failure rate.

COITUS INTERRUPTUS OR WITHDRAWAL

This is the act of withdrawing the penis during sexual intercourse before ejaculation takes place. It has a high failure rate and is not a recommended method.

BREAST FEEDING

Another "method" with a high failure rate is breast feeding. Thought by many to be a natural way to plan families, it is only likely to work if the baby is being breast fed at least 5 times in a 24-hour period and is not being fed solid foods, and the woman has not started to menstruate.

FEMALE STERILIZATION AND VASECTOMY

Permanent contraception methods such as sterilization and vasectomy continue to be popular among both men and women. A woman is sterilized by blocking both fallopian tubes (through which sperm reach the egg). This tubal ligation is a surgical procedure which is now commonly carried out by endoscopy using an optical instrument through an abdominal incision. A vasectomy blocks the tube (the vas deferens) that carries the sperm from the testis to the semen. This is a simple operation done via a small incision in the front of the scrotum.

While the success rate in reversing these methods can be as high as 70 percent, they are not recommended birth control methods for people who may later want children.

Circumcision

The surgical process of cutting away all or part of the foreskin of the penis is called circumcision. The operation of removing part or all of the female genitalia is known as female circumcision, clitoridectomy or female genital mutilation.

Circumcision of boys is done before or at puberty or, in some Arab peoples, just before marriage. A rite since the time of the ancient Egyptians, circumcision is today an important religious ritual in Islamic and Jewish communities worldwide. It became popular in many industrialized cultures in the early 1900s in the belief that it promoted good hygiene and also discouraged masturbation.

Generally, the arguments in favor of circumcision are that there is less likelihood of penile cancer, urinary tract infections or sexually transmitted diseases; the penis is easier to clean; the rest of the males in the family are circumcised; or problems with the foreskin. The arguments against are that it is an unnecessary, painful operation with the possibility of complications, the sensation at the tip of the penis (the glans) is diminished, it is just as easy to clean under a foreskin as without, and that problems with the foreskin are most often caused by adults forcibly attempting to retract it in a young boy before it is ready.

Female circumcision has been performed for centuries in parts of Africa, the Middle East and Southeast Asia with the purpose of preserving virginity, improving hygiene or as a religious ritual. It is usually performed by non-medical practitioners in non-medical settings and has a high risk of complications and long-term consequences including sterility.

The procedure can involve infibulation which is the removal of the clitoris, the labia minora and the anterior two-thirds of the labia majora with a join made leaving a small opening. It can also involve the introduction of corrosive substances to cause bleeding for the purpose of narrowing the vagina; or any or all of the above. Even when the wounds heal normally, urination and sexual intercourse can be painful and unreleased menstrual blood can cause problems. In societies where clitoridectomy is practised, male circumcision is usually also practised and it is considered to be an important religious ritual or ethnic tradition.

The World Health Organization, the United Nations Population Fund and the United Nations Children's Fund have drawn up a plan to reduce female genital mutilation, with the aim of eliminating it completely. Strong lobby groups against female genital mutilation exist in many industrialized countries who view it as a major health risk and also as the subjugation of women so that they will never enjoy sexual relations.

Hormone replacement therapy

At menopause, the ovaries start to regress and women produce less of their reproductive hormones, estrogen and progesterone. The decrease in the quantity of these hormones can cause a variety of symptoms, which include hot flashes (sudden feeling of heat, usually over the face and neck, often with redness and sweating), headaches, vaginal dryness, anxiety and sleep problems. The hot flashes are probably the result of sharp surges in pituitary hormones as a result of falling ovarian hormone levels while vaginal dryness is a direct result of estrogen withdrawal.

Estrogen loss increases the risk of other health problems such as heart disease. Loss of bone substance also increases after menopause, and can lead to osteoporosis where bones lose calcium and are easily susceptible to fracture.

Most of the problems associated with menopause can be prevented by hormone replacement therapy (HRT). HRT involves supplementing the patient's hormone levels with doses of estrogen, often in combination with progesterone. HRT can be very effective in relieving symptoms such as hot flashes and vaginal dryness, while also aiding in reducing the risk of osteoporosis.

In some cases HRT is taken cyclically to allow for monthly bleeding, much like a menstrual cycle, and in other cases it is taken without pause, which will cause regular bleeding to cease. The HRT regimen can be tailored to suit each patient, depending on her symptoms and other health problems.

HRT can also be useful in some cases of incontinence, where there is a loss of bladder control leading to leakage of urine. This is known as "stress incontinence," where increases in abdominal pressure can force urine out, for instance, while coughing or sneezing. This is may be more common in postmenopausal women, and is often caused by a weakening of the pelvic floor muscles. Estrogen treatment may be helpful, as may pelvic muscle exercises and other medications. Changes in the lower urinary tract soon after menopause can also predispose a woman to cystitis (bladder infection).

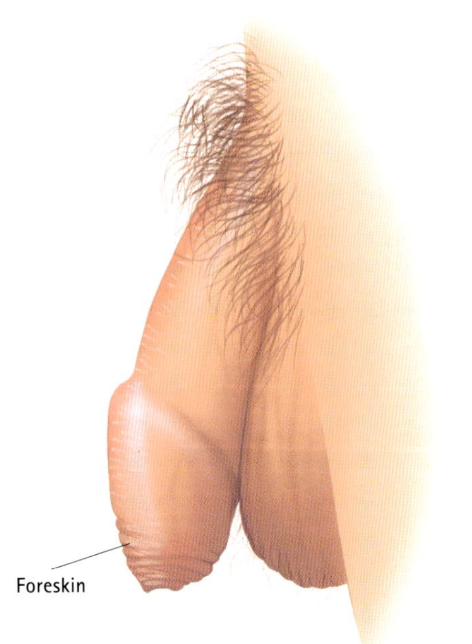

Uncircumcised penis

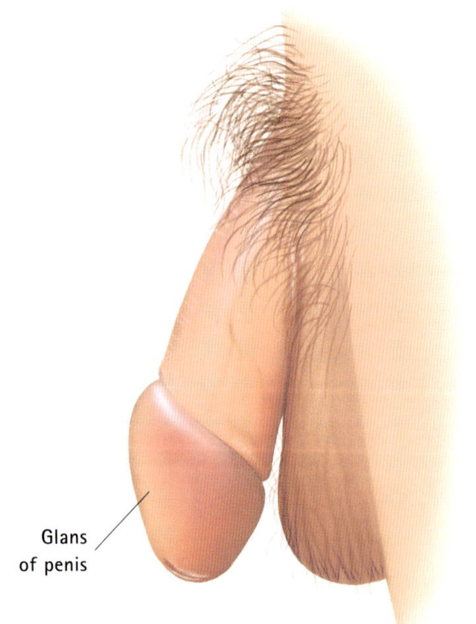

Circumcised penis

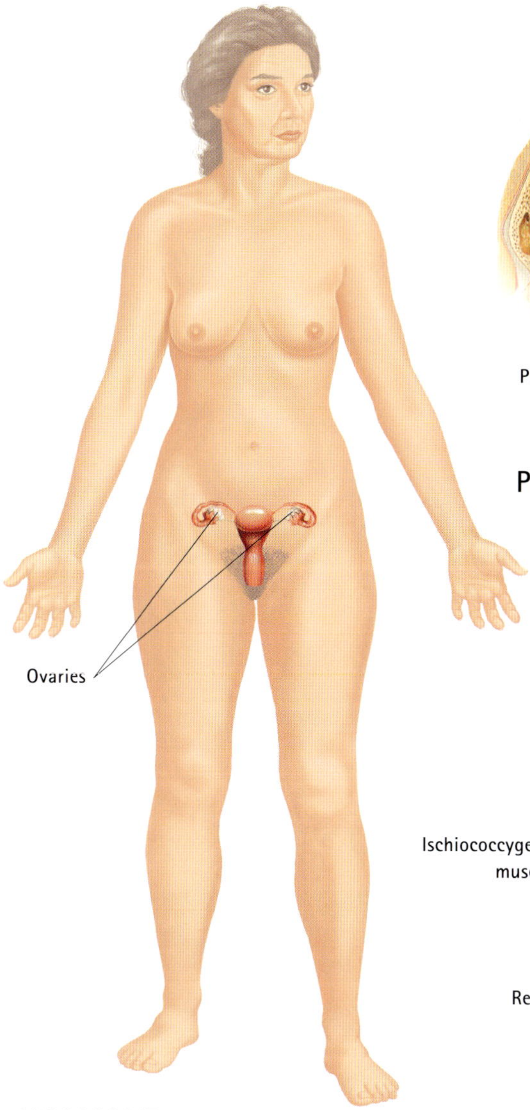

HORMONE REPLACEMENT THERAPY

At menopause, the ovaries produce less estrogen and progesterone. A woman may choose to supplement her hormone levels with hormone replacement therapy. HRT can reduce the risk of osteoporosis and bladder problems such as incontinence.

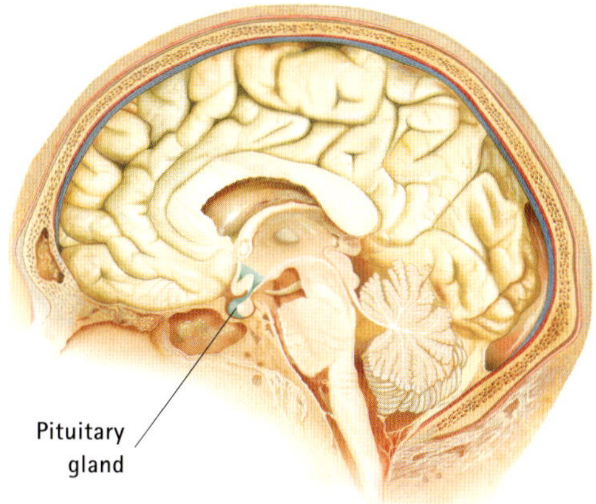

Pituitary hormones

Surges of pituitary hormones as a result of falling ovarian hormone levels cause the hot flashes often experienced during menopause.

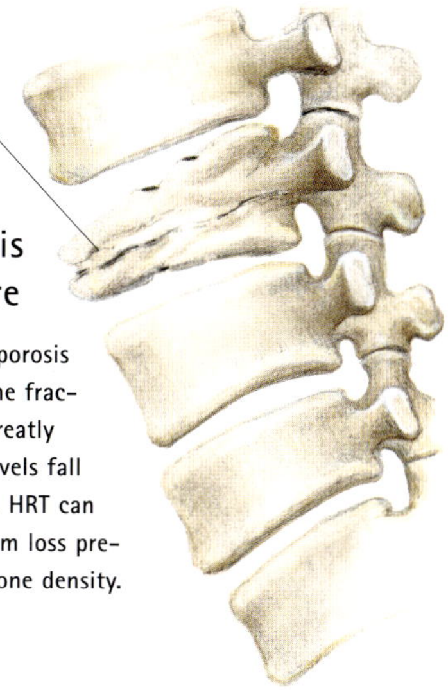

Osteoporosis and fracture

The risk of osteoporosis and resulting bone fractures increases greatly when estrogen levels fall after menopause. HRT can slow down calcium loss preventing loss of bone density.

Incontinence

Weakening of the pelvic floor muscles is common after menopause and may result in incontinence. Estrogen treatment can improve symptoms.

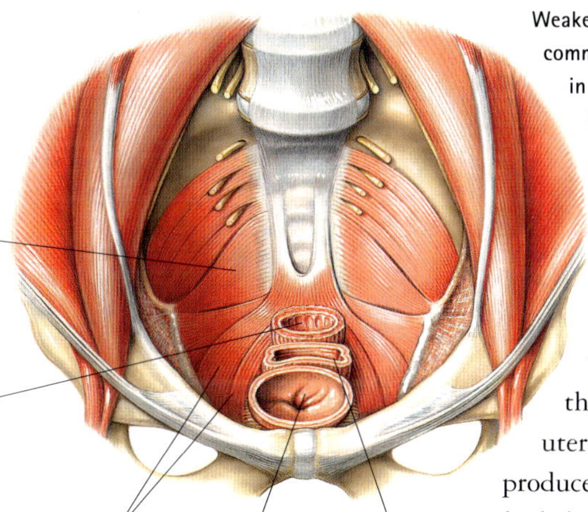

This causes frequent and sometimes painful urination and can be helped by HRT.

HRT can lead to a number of unwanted side effects. Women may experience breast tenderness and swelling, unwanted vaginal bleeding, nausea, fluid retention and a degree of weight gain.

The levels of estrogen and progesterone used in HRT are lower than those used in the oral contraceptive pill. Thus complications which have been associated with taking the pill (particularly the higher doses used in early pill preparations), such as increased blood pressure and increased risk of pulmonary embolism, are not considered to be problems with HRT. However HRT may not be considered appropriate in people who have already had these problems.

Estrogen administered on its own has been shown to increase the risk of uterine (endometrial) cancer. Combination therapy (estrogen and progesterone) is not associated with an increased risk of uterine cancer, so it is the preferred treatment for women who have not had a hysterectomy. HRT is not recommended in certain conditions. For instance, many breast cancers are sensitive to estrogen, which stimulates their growth, so HRT should not be used in these cases.

A full breast examination and mammogram is therefore recommended before HRT treatment begins. The recent Women's Health Initiative (WHI) study showed that HRT was of no value in preventing heart disease and that there was a significant risk of breast cancer with long-term use of HRT. Short-term use of HRT (defined as less than 2 years) appears to be safe.

In vitro fertilization

In vitro fertilization (IVF)—the fertilization of an ovum ("egg") in vitro ("in glass") and the return of the resulting embryo to the woman's uterus—is the pioneer procedure that produced what were originally called "test tube babies." This technique was used for the first time in the 1970s to offer hope to women who were infertile because something interfered with the passage of the ovum from the ovary to the uterus via the fallopian tube. It is now available to treat infertility arising from most causes.

IVF begins when the woman takes ovary-stimulating drugs for about 10 days. Then, ultrasound checks determine the number and size of the ovarian follicles in which eggs are developing. The eggs are collected using a trans-vaginal probe when the woman is under local or general anesthetic.

The man will be required to masturbate to produce a semen sample. A small amount of this semen will be processed to remove debris and then added to the dish that holds an egg. Fertilization can take up to 18 hours, and 1–5 days later the embryos—which at this stage are technically known as blastocysts—will be ready for implanting into the woman's uterus.

The technique whereby a single sperm is injected into the egg has helped revolutionize

IVF treatment for men with very low sperm counts. Generally up to three fertilized eggs are transferred to the uterus to increase the chances of pregnancy, and most transfers are done on day two. The transfer is done using a fine catheter and the woman does not need to be sedated. The woman may be prescribed hormones to increase the chances that the fertilized eggs will successfully implant. Only between 15 and 20 percent of embryos created in this artificial environment are truly viable—that is, have the potential to continue their development.

Lithotripsy

Lithotripsy is a procedure used to shatter stones (calculi) that have formed in the bladder, gallbladder, ureter or kidney. Extracorporeal shockwave lithotripsy is a common technique that creates shock waves outside the body, which are targeted at the stone to break it up. This technique is generally used on stones no more than ⅝ inch (1.5 centimeters) in diameter and some sort of anesthesia is given. The patient may be positioned in water during the procedure. X-rays or ultrasound are used to ensure the accurate location of the stone.

While surgical stone removal once required a lengthy stay in hospital, lithotripsy may take only 45 minutes and can sometimes be performed on an out-patient basis. Depending on the size of the stone and the strength and duration of the shockwaves needed, anesthesia may be used to alleviate pain. It may take between 800 and 2000 shocks, for example, to break up a kidney stone. Stone fragments are sometimes passed from the body with the aid of a catheter inserted in the ureter.

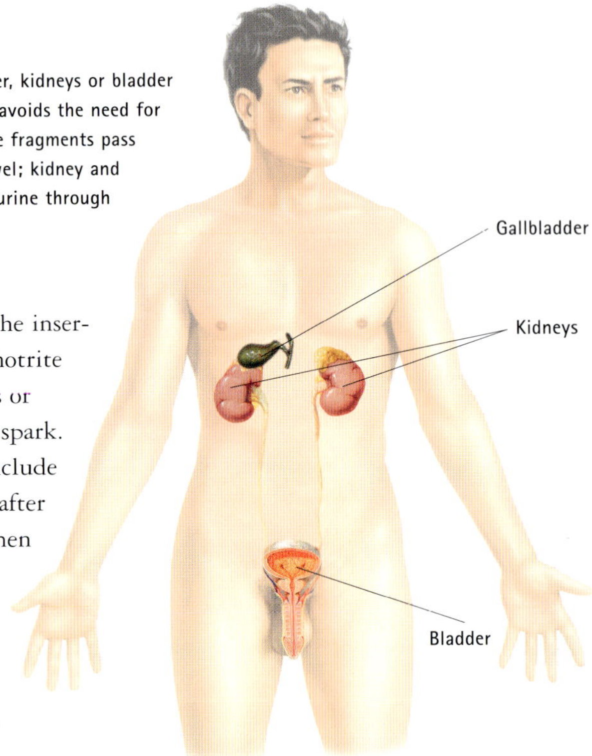

Lithotripsy

Stones that have formed in the gallbladder, kidneys or bladder can be shattered using shock waves. This avoids the need for invasive surgery. Once shattered, gallstone fragments pass out of the body via the bile duct and bowel; kidney and bladder stone fragments are passed with urine through the ureter.

Another type of lithotripsy is the insertion of an instrument called a lithotrite through the ureter which shatters or crushes stones using an electrical spark.

Complications of lithotripsy include blood in the urine for a few days after treatment, bruising on the abdomen or back, and pain or discomfort as stones pass out of the body in the urine. More than one treatment may be required if the stone is not completely shattered the first time.

Catheterization

A catheter is a flexible, hollow tube that is introduced into a cavity through a narrow opening, in order to discharge the fluid in the cavity or to unblock a vessel. It may be made of glass, metal, hard or soft rubber, rubberized silk, or plastic. A catheter may be inserted into the bladder temporarily during an operation, or permanently after a brain or spinal injury to relieve an underactive or overactive bladder, or because of paralysis.

A catheter may also be inserted into an artery or vein or directly into the heart to inject drugs, to measure blood flow or pressure, to use in the diagnosis of congenital heart disease, to explore narrow passages, or to pass electrodes into the heart so that the heartbeat can be restored or made regular.

Angioplasty

Angioplasty is the repair of blood vessels affected by disease (usually atherosclerosis). It may be performed in the blood vessels supplying the heart or in the arteries of the limbs, brain or kidney.

Angioplasty may be surgical or non-surgical. In surgical angioplasty, segments of the affected artery are removed and replaced with tissue from a vein from elsewhere in the body, or synthetic tissue. This procedure provides an alternative to bypass surgery.

Balloon angioplasty is a non-surgical method of removing atherosclerotic plaque, the fibrous and fatty deposits on the walls of blocked arteries. A catheter with a balloon-like tip is threaded up from the arm or groin through the artery until it reaches the blocked area. The balloon is then inflated many times, flattening the plaque and widening the blood vessel; the balloon is

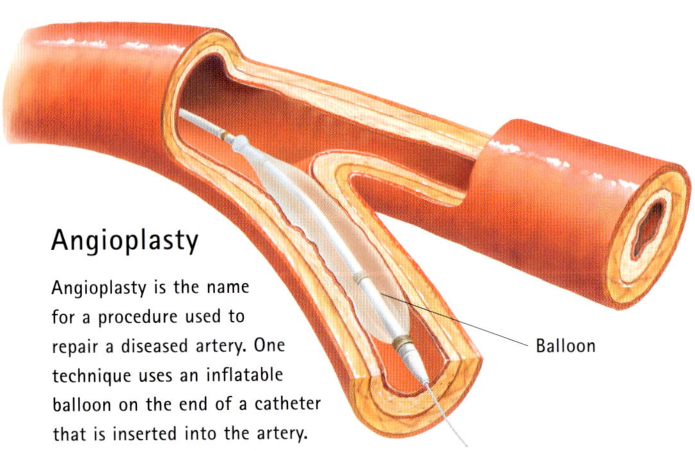

Angioplasty

Angioplasty is the name for a procedure used to repair a diseased artery. One technique uses an inflatable balloon on the end of a catheter that is inserted into the artery. The balloon expands and compresses the blockage.

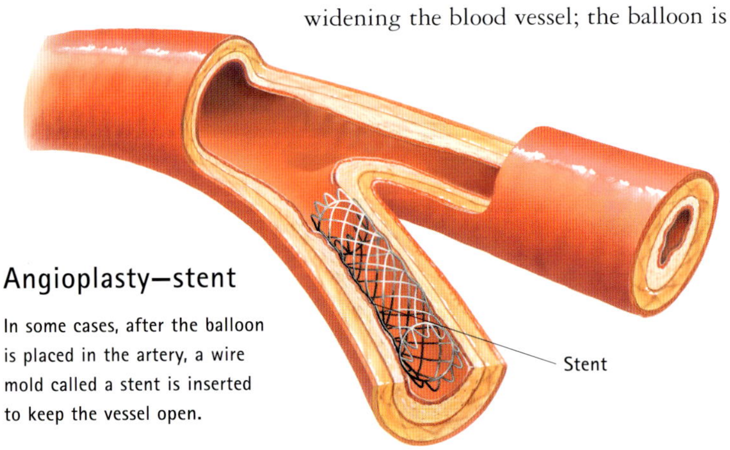

Angioplasty—stent

In some cases, after the balloon is placed in the artery, a wire mold called a stent is inserted to keep the vessel open.

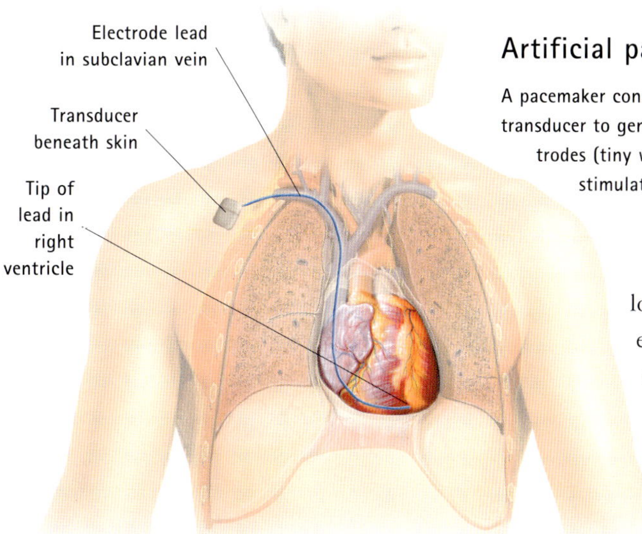

Artificial pacemaker

A pacemaker consists of two main components: a transducer to generate electrical impulses, and electrodes (tiny wires) to make contact with and stimulate the heart muscle.

then removed. Often a wire mold called a stent is then inserted into the vessel to keep the vessel open. This technique is especially suitable when only one vessel is blocked.

Laser angioplasty involves a similar technique, except that the catheter has a camera lens on its tip connected to a screen, allowing the physician to spot areas where plaque has built up. A laser fixed to the tip is then used to destroy the plaque.

Pacemaker

This term is often used in reference to the heart, although other organs (such as the uterus and bowel) may also have specialized tissue which controls rhythmic contraction. There is a naturally occurring pacemaker, called the sinoatrial node, which directly controls the rhythmic contraction of the heart. This node is located in the wall of the right atrium near the superior vena cava and has its own rhythmic cycle of electrical activity. This activity is transmitted to surrounding heart muscle and through specialized conducting cells to all parts of the heart, thus initiating heart muscle contraction. The electrical rhythm of the sinoatrial node may be slowed by the vagus nerve, or quickened by the sympathetic cardiac nerves and circulating hormones such as epinephrine (adrenaline).

If the heart is unable to maintain an adequate rhythm due to disease, then artificial pacemakers may be used. They give the heart small stimulant shocks by delivering electrical impulses at a pre-determined rate. These impulses trigger contractions of the heart muscle which cause blood to pump through the heart.

Artificial pacemakers usually consist of two main components: a transducer for generating electrical impulses and tiny wires, or electrodes, which are fed into veins and make contact with the heart muscle. The transducer is a small device weighing about 1 ounce (28 grams) and powered by a lithium battery that may not need changing for at least five years. The transducer is usually implanted beneath the skin just below the collar bone via a small incision. This is a minor procedure carried out using mild sedation and local anesthetic.

Transducers are can usually be encased in titanium, which usually does not irritate the body. Alternatively, the transducer may be worn on a belt around the body.

Electrical impulses from the transducer travel along the electrodes to the heart muscle at a rate that may be preset or controlled externally by a remote switch. Depending on the condition of the heart, the transducer may monitor the heart's rate of contraction and send out electrical impulses only when it beats abnormally. In this way, pacemakers can be used to stimulate a heart that beats too slowly due to such problems as blocked arteries, metabolic abnormalities or the side effects of certain medications. They can also help to stabilize and slow down a heart that beats too fast or to re-establish the heart's rhythm after cardiac arrest.

Defibrillation

Fibrillation is rapid, irregular twitching of muscle fibers. In the heart it may be caused by heart disease, such as coronary artery disease, by drugs (such as digitalis), or by electrocution. Defibrillation is the term used for stopping fibrillation of the heart muscle.

If fibrillation affects the lower chambers (ventricles) of the heart, it causes cardiac arrest, which will rapidly lead to death because a heart that is fibrillating will pump little or no blood around the body's circulatory system. Normal heart contractions can often be restored by electric shocks from a machine called a cardiac defibrillator. Cardiac defibrillation is an emergency procedure, and is combined with cardiopulmonary resuscitation (CPR). If cardiac arrest and defibrillation occur in a hospital, the survival rate is high; elsewhere it is very low. Defibrillation is most effective when performed soon after cardiac arrest.

There are three types of defibrillators: manual defibrillator; automated external defibrillator (AED), which can read a patient's heart rhythm and deliver the shock automatically; and implantable cardiac defibrillators (ICDs), which are surgically implanted and are suitable for people with recurrent fibrillation.

Life support

Life support is the term given to the efforts to maintain vital body functions, such as respiration, circulation and fluid and nutrient intake, in comatose patients who cannot maintain these functions naturally. Life support involves the use of mechanical ventilation, physical therapy, intravenous fluids, and round-the-clock nursing care in an intensive care unit in hospital.

Ethical questions have been raised regarding the quality of life of people kept alive by artificial life support. When the patient is unlikely to make any recovery, physicians—with the cooperation of the patient's relatives—may elect not to continue life support, especially where the patient has been declared brain dead. Some individuals draw up what is known as a "living will" in which they specify how they wish to be treated if accident or illness leaves them in a vegetative state.

Dialysis

Dialysis is a method of removing toxic substances (impurities or wastes) from the blood when the kidneys cannot do so. This is usually because of kidney disease, but dialysis may sometimes be used to quickly remove drugs or poisons from the bloodstream in people who have been poisoned or who have overdosed on drugs. If the kidneys are not able to fulfil over 10 percent of their normal function, it is considered end-stage kidney disease and dialysis is usually necessary. There are two types of dialysis: peritoneal dialysis and hemodialysis.

Peritoneal dialysis uses the sufferer's abdominal peritoneal membrane to act as a dialysis mechanism. It involves filling up the abdominal cavity (through a catheter) with a special solution, using the peritoneal membrane inside the abdomen as the semi-permeable membrane. The fluid is allowed to absorb wastes for several hours, and then the waste-filled fluid is exchanged for a fresh batch of solution.

Hemodialysis involves filtering the blood slowly through an artificial kidney machine called a dialyzer. Inside the dialyzer, blood is run through tubes with semipermeable membranes, and the tubes are bathed with solutions that will remove small soluble molecules (such as urea) from the blood. The purified blood is then fed back into one of the patient's veins.

For hemodialysis to be successful, there must be adequate access to the circulation. A normal vein is not big enough to carry a wide bore IV (intravenous) line, so special types of arterial and venous access have to be constructed. A common method is to surgically join an artery and vein together under the skin; this is called an A (arterio-venous) fistula. The increased blood volume stretches the elastic vein to allow a larger volume of blood flow into which a wide bore IV line can be inserted.

In people whose veins are not very suitable for an AV fistula, a graft from an artery to a vein may be used. The graft may come from the person's own saphenous vein (in the leg), or it may be a synthetic graft.

Usually, dialysis needs to be performed three times a week for periods of four to six hours each time. Dialysis may take place in a hospital, at a special dialysis center, or at the patient's home. It needs to be continued until the kidneys recover their normal function completely. In the case of chronic renal failure, this usually means for the rest of the person's life or until a kidney transplant is performed.

Paracentesis

Paracentesis (also known as abdominal tap or peritoneal tap) is a medical procedure which involves removing fluid from a body cavity, usually from the abdomen. It can be performed in a doctor's treatment room or a hospital. A specialized needle is inserted through the body wall, after the area has first been numbed with a local anesthetic, and fluid is then drawn off. An incision may be needed to assist insertion of the needle.

Paracentesis is commonly used to sample excessive abdominal fluid (ascites) to test for what is causing it, check for internal bleeding or relieve the effects of fluid build-up.

Blood transfusion

Blood transfusion is the transfer of blood from one person to another. A transfusion may be required in serious cases of anemia, either as a result of disease or from a loss of blood, or it may be part of the treatment of acute shock. Blood for transfusion usually comes from a donor and is stored in a blood bank (though the bank may store your own blood to be given back to you later for elective surgery). The donor's blood is screened for HIV and other viruses prior to the transfusion. Blood is cross-matched by the bank to prevent incompatibility between the donor's and the recipient's blood. Incompatibility can cause a transfusion reaction, which in severe cases can be fatal.

An exchange transfusion is the complete replacement of a person's blood. It is sometimes necessary in newborn babies or for people suffering severe poisoning.

Plasmapheresis

Plasmapheresis is a method in which plasma is passed through a filtration device that removes certain antibodies and abnormal protein elements and returns the plasma, free of antibodies and protein elements, to the patient. It is used for treating certain autoimmune diseases (caused by antibodies attacking the body's own tissue), such as systemic lupus erythematosus (SLE), myasthenia gravis and also in multiple myeloma.

Anesthesia

Anesthesia is loss of sensation, particularly to pain. Anesthetics are drugs that halt the sensation of pain through the body's nervous system. The nervous system is made up of neurons (nerve cells) which send information to the brain. The brain then processes this information and relays it to the muscles via motor neurons. Anesthetics act in various ways, such as affecting certain parts of nerves (called axons) and interfering with transmission between neurons.

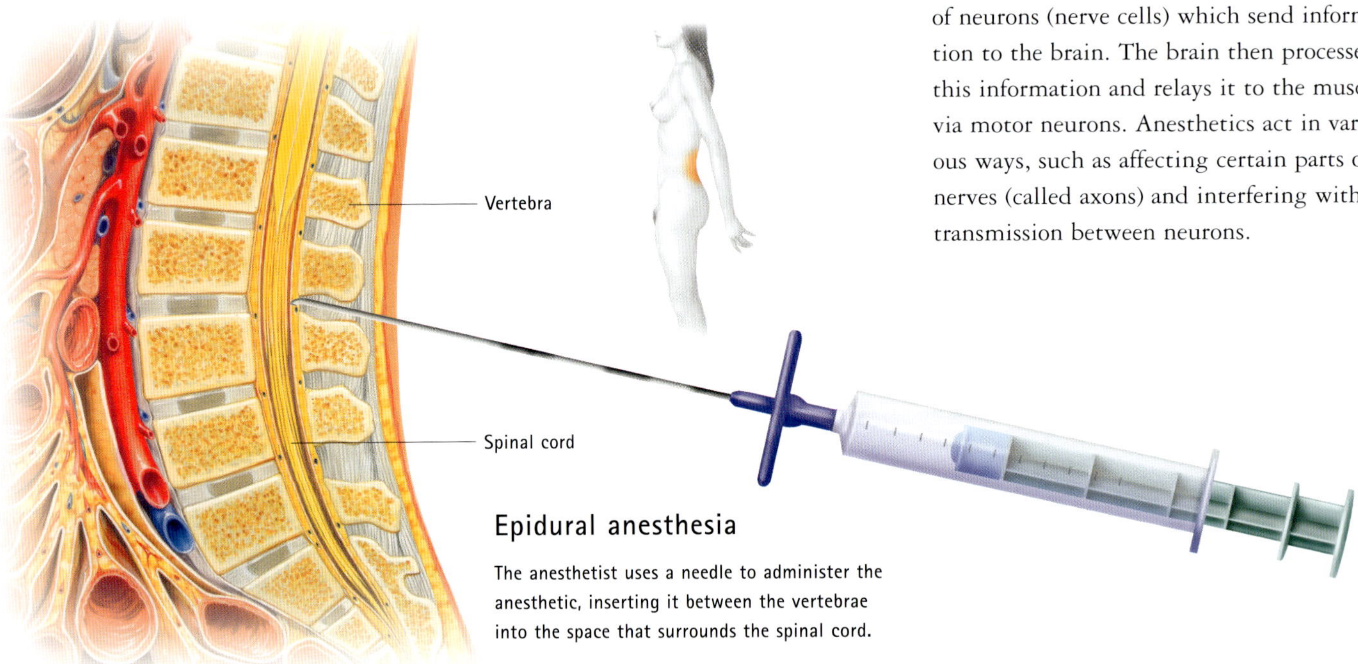

Epidural anesthesia

The anesthetist uses a needle to administer the anesthetic, inserting it between the vertebrae into the space that surrounds the spinal cord.

TYPES OF ANESTHETICS

There are three main types of drug-induced anesthesia: local anesthesia, which involves the numbing of a local nerve and can be administered in the form of a cream, injection or eye drops; regional anesthesia, which is the numbing of larger areas, for example an entire limb, and is done by a series of local anesthetic injections around a nerve or number of nerves; and general anesthesia, which renders the person totally unconscious and involves the inhalation of gas, the injection of drugs into the bloodstream or a combination of both.

When local anesthetics are applied directly to mucous membranes they are known as topical anesthetics—teething gels being an example. When a local anesthetic is injected into the space just outside the membranous sheath of the lumbar spinal cord (dura), it removes sensation from the lower part of the body, either partially or completely. This form of regional anesthesia is known as a lumbar epidural injection or epidural anesthesia and is used in childbirth. Spinal anesthesia is an injection into the cerebrospinal fluid in the lower spine.

Acupuncture uses fine needles inserted into key areas of the body to stimulate lines of energy and can provide relief, though how it works is still poorly understood.

Hypnosis can also be used to reduce the amount of pain a person is experiencing. Rather than anesthetizing, hypnosis is most useful in relaxing the patient prior to anesthesia and in improving the effect of the drugs, as well as reducing the dosage required. Acupuncture and hypnosis may reduce the need for other analgesia in childbirth. In the treatment of intractable pain, hypnosis is a useful tool and it is also sometimes used in dental treatment.

TENS (transcutaneous electrical nerve stimulation) is the application of a gentle current to the surface of the skin. It is a form of pain relief which is used for back, neck and joint pain, as well as arthritis, migraines and the pain of childbirth.

Traction

Traction is the application of a pulling force to a part of the body, a technique used for aligning fractured bones to allow proper healing, and also used in physical therapy to overcome muscle tension, or to relieve pressure on bulging or herniated disks of the spine. Traction is generally applied by using weights to exert a sustained pull in one direction for a set period of time.

If applied to the spine (lumbar traction), the therapist may recline the patient and attach a fixed harness to restrain the upper body. A second harness is then fitted around the hips or the upper thighs and weights or a machine is used to exert a carefully calculated degree of pull that is intended to straighten the spine and relieve pressure from the vertebrae on the disks.

In fractures of major bones such as the femur or pelvis, the patient may be confined to bed with traction being applied by weights attached to the limb or a cast to ensure the bones knit in exact alignment. Traction may be applied for for half an hour in physical therapy to several months in the case of severely broken or shattered bones.

Some forms of exercise, notably yoga, use postures which place muscles under traction to properly align the spine and extend the range of available joint movement.

Light therapy and melatonin

Produced by the pineal gland in the brain, the hormone melatonin has been linked to the regulation of sleep. Light therapy suppresses the secretion of melatonin and alters the body's circadian rhythm to produce an antidepressant effect.

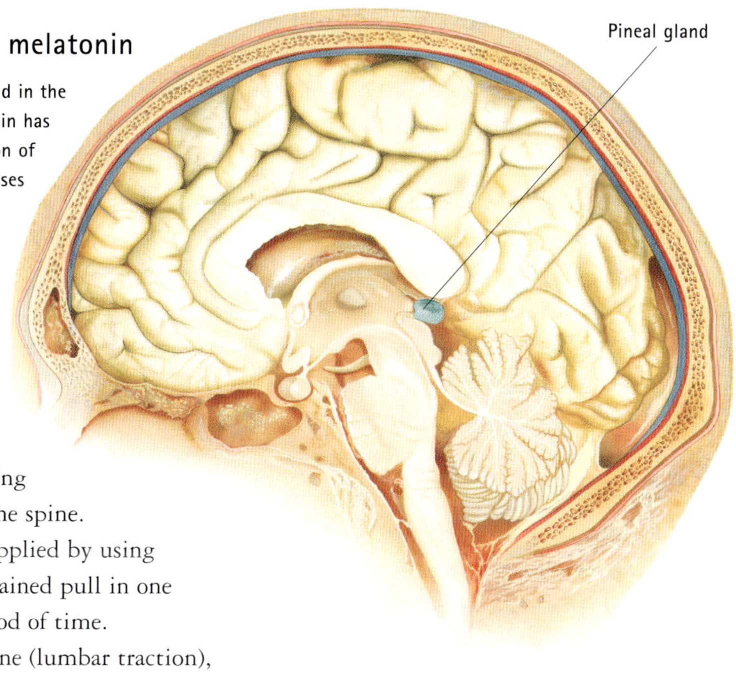

Electroconvulsive therapy

Electroconvulsive therapy (ECT) is a psychiatric treatment that involves bringing on a seizure by administering an electric shock to the brain. The patient is first anesthetized and given a muscle relaxant, then an electrode is placed on each temple and an electric current is passed through the brain between the two electrodes.

The patient may experience memory disturbance immediately after treatment, but this usually passes with time. Exactly how or why ECT works is still unclear but it is known to be effective in the treatment of certain types of depression.

ECT is probably the most controversial psychiatric treatment in use today, the controversy revolving around its effectiveness versus its side effects. The major side effect is the risk of prolonged memory loss and confusion in some patients.

Although ECT was widely used in the 1940s and 50s to treat cases of severe psychiatric illness, it was subsequently found to be relatively ineffective in many cases, particularly those involving psychosis. Today its use is more limited but it is still being used, usually in conjunction with medication, to successfully treat cases of severe depression and sometimes to interrupt manic episodes. Generally, a number of treatments are required over a period of several weeks for ECT to be effective.

Light therapy

Since 1981, intense artificial light therapy has been used in the USA and Europe for the treatment of seasonal affective disorder (SAD) or winter depression. Light therapy affects the body's circadian rhythms by lengthening the day with bright light.

Exposure to the light suppresses secretion of the night-time hormone melatonin and may sometimes enhance the effectiveness of serotonin and other neurotransmitters, producing an antidepressant effect in those whose depression is caused by a sensitivity to the reduction in sunlight.

Lamps used for this therapy produce a high intensity light, brighter than normal lighting (eight times brighter than normal lighting) and equal to standing outside on a sunny spring day. The individual with SAD sits in front of this lamp, usually for half an hour upon first waking. It is not necessary to stare at the lamp, but the patient may be advised to look at the lamp at regular intervals whilst carrying on with activities such as reading or watching television. Treatment for SAD commences in early autumn as the days begin to shorten and ends in spring when longer hours of daylight diminish the symptoms of SAD.

Light therapy is also used to treat sleep disorders, since exposure to light in the morning advances the circadian rhythm. This means that a person who cannot fall asleep until very late at night can fall asleep earlier. Exposure to light in the night delays the circadian rhythm, normalizing the sleep of those who are fatigued in the early evening and then wake early in the morning.

Light therapy can help shift workers and jet-lag sufferers to reset their body clocks. It may give SAD sufferers relief from irritability, fatigue, low energy levels and weight gain. With regular light therapy treatment symptoms may disappear and the length of the sessions may be reduced.

Ultraviolet therapy

Ultraviolet (UV) therapy is a form of light therapy (or phototherapy) used to treat skin disorders. The patient stands undressed in a specially designed cabinet containing fluorescent light tubes. Ultraviolet light is administered as either longwave UV light energy (UVA) or shortwave UV light energy (UVB). One form of this treatment, known as PUVA, combines UVA light with psoralen medication, which renders the skin more responsive to the therapy. Ultraviolet therapy is principally used in the treatment of psoriasis, as well as for dermatitis, vitiligo and cutaneous T cell lymphoma.

Radiation therapy

Radiation therapy (or radiotherapy) is the treatment of disorders using radiation. It is most widely used in the treatment of many different types of cancer. High-energy rays are used to damage cancer cells and stop them from growing and dividing. Similar to surgery, radiation therapy is a local treatment; it affects cancer cells only in the area which is being treated.

It may be used with other forms of treatment, like chemotherapy and surgery. Often it is used to shrink a tumor, which is then removed during surgery. It may also be used to provide a temporary relief of symptoms, or to treat malignancies which are not accessible to surgery.

The machinery used for radiation therapy is similar to x-ray equipment, but it contains a source of high-energy radiation, such as radium or a radioactive isotope of cobalt. Treatment is usually given on an outpatient basis in a hospital over several weeks. Patients are not radioactive during or after the treatment. Alternatively, the radiation can be administered via an implant, a small container of radioactive material placed directly into or near the tumor. In a few cases, some patients may receive both kinds of radiation therapy.

Radioimmunotherapy is a radiation therapy in which radioactive particles (also called radionuclides) are attached to antibodies. These antibodies are then introduced into the body, travel to tumor cells, attach to them and kill them. This technique has the advantage over other types of radiation therapy in that normal (non-cancer) cells are unaffected.

The side effects depend on the treatment dose and the part of the body that is treated. Tiredness, skin reactions (such as a rash or redness) in the treated area, and loss of appetite are the most common side effects. Sometimes the production of white blood cells may be suppressed. The side effects of radiation therapy can usually be treated or controlled and are not permanent.

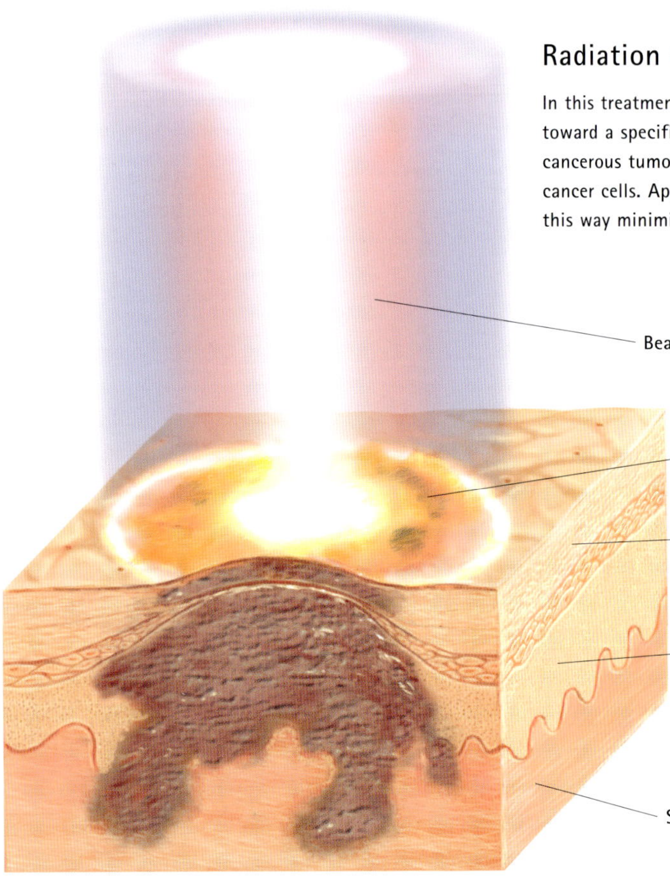

Radiation therapy

In this treatment, high energy rays are directed toward a specific area of the body (such as a cancerous tumor) in order to target damaged cancer cells. Applying localized radiation in this way minimizes damage to normal tissues.

Cryosurgery

Cryosurgery is a medical technique in which extreme cold is used to destroy tissue. A part of the body is rapidly cooled to minus 76°F (-60°C). The targeted tissue freezes and ice crystals form, breaking up and destroying the cell structure. Freezing can also stimulate the immune system to produce antibodies that attack diseased cells.

Cryosurgery is used for destroying skin cancers, warts, hemorrhoids, cataracts and difficult-to-reach internal tumors (such as in the kidney or brain). It is less invasive, less traumatic and involves less blood loss than traditional surgery. When treating growths on internal organs, liquid nitrogen is delivered through a small incision, thus avoiding major surgery.

Contact lenses

Contact lenses are artificial lenses worn on the eyes' surface to correct vision defects, commonly astigmatism, aphakia (absence of the eye's crystalline lens) and myopia (short or near sightedness). They are a popular alternative to glasses because in many cases they provide better vision, and some people think they are more attractive. Today's lenses are also comfortable and easy to wear and care for. However, they are not suitable for everyone: elderly people and those with arthritis often find them difficult to use.

The first lenses were made of glass in 1887, but today's lenses are made of plastic. They were originally made on a mold taken from an impression of the eye; now a measurement of the curvature of the cornea is made and the plastic lens sits on a cushion of tears covering the iris and the pupil. Lenses need to be individually prescribed and must be disinfected and cleaned at regular intervals. People who wear lenses require more frequent eye examinations than those who wear glasses.

There are two types of plastic lenses available: soft, which are the most common, and rigid. Soft lenses can be tinted and are quickly adapted to and comfortable to wear. They include disposable lenses, which are designed to be replaced every two to four weeks, and extended wear lenses which can be kept in the eye for up to thirty days. These are specially designed to allow a large amount of oxygen to pass to the eye and are not suitable for all. Rigid lenses, while they give the wearer better vision, require more adaptation and are typically less comfortable. On the other hand, they last longer.

Dental procedures and therapies

Dental procedures may be undertaken for cosmetic reasons, to combat tooth decay and disease, or to improve the functioning of the teeth and jaw.

SEE ALSO *Teeth on page 180*

Fillings

Fillings are metal, porcelain or plastic material used to replace decayed parts of teeth and to halt the growth of dental cavities. The process of filling a tooth may begin by drilling into the cavity to remove the decayed matter. Smaller cavities may not require drilling. The hole is then filled with gold, silver alloy, porcelain or plastic, depending on the desired effect. Plastic and porcelain look more like the natural tooth and may be preferred for fillings in front teeth. Metal fillings are usually stronger. Fillings may be done with the aid of local or general anesthetic, or nitrous oxide gas.

Root canal therapy

Root canal therapy is a dental procedure in which the center, or the pulp, of a dead or infected tooth is removed. The pulp includes nerves, blood vessels and lymphatic tissue which pass through the root canal. Root canal therapy is usually needed to avoid severe toothache and abscesses that can result from infection or trauma such as a blow to the mouth.

It may save a tooth that would otherwise be extracted. The empty tooth is usually filled with a plastic substance. A crown may be fitted to improve the strength of the tooth, which may become weak as a result of the treatment.

Orthodontics

Orthodontics is a branch of dentistry which is concerned with the diagnosis, prevention and correction of misaligned teeth and jaws. This procedure involves the fitting of braces, plates, retainers and other dental appliances to reposition and straighten teeth. The result is improved appearance, function and dental health.

Root canal therapy

A damaged tooth may be saved by root canal therapy. First the tooth is opened up to allow access to the abscess and all pulp is removed (a). Then the abscess and canals are cleaned and sterilized (b). Finally the canals are sealed with filling (c).

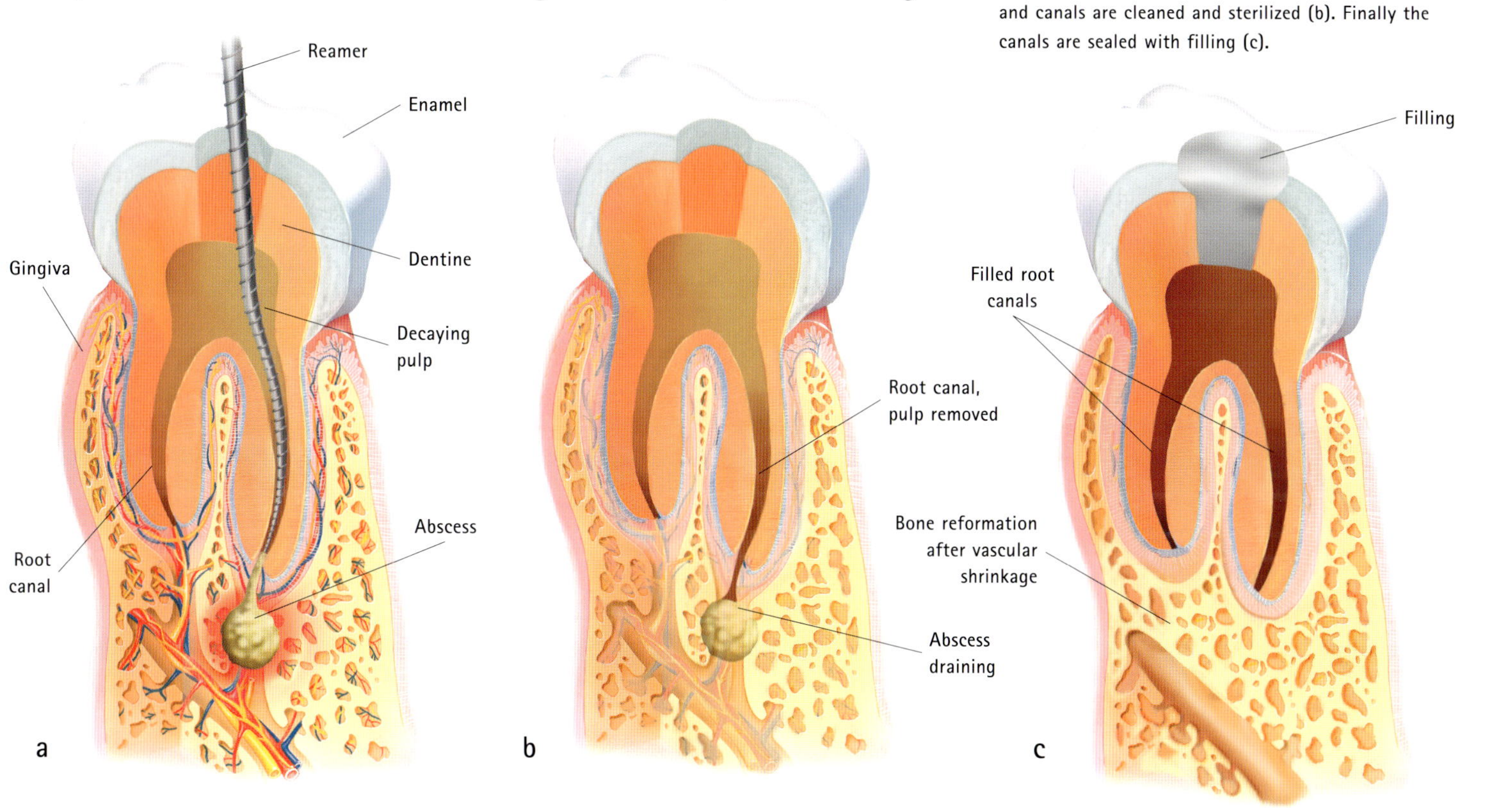

Orthodontics

The misalignment of teeth and jaws can be corrected by using orthodontic appliances. Many teenagers are fitted with braces to rearrange the position of their teeth within the jaw to obtain a more favorable even line of teeth, improving bite and appearance.

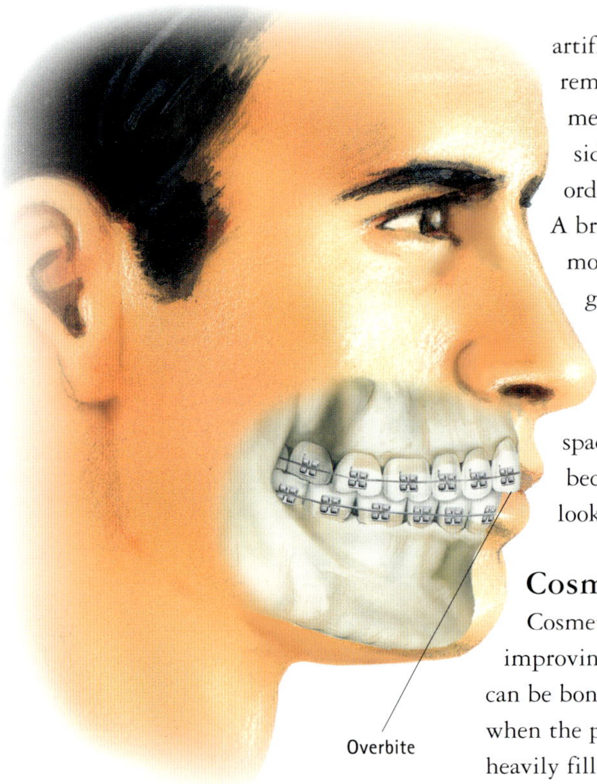

Most orthodontic procedures are performed on children whose jaws, unlike those of adults, are not fully developed. This early treatment makes it easier to manipulate the teeth and jaw and prevent malocclusion, or a bad bite, from developing or worsening.

When upper and lower teeth do not meet correctly, problems with biting and chewing can arise. Teeth may wear unevenly or excessively due to grinding or clenching. They may also become overcrowded, making cleaning difficult and increasing the risk of inflammation, gum disease and the loss of tooth. In some severe cases, where the teeth and/or jaw have developed incorrectly, or where they have suffered trauma, results can be infection, difficulty with speech and poor nutrition.

Aside from the physical aspects of poor dental development, many orthodontic procedures are carried out for cosmetic reasons as a means of improving self-confidence.

Common orthodontic problems

A common type of malocclusion is overbite, where the upper jaw juts outward past the bottom jaw, making teeth more prone to damage. An underbite is where the lower jaw protrudes further than the upper jaw and can wear down the front teeth. With a normal bite, the top teeth close over the bottom teeth slightly.

In some instances the top teeth may completely cover the bottom teeth. This is known as a deep bite and can damage the gum behind the top front teeth. Cross-bite is when the upper jaw is too narrow, causing the bottom jaw to swing to one side so that the teeth can meet.

An open bite occurs when some teeth don't meet and is a common cause of eating and speech problems. Orthodontic procedures may also be recommended in order to fill gaps made by missing teeth or where teeth or jaw are crooked or misshapen due to such things as thumb sucking in infancy.

Treatment

The orthodontist assesses dental problems with x-rays to decide how teeth and bone can be manipulated. Treatment may involve the extraction of teeth or surgical manipulation of the jaw. With children, assessment usually occurs at 8–10 years of age when most adult teeth have appeared. Treatment, usually involving braces, is a slow process during which teeth are gradually moved over a period of about 18–24 months. It is slower and perhaps a little more painful for adults as the fully developed adult jaw is more difficult to manipulate.

Braces usually consist of brackets fitted over individual teeth and connected by wires. Rubber bands attached to the braces, or an appliance fitted to the head, may also be needed. Every 4–8 weeks the orthodontist adjusts the wires. Traditional designs are made of stainless steel, although less conspicuous clear ceramic types are now available. When the braces are taken off, a removable appliance called a retainer, or plate, is worn for a certain period to help keep the teeth in the correct position.

Crown and bridge

These are two dental procedures, often used in combination, to fill a space created by one or more missing teeth. A crown is an artificial tooth or covering for the remains of a natural tooth, made of metal or porcelain. The teeth either side of a gap may be crowned in order to act as an anchor for a bridge. A bridge consists of false teeth on a mounting, and is often made of gold or porcelain on gold. It functions as normal teeth would and may prevent adjacent teeth from moving into an empty space. The crowns and bridge become one solid piece designed to look like individual natural teeth.

Cosmetic dentistry

Cosmetic dentistry involves repairing or improving the appearance of teeth. Teeth can be bonded with a tooth-colored plastic when the problem is chipped, stained or heavily filled front teeth or gaps that need closing up. Porcelain veneers can also improve the appearance of the teeth; and teeth can be whitened using a bleach. A missing front tooth can be replaced with a partial denture, bridgework or an implant. One or more missing teeth elsewhere may be replaced by a partial denture, either removable or fixed. Implants are artificial teeth, requiring surgery which can only be performed on people with healthy gums and adequate bone; implants require a commitment to meticulous ongoing hygiene.

Dentures

Dentures, also called (plates or false teeth, are composite artificial replacements of teeth and gums. A complete denture is required when all the teeth in the upper or lower jaw have been removed. Partial dentures are used to replace one or more missing teeth. Complete dentures are usually removable; partial dentures may be removable or fixed. An overdenture is a partial denture that takes support from any remaining teeth roots and gum near the teeth that are missing.

Titanium implants can be surgically inserted into the jawbone; replacement teeth are attached later, providing a fixed denture. This prevents resorption, which is the shrinkage of the jawbone with loss of nerve and surrounding tissue, a common situation where all the teeth have been removed.

Surgical procedures

Manual operations can be the most effective method of treatment for a variety of diseases and injuries and can also be used to correct or improve the functioning of the body. Surgical procedures are continually being improved and refined.

Plastic surgery

Plastic or reconstructive surgery is a versatile specialty which includes the correction of disfigurement, restoring impaired function and improving physical appearance (cosmetic surgery). Reconstructive surgery is performed on abnormalities caused by congenital defects, developmental abnormalities, trauma, infection, tumors or disease. One of the possible sources of the term is the Greek word *plastikos*, which means "to mold or give form."

The driving force behind most developments in plastic surgery was war. World War I (1914–18) resulted in outstanding developments in plastic surgery. At this time doctors were required to treat many extensive facial and head injuries, ranging from shattered jaws and blown-off noses to gaping skull wounds. At first this specialty did not have any formal training; then in the 1930s the American Society of Plastic and Reconstructive Surgeons was launched. In the 1940s, during World War II, the skills of plastic surgeons were again in great demand.

In the 1960s silicone emerged as a tool for plastic surgeons. Initially it was used to treat skin imperfections. However, in 1962 it was first used by an American surgeon, Thomas Cronin, in a breast implant device. It was not until 1990 that concerns about silicone implants came to a head and patients began to sue manufacturers, leading to the ban of silicone breast implants in the USA in 1992.

Plastic surgery repairs or reshapes tissue structures, removes tissues and grafts or transfers tissue. Grafting is often used to treat cases of trauma such as severe burns, injuries from automobile accidents and gunshot wounds. If the circumstances are favorable, plastic surgeons may also reattach a severed body part. Congenital deformities—commonly cleft lip and palate—can also be corrected by plastic surgery.

Transplant surgery

Organ, tissue and limb transplant surgery, or grafts, means permanently transferring a tissue or organ from one part of the body to another, or from one person to another. The first records come from the sixth century BC, when Hindu doctors used skin flaps from the patient's own arm to repair damage to the nose. Kidneys were the first organs used in person-to-person transplants and are now the organ that is most commonly transplanted.

Current surgical techniques also make it possible to graft corneas, teeth, heart, lungs, liver, pancreas, cartilage, bone marrow, brain tissue, blood and skin to one person from another. More recently hip and knee joints have been transplanted, and even hands. Transplants of animal donor tissues to humans (xenotransplants) have also been made, and have included chimpanzee and baboon hearts, and heart valves and fetal brain cells from pigs. Some of these are highly experimental and are very cotentious, largely due to the possibility of serious viral infections.

Autografts, where donor and host are the same person, pose no threat of tissue rejection, nor do grafts from one identical twin to another. A graft from any other source will stimulate rejection by the host because of the differences in histocompatibility antigens on the cell surfaces of donor and host tissues, which can quickly kill the implant. Immunosuppressive drugs are used to counter this reaction.

Organ grafts are usually a long-term solution undertaken to sustain life in a patient who is chronically ill. Skin grafts may be short-term measures to prevent fluid loss and infection at burn sites while host tissue regenerates.

People who have decided that in the event of their death their organs can be used as transplants are referred to as organ donors. This information may be recorded on a card or driver's license. Organs are removed after death and transported to hosts, patients on a waiting list in priority of need. Transplant organs go to the patient on the waiting list who is nearest death, priority being given to those in whom earlier transplants have failed.

Transplant surgery is a very costly and uncertain procedure not funded by many private health insurers, and ethical questions stimulate much debate about the procedure. A black market in organs exists in some developing countries.

Prosthetic surgery

In prosthetic surgery an artificial body part is surgically implanted as a substitute for an existing diseased body part. Prosthetic

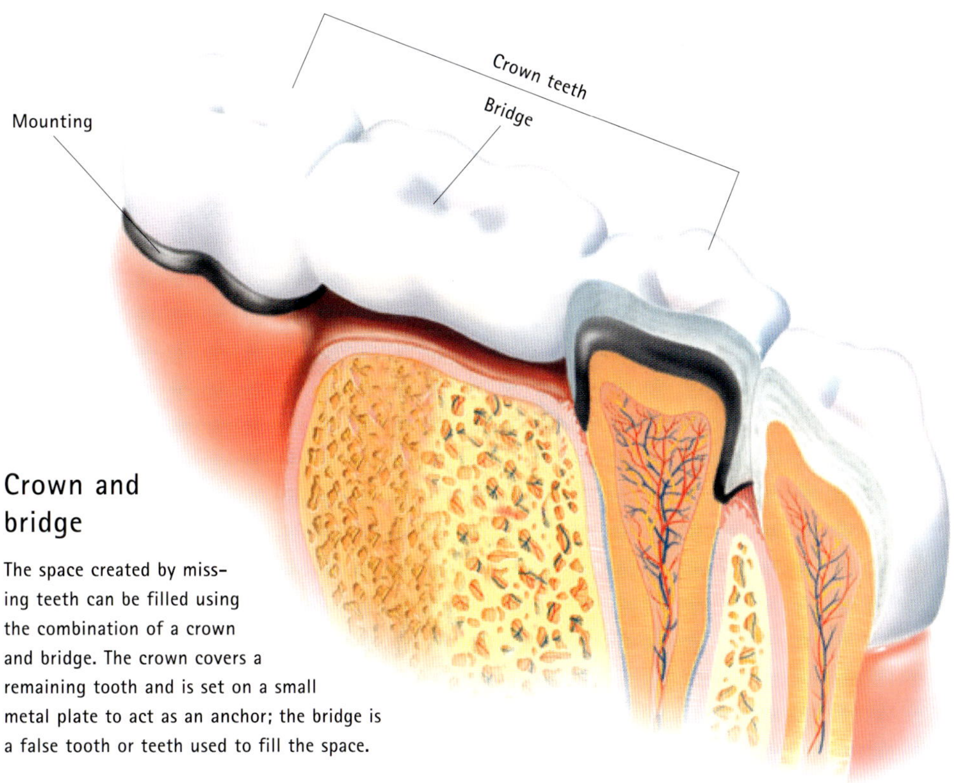

Crown and bridge

The space created by missing teeth can be filled using the combination of a crown and bridge. The crown covers a remaining tooth and is set on a small metal plate to act as an anchor; the bridge is a false tooth or teeth used to fill the space.

replacement of various joints in the body is becoming increasingly common. The hip and knee joints are the most commonly replaced but there are now prostheses for most joints in the body.

Diseased heart valves may be replaced with prosthetic valves. The choice of artificial valve rests with the surgeon, but the Starr-Edwards ball and cage valve is still probably the most widely used. The principle complication of prosthetic valve replacement is thromboembolism. Infective endocarditis and hemolysis due to trauma to red blood cells may also occur.

Microsurgery

Microsurgery is a surgical method in which a surgeon uses a microscope and specially adapted hand-held miniaturized instrments to operate on tiny structures, such as very small blood vessels and nerves. It is used for operations that require extreme delicacy, such as plastic surgery, reconstruction and transplantation. Microsurgery is quite frequently used in surgery of the ears, the eyes, or of the brain.

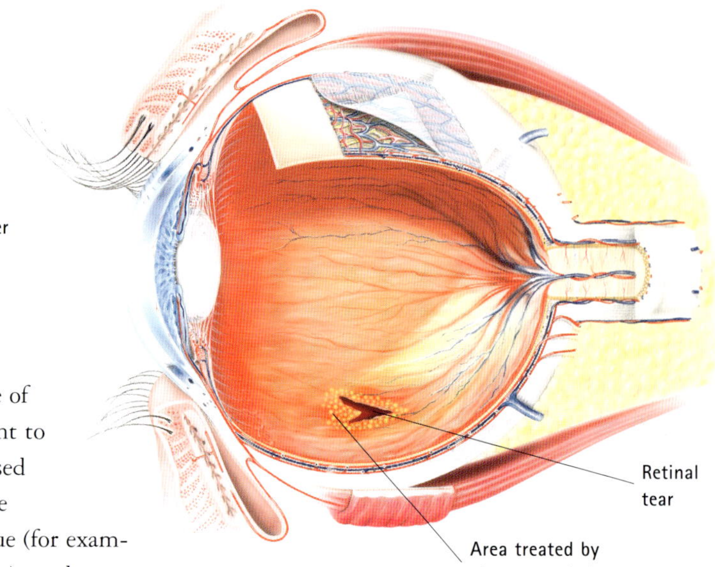

Photocoagulation

In cases of retinal detachment, photocoagulation can be used to "weld" the edges of the torn retina to the deeper layers of the eyeball and prevent further detachment.

Electrosurgery

Electrosurgery is the use of repeated electrical current to destroy healthy or diseased tissue. It is used to make incisions in healthy tissue (for example, to excise skin lesions), to destroy diseased tissue such as tumors, or to coagulate bleeding blood vessels during surgery.

There are several electrosurgery techniques, all of which use the principle that as electrical current is converted into heat by tissue resistance, it destroys tissue.

Electrosurgery has largely been replaced by laser surgery, which is more accurate and can be used in areas where electrosurgery cannot, for example in eye surgery.

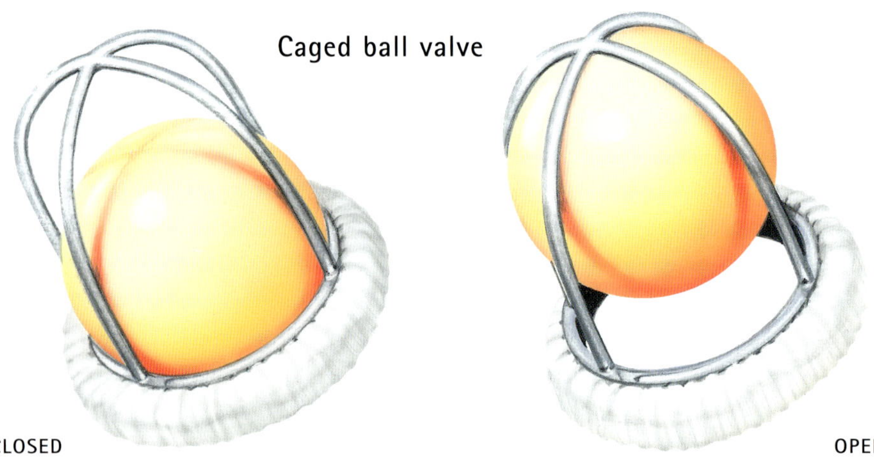

Valve replacement

Artificial valves can be used to replace heart valves that have been damaged by disease. These artificial valves are made from metal and plastic and may use a caged-ball or tilting disc mechanism.

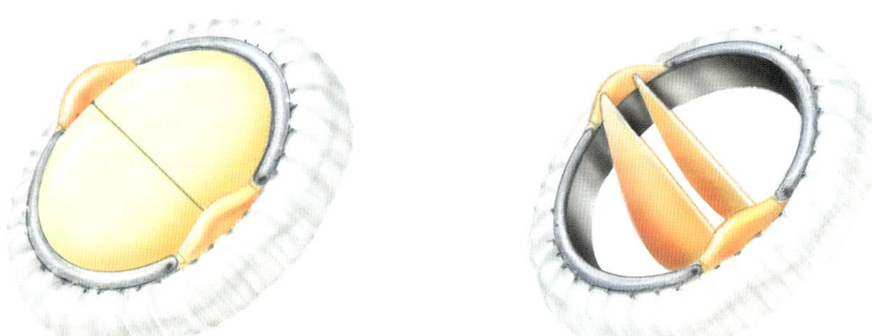

Photocoagulation

Photocoagulation is a form of laser surgery to correct certain eye disorders. An intense beam of light is aimed into the eye to destroy damaged blood vessels and seal leaking ones. Photocoagulation takes about 30 minutes, is relatively painless and is performed by an ophthalmologist.

Cholecystectomy

Cholecystectomy is the surgical removal of the gallbladder. It is a common treatment in severe cases of pain due to gallstones or acute cholecystitis.

Traditionally, cholecystectomy involved removal of the gallbladder through a surgical incision in the abdomen under general anesthesia. However, laparoscopic cholecystectomy has now largely replaced it. In this procedure, the gallbladder is surgically removed through a small incision in the abdomen with the aid of a fiberoptic tube called a laparoscope. It can be performed under local anesthetic, and greatly reduces the hospital stay and recovery time, although there can often be more complications.

Gastrectomy

Gastrectomy is the surgical removal of part or (in some cases) all of the stomach. It is usually used for the treatment of gastric cancer, and gastric ulcers that do not respond to any other type of medical treatment. Partial gastrectomy involves the removal of the lower parts of the stomach, while keeping other parts (the fundus and cardiac parts) intact.

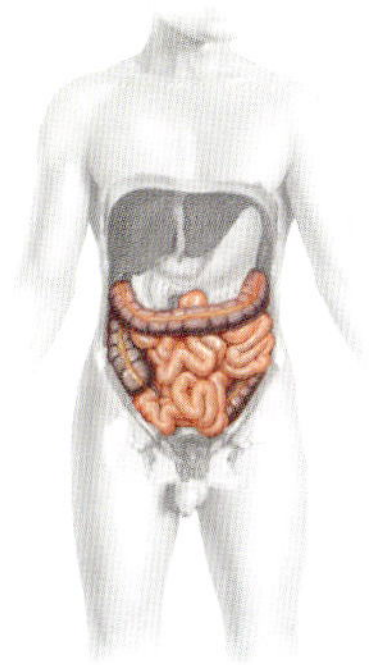

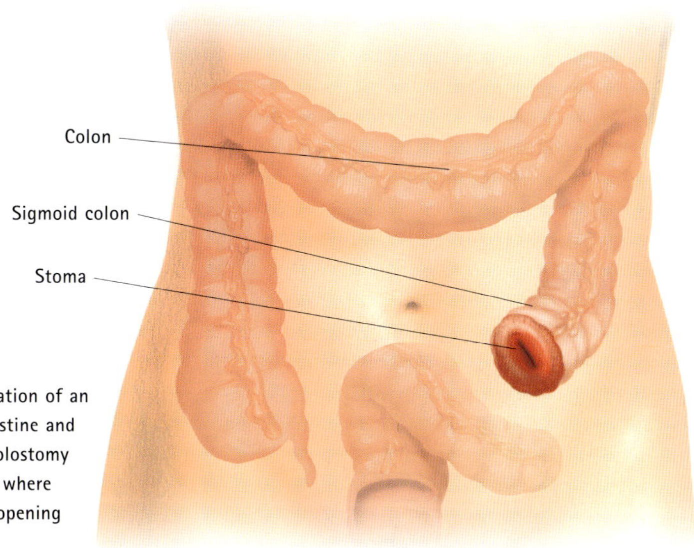

Colostomy

A colostomy is the surgical creation of an opening between the large intestine and abdominal wall. A removable colostomy bag is attached to the opening where the colon meets the skin. This opening is known as a stoma.

Colostomy

A colostomy is a surgical procedure in which the interior of the bowel is opened and brought to the surface of the body. The resulting opening is called a stoma. Colostomies may be used to relieve pressure in an obstructed bowel, to divert the stream of feces in preparation of the bowel for surgery, to allow for removal of feces from the lower bowel when a portion is removed, and to protect surgically repaired bowel further down the gut. Colostomies may be temporary or permanent and may be performed by opening one side or one end of the bowel onto the skin surface.

The most common type of permanent colostomy involves the sigmoid part of the colon, which is located in the pelvis. This is usually performed at the time of removal of the rectum for cancer. Usually patients can eat the same types of food which they enjoyed before colostomy, except that fruits may cause diarrhea.

Sterilization

Sterilization is a permanent surgical form of contraception involving vasectomy in men and tubal ligation or tubal sterilization in women, the latter often referred to as tying of the tubes. These procedures are chosen by couples who do not want to have more or any children, and who prefer to avoid drug-based forms of birth control or those that carry a higher risk of pregnancy. It is estimated that as many as 100 million couples worldwide have opted for sterilization as a method of birth control. More female sterilizations have been done than male, but demand is high for both.

The advantages of permanent protection against pregnancy, with no effect on sexual pleasure and no lasting side effects, may outweigh any costs or the possibility of later regrets. As both of these operations should be regarded as permanent, it is a good thought for anyone contemplating sterilization to receive some counseling.

While reversal of these operations is possible, the success rates differ. Although microsurgery is improving all the time, sterilization is not a procedure that should be undertaken if the individual has any thoughts of producing children later in life.

Vasectomy

Vasectomy is a surgical operation which involves cutting the two tubes (each called the ductus deferens or vas deferens) which run from the testes to the urethra. This prevents sperm reaching the seminal vesicles where they are held prior to ejaculation through the urethra during orgasm.

Vasectomy is a simple procedure that usually requires only a local anesthetic. There is some local bruising and discomfort but recovery takes only a few days. The testes continue to produce sperm but they die and degenerate in the ductus deferens or in the epididymis.

The other secretions which make up the seminal fluid are still produced and thus the man continues to ejaculate. After the operation it takes 16–20 ejaculations to expel all the stored sperm and for the man to be declared infertile.

Reversal of the operation requires a general anesthetic and takes up to 2 hours, with recovery time related to the period spent anesthetized. About 60 percent of reversals succeed, enabling their subjects to father children. The best chances of success come with reversals performed less than 5 years after the original vasectomy and where the female partner is under 30 years of age and normally fertile.

Factors mitigating against success include the growth of fibrous tissue in the spermatic tubes and the production of antibodies to sperm after the vasectomy. Sperm may be frozen prior to vasectomy, for future use.

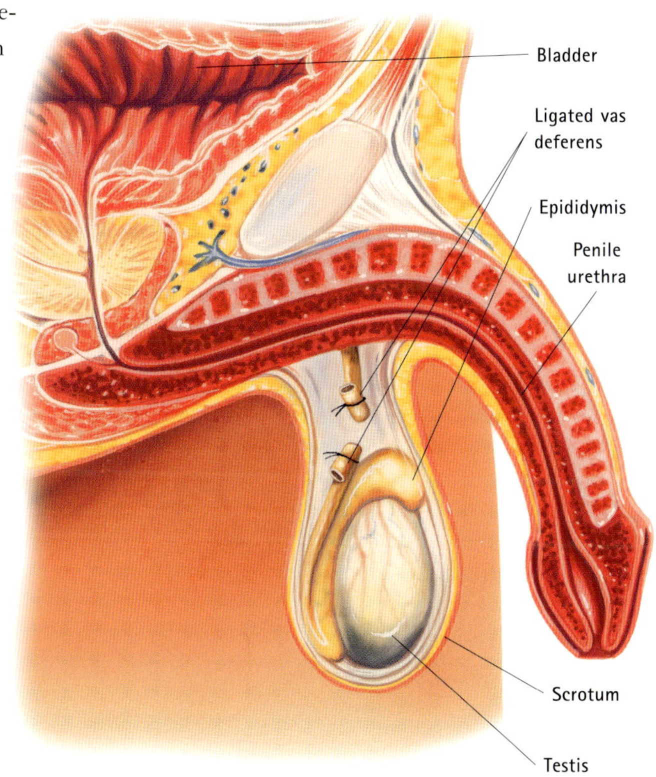

Vasectomy

Usually performed under local anesthetic, a vasectomy involves cutting the vas deferens tube that carries sperm from the testes to the urethra. Sterilization can be reversed if the tubes are rejoined, but this is not always possible.

Vasectomy is the most effective form of contraception for men and may be done voluntarily for this purpose. It may also be done for prophylactic reasons where there is infection of the epididymis or testes.

Laryngectomy

This operation involves the surgical removal of the voice box (larynx), and is most commonly performed when the larynx has been irretrievably damaged by injury or invaded by cancer. The patients most at risk of needing a laryngectomy include heavy smokers and drinkers.

Air travels past the vocal cords while exhaling, causing them to vibrate and cause sounds. However, after a laryngectomy the passage of air between lungs and mouth cannot occur as the connection between the mouth and windpipe no longer exists.

The operation leaves an opening from the windpipe to the outside. This permanent opening, or stoma, allows air to pass directly into the trachea and on to the lungs. Sometimes a tracheostomy tube is inserted in the stoma temporarily to help it stay open during the healing process.

It is important that the stoma does not become blocked as it is the only existing air passage. In most cases the laryngectomy procedure does not affect the swallowing of food and liquid as the connection between mouth and esophagus remains. After a laryngectomy, laryngeal speech is no longer possible. In the past, patients who underwent the procedure were taught to "speak" again by trapping air in the eso-phagus. Patients now often speak through the use of electronic devices that are held against the throat. Artificial vocal cords may also be surgically implanted.

Tracheostomy

Tracheostomy is a surgical procedure in which an opening in the front of the windpipe (the trachea) is created to maintain a clear airway. A silicon tracheostomy tube is then placed in the opening. A tracheostomy is performed on someone who has had a mouth or chest injury or who has undergone a major operation, such as lung surgery. It is also performed on someone who needs to be on a mechanical ventilator for an extended period of time. An emergency tracheostomy may be needed after an accident or injury if the normal airway is blocked by swelling or by blood.

Female procedures

A variety of procedures are available to treat disorders or conditions that affect women. These procedures focus on treating the breasts and the female reproductive and urinary organs.

SEE ALSO *Female reproductive organs on page 303; Female reproductive system on page 104; Urinary system on page 98*

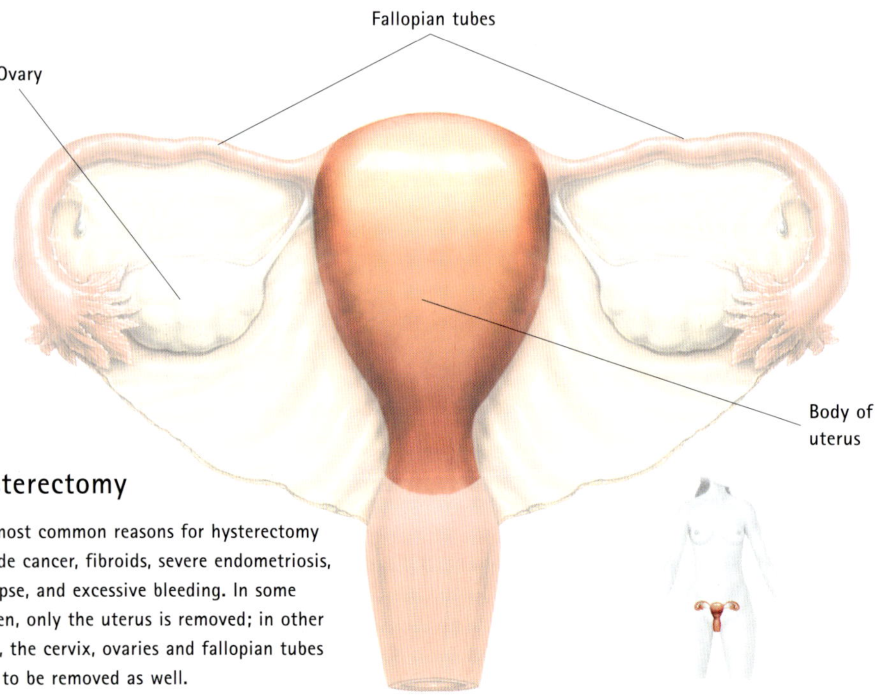

Hysterectomy

The most common reasons for hysterectomy include cancer, fibroids, severe endometriosis, prolapse, and excessive bleeding. In some women, only the uterus is removed; in other cases, the cervix, ovaries and fallopian tubes need to be removed as well.

Hysterectomy

Hysterectomy is the surgical removal of the uterus and it is the second most common operation performed on women in the industrialized world after dilation and curettage. It may be performed for a number of reasons. These include cancer of the uterus, the cervix or the ovaries; benign tumors, such as large fibroids; extreme cases of endometriosis; a severe prolapse of the uterus; excessive blood loss that is not responding to treatment; and, very rarely, after childbirth or gynecological surgery.

There are many forms of this operation. A total hysterectomy is the removal of the entire uterus and the cervix. A subtotal hysterectomy removes the uterus but not the cervix. A radical hysterectomy removes the uterus and the associated lymph glands in the pelvis. A hysterosalpingo-oophorectomy removes the uterus, including the ovaries and tubes on both sides.

The operation can be performed through an abdominal incision or through the vagina; the latter method is now more common. A laparoscope, a narrow tube with a fiber-optic light on one end, is often used in the operation. It is inserted through a small incision just below the navel and enables the surgeon to view the reproductive organs and to operate using small instruments guided by the laparoscope.

Risks include infections and damage to other organs. An abdominal hysterectomy is major surgery which will require post-operative painkillers and a convalescence period of around six weeks. Vaginal hysterectomy causes less post-operative pain and requires a shorter convalescence. Laparoscopic hysterectomy, on the other hand, has more complications but has a very short recovery time.

Once the woman has recovered from the operation, hysterectomy will not affect her ability to have sexual intercourse and may even improve her quality of life. Side effects include no more menstrual periods and no risk of pregnancy. A woman who must have a hysterectomy before she has reached menopause will need hormone replacement therapy, if her ovaries are not left in place.

Fewer hysterectomies are performed now than during the 1970s, when there was some controversy over unnecessary operations.

Oophorectomy

Oophorectomy means surgical removal of one or both ovaries (unilateral or bilateral oophorectomy). Ovaries are removed for a variety of reasons, including extensive or very large cysts, or ovarian cancer. Oophorectomy may sometimes accompany hysterectomy, even though the ovaries are not diseased. It has been argued that removing the ovaries avoids possible future problems of cysts or ovarian cancer.

However, the advantage of leaving at least one ovary in premenopausal women is the avoidance of symptoms of premature menopause and the possible need for hormone replacement therapy.

Myomectomy

Myomectomy is an operation to remove a fibroid, or tumor, formed of muscle tissue in the uterus. It is performed instead of a hysterectomy when it is important to preserve the woman's fertility. About one quarter of women who would formerly have had a hysterectomy now have a myomectomy, which leaves the uterus in place and is not a major operation.

This operation can be done by vaginal incision, or abdominal incision, which may mean that a cesarean is necessary should the woman later decide to have children.

Dilation and curettage

Often called D & C, dilation and curettage is the most commonly performed gynecological procedure. It involves the surgical opening (dilation) of the cervix (the neck of the uterus), and the removal of the contents of the uterus with a curette, an instrument with a long handle and an end shaped like a spoon. This is used to obtain tissue from the uterus lining (endometrium) for examination or to remove fragments of placenta after a miscarriage. The procedure can be performed under a general or a local anesthetic and has few side or after effects.

Abortion

An abortion is a pregnancy that ends prematurely with the loss of the embryo or fetus, either spontaneously or by artificial induction. The word "abortion" refers to artificial induction, while a spontaneous abortion is generally called miscarriage.

Abortion is an issue that raises strong passions. Those supporting abortion argue that it should be made legally available to women who choose to have it, and that refusing to do so often leads to women trying various unsafe ways of aborting the fetus themselves. Those against abortion argue that as soon as an embryo is conceived it is a life, and that to abort it is to take this life.

The decision to have an abortion is never free of conflict, and counseling is very important. When antenatal testing reveals that the fetus has a lethal abnormality such as anencephaly, or a defect that can be a major disability such as spina bifida with hydrocephalus, termination is acceptable to many parents. Some women choose not to have an antenatal test if their beliefs would not allow them to have an abortion; or they may decide not to abort a fetus even if it has a high risk of a birth defect.

In any such situation, difficulties can arise and counseling is advisable, and imperative when defects with less predictable outcomes are detected.

TYPES OF ABORTIONS

There are several methods of performing an abortion. The method used generally depends on how far the pregnancy is advanced, with the duration of the pregnancy being calculated from the date of the woman's last menstrual period.

Surgical evacuation—removing the contents of the uterus through the vagina—is the method used in over 95 percent of abortions and almost always for pregnancies of less than 12 weeks. The technique is known as suction aspiration or suction curetting. The cervix is dilated and the lining of the uterus, containing the embryo, is drawn out by suction. A curette may also be used to scrape the uterine walls.

For pregnancies over 12 weeks and under 20 weeks, dilation and evacuation (D&E) is used. When the cervix is dilated enough, forceps and a combination of suction and curettage remove the fetus. This can be done under general or local anesthetic.

A hysterotomy, the cutting open of the uterus to remove the embryo or fetus, is the same procedure as a cesarean delivery and is used only when the pregnancy must be terminated quickly in order to save the mother or when other methods are impossible.

Drugs such as mifepristone (RU-486) and prostaglandins are used in some cases, especially after 16 weeks of pregnancy, though mifepristone may be used shortly after conception. Taken as a tablet, it is followed 48 hours later with a prostaglandin and the abortion starts soon after this. In over 95 percent of cases it is completed quickly with minimal blood loss. If unsuccessful, surgical termination is necessary.

COMPLICATIONS

Induced abortion before the twelfth week of the pregnancy and performed by a skilled doctor in a safe environment is a safe procedure. However, as with any surgical procedure, there are possible complications and

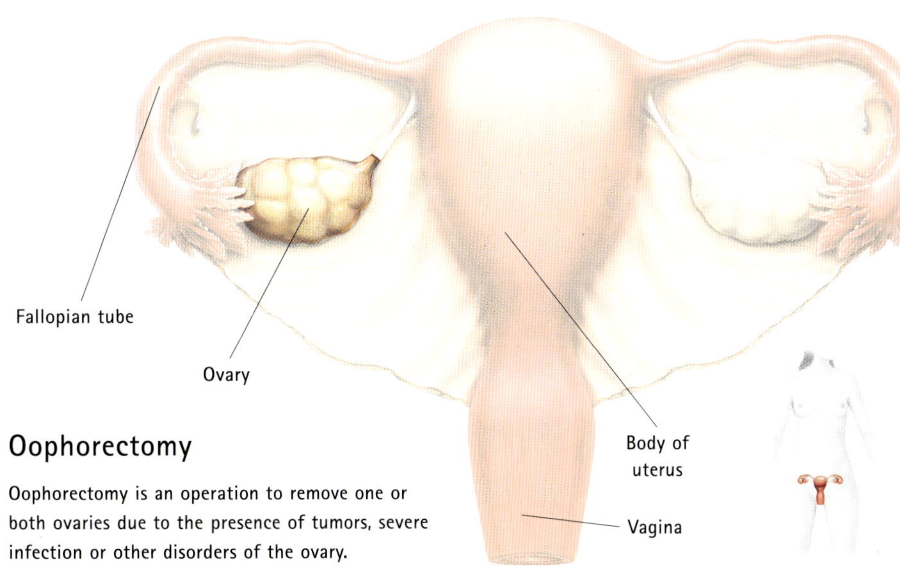

Oophorectomy

Oophorectomy is an operation to remove one or both ovaries due to the presence of tumors, severe infection or other disorders of the ovary.

some risks. The probability of complications increases with the duration of the pregnancy.

Complications include: infection (many clinics routinely prescribe antibiotics to reduce this risk and women are advised of the symptoms, as immediate medical attention is needed); incomplete abortion (this will require a further curettage); hemorrhage; perforation of the uterus (a rare complication when the operation is performed by a skilled doctor); and tearing of the cervix (which will require stitching). Occasionally, the abortion is unsuccessful and the pregnancy continues. Bleeding, similar to a menstrual period, is normal within a few days of the abortion and a checkup is part of the normal procedure.

Many women find that an abortion can have psychological repercussions. These are less common when counseling is part of the procedure.

Artificial insemination

The introduction of semen into the vagina or cervix of a female by any method other than sexual intercourse is known as artificial insemination. In this procedure, a solution is used to impregnate women whose partners are sterile or impotent. Apart from sterility or impotence problems in the male, artificial insemination can also be an option when the female's cervical mucus is unreceptive, or if the infertility is unexplained.

Semen may be freshly obtained or it may be frozen. It may be from the partner (if he is not impotent) or from some other male donor (if the partner is sterile), in which case it may be described as donor insemination.

If freshly obtained from the partner, he needs to masturbate and ejaculate in the hour before the insemination is scheduled to take place. Frozen sperm can be kept for long periods and used, for example, if the partner is about to undergo medical treatment, often for cancer, which could render him infertile.

At around the time of ovulation, determined by checking her temperature or cervical mucus, the woman goes through a simple procedure in which the sperm is introduced by a syringe into her vagina or cervix. The technique has been reasonably successful, with 50 to 65 percent achieving conception and pregnancy.

Tubal sterilization

Tubal sterilization, also called tubal ligation, is the surgical closing of the fallopian (uterine) tubes to prevent future pregnancies. The fallopian tubes allow passage of sperm cells to the ovum and movement of the fertilized ovum to the body of the uterus for implantation and development into a baby. Removal of a length of the fallopian tube, combined with clipping or tying of the cut ends, will prevent sperm reaching the ovum, thus sterilizing the woman.

The procedure is performed in a hospital under general anesthia by a gynecologist usually via a laparoscope (an optical instrument) inserted through an incision near the navel and into the abdominal cavity, or through the vaginal vault. The tubes, once found, are blocked by rings or clips, or by burning with diathermy or laser. A minilaparotomy involves a small incision just below the pubic hairline.

After sterilization, the ovaries will continue to release ova, but sperm will not be able to reach them. The menstrual cycle is not affected by the procedure, with periods and menopause still occurring. There are usually no side effects after recovery from the operation, although some women have reported heavier than usual periods, strong cramps, recurrent abdominal pain and pelvic infections. Recovery is quick, allowing the patient to return home the same day as the procedure. Strenuous exercise is best avoided for a few days, after which time work may be resumed.

The failure rate of tubal ligation is around two per 1,000. This may be due to tubes joining themselves back together and, on rare occasions, may result in a potentially dangerous ectopic pregnancy where a fertilized ovum embeds in a fallopian tube or somewhere else outside the uterus. This procedure may be reversed (with difficulty) by reparative microsurgery; it is suitable only for women who are certain they wish to permanently prevent future pregnancies.

Mastectomy

Mastectomy is the surgical removal of the breast to treat diseased breast tissue, usually cancer. There are many ways to remove diseased breast tissue, depending on the type of tumor, its size, how fast it has grown, how widely the cancerous cells have spread, and the patient's general health.

Lumpectomy is the surgical removal of a lump; simple mastectomy is the removal of the breast. Radical mastectomy is the removal of the breast and the surrounding lymph nodes, muscles, fatty tissue and skin. Modified mastectomy is the removal of the breast and part of the muscles. A breast implant (also called prosthesis) can be inserted during surgery (except in the case of radical mastectomy).

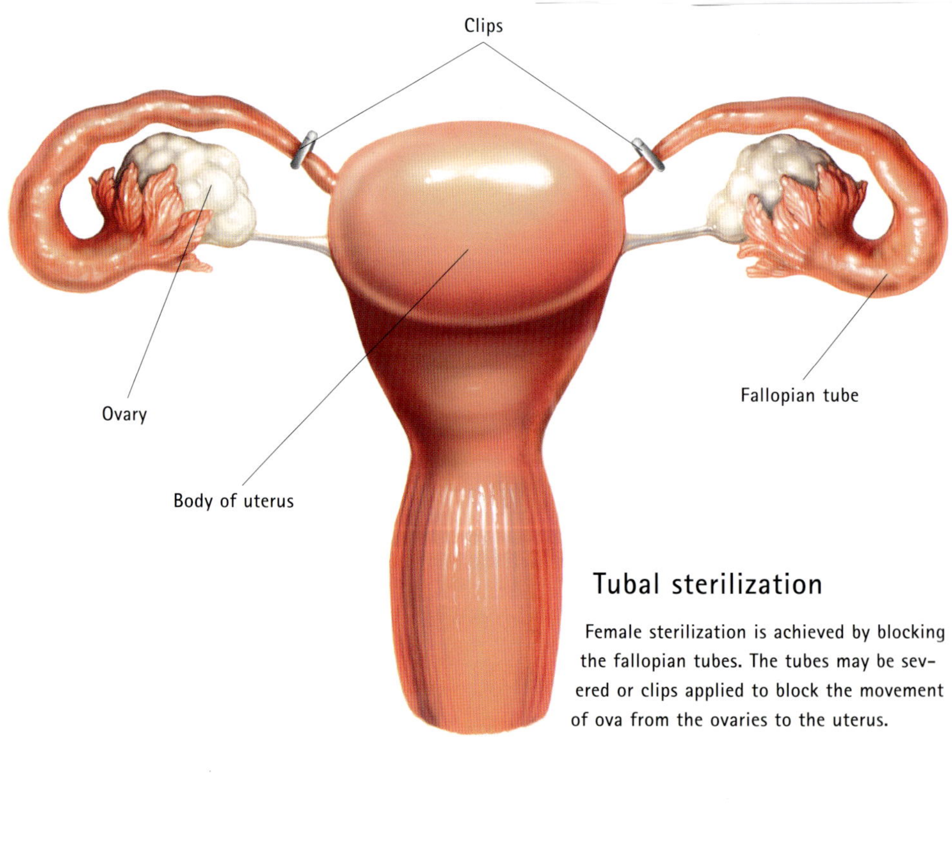

Tubal sterilization

Female sterilization is achieved by blocking the fallopian tubes. The tubes may be severed or clips applied to block the movement of ova from the ovaries to the uterus.

Symptoms table

This section provides information on common symptoms and associated diseases. You may find it useful to refer to this symptoms table if you would like to find out more information on a particular symptom, and what it may mean.

This symptoms table does not replace your doctor, and should not be used as a guide to diagnosing medical conditions in yourself or a family member. Refer to your doctor or other appropriately qualified health professionals if you believe you or another person may have one of the diseases described.

How to use this table

Symptoms are listed alphabetically, and usually contain the name of the affected body part. For example, if you would like information on abdominal problems, refer to Abdominal distension to find entries on distension and other problems relating to the abdomen.

Each entry includes a list of descriptions of various ways in which the symptom may be experienced. Look down the Characteristic symptom column until you find a description that best matches the symptom in which you are interested.

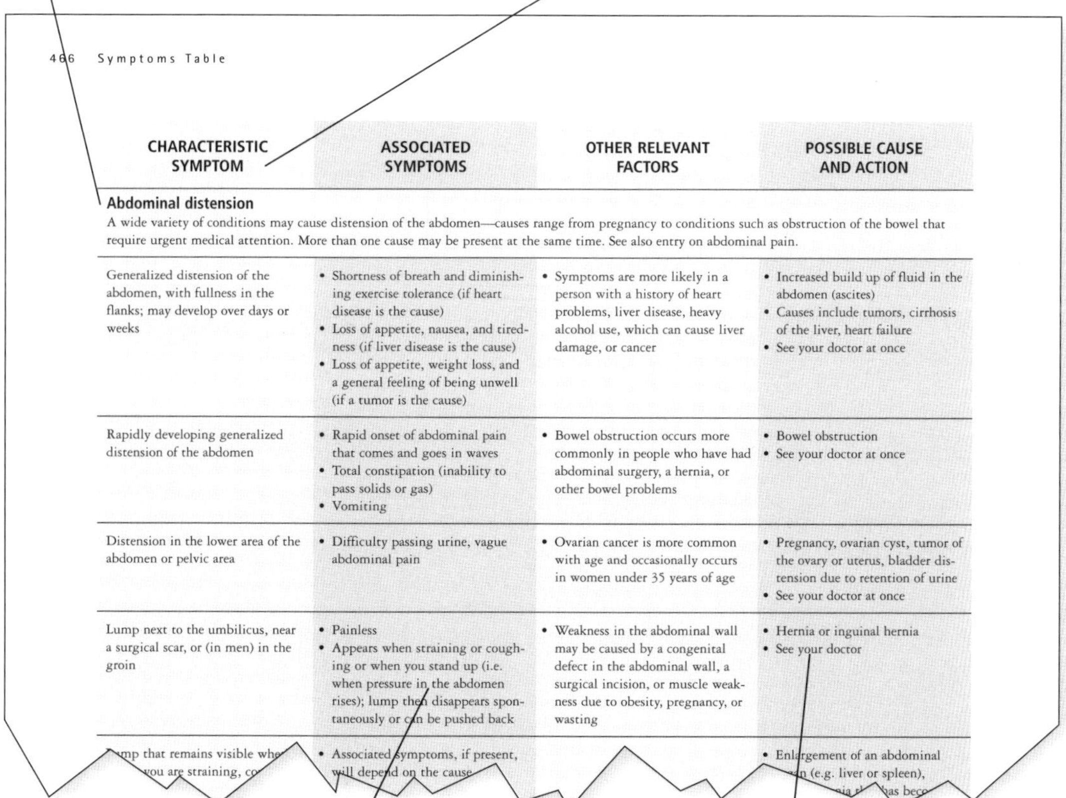

466 Symptoms Table

CHARACTERISTIC SYMPTOM	ASSOCIATED SYMPTOMS	OTHER RELEVANT FACTORS	POSSIBLE CAUSE AND ACTION
Abdominal distension A wide variety of conditions may cause distension of the abdomen—causes range from pregnancy to conditions such as obstruction of the bowel that require urgent medical attention. More than one cause may be present at the same time. See also entry on abdominal pain.			
Generalized distension of the abdomen, with fullness in the flanks; may develop over days or weeks	• Shortness of breath and diminishing exercise tolerance (if heart disease is the cause) • Loss of appetite, nausea, and tiredness (if liver disease is the cause) • Loss of appetite, weight loss, and a general feeling of being unwell (if a tumor is the cause)	• Symptoms are more likely in a person with a history of heart problems, liver disease, heavy alcohol use, which can cause liver damage, or cancer	• Increased build up of fluid in the abdomen (ascites) • Causes include tumors, cirrhosis of the liver, heart failure • See your doctor at once
Rapidly developing generalized distension of the abdomen	• Rapid onset of abdominal pain that comes and goes in waves • Total constipation (inability to pass solids or gas) • Vomiting	• Bowel obstruction occurs more commonly in people who have had abdominal surgery, a hernia, or other bowel problems	• Bowel obstruction • See your doctor at once
Distension in the lower area of the abdomen or pelvic area	• Difficulty passing urine, vague abdominal pain	• Ovarian cancer is more common with age and occasionally occurs in women under 35 years of age	• Pregnancy, ovarian cyst, tumor of the ovary or uterus, bladder distension due to retention of urine • See your doctor at once
Lump next to the umbilicus, near a surgical scar, or (in men) in the groin	• Painless • Appears when straining or coughing or when you stand up (i.e. when pressure in the abdomen rises); lump then disappears spontaneously or can be pushed back	• Weakness in the abdominal wall may be caused by a congenital defect in the abdominal wall, a surgical incision, or muscle weakness due to obesity, pregnancy, or wasting	• Hernia or inguinal hernia • See your doctor
...mp that remains visible wh... you are straining, c...	• Associated symptoms, if present, will depend on the cause...		• Enlargement of an abdominal ...n (e.g. liver or spleen), ...

There is also a list of other symptoms that may be associated with the main symptom. Symptoms listed in the Associated symptoms column may not all occur at once, and not all of them must be present for the medical problem suggested under Possible cause to be a likely explanation for the symptoms.

Diseases mentioned in the Possible cause column suggest problems that may cause the symptoms described, but do not represent definitive diagnoses. Not all disease possibilities are listed, especially those that are less common.

LIST OF SYMPTOM ENTRIES

- Abdominal distension
- Abdominal pain
- Arm and shoulder problems
- Back deformity
- Back pain
- Blackouts, including fainting
- Bones, fracture/brittleness
- Bowel, bleeding from
- Breast lump
- Breathing problems
- Bruising
- Chest pain
- Constipation
- Cough
- Cramp
- Diarrhea
- Dizziness
- Ear pain (earache)
- Eye problems
- Eyesight problems
- Facial pain
- Fever
- Foot and ankle problems
- Genital pain
- "Glands," swollen
- Gums, bleeding
- Hand and wrist problems
- Headache
- Hearing loss (deafness)
- Heartburn
- Hip problems
- Knee problems
- Leg problems
- Menstrual problems
- Nausea and vomiting
- Neck problems
- Numbness and tingling
- Palpitations
- Seizures (known as "fits," including convulsions)
- Sexual function problems
- Skin problems
- Urination, frequent
- Urination, painful
- Vaginal problems
- Weakness and paralysis
- Wounds or sores, non-healing

You can read more about many of these symptoms, diseases and illnesses in the main section of this book.
The symptoms appear in alphabetical order and are listed above.

CHARACTERISTIC SYMPTOM	ASSOCIATED SYMPTOMS	OTHER RELEVANT FACTORS	POSSIBLE CAUSE AND ACTION

Abdominal distension

A wide variety of conditions may cause distension of the abdomen—causes range from pregnancy to conditions such as obstruction of the bowel that require urgent medical attention. More than one cause may be present at the same time. See also entry on abdominal pain.

CHARACTERISTIC SYMPTOM	ASSOCIATED SYMPTOMS	OTHER RELEVANT FACTORS	POSSIBLE CAUSE AND ACTION
Generalized distension of the abdomen, with fullness in the flanks; may develop over days or weeks	• Shortness of breath and diminishing exercise tolerance (if heart disease is the cause) • Loss of appetite, nausea, and tiredness (if liver disease is the cause) • Loss of appetite, weight loss, and a general feeling of being unwell (if a tumor is the cause)	• Symptoms are more likely in a person with a history of heart problems, liver disease, heavy alcohol use, which can cause liver damage, or cancer	• Increased build up of fluid in the abdomen (ascites) • Causes include tumors, cirrhosis of the liver, heart failure • See your doctor at once
Rapidly developing generalized distension of the abdomen	• Rapid onset of abdominal pain that comes and goes in waves • Total constipation (inability to pass solids or gas) • Vomiting	• Bowel obstruction occurs more commonly in people who have had abdominal surgery, a hernia, or other bowel problems	• Bowel obstruction • See your doctor at once
Distension in the lower area of the abdomen or pelvic area	• Difficulty passing urine, vague abdominal pain	• Ovarian cancer is more common with age and occasionally occurs in women under 35 years of age	• Pregnancy, ovarian cyst, tumor of the ovary or uterus, bladder distension due to retention of urine • See your doctor at once
Lump next to the umbilicus, near a surgical scar, or (in men) in the groin	• Painless • Appears when straining or coughing or when you stand up (i.e. when pressure in the abdomen rises); lump then disappears spontaneously or can be pushed back	• Weakness in the abdominal wall may be caused by a congenital defect in the abdominal wall, a surgical incision, or muscle weakness due to obesity, pregnancy, or wasting	• Hernia or inguinal hernia • See your doctor
Lump that remains visible whether or not you are straining, coughing, or standing	• Associated symptoms, if present, will depend on the cause		• Enlargement of an abdominal organ (e.g. liver or spleen), tumor, hernia that has become trapped • See your doctor at once

Abdominal pain

Abdominal pain may be related to any abdominal organ, and the type of pain will vary widely, depending on the cause. See your doctor or emergency room immediately if abdominal pain is severe or of sudden onset, since many common causes require immediate surgical intervention. See also entry on abdominal distension.

CHARACTERISTIC SYMPTOM	ASSOCIATED SYMPTOMS	OTHER RELEVANT FACTORS	POSSIBLE CAUSE AND ACTION
Episodes of pain in the upper abdomen, typically central and characterized as a burning sensation or heartburn	• Bitter fluid may return into the mouth from the stomach • Symptoms may be brought on by stooping, straining, or lying down	• Symptoms may be alleviated by antacids	• Esophagitis caused by reflux • See your doctor
Episodes of pain in the upper abdomen, typically central but may also be on the right or left; may be characterized as a burning or gnawing sensation	• Pain is typically related to food, and may either occur before meals and be relieved by eating, antacids, or vomiting, or may occur shortly after meals, sometimes making the person reluctant to eat	• The incidence of peptic ulcer is highest among people aged 30–50 years; may occur in teenagers and rarely in children • Ulcers are most frequently caused by a *Helicobacter pylori* infection or by the use of non-steroidal anti-inflammatory drugs	• Peptic ulcer • See your doctor

CHARACTERISTIC SYMPTOM	ASSOCIATED SYMPTOMS	OTHER RELEVANT FACTORS	POSSIBLE CAUSE AND ACTION
Episodes of colicky or constant pain in the upper abdomen, typically central or on the right side and of rapid onset; pain may radiate around the right side of the body into the shoulder and typically lasts for hours at a time	• Nausea • Sometimes vomiting • Sometimes yellowness of the skin may ensue (jaundice), possibly with darkened urine and pale stools • This pain is not relieved by vomiting or antacids	• An episode may follow eating a fatty meal • Occurs in adults; women on oral contraceptives are at higher risk • May have a family history of gallstones	• Gallstones • See your doctor
Upper abdominal pain, gradually worsening and typically constant over hours or days before subsiding; pain may radiate directly through to the back	• Nausea, vomiting • May be aggravated by food, drink, or vomiting • Typically aggravated by lying down	• Symptoms may follow a bout of heavy drinking • May have background of alcohol use or gallstones • May be alleviated by sitting up and leaning forward	• Sudden-onset (acute) pancreatitis • See your doctor at once
Persistent upper abdominal pain, typically characterized as a constant background pain; pain may radiate directly through to the back	• May later develop weight loss or pale stools • Pain may be aggravated by lying down	• May have background of alcohol use or gallstones • Pain may be alleviated by sitting up and leaning forward	• Long-term (chronic) pancreatitis or pancreatic tumor • See your doctor
Constant discomfort on the right side of upper abdomen, typically felt as a dull ache	• Nausea, loss of appetite, tiredness, fever, eventually jaundice • Other symptoms depend on the cause • Shortness of breath, diminishing exercise tolerance, and ankle swelling may occur with a heart problem • Loss of appetite, weight loss, and a general feeling of being unwell may occur with a tumor	• Liver symptoms are more likely to occur in people with a history of heart disease, heavy alcohol users, cancer, recent intravenous drug use or blood transfusion, overseas travel, recent contact with a person with a similar illness	• Liver pain due to heart failure (causing swelling of the liver), viral hepatitis, hepatitis caused by heavy alcohol use, liver tumor • See your doctor
Rapid onset of diffuse cramping pains, typically centered around the central abdomen	• Diarrhea, nausea, vomiting, fever	• Symptoms may follow overseas travel or eating carelessly prepared food	• Gastroenteritis or food poisoning • Drink plenty of fluids • See your doctor if you are worried or if symptoms persist
Rapid onset of pain on the right side of the lower abdomen; the pain may either be constant or occur in waves; pain typically starts around the middle of the abdomen, then moves to the right lower abdomen	• Loss of appetite, nausea, vomiting, low fever	• Most common in children or young adults	• Appendicitis • This is a medical emergency • Go to the emergency room or call an ambulance immediately
Pains in the lower abdomen that may be cramping, occur in waves, or vague discomfort	• Change in bowel habit • May have blood or mucus in the stools • Pain may be worse before defecation and alleviated by defecation or passing wind	• Cancer is a common cause in patients over 50, but can also occur in younger people • In older patients (over 70 years of age), bowel obstruction is one of the most common causes of abdominal pain requiring surgical intervention	• Inflammatory bowel disease or tumor of the colon • See your doctor • If severe pain, abdominal distension and total constipation (inability to pass solids or gas) occurs, this is a medical emergency • Go to the emergency room or call an ambulance immediately
Sudden severe pain on one flank, typically radiating from the loins around to the front of the abdomen and down to the groin	• May have a change in the color of your urine or blood in the urine • May be associated with nausea	• May have a predisposing factor such as gout • Passing a stone in the urine may cause severe pain	• Kidney stone • This is a medical emergency • Go to the emergency room or call an ambulance immediately

CHARACTERISTIC SYMPTOM	ASSOCIATED SYMPTOMS	OTHER RELEVANT FACTORS	POSSIBLE CAUSE AND ACTION
Abdominal pain (cont.)			
Loin pain, typically a dull ache	• Fever • Passing urine more frequently • Painful urination	• May be a predisposing condition like an abnormality in the urinary tract or a disorder that impairs immune defenses • In women and infants, may occur without a predisposing factor	• Pyelonephritis (kidney infection) • See your doctor at once
Severe, sharp, well localized abdominal pain; pain may be of abrupt onset, but sometimes is more gradual	• Pain aggravated by movement or coughing	• Predisposing conditions include peptic ulcer, abdominal aneurysm, hernia	• Peritonitis caused by appendicitis, perforation of a peptic ulcer, complication of abdominal aneurysm, strangulation or twisting of the bowel • This is a medical emergency. • Go to the emergency room or call an ambulance immediately
Bouts of severe abdominal pain in an infant	• Screaming and drawing up the legs during the bouts of pain, vomiting, red stools	• Intussusception is most common during the first two years of life	• Intussusception • This is a medical emergency. • Go to the emergency room or call an ambulance immediately

Arm and shoulder problems

Symptoms in the arm or shoulder often result from overuse, injury, and falls. Sporting activities and occupational tasks are common causes of problems in this area. Pain in the arm can also be a symptom of a heart attack, so see your doctor immediately if you experience unexplained pain. See also entry on chest pain.

CHARACTERISTIC SYMPTOM	ASSOCIATED SYMPTOMS	OTHER RELEVANT FACTORS	POSSIBLE CAUSE AND ACTION
Pain in the shoulder with movement, especially lifting the arm over the head; may occur following an injury or activity involving repetitive movements	• Tenderness and swelling in the area, weakness of the arm • Symptoms are aggravated by further movement • Pushing and twisting movements (e.g. reaching behind the back, dressing) are often painful	• Lifting, twisting and sporting activities often cause these problems	• Tendinitis • Rest and apply ice • See your doctor if you are worried or if pain persists
Sudden pain on the top of the shoulder; may be worse upon waking	• May be some redness and tenderness, limited movement • Symptoms are usually aggravated by raising arm out to the side or twisting movement	• Can occur with overuse, infections, arthritis, injuries	• Bursitis • See your doctor for advice
Gradually worsening elbow pain that occurs with repeated movements	• Pain may spread down into the muscles in the forearm • May be tender to touch • Symptoms are aggravated by further movement and alleviated by rest	• More likely to occur in those with weak wrist and arm muscles, and in tennis players using an incorrect backhand stroke	• Tennis elbow • Rest and apply ice • See your doctor if you are worried or if pain persists
Increasing shoulder pain and stiffness that has worsened over several weeks	• Limited movement in all directions • Symptoms are aggravated by movement in most directions, especially away from the body	• Often occurs in older people, especially following an injury or shoulder/chest surgery	• Frozen shoulder • See your doctor
Intense arm pain following a fall or direct blow to the arm	• Arm or shoulder may be deformed • Symptoms are aggravated by an attempt at movement		• Fracture or dislocation • See your doctor at once

CHARACTERISTIC SYMPTOM	ASSOCIATED SYMPTOMS	OTHER RELEVANT FACTORS	POSSIBLE CAUSE AND ACTION
Persistent, severe pain radiating from the chest down the left arm	• Shortness of breath, nausea, sweating • Feeling of chest being squeezed • Pain does not change with arm movements	• Increased likelihood in a person who has previously had a heart attack or with a history of coronary heart disease	• Myocardial infarction (heart attack) • This is a medical emergency. • Go to the emergency room or call an ambulance immediately

Back deformity

Abnormalities of the spine may occur due to an acquired disease, as a developmental problem in growing children, or as part of one of the syndromes of multiple anatomical abnormalities that are caused by genetic defects. Any unusual appearance of the spine should be checked by a doctor.

CHARACTERISTIC SYMPTOM	ASSOCIATED SYMPTOMS	OTHER RELEVANT FACTORS	POSSIBLE CAUSE AND ACTION
Sideways curvature of the spine (scoliosis) in an adolescent, visible while sitting erect and straight; spine is S-shaped or corkscrew shaped	• Postural problems, pain, fatigue of back muscles, visible disfigurement in severe cases	• The cause of idiopathic scoliosis is unknown, but it is thought to be due to complex inherited causes • The condition is twice as common in females as males, and usually appears at around puberty	• Idiopathic scoliosis
Exaggeration of the normal outward curve of the spine at the level of the rib cage (kyphosis)	• Poor posture, fatigue or pain in back muscles	• Abnormal curvature may first appear in a young growing child	• Exaggerated kyphosis ("hunchback"), due to injury or a developmental abnormality • See your doctor
Exaggeration of the normal inward curve of the spine at the lower back (lordosis)	• Poor posture, fatigue or pain in back muscles	• Abnormal curvature may first appear in a young growing child	• Exaggerated lordosis ("swayback"), due to injury or a developmental abnormality • See your doctor
Sideways curvature of the spine (scoliosis) which worsens over time	• Difficulty running and jumping, muscle weakness, enlarged calf muscles, breathing problems • Children use their arms to pull themselves up from the ground after sitting or lying down • Spinal curvature may become painful over time	• Duchenne muscular dystrophy is a genetic disorder affecting only boys • The condition is present at birth, but signs only appear later between 3–5 years of age	• Duchenne muscular dystrophy • See your doctor
Gradual development of a exaggerated outward curvature (kyphosis) or sideways curvature of the spine (scoliosis) in an elderly person	• Multiple fractures of the vertebrae may result in "dowager's hump" • New vertebral fractures may be very painful, but pain usually subsides within weeks to months	• Osteoporosis is more common in women than men	• Osteoporosis • See your doctor

Back pain

The majority of cases of backache relate to the lower back. Low back pain is one of the most common symptoms that people experience. In adults, a majority of cases are based on a mechanical problem relating to posture, muscle strain, arthritis, or disk problems. Sometimes, no physical cause is found. In some circumstances, prompt and adequate early treatment of acute mechanical back pain may help to avoid progression to ongoing long-term pain, so do not hesitate to see your doctor. Back pain that is progressive, severe, persistent, or associated with neurological symptoms such as weakness, numbness, or tingling sensations always warrants a visit to a doctor. See also entry on numbness and tingling.

CHARACTERISTIC SYMPTOM	ASSOCIATED SYMPTOMS	OTHER RELEVANT FACTORS	POSSIBLE CAUSE AND ACTION
Sudden episodes of pain, usually in the lower back; often described as sharp and/or shooting in character; often radiates down into the buttock and leg	• Neurological symptoms may develop, such as tingling or some loss of sensation in the buttock or part of the leg • Pain aggravated by bending (especially to the side), coughing/sneezing, laughing	• Pain may have been precipitated by trauma or an unusual movement • The pain may spontaneously go away for a time • Pain may be alleviated by lying in the fetal position	• Intervertebral disk problem (e.g. herniated disk or disk protrusion) • See your doctor at once

CHARACTERISTIC SYMPTOM	ASSOCIATED SYMPTOMS	OTHER RELEVANT FACTORS	POSSIBLE CAUSE AND ACTION
Back pain (cont.)			
Gradual onset of pain, usually in the lower back	• Similar symptoms in other joints, stiffness in the mornings • Stiffness tends to gradually ease as the day progresses, pain tends to be worse with activity and eased by resting	• Not precipitated by physical activity or trauma • Osteoarthritis is more common with age; other forms of arthritis may occur in younger people	• Arthritis (e.g. osteoarthritis or rheumatoid arthritis) • See your doctor
Dull aching pain; may radiate to one side; most commonly in the middle or lower spine	• Worse with exertion, posture, movement, coughing	• May result from severe trauma, or minor or unnoticed trauma in a person with osteoporosis or a bone abnormality	• Vertebral fracture • See your doctor at once
Lower back ache, sometimes throbbing in nature	• May have stiffness, rash, other joints affected, or symptoms from the gastrointestinal, urinary, or genital tract	• Often in young adults or people with inflammatory disease of the bowel • Characterized by being worse with rest, and alleviated by activity; this contrasts with mechanical causes of back pain	• An inflammatory joint disorder • See your doctor
Unremitting continuous and possibly progressive pain, usually of gradual onset; may be deep, boring pain	• Weight loss, loss of appetite, malaise, fever, chills, nervous system problems • Posture or motion typically have little effect on the pain	• Pathological lesions are rarely the cause of low back pain in adults; they are a more common underlying cause of pain higher up in the spine or low back pain in a child	• Cancer metastasis, cancer of the blood (e.g. leukemia), inflammation, or infection • See your doctor at once

Blackouts, including fainting

Fainting results from a transient loss of consciousness. Lapses in awareness or falling to the ground suddenly without warning may also occur without loss of consciousness. Faints are more common in elderly people, who are also at greatest risk of injury from the resulting falls. In about 50 percent of cases, no cause is found.

CHARACTERISTIC SYMPTOM	ASSOCIATED SYMPTOMS	OTHER RELEVANT FACTORS	POSSIBLE CAUSE AND ACTION
Transient loss of consciousness often without warning; may occur when upright or when lying down	• Lightheadedness, palpitations, chest pain, or breathlessness preceding loss of consciousness • While unconscious, the person may appear pallid • On waking, the person is usually well oriented and quickly feels well again	• In some cases, fainting is precipitated by exercise • The risk is increased if the person is also dehydrated	• Heart disease • See your doctor at once
Falling to the ground suddenly without warning; may occur with or without loss of consciousness	• Sudden onset of persistent nervous system problems (e.g. blindness, weakness, pins and needles, numbness, difficulties with speech, lack of muscle coordination)	• More common with advancing age • Risk increased in people with hypertension, lipid abnormalities (e.g. high cholesterol), diabetes	• Stroke • This is a medical emergency • Go to the emergency room or call an ambulance at once
Sudden transient loss of consciousness, or falling to the ground, without warning	• There may be a sudden onset of transient neurological problems such as loss of vision, weakness, pins and needles, numbness, difficulties with speech, lack of muscle coordination, vertigo • These symptoms spontaneously resolve	• May have a history of cardiovascular disease or osteoarthritis of the neck • May have been precipitated by a particular movement of the head • More common with advancing age • Risk increased in people with hypertension, lipid abnormalities (e.g. high cholesterol), diabetes	• Transient ischemic attack • This is a medical emergency • Go to the emergency room or call an ambulance at once

CHARACTERISTIC SYMPTOM	ASSOCIATED SYMPTOMS	OTHER RELEVANT FACTORS	POSSIBLE CAUSE AND ACTION
Loss of consciousness, generally prolonged	• Before losing consciousness may experience confusion, sweating, possibly headache, faintness, and weakness	• Person on medication to treat diabetes • May have missed a meal, exercised more than anticipated, or recently altered their diabetic medication • Loss of consciousness can usually be prevented by eating glucose candy or jelly beans and taking sugary drinks (e g. fruit juice) as soon as the attack is recognized	• Hypoglycemia (low blood sugar) • This is a medical emergency • Go to the emergency room or call an ambulance at once
Bones, fracture/brittleness Suspected fractures always require immediate medical attention, especially when there is substantial blood loss, symptoms of nerve damage, or any suggestion of damage to internal organs. If the injury may affect the spinal column—for example following a direct blow to the neck or back, falls from a height, or high-speed accidents—great care must be taken to immobilize the spine until an ambulance arrives to transport the person to hospital.			
Pain with loss of limb function and possibly also change in shape	• The skin overlying the fracture may be broken (open fracture) or intact (closed fracture) • There may be substantial blood loss, or damage to nerves or internal organs	• Fractures may be caused by a heavy blow hitting or crushing a bone (direct force) or twisting, bending, or compressing a bone (indirect force) • In adults, the break is usually through the full thickness of the bone • In children, the bones are more springy and incomplete ("greenstick") fractures are common	• Fracture • See your doctor at once • First aid involves controlling bleeding, covering any open wounds, and supporting the limb in its most comfortable position
Broken bone following repeated activity that stresses a particular site (usually a limb)	• Symptoms usually include pain that initially starts after the exercise or activity, then begins to occur during the activity, and later is present at other times as well • There is no deformity and the limb functions normally	• Usually occurs in otherwise healthy, active people	• Stress fracture • See your doctor at once
Broken bone following slight trauma	• Pain with loss of limb function and possible deformity • Elderly women with long-term multiple fractures of the vertebrae may develop a hump at the top of the back • Pain is aggravated by any movement of the affected part	• Pain or deformity of the bone before the fracture occurred suggests a pre-existing abnormality of the bone • Fractures of the hip, wrist, and vertebrae are most commonly associated with osteoporosis • Osteoporosis is more common in women than men • Factors that increase risk of developing osteoporosis include inherited factors, hormonal factors, calcium deficiency, alcohol abuse, smoking, lack of exercise, or drug treatment	• Fracture due to an abnormality making the bone more fragile (pathological fracture) • Causes include osteoporosis, cancer with metastasis, blood cell abnormalities • See your doctor at once

CHARACTERISTIC SYMPTOM	ASSOCIATED SYMPTOMS	OTHER RELEVANT FACTORS	POSSIBLE CAUSE AND ACTION

Bowel, bleeding from

Bleeding from the bowel always warrants a visit to the doctor to determine the cause and, if blood loss has been substantial, to support the circulation. The most common cause is hemorrhoids, but more serious conditions such as bowel cancer are common and may be responsible for a bleed of any description. Serious pathologies need to be excluded by a doctor even if a minor ailment such as hemorrhoids is present. Some of the important causes in adults, such as hemorrhoids and rectal cancer, do not occur in children.

This table deals with bleeding through the rectum. Bleeding from the upper gastrointestinal tract often causes vomiting of blood, which may either be fresh blood or darkened blood resembling coffee grounds.

CHARACTERISTIC SYMPTOM	ASSOCIATED SYMPTOMS	OTHER RELEVANT FACTORS	POSSIBLE CAUSE AND ACTION
Squirt or drip of fresh blood in toilet or on toilet paper after defecation	• Hemorrhoids are usually painless unless there are complications • A sensation of unsatisfied defecation or urgency signals a problem inside the rectum	• Straining to defecate predisposes to hemorrhoids	• Hemorrhoids, polyps, tumors • See your doctor
Small amount of fresh blood on toilet paper after defecation with sharp pain in the anus during or after defecating		• Straining to defecate may predispose to anal injuries	• Anal tear or break in tissue (fissure), tumor • See your doctor
Fresh blood mixed with stool or dark blood in stool, with or without blood clots	• No other symptoms or there may be diarrhea, constipation, abdominal pain, abdominal distension, weight loss, loss of appetite, general feeling of being unwell, symptoms of anemia (tiredness, lack of energy, breathlessness on exertion, pallor of skin, gums, or fingernails) • Mucus may be passed with stools	• More likely in a person with personal or family history of polyps or bowel cancer • Conditions that predispose to bowel cancer include benign polyps, ulcerative colitis	• Bowel tumor, polyp, other problems of rectum or bowel • See your doctor
Blood in feces or red-colored stool in an infant	• Unwell with vomiting and bouts of abdominal pain and screaming	• Most common in children under 2 years	• Intussusception • This is a medical emergency • Go to the emergency room or call an ambulance at once
Episodes of bloody diarrhea	• Abdominal pain, aches, general feeling of being unwell, weight loss, fever, symptoms of anemia (tiredness, lack of energy, breathlessness on exertion, pallor of skin, gums, or fingernails) • Mucus may be passed in stool	• Inflammatory bowel disease is most common in Jewish people (3–6 times higher than for non-Jewish racial or ethnic groups), and is higher among white racial groups than in black people or Asians	• An inflammatory bowel disease such as Crohn's disease or ulcerative colitis • See your doctor at once
Bloody diarrhea of sudden onset	• Abdominal pain	• It commences most commonly at age 15–35 • The cause of inflammatory bowel disease is unknown • Most common in elderly people with a history of other cardiovascular disease, but no contact with dysentery	• Loss of blood supply to a portion of the bowel (ischemic colitis) • This is a medical emergency • Go to the emergency room or call an ambulance at once
Heavy bleeding from the rectum	• Lightheadedness or fainting, confusion, severe thirst, feeling cold and clammy, pallor of skin, gums, and nails	• More common in men than in women • Heavy bleeding is a rare complication of diverticulitis, which is usually associated with invisible or light bleeding	• Diverticulitis • This is a medical emergency • Go to the emergency room or call an ambulance at once

CHARACTERISTIC SYMPTOM	ASSOCIATED SYMPTOMS	OTHER RELEVANT FACTORS	POSSIBLE CAUSE AND ACTION
Melena (black tarry stools)	• May also vomit blood, either as fresh blood or darkened blood resembling coffee grounds	• Peptic ulcer can occur in any age group, but is relatively common in elderly people taking non-steroidal anti-inflammatory drugs for painful inflammatory conditions such as osteoarthritis or rheumatoid arthritis	• Bleeding from the esophagus, stomach, or duodenum, caused by a peptic ulcer, or gastritis, or dilated veins in the esophagus • This is a medical emergency • Go to the emergency room or call an ambulance at once

Breast lump

Most women will detect a lump in their breast at some time in their life. Most of these lumps are found to be benign on further assessment. Breast cancer has been found responsible for up to 25 percent of breast lumps and remains the most common cancer in women.

CHARACTERISTIC SYMPTOM	ASSOCIATED SYMPTOMS	OTHER RELEVANT FACTORS	POSSIBLE CAUSE AND ACTION
General breast lumpiness, tender breast lumps usually affecting both breasts, but occasionally a single lump only; lumps are soft, round, and smooth; pain and tenderness may occur during menstrual cycles, usually of both breasts; lumps may come and go with menstrual cycles	• Pain and tenderness worse premenstrually	• Usually affects premenopausal women • Symptoms usually diminish or disappear after menopause • Sometimes symptoms may be alleviated by reducing intake of coffee, tea, and chocolate	• Fibrocystic disease of the breast • See your doctor • If fibrocystic disease is diagnosed a well-fitted bra may help provide adequate support • Vitamin E supplementation may reduce symptoms
Firm, elastic breast lump that moves freely; usually a single, smooth lump approximately 1½ inches (3–4 centimeters) in diameter	• Usually painless	• Usually occurs in women aged 20–40 years	• Fibroadenoma • See your doctor
Hard or irregular breast lump which may be fixed in one position or attached to nearby tissues; usually only one breast is affected	• No other symptoms or pain (5 percent of patients), bloody or non-milky discharge from nipple, dimpling of the skin, tenderness or lumps under the armpit	• More common in women than in men • Most common after 40 years of age but may occur much earlier in women with a genetic predisposition	• Breast cancer • See your doctor at once
Firm, single, irregular lump, which may be attached to the skin	• Lump is painful and tender	• Gradual onset (over months) after breast trauma; the trauma involved may be subtle	• Abnormality in fat tissue • See your doctor
Single, tender breast lump that develops rapidly in a women who is breast feeding	• Pain and tenderness of lump, redness of the overlying skin, fever, chills	• Occurs during lactation	• Breast abscess • See your doctor

Breathing problems

Shortness of breath can be a gradually developing phenomenon or may be rapidly developing and life-threatening. Some of the more prominent causes of each are described in this section.

CHARACTERISTIC SYMPTOM	ASSOCIATED SYMPTOMS	OTHER RELEVANT FACTORS	POSSIBLE CAUSE AND ACTION
Long-term or repeated shortness of breath on exertion; onset gradual	• Shortness of breath on lying flat, waking up at night with shortness of breath that is relieved by sitting up, swelling of the legs, cough, wheeze may be present, gradual decline in exercise capacity, shortness of breath at rest	• Chronic heart failure is most common in the middle-aged and elderly, and rarely occurs in younger people • Heart failure may follow rheumatic fever, heart attack/coronary artery disease, hypertension, arrhythmia, or heart valve abnormalities	• Chronic heart failure • See your doctor at once

CHARACTERISTIC SYMPTOM	ASSOCIATED SYMPTOMS	OTHER RELEVANT FACTORS	POSSIBLE CAUSE AND ACTION
Breathing problems (cont.)			
Long-term or repeated shortness of breath on exertion; may progress gradually over many years	• Wheezing • Cough with sputum, often worse in the morning • Gradual decline in exercise capacity • Sometimes alleviated by leaning forward or lying on stomach with head down	• Chronic obstructive lung disease is very much more common in smokers and in the elderly	• Chronic obstructive lung disease, bronchitis, emphysema • See your doctor
Long-term or repeated shortness of breath on exertion, occurring in episodes with no symptoms at other times	• Wheezing, persistent cough, usually without sputum • Symptoms may be triggered by exposure to cigarette smoke, air pollution, pollens, or pets, cold air, exercise, respiratory infections	• History of hay fever or eczema; family history of asthma, eczema, or allergies • Usually more frequent in winter, following a common cold, or when pollen levels are high	• Asthma • See your doctor • If breathing is difficult, or you suspect oxygen shortage, this is a medical emergency • Go to the emergency room or call an ambulance at once
Acute shortness of breath at rest; onset may be sudden	• Chest pain if the episode is precipitated by a heart attack • May have a background history or recent worsening of shortness of breath on exertion and/or when lying flat, and/or waking up short of breath at night	• Pulmonary edema is more likely in a person with a history of this problem • May be caused by cardiac disease, lung infections, shock (e.g. cardiac surgery, serious infections, inhaled toxins)	• Pulmonary edema • This is a medical emergency • Go to the emergency room or call an ambulance at once
Sudden onset shortness of breath at rest	• Fever, chills, pain around the chest that is aggravated by breathing in	• Pneumonia is a relatively common condition in all age groups, but may be more likely with an inhaled foreign body, pre-existing lung diseases, immune suppression (e g. other disease, medications)	• Pneumonia • See your doctor at once
Acute shortness of breath at rest; sudden onset	• Chest pain	• May occur spontaneously or as a result of chest trauma • Most often occurs in young, healthy people without a prior lung problem • Predisposing factors include cystic fibrosis, chronic obstructive lung disease	• Pneumothorax • This is a medical emergency • Go to the emergency room or call an ambulance at once
Acute shortness of breath at rest	• Chest pain, dizziness, faintness, loss of consciousness, rapidly turning blue	• Predisposing factors to an embolus include the use of oral contraceptives in young women, recent surgery, recent childbirth, a long period of bed rest or immobility, history of a blood clot in the leg, abnormal heart rhythms	• Pulmonary embolism • This is a medical emergency • Go to the emergency room or call an ambulance at once
Very sudden shortness of breath at rest with choking; occurring while eating or where there is a high probability of a toddler inhaling a small object	• Severe distress with inability to take in air despite desperate efforts, collapse, blue skin and gums, loss of consciousness (in complete obstruction) • Rapid labored breathing with loud harsh noise while breathing in, gagging (in partial blockage of the airway)	• Inhalation is probable if onset while eating • In adults, this is more likely to occur if a person is intoxicated or has a reduced level of consciousness/awareness • Toddlers commonly inhale other objects	• Upper airway obstruction due to inhaled foreign body such as a piece of food • This is a medical emergency • Go to the emergency room or call an ambulance at once

CHARACTERISTIC SYMPTOM	ASSOCIATED SYMPTOMS	OTHER RELEVANT FACTORS	POSSIBLE CAUSE AND ACTION
Sudden shortness of breath at rest with choking; occurring when inhalation of food or a small object is unlikely	• Severe distress with inability to take in air despite desperate efforts, collapse, blue skin and gums, loss of consciousness (in complete obstruction) • Rapid labored breathing with loud harsh noise while breathing in, gagging (in partial blockage of the airway)	• Inflammation of the airways may be caused by an insect bite or sting, an allergic reaction to food or a new medication, inhalation of hot gases or other toxins • Eating peanuts is a relatively common cause of severe reaction, which may occur in a person without a known history of peanut allergy	• Upper airway obstruction due to swelling of structures in the throat • This is a medical emergency • Go to the emergency room or call an ambulance at once

Bruising

Bruising is a normal reaction to injury and is caused by blood leaking from damaged blood vessels under the skin. Bruises start off as a red mark, quickly turn blue then fade into a green or yellow color as they heal. Frequent or severe bruising after a minor injury could be a sign of an underlying disorder of the blood vessels or the ability of the blood to clot. Any unusual bruising should be checked by a doctor.

CHARACTERISTIC SYMPTOM	ASSOCIATED SYMPTOMS	OTHER RELEVANT FACTORS	POSSIBLE CAUSE AND ACTION
Bruise confined to a single area	• Pain, tenderness, may be a lump under the bruise if severe	• Bruising occurs in healthy people, usually after injury • Women tend to bruise more easily then men • Bruising is often more extensive in the elderly	• Bruise resulting from minor injury • Placing an ice pack over the injured site may reduce bruising • See your doctor if you are worried or if symptoms persist
Extensive dark purple bruises in an elderly person, often on arms or legs	• Pain, tenderness, may be a lump under the bruise if severe	• Usually a history of minor injury • Occurs in the elderly due to weakening of the blood vessels and skin	• Senile purpura (bruising due to old age) • Creams available from your pharmacist can speed healing of the bruise • See your doctor if you are worried or if symptoms persist
Slow healing bruises	• Slow healing wounds, frequent infections, tingling and numbness in the hands or feet	• May occur in people with a history of poorly controlled diabetes	• Diabetes • See your doctor
Bruising following minor injuries	• May be prolonged bleeding after cuts	• Medications that may be associated with increased bruising include aspirin, cortisone-type drugs (corticosteroids), non-steroidal anti-inflammatory drugs, blood thinning drugs (anticoagulants)	• Medication-related bruising • See your doctor
Development of bruising and tiny red dots on the skin	• Nosebleeds or bleeding from gums in severe cases • Heavy menstrual bleeding in young women	• Most common in children from 2– 4 years of age and young women • Illness may be prolonged in adults but is usually self-limiting in children without treatment	• Thrombocytopenia • See your doctor
Scattered bruises of varying ages, appearing over a period of weeks	• Severe fatigue, loss of appetite, pallor, episodes of fever, bone pain	• May be a history of frequent infections	• Leukemia • See your doctor at once

CHARACTERISTIC SYMPTOM	ASSOCIATED SYMPTOMS	OTHER RELEVANT FACTORS	POSSIBLE CAUSE AND ACTION
Bruising (cont.)			
Frequent, often severe bruising without significant injury; bruises may be firm and hard due to blood clotting under the skin	• Other signs of bleeding may be also present such as painful or swollen joints, excessive bleeding following cuts or dental extractions, blood in the urine • Symptoms of brain hemorrhage include irritability, drowsiness, headache, confusion, nausea, vomiting and double vision • Other internal bleeding may result in breathing problems, weakness, pallor or loss of consciousness	• Hemophilia occurs only in males • The worldwide incidence is one in 10,000 males • Hemophilia is an inherited disorder, so there may be family history of a bleeding tendency	• Hemophilia or other bleeding disorder • See your doctor • If the person is bleeding or shows signs of internal bleeding, this is a medical emergency • Go to the emergency room or call an ambulance at once

Chest pain

Chest pain typically arises from the heart, lungs, chest wall, or esophagus. Occasionally, chest pain is due to disease of upper abdominal structures such as the stomach or gallbladder. Any chest pain warrants full examination by a doctor.

CHARACTERISTIC SYMPTOM	ASSOCIATED SYMPTOMS	OTHER RELEVANT FACTORS	POSSIBLE CAUSE AND ACTION
Repeated episodes of central chest heaviness, pressure, tightness, or discomfort occurring on exertion and relieved by rest; often the pain radiates into the neck and/or left arm, or the arm may feel numb or tingly; attacks usually last a few minutes, sometimes up to 15 minutes	• Shortness of breath or anxiety may be present	• Angina pectoris occurs most frequently in the middle-aged or elderly • Alleviated by rest or nitrate tablet taken under the tongue • Rarely occurs in children with congenital heart problems	• Angina pectoris • See your doctor
Central burning pain radiating up the throat	• Bitter taste in the mouth or return of bitter fluids into the mouth from the stomach	• Heartburn occurs in adults and children and is a symptoms of gastroesophageal reflux • Attacks are often provoked by lying flat in bed at night, or bending over	• Heartburn • Avoid large meals, eating within a few hours of bedtime and any specific foods which provoke symptoms • See your doctor
Dull cramp-like central chest pain which occurs only after swallowing; the pain can be quite severe	• Difficulty swallowing	• Esophageal spasm may occur as a result of gastroesophageal reflux	• Esophageal spasm (a disturbance of the contractions of the esophagus which usually push food through to the stomach) • See your doctor
Severe central chest heaviness, pressure, tightness, or discomfort that usually lasts more than 30 minutes; often the pain radiates into the neck and/or left arm; onset at rest or on exercise but without relief after resting	• Shortness of breath, faintness, sweating, cold, nausea or vomiting, palpitations • Symptoms may be relatively mild in a person with a past history of heart disease	• Myocardial infarction most commonly occurs in middle-aged or older people	• Myocardial infarction (heart attack) • This is a medical emergency • Go to the emergency room or call an ambulance at once
Sharp, stabbing chest pain within a well-defined area, occurring on or worsened by breathing in	• Fever, cough with sputum that may contain pus, shortness of breath • Pain is aggravated by taking a deep breath, or by coughing	• Pneumonia may result from a bacterial infection in previously healthy people, or due to lowering of resistance resulting from another illness	• Pneumonia • See your doctor at once

CHARACTERISTIC SYMPTOM	ASSOCIATED SYMPTOMS	OTHER RELEVANT FACTORS	POSSIBLE CAUSE AND ACTION
Sharp, stabbing chest pain within a well-defined area, worse on breathing in; may affect one or both sides of the chest	• Shortness of breath • Coughing up blood • Pain is aggravated by taking a deep breath, or by coughing	• Predisposing factors include a period of immobilization or bed rest, recent surgical procedure, recent childbirth, history of a blood clot in the leg, the use of oral contraceptives	• Loss of blood supply to an area of lung tissue (pulmonary infarct), usually due to blockage of blood vessels by a clot • See your doctor at once
Sudden onset of sharp, localized pain on one side of the chest, worse on breathing in	• Shortness of breath • Pain is aggravated by taking a deep breath, or by coughing	• Pneumothorax may occur in adults and children, either spontaneously or as a result of injury to the chest • Most often occurs in young healthy people without a prior lung problem • Predisposing factors include cystic fibrosis, chronic obstructive lung disease	• Pneumothorax • See your doctor at once
Dull aching localized pain, usually on one side, and worse with movement (turning, twisting, bending), straining (e.g. lifting), possibly by coughing or deep breathing in	• Tenderness of the chest wall	• May follow trauma or unusual exertion	• Injury to the chest wall • See your doctor
Constant burning pain on one side of the chest followed by rash	• Painful blistering rash appears in the painful area a few days after the initial pain	• Shingles mainly occurs in adults over 60 years, and is due to reactivation of the same virus responsible for chickenpox • In children, chickenpox frequently causes a painful blistering rash on the chest	• Shingles • See your doctor at once
Sudden, severe central chest pain	• Severe shortness of breath, fainting • Skin may be cold, clammy, pale	• Predisposing factors include a period of immobilization or bed rest, recent surgical procedure, recent childbirth, history of a blood clot in the leg, use of oral contraceptives	• Pulmonary embolism • This is a medical emergency • Go to the emergency room or call an ambulance at once
Sudden onset of severe, tearing pain of the front or back of the chest that moves as time progresses; the pain often radiates through to the back or may be felt predominantly in the back of the chest; pain may also involve the arms, neck, trunk, or legs	• Shortness of breath, fainting • Complications may cause symptoms in other bodily systems (e.g. paralysis of the legs, mental disturbances, bloody diarrhea)	• Most often occurs in middle-aged or elderly men with hypertension	• Tearing along the inside wall of the aorta (dissecting aneurysm) • This is a medical emergency • Go to the emergency room or call an ambulance at once
Central chest tightness or discomfort	• Shortness of breath, wheezing • May be exacerbated by exposure to pets or cigarette smoke	• Asthma occurs more commonly in adults or children with a history of hay fever or eczema, or a family history of asthma, eczema, or allergies • Usually more frequent in winter, following a common cold or when pollen levels are high • May be exacerbated by exposure to pets or cigarette smoke	• Asthma • See your doctor at once

CHARACTERISTIC SYMPTOM	ASSOCIATED SYMPTOMS	OTHER RELEVANT FACTORS	POSSIBLE CAUSE AND ACTION

Constipation

Constipation is the most common symptom relating to the gastrointestinal tract. Different people have different normal bowel habits—for a particular person it may be usual to pass anywhere between three stools per week and three per day. In most cases, constipation can be thought of in terms of a change away from an individual's usual pattern. Sometimes a person may by troubled by very hard stools or difficulty in expelling stools rather than a reduction in frequency.

Frequently the cause of constipation relates to lifestyle factors. Laxatives are not usually needed and can themselves cause constipation if used inappropriately, so it is preferable to seek medical advice before taking them. Occasionally serious underlying pathology is responsible, or constipation is a manifestation of disease in another bodily system. Prescription and over-the-counter medicines commonly cause or aggravate constipation.

CHARACTERISTIC SYMPTOM	ASSOCIATED SYMPTOMS	OTHER RELEVANT FACTORS	POSSIBLE CAUSE AND ACTION
Stools are dry and hard and may require straining to expel; the condition may be long-standing	• Vague discomfort in the abdomen • Children may also have a sense of bloating, irritability, soiling of underwear	• Most common in adults; may occur in children • Simple constipation is commonly due to a low-fiber diet, inadequate fluid intake, lack of exercise • May be triggered by any upset of eating or bowel routine (e.g. travel, recent bed rest or immobilization, recent dietary changes), ignoring the urge to have a bowel movement, recently toilet-trained children	• Simple constipation • Increase your dietary fiber intake from vegetables, fruits, and whole grains, avoid excessive intake of fatty or sugary foods, engage in some form of physical activity • See your doctor if you are worried or if symptoms persist, or if constipation occurs in a child
Inconsistent frequency and/or character of stools; may be long-standing	• May have a sense of bloating or vague abdominal discomfort	• Most commonly young adults, but may occur at any age	• Irritable bowel syndrome • See your doctor
Recent change from usual bowel habit or worsening of any pre-existing problem; may experience an early morning rush to go to the toilet	• Vague abdominal discomfort or crampy abdominal pains, blood or mucus in the stools, weight loss, loss of appetite, general feeling of being unwell • Laxative use less effective than previously	• More common with advancing age • Usually occurs after 40 years of age, but occasionally may occur at younger age • Predisposing conditions for bowel cancer include benign polyps, familial polyposis (an inherited condition in which many polyps form in the bowel at puberty), ulcerative colitis	• Bowel tumor • See your doctor
Passage of stools causes pain, so is consciously avoided by the person	• May be associated with urgent or painful desire to pass stools or fresh blood on the toilet paper	• May occur at any age • If an urgent or painful desire to pass stools is present, this points toward a potentially serious rectal lesion	• Anal tear/fissure, hemorrhoids, tumor of the anus or rectum • See your doctor
Stools are small, hard and are passed infrequently (e.g. once every few days); may be a recurrent problem	• Soiling of underwear	• Most likely in the elderly person • May be triggered by periods of bed rest or immobilization, poor fluid intake	• Feces wedged firmly in the rectum • See your doctor

Cough

Coughing is the body's way of clearing the airways of secretions or foreign material. A cough may be caused by any irritation or inflammation of the airways or lungs. Any persistent cough should be assessed by a doctor.

CHARACTERISTIC SYMPTOM	ASSOCIATED SYMPTOMS	OTHER RELEVANT FACTORS	POSSIBLE CAUSE AND ACTION
Rapid onset of cough that is dry or productive (produces phlegm)	• Runny nose, sore throat, fever, fatigue • If sputum is present it is usually clear or white	• May follow recent contact with person with cough or cold • Often occurs in winter	• Infectious cough, often caused by a virus which causes irritation of the lining of the windpipe and the production of mucus • Usually self-limiting without treatment • See your doctor if you are worried or if symptoms persist

CHARACTERISTIC SYMPTOM	ASSOCIATED SYMPTOMS	OTHER RELEVANT FACTORS	POSSIBLE CAUSE AND ACTION
Acute onset of moist cough productive of phlegm	• Fever, shortness of breath, fatigue, sharp chest pain on breathing in due to inflammation of the pleura (pleurisy) • Sputum produced is yellow or green and may contain some blood • Cough may be aggravated by cold air, smoky environment, or exercise	• May follow recent cold or flu • More common in smokers	• Bronchitis or pneumonia caused by an infection • If the cause is bacterial, you may require treatment with antibiotics • See your doctor
Episodic dry cough with tight feeling in chest, often occurring at night or after exercise	• Wheeze, shortness of breath • May be exacerbated by exposure to pets or cigarette smoke	• History of hay fever or eczema • Family history of asthma, eczema, or allergies • Usually more frequent in winter, following a common cold, or when pollen levels are high	• Asthma • Any episodic cough, especially if associated with wheeze, should be assessed by a doctor • Management of acute episodes of asthma usually requires inhaled medications to open the airways (bronchodilator or "reliever" medication) • Continuous preventative medication may be required even when cough is not present • See your doctor if having difficulty breathing
Sudden onset of cough which persists, without other explanation (e.g. signs of a cold infection)	• Occasional wheeze	• History of choking • More common in children • Small objects such as peanuts or beads are commonly responsible	• Inhalation of a foreign body • See your doctor at once
Chronic cough, often productive of phlegm	• Shortness of breath on exertion, wheezing	• History of smoking or exposure to cigarette smoke	• Chronic bronchitis or chronic lung disease • See your doctor • Any change in the nature of a chronic "smoker's cough" should be checked by a doctor at once to rule out cancer of the airways (lung cancer)

Cramp

A cramp is a sudden, prolonged and painful contraction of one or a group of muscles. Although the term "cramp" is often used to describe abdominal pain (abdominal cramps) and period pain (menstrual cramps), this table will focus on cramps of voluntary muscles of the body. Muscular cramps occur very commonly and are usually benign. Occasionally they may indicate a serious underlying disease process so it is important to have any frequent or persistent cramps assessed by a doctor.

CHARACTERISTIC SYMPTOM	ASSOCIATED SYMPTOMS	OTHER RELEVANT FACTORS	POSSIBLE CAUSE AND ACTION
Sudden, prolonged and painful contraction of a small group of muscles	• If severe, tenderness and weakness may remain after the cramp has resolved	• Most common in the elderly, often occurs in the calves	• Benign cramps • See your doctor
Sudden, prolonged and painful contraction of muscles		• May occur after insufficient warm-up before strenuous exercise	• Benign cramps induced by exercise • Thorough warm-up and stretching before exercise can help prevent cramps • See your doctor if you are worried or if symptoms persist

CHARACTERISTIC SYMPTOM	ASSOCIATED SYMPTOMS	OTHER RELEVANT FACTORS	POSSIBLE CAUSE AND ACTION
Cramp (cont.)			
Very prolonged painful contraction of muscles, usually in the extremities	• Occur during strenuous exercise involving excessive sweating	• Sometimes known as "miner's cramps" or "stoker's cramps"	• Heat cramps • Replacement of depleted salts and water by consuming electrolyte enriched drinks may relieve symptoms • See your doctor if you are worried or if symptoms persist
Frequent generalized cramps in muscle groups anywhere in the body	• May be a sign of underlying neurological disease or electrolyte imbalance • Motor neuron disease is relatively rare, occurring in less than 0.01percent of people; more common in men than women, and occurs most often at around 60 years of age		• Motor neuron disease • See your doctor

Diarrhea

Diarrhea refers to the passing of frequent and abnormally loose or watery bowel actions. Diarrhea is often the result of gastrointestinal infection and lasts only a few days. Sometimes diarrhea can be caused by a more serious disease process. The consistency, color, and frequency of the diarrhea will give valuable clues with regard to the underlying cause. Any prolonged or unusual diarrhea should be assessed by a doctor to rule out an underlying medical condition.

CHARACTERISTIC SYMPTOM	ASSOCIATED SYMPTOMS	OTHER RELEVANT FACTORS	POSSIBLE CAUSE AND ACTION
Abrupt onset of watery diarrhea; mucus often present but rarely blood; lasts 1–3 days	• Vomiting, mild abdominal pain, low-grade fever • May follow an upper respiratory tract infection • Eating and drinking may aggravate diarrhea	• Very common in children, especially in winter • Highly contagious, spread by fecal–oral transmission	• Viral gastroenteritis • Oral rehydration is usually sufficient treatment • Medications to reduce nausea and vomiting, and acetaminophen (paracetamol) can be helpful • Contact your doctor for advice
Abrupt onset of watery diarrhea, moderate in quantity, lasting 1–3 days	• Generally painless, although mild abdominal cramps may be present • Low-grade fever or vomiting may be present	• May be history of ingesting poorly prepared food or travel to South America, Africa or Asia	• A bacterial infection • See your doctor at once
Abrupt onset of watery diarrhea, copious in quantity, lasting 2–7 days	• Effortless vomiting usually follows the onset of diarrhea, nausea usually absent • Dehydration, muscle cramps • May cause serious, possibly fatal dehydration within hours	• Most common in developing countries including India, Bangladesh, African and Latin American countries • Often occurs in epidemics • Usually spread by drinking contaminated water • More common and serious in children	• Cholera • This is a medical emergency • Go to the emergency room or call an ambulance at once • Where medical help is unavailable, dehydration may be alleviated by maintaining fluid intake with solutions containing glucose and salt, or by breast feeding in babies
Abrupt onset of diarrhea containing blood or mucus, with an offensive odor	• High fever, aches and pains, nausea and vomiting, moderate to severe abdominal cramps, frequent urge to pass stool with little result • Eating and drinking may aggravate diarrhea	• May be a history of travel to a developing country or simultaneous illness in people who have shared contaminated food	• Food poisoning • Antidiarrheal agents are not recommended for this type of diarrhea • See your doctor at once

CHARACTERISTIC SYMPTOM	ASSOCIATED SYMPTOMS	OTHER RELEVANT FACTORS	POSSIBLE CAUSE AND ACTION
Abrupt onset of watery diarrhea; symptoms may persist for weeks	• Mild abdominal pain, bloating, flatulence, malabsorption (passage of larger volume stools than normal due to inability to absorb food), low-grade fever • Eating and drinking may aggravate diarrhea	• Often a history of travel • Common in children	• Infection with protozoal organism called giardia (giardiasis) • See your doctor
Abrupt onset of diarrhea while taking antibiotic medication	• Range from mild abdominal pain and bloating to severe abdominal cramps, fever and bloody diarrhea	• May be simply a side effect caused by overgrowth of gastrointestinal flora or due to colitis caused by overgrowth of the bacteria *Clostridium difficile*	• Antibiotic-related diarrhea • See your doctor
Chronic intermittent diarrhea with mucus, sometimes alternating with constipation	• Abdominal pain, flatulence, bloating • Emotional stress can aggravate symptoms	• Most common in young to middle-aged women • Increasing dietary fiber can improve symptoms	• Irritable bowel syndrome • See your doctor
Watery diarrhea; no mucus or blood	• Recent constipation, bloating and abdominal discomfort • No fever or other signs of infection • Fecal incontinence may occur	• May be a history of chronic constipation • Watery stool leaks around the compacted feces	• Chronic constipation with overflow diarrhea • See your doctor
Loose, bulky, offensively smelling stools	• Abdominal bloating and discomfort, flatulence, weight loss, signs of malnutrition or poor growth • Worse with certain types of food such as dairy or wheat products	• May be a history of gallbladder disease, disorders of the pancreas	• Poor absorption of food from the bowel due to lactose intolerance or celiac disease • Malabsorption • See your doctor
Gradual onset of frequent, loose stools containing blood and mucus	• Abdominal pain, weight loss, abdominal tenderness, bloating and passage of larger volume stools than normal due to inability to absorb food • May be associated arthritis or skin lesions	• There may be a family history of inflammatory bowel disease	• Inflammatory bowel disease • Ulcerative colitis and Crohn's disease are the most common of these • See your doctor

Dizziness

Dizziness is a commonly reported symptom. The term "dizziness" is used to describe a variety of sensations such as lightheadedness, faintness, spinning or giddiness. Dizziness usually means either faintness (the feeling that precedes a faint) or vertigo (a false sense of movement). It is often a brief sensation, however persistent or frequent episodes of dizziness occasionally herald a serious medical condition so it is important to have these assessed by a doctor.

CHARACTERISTIC SYMPTOM	ASSOCIATED SYMPTOMS	OTHER RELEVANT FACTORS	POSSIBLE CAUSE AND ACTION
Faintness	• Lightheadedness, visual blurring, loss of vision, heaviness of limbs, loss of consciousness	Usually a brief episode	• Faint caused by reduced blood supply to the brain • Causes include low blood pressure on standing, disturbance in heartbeat rhythm, dehydration • See your doctor at once
Sensation of movement or spinning	• Nausea	• Commonly occurs while traveling in cars or on boats	• Motion sickness • Medication to relieve nausea and vomiting can relieve symptoms • See your doctor if you are worried or if symptoms persist

CHARACTERISTIC SYMPTOM	ASSOCIATED SYMPTOMS	OTHER RELEVANT FACTORS	POSSIBLE CAUSE AND ACTION
Dizziness (cont.)			
Sudden feeling of dizziness or spinning lasting a few seconds	• Nausea	• Related to the blockage of fluid in the balance centers of the inner ear sometimes by tiny "stones"	• Benign paroxysmal positional vertigo • Your doctor can show you simple head movements to unblock the inner ear • Antiemetic medication can relieve symptoms • See your doctor
Severe episodes of dizziness with a sensation of spinning or linear movement	• Nausea, tendency to fall toward the side of the abnormality, rapid involuntary movements of the eyes	• May occur only once or lead to recurrent episodes of vertigo	• Acute labyrinthitis caused by infection, poor blood supply to the inner ear or toxicity (drugs or alcohol) • See your doctor
Recurrent episodes of dizziness lasting 20 minutes to 2 hours	• Ringing in the ears (tinnitus), feeling of fullness of the ears, balance problems, nausea, hearing loss	• May be caused by fluid in the canals of the inner ear	• Menière's disease • Full medical assessment is required • Investigations to rule out other causes of symptoms such as acoustic neuroma may be required • Diuretics can help prevent attacks, antiemetics can provide symptomatic relief • See your doctor

Ear pain (earache)

Ear pain is usually due to increased pressure in the middle ear or infection of the ear or surrounding structures. In addition to significant distress and discomfort, ear infections can lead to serious complications such as perforation of the eardrum or spread of an infection to the covering of the brain (meningitis). It is important to seek a doctor's advice, since antibiotic therapy may be needed.

CHARACTERISTIC SYMPTOM	ASSOCIATED SYMPTOMS	OTHER RELEVANT FACTORS	POSSIBLE CAUSE AND ACTION
Dull, deep ear pain, feeling of fullness or pressure	• Fever, hearing loss, feeling of pressure in the ear, discharge if perforation of eardrum occurs, dizziness may occur if inner ear is involved	• Swelling and blockage of the tubes draining the middle ear results in increased pressure • Swallowing can relieve the pressure in the middle ear	• Fluid and pressure build-up in the middle ear • Treatment with decongestant medication may help improve the drainage from the middle ear • See your doctor if you are worried or if symptoms persist
Rapid onset of severe, deep ear pain	• Fever, swelling and tenderness over the mastoid (bony process behind the ear containing air cells	• Fever may be the only symptom in young children • Spontaneous perforation of the eardrum will relieve pain but prolong hearing loss	• Infection of the middle ear (otitis media) caused by a virus or bacteria • See your doctor
Rapid onset of severe pain behind the ear	• Fever, swelling and tenderness over the mastoid (bony process behind the ear containing air cells that are linked to the middle ear)	• May be a history of otitis media as infection can spread from the middle ear to the mastoid • Meningitis is a complication of mastoid infection	• Bacterial infection of the mastoid air cells (mastoiditis) • See your doctor at once
Itching progressing to throbbing pain of ear	• Swelling of external ear canal, foul smelling discharge, hearing loss, enlarged glands in the neck	• Trauma to the ear canal, swimming, humid conditions, skin disorders such as eczema can predispose to infection	• Bacterial infection of the external ear canal (otitis externa) • See your doctor. Treatment with antibiotic and anti-inflammatory drops may be required • Careful cleansing of the ear canal as instructed by a doctor will aid healing

CHARACTERISTIC SYMPTOM	ASSOCIATED SYMPTOMS	OTHER RELEVANT FACTORS	POSSIBLE CAUSE AND ACTION
Eye problems Mild redness and irritation of the eye is common and usually has a simple underlying cause. Symptoms to beware of include sudden or severe eye pain and changes in vision, especially when associated with a red eye. If in doubt, it is safer to have a doctor check the eye early to prevent any long-term complications. See also entry on eyesight problems.			
Sore, dry, gritty eyes	• May be mild redness, normal vision • Tiredness and the wearing of contact lenses can increase eye irritation	• More common in the elderly • Some medications can predispose to dry eyes (diuretics, decongestants, sleeping pills)	• Dry eyes due to reduced production of tears. • Artificial lubricants (or tears) can relieve symptoms • See your doctor
Sore, red eye, yellow sticky discharge, affecting one or both eyes	• Sensitivity to light, itching, burning, gritty feeling on moving the eyes, normal vision	• Highly contagious, often spreads from one eye to the other • Rubbing eyes increases itching and discharge	• Conjunctivitis caused by a bacteria or virus • Gently cleanse eyes with cool water and avoid wearing contact lenses • See your doctor if you are worried or if symptoms persist
Rapid onset of red, painful eye	• Blurred vision, sensitivity to light, increased tear production, small pupil	• More common in people with joint diseases such as ankylosing spondylitis or Behçet's disease or other autoimmune diseases	• Iritis • See your doctor at once
Rapid onset of red, extremely painful eye, on one side	• Nausea, vomiting, and blurred vision, cornea may be hazy due to swelling, pupil becomes fixed and dilated in severe cases • Episode may be preceded by vision disturbances such as halos around lights	• May be a family history of glaucoma • More common in older people with diabetes or hypertension • Some medications can make glaucoma worse, such as some antidepressants or steroids	• Glaucoma • See your doctor at once
Rapid onset of eye pain, oversensitivity to light and blurred vision on one side	• May be a visible white spot on the clear surface of the eye • Symptoms worsen at night when the pupil dilates	• May be a history of trauma to the eye such as a scratch	• Corneal ulceration due to infection, exposure or trauma • See your doctor at once
Eyesight problems Disturbances of vision are very common. Whether vision is just blurred or there is complete or partial loss of eyesight, visual disturbance is a very distressing and disabling symptom. The eyes are extremely delicate organs and can be damaged by many diseases involving other parts of the body. This section will concentrate on conditions localized to the eye. See also entries on eye problems.			
Gradual onset of distorted vision, blurred vision especially when reading	• Central vision is affected but peripheral vision remains intact	• Most common cause of loss of vision in the elderly	• Macular degeneration • Have your vision tested as you may need glasses • See your doctor if you are worried or if symptoms persist
Gradual onset of cloudy, foggy vision, distorted color, halos around lights	• Pupil of the eye may appear milky instead of black • Frequent changes in eyeglass prescription	• Cataracts are frequently a complication of diabetes • Vision is worse at night with glare from bright lights	• Cataracts • See your doctor
Loss of parts of peripheral vision (visual field defects)	• Vision loss is the first symptom	• Risk factors include older age, black race, diabetes, hypertension, nearsightedness, and a family history of glaucoma • Those people at risk should be tested for glaucoma • Early detection and treatment is the only way to prevent permanent vision loss	• Glaucoma • See your doctor at once

CHARACTERISTIC SYMPTOM	ASSOCIATED SYMPTOMS	OTHER RELEVANT FACTORS	POSSIBLE CAUSE AND ACTION
Eyesight problems (cont.)			
Rapid development of hazy vision in one or both eyes lasting days to weeks	• Distortion of color, eye movements may be painful • Pupil is less reactive to light on the affected side	• Leads to an increased risk of developing multiple sclerosis	• Inflammation of the optic nerve (optic neuritis) • See your doctor at once
Rapid loss of vision in one eye	• Recent vision disturbances such as "floaters" (dark spots or shapes floating in the field of vision) due to previous small bleeds	• Common in diabetics with disease of the blood vessels of the retina	• Hemorrhage into the fluid in the eye (vitreous hemorrhage) • See your doctor at once
Sudden loss of vision on one side	• Headache, tenderness of the temple area, pain in the jaw on chewing, fever • May be associated with muscle and joint pains and fatigue	• Rare disease occurring in the elderly • More common in women	• Temporal arteritis • See your doctor at once
Sudden, painless deterioration of vision in one eye	• Vision loss may be moderate to severe (unable to count fingers)	• More common in people with glaucoma, atherosclerosis and high blood pressure	• Blockage of the central retinal vein • See your doctor at once
Sudden, complete, painless loss of vision in one eye	• Pupil does not react to light on one side	• More common in people with carotid artery atherosclerosis • Vision loss is often permanent	• Blockage of the central retinal artery • This is a medical emergency • Go to the emergency room or call an ambulance at once
Sudden painless loss of vision in one eye, like a curtain falling down	• Recent visual disturbances such as flashing lights or spots before the eyes • Pressure on eye may aggravate vision loss	• May be a history of trauma, glaucoma, cataract surgery, or nearsightedness	• Detached retina • This is a medical emergency • Go to the emergency room or call an ambulance at once

Facial pain

Facial pain can be a severe and distressing symptom. Facial neuralgia (recurrent stabbing, burning facial pain in a particular area of the face) may be severe, and do not respond well to simple painkillers. Any worrying facial pain should be assessed by a doctor.

CHARACTERISTIC SYMPTOM	ASSOCIATED SYMPTOMS	OTHER RELEVANT FACTORS	POSSIBLE CAUSE AND ACTION
Aching pain in the forehead, cheeks and upper jaw	• Feeling of pressure, nasal discharge, discharge from the back of the nose into throat, fever, tenderness of the forehead and facial bones	• May be a history of upper respiratory tract infection	• Sinus congestion or sinus infection (sinusitis) • Try decongestant medications from your pharmacist • See your doctor if you are worried or if symptoms persist
Episodes of sudden excruciating pain in the lips, gum, cheek or chin on one side of the face, lasting seconds to minutes	• Episodes occur regularly, day or night for several weeks	• Usually occurs in the elderly • More common in women than men • There is usually no known cause	• Tic douloureux (trigeminal neuralgia) • See your doctor at once
Episodes of aching or burning pain on one side of the face	• Numbness, tingling in the same area	• May be a history of shingles	• Pain following shingles (post-zoster neuralgia) • See your doctor at once
Episodes of burning, stabbing pain, usually on one side of the face lasting up to an hour	• Numbness or tingling, may be followed by a headache	• May be a history of migraine • More common in women	• Migrainous neuralgia • See your doctor at once
Severe aching pain from the side of the face, in front of the ear on one side	• Tenderness over upper jaw joint (temporomandibular joint), malocclusion of teeth	• Usually occurs in elderly women • May be a history of rheumatoid arthritis	• Arthritis of the temporomandibular joint • See your doctor

CHARACTERISTIC SYMPTOM	ASSOCIATED SYMPTOMS	OTHER RELEVANT FACTORS	POSSIBLE CAUSE AND ACTION

Fever

Fever is defined as a body temperature higher than the normal 98.6°F (37°C) taken orally. Mild or short-term temperature rises are common with minor infections, but high or sustained fever can signal a potentially dangerous and serious infection. Fever is often accompanied by other symptoms, as outlined below, which may help identify the cause.

The first table that follows applies to fever in adults and children of all ages. The second table deals with some additional symptoms specific to infants and children.

Fever in adults and children of all ages

CHARACTERISTIC SYMPTOM	ASSOCIATED SYMPTOMS	OTHER RELEVANT FACTORS	POSSIBLE CAUSE AND ACTION
Fever—body temperature above 98.6°F (37°C)	• Flushed face, hot skin, sore throat, mild headache	• May follow exposure to people with similar symptoms	• Viral infection (e g. cold or flu) • Take decongestant and fever-reducing medication • For any fever it is advisable to: – remove excess layers of clothing – drink plenty of fluids – take fever-reducing medication – have a lukewarm bath – check temperature every 4–6 hours • See your doctor if: – temperature remains at 102.2°F (39°C) or above in adults or 101.3°F (38.5°C) in children – fever lasts longer than 48 hours – the patient is pregnant – the patient is a child • In children, use a non-aspirin fever-reducing medication formulated for children • See your doctor if the fever does not respond quickly to medication or if the child has a sore throat or painful ear
Fever	• Aches, chills, nausea, vomiting, cramping, diarrhea	• Affects all age groups, but may cause life-threatening dehydration in the very young, the very ill, and the elderly	• Gastroenteritis or other viral infection • Rest and follow standard advice for any fever: – remove excess layers of clothing – drink plenty of fluids – take a fever-reducing medication – have a lukewarm bath – check temperature every 4–6 hours • Use anti-diarrhea and anti-vomiting medications as advised by your pharmacist or doctor • See your doctor if vomiting lasts longer than 12 hours or if there is bloody diarrhea • See your doctor if you are worried or if symptoms persist

CHARACTERISTIC SYMPTOM	ASSOCIATED SYMPTOMS	OTHER RELEVANT FACTORS	POSSIBLE CAUSE AND ACTION
Fever (cont.)			
Fever	• Severe headache, neck stiffness, drowsiness, vomiting, sensitivity to light	• May be confused and unable to respond well to questioning	• An infection in the area around the brain (meningitis) • This is a medical emergency • See your doctor or emergency department
Sudden onset of fever with simultaneous sore throat	• Headache	• Tends to occur during the colder months and can be precipitated by stress, overwork, exhaustion, and when the body's immune system is fighting other infections	• Strep throat (streptococcal infection) • Rest and follow standard advice for any fever: – remove excess layers of clothing – drink plenty of fluids – take a fever-reducing medication – have a lukewarm bath – check temperature every 4–6 hours • See your doctor
Fever	• Ear pain, hearing loss, feeling of fullness or fluid in the ear	• Fever may be the predominant sign in a child too young to indicate other symptoms	• Middle ear infection (otitis media) or outer ear infection (otitis externa) • Ear infections can lead to more serious problems if not treated, so see your doctor for assessment • Antibiotic treatment may be required • See your doctor
Fever	• Pain with urination or low back pain, tenderness on both sides of the lower back		• Kidney infection • You may require treatment with antibiotics • See your doctor at once
Fever	• Open sore or wound that is red • Red streaking on the arms or legs originating near the wound • Surrounding skin may be tender and hot; there may also be localized swelling		• Blood poisoning as a result of infection of the skin or lymphatic system • See your doctor at once
Fever in adults and children of all ages			
Fever in a child aged under 3 months	• Lethargy, pale skin, irritability		• Fever in a baby should always be investigated to rule out serious infection • See your doctor at once

CHARACTERISTIC SYMPTOM	ASSOCIATED SYMPTOMS	OTHER RELEVANT FACTORS	POSSIBLE CAUSE AND ACTION
High fever over 101.3°F (38.5°C)	• Barking cough		• Croup • See a doctor as soon as possible • For any fever it is advisable to: – remove excess layers of clothing – drink plenty of fluids – take a fever-reducing medication – have a lukewarm bath – check temperature every 4–6 hours • See your doctor at once
Fever	• Blisters over face, back, neck, and chest	• Occurs most commonly in children • May occur in adult not previously infected • Follows recent contact with person with chickenpox • Highly infectious, so common among childcare and hospital workers • Vaccination to prevent is now available in some countries	• Chickenpox • Chickenpox infection may be severe in adults, so seek medical advice at once • Take children with suspected chickenpox for assessment by a doctor, and keep child away from others who have not been infected • See your doctor • Chickenpox is highly contagious, so warn the doctor's office of the possibility of infection before attending
High fever, above 102°F (39°C)	• Seizure or convulsion may be triggered by high fever	• Three percent of children have at least one febrile convulsion • Cooling a feverish child in a lukewarm bath can help prevent a convulsion • A rapid rise in body temperature is more likely to cause seizures than a slow rise to the same temperature	• Febrile seizure • This requires immediate action • Ensure the airway is clear and turn child on to the side • Remove clothing and bathe or sponge with lukewarm water after the seizure has finished • See your doctor soon

Foot and ankle problems

Most people experience occasional problems with the foot or ankle. Ankle sprains are one of the most common musculoskeletal injuries, while degenerative disorders such as gout and heel spurs may often affect people as they age.

CHARACTERISTIC SYMPTOM	ASSOCIATED SYMPTOMS	OTHER RELEVANT FACTORS	POSSIBLE CAUSE AND ACTION
Pain and swelling, usually on one side of the ankle, following twisting injury or fall	• Bruising, warmth • May be aggravated by walking, but pain does not stop walking	• Most commonly affects the outer side of the ankle	• Injury to the ligaments in the ankle (ankle sprain) • Rest, apply ice, elevate the foot, use a compression bandage • See your doctor
Severe pain around the ankle following a fall, twisting injury, or direct blow to the ankle	• Swelling of the ankle and possibly foot and toes, throbbing, warmth, bruising, inability to walk		• Fracture or severe ligament sprain • Elevate the leg, apply ice • See your doctor at once
Pain under the heel and arch of the foot	• May be stiffness in the heel • Aggravated by prolonged walking or running	• Can lead to development of a heel spur (see directly below)	• Plantar fasciitis • See your doctor
Sharp pain under the heel	• May be some swelling • Aggravated by walking, pressing on the heel	• May be more common in those with flat feet	• Excess growth of bone at the heel (spur) • See your doctor

CHARACTERISTIC SYMPTOM	ASSOCIATED SYMPTOMS	OTHER RELEVANT FACTORS	POSSIBLE CAUSE AND ACTION
Foot and ankle problems (cont.)			
Sudden onset of severe pain in the big toe, foot, or ankle; pain often begins at night and may last for several days	• Swelling, skin over the affected area is usually red, shiny, and very tender to touch • May be fever and chills	• Most commonly affects the big toe, but can affect many other joints in the body • Usually affects one or two joints at a time, and attacks recur	• Gout • See your doctor
Dull ache or pain at the back of the heel that travels up the back of the ankle and lower calf	• Mild swelling, tenderness, warmth at the back of the ankle • Aggravated by running, jumping, walking, bicycling especially when commencing the activity	• More common in those with flat feet or tight muscles in the calf	• Inflammation of the Achilles tendon at the back of the ankle (tendinitis) • Avoid activities that cause pain • See your doctor for advice

Genital pain

Pain in the genital area is usually the result of an infection, or may occur following injury. Sexually transmitted diseases are a common cause of infection, so it is important to practice safe sex to prevent further spread of these conditions, and consult your doctor for treatment. Any pain with sexual intercourse should be investigated by a doctor. See also entry on vaginal problems.

CHARACTERISTIC SYMPTOM	ASSOCIATED SYMPTOMS	OTHER RELEVANT FACTORS	POSSIBLE CAUSE AND ACTION
Painful blisters on the genitals, which may break, weep, and form sores; the first occurrence of blisters may last from 5 days to several weeks; further episodes are usually shorter and less severe, and usually occur less frequently	• Itchy, tingling sensation, swollen and tender lymph nodes in the groin, pain on urination, discharge from the urethra in men and vagina in women • Weakness or constipation may occur • Erectile dysfunction may occur in men	• Triggers for an episode may be stress, illness, sexual intercourse, menstruation • If a pregnant woman has an attack of genital herpes toward the end of her pregnancy, there is a risk of passing it on to the baby and causing serious problems such as brain damage; cesarean section may overcome this problem	• Genital herpes • Wear cotton underwear and loose clothing • See your doctor at once • Avoid sexual contact until medical advice is obtained
In women, tender swelling of the vaginal opening or swelling of one labia	• May be hot and painful to touch		• Infection or abscess in a Bartholin's gland, or an infection of the labia • See your doctor at once
Burning sensation or pain with urination	• Frequent urge to urinate, even when bladder is empty • Urinating only small amounts • Burning sensation in lower abdomen, urine with a strong odor	• Cystitis is more common in women than in men • Infection may be triggered by sexual contact, use of diaphragm for birth control or use of urinary catheters	• Cystitis • Drink plenty of water • See your doctor
In men, tender and swollen tip of the penis	• May be aggravated by sexual contact, pressure from tight clothing		• Infection of the head of the penis (balanitis) • See your doctor at once
In men, pain with ejaculation	• Blood in the semen, tenderness with bowel movements, or pain behind the penis or scrotum	• Sudden-onset prostatitis caused by a bacterial infection is most frequent in young men • Prostatitis may be triggered by the use of urinary catheters	• Prostatitis • See your doctor at once
In men, pain during urination	• Pus or mucus may be visible in urine	• Urethritis is the most common form of sexually transmitted disease • Occurs most frequently in sexually active young men	• Urethritis, caused by sexually transmitted infection • See your doctor at once
In women, mild pain or discomfort while urinating	• Vaginal discharge, abdominal pain, rectal pain, sore throat	• Chlamydia is one of the most common sexually transmitted infections, and is most common in young, sexually active people	• Chlamydia • See your doctor at once • Avoid sexual contact until medical advice is obtained

CHARACTERISTIC SYMPTOM	ASSOCIATED SYMPTOMS	OTHER RELEVANT FACTORS	POSSIBLE CAUSE AND ACTION
Painful urination in men or women	• Inflamed genitals after sexual intercourse • In males, discharge of pus from the penis • In females, vaginal discharge, urge to urinate frequently, abnormal menstrual bleeding • The infection may spread from the genitals to the urethra, rectum, conjunctiva, pharynx or cervix	• The incidence of gonorrhea is much higher in the USA than in other industrialized countries • Potential complications include inflammation of reproductive organs, peritonitis, inflammation around the liver, inflammation of the Bartholin's gland in women and epididymitis or abscess around the urethra in men, arthritis, dermatitis, endocarditis, meningitis, myocarditis or hepatitis	• Gonorrhea • See your doctor at once
Pain and tenderness in the genital area	• Bruising, possibly some discharge, urinary tract infection • Person may appear irritable or fearful		• Sexual abuse or rape • See your doctor at once
Pain with sexual intercourse in women	• Bleeding after sexual intercourse, watery discharge from vagina, painful bowel movements, frequent urge to urinate	• Vaginal tumors are rare • Most chlamydial infections do not cause any symptoms, so transmission is very common and easy	• Tumor of the vagina • See your doctor at once

"Glands," swollen

The term "swollen glands" usually refers to swelling of the lymph nodes (also known as lymph glands) in the neck, armpit, or groin. There are many causes of swollen lymph nodes, ranging from mild infections to serious disorders such as cancer. Swelling often goes down as the infection resolves, but if your glands stay swollen for more than two weeks consult your doctor.

CHARACTERISTIC SYMPTOM	ASSOCIATED SYMPTOMS	OTHER RELEVANT FACTORS	POSSIBLE CAUSE AND ACTION
Swollen lymph nodes in the neck	• Fever, fatigue, sore throat, headache	• Infectious mononucleosis occurs most commonly in young adults • The virus is contagious and is passed via kissing, coughing, and sneezing; sometimes known as "the kissing disease"	• Infectious mononucleosis (glandular fever) • See your doctor
Swollen, tender lymph nodes in the neck	• Sore throat, headache, fever, bad breath, white spots on the tonsils		• Streptococcal infection ("strep throat") • See your doctor
Enlarged, non-tender lymph nodes throughout the body	• Fatigue, night sweats, weight loss, fever, severe itching all over the body • Lymph nodes enlarge slowly and are usually painless	• Non-Hodgkin's lymphoma is a relatively common cancer, and occurs most frequently in children and young adults • Hodgkin's disease occurs most commonly in young adults or in people over 50 years, and is more prevalent in men than women	• Cancer of the lymphatic system (lymphoma) (e.g. Hodgkin's lymphoma or non-Hodgkin's lymphoma) • See your doctor at once
Swollen lymph nodes in the neck, groin, or armpit	• Weight loss, fatigue, tendency to bruise or bleed easily, loss of appetite • Anemia is common during the early stages of leukemia	• Leukemia may increase susceptibility to infections (e.g. pneumonia, tonsillitis) • Leukemia is most common in children and young adults	• Leukemia • See your doctor at once

CHARACTERISTIC SYMPTOM	ASSOCIATED SYMPTOMS	OTHER RELEVANT FACTORS	POSSIBLE CAUSE AND ACTION
"Glands," swollen (cont.)			
Swollen, inflamed lymph nodes just above the angle of the jaw, on one or both sides of the face	• Fever, fatigue, swelling of the lymph nodes under the tongue, testicular swelling in males, or abdominal pain	• Mumps is contagious, but less so than other infections such as measles or chickenpox • It is a preventable through immunization • In countries where mumps vaccination is widely practiced, mumps occurs most frequently in adults • Prior to routine vaccination it occurred most commonly in children • Mumps is spread by close contact	• Mumps • See your doctor

Gums, bleeding

Bleeding gums are usually a sign of gum disease, most often caused by poor oral hygiene. Smokers are more than twice as likely as non-smokers to develop gum disease, while other at-risk groups include people with diabetes, leukemia, and Crohn's disease.

CHARACTERISTIC SYMPTOM	ASSOCIATED SYMPTOMS	OTHER RELEVANT FACTORS	POSSIBLE CAUSE AND ACTION
Swollen, red gums that bleed easily, e.g. with brushing of the teeth or with eating	• Gums are tender when touched or when chewing food	• The hormonal changes of pregnancy can worsen the condition • Gingivitis is most commonly caused by poor oral hygiene, including inadequate brushing and flossing of the teeth	• Gingivitis • See your doctor or dentist
Bleeding, red gums	• Pus around the teeth, bad taste in the mouth, halitosis (bad breath) • Periodontitis may lead eventually to deepening of the pockets around the teeth, loose teeth, and loss of teeth	• Periodontitis occurs when gingivitis extends to the supporting structures of the teeth	• Periodontitis • See your dentist
Sudden onset of painful gums that bleed easily	• Fatigue, bad breath, excess saliva, appearance of gray-white mucus covering the gums	• Poor oral hygiene predisposes to this condition, known as trench mouth; the term comes from World War I when many soldiers in the trenches developed the infection	• Bacterial infection of the mouth (Vincent's angina, Vincent's disease known as trench mouth) • See your dentist
Tendency to bleed easily from the gums and nose	• Swollen lymph nodes in the neck, groin or armpit, weight loss, fatigue, tendency to bruise easily, loss of appetite	• Increased susceptibility to infections (e.g. tonsillitis, pneumonia)	• Leukemia or serious infection (e.g. AIDS) • See your doctor at once
Extremely sore, swollen gums that bleed easily	• Earaches, symptoms similar to those of sinusitis, nosebleeds, fever, weight loss, cough, fatigue	• Wegener's granulomatosis is a rare disease occurring most commonly in white people • It occurs at any age, most common around 40 years	• Wegener's granulomatosis, a rare, serious disease • See your doctor at once

Hand and wrist problems

Problems in the wrist or hand can be caused by a variety of conditions, including overuse, injury and falls. The hands are also a common site for developing arthritis.

CHARACTERISTIC SYMPTOM	ASSOCIATED SYMPTOMS	OTHER RELEVANT FACTORS	POSSIBLE CAUSE AND ACTION
Pain in the wrist or hand following repeated movements	• Aggravated by continuing the repeated movements; alleviated by rest	• Inflammation of the tendons in the wrist or hand (tendinitis) may be caused by overuse • Healing is often delayed in people with arthritis, diabetes, or gout	• Tendinitis • Try anti-inflammatory medication as advised by your pharmacist • See your doctor

CHARACTERISTIC SYMPTOM	ASSOCIATED SYMPTOMS	OTHER RELEVANT FACTORS	POSSIBLE CAUSE AND ACTION
Numbness, tingling, burning in the hand, wrist pain that shoots into the palm of the hand; may be worse at night	• Aggravated by flexing the wrist, making a fist • Weakness, may be mild swelling	• More common in women • Associated with occupations that involve repeated forceful movements of the wrist (e.g. using a screwdriver)	• Carpal tunnel syndrome • See your doctor
Pain, swelling, stiffness in the wrist and/or finger joints, often worse after periods of inactivity	• Affected joints may feel hot, possible chills or fever • May be aggravated by movement • Can progress to cause deformities of the hand	• Usually affects both hands/wrists at the same time • May have other joints that are affected • Affects more women than men	• Rheumatoid arthritis • See your doctor at once
Temporary, patchy, red and white discoloration of the fingers, usually following exposure to cold, may last for minutes or hours	• May be associated numbness, tingling, burning, feeling of pins and needles	• Can also affect the feet • Most common in young women • More likely in smokers • Alleviated by warming the hands	• Raynaud's disease • See your doctor

Headache

Headache is the term used to describe any form of pain or discomfort in the head. It is an extremely common problem and one that most people have experienced. While most headaches are minor and easily treated with pain relievers, some warrant further medical investigation, and occasionally signal a more serious problem.

CHARACTERISTIC SYMPTOM	ASSOCIATED SYMPTOMS	OTHER RELEVANT FACTORS	POSSIBLE CAUSE AND ACTION
Dull, non-throbbing pain that feels like a vise around the head, squeezing both temples, extending into the neck	• Scalp or neck tenderness, tight or tender neck and shoulder muscles • Symptoms may start after working in one position for several hours, or after driving; may be related to stress or anxiety		• Tension headache • Try relaxation techniques • Heat may help to relax neck and shoulder muscles, and analgesics may help the pain • See your doctor if you are worried or if symptoms persist
Intense, throbbing, one-sided headache, may be centered around the eye; pain may last from a few hours to several days	• Vomiting and nausea may occur • In some people, the headache is preceded by a warning sign (aura), which may include visual disturbances such as flashing lights or spots • Oversensitivity to light, odors, or sound may be experienced	• Migraine may be triggered by certain foods (e.g. cheese, strawberries, chocolate) • In women, migraine may be associated with the menstrual cycle	• Migraine • Take analgesics such as acetaminophen (paracetamol) at once on onset of symptoms, and lie down • See your doctor if you are worried or if symptoms persist
Throbbing pain in the front of the head and around the eyes	• Feeling of pressure around the eyes and nose, thick nasal discharge • May worsen with bending forward; may follow a recent cold or episode of hay fever	• Occurs with viral infections, allergies, deep sea diving, or dental infections	• Sinusitis • Decongestants may be helpful to relieve symptoms • See your doctor if you are worried or if symptoms persist
Severe headache	• Stiff neck, vomiting, fever, drowsiness, delirium, unconscious, or have convulsions • May be worsened by exposure to bright lights	• A child may be difficult to wake and have an unusual high-pitched moaning cry	• Meningitis • This is a medical emergency • Go to the emergency room or call an ambulance at once
Severe, piercing pain in and around one eye, lasting 30 minutes to several hours; headaches may occur one or more times a day for a period of weeks or months	• The affected eye may be bloodshot and watery • There may be associated nasal congestion and facial flushing • Pain often occurs at night	• Occurs much more frequently in men than in women	• Cluster headache • See your doctor

CHARACTERISTIC SYMPTOM	ASSOCIATED SYMPTOMS	OTHER RELEVANT FACTORS	POSSIBLE CAUSE AND ACTION
Headache (cont.)			
Persistent, throbbing headache that begins first thing in the morning, and may lessen during the day	• Vomiting, nausea, fatigue, blurred vision, weakness • Headache may be unlike any other headache the person has had before • May be worsened by changing positions (e.g. moving from lying down to standing	• Hypertension causing these symptoms is relatively rare • Brain tumors are rare, and may be inherited or associated with exposure to ionizing radiation	• Severe hypertension or brain tumor • This is a medical emergency • Go to the emergency room or call an ambulance at once
Severe headache that begins suddenly	• Vomiting, limb weakness, double vision, slurred speech, difficulty swallowing, loss of consciousness		• Cerebral hemorrhage or aneurysm • This is a medical emergency • Go to the emergency room or call an ambulance at once

Hearing loss (deafness)

Significant loss of hearing is common, and can result from damage to the ear, disease of the ear or changes with age that damage the delicate structures enabling us to hear. Hearing loss is due to disturbances in the external or middle ear or abnormalities in the inner ear or neuronal (nerve) pathways. Any persistent loss of hearing should be assessed by a doctor to determine the type of hearing loss, possible causes and whether treatment is available which will restore hearing.

CHARACTERISTIC SYMPTOM	ASSOCIATED SYMPTOMS	OTHER RELEVANT FACTORS	POSSIBLE CAUSE AND ACTION
Intermittent hearing loss in one or both ears	• Usually no associated symptoms or pain	• May have a history of ear wax blockage • Wax-softening drops may relieve hearing loss	• Ear wax blockage (ceruminosis) • See your doctor • Do not try to remove the blockage yourself, as you may damage the eardrum or small bones in the ear • Your doctor has special instruments to do this safely
Sudden hearing loss in one or both ears	• Fever, symptoms of a cold, discomfort ranging from a feeling of pressure in the ear to persistent, severe ache in one or both ears • Nausea and vomiting, dizziness, or ringing in the ears	• Most commonly occurs following a cold • Most common in children aged 3 months to 3 years, due to narrow eustachian tubes (tubes that allows pressure between the mouth or nose and ears to be equalized) • Eustachian tubes that are not fully developed may block easily with inflammation of the nose or throat, or with allergies • Decongestants can aid drainage from the inner ear	• Otitis media (viral or bacterial ear infection) commonly caused by a cold virus • Symptoms are caused by build-up of fluid • Perforation of the eardrum is a potential complication, which can lead to more prolonged but usually temporary hearing loss • See your doctor • See your doctor at once if the patient is a child
Recurrent episodes of hearing loss, mainly low tone sounds, usually lasting 20 minutes to several hours	• Dizziness, ringing, rushing or buzzing sound in the ears (tinnitus), nausea, feeling of movement, or dizziness (vertigo) • Occurs intermittently	• Affects only one ear in the majority of people with the disease • May be caused by fluid in the canals of the inner ear • Reducing dietary salt, caffeine, alcohol may help control episodes	• Menière's disease • See your doctor • Antiemetics may provide symptomatic relief
Gradual onset of hearing loss on one side	• May have facial weakness on the same side	• Acoustic neuromas grow slowly and are more common in older people	• Acoustic neuroma (a benign tumor of nerve cells) • Tumors may grow large enough to put pressure on other structures • See your doctor

CHARACTERISTIC SYMPTOM	ASSOCIATED SYMPTOMS	OTHER RELEVANT FACTORS	POSSIBLE CAUSE AND ACTION
Hearing loss in childhood	• Delayed language development	• Infection during pregnancy (e.g. rubella or cytomegalovirus) can cause congenital hearing loss in children • Repeated ear infections may cause deafness in children • Meningitis can lead to hearing loss in one or both ears • Any child with suspected hearing loss or delayed language development should have their hearing tested	• Congenital deafness, meningitis, or otitis media • See your doctor
Heartburn Heartburn does not involve the heart, but is a traditional name given to a symptom of a digestive problem that can often be relieved by indigestion medications. However, it is important to make sure the chest pain is not caused by angina or a heart attack.			
Painful, burning sensation in the chest, behind the breast bone, which may rise up to the throat	• Bitter taste in the mouth • Large meals, fatty or spicy foods may cause symptoms • Lying down or bending may worsen symptoms • Smoking or alcohol may aggravate symptoms • Tight clothing or belts may make symptoms worse	• Typically occurs after food • Heartburn may occur during pregnancy but usually resolves after the baby is born	• Back-washing of food and stomach acid upward into the esophagus (gastroesophageal reflux) • Take an antacid as advised by your pharmacist or doctor • Avoid foods that seem to cause the symptoms, and do not eat within 2 hours of going to bed • Quit smoking and reduce alcohol intake • Raise the head end of the bed 4 inches (10 centimeters) • Lose weight if you are overweight • See your doctor if you are worried or if symptoms persist
Intermittent pain behind the breast bone, which at first may not be easily distinguished from heartburn	• Worse with exercise, relieved by resting or rapidly-acting nitrate drugs	• Most common in those with previous history of coronary heart disease	• Angina pectoris, a form of coronary heart disease • See your doctor at once • Failure to treat the cause of angina may result in a heart attack
Intense chest pain, which may initially be mistaken for severe heartburn	• Pain spreading to left arm or both arms; pain in jaw; feeling of chest being squeezed	• Increased likelihood in a person who has previously had a heart attack or with a history of coronary heart disease	• Heart attack (myocardial infarction) due to sudden loss of blood supply to a section of the heart muscle due to blockage of the coronary arteries supplying the heart muscle • This is a medical emergency • Go to the emergency room or call an ambulance at once
Hip problems Hip problems often occur following a fall, especially in the elderly, or because of arthritis. Other causes of hip pain and stiffness include frequent running or problems with the cartilage in the hip joint.			
Intense hip pain following a fall	• Leg may be held in an abnormal position, may develop swelling • Aggravated by standing, straightening the leg, lifting the leg	• More likely to occur in the elderly	• Hip fracture • See your doctor at once

CHARACTERISTIC SYMPTOM	ASSOCIATED SYMPTOMS	OTHER RELEVANT FACTORS	POSSIBLE CAUSE AND ACTION
Hip problems (cont.)			
Stiffness and pain in one or both hips	• May have swelling and redness around the joints, stiffness and pain in other joints • Stiffness often aggravated by long periods in one position • Pain aggravated by lots of walking or standing	• More likely in older people	• Arthritis • See your doctor for advice
In infants, clicking of the hip	• May be some pain when the hip is stretched, movement may be limited	• More common in girls than boys, also more common in babies born breech (buttocks first), or in those with a relative who has the same disorder	• Congenital dislocation of the hip • See your doctor
In teenagers, stiffness in the hip, pain, and limping	• May also have pain in the knee or thigh • Affected leg may be twisted outward • Aggravated by walking	• More common in overweight teens • Affects boys more than girls	• Dislocation of the top of the thigh bone (slipped capital femoral epiphysis) • See your doctor at once
In children, gradual onset of hip pain and stiffness; symptoms progress slowly	• Limping, wasting of thigh muscles, limited movements • Aggravated by walking	• Most common in 5–10 year olds • Affects boys more than girls	• Degeneration of the top of the thigh bone (Perthes' disease) • See your doctor
Shooting or burning pain in the back of one hip or buttock; pain may travel down the back of one leg	• May also have low back pain, numbness or tingling in the foot • Aggravated by coughing, sneezing, bending, lifting	• More common in people with stiff backs or past back injury	• Sciatica • See your doctor

Knee problems

Knee pain is a common symptom in all age groups. Problems range from mild pain under the kneecap to ligament tears requiring surgery. The knee is also a common site for developing arthritis.

CHARACTERISTIC SYMPTOM	ASSOCIATED SYMPTOMS	OTHER RELEVANT FACTORS	POSSIBLE CAUSE AND ACTION
Intermittent pain inside the knee or along one side of the knee, often starts following a twisting injury	• Knee may lock, may feel blocked and unable to straighten it fully, may have clicking of the knee; some swelling may be present • Aggravated by squatting or twisting	• More commonly occurs on the medial (inside) part of the knee	• Tearing of the cartilage in the knee (meniscus) • Apply ice and rest • See your doctor
Knee pain following a fall, twisting injury, hyperextension (knee forced straight) injury, or direct blow to the knee	• Popping sound at the time of injury, swelling that develops soon after the injury, giving way of the knee	• Common sporting injury and one of the most serious, often requiring surgery and extensive rehabilitation	• Tearing of one of the ligaments running through the knee joint from front to back (anterior cruciate ligament) • See your doctor
Long-term aching and stiffness in the knee that has become worse over a period of months	• Limited movement; the person may be unable to bend or straighten fully • Pain is alleviated by rest, aggravated by a lot of activity • Stiffness is often worse in the morning	• More common in people over 50 years of age	• Osteoarthritis • See your doctor
Pain along the inner or outer knee, usually following a direct force to one side of the knee while the foot remains planted on the ground	• Tenderness on one side of the knee	• Often occurs during sports such as soccer, football, skiing	• Damage to one or more of the ligaments on the inside or outside of the knee (collateral ligaments) • See your doctor

CHARACTERISTIC SYMPTOM	ASSOCIATED SYMPTOMS	OTHER RELEVANT FACTORS	POSSIBLE CAUSE AND ACTION
Red, swollen knee	• Constant ache, swelling, fever, generally feeling unwell • May be aggravated by movement	• Infection may spread through the blood, through a penetrating injury	• Osteomyelitis or joint infection (septic arthritis) • See your doctor at once

Leg problems

Leg problems can arise from a variety of conditions, from simple muscle strains to fractures and serious circulation disorders. See also entries on knee problems, hip problems, and foot and ankle problems.

CHARACTERISTIC SYMPTOM	ASSOCIATED SYMPTOMS	OTHER RELEVANT FACTORS	POSSIBLE CAUSE AND ACTION
Sudden pain in the leg associated with quick movement of the leg (e.g. kicking, sprinting, change of direction)	• Swelling, bruising • Pain when stretching or bending the leg, but can still move it	• Common sporting injury, often affects hamstrings (muscles at back of thigh), quadriceps (muscles on the front of thigh), and calf muscles	• Muscle strain or tear • Rest and apply ice to the area • See your doctor
Severe, constant leg pain following an injury, fall, or after a direct blow to the leg	• Swelling, may be some deformity of the leg • Aggravated by attempts to walk or move the leg		• Fracture • This is a medical emergency • Go to the emergency room or call an ambulance at once
Intermittent pain over the front of the shin	• May be some pain when the shin is pressed • Aggravated by repetitive motion (e.g. running, bicycling, jumping, walking up and down hills)	• More common in those with flat feet, bow legs, knock knees	• Shin splints • See your doctor
Cramping pain in the calves, feet, or hips while walking	• May develop numbness or tingling in the feet • Aggravated by walking, especially quickly or up hills, usually alleviated by rest	• Risk is increased in people with abnormal cholesterol or triglyceride levels, diabetes, hypertension, men over 55 years old, women over 65 years old, those with a family history of cardiovascular disease, cigarette smokers, and those with obesity or little physical activity	• Peripheral vascular disease • See your doctor
Shooting or burning pain in the buttock and down the back of one leg	• May also have low back pain, numbness or tingling in the foot • Aggravated by coughing, sneezing, bending, lifting	• More common in people with stiff backs or past back injury	• Sciatica • See your doctor
Pain and swelling in the back of the calf	• Warmth, pain when touched	• Most common following surgery or after long air flights or bus trips	• Deep venous thrombosis • This is a medical emergency • Go to the emergency room or call an ambulance at once

Menstrual problems

Most menstrual problems warrant a full medical investigation, since they may indicate the presence of a disease that requires treatment. Although period pain or premenstrual syndrome may respond to simple self-treatment, it is important to ask your doctor's advice if symptoms persist or worsen.

CHARACTERISTIC SYMPTOM	ASSOCIATED SYMPTOMS	OTHER RELEVANT FACTORS	POSSIBLE CAUSE AND ACTION
Temporary emotional instability just prior to menstrual period	• Bloating or discomfort in lower abdomen, irritability, depression, tearfulness, inability to concentrate, sleep disturbances, fatigue, lethargy • Symptoms usually disappear when menstruation begins • Caffeine may worsen irritability	• Approximately one-third of fertile women experience some premenstrual symptoms • The full premenstrual syndrome occurs in about 3–10 percent of fertile women	• Premenstrual syndrome (PMS) • There is no standard treatment • Ask your pharmacist's advice on over-the-counter medication for bloating or pain • Vitamin B_6 supplements may help ease the symptoms • See your doctor if you are worried or if symptoms persist

CHARACTERISTIC SYMPTOM	ASSOCIATED SYMPTOMS	OTHER RELEVANT FACTORS	POSSIBLE CAUSE AND ACTION
Menstrual problems (cont.)			
Mild to moderate cramping pain during menstrual period	• Pain may be aggravated by flatulence or constipation	• Period pain sufficiently severe to cause missed school or work days is common in teenagers and young women • Pain may be alleviated by heat applied to the lower abdomen (e.g. a hot water bottle or bath) • Severe symptoms may suggest endometriosis	• Period pain (menstruation) • Try analgesic or non-steroidal anti-inflammatory drugs as recommended by your pharmacist • See your doctor if you are worried or if symptoms persist
Gradual onset of more pain than usual during and just before menstrual period	• Low back pain, period pain lasting more than 2–3 days and starting before the onset of bleeding • Spotting of small amounts of blood for 1–3 days prior to onset of period • Menstrual bleeding may be heavier than usual • Pain in pelvic area may worsen during sexual contact	• Endometriosis occurs in approximately 5–10 percent of women, and is more likely in women with a mother or sister with the disease, and in women who have never become pregnant • Pregnancy may temporarily resolve the problem, though symptoms may recur months or years later • Endometriosis may result in infertility	• Endometriosis • See your doctor
More pain than usual during and just before menstrual period	• Fever, vaginal discharge with offensive odor, abnormal vaginal bleeding, abdominal pain, pain during urination • Onset of symptoms is usually gradual when caused by an intrauterine device (IUD)	• Pelvic inflammatory disease occurs almost exclusively in sexually active women • Pelvic inflammatory disease may result from infections (usually sexually transmitted infections), uterine surgery (e.g. dilation and curettage, insertion of IUD, cesarean section), or childbirth	• Pelvic inflammatory disease • See your doctor at once
Increased volume and length of menstrual bleeding in women with an intrauterine device (IUD)	• Spotting of blood between menstrual periods, increased pain during periods	• IUD may change the pattern of menstrual bleeding	• IUD related adverse effect • See your doctor
Excessive menstrual bleeding	• Pain during menstrual bleeding, longer than usual menstrual periods	• Fibroids are most common in women over 35 years old or who have had several pregnancies	• Uterine fibroids • See your doctor
Irregularity or cessation of menstrual periods in a woman who is not pregnant	• Fatigue or lethargy may occur with thyroid disease	• A menstrual period may occasionally be missed in some women during the use of oral contraceptives • Excessive exercise or weight loss (e.g. during athletic training or anorexia nervosa) may cause cessation of menstrual periods	• Hormonal abnormality due to an ovarian problem, oral contraceptive use or a thyroid problem • See your doctor
Cessation of menstrual periods in sexually active women	• Breast tenderness, abdominal bloating or feeling of fullness, nausea	• All methods of contraception carry a slight chance of failure leading to pregnancy	• Pregnancy • Use pregnancy test kit—if positive, see your doctor • See your doctor if you are worried or if symptoms persist
Cessation of menstrual periods in women aged over 35 years	• Irritability, hot flashes (flushes)	• The onset of menopause most commonly occurs between the ages of 40–55 • Early menopause may occur from 35 years, or younger in rare cases	• Menopause • See your doctor

CHARACTERISTIC SYMPTOM	ASSOCIATED SYMPTOMS	OTHER RELEVANT FACTORS	POSSIBLE CAUSE AND ACTION
Recommencement of menstrual bleeding in a woman who has already gone through menopause	• Abdominal swelling or discomfort, vaginal discharge	• Some hormonal medications may cause uterine bleeding	• Uterine tumor or vaginal infection • See your doctor

Nausea and vomiting

Nausea and vomiting occur with many medical conditions, and the cause is not always obvious. Since some conditions that may cause these symptoms are potentially serious, it is advisable to consult a doctor if the problem does not resolve quickly. If symptoms recur, and/or are accompanied by any other unusual symptoms, you may need medical tests to find the problem. When a person vomits blood or has severe pain, the situation should be treated as an emergency and a doctor consulted at once.

CHARACTERISTIC SYMPTOM	ASSOCIATED SYMPTOMS	OTHER RELEVANT FACTORS	POSSIBLE CAUSE AND ACTION
Nausea and vomiting after eating	• Diarrhea may follow • Unable to tolerate food or liquids	• Symptoms occur after eating food that may have been kept too long or at incorrect temperature such as hot food kept warm several hours, or cold food that has been kept at room temperature or uncovered for several hours	• Bacterial contamination of food (food poisoning) • Take frequent small amounts of fluid, if tolerated • Typical cases of food poisoning will usually pass in under 12 hours • See your doctor if the person is severely ill and unable to drink fluids, if you are worried, or if symptoms persist
Intermittent nausea and vomiting	• Burning pain high in the abdomen • Worse after eating, especially spicy foods • Bland foods may relieve pain	• Use of anti-inflammatory medications for pain (prescription or non-prescription) may damage stomach lining • Ulcers are commonly caused by a bacterial infection and require antibiotics	• Gastritis, or ulcer of stomach or esophagus • If symptoms are mild and not persistent, use an antacid (your pharmacist may advise you on a suitable choice) • See your doctor
Recent onset nausea and vomiting	• Fever and cold or flu symptoms, diarrhea • Inability to tolerate food or liquids		• Viral gastroenteritis • Rest and take frequent small amounts of fluids if tolerated (e.g. diluted soft drink or an electrolyte sachet from your pharmacist) • Your doctor or pharmacist may advise you further about treating specific symptoms • See your doctor if the person is unable to tolerate fluids, if you are worried, or if symptoms persist
Nausea and vomiting with intermittent severe pain	• Pain in the upper right abdomen, fever • Pain may worsen after eating greasy foods		• Gallbladder inflammation or gallstones • See your doctor if you are worried or if symptoms persist
Nausea and vomiting with steady worsening pain	• Recent onset abdominal pain in middle or lower right, fever	• Pain may begin as dull discomfort centrally and become more severe and localized to the right side	• Appendicitis or a bowel obstruction • This is a medical emergency • Go to the emergency room or call an ambulance at once
Persistent nausea and vomiting over more than a week in women of child-bearing age	• Missed menstrual period • Certain foods or smells may worsen symptoms • Symptoms may be consistently worse at certain times of day	• Non-predictable; a woman may experience morning sickness with one pregnancy but not a subsequent pregnancy	• "Morning sickness" of pregnancy • See your doctor • Avoid an empty stomach by eating frequent small meals • Nibbling dry crackers between meals and before getting out of bed may help

CHARACTERISTIC SYMPTOM	ASSOCIATED SYMPTOMS	OTHER RELEVANT FACTORS	POSSIBLE CAUSE AND ACTION
Nausea and vomiting (cont.)			
Vomiting in a baby or young child	• Crying, irritability, or quietness, or an inability to become interested in toys	• Child under 2 vomiting for more than 6 hours, or child over 2 vomiting for more than 12 hours	• Viral infections are a common cause • Children may rapidly become dehydrated • See your doctor • If you suspect severe dehydration, go to the emergency room or call an ambulance at once
Vomiting in a baby or young child	• Uncontrollable crying, dark red diarrhea, unable to keep down any fluids	• Obstruction is relatively rare	• Intestinal obstruction • This is an emergency • Visit your doctor or the emergency room at once
Vomiting in a baby	• Forceful expulsion of stomach contents, persistent vomiting	• 20 percent of healthy babies vomit or regurgitate frequently enough to worry parents and cause them to seek medical advice • Approximately 7 percent of babies show more severe symptoms suggesting gastroesophageal reflux disease	• Stomach obstruction or reflux • Ask your doctor's advice to confirm the cause

Neck problems

Symptoms involving the neck may result from a wide variety of conditions. Infections in the body will often lead to swelling of the neck glands, while poor posture and arthritis can cause neck pain and stiffness.

CHARACTERISTIC SYMPTOM	ASSOCIATED SYMPTOMS	OTHER RELEVANT FACTORS	POSSIBLE CAUSE AND ACTION
Neck stiffness that is present after sleep or periods of inactivity, gradually worsens over time	• Pain and limitation of movement, spine may be tender to touch • Aggravated by periods of inactivity or following exercise	• Most common in those aged over 40 • May be alleviated by moving the neck gently	• Osteoarthritis • See your doctor
Intense neck pain that radiates into the shoulders and possibly down the arms	• Tingling or numbness in the hands, arm weakness • May be aggravated by neck movements, sneezing, or coughing	• May follow an injury or begin after regular daily activities	• Vertebral disk injury causing pressure on a spinal nerve • See your doctor
Neck stiffness with a severe headache	• Vomiting, fever, drowsiness, may become delirious, unconscious, or have convulsions • Exposure to bright lights increases pain	• A child may be difficult to wake or have a high-pitched cry	• Meningitis • This is a medical emergency • Go to the emergency room or call an ambulance at once

Numbness and tingling

A feeling of numbness or tingling usually results from a malfunction in part of the body's nervous system. The symptoms may be caused by an isolated problem in one nerve, or may be part of a more serious degenerative disease. See also entry on weakness and paralysis.

CHARACTERISTIC SYMPTOM	ASSOCIATED SYMPTOMS	OTHER RELEVANT FACTORS	POSSIBLE CAUSE AND ACTION
Numbness or tingling in one arm or one leg	• Neck or back pain, weakness of the affected limb • In serious cases, may have difficulty urinating • Aggravated by sitting, bending forward, sneezing, coughing	• The precise location of the symptoms defines which part of the back or neck is affected	• Pressure on a nerve caused by swelling of a ruptured or bulging vertebral disk in the spine • See your doctor at once

CHARACTERISTIC SYMPTOM	ASSOCIATED SYMPTOMS	OTHER RELEVANT FACTORS	POSSIBLE CAUSE AND ACTION
Tingling or numbness in the arms, legs, trunk, or face	• Loss of strength or dexterity, vision disturbances, dizziness, unusual tiredness, difficulty walking, trembling, loss of bladder control • Aggravated by very warm weather, hot bath, fever	• More common among people who have lived in a temperate climate up to age 10 • Occurs much less commonly in those whose childhood was spent in a tropical climate, and extremely rare at the equator	• Multiple sclerosis • See your doctor at once
Numbness or tingling on one side of the body; symptoms usually start suddenly	• Weakness in hands or feet, confusion, dizziness, partial loss of vision or hearing, slurred speech, inability to recognize parts of the body, unusual movements, fainting	• More common with advancing age • Risk increased in people with cardiovascular disease (e.g. hypertension, coronary heart disease), lipid abnormalities (e.g. high cholesterol), diabetes	• Transient ischemic attack or stroke • This is a medical emergency • Go to the emergency room or call an ambulance at once

Palpitations

Palpitations is the term used to describe an uncomfortable awareness of your heartbeat. The palpitations may take the form of fluttering, throbbing, pounding, or racing in the chest. The heart may feel as though it is beating irregularly. Palpitations may be harmless, but in certain cases they signal underlying heart disease.

CHARACTERISTIC SYMPTOM	ASSOCIATED SYMPTOMS	OTHER RELEVANT FACTORS	POSSIBLE CAUSE AND ACTION
Recurrent fluttering, racing, pounding, thumping in the chest; may have feeling of a strong pulse in the neck	• Chest discomfort, weakness, dizziness, shortness of breath	• There are many types of variation from normal heartbeat rhythm, some of which are serious • Arrhythmia is most commonly caused by heart disease, but may also occur with caffeine use, excessive alcohol, vigorous exercise	• Arrhythmia • See your doctor at once
Temporary racing, pounding, thumping in the chest; usually lasts for 10–20 minutes	• Trembling, dizziness, shortness of breath, feeling of choking, nausea, diarrhea, an out-of-body sensation, tingling in the hands, chills, fear of dying • May be aggravated by stress, and the fear of further attacks	• Women are 2–3 times more likely than men to have these attacks	• Panic attack • See your doctor if you are worried or if symptoms persist
Racing heartbeat	• Shortness of breath on exertion, tiring easily, swelling in the legs and abdomen • May be sudden fever and flu-like symptoms	• Can occur as the result of infection, or in association with many diseases, or may have no identifiable cause	• Cardiomyopathy (disease of the heart muscle) • See your doctor at once
Awareness of forceful heartbeats, especially when lying on the left side	• Shortness of breath on exertion, swelling of the legs, chest pain, dizziness	• More common in those who have had rheumatic fever	• Heart valve disorder • See your doctor for advice
Sudden, heavy pounding or thumping in the chest	• Pain in the middle of the chest that may spread down the left arm, sweating, shortness of breath, faintness, anxiousness, sense of impending doom • Symptoms are not alleviated by rest	• Increased likelihood in a person who has previously had a heart attack or with a history of coronary heart disease	• Myocardial infarction (heart attack) • This is a medical emergency • Go to the emergency room or call an ambulance at once

CHARACTERISTIC SYMPTOM	ASSOCIATED SYMPTOMS	OTHER RELEVANT FACTORS	POSSIBLE CAUSE AND ACTION

Seizures (known as "fits," including convulsions)

Seizures result from an abrupt episode of abnormal electrical activity within the brain. There are many different types of seizures and many possible causes. Seizures may be generalized (generalized tonic-clonic convulsion also known as grand mal seizure) or localized to a particular part of the body (focal convulsion). Some seizures manifest as a brief aura followed by loss of awareness of surroundings. Any seizure warrants assessment by a doctor and often full medical investigation.

CHARACTERISTIC SYMPTOM	ASSOCIATED SYMPTOMS	OTHER RELEVANT FACTORS	POSSIBLE CAUSE AND ACTION
Repeated episodes of a generalized tonic-clonic seizure (grand mal seizure); begins with stiffness of limbs and jaw locking (tonic phase) followed by jerking of limbs (clonic phase) then a period of drowsiness and confusion (postictal phase)	• Urinary incontinence during fit • Aggravated by sleep deprivation, flickering lights, hyperventilation	• No fever or current illness	• Epilepsy • See your doctor at once
Repeated episodes of seizures that involve disturbances in the senses (sensory seizures)	• May have preceding aura involving visual and auditory hallucinations or distortions of taste and smell, followed by period of altered awareness sometimes associated with lipsmacking or repetitive movements (automatisms) • May be brought on by sleep deprivation, flickering lights, hyperventilation		• Temporal lobe epilepsy • See your doctor at once
Brief generalized seizure in child under 5 years	• Fever • No signs of infection of the brain (encephalitis) or covering of the brain (meningitis), no history of epilepsy • Rapid rise in temperature	• 3 percent of children have at least one febrile convulsion	• Simple febrile seizure of childhood • See your doctor at once
Generalized or focal convulsion	• Headache, drowsiness, neck stiffness, oversensitivity to light, fever • May have preceding febrile illness	• May occur in previously healthy person	• Meningitis or encephalitis • This is a medical emergency • Go to the emergency room or call an ambulance at once
Isolated generalized or focal convulsion	• Headache, nervous system abnormalities, decreased consciousness, newly developed squint • May follow head injury	• May indicate raised pressure within the confined space of the skull	• Brain tumor, abscess, or cerebral hemorrhage • This is a medical emergency • Go to the emergency room or call an ambulance at once
Repeated episodes of focal seizures (localized to a particular part of the body)	• Involuntary movements may occur in a single limb, one side of the body, or involve eyes deviating to one side	• Indicates a localized lesion within the brain triggering the seizures	• Head injury is the most common cause in young adults, while cerebrovascular accidents (strokes) are the most common cause in the elderly • Congenital malformations of the brain, early meningitis, or perinatal brain damage are common causes in children • This is a medical emergency • Go to the emergency room or call an ambulance at once

CHARACTERISTIC SYMPTOM	ASSOCIATED SYMPTOMS	OTHER RELEVANT FACTORS	POSSIBLE CAUSE AND ACTION

Sexual function problems

Several problems limit sexual pleasure or the proper function of sex organs. See your doctor, since many problems affecting sexual function may be treated. Untreated sex problems can lead to relationship problems, depression, and anxiety. See also entries on genital pain and vaginal problems.

CHARACTERISTIC SYMPTOM	ASSOCIATED SYMPTOMS	OTHER RELEVANT FACTORS	POSSIBLE CAUSE AND ACTION
In males, ejaculation before or immediately after intercourse begins	• Anxiety, frustration, depression • May be aggravated by further worry about it happening	• More common in young men • Physical causes are rare, and most cases have a psychological cause	• Premature ejaculation • See your doctor if you are worried or if symptoms persist
In males, inability to have or keep an erection sufficient for sexual intercourse	• Anxiety, frustration, depression	• More likely as men get older • Physical disorders are the main cause, especially in men aged over 50	• Erectile dysfunction (impotence) • See your doctor
In males, pain during sexual contact	• May be redness or rash, or other symptoms of infection such as fever • Aggravated by continued sexual contact		• Infection (e.g. prostate, testes, or urethra), allergic reaction to spermicide • See your doctor
Lack of sexual desire or inability to experience sexual pleasure	• Anxiety, frustration • May be aggravated by stress, fatigue, anxiety, relationship problems	• More common in women than men • Physical and psychological causes can lead to this problem	• Arousal dysfunction • See your doctor
In females, pain during intercourse; pain may be in the vaginal area or deeper in the pelvis	• Vaginal discharge, itching, dryness • May be aggravated by continued sexual contact	• Pelvic inflammatory disease and infections are most common among young, sexually active women • Endometriosis is most common among women with a family history of the disease, and in women who have never been pregnant • Hormonal imbalances causing vaginal dryness are more common following menopause	• Pelvic inflammatory disease, infections, hormonal imbalance, endometriosis • See your doctor
In females, inability to have intercourse due to contraction of the vaginal muscles	• Fear, anxiety, pain	• This is an involuntary response, outside the woman's control	• Vaginismus • See your doctor

Skin problems

The skin can show a very wide range of noticeable changes. It is important to check your skin regularly and report any changes to your doctor, since it is often difficult to tell the difference between significant changes (e.g. early skin cancers or eruptions due to other diseases) and unimportant ones, by appearance alone.

CHARACTERISTIC SYMPTOM	ASSOCIATED SYMPTOMS	OTHER RELEVANT FACTORS	POSSIBLE CAUSE AND ACTION
A new, growing or changing, brown or blue-black pigmented lesion, usually irregular or asymmetric in shape and color; often over ¼ inch (0.5 centimeter) in diameter; may be bleeding or ulcerated	• Usually painless • May be itchy • Surrounding skin may be inflamed	• Melanoma occurs in all adult age groups but is rare in pre-pubescent children • Risk is increased in people with fair skin or hair, many freckles or moles, or moles of unusual appearance • Up to 50 percent of melanomas develop from moles • Sun exposure, especially before age 10, may predispose to melanoma	• Melanoma • See your doctor at once
Sharply defined white (depigmented) patches of skin		• Vitiligo may occur in people with a family history of the disease or in those with immune disorders	• Vitiligo • See your doctor

CHARACTERISTIC SYMPTOM	ASSOCIATED SYMPTOMS	OTHER RELEVANT FACTORS	POSSIBLE CAUSE AND ACTION
Skin problems (cont.)			
Rash of pimples and pustules on face, chest, and back	• Inflamed, raised, red spots, excessive oiliness, blocked pores (whiteheads, blackheads), scarring • May be exacerbated by some foods or medications	• Acne is most common in teenagers but also occurs in 10–20 percent of adults	• Acne (acne vulgaris) • See your doctor
Unusual growth or ulcerated, raised lump on the face	• May be itchy or painful	• Incidence increases with age • Skin cancers are most common on the face, but can develop on other sun-exposed areas of the body • Growth rate depends on the type of cancer; the growth may develop over a month or so, or slowly over many months	• Skin cancer • See your doctor
Small (pinpoint or pin head size), raised, round, pink or pearly, shiny bumps with pits in the center		• Molluscum contagiosum is most common in children • Contacts (e.g. family members or friends) may also be affected • Commonly occurs on the face, eyelids, or genitals, but may develop on any area	• Molluscum contagiosum • See your doctor
Small, red, warm, tender bump around a hair follicle, that develops suddenly	• Painful	• Boils may occur singly, or several may appear at the same time • Occasionally multiple boils in the same area result in inflammation of the whole area • Conditions that may predispose to boils include scratching of the skin, which allows bacteria to enter, illnesses that lower the body's resistance (e.g. diabetes)	• Boil • See your doctor
Red bumps or elevated red patches that appear suddenly, each of which lasts from a few hours to 2 days; may be white in the center	• Itching and tingling or a pricking sensation • Swelling around the mouth or throat, difficulty breathing • Scratching may worsen inflammation and may cause open, weeping sores	• Hives most often appear on the arms, legs, or waist, but any part of the body may be affected • Common causes include food allergies, exposure to dusts, medicines, infections, heat or cold	• Hives (urticaria) • If needed, calamine or other soothing lotions as recommended by your pharmacist may relieve itching • See your doctor at once
Sudden appearance of bright red or dark red-blue, tender, deep-seated bumps or raised areas about 1–2 inches (2–5 centimeters) in diameter; usually on the front of both legs, occasionally on the outer forearms	• Lumps are painful • Fever, feeling of being generally unwell, joint pains, sore throat	• Most often affects 20–30 year olds, more commonly females • Erythema nodosum may occurs as a symptom of infection, drug reaction, or an underlying illness	• Skin eruption caused by inflammation within the skin (erythema nodosum) • See your doctor
Rashes			
A tender, red, warm, swollen area of skin with an undefined border	• Fever	• May occur where skin is broken (e.g. a cut or scratch)	• Cellulitis • See your doctor

CHARACTERISTIC SYMPTOM	ASSOCIATED SYMPTOMS	OTHER RELEVANT FACTORS	POSSIBLE CAUSE AND ACTION
Small, purplish-red, bruise-like spots, may be flat or slightly raised	• Associated symptoms, if present, will depend on the underlying cause	• Purpura may be due to bruising, inflammation of capillaries, the use of cortisone-type medications (e.g. ointments or oral medications), diseases affecting the platelets, or serious infections	• Bleeding into the skin (purpura) due to medical condition affecting the blood or blood vessels • See your doctor at once
Red rash with tiny, fluid-filled blisters; lesions tend to be dry and fragmented; may be swollen, scaly, or develop painful cracks; the margins of the rash are often ill defined	• Severe itching • Skin may be dry in general • Other symptoms of allergies • Itching is exacerbated by changes in temperature, mood, and contact with irritating materials	• Symptoms may commence at any age, may occur intermittently or long-term, and may fluctuate in severity • Allergic dermatitis is most common in people with a family history of allergic diseases • May occur with other allergic conditions (e.g. asthma or hay fever)	• Eczema or dermatitis • See your doctor
Red rash with tiny, fluid-filled blisters (vesicles) in an area exposed to an irritating substance; may be swollen or scaly	• May be itchy or sore • Scratching may worsen inflammation and may cause open, weeping sores	• Contact dermatitis may occur following exposure to clothing, cosmetics, household detergents, occupational exposure to oils, petroleum-based products, solvents, paint, cement, rubber, resins, plants, or medicines that are applied directly to the skin • Contact dermatitis may occur on the shoulders, neck, and scalp if the irritant is in the form of dust	• Contact dermatitis • Try to identify the cause by eliminating suspected substances • Avoid contact with the substance by wearing protective clothing • See your doctor if you are worried or if symptoms persist
Red rash or ring-like area; may be scaly	• Itchy • If the area affected is the scalp, hairs within the affected area tend to be broken	• May affect the nails, feet, hands, groin, trunk, or scalp • Ringworm may follow contact with pets (e.g. dogs, cats, or horses) • Usually chronic	• Fungal infection (e.g. ringworm) • See your doctor
Reddish plaques covered with silvery scales; the margins are well defined	• Usually not itchy • Arthritis may occur • Trauma, infections, or emotional upsets may predispose to symptoms	• May affect any age group, but uncommon before 10 years of age and most common at 15–30 years of age • Usually develops gradually	• Psoriasis • See your doctor
Rapid onset of well defined, reddish, slightly scaly patches on the trunk; usually a single patch precedes the development of others by a week or so	• Patches may be slightly itchy • May also affect the arms and legs	• The cause of pityriasis rosea is unknown • Symptoms occur most commonly in spring and autumn • Any age group may be affected, most commonly young adults	• Pityriasis rosea • See your doctor
Sudden development of small, round, red, target-like patches and bumps that are darker in the center than the outside of the lesion; may have blistering	• Feeling of being generally unwell, fever, sore throat, diarrhea	• Attacks may be triggered by medications, infections, cancer, pregnancy • Most commonly affects the back of the hands and forearms in a symmetrical fashion; may affect other areas	• Erythema multiforme, an inflammatory disease of the skin, which is usually a reaction to infection or medication • See your doctor

CHARACTERISTIC SYMPTOM	ASSOCIATED SYMPTOMS	OTHER RELEVANT FACTORS	POSSIBLE CAUSE AND ACTION
Skin problems–rashes (cont.)			
Flat, blotchy, red rashes, which begin 4 days after symptoms of a cold in a child; rashes may join up to form one larger red area	• Before the rash develops there may be a general feeling of being unwell, loss of appetite, fever, cough, runny nose, red watering eyes	• Measles is preventable by vaccination • Measles tends to be more severe in adults than in children	• Measles • Rest in bed, and avoid contact with others, especially pregnant women • See your doctor • Measles is highly contagious, warn the doctor's office of symptoms before you attend
Scaly rash across the nose and cheeks or forehead; spots are round or butterfly-shaped, with well defined margins	• Arthritis, joint pain, fever, hair loss, kidney problems	• Usually long-term • Onset is gradual • Women are twice as likely as males to develop lupus erythematosus • More common in African races (e.g. Afro-Americans) than white races	• Lupus erythematosus • See your doctor
Bright red rash on the cheeks ("slapped cheek"); after a day or two, rash typically also appears on the forearms and thighs	• Fever, general feeling of being unwell • Arthritis may occur in adults	• Most common in children aged 3–12 years • Infection during pregnancy may cause fetal damage	• Fifth disease (erythema infectiosum), an infectious viral disease • Avoid contact with other people, especially pregnant women • See your doctor • Fifth disease is infectious, so warn the doctor's office of symptoms before you attend
Blistering conditions			
Groups of bright red, tiny, fluid-filled blisters, which rupture and form crusts	• Rash is itchy • Fever (before blisters appear), feeling of being generally unwell • Adults may also experience aches and pains, headaches, serious nerve damage (rare)	• Rash mostly affects the trunk and face • Children are most often affected, especially between 2–8 years of age; condition is usually mild, but occasionally can be fatal • A person may be affected at any age if they have not previously had the condition • Severe illness occurs more often in adults	• Chickenpox • Rest, use calamine lotion or other medications recommended by your doctor or pharmacist to relieve itching, acetaminophen (paracetamol) for fever, daily bathing • Avoid scratching spots • See your doctor • Chickenpox is highly contagious, so warn the doctor's office of symptoms before you attend
Groups of tiny, fluid-filled blisters in a band-like distribution on one side of the body; the vesicles usually rupture and crust over	• Severe pain in the area of the rash usually begins 1–2 days before the skin lesions	• Mostly occurs in adults • May recur	• Shingles • See your doctor at once
Small and larger blisters around the face and ears in a child	• Often itchy • Scratching can lead to further spread of the lesions	• May affect adults but more common in children	• Impetigo • Highly contagious • Other members of the household should avoid unnecessary contact with toweling or napkins that come into contact with the lesions • See your doctor

CHARACTERISTIC SYMPTOM	ASSOCIATED SYMPTOMS	OTHER RELEVANT FACTORS	POSSIBLE CAUSE AND ACTION

Urination, frequent

Increase in the production or frequency of urination should be assessed by a doctor to ensure the cause is not potentially serious. Most common causes for these symptoms can be overcome or controlled, so there is no need to tolerate these symptoms without asking medical advice. See also entries on urination, painful.

CHARACTERISTIC SYMPTOM	ASSOCIATED SYMPTOMS	OTHER RELEVANT FACTORS	POSSIBLE CAUSE AND ACTION
Involuntary leaking of urine	• Occurs with coughing, sneezing, or exercise • May be worse during a bladder infection (cystitis), after drinking more fluids than usual, or in cold weather	• Common in women after childbirth or with ageing	• Stress incontinence (weakness of bladder muscles causing leakage of urine) • See your doctor • Exercises may help strengthen the surrounding muscles • Severe cases may require surgery
Producing more urine than usual	• Discolored urine • Waking at night to urinate • Puffy swelling of extremities • Generally feeling unwell	• May occur with high blood pressure	• Kidney disease • See your doctor at once
Producing more urine than usual	• Excessive thirst, frequent urination, increased appetite, weight loss, nausea, blurred vision • In women, frequent vaginal infections; in men, impotence; recurring yeast infections in both sexes	• Diabetes is more common in people with obesity, hypertension, or a family history of diabetes	• Diabetes • See your doctor at once
In men, waking several times during the night to urinate	• Difficulty starting a urine stream • Urine dribbling after urinating	• More common with ageing	• Prostate problems (e.g. enlargement, prostatitis, or prostate cancer) • See your doctor at once.
Discomfort or burning pain on urination	• Pain under scrotum, difficulty urinating		• Kidney infection or kidney stones • See your doctor at once

Urination, painful

Any new occurrence of pain when urinating should be fully investigated by a doctor. For intermittent problems with which you are already familiar, like cystitis or genital herpes, your doctor or pharmacist can give advice on how to manage the problem when you recognize a new episode. See also entry on urination, frequent.

CHARACTERISTIC SYMPTOM	ASSOCIATED SYMPTOMS	OTHER RELEVANT FACTORS	POSSIBLE CAUSE AND ACTION
Burning pain on urination	• Frequent urge to urinate, even when bladder is empty • Urinating only small amounts • Burning sensation in lower abdomen, urine with a strong odor • Lack of adequate fluid intake may prolong symptoms • Sexual contact may aggravate pain	• Cystitis is more common in women than in men; infection may be triggered by sexual contact, use of diaphragm for birth control, or use of urinary catheters	• Cystitis • Drink plenty of water • See your doctor
Painful urination	• Pain under scrotum, difficulty urinating	• More common with ageing	• Prostate problems (e g. prostatitis or prostate cancer) • See your doctor at once
Painful urination in men or women	• Inflamed genitals after sexual intercourse • In males, pus discharge from penis • In females, vaginal discharge, urge to urinate frequently, abnormal menstrual bleeding • The infection may spread from the genitals to the urethra, rectum, conjunctiva, pharynx, or cervix	• Complications include inflammation of reproductive organs, peritonitis, inflammation around the liver, inflammation of the Bartholin's gland in women and epididymitis or abscess around the urethra in men, arthritis, dermatitis, endocarditis, meningitis, myocarditis, or hepatitis	• Gonorrhea • See your doctor at once

CHARACTERISTIC SYMPTOM	ASSOCIATED SYMPTOMS	OTHER RELEVANT FACTORS	POSSIBLE CAUSE AND ACTION
Urination, painful (cont.)			
Burning pain on urination	• Blisters or sores on external genital areas (may later become scabby) • Sore, raw-feeling area inside vagina or on labia (women) • Vaginal discharge (women) • Discharge from infected sores • Burning pain in lower abdomen	• New sores after others heal • Sores may become infected • First infection may cause flu-like symptoms • Urinating in a warm bath may alleviate scalding sensation • Outbreaks may be triggered by other viruses or stress	• Genital herpes • Wear cotton underwear and loose clothing • See your doctor at once • Avoid sexual contact until you obtain medical advice

Vaginal problems

Some vaginal problems may be successfully treated with medications available from pharmacists. If you are unsure of the cause or if symptoms persist, see your doctor or sexual health clinic for a full sexual health check-up and to ensure possible infections do not result in fertility problems or other complications. See also entry on genital pain.

CHARACTERISTIC SYMPTOM	ASSOCIATED SYMPTOMS	OTHER RELEVANT FACTORS	POSSIBLE CAUSE AND ACTION
Thick, white discharge forming clumps	• In women, the vulva may be swollen and red, and there may be a thick, white discharge • Sexual intercourse may increase discomfort • Wearing tight clothing or synthetic underwear may worsen symptoms	• Yeast infections can be a side effect of taking antibiotics, or may be triggered by stress, pregnancy, or use of the contraceptive pill • Candidiasis is most likely in women with diabetes or as a side effect of some antibiotics	• Yeast infection (candidiasis) • Wear cotton underwear and loose clothing to allow air to the area • See your doctor
Greenish-yellow discharge with unpleasant smell		• Recent sexual contact or new sexual partner in the last month	• Infection such as bacterial vaginosis or trichomoniasis (a parasitic infection) • See your doctor
Greenish-yellow discharge with unpleasant smell	• Pain in the lower abdomen, fever		• Pelvic inflammatory disease • See your doctor at once
Yellow discharge, may be thick like mucus	• Cervix bleeds when touched or scraped, abnormal menstrual bleeding, abdominal pain, fever, pain when urinating, and pain and swelling of one or both labia	• May follow recent sexual contact with a new partner • The incidence of gonorrhea is much higher in the USA than in other industrialized countries • Potential complications include inflammation of reproductive organs, peritonitis, inflammation around the liver, inflammation of the Bartholin's gland in women and epididymitis or abscess around the urethra in men, arthritis, dermatitis, endocarditis, meningitis, myocarditis, or hepatitis	• Gonorrhea • See your doctor or a sexual health clinic as soon as possible • Antibiotic treatment is important to prevent serious complications

Weakness and paralysis

Weakness and paralysis are usually caused by disorders in the nervous system. Symptoms may involve the entire body, or be limited to one part such as an arm or leg. Paralysis (loss of muscle function) is a serious symptom and should be investigated by a doctor. See also entry on numbness and tingling.

CHARACTERISTIC SYMPTOM	ASSOCIATED SYMPTOMS	OTHER RELEVANT FACTORS	POSSIBLE CAUSE AND ACTION
Weakness or paralysis of the arms or legs	• Progressive numbness in the arms or legs, back or neck pain • Bladder, bowel, and sexual functions may be affected • May be aggravated by moving the neck or back	• Can occur following injury (e.g. broken neck or back), or due to a tumor, disease, or infection	• Spinal cord damage • See your doctor at once

CHARACTERISTIC SYMPTOM	ASSOCIATED SYMPTOMS	OTHER RELEVANT FACTORS	POSSIBLE CAUSE AND ACTION
Weakness in one arm or one leg	• Neck or back pain, numbness and tingling of the affected limb • Aggravated by bending, sitting, coughing, sneezing	• The precise location of the symptoms defines which part of the neck or back is affected	• Ruptured vertebral disk causing nerve compression • See your doctor at once
Weakness or paralysis on one side of the body, usually starts suddenly	• Tingling, confusion, dizziness, partial loss of hearing, slurred speech, inability to recognize parts of the body, unusual movements, fainting	• Common with advancing age • Risk increased in people with cardiovascular disease (e.g. hypertension, coronary heart disease), lipid abnormalities (e.g. high cholesterol), diabetes	• Transient ischemic attack or stroke • See your doctor at once
Profound weakness in both legs, then progresses upward to both arms	• Tingling, numbness	• In the majority of cases, symptoms begin 3–21 days after a mild infection or surgery	• Guillain-Barré syndrome • This is a medical emergency • Go to the emergency room or call an ambulance at once
In males, progressive muscle weakness throughout the body, usually beginning in the muscles of the pelvis	• Muscles often enlarge • May also have trouble climbing stairs or getting out of a chair, frequent falls	• Usually first occurs in boys aged 3–7 years	• Muscular dystrophy • See your doctor at once
Weakness or paralysis on one side of the body	• Constant headache, poor balance and coordination, dizziness, double vision, loss of sensation, loss of hearing	• Most common in people with cancer in another part of the body	• Brain tumor • See your doctor at once

Wounds or sores, non-healing

Any wound that takes longer than normal to heal warrants investigation by your doctor, as slow-healing wounds may indicate an infection or significant problem affecting the whole body.

CHARACTERISTIC SYMPTOM	ASSOCIATED SYMPTOMS	OTHER RELEVANT FACTORS	POSSIBLE CAUSE AND ACTION
Area where normal skin surface has been lost and fails to heal (ulcer), especially if overgrown or sealed-looking at the edges	• May be itchy and/or painful	• Typically occurs in sun-exposed areas of the skin	• Skin cancer • See your doctor at once
Area where normal skin surface has been lost and fails to heal (ulcer) on the lower leg	• Long history of discomfort, swelling, and skin changes of the lower leg	• History of disease of the leg veins (e.g. abnormal clotting or deep venous thrombosis) • Affects women more often than men	• Persistent ulcer that does not heal due to long-term, abnormally high pressure in the leg veins (venous ulcer) • See your doctor at once
Area where normal skin surface has been lost and fails to heal (ulcer) on the tips of the toes or fingers or in areas subject to pressure	• Painful • Discomfort is aggravated by pressure to the area	• History of disease in the arteries supplying blood to the affected area, or an injury affecting blood supply to the area	• Persistent ulcer that does not heal due to severe reduction of blood supply to the area (ischemic ulcer) • See your doctor at once
Area where normal skin surface has been lost and fails to heal (ulcer) in an area that suffers repeated trauma or pressure	• Painless • Numbness of the surrounding skin	• History of injury or disorder of the nerves supplying the affected area • The most common underlying cause is diabetes	• Persistent ulcer caused by repeated trauma to an area where the individual is unable to sense pain (neuropathic ulcer) • See your doctor at once
Area where normal skin surface has been lost and fails to heal (ulcer) on foot or lower leg in a person with diabetes		• Embedded foreign bodies of which the patient was unaware are commonly found in people with diabetic ulcers • Close control of blood sugar may help minimize the risk of foot ulcers	• Diabetic ulcer • See your doctor at once • Keep feet clean and dry at all times and wear properly fitted shoe • Inspect feet daily for callus, infection, abrasions, or blisters

The Time of Your Life

DEVELOPMENTAL MILESTONES AND PREVENTIVE HEALTH ISSUES FOR WOMEN

WORLD HEALTH ORGANIZATION RECOMMENDED IMMUNIZATION SCHEDULE

This table shows a typical immunization schedule. While it highlights the main immunization recommendations for many regions, the schedule will vary from country to country, and dosing intervals and frequencies may be different. Furthermore, some high-risk areas will require additional vaccinations not listed here. Please consult your local health authority for the appropriate immunization schedule for your country. Certain immunizations are effective for a limited time and booster shots are required throughout life to maintain immunity.

DISEASE	Birth	2 months	4 months	6 months	12 months	12–18 months	4 years	10–13 years	15–19 years	50 years	65 + years
Hepatitis B	•	•	•	•				•*			
Diphtheria		•	•			•	•				
Tetanus		•	•			•	•		•	•	
Pertussis (whooping cough)		•	•			•	•				
Hemophilus influenzae B		•	•		•						
Polio		•	•	•			•		•		
Measles					•		•				
Mumps					•		•				
Rubella (German measles)					•		•				
Chickenpox (optional)						•**					
Pneumococcal infection											•***
Influenza											•****

* give 3 doses if not given as an infant. Second dose 1 month after first, third dose 5 months after second dose.

** can be given anytime after 12 months of age

*** give every 5 years from age 65

**** give every year from age 65

Birth–6 months

- At birth, babies have the ability to see about 8–12 inches (20–30 centimeters) and have fully developed hearing. Females may have some discharge or spotting from the vagina. Labia may be swollen.
- Within a few days of birth, babies develop sense of taste and respond to their mother's voice and smell.
- First smile at 4–6 weeks.
- By 6 months, most babies can hold objects placed in hand, focus in all directions, roll over and lift head and shoulders when lying on their stomach.

HEALTH CHECK

- Apgar score at 1 minute and 5 minutes after birth to assess color, heart rate, breathing, responsiveness, muscle tone. Length, weight and head circumference measured, plus thorough physical examination in first 12 hours after birth.
- Screening test at 4–5 days of age to detect presence of any rare metabolic diseases and some inherited diseases.
- Thorough physical examination at 6–8 weeks.
- Immunization at birth, 2 and 4 months. (See Schedule)

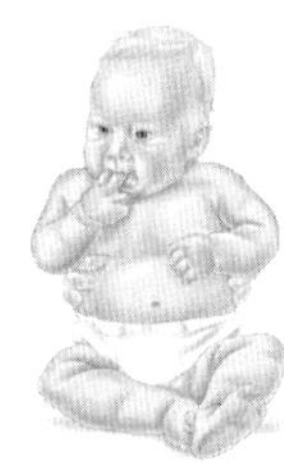

6–12 months

- First teeth appear by 6–8 months. Most babies can now sit with some support.
- By 9 months, babies are usually crawling, and may wave and clap hands. They may imitate sounds and will respond to their own name.
- By 12 months of age, many babies will be walking by holding on to furniture, standing alone for a few seconds at a time and saying 1 or 2 single words.

HEALTH CHECK

- Immunization at 6 months and 12 months. (See Schedule)
- Physical and developmental examination may be done at 6–8 months.

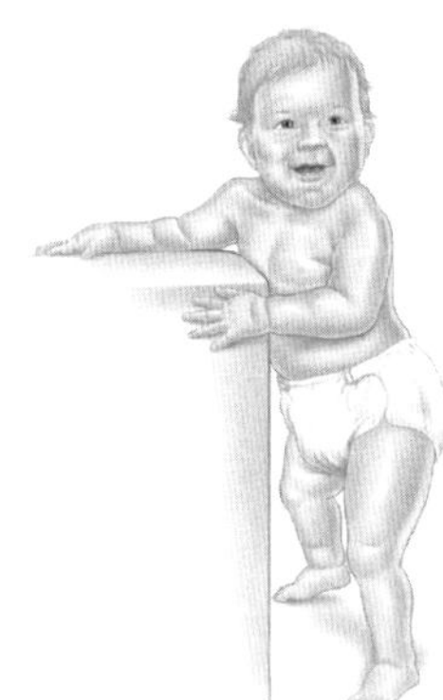

20–30 years

- Fertility peaks in the mid-20s, and this decade is the most likely time for childbearing to begin.
- Awareness of sexually transmitted diseases (STDs) and use of contraception is essential in all sexually active females.

HEALTH CHECK

- Physical examination every 2 years, including blood pressure, height and weight. From age 18, all women should have a Pap smear every 2 years.
- Breast self-examination should be done every month for life.
- Certain immunization boosters are necessary every 10 years. (See Schedule)

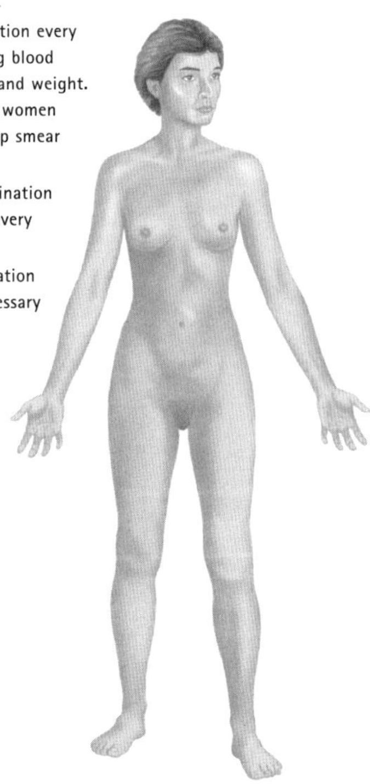

30–40 years

- Fertility declines, and risk of fetal abnormalities in the older pregnant woman increases.
- Bone density starts to decline.

HEALTH CHECK

- Physical examination every 2 years, including blood pressure, height and weight. Pap smears continue every 2 years.
- Breast self-examination should be done every month.
- Cholesterol should be checked every 5 years.

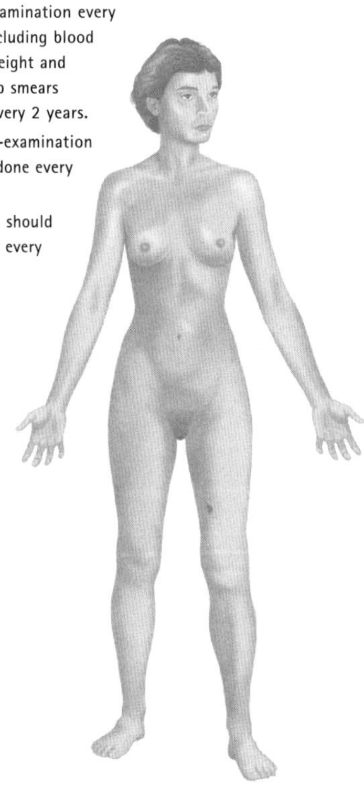

40–50 years

- Menopause usually commences, signifying the end of menstruation and fertility. Women may suffer hot flashes/flushes and other symptoms due to the decline in estrogen levels.
- Calcium and mineral content of bones decreases significantly.

HEALTH CHECK

- Physical examination every year, including blood pressure, height and weight, and clinical examination of breasts. Gynecological exam, including internal pelvic assessment every year. Pap smears should also now be every year. Rectal examination and urine tests yearly.
- Breast self-examination should be done every month.
- Mammograms may be done every 2 years. Eye test, including glaucoma screen, every 2 years.

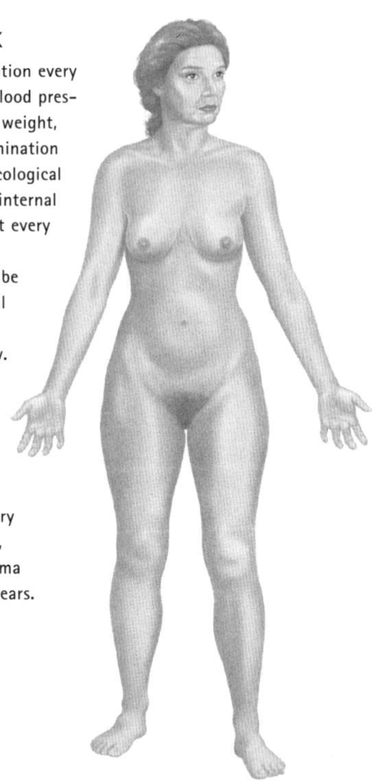

1–5 years

- The preschool years are a time of rapid development, and a time when a child's individuality becomes noticeable.
- At 18 months, she can walk well, stack blocks, throw a ball and push toys around the room.
- By her second birthday, she can run, walk up and down stairs and feed herself with a spoon. 2 year olds are easily frustrated and liable to throw temper tantrums. Vocabulary is expanding dramatically.
- 3–4 year olds learn to use simple sentences, and play involves more interaction with playmates. Girls are usually fully toilet-trained by this age.
- By age 4, girls have grown taller and appear slimmer, due to losing fat and gaining muscle. Most height increase involves the legs.

HEALTH CHECK

- Immunization at 12–18 months, and 4–5 years prior to school entry. (See Schedule)
- Development will be continually monitored at doctor's visits.

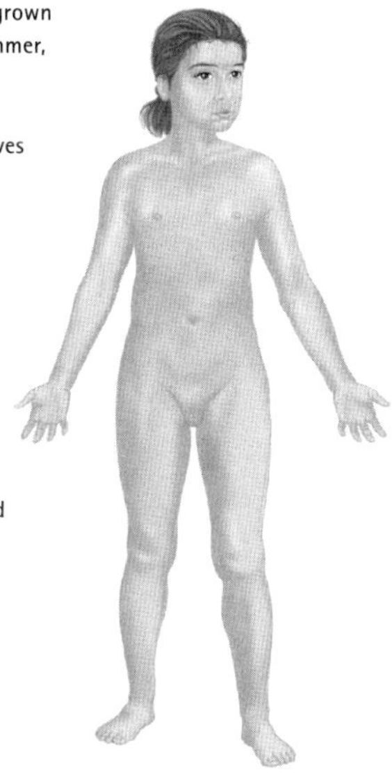

6–11 years

- These years are a time of rapid physical, intellectual and psychological growth.
- School begins at age 5–6, and girls start to develop close friendships, usually with other girls.
- Baby teeth begin to fall out around the sixth year.
- Reading and writing skills are developed and refined.
- For most girls, puberty begins after the age of 10. Signs of puberty include development of breasts and pubic hair, and a marked increase in physical size (growth spurt).
- Children of this age usually have the co-ordination and balance of an adult.

HEALTH CHECK

- Children who experience difficulties with schoolwork may be tested for attention deficit disorder (ADD) or learning problems such as dyslexia.
- Immunization boosters may be necessary. (See Schedule)

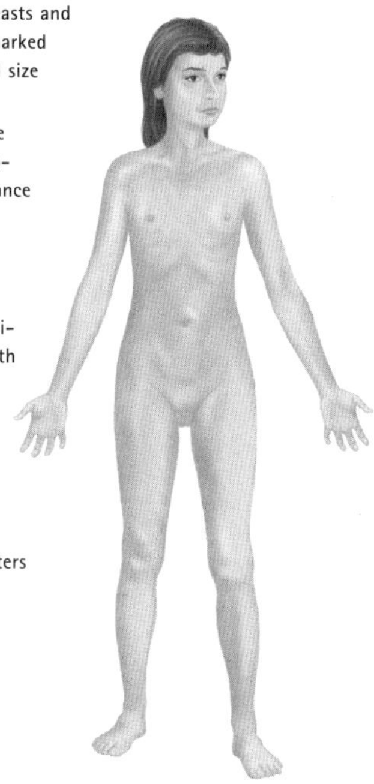

12–19 years

- Menstruation usually begins at 11–14 years of age, after which growth may slow down. Body fat increases around the hips and thighs, sweat glands develop further, and hormonal changes can start to cause skin problems.
- Eating disorders sometimes affect young women in this age group.

HEALTH CHECK

- Immunization at 10–13 years and 15–19 years. (See Schedule)
- Scoliosis screening is carried out at many schools. More girls than boys develop scoliosis—a sideways curvature of the spine.

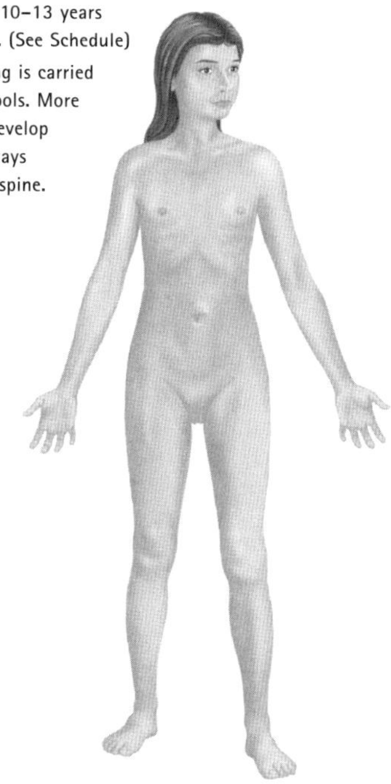

50–65 years

- Following menopause, bone density falls significantly. Lean body mass declines, and the metabolic rate decreases.

HEALTH CHECK

- Yearly screening continues, including blood pressure, height and weight, and clinical examination of breasts. Gynecological exam, including internal pelvic assessment and Pap smear every year. Rectal examination, urine tests and skin checks yearly. Eye test, including glaucoma screen, every 2 years.
- Additional annual screening following menopause is also necessary, including assessment of heart function, bone density and stools.
- Mammograms are recommended yearly.
- Sigmoidoscopy (visual examination of the rectum and lower colon) every 3–5 years is advised, and cholesterol screening continues every 5 years.

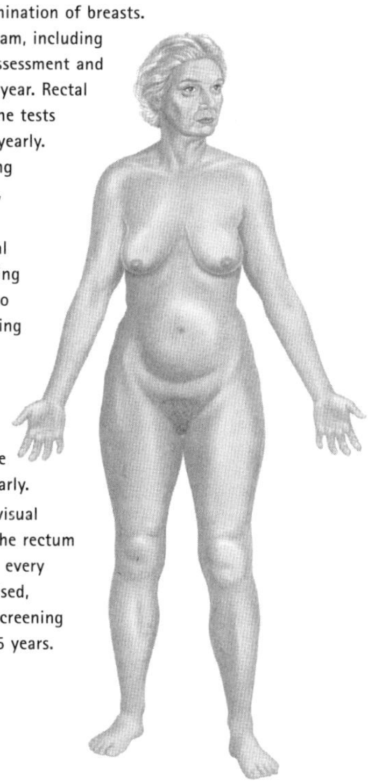

65–85 years

- By 65–70 years of age, a woman has half the bone density she had at age 30.
- There may be a decline in organ function, including the brain. Many older women will continue to be very active, but physical capabilities may become limited or more difficult.

HEALTH CHECK

- Yearly screening continues. Physical examination should include blood pressure, height and weight, and clinical examination of breasts. A gynecological examination, including internal pelvic exam and Pap smear, and rectal examination and urine tests are also advised. Yearly mammograms, as well as heart, bone, stool and cholesterol tests are recommended. Eye test, including glaucoma screen, every 2 years.
- Monthly breast self-examination should continue.
- Annual vaccination against influenza is advised.
- Pneumococcal immunization every 5 years.

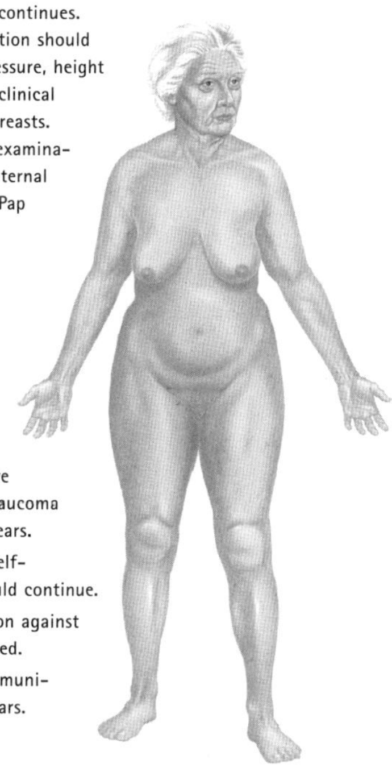

85+ years

- Loss of bone density continues. About 30 percent of all women reaching 90 years of age will suffer a hip fracture due to weakened bones.
- The incidence of dementia continues to increase. Lapses of memory and difficulty learning new information are noticed in those affected.

HEALTH CHECK

- Yearly screening continues. Physical examination should include full gynecological examination, including Pap smear, rectal examination, blood pressure, height and weight, skin checks and clinical examination of breasts and urine tests.
- Blood tests for cholesterol and sugar.
- Yearly mammograms, as well as heart, bone density and stool tests are recommended.
- Pneumococcal and influenza immunizations continue.
- Eye test, including glaucoma screen, every 2 years.
- Monthly breast self-examination should continue.

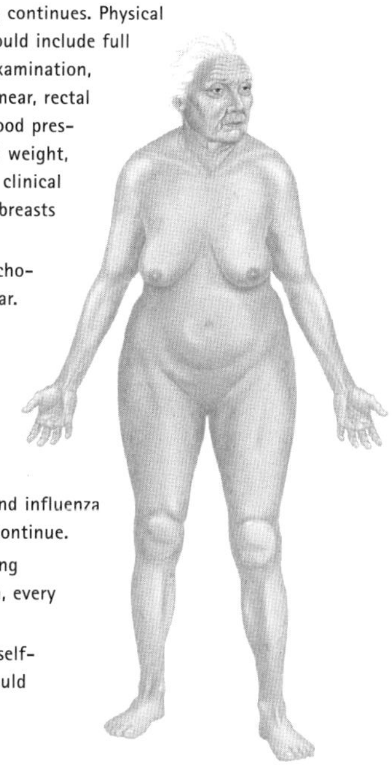

The Time of Your Life

DEVELOPMENTAL MILESTONES AND PREVENTIVE HEALTH ISSUES FOR MEN

WORLD HEALTH ORGANIZATION RECOMMENDED IMMUNIZATION SCHEDULE

This table shows a typical immunization schedule. While it highlights the main immunization recommendations for many regions, the schedule will vary from country to country, and dosing intervals and frequencies may be different. Furthermore, some high-risk areas will require additional vaccinations not listed here. Please consult your local health authority for the appropriate immunization schedule for your country. Certain immunizations are effective for a limited time and booster shots are required throughout life to maintain immunity.

DISEASE	Birth	2 months	4 months	6 months	12 months	12–18 months	4 years	10–13 years	15–19 years	50 years	65 + years
Hepatitis B	•	•	•	•				•*			
Diphtheria		•	•			•	•				
Tetanus		•	•			•	•		•	•	
Pertussis (whooping cough)		•	•			•	•				
Hemophilus influenzae B		•	•		•						
Polio		•	•	•			•		•		
Measles					•		•				
Mumps					•		•				
Rubella (German measles)					•		•				
Chickenpox (optional)						•**					
Pneumococcal infection											•***
Influenza											•****

* give 3 doses if not given as an infant. Second dose 1 month after first, third dose 5 months after second dose.

** can be given anytime after 12 months of age

*** give every 5 years from age 65

**** give every year from age 65

Birth–6 months

- At birth, babies have the ability to see about 8–12 inches (20–30 centimeters) and have fully developed hearing. Both testes should be present in the scrotum.
- Within a few days of birth, babies develop sense of taste and respond to their mother's voice and smell.
- First smile is usually at 4–6 weeks.
- By 6 months, most babies can hold objects placed in hand, focus in all directions, roll over and lift head and shoulders when lying on their stomach.

HEALTH CHECK

- Apgar score at 1 minute and 5 minutes after birth to assess color, heart rate, breathing, responsiveness, muscle tone. Length, weight and head circumference measured, plus thorough physical examination in first 12 hours after birth.
- Screening test at 4–5 days of age to detect presence of any rare metabolic diseases and some inherited diseases.
- Thorough physical examination at 6–8 weeks.
- Immunization at birth, 2 and 4 months. (See Schedule)

6–12 months

- First teeth usually appear between 6–8 months. Most babies can now sit with some support.
- By 9 months, babies are usually crawling, and may wave and clap hands. They may imitate sounds and will respond to their own name.
- By 12 months of age, most babies have tripled their birth weight. Many babies will be walking by holding on to furniture, standing alone for a few seconds at a time and saying 1 or 2 single words.

HEALTH CHECK

- Immunization at 6 months and 12 months. (See Schedule)
- Detailed physical and developmental examination may be done at 6–8 months.

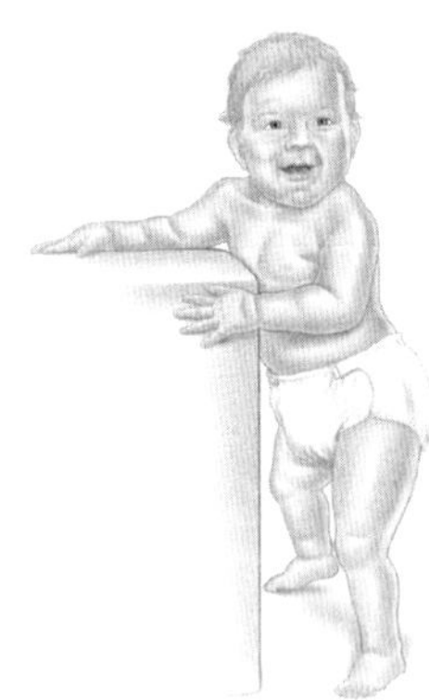

20–30 years

- Bones continue to broaden until the age of 20 in most males.
- Awareness of sexually transmitted diseases (STDs) and use of contraception is essential in all sexually active males.

HEALTH CHECK

- Regular physical examinations are recommended every 2 years, including height, weight, and blood pressure assessment.
- Testes should be clinically examined every 2 years, and self-examination is recommended every month.
- Certain immunization boosters are necessary every 10 years. (See Schedule)

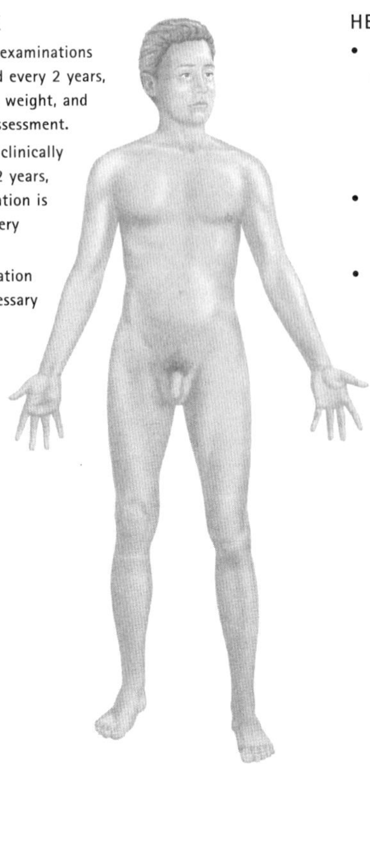

30–40 years

- Distribution of body fat may begin to change, becoming more prevalent around the abdomen.
- 30 percent of males in their 30s may notice some hair loss.

HEALTH CHECK

- Regular physical exams should continue every 2 years, including height, weight, blood pressure, testicular examination.
- Cholesterol screening every 5 years is advised.
- Self-examination of testes is recommended every month.

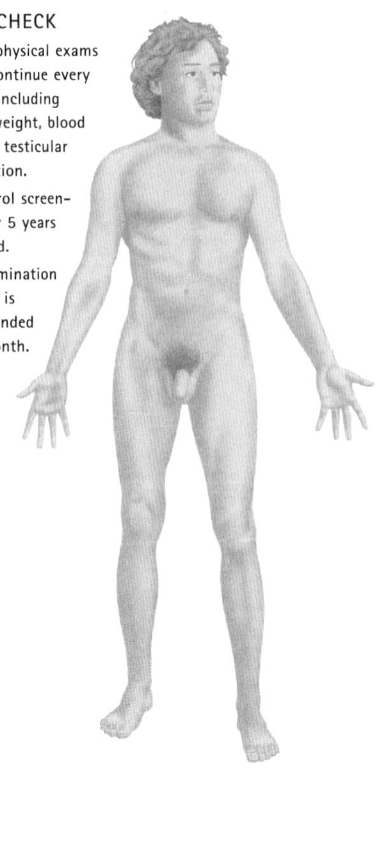

40–50 years

- Metabolism may slow, causing a tendency to gain weight.
- Prostate enlargement usually begins around age 45.

HEALTH CHECK

- Physical examination every year, including height, weight, blood pressure, and clinical examination of testes. Rectal examination and urine tests yearly.
- Cholesterol screening should become more frequent, and may be tested every 1–2 years.
- Self-examination of testes is recommended every month.
- Eye exam, including glaucoma screening, is recommended every 2 years.

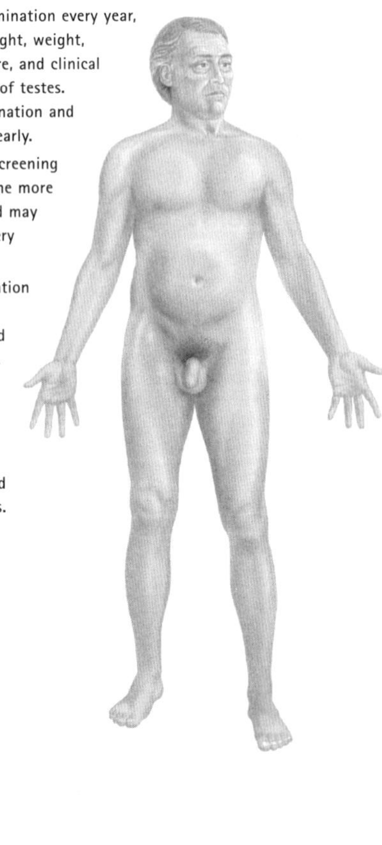

1–5 years

- The preschool years are a time of rapid development, and a time when a child's individuality becomes noticeable.
- At 18 months, he can walk well, stack blocks, throw a ball and push toys around the room.
- By his second birthday, he can run, walk up and down stairs, and feed himself with a spoon. 2 year olds are easily frustrated and liable to throw temper tantrums. Vocabulary is expanding dramatically.
- 3–4 year olds learn to use simple sentences, and play involves more interaction with playmates.
- Toilet-training in boys is usually complete by this time, although some boys may continue to have problems particularly at night.
- By age 4, boys have grown taller and appear slimmer, due to losing fat and gaining muscle. Most height increase involves the legs.

HEALTH CHECK

- Immunization at 18 months, and 4–5 years prior to school entry. (See Schedule)
- Development will be continually monitored at doctor's visits.

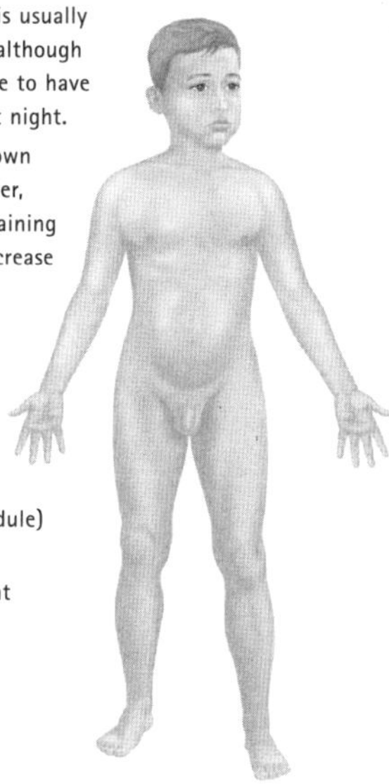

6–11 years

- These years are a time of rapid physical, intellectual and psychological growth.
- School begins at age 5–6, and boys start to develop close friendships, usually with other boys.
- Baby teeth begin to fall out around the sixth year.
- Reading and writing skills are developed and refined.
- Children of this age usually have the coordination and balance of an adult.

HEALTH CHECK

- Children who experience difficulties with schoolwork may be tested for attention deficit disorder (ADD) or learning problems such as dyslexia.
- Immunization boosters may be necessary. (See Schedule)

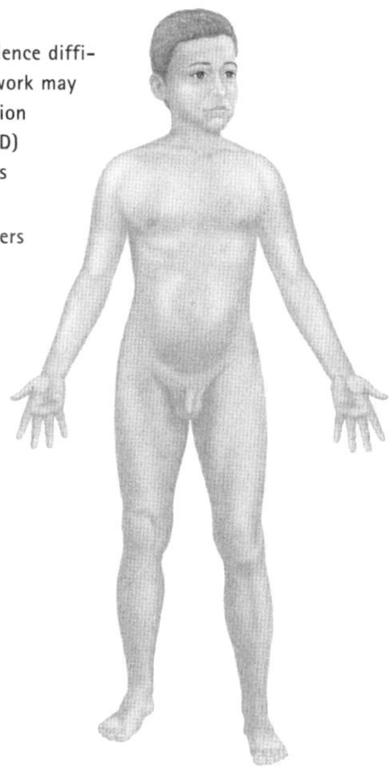

12–19 years

- Puberty in boys normally begins at around 11–12 years. The testes and penis grow larger and the skin of the scrotum darkens. Pubic hair begins to appear.
- At the age of 12–13, boys have a growth spurt, growing rapidly in weight and height. The chest and shoulders become broader.
- At 13 or 14 the voice box begins to grow and the voice starts change. Many boys are able to ejaculate by this age.
- Underarm and facial hair, as well as sweat glands, typically begin to appear between 13 and 15 years of age.
- Height continues to increase until around 17–18 years of age.

HEALTH CHECK

- Immunization at 10–13 years and 15–19 years. (See Schedule)
- Screening for scoliosis (sideways curvature of the spine) is carried out at many schools.

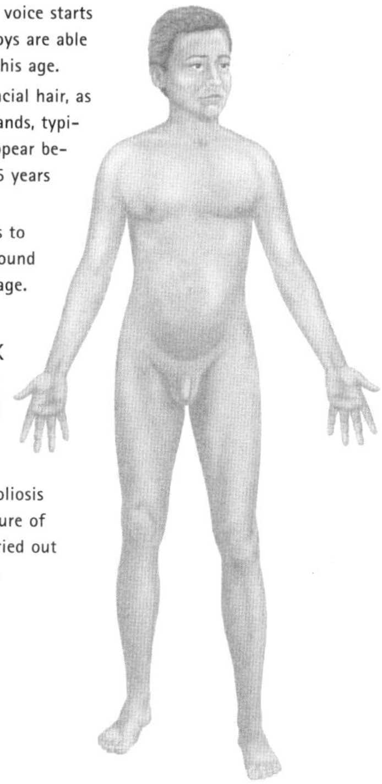

50–65 years

- Prostate enlargement continues. Bone density falls.
- Hair loss is evident in 50 percent of men aged over 50.

HEALTH CHECK

- Yearly screening continues, including height, weight, blood pressure, clinical examination of testes, urine tests, skin checks and stool test. Self-examination of testes is recommended every month.
- Annual rectal examination and blood tests for prostate-specific antigen (marker of prostate cancer).
- Sigmoidoscopy (visual study of the rectum and lower colon) every 3–5 years is advised.
- Eye exam, including glaucoma screening, is recommended every 2 years.
- Stress test and electrocardiogram (EKG) may be necessary, depending on overall health and family history of heart disease.

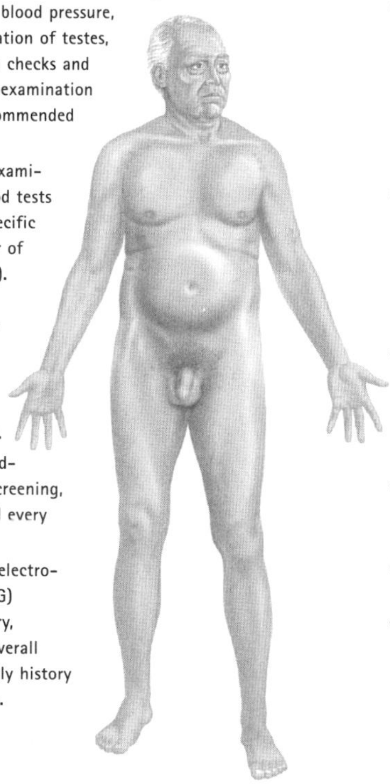

65–85 years

- Many older men will continue to be very active, but physical capabilities may become limited or more difficult. There may be a decline in organ function, including the brain.

HEALTH CHECK

- Yearly screening continues. Physical examination should include clinical examination of the testes and prostate, cholesterol measurements, height, weight, blood pressure, skin checks, heart tests and urine tests.
- Annual rectal examination, blood tests for prostate-specific antigen, and annual stool tests are necessary.
- Eye exam, including glaucoma screening, is recommended every 2 years.
- Yearly vaccination against influenza is advised.
- Pneumococcal immunization every 5 years.
- Self-examination of testes is recommended every month.

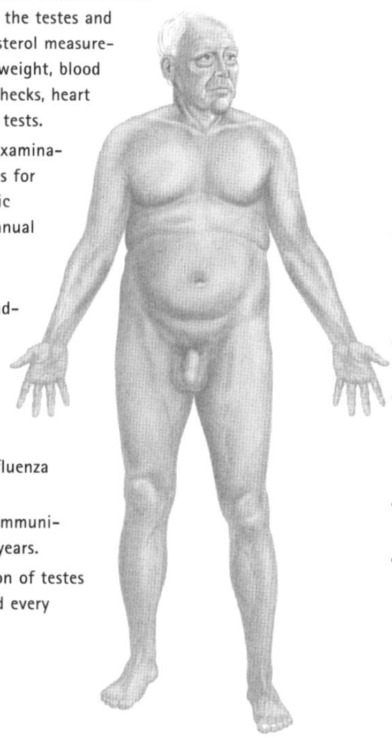

85+ years

- The incidence of dementia increases with age, affecting about 1 in 5 men over 85 years.
- Although age-related loss of bone density begins earlier and proceeds more rapidly in women, it is also a significant health problem in elderly men.
- Most elderly men remain independent.

HEALTH CHECK

- Yearly screening continues. Physical examination should include blood pressure check, rectal examination, height, weight, skin checks, heart tests and urine tests.
- Blood tests for cholesterol, sugar and prostate-specific antigen.
- Annual stool test for blood.
- Pneumococcal and influenza vaccinations continue.
- Tests for bone density are recommended.
- Self-examination of testes continues every month.

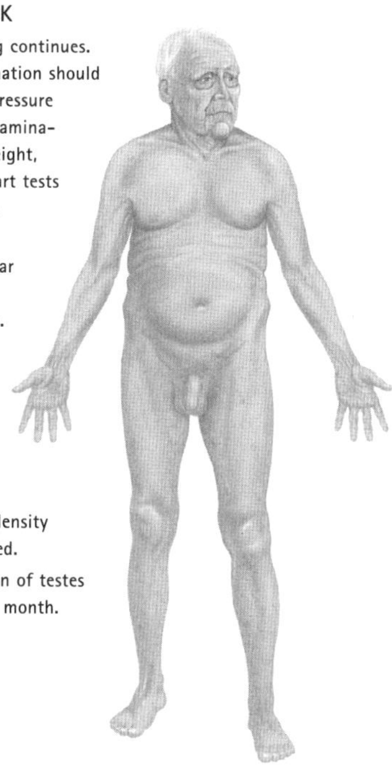

Index

Bold text denotes the main entry for a topic, *italicised numerals* denote illustrations for entries.

A

I

M

N

Q

R

T

U

V

W

X

Y

Z